FIFTH EDITION

Integrated Arithmetic and Basic Algebra

BILL E. JORDAN
Seminole State College of Florida

WILLIAM P. PALOW
Miami Dade College

PEARSON

Boston   Columbus   Indianapolis   New York   San Francisco   Upper Saddle River
Amsterdam   Cape Town   Dubai   London   Madrid   Milan   Munich   Paris   Montréal   Toronto
Delhi   Mexico City   São Paulo   Sydney   Hong Kong   Seoul   Singapore   Taipei   Tokyo

Editorial Director	Christine Hoag
Editor in Chief	Maureen O'Connor
Executive Content Editor	Kari Heen
Senior Content Editor	Lauren Morse
Associate Content Editor	Katherine Minton
Editorial Assistant	Rachel Haskell
Senior Managing Editor	Karen Wernholm
Senior Production Project Manager	Beth Houston
Digital Assets Manager	Marianne Groth
Associate Media Producer	Stephanie Green
Content Development Manager	Rebecca E. Williams
QA Manager Assessment Content	Marty Wright
Associate Marketing Manager	Alicia Frankel
Marketing Assistant	Ashley Bryan
Rights and Permissions Advisor	Michael Joyce
Manufacturing Buyer	Linda Cox
Design Manager	Andrea Nix
Senior Designer Specialist	Heather Scott
Text Design	Tamara Newnam
Composition	PreMediaGlobal
Cover Image	Diane Diederich/iStockphoto

Many of the designations used by manufacturers and sellers to distinguish their products are claimed as trademarks. Where those designations appear in this book, and Pearson Education was aware of a trademark claim, the designations have been printed in initial caps or all caps.

Library of Congress Cataloging-in-Publication Data

Jordan, Bill E.
    Integrated arithmetic and basic algebra/Bill E. Jordan, William P. Palow.—5th ed.
        p. cm.
    Includes index.
    ISBN 13: 978-0-321-74738-9
    ISBN 10: 0-321-74738-0
    1. Arithmetic. 2. Algebra.   I. Palow. William P.   II. Title.

QA107.2.J67 2013

513'.12—dc22   2011005565

1 2 3 4 5 6 7 8 9 10—CRK—15 14 13 12 11   ISBN 13: 978-0-321-74738-9
                                                 ISBN 10: 0-321-74738-0

Contents

Preface

Benefits of the Integrated Approach

The Integrated Approach   Why use an "integrated approach" for a standard prealgebra/beginning algebra or a beginning algebra course? Our goal is for students to have a better understanding of both courses and progress more easily through the college math curriculum. This idea was generated from our long teaching experience. We watched our students perform reasonably well in arithmetic courses until they progressed to algebra, when they became far more frustrated and often saw algebra as an alien subject. Our exciting challenge was to make explicit connections so students could overcome this gap by learning the concepts in arithmetic and algebra simultaneously, when possible.

Mainstream Coverage   This is a mainstream approach to teaching prealgebra/beginning algebra in that the usual topics and sequence of topics for the course are for the most part standard and natural. What makes this book unique, and will give pause to anyone who wearily says "all textbooks are the same," is the parallelism of the pedagogical approach. For example, operations with rational numbers and rational expressions are presented together rather than separately, emphasizing for students the logical connections between arithmetic and algebra.

Proven Student Success   We are proud that our first four editions have helped so many students find success where they previously met frustration, and we thank our many ardent supporters for giving our book a try and giving us suggestions, both large and small, to make it a more polished teaching and learning tool. This text takes the burden off of students to make connections between basic operations and thus allows them to make connections on higher levels. That in turn leaves them better prepared for intermediate algebra and later math courses. Studies performed at Seminole Community College, Midlands Technical College, and Miami-Dade College substantiated higher student performance using this text. An added benefit is the improvement in students' attitudes: they are sometimes reluctant to be seen taking a course focused only on arithmetic, but that onus is removed when they can say they are enrolled in an algebra course.

Modern Pedagogy   Our book is not limited to its integrated approach, and we have readily embraced new pedagogical practices emanating from conferences and professional organizations in addition to enrichment techniques learned from our many years of teaching. Thus, we emphasize real-world relevancy, critical thinking, collaboration, and a determined progression from skills to conceptual understanding to make college students better thinkers and problem solvers. Also, once concepts are introduced, they are reused and developed throughout the book so that students will continue practicing concepts and deepen their understanding of them. Finally, geometry is integrated into examples and exercises throughout the book rather than relegated to a separate section, reinforcing the notion that geometry and algebra support each other in problem solving.

Hallmark Features

Preparation

The integrated approach allows students to see the big picture of math through the often parallel relationship of arithmetic and algebra.

The Annotated Instructor's Edition shows all answers next to the exercises and contains helpful teaching tips in the margin.

Prerequisite Skills needed to be successful are listed at the end of section objectives, and include text references to allow students to strengthen these skills.

Getting Ready Exercises open each section and prepare students for the upcoming material.

Examples include explanations next to each step describing what students *will* do to get to the next step, rather than showing what *was* done to get to that step as in most texts.

Support

Be Careful notes appear throughout the text to warn students about common errors.

Study Tip boxes are positioned in Chapters 1 and 2 to help students improve their study habits.

Note boxes call out additional information, explanations, or observations about the concept just discussed.

Practice

Practice Exercises follow examples to allow students to attempt solving similar problems immediately after the concept is presented.

Additional Practice Exercises double the number of Practice Exercises as an extra resource.

Exercise sets provide students with ample opportunities for practice. All exercise sets include **Writing Exercises, Challenge Exercises,** and **Group Activities**.

Mastery

Chapter summaries are presented in a concise, easy-to-read table format that offers an explanation and example for each key concept.

Chapter Review Exercises contain additional exercises keyed to the chapter's sections.

Chapter Tests simulate a class test to give students extra practice and experience with testing conditions.

New in the Fifth Edition

General

- **Approximately 700 new or revised exercises** have been added to the text.

- **NEW design** enhances the presentation of the material.

- **NEW** *Instructor's Resource Manual with Tests* has been added to the list of instructor resources.

- **ENHANCED** *Worksheets for Classroom and Lab Practice* are available for students to use and now include "Getting Ready" exercises to correspond to those found in the text.

Content Changes

- Revised and reorganized Chapter R.

- Combined Sections R.3 (Multiplication of Whole Numbers) with R.4 (Division of Whole Numbers).

- Added reducing fractions to lowest terms to Section R.4 (formerly R.5).

- Combined the Study Tips so students see them at the start of the course in Chapters 1 and 2.

- Altered the charts for application problems in Sections 3.5 and 3.6.

- Moved Section 3.5, Solving Linear Inequalities, to the end of the chapter so it is now Section 3.8.

- Labeled the graphs of linear equations and inequalities in Chapter 4.

- Eliminated Section 6.1 by removing Objectives A and B, moving Objective C (Represent a rational number as division and change a rational number to a decimal) to Section R.6, moving Objective D (Changing improper fractions to mixed numbers and mixed number into improper fractions) to Section R.4, and moving Objective E (Graph rational numbers) to Section 6.2.

- Labeled the graphs of systems of linear equations and inequalities in Chapter 9.

Ancillaries

The following ancillaries are available to help both instructors and students use the text more effectively:

STUDENT SUPPLEMENTS	INSTRUCTOR SUPPLEMENTS
Student's Solutions Manual	**Annotated Instructor's Edition**
• Contains solutions to odd-numbered exercises as well as all Chapter Review and Chapter Test problems. ISBN-10: 0-321-75927-3 ISBN-13: 978-0-321-75927-6	• Includes all answers to exercises printed in red beneath each problem as well as Teaching Tips in the margin. ISBN-10: 0-321-75923-0 ISBN-13: 978-0-321-75923-8
ENHANCED Worksheets for Classroom or Lab Practice	**Instructor's Solutions Manual (download only)**
These lab- and classroom-friendly workbooks offer the following resources for every section of the text: • A list of learning objectives • NEW! Getting Ready exercises matched to those in the main text to prepare students for the upcoming material • Vocabulary practice problems • Extra practice exercises with ample space for students to show their work ISBN-10: 0-321-75924-9 ISBN-13: 978-0-321-75924-5	• Contains solutions to even-numbered exercises from the text, as well as for group projects. • Available in MyMathLab® and on the Instructor's Resource Center.
	PowerPoint® Slides
	• Include explanations of objectives, definitions from the text, as well as examples from each section of the textbook. • Available for download in MyMathLab® and on the Instructor's Resource Center.
Video Resources	**Instructor's Resource Manual with Tests (download only)**
• Present lectures for each section in Chapters 1-9, and present overview lectures for Chapters R, 10, and 11. • Also available in MyMathLab®. ISBN-10: 0-321-78630-0 ISBN-13: 978-0-321-78630-2	• Includes resources designed to help both new and adjunct faculty with course preparation and classroom management. • Offers helpful teaching tips. • Contains two multiple-choice tests per chapter; three free-response tests per chapter; chapter overviews and teaching notes. • Available in MyMathLab® and on the Instructor's Resource Center.
	TestGen® Test Bank
	TestGen® (www.pearsoned.com/testgen) enables instructors to build, edit, print, and administer tests using a computerized bank of questions developed to cover all the objectives of the text. TestGen is algorithmically based, allowing instructors to create multiple but equivalent versions of the same question or test with the click of a button. Instructors can also modify test bank questions or add new questions. The software and testbank are available for download from Pearson Education's online catalog.

Media Supplements

MyMathLab® Online Course (access code required)

MyMathLab delivers **proven results** in helping individual students succeed.

- MyMathLab has a consistently positive impact on the quality of learning in higher education math instruction. MyMathLab can be successfully implemented in any environment—lab-based, hybrid, fully online or traditional—and demonstrates the quantifiable difference that integrated usage has on student retention, subsequent success, and overall achievement.
- MyMathLab's comprehensive online gradebook automatically tracks your students results on tests, quizzes, homework, and in the study plan. You can use the gradebook to quickly intervene if your students have trouble or to provide positive feedback on a job well done. The data within MyMathLab is easily exported to a variety of spreadsheet programs, such as Microsoft Excel. You can determine which points of data you want to export, and then analyze the results to determine success.

MyMathLab provides **engaging experiences** that personalize, stimulate, and measure learning for each student.

- **Tutorial Exercises:** The homework and practice exercises in MyMathLab are correlated to the exercises in the textbook, and they regenerate algorithmically to give students unlimited opportunity for practice and mastery. The software offers immediate, helpful feedback when students enter incorrect answers.
- **Multimedia Learning Aids:** Exercises include guided solutions, sample problems, animations, videos, and eText clips for extra help at point-of-use.
- **Expert Tutoring:** Although many students describe the whole of MyMathLab as "like having your own personal tutor," students using MyMathLab do have access to live tutoring from Pearson, from qualified math and statistics instructors who provide tutoring sessions for students via MyMathLab.

MyMathLab Standard allows you to build your course your way, offering maximum flexibility and complete control over all aspects of assignment creation. Starting with a clean slate lets you choose the exact quantity and type of problems you want to include for your students. You can also select from pre-built assignments to give you a starting point.

Ready-to-Go MyMathLab comes with assignments pre-built and pre-assigned, reducing start-up time. You can always edit individual assignments as needed throughout the semester.

And, MyMathLab comes from a **trusted partner** with educational expertise and an eye on the future.

> Knowing that you are using a Pearson product means knowing that you are using quality content. That means that our eTexts are accurate, that our assessment tools work, and that our questions are error-free. And whether you are just getting started with MyMathLab, or have a question along the way, we're here to help you learn about our technologies and how to incorporate them into your course.

To learn more about how MyMathLab combines proven learning applications with powerful assessment, visit **www.mymathlab.com** or contact your Pearson representative.

MyMathLab®Plus/MyStatLab™ Plus

MyLabsPlus combines proven results and engaging experiences from MyMathLab® and MyStatLab™ with convenient management tools and a dedicated services team.

Designed to support growing math and statistics programs, it includes additional features such as:

- **Batch Enrollment:** Your school can create the login name and password for every student and instructor, so everyone can be ready to start class on the first day. Automation of this process is also possible through integration with your school's Student Information System.
- **Login from your campus portal:** You and your students can link directly from your campus portal into your MyLabsPlus courses. A Pearson service team works with your institution to create a single sign-on experience for instructors and students.
- **Advanced Reporting:** MyLabsPlus's advanced reporting allows instructors to review and analyze students' strengths and weaknesses by tracking their performance on tests, assignments, and tutorials. Administrators can review grades and assignments across all courses on your MyLabsPlus campus for a broad overview of program performance.
- **24/7 Support:** Students and instructors receive 24/7 support, 365 days a year, by email or online chat.

MyLabsPlus is available to qualified adopters. For more information, visit our website at www.mylabsplus.com or contact your Pearson representative.

MathXL® Online Course (access code required)

MathXL® is the homework and assessment engine that runs MyMathLab. (MyMathLab is MathXL plus a learning management system.) With MathXL, instructors can:

- Create, edit, and assign online homework and tests using algorithmically generated exercises correlated at the objective level to the textbook.
- Create and assign their own online exercises and import TestGen tests for added flexibility.
- Maintain records of all student work tracked in MathXL's online gradebook.

With MathXL, students can:

- Take chapter tests in MathXL and receive personalized study plans and/or personalized homework assignments based on their test results.
- Use the study plan and/or the homework to link directly to tutorial exercises for the objectives they need to study.
- Access supplemental animations and video clips directly from selected exercises.

MathXL is available to qualified adopters. For more information, visit our website at www.mathxl.com, or contact your Pearson representative.

Acknowledgments

We would like to express our appreciation to the following very special people at Pearson for their guidance and, most of all, for being so darn nice and cooperative: Maureen O'Connor (our Editor in Chief), Kari Heen (Executive Content Editor), Lauren Morse (Senior Content Editor), Katherine Minton (Associate Content Editor), Rachel Haskell (Editorial Assistant), Beth Houston (Production Supervisor), Michelle Renda (Executive Marketing Manager), Ashley Bryan (Marketing Assistant), Stephanie Green (Associate Media Producer), Rebecca Williams (Content Development Manager), and Heather Scott (Senior Design Specialist). In addition, we would also like to thank Trish O'Kane at PreMediaGlobal (Project Manager); Laurie Semarne and Steve Ouellette for accuracy checking the manuscript; Elizabeth Morrison and Gary Williams for accuracy checking pages; Cheryl Cantwell, Helen Medley, and Andrea Roberts for their work on the solutions manuals; Beverly Fusfield and John Morin for their work on the *Instructor's Resource Manual with Tests*; and Carrie Green and Diane Cook for their work on the *Worksheets for Classroom or Lab Practice*.

In addition, we would especially like to thank the following reviewers, whose contributions were invaluable in producing a text of this quality:

Sandra Belcher, *Midwestern State University*

Dixielee Blackinton, *Weber State University*

Laurie E. Braga, *Loyola University*

Stephanie Clute, *Sam Houston State University*

Jonathan Cornick, *Queensborough Community College*

Arianna S. Derrick, *Midlands Technical College*

Kathy Garrison, *Clayton College and State University*

Tim Hagopian, *Worcester State College*

Mary Lou Hammond, *Spokane Community College*

Jon Harper, *Weber State University*

Jim Hodge, *Mountain State University*

Sandee D. House, *Georgia Perimeter College*

Kay L. Keagy, *Community College of Allegheny County, Allegheny Campus*

Roberta Lacefield, *Waycross College*

Kevin Lawson, *Teach Educational Consultants*

Randal Leichty, *Indiana Institute of Technology*

Elizabeth Morrison, *Valencia Community College, West Campus*

Bronte Overby, *Patrick Henry Community College*

Ellyn Stewart, *Midlands Technical College*

Alicia Taylor, *Shelton State Community College*

Mike Totoro, *Nassau Community College*

Shirley Treadway, *Eastern Illinois University*

Dr. Amit Trehan, *Montgomery College*

Tony Vavra, *West Virginia Northern Community College*

We would also like to thank our colleagues at Seminole State College of Florida and Miami Dade College for allowing us to create the Integrated Arithmetic and Algebra courses at our schools and to class test the materials.

This book has been written with the characteristics and needs of the beginning mathematics student in mind. It is based on the research and experience of authors who have each taught for over 30 years and who have received numerous awards for teaching excellence. We have tried to combine our experience with current learning theory and practice to produce a book that we hope will make a difference. It is our belief that the beginning mathematics student needs a fresh way of viewing mathematics so that they are not looking at the same old material in the same old way. We hope that your experiences with this text are as pleasant as ours have been.

B. E. Jordan

W. P. Palow

Basic Ideas

Chapter R covers the operations of addition, subtraction, multiplication, and division of whole numbers and decimals. Fractions and application problems are briefly introduced. In addition, we will learn *when* to add, subtract, multiply, or divide so we can combine this ability with knowing *how* to carry out these operations.

This chapter, and indeed the whole book, views mathematics as consisting of three components: **concepts, algorithms**, and **applications**. The ideas of mathematics are called *concepts*. A process for carrying out an operation is an *algorithm*. An *application* is a situation, usually posed in words, in which a concept is recognized and an algorithm applied to calculate a result.

If you feel you have already mastered the material in Chapter R, you may go directly to Chapter 1.

Section R.1   Reading and Writing Numerals

OBJECTIVES *When you complete this section, you will be able to:*

Ⓐ Read and write numerals that represent whole numbers.
Ⓑ Write whole numbers in expanded notation.
Ⓒ Write the word name for a standard numeral.

PREREQUISITE SKILLS *none*

GETTING READY FOR SECTION R.1

No prerequisite skills are called for, so no Getting Ready exercises are presented.

Introduction

A **number** is a concept or idea that exists only in our minds. A **numeral** is a symbol that represents a number. In Sections R.1–R.3, we will discuss properties and operations with **Whole Numbers**.

> **Definition** Whole Numbers
>
> The numbers $0, 1, 2, 3, \ldots$ are called the whole numbers.

The three dots, $\ldots$, mean that the pattern established by the previous numbers continues forever.

OBJECTIVE ## Reading and writing numerals that represent whole numbers

In the Middle Ages India was called Hindu.

The development of the Hindu–Arabic system of numerals with a zero gave mathematicians a powerful tool to develop the present-day subject of arithmetic. The Hindu–Arabic enumeration system uses 10 symbols called **digits**—0, 1, 2, 3, 4, 5, 6, 7, 8, and 9. The meaning of each digit comes from counting and tells us "how many." We associate each of these symbols with the corresponding number of objects in a **set**, or collection of things. Sets may be designated by writing the **members**, or things in the set, between two symbols called **braces**, $\{\ \}$. Set A = $\{a, b, c\}$ has three members and is thus associated with the number 3.

The most important thing the Hindu–Arabic system gives us is the concept of **place value**, in which the value of a digit depends on its left–right location. As we go to the left, we find the value of each place by multiplying the value of the place before it by 10. The **value of a digit** is found by multiplying the digit by its place value. The **value of the numeral** is found by adding the individual amounts represented by each digit of the numeral.

Place values allow construction of numerals that represent any number. Some of the place values in our system of numeration are shown in the following table:

	,				,				,			
B	H	T	M	H	T	T	H	T	O
i	u	e	i	u	e	h	u	e	n
l	n	n	l	n	n	o	n	n	e
l	d	M	l	d	T	o	d	s	s
i	r	i	i	r	h	u	r		
o	e	l	o	e	o	s	e		
n	d	l	n	d	u	a	d		
s	M	i	s	T	s	n	s		
	i	o		h	a	d			
	l	n		o	n	s			
	l	s		u	d				
	i			s	s				
	o			a					
	n			n					
	s			d					
				s					

In Europe, the comma is used in the same way that we use a decimal point in the United States; where we use a comma, they use a space or a period. For example 42,345.97 would be written as 42 345,97 or 42.345,97.

Notice that the place values are separated into groups of three by a comma so that a numeral is easier to read. (Note: Commas are often omitted on numbers from 1000 to 9999.) As shown above, the place values start on the far right with a value of 1, or one unit. As we go from right to left, each place value is found by multiplying the one before it by 10. Thus, we obtain one (10), ten (1×10), one hundred $(1 \times 10 \times 10)$, one thousand $(1 \times 10 \times 10 \times 10)$, ten thousand $(1 \times 10 \times 10 \times 10 \times 10)$, one hundred thousand $(1 \times 10 \times 10 \times 10 \times 10 \times 10)$, one million $(1 \times 10 \times 10 \times 10 \times 10 \times 10 \times 10)$, ten million $(1 \times 10 \times 10 \times 10 \times 10 \times 10 \times 10 \times 10)$, and so on.

Example 1   Give the place value of the underlined digit.

a. 64,3̲89

Solution

3rd place, so
$1 \cdot 10 \cdot 10 =$ one hundred.

b. 4̲0,934

Solution

4th place, so
$1 \cdot 10 \cdot 10 \cdot 10 =$ one thousand.

c. 9,487,25̲4

Solution

1st place, so one.

PRACTICE EXERCISE 1

Give the place value of the underlined digit.

a. 98,5̲46

b. 3̲78,035

c. 956,0̲43

Answers: **Practice Exercise 1: a.** ten **b.** ten thousand **c.** one hundred

The digit in a numeral tells us how many of each of the values we have.

| **Example 2** | **Give the meaning of the underlined digit in each case.** |

a. 46,5̲03

b. 6,09̲4,321

Solution

The 5 is in the 3rd place (hundreds), so it represents 5 hundreds, $5 \cdot 100 = 500$.

Solution

The 9 is in the 5th place (ten thousands), so it represents 9 ten thousands, or $9 \cdot 10,000 = 90,000$.

PRACTICE EXERCISE 2

Give the meaning of the underlined digit in each case.

a. 143̲,706

b. 97̲,890,462

OBJECTIVE **B** Writing whole numbers in expanded notation

As mentioned before, the value represented by a numeral is found by adding the individual amounts represented by the various digits. For example, $15,486 = 1 \cdot 10,000 + 5 \cdot 1000 + 4 \cdot 100 + 8 \cdot 10 + 6 \cdot 1 = 10,000 + 5000 + 400 + 80 + 6$. A numeral written in this form is said to be in **expanded notation**.

| **Example 3** | **Write in expanded notation.** |

a. 64,253

b. 205,650

Solution

$64,253 = 6 \cdot 10,000 + 4 \cdot 1000 + 2 \cdot 100$
$+ 5 \cdot 10 + 3 \cdot 1$
$= 60,000 + 4000 + 200 + 50 + 3$

Solution

$205,650 = 2 \cdot 100,000 + 0 \cdot 10,000$
$+ 5 \cdot 1000 + 6 \cdot 100 + 5 \cdot 10 + 0 \cdot 1$
$= 200,000 + 5000 + 600 + 50$

PRACTICE EXERCISE 3

Write in expanded notation.

a. 4819

b. 214,804

If you need more practice, do the following Additional Practice Exercises.

Additional Practice Exercise 3   Write the following in expanded notation:

a. 56,149

b. 2,830,456

OBJECTIVE **C** Writing the word name for a standard numeral

The numerals that we use to represent Hindu–Arabic numbers are said to be in **standard notation** or to be **standard numerals**. Thus, 15,486 is in standard notation, or is called a *standard numeral*.

Answers: Practice Exercise 2: a. 3000 **b.** 7,000,000 **Practice Exercise 3: a.** 4000 + 800 + 10 + 9
b. 200,000 + 10,000 + 4000 + 800 + 4 **Additional Practice 3: a.** 50,000 + 6000 + 100 + 40 + 9
b. 2,000,000 + 800,000 + 30,000 + 400 + 50 + 6

Using the concept of place value, the standard numeral 6,204,893 means 6 millions and 2 hundred thousands and 0 ten thousands and 4 thousands and 8 hundreds and 9 tens and 3 ones. However, this is not the way the numeral is properly read aloud or written. The numeral is written or read as six million, two hundred four thousand, eight hundred ninety-three.

When writing out numerals, the numerals twenty-one through twenty-nine, thirty-one through thirty-nine, forty-one through forty-nine, and so on to ninety-nine are hyphenated.

The words that we say while reading the numeral for a number are called its **word name**. The word name is a third type of numeral, or symbol for the number.

Be Careful

Whole numbers do not have the word "and" in their names.

Recall that the place values are separated into groups of three. From right to left, these groups are the units, thousands, millions, billions, and so on. As we read a numeral, we say the numeral within a group, then name the group. For example, in the numeral 414,678, we say, "four hundred fourteen," then we say, "thousands," because 414 is in the thousands group. We do not name the units group, but simply say, "six hundred seventy-eight."

| **Example 4** | **Give the word name for the standard numeral.** |

Note

In Example 4a, we do not say, "zero ten thousands." We leave out any mention of ten thousands. When reading a numeral that has the digit 0, leave out the place value of the 0.

a. 18,704,234

Solution

Eighteen million, seven hundred four thousand, two hundred thirty-four

b. 4,914,042

Solution

Four million, nine hundred fourteen thousand, forty-two

PRACTICE EXERCISE 4

Write the word name for each standard numeral.

a. 6,722,414

b. 7,304,056

Given a word name, we will be able to write the standard numeral.

| **Example 5** | **Express each of the following as a standard numeral:** |

a. Six thousand, nine hundred six

Solution

6906

b. One hundred million, eighty-six thousand, three hundred five

Solution

100,086,305

Answers: Practice Exercise 4: a. Six million, seven hundred twenty-two thousand, four hundred fourteen **b.** Seven million, three hundred four thousand, fifty-six

PRACTICE EXERCISE 5

Express each of the following as a standard numeral:

a. Twelve thousand, five hundred seven

b. Eight million, nine hundred thirteen thousand, seven hundred thirty-two

For Extra Help

Exercise Set R.1   MyMathLab®

Give the place value of the underlined digit in each of the following. (See Example 1.)

1. 8<u>6</u>41

2. 9,<u>7</u>64,534

3. 85,<u>7</u>36,394

4. 7,563,0<u>8</u>9

5. 246,576,83<u>4</u>

6. <u>3</u>4,255,768

What is the meaning of the underlined digit in each of the following numerals? (See Example 2.)

7. <u>6</u>0,171

8. 338,<u>4</u>20

9. 10<u>5</u>,268

10. <u>2</u>81,093

11. 34,<u>9</u>20,385

12. 10,233,<u>4</u>58

Write each of the following in expanded form. (See Example 3.)

13. 397

14. 818

15. 3033

16. 740,992

17. 409,135

18. 1,017,819

Write the word name for each of the following standard numerals. (See Example 4.)

19. 12

20. 24

21. 905

22. 970

23. 7149

24. 2013

25. 8902

26. 5003

27. 30,209

28. 402,668

29. 4,078,074

30. 9,203,441

31. 29,756,011

32. 17,104,869

Answers: Practice Exercise 5:   a. 12,507   **b.** 8,913,732

Write each of the following as a standard numeral. (See Example 5.)

33. Four hundred seven

34. Eight thousand, eighty-one

35. Fourteen thousand, seventy-three

36. One hundred five thousand, two hundred twenty-seven

37. Nine hundred two thousand, four hundred sixty

38. Forty-five million, two hundred thousand, six hundred one

Challenge Exercises (39 and 40)

39. Some enumeration systems in the past were repetitive. If you want to represent three of something, you repeat the symbol for 1 three times. If you want to represent 30 things, you write the symbol for 10 three times, and so on. In ancient Egypt, the symbol for 1 was / and the symbol for 10 was ∩. How would you write the numeral for 25 in the Egyptian system?

40. Some systems of enumeration are multiplicative. If you want to represent 30 things, you write the symbol for 3 followed by the symbol for 10. In China and Japan the symbol for 3 is 三, and the symbol for 10 is 十. Thus, 30 is symbolized 三十. If the symbol for 8 is 八 and the symbol for 6 is 六, how do the Chinese write 860?

Writing Exercises (41 and 42)

41. Suppose that you are in a culture that counts like this: 1, 2, 3, many. (That is, there are no distinct numerals beyond 3.) Being a large landowner, you have many, many sheep. How could you keep track of the sheep so that you would know if any disappeared overnight? (*Hint:* Try matching sheep with objects.)

42. At least two ancient cultures had enumeration systems based on 20 rather than on 10. What do you think is the reason for using base 20?

Section R.2   Addition and Subtraction of Whole Numbers

OBJECTIVES *When you complete this section, you will be able to:*

A Find the sum of two digits from memory.
B Add whole numbers.
C Use combinations of numbers to add columns of whole numbers.
D Find the difference of two digits from memory.
E Subtract whole numbers without regrouping (borrowing).
F Subtract whole numbers with regrouping (borrowing).
G Solve application problems using subtraction.

PREREQUISITE SKILLS *Before beginning this section, you should be able to:*

Construct standard numerals for whole numbers, using place value. (Section R.1)

GETTING READY FOR SECTION R.2

Construct standard numerals for each of the following whole numbers from word notation:

1. Eight hundred five   **2.** Seventy thousand, forty-three

3. Five million, four hundred thirty-three thousand, six hundred one

Introduction

Note

To conform with standard usage, we will refer to numerals as numbers from this point forward.

In Section R.1, we discussed numbers. In this section, we begin our study of operations on numbers. The four basic operations of arithmetic are **addition, subtraction, multiplication**, and **division**. We begin with addition.

The concept of addition may be thought of as putting two nonoverlapping sets together to form a combined set. The number of elements in the combined set is the sum of the numbers of elements in the individual sets.

OBJECTIVE Finding the sum of any two digits from memory

In order to discuss several numbers at the same time, we introduce the concept of a **variable**. A variable is a symbol, usually a letter, that stands for the numbers of a specific set. In this case, the specified set is the set of whole numbers. Any letter may be used as a variable—for example, a, b, c, and so on. We give a formal definition of addition.

> **Definition** The Sum of Whole Numbers
>
> The sum of any two whole numbers a and b is the whole number c such that $a + b = c$. The numbers a and b are called **addends**, and the number c is called the **sum**.

For example, the sum of 2 and 4 is represented as $2 + 4 = 6$. That is, 2 (addend) + 4 (addend) = 6 (sum).

Answers: Getting Ready: 1. 805 **2.** 70,043 **3.** 5,433,601

There are certain sums that we must memorize before we can add anything but the smallest whole numbers. The best way to organize these sums is to put them in a table. This way, we can easily find the appropriate sum and look for some important patterns.

Table of Sums for Two Digits

+	0	1	2	3	4	5	6	7	8	9
0	0	1	2	3	4	5	6	7	8	9
1	1	2	3	4	5	6	7	8	9	10
2	2	3	4	5	6	7	8	9	10	11
3	3	4	5	6	7	8	9	10	11	12
4	4	5	6	7	8	9	10	11	12	13
5	5	6	7	8	9	10	11	12	13	14
6	6	7	8	9	10	11	12	13	14	15
7	7	8	9	10	11	12	13	14	15	16
8	8	9	10	11	12	13	14	15	16	17
9	9	10	11	12	13	14	15	16	17	18

A line of numbers going left and right is called a **row** and a line of numbers going up and down is called a **column**. The first addend in the sum comes from the first column, and the second addend comes from the first row. If we choose 3 in the first column, and 5 in the first row, the place where the corresponding column and row intersect is the sum of 3 and 5, which is 8. That is, $3 + 5 = 8$.

OBJECTIVE Adding whole numbers

After we have mastered the addition facts, we may use them to add any two whole numbers. The process of addition with standard numerals is known as the **addition algorithm**. This process involves lining up the corresponding place values to ensure like values are added to like values as digits are added from right to left.

If the sum of the digits in a place value is greater than or equal to 10, we **regroup**. To use a simple example, we add 8 and 6. We get more units than we can symbolize in the units place, since the largest single digit we can use is 9. Therefore, we take 2 units from the 6 and add them to the 8. This gives us one group of 10. The remaining 4 units from the 6 are left in the units place, and the one group of 10 is "carried" to the tens place value by putting a 1 above the digits already there. This process is called **regrouping** or **carrying** and is illustrated by parts b and c of Example 1.

> **Procedure: Addition Algorithm**
>
> To add whole numbers, line up the corresponding place values and add from right to left. If the sum in any corresponding place value is 10 or greater, then carry the excess to the next larger place value.

Example 1 **Add the following whole numbers:**

a. 154 and 21

Solution

We line up place values. Units are over units, tens over tens, and so on.

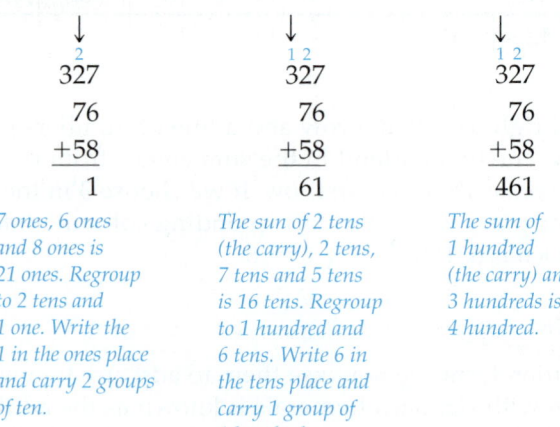

Add units.	Add tens.	Add hundreds.
↓	↓	↓

154	154	154
+21	+21	+21
5	75	175

4 units and 1 unit is 5 units.

5 tens and 2 tens is 7 tens, or 70.

1 hundred and 0 hundred is 1 hundred.

b. 706,469 and 514,088

Solution

$$\begin{array}{r} {\scriptstyle 1\ \ 11} \\ 706{,}469 \\ +514{,}088 \\ \hline 1{,}220{,}557 \end{array}$$

Starting with the units column, we add each column in turn and carry when necessary. For example, 9 ones and 8 ones is 17 ones. Write 7 in the ones place and carry 1 group of 10. 1 ten, 6 tens and 8 tens is 15 tens. Write 5 in the tens place and carry 1 group of 100. Continue until all the digits are added.

c. 327, 76, 58

Solution

↓	↓	↓
²	¹ ²	¹ ²
327	327	327
76	76	76
+58	+58	+58
1	61	461

7 ones, 6 ones and 8 ones is 21 ones. Regroup to 2 tens and 1 one. Write the 1 in the ones place and carry 2 groups of ten.

The sum of 2 tens (the carry), 2 tens, 7 tens and 5 tens is 16 tens. Regroup to 1 hundred and 6 tens. Write 6 in the tens place and carry 1 group of 1 hundred.

The sum of 1 hundred (the carry) and 3 hundreds is 4 hundred.

PRACTICE EXERCISE 1

Add each of the following:

a. 99 and 15   **b.** 140,356 and 92,076   **c.** 539, 42, and 78

OBJECTIVE **C** Using combinations of numbers to add columns of whole numbers

There are techniques that can help in the adding process. One of these is to look for **combinations** of digits that when added give easy sums. We can find various combinations of digits whose sums are 5, 10, or 15, and so forth. For example, $1 + 4 = 5, 8 + 2 = 10, 7 + 3 = 10,$ or $7 + 8 = 15.$

Example 2   Add the following:

a. $82 + 73 + 17$

b. $149, 265, 321$ and 785

Solution

$$\begin{array}{r} 1 \\ 82 \\ 73 \\ +17 \\ \hline 172 \end{array}$$

Ones Column
Combine 3 and 7 for a partial sum of 10.
Add 10 and 2 to get 12.
Write 2 ones and carry 1 ten.

Tens Column
Combine 7 and 8 to get 15.
Add 1 and 15 and 1 to get 17.

Solution

$$\begin{array}{r} 2\,2 \\ 149 \\ 265 \\ 321 \\ 785 \\ \hline 1520 \end{array}$$

Ones Column
$9 + 1 = 10$
$5 + 5 = 10$
$10 + 10 = 20$
Put 0 ones, carry 2 tens.

Tens Column
$4 + 6 = 10$
$2 + 8 = 10,$
$10 + 10 + 2 = 22$
Put 2 tens, carry 2 hundreds.

Hundreds Column
$2 + 1 + 2 = 5$
$3 + 7 = 10,$
$5 + 10 = 15$
Put 5 hundreds, carry 1 thousand.

PRACTICE EXERCISE 2

Add each of the following columns of numbers:

a.
$$\begin{array}{r} 39 \\ 82 \\ 46 \\ +71 \end{array}$$

b.
$$\begin{array}{r} 963 \\ 148 \\ 252 \\ +377 \end{array}$$

If you cannot find combinations that result in 5, 10, 15, and so forth, add in the usual manner as in Example 1.

OBJECTIVE **Finding the difference of two digits from memory**

Subtraction may be thought of as a taking away of objects from a set, or the reduction of the number of objects in a set.

Definition Difference of Whole Numbers

The difference of two whole numbers **a** and **b** is the whole number **c** such that **b** + **c** = **a**. We call **a** the **minuend**, **b** the **subtrahend**, and **c**, or **a** − **b**, the **difference**.

In the sentence $a - b = c$, b is subtracted from a to get c. Thus, $9 - 7 = 2$ becomes

$$\begin{array}{r} 9 \\ -7 \\ \hline 2 \end{array}$$

Minuend.
Subtrahend.
Difference.

If $a - b = c$, then $b + c = a$. In other words, $a - b$ is that number which you add to b to get a. This discussion also leads to a very important conclusion: For every subtraction statement, such as $9 - 3 = 6$ $(a - b = c)$, there is a related addition statement, $6 + 3 = 9$ $(c + b = a)$.

If we have mastered the addition facts such as those given in the table on page 9, it is unnecessary to memorize subtraction facts. For example, consider $12 - 8$. This may be thought of as, "What number plus 8 equals 12?" From the addition facts table on page 9, we know that $4 + 8 = 12$. Thus, $12 - 8 = 4$. Since we can apply this logic to any difference, there is no need for a subtraction fact table.

Example 3 Use a related addition statement to find the indicated difference.

a. $9 - 4$

Solution

The relationship may be thought of as "What number plus 4 equals 9?" Since $5 + 4 = 9, 9 - 4 = 5$.

b. $15 - 6$

Solution

The relationship may be thought of as "What number plus 6 equals 15?" Since $9 + 6 = 15, 15 - 6 = 9$.

PRACTICE EXERCISE 3

Find the following differences:

a. $9 - 2$ (What number is added to 2 to get 9?)

b. $8 - 3$ (What number is added to 3 to get 8?)

Two ideas are basic to subtraction. The first is that zero subtracted from a number yields the original number; for example, $5 - 0 = 5$. The second is that any number subtracted from itself yields zero; for example, $5 - 5 = 0$.

OBJECTIVE Subtracting whole numbers without regrouping (borrowing)

Next, we consider the process, or algorithm, for subtraction of whole numbers. First, we show subtraction that requires no **borrowing**, or regrouping. Since there is a related addition statement for each subtraction statement, we can easily check our differences to see if they are correct. That is, if $8 - 5 = 3$, then $3 + 5 = 8$.

Note

Place values *must* be lined up because we may subtract only things that are alike. That is, ones from ones, tens from tens, hundreds from hundreds, and so on.

Procedure: Subtraction without Regrouping (borrowing)

To subtract whole numbers without regrouping, line up the corresponding place values and subtract from right to left.

Example 4 Subtract and check the results.

a. 23 from 37

Solution

Line up the place values.

$$
\begin{array}{r} 37 \\ -23 \\ \hline 4 \end{array}
\qquad
\begin{array}{r} 37 \\ -23 \\ \hline 14 \end{array}
\qquad
\begin{array}{r} 37 \\ -23 \\ \hline 14 \\ \hline 37 \end{array}
$$

7 ones minus 3 ones is 4 ones.

3 tens minus 2 tens is 1 ten.

CHECK: Add 23 and 14
This sum is 37 which is the minuend. Hence, the subtraction is correct.

b. 365 from 2599

Solution

$$
\begin{array}{r} 2599 \\ -365 \\ \hline +2234 \\ \hline 2599 \end{array}
$$

CHECK: Add 365 and 2234.
This sum is 2599 which is the minuend. Therefore, the subtraction is correct.

Answers: Practice Exercise 3: a. 7 **b.** 5

PRACTICE EXERCISE 4

Subtract the following:

a. $77 - 61$ **b.** $279 - 76$ **c.** $982 - 351$ **d.** $775 - 34$

OBJECTIVE Subtracting whole numbers with regrouping (borrowing)

We are ready to expand our discussion of subtraction to problems that require regrouping, or borrowing, to find the differences.

> **Procedure: Subtraction of Whole Numbers with Regrouping (borrowing)**
>
> Line up corresponding place values. If a digit in the subtrahend is larger than the digit in the minuend in the same column, borrow one place value from the next digit to the left in the minuend. The one place value that is borrowed is placed in front of the digit in the minuend from which the larger digit in the subtrahend is being subtracted, and the digit from which the place value was borrowed is decreased by one. Subtract from right to left.

Example 5 | **Subtract.**

a. 36 from 52

Solution

$$\begin{array}{r} 52 \\ -36 \end{array}$$ *We cannot subtract 6 ones from 2 ones, so we regroup by taking 1 ten from the 5 tens, which gives us 4 tens and 12 ones.*

$$\begin{array}{r} ^{4\,12}52 \\ -36 \\ \hline 6 \end{array}$$ *Regrouping allows us to subtract 6 ones from 12 ones to get 6 ones.*

$$\begin{array}{r} ^{4\,12}52 \\ -36 \\ \hline 16 \end{array}$$ *4 tens minus 3 tens is 1 ten.*

Therefore, 36 from 52 is 16.   CHECK: $16 + 36 = 52$.

b. 76 from 494

Solution

$$\begin{array}{r} 494 \\ -76 \end{array}$$ *We cannot subtract 6 ones from 4 ones, so we borrow 1 ten from the 9 tens in the next column. This gives us 14 ones and 8 tens.*

$$\begin{array}{r} ^{8\,14}494 \\ -76 \\ \hline 8 \end{array}$$ *Six ones from 14 ones gives us 8 ones.*

Answers: Practice Exercise 4: a. 16 **b.** 203 **c.** 631 **d.** 741

494 Seven tens from 8 tens gives us 1 ten.
−76
‾‾‾‾
18

494 Zero hundreds from 4 hundreds gives us 4 hundreds.
−76
‾‾‾‾
418

Therefore, 76 from 494 is 418.　CHECK: 418 + 76 = 494.

c. 5003 minus 981

Solution

5003
−981 1 unit from 3 units gives us 2 units.
‾‾‾‾
2

5003 We cannot subtract 8 tens from 0 tens, so we borrow 1 hundred from the hundreds column. But there
−981 are no hundreds, so we must go to the thousands column. From 5 thousands we borrow 1 thousand,
‾‾‾‾ leaving 4 thousands. We rewrite the 1 thousand we borrowed as 10 hundreds.
2

5003
−981 We still need to borrow from the tens column, so we borrow 1 hundred from the hundreds place,
‾‾‾‾ leaving 9 hundreds. We rewrite the hundred we borrowed as 10 tens and put it in the tens column.
22   Now we can subtract 8 tens from 10 tens and get 2 tens.

5003
−981 We subtract 9 hundreds from 9 hundreds and get 0 hundreds.
‾‾‾‾
022

5003 Finally, we subtract the unwritten 0 thousands from 4 thousands and obtain 4 thousands.
−981
‾‾‾‾
4022

Therefore, 5003 minus 981 is 4022.　CHECK: 4022 + 981 = 5003.

PRACTICE EXERCISE 5

Subtract. Each of the following involves at least one borrowing:

a. 360 − 92　　　　**b.** 547 − 277　　　　**c.** 1308 − 163　　　　**d.** 9007 − 328

OBJECTIVE Solving application problems using subtraction

It is often necessary to add and/or subtract whole numbers when solving application problems.

Answers:　**Practice Exercise 5:**　**a.** 268　**b.** 270　**c.** 1145　**d.** 8679

| **Example 6** | **Solve. Some may contain unnecessary information.** |

a. On Black Monday, a stock was selling for $15 per share. By Monday of the following week, it was down to $8 per share. How much value did the stock lose in that week?

Solution

We may think of this problem as asking, "What number must we add to $8 to get $15?" This we recognize from Example 3 as a related addition statement for subtraction. Therefore, $15 − $8 = $7.

b. Professor Jones has 28 students in her integrated arithmetic and algebra class. There are 13 men in the class. How many women are in the class?

Solution

The class is composed only of men and women. Therefore, the total number of students in the class (28) minus the number of men (13) equals the unknown number of women. In symbols, $28 - 13 = 15$.

c. At the beginning of training camp in August, the Bulldogs pro football team had 60 players report. By September 15, they were down to 44 players on the roster. How many players were taken off the roster?

Solution

Notice that the dates have nothing to do with the answer. They are extra information and should be ignored. The team started with 60 players and had to take away some players to get down to 44. "Take away" suggests subtraction. Therefore, $60 - 44 = 16$.

PRACTICE EXERCISE 6

Solve.

a. North Campus has 237 faculty members. If 148 are women, how many are men?

b. Angela has been working all summer and saving to go back to school in the fall. She now has $2300, but it will cost $5000 for the academic year. How much must she borrow?

c. One morning, the temperature in Calgary, Canada, was 75°F at 8:00 A.M. During the day, a cold front came in from the Arctic Circle. By 5:00 P.M., the temperature was 35°F. How much did the temperature drop?

Exercise Set R.2   For Extra Help   MyMathLab®

Add each of the following. (See Examples 1 and 2.)

1. 423	**2.** 235	**3.** 903	**4.** 607
+445	+142	+27	+89

5. 9538	**6.** 4847	**7.** 42,807	**8.** 29,015
+203	+608	+37,155	+50,474

Answers: Practice Exercise 6: a. 89 men **b.** $2700 **c.** 40°

9. 26
    42
    +19

10. 92
    46
    +67

11. 599
    290
    656
    +179

12. 536
    920
    732
    +357

13. 4189
    60
    157
    4
    +2119

14. 63
    4219
    4
    622
    +8965

Subtract. Check each difference as in Example 8. (See Examples 3–5.)

15. $252 - 41$

16. $665 - 51$

17. $1974 - 923$

18. $8717 - 372$

19. $63 - 25$

20. $75 - 38$

21. $732 - 279$

22. $655 - 277$

23. $4102 - 3226$

24. $6023 - 2635$

25. $85 - 47$

26. $56 - 19$

27. $808 - 438$

28. $709 - 256$

29. $3005 - 1132$

30. $8009 - 6679$

Solve. Some of the exercises may contain extra information or involve addition as well as subtraction. (See Example 6.)

31. On payday, John Lee had $1500 in his checking account. After paying the bills, he had $230 left. How much did he spend on bills?

32. Freshman English started the term with 32 students. At the end of the term, 24 people were left in the class. How many students dropped the course?

33. Last year, Diana bought a new car. After the appropriate negotiations, she paid $18,000 for the car. Now the car has nagging electrical problems, and she wants to trade it in. The dealer says the car is worth $15,000. How much value has the car lost since she bought it?

34. In order to get a B in beginning algebra, Hank needs 480 out of 600 points. He now has 395 points after taking four tests. How much must he get on the fifth and last test to get the B?

Challenge Exercises (35–37)

35. The algorithm for adding in base 5 is the same as in base 10, which is our numeral system. For example, $3 + 2 = 10_{\text{five}}$, $3 + 3 = 11_{\text{five}}$, and $3 + 4 = 12_{\text{five}}$. What is the sum of $4 + 4$ in base 5? (*Hint:* 12_{five} means one five and two units.)

36. During the lifetime of a 30-year mortgage, you pay $275,000 in principal and interest. If the principal is $55,000, how much interest is paid?

37. Last year, a small business had $90,000 in sales. The expenses for doing business that year were as follows: (a) rent—$24,000, (b) electricity—$6000, and (c) personnel—$48,000. How much profit was made?

Writing Exercises (38–41)

38. Suppose we did not have a digit zero. How would we write numerals like 204?

39. The Chinese added on an abacus, called a *suan-pan*. This instrument consists of a frame with rods running from one side to the opposite side. Each rod represents a place value. Some people say that an expert on the suan-pan can add as fast as you can say the addends.

Explain how you could add 213 and 152 on the abacus.

Other cultures used the abacus as well. These include the Egyptian, Japanese, Mayan, and Roman civilizations.

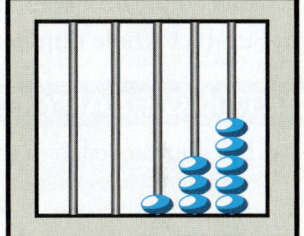

The number 135 shown on a suan-pan.

40. Suppose you were offered $20 an hour to teach a 6- or 7-year-old child how to subtract. How would you go about it? Where would you start?

41. Suppose you were offered $30 an hour to teach an adult how to subtract. Where would you start? What would you cover? Would you teach an adult any differently than you would teach a child?

Section R.3 — Multiplication and Division of Whole Numbers

OBJECTIVES *When you complete this section, you will be able to:*

A Find the product of two numbers.
B Know the multiplication facts through 12 by 12.
C Use an algorithm to multiply in standard notation.
D Recognize and apply the relationship between multiplication and division.
E Recognize that division by zero is not possible.
F Divide whole numbers.

PREREQUISITE SKILLS *Before beginning this section, you should be able to:*

a. Add whole numbers. (Section R.2)
b. Subtract whole numbers. (Section R.2)

GETTING READY FOR SECTION R.3

Add or subtract each of the following whole numbers:

1.
$$\begin{array}{r} 708 \\ +465 \\ \hline \end{array}$$

2. $43 + 43 + 43$

3.
$$\begin{array}{r} 54 \\ -37 \\ \hline \end{array}$$

4.
$$\begin{array}{r} 891 \\ -723 \\ \hline \end{array}$$

Introduction

So far, we have studied the addition and subtraction of whole numbers. In this section, we consider the operation of multiplication. Multiplication may be indicated in three ways: a raised dot ($2 \cdot 3 = 6$), parentheses ($2(3) = 6$), and with a "$\times$" ($2 \times 3 = 6$).

Multiplication may be thought of as repeated addition of the same number. To multiply two numbers, we use the first number as an addend the number of times specified by the second number. For example, $2 \cdot 3 = 2 + 2 + 2 = 6$.

OBJECTIVE Finding the product of two numbers

Definition The Product of Whole Numbers

The **product** of two whole numbers **a** and **b** is a whole number **c** that is the result of multiplying the **factor a** by the **factor b**. That is, $a \cdot b = c$.

In a typical multiplication statement, we have (factor) $\times$ (factor) $=$ (product). For example, 2 (factor) $\times$ 4 (factor) $=$ 8 (product). We need to memorize some basic products so that we have them at our fingertips when we need them. If you ever forget a product, you can always calculate it by using the concept of repeated addition. For example, $9 \times 8 = 9 + 9 + 9 + 9 + 9 + 9 + 9 + 9 = 72$.

Answers: Getting Ready: 1. 1173 **2.** 129 **3.** 17 **4.** 168

OBJECTIVE Knowing the multiplication facts through 12 × 12

Table of Products for Two Numbers

×	0	1	2	3	4	5	6	7	8	9	10	11	12
0	0	0	0	0	0	0	0	0	0	0	0	0	0
1	0	1	2	3	4	5	6	7	8	9	10	11	12
2	0	2	4	6	8	10	12	14	16	18	20	22	24
3	0	3	6	9	12	15	18	21	24	27	30	33	36
4	0	4	8	12	16	20	24	28	32	36	40	44	48
5	0	5	10	15	20	25	30	35	40	45	50	55	60
6	0	6	12	18	24	30	36	42	48	54	60	66	72
7	0	7	14	21	28	35	42	49	56	63	70	77	84
8	0	8	16	24	32	40	48	56	64	72	80	88	96
9	0	9	18	27	36	45	54	63	72	81	90	99	108
10	0	10	20	30	40	50	60	70	80	90	100	110	120
11	0	11	22	33	44	55	66	77	88	99	110	121	132
12	0	12	24	36	48	60	72	84	96	108	120	132	144

As in the addition fact table, the lines of numbers going up and down are called columns, and the lines of numbers going right and left are called rows. We find a product by choosing one factor in the first column and the second factor in the first row. Then we find where the row and column meet. For example, choose 4 in the first column and 6 in the first row. Run your finger down the first column until you reach 4. Move your finger to the right until you reach the column under 6 in the first row. You will find the number 24. Therefore, $4 \cdot 6 = 24$.

The **multiples** of a number are the results of multiplying the number by each of the whole numbers, starting with zero. Therefore, each column shows the multiples of the entry at the top of the column, and each row shows the multiples of the number at the beginning of the row. For example, the multiples of 8 are the following: 0, 8, 16, 24, 32, 40, 48, and so on. Thus, we have a handy summary of the first 13 multiples of the numbers 0 through 12.

OBJECTIVE Using an algorithm to multiply in standard notation

When multiplying vertically, the top number is the **multiplicand** and the bottom number is the **multiplier**. The process we normally use for multiplication is called the **multiplication algorithm**.

> **Procedure: Multiplication Algorithm**
>
> To find the product of two whole numbers, follow these steps:
>
> **1.** Line up the place values of the two factors.
>
> **2.** Starting from the right in the multiplier, multiply each digit of the multiplicand by that digit (carrying as necessary) and write the product directly beneath the multiplier.
>
> **3.** Multiply each digit of the multiplicand by the tens digit of the multiplier (carrying as necessary) and place the product beneath the previous product, but indented one place to the left.
>
> **4.** Continue until you have multiplied the multiplicand by all the digits of the multiplier.
>
> **5.** Sum the results.

Example 1   **Multiply 607 by 52, using the multiplication algorithm.**

Solution

Read from left to right and down.

$$\begin{array}{r} 607 \\ \times 52 \\ \hline 4 \end{array}$$

607   *Multiplicand.*
×52   *Multiplier.*
4   *$2 \cdot 7 = 14$. Write 4, carry 1 ten.*

$$\begin{array}{r} \overset{1}{607} \\ \times 52 \\ \hline 1214 \end{array}$$

$2 \cdot 600 = 1200$.
Write 2 in the hundreds place.
Write 1 in the thousands place.

$$\begin{array}{r} \overset{31}{607} \\ \times 52 \\ \hline 1214 \\ 35 \end{array}$$

Carry 3 hundreds.
$50 \cdot 0 = 0$.
$0 + 300 = 300$.
Put 3 in the hundreds column.

$$\begin{array}{r} \overset{1}{607} \\ \times 52 \\ \hline 14 \end{array}$$

Carry 1 ten.
$2 \cdot 0 = 0, 0 + 10 = 10$. Write 1 in the tens column.

$$\begin{array}{r} \overset{1}{607} \\ \times 52 \\ \hline 1214 \\ 5 \end{array}$$

$50 \cdot 7 = 350$.
Write 5 in tens column. Leave the ones place blank for 0 and carry 300.

$$\begin{array}{r} \overset{31}{607} \\ \times 52 \\ \hline 1214 \\ 3035 \\ \hline 31{,}564 \end{array}$$

$50 \cdot 600 = 3000$.
Put 0 in the thousands place and 3 in the ten thousands place.
Add to get product.

Therefore, 607 multiplied by 52 is 31,564.

PRACTICE EXERCISE 1

Multiply.

a. $\begin{array}{r} 532 \\ \times 48 \\ \hline \end{array}$

b. $\begin{array}{r} 857 \\ \times 602 \\ \hline \end{array}$

Application problems often involve multiplication.

Example 2   **Solve the following:**

a. The maximum seating capacity of a commuter train car is 64 passengers. What is the largest number of seated passengers in a six-car train?

Solution

We have six groups with 64 members each; therefore, we multiply 64 by 6. Using the multiplication algorithm, we have $6 \cdot 64 = 384$.

b. There are 22 rows of 14 acoustic tiles on the ceiling of classroom 2206. What is the total number of tiles in the ceiling?

Solution

We have 22 groups of 14 members each; therefore, we multiply 22 by 14. Using the multiplication algorithm, we have

$$\begin{array}{r} 22 \\ \times 14 \\ \hline 88 \\ 22 \\ \hline 308 \text{ tiles} \end{array}$$

Answers: Practice Exercise 1: a. 25,536 **b.** 515,914

PRACTICE EXERCISE 2

Solve the following:

a. A word processor is set for 58 lines per page. If you type five pages, how many lines will there be?

b. The Agricultural and Mechanical University marching band has 12 rows of 14 musicians each. How many musicians are in the marching band?

Now we consider the operation of division. There are several ways we can symbolize division. All of them mean the same, but are used in different situations. Consider 15 divided by 3. In each of the four cases,

$$15 \div 3 = 5, \qquad 15/3 = 5, \qquad \frac{15}{3} = 5, \qquad 3\overline{)15}^{\,5}$$

we call 3 the **divisor**, 15 the **dividend**, and 5 the **quotient**.

OBJECTIVE Recognizing and applying the relationship between multiplication and division

There is a very close relationship between multiplication and division. When we say $5 \cdot 3 = 15$, we mean that there are three groups of 5 in 15. On the other hand, $15 \div 5$ means "How many groups of 5 are in 15?" We know the answer from the multiplication. There are three groups of 5 in 15. Therefore, $15 \div 5 = 3$.

Each division has a related multiplication:

If $\frac{20}{4} = 5$, then $4 \cdot 5 = 20$. Since $\frac{20}{5} = 4$, then $5 \cdot 4 = 20$.

Each multiplication has two related divisions:

If $3 \cdot 6 = 18$, then both 3 and 6 will divide into 18.

That is, $\frac{18}{3} = 6$ and $\frac{18}{6} = 3$.

OBJECTIVE Recognizing that division by zero is not possible

A very special case arises if we consider **zero as a divisor**. What does $2 \div 0$ mean if it refers to how many groups of 0 there are in 2? Consider some popular answers.

$2 \div 0 = 0$?	The related multiplication is $2 = 0 \cdot 0 = 0$. Does $2 = 0$? No, so $2 \div 0 \neq 0$.
$2 \div 0 = 2$?	The related multiplication is $2 = 2 \cdot 0 = 0$. Does $2 = 0$? No, so $2 \div 0 \neq 2$.

Regardless of what number we let equal $2 \div 0$, the related multiplication will always result in the statement $2 = 0$, which is impossible. Therefore, we know that **division by zero is impossible**, or **undefined**. Now we can make a formal definition of division.

> ### Definition Division of Whole Numbers
> The **quotient** of two whole numbers, **a** and **b** (**b** not zero), is a whole number **c** such that $b \cdot c = a$. The number **a** is called the dividend, and the number **b** is called the divisor, and the number **c** is called the quotient. Using variables, if $a \div b = c$, then $b \cdot c = a$.

Substituting numbers for the variables, if $30 \div 6 = 5$, then $6 \cdot 5 = 30$.

Answers: Practice Exercise 2: a. 290 **b.** 168

OBJECTIVE Dividing whole numbers

Since multiples are so important to the division process, you should go back and review the multiplication table on page 19.

One process for division uses multiples of the divisor. For example, the multiples of 7 are 7, 14, 21, 28, 35, and so on, and the multiples of 6 are 6, 12, 18, 24, 30, 36, and so on. We will be looking for the largest multiple of the divisor that is less than or equal to the dividend. The quotient is the number you would multiply by the divisor in order to get the multiple previously mentioned. If there is no multiple of the divisor that equals the dividend exactly, then find the largest multiple of the divisor that is less than the dividend. The difference between the dividend and this multiple is the **remainder**. To check division, divisor × quotient + remainder = dividend.

Example 3   Find the quotient by using multiples, and check the answer.

a. $21 \div 7$

Solution

$$\begin{array}{r} 3 \\ 7\overline{)21} \\ -21 \\ \hline 0 \end{array}$$

The smallest multiple of 7 less than or equal to 21 is $3 \cdot 7 = 21$. So the quotient is 3.

Subtract $3 \cdot 7 = 21$.

Remainder.

CHECK:

$7 \cdot 3 = 21$

b. $27 \div 6$

Solution

$$\begin{array}{r} 4\ \text{R}\ 3 \\ 6\overline{)27} \\ -24 \\ \hline 3 \end{array}$$

The largest multiple of 6 less than or equal to 27 is $4 \cdot 6 = 24$. So the quotient is 4.

Subtract $4 \cdot 6 = 24$.

Remainder.

CHECK:

$(6 \cdot 4) + 3 = 24 + 3 = 27$

PRACTICE EXERCISE 3

Find the quotient by using multiples, and check by using a related multiplication.

a. $45 \div 9$

b. $35 \div 8$

Now we look at division of whole numbers. Be sure to follow the steps in the number that they are ordered and not the vertical position of the step.

Example 4   Find the quotient.

a. 725 and 5

Solution

$$\begin{array}{r} 1 \\ 5\overline{)725} \\ -5 \\ \hline 2 \end{array}$$

2. There is one 5 in 7. Put 1 in the hundreds place.
1. How many 5's are there in 7?
3. Multiply 1 times 5.
4. Subtract 5 from 7.

$$\begin{array}{r} 14 \\ 5\overline{)725} \\ -5 \\ \hline 22 \\ -20 \\ \hline 2 \end{array}$$

6. There are four 5's in 22. Put 4 in the tens place.
5. Put the 2 in the tens place. This is called **bringing down**. How many 5's are in 22?
7. Multiply 4 times 5.
8. Subtract 20 from 22.

$$\begin{array}{r} 145 \\ 5\overline{)725} \\ -5 \\ \hline 22 \\ -20 \\ \hline 25 \\ -25 \\ \hline 0 \end{array}$$

10. There are five 5's in 25. Put 5 in the ones place.
9. Bring down the 5. How many 5's are in 25?
11. Multiply 5 times 5.
12. Subtract 25 from 25. Remainder is 0.

Note

In division, we must be careful to line up corresponding columns that represent the same place values. We thus avoid making careless errors.

CHECK:

$5 \cdot 145 = 725$

Therefore, there are 145 fives in 725; or 725 divided into 5 parts has 145 in each part with none left over.

---

b. 12 divided into 456

c. How many 86's are in 3893?

Solution

$12\overline{)456}$ *2. There are no 12's in 4. Leave the hundreds place blank.*

1. How many 12's are in 4?

$\begin{array}{r} 3 \\ 12\overline{)456} \\ -36 \\ \hline 9 \end{array}$ *4. There are three 12's in 45. Put 3 in the tens place.*

3. How many 12's are in 45?

5. Multiply 3 times 12.

6. Subtract 36 from 45.

$\begin{array}{r} 38 \\ 12\overline{)456} \\ -36 \\ \hline 96 \\ -96 \\ \hline 0 \end{array}$ *9. There are eight 12's in 96. 8 is written in the ones place.*

7. Put the 6 in the ones place. (Bring down the 6.)

8. How many 12's are in 96?

10. Multiply 12 times 8.

11. Subtract 96 from 96.

12. Remainder.

Therefore, $456 \div 12 = 38$.

CHECK:

$12 \cdot 38 = 456$

Solution

$86\overline{)3893}$ *1. There are no 86's in 3. Leave the thousands place blank.*

$86\overline{)3893}$ *2. How many 86's are in 38?*

3. There are no 86's in 38. Leave the hundreds place blank.

$\begin{array}{r} 4 \\ 86\overline{)3893} \\ -344 \\ \hline 45 \end{array}$ *5. 86 is close to 90, and 389 is close to 390, so we guess 4.*

4. How many 86's are in 389?

6. Multiply 86 times 4.

7. Subtract 344 from 389.

$\begin{array}{r} 45 \\ 86\overline{)3893} \\ -344 \\ \hline 453 \\ -430 \\ \hline 23 \end{array}$ *10. 86 is close to 90, and 453 is close to 450, so we guess 5.*

8. Bring down the 3.

9. How many 86's are in 453?

11. Multiply 86 times 5.

12. Subtract 430 from 453.

13. Remainder.

Therefore, $3893 \div 86 = 45\,R\,23$.

CHECK:

$(86 \cdot 45) + 23 = 3870 + 23 = 3893$

Note

Sometimes we may make a mistake and end up with a remainder greater than the divisor. If this happens, just increase the last digit in the quotient until the resulting multiplication yields a remainder less than the divisor.

PRACTICE EXERCISE 4

Find each of the quotients:

a. $493 \div 4$ **b.** $873 \div 9$ **c.** $3185 \div 49$ **d.** $1652 \div 15$

We now consider the solution of application problems involving division.

Answers: **Practice Exercise 4:** **a.** 123 R 1 **b.** 97 **c.** 65 **d.** 110 R 2

Example 5 | **Solve.**

a. One version of the game of dominoes requires that all 28 game pieces be drawn at the beginning of play. If three people are playing, how many pieces does each person get?

Solution

The dominoes are distributed to each player in turn until there are not enough dominoes for each player to receive one. They will go around 9 times with 1 left over. That is, $28 \div 3 = 9 \text{ R } 1$. So, each player receives 9 pieces, and the 1 left over is put aside and not used for play.

b. The area of a standard parking space is 162 square feet (18 feet deep by 9 feet wide). How many of these spaces can we get into a rectangular lot whose area is 6156 square feet (36 feet by 171 feet)?

Solution

To find the number of parking spaces, divide the area of one space into the total area available. In other words, divide 162 into 6156.

$$
\begin{array}{r}
38 \\
162\overline{)6156} \\
-486 \\
\hline
1296 \\
-1296 \\
\hline
\end{array}
$$

Therefore, there is room for 38 parking spaces.

PRACTICE EXERCISE 5

a. Michael earns $25 each time he mows a lawn. How many lawns would he have to mow to earn the $300 needed to buy a used computer?

b. It is 700 miles from Chicago to Kansas City. If Laura's car gets 35 miles to the gallon on the highway, how many gallons of gasoline would she use on the trip?

Exercise Set R.3   For Extra Help   MyMathLab®

Multiply each of the following. (See Example 1.)

1. 4(17)   **2.** 6(14)   **3.** 5(46)   **4.** 3(58)

5. 2(79)   **6.** 9(87)   **7.** 6(299)   **8.** 7(541)

Multiply each of the following. (See Example 1.)

9.   65   **10.**   45   **11.**   27   **12.**   88
      ×8   ×6   ×38   ×42

Answers: Practice Exercise 5: a. 12 lawns **b.** 20 gallons

13. 593
    ×20

14. 336
    ×70

15. 102
    ×86

16. 250
    ×66

17. 508
    ×204

18. 709
    ×306

19. 656
    ×331

20. 591
    ×227

Solve each problem. You may need to do more than just multiply. Be careful—there may be extra information. (See Example 2.)

21. The new dormitory at University of the North needs to order chairs for 230 rooms. If they need three chairs per room, how many chairs should they order?

22. The main library has 60 tables with 16 chairs each in the undergraduate study area. How many students can be seated at one time?

23. Alice is in the real estate business. Last year, she sold eight tract houses for a builder. If her commission was $4200 per house, how much did she make?

24. Including cement, blocks, and labor, it costs $4 a block to build a wall. If the wall is 16 blocks high and 263 blocks long, how many blocks are in the wall?

25. Looking up at a downtown office building, Carlos counted eight windows for each of the 12 floors of the building. How many windows did he count?

26. The committee for the annual Easter egg hunt ordered 150 dozen eggs for the big event this year. If 800 children show up, will this be enough eggs to have at least 1 per child?

27. A computer spreadsheet has 15 columns across the paper and 35 rows down the paper. How many possible entries can there be?

28. The northwest section of a large city is laid out in streets and avenues. The city stretches 249 blocks north from Main Street and goes 120 blocks west from Prime Avenue. How many city blocks are there?

Divide by using multiples of the divisor and check by using related multiplication. (See Example 3.)

29. $72 \div 9$

30. $64 \div 8$

31. $86 \div 12$

32. $93 \div 11$

33. $57 \div 6$

34. $37 \div 4$

35. $96 \div 10$

36. $72 \div 11$

Divide and check by using the related multiplication. (See Example 4.)

37. $4\overline{)72}$

38. $7\overline{)91}$

39. $5\overline{)594}$

40. $6\overline{)249}$

41. $14\overline{)298}$

42. $23\overline{)534}$

43. $67\overline{)979}$

44. $93\overline{)788}$

45. $237\overline{)5430}$

46. $385\overline{)4964}$

Solve by performing the necessary operation(s). There may be more than one operation needed, and there may be more information than needed. (See Example 5.)

47. Hank wishes to serve a snack at a nursery school. He has only 4-ounce cups. How many children can he serve from a half gallon (64-oz) jug of orange juice?

48. Mass transit needs to move 30,000 people to and from a downtown parade. With standing room only, one car of the train will hold 200 people. How many train cars do they need to carry the 30,000 people downtown?

49. Four students are going to share equally in the expenses of a trip to New Orleans for the Mardi Gras celebration. If it costs $240 for gasoline, tolls, and parking, how much should each student pay?

50. Five children inherited $12,000 from their mother. The estate is to be shared equally. How much will each child get?

51. Mary is selling flowers for Mother's Day. If she buys 1 gross (144) of flowers, how many bunches of six flowers each can she make?

52. An automobile dealer has ordered 117 new cars for the beginning of the model year. If the delivery rigs hold 9 cars each, how many loads will it take to deliver all of the new cars?

53. Seven students decided to have lunch at a pizza parlor. They shared three pizzas, which along with the drinks cost $35. How much did each student have to pay?

54. The four Diaz brothers have a large paper route of 136 customers. If they split the route equally, how many papers does each brother have to deliver?

55. Winton is a fat cat. If he eats 6 ounces of dry cat food a day, how long will a 2-pound (1 lb = 16 oz) bag last?

56. If 2 quarts (1 qt = 32 oz) of soda are divided equally between 4 people, how many ounces does each receive?

Challenge Exercises (57–61)

57. Alpha Club wishes to serve doughnuts at its next meeting. It estimates that each of the 35 expected members will eat two doughnuts. Will five dozen doughnuts be enough?

58. Mario has three 300-calorie workouts per week. Will this be enough to burn off his snack of two scoops of ice cream (180 calories each) and 4 ounces of potato chips (150 calories per ounce)?

59. There are five people in George's family: two adults, and three children ages 3, 7, and 11. The family decided to visit an amusement park and buy the three-day pass. Adult passes are $120 per adult, and children's passes are $80 per child if the child is 9 years old or younger. How much admission cost will the family have to pay?

60. David is going to have a midnight study break with a snack. He eats 6 ounces of potato chips at 160 calories per ounce, and five cookies at 120 calories each, and drinks two sodas at 350 calories each. How many calories does he consume?

61. Rosie is making a counted cross-stitch picture having a design that is 110 by 88 stitches. If she is stitching on material that requires 11 stitches per inch, what size frame should she buy?

62. A homeowner wants to fertilize his lawn. His lot measures 120 feet by 110 feet, and his house covers an area of 3500 square feet. If a bag of fertilizer covers 8000 square feet, how many bags of fertilizer must the homeowner buy?

Writing Exercises (63–67)

63. The Romans did not have algorithms for multiplication. However, they did multiply numbers. How do you suppose they did it?

64. All computers and almost all calculators can perform only one operation—addition. Subtraction and multiplication are done through addition. How do you think this is possible?

65. How are multiplication and division related?

66. How do 0 ÷ 3 and 3 ÷ 0 differ?

67. How are subtraction, multiplication, and division all related to addition?

Section R.4 A Brief Introduction to Fractions

OBJECTIVES *When you complete this section, you will be able to:*

A Write a fraction that represents a given part of a whole (unit).
B Identify the part of the whole (unit) that is represented by a fraction.
C Change an improper fraction into a mixed number and a mixed number into an improper fraction.
D Reduce a fraction to lowest terms.
E Multiply two fractions.
F Divide two fractions.
G Add and subtract fractions with the same denominators.

PREREQUISITE SKILLS *Before beginning this section, you should be able to:*

a. Add and subtract whole numbers. (Section R.2)
b. Multiply and divide whole numbers. (Section R.3)
c. Recognize situations that call for addition, subtraction, multiplication, or division. (Sections R.1–R.3)
d. Know the names of the results of the four basic operations of arithmetic. (Sections R.1–R.3)

GETTING READY FOR SECTION R.4

Add or subtract the following whole numbers:

1. $8 + 7 + 3$ **2.** $54 - 39$

Multiply or divide the following whole numbers:

3. $(6)(4)$ **4.** $14 \div 2$

5. Match one of the terms *sum*, *difference*, *product*, and *quotient* with the results of each of the following operations:

a. $(7)(4)$ **b.** $5 + 4$ **c.** $8 \div 4$ **d.** $9 - 2$

Introduction

In this section, we will give a very brief introduction to fractions. We will discuss how to multiply, divide, add, and subtract two fractions. We need just the basic concepts in order to develop additional ideas in later sections of Chapter R and in Chapter 2. A thorough discussion of fractions occurs in Chapters 6 and 7.

OBJECTIVE Writing a fraction that represents a given part of the whole

Fractions can be used to indicate division, but in this section we look at fractions from a different point of view. A **fraction** is a number in the form $\frac{a}{b}$ with $b \neq 0$. The number a is called the **numerator**, and b is the **denominator**. If an object (unit) is divided into a number of equal parts and we select some of those parts, we can represent the part of the object selected by using a fraction. The numerator represents the number of equal parts selected, and the denominator represents the number of equal parts into which the object has been divided.

Example 1 **Using a fraction, represent the part of the whole region that is shaded.**

a.
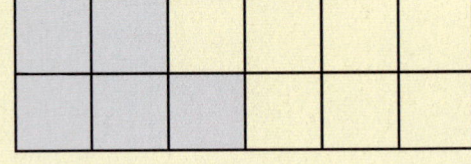
1 unit

Three of the ten equal parts are shaded. Therefore, the numerator is 3 and the denominator is 10. So the fraction is written as $\frac{3}{10}$ and is read as "three-tenths."

b.

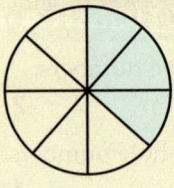
1 unit

Two of the four equal parts are shaded. Therefore, the numerator is 2 and the denominator is 4. So the fraction is written as $\frac{2}{4}$ and is read as "two-fourths."

PRACTICE EXERCISE 1

Using a fraction, represent the part of the whole region that is shaded and write the fraction in words.

a.

1 unit

b.

1 unit

OBJECTIVE Ⓑ **Identifying the part of the whole that is represented by a fraction**

The opposite of representing part of a whole by a fraction is identifying the part of the whole that is represented by a fraction.

Example 2 **Shade the part of the whole region that is represented by the given fraction.**

a. $\frac{1}{4}$
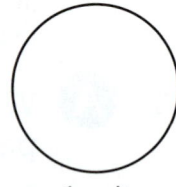
1 unit

The denominator, 4, tells us to divide the region into four equal parts. The numerator, 1, tells us to shade any one of the four smaller regions.

b. $\frac{5}{8}$
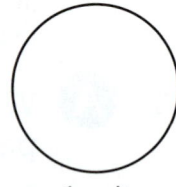
1 unit

The denominator tells us to divide the circle into eight equal regions. The numerator tells us to shade any five of the regions.

Solution
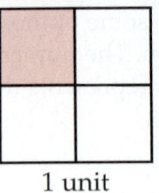
1 unit

Solution
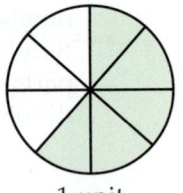
1 unit

Answers: Practice Exercise 1: a. $\frac{5}{12}$, five-twelfths **b.** $\frac{3}{8}$, three-eighths

PRACTICE EXERCISE 2

Shade the part of the whole that is represented by each of the following fractions:

a. $\frac{7}{15}$

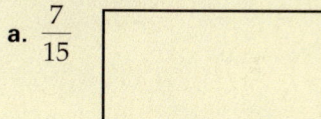

1 unit

b. $\frac{2}{3}$

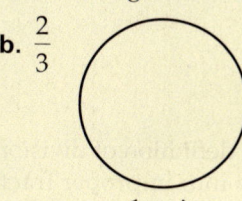

1 unit

OBJECTIVE Changing an improper fraction into a mixed number and a mixed number into an improper fraction

A fraction whose numerator is greater than or equal to its denominator is called **improper**. A **mixed number** is greater than 1 and has a whole-number part and a fractional part. Remember from Section R.3 that when dividing whole numbers, the remainder can be put over the divisor. If the dividend is larger than the divisor, this procedure results in a mixed number. Therefore, this gives us a procedure for changing improper fractions into mixed numbers.

Example 3   **Change each of the improper fractions into mixed numbers:**

a. $\frac{11}{4}$

Solution

$\frac{11}{4}$ means $11 \div 4$. We will divide 4 into 11 and put the remainder over 4.

$$
\begin{array}{r}
2 \\
4\overline{)11} \\
-8 \\
\hline
3
\end{array}
$$ 3 is the remainder.

Therefore, $\frac{11}{4} = 2\frac{3}{4}$.

b. $\frac{69}{5}$

Solution

$\frac{69}{5}$ means $69 \div 5$. We will divide 5 into 69 and put the remainder over 5.

$$
\begin{array}{r}
13 \\
5\overline{)69} \\
-5 \\
\hline
19 \\
-15 \\
\hline
4
\end{array}
$$ The remainder is 4.

Therefore, $\frac{69}{5} = 13\frac{4}{5}$.

Answers: Practice Exercise 2: a. 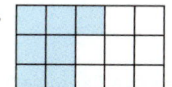   **b.**

PRACTICE EXERCISE 3

Change each of the improper fractions into a mixed number:

a. $\dfrac{23}{6}$

b. $\dfrac{30}{13}$

The definition of division provides us with an easy method for converting mixed numbers into improper fractions. In Example 3a, the improper fraction $\frac{11}{4}$ was converted into the mixed number $2\frac{3}{4}$ by division. The procedure for checking the answer to a division problem is to multiply the divisor, 4, by the quotient, 2, and add the remainder, 3. This results in the dividend ($4 \cdot 2 + 3 = 11$). Note that if $\frac{11}{4}$ is written as the mixed number $2\frac{3}{4}$, the divisor, 4, is the denominator of the fractional part. The quotient, 2, is the whole part of the mixed number, and the remainder, 3, is the numerator of the fractional part. Therefore, to change a mixed number to an improper fraction, we multiply the denominator of the fractional part by the whole part and add the numerator of the fractional part. This result is then written as the numerator, and the denominator is the denominator of the fractional part. In symbols, $2\frac{3}{4} = \frac{4 \cdot 2 + 3}{4} = \frac{11}{4}$.

Example 4   **Change each of the mixed numbers into improper fractions:**

a. $4\dfrac{2}{3}$

b. $5\dfrac{7}{8}$

Solution

Multiply the denominator, 3, by the whole part, 4, and add the numerator, 2. Put the result over the denominator, 3. Therefore, $4\frac{2}{3} = \frac{3 \cdot 4 + 2}{3} = \frac{14}{3}$.

Solution

Multiply the denominator, 8, by the whole part, 5, and add the numerator, 7. Therefore, $5\frac{7}{8} = \frac{8 \cdot 5 + 7}{8} = \frac{47}{8}$.

PRACTICE EXERCISE 4

Change each of the mixed numbers into improper fractions:

a. $6\dfrac{3}{4}$

b. $11\dfrac{4}{9}$

OBJECTIVE Reducing a fraction to lowest terms

In Example 1b, we shaded $\frac{2}{4}$ of a circle. Note that $\frac{2}{4}$ of the circle is also $\frac{1}{2}$ of the circle. When $\frac{2}{4}$ is rewritten as $\frac{1}{2}$, we say that the fraction has been **reduced to lowest terms**.

Note

In Chapter 6, we will give a definition of lowest terms and a procedure for reducing fractions to lowest terms that are more consistent with the algebraic approach.

Definition Reduced to Lowest Terms

A fraction is reduced to lowest terms if no whole number (other than 1) divides evenly into both the numerator and denominator.

Answers: Practice Exercise 3: a. $3\dfrac{5}{6}$ **b.** $2\dfrac{4}{13}$ **Practice Exercise 4: a.** $\dfrac{27}{4}$ **b.** $\dfrac{103}{9}$

To reduce a fraction to lowest terms, divide both the numerator and denominator by the largest whole number that divides evenly into both.

Example 5   Reduce to lowest terms.

a. $\dfrac{8}{12}$

b. $\dfrac{9}{12}$

Solution

The largest number that divides evenly into 8 and 12 is 4, so divide both the numerator and denominator by 4.

$$\frac{8}{12} = \frac{8 \div 4}{12 \div 4} = \frac{2}{3}$$

Solution

The largest number that divides evenly into 9 and 12 is 3, so divide both the numerator and denominator by 3.

$$\frac{9}{12} = \frac{9 \div 3}{12 \div 3} = \frac{3}{4}$$

PRACTICE EXERCISE 5

Reduce to lowest terms.

a. $\dfrac{3}{15}$

b. $\dfrac{12}{16}$

OBJECTIVE Multiplying two fractions

Multiplication of fractions can be represented by shading regions. We can represent $\frac{1}{5}$ by dividing a region into five equal parts and shading one of those parts, as shown in the following diagram:

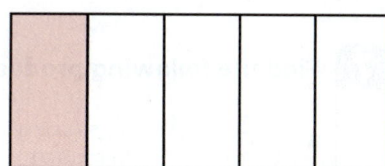

Suppose we now wish to further shade $\frac{1}{2}$ of the $\frac{1}{5}$ that is already shaded. We divide each of the five parts into two equal parts and further shade one of the two shaded parts.

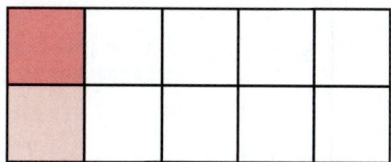

By dividing each of the five equal parts into two equal parts, the entire region has been divided into ten equal parts. Therefore, the one part that was further shaded represents $\frac{1}{10}$ of the entire region. In words, we would say that $\frac{1}{2}$ of $\frac{1}{5}$ is $\frac{1}{10}$. When performing operations with fractions, the word "of" is used to indicate multiplication. Consequently, $\frac{1}{2} \cdot \frac{1}{5} = \frac{1}{10}$. Note, this is the same as $\frac{1}{2} \cdot \frac{1}{5} = \frac{1 \cdot 1}{2 \cdot 5} = \frac{1}{10}$. Let us repeat this procedure with a more complicated example.

Answers: Practice Exercise 5: **a.** $\dfrac{1}{5}$ **b.** $\dfrac{3}{4}$

Example 6   Represent $\frac{2}{3}$ of $\frac{2}{5}$ by using shaded regions and give the result as a fractional part of the entire region.

Solution

The fraction $\frac{2}{5}$ means divide the region (unit) into five equal parts and shade two.

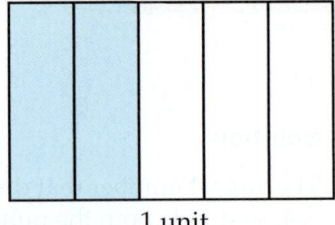

1 unit

To find $\frac{2}{3}$ of the two parts that are shaded, we will divide each of the five parts into three equal parts. Then we further shade two parts of each of the three-part sections that were already shaded.

Notice that the entire figure is now divided into fifteen equal parts, of which four have been further shaded. This represents $\frac{4}{15}$ of the entire figure. Therefore, $\frac{2}{3}$ of $\frac{2}{5}$ is $\frac{4}{15}$. As multiplication, $\frac{2}{3} \cdot \frac{2}{5} = \frac{4}{15}$. Note that this can be computed as $\frac{2}{3} \cdot \frac{2}{5} = \frac{2 \cdot 2}{3 \cdot 5} = \frac{4}{15}$.

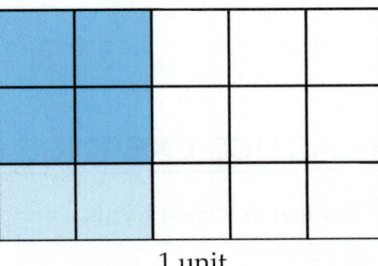

1 unit

Based on the preceding examples, we are now ready to state the procedure for multiplication of fractions.

Rule: Multiplication of Fractions

For all numbers a, b, c, and d, with b and $d \neq 0$, $\dfrac{a}{b} \cdot \dfrac{c}{d} = \dfrac{a \cdot c}{b \cdot d}$. In words, to multiply two fractions, find the product of the numerators divided by the product of the denominators.

Example 7   Find the following products:

a. $\dfrac{3}{5} \cdot \dfrac{2}{7} =$ *Find the product of the numerators divided by the product of the denominators.*

$\dfrac{3 \cdot 2}{5 \cdot 7} =$ *Multiply.*

$\dfrac{6}{35}$ *Product.*

b. $\dfrac{2}{3} \cdot \dfrac{3}{4} =$ *Find the product of the numerators divided by the product of the denominators.*

$\dfrac{2 \cdot 3}{3 \cdot 4} =$ *Multiply.*

$\dfrac{6}{12}$ *Reduce to lowest terms.*

$\dfrac{1}{2}$ *Product.*

c. $8 \cdot \dfrac{3}{4} =$ $8 = \frac{8}{1}.$

$\dfrac{8}{1} \cdot \dfrac{3}{4} =$ *The product of the numerators divided by the product of the denominators.*

$\dfrac{8 \cdot 3}{1 \cdot 4} =$ *Multiply.*

$\dfrac{24}{4} =$ $24 \div 4 = 6.$

6 *Product.*

PRACTICE EXERCISE 7

Find the following products:

a. $\dfrac{2}{3} \cdot \dfrac{4}{9}$

b. $\dfrac{1}{3} \cdot \dfrac{3}{7}$

c. $9 \cdot \dfrac{2}{3}$

Before we discuss division of fractions, we need to introduce the **multiplicative inverse**, also called the **reciprocal**. Two numbers are multiplicative inverses if their product is 1. For example, the multiplicative inverse (reciprocal) of $\frac{2}{3}$ is $\frac{3}{2}$ since $\frac{2}{3} \cdot \frac{3}{2} = 1$. The multiplicative inverse of 2 is $\frac{1}{2}$ since $2 \cdot \frac{1}{2} = 1$. In general, for a and b with neither a nor b equal to 0, the multiplicative inverse of $\frac{a}{b}$ is $\frac{b}{a}$ since $\frac{a}{b} \cdot \frac{b}{a} = 1$. To find the multiplicative inverse of a number, we interchange the numerator and the denominator. This procedure is sometimes referred to as "inverting the fraction."

OBJECTIVE (F) Dividing two fractions

Suppose a region is divided into two equal parts so that each part is $\frac{1}{2}$ of the original region, as shown in the following diagram:

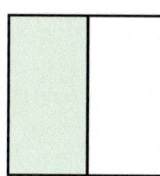

Now take the $\frac{1}{2}$ that is shaded and *divide* it into two equal parts.

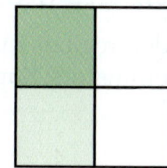

From the figure, we can see that one of the two equal shaded parts is $\frac{1}{4}$ of the original region. In symbols, $\frac{1}{2} \div 2 = \frac{1}{4}$. Notice that $\frac{1}{2} \cdot \frac{1}{2} = \frac{1}{4}$ also. Consequently, $\frac{1}{2} \div 2 = \frac{1}{2} \cdot \frac{1}{2}$. Without shading parts of a figure, we can see that $6 \div 3 = 2$ and $6 \cdot \frac{1}{3} = 2$, so $6 \div 3 = 6 \cdot \frac{1}{3}$.

Answers: Practice Exercise 7: a. $\dfrac{8}{27}$ **b.** $\dfrac{1}{7}$ **c.** 6

This suggests that dividing by a whole number, except for 0, is the same as multiplying by its multiplicative inverse (reciprocal).

The preceding suggests the following rule for the division of fractions:

Rule: Division of Fractions

For all numbers a, b, c, and d, with b, c, and $d \neq 0$, $\dfrac{a}{b} \div \dfrac{c}{d} = \dfrac{a}{b} \cdot \dfrac{d}{c}$. In words, to divide by a number, multiply by its multiplicative inverse (reciprocal).

This rule is usually stated more loosely as "When dividing two fractions, invert the fraction on the right and change the operation to multiplication." It is this form of the rule that we apply in the following examples:

Example 8 **Find the following quotients:**

a. $\dfrac{2}{3} \div \dfrac{3}{5} =$ *Invert the fraction on the right and change the operation to multiplication.*

$\dfrac{2}{3} \cdot \dfrac{5}{3} =$ *Multiply numerator times numerator and denominator times denominator.*

$\dfrac{2 \cdot 5}{3 \cdot 3} =$ *Multiply.*

$\dfrac{10}{9}$ *Quotient.*

b. $\dfrac{3}{7} \div 8$ $8 = \frac{8}{1}$.

$\dfrac{3}{7} \div \dfrac{8}{1} =$ *Invert the fraction on the right and change the operation to multiplication.*

$\dfrac{3}{7} \cdot \dfrac{1}{8} =$ *Multiply numerator times numerator and denominator times denominator.*

$\dfrac{3 \cdot 1}{7 \cdot 8} =$ *Multiply.*

$\dfrac{3}{56} =$ *Quotient.*

PRACTICE EXERCISE 8

Find the following quotients:

a. $\dfrac{3}{4} \div \dfrac{2}{7}$

b. $\dfrac{2}{9} \div 5$

OBJECTIVE Adding and subtracting fractions with the same denominator

Shaded regions may also be used to illustrate the addition of fractions with the same denominators. Suppose $\frac{3}{8}$ of a region is shaded.

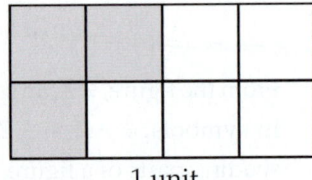

1 unit

Answers: Practice Exercise 8: a. $\frac{21}{8}$ **b.** $\frac{2}{45}$

Now shade another $\frac{4}{8}$ of the same region.

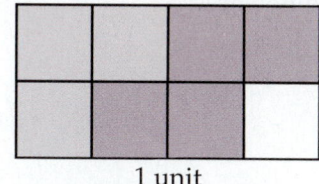

1 unit

Observe that $\frac{7}{8}$ of the region is now shaded. In other words, $\frac{3}{8} + \frac{4}{8} = \frac{3+4}{8} = \frac{7}{8}$. Note that the denominators of the fractions are the same and only the numerators are added. This leads to the following rule for adding fractions with like denominators:

> ### Rule: Addition of Fractions with the Same Denominators
>
> For all numbers a, b, and c, with $c \neq 0$, $\dfrac{a}{c} + \dfrac{b}{c} = \dfrac{a+b}{c}$. In words, to add two or more fractions with the same denominator, add the numerators and place the sum over the same denominator.

It can easily be shown that a similar rule is used to subtract fractions. The rule is given in the following box:

> ### Rule: Subtraction of Fractions with the Same Denominators
>
> For all numbers a, b, and c, with $c \neq 0$, $\dfrac{a}{c} - \dfrac{b}{c} = \dfrac{a-b}{c}$. In words, to subtract two fractions with the same denominators, subtract the numerators and place the difference over the same denominator.

Example 9   Find the following sums or differences:

a. $\dfrac{1}{5} + \dfrac{3}{5} =$ *The denominators are the same, so add the numerators and place the sum over the same denominator.*

$\dfrac{1+3}{5} =$ *Add 1 and 3.*

$\dfrac{4}{5}$ *Sum.*

b. $\dfrac{7}{9} - \dfrac{2}{9} =$ *The denominators are the same, so subtract the numerators and place the difference over the same denominator.*

$\dfrac{7-2}{9} =$ *Subtract 7 and 2.*

$\dfrac{5}{9}$ *Difference.*

c. $\dfrac{2}{13} + \dfrac{4}{13} + \dfrac{1}{13} =$ *The denominators are the same, so add the numerators and place the sum over the same denominator.*

$\dfrac{2+4+1}{13} =$ *Add 2, 4, and 1.*

$\dfrac{7}{13}$ *Sum.*

d. $\dfrac{2}{11} + \dfrac{5}{11} - \dfrac{4}{11} =$ *The denominators are the same, so add and subtract the numerators and place the sum and difference over the same denominator.*

$\dfrac{2+5-4}{11} =$ *Add 2 and 5; then subtract 4.*

$\dfrac{3}{11}$ *Answer.*

PRACTICE EXERCISE 9

Find the following sums or differences:

a. $\dfrac{4}{11} + \dfrac{2}{11}$

b. $\dfrac{11}{15} - \dfrac{4}{15}$

c. $\dfrac{5}{17} + \dfrac{3}{17} + \dfrac{6}{17}$

d. $\dfrac{3}{10} + \dfrac{5}{10} - \dfrac{1}{10}$

Example 10 | **Solve the following:**

a. To make a circuit board requires $\frac{2}{15}$ of an ounce of silver. How many circuit boards could be made using 1 pound (1 pound = 16 ounces) of silver?

Solution

To find the number of circuit boards, divide the number of ounces of silver available by the number of ounces needed to make one circuit board.

$16 \div \dfrac{2}{15} =$ *Invert $\frac{2}{15}$ and multiply.*

$16 \cdot \dfrac{15}{2} =$ *Multiply.*

$\dfrac{16 \cdot 15}{2} =$ *$16 \cdot 15 = 240$.*

$\dfrac{240}{2} =$ *$240 \div 2 = 120$.*

120 *There is enough silver for 120 circuit boards.*

b. A recipe for a cake calls for $\frac{2}{3}$ of a cup of chopped nuts. Find the number of cups of nuts needed for nine such cakes.

Solution

To find the number of cups of nuts needed for nine cakes, multiply the number of cups per cake by the number of cakes. Hence,

$\dfrac{2}{3} \cdot 9 =$ *Think of 9 as $\frac{9}{1}$.*

$\dfrac{2}{3} \cdot \dfrac{9}{1} =$ *Multiply the fractions.*

$\dfrac{2 \cdot 9}{3 \cdot 1} =$ *Multiply.*

$\dfrac{18}{3} =$ *Divide.*

6 *Therefore, 6 cups of nuts are required for the nine cakes.*

---

c. The water level rose $\frac{5}{8}$ of an inch on Thursday and $\frac{3}{8}$ of an inch on Friday. How much did it rise in two days?

Solution

To find the total amount it rose, we add the amount it rose on Thursday to the amount it rose on Friday.

$\frac{5}{8} + \frac{3}{8} =$ *Add the numerators and keep the common denominator.*

$\frac{5 + 3}{8} =$ *Add.*

$\frac{8}{8} =$ *Divide.*

1 *Therefore, the total amount it rose was 1 inch.*

d. The battery on a cell phone was at $\frac{11}{15}$ of a full charge. After 3 hours of usage, the battery's charge was reduced by $\frac{7}{15}$ of a charge. How much charge was remaining?

Solution

To find the battery's remaining charge, we reduce, or subtract, the expended charge from the beginning charge.

$\frac{11}{15} - \frac{7}{15} =$ *Subtract the numerators and keep the common denominator.*

$\frac{11 - 7}{15} =$ *Subtract.*

$\frac{4}{15}$ *Therefore, the remaining charge was $\frac{4}{15}$ of a full charge.*

PRACTICE EXERCISE 10

Solve the following:

a. A baby bottle holds $\frac{1}{8}$ of a quart of formula. If a case of formula contains 16 quarts, how many times can the bottle be filled from a case of formula?

b. A pharmacist fills a prescription for a high blood pressure medication. She uses $\frac{2}{3}$ of the pills in a bottle that is $\frac{2}{5}$ full. What part of a full bottle did she use?

c. It took Jennifer $\frac{3}{4}$ of an hour to complete her English homework and $\frac{5}{4}$ of an hour to complete her mathematics homework. How much time did she take to complete these two sets of homework?

d. At the beginning of a bicycle race, Jane had $\frac{15}{17}$ of a bottle of water. After riding the 10-kilometer race she had $\frac{7}{17}$ of a bottle left. How much water did she drink during the race?

Answers: Practice Exercise 10: a. 128 times **b.** $\frac{4}{15}$ **c.** 2 hr **d.** $\frac{8}{17}$ of a bottle

For Extra Help
Exercise Set R.4   MyMathLab®

Using a fraction, represent the part of the whole region that is shaded. (See Example 1.)

1.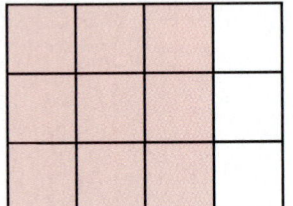

2.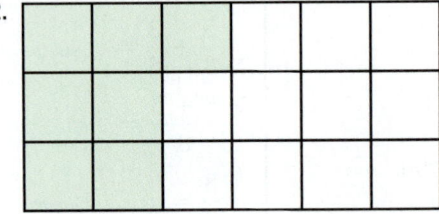

3.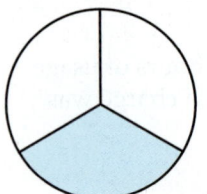

4.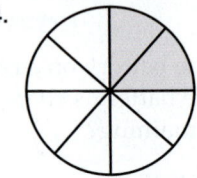

5.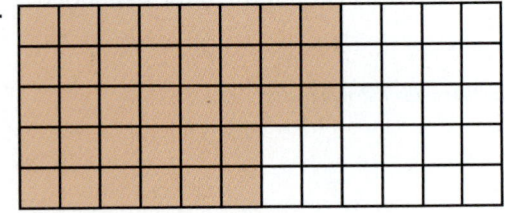

6.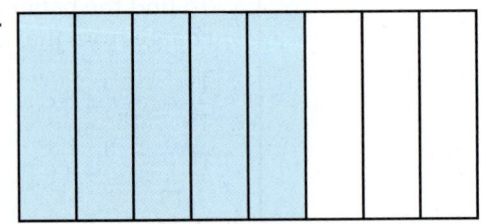

Shade the part of the whole that is represented by the given fraction. (See Example 2.)

7. $\dfrac{3}{4}$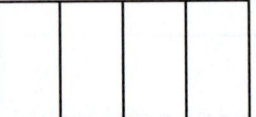

8. $\dfrac{4}{5}$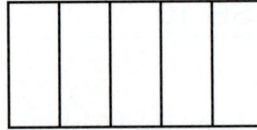

9. $\dfrac{3}{8}$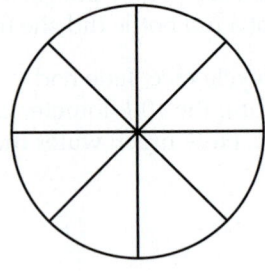

10. $\dfrac{2}{4}$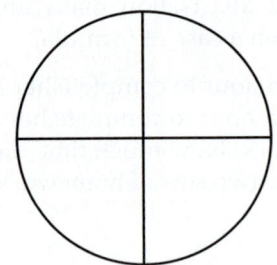

Change the improper fractions into mixed numbers. (See Example 3.)

11. $\dfrac{13}{5}$

12. $\dfrac{19}{6}$

13. $\dfrac{36}{7}$

14. $\dfrac{44}{9}$

15. $\dfrac{47}{13}$

16. $\dfrac{57}{22}$

17. $\dfrac{63}{29}$

18. $\dfrac{71}{34}$

Change the mixed numbers into improper fractions. (See Example 4.)

19. $4\dfrac{2}{5}$

20. $3\dfrac{3}{4}$

21. $5\dfrac{4}{9}$

22. $7\dfrac{3}{8}$

23. $6\dfrac{5}{12}$

24. $2\dfrac{8}{15}$

25. $12\dfrac{5}{6}$

26. $14\dfrac{7}{8}$

27. $14\dfrac{21}{23}$

28. $18\dfrac{31}{40}$

Reduce to lowest terms. (See Example 5.)

29. $\dfrac{6}{8}$

30. $\dfrac{6}{9}$

31. $\dfrac{16}{20}$

32. $\dfrac{18}{24}$

33. $\dfrac{16}{24}$

34. $\dfrac{20}{28}$

Find the following products. (See Example 7.)

35. $\dfrac{1}{3} \cdot \dfrac{1}{5}$

36. $\dfrac{1}{4} \cdot \dfrac{1}{6}$

37. $\dfrac{2}{5} \cdot \dfrac{4}{9}$

38. $\dfrac{3}{7} \cdot \dfrac{1}{4}$

39. $\dfrac{4}{7} \cdot \dfrac{3}{8}$

40. $\dfrac{5}{8} \cdot \dfrac{3}{10}$

41. $\dfrac{5}{12} \cdot \dfrac{7}{15}$

42. $\dfrac{4}{13} \cdot \dfrac{3}{10}$

43. $6 \cdot \dfrac{1}{3}$

44. $8 \cdot \dfrac{1}{4}$

45. $\dfrac{3}{5} \cdot 10$

46. $\dfrac{5}{6} \cdot 12$

Find the following quotients. (See Example 8.)

47. $3 \div \dfrac{1}{2}$

48. $\dfrac{1}{2} \div 3$

49. $\dfrac{3}{5} \div \dfrac{7}{4}$

50. $\dfrac{4}{9} \div \dfrac{3}{2}$

51. $\dfrac{1}{2} \div \dfrac{5}{9}$

52. $\dfrac{1}{4} \div \dfrac{4}{7}$

53. $\dfrac{7}{12} \div \dfrac{3}{4}$

54. $\dfrac{7}{8} \div \dfrac{3}{4}$

55. $\dfrac{2}{5} \div 3$

56. $\dfrac{5}{8} \div 6$

57. $7 \div \dfrac{7}{8}$

58. $8 \div \dfrac{8}{11}$

Find the following sums and differences. (See Example 9.)

59. $\dfrac{3}{7} + \dfrac{2}{7}$

60. $\dfrac{2}{15} + \dfrac{2}{15}$

61. $\dfrac{6}{17} + \dfrac{9}{17}$

62. $\dfrac{3}{19} + \dfrac{11}{19}$

63. $\dfrac{4}{5} - \dfrac{2}{5}$

64. $\dfrac{7}{9} - \dfrac{2}{9}$

65. $\dfrac{9}{13} - \dfrac{4}{13}$

66. $\dfrac{9}{11} - \dfrac{5}{11}$

67. $\dfrac{2}{7} + \dfrac{1}{7} + \dfrac{3}{7}$

68. $\dfrac{4}{23} + \dfrac{8}{23} + \dfrac{3}{23}$

69. $\dfrac{7}{13} + \dfrac{4}{13} - \dfrac{2}{13}$

70. $\dfrac{11}{15} + \dfrac{4}{15} - \dfrac{7}{15}$

Answer the following. (See Example 10.)

71. If 10 sugar bowls hold $\frac{1}{2}$ cup of sugar each, how much sugar does it take to fill them all?

72. Tia works part time. She works $\frac{2}{3}$ of a work day. How many days will she have to work in order to work the equivalent of 8 full days?

73. A pizza is cut into eight equal pieces. If Sue eats one piece, John eats two pieces, and Ellen eats two pieces, what part of the pizza has been eaten?

74. Allen, Jose, and Ismel take a trip. If Allen has driven $\frac{4}{13}$ of the trip, Jose has driven $\frac{5}{13}$ of the trip, and Ismel has driven $\frac{2}{13}$ of the trip, what part of the trip has been completed?

75. John needs eight pieces of lumber, each $\frac{7}{12}$ of a foot long. How many feet of lumber does he need?

76. If Candido spends $\frac{11}{8}$ of an hour on his algebra homework each day, how many hours does he spend on his algebra homework per week (7 days)?

77. A board 12 feet long is to be cut into pieces that are $\frac{2}{3}$ of a foot each. How many pieces will there be?

78. If a tank on a lawnmower holds $\frac{1}{4}$ gallon of gasoline, how many times can the tank be filled from a 5-gallon can?

79. If Elena drives to campus, it takes $\frac{13}{15}$ of an hour. If she catches the Metro rail, it takes $\frac{9}{15}$ of an hour to reach campus. How much longer does it take for her to drive?

80. Juan has $\frac{3}{13}$ of the amount on his phone debit card left after talking to his girlfriend. If he started with $\frac{9}{13}$ of the value of his phone card, how much of the card did he spend on the phone call?

Writing Exercises (81–84)

81. Name three types of documents in which you have seen fractions used.

82. Write a problem that requires the multiplication of fractions.

83. Write a problem that requires the division of fractions.

84. Write a problem that requires the addition of fractions with the same denominators.

Section R.5 — Addition and Subtraction of Decimal Numerals

OBJECTIVES *When you complete this section, you will be able to:*

A Recognize, read, and write decimals in standard notation.
B Determine which decimal is larger or largest.
C Add decimals.
D Subtract decimals.

PREREQUISITE SKILLS *Before beginning this section, you should be able to:*

a. Add and subtract whole numbers. (Section R.2)
b. Multiply and divide fractions. (Section R.4)
c. Name the results of addition, subtraction, multiplication, and division. (Section R.4)
d. Recognize situations that call for additions or subtractions. (Sections R.1–R.3)

GETTING READY FOR SECTION R.5

Add or subtract the following whole numbers, as indicated:

1. $285 + 373$

2.
$$\begin{array}{r} 907 \\ 35 \\ +1357 \end{array}$$

3. $1854 - 395$

Multiply or divide the following numbers as indicated:

4. $\dfrac{1}{10} \cdot \dfrac{1}{100}$

5. $\dfrac{1000}{10}$

6. Give the names of the results in exercises 1 through 5: sum, difference, product, or quotient.

7. The flight from Houston to Fort Hood, Texas, takes 55 minutes. Josh has been in the air for 30 minutes. How much longer is it until touchdown?

Introduction

In Section R.4, we introduced and discussed fractions. In this section, we will be working with special fractions that have whole numbers in the numerator and 10, 100, 1000, 10,000, 100,000, and so on as denominators. In order to see where these fractions come from, we need to go back to the idea of place value in the Hindu–Arabic system.

OBJECTIVE **Recognizing, reading, and writing decimals in standard notation**

Whole numbers have the following pattern of place values:

one hundred thousand	ten thousand	one thousand	one hundred	ten	one
← 100,000	10,000	1000	100	10	1 →

If we start on the right and go to the left, we see that the place values increase by a factor of 10. If we multiply a place value by 10, we get the next place value to the left.

Answers: Getting Ready: 1. 658 **2.** 2299 **3.** 1459 **4.** $\dfrac{1}{1000}$ **5.** 100 **6.** sum, sum, difference, product, quotient, respectively **7.** 25 minutes

Now we consider the pattern in the opposite direction (from left to right). The opposite operation of multiplication is division, so if we multiply by 10 going in one direction, then we divide by 10 going in the other direction. One way to divide by 10 is to multiply by $\frac{1}{10}$. Therefore, if we multiply a place value by $\frac{1}{10}$, we get the next place value to the right. For example:

$$\frac{1}{10} \cdot \frac{1000}{1} = \frac{1000}{10} = 100;$$

$$\frac{1}{10} \cdot \frac{100}{1} = \frac{100}{10} = 10;$$

$$\frac{1}{10} \cdot \frac{10}{1} = \frac{10}{10} = 1;$$

$$\frac{1}{10} \cdot 1 = \frac{1}{10} \cdot \frac{1}{1} = \frac{1}{10}$$

In Europe, a comma is used instead of a decimal point.

The last product yielded the fraction $\frac{1}{10}$, one of those special fractions we talked about earlier. We represent these special fractions with a **decimal point**. The first place value to the right of the ones place is $\frac{1}{10}$, so we represent $\frac{1}{10}$ as .1. Since both symbols $\frac{1}{10}$ and .1 represent the same number, they are both read as "one tenth." To help avoid confusion, we shall call fractions of the form $\frac{1}{10}$ **ratio fractions** and fractions of the form .1 **decimal fractions**.

Consider the second place value to the right of the decimal point. If we continue our process of multiplying by $\frac{1}{10}$, the value of this next place would be $\frac{1}{10} \cdot \frac{1}{10} = \frac{1}{100}$, symbolized as .01 and read as "one hundredth." This process of multiplying a place value by $\frac{1}{10}$ is continued forever as we go to the right. We summarize the first few of these decimal fractions next.

Ratio Fraction	Decimal Fraction	Name
$\frac{1}{10}$	.1	one tenth
$\frac{1}{100}$	.01	one hundredth
$\frac{1}{1000}$	.001	one thousandth
$\frac{1}{10,000}$	.0001	one ten-thousandth
$\frac{1}{100,000}$	.00001	one hundred-thousandth
$\frac{1}{1,000,000}$	.000001	one millionth

Note that the decimal fraction is found by writing the same number of decimal places as we have 0's in the denominator of the corresponding ratio fraction.

Place values to the right of the decimal are as follows:

.	Tenths	Hundredths	Thousandths	Ten-Thousandths	Hundred-Thousandths	Millionths

Note

The only time "and" should be used in a numeral is to indicate a decimal point.

A decimal numeral is read as though it were a whole number, and the place value of the last digit on the right is then given. Decimal numerals may include whole numbers as well as decimal fractions. We call these numerals **decimal mixed numerals**. We read the whole number, then read the decimal point as "and," then read the decimal part of the numeral.

Example 1   Read as a decimal fraction.

a. .408

Solution

Starting at the decimal point, the place values are tenths, hundredths, thousandths. Thus, 8 is in the thousandths place, so we read "four hundred eight thousandths."

b. .0169

Solution

Starting from the decimal point, the place values are tenths, hundredths, thousandths, and ten-thousandths. Thus, 9 is in the ten-thousandths place, so we read "one hundred sixty-nine ten-thousandths."

c. 6.485

Solution

We read six (whole-number portion) **and** (decimal point) four hundred eighty-five thousandths (decimal portion).

d. 305.14

Solution

We read three hundred five (whole-number portion) **and** (decimal point) fourteen hundredths (decimal portion).

PRACTICE EXERCISE 1

Read each of the following. Then write the word name.

a. .702   **b.** .4083   **c.** 69.12   **d.** 524.035

If you need more practice, do the following Additional Practice Exercises.

Additional Practice Exercise 1 Read and write the word name for each.

a. .862   **b.** 2.83   **c.** 68.469

OBJECTIVE Determining which decimal number is larger or largest

As in whole numbers, it is useful to know which decimal number is larger. We can easily tell which number is larger by comparing the digits in corresponding decimal places. In doing this it is convenient to use the symbol > for "greater than" and the symbol < for "less than."

| **Example 2** | **Which is the larger or largest number?** |

a. 38.547 or 206.1

Solution

We can write each numeral with the same number of digits by writing zeroes both to the left and to the right of the decimal point until each numeral has the same number of place values filled. So 38.547 becomes 038.547 and 206.1 becomes 206.100.

0 3 8 . 5 4 7   *Comparing digits from left to right, 2 is greater than 0, which means that 200 is*
⇕ ⇕ ⇕ ⇕ ⇕ ⇕ ⇕   *greater than 0 hundreds. Therefore, 206.1 is greater than 38.547, and this is*
2 0 6 . 1 0 0   *written as 206.1 > 38.547.*

b. 17.637 or 17.652

Solution

Since these numerals have the same number of digits both left and right of the decimal point, we go from left to right, checking each place value.

1 7 . 6 3 7   *The tens places have the same value (1); the units places have the same value (7);*
⇕ ⇕ ⇕ ⇕ ⇕ ⇕   *the tenths places have the same value (6). The hundredths places are different.*
1 7 . 6 5 2   *Since 5 is larger than 3, 17.652 > 17.637.*

c. 2.31, 2.504, or 2.5

Solution

2.310   Line up the place values.
2.504   Write as many zeroes as necessary to give each numeral the same number of
2.500   place values.

2 . 3 1 0   *From left to right, the units digits are the same (2). In the tenths places, 5 > 3, so*
⇕ ⇕ ⇕ ⇕ ⇕   *2.31 is the smaller number. Since the same digit, 5, is in both 2.504 and 2.5, we go*
2 . 5 0 4   *to the hundredths digits. The hundredths digits are the same (0) in 2.504 and 2.500,*
⇕ ⇕ ⇕ ⇕ ⇕   *so we go to the thousandths digits, 4 > 0. Therefore, 2.504 is the largest number.*
2 . 5 0 0

Answers: Practice Exercise 1: a. seven hundred two thousandths **b.** four thousand eighty-three ten-thousandths **c.** sixty-nine and twelve hundredths **d.** five hundred twenty-four and thirty-five thousandths **Additional Practice 1: a.** eight hundred sixty-two thousandths **b.** two and eighty-three hundredths **c.** sixty-eight and four hundred sixty-nine thousandths

PRACTICE EXERCISE 2

Decide which of the following decimals in each pair is larger or largest:

a. .38, .308

b. 76.085, 7.9965

c. 554.03, 554.44

d. 7.022, 7.7922, 7.124

If you need more practice, do the following Additional Practice Exercises.

Additional Practice Exercise 2 Which is larger?

a. .4157, .4152

b. 67.602, 6.7602

c. 99.056, 99.048

OBJECTIVE Adding decimals

All whole numbers have an understood, but unwritten, decimal point to the right of the numeral. For example, 15 = 15. and so on. Consequently, all whole numbers are also decimals. Therefore, the addition of decimals is very similar to the addition of whole numbers.

We can add only digits that are in the same place values. Combinations of decimal fractions and/or decimal mixed numerals may be added by putting them in vertical form and lining up the decimal points so that digits with the same place value are in columns. We then add the columns just as with whole numbers and bring the decimal point straight down. This process is summarized next.

> **Procedure: Addition Algorithm for Decimals**
>
> 1. Line up the decimals so corresponding place values are lined up in the same columns. Write zeroes so that all addends have the same number of decimal places.
>
> 2. Add the columns as in whole numbers.
>
> 3. Bring the decimal point straight down.

Once we have lined up the place values by lining up the decimal points, addition of decimals follows the same pattern as addition of whole numbers.

Example 3 Add.

a. .24 and .5

Solution

$$
\begin{array}{r}
.24 \\
+.50 \\
\hline
4
\end{array}
$$

.24 Addend.

+.50 Addend.

4 Sum.

4 hundredths and
0 hundredths is
4 hundredths.

$$
\begin{array}{r}
.24 \\
+.50 \\
\hline
.74
\end{array}
$$

2 tenths and 5 tenths is 7 tenths. Write the decimal point below those in the addends.

b. .16 and .23

Solution

$$
\begin{array}{r}
.16 \\
+.23 \\
\hline
9
\end{array}
$$

.16 Addend.

+.23 Addend.

9 Sum.

6 hundredths and
3 hundredths is
9 hundredths.

$$
\begin{array}{r}
.16 \\
+.23 \\
\hline
.39
\end{array}
$$

1 tenth and 2 tenths is 3 tenths. Write the decimal point below those in the addends.

Answers: Practice Exercise 2: a. .38 **b.** 76.085 **c.** 554.44 **d.** 7.7922 **Additional Practice 2: a.** .4157 **b.** 67.602 **c.** 99.056

c. 43.834 and 35.16

d. 89.437 and 62.7381

Solution

Line up decimal points and write zeroes to make the number of digits equal.

$$\begin{array}{r} 43.834 \\ +35.160 \\ \hline 78.994 \end{array}$$

Solution

Line up decimal points and write zeroes to make the number of digits equal.

$$\begin{array}{r} {\scriptstyle 11\,1\quad1} \\ 89.4370 \\ +62.7381 \\ \hline 152.1751 \end{array}$$

e. Add: 616.57 + 92.372 + 484.13 + 18.059.

Solution

Line up the decimal points to match place values.

$$\begin{array}{r} {\scriptstyle 2\,21\;21} \\ 616.570 \\ 092.372 \\ 484.130 \\ +018.059 \\ \hline 1211.131 \end{array}$$

PRACTICE EXERCISE 3

Add each of the following:

a. .48 and .3

b. .25 and .73

c. .061 and .52

d. 74.26 + 64.297

e. 50.076 + 16.77

f. 1.8 + 77.1 + 32.5 + 206.26

OBJECTIVE Subtracting decimals

The process for subtraction of decimals is exactly the same as that for subtraction of whole numbers, except that, as in addition of decimals, place values must be lined up by lining up the decimal points of the minuend and subtrahend.

> **Procedure: Subtraction Algorithm for Decimals**
>
> 1. Line up the decimals so corresponding place values are lined up in the same columns. Write zeroes so that the minuend and subtrahend have the same number of decimal places.
>
> 2. Subtract the columns as in whole numbers.
>
> 3. Bring the decimal point straight down from the subtrahend.

Answers: Practice Exercise 3: a. .78 **b.** .98 **c.** .581 **d.** 138.557 **e.** 66.846 **f.** 317.66

Once we have lined up the place values by lining up the decimal points, subtraction of decimals follows the same pattern as subtraction of whole numbers.

Example 4 **Subtract each of the following:**

a. .3 from .7.

Solution

Line up decimal points and subtract.

$$\begin{array}{r} .7 \\ -.3 \\ \hline .4 \end{array}$$

Minuend.

Subtrahend.

Difference.

7 tenths minus 3 tenths equals 4 tenths. Write the decimal point below the decimal point in the subtrahend.

b. .2 from .35.

Solution

Line up decimal points. Write 0 after 2 and subtract.

$$\begin{array}{r} .35 \\ -.20 \\ \hline 5 \end{array}$$

Minuend.

Subtrahend.

Difference.

5 hundredths minus 0 hundredths equals 5 hundredths.

$$\begin{array}{r} .35 \\ -.20 \\ \hline .15 \end{array}$$

3 tenths minus 2 tenths equals 1 tenth. Write the decimal point below the decimal point in the subtrahend.

c. 54.32 from 70.86.

Solution

Line up decimal points and subtract.

$$\begin{array}{r} 70.86 \\ -54.32 \\ \hline .54 \end{array}$$

2 hundredths from 6 hundredths is 4 hundredths and 3 tenths from 8 tenths is 5 tenths. Bring the decimal point down.

$$\begin{array}{r} \overset{6\ 10}{70.86} \\ -54.32 \\ \hline 6.54 \end{array}$$

We cannot take 4 ones from 0 ones so we borrow 1 ten from 7 tens leaving 6 tens and change the 1 ten that we borrowed into 10 ones, 4 ones from 10 ones is 6 ones.

$$\begin{array}{r} \overset{6\ 10}{70.86} \\ -54.32 \\ \hline 16.54 \end{array}$$

5 tens from 6 tens is one ten.

d. 21.43 from 853.02.

Solution

Line up decimal points and subtract.

$$\begin{array}{r} \overset{9}{\underset{}{}}\,\overset{2\ 10^{12}}{853.02} \\ -21.43 \\ \hline 9 \end{array}$$

We cannot take 3 from 2, so we try to borrow from 0. Zero doesn't have anything, so we go to 3. Borrow 1 one from 3 ones leaving 2 ones. Change the 1 one to 10 tenths. Borrow 1 tenth from the 10 tenths. Change 1 tenth to 10 hundredths and add to 2 hundredths, giving 12 hundredths. 3 hundredths from 12 hundredths = 9 hundredths.

$$\begin{array}{r} \overset{9}{}\,\overset{2\ 10^{12}}{853.02} \\ -21.43 \\ \hline .59 \end{array}$$

4 tenths from 9 tenths = 5 tenths. Being the decimal point down.

$$\begin{array}{r} \overset{9}{}\,\overset{2\ 10^{12}}{853.02} \\ -21.43 \\ \hline 1.59 \end{array}$$

1 one from 2 ones is 1 one.

$$\begin{array}{r} \overset{9}{}\,\overset{2\ 10^{12}}{853.02} \\ -021.43 \\ \hline 831.59 \end{array}$$

2 tens from 5 tens is 3 tens, and 0 hundreds from 8 hundreds is 8 hundreds.

e. .24 from .8.

Solution

Line up decimal points. Annex 0 after 8 and subtract.

$$\begin{array}{r} {\scriptstyle 710} \\ .80 \\ -.24 \\ \hline 6 \end{array}$$ Minuend
Subtrahend
Difference

Cannot take 4 hundredths from 0 hundredths. Borrow 1 tenth and express as 10 hundredths. 4 hundredths from 10 hundredths is 6 hundredths.

$$\begin{array}{r} {\scriptstyle 710} \\ .80 \\ -.24 \\ \hline .56 \end{array}$$ *7 tenths minus 2 tenths is 5 tenths. Bring the decimal point down.*

PRACTICE EXERCISE 4

Subtract each of the following:

a. $.45 - .24$

b. $.873 - .6$

c. $80.80 - 61.23$

d. $16 - .0683$

e. $22.34 - 15$

f. $873.08 - 45.13$

When writing a check, we must write both a decimal numeral and the word name for the dollar amount to be paid. This way the bank can make sure of the correct amount to be paid. See the following example:

Example 5   **Write a check for $85.59.**

Solution

Note

The number of pennies is *always* written as a fraction of a dollar. That is, the 59 cents is written $\frac{59}{100}$ Dollars.

Example 6

Bryan needed some new clothes because cold weather was coming. He bought a pair of jeans for $27.99, a sweatshirt for $29.95, and a pair of shoes for $59.99. If he had a gift certificate for $20.00, how much would he have to pay?

Solution

First we add all of the costs.

$$\begin{array}{r} {\scriptstyle 2\ 2\ 2} \\ \$27.99 \\ \$29.95 \\ +\$59.99 \\ \hline \$117.93 \end{array}$$ Jeans.
Sweatshirt.
Shoes.
Sum.

Then we subtract the amount of the gift certificate.

$$\begin{array}{r} \$117.93 \\ -\$20.00 \\ \hline \$97.93 \end{array}$$ Sum.
Gift certificate.
Difference.

Therefore, Bryan would have to pay $97.93.

Answers: Practice Exercise 4: a. .21 **b.** .273 **c.** 19.57 **d.** 15.9317 **e.** 7.34 **f.** 827.95

PRACTICE EXERCISES 5 AND 6

5. Write the decimal numeral and the word name as they would appear on a check for the difference of $154.78 and $68.91.

6. Donna went to the discount pharmacy. She bought a toothbrush for $2.99, deodorant for $2.79, a hairbrush for $5.54, a bottle of shampoo for $4.69, and some vitamins for $7.49. At the same time, she got a refund for a $12.45 calculator that did not work. How much did she have to pay?

For Extra Help

Exercise Set R.5 MyMathLab®

Read each of the following, and then write the word name. (See Example 1.)

1. .6

2. .3

3. .84

4. .39

5. .763

6. .499

7. 27.45

8. 93.54

9. 68.607

10. 17.095

Identify the place value of the underlined digit. (See Objective A.)

11. 8.7$\underline{4}$

12. 9.0$\underline{5}$

13. .04$\underline{7}$

14. .094$\underline{7}$

15. 12.$\underline{0}$18

16. 34.87$\underline{9}$2

Decide which decimal numeral in each exercise represents the larger or largest number. (See Example 2.)

17. .520, 52.0

18. .680, 6.80

19. 18.40, 18,043

20. 61.023, 61,000

21. 298.20, 289.20

22. 840.67, 804.67

23. .66499, .66399

24. .89532, .89668

25. 1.01221, 1.011, 1.1022

26. 35,071, 3507, 350.07

Add each of the following. (See Example 3.)

27. .5 + .9

28. .7 + .4

29. .17 + .54

30. .38 + .27

31. 4.37 + 1.69

32. 8.56 + 3.23

33. 78.077 + 41.42

34. 23.98 + 630.034

35. 383.58 + 807.03

36. 573.06 + 12.982

37. 124.97 + 81.3

38. 99.46 + 566.7

Answers: Practice Exercises 5 and 6: 5. $85.87 Eighty-five and $\frac{87}{100}$ dollars **6.** $11.05

Subtract each of the following. (See Example 4.)

39. .8 − .2

40. .9 − .4

41. .75 − .17

42. .56 − .39

43. 68.63 − 39.06

44. 75.64 − 68.45

45. 48.76 − 13.5

46. 89.95 − 14.3

47. 212.16 − 98.204

48. 634.45 − 27.022

49. 20 − .8369

50. 42 − .1895

51. 70.34 − 38

52. 60.95 − 33

Add.

53. 67.46 + 117.85 + 35.19

54. 135.31 + 87.62 + 43.77

55. 346.48 + 73.373 + 686.45 + 52.05

56. 79.626 + 84.84 + 431.28 + 5.422

Write the word name for each of the following money amounts as it would appear on a check. (See Example 5.)

57. $706.45

58. $908.44

59. $305.02

60. $602.30

Solve each of the following. (See Example 6.)

61. At one time, the city of Houston was owed $4.05 million in unpaid parking tickets, and the city of New Orleans was owed $4.24 million in unpaid parking tickets. How much was owed to the two cities in unpaid parking tickets? (*Source: Miami Herald*)

62. At the same time, the cities of Anaheim and San Diego were owed $1.65 million and $.55 million in unpaid parking tickets. How much was owed to the two cities in unpaid parking tickets? (*Source: Miami Herald*)

63. Driving into the parking garage, Diana saw a sign that said "Clearance 7.6 ft." If her van is 6.8 feet tall, how much clearance does she have?

64. Anita and Alice are long-distance truck drivers. They noticed that an overpass on the interstate highway had a clearance of 16.25 feet and the height of their tractor trailer is 14.50 feet. What was the clearance between the overpass and their rig?

65. On long driving trips, Bill likes to stop every couple of hours. On the first part of his trip, he drove 129.4 miles, and on the rest of the trip he drove 168.7 miles. How many miles was the total trip?

66. In traveling from New York to San Francisco one way one stop, Anne paid $548.00 for airfare and $12.45 for the airport shuttle. How much did she pay for this transportation?

67. A gallon of gasoline costs $3.29 on the turnpike and $3.13 off the turnpike. How much cheaper is gasoline off the turnpike?

68. Two college professors were going to the same mathematics meeting in Baltimore. One professor bought her ticket a month ahead of time and paid $249.95, while her colleague bought a ticket at the last minute when the same passage was reduced to $168.40. What was the difference in fares?

69. Andy was doing some comparison shopping. He found a portable radio with earphones for $34.95 at one store and the same item for $37.98 at a second store. How much did he save by buying the cheaper radio?

70. Mom wants to buy Miguel a video game for his birthday. At the toy store she found one Miguel wants for $33.49, and at the electronics store she saw the same game for $39.99. How much more would she pay if she bought the video game at the electronics store?

71. Robin fills up her gas tank every Friday afternoon after being paid. This week she put in 13.883 gallons, and last week she put in 15.088 gallons. How much gasoline did she use for the two weeks?

72. Jackie is taking her family of two adults and two children to the movies. If adult tickets are $8.50 each and children's tickets are $6.50 each, how much will it cost to take the family to the movies?

Challenge Exercises (73 and 74)

73. Gasoline sells for 2.25\frac{9}{10}$ per gallon. How much is this in cents?

74. The odometer tells you how far a car has been driven. If the odometer reads 115,436.8 miles at the beginning of a trip and 116,123.5 miles at the end of the trip, how many miles were driven?

Writing Exercises (75–78)

75. Statistics say that, on the average, 1.8 people ride in an automobile. If you own a new car and do not want it to be hit by people opening car doors, which side of a parked car would you park on? Why?

76. The metric system is composed of all decimal numbers. The United States is the only major country in the world that has not adopted the metric system. Why do you suppose that is true?

77. How is adding decimals like adding whole numbers?

78. Does $7.2 = 7.20 = 7.200$? Why or why not?

Section R.6 — Multiplication and Division of Decimal Numerals

OBJECTIVES *When you complete this section, you will be able to:*

A Multiply decimals.
B Divide decimals.
C Round decimals.
D Divide decimals and round the quotient.
E Represent a fraction as division and rewrite a fraction as a decimal.

PREREQUISITE SKILLS *Before beginning this section, you should be able to:*

a. Add, subtract, multiply, and divide whole numbers. (Sections R.1–R.3)
b. Multiply fractions. (Section R.4)
c. Recognize situations that call for multiplication or division. (Section R.3)

GETTING READY FOR SECTION R.6

Multiply each of the following whole numbers:

1.
$$\begin{array}{r} 28 \\ \times\ 13 \\ \hline \end{array}$$

2.
$$\begin{array}{r} 416 \\ \times\ 7 \\ \hline \end{array}$$

Divide each of the following whole numbers:

3. $120 \div 8$ **4.** $527 \div 17$

Find the product.

5. $\dfrac{1}{10} \cdot \dfrac{1}{100}$

Introduction

As we discussed in the previous section, decimal fractions are a special type of ratio fraction that can easily be written by using a decimal point. Recall that $.1 = \frac{1}{10}$; $.01 = \frac{1}{100}$; $.001 = \frac{1}{1000}$; and so on. Since decimal fractions are ratio fractions, it is reasonable to expect that the multiplication of decimal fractions should be based on the multiplication of ratio fractions.

OBJECTIVE **Multiplying decimals**

The numerator of a ratio fraction associated with a decimal is a whole number. A process you can use to multiply decimals is to multiply the whole numbers in the numerators, then find the denominator by multiplying the place values of the last digits of the equivalent decimal fractions. For example, if we multiply $(.7)(.31)$ using ratio fractions, we have $(.7)(.31) = \frac{7}{10} \cdot \frac{31}{100} = \frac{7 \cdot 31}{10 \cdot 100} = \frac{217}{1000}$. The 217 comes from multiplying the whole numbers 7 and 31. The thousandths place results from multiplying the place value of .7, which is $\frac{1}{10}$, with the place value of the 1 in .31, which is $\frac{1}{100}$. This process of decimal multiplication can be broken down into two parts, as illustrated by Example 1.

| Example 1 | **Multiply.** |

Note

This example illustrates an interesting fact about the multiplication of decimals. In whole numbers, the product is always larger than any factor. In decimal multiplication, the product can be smaller than either factor. For example, $(0.1)(0.2) = 0.02$ which is smaller than 0.1 or 0.2.

a. 1.4 by .12

Solution

$$
\begin{array}{r}
1.4 \\
\times\ .12 \\
\hline
28 \\
14 \\
\hline
.168
\end{array}
$$

First, multiply the numbers, disregarding the decimal point: $14 \cdot 12 = 168$. This is the same as multiplying numerator by numerator of the associated fractions. Second, multiply the place values of the last digits of the associated fractions: $\frac{1}{10} \cdot \frac{1}{100} = \frac{1}{1000}$, or in decimal form, $.1 \cdot .01 = .001$. The last digit of the product, 8, has to be in the thousandths place, so we have .168. Hence, $1.4(.12) = .168$.

b. 82.47 by .23

Solution

Carry out the standard process for multiplying whole numbers.

$$
\begin{array}{r}
\overset{\scriptstyle 1}{}\overset{\scriptstyle 1\ 2}{} \\
82.47 \\
\times\ \ .23 \\
\hline
24741 \\
16494 \\
\hline
18.9681
\end{array}
$$

We have $\frac{1}{100} \cdot \frac{1}{100} = \frac{1}{10,000}$, or in decimal form, $.01 \cdot .01 = .0001$. The last digit, 1, must be in the ten-thousandths place, so we place the decimal between the 8 and the 9. Hence, $82.47(.23) = 18.9681$.

Before we continue, let us make an observation about multiplying fractions. Consider

$$ \frac{1}{100} \cdot \frac{1}{1000} = \frac{1}{100,000}. $$

As decimals,

$$ (.01)(.001) = .00001. $$

Number of decimal places is

$$ 2 + 3 = 5. $$

From this observation, we see that the sum of the number of decimal places in all of the factors is the number of decimal places in the product. We use this result in the following algorithm:

Procedure: Multiplication Algorithm for Decimals

To multiply decimal numerals, follow this procedure:

1. Ignore the decimal points and multiply as in whole numbers.

2. Count the number of decimal places to the right of the decimal point in each factor.

3. Add the number of decimal places in all of the factors.

4. Place the decimal point in the product by counting from right to left the number of decimal places found in step 3. If necessary, write zeroes.

We now show some examples of applying the multiplication algorithm.

| **Example 2** | **Multiply by using the multiplication algorithm for decimals.** |

a. 1.8 by .9

Solution

Treat the decimal numerals as whole numbers and multiply.

$$\begin{array}{r} \overset{7}{1.8} \\ \times .9 \\ \hline 162 \end{array}$$

1.8   Factor.
×.9   Factor.
162   Whole-number product.

We add the number of decimal places in both factors and put the decimal point this number of places to the left in the whole-number product, starting with the digit on the right.

$$\begin{array}{r} \overset{7}{1.8} \\ \times .9 \\ \hline 1.62 \end{array}$$

1.8   1 decimal place.
×.9   1 decimal place.
1.62   Product. 1 + 1 = 2 decimal places.

b. 2.38 by 6

$$\begin{array}{r} \overset{2\ 4}{2.38} \\ \times 6 \\ \hline 14.28 \end{array}$$

2.38   Factor, 2 decimal places.
×6   Factor, 0 decimal places.
14.28 Product. 2 + 0 = 2 decimal places.

c. 29.06 by 4.3

Solution

Treat the decimal numerals as whole numbers and multiply.

$$\begin{array}{r} \overset{3\ \ 2}{\underset{2\ \ 1}{29.06}} \\ \times 4.3 \\ \hline 8718 \\ 11624\ \ \\ \hline 124.958 \end{array}$$

29.06   Factor, 2 decimal places.
×4.3   Factor, 1 decimal place.

124.958 Product. 2 + 1 = 3 decimal places.

PRACTICE EXERCISE 2

Multiply each of the following by using the multiplication algorithm for decimals:

a. 4.3(.7)   **b.** 9(5.03)   **c.** 6.3(4.12)

Additional Practice Exercise 2   Multiply each of the following by using the multiplication algorithm for decimals:

d. 6.2(.4)   **e.** (23.04)(7)   **f.** 6.07(19.457)

Answers: **Practice Exercise 2: a.** 3.01 **b.** 45.27 **c.** 25.956   **Additional Practice Exercise 2: d.** 2.48
e. 161.28 **f.** 118.10399

OBJECTIVE **B** Dividing decimals

The process for division of decimals is also based on ratio fractions. We have defined ratio fractions as having meaning only when the denominator is a natural number $\{1, 2, 3, \ldots\}$. Therefore, the process for division of decimals depends on the divisor being a natural number. Hence, part of the process for division of decimals is to do what is necessary to make the divisor a natural number.

In the next example, bear in mind that $10 \times .3$ is $10 \times \frac{3}{10} = 3$. Hence, the decimal point appears to have "moved" one place to the right upon multiplication by 10. Multiplication by 100 "moves" the decimal point two places to the right, multiplication by 1000 "moves" the decimal point three places to the right, and so on. Let us do an example.

Example 3   Find the quotient.

a. $1.2 \div .3$

Solution

$$1.2 \div .3 = \frac{1.2}{.3}$$

Multiply the numerator and the denominator by 10 to get 3 in the denominator. This is the same as multiplying by $\frac{10}{10}$.

$$= \frac{1.2 \cdot 10}{.3 \cdot 10}$$

$$= \frac{12}{3} = 4$$

Note that the decimal in both the divisor and dividend was moved one place to the right.

b. $7.5 \div .25$

Solution

$$7.5 \div .25 = \frac{7.5}{.25}$$

Multiply by $\frac{100}{100} = 1$ to make the denominator a natural number.

$$\frac{7.5}{.25} \cdot \frac{100}{100} = \frac{7.5 \cdot 100}{.25 \cdot 100}$$

Note that the decimal in both the numerator and denominator was moved two places to the right.

$$= \frac{750}{25} = 30$$

Example 3a can be written as $.3\overline{)1.2}$ and 3b as $.25\overline{)7.5}$.

The results of Example 3 lead to the algorithm for division of decimals.

> ### Procedure: Division Algorithm for Decimals
>
> To divide decimal numerals, follow these steps:
>
> 1. Make the divisor a whole number by moving the decimal point; that is, multiply the divisor by whatever it takes to make the divisor a whole number. Move the decimal point in the dividend the same number of places it was moved in the divisor. If the divisor is already a whole number, go to step 2.
>
> 2. Place the decimal point in the quotient directly above the decimal point in the dividend.
>
> 3. Carry out the division as in whole-number division.

Example 4 Find the quotient.

a. .045 ÷ .5

Solution

$$.5\overline{)\,.045}$$

Multiply the divisor and dividend by 10 to make the divisor a whole number. This will, in effect, move the decimal point one place to the right in both the divisor and the dividend. Write the decimal point in the quotient above the decimal point in the dividend.

$$\begin{array}{r} .09 \\ 5\overline{)\,.45} \\ -45 \\ \hline 0 \end{array}$$

Proceed as in whole-number division. Since 5 will not divide 4, write 0 in the value place above the 4. We see that 5 divides 45 exactly 9 times. Multiply and subtract.

CHECK:

$(.09)(.5) = .045$

b. Divide 8 into 7.2.

Solution

$$\begin{array}{r} 0.9 \\ 8\overline{)\,7.2} \\ -72 \\ \hline 0 \end{array}$$

Since the divisor is already a whole number, there is no need to move the decimal to make it a whole number. This means that the decimal point in the dividend stays where it is and the decimal point in the quotient goes right above it.

Proceed as in whole-number division.

CHECK:

$(8)(0.9) = 7.2$

c. 13.5 ÷ .15

Solution

$$.15\overline{)\,13.5}$$

Multiply the divisor and dividend by 100 to make the divisor a whole number. This will, in effect, move the decimal point two places to the right in both the divisor and the dividend. Write 0's as needed to fill in the place values. Write the decimal point in the quotient above the decimal point in the dividend.

$$\begin{array}{r} 90. \\ 15\overline{)\,1350.} \\ -135 \\ \hline 0 \end{array}$$

Proceed as in whole-number division.

CHECK:

$(90)(.15) = 13.5$

PRACTICE EXERCISE 4

Find the quotient.

a. .72 ÷ .8   **b.** 4.8 ÷ 6   **c.** 16.1 ÷ .23

OBJECTIVE Rounding decimals

As we know from division of whole numbers, the remainder is not always 0. If the remainder is not 0, we usually continue dividing for a given number of decimal places, then round.

The process of rounding involves writing a numeral to the nearest desired place value. If we wanted to round .147 to the nearest hundredth, we go to the *thousandths* place,

---

Answers: Practice Exercise 4: **a.** .9 **b.** .8 **c.** 70

see a 7, and add 1 to the 4 in the *hundredths* place. That is, .147 is .15 to the *nearest hundredth*.

On the other hand, if we wanted to round .147 to the nearest tenth, we would go to the *hundredths* place, see a 4, and drop it and all other digits to the right of it. So .147 to the *nearest tenth* is .1.

Note

The last digit should be in the place value to which you are rounding.

Procedure: Rounding to the Nearest Place Value

To round to the nearest place value, do the following:

1. Go to the place value to the right of the place value under consideration.

2. If this digit is 5 or more, add 1 to the digit under consideration and drop all digits that follow the digit under consideration.

3. If this digit is 4 or less, drop it and all that follow it. Keep the digit under consideration as is.

Example 5   **Round the given number to the indicated place value.**

a. .367, nearest tenth

Solution:

The place value to the right of the tenths place has a 6. Since 6 is greater than 5, we add 1 to the 3 in the tenths place and drop all digits that follow the tenths place. Thus, .367 rounded to the nearest tenth is .4.

b. 12.573, nearest hundredth

Solution:

The place value to the right of the hundredths place has a 3. Since 3 is less than 5, we drop the 3 and all digits that follow. Thus, 12.573 rounded to the nearest hundredth is 12.57.

PRACTICE EXERCISE 5

Round to the nearest tenth.
a. 15.082

b. 85.349

Round to the nearest hundredth.
c. 65.6805

d. 1.71518

Round to the nearest thousandth.
e. 25.0963

f. .00499

OBJECTIVE Dividing decimals and rounding the quotient

When dividing, the results can often continue for many decimal places (or forever) making it necessary to round the answer to a specified place value.

Answers: **Practice Exercise 5:** **a.** 15.1 **b.** 85.3 **c.** 65.68 **d.** 1.72 **e.** 25.096 **f.** .005

Example 6   **Divide and round the answer to the nearest hundredth.**

a. 4.21 by 5

Solution

$$5\overline{)4.21}$$

The divisor is already a whole number, so write the decimal point in the quotient above the decimal point in the dividend.

$$
\begin{array}{r}
0.842 \\
5\overline{)4.210} \\
\underline{-40} \\
21 \\
\underline{-20} \\
10 \\
\underline{-10} \\
0
\end{array}
$$

To round to the nearest hundredth, we need the digit in the thousandths place. The thousandths place has a 2, and 2 is less than 5. We drop the 2. Hence, our rounded quotient is .84.

Answer: .84

b. .625 by .21

Solution

$$.21\overline{)\,.625}$$

Move the decimal point two places to the right in both divisor and dividend. Write the decimal point in the quotient above the decimal point in the dividend. Proceed as in division of whole numbers.

$$
\begin{array}{r}
2.976 \\
21\overline{)62.500} \\
\underline{-42} \\
205 \\
\underline{-189} \\
160 \\
\underline{-147} \\
130 \\
\underline{-126} \\
4
\end{array}
$$

Since we are rounding to the nearest hundredth, we stop at the thousandths place in the quotient. The thousandths place is 6, and 6 is greater than 5. We add 1 to the 7 in the hundredths place and drop digits to the right of the 7. Hence, our rounded quotient is 2.98.

Answer: 2.98

c. 79 by .15

Solution

$$.15\overline{)79}$$

Move the decimal point two places to the right in both divisor and dividend. Write the decimal point in the quotient above the decimal point in the dividend. Proceed as in division of whole numbers.

$$
\begin{array}{r}
526.666\ldots \\
15\overline{)7900.} \\
\underline{-75} \\
40 \\
\underline{-30} \\
100 \\
\underline{-90} \\
100 \\
\underline{-90} \\
10,\ \text{etc.}
\end{array}
$$

The quotient is a repeating decimal. Since we are rounding to the nearest hundredth, we stop at the thousandths place in the quotient. The thousandths place has a 6, and 6 is greater than 5. We add 1 to the 6 in the hundredths place and drop digits to the right of the resulting 7. Hence, our rounded quotient is 526.67.

Answer: 526.67

Note

If we round a number, the original number is not equal to the rounded number. For example, 5.467 rounded to the nearest tenth is 5.5. However, $5.5 \neq 5.467$. Thus, we use a special symbol which means "approximately equal to." That is, $5.5 \approx 5.467$.

PRACTICE EXERCISE 6

Divide each of the following. If a quotient goes beyond the hundredths place, round to the nearest hundredth.

a. 21.5 ÷ 9 **b.** 2.4 ÷ .7 **c.** .234 ÷ .08 **d.** 65 ÷ .32

OBJECTIVE **Representing a fraction as division and rewriting a fraction as a decimal**

Any fraction represents division, where the numerator is divided by the denominator. For example, $\frac{3}{4} = 3 \div 4$. In general, $\frac{a}{b} = a \div b$. This interpretation allows us to write a fraction as a decimal.

Example 7 **Change the rational numbers into decimals. If necessary, round the answer to the nearest hundredth.**

a. $\frac{3}{4}$

Solution

$\frac{3}{4}$ means $3 \div 4$. Since $4 > 3$, it is necessary to put a decimal point and some zeroes after the 3.

$$
\begin{array}{r}
.75 \\
4\overline{)3.00} \\
-28 \\
\hline
20 \\
-20 \\
\hline
0
\end{array}
$$

Therefore, $\frac{3}{4} = .75$.

b. $\frac{9}{16}$

Solution

$\frac{9}{16}$ means $9 \div 16$. As in Example 7a, we will need to put a decimal point after the 9, followed by one or more zeroes.

$$
\begin{array}{r}
.562 = .56 \\
16\overline{)9.000} \\
-80 \\
\hline
100 \\
-96 \\
\hline
40 \\
32
\end{array}
$$

In order to round to the nearest hundredth, we need an answer to the thousandth.

c. $\frac{12}{5}$

Solution

$\frac{12}{5}$ means $12 \div 5$. Since 5 will not divide into 12 evenly, put a decimal point after the 12 followed by one or more zeroes.

$$
\begin{array}{r}
2.4 \\
5\overline{)12.0} \\
-10 \\
\hline
20 \\
-20 \\
\hline
0
\end{array}
$$

d. $\frac{24}{7}$

Solution

$\frac{24}{7}$ means $24 \div 7$. As in Example 7c, place a decimal point and as many zeroes as needed after 24.

$$
\begin{array}{r}
3.428 = 3.43 \\
7\overline{)24.000} \\
-21 \\
\hline
30 \\
-28 \\
\hline
20 \\
-14 \\
\hline
60 \\
-56
\end{array}
$$

In order to round to the nearest hundredth, we need the answer to the thousandth.

Answers: **Practice Exercise 6:** **a.** 2.39 **b.** 3.43 **c.** 2.93 **d.** 203.13

PRACTICE EXERCISE 7

Write each of the rational numbers as a decimal. If necessary, round to the nearest hundredth.

a. $\dfrac{3}{8}$ **b.** $\dfrac{12}{17}$ **c.** $\dfrac{15}{6}$ **d.** $\dfrac{29}{9}$

We see applications of decimals everywhere. Our money system is based on the decimal system, and most of our everyday calculations are done in decimals. Most calculators work only with decimals.

Example 8 Solve each of the following:

a. A package of cookies weighs 1.25 pounds. How much will eight packages of cookies weigh?

Solution

Since there are eight packages with 1.25 pounds in each package, this is a case of multiplication.

$$\begin{array}{r} 1.25 \\ \underline{\times 8} \\ 10.00 \end{array}$$ The cookies weigh 10 pounds.

b. Dad took his family of four out to dinner to celebrate his daughter's 13th birthday. If the meal cost $72.12, what was the average cost per person?

Solution

The average cost per person is like sharing the cost among four people. Therefore, this is a case of division.

$$4)\overline{\$72.12}\quad \$18.03$$ The meals cost $18.03 per person.

PRACTICE EXERCISE 8

Solve each problem. Round to the nearest hundredth where necessary.

a. A shipping container has a 32,580-kilogram capacity. If there are 2.2 pounds per kilogram, what is the capacity in pounds?

b. If a 5-pound bag of potatoes costs $3.49, find, to the nearest cent, how much 1 pound of potatoes costs.

Exercise Set R.6

For Extra Help

MyMathLab®

Multiply. (See Example 2.)

1. 7(8.1) **2.** 5.2(6) **3.** 3.2(4.4) **4.** 6.5(3.3)

5. 1.8(.52) **6.** .15(2.6) **7.** 3.9(6) **8.** 4.8(5)

9. 5.1(8.4) **10.** 3.2(8.3) **11.** .3(.89) **12.** .6(.74)

13. 2.59(.15) **14.** 3.34(.07) **15.** 5.27(.603) **16.** (23.9).834

Round to the indicated place value. (See Example 5.)

17. .818 nearest hundredth **18.** .974 nearest hundredth

19. 8.428 nearest tenth **20.** 2.675 nearest tenth

21. 64.708 nearest one **22.** 5.607 nearest one

Answers: Practice Exercise 7: a. .38 **b.** .71 **c.** 2.5 **d.** 3.22 **Practice Exercise 8: a.** 71,676 pounds **b.** $.70

23. .00684 nearest thousandth

24. 75.080691 nearest thousandth

25. 4.9899 nearest tenth

26. .9999 nearest hundredth

Divide. If a quotient goes beyond the hundredths place, round to the nearest hundredth. (See Examples 4 and 6.)

27. $5.6 \div 8$

28. $7.2 \div 9$

29. $.50 \div 6$

30. $.101 \div 1.1$

31. $.28 \div .4$

32. $.84 \div .8$

33. $20.4 \div .9$

34. $35.3 \div .7$

35. $97.43 \div .066$

36. $61.16 \div .02$

37. $5.23 \div .143$

38. $6.81 \div .257$

Change the fractions into decimals. If necessary, round answers to the nearest hundredth. (See Example 7.)

39. $\dfrac{1}{4}$

40. $\dfrac{3}{6}$

41. $\dfrac{7}{8}$

42. $\dfrac{5}{8}$

43. $\dfrac{9}{16}$

44. $\dfrac{10}{17}$

45. $\dfrac{2}{3}$

46. $\dfrac{4}{9}$

47. $\dfrac{9}{4}$

48. $\dfrac{13}{4}$

49. $\dfrac{22}{5}$

50. $\dfrac{27}{5}$

51. $\dfrac{33}{25}$

52. $\dfrac{27}{20}$

53. $\dfrac{31}{14}$

54. $\dfrac{32}{15}$

Solve. Round to the nearest hundredth where necessary. (See Example 8.)

55. There are about 1.6 kilometers in 1 mile. If you are traveling at 60 miles per hour, how fast are you going in kilometers per hour?

56. Five people have formed a lottery pool. If they win the $6.7 million jackpot, how much will each person receive?

57. An author has to write 250 pages in 60 days. What is the average number of pages she needs to write each day?

58. An airline pilot has a schedule to meet. At what rate must she fly if she has 4300 miles to travel in 8 hours?

59. Arthur Chung has pledged to read five books that together have a total of 1089 pages. If he has pledges for 3 cents per page, how much can he raise in the read-a-thon?

60. A student sold 650 copies of the campus humor magazine. If she makes $.48 a copy, how much has she earned for her activities?

61. A railroad club has its own private railroad car. It costs $32,000 to have the local railroad hook up and pull the car to a railroad convention. If there are 28 people going on a trip, what is the cost per person?

62. Carmen wants to drive her car out west on vacation. She estimates that she will drive about 5500 miles. If her car gets 32 miles to the gallon of gasoline, how many gallons will she need?

Challenge Exercises (63 and 64)

63. If the sales tax rate in a certain state is 6 cents on the dollar, how much is the sales tax on $8400?

64. In an 8-hour period the temperature outside fell from 42°F to 23°F. Assuming that the decline was constant, on the average how many degrees per hour did the temperature fall?

Writing Exercises (65 and 66)

65. Some people say that with decimal fractions we have no need for any other kind of fractions. Do you think that this is true? Why or why not?

66. Why do we line up the decimal points when adding decimals, but not when multiplying them?

Linear Measurement in the American and Metric Systems

OBJECTIVES *When you complete this section, you will be able to:*

A Convert linear units within the American system of measurement.
B Convert linear units within the metric system of measurement.

PREREQUISITE SKILLS *Before beginning this section, you should be able to:*

a. Multiply and divide decimals (Section R.6).
b. Multiply by fractions whose numerator or denominator is one. (Section R.4).

GETTING READY FOR SECTION R.7

Perform each indicated operation.

1. $2.4 \cdot 12$ **2.** $6.28 \cdot 100$ **3.** $5.78 \div 10$ **4.** $763 \div 100$

Multiply the following:

5. $8 \cdot \dfrac{12}{1}$ **6.** $216 \cdot \dfrac{1}{36}$ **7.** $10,560 \cdot \dfrac{1}{5280}$ **8.** $3.4 \cdot \dfrac{36}{1}$

OBJECTIVE Converting linear units within the American system of measurement

In the American system of linear measure, the units are inch, foot, yard, rod, furlong, and mile. The rod and furlong are not often used in everyday situations and will not be discussed here. We often abbreviate the names of the units. We abbreviate inch as "in.," foot as "ft," yard as "yd," and mile as "mi." It is possible to measure the length of an object and get different numerical answers because we used different units. In order to change from one unit of measure to another, we must know the conversion factors. The conversion factors for the American system are given in the following table:

Conversion Factors: American System of Linear Measure	
1 ft = 12 in.	1 yd = 3 ft
1 yd = 36 in.	1 mi = 5280 ft
1 mi = 1760 yd	

There are several methods for converting from one unit to another. One method is used extensively in the sciences and is the method that we will use. It is called the **unit-cancellation, unit analysis**, or **factor-label** method. When using this method, we multiply the expression to be converted by the ratio of the conversion factors in which the numerator is the unit into which we wish to convert and the denominator is the unit from which we are converting. For example, if we want to convert a given number of feet to inches, we multiply the given number of feet by the ratio $\frac{12 \text{ in.}}{1 \text{ ft}}$. We then "cancel" the units. The word *cancel* is often used to indicate division, but will not be used as such in this book.

Answers: Getting Ready: 1. 28.8 **2.** 628 **3.** .578 **4.** 7.63 **5.** 96 **6.** 6 **7.** 2 **8.** 122.4

Example 1   **Convert the following into the indicated unit:**

a. 4 ft = _____ in.   Multiply 4 ft by the ratio $\frac{12 \text{ in.}}{1 \text{ ft}}$.

$4 \text{ ft} \cdot \frac{12 \text{ in.}}{1 \text{ ft}} =$ Divide the ft.

$4 \cancel{\text{ft}} \cdot \frac{12 \text{ in.}}{1 \cancel{\text{ft}}} =$ 12 in. ÷ 1 = 12 in.

$4 \cdot 12 \text{ in.} =$ Multiply.

48 in.   Therefore, 4 ft = 48 in.

c. 2.4 mi = _____ ft   Multiply 2.4 mi by $\frac{5280 \text{ ft}}{1 \text{ mi}}$.

$2.4 \text{ mi} \cdot \frac{5280 \text{ ft}}{1 \text{ mi}} =$ Divide the mi.

$2.4 \cancel{\text{mi}} \cdot \frac{5280 \text{ ft}}{1 \cancel{\text{mi}}} =$ 5280 ft ÷ 1 = 5280 ft.

$2.4 \cdot 5280 \text{ ft} =$ Multiply.

12,672 ft   Therefore, 2.4 mi = 12,672 ft.

b. 72 in. = _____ yd   Multiply 72 in. by $\frac{1 \text{ yd}}{36 \text{ in.}}$.

$72 \text{ in.} \cdot \frac{1 \text{ yd}}{36 \text{ in.}} =$ Divide the in.

$72 \cancel{\text{in.}} \cdot \frac{1 \text{ yd}}{36 \cancel{\text{in.}}} =$ Think of 72 as $\frac{72}{1}$.

$\frac{72}{1} \cdot \frac{1 \text{ yd}}{36} =$ Multiply.

$\frac{72 \text{ yd}}{36} =$ Divide 72 by 36.
   72 ÷ 36 = 2.

2 yd   Therefore, 72 in. = 2 yd.

PRACTICE EXERCISE 1

Convert each of the following to the indicated unit:

a. 4 yd = ____ ft   **b.** 48 in. = ____ ft   **c.** 3 ft = ____ in.   **d.** 15,840 ft = ____ mi

If you need more practice, do the following Additional Practice Exercises.

Additional Practice Exercise 1   Convert each of the following to the indicated unit:

a. 180 in. = ____ yd   **b.** 2 yd = ____ ft   **c.** 1.2 mi = ____ ft   **d.** 36 ft = ____ yd

OBJECTIVE **B** ## Converting linear units within the metric system of measurement

One of the greatest advantages of the metric system is that conversions from one linear unit of measure to another are accomplished by powers of 10. When multiplying or dividing by a power of 10, we simply move the decimal point the appropriate number of places. Study the following for a pattern that gives the number of places the decimal point is moved when multiplying or dividing by a power of 10:

Example 2   **Find the following products:**

a. $2.763 \times 100 = 276.3$ Decimal point moved right two places.

$43.91 \times 1000 = 43,910$ Decimal point moved right three places.

$0.0062 \times 10 = 0.062$ Decimal point moved right one place.

$0.000123 \times 10,000 = 1.23$ Decimal point moved right four places.

b. $467.91 \div 10 = 46.791$ Decimal point moved left one place.

$5.78 \div 1000 = 0.00578$ Decimal point moved left three places.

$0.0142 \div 100 = 0.000142$ Decimal point moved left two places.

$100.93 \div 10,000 = 0.010093$ Decimal point moved left four places.

Answers: **Practice Exercise 1:** **a.** 12 **b.** 4 **c.** 36 **d.** 3   **Additional Practice 1:** **a.** 5 **b.** 6 **c.** 6336 **d.** 12

Based on the preceding example, we make the following observations:

Multiplication and Division by Powers of 10

a. When multiplying by a power of 10, move the decimal point to the right the same number of places as there are zeroes in the power of 10.

b. When dividing by a power of 10, move the decimal point to the left the same number of places as there are zeroes in the power of 10.

The basic unit of linear measure in the metric system is the meter, which is abbreviated as "m." The basic unit is then prefixed by one of the following, which tell how many meters or what part of a meter the unit is measuring:

Partial List of Metric Prefixes

kilo- (k) = 1000 units

hecto- (h) = 100 units

deca- (dk or da) = 10 units

deci- (d) = $\frac{1}{10}$ of a unit

centi- (c) = $\frac{1}{100}$ of a unit

milli- (m) = $\frac{1}{1000}$ of a unit

Consequently, "km" means kilometer and equals 1000 meters, "dm" means decimeter and equals $\frac{1}{10}$ of a meter, and "mm" means millimeter and equals $\frac{1}{1000}$ of a meter. The most commonly used units are the kilometer, which is used in the way the mile is used in the American system; the meter, which is used in the way feet or yards are used in the American system; and the centimeter and millimeter, which are used in the way the inch is used in the American system.

Conversions within the metric system can be done by the factor-label method, but there is a much easier way. Since the metric system is based on powers of 10, and multiplying or dividing by powers of 10 simply moves the decimal point, all we have to do is determine how many places and in which direction to move the decimal point. One way of doing this is to use the accompanying diagram. The units on the line correspond to place values in our base-10 number system.

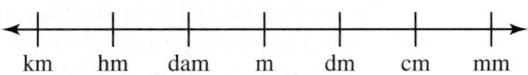

Using the proceeding diagram, we convert within the metric system as follows.

Procedure: Converting Within the Metric System

1. Locate the unit given and count the number of places, left or right, to the unit into which you are converting.

2. Move the decimal point the number of place values and in the same direction as found in step 1.

Example 3   **Convert each of the following into the indicated unit:**

a. 3.6 m = _____ cm

Solution

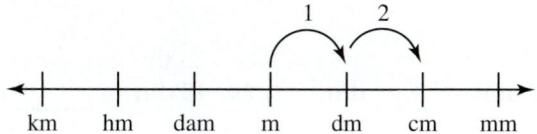

Since cm is two units to the right of m, move the decimal point two places to the right. Therefore, 3.6 m = 360 cm.

b. 7354 dm = _____ hm

Solution

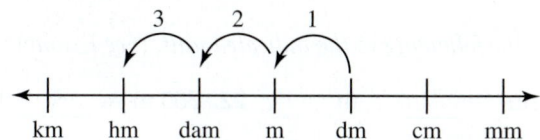

Since hm is three units to the left of dm, move the decimal point three places to the left. Therefore, 7354 dm = 7.354 hm.

c. 0.0178 km = _____ cm

Solution

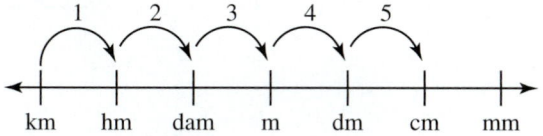

Since cm is five units to the right of km, move the decimal point five places to the right. Therefore, 0.0178 km = 1780 cm. Note: It was necessary to write a 0 after the 8 in order to be able to move the decimal point five places.

d. 9.3 cm = _____ dam

Solution

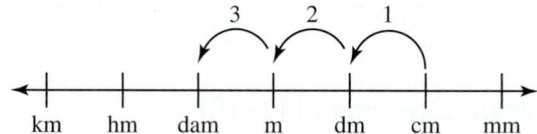

Since dam is three units to the left of cm, move the decimal point three places to the left. Therefore, 9.3 cm = 0.0093 dam. Note: It was necessary to write two 0s before the 9 in order to be able to move the decimal point three places.

PRACTICE EXERCISE 3

Convert each of the following to the indicated unit:

a. 76.323 dam = _____ dm

b. 5824 m = _____ hm

c. 2.45 km = _____ dm

d. 2.9 cm = _____ dam

If you need more practice, do the following Additional Practice Exercises.

Additional Practice Exercise 3   Convert each of the following to the indicated unit:

a. 4.478 hm = _____ m

b. 8345 dm = _____ km

c. 6.2 hm = _____ dm

d. 0.063 cm = _____ dm

Exercise Set R.7   For Extra Help   MyMathLab®

Convert the following to the indicated unit. (See Example 1.)

1. 2 ft = _____ in.

2. 36 in. = _____ ft

3. 9 ft = _____ yd

4. 4 yd = _____ ft

5. 5 mi = _____ ft

6. 15,840 ft = _____ mi

7. 2 mi = _____ yd

8. 3520 yd = _____ mi

9. 60 in. = _____ ft　　**10.** 8 ft = _____ in.　　**11.** 24 ft = _____ yd　　**12.** 15 yd = _____ ft

13. 288 in. = _____ yd　　**14.** 4.5 yd = _____ in.　　**15.** 255 yd = _____ ft　　**16.** 324 in. = _____ yd

17. 5.25 mi = _____ yd　　**18.** 2.5 yd = _____ in.　　**19.** 31,680 ft = _____ mi　　**20.** 7040 yd = _____ mi

Convert the following to the indicated unit. (See Example 3.)

21. 3 hm = _____ m　　**22.** 500 m = _____ hm　　**23.** 400 mm = _____ dm　　**24.** 15 dm = _____ mm

25. 8 dam = _____ cm　　**26.** 12,000 cm = _____ dam　　**27.** 1.4 km = _____ m　　**28.** 500 m = _____ km

29. .7 m = _____ mm　　**30.** 80 mm = _____ m　　**31.** 458 dm = _____ dam　　**32.** 56 dam = _____ cm

33. 10,000 m = _____ km　　**34.** 7.89 km = _____ m　　**35.** 8.4 dm = _____ m　　**36.** 6.97 hm = _____ km

37. .009 km = _____ m　　**38.** 8.3 m = _____ mm　　**39.** 16.2 dm = _____ hm　　**40.** .005 m = _____ mm

Challenge Exercises (41–48)

Convert the following to the indicated unit. (See Examples 1 and 3.)

41. 540 in. = _____ yd　　**42.** 243 yd = _____ in.　　**43.** 6.7 mi = _____ ft

44. 18.75 yd = _____ in.　　**45.** 5280 mm = _____ km　　**46.** .0046 hm = _____ cm

47. .0000001 m = _____ mm　　**48.** 8903 km = _____ dm

Writing Exercises (49–51)

49. List the conversion factors in the American system for measuring weight.

50. List the conversion factors in the American system for measuring dry volume.

51. List the conversion factors in the American system for measuring liquid volume.

Chapter R Summary

Concept/Procedure	Example
Numerals [Section R.1]	
• Digits are the symbols 0, 1, 2, 3, 4, 5, 6, 7, 8, and 9.	**Example 1:**
• A numeral is a symbol composed of digits used with place value.	Give the meaning of the underlined digit in 73,<u>4</u>60.
• Determining place value involves assigning the digits of a numeral values starting at the right and going left from one, then ten, one hundred, one thousand, ten thousand, one hundred thousand, one million, and so on.	Solution: The 4 is in the 3rd place, so it represents 4 hundreds, or $4 \cdot 100 = 400$.

Concept/Procedure	Example
Notation [Section R.1]	
• A numeral is in expanded notation when it is formed by writing the sum of the products of each digit multiplied by its place value.	**Example 2:** Write 56,201 in expanded notation.
• Standard notation is the way we normally write numerals.	Solution: $5 \cdot 10{,}000 + 6 \cdot 1000 + 2 \cdot 100 + 0 \cdot 10 + 1 \cdot 1$ $= 50{,}000 + 6000 + 200 + 1$
• The word name of a numeral consists of the words we say when we read aloud a numeral written in standard notation.	**Example 3:** Express four hundred eighty thousand, six hundred nine as a standard numeral. Solution: 480,609

Concept/Procedure	Example
Addition of Whole Numbers [Section R.2]	
• $a + b = c$, a is an addend, b is an addend, and c is the sum of a and b.	

Concept/Procedure	Example
Addition Algorithm [Section R.2]	
• To add whole numbers, line up the corresponding place values and add from right to left. If the sum in any corresponding place values is 10 or greater, then carry the excess to the next larger place value.	**Example 4:** Add: $\begin{array}{r} \overset{1}{5}37 \\ +492 \\ \hline 1029 \end{array}$

Concept/Procedure	Example
Difference of Whole Numbers [Section R.2]	
• If $a - b = c$, then $b + c = a$, a is the minuend, b is the subtrahend, and c is the difference.	

Concept/Procedure	Example
Subtraction Algorithm [Section R.2]	
• To subtract two whole numbers, line up the corresponding place values and subtract from right to left. If the digit being subtracted is larger than the digit it is being subtracted from, then borrow one from the digit in the next-larger place value.	**Example 5:** Subtract: $\begin{array}{r} \overset{5\,12\,15}{6{,}635} \\ -586 \\ \hline 6{,}049 \end{array}$

Concept/Procedure	Example
Product of Whole Numbers [Section R.3] • $a \cdot b = c$, a is a factor, b is a factor, and c is the product.	

Multiplication Algorithm [Section R.3]

• To multiply two whole numbers, follow these steps:

1. Line up the place values, multiply each digit of the multiplicand by the units digit of the multiplier (carrying as necessary), and place the product directly beneath the multiplier.
2. Multiply each digit of the multiplicand by the tens digit of the multiplier (carrying as necessary) and place the product beneath the previous product, but indented one place to the left.
3. Continue until you have multiplied the multiplicand by all the digits of the multiplier.
4. Add the results.

Example 6:

Multiply:

$$\begin{array}{r} \overset{1}{\overset{24}{748}} \\ \times 26 \\ \hline 4488 \\ 1496 \\ \hline 19,448 \end{array}$$

748   Multiplicand
×26   Multiplier
4488   Multiply 748 by 6.
1496   Multiply 748 by 2, but indent one place to the left.
19,448   Add the results.

Division by Zero Is Impossible [Section R.3]

• $a \div 0$ is undefined.

Quotient of Whole Numbers [Section R.3]

• $a \div b = c$, if $b \cdot c = a$, b is the divisor, a is the dividend, and c is the quotient. In division, we say, "How many b's are in a?" Division may be checked with multiplication.

Division of Whole Numbers [Section R.3]

• See Examples 5 and 6.

Example 7:

Divide:

$$\begin{array}{r} 133 \\ 28\overline{)3741} \\ -28 \\ \hline 94 \\ -84 \\ \hline 101 \\ -84 \\ \hline 17 \end{array}$$

1. 28 goes into 37 1 time. Put 1 above 7.
2. Multiply 1 times 28 and put the product beneath 37.
3. Subtract 28 from 37 and bring down the 4.
4. 28 goes into 94 3 times. Put 3 above 4.
5. Multiply 3 times 28 and put the result beneath 94.
6. Subtract 84 from 94 and bring down the 1.
7. 28 goes into 101 3 times. Put 3 above 1.
8. Multiply 3 times 28 and put the result beneath 101.
9. Subtract 84 from 101. Since there is nothing else to bring down, and 17 < 28, 17 is the reminder.

Therefore, $3741 \div 28 = 133 \text{ R } 17$

Concept/Procedure	Example
Ratio Fractions [Section R.4] • $\frac{a}{b}$ The top number, a, is the numerator, and the bottom number, b, is the denominator (not zero); a form of division.	
Proper and Improper Fractions: [Section R.4] a. A fraction is **proper** if the numerator is less than the denominator. b. A fraction is **improper** if the numerator is greater than or equal to the denominator.	
Mixed Number: [Section R.4] • A number is mixed if it is greater than 1 and has a whole part and a fractional part.	
Changing a Mixed Number into an Improper Fraction, and Vice Versa: [Section R.4] a. To change a mixed number into an improper fraction, multiply the denominator of the fractional part by the whole part and add the numerator of the fractional part. This is the numerator of the improper fraction, and the denominator is the denominator of the fractional part. b. To change an improper fraction into a mixed number, divide the denominator into the numerator. The number of times the denominator divides into the numerator is the whole part, and the number left over is the numerator of the fractional part.	Example 8: Convert $5\frac{3}{8}$ into an improper fraction. Solution: Multiply the denominator, 8, by the whole part, 5, and add the numerator, 3. Put the result over the denominator, 8. Therefore, $5\frac{3}{8} = \frac{8 \cdot 5 + 3}{8} = \frac{43}{8}$. Example 9: Convert $\frac{49}{6}$ into a mixed number. Solution: Divide 6 into 49 and put the remainder over 6. $$\begin{array}{r} 8 \\ 6\overline{)49} \\ -48 \\ \hline 1 \end{array}$$ Therefore, $\frac{49}{6} = 8\frac{1}{6}$.
Reducing Fractions to Lowest Terms [Section R.4] • To reduce a fraction to lowest terms, divide both the numerator and denominator by the largest whole number that divides evenly into both.	Example 10: Reduce $\frac{10}{15}$ to lowest terms. $\frac{10}{15} = \frac{10 \div 5}{15 \div 5} = \frac{2}{3}$

Concept/Procedure	Example

Multiplication of Fractions [Section R.4]

- $\frac{a}{b} \times \frac{c}{d} = \frac{a \cdot c}{b \cdot d}$ For all numbers a, b, c, d (with b and d not zero), to multiply two fractions, find the product of the numerators divided by the product of the denominators.

Example 11:

Multiply: $\dfrac{3}{5} \cdot \dfrac{2}{7} = \dfrac{3 \cdot 2}{5 \cdot 7} = \dfrac{6}{35}$ Multiply numerator by numerator and denominator by denominator.

Quotient of Ratio Fractions [Section R.4]

- $\frac{a}{b} \div \frac{c}{d} = \frac{a}{b} \times \frac{d}{c}$ For all numbers a, b, c, d (with $b, c,$ and d not zero), to divide a fraction by a fraction, multiply the first fraction by the multiplicative inverse (reciprocal) of the second fraction.

Example 12:

Divide: $\dfrac{7}{8} \div \dfrac{4}{3} = \dfrac{7}{8} \cdot \dfrac{3}{4} = \dfrac{7 \cdot 3}{8 \cdot 4} = \dfrac{21}{32}$ Invert the divisor and multiply.

Sum of Ratio Fractions with the Same Denominators [Section R.4]

- $\frac{a}{c} + \frac{b}{c} = \frac{a + b}{c}$ For all numbers $a, b,$ and c, with c not zero, to add two or more fractions with the same denominators, add the numerators and place the sum over the same denominator.

Example 13:

Add: $\dfrac{4}{11} + \dfrac{6}{11} = \dfrac{4 + 6}{11} = \dfrac{10}{11}$ Add the numerators and put the result over the denominator.

Difference of Ratio Fractions with the Same Denominators [Section R.4]

- $\frac{a}{c} - \frac{b}{c} = \frac{a - b}{c}$ For c not zero, to subtract two fractions with the same denominator, subtract the numerators and place the difference over the same denominator.

Example 14:

Subtract: $\dfrac{13}{17} - \dfrac{5}{17} = \dfrac{13 - 5}{17} = \dfrac{8}{17}$ Subtract the numerators and put the result over the denominator.

Decimal Point [Section R.5]

- A symbol used to denote fractions with denominators of 10, 100, 1000, and so on that are written in decimal form.

Decimal Fraction [Section R.5]

- A fraction with a denominator of 10, 100, 1000, and so on that is written in the form .1, .01, .001, .0001, and so on.

Decimal Mixed Numeral [Section R.5]

- Numerals in decimal notation that include both a whole number and a decimal fraction are decimal mixed numerals.

Concept/Procedure	Example
Reading a Decimal [Section R.5] • Read the whole number; then read the decimal as "and"; finally, read the decimal part as a whole number followed by the place value of the last digit.	Example 15: Write how 258.62 would be read. Solution: Two hundred fifty-eight and sixty-two hundredths.
Comparing Decimals [Section R.5] • Compare decimals by comparing corresponding place values from left to right.	

Addition and Subtraction of Decimals [Section R.5]
 	
1. Write the numerals with corresponding place values lined up in the same columns. Add zeros as needed.
2. Add or subtract the columns as in whole numbers.
3. Bring the decimal straight down.
	Example 16:
	Add 136.7, 23.48, and 1258.2.
	
	Solution: 0136.70
	    0023.48
	    +1258.20
	    1418.38

Multiplication Rule for Decimals [Section R.6]
 	
1. Ignore the decimal points and multiply as in whole numbers.
2. Count the number of decimal value places in each of the factors.
3. Find the sum of the decimal value places in the factors.
4. Place this number of decimal value places in the product, counting from right to left.
	Example 17:
	Multiply 35.18 by 5.7.
	
	Solution: 35.18 Two decimal places.
	    ×5.7 One decimal place.
	    24626
	    17590
	    200.526 Product has 2 + 1 = 3 decimal places.

Quotient of Decimals [Section R.6] • In $a \div b = c$, if $b \cdot c = a$, b is the divisor, a is the dividend, and c is the quotient. Division may be checked with multiplication.	

Concept/Procedure	Example

Rounding to the Nearest Place Value [Section R.6]

- Writing a number to the nearest desired value place is called rounding.
- To round to the nearest place value, do the following:
  1. Go to the first value place to the right of the place under consideration.
  2. If this digit is 5 or more, add 1 to the digit under consideration and drop the digits that follow.
  3. If this digit is 4 or less, drop it and all digits that follow. Keep the digit under consideration.

Example 18:

Round 241.345 to the nearest tenth.

Solution: Since the hundredths digit is 4, drop the 4 and all digits following, so that 241.345 = 241.3 to the nearest tenth.

Division Rule for Decimals [Section R.6]

- In dividing decimal numerals, apply these rules:
  1. Make the divisor a whole number by moving the decimal point the required number of places to the right. Then move the decimal point the same number of places to the right in the dividend. If the divisor is already a whole number, go to step 2.
  2. Place the decimal point in the quotient directly above the decimal point in the dividend.
  3. Carry out the whole-number division.

Example 19:

Find the quotient. $1.512 \div .12$

Solution: $.12\overline{)1.512}$ Multiply the divisor and dividend by 100 to make the divisor a whole number.

$$
\begin{array}{r}
12.6 \\
12\overline{)151.2} \\
\underline{12} \\
31 \\
\underline{24} \\
72 \\
\underline{72} \\
0
\end{array}
$$

Bring the decimal straight up and proceed as in whole-number division.

Changing a Rational Number into a Decimal: [Section R.6]

- To change a rational number into a decimal, divide the denominator into the numerator, and if necessary, round the quotient to the indicated number of decimal places.

Example 20: Convert $\dfrac{3}{7}$ into a decimal and round to the nearest hundredth.

Solution: Divide the denominator into the numerator to the nearest thousandths and then round to the nearest hundredth.

$$
\begin{array}{r}
.428 \\
7\overline{)3.000} \\
\underline{-28} \\
20 \\
\underline{-14} \\
60 \\
\underline{-56} \\
4
\end{array}
$$

Therefore, $\dfrac{3}{7} = 0.43$ to the nearest hundredth.

Concept/Procedure	Example

Conversion Factors—American System [Section R.7]

- 1 ft = 12 in.
- 1 yd = 3 ft
- 1 yd = 36 in.
- 1 mi = 5280 ft
- 1 mi = 1760 yd

Conversions by the Factor-Label Method [Section R.7]

- To convert from one unit to another using the factor-label method, multiply the expression to be converted by the ratio of the conversion factors in which the numerator is the unit into which we wish to convert and the denominator is the unit from which we are converting.

Example 21:

Convert 252 inches to yards.

Solution: 252 inches = $\quad$ Multiply 252 inches by $\dfrac{1 \text{ yard}}{36 \text{ inches}}$.

$252 \text{ in.} \cdot \dfrac{1 \text{ yd}}{36 \text{ in.}} = \quad$ Divide the inches.

$252 \text{ in.} \cdot \dfrac{1 \text{ yd}}{36 \text{ in.}} = \quad$ Think of 252 as $\dfrac{252}{1}$.

$\dfrac{252}{1} \cdot \dfrac{1 \text{ yd}}{36} = \quad$ Multiply.

$\dfrac{252 \text{ yd}}{36} = \quad$ Divide 36 into 252.

$7 \text{ yd} \qquad$ Therefore, 252 inches = 7 yards.

Metric Prefixes [Section R.7]

- kilo- (k) = 1000 units
- hecto- (h) = 100 units
- deca- (da or dk) = 10 units
- deci- (d) = $\frac{1}{10}$ of a unit
- centi- (c) = $\frac{1}{100}$ of a unit
- milli- (m) = $\frac{1}{1000}$ of a unit

Converting within the Metric System [Section R.7]

- Using the diagram that follows, locate the unit given and count the number of places, left or right, to the unit into which we are converting. Then move the decimal point this number of place values in the same direction.

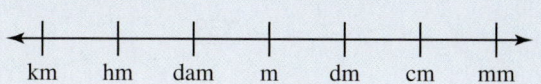

Example 22:

Convert 6458 cm into m.

Solution:

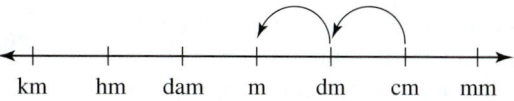

Since m is two places to the left of cm, move the decimal point two places to the left.

Therefore, 6458 cm = 64.58 m.

Chapter R Review Exercises

Write the word name for each standard numeral. [Section R.1]

1. 6051

2. 30,459

Write as a standard numeral. [Section R.1]

3. Thirty-seven thousand, two hundred four

Give the place value of the underlined digit for each of the following: [Section R.1]

4. 7̲2,891

5. 17̲2,808,369

What is the meaning of the designated digit in each of the following numerals? [Section R.1]

6. 4 in 107,461,077

7. 0 in 8208

Write in expanded notation. [Section R.1]

8. 396,071

Add each of the following: [Section R.2]

9.   54
      +45

10.   58
       +27

11. 583
       +45

12. 5409
       +747

13. 652,428
       +76,923

14. 5541
          27
        7019
        1510
        +286

Solve the application problems. [Section R.2]

15. A football team scored 14 points in the first quarter, 7 points in the second quarter, 0 points in the third quarter, and 21 points in the fourth quarter. How many points did this team score in the four quarters of the game?

16. In 1981, Kay and Dani planted a 6-foot-tall oak tree. Since then, it has grown 55 feet. How tall is it now?

17. Last year, the parks department planted 18,000 trees. This year they planted 27,000 trees. How many trees have they planted in the last 2 years?

18. A TV cable is 8 feet long. Twelve more feet of cable are needed to reach the TV. How long must the cable be?

Subtract each of the following: [Section R.2]

19. $729 - 41$

20. $623 - 544$

21. $82,501 - 7345$

Solve the application problems. Some of them may contain extra information or involve addition as well as subtraction. [Section R.2]

22. Marti has 350 miles to drive from her hometown to the university she attends. If she has driven 168 miles, how much farther does she have to go?

23. Dr. Lopez is cleaning her office. She has 164 books, but 28 of them will not fit on the shelves. How many will she have left if she gives the 28 extra books to the campus library?

24. Fran and Dan were sitting in the student union, comparing how many relatives each has. Fran has 13 first cousins, and Dan has 47 first cousins. How many more first cousins does Dan have than Fran?

25. Cesar got his paycheck today. Out of his salary of $875, the following amounts were withheld: $175 for income tax, $70 for FICA, $53 for retirement, and $71 for health insurance. How much does he actually take home?

Multiply. [Section R.3]

26.   65
       ×7

27.   76
       ×39

28. 306
$\times$42

29. 671
$\times$30

30. 415
$\times$108

Solve each of the exercises. You may need to do more than just multiply. There may be extra information as well. [Section R.3]

31. As part of her financial-aid package, Penny got a job working in the chemistry lab. She washed 15 loads of 24 test tubes in the dishwasher. How many test tubes did she wash?

32. In the Tournament of Roses parade, one band marched 13 members across and 15 rows deep. How many band members were there?

33. Juan bought five shirts at $35 each and three pairs of pants at $47 each. How much did he spend on shirts and pants?

Divide by using the standard process of long division and check by using the related multiplication. [Section R.3]

34. $481 \div 13$

35. $625 \div 7$

36. $18\overline{)594}$

37. $21\overline{)9542}$

Solve the problems. There may be more than one operation, and there may be more information than needed. [Section R.3]

38. The county will pave a gravel road in front of the property of eight landowners if the landowners will pay the cost of $42,000. If they shared the cost equally, how much would each landowner have to pay?

39. There are 1050 ROTC cadets at Land Grant University. How many classes of 30 students each can be formed?

Convert the improper fractions into mixed numbers. [Section R.4]

40. $\dfrac{17}{8}$

41. $\dfrac{35}{6}$

42. $\dfrac{53}{21}$

43. $\dfrac{81}{33}$

Change the mixed numbers into improper fractions. [Section R.4]

44. $5\dfrac{4}{7}$

45. $6\dfrac{7}{9}$

46. $3\dfrac{7}{18}$

47. $8\dfrac{5}{23}$

Multiply. [Section R.4]

48. $\dfrac{3}{4} \cdot \dfrac{5}{2}$

Divide. [Section R.4]

49. $\dfrac{6}{7} \div \dfrac{11}{9}$

Add. [Section R.4]

50. $\dfrac{3}{11} + \dfrac{4}{11}$

Subtract. [Section R.4]

51. $\dfrac{8}{9} - \dfrac{4}{9}$

Add. [Section R.5]

52. $1.036 + .093$

53. $79.149 + 118$

Subtract. [Section R.5]

54. $7.58 - 5.38$

55. $201.9 - 55.09$

56. $671.26 - 167.52$

Add the following: [Section R.5]

57. $16.77 + 371.16 + 58.55 + 21.462$

Solve. Each exercise may require addition or subtraction or both. [Section R.5]

58. Jill bought 1.82 pounds of pork chops and 1.99 pounds of spareribs for a family barbecue. How many pounds of pork did she buy?

59. Last month the gas bill included a $10.67 "energy" charge, a $35.21 cost of gas charge, a $6.60 "gas adjustment" charge, and a $10.27 utility tax charge. How much was the total bill?

60. Jose Canseco had a batting average of .269 during the regular baseball season. He had a .357 batting average for the World Series. How much higher was his batting average during the series?

Multiply. [Section R.6]

61. .04(.85)   **62.** 6.78(5.7)

Round to the indicated place value. [Section R.6]

63. 38.689, nearest hundredth

64. 73.49, nearest tenth

Divide. Round to the nearest hundredth if necessary. [Section R.6]

65. 2.7 ÷ .9   **66.** 19.3 ÷ .62

67. 78.53 ÷ 2.55

Change the fractions into decimals. If necessary, round the answer to the nearest hundredth. [Section 6.1]

68. $\dfrac{5}{16}$ **69.** $\dfrac{7}{9}$

70. $\dfrac{11}{13}$ **71.** $\dfrac{12}{19}$

72. $\dfrac{11}{5}$ **73.** $\dfrac{17}{4}$

Solve each of the problems. If necessary, round to the nearest hundredth. [Section R.6]

74. Little League baseball cleats cost $37.54 per pair. How much would a sponsor have to pay to buy shoes for his team of 12 children?

75. There are 355 milliliters (ml) in 12 ounces (oz) of cola. How many ml are in 1 oz?

76. Karol sent three pairs of pants and five blouses to the dry cleaner. It cost $3.95 per pair of pants and $4.49 per blouse. How much was her total dry-cleaning bill?

Convert the following to each indicated unit, using the factor-label method where appropriate: [Section R.7]

77. 5 ft = _____ in.   **78.** 1.2 yd = _____ ft

79. 90 in. = _____ ft   **80.** 3.7 m = _____ mm

81. 18 km = _____ cm   **82.** .89 hm = _____ dm

83. $\dfrac{1}{4}$ yd = _____ in.   **84.** 16,896 ft = _____ mi

85. 9215 mm = _____ m   **86.** 43.6 hm = _____ dm

Chapter R   Test

1. Give the place value of the underlined digit in 98<u>3</u>,614.

2. Write 82,063 in expanded notation.

Add.

3. 728
    +674

4. 88,913
    +7,108

5.   907
    5831
     412
    6053
    +6980

6. Clara collects baseball cards. She has 143 from 1990, 208 from 1991, and 182 from 1992. How many cards does she currently have?

Subtract.

7. 11,683
−5,702

8. 590,042
−72,672

9. Angela planted 1050 seeds for her biology project. Only 964, however, sprouted. How many seeds did not sprout?

10. Change $\frac{29}{13}$ to a mixed number.

11. Change $6\frac{4}{7}$ to an improper fraction.

Multiply.

12. 749
×8

13. 805
×67

14. There are 38 students in each of the 17 classes of integrated algebra and arithmetic. Each student needs a #2 pencil. How many pencils are needed?

Divide and, if necessary, round to the nearest hundredth.

15. $9\overline{)855}$

16. $26\overline{)966}$

Multiply.

17. $\frac{2}{5} \cdot 9$

Subtract.

18. $\frac{11}{12} - \frac{6}{12}$

19. Which is larger, 70.2259 or 80.967?

Add.

20. 68.52 + 469.514

Subtract.

21. 60.38 − 45.726

22. Bonnie needs 8.5 pounds of hamburger meat to make hamburgers for her friends after the football game. She already has 2.3 pounds in the freezer. How many more pounds of meat does she need?

Multiply.

23. 8.25(.47)

24. 5.26(.363)

Round to the nearest hundredth.

25. 3.6634

26. 64.999

Divide. Round to the nearest hundredth.

27. 27.283 ÷ 13

28. 843 ÷ .9

29. Write $\frac{5}{16}$ as a decimal rounded to the nearest hundredth.

Convert the following to each indicated unit:

30. 42 ft = _____ yd

31. 5.5 ft = _____ in.

32. 7.6 hm = _____ cm

33. .09 m = _____ mm

77

1

Adding and Subtracting Integers and Polynomials

Chapter Outline

We begin the chapter by introducing exponents and variables. Variables allow us to generalize the concepts of arithmetic and expand on these concepts. This is followed by the order of operations on whole numbers needed to do the sections on geometry and signed numbers. Next, we discuss addition and subtraction of signed numbers and show how these operations are used to combine like terms and polynomials.

Much of arithmetic is the study of performing operations on numbers. Much of algebra is performing these same operations on polynomials.

Section 1.1 Variables, Exponents, and Order of Operations

OBJECTIVES *When you complete this section, you will be able to:*

A Identify variables and constants.
B Read and evaluate expressions raised to powers.
C Simplify expressions involving more than one operation.
D Evaluate expressions containing variables, when given the value(s) of the variable(s).

PREREQUISITE SKILLS *Before beginning this section you should be able to:*

a. Add, subtract, multiply, and divide whole numbers. (Sections R.2 and R.3)

GETTING READY FOR SECTION 1.1

Add, subtract, multiply, or divide the following whole numbers:

1. $19 + 43$ **2.** $65 - 28$ **3.** $6 \cdot 6 \cdot 6$ **4.** $\dfrac{68}{17}$

Introduction

In this book, we will discuss some algebra, much of which is arithmetic from a more general point of view. This more general point of view is made possible by the use of **variables**. You have previously used variables in formulas for the perimeters, areas, or volumes of geometric figures. For example, in the formula for the circumference of a circle, $C = 2\pi r$, both C and r are variables. When given a value for r, we can find a value for C. Remember that π is a **constant** whose approximate value is 3.14.

OBJECTIVE Identifying variables and constants

> **Definition** Variable and Constant
> A **variable** is a symbol, usually a letter, that is used to represent a number.
> A **constant** is a symbol (a numeral) whose value does not change.

Variables are to mathematics as pronouns are to a language. Pronouns take the place of nouns, and variables take the place of numbers.

Constants are used in a variety of ways in mathematics. Often they are given special names, depending on how they are used.

Example 1 **Identify the constant and the variable(s) in each of the following:**

a. $3x$ The constant is 3 and the variable is x.

b. $4xy$ The constant is 4 and the variables are x and y.

c. n This means $1 \cdot n$, so the constant is 1 and the variable is n.

d. 5 The constant is 5 and there is no variable.

Answers: **Getting Ready:** **1.** 62 **2.** 37 **3.** 216 **4.** 4

PRACTICE EXERCISE 1

Identify the constant and the variable(s) in each of the following:

a. $6y$ Constant = _____ , variable = _____ .

b. $5xy$ Constant = _____ , variables = _____ and _____ .

c. z Constant = _____ , variable = _____ .

d. 8 Constant = _____ , variable = _____ .

In arithmetic, multiplication is a short way of performing repeated addition of the same number. For example, $4 \cdot 3$ means $3 + 3 + 3 + 3$. The "4" tells us how many 3s are being added. Likewise, there is a short way of indicating repeated multiplication of the same number by the use of **exponents**. For example, $3 \cdot 3 \cdot 3 \cdot 3$ is written as 3^4, where 4 is the **exponent** and 3 is the **base** of the exponent. Since each of the numbers being multiplied is called a **factor**, the exponent tells us how many times the base is used as a factor.

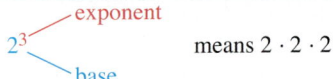

2^3 means $2 \cdot 2 \cdot 2$

OBJECTIVE **B** **Reading and evaluating expressions raised to powers**

In reading expressions that have exponents, the base is read first and then the exponent is read as a power. The preceding expression, 2^3, is read "two to the third power." The power 2 is most often read as "squared," and the power 3 is most often read as "cubed." Consequently, 2^3 could also be read as "two cubed." The reasons for reading the second power as squared and the third power as cubed will be made clear after the sections on geometry. There are no special names for powers greater than three.

Since exponents indicate multiplication, constants raised to powers can be evaluated.

Example 2

	Exponential Expression	Read as	Meaning	Value
a.	5^2	5 squared, or 5 to the second power	$5 \cdot 5$	25
b.	6^3	6 cubed, or 6 to the third power	$6 \cdot 6 \cdot 6$	216
c.	8^5	8 to the fifth power	$8 \cdot 8 \cdot 8 \cdot 8 \cdot 8$	32,768
d.	$\left(\dfrac{2}{3}\right)^4$	Two-thirds to the fourth power	$\dfrac{2}{3} \cdot \dfrac{2}{3} \cdot \dfrac{2}{3} \cdot \dfrac{2}{3}$	$\dfrac{16}{81}$

PRACTICE EXERCISE 2

Complete the following table:

	Exponential Expression	Read as	Meaning	Value
a.	4^2			
b.	5^3			
c.	3^6			
d.	$\left(\dfrac{1}{2}\right)^5$			

Raising variables to powers has the exact same meaning as raising constants to powers. However, great care must be taken in identifying the base of the exponent.

Example 3   **Write each of the following as repeated multiplication:**

a. $x^4 = x \cdot x \cdot x \cdot x$

b. $3x^5 = 3 \cdot x \cdot x \cdot x \cdot x \cdot x$

NOTE: In part b, 3 is *not* raised to the fifth power; only *x* is.

c. $n^3 \cdot y^2 = n \cdot n \cdot n \cdot y \cdot y$

d. $3^3 \cdot x^5 = 3 \cdot 3 \cdot 3 \cdot x \cdot x \cdot x \cdot x \cdot x$

e. $(3a)^3 = (3a)(3a)(3a)$

NOTE: The parentheses are used to indicate that the base is 3*a*.

PRACTICE EXERCISE 3

Write each of the following as repeated multiplication:

a. $z^3 =$ **b.** $4y^3 =$ **c.** $n^4y^5 =$

d. $7^2x^6 =$ **e.** $(2y)^5 =$

Example 4   **Write each of the following by using exponents:**

a. $a \cdot a \cdot a \cdot a \cdot a = a^5$

b. $5 \cdot r \cdot r = 5r^2$

c. $n \cdot n \cdot n \cdot p \cdot p \cdot q \cdot q \cdot q = n^3 p^2 q^3$

d. $8 \cdot 8 \cdot 8 \cdot 8 \cdot y \cdot y \cdot y \cdot z \cdot z = 8^4 y^3 z^2$

e. $(8z)(8z)(8z)(8z) = (8z)^4$

Answers: Practice Exercise 2: a. 4 squared or 4 to the second power, $4 \cdot 4$, 16 **b.** 5 cubed or 5 to the third power, $5 \cdot 5 \cdot 5$, 125 **c.** 3 to the sixth power, $3 \cdot 3 \cdot 3 \cdot 3 \cdot 3 \cdot 3$, 729 **d.** One-half to the fifth power, $\dfrac{1}{2} \cdot \dfrac{1}{2} \cdot \dfrac{1}{2} \cdot \dfrac{1}{2} \cdot \dfrac{1}{2}$, $\dfrac{1}{32}$ **Practice Exercise 3: a.** $z \cdot z \cdot z$ **b.** $4 \cdot y \cdot y \cdot y$ **c.** $n \cdot n \cdot n \cdot n \cdot y \cdot y \cdot y \cdot y \cdot y$ **d.** $7 \cdot 7 \cdot x \cdot x \cdot x \cdot x \cdot x \cdot x$ **e.** $(2y)(2y)(2y)(2y)(2y)$

PRACTICE EXERCISE 4

Write each of the following by using exponents:

a. $a \cdot a \cdot a \cdot a \cdot a =$

b. $8 \cdot y \cdot y \cdot y \cdot y \cdot y =$

c. $x \cdot x \cdot x \cdot x \cdot x \cdot y \cdot y \cdot y \cdot z \cdot z =$

d. $4 \cdot 4 \cdot 4 \cdot w \cdot w \cdot w \cdot w =$

e. $(6w)(6w)(6w) =$

OBJECTIVE Simplifying expressions involving more than one operation

In mathematics, we have the operations of addition, subtraction, multiplication, and division. We just introduced raising to powers, and later we will discuss finding roots. Suppose a problem has more than one operation. Which one do we do first? For example, if we add first, then $2 + 3 \cdot 4 = 5 \cdot 4 = 20$; but if we multiply first, then $2 + 3 \cdot 4 = 2 + 12 = 14$. Both answers cannot be correct, so which one is correct?

Consider the following situations:

a. You are shopping for some clothes. You buy a shirt for \$30 and four pairs of socks for \$2 per pair. The amount you would pay is $30 + 4 \cdot 2 = \$38$. Did you add or multiply first?

b. An item is marked "three for \$12." How much would you pay for two of these items? The amount you would pay is $12 \div 3 \cdot 2 = \$8$. Did you divide or multiply first?

c. Three friends are buying soft drinks for a party. They purchase five cartons at \$6 per carton. If they pay equal amounts, how much does each pay? The amount each pays is $5 \cdot 6 \div 3 = \$10$. Do you multiply or divide first?

d. You purchase a jacket for \$30 and two shirts marked at \$20 each, but with a tag for \$5 off on each. You also return a pair of pants originally marked \$15, but purchased with a \$3 discount. How much money do you owe? You owe $30 + 2(20 - 5) - (15 - 3) = \$48$. In what order would you perform these calculations?

In situation a, you multiply before adding. In situation b, you divide before multiplying, but in situation c you multiply before dividing. In situation d, you simplify inside the parentheses first, then multiply by 2, and finally add and subtract in order from left to right.

It appears that we need some rules as to which order to perform operations in problems that involve more than one operation. Hence, we have the following order of operations:

Procedure: Order of Operations

If an expression contains more than one operation, operations are to be performed in the following order:

1. If parentheses or other inclusion symbols (braces or brackets) are present, begin within the innermost and work outward, using the order in steps 3–5 in doing so.

2. If an implied grouping like a fraction bar is present, simplify above and below the fraction bar separately in the order given by steps 3–5.

3. Evaluate all indicated powers.

4. Perform all multiplications or divisions in the order in which they occur as you work from left to right.

5. Then perform all additions or subtractions in the order in which they occur as you work from left to right.

Answers: Practice Exercise 4: a. a^5 **b.** $8y^5$ **c.** $x^5 y^3 z^2$ **d.** $4^3 w^4$ **e.** $(6w)^3$

Example 5   **Simplify the following, using the order of operations:**

a. $4 \cdot 5 - 8 =$ Multiply before subtracting, so multiply 4 and 5.
   $20 - 8 =$ Subtract 8 from 20.
   12 Answer.

b. $5 + 12 \div 4 =$ Divide before adding, so divide 12 by 4.
   $5 + 3 =$ Add 5 and 3.
   8 Answer.

c. $24 \div 6 \cdot 2$ Multiply or divide left to right, so divide 24 by 6.
   $4 \cdot 2$ Multiply 4 and 2.
   8 Answer.

d. $2 + 3 \cdot 4^2 =$ Raise to powers first, so square 4.
   $2 + 3 \cdot 16 =$ Multiply before adding, so multiply 3 and 16.
   $2 + 48 =$ Add 2 and 48.
   50 Answer.

e. $5(6 - 3) + 2 =$ Perform operations inside parentheses first, so subtract 3 from 6.
   $5(3) + 2 =$ Multiply before adding, so multiply 5 and 3.
   $15 + 2 =$ Add 15 and 2.
   17 Answer.

f. $15 - 2(3 + 2^2) =$ Perform operations inside parentheses first, so square 2.
   $15 - 2(3 + 4) =$ There is still an operation inside parentheses, so add 3 and 4.
   $15 - 2(7) =$ Multiply before adding, so multiply 2 and 7.
   $15 - 14 =$ Subtract 14 from 15.
   1 Answer.

g. $15 + 3 \cdot 8 \div 12 =$ Multiply 3 and 8.
   $15 + 24 \div 12 =$ Divide 24 by 12.
   $15 + 2 =$ Add 15 and 2.
   17 Answer.

h. $\dfrac{4 + 3 \cdot 2^2}{2 + 3 \cdot 2} =$ Simplify above and below the fraction bar separately.
Square 2 in the numerator and multiply in the denominator.

   $\dfrac{4 + 3 \cdot 4}{2 + 6} =$ Multiply in the numerator and add in the denominator.

   $\dfrac{4 + 12}{8} =$ Add 4 and 12.

   $\dfrac{16}{8} =$ Divide 8 into 16.

   2 Answer.

PRACTICE EXERCISE 5

Simplify the following, using the order of operations:

a. $9 + 3 \cdot 6$

b. $12 - 10 \div 5 \cdot 3$

c. $20 \div 5 \cdot 2$

d. $5 + 4 \cdot 3^3$

e. $6(12 - 2^2) + 4$

f. $15 - 20 \div 4 \cdot 2 - 3$

g. $4 + 3(6 + 5 \cdot 3^2) - 10$

h. $\dfrac{8 + 15 \cdot 2}{2 \cdot 6 + 7}$

If you need more practice, do the following Additional Practice Exercises.

Additional Practice Exercise 5   Simplify the following, using the order of operations:

a. $5 + 4(6)$

b. $24 - 21 \div 3$

c. $4 + 2 \cdot 3^3$

d. $36 \div 6 \cdot 2$

e. $7(25 - 15 \div 3) - 8$

f. $6 + 2(9 + 3 \cdot 7)$

g. $24 - 15 \div 3 \cdot 2^2$

h. $\dfrac{6 + 12 \div 3}{2 \cdot 3 - 1}$

Answers: **Practice Exercise 5:**   **a.** 27 **b.** 6 **c.** 8 **d.** 113 **e.** 52 **f.** 2 **g.** 147 **h.** 2   **Additional Practice 5:**   **a.** 29
b. 17 **c.** 58 **d.** 12 **e.** 132 **f.** 66 **g.** 4 **h.** 2

OBJECTIVE Evaluating expressions containing variables

The same order of operations applies when working with expressions containing variables. If a variable and a constant, or two variables, are written together with no operation sign, the operation is understood to be multiplication. For example, $3x$ means $3 \cdot x$ and xy means $x \cdot y$. We substitute the value(s) for the variable(s) by placing the value in parentheses and simplify by performing the correct order of operations.

Example 6 **Let $x = 2$ and $y = 5$ and evaluate.**

Note

Remember that $3x + 4y$ means $3 \cdot x + 3 \cdot y$.

a. $3x + 4y =$ Substitute 2 for x and 5 for y.

$3(2) + 4(5) =$ Multiply in order from left to right, so multiply $3(2)$ and $4(5)$.

$6 + 20 =$ Add 6 and 20.

26 Answer.

b. $2x^2 + 3y^3 =$ Substitute 2 for x and 5 for y.

$2(2)^2 + 3(5)^3$ First, raise to powers.

$2(4) + 3(125) =$ Multiply in order from left to right.

$8 + 375 =$ Add.

383 Answer.

c. $5x^2y^2 =$ Substitute 2 for x and 5 for y.

$5(2)^2(5)^2 =$ Raise to powers.

$5 \cdot 4 \cdot 25 =$ Multiply in order from left to right.

$20 \cdot 25 =$ Multiply.

500 Answer.

d. $10x^2 \div y =$ Substitute 2 for x and 5 for y.

$10(2)^2 \div 5 =$ Raise to powers.

$10(4) \div 5 =$ Multiply and divide in order from left to right, so multiply 10 and 4.

$40 \div 5 =$ Divide.

8 Answer.

PRACTICE EXERCISE 6

Let $x = 3$ and $y = 4$ and evaluate.

a. $4x + 3y$ **b.** $2x^2 + 5y$ **c.** $2x^3y^2$ **d.** $(8x^3) \div (2y)$

If you need more practice, do the following Additional Practice Exercises.

Additional Practice Exercise 6 Let $x = 4$ and $y = 5$ and evaluate.

a. $2x^2 + y^2$ **b.** $3x^2y^2$ **c.** $3x^3y + 4y^2$ **d.** $(5x^3) \div (2y)$

Answers: Practice Exercise 6: **a.** 24 **b.** 38 **c.** 864 **d.** 27 Additional Practice 6: **a.** 57 **b.** 1200 **c.** 1060 **d.** 32

Order of operations occurs in everyday situations.

| **Example 7** | **Write an expression for each of the following and evaluate:** |

a. Tara buys living room furniture by paying $500 down and $150 for 18 months. Find the total amount that she pays.

Solution

The amount that she pays can be represented by $500 + 150 \cdot 18$.

$$
\begin{aligned}
\text{Amount paid} &= 500 + 150 \cdot 18 && \text{Multiply before adding.}\\
&= 500 + 2700 && \text{Add.}\\
&= 3200 && \text{Therefore, she pays \$3200 for the living room set.}
\end{aligned}
$$

b. Jonathon leased a car with the following terms: He paid $1000 down, $199 per month for 2 years, and 12 cents per mile for all mileage over 24,000. When he returned the car, it had 32,000 miles on the odometer. Find the total amount that he paid for the lease.

Solution

The amount that he paid can be represented by $1000 + 199 \cdot 2 \cdot 12 + .12(32{,}000 - 24{,}000)$. Note that 12 cents was changed to $.12$, since all the other amounts were given in dollars.

$$
\begin{aligned}
\text{Amount paid} &= 1000 + 199 \cdot 2 \cdot 12 + .12(32{,}000 - 24{,}000) && \text{Simplify inside the parentheses.}\\
&= 1000 + 199 \cdot 2 \cdot 12 + .12(8000) && \text{Multiply 199 and 2.}\\
&= 1000 + 398 \cdot 12 + .12(8000) && \text{Multiply } 398 \cdot 12.\\
&= 1000 + 4776 + .12(8000) && \text{Multiply .12 and 8000.}\\
&= 1000 + 4776 + 960 && \text{Add.}\\
&= 6736 && \text{Therefore, the lease cost \$6736.}
\end{aligned}
$$

PRACTICE EXERCISE 7

For each of the following, write an expression similar to those in Example 7 and evaluate:

a. Wolfgang bought a car by paying $1200 down and $350 per month for 4 years. What is the total amount that he paid for the car?

b. Five people agree to split the cost of some take-out food. Three of the orders cost $4.99 each, and two orders cost $6.99 each. They also purchased two 2-liter soft drinks that cost $1.29 each. To the nearest cent (hundredths place), how much did each pay?

Study Tip 1 Motivation—The Key to Success

The attitude with which you approach a course is the single most important factor in determining your success in that course. Begin with a positive attitude and a belief that you can be successful. Mathematics can be learned and understood. We have done our very best to take the mysteries out of mathematics. But for you to reach your goals, you need to make a commitment right now to attend class, to study as we suggest, to do your homework, and to give this course your very best effort. With the proper determination, you can master this course.

Answers: Practice Exercise 7: a. $1200 + 350(12)(4) = \$18{,}000$ **b.** $\dfrac{3(4.99) + 2(6.99) + 2(1.29)}{5} = \$6.31$

For Extra Help

Exercise Set 1.1   MyMathLab®

Write a statement indicating how each of the following would be read, and evaluate. (See Example 2.)

1. 8^2

2. 4^2

3. 5^3

4. 8^3

5. 9^4

6. 2^4

7. 2^5

8. 10^5

9. 10^6

10. 3^7

11. $\left(\dfrac{4}{5}\right)^2$

12. $\left(\dfrac{3}{4}\right)^2$

13. $(1.3)^3$

14. $(2.1)^3$

Write each of the following as multiplication. (See Example 3.)

15. a^4

16. c^6

17. a^2b^3

18. r^2s^4

19. $6b^4$

20. $3r^5$

21. $(4a)^4$

22. $(5z)^3$

Write each of the following by using exponents:

23. $p \cdot p \cdot p \cdot p$

24. $r \cdot r \cdot r \cdot r \cdot r$

25. $b \cdot b \cdot b \cdot b \cdot d \cdot d \cdot d$

26. $t \cdot t \cdot t \cdot t \cdot t \cdot f \cdot f \cdot f \cdot f$

27. $3 \cdot x \cdot x \cdot x \cdot x$

28. $5 \cdot y \cdot y \cdot y \cdot y \cdot y \cdot y \cdot y$

29. $(4r)(4r)(4r)(4r)(4r)(4r)$

30. $(6w)(6w)(6w)(6w)(6w)$

Simplify the following by using the order of operations. (See Example 5.)

31. $7 - 3 + 5$

32. $8 + 6 - 4$

33. $5 - 6 \div 2$

34. $8 + 15 \div 3$

35. $7 + 2 \cdot 6$

36. $10 - 2 \cdot 4$

37. $8 + 3 \cdot 4^2$

38. $25 - 3 \cdot 2^3$

39. $2^3 + 7 \cdot 3$

40. $3^4 + 12 \div 3$

41. $24 - 3(8 - 2)$

42. $25 - 5(12 - 7)$

43. $10 + (13 - 9) \div 2$

44. $15 + (24 - 6) \div 3$

45. $2^3 \cdot 6^2$

46. $2^5 \cdot 4^2$

47. $36 \div 9 \cdot 2 - 4$

48. $3 + 48 \div 6 \cdot 2$

49. $\left(\dfrac{1}{2}\right)^3 \cdot \left(\dfrac{5}{3}\right)^2$

50. $\left(\dfrac{4}{3}\right)^2 \cdot \left(\dfrac{1}{3}\right)^3$

51. $2^3 + 3 \cdot 4^2$

52. $3^4 - 4 \cdot 2^3$

53. $26 - 18 \div 3^2$

54. $32 - 24 \div 2^3$

55. $3^4 - 45 \div 3^2$

56. $5^3 - 4 \cdot 3^3$

57. $22 + 18 \div 3 - 12$

58. $31 - 24 \div 6 + 8$

59. $32 - 4^2 + (8 - 3)$

60. $9 - 2^2 + (11 - 7)$

61. $45 \div (18 - 3^2)$

62. $36 \div (22 - 2^2)$

63. $3(16 - 8) \div 3$

64. $4(21 - 6) \div 20$

65. $4(2^4) - 32$

66. $5(3^3) - 45$

67. $6(4^2 - 8) \div 12$

68. $5(6^2 - 28) \div 10$

69. $17 - 9 + 8 \cdot 4 \div 16$

70. $29 - 12 \cdot 5 \div 30$

71. $48 \div 6 \cdot 5 - 16$

72. $64 \div 8 \cdot 4 - 17$

73. $5 + 35 \div 7 \cdot 3 - 16$

74. $14 + 81 \div 9 \cdot 2 - 26$

75. $200 - 84 \div 4 \cdot 2^3 + 29$

76. $225 - 48 \div 8 \cdot 3^2 - 17$

77. $\dfrac{8 + 2 \cdot 3}{7}$

78. $\dfrac{4 + 7 \cdot 5}{13}$

79. $\dfrac{9 - 12 \div 4}{8 \div 4}$

80. $\dfrac{12 - 16 \div 8}{15 \div 3}$

81. $\dfrac{6^2 + 8 \cdot 2^2}{4^2 + 1}$

82. $\dfrac{8^2 + 6 \cdot 4^2}{6^2 + 4 \cdot 11}$

Calculator Exercises (83–90, Optional)

Evaluate the following by using a calculator:

83. 3^7

84. 12^3

85. $(7.32)^2$

86. $4^4 \cdot 6^5$

87. $(4.13)(8.95) + (2.6)^2$

88. $20.48 \div 5.12 - 3.15$

89. $51.23(6.14^2 - 23.46) - 17.3^2$

90. $8.1^3 \div 2.7^2 - 4.7 \div 3.1^3$

Let $x = 2$ and $y = 3$ and evaluate. (See Example 6.)

91. $3x + 4y$

92. $5x + 3y$

93. $3x^4$

94. $5y^3$

95. $x^2 y$

96. xy^4

97. $5x^2 y^2$

98. $7x^3 y$

99. $3x^2 + 4y^2$

100. $6x^3 - 2y^2$

101. $9x^2 \div y^2$

102. $18x^3 \div y^2$

103. $\dfrac{x^2 + 2y^2}{2x^2 + 3}$

104. $\dfrac{3x^2 + y^2}{2x^2 - 1}$

Challenge Exercises (105–108)

Find the value of each of the following when $x = \frac{1}{3}$ and $y = \frac{3}{5}$:

105. xy^2

106. $x^3 y$

107. $9x^2 y^2$

108. $25x^2 y^3$

Calculator Exercises (109–114, Optional)

Find the value of the following for $x = 2.4$ and $y = 3.6$:

109. $1.7x + 2.6y$

110. $4.2y - 1.2x$

111. $1.6x^2 + 3.7y^3$

112. $3.3x^3 + 8.7y^2$

113. $6x^3 \div 3y$

114. $1.2x^4 \div (.6y^2)$

Write an expression for each of the following and evaluate. (See Example 7.)

115. Sadie bought a sofa by paying $150 down and $28 per month for 18 months. Find the cost.

116. Diane bought a refrigerator by paying $225 down and $48 per month for 12 months. Find the cost.

117. Deon buys a used car by paying $600 down and $190 per month for 24 months. What is the total amount Deon paid for the car?

118. An automobile lease company offers a plan where the customer pays $1000 down and $399 per month for 24 months. Find the total cost of leasing the car.

119. Under the property settlement terms of his divorce, Manuel is to pay his ex-wife $3000 per month for the first 4 years and then $3500 per month for the next 6 years. How much will Manuel pay in property settlement?

120. Harold buys a lot with the following terms: He is to pay $250 per month for the first 5 years and $350 per month for the next 5 years. Find the total amount Harold will pay for the lot.

121. Five fraternity brothers agree to split the cost equally for three pizzas that cost $10.99 each and five soft drinks that cost $1.29 cents each. Find the amount (to the nearest cent) that each pays.

122. Six people agree to split the cost equally for some Chinese takeout. If two orders cost $6.99 each, three orders cost $7.50 each, and one order costs $8.50 and they also purchase three bottles of soft drinks that cost $1.49 each, find the amount (to the nearest cent) that each pays.

123. Ezra leased a car with the following terms: He agreed to pay $800 down, $299 per month for 3 years, and 15 cents per mile for all mileage in excess of 45,000. When Ezra returned the car at the end of the lease period, it had 58,000 miles on the odometer. Find the total amount Ezra paid for the lease.

124. Candace leased a car with the following terms: She agreed to pay $1500 down, $499 per month for 24 months, and 20 cents per mile for all mileage in excess of 24,000. When she returned the car at the end of the lease period, it had 32,000 miles on the odometer. Find the total amount Candace paid for the lease.

Challenge Exercises (125–128)

Simplify the following by using the order of operations:

125. $6^2 + 2^3(4 \cdot 5^2 - 4 \cdot 5) \div 20 + 2$

126. $4^3 + 3 \cdot 5(36 \div 3^2 + 7 \cdot 4) \div 8 - 3^2$

Evaluate each of the following for $x = 3, y = 4$, and $z = 6$:

127. $x^2y \div 2z + 3y^3 \div xy^2 - z^2$

128. $y^2z^2 \div x^2y \div (2y)(4x^2y)$

Writing Exercises (129–132)

129. How do constants and variables differ in meaning?

130. Give three different ways of indicating multiplication.

131. When a number is written with an exponent, what is the meaning of the exponent?

132. Why is an agreement on order of operations necessary?

Critical-Thinking Exercises (133 and 134)

133. Compare the meaning of multiplication to the meaning of raising a number to a power.

134. Write a word problem that could represent the expression $3 \cdot 2 + 2(20 - 5)$.

Writing Exercises or Group Activities (135 and 136)

If these exercises are done as a group project, each group should write at least three exercises of each type. Then exchange exercises with another group and solve. If the exercises are done as writing exercises, each student should both write and solve one of each.

135. Write an application problem similar to Exercises 115–124 that involves at least two operations.

136. Write an application problem similar to Exercises 115–124 that involves at least three operations.

OBJECTIVES *When you complete this section, you will be able to:*

A Find the perimeters of geometric figures including triangles, squares, and rectangles, and the circumference of a circle.

B Solve application problems involving perimeters of geometric figures.

PREREQUISITE SKILLS *Before beginning this section, you should be able to:*

a. Add, multiply, and divide whole numbers (Sections R.2 and R.3).

b. Substitute values for variables and evaluate by using the order of operations (Section 1.1).

GETTING READY FOR SECTION 1.2

Add, multiply, or divide the following whole numbers:

1. $28 + 32 + 14$ **2.** $2(3.14)(8)$ **3.** $2(12) + 2(15)$ **4.** $\dfrac{22 \cdot 14}{7}$

Substitute the given value(s) of the variable(s) and evaluate by using the order of operations.

5. $2l + 2w; l = 6, w = 3$ **6.** $2\pi r: \pi \approx \dfrac{22}{7}, r = 21$

Introduction

One concept that involves unit conversions, the evaluation of expressions for specific values of the variable(s), and the order of operations is the evaluation of geometric formulas.

The **perimeter** of a geometric figure is the distance around it. Think of the figure as being made of string. If we cut the string and stretch it out, the total length of the string is the same as the perimeter of the figure.

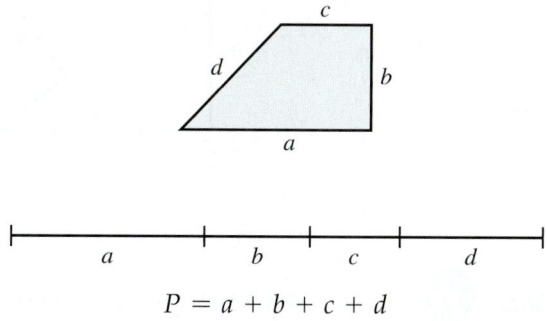

$$P = a + b + c + d$$

Consequently, to find the perimeter of a figure, we find the sum of the lengths of the sides. Since the perimeter is a measure of distance, it is expressed as a linear unit such as inch, foot, mile, centimeter, or meter. If the sides are not given in the same units (e.g., feet, inches), then we must convert to the same unit before doing any calculations.

Be Careful

When substituting into geometric formulas, be sure all the measurements are in the same units. For example, one side cannot be in feet and another in inches.

Definition Perimeter of Geometric Figures

If $a, b, c, \ldots$ are the lengths of the sides of a geometric figure, then the perimeter of the geometric figure is the sum of the lengths of the sides. That is,

$$P = a + b + c + \cdots.$$

Answers: Getting Ready: 1. 74 **2.** 50.24 **3.** 54 **4.** 44 **5.** 18 **6.** 132

Example 1   **Find the perimeter of each of the following:**

a.

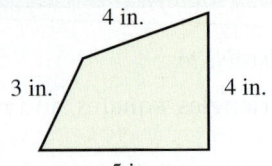

To find the perimeter, add the lengths of the sides. Hence,

$P = 3$ in. $+ 5$ in. $+ 4$ in. $+ 4$ in. $= 16$ inches.

b.

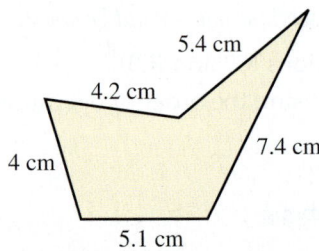

To find the perimeter, add the lengths of the sides. Hence,
$P = 4$ cm $+ 5.1$ cm $+ 7.4$ cm $+ 5.4$ cm $+ 4.2$ cm $=$
26.1 centimeters.

c. Express the answer in inches.

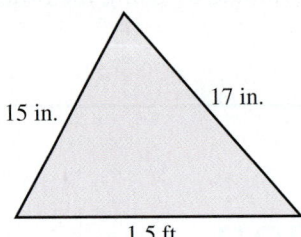

First, we need to change all measurements into the same unit. Let's change 1.5 ft into inches. 1.5 ft $= 1.5$ ft $\cdot \frac{12\,\text{in.}}{1\,\text{ft}} = 18$ in. To find the perimeter, we add the lengths of all the sides. Hence, $P = 18$ in. $+ 17$ in. $+ 15$ in. $= 50$ inches.

PRACTICE EXERCISE 1

Find the perimeter of each of the following:

a.

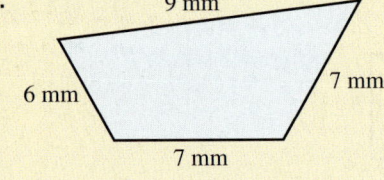

b.

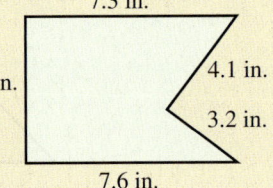

c. Express the answer in feet.

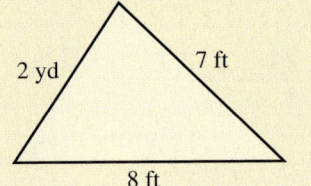

OBJECTIVE **A** Finding the perimeters of triangles, rectangles, and squares, and the circumference of a circle

Some special geometric figures have formulas for finding their perimeters. Following are examples of some common geometric figures, their characteristics, and the formula for calculating the perimeter (circumference, in the case of the circle) of each:

Figure	Characteristics	Formula for Perimeter
Triangle 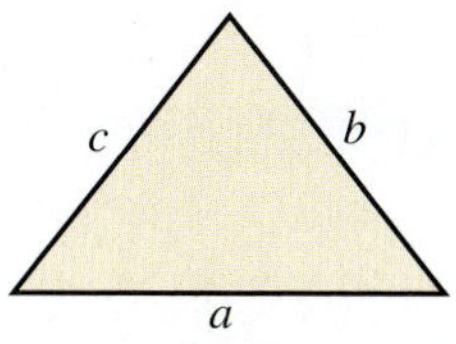	Three sides of lengths a, b, and c.	$P = a + b + c$
Rectangle 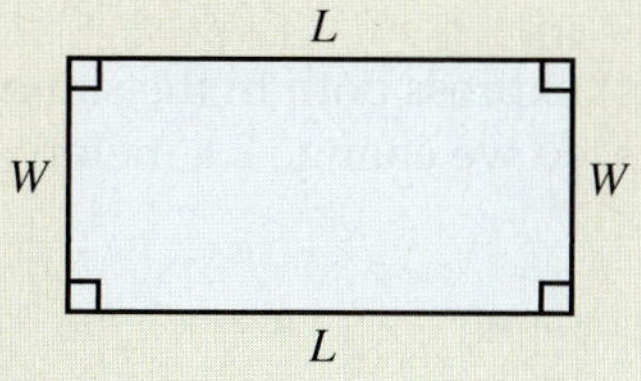	Four sides; all angles are right angles (90°); opposite sides are equal in length. The two longest sides are usually called the **length**, the two shortest sides the **width**.	$P = 2L + 2W$
Square 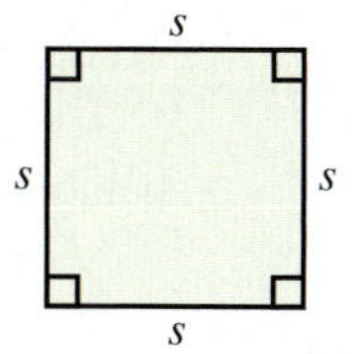	Four sides of equal length and four right angles. The length of each side is s.	$P = 4s$
Circle 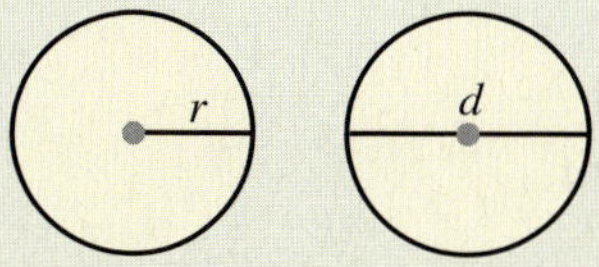	The set of all points in a plane that are an equal distance from a given point in the plane. The given point is called the **center** of the circle, and the given distance is the **radius**. The distance around the circle is called the **circumference** rather than the perimeter. A line segment with endpoints on the circle and passing through the center is called a **diameter**. Hence, $d = 2r$, where d is the diameter and r is the radius.	**Formulas for Circumference** $C = \pi d$ or $C = 2\pi r$ $\pi \approx 3.14$ or $\dfrac{22}{7}$ The symbol $\approx$ means "is approximately equal to." **Note:** Many calculators have a $\boxed{\pi}$ key. If you use a $\boxed{\pi}$ key, your results may differ slightly from those in the text.

The formulas for the circumference of a circle involve the constant π, which is the ratio of the circumference of any circle to its diameter and whose approximate value is 3.14 or $\frac{22}{7}$. There is no exact decimal or fractional representation of π. Better approximations are 3.142 and $\frac{355}{113}$. Since there is no exact decimal or fractional representation for π, answers obtained by substituting values for π are approximate answers. When answers are left in terms of π, the answers are exact.

When evaluating geometric formulas, we suggest the following procedure:

Procedure: Evaluating Geometric Formulas

1. Write the appropriate formula.

2. Substitute the known quantities. (Make sure the units are the same.)

3. Evaluate by using the order of operations.

Example 2   **Find the perimeter of each of the following:**

a.

9 in.   11 in.   7 in.

$P = a + b + c$ Formula for the perimeter of a triangle. Substitute for a, b, and c.

$P = 7 \text{ in.} + 9 \text{ in.} + 11 \text{ in.}$ Add.

$P = 27 \text{ in.}$ Therefore, the perimeter is 27 inches.

b. A rectangle whose length is 2.4 meters and whose width is 93 centimeters. Express the answer in centimeters.

Since the sides are given in different units, we must first express both in the same unit. We are asked to express the answer in centimeters, so we change 2.4 meters into centimeters. 2.4 meters = 240 centimeters.

$P = 2L + 2W$ Formula for the perimeter of a rectangle. Substitute for L and W.

$P = 2(240 \text{ cm}) + 2(93 \text{ cm})$ Multiply before adding.

$P = 480 \text{ cm} + 186 \text{ cm}$ Add.

$P = 666 \text{ cm}$ Therefore, the perimeter is 666 centimeters.

c.

3.2 ft   3.2 ft   3.2 ft   3.2 ft

$P = 4s$ Formula for the perimeter of a square. Substitute for s.

$P = 4(3.2 \text{ ft})$ Multiply.

$P = 12.8 \text{ ft}$ Therefore, the perimeter is 12.8 feet.

d. Find the circumference of the circle shown. Use $\pi \approx 3.14$.

8 ft

$C = \pi d$ Formula for the circumference of a circle. Substitute for π and d.

$C \approx (3.14)(8 \text{ ft})$ Multiply.

$C \approx 25.12 \text{ ft}$ Therefore, the circumference is approximately 25.12 feet.

Note

The exact answer to Example 2d is $C = \pi(8 \text{ ft}) = 8\pi$ ft.

e. Find the circumference of a circle with radius $\frac{9}{5}$ feet. Use $\pi \approx \frac{22}{7}$.

$C = 2\pi r$ Formula for the circumference of a circle. Substitute for π and r.

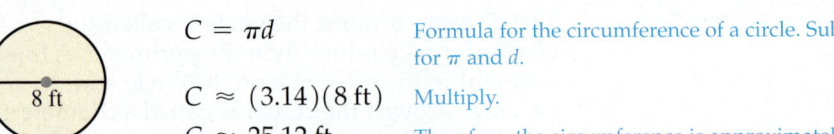

$C \approx 2\left(\frac{22}{7}\right)\left(\frac{9}{5} \text{ft}\right)$ Multiply.

$C \approx \frac{396}{35} \text{ft}$ Therefore, the circumference is approximately $\frac{396}{35}$ feet.

Note

The exact answer to Example 2e is $C = 2\pi\left(\frac{9}{5} \text{ft}\right) = \frac{18}{5}\pi$ ft.

PRACTICE EXERCISE 2

Find the perimeter of each of the following:

a. A triangle whose sides are 2 feet, 18 inches, and 15 inches. Express the answer in inches.

b.

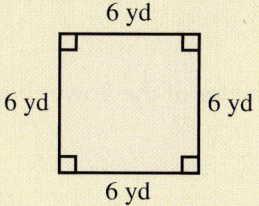

c.

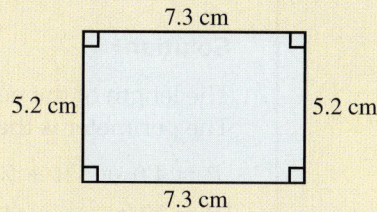

Find the circumference of the following circles. Give both approximate and exact answers.

d. Diameter of 5.6 yards. Use $\pi \approx 3.14$ and round the answer to the nearest tenth of a yard.

e. Use $\pi \approx \dfrac{22}{7}$.

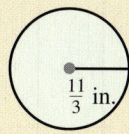

If you need more practice, do the following Additional Practice Exercises.

Additional Practice Exercise 2 Find the perimeter of each of the following:

a. A triangle whose sides are 3 inches, 5 inches, and 6 inches.

b.

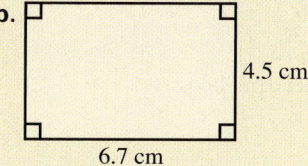

c.

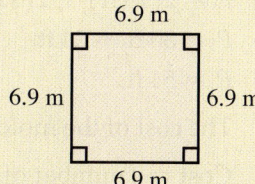

Find the circumference of the following circles. Give both approximate and exact answers.

d. Use $\pi \approx 3.14$ and round the answer to the nearest tenth of a meter.

e. Radius of $\dfrac{5}{3}$ feet. Use $\pi \approx \dfrac{22}{7}$.

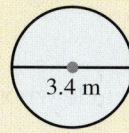

OBJECTIVE **B** Solving application problems involving perimeters of geometric figures

The use of these geometric formulas occurs in many everyday situations. In order to solve these problems, we must determine the appropriate formula and the value(s) of the variable(s), substitute for the variables, and simplify. Often there are additional factors that must be taken into consideration. We illustrate with some examples.

Answers: Practice Exercise 2: a. 57 in. **b.** 24 yd **c.** 25 cm **d.** 17.6 yd, 5.6π yd **e.** $\dfrac{484}{21}$ in., $\dfrac{22}{3}\pi$ in.
Additional Practice 2: a. 14 in. **b.** 22.4 cm **c.** 27.6 m **d.** 10.7 m, 3.4π m **e.** $\dfrac{220}{21}$ ft, $\dfrac{10}{3}\pi$ ft

Example 3   Applications involving perimeter

a. An irregularly shaped flower bed has sides whose lengths are 4 feet, 5 feet, 7 feet, 6 feet, and 8 feet. The flower bed is to be enclosed with landscape timbers. How many feet of timbers are needed?

Solution

The length of the needed timbers is the same as the perimeter of the flower bed. The perimeter is the sum of the lengths of all the sides.

$P = 4\text{ ft} + 5\text{ ft} + 7\text{ ft} + 6\text{ ft} + 8\text{ ft}$

$P = 30\text{ ft}$ It would require 30 feet of timbers to enclose the flower bed.

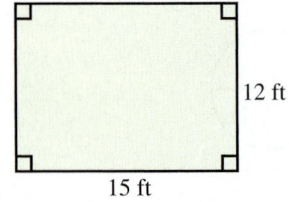
12 ft
15 ft

b. A carpenter needs to put molding around the ceiling of a rectangular room whose length is 15 feet and whose width is 12 feet. How many feet of molding does he need? If the molding costs $2.59 per foot, what is the cost of the molding? See the figure in the margin.

Solution

Finding the amount of molding needed is the same as finding the perimeter of a rectangle whose length is 15 feet and whose width is 12 feet.

$P = 2L + 2W$ Formula for the perimeter of a rectangle.

$P = 2(15\text{ ft}) + 2(12\text{ ft})$ Substitute for L and W.

$P = 30\text{ ft} + 24\text{ ft}$ Multiply before adding.

$P = 54\text{ ft}$ Therefore, the carpenter needs 54 feet of molding.

The cost of the molding is:

Cost = (number of feet) $\times$ (cost per foot)

= (54)(2.59)

= $139.86

3 ft

c. An architect wishes to put a border around a circular fountain whose radius is 3 feet. How much of the border material will she need? See the figure in the margin.

Solution

This is the same as finding the circumference of a circle whose radius is 3 feet.

$C = 2\pi r$ Formula for the circumference of a circle.

$C \approx 2(3.14)(3\text{ ft})$ Substitute for π and r.

$C \approx 18.84\text{ ft}$ Therefore, the architect needs approximately 18.84 feet of border material.

PRACTICE EXERCISE 3

Solve each problem. The figures for some are in the margin.

a. A surveyor finds the lengths of the sides of a field with four sides to be 180 meters, 192 meters, 223 meters, and 173 meters. Find the perimeter of the field.

b. A bicycle tire is 26 inches in diameter. How far does the bicycle travel in 10 revolutions of the wheel? Use 3.14 for π.

c. A drainage ditch is dug around a square field, each of whose sides is 180 feet long. If Charles is paid $.60 per foot for digging the ditch, how much money does he earn?

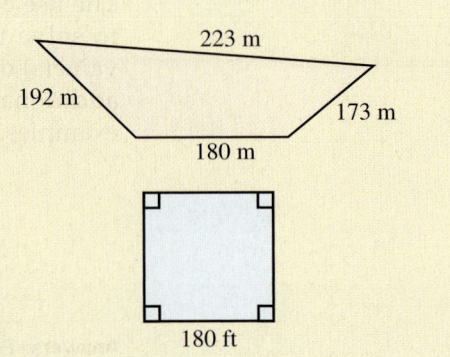
223 m
192 m
173 m
180 m
180 ft

Answers: Practice Exercise 3: a. 768 m **b.** 816.4 in. **c.** $432

Study Tip 2 Developing a Positive Approach

A positive approach can help ensure success in your math courses. Your approach should include the following:

1. Recognize that your degree of success depends upon you and you alone.
2. Make an all-out effort to do well in the course.
3. Work hard enough to do much better than just pass. Set high, but realistic, goals for yourself.

4. Make a commitment to overcome *any* setbacks, personal or otherwise, and work hard until the very end of the course.

Commitment and determination can go a long way toward guaranteeing that you will be successful. You can do it!

Exercise Set 1.2

For Extra Help

MyMathLab®

Find the perimeter of each of the following shapes. (See Example 1.)

1.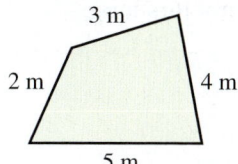
3 m
2 m
4 m
5 m

2.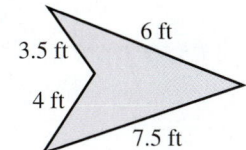
3.5 ft
6 ft
4 ft
7.5 ft

3.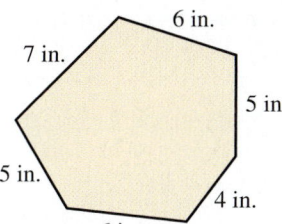
6 in.
7 in.
5 in.
5 in.
4 in.
6 in.

4.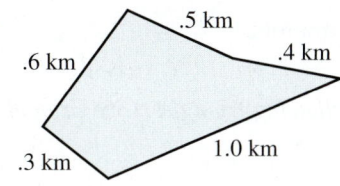
.5 km
.6 km
.4 km
.3 km
1.0 km

5. Express the answer in inches.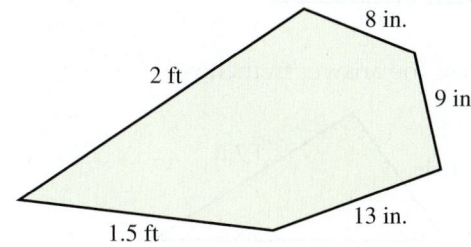
8 in.
2 ft
9 in.
13 in.
1.5 ft

6. Express the answer in decimeters.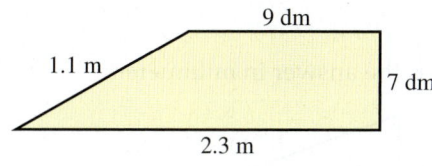
9 dm
1.1 m
7 dm
2.3 m

Find the perimeter of each of the following rectangles. (See Example 2b.)

7.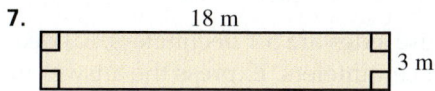
18 m
3 m

8.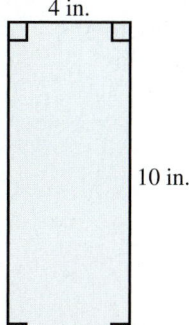
4 in.
10 in.

9.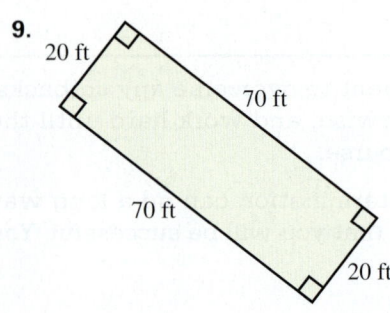

10. Express the answer in feet.

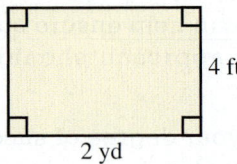

11. Length of 4 yards and width of 6 feet. Express the answer in feet.

12. Length of 2 feet and width of 19 inches. Express the answer in inches.

13. Length of 0.63 decimeters and width of 4.2 centimeters. Express the answer in decimeters.

14. Length of 5.7 meters and width of 650 centimeters. Express the answer in meters.

Find the perimeter of each of the following squares. (See Example 2c.)

15.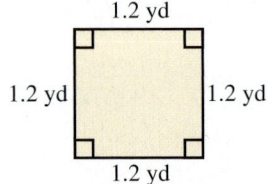

16. Each of whose sides is 9 decimeters long.

17. Each side is 14 centimeters.

18. Each side is 7 feet.

Find the perimeter of the following triangles. (See Example 2a.)

19.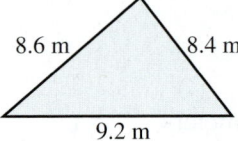

20.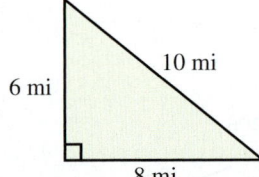

21. Express the answer in millimeters.

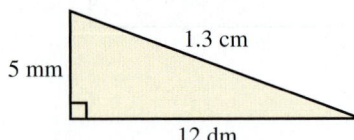

22. Express the answer in inches.

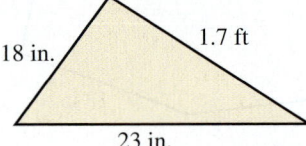

23. A triangle whose sides are 6 inches, 7 inches, and 9 inches in length.

24. A triangle whose sides are 12 kilometers, 14 kilometers, and 18 kilometers long.

25. A triangle whose sides are 43 decimeters, 2.7 meters, and 3.2 meters. Express the answer in meters.

26. A triangle whose sides are 5.1 decimeters, 6.3 decimeters, and 42 centimeters. Express the answer in decimeters.

Find the exact value (express answers in terms of π) and the approximate value (use π ≈ 3.14) of the circumference of each of the following circles. If necessary, round answers to the nearest hundredth. (See Examples 2d and 2e.)

27.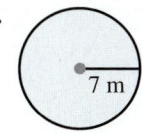

    7 m

28.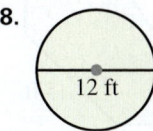

    12 ft

29. A circle whose diameter is 5.5 feet.

30. A circle with a radius of 9.8 centimeters.

Find the exact and approximate circumference of each of the following circles. Use π ≈ $\frac{22}{7}$.

31. A circle with a radius of $\frac{4}{3}$ inches.

32. A circle with a diameter of $\frac{3}{5}$ feet.

Challenge Exercises (33–36)

Find the perimeter of the following.

33.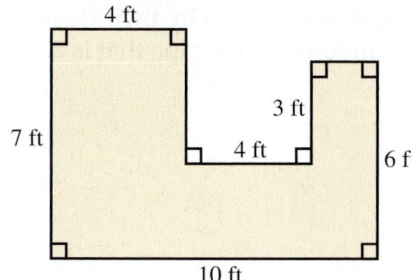

34. Use π ≈ 3.14.

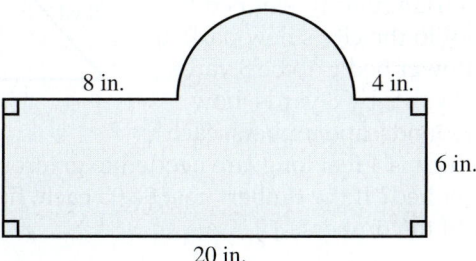

35. Use π ≈ 3.14.

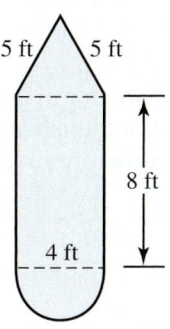

36. Use π ≈ 3.14.

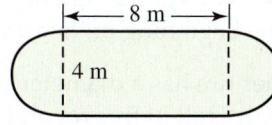

Solve each problem. If necessary, use π ≈ 3.14 and round your answers to the nearest hundredth of a unit. (See Example 3.)

37. A carpenter wants to make square window frame. If the window measures 52 inches on each of the four sides, how much material does he need to make the frame?

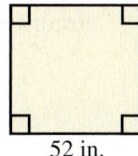

52 in.

38. A rectangular football field measures 120 yards long by 150 feet wide. If the school band is to march all the way around the edge of the field, how far will they march? Express the answer in yards.

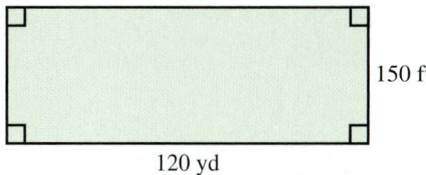

150 ft

120 yd

39. A police officer wishes to tape off an irregularly shaped crime scene in the shape of the given figure. How much tape will the officer need to go all the way around the scene?

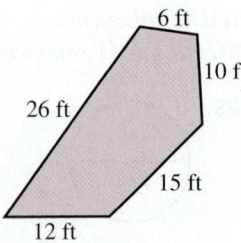

40. Magna has a square garden plot that measures 18 feet on each side. If fencing costs $.50 a linear foot, how much does it cost to fence her garden?

41. A local sandwich shop wishes to put up a sign with a fluorescent light border. The shape and size of the sign are given in the accompanying diagram. If the fluorescent tubing costs $5.00 per foot, how much does the border around the sign cost?

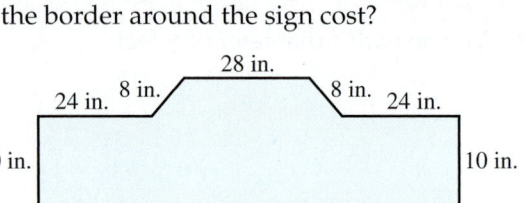

42. A lawn sprinkler rotates on a shaft so that it wets a circular region of the lawn. If the water sprays out a distance of 50 feet from the sprinkler, what is the circumference of the circle of wet grass?

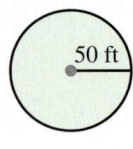

43. A landscape architect is designing a triangular flower bed to be put in the city's new park. If the flower bed is to be 8 yards by 12 yards by 8 yards, how many landscape timbers, each of which is 4 feet long, are needed to go around the flower bed? If the timbers cost $4.95 each, find the cost of the timbers.

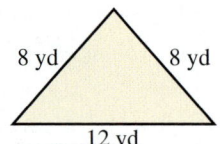

44. Plumbers measure water pipes by their diameter. What is the circumference of a pipe that is 2.5 inches in diameter?

45. The runway at a certain airport is 500 yards long and 60 yards wide. How many runway lights are needed if lights are placed every 20 yards at both of the longer sides of the runway? The first light is placed at the beginning of the runway.

46. The diameter of the bore of the barrels of the largest guns used on a battleship is 16 inches. What is the circumference of the bore of these barrels?

47. A boat trailer tire has a diameter of 24 inches. How far will the tire roll in five revolutions?

48. A beach ball has a diameter of 18 inches. If the wind is blowing the ball along the beach and the ball makes 20 revolutions before Juan can catch it, how far has it traveled?

Writing Exercise or Group Project (49 and 50)

If these exercises are done as a group project, each group should write at least one exercise. The groups should then exchange exercises with other groups and solve the exercise(s) that they receive.

49. Write at least one application problem involving the perimeter of a rectangle.

50. Write at least one application problem involving the circumference of a circle.

Section 1.3 Areas of Geometric Figures

OBJECTIVES *When you complete this section, you will be able to:*

A Find the areas of rectangles, squares, triangles, parallelograms, trapezoids, and circles.

B Solve application problems involving area.

PREREQUISITE SKILLS *Before beginning this section, you should be able to:*

a. Multiply and divide whole numbers and decimals (Sections R.3 and R.6).

b. Substitute values for variables and evaluate by using the order of operations (Section 1.1).

c. Change units of linear measure (Section 1.2).

GETTING READY FOR SECTION 1.3

Multiply or divide the following whole numbers and/or decimals:

1. $(15)(6)$　　　　**2.** $(3.14)(25)$　　　　**3.** $\dfrac{(4.2)(9)}{2}$

Substitute the given value(s) of the variable(s) and evaluate by using the order of operations.

4. bh: $b = 8.3, h = 3.1$　　**5.** $\dfrac{h(B + b)}{2}$: $h = 7, B = 13, b = 5$　　**6.** πr^2, $\pi \approx 3.14, r = 4$

Introduction

In the previous section, we found the perimeter, or distance around a geometric figure. In this section, we find the size of the surface of a geometric figure.

Developing the concept of area

The units used for measuring perimeter are linear units that measure distances. Suppose we want to carpet the floor of a room that is 14 feet long and 10 feet wide. We could not measure the amount of carpet needed by using linear units, since the size of the floor cannot be expressed as a distance. The floor of the room is a surface. The measure of the size of a surface is called its **area**. The units used to measure area are **square units**. A square unit is a square each of whose sides is one linear unit. The following are some examples of square units:

Area units of measure

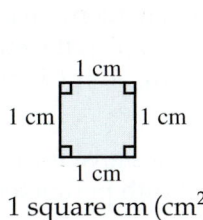

1 square cm (cm²)

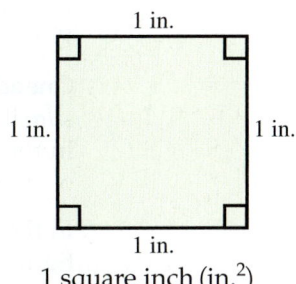

1 square inch (in.²)

Instead of writing "square cm" or "square in.," we will use the notation **cm²** and **in.²**. Thus, ft² is read **square foot** and m² is read **square meter**. This is why the exponent "2" is often read as **squared**.

Answers: Getting Ready: 1. 90 **2.** 78.5 **3.** 18.9 **4.** 25.73 **5.** 63 **6.** 50.24

OBJECTIVE Finding the areas of rectangles, squares, triangles, parallelograms, trapezoids, and circles

In finding the area of a surface, we find the number of square units that are contained within the surface. For example, in the 10-foot-by-14-foot room just mentioned, the area of the floor is the same as the number of square feet of carpet needed to cover the floor. We will develop the formula for the area of a rectangle to reinforce the concept of area and then give the formulas for other geometric figures.

| **Example 1** | **Find the area of a rectangle 5 centimeters long and 3 centimeters wide.** |

Solution

We begin by drawing the rectangle.

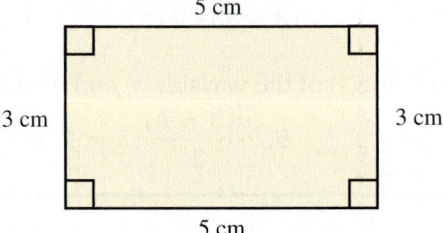

We now draw horizontal and vertical lines 1 centimeter apart.

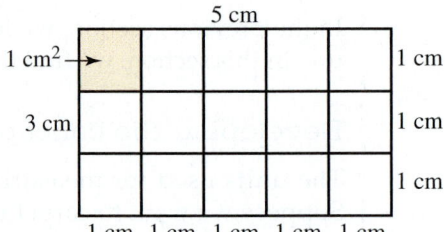

We have created a number of squares within the rectangle. The length of each side of each square is 1 centimeter, so each square is 1 square centimeter. Each row contains 5 squares, and there are 3 rows of squares. Therefore, there are $5 \cdot 3 = 15$ squares. Consequently, the area of the rectangle is 15 square centimeters. Notice that this is the same as multiplying the length (5 centimeters) times the width (3 centimeters). Therefore, the area of a rectangle with length L and width W is $A = LW$.

One advantage of the unit^2 notation is that we will automatically get the correct unit of measure if we write the unit as well as the numerical value when performing computations. In the rectangle just shown, $A = (5\,\text{cm})(3\,\text{cm}) = (5 \cdot 3)(\text{cm} \cdot \text{cm}) = 15\,\text{cm}^2$. (Think of cm $\cdot$ cm as cm^2.)

In the chart that follows, we give formulas for the areas of some common geometric figures. As was the case with perimeters, the dimensions must be given in the same units.

Figure	Characteristics	Formula
Rectangle 	Four sides, all angles are right angles ($90°$), opposite sides are equal in length. The two longest sides are usually called the *length*, and the two shortest sides are called the *width*.	$A = LW$
Square 	Four sides of equal length and four right angles. The length of each side is s.	$A = s^2$
Triangle 	Any side can be used as the base b, and the perpendicular (forms a right angle) from the opposite vertex (point where two sides intersect) to that side is the height h.	$A = \dfrac{bh}{2}$
Parallelogram 	Has four sides, and opposite sides are parallel (do not intersect and remain the same distance apart) and are equal in length. Any side can be used as the base b, and the perpendicular to that side from the opposite vertex is the height h.	$A = bh$
Trapezoid 	Has four sides in which one pair is parallel. The parallel sides are called the *bases*. The longer base is designated as B and the shorter base b. The distance between the parallel sides is the height h.	$A = \dfrac{h(B + b)}{2}$
Circle 	Characteristics given earlier when perimeter was discussed (see table on page 91).	$A = \pi r^2$

We use the same procedure for finding areas that we used in the last section for finding perimeters: 1. Write the formula; 2. substitute; 3. evaluate.

Example 2   **Find the area of each of the following:**

a.

$A = LW$ Substitute 4 in. for L and 2 in. for W.

$A = (4\text{ in.})(2\text{ in.})$ Multiply.

$A = 8\text{ in.}^2$ Therefore, the area is 8 square inches.

[rectangle: 2 in. height, 4 in. base]

b.

$A = s^2$ Substitute 7 ft for s.

$A = (7\text{ ft})^2$ Raise to the power.

$A = 49\text{ ft}^2$ Therefore, the area is 49 square feet.

[square: 7 ft, 7 ft]

c. Express the answer in square inches.

Change 2 ft into inches. $2\text{ ft} = 2\text{ ft} \cdot \dfrac{12\text{ in}}{1\text{ ft}} = 24\text{ in.}$

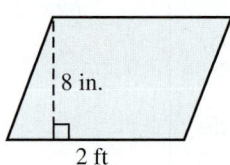

$A = bh$ Substitute 24 in. for b and 8 in. for h.

$A = (24\text{ in.})(8\text{ in.})$ Multiply.

$A = 192\text{ in.}^2$ Therefore, the area is 192 square inches.

d.

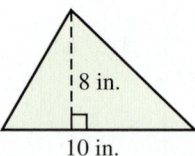

$A = \dfrac{bh}{2}$ Substitute for b and h.

$A = \dfrac{(10\text{ in.})(8\text{ in.})}{2}$ Multiply in the numerator.

$A = \dfrac{80\text{ in.}^2}{2}$ Divide.

$A = 40\text{ in.}^2$ Therefore, the area is 40 square inches.

e. Express the answer in square centimeters.

Change .3 dm into centimeters. $.3\text{ dm} = 3\text{ cm}$

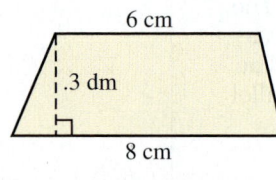

$A = \dfrac{h(B + b)}{2}$ Substitute for h, B, and b.

$A = \dfrac{(3\text{ cm})(8\text{ cm} + 6\text{ cm})}{2}$ Add inside parentheses.

$A = \dfrac{(3\text{ cm})(14\text{ cm})}{2}$ Multiply in the numerator.

$A = \dfrac{42\text{ cm}^2}{2}$ Divide.

$A = 21\text{ cm}^2$ Therefore, the area is 21 square centimeters.

f. Circle with diameter of 8.2 inches. Round the answer to the nearest hundredth of a square inch. Use $\pi \approx 3.14$.

Since the diameter is 8.2 in., the radius is $\dfrac{8.2}{2} = 4.1\text{ in.}$

$A = \pi r^2$ Substitute for r.

$A = \pi(4.1\text{ in.})^2$ Square 4.1 in. and substitute for π.

$A \approx (3.14)(16.81\text{ in.}^2)$ Multiply.

$A \approx 52.7834\text{ in.}^2$ Round to the nearest hundredth of an in^2.

$A \approx 52.78\text{ in.}^2$ Therefore, the area is approximately 52.78 square inches.

Note

The exact answer to Example 2f is $A = \pi r^2 = \pi(4.1)^2 = 16.81\pi\text{ in}^2$.

PRACTICE EXERCISE 2

a. Find the area of the following rectangle. Express the answer in square centimeters.

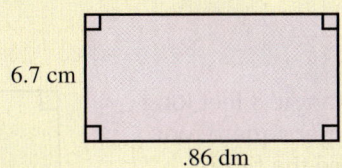

6.7 cm, .86 dm

b. Find the area of the following square:

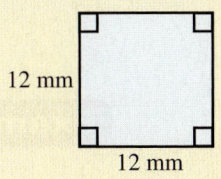

12 mm, 12 mm

c. Find the area of the following parallelogram:

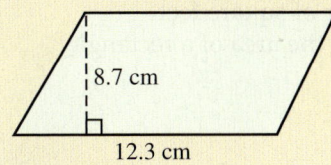

8.7 cm, 12.3 cm

d. Find the area of the following triangle:

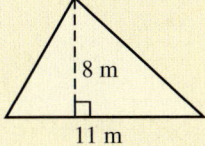

8 m, 11 m

e. Find the area of the following trapezoid. Express the answer in square inches.

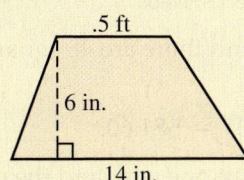

.5 ft, 6 in., 14 in.

f. Find the area of the following circle. Use $\pi = 3.14$.

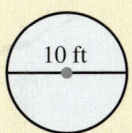

10 ft

If you need more practice, do the following Additional Practice Exercises.

Additional Practice Exercise 2

a. Find the area of the following rectangle. Express the answer in square feet.

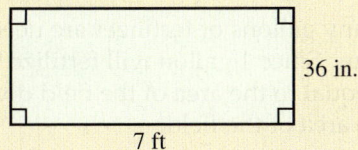

36 in., 7 ft

b. Find the area of the following square:

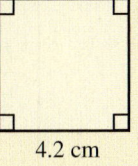

4.2 cm

c. Find the area of the following parallelogram. Express the answer in square centimeters.

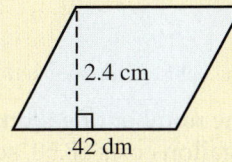

2.4 cm, .42 dm

d. Find the area of a triangle with a base of 4.2 centimeters and height of 5 centimeters.

e. Find the area of the following trapezoid:

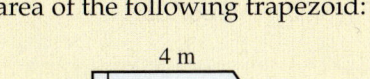
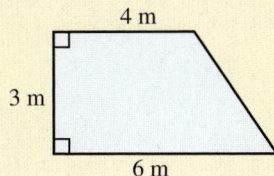

4 m, 3 m, 6 m

f. Find the area of a circle with diameter of 12.8 feet. Round the answer to the nearest hundredth. Use $\pi \approx 3.14$.

Answers: Practice Exercise 2: a. 57.62 cm² **b.** 144 mm² **c.** 107.01 cm² **d.** 44 m² **e.** 60 in.² **f.** 78.5 ft² **Additional Practice 2: a.** 21 ft²
b. 17.64 cm² **c.** 10.08 cm² **d.** 10.5 cm² **e.** 15 m² **f.** 128.61 ft²

OBJECTIVE Solving application problems involving area

There are many applications of areas of geometric figures.

Example 3

a. A piece of wood is in the shape of a rectangle 8 feet long and 6 feet wide. If the wood costs $1.70 per square foot, what is the cost of the piece of wood? See the figure to the right.

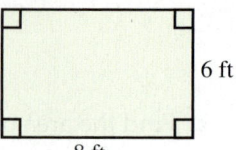

6 ft

8 ft

Strategy We must first find the number of square feet of wood in the piece. This is the same as the area of a rectangle 6 feet long and 8 feet wide.

Solution

$A = LW$ Substitute for L and W.

$A = (8 \text{ ft})(6 \text{ ft})$ Multiply.

$A = 48 \text{ ft}^2$ Therefore, there are 48 square feet of wood.

Since the wood costs $1.70 per square foot and there are 48 square feet, multiply 48 and $1.70. Therefore,

$$\text{cost} = (48)(1.70) = \$81.60.$$

b. An irrigation pipe 150 feet long is anchored at one end, and the other end is mounted on wheels. Consequently, it sweeps out a circle when in operation. Suppose the field is planted in potatoes. The potatoes can be fertilized by putting liquid fertilizer into the irrigation system. When diluted, 1 gallon of fertilizer will fertilize 900 square feet of potatoes and costs $8.50. How much does it cost to fertilize the field? See the figure in the margin.

150 ft

Strategy We must determine how many gallons of fertilizer are needed to fertilize a circular field 150 feet in radius. Since 1 gallon will fertilize 900 square feet, the number of gallons needed is equal to the area of the field divided by 900. Therefore, we must first determine the area of the field.

Solution

$A = \pi r^2$ Substitute for π and r.

$A \approx (3.14)(150 \text{ ft})^2$ Square 150 ft.

$A \approx (3.14)(22{,}500 \text{ ft}^2)$ Multiply.

$A \approx 70{,}650 \text{ ft}^2$ Therefore, the area of the field is approximately 70,650 square feet.

Since 1 gallon will fertilize 900 square feet, the number of gallons of fertilizer needed is $70{,}650 \div 900 = 78.5$ gallons. Each gallon costs $8.50, so

$$\text{cost} = (78.5)(8.50) = \$667.25.$$

PRACTICE EXERCISE 3

Solve each problem. The figures are in the margin.

a. How many square feet of sod will it take to cover a rectangular yard 75 feet long and 50 feet wide?

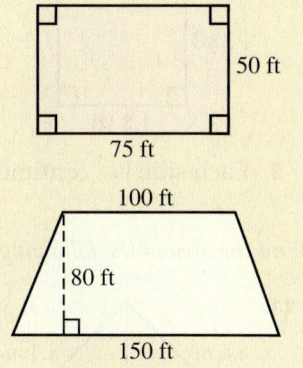

b. A field is in the shape of a trapezoid whose parallel sides are 150 feet and 100 feet long. The height of the trapezoid is 80 feet. A bag of grass seed costs $6.00 and will plant 500 square feet. How much will the seed to plant the field cost?

Study Tip 3 Mathematics Is Sequential Learning

Mathematical knowledge is sequential. New concepts and principles build upon previously learned concepts and principles. The material you learn each day will depend upon material that you learned prior to that day. Therefore, you must do all of your homework *before* each class meeting. If you are absent from class, you must first learn the material that was covered while you were absent. Otherwise, you lack the foundation on which the next day's lesson is based.

Exercise Set 1.3 For Extra Help MyMathLab®

Find the area of each of the following rectangles. (See Example 2a.)

1.   18 m   3 m

2.   4 in.   10 in.

3.   20 ft   70 ft   70 ft   20 ft

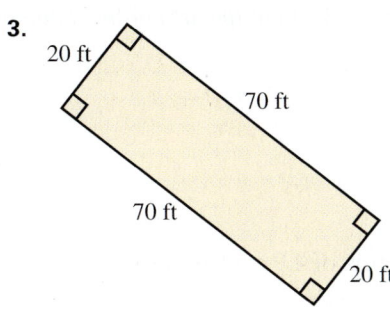

4.   4 ft   6 ft

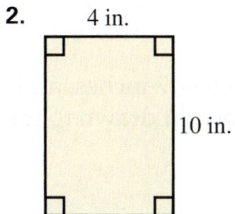

5. Length of 3 yards and width of 5 feet. Express the answer in square feet.

6. Length of 2 feet and width of 8 inches. Express the answer in square inches.

Find the area of each of the following squares. (See Example 2b.)

7.

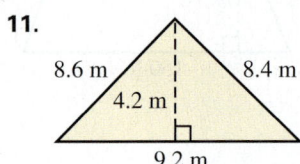

1.2 yd (top)
1.2 yd (left)   1.2 yd (right)
1.2 yd (bottom)

8. Each side is 9 decimeters long.

9. Each side is $\frac{5}{7}$ centimeter.

10. Each side is $\frac{5}{8}$ feet.

Find the area of the following triangles. (See Example 2d.)

11.

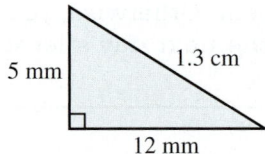

8.6 m   8.4 m
4.2 m
9.2 m

12. Express answer in square inches.

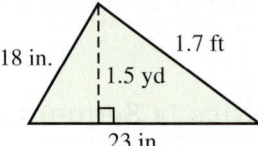

18 in.   1.7 ft
1.5 yd
23 in.

13. Express the answer in square millimeters.

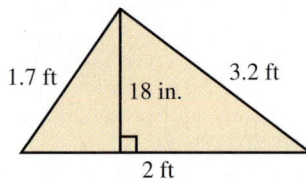

5 mm   1.3 cm
12 mm

14.

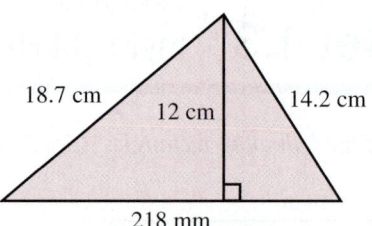

6 mi   10 mi
8 mi

15. Express the answer in square inches.

1.7 ft   3.2 ft
18 in.
2 ft

16. Express the answer in square centimeters.

18.7 cm   14.2 cm
12 cm
218 mm

17. A triangle whose sides are 6 inches, 7 inches, and 8 inches in length, and whose height drawn to the 6-inch side is 6.8 inches.

18. A triangle whose sides are 12 kilometers, 14 kilometers, and 16 kilometers long, and whose height drawn to the 14-kilometer side is 11.6 kilometers.

Find the exact value (express the answers in terms of π) and the approximate value (use $\pi \approx 3.14$) of the area of the following circles. Round approximate answers to the nearest tenth of a unit. (See Example 2f.)

19.

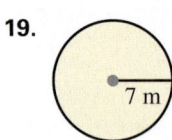

7 m

20.

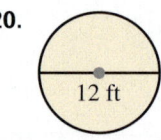

12 ft

21. A circle with a diameter of 13 feet.

22. A circle with a radius of 4.9 centimeters.

Find the area of each of the following circles. Use $\pi \approx \dfrac{22}{7}$.

23. A circle with a radius of $\dfrac{5}{3}$ inches.

24. A circle with a diameter of $\dfrac{3}{5}$ feet.

Find the area of each of the following parallelograms. (See Example 2c.)

25.

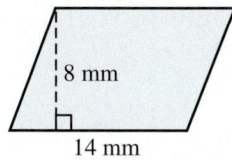

8 mm

14 mm

26.

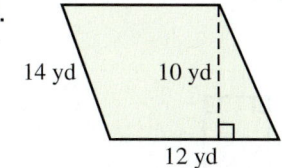

14 yd   10 yd

12 yd

27.

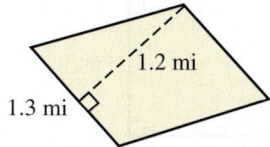

1.2 mi

1.3 mi

28.

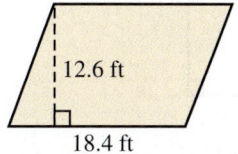

12.6 ft

18.4 ft

29. A parallelogram whose base is 31 millimeters and whose height is 2.3 centimeters. Express the answer in square centimeters.

30. A parallelogram whose base is 7 feet and whose height is 2 yards. Express the answer in square feet.

Find the areas of the following trapezoids. (See Example 2e.)

31.

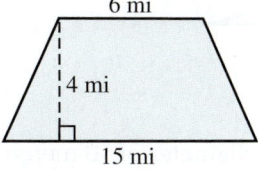

6 mi

4 mi

15 mi

32.

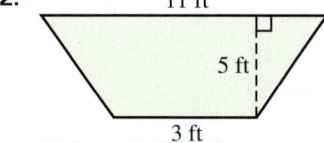

11 ft

5 ft

3 ft

33. Express the answer in square inches.

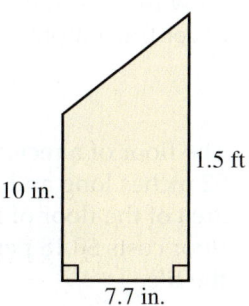

1.5 ft

10 in.

7.7 in.

34. A trapezoid with a large base of 11.3 centimeters, a small base of 9.7 centimeters, and an altitude of 6.5 centimeters.

35. Express the answer in square decimeters.

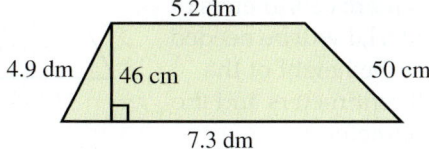

5.2 dm

4.9 dm   46 cm   50 cm

7.3 dm

36. Express the answer in square feet.

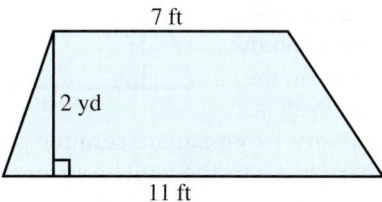

7 ft

2 yd

11 ft

Challenge Exercises (37–40)

Find the area of each of the following:

37.

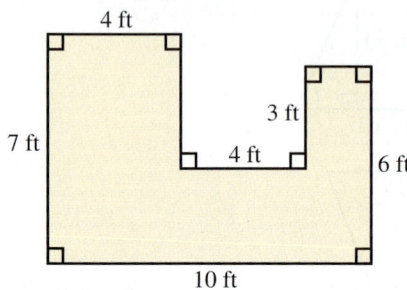

38. Use $\pi \approx 3.14$.

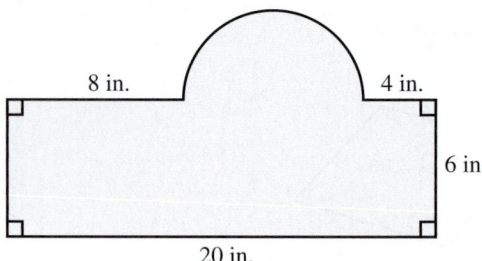

39. Find the area inside the square and outside the circle. Use $\pi \approx 3.14$.

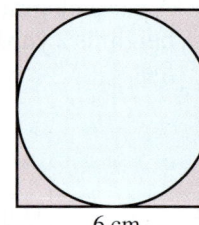

40.

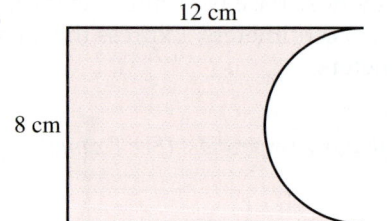

Solve the following. (See Example 3.)

41. A university wants to install artificial turf in its football stadium. The football field, including the end zones, is rectangular and measures 150 yards long and 360 feet wide. How many square yards of artificial turf are needed?

42. A circular fountain has a diameter of 10 meters. How many square-meters of tile are needed to cover the bottom of the fountain? Use $\pi \approx 3.14$.

43. A certain type of plant needs 1 square yard of ground to grow in. If a field is 200 yards long and 150 yards wide, how many plants can be grown in the field? If each plant costs $2.95, find the cost of the plants.

44. The floor of a rectangular shower stall measures 52 inches long and 32 inches wide. What is the area of the floor of the shower stall? If tile for the floor costs $0.18 per square inch, find the cost of the tile.

45. A business concern wants to build a decorative wall in the shape of a trapezoid that measures 25 feet on the top base and 40 feet on the bottom base. If the wall is 8 feet high, how many 1-foot-square ceramic pieces of tile are necessary to cover the wall? Assume no waste. If each piece of tile costs $2.59, find the cost of the tiles.

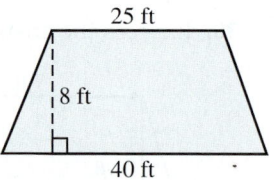

46. The airport authority wants to design a triangular caution sign. How many square centimeters of reflective material will be needed for the sign if the height of the triangle is 90 centimeters and the base is 10 decimeters?

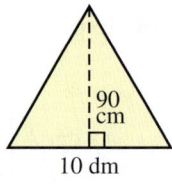

47. A family wishes to put an addition on their house. The roof of the addition is rectangular and measures 25 feet long and 12 feet wide. If roofing costs $3.50 a square foot, how much will the material for the new roof cost?

48. A ceiling, which measures 20 feet long by 17 feet wide, is to be covered in acoustic ceiling tiles with an area of 1 square foot each. How many 1-foot-square tiles are needed? If the tiles cost $.49 each, find the cost of the tile.

49. A room is 12 feet long and 18 feet wide. If carpet costs $18.95 per square yard, find the cost of the carpet needed to carpet the floor of the room.

50. It costs the Department of Transportation $.43 per square foot for the material used in making traffic signs. Each sign is rectangular with a length of 2 feet and a width of 1.5 feet. Find the cost of the material to make 400 such signs.

Challenge Exercises (51–55)

51. An artist wishes to make one wall of his studio out of glass brick to capture the natural light. The wall measures 26 feet long and 12 feet high. If the glass blocks are 6-inch squares and cost $1.25 each, how much will the blocks for the wall cost?

52. The side of an A-frame house is in the shape of a triangle and has a rectangular window as shown in the figure. The window is 2.6 feet by 5.2 feet. If the wall is to be stuccoed, find the area to be stuccoed.

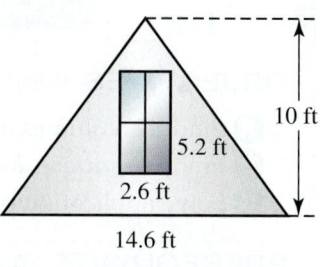

53. A house is situated on a trapezoidal lot as shown. If the area around the house is to be sodded, find the area to be sodded.

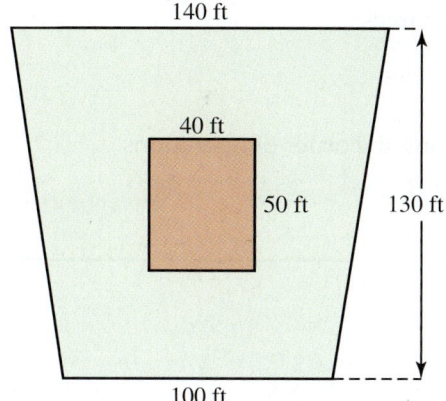

54. The front of the storage shed shown in the figure is to be painted. The door, which is metal and is 2.4 ft by 6.5 ft, will not be painted. Find the area to be painted.

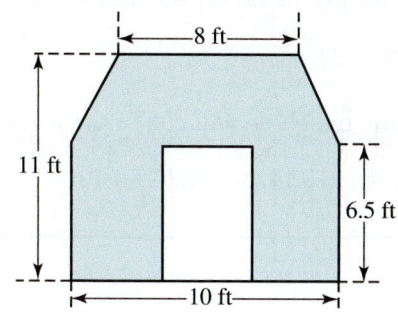

55. A very large washer used in heavy equipment has an outer diameter of 4 inches, and the diameter of the hole is 0.5 inches. What is the area of the washer? Use $\pi \approx 3.14$.

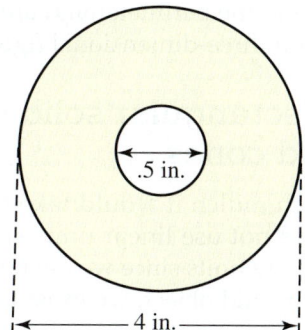

Writing Exercises or Group Project (56 and 57)

If these exercises are done as a group project, each group should write at least one exercise. The groups should then exchange exercises with other groups and solve the exercise(s) that they receive.

56. Write at least one application problem involving the area of a rectangle.

57. Write at least one application problem involving the area of a circle.

Group Project

58. Look around your campus or local community and find examples of at least five of the geometric figures that we studied in this section. Find the area of each.

Section 1.4   Volumes and Surface Areas of Geometric Figures

OBJECTIVES *When you complete this section, you will be able to:*

Ⓐ Find the volumes of rectangular solids, cubes, right circular cylinders, and cones.
Ⓑ Find the surface area of rectangular solids, cubes, and right circular cylinders.
Ⓒ Solve application problems involving volumes and surface areas.

PREREQUISITE SKILLS *Before beginning this section, you should be able to:*

a. Add, multiply, and divide whole numbers and decimals (Sections R.2, R.3, R.5, and R.6).
b. Substitute values for variables and evaluate by using the order of operations (Section 1.1).

GETTING READY FOR SECTION 1.4

Add, multiply, or divide the following whole numbers and/or decimals:

1. $(7)(9)(12)$ 　　　　　　　　 **2.** $\dfrac{4(3.14)(27)}{3}$

Substitute the given value(s) of the variable(s) and evaluate by using the order of operations.

3. $\pi r^2 h$: $\pi \approx 3.14, r = 4, h = 6$ 　　　　 **4.** $\dfrac{4}{3}\pi r^3$: $\pi \approx 3.14, r = 3$

Introduction

In the previous sections, we found the perimeter and area of two-dimensional figures. In this section, we concentrate on three-dimensional figures.

OBJECTIVE Ⓐ **Finding volumes of rectangular solids, cubes, right circular cylinders, and cones**

If we wanted to find how much mulch it would take to fill a box 4 feet long, 2 feet wide, and 3 feet high, we could not use linear units since we are not finding a distance, and we could not use square units since we are not finding the size of a surface. In order to find the volume of a solid object, we must use another type of unit called a **cubic unit**.

The word cube is often misused in everyday English. Have you ever seen an ice cube that is actually a cube?

Imagine a box, including the top, each of whose faces is a square. This type of figure is a **cube**. A cubic unit is a cube each of whose faces is a square whose sides are one unit each. For example, the following figure is a cubic inch (in.^3):

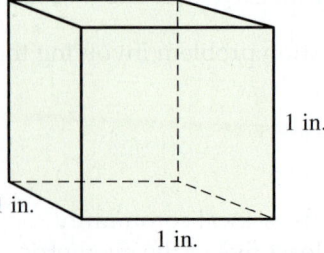

1 in.
1 in.
1 in.

When finding the **volume** of a solid, we are finding the number of cubic units it takes to fill the solid. This may be a little confusing, since the volumes we encounter in everyday situations are usually liquid measures of volume like fluid ounces and gallons.

Answers: Getting Ready: 1. 756 **2.** 113.04 **3.** 301.44 **4.** 113.04

Volume of a rectangular solid

A **rectangular solid** is a geometric figure with six faces, each of which is a rectangle. A box is an example of a rectangular solid. See the following figure:

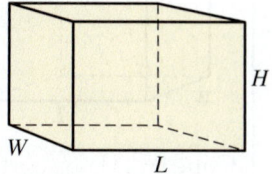

We will develop the formula for the volume of a rectangular solid and then give the formulas for the volumes of other types of solids. Let us find the volume of a rectangular solid that is 4 feet long, 3 feet wide, and 2 feet high. We are finding the number of cubic feet necessary to fill the solid. First we calculate how many cubic feet (ft^3) it would take to cover the bottom of the rectangular solid.

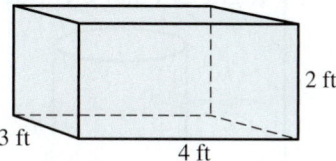

We mark off units of one foot each along the length and width of the base of the solid and draw horizontal lines at these units.

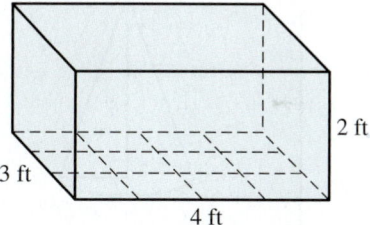

The lines divide the base into $3 \cdot 4 = 12$ unit squares each of whose sides is 1 foot. Since each face of a cube is a square, we can place 1 cubic foot on top of each of these squares, forming one layer of 12 cubic feet on the base.

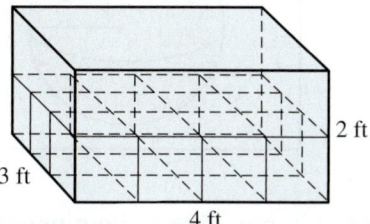

Since the height is 2 feet, it would take another layer of cubes identical to this layer to completely fill the rectangular solid. Therefore, it would take $12 \cdot 2 = 24 \ ft^3$ to fill the rectangular solid. Notice the length (4) times the width (3) gave us the number of cubes in the first layer (12). The height (2) gave us the number of layers. Therefore, the volume is $3 \cdot 4 \cdot 2 = 24 \ ft^3$. Consequently, the volume of a rectangular solid is $V = LWH$. Following are some common geometric solids and the formulas for their volumes:

Type of Solid	Characteristics	Formula
Rectangular Solid	Six faces all of which are rectangles.	$V = LWH$
Cube	A rectangular solid in which each face is a square. The intersection of any two faces is called an **edge** and is denoted by e. All edges are of the same length.	$V = e^3$
Right Circular Cylinder	Looks like a tin can. The top and bottom are circles, and the "sides" are perpendicular to the top and bottom. The radius of the base is denoted by r, and the distance between the top and bottom is the height h.	$V = \pi r^2 h$ Note that this is the area of the base times the height.
Cone	The bottom, called the **base**, is a circle, and the figure comes to a single point at the top, called the **vertex**. The radius of the base is denoted by r. The perpendicular distance between the vertex and the base is called the *height* and is denoted by h.	$V = \dfrac{1}{3}\pi r^2 h$ or $\dfrac{\pi r^2 h}{3}$ Note that this is $\frac{1}{3}$ the area of the base times the height and the volume is $\frac{1}{3}$ the volume of a right circular cylinder with the same base and height.
Sphere	The set of all points in space that are equidistant from a given point. The given point is the **center**, and the given distance is the **radius**.	$V = \dfrac{4}{3}\pi r^3$

Example 1   Find the volumes of the given geometric solids. If necessary, round answers to the nearest hundredth.

a.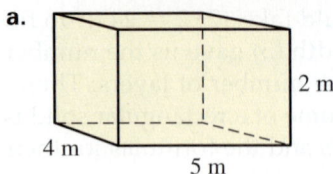

Solution

$V = LWH$ Formula for the volume of a rectangular solid. Substitute for L, W, and H.

$V = (5\,\text{m})(4\,\text{m})(2\,\text{m})$ Multiply.

$V = 40\,\text{m}^3$ Therefore, the volume is 40 cubic meters.

b.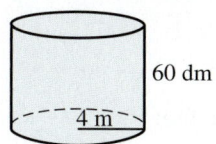

$V = e^3$ — Formula for the volume of a cube. Substitute for e.

$V = (4 \text{ ft})^3$ — Raise to the power.

$V = 64 \text{ ft}^3$ — Therefore, the volume is 64 cubic ft.

c. Express the answer in cubic meters. Use $\pi \approx 3.14$.

First, change 60 decimeters into meters. $60 \text{ dm} = 6 \text{ m}$.

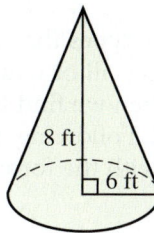

$V = \pi r^2 h$ — Substitute for $\pi, r,$ and h.

$V \approx (3.14)(4 \text{ m})^2(6 \text{ m})$ — Raise to powers first.

$V \approx (3.14)(16 \text{ m}^2)(6 \text{ m})$ — Multiply.

$V \approx 301.44 \text{ m}^3$ — Therefore, the volume is approximately 301.44 m³.

d. Right circular cylinder where the radius of the base is 2.1 meters and the height is 3.2 m. Round the answer to the nearest hundredth of a cubic meter. Use $\pi \approx 3.14$.

$V = \pi r^2 h$ — Substitute for $\pi, r,$ and h.

$V \approx (3.14)(2.1 \text{ m})^2(3.2 \text{ m})$ — Raise to powers first.

$V \approx (3.14)(4.41 \text{ m}^2)(3.2 \text{ m})$ — Multiply.

$V \approx 44.31168 \text{ m}^3$ — Round to the nearest hundredth.

$V \approx 44.31 \text{ m}^3$ — Therefore, the volume is 44.31 cubic meters to the nearest hundredth.

Note

The exact volumes for Examples 1c and d are:

$$V = \pi r^2 h = \pi(4 \text{ m})^2(6 \text{ m}) = \pi(16 \text{ m}^2)(6 \text{ m}) = 96\pi \text{ m}^3$$

$$V = \pi r^2 h = \pi(2.1 \text{ m})^2(3.2 \text{ m}) = \pi(4.41 \text{ m}^2)(3.2 \text{ m}) = 14.112\pi \text{ m}^3$$

e. Find the exact volume of the cone. (Express the answer in terms of π.)

$V = \dfrac{\pi r^2 h}{3}$ — Substitute for r and h.

$V = \dfrac{\pi(6 \text{ ft})^2(8 \text{ ft})}{3}$ — Raise to powers first.

$V = \dfrac{\pi(36 \text{ ft}^2)(8 \text{ ft})}{3}$ — Multiply in the numerator.

$V = \dfrac{\pi(288 \text{ ft}^3)}{3}$ — Divide.

$V = 96\pi \text{ ft}^3$ — Therefore, the exact volume is 96π ft³.

PRACTICE EXERCISE 1

Find the volume of the following geometric solids:

a. Rectangular solid with length of 13 millimeters, width of 2 millimeters, and height of 24 millimeters.

b.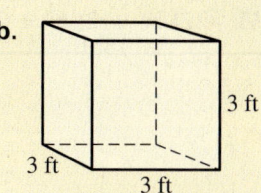

(Continued)

Answers: Practice Exercise 1: a. 624 mm³ **b.** 27 ft³

c. Right circular cylinder for which the radius of the base is $\frac{2}{3}$ inch and the height is $\frac{5}{3}$ inch. Use $\pi \approx \frac{22}{7}$. What is the exact volume?

d. Find the exact volume (in terms of π) and in cubic feet.

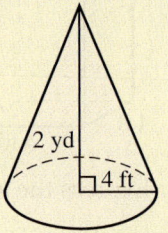

2 yd

4 ft

If you need more practice, do the following Additional Practice Exercises.

Additional Practice Exercise 1

a. Find the volume of the following rectangular solid: Length of 8 inches, width of 4 inches, and height of 5 inches.

b. Find the volume of the following cube:

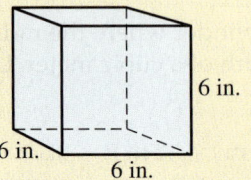

6 in.

6 in.

6 in.

c. Find the approximate and exact volume of the following right circular cylinder: Radius of the base is 5.1 feet, and the height is 2.4 feet. For approximate, use $\pi \approx 3.14$ and round to the nearest tenth.

d. Find the exact volume of the following cone: (Express the answer in terms of π.)

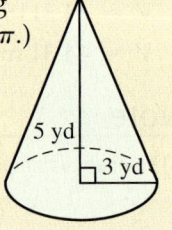

5 yd

3 yd

OBJECTIVE Finding the surface area of rectangular solids, cubes, and right circular cylinders

Three-dimensional figures are made up of surfaces that often are of the types that we studied in Section 1.3. For example, a rectangular solid has six surfaces all of which are rectangles, and a cube has six surfaces all of which are squares. When we find the areas of all these surfaces, we have found the **surface area** of the solid. Following are some of the solids for which we previously found the volume, along with the formulas for their surface areas:

Type of Solid	Rectangular Solid	Cube	Right Circular Cylinder
	H / W / L	e / e / e	h / r
Formula for Surface Area	$SA = 2LW + 2LH + 2WH$	$SA = 6e^2$	$SA = 2\pi r^2 + 2\pi rh$

Answers: Practice Exercise 1: c. $\frac{440}{189}$ or $2\frac{62}{189}$ in.^{3}, $\frac{20}{27}\pi$ in.^{3} **d.** 32π ft^3

Additional Practice 1: a. 160 in.^{3} **b.** 216 in.^{3} **c.** 196.0 ft^3, 62.424π ft^3 **d.** 15π yd^3

Example 2   **Find the surface area of each of the following:**

a.

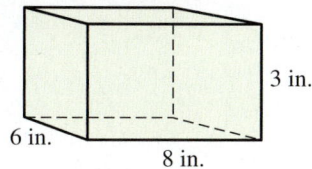

3 in.

6 in.

8 in.

Solution

$SA = 2LW + 2LH + 2WH$ — Substitute.

$SA = 2(8\text{ in.})(6\text{ in.}) + 2(8\text{ in.})(3\text{ in.}) + 2(6\text{ in.})(3\text{ in.})$ — Multiply.

$SA = 96\text{ in.}^2 + 48\text{ in.}^2 + 36\text{ in.}^2$ — Add.

$SA = 180\text{ in.}^2$ — Answer.

b.

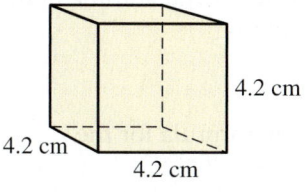

4.2 cm

4.2 cm

4.2 cm

Solution

$SA = 6e^2$ — Substitute.

$SA = 6(4.2\text{ cm})^2$ — Evaluate $(4.2)^2$

$SA = 6(17.64\text{ cm}^2)$ — Multiply.

$SA = 105.84\text{ cm}^2$ — Answer.

c. Leave answer in terms of π (exact answer).

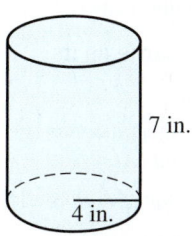

7 in.

4 in.

Solution

$SA = 2\pi r^2 + 2\pi rh$ — Substitute.

$SA \approx 2(\pi)(4\text{ in.})^2 + 2(\pi)(4\text{ in.})(7\text{ in.})$ — Square 4.

$SA \approx 2(\pi)(16\text{ in.}^2) + 2(\pi)(4\text{ in.})(7\text{ in.})$ — Multiply.

$SA \approx 32\pi\text{ in.}^2 + 56\pi\text{ in.}^2$ — Add.

$SA \approx 88\pi\text{ in.}^2$ — Answer.

PRACTICE EXERCISE 2

Find the surface area of each of the following:

a.

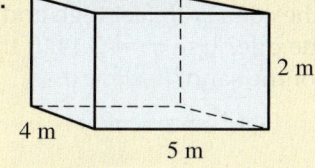

2 m

4 m

5 m

b.

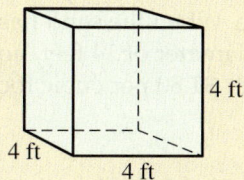

4 ft

4 ft

4 ft

c. Leave the answer in terms of π (exact answer).

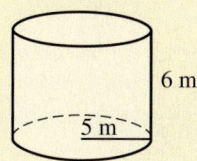

6 m

5 m

If you need more practice, do the following Additional Practice Exercises.

Additional Practice Exercise 2   Find the surface area of each of the following:

a.

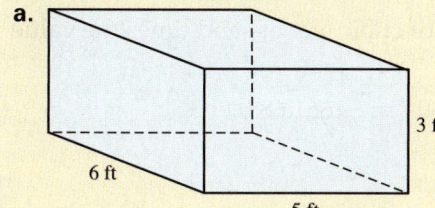

3 ft

6 ft

5 ft

b.

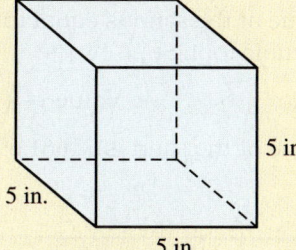

5 in.

5 in.

5 in.

c. Leave the answer in terms of π (exact answer).

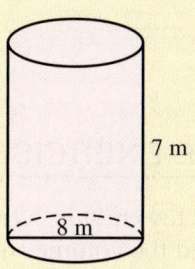

7 m

8 m

Answers: **Practice Exercise 2: a.** 76 m^2 **b.** 96 ft^2 **c.** 110π m^2 **Additional Practice 2: a.** 126 ft^2 **b.** 150 in.^{2} **c.** 88π m^2

OBJECTIVE Solving application problems involving volumes and surface areas

Applications involving volume and surface area often occur.

Example 3 | **Solve the following:**

a. A toolbox manufacturer makes toolboxes in the shape of rectangular solids that are 48 inches long, 16 inches wide, and 22 inches high. What is the volume of these boxes, and how much material does it take to make a toolbox?

Solution

Since the toolbox is in the shape of a rectangular solid, the formula for the volume is

$V = LWH$ Substitute for L, W, and H.

$V = (48 \text{ in.})(16 \text{ in.})(22 \text{ in.})$ Multiply.

$V = 16,896 \text{ in.}^3$ Therefore, the volume of the box is 16,896 cubic inches.

The amount of material used in manufacturing the toolbox is the same as its surface area. The formula is

$SA = 2LW + 2LH + 2WH$ Substitute for L, W, and H.

$SA = 2(48 \text{ in.})(16 \text{ in.}) + 2(48 \text{ in.})(22 \text{ in.}) + 2(16 \text{ in.})(22 \text{ in.})$ Multiply before adding.

$SA = 1536 \text{ in.}^2 + 2112 \text{ in.}^2 + 704 \text{ in.}^2$ Add.

$SA = 4352 \text{ in.}^2$ Therefore, it takes 4352 square inches of material to make a box.

b. As sand comes off a conveyor belt, it forms a pile that is in the shape of a right circular cone. When the sand has accumulated until the cone is 18 feet high and has a base diameter of 14 feet, how much sand is in the pile? Use $\pi \approx 3.14$. If the sand is worth $1.80 per cubic foot, what is the value of the sand?

Solution

The amount of sand in the pile is the same as the volume. Therefore, the formula is

$V = \frac{1}{3}\pi r^2 h$ Substitute for π, r, and h. Note the radius is 7 ft.

$V \approx \frac{1}{3}(3.14)(7 \text{ ft})^2(18 \text{ ft})$ Square 7 ft.

$V \approx \frac{1}{3}(3.14)(49 \text{ ft}^2)(18 \text{ ft})$ Multiply.

$V \approx 923.16 \text{ ft}^3$ Therefore, there are approximately 923.16 cubic feet of sand in the pile.

The value of the sand is equal to the number of cubic feet of sand times the value of one cubic foot. So,

$$\text{Value} = (923.16)(\$1.80) = \$1661.688$$

The value of the sand is $1661.69.

PRACTICE EXERCISE 3

a. A child's toy chest is in the shape of a rectangular solid that is 32 inches long, 14 inches wide, and 18 inches high. Find the volume and surface area of the toy chest. If the material used to make the chest cost $0.08 per square inch, find the cost of the material.

b. A drum is in the shape of a right circular cylinder and is 40 inches high and 18 inches in diameter. Find the volume and surface area of the drum. Use $\pi \approx 3.14$.

Answers: Practice Exercise 3: a. $V = 8064 \text{ in.}^3$, $SA = 2552 \text{ in.}^2$, cost = $204.16 **b.** $V = 10,173.6 \text{ in.}^3$, $SA = 2769.48 \text{ in.}^2$

Study Tip 4 Reading Mathematics Critically

Mathematics is read differently than other disciplines. In a novel or a story, a whole chapter may contribute only a single idea to the plot. In mathematics, however, every single word or symbol has a precise meaning, and each may contribute one or more ideas to the overall subject. Consequently, you must read slowly and carefully so that you understand the meaning of *each* word and symbol.

You must be able to tell the difference between expressions that look very much alike, but have different meanings. For example, "the difference of 8 and 5, written $8 - 5$" is very different from "the difference of 5 and 8, written $5 - 8$."

Read with pencil and paper available. If you do not understand what the author did in getting from one step to the next, take the time to work it out for yourself.

For Extra Help

Exercise Set 1.4   MyMathLab®

Find the volume and surface area of the given rectangular solids. (See Examples 1a and 2a.)

1.

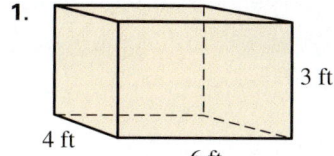

3 ft
4 ft
6 ft

2. Express the volume in cubic centimeters and the surface area in square centimeters.

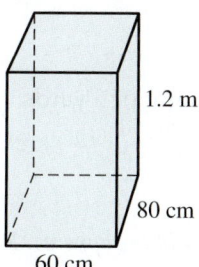

1.2 m
80 cm
60 cm

3. Express the volume in cubic yards and the surface area in square yards.

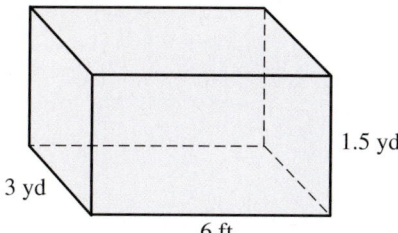

1.5 yd
3 yd
6 ft

4. Express the volume in cubic meters and the surface area in square meters.

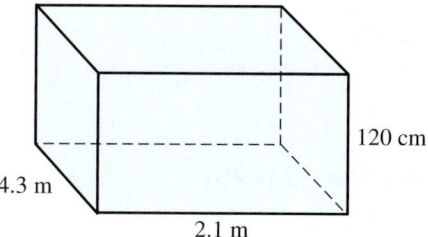

120 cm
4.3 m
2.1 m

5. Length of 8 meters, width of 5.2 meters, and height of 6.5 meters.

6. Length of 8 feet, width of 7 feet, and height of 4 yards. Express the volume in cubic feet and the surface area in square feet.

Find the exact and approximate volume and surface area of the following cubes. (See Examples 1b and 2b.)

7.

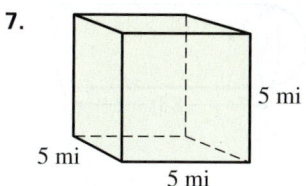

5 mi
5 mi
5 mi

8. Length of each edge is 2.3 kilometers.

9. Cube each of whose edges is $\frac{3}{5}$ centimeter.

10. Cube each of whose edges is $\frac{4}{7}$ feet.

Find the exact and approximate volume and surface area of the given right circular cylinders. Round approximate answers to the nearest tenth. Use π ≈ 3.14. (See Examples 1c and 2c.)

11.

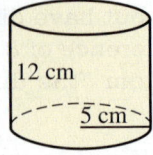

12 cm

5 cm

12.

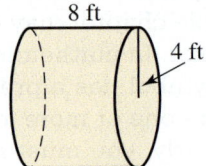

8 ft

4 ft

13. Radius of the base is 4.2 meters, and the height is 5.3 meters.

14. Radius of the base is 2.3 centimeters, and the height is 8.5 centimeters.

Find the exact and approximate volumes of the given cones. Round the approximate answers to the nearest tenth. (See Example 1e.)

15.

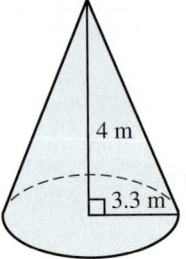

4 m

3.3 m

16.

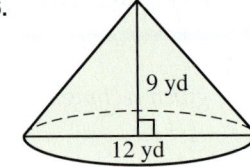

9 yd

12 yd

17. Base radius of 18 inches and height of 6 yards. Express the answer in cubic yards.

18. Base radius of 3 meters and height of 450 centimeters. Express the answer in cubic meters.

Find the exact and approximate volumes of the spheres. Round the approximate answers to the nearest hundredth if necessary.

19.

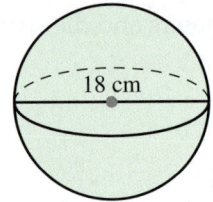

18 cm

20. Find the volume of a sphere whose radius is 6 inches.

Challenge Exercise (21–25)

Find the volumes of the given cones:

21. Base radius of $\frac{5}{7}$ inch and height of 14 inches. Use $\pi \approx \frac{22}{7}$.

22. Base diameter of 14 feet and height of $\frac{23}{11}$ feet. Use $\pi \approx \frac{22}{7}$.

23. Find the volume of the figure shown. Express the answer in terms of π.

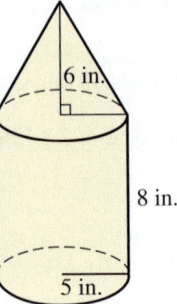

6 in.

8 in.

5 in.

24. Find the volume inside the cylinder, but outside the cone. Use $\pi \approx 3.14$.

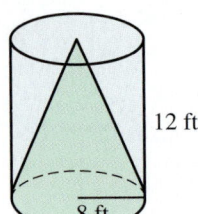

12 ft

8 ft

25. Find the volume of the propane tank. Express the answer in terms of π.

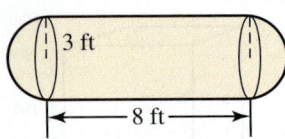

3 ft

8 ft

Answer the following. (See Example 3.)

26. A cereal box is 11 inches high, 7 inches long, and 2 inches wide. How many cubic inches of cereal can it hold?

27. The body of a dump truck used for hauling dirt and sand is in the shape of a rectangular solid and measures 8 feet by 12 feet by 4 feet. How many cubic yards of sand will it hold? Round the answer to the nearest hundredth of a cubic yard.

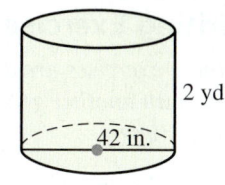

28. A building contractor wishes to put a basement in a new office building on campus. The basement is to measure 12 feet deep, and the building is to measure 90 feet long by 40 feet wide. How many cubic yards of dirt must be removed before the basement can be built?

29. A storage tank is in the shape of a right circular cylinder that has a base diameter of 42 inches and a height of 2 yards. Find the volume in cubic inches and express the answer in terms of π.

30. The men's size 7 regulation basketball has a diameter of 9.39 inches. Find the volume. Use $\pi \approx 3.14$ and round the answer to the nearest tenth.

31. The women's size 6 regulation basketball has a diameter of 9.07 inches. Find the volume. Use $\pi \approx 3.14$ and round the answer to the nearest tenth.

32. John and Kay wish to fill their swimming pool. Their pool is shaped like a rectangular solid measuring 5 yards wide by 40 feet long with the water at a depth of 6 feet. If water costs $0.75 per 100 cubic feet, how much would it cost to fill the pool?

33. Alfredo is estimating the cost of an addition to his home. He will build two rectangular-shaped rooms with a combined area that measures 60 feet by 18 feet. If the concrete slab (foundation) must be 8 inches thick and the concrete costs $27.00 a cubic yard, how much will the concrete for the slab cost?

34. A carpenter is building a wooden toy chest in the shape of a rectangular solid. The chest is to be 2 feet wide, 3 feet long, and 1.5 feet high. If the wood costs $5.50 per square foot, find the cost of the wood needed to build the chest.

35. A manufacturer of cardboard boxes has received an order for 500 boxes that are to be 18 inches wide, 24 inches long, and 20 inches high. How much cardboard will she need in order to manufacture these boxes?

36. Two walls of a room are 10 feet long, and the other two are 16 feet long. The ceiling is 12 feet high. If a gallon of paint will cover 500 square feet, find (to the nearest hundredth of a gallon) the number of gallons it will take to paint the room.

37. Two walls of a room are 4 yards long, and the other two are 5 yards long. The ceiling is 10 feet high. If it costs $.80 per square foot for plastering, find the cost of plastering the walls and ceiling.

38. As part of a mathematics exhibit at the science center, a cube is constructed so that the length of each edge is 15 inches. The material used in making the cube costs $0.12 per square inch. Find the cost of the material.

39. A cardboard box is in the shape of a cube with each edge 2 feet long. The box is to be wrapped with wrapping paper that costs $.35 per square foot. Find the cost of the paper.

40. A container of raisins is in the shape of a right circular cylinder with a radius of 2 inches and height of 6 inches. What is the surface area of the container? Use $\pi \approx 3.14$.

41. A can of premium chunk white chicken is in the shape of a right circular cylinder and is 4 inches in diameter and 2 inches high. What is the surface area of the can? Use $\pi \approx 3.14$.

42. A cereal container is in the shape of a right circular cylinder and has a radius of 3 inches and a height of 10 inches. The cardboard used to make the box costs $.0002 per square inch. Find the cost of the material needed for 6000 containers. Use $\pi \approx 3.14$.

43. A can manufacturer receives an order for 10,000 cans with a radius of 5 centimeters and a height of 15 centimeters. If the material used in constructing the cans costs $.0002 per square centimeter, find the cost of the material used in filling the order. Use $\pi \approx 3.14$.

Challenge Exercise

44. A gasoline-storage tank at a service station is in the shape of a right circular cylinder with a diameter of 6 feet and a height of 12 feet. If there are 231 cubic inches in a gallon and the gasoline costs the service station owner $2.59 per gallon, how much will it cost to fill the tank? Use $\pi \approx 3.14$ and round the number of gallons to the nearest whole gallon.

Writing Exercises or Group Projects (45–48)

If these exercises are done as a group project, each group should write at least one exercise of each type. Then exchange exercises with another group and solve. If the exercises are done as writing exercises, each student should both write and solve one of each type.

45. Write an application problem involving the volume of a rectangular solid.

46. Write an application problem involving the volume of a right circular cylinder.

47. Write an application problem involving the surface area of a rectangular solid.

48. Write an application problem involving the surface area of a right circular cylinder.

Group Project

49. Look around your campus or local community and find examples of at least five of the geometric figures that we studied in this section. Find the volume and surface area (except for a cone) of each.

Section 1.5

Introduction to Integers

OBJECTIVES *When you complete this section, you will be able to:*

A Identify natural numbers, whole numbers, and integers.
B Graph natural numbers, whole numbers, and integers.
C Determine order relations for integers.
D Find the negative (opposite) of an integer.
E Find the absolute value of an integer.

PREREQUISITE SKILLS *None*

GETTING READY FOR SECTION 1.5

No prerequisite skills, so no getting ready exercises.

Introduction

A **set** is any collection of objects, and the objects that make up the set are called the **elements** of the set. In mathematics, the sets that we are most often concerned with are sets whose elements are numbers. Sets are usually named by a capital letter and are indicated by enclosing its elements in braces and separating them by commas. If the elements of a set are listed individually, then the set is said to be listed by **roster**. For example, $\{a, b, c\}$ is read "the set whose elements are a, b, and c," and the elements are listed by roster. If all the elements of set A are also elements of set B, then set A is a **subset** of set B. For example, if set A $= \{1, 2\}$ and set B $= \{1, 2, 3, 4\}$, then set A is a subset of set B.

OBJECTIVE Identifying natural numbers, whole numbers, and integers

Historically, numbers have been invented as they were needed. Our ancestors probably kept track of "how many" by using tally marks or stones, one for each object being "counted." No doubt, the first numbers needed were those that indicated how many. We call these the **natural**, or **counting**, numbers.

> **Definition** Set of Natural Numbers
> $$N = \{1, 2, 3, 4, 5, \dots\}$$

If the set of natural numbers is expanded to include 0, we get another set of numbers called the **whole numbers**. Note that the natural numbers are a subset of the whole numbers, since all the elements of the set of natural numbers are also elements of the set of whole numbers.

> **Definition** Set of Whole Numbers
> $$W = \{0, 1, 2, 3, 4, \dots\}$$

If limited to just the whole numbers, we would have no way of representing 10° below zero or a loss of $150. Consequently, **negative numbers** were invented to help us make these representations.

Example 1

a. If $+10°$ represents $10°$ above $0°$, then $-10°$ (read "negative 10") represents $10°$ below $0°$.

b. If $+45$ feet represents 45 feet above sea level, then -45 feet represents 45 feet below sea level.

c. If $+\$250$ represents a deposit of $\$250$, then $-\$250$ represents a withdrawal of $\$250$.

PRACTICE EXERCISE 1

a. If $+3$ yards represents a gain of 3 yards, then _____ represents a loss of 3 yards.

b. If $-\$100$ represents a withdrawal of $\$100$, then _____ represents a deposit of $\$100$.

c. If $+12°$ represents a rise of $12°$ in temperature, then _____ represents a drop of $12°$ in temperature.

If you need additional practice, do the following Additional Practice Exercises.

Additional Practice Exercise 1

a. If $+23$ represents going up 23 floors in an elevator, how would you represent going down 23 floors?

b. If $+\$250$ means receiving $\$250$, what would $-\$250$ represent?

c. If $+33$ represents going 33 miles north, how would you represent going 33 miles south?

If the set of whole numbers is expanded to include the negatives of the natural numbers, we get a new set of numbers called the **integers**. Note that both the natural numbers and the whole numbers are subsets of the integers.

> **Definition** Set of Integers
> $$I = \{\ldots, -3, -2, -1, 0, +1, +2, +3, \ldots\}$$

OBJECTIVE **B** Graphing natural numbers, whole numbers, and integers

Positive and negative numbers are often referred to as signed numbers, with 0 being neither positive nor negative. Another way of representing the integers is by using the number line. The number line is like a thermometer that is horizontal instead of vertical. To construct a number line, draw a horizontal line and choose an arbitrary point that is labeled "0." Choose a unit of length and mark off the line both to the left and right of 0 in terms of this unit. Label the units to the right of 0 with the positive integers and the units to the left with the negative integers. Positive numbers are usually written without the $+$ sign. For example, $+3$ is written as 3 and $+6$ is written as 6. The resulting number line should look something like this:

In this manner, any integer can be paired with a point on the number line. This procedure is called **graphing**.

Example 2 | **Graph the integers {− 4, − 2, 0, 3}**

We put a dot on the number line at the location of each of the integers $-4, -2, 0, 3$.

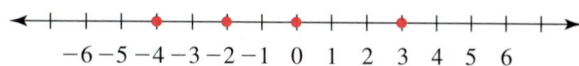

PRACTICE EXERCISE 2

Graph the following sets of integers:

a. $\{-5, -2, 3, 6\}$

b. $\{-4, -3, -1, 1, 2, 3\}$

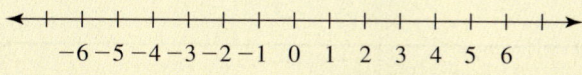 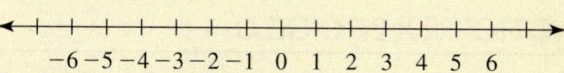

If you need more practice, do the following Additional Practice Exercises.

Additional Practice Exercise 2 Draw a number line and graph each of the following:

a. $\{-3, -1, 0, 2, 4\}$

b. $\{-4, -2, 3, 5, 6\}$

Note: There are points on the number line that do not have integers assigned to them. Consequently, other types of numbers will be needed in order to assign a number to every point on the number line. These types of numbers will be discussed later in the text.

OBJECTIVE Determining order relations for integers

The integers possess the property of being **ordered**. By this, we mean that when given two integers, it is possible to determine whether they are equal or which is the larger or smaller of the two.

To show the order relationships of numbers, we use the following symbols:

Definition Order Symbols

$a = b$ Read "a is equal to b." This means that the number represented by a has the same value as the number represented by b. For example, $5 = 5$.

$a \neq b$ Read "a is not equal to b." This means that the number represented by a does not have the same value as the number represented by b. For example, $-3 \neq 4$.

$a < b$ Read "a is less than b." This means that the number represented by a is to the left of the number represented by b on the number line. For example, $2 < 3$ because 2 is to the left of 3 on the number line.

$a \leq b$ Read "a is less than or equal to b." This means that the number represented by a is either less than the number represented by b or equal to the number represented by b. For example, $3 \leq 4, 5 \leq 5$.

$a > b$ Read "a is greater than b." This means that the number represented by a is to the right of the number represented by b on the number line. For example, $5 > 1$ because 5 is to the right of 1 on the number line.

$a \geq b$ Read "a is greater than or equal to b." This means that the number represented by a is either greater than the number represented by b or equal to the number represented by b. For example, $4 \geq 2, 7 \geq 7$.

Answers: Practice Exercise 2: a.

$-6\ -5\ -4\ -3\ -2\ -1\ 0\ 1\ 2\ 3\ 4\ 5\ 6$

b.

$-6\ -5\ -4\ -3\ -2\ -1\ 0\ 1\ 2\ 3\ 4\ 5\ 6$

Additional Practice 2: a.

$-6\ -5\ -4\ -3\ -2\ -1\ 0\ 1\ 2\ 3\ 4\ 5\ 6$

b.

$-6\ -5\ -4\ -3\ -2\ -1\ 0\ 1\ 2\ 3\ 4\ 5\ 6$

Remember, the smaller number is always to the left of the larger on the number line. Or equivalently, the larger is always to the right of the smaller.

Look back at the graph of Example 2 when doing Example 3.

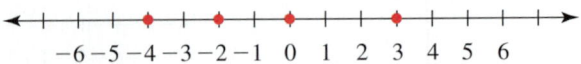

Example 3

a. 0 is to the left of 3. Therefore, $0 < 3$ or, equivalently, $3 > 0$. We could also write $0 \leq 3$ or $3 \geq 0$.

b. -2 is to the left of 0. Therefore, $-2 < 0$ or $0 > -2$ **c.** -4 is to the left of -2. Therefore, $-4 < -2$.

PRACTICE EXERCISE 3

Insert the proper symbol ($=$, $<$, or $>$) in order to make each of the following true:

a. -2 _____ 3   **b.** 4 _____ -5 **c.** -5 _____ -8

If you need more practice, do the following Additional Practice Exercises.

Additional Practice Exercise 3   Insert the proper symbol ($=$, $<$, or $>$) in order to make each of the following true:

a. 8 _____ 8   **b.** 4 _____ $2 + 5$ **c.** -9 _____ -5

OBJECTIVE Finding the negative (opposite) of an integer

From the accompanying diagram, you can see that each integer, except 0, can be paired with another integer the same distance from 0, but on the opposite side. Each such integer is called the **opposite**, or **negative**, of the other. Zero is its own opposite.

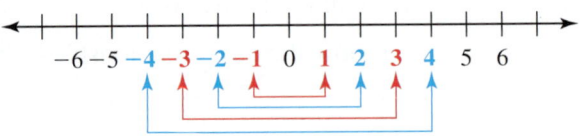

The negative of 3 is -3, the negative of -4 is 4, and so on. The negative of a number is usually indicated by placing a negative sign before the number. Consequently, the negative of -4 is written $-(-4)$ and equals 4, since 4 is the negative, or opposite, of -4. Likewise, $-(-6) = 6$ and $-(-2) = 2$. This leads us to an important characteristic of signed numbers.

> **Rule: The Opposite of a Negative**
>
> $-(-x) = x$ for all values of x.

OBJECTIVE Finding the absolute value of an integer

Another important concept associated with signed numbers is that of **absolute value**.

> **Definition** Absolute Value
> The absolute value of a number is defined as the distance from 0 to the number.

Absolute value is indicated by placing the number whose absolute value you wish to find between vertical bars. Since absolute value is a distance, it can never be negative.

Answers: Practice Exercise 3: a. < b. > c. >   Additional Practice 3: a. = b. < c. <

Example 4 Find the indicated absolute values.

a. $|4| = 4$ The distance from 0 to 4 is 4.

b. $|-4| = 4$ The distance from 0 to -4 is 4.

c. $|0| = 0$ The distance from 0 to 0 is 0.

d. $|-8| = 8$ The distance from 0 to -8 is 8.

e. $|8| = 8$ The distance from 0 to 8 is 8.

As stated earlier, the absolute value of a number is never negative. As long as we are finding absolute values of constants, there is no problem, because it is obvious whether the constant is positive, negative, or zero. The difficulty occurs when we are asked to represent the absolute values of variables. Since a variable may be replaced by any number, the variable x could represent either a positive number, a negative number, or 0. Consequently, $x < 0, x > 0$, or $x = 0$, depending on what we choose to substitute for x. Therefore, x is not necessarily positive, nor is $-x$ necessarily negative. If x is positive, then $-x$ is negative; in symbols, if $x > 0$, then $-x < 0$. In words, the opposite of a positive number is a negative number. If x is negative, then $-x$ is positive; that is, if $x < 0, -x > 0$. In words, the opposite of a negative number is a positive number. For example, if $x = -3$, then $-x = -(-3) = 3$. So when asked to represent $|x|$, we have to consider three cases: one case when $x > 0$, another when $x < 0$, and another when $x = 0$. The case where $x = 0$ is usually combined with the case where $x > 0$.

> **Rule: Absolute Value of x**
>
> **a.** If $x \geq 0, |x| = x$.
>
> **b.** If $x < 0, |x| = -x$.
>
> This means that if x is positive or 0, then x is its own absolute value. If x is negative, the opposite of x is its absolute value.

This says that the absolute value of a positive number is the same positive number ($|6| = 6$) and the absolute value of a negative number is the negative, or opposite, of that negative number ($|-5| = -(-5) = 5$).

Example 5 Find the following by using the definition of absolute value:

a. $|2| = 2$ Since $2 \geq 0, |x| = x$.

b. $|-8| = -(-8) = 8$ Since $-8 < 0, |x| = -x$.

c. $|0| = 0$ Since $0 \geq 0, |x| = x$.

d. $|n| = \begin{cases} n \text{ for } n \geq 0 \\ -n \text{ for } n < 0 \end{cases}$ We must include all possibilities, since we do not know if n is positive, negative, or zero.

e. $-|10| = -10$ The "$-$" is outside the absolute value marks, so first find the absolute value of 10 and then take its negative.

f. $-|-15| = -15$ The "$-$" is outside the absolute value marks, so first find the absolute value of -15 and then take its negative.

Note

Don't confuse $-(-15)$ with $-|-15|$; $-(-15) = 15$ and $-|-15| = -(+15) = -15$.

PRACTICE EXERCISE 5

Find the value of each of the following:

a. $|9| = $ _____

b. $|-10| = $ _____

c. $|13| = $ _____

d. $|y| = $ _____

e. $-|18| = $ _____

f. $-|-23| = $ _____

If you need more practice, do the following Additional Practice Exercises.

Additional Practice Exercise 5   Find the value of each of the following:

a. $|4| = $ _____

b. $|-8| = $ _____

c. $|-25| = $ _____

d. $|z| = $ _____

e. $-|15| = $ _____

f. $-|-8| = $ _____

When evaluating expressions containing absolute values, we evaluate inside the absolute value bars first much like we do when parentheses are present. After evaluating the expression inside the absolute value bars (if necessary), we then find all absolute values before performing any operations.

| **Example 6** | **Evaluate the following:** |

a. $|10 - 4| = $ Simplify inside the absolute value bars first.

$\quad |6| = $ Find the absolute value of 6.

$\quad 6$ Answer.

b. $|-5| + |4| = $ Find the absolute values of -5 and 4.

$\quad 5 + 4 = $ Add 5 and 4.

$\quad 9$ Answer.

c. $|-6| - |-5| = $ Find the absolute values of -6 and -5.

$\quad 6 - 5 = $ Subtract

$\quad 1$ Answer.

d. $|12 - 8| + |14 + 5| = $ Evaluate inside the absolute value bars first.

$\quad |4| + |19| = $ Find the absolute values of 4 and 19.

$\quad 4 + 19 = $ Add.

$\quad 23$ Answer.

PRACTICE EXERCISE 6

Evaluate the following.

a. $|15 - 6|$

b. $|-7| + |4|$

c. $|-7| - |-6|$

d. $|8 - 2| - |6 - 3|$

If you need more practice, do the following Additional Practice Exercises.

Additional Practice Exercise 6   Evaluate the following:

a. $|12 - 9|$

b. $|4| + |-9|$

c. $|-10| - |-6|$

d. $|15 - 4| - |-3 - 5|$

Answers: Practice Exercise 5: a. 9 **b.** 10 **c.** 13 **d.** y if $y \geq 0$, $-y$ if $y < 0$ **e.** -18 **f.** -23 **Additional Practice 5: a.** 4 **b.** 8 **c.** 25 **d.** z if $z \geq 0$, $-z$ if $z < 0$ **e.** -15 **f.** -8 **Practice Exercise 6: a.** 9 **b.** 11 **c.** 1 **d.** 3   **Additional Practice 6: a.** 3 **b.** 13 **c.** 4 **d.** 3

Exercise Set 1.5   MyMathLab®

For Extra Help

Answer each of the following by using signed numbers. (See Example 1.)

1. If +8 represents 8 units to the right, how do you represent 8 units to the left?

2. If −80 feet represents a scuba diver descending 80 feet, how do you represent a scuba diver ascending 50 feet?

3. If +12 pounds represents gaining 12 pounds, what would −10 pounds represent?

4. If −$25 represents losing $25, what would +$55 represent?

5. If +250 miles represents 250 miles east, what would −250 miles represent?

6. If +5 represents gaining 5 points on an exam, what would −5 represent?

Draw a number line for each of the following and graph the indicated integers. (See Example 2.)

7. $\{-5, -4, -1, 0\}$

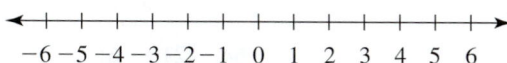

8. $\{-3, -1, 2, 3\}$

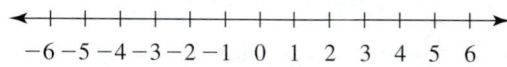

9. $\{-4, -2, 3, 5, 6\}$

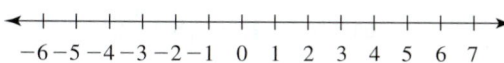

10. $\{-8, -3, 0, 2, 6\}$

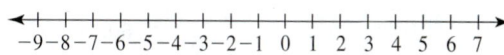

For Exercises 11–13, give three examples of each of the following. (See Objective A.)

11. Integers that are not natural numbers.

12. Integers that are not whole numbers.

13. Numbers that are both integers and natural numbers.

14. Given the set $\{8, -.43, 0, -4, \frac{2}{3}\}$, give the elements that are

   a. natural numbers.
   b. whole numbers.
   c. integers.
   d. not a natural number, nor a whole number, nor an integer.

15. Given the set $\{-6.2, 5, \frac{7}{10}, 0, -10\}$, give the elements that are

   a. natural numbers.
   b. whole numbers.
   c. integers.
   d. not a natural number, nor a whole number, nor an integer.

Answer the following as **always true,** *sometimes true, or* **never true.** *(See Objective A.)*

16. A natural number is an integer.

17. An integer is a whole number.

18. A whole number is a natural number.

19. A whole number is an integer.

20. An integer is a natural number.

21. A natural number is a whole number.

Find the negative (opposite) of each of the following. (See Objectives D and E.)

22. 47   **23.** 17   **24.** -12 **25.** -15

26. $|22|$ **27.** $|16|$ **28.** $|-6|$ **29.** $|-8|$

30. $-|7|$ **31.** $-|15|$ **32.** $-|-23|$ **33.** $-|-12|$

Insert the proper symbol $=$, $<$, or $>$ in order to make each of the following true. (See Example 3.)

34. 8 _____ 10   **35.** 29 _____ 12   **36.** 14 _____ -3 **37.** -10 _____ 8

38. -8 _____ -4 **39.** -3 _____ -7 **40.** $|6|$ _____ 6   **41.** $|-5|$ _____ 5

42. $-|-4|$ _____ 4   **43.** -7 _____ $-|-7|$ **44.** $|10|$ _____ $|-10|$ **45.** $-|-10|$ _____ $-(-10)$

46. $|24-18|$ _____ $|8-1|$ **47.** $|32-15|$ _____ $|13-5|$ **48.** $-(-5)$ _____ 5

49. 8 _____ $-(-8)$ **50.** -10 _____ $-(-10)$

Evaluate the following. (See Example 6.)

51. $|14-6|$ **52.** $|15-8|$ **53.** $|7-2|$ **54.** $|21-6|$

55. $|12|-|6|$ **56.** $|8|-|3|$ **57.** $|-4|+|6|$ **58.** $|-7|+|10|$

59. $|-7|-|-5|$ **60.** $|-9|-|-8|$ **61.** $|6+4|+|8-2|$

62. $|9+3|+|14-10|$ **63.** $|5+7|-|14-10|$ **64.** $|12-8|+|15-4|$

Challenge Exercises (65–68)

Evaluate each of the following:

65. $2|-4|+3|-5|-2\cdot3^2$ **66.** $4|5|-2|-3|+3\cdot2^3$

67. $5|4^2+3\cdot2|-4^2\cdot2$ **68.** $6|5\cdot6^2-8\cdot7|-8\cdot2^3$

Writing Exercises (69–73)

69. Give three examples of numbers that are not integers, and explain why they are not integers.

70. Why is any positive number greater than any negative number?

71. Does $-(-x)$ always represent a positive number? Why or why not?

72. What does $\neq$ mean?

73. Why is the absolute value of a number always nonnegative?

Group Project

74. In this section, examples of the occurrence of negative numbers were given, including the withdrawal of money and a drop in temperature. Give two more examples where negative numbers occur, other than those given in the text.

Section 1.6   Addition of Integers

OBJECTIVES *When you complete this section, you will be able to:*

Ⓐ Add integers with the same and with opposite signs.

Ⓑ Translate verbal expressions into mathematical expressions and simplify.

Ⓒ Identify properties of addition.

PREREQUISITE SKILLS *Before beginning this section, you should be able to:*

a. Add and subtract whole numbers and decimals (Section R.2 and R.5).

b. Graph integers by using the number line (Section 1.5).

c. Add fractions with common denominators (Section R.4).

GETTING READY FOR SECTION 1.6

Add or subtract the following whole numbers and decimals.

1. $15 + 6$ **2.** $12 - 9$ **3.** $8.2 - 3.7$

4. Graph the following set of integers on the number line. $\{-3, -1, 2, 4\}$

Introduction

The addition of integers occurs in many everyday situations. If the temperature rises 8° from 9:00 to 10:00 A.M. and then rises 5° from 10:00 to 11:00, there has been a total change in temperature of $8° + 5° = 13°$ from 9:00 to 11:00. If a football team gains 8 yards $(+8)$ on first down and loses 3 yards (-3) on second down, then there is a net gain on the two downs of $(+8) + (-3) = 5$ yards. If a gambler loses $20 (-20) on the first hand and wins $8 $(+8)$ on the second hand, there is a total loss of $(-\$20) + \$8 = -\$12$ on the two hands. If a scuba diver descends 30 feet (-30) and then descends another 20 feet (-20), then she has descended a total of $(-30) + (-20) = -50$ feet.

Calculator Exploration Activity (Optional)

Using a calculator, complete the following: Be sure to use the minus $(-)$ key and not the subtraction key to represent negative numbers.

Column A	Column B
$3 + 3 =$	$6 + (-2) =$
$4 + 5 =$	$-6 + 2 =$
$6 + 2 =$	$4 + (-6) =$
$(-3) + (-4) =$	$5 + (-2) =$
$(-5) + (-7) =$	$-8 + 5 =$
$(-2) + (-6) =$	$-4 + 9 =$

On the basis of your answers to column A, answer the following:

1. When adding two numbers with the same sign, do you add or subtract the absolute values of the numbers being added?

2. If the signs of the numbers being added are the same, how does the sign of the answer compare with the signs of the numbers being added?

3. On the basis of your answers to 1 and 2, write a rule for adding two numbers with the same sign.

On the basis of your answers to column B, answer the following:

4. When adding two numbers with the opposite signs, do you add or subtract the absolute values of the numbers being added?

5. If the signs of the numbers being added are opposites, how does the sign of the answer compare with the signs of the numbers being added?

6. On the basis of your answers to 4 and 5, write a rule for adding two numbers with opposite signs.

Following are some justifications for the conclusions you should have reached in questions 1–6:

Answers: Getting Ready: 1. 21 **2.** 3 **3.** 4.5 **4.**

$$\xleftarrow{\quad} \overset{\bullet}{-6} \; \overset{}{-5} \; \overset{}{-4} \; \overset{\bullet}{-3} \; \overset{}{-2} \; \overset{\bullet}{-1} \; \overset{}{0} \; \overset{}{1} \; \overset{\bullet}{2} \; \overset{}{3} \; \overset{\bullet}{4} \; \overset{}{5} \; \overset{}{6} \xrightarrow{\quad}$$

Calculator Exploration Activity: 1. add **2.** same **3.** Add the absolute values of the numbers being added and give the answer the same sign. **4.** subtract **5.** Same as the sign of the number with the larger absolute value. **6.** Subtract the absolute values of the numbers being added and give the answer the sign of the number with the larger absolute value.

OBJECTIVE **Adding integers with the same and with opposite signs**

When adding integers, it is convenient to think of them as directed numbers. In the examples that follow, the number line will be used to show the addition of integers with direction to the right as positive and direction to the left as negative. We will always begin at 0. The operation sign + is like a verb, since it is telling you what action to perform. Think of the action as "and then go." Do not confuse the addition sign with the + sign indicating a positive number. If a signed number follows an operation sign $(+, -, \times, \div)$, the signed number is placed inside parentheses.

Example 1	**Find the following sums by using the number line:**

Note

$(+3) + (+2) = 5$ can be thought of as a gain of 3 followed by a gain of 2 for a net gain of 5.

a. $(+3) + (+2)$ (This is usually written as $3 + 2$, since a number written without a sign is assumed to be positive.)

This means begin at 0 and go 3 units to the right (because 3 is positive). Then go 2 more to the right (because 2 is also positive). You should now be at $+5$, showing that $(+3) + (+2) = +5$ or $3 + 2 = 5$. Using the number line, it appears as follows:

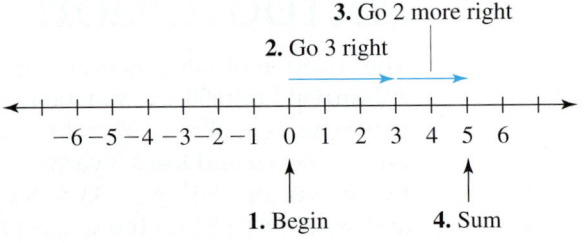

Note

$6 + (-2) = 4$ can be thought of as a gain of 6 followed by a loss of 2 for a net gain of 4.

b. $(+6) + (-2)$ (Usually written as $6 + (-2)$ or $6 - 2$.)

This means begin at 0 and go 6 units to the right. Then go 2 units to the left. You should now be at 4. Therefore, $6 + (-2) = 4$, as the following illustrates:

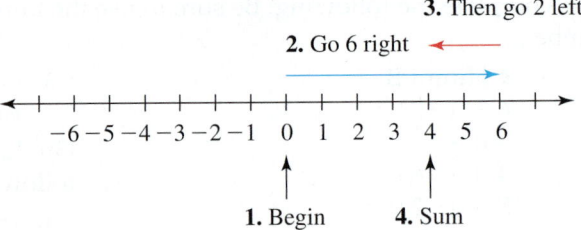

Note

$-6 + 4 = -2$ can be thought of as a loss of 6 followed by a gain of 4 for a net loss of 2.

c. $-6 + (+4)$ (Usually written as $-6 + 4$.)

This means begin at 0 and go 6 units to the left. Then go 4 units to the right. You should now be at -2. Therefore, $-6 + 4 = -2$, as the following illustrates:

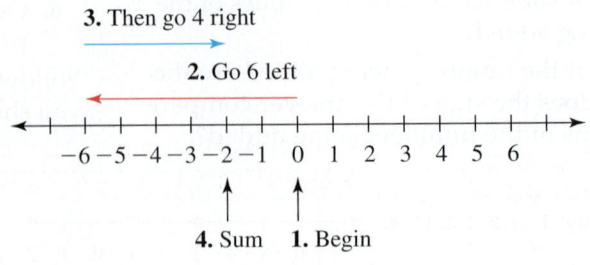

Note

$(-4) + (-2) = -6$ can be thought of as a loss of 4 followed by a loss of 2 for a net loss of 6.

d. $(-4) + (-2)$

This means begin at 0 and go 4 units left. Then go 2 more units left. You should now be at -6. Therefore, $(-4) + (-2) = -6$, as the following illustrates:

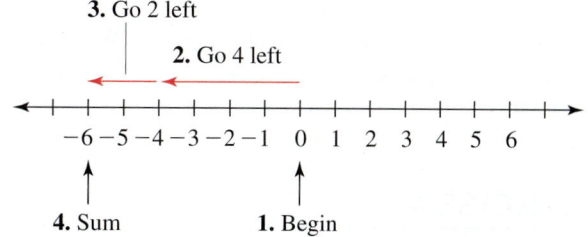

PRACTICE EXERCISE 1

Use the number line to find the following sums:

a. $2 + 4$

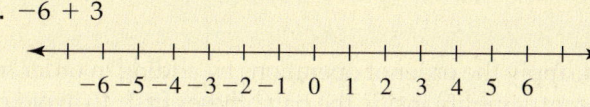

b. $5 + (-4)$

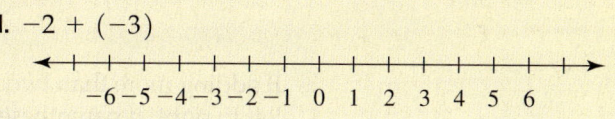

c. $-6 + 3$

d. $-2 + (-3)$

Following is a table summarizing the results of Example 1:

Problem	Signs of Addends	Add or Subtract the Absolute Values of the Addends?	Sign of the Answer
$(+3) + (+2) = 5$	same (both +)	add	positive—the same as sign of the addends
$(+6) + (-2) = 4$	opposite	subtract	positive—the same as addend with the larger absolute value
$-6 + (+4) = -2$	opposite	subtract	negative—the same as addend with the larger absolute value
$(-4) + (-2) = -6$	same (both −)	add	negative—the same as sign of the addends

Study Tip

One way to remember whether to add or subtract the absolute values is to remember what is happening on the number line. If you move in the same direction, add the absolute values. If you move in the opposite direction, subtract the absolute values.

Look back at the situations in the introduction and see that the results are consistent. This brings us to a general rule for adding signed numbers.

Rule: Adding Signed Numbers

a. To add numbers with the same sign, add the absolute values of the numbers and give the answer the same sign as the numbers being added.

b. To add numbers with opposite signs, subtract the absolute values of the numbers and give the answer the sign of the number with the larger absolute value.

Answers: Practice Exercise 1: a. 6 **b.** 1 **c.** −3 **d.** −5

Example 2   Find the following sums:

a. $7 + 5 =$ *Signs are the same, so add the absolute values.*

$\quad$ 12   *Both numbers are positive, so sum is positive.*

b. $6 + (-13) =$ *Signs are opposite, so subtract the absolute values.*

$\quad$ -7 *-13 has the larger absolute value, so the sum is negative.*

c. $-6 + (-7) =$ *Signs are the same, so add the absolute values.*

$\quad$ -13 *Both numbers are negative, so the sum is negative.*

d. $-14 + 5 =$ *Signs are opposite, so subtract the absolute values.*

$\quad$ -9 *-14 has the larger absolute value, so the sum is negative.*

PRACTICE EXERCISE 2

Find the following sums:

a. $9 + 7$ $\qquad$ **b.** $-12 + 7$ $\qquad$ **c.** $-8 + (-5)$ $\qquad$ **d.** $9 + (-16)$

If you need more practice, do the following Additional Practice Exercises.

Additional Practice Exercise 2   Find the following sums:

a. $7 + 12$ $\qquad$ **b.** $-15 + 9$ $\qquad$ **c.** $-12 + (-15)$ $\qquad$ **d.** $13 + (-13)$

If adding more than two numbers, apply the order of operations by adding in order from left to right. If parentheses are present, simplify inside the parentheses first. To avoid confusion, brackets are often used to show grouping when parentheses are already present.

Example 3   Find the following sums:

a. $4 + (-8) + 7 =$ *First add 4 and -8.*

$\quad$ $-4 + 7 =$ *Now add -4 and 7.*

$\quad$ 3   *Sum.*

b. $-24 + 16 + (-18) =$ *First add -24 and 16.*

$\quad$ $-8 + (-18) =$ *Now add -8 and -18.*

$\quad$ -26 *Sum.*

c. $26 + [17 + (-13)] =$ *Perform operations inside brackets first, so add 17 and -13.*

$\quad$ $26 + 4 =$ *Add 26 and 4.*

$\quad$ 30   *Sum.*

d. $(-52 + 34) + [-21 + (-8)] =$ *Perform operations inside parentheses first.*

$\quad$ $-18 + (-29) =$ *Add -18 and -29.*

$\quad$ -47 *Sum.*

PRACTICE EXERCISE 3

Find the following sums:

a. $-10 + 8 + (-15)$ $\qquad\qquad\qquad\qquad$ **b.** $32 + [-18 + (-21)]$

c. $-9 + [15 + (-6)]$ $\qquad\qquad\qquad\qquad$ **d.** $(-45 + 34) + [36 + (-22)]$

If you need more practice, do the following Additional Practice Exercises.

Additional Practice Exercise 3   Find the following sums:

a. $-14 + 9 + (-16)$ $\qquad\qquad\qquad\qquad$ **b.** $24 + [(-9) + (-22)]$

c. $18 + [27 + (-17)]$ $\qquad\qquad\qquad\qquad$ **d.** $(-23 + 52) + (-45 + 33)$

Answers: Practice Exercise 2: a. 16 **b.** -5 **c.** -13 **d.** -7 **Additional Practice 2: a.** 19 **b.** -6 **c.** -27 **d.** 0   **Practice Exercise 3: a.** -17 **b.** -7 **c.** 0 **d.** 3   **Additional Practice 3: a.** -21 **b.** -7 **c.** 28 **d.** 17

OBJECTIVE (B) Translating verbal expressions into mathematical expressions and simplifying

Often, it is necessary to translate expressions written in English into mathematical expressions. There are many expressions in English that indicate addition. For example, $2 + 3$ can be written as 2 plus 3, the sum of 2 and 3, 3 more than 2, 2 increased by 3, and 3 added to 2.

Example 4   **Write a numerical expression for each of the following and evaluate:**

a. The sum of -7 and -2

Solution

Sum means add, so this means add -7 and -2.

$-7 + (-2) =$ Signs are the same, so add the absolute values.
-9 Both signs are negative, so the sum is negative.

b. -10 increased by 4

Solution

This implies that you begin with -10 and increase it by 4, which means add 4 to -10.

$-10 + 4 =$ Signs are opposite, so subtract the absolute values.
-6 -10 has the larger absolute value, so the sum is negative.

c. 6 more than -2

Solution

This implies that you begin with -2 and you need 6 more, which means add -2 and 6.

$-2 + 6 =$ Signs are opposite, so subtract the absolute values.
4 6 has the larger absolute value, so the sum is positive.

d. -3 added to 5

Solution

This implies that you begin with 5 and -3 is added to 5.

$5 + (-3) =$ Signs are opposite, so subtract the absolute values.
2 5 has the larger absolute value, so the sum is positive.

PRACTICE EXERCISE 4

Write a numerical expression for each of the following and evaluate:

a. The sum of -12 and -5

b. -8 increased by 14

c. 5 more than -13

d. -6 added to 4

If you need more practice, do the following Additional Practice Exercises.

Additional Practice Exercise 4   Write a numerical expression for each of the following and evaluate:

a. The sum of 8 and -3

b. -6 increased by 4

c. 9 more than -23

d. -6 added to -5

Answers: Practice Exercise 4: a. $-12 + (-5) = -17$ **b.** $-8 + 14 = 6$ **c.** $-13 + 5 = -8$ **d.** $4 + (-6) = -2$
Additional Practice 4: a. $8 + (-3) = 5$ **b.** $-6 + 4 = -2$ **c.** $-23 + 9 = -14$ **d.** $-5 + (-6) = -11$

OBJECTIVE Identifying properties of addition

The operation of addition has several properties that are of such importance that they are given names for easy reference. You may have already noticed that the order in which the numbers are added makes no difference. For example, $2 + 4 = 4 + 2$, and $5 + 9 = 9 + 5$. The same is true for the addition of integers. This relationship is known as the **commutative property of addition**. It says that the order in which the numbers are added does not matter since the sum is the same. We use this property every day. Consider shopping. Without the commutative property of addition, the bill would be determined by the order in which the items were rung up.

Study Tip

One way to remember the commutative property is to think how each day you commute from home to school. You take the same route each way, but travel in the opposite direction. Similarly, the commutative property involves the same numbers (routes), but in the opposite order (direction).

Example 5 **Simplify on both sides of the equal sign:**

a. $-4 + 7 = 7 + (-4)$ Simplify each side.
   $3 = 3$ Therefore, both sides are equal.

b. $-5 + (-8) = -8 + (-5)$ Simplify each side.
   $-13 = -13$ Therefore, both sides are equal.

If three or more numbers are added, the manner in which they are grouped makes no difference. For example, $2 + (3 + 4) = (2 + 3) + 4$. So it does not matter which we add first, the "3" and the "4" or the "2" and the "3." This relationship is called the **associative property of addition**. It says the manner in which the numbers are grouped does not matter, since the sum is the same.

Example 6 **Simplify on both sides of the equal sign:**

a. $4 + (5 + 8) = (4 + 5) + 8$ Simplify each side by adding inside the parentheses.
   $4 + 13 = 9 + 8$ Add.
   $17 = 17$ Therefore, both sides are equal.

b. $-6 + [(-4) + 9] = [-6 + (-4)] + 9$ Simplify inside brackets first.
   $-6 + 5 = -10 + 9$ Add.
   $-1 = -1$ Therefore, both sides are equal.

Note

By combining both the commutative and associative properties, we can often make addition easier by mentally changing the order and grouping. For example,

$$-1 + [14 + (-9)] = [-1 + (-9)] + 14 = -10 + 14 = 4.$$

The number 0 is called the **additive identity**, or the identity element, for addition. Adding 0 to any number gives a result identical to the original number. If the sum of two numbers is 0, the two numbers are **additive inverses** of each other. In Section 1.5 these were called *negatives*, or *opposites*.

Example 7

a. The additive inverse of 5 is -5 because $5 + (-5) = 0$.

b. The additive inverse of -14 is 14 because $-14 + 14 = 0$.

c. 0 is its own additive inverse because $0 + 0 = 0$.

Following is a summary of the properties of addition:

> **Definition** Properties of Addition
>
> All real numbers a, b, and c satisfy the following properties:
>
> **Commutative Property of Addition**
> $a + b = b + a$ Addition may be performed in any order.
>
> **Associative Property of Addition**
> $a + (b + c) = (a + b) + c$ Addition may be grouped in any manner.
>
> **Additive Inverse**
> Every real number a has an additive inverse $-a$ so that $a + (-a) = (-a) + a = 0$.
>
> **Additive Identity**
> The real number 0 is called the additive identity because $a + 0 = 0 + a = a$. The sum of any number and 0 is identical to the number.

Example 8   **Give the name of the property illustrated by each of the following:**

a. $3 + (-5) = -5 + 3$ *Order is different. Commutative property of addition.*

b. $4 + (-6 + 2) = [4 + (-6)] + 2$ *Order is the same and grouping is different. Associative property of addition.*

c. $5 + (8 + 4) = 5 + (4 + 8)$ *Grouping is the same and order is different. Commutative property of addition.*

d. $5[6 + (-6)] = 5(0)$ $6 + (-6) = 0$. *Additive inverse property.*

e. $-6 + 0 = -6$ *Additive identity property.*

PRACTICE EXERCISE 8

Give the name of the property illustrated by each of the following:

a. $-8 + 2 = 2 + (-8)$

b. $7 + (-6 + 3) = [7 + (-6)] + 3$

c. $4(6 + 3) = 4(3 + 6)$

d. $-9(-3 + 3) = -9(0)$

e. $5 + 0 = 5$

Answers: Practice Exercise 8: a. commutative property of addition. **b.** associative property of addition **c.** commutative property of addition **d.** additive inverse **e.** additive identity

Study Tip 5 Memorizing and Understanding Mathematics

Some things in mathematics must be memorized. Symbols, definitions, rules, and algorithms have to be both memorized and understood. We all forget, so those things that must be memorized also need to be practiced frequently. We can do this by writing definitions, rules, and so on, on note cards and reading them while we wait for a bus or train or between classes. However, you should realize that you cannot memorize everything in a mathematics course. You may be able to memorize enough for one unit exam, but there is always a final exam.

Psychologists tell us that we learn best when what we are learning has meaning for us. That is, we can learn more and keep it longer when we understand the material. The more we review, think about, and see how things fit together, the more meaning these things have for us and the better we can understand and apply them.

In mathematics, there is a reason for everything that we do, and we must know and understand that reason. Mathematics is not learned just by doing problems. It is learned by doing the problems and understanding why we did them the way we did. The "why" of mathematics is as important, if not more important, than the "how" of mathematics. In general, mathematics is to be *understood*, not memorized.

Exercise Set 1.6

For Extra Help

MyMathLab®

Find the following sums. (See Examples 2 and 3.)

1. $2 + 16$

2. $9 + 5$

3. $-2 + 13$

4. $-5 + 16$

5. $7 + (-14)$

6. $8 + (-18)$

7. $17 + (-8)$

8. $14 + (-5)$

9. $-17 + 6$

10. $-21 + 3$

11. $-7 + (-4)$

12. $-6 + (-3)$

13. $13 + 35$

14. $32 + 43$

15. $-24 + 35$

16. $-54 + 67$

17. $-25 + 19$

18. $-42 + 37$

19. $63 + (-34)$

20. $72 + (-54)$

21. $4.3 + (-5.7)$

22. $-6.7 + 5.3$

23. $-9.2 + 4.5$

24. $7.6 + (-4.1)$

25. $8 + (-5) + 4$

26. $9 + (-6) + (-6)$

27. $-8 + 10 + (-7)$

28. $-6 + 12 + (-4)$

29. $13 + (-17) + (-21)$

30. $21 + (-16) + (-32)$

31. $-16 + (-23) + (-31)$

32. $-25 + (-32) + (-47)$

33. $[9 + (-3)] + (-8 + 5)$

34. $[7 + (-4)] + (-6 + 9)$

35. $(-13 + 9) + (-21 + 14)$

36. $(-17 + 8) + (-27 + 17)$

37. $[-4 + (-7)] + (-10 + 4)$

38. $[-8 + (-5)] + (-13 + 7)$

39. $[16 + (-32)] + [17 + (-6)]$

40. $[-25 + 43] + [37 + (-21)]$

Challenge Exercises (41–46)

Find the following sums of fractions:

41. $\dfrac{7}{11} + \dfrac{-3}{11}$

42. $\dfrac{-5}{29} + \dfrac{2}{29}$

43. $\dfrac{-7}{17} + \dfrac{3}{17}$

44. $\dfrac{-8}{19} + \dfrac{5}{19}$

45. $-\dfrac{2}{11} + \left(-\dfrac{5}{11}\right)$

46. $\dfrac{-4}{13} + \left(-\dfrac{2}{13}\right)$

▦ Calculator Exercises (47–52, Optional)

Find the following sums using a calculator:

47. $14.62 + (-33.21)$

48. $-67.43 + 49.62$

49. $-896.456 + (-45.7689)$

50. $-985.35 + (-83678.587)$

51. $(-765.45 + 65.78) + [473.21 + (-860.32)]$

52. $[-75849.359 + (-94327.432)] + [79235.937 + (-27807.857)]$

Give the name of the property illustrated by each of the following. (See Example 8.)

53. $6 + (-2) = -2 + 6$

54. $5 + (6 + 4) = (5 + 6) + 4$

55. $-8 + 8 = 0$

56. $0 + (-21) = -21$

57. $7(5 + (-3)) = 7(-3 + 5)$

58. $6[4 + (-2)] = 6[-2 + 4]$

59. $0 + (6 + 4) = 6 + 4$

60. $0 + 8 = 8 + 0$

61. $12 + (-4 + 9) = 12 + [9 + (-4)]$

62. $-5(-4 + 4) = -5(0)$

63. $(3 + 5) + (-7) = 3 + [5 + (-7)]$

64. $(-10 + 10) + 7 = 0 + 7$

Complete each of the following by using the given property. (See Examples 5–7.)

65. $-5 + 6 =$ _____   commutative for addition

66. $-4 + (9 + 12) =$ _____   associative for addition

67. $0 + 34 =$ _____   additive identity

68. $-4 + (9 + 12) =$ _____   commutative for addition

69. $-14 + 14 =$ _____   additive inverse

70. $45 + 0 =$ _____   commutative for addition

71. $(8 + 3) + 5 =$ _____   associative for addition

72. $7(5 + 8) =$ _____   commutative for addition

Write a numerical expression for each of the following and evaluate. (See Example 4.)

73. The sum of -9 and 5

74. The sum of -11 and 8

75. The sum of $-3, 6,$ and -7

76. The sum of $-6, -4,$ and 10

77. 6 increased by 15

78. 8 increased by 11

79. -18 increased by 14

80. -12 increased by 7

81. 23 more than 36

82. 35 more than 49

83. 42 more than -30

84. 36 more than -18

85. 6 added to -14

86. 5 added to -12

87. -6 added to -11

88. -12 added to -5

Problems 89–96 involve changes in temperature. Write each as an addition problem and solve.

	Beginning Temp.	Change	Addition	New Temperature
89.	78°	Drops 15°		
91.	18°	Drops 28°		
93.	−5°	Rises 15°		
95.	−14°	Drops 12°		

	Beginning Temp.	Change	Addition	New Temperature
90.	92°	Drops 25°		
92.	12°	Drops 19°		
94.	−12°	Rises 24°		
96.	−21°	Drops 7°		

Problems 97–100 involve a scuba diver. Express each as an addition problem and solve.

	Beginning Depth	Change in Depth	Addition	New Depth
97.	−43 feet	Ascends 27 feet		
99.	−53 feet	Descends 22 feet		

	Beginning Depth	Change in Depth	Addition	New Depth
98.	−68 feet	Ascends 39 feet		
100.	−42 feet	Descends 43 feet		

Solve each of the following word problems by using addition of integers:

Use the following information for Exercises 101–104. In Europe, the first floor of a building is called the ground floor, what we call the second floor is called the first floor, and so on. Consequently, we could think of the ground floor as being the "0" floor. Suppose a European hotel has a 30th floor and four levels of underground parking garages. Represent each of the following as an addition problem and solve. See the figure in the margin.

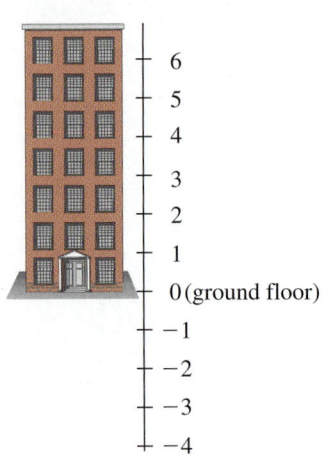

101. If an elevator in a European hotel starts on the 27th floor and goes down 29 levels, where does it stop?

102. Suppose the elevator in Exercise 101 begins on the 16th floor and goes down 19 levels. Where does it stop?

103. Suppose the elevator in Exercise 101 begins on the third parking level below ground and goes up 23 levels. Where does it stop?

104. Suppose the elevator in Exercise 101 begins on the second parking level below ground and goes up 14 levels. Where does it stop?

105. Hernando has a balance of $130 in his checking account. He makes a deposit of $230. What is his new balance?

106. Farshad has a balance of $289 in his checking account. He writes a check for $146. What is his new balance?

107. Sara has a balance of $319 in her checking account. She makes a deposit of $132 and then writes a check for $350. What is her new balance?

108. Mira has a balance of $293 in her checking account. She writes a check for $167 and later makes a deposit of $250. What is her new balance?

109. On 4/8/10 the Dow Jones Industrial Average closed at 10,967.8. It rose 29.55 points on 4/9/10. What was the closing average on 4/9/10?

110. On 4/8/10 the New York Stock Exchange Composite was 7609.87. It rose 19.12 points on 4/9/10. What was the closing composite on 4/9/10?

111. A hiker begins at the rim of the Grand Canyon and descends 1500 feet into the canyon. He then turns around and ascends 850 feet. How far is the hiker from the rim?

112. The balance in a credit card account is $1396 before a payment of $450 is made. Find the new balance

113. A group of explorers is located in Death Valley and is 32 feet below sea level. An airplane drops supplies from a height of 1800 feet above sea level. How far did the supplies fall?

114. On successive plays, a running back gained 5 yards, lost 2 yards, and gained 8 yards. If a net gain of 10 yards is needed to earn a first down, did the team earn a first down? Why?

Challenge Exercises (115–118)

Evaluate each of the following:

115. $4^2(-7 + 2 \cdot 5) \div 3 \cdot 4$

116. $(-3 + 6 \cdot 3^2) \div 17 \cdot 6$

117. $3^2 \cdot 4(-12 + 4^2) \div 2^2 \cdot 3^2$

118. $48 \div 2^2 \cdot 3(-18 + 2 \cdot 4^2)$

Writing Exercise

119. Why is 0 its own additive inverse (opposite)?

Critical-Thinking Exercises (120 and 121)

120. In your own words, explain why you add the absolute values of two numbers with like signs when finding their sum.

121. In your own words, explain why you subtract the absolute values of two numbers with unlike signs when finding their sum.

Group Projects (122 and 123)

122. In the introduction, several examples were given where the addition of integers occurred in everyday situations. Give three additional examples not found in the text.

123. Interview an engineer, a physics or chemistry instructor, or another professional and make a list of at least three instances in which they have used addition of signed numbers.

OBJECTIVES *When you complete this section, you will be able to:*

A Subtract integers.
B Combine like terms.

PREREQUISITE SKILLS *Before beginning this section you should be able to:*

a. Add and subtract whole numbers and decimals (Sections R.2 and R.5).
b. Add integers (Section 1.6).
c. Find the negative (opposite) of a negative number (Section 1.5).

GETTING READY FOR SECTION 1.7

Add or subtract the following whole numbers and decimals:

1. $4.7 + 7.8$ **2.** $21 - 14$

Add the following integers:

3. $-5 + 8$ **4.** $4 + (-12)$ **5.** $-6 + (-9)$

6. Find the negative (opposite) of -6.

Introduction

In this section, we will see that we already know how to subtract signed numbers.

Calculator Exploration Activity (Optional)

Complete the following by using a calculator:

Column A	Column B
a. $4 - (+3) =$	$4 + (-3) =$
b. $-6 - (+5) =$	$-6 + (-5) =$
c. $7 - (-3) =$	$7 + (+3) =$
d. $-8 - (-2) =$	$-8 + (+2) =$

By looking at the corresponding lines in columns A and B, answer the following:

Subtracting a number is the same as _____ its _____.

Following are some real-world situations that support the preceding conclusion:

OBJECTIVE Subtracting integers

As with the addition of integers, the subtraction of integers occurs in many everyday situations. Look for a pattern in the following examples using temperature:

Example 1

a. If the temperature changes from $50°$ to $70°$, what is the change in temperature? We know the temperature has risen $20°$, which can be represented as $+20°$, but how do we find it mathematically? The change is (the new temperature) − (the old temperature). Therefore, $70° - (+50°) = +20°$ represents the change in temperature. The positive indicates the temperature has risen by 20°. Note that $70° - (+50°) = 70° + (-50°) = 20°$.

b. If the temperature changes from 60° to 50°, what is the change in temperature? We know the temperature has dropped 10°, which can be represented by $-10°$. The change can be found mathematically by taking (the new temperature) $-$ (the old temperature). Therefore, the change would be $50° - (+60°) = -10°$, where the negative indicates that the temperature has dropped. Note that $50° - (+60°) = 50° + (-60)° = -10°$.

c. If the temperature changes from $-10°$ to 20°, what is the change in temperature? We know the temperature has risen 30°, which can be represented by $+30°$. Mathematically, the change would be $20° - (-10°) = +30°$, where the positive indicates that the temperature has risen. Note that $20° - (-10°) = 20° + (+10)° = +30°$.

d. Likewise, if the temperature changes from $-20°$ to $-10°$, how much has it changed? We know it has risen 10°. Mathematically, the change would be $(-10°) - (-20°) = +10°$. The positive indicates the temperature has risen by 10°. Notice $(-10°) - (-20°) = -10° + (+20°) = +10°$.

Following is a table summarizing the results of Example 1:

Temperature Change	Written as Subtraction	Written as Addition
From 50° to 70°	$70° - (+50°) = 20°$	$70° + (-50°) = 20°$
From 60° to 50°	$50° - (+60°) = -10°$	$50° + (-60°) = -10°$
From $-10°$ to 20°	$20° - (-10°) = 30°$	$20° + (+10°) = 30°$
From $-20°$ to $-10°$	$-10° - (-20°) = 10°$	$-10° + (+20°) = 10°$

From the preceding examples, it seems there is a very close relationship between subtraction and addition, since each subtraction problem can be rewritten as addition. You will notice that the procedure is to change the sign of the number being subtracted and change the operation from subtraction to addition. This leads to the following procedure for subtracting integers:

Procedure for Subtracting

$$a - b = a + (-b)$$

To subtract a number, add its negative (opposite). Remember, a and b can represent either positive or negative numbers.

Example 2   **Rewrite each of the following as addition and evaluate:**

a. $7 - 5 =$ $7 - 5$ means $7 - (+5)$, so change the subtraction to addition and change $+5$ to -5.

$7 + (-5) =$ Add.

2 Difference.

b. $-6 - 2 =$ $-6 - 2$ means $-6 - (+2)$, so change the subtraction to addition and change $+2$ to -2.

$-6 + (-2) =$ Add.

-8 Difference.

Note

We can think of this as subtracting a debt of $5 from a balance of $8. Subtracting a debt is the same as making a payment.

c. $8 - (-5) =$ Change the subtraction to addition and change -5 to $+5$.

$8 + 5 =$ It is not necessary to write $+(+5)$ since 5 means $+5$.

13 Difference.

d. $-12 - (-8) =$ Change the subtraction to addition and change -8 to $+8$.

$-12 + 8 =$ Add.

-4 Difference.

PRACTICE EXERCISE 2

Rewrite each of the following as addition and evaluate:

a. $12 - 8$

b. $-7 - 5$

c. $9 - (-4)$

d. $-14 - (-6)$

If you need more practice, do the following Additional Practice Exercises.

Additional Practice Exercise 2   Evaluate each of the following:

a. $8 - 13$

b. $-3 - 9$

c. $5 - (-12)$

d. $-16 - (-9)$

Note

Since $a - b = a + (-b)$, we will no longer write two signs when adding or subtracting, except in the case of subtracting a negative number. **We will assume the operation is addition and the sign is the sign of the number following it.** For example, $6 - 4 = 6 + (-4) = 2$ and $-8 - 5 = -8 + (-5) = -13$.

If there are more than two numbers, simplify inside any parentheses or brackets first from the innermost to the outermost, then add or subtract in order from left to right, as in the following examples:

Example 3   **Evaluate. Remember the order of operations.**

a. $8 - 10 + 5 =$ Add 8 and -10. $(8 - 10 = 8 + (-10))$

$-2 + 5 =$ Add -2 and 5. $(-2 + 5 = -2 + (+5))$

3 Sum.

b. $-14 - 7 + 6 - (-4) =$ Add -14 and -7. $(-14 - 7 = -14 + (-7))$

$-21 + 6 - (-4) =$ Add -21 and 6. $(-21 + 6 = -21 + (+6))$

$-15 - (-4) =$ Change $-(-4)$ to $+4$.

$-15 + 4 =$ Add -15 and 4. $(-15 + 4 = -15 + (+4))$

-11 Sum.

c. $14 - (9 - 21) =$ Simplify inside parentheses first, so add 9 and -21.

$14 - (-12) =$ Change $-(-12)$ to $+12$.

$14 + 12 =$ Add 14 and 12.

26 Sum.

d. $(-16 - 8) - (18 - 9) =$ Simplify inside parentheses first.

$-24 - 9 =$ Add -24 and -9.

-33 Sum.

Answers: Practice Exercise 2: a. $12 + (-8) = 4$ **b.** $-7 + (-5) = -12$ **c.** $9 + 4 = 13$ **d.** $-14 + 6 = -8$
Additional Practice 2: a. -5 **b.** -12 **c.** 17 **d.** -7

e. $5 - [7 - (-6 - 5)] =$ Simplify inside parentheses first.

$\quad 5 - [7 - (-11)] =$ Change $-(-11)$ to $+11$.

$\quad 5 - [7 + 11] =$ Simplify inside brackets next.

$\quad 5 - 18 =$ Add 5 and -18.

$\quad -13$ Sum.

Note

In Example 3e, we used brackets and parentheses to show grouping, since it is confusing to have parentheses inside parentheses.

PRACTICE EXERCISE 3

Evaluate the following:

a. $-7 + 10 - 8$

b. $10 - (-6) + 4 - 12$

c. $17 - (-5 - 8)$

d. $(8 - 16) - (12 - 21)$

e. $8 - [-9 - (-24 + 18)]$

If you need more practice, do the following Additional Practice Exercises.

Additional Practice Exercise 3   Evaluate the following:

a. $-9 + 7 - 3$

b. $14 - 6 - (-7) + 5$

c. $15 - (-11 + 7)$

d. $(12 - 19) - (-6 + 13)$

d. $5 - [6 - (-15 + 9)]$

OBJECTIVE B Combining like terms

Addition and subtraction of integers is used to combine like terms. A **term** is a number, variable, or product and/or quotient of numbers and variables raised to powers. Examples of terms are $5, x, 5x, -3x^2, 2x^3y^2, -7x^2y^3z^4, \frac{x}{y}$, and $4x^{-2}$.

Distributive property

Before we can discuss adding terms, we need another number property. Let us simplify $3(5 + 2)$ in two different ways. If we follow the order of operations, we get $3(5 + 2) = 3(7) = 21$. If we find $3 \cdot 5 + 3 \cdot 2$, we get $15 + 6 = 21$. Therefore, $3(5 + 2) = 3 \cdot 5 + 3 \cdot 2$. This example can be generalized into the **distributive property** stated as follows:

> **Definition** Distributive Property of Multiplication over Addition
>
> For all numbers a, b, and c, $a(b + c) = a \cdot b + a \cdot c$.

The distributive property states that it does not matter whether we add inside the parentheses and then multiply or multiply first and then add. We will not discuss multiplication of signed numbers until a later section, so Example 4 will be limited to natural numbers only. The distributive property holds, however, for all numbers.

Answers: Practice Exercise 3: a. -5 **b.** 8 **c.** 30 **d.** 1 **e.** 11 **Additional Practice 3: a.** -5 **b.** 20 **c.** 19 **d.** -14 **e.** -7

Example 4 **Verify the distributive property for each of the following by evaluating both sides and showing that they are equal:**

a. $2(3 + 5) = 2 \cdot 3 + 2 \cdot 5$ *$a(b + c) = ab + ac$. Simplify by using the order of operations.*
$$2(8) = 6 + 10$$
$$16 = 16$$

b. $5(2 + 8) = 5 \cdot 2 + 5 \cdot 8$ *$a(b + c) = ab + ac$. Simplify by using the order of operations.*
$$5(10) = 10 + 40$$
$$50 = 50$$

PRACTICE EXERCISE 4

Verify the distributive property for each of the following. Rewrite each expression as in Example 4, then evaluate each side and show they are equal.

a. $3(4 + 2)$

b. $5(6 + 3)$

We need to be able to recognize the distributive property when we see it in various forms. Instead of writing it as $a(b + c) = ab + ac$, another form of the distributive property is $(b + c)a = ba + ca$. This equation may be "turned around" and written as $ba + ca = (b + c)a$. For example, $3 \cdot 4 + 5 \cdot 4 = (3 + 5) \cdot 4$. It is this form that we will use in Example 5.

Terms that have the same variables with the same exponents on these variables are called **like terms**. Examples of like terms are $3x$ and $5x$, $6y$ and $9y$, $11z^2$ and $9z^2$, $3x^2y$ and $7x^2y$. The terms $4x$ and $3y$ are not like terms, because the variables are not the same. The terms $3x$ and $3x^2$ are not like terms, because the exponents on the variables are not the same. In a term, the constant preceding the variable(s) is called the **numerical coefficient**.

We can approach the addition of like terms from an intuitive point of view, as follows: If we have 2 apples and we buy 7 apples, then we have 2 apples + 7 apples = 9 apples. If we let a represent apple, this becomes $2a + 7a = 9a$. Similarly, if we have 3 pears and 4 bananas and we buy 6 pears and 8 bananas, we have 3 pears + 4 bananas + 6 pears + 8 bananas = 9 pears + 12 bananas. If we let p represent pear and b represent banana, this becomes $3p + 4b + 6p + 8b = 9p + 12b$. In other words, we can add the number of pears to the number of pears and the number of bananas to the number of bananas, but we cannot add the number of pears to the number of bananas.

The mathematical basis for adding like terms lies in the distributive property in the form $ba + ca = (b + c)a$, as in the following examples:

Example 5 **Find the following sums:**

a. $2x + 4x =$ *Apply $ba + ca = (b + c)a$.*
$(2 + 4)x =$ *Now add 2 and 4.*
$6x$ *Therefore, $2x + 4x = 6x$.*

b. $5xy - 3xy =$ *Apply $ba + ca = (b + c)a$.*
$(5 - 3)xy =$ *Now add 5 and -3.*
$2xy$ *Therefore, $5xy - 3xy = 2xy$.*

c. $3x^2 + 8x^2 =$ *Apply $ba + ca = (b + c)a$.*
$(3 + 8)x^2 =$ *Now add 3 and 8.*
$11x^2$ *Therefore, $3x^2 + 8x^2 = 11x^2$.*

d. $3x + 7y$ *The distributive property does not apply, so this expression cannot be simplified.*

Answers: Practice Exercise 4: a. $3(4 + 2) = 3 \cdot 4 + 3 \cdot 2$ $3(6) = 12 + 6$ $18 = 18$ **b.** $5(6 + 3) = 5 \cdot 6 + 5 \cdot 3$ $5(9) = 30 + 15$ $45 = 45$

In Examples 5a–c, we were adding like terms. In each case, the variable was placed outside the parentheses with the coefficients inside and the coefficients were then added. This leads to the following procedure for adding like terms:

Rule: Addition of Like Terms

To add like terms, add the numerical coefficients and leave the variable portion unchanged.

The preceding rule greatly simplifies the addition of like terms, since we no longer have to demonstrate that we have applied the distributive property, but the property is still being applied.

Example 6 | **Simplify the following, if possible, by adding like terms:**

a. $6x + 2x =$ These are like terms, so add the coefficients, 6 and 2.
  $8x$ Leave the variable portion unchanged.

b. $4x - 7x =$ These are like terms, so add the coefficients, 4 and -7.
  $-3x$ Leave the variable portion unchanged.

c. $5x^2 + x^2 =$ These are like terms, so add the coefficients, 5 and 1.
  $6x^2$ Leave the variable portion unchanged.

d. $6xy - 2xy + 8 =$ $6xy$ and $-2xy$ are like terms, so add them.
  $4xy + 8$

e. $5x + 4y - 2x + 6y =$ $5x$ and $-2x$ are like terms, and $4y$ and $6y$ are
  $3x + 10y$ also like terms, so add each pair.

Note

Some people prefer to use the commutative and associative properties to rearrange the terms so that the like terms are together before adding. Example 6e could be rewritten as follows: $5x + 4y - 2x + 6y = 5x - 2x + 4y + 6y = 3x + 10y$. Also, some prefer to add vertically by placing like terms underneath each other as follows:

$$\begin{array}{r} 5x + 4y \\ -2x + 6y \\ \hline 3x + 10y \end{array}$$

f. $3x^2 + 4x - 7x^2 + 5 + 3x =$ $3x^2$ and $-7x^2$ are like terms, and so are $4x$ and $3x$.
  $-4x^2 + 7x + 5$ Add each pair.

As in Example 6e, the terms could be rearranged before adding. So,
$3x^2 + 4x - 7x^2 + 5 + 3x = 3x^2 - 7x^2 + 4x + 3x + 5 = -4x^2 + 7x + 5$.

g. $2x - (-5x) =$ Change to addition.
  $2x + 5x =$ Add like terms.
  $7x$

h. $3x + 4y - 3x^2$ There are no like terms, so this expression cannot be simplified.

PRACTICE EXERCISE 6

Simplify the following, if possible:

a. $7x + 5x$

b. $4xy + 8xy$

c. $6y^2 + y^2$

d. $5x - 9x + 3x$

e. $4x^2y - 5 + 8x^2y$

f. $7z + 6x - 3z + 9x$

g. $4x - 5x^2 - 6x + 4 - 3x$

h. $4x - (-7x)$

i. $6a + 7b + 5a^2$

If you need more practice, do the following Additional Practice Exercises.

Additional Practice Exercise 6   Simplify the following, if possible:

a. $4y + 6y$

b. $5ab + 9ab$

c. $7x - 4x + x$

d. $5z^2 + 9z^2$

e. $9x^2y - 5x + 3x^2y$

f. $6y + 2z - 9y - 7z$

g. $7x - 6x^2 - 9x + 6 + 7x^2$

h. $7y - (-5y)$

i. $7r^2 + 3s - 5t$

Often, problems present themselves to us in the form of words rather than symbols. When this happens it is necessary to translate from English to mathematics. There are many ways of saying "subtract" in English. For example, $6 - 2$ can be expressed as the difference of 6 and 2, 2 less than 6, 2 subtracted from 6, 6 minus 2, 6 decreased by 2, and 6 less 2.

Example 7   Write each of the following as a subtraction problem and evaluate:

a. The difference of 7 and 12

Solution

The difference of 7 and 12 means subtract 12 from 7, which is written as $7 - 12$.

$7 - 12 =$ Add 7 and -12.

-5 Therefore, the difference of 7 and 12 is -5.

b. $8a$ less than $2a$

Solution

$8a$ less than $2a$ means subtract $8a$ from $2a$, which is written as $2a - 8a$.

$2a - 8a =$ Add the coefficients 2 and -8.

$-6a$ Therefore, $8a$ less than $2a$ is $-6a$.

c. The difference of 4 and -5 added to 10

Solution

The difference of 4 and -5 means subtract -5 from 4, which is $4 - (-5)$. This entire difference is added to 10, which is written as $10 + [4 - (-5)]$. The brackets around $4 - (-5)$ are necessary to show that we are adding the difference of 4 and -5 to 10. Without the brackets we would have $10 + 4 - (-5)$, which means we

---

Answers: **Practice Exercise 6: a.** $12x$ **b.** $12xy$ **c.** $7y^2$ **d.** $-x$ **e.** $12x^2y - 5$ **f.** $4z + 15x$ **g.** $-5x^2 - 5x + 4$ **h.** $11x$ **i.** cannot be simplified   **Additional Practice 6: a.** $10y$ **b.** $14ab$ **c.** $4x$ **d.** $14z^2$ **e.** $12x^2y - 5x$ **f.** $-3y - 5z$ **g.** $x^2 - 2x + 6$ **h.** $12y$ **i.** cannot be simplified

are adding 10 and 4 and then finding the difference between that sum and -5. So we have:

$10 + [4 - (-5)] =$ Rewrite $4 - (-5)$ as $4 + 5$.

$10 + [4 + 5] =$ Add 4 and 5. Drop the brackets.

$10 + 9 =$ Add 10 and 9.

19 Therefore, the difference of 4 and -5 added to 10 is 19.

d. The sum of $-8x$ and $6x$ subtracted from the difference of $-3x$ and $2x$.

Solution

The sum of $-8x$ and $6x$ means add $-8x$ and $6x$, which is written as $-8x + 6x$. The difference of $-3x$ and $2x$ means subtract $2x$ from $-3x$ and is written as $-3x - 2x$. We need to subtract $-8x + 6x$ from $-3x - 2x$. This is written as $(-3x - 2x) - (-8x + 6x)$. The parentheses are necessary to show we are subtracting the *sum* of $-8x$ and $6x$ (in parentheses) from the *difference* of $-3x$ and $2x$ (in parentheses). So we have

$(-3x - 2x) - (-8x + 6x) =$ Simplify inside the parentheses first.

$(-5x) - (-2x) =$ Rewrite $(-5x) - (-2x)$ as $-5x + 2x$.

$-5x + 2x =$ Add.

$-3x.$ Therefore, the sum of $-8x$ and $6x$ subtracted from the difference of $-3x$ and $2x$ is $-3x$.

PRACTICE EXERCISE 7

Write each of the following as a subtraction problem and evaluate:

a. The difference of -5 and -9

b. $10x$ less than $-3x$

c. The difference of 3 and -1 added to -3

d. The sum $5b$ and $-8b$ subtracted from the difference of $-4b$ and $6b$

If you need more practice, do the following Additional Practice Exercises.

Additional Practice Exercise 7

Write each of the following as a subtraction problem and evaluate:

a. The difference of 4 and -7

b. $6y$ less than $2y$

c. The difference of -4 and -8 added to 3

d. The sum of $-5m$ and $2m$ subtracted from the difference of $4m$ and $-8m$

For Extra Help

Exercise Set 1.7   MyMathLab®

Evaluate each of the following. (See Examples 2 and 3.)

1. $9 - 3$

2. $11 - 7$

3. $6 - 12$

4. $2 - 17$

5. $8 - (-4)$

6. $6 - (-7)$

7. $-5 - (-9)$

8. $-7 - (-12)$

9. $-14 - (-6)$

Answers: Practice Exercise 7: a. $-5 - (-9) = 4$ **b.** $-3x - 10x = -13x$ **c.** $-3 + [3 - (-1)] = 1$ **d.** $(-4b - 6b) - [5b + (-8b)] = -7b$
Additional Practice 7: a. $4 - (-7) = 11$ **b.** $2y - 6y = -4y$ **c.** $3 + [-4 - (-8)] = 7$ **d.** $[4m - (-8m)] - (-5m + 2m) = 15m$

10. $-9 - (-3)$

11. $7.5 - 9.2$

12. $4.6 - 10.3$

13. $-7 + 4 - 6$

14. $-5 + 9 - 8$

15. $-13 - 6 + 7$

16. $-17 - 7 + 4$

17. $21 - (-8) - 4$

18. $14 - (-6) - 9$

19. $10.6 - 5.1 + 3.3$

20. $11.4 - 2.1 + 6.3$

21. $8 - (6 - 2)$

22. $12 - (7 - 5)$

23. $9 - (5 - 8)$

24. $7 - (3 - 8)$

25. $(16 - 12) - (8 - 5)$

26. $(26 - 17) - (7 - 3)$

27. $(14 - 23) - (31 - 26)$

28. $(16 - 28) - (34 - 27)$

29. $(-12 + 24) - (-19 - 8)$

30. $(-14 + 29) - (-24 - 9)$

31. $5 - [3 - (2 - 8)]$

32. $6 - [4 - (3 - 9)]$

33. $-13 + [24 - (-9 + 3)]$

34. $-25 + [37 - (-8 + 7)]$

35. $[(33 - 24) - 12] - 23$

36. $[(46 - 39) - 32] - 44$

Challenge Exercises (37–46)

Evaluate each of the following:

37. $\dfrac{-11}{8} + \dfrac{5}{8}$

38. $-\dfrac{5}{6} + \dfrac{7}{6}$

39. $-\dfrac{15}{20} - \dfrac{8}{20}$

40. $-\dfrac{14}{21} - \dfrac{12}{21}$

41. $\dfrac{11}{17} - \dfrac{18}{17}$

42. $\dfrac{6}{13} - \dfrac{10}{13}$

43. $\dfrac{20}{24} - \left(-\dfrac{9}{24}\right)$

44. $\dfrac{11}{19} - \left(-\dfrac{3}{19}\right)$

45. $\left(\dfrac{7}{13} - \dfrac{5}{13}\right) - \left(\dfrac{10}{13} - \dfrac{2}{13}\right)$

46. $\left(\dfrac{10}{17} - \dfrac{14}{17}\right) - \left(\dfrac{21}{17} - \dfrac{12}{17}\right)$

▦ Calculator Exercises (47–50, Optional)

Find the following differences, using a calculator when necessary:

47. $7564.968 - (-8573.3547)$

48. $6503.954 - (-2759.8735)$

49. $56.45 - [45.93 - (45.98 - 65.43)]$

50. $87.4 - [8.94 - (5.7 - 87.64)]$

Simplify the following by adding like terms. (See Examples 5 and 6.)

51. $4x + 7x$

52. $6x + 9x$

53. $-3y + 9y$

54. $-7y + 4y$

55. $9ab - 2ab$

56. $9xy - 7xy$

57. $-4x^2 - 3x^2$

58. $-5y^2 - 4y^2$

59. $12z - 8z + 3$

60. $15z - 9z + 7$

61. $-8y - 2y + 3$

62. $-10a - 3a + 6$

63. $3x - 8x - 6x$

64. $4z - 5z - 2z$

65. $-4r + 7r - 8r$

66. $-7b + 2b - 3b$

67. $8xy + 6x - 2xy + x$

68. $9xy + 5x - 3xy + x$

69. $3x^2 - 2y^2 - 6x^2 - 5x^2$

70. $4x^2 + 5y^2 - 6x^2 + 3y^2$

71. $9z - 5 + 3x + 3z - 2x - 7$

72. $8y - 4k + 2 - 4y + 3 - 9k$

73. $9b^2 - 8b + 2 - 2b^2 - 2b - 2$

74. $4d^2 - 4 + 4d + 4d^2 + 4 - 3d$

75. $m^2r + 3r - 5mr^2 + 3r + 5mr^2$

76. $n^2t - 8t + 6tn^2 - 2t - n^2t$

Write each of the following as a subtraction problem and evaluate. (See Example 7.)

77. The difference of 7 and −5

78. The difference of 9 and −5

79. The difference of −10*a* and 6*a*

80. The difference of −14*a* and 9*a*

81. 5 less than 8

82. 7 less than 12

83. 8*m* less than −5*m*

84. 13*n* less than −8*n*

85. 6 less 9

86. 7 less 12

87. −13 subtracted from 9

88. −21 subtracted from 11

89. 32*r* subtracted from 26*r*

90. 29*r* subtracted from 21*r*

91. −3*x* decreased by 5*x*

92. −5*y* decreased by 3*y*

93. 12*a* decreased by 19*a*

94. 7*r* decreased by 13*r*

95. The difference of 6 and −5 added to 14

96. The difference of 8 and −3 added to 23

97. 15*ab* added to the difference of 4*ab* and −6*ab*

98. 7*mn* added to the difference of 9*mn* and −5*mn*

99. The sum of −5 and 6 subtracted from the difference of 9 and −4

100. The sum of 7 and −3 subtracted from the difference of −7 and 5

101. The difference of −5*y* and −2*y* added to the difference of 7*y* and −3*y*

102. The difference of 4*y* and 9*y* added to the difference of −5*y* and 8*y*

Exercises 103–106 involve a scuba diver. Write as subtraction and complete the table.

	Previous Depth	Present Depth	Subtraction	Change in Depth
103.	−18 ft	−62 ft		
104.	−22 ft	−37 ft		
105.	−34 ft	−15 ft		
106.	−84 ft	−49 ft		

Write each as subtraction and answer the following:

107. A submarine is submerged at a depth of 145 feet. An airplane searching for the sub flies over at a height of 350 feet. What is the distance between them?

108. If the submarine in Exercise 107 is submerged at a depth of 245 feet and the airplane is at an altitude of 435 feet, what is the distance between them?

109. If a company loses $28,000 in its first year of operation and has a profit of $86,000 the following year, how much greater was the company's earnings the second year?

110. If a company has a profit of $77,500 dollars one year and a loss of $7300 the next, how much less did the company earn the second year than it did the first?

111. A mine shaft began at 500 feet above sea level and ended at 300 feet below sea level. Suppose a miner is 250 feet above sea level and descends to 125 feet below sea level. What is his change in altitude? Remember, change = new altitude − old altitude.

112. Suppose the miner in Exercise 111 is 210 feet below sea level and ascends to 109 feet below sea level. What is his change in altitude?

113. An airplane is flying over Death Valley at a height of 2500 feet when it flies over a hiker whose elevation is 187 feet below sea level. What is the distance between the airplane and the hiker?

114. The highest point in Asia is the peak of Mount Everest at an elevation of 8848 meters and the lowest point is the Dead Sea at 418 meters below sea level. What is the difference between the elevations of Mount Everest and the Dead Sea?

115. The daytime temperature on the moon can get as high as 100°C, and the nighttime temperature as low as −173°C. What is the difference between the daytime high and the evening low?

116. On the same day in February, the low temperature in East Yellowstone, Montana, was −36°F and the low temperature in Orlando, Florida, was 52°F. How much higher was the low temperature in Orlando than in East Yellowstone?

John operates a small part-time business from his home. The accompanying graph shows the daily profit or loss for his business. Use the graph to answer Exercises 117 and 118.

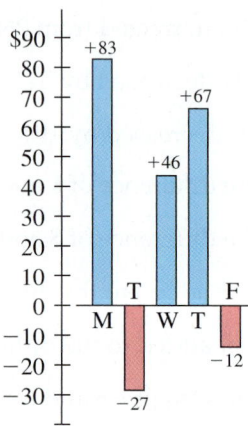

117. What was the total profit/loss for the week?

118. How much more did the company earn on Monday than on Tuesday?

Challenge Exercises (119–128)

Evaluate the following:

119. $5^2[6^2 \cdot 2 - (-4)] \div 19$

120. $2^4[48 - (-12) - 32] \div 7 \cdot 4$

121. $5^2\{6 - 2[10 - 3(-9 + 12)]\}$

122. $2^3\{24 - 3[12 - 2(-7 + 11)]\}$

Simplify the following:

123. $\dfrac{4}{5}x^2y + \dfrac{3}{7}xy^2 - \dfrac{2}{5}x^2y + \dfrac{2}{7}xy^2$

124. $\dfrac{7}{9}ab^2 + \dfrac{3}{11}a^2b^2 - \dfrac{5}{9}ab^2 - \dfrac{5}{11}a^2b^2$

125. $\dfrac{3}{13}m^3n^2 - \dfrac{5}{9}mn^2 - \dfrac{7}{13}m^3n^2 - \dfrac{2}{9}mn^2$

126. $\dfrac{4}{17}a^4b^3 - \dfrac{5}{19}a^2b - \dfrac{8}{17}a^4b^3 - \dfrac{3}{19}a^2b$

127. $\dfrac{6}{23}x^3y - \dfrac{16}{29}xy^2 - \dfrac{17}{23}x^3y - \dfrac{7}{29}xy^2$

128. $\dfrac{16}{17}ab^2 + \dfrac{5}{11}a^2b - \dfrac{9}{17}ab^2 - \dfrac{7}{11}a^2b$

Writing Exercises (129 and 130)

129. Is subtraction commutative? Why or why not? (*Hint:* Try some examples.)

130. Is subtraction associative? Why or why not? (*Hint:* Try some examples.)

Writing Exercise or Group Project

If this exercise is done as a group project, each group should write at least two exercises. Then exchange with another group and solve. If the exercise is done as a writing exercise, each student should write and solve one exercise.

131. Write an application problem involving the subtraction of integers, at least one of which is negative.

OBJECTIVES *When you complete this section, you will be able to:*

Ⓐ Identify types of polynomials.
Ⓑ Find the degree of a term.
Ⓒ Write polynomials in descending order and find their degrees.
Ⓓ Evaluate polynomials for a specified value of the variable.

PREREQUISITE SKILLS *Before beginning this section, you should be able to:*

a. Substitute for variables and evaluate by using the order of operations (Section 1.1).
b. Add like terms (Section 1.7).
c. Translate expressions in English into mathematics and simplify (Sections 1.6 and 1.7).

GETTING READY FOR SECTION 1.8

Substitute the given value for the variable and evaluate by using the order of operations.
1. $3x - 8: x = 2$ **2.** $x^2 + 6x - 4: x = 3$

Add the following like terms:
3. $4y + 5 - 2y - 6$ **4.** $2x^2 - 4x - 6 + 4x^2 + 7x + 3$

Translate the following expressions in English into mathematics and simplify:
5. $6x$ more than $-3x$ **6.** $8a$ less than $-2a$

Introduction

In arithmetic, operations are performed on numbers. In algebra, operations are performed on algebraic expressions. An **algebraic expression** is any constant, variable, or combination of constants and variables using the operations of addition, subtraction, multiplication, division, raising to powers, or taking of roots (which will be discussed later). Examples of algebraic expressions are

$$5, \ -3, \ x, \ 3x, \ x^3, \ -4x^4y, \ 4x^2 - 5x + 2, \ \frac{2x}{y}, \ \frac{2x - 5}{3x + 5}, \ \frac{2x^3 - 4y^2}{4y^4 + 3x^2}$$

Definition of a term

We defined *term* in Section 1.7, but will repeat the definition here. A **term** is a number, variable, or product and/or quotient of numbers and variables raised to powers. Examples of terms are $5, x, 5x, -3x^2, 2x^3y^2, \frac{x}{y}$, and $4x^{-2}$. An expression like $3x + 2$ is not a term, since addition or subtraction is not allowed in a term. This is actually the sum of two terms, $3x$ and 2. We have not used 0 or negative integers as powers, but we do so in Chapter 2, where we show that $x^0 = 1$ if $x \neq 0$. We mention this now because we will need it when we discuss the degree of a polynomial later in this section.

Definition of a polynomial

A **polynomial** is the sum or difference of a **finite** number of terms in which there is no variable as a divisor and the exponents on the variables are whole numbers. A set is finite if it is possible to represent the number of elements in the set with a whole number. Examples of polynomials are $x, 2x + 3, 5x^2 + 2x - 4, 2x^3 + 2x^2 - 4x + 6,$

Answers: Getting Ready: 1. -2 **2.** 23 **3.** $2y - 1$ **4.** $6x^2 + 3x - 3$ **5.** $3x$ **6.** $-10a$

$\frac{2}{3}x^2 + \frac{4}{5}x - 4$, and 3. Expressions like $3x^{-1} + 2$, $2x^{1/2} + 5x$, and $\frac{3}{x} + 6$ are not polynomials, because exponents on variables must be whole numbers and division by variables is not allowed in a polynomial.

OBJECTIVE **Identifying types of polynomials**

Polynomials can also be classified by the number of terms they contain, as illustrated in the following table:

Name	Number of Terms	Examples	Terms
Monomial	One term	4 x $-5x^2$	4 x $-5x^2$
Binomial	Two terms	$3x - 2$ $4x^3 - 3x$	$3x, -2$ $4x^3, -3x$
Trinomial	Three terms	$y^2 - 3y + 5$ $6x^3 + 2x^2 - 4x$	$y^2, -3y, 5$ $6x^3, 2x^2, -4x$

Note

Polynomials with more than three terms are not given special names. You will be asked to suggest names for some of these in the writing exercises.

Example 1 **Determine whether each expression is a polynomial. If it is a polynomial, determine the number of terms. Identify any monomials, binomials, or trinomials.**

a. $2x + 4$ *This is a polynomial with two terms, $2x$ and 4. Therefore, this is a binomial.*

b. $3x^3$ *This is a polynomial with only one term. Therefore, this is a monomial.*

c. $2y^3 + 4y - 2$ *This is a polynomial with three terms, $2y^3$, $4y$, and -2. Therefore, this is a trinomial.*

d. $6x^2 - 4x^4 + 5x - 4$ *This is a polynomial with four terms. Therefore, this is none of these.*

e. -4 *This is a polynomial with one term. Therefore, this is a monomial.*

f. $\frac{5}{y} - 10$ *We have division by a variable. Therefore, this is not a polynomial.*

PRACTICE EXERCISE 1

Determine whether each expression is a polynomial. If it is a polynomial, determine the number of terms. Identify any monomials, binomials, or trinomials.

a. $2y^2 + 4$ **b.** $8x^4$ **c.** $-3x^2 - 6x^5 - 7x$

d. $2x^4 - 2x - x^2 + 9$ **e.** 7   **f.** $\frac{3}{x} + 5$

Answers: Practice Exercise 1: a. 2, binomial **b.** 1, monomial **c.** 3, trinomial **d.** 4 **e.** 1, monomial **f.** not a polynomial

If you need more practice, do the following Additional Practice Exercises.

Additional Practice Exercise 1 Determine whether each expression is a polynomial. If it is a polynomial, determine the number of terms. Identify any monomials, binomials, or trinomials.

a. $-3z^4 - 6z^2$ **b.** $12m^6$ **c.** $4 + 6x - 9x^2$

d. $-9n^4 - 2n^3 + 9n - 5$ **e.** -15 **f.** $4y^{-2} + 5$

OBJECTIVE Finding the degree of a term

Recall from Section 1.7 that the constant written before the variable(s) is called the numerical coefficient or simply the *coefficient*. For example, in the term $3x$, the numerical coefficient is 3. The coefficient includes the sign of the number. If a monomial has only one variable, the exponent on the variable is called the **degree** of the monomial. If there is more than one variable, the degree of the monomial is the sum of all the exponents on all the variables.

Example 2 | **Give the coefficient, variable, and degree of each of the following monomials:**

Be Careful

In Example 2e, the degree of -3 is 0, since $-3 = -3x^0$. It is often mistakenly given as degree of 1. Remember, the degree of a term is the exponent of the variable, so $-3x$ has a degree of 1. Thus, -3 and $-3x$ could not have the same degree.

	Monomial	Coefficient	Variable(s)	Degree
a.	$4x^3$	4	x	3
b.	$-8z^6$	-8	z	6
c.	y	$1 (y = 1y^1)$	y	1
d.	$2m^3n^4$	2	m, n	$3 + 4 = 7$
e.	-3	-3	none	0, since $-3 = -3x^0$

PRACTICE EXERCISE 2

Give the coefficient, variable(s), and degree of each of the following:

a. $-8x^3$ coefficient = _____ variable(s) = _____ degree = _____

b. $-x$ coefficient = _____ variable(s) = _____ degree = _____

c. $2x^2y^4$ coefficient = _____ variable(s) = _____ degree = _____

d. -5 coefficient = _____ variable(s) = _____ degree = _____

OBJECTIVE Writing polynomials in descending order and finding their degrees

A polynomial of one variable is written in **descending order** when the term of highest degree is written first, the term of next highest degree second, and so on. For example,

$3x^2 - 2x + 4$ is in descending order, but $2x - 3x^3 - 5$ is not, because the term of highest degree is not written first. The **degree of a polynomial** is the same as the degree of the term of the polynomial with the highest degree. See the chart in Example 3.

Example 3 Write each polynomial in descending order and find the degree of the polynomial:

Polynomial	Descending Order	Terms	Degree of Terms	Degree of Polynomial
$3 + 4x$	$4x + 3$	$4x, 3$	$1, 0$	1
$2x - 4 + x^2$	$x^2 + 2x - 4$	$x^2, 2x, -4$	$2, 1, 0$	2
$4x - 5x^3 - 7x^4 + 2$	$-7x^4 - 5x^3 + 4x + 2$	$-7x^4, -5x^3, 4x, 2$	$4, 3, 1, 0$	4

PRACTICE EXERCISE 3

Write each of the following polynomials in descending order and give its degree:

Polynomial	**Descending Order**	**Degree**
a. $-5 + 3x$	_____	_____
b. $-5 + 2x^2 + 8x$	_____	_____
c. $6x - 7x^4 - 8x^2 + x^3$	_____	_____

If you need more practice, do the following Additional Practice Exercises.

Additional Practice Exercise 3   Write each of the following in descending order and give its degree:

a. $3 - 7x$ **b.** $7 - 3x + 5x^2$ **c.** $5y - 4y^2 + 6y^4 - y^3$

OBJECTIVE **Evaluating polynomials for a specified value of the variable**

Since polynomials contain variables and variables represent numbers, it is possible to evaluate a polynomial when given value(s) for its variable(s). Remember to keep the order of operations in mind when evaluating the polynomial. For convenience in evaluating, polynomials are often written with a special notation. The polynomial is given a name, usually a letter, and then its variable is specified in parentheses. For example, we might write $P(x) = x^2 + 2x - 3$, where P is the name we have given the polynomial and x is the variable. When we wish to assign the variable a value, the value replaces the variable inside the parentheses and is substituted for the variable throughout the expression. If we wish to evaluate $P(x) = x^2 + 2x - 3$ for $x = 3$, we write $P(3) = 3^2 + 2(3) - 3$ and then simplify. So, $P(3) = 9 + 6 - 3 = 12$. $P(x)$ is read "P of x" and means the value of P for a specific value of x. Consequently, $P(3) = 12$ means that the polynomial named P has a value of 12 when x has a value of 3.

Be Careful

Parentheses often mean multiply, but $P(x)$ does *not* mean multiply P and x.

Example 4 **Evaluate the following polynomials for the given value of the variable:**

a. $P(x) = 2x + 4; x = 5$ Substitute 5 for x.

$P(5) = 2(5) + 4$ Multiply 2 and 5.

$P(5) = 10 + 4$ Add 10 and 4.

$P(5) = 14$ Therefore, the value of P when x is 5 is 14.

b. $P(t) = t^2 + 3t - 4; t = 2$ Substitute 2 for t.

$P(2) = (2)^2 + 3(2) - 4$ Raise to powers first, so square 2.

$P(2) = 4 + 3(2) - 4$ Multiply 3 and 2.

$P(2) = 4 + 6 - 4$ Add in order left to right.

$P(2) = 10 - 4$ Add.

$P(2) = 6$ Therefore, the value of P when t is 2 is 6.

PRACTICE EXERCISE 4

Evaluate each of the following polynomials for the given value of the variable:

a. $P(x) = 2x + 5; x = 2$

b. $P(y) = y^2 + 3y - 6; y = 4$

If you need more practice, do the following Additional Practice Exercises.

Additional Practice Exercise 4 Evaluate each of the following for the given value of the variable:

a. $P(x) = 3x - 7; x = 2$

b. $P(t) = t^2 + 4t - 4; t = 6$

Evaluations of polynomials occur in many situations in the sciences and in the real world.

Example 5 **Solve the following:**

A particle moves along a straight line such that the distance from the starting point is given by $s(t) = t^2 + 3t + 2$, where $s(t)$ is the distance after t seconds. Find the location of the particle after 3 seconds by finding $s(3)$.

$s(t) = t^2 + 3t + 2$ To find $s(3)$, replace t with 3.

$s(3) = 3^2 + 3(3) + 2$ Square 3.

$s(3) = 9 + 3(3) + 2$ Multiply $3(3)$.

$s(3) = 9 + 9 + 2$ Add in order left to right.

$s(3) = 20$ Therefore, the particle is 20 units from the starting point after 3 seconds.

PRACTICE EXERCISE 5

a. If x units of a product are sold, the revenue in dollars is given by $R(x) = 1000x + .2x^2$. Find the revenue when 25 units are sold by finding $R(25)$.

If you need more practice, do the following Additional Practice Exercise.

Additional Practice Exercise 5

a. If n is the number of sides of a polygon, the number of diagonals is given by $N(n) = \dfrac{n(n-3)}{2}$. Find the number of diagonals for a polygon with 6 sides by finding $N(6)$.

The polynomial plays a central role in all of algebra. A great deal of the remainder of this book will be spent performing operations on polynomials.

Answers: Practice Exercise 4: a. 9 **b.** 22 **Additional Practice 4: a.** -1 **b.** 56 **Practice Exercise 5: a.** \$25,125 **Additional Practice 5: a.** 9

Since polynomials are made of terms, adding or subtracting polynomials is very similar to adding like terms. Parentheses are used to show which polynomials are being added or subtracted, so we must remove the parentheses and combine like terms. If there is no sign preceding the parentheses, there is an understood positive sign and we do not change any signs. We found the negative of an integer by changing its sign. In the same manner, we will find the negative of a polynomial by changing *all* of its signs.

Example 6	**Remove the parentheses from each of the following:**

a. $(2x - 3) = 2x - 3$ There is an understood + sign before the parentheses, so do not change any signs.

b. $-(3x - 6) = -3x + 6$ There is a − sign before the parentheses, so change all the signs of the polynomial.

c. $(-4x + 2) = -4x + 2$ There is an understood + sign before the parentheses, so do not change any signs.

d. $-(2x^2 + 4x - 5) = -2x^2 - 4x + 5$ There is a − sign before the parentheses, so change all the signs of the polynomial.

To add or subtract polynomials, we remove the parentheses and add like terms.

PRACTICE EXERCISE 6

Remove the parentheses from each of the following:

a. $(5a - 8)$

b. $-(3c^2 - 5c + 7)$

Example 7	**Find the following sums or differences:**

a. $(2x + 4) + (3x - 6) =$ Remove the parentheses. Do not change signs.

$2x + 4 + 3x - 6 =$ Add like terms.

$5x - 2$ Sum.

b. $(2x^2 - 6x + 4) + (x^2 - 3x - 7) =$ Remove parentheses.

$2x^2 - 6x + 4 + x^2 - 3x - 7 =$ Add like terms.

$3x^2 - 9x - 3$ Sum.

c. $(2x^2 - 6x + 3) - (x^2 - 2x + 7) =$ Remove parentheses and change all the signs of the second polynomial.

$2x^2 - 6x + 3 - x^2 + 2x - 7 =$ Add like terms.

$x^2 - 4x - 4$ Difference.

d. $(x^2 + 6x - 5) - (-2x^2 + 7x - 3) =$ Remove parentheses and change all the signs of the second polynomial.

$x^2 + 6x - 5 + 2x^2 - 7x + 3 =$ Add like terms.

$3x^2 - x - 2$ Difference.

PRACTICE EXERCISE 7

Add or subtract the following polynomials:

a. $(4x + 6) + (3x + 5)$

b. $(x^2 - 5x + 4) + (2x^2 - 3x - 6)$

c. $(3x^2 - 7x - 4) - (x^2 + 5x - 6)$

d. $(2x^2 + 4x - 7) - (-3x^2 - 8x + 3)$

Answers: Practice Exercise 6: a. $5a - 8$ **b.** $-3c^2 + 5c - 7$ **Practice Exercise 7: a.** $7x + 11$ **b.** $3x^2 - 8x - 2$ **c.** $2x^2 - 12x + 2$ **d.** $5x^2 + 12x - 10$

If you need more practice, do the following Additional Practice Exercises.

Additional Practice Exercise 7 Add or subtract the following polynomials:

a. $(5x - 3) + (-3x + 6)$

b. $(x^2 - 4x + 2) + (x^2 + 6x - 5)$

c. $(x^2 + x - 7) - (x^2 - 8x - 8)$

d. $(3x^2 + 2x - 9) - (-x^2 + 2x + 3)$

As was the case with integers, addition and subtraction of polynomials is often expressed in words.

| **Example 8** | **Write each of the following as addition and/or subtraction and simplify:** |

a. Find the sum of $2x^2 + 3x - 6$ and $4x^2 - 5x + 2$.

Solution

Since *sum* means add, this is written as $(2x^2 + 3x - 6) + (4x^2 - 5x + 2)$.

$(2x^2 + 3x - 6) + (4x^2 - 5x + 2) =$ Remove the parentheses.

$2x^2 + 3x - 6 + 4x^2 - 5x + 2 =$ Add like terms.

$6x^2 - 2x - 4$ Sum.

b. Subtract $3y^2 - 4y + 3$ from $y^2 - 5y + 7$.

Solution

This implies that we begin with $y^2 - 5y + 7$ and subtract $3y^2 - 4y + 3$ from it. This is written as $(y^2 - 5y + 7) - (3y^2 - 4y + 3)$.

$(y^2 - 5y + 7) - (3y^2 - 4y + 3) =$ Remove parentheses.

$y^2 - 5y + 7 - 3y^2 + 4y - 3 =$ Add like terms.

$-2y^2 - y + 4$ Difference.

c. From the sum of $4x + 5$ and $2x - 8$, subtract $7x - 1$.

Solution

This means that we first add $4x + 5$ and $2x - 8$ and, from this sum, subtract $7x - 1$. This is written as $[(4x + 5) + (2x - 8)] - (7x - 1)$.

$[(4x + 5) + (2x - 8)] - (7x - 1) =$ Remove the parentheses inside the brackets.

$[4x + 5 + 2x - 8] - (7x - 1) =$ Add like terms inside the brackets.

$[6x - 3] - (7x - 1) =$ Remove the parentheses and brackets.

$6x - 3 - 7x + 1 =$ Add like terms.

$-x - 2$ Answer.

PRACTICE EXERCISE 8

Write each of the following as addition and/or subtraction and simplify:

a. Find the sum of $5z^2 - 7z + 4$ and $2z^2 + 4z - 6$.

b. Subtract $4a^2 - 5a - 5$ from $2a^2 + 7a - 8$.

c. From the sum of $5a + 2$ and $-2a - 5$, subtract $8a - 1$.

(Continued)

Answers: Additional Practice 7: a. $2x + 3$ **b.** $2x^2 + 2x - 3$ **c.** $9x + 1$ **d.** $4x^2 - 12$ **Practice Exercise 8: a.** $(5z^2 - 7z + 4) + (2z^2 + 4z - 6) =$ $7z^2 - 3z - 2$ **b.** $(2a^2 + 7a - 8) - (4a^2 - 5a - 5) = -2a^2 + 12a - 3$ **c.** $[(5a + 2) + (-2a - 5)] - (8a - 1) = -5a - 2$

If you need more practice, do the following Additional Practice Exercises.

Additional Practice Exercise 8   Write each of the following as addition and/or subtraction, and simplify:

a. Find the sum of $6y^2 - 2y + 8$ and $3y^2 + 8y - 9$.

b. Subtract $8a^2 - 3a - 7$ from $2a^2 + 5a - 9$.

c. From the sum of $8x - 3$ and $-5x + 7$, subtract $3x + 2$.

Optional: Adding and Subtracting Polynomials Vertically

Addition and subtraction of polynomials may be done vertically as well as horizontally. The procedure is similar to the way whole numbers and decimals are added. We place digits with the same place value underneath each other and add. In adding polynomials vertically, we put them in descending order, then place like terms underneath each other and add. Of course, there is no regrouping (carrying) when adding polynomials.

Example 9   Add the following vertically:

a. 374 and 68

Solution

Place the digits with the same place value underneath each other and add.

$$
\begin{array}{r}
374 \\
68 \\
\hline
442
\end{array}
$$

b. Add $13p^2 + 3p - 4$ and $7p^2 - 4p - 4$.

Solution

Place like terms underneath each other and add.

$$
\begin{array}{r}
13p^2 + 3p - 4 \\
7p^2 - 4p - 4 \\
\hline
20p^2 - p - 8
\end{array}
$$

c. Add $9w^2 + 12w - 14$ and $-5w^2 + 12$.

Solution

Place like terms underneath each other and add.

$$
\begin{array}{r}
9w^2 + 12w - 14 \\
-5w^2 + 12 \\
\hline
4w^2 + 12w - 2
\end{array}
$$

d. Add $3x^3 - 5x + 7$ and $2x^2 + 4x - 9$.

Solution

Place like terms underneath each other and add.

$$
\begin{array}{r}
3x^3 - 5x + 7 \\
2x^2 + 4x - 9 \\
\hline
3x^3 + 2x^2 - x - 2
\end{array}
$$

Remember, when subtracting polynomials we add the negative of the polynomial being subtracted. Consequently, all its signs are changed. When subtracting vertically, the polynomial being subtracted is the one on the bottom, so we change the signs of the polynomial on the bottom and add.

Example 10   Subtract the following vertically:

a. Subtract $3y^2 - 8y - 10$ from $5y^2 - 9y + 3$.

Solution

Written horizontally, this is $(5y^2 - 9y + 3) - (3y^2 - 8y - 10)$, so it is necessary to change the signs of $3y^2 - 8y - 10$. Written vertically, it appears as follows:

Subtract:

$$
\begin{array}{r}
5y^2 - 9y + 3 \\
3y^2 - 8y - 10 \\
\hline
\end{array}
$$

Change the signs of the polynomial on the bottom, since it is the polynomial being subtracted, and change the operation to addition.

Answers: Additional Practice 8: a. $(6y^2 - 2y + 8) + (3y^2 + 8y - 9) = 9y^2 + 6y - 1$ **b.** $(2a^2 + 5a - 9) - (8a^2 - 3a - 7) = -6a^2 + 8a - 2$
c. $[(8x - 3) + (-5x + 7)] - (3x + 2) = 2$

Add:

$$5y^2 - 9y + 3$$
$$-3y^2 + 8y + 10$$
$$\overline{2y^2 - y + 13}$$

b. Find the difference of $6x^2 - 11x + 9$ and $3x^2 + 6$.

Note

This is the same as saying, "Subtract $3x^2 + 6$ from $6x^2 - 11x + 9$."

Solution

Just like with whole numbers, *difference* means subtract the second expression from the first.

Subtract:

$$6x^2 - 11x + 9$$
$$3x^2 \qquad + 6$$

Change the signs of the bottom polynomial and add.

Add:

$$6x^2 - 11x + 9$$
$$-3x^2 \qquad -6$$
$$\overline{3x^2 - 11x + 3}$$

Note

This is the same as saying, "Find the difference of $3x^3 - 7x^2 + 5$ and $4x^2 - 6x - 8$."

c. Subtract $4x^2 - 6x - 8$ from $3x^3 - 7x^2 + 5$.

Solution

Subtract:

$$3x^3 - 7x^2 \qquad + 5$$
$$\qquad 4x^2 - 6x - 8$$

Change the signs of the bottom polynomial and add.

Add:

$$3x^3 - 7x^2 \qquad + 5$$
$$\qquad -4x^2 + 6x + 8$$
$$\overline{3x^3 - 11x^2 + 6x + 13}$$

PRACTICE EXERCISE 10

Add or subtract the following vertically:

a. Add $5x^2 + 7x - 3$ and $3x^2 - 8x + 5$.

b. Add $10x^2 + 9x - 6$ and $-4x^2 - 9$.

c. Add $3x^3 - 4x + 5$ and $2x^2 + 6x - 9$.

d. Subtract $6x^2 + 7x - 9$ from $2x^2 + 5x + 7$.

e. Find the difference of $8y^2 - 9y + 3$ and $5y^2 + 9$.

f. Subtract $7x^2 - 8x + 2$ from $8x^3 - 9x^2 + 4$.

Study Tip 6 Making Sure: Confidence Building

We feel good about what we have done when we know that it is correct. We can tell if our work is accurate by checking our answers. Sometimes, the answer is in the back of the book, but most of the time we need to make our own check. The check should be different than reworking the exercise. For example, the check for subtraction is addition: $8 - 5 = 3$ if $3 + 5 = 8$. Also, the check for division is multiplication: $54 \div 9 = 6$ if $6 \cdot 9 = 54$. Always check your work!

Answers: Practice Exercise 10: a. $8x^2 - x + 2$ **b.** $6x^2 + 9x - 15$ **c.** $3x^3 + 2x^2 + 2x - 4$ **d.** $-4x^2 - 2x + 16$ **e.** $3y^2 - 9y - 6$
f. $8x^3 - 16x^2 + 8x + 2$

Exercise Set 1.8

For Extra Help

MyMathLab®

Determine whether each expression is a polynomial. If it is a polynomial, determine the number of terms. Identify any monomials, binomials, or trinomials. (See Example 1.)

1. $2x^2$

2. $3y^3$

3. $x^2 - 3x + 4$

4. $x^2 - 5x + 2$

5. $5 + 8x$

6. $3 - 7x$

7. $2x - 2x^2 + 4$

8. $4 - 3x + 5x^2$

9. $\dfrac{2x^2}{y^2} - 4x$

10. $\dfrac{3y^3}{2x^2} - 2y$

11. 5

12. -7

13. $x^3 + 3x - 4x^2 - 3$

14. $x^4 - 3x + 2x^3 + 4$

Give the coefficient, variable(s), and degree of each of the following. (See Example 2.)

	Coefficient	Variable(s)	Degree		Coefficient	Variable(s)	Degree
15. $3x$	____	____	____	**16.** $2y$	____	____	____
17. x^6	____	____	____	**18.** y^4	____	____	____
19. $-5x$	____	____	____	**20.** $-7z$	____	____	____
21. $4x^3$	____	____	____	**22.** $7y^5$	____	____	____
23. $-x^3$	____	____	____	**24.** $-y^5$	____	____	____
25. $10x^2y^4$	____	____	____	**26.** $12x^5y^2$	____	____	____
27. $-22a^6b^5$	____	____	____	**28.** $-16y^4z^6$	____	____	____

Write each of the following in descending order and give the degree of the polynomial. (See Example 3.)

	Descending Order	Degree		Descending Order	Degree
29. $4 - 5x$	____	____	**30.** $5 + 2y$	____	____
31. $3x - 4 + 2x^2$	____	____	**32.** $5x - 6 + 4x^2$	____	____
33. $3x - 4x^2 - x^4 + x^3$	____	____	**34.** $2y + 3y^3 - y^4 + 5$	____	____
35. $b + 6b^4$	____	____	**36.** $4 + 3x^2$	____	____

Evaluate each of the following for the indicated value of the variable. (See Example 4.)

37. $P(x) = 3x - 9; x = 3$

38. $P(y) = 4y - 3; y = 2$

39. $P(z) = 3z^2 - 8; z = 1$

40. $P(x) = 2x^2 - 5; x = 1$

41. $P(x) = x^2 + 3x - 5; x = 1$

42. $P(x) = x^2 + 4x - 3; x = 5$

43. $P(s) = 2s^2 + 5s - 7; s = 4$

44. $P(s) = 3s^2 + 6s - 12; s = 5$

45. $P(x) = x^3 + 2x^2 + 3x - 3; x = 2$

46. $P(x) = x^3 + 4x^2 + 2x + 5; x = 3$

If an object is dropped from a resting position, the distance (in feet) it falls is given by $s(t)=16t^2$. Use this to answer Exercises 47 and 48. (See Example 5.)

47. While hunting in rural Georgia, a hunter discovers an old homesite with an open well. In order to estimate the depth of the well, he drops a stone into it and observes that it takes $t = 2.5$ seconds for the stone to hit the water. Find the depth of the well by finding $s(2.5)$.

48. If a baseball is dropped from the top of the Sears Tower, find how far it has fallen after 3 seconds by finding $s(3)$.

On Mars, if an object is dropped from a height of 500 meters, its distance above the ground is given by $s(t) = -3.72t^2 + 500$, where t is in seconds and s is in meters. Use this to answer Exercises 49 and 50. (See Example 5.)

49. Find the distance above the ground after 4 seconds by finding $s(4)$.

50. Find the distance above the ground after 10 seconds by finding $s(10)$.

A particle moves along a line such that the distance from the starting point is given by $s(t) = t^3 + 2t^2 + 3t + 5$, with t measured in seconds. Use this to answer Exercises 51 and 52. (See Example 5.)

51. Find the location of the particle after 2 seconds by finding $s(2)$.

52. Find the location of the particle after 5 seconds by finding $s(5)$.

A furniture manufacturing company finds that the cost in dollars of manufacturing x sofas is given by $C(x) = 3500 + 23x + .4x^2$. Use this to answer Exercises 53 and 54. (See Example 5.)

53. Find the cost of producing 100 sofas by finding $C(100)$.

54. Find the cost of producing 500 sofas by finding $C(500)$.

Find the following sums and differences of polynomials. (See Example 7.)

55. $(6x - 5y) + (3x + 2y)$

56. $(3x + 8y) + (7x - 2y)$

57. $(5x - 4) - (3x - 6)$

58. $(4y + 7) - (6y - 2)$

59. $(4y^2 + 6y - 3) + (2y^2 - 3y - 6)$

60. $(3x^2 - 5x + 4) + (4x^2 - 5x + 8)$

61. $(2m^2c^2 + 5mc - 9c^2 + 5) + (3m^2c^2 - mc - 2)$

62. $(3q^2d^2 + 6qd - 4d^2 + 4) + (3q^2d^2 - qd - 3)$

63. $(4r^2 + 3r - 5) - (9r^2 - 3r + 5)$

64. $(2k^2 + 9k - 5) - (9k^2 - 9k + 5)$

65. $(8rp^2 - 5rp) - (r^3 + 4rp) + (4rp - 7rp^2)$

66. $(-5m^2t + 9) - (4mt^2 + 9m - 9) + (6mt^2 + 5m^2t)$

67. $(4h^2 + 7h - 4) - (3h^2 - 4h - 5) + (-11h - 10)$

68. $(3p^2 - 9p + 7) + (3p^2 + 5p + 6) - (4p^2 + 13)$

Represent each of the following by using addition and/or subtraction and simplify. (See Example 8.)

69. Add $7x^2 + 9x - 1$ and $4x^2 - 3x + 4$.

70. Add $4x^2 + 9x - 2$ and $9x^2 - 3x - 7$.

71. Subtract $6x^2 + 7x - 3$ from $3x^2 - 3x + 1$.

72. Subtract $2z^2 + z - 3$ from $4z^2 + 7z - 4$.

73. Find the sum of $3x + 13$ and $5x - 8$.

74. Find the sum of $5x + 16$ and $7x - 3$.

75. Find the difference of $5a + 2$ and $2a - 5$.

76. Find the difference of $7r + 3$ and $5r - 4$.

77. Find the sum of $x^2 - 3x + 4$ and $2x^2 + 4x - 6$.

78. Find the sum of $y^2 - 7y + 3$ and $3y^2 + 6y - 8$.

79. Find the difference of $z^2 + 7z - 9$ and $z^2 - 5z + 2$.

80. Find the difference of $w^2 - 8w + 4$ and $w^2 - 7w - 9$.

81. From the sum of $2x^2 - 7x - 2$ and $x^2 + 5$, subtract $2x^2 + 4x - 3$.

82. From the sum of $3y^2 + 8y - 6$ and $2y^2 - 9$, subtract $4y^2 - 7y - 4$.

83. To the difference of $3x - 7$ and $4x + 2$, add $5x - 9$.

84. To the difference of $4x + 2$ and $6x - 3$, add $9x + 5$.

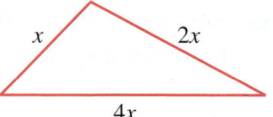

Recall that the perimeter of a triangle is found by adding the lengths of all three sides. For example, if the sides of a triangle are x inches, 4x inches, and 2x inches, the perimeter is $x + 4x + 2x = 7x$ *inches. Find the following:*

85. Find the perimeter of a triangle whose sides are $2x$ centimeter, $5x$ centimeter, and x centimeter.

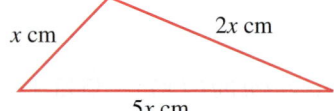

86. Find the perimeter of a triangle whose sides are x feet, $6x$ feet, and $4x$ feet.

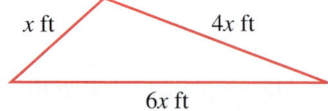

87. The lengths of the sides of a triangular sign are x inches, $(2x - 3)$ inches, and $(5x + 4)$ inches. What is the perimeter of the sign?

88. The lengths of the sides of a triangular piece of land are x feet, $(3x - 2)$ feet, and $(7x + 1)$ feet. How many feet of fencing would it take to enclose it?

A quadrilateral is a figure with four sides. The perimeter of a quadrilateral is found by adding the lengths of all four sides. For example, if the sides are 2x feet, 3x feet, x feet, and 2x feet, the perimeter is $2x + 3x + x + 2x = 8x$ *feet.*

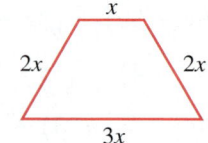

89. Find the perimeter of a quadrilateral whose sides are x inches, $3x$ inches, $5x$ inches, and $7x$ inches.

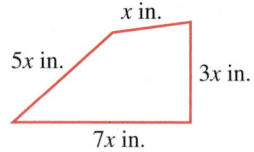

90. Find the perimeter of a quadrilateral whose sides are x meters, $4x$ meters, $2x$ meters, and $8x$ meters.

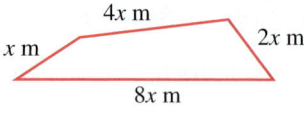

91. If the dimensions of a yard are x feet, $(2x + 3)$ feet, $(x + 4)$ feet, and $(3x + 2)$ feet, how many feet of fencing would it take to enclose it?

92. If the dimensions of a patio are x feet, $(3x - 1)$ feet, $(4x + 2)$ feet, and $(5x - 7)$ feet, what is the perimeter of the patio?

Represent each of the following by using variables. Make a drawing for Exercises 96–98 similar to those given for Exercises 93 and 95.

93. John is building a piece of furniture. He needs a board that is x feet long and another that is $(2x - 2)$ feet long. What is the total length of the two boards needed?

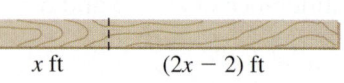

94. If John has x dollars and Mary has $(3x - 5)$ dollars, how much money do they have combined?

95. Christine is making a shelf for a closet and has a board $(2x - 5)$ feet long. She has to cut off a piece $(x - 3)$ feet long. How long is the remaining piece?

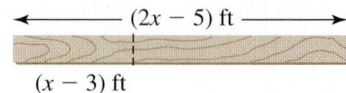

$(2x - 5)$ ft

$(x - 3)$ ft

96. A carpenter is installing baseboard and has a board $(3y + 2)$ feet long. He has to cut off a piece $(y + 4)$ feet long. How long is the remaining piece?

97. If Chan climbs $(3x + 3)$ meters and then climbs $(2x - 5)$ meters more, how high has he climbed altogether?

98. If Yolanda has driven $(5x - 6)$ miles and then drives $(3x + 2)$ miles farther, what is the total distance she has driven?

Challenge Exercises (99–106)

99. $\left(\dfrac{5}{7}x + \dfrac{3}{11}y\right) + \left(\dfrac{1}{7}x - \dfrac{4}{11}y\right)$

100. $\left(\dfrac{4}{13}a - \dfrac{3}{7}b\right) + \left(\dfrac{5}{13}a - \dfrac{2}{7}b\right)$

101. $\left(\dfrac{9}{13}r - \dfrac{6}{17}m\right) - \left(\dfrac{7}{13}r + \dfrac{10}{17}m\right)$

102. $\left(\dfrac{3}{7}m - \dfrac{4}{5}n\right) - \left(\dfrac{5}{7}m - \dfrac{1}{5}n\right)$

103. $\left(\dfrac{2}{3}a^2 - \dfrac{3}{5}a + \dfrac{4}{15}\right) + \left(-\dfrac{1}{3}a^2 + \dfrac{1}{5}a + \dfrac{3}{15}\right)$

104. $\left(\dfrac{4}{9}x^2 + \dfrac{2}{11}x - \dfrac{9}{19}\right) + \left(\dfrac{4}{9}x^2 - \dfrac{7}{11}x + \dfrac{4}{19}\right)$

105. $\left(\dfrac{8}{23}y^2 - \dfrac{8}{19}y + \dfrac{6}{13}\right) - \left(\dfrac{12}{23}y^2 - \dfrac{7}{19}y + \dfrac{2}{13}\right)$

106. $\left(\dfrac{10}{17}y^2 + \dfrac{16}{29}y - \dfrac{5}{31}\right) - \left(\dfrac{6}{17}y^2 + \dfrac{23}{29}y + \dfrac{16}{31}\right)$

Writing Exercises (107–113)

107. How does an algebraic expression differ from a polynomial?

108. Give three examples of algebraic expressions that are not polynomials.

109. If special names were given to polynomials of more than three terms, what suggestions would you have for the name of a polynomial of four terms? Five terms?

110. How do you find the degree of a monomial? A polynomial with two or more terms?

111. What is meant by writing a polynomial in descending order?

112. Why does 2^4y^2 have degree 2 and x^4y^2 have degree 6?

113. How does a term differ from a monomial?

Critical-Thinking Exercises (114–116)

114. Are all terms monomials? Explain. Are all monomials terms? Explain.

115. Do the following: (1) Choose any number; (2) triple the number; (3) add 1; (4) subtract the original number; (5) repeat steps 3 and 4; (6) repeat steps 3 and 4 again. The result is always 3; write a paragraph explaining why.

116. Write a riddle similar to the one in Exercise 115.

Chapter 1   Summary

Concept/Procedure	Example

Definition of Variables and Constants: [Section 1.1]

- A variable is a symbol, usually a letter, that is used to represent a number.
- A constant is a symbol, a number, whose value does not change.

Definition of Natural Number Exponents: [Section 1.1]

- a^n means multiply n factors of a.

Order of Operations: [Section 1.1]

- If an expression contains more than one operation, operations are to be performed in the following order:

  1. If parentheses or other inclusion symbols (braces or brackets) are present, begin within the innermost and work outward, using the order in steps 3–5 in doing so.
  2. If an implied grouping like a fraction bar is present, simplify above and below the fraction bar separately in the order given by steps 3–5.
  3. Evaluate all indicated powers.
  4. Perform all multiplications or divisions in the order in which they occur as you work from left to right.
  5. Then perform all additions or subtractions in the order in which they occur as you work from left to right.

Example 1:

Simplify: $12 + 36 \div 3 \cdot 2 - 15 =$ Divide 3 into 36, since division came before multiplication.

$\qquad\qquad 12 + 12 \cdot 2 - 15 =$ Multiply 12 and 2, since multiplication is done before addition or subtraction.

$\qquad\qquad\qquad 12 + 24 - 15 =$ Add and subtract in order from left to right.

$\qquad\qquad\qquad\qquad 21$ Answer.

Evaluating Expressions Containing Variables: [Section 1.1]

- To evaluate an expression containing variable(s), replace the variable with the value(s) of the variable(s), using parentheses, and simplify, using the order of operations.

Example 2:

Evaluate $6x^2 \div y$ when $x = 3$ and $y = 2$.

Solution: $6x^2 \div y =$ Replace x with 3 and y with 2.

$\qquad 6(3)^2 \div (2) =$ Square 3.

$\qquad 6 \cdot 9 \div 2 =$ Multiply 6 and 9.

$\qquad 54 \div 2 =$ Divide.

$\qquad 27$ Answer.

Definition of Perimeter: [Section 1.2]

- The perimeter of a geometric figure is the distance around the figure. The distance around a circle is the circumference. Perimeter and circumference are expressed in linear units.

Concept/Procedure	Example

Perimeter Formulas: [Section 1.2]

Figure	Formula
Triangle	$P = a + b + c$
Rectangle	$P = 2L + 2W$
Square	$P = 4s$
Circle	$C = 2\pi r$ or $C = \pi d$

Example 3:

Find the perimeter of a rectangle whose length is 8 cm and whose width is 5 cm.

Solution: $P = 2l + 2w$ Substitute 8 cm for the length and 5 cm for the width.

$$P = 2(8\text{ cm}) + 2(5\text{ cm})$$ Multiply.

$$P = 16\text{ cm} + 10\text{ cm}$$ Add.

$$P = 26\text{ cm}$$ Therefore, the perimeter is 26 cm.

Definition of Area: [Section 1.3]

- The area is the measure of the size of a surface and is expressed in square units.

Area Formulas: [Section 1.3]

Figure	Formula
Rectangle	$A = LW$
Square	$A = s^2$
Triangle	$A = \dfrac{bh}{2}$
Parallelogram	$A = bh$
Trapezoid	$A = \dfrac{h(B + b)}{2}$
Circle	$A = \pi r^2$

Example 4:

Find the area of a triangle whose base is 6 ft and whose height is 5 ft.

Solution: $A = \dfrac{bh}{2}$ Substitute 6 ft for the base and 5 ft for the height.

$$A = \frac{(6\text{ ft})(5\text{ ft})}{2}$$ Multiply in the numerator.

$$A = \frac{30\ \text{ft}^2}{2}$$ Divide.

$$A = 15\text{ ft}^2$$ Therefore, the area is 15 square feet.

Definition of Volume: [Section 1.4]

- The volume of a geometric solid is a measure of how much it will hold and is expressed in cubic units.

Volume Formulas: [Section 1.4]

Solid	Formula
Rectangular Solid	$V = LWH$
Cube	$V = e^3$
Right Circular Cylinder	$V = \pi r^2 h$
Cone	$V = \dfrac{\pi r^2 h}{3}$
Sphere	$V = \dfrac{4}{3}\pi r^3$

- In general, volume equals the area of the base · height, or if it comes to a point, area equals $\frac{1}{3}$ (area of base) (height).

Example 5:

Find the volume of a right circular cylinder whose height is 6 m and the radius of whose base is 3 m.

Solution: $V = \pi r^2 h$ Substitute 3.14 for π, 3 m for r, and 6 m for h.

$$V \approx (3.14)(3\text{ m})^2(6\text{ m})$$ Square 3 m.

$$V \approx 3.14(9\text{ m}^2)(6\text{ m})$$ Multiply from left to right.

$$V \approx 169.56\text{ m}^3$$ Therefore, the volume is 169.56 cubic meters.

Exact volume is $V = \pi(3\text{m})^2(6\text{m}) = \pi(9\text{m}^2)(6\text{m}) = 54\pi\text{ m}^3$

Concept/Procedure	Example

Definition of Surface Area: [Section 1.4]

- The surface area of a three-dimensional object is the sum of the areas of all its surfaces and is expressed in square units.

Surface Area Formulas: [Section 1.4]

Solid	Formula
Rectangular Solid	$SA = 2LW + 2LH + 2WH$
Cube	$SA = 6e^2$
Right Circular Cylinder	$SA = 2\pi r^2 + 2\pi rh$

Example 6:

Find the surface area of a rectangular solid with length 9 yd, width 3 yd, and height 5 yd.

Solution: $SA = 2LW + 2LH + 2WH$ Substitute 9 yd for L, 3 yd for W, and 5 yd for H.

$SA = 2(9\,\text{yd})(3\,\text{yd}) + 2(9\,\text{yd})(5\,\text{yd})$ Multiply.
$\quad\quad + 2(3\,\text{yd})(5\,\text{yd})$

$SA = 54\,\text{yd}^2 + 90\,\text{yd}^2 + 30\,\text{yd}^2$ Add.

$SA = 174\,\text{yd}^2$ Therefore, the surface area is 174 square yards.

Types of Numbers: [Section 1.5]

- Natural Numbers $= \{1, 2, 3, \ldots\}$
- Whole Numbers $= \{0, 1, 2, 3, \ldots\}$
- Integers $= \{\ldots, -3, -2, -1, 0, 1, 2, 3, \ldots\}$

Order Symbols: [Section 1.5]

- $=$ equals
- $\neq$ does not equal
- $<$ is less than
- $\leq$ is less than or equal to
- $>$ is greater than
- $\geq$ is greater than or equal to

Example 7:

$4 = 4$

$4 \neq 6$

$4 < 5$

$4 \leq 4$

$4 > 2$

$4 \geq 2$

Negative of a Negative: [Section 1.5]

- $-(-x) = x$ for all x.

Example 8:

$-(-6) = 6$

Absolute Value: [Section 1.5]

- The absolute value of a number is its distance from 0.
- $|x| = x$ if $x \geq 0$ and $|x| = -x$ if $x < 0$.

Example 9:

$|5| = 5$

$|-7| = 7$

Concept/Procedure	Example
Rules for Addition of Integers: [Section 1.6] • To add integers with the same sign, add their absolute values and give the answer the same sign as the numbers being added. • To add integers with opposite signs, subtract their absolute values and give the answer the sign of the number with the larger absolute value.	**Example 10:** Add: $-13 + (-9) = -22$ Signs are the same; add the absolute values and give the answer the same sign. $-11 + 5 = -6$ Signs are opposite; subtract absolute values and give answer the sign of the number with the larger absolute value.
Properties of Addition: [Section 1.6] • Commutative: $a + b = b + a$ Addition may be performed in any order. • Associative: $a + (b + c) = (a + b) + c$ Addition may be grouped in any manner. • Inverse: For every integer a, there exists an integer $-a$ (read "the opposite of a") such that $a + (-a) = -a + a = 0$. • Identity: The integer 0 is the additive identity, since $0 + a = a + 0 = a$.	**Example 11:** $-2 + 3 = 3 + (-2)$ $-4 + (2 + 5) = (-4 + 2) + 5$ $7 + (-7) = 0$ $4 + 0 = 4$
Rule for Subtraction: [Section 1.7] • $a - b = a + (-b)$ To subtract a number, add its inverse (opposite).	**Example 12:** Subtract: $-15 - (-8) = -15 + 8 = -7$ To subtract -8, add $+8$.
Definition of Term: [Sections 1.7, 1.8] • A term is a number, variable, or product and/or quotient of numbers and variables raised to powers.	**Example 13:** The expressions 3, x, $3x$, $3x^4$ and $\frac{3x^5}{y^2}$ are terms. The expression $2x^2 + 5$ is not a term.
Definition of Like Terms: [Sections 1.7, 1.8] • Like terms are terms that have the same variables with the same exponents on these variables.	**Example 14:** The terms $4x$ and $6x$ are like terms, the terms $4x$ and $6y$ are not like terms.
Distributive Property of Multiplication over Addition: [Section 1.7] • For all numbers $a, b,$ and $c, a(b + c) = ab + ac$.	**Example 15:** $2(4 + 7) = 2 \cdot 4 + 2 \cdot 7 = 8 + 14 = 22$
Addition of Like Terms: [Section 1.7] • To add like terms, add the numerical coefficients and leave the variable portion unchanged.	**Example 16:** Simplify: $7x + 2y + 8 - 5y + 4x - 5 =$ Add like terms. $11x - 3y + 3$ Answer.

Concept/Procedure	Example
Numerical Coefficient: [Section 1.8] • The numerical coefficient of a term is the number factor of the term.	Example 17: The numerical coefficient of $-6x^2y$ is -6.
Degree of a Term: [Section 1.8] • The degree of a term is the exponent on the variable if there is only one variable and the sum of the exponents on the variables if there is more than one variable.	Example 18: Find the degree of $6x^2$. Solution: The degree is the exponent of x which is 2.
Definition of Polynomial: [Section 1.8] • A polynomial is the sum of a finite number of terms that do not have a variable as a divisor and the exponents on the variables are whole numbers.	Example 19: The expression $3x^2 + 2x - 5$ is a polynomial and the expression $3x + \frac{4}{y}$ is not a polynomial.
Types of Polynomials: [Section 1.8] • A monomial is a polynomial with one term. • A binomial is a polynomial with two terms. • A trinomial is a polynomial with three terms.	Example 20: Given $-2x + 3x^3 + 5$. Is it a monomial, binomial, or trinomial? Solution: Three terms; therefore, it is a trinomial.
Descending Order: [Section 1.8] • A polynomial is written in descending order if the term of the highest degree is written first, the term of the second highest degree is written second, and so on.	Example 21: Write $-2x + 3x^3 + 5$ in descending order. Solution: $3x^3 - 2x + 5$
Degree of a Polynomial: [Section 1.8] • The degree of a polynomial is the same as the degree of the term of the polynomial with the highest degree.	Example 22: What is the degree of $3x^3 - 2x + 5$? Solution: Third, because the degree of the term of highest degree is 3.
Addition of Polynomials: [Section 1.8] • To add polynomials, add the like terms and leave the sum in descending order.	Example 23: Add: $(2a^2 - 6a - 9) + (4a^2 + 2a + 3) =$ Remove the parentheses. $2a^2 - 6a - 9 + 4a^2 + 2a + 3 =$ Add like terms. $6a^2 - 4a - 6$ Answer.
Subtraction of Polynomials: [Section 1.8] • To subtract a polynomial, add the negative of the polynomial.	Example 24: Subtract: $(4r^2 - 2r + 8) - (-3r^2 + 4r - 9) =$ Remove the parentheses $4r^2 - 2r + 8 + 3r^2 - 4r + 9 =$ Add like terms. $7r^2 - 6r + 17$ Answer.

Chapter 1   Review Exercises

Write each expression by using exponents. [Section 1.1]

1. $9 \cdot 9 \cdot 9 \cdot 9 \cdot 9$

2. $(-2)(-2)(-2)(-2)$

3. $3 \cdot 3 \cdot 3 \cdot 5 \cdot 5 \cdot 5 \cdot 5$

4. $12 \cdot 9 \cdot 9 \cdot 9 \cdot 2 \cdot 2$

5. $3 \cdot 3 \cdot 3 \cdot a \cdot a \cdot a$

6. $7 \cdot 7 \cdot 7 \cdot r \cdot r \cdot r \cdot r \cdot s \cdot s$

Evaluate each expression. [Section 1.1]

7. $2^3 \cdot 3^2$

8. $5 \cdot 2^4$

9. $3^2 \cdot 5^2 \cdot 2^2$

10. $10^2 \cdot 2^2 \cdot 5^3$

Find the value of each expression when $x = 3$ and $y = 2$. [Section 1.1]

11. $6x + 2y$

12. $x^2 y^2$

13. $4xy^5$

14. $6x^2 - 5y^3$

15. $8x^2 \div y$

16. $\dfrac{2x^2 + 3y^2}{3x^2 + 3}$

Evaluate each expression, using the order of operations. [Section 1.1]

17. $36 - 12 \div 6$

18. $2^4 + 8 \div 4$

19. $29 - 4(7 - 2)$

20. $5 + 2 \cdot 3^2$

21. $4^3 - 2^3 \cdot 3 \div 2^2$

22. $64 \div (3^3 - 11)$

23. $8 + 48 \div 8 \cdot 3 + 5$

24. $12 + 8 \cdot 9 \div 3 - 9$

Find the perimeter of each of the following. [Section 1.2]

25.

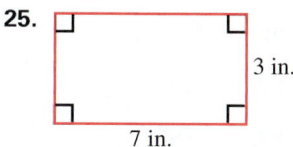

3 in.

7 in.

26.

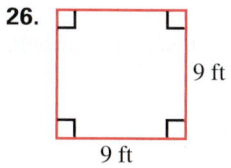

9 ft

9 ft

27. Express the answer in inches.

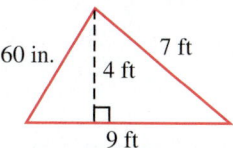

60 in.   7 ft   4 ft   9 ft

28. Find the exact and approximate circumference. Use $\pi \approx 3.14$.

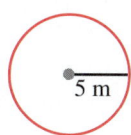

5 m

Find the area of each of the following. [Section 1.3]

29.

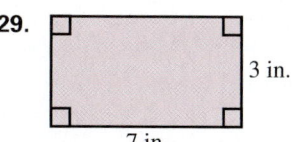

3 in.

7 in.

30.

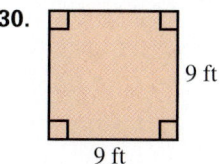

9 ft

9 ft

31.

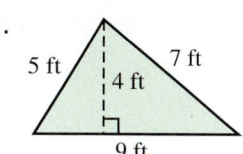

5 ft   7 ft   4 ft   9 ft

32. Find the exact and approximate area. Use $\pi \approx 3.14$.

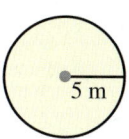

5 m

33. Express the answer in yd^2.

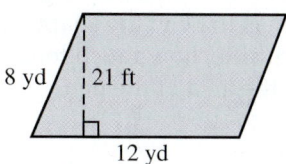

8 yd   21 ft   12 yd

34. Express the answer in cm^2.

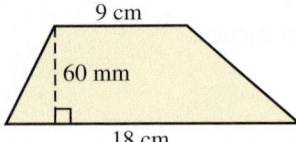

Find the volume and surface area of each geometric figure. [Section 1.4]

35.

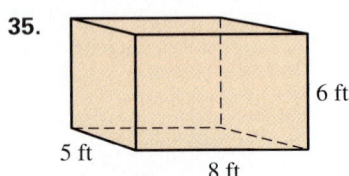

36.

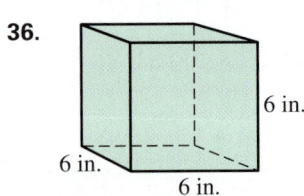

37. Express the answer in terms of π.

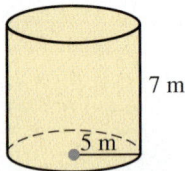

38. Find the volume only. Express the answer in terms of π.

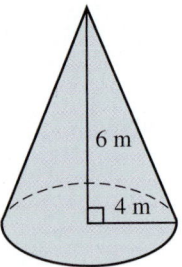

Solve the problems. [Sections 1.2–1.4]

39. A rancher is planning to fence in a holding pen for his cattle. If the pen is to be in the shape of a square 250 feet on each side, how much fencing will he need? What is the area of the pen?

40. An irrigation system consists of a pipe that is anchored on one end to remain stationary, and the other end is on wheels. When the system is turned on, the pipe rotates about the stationary end and the end on wheels sweeps out a circle. If the pipe is 150 feet long, how far do the wheels travel in one rotation? Use $\pi \approx 3.14$.

41. A rectangular bedspread is 120 inches long and 90 inches wide. Pat is going to sew a lace border around the bedspread. If the lace costs $1.50 per yard, how much will the lace cost?

42. Find the cost of the carpet needed to cover the floor of a room that is 14 yards long and 10 yards wide if the carpet costs $23.50 per square yard.

43. A window is in the shape of a square each of whose sides is 52 inches long. What is the area of the window?

44. A sign for an advertisement is in the shape of a trapezoid. If the parallel sides are 10 inches and 6 inches long and the height of the sign is 8 inches, find the area of the sign.

45. Find the area watered in one revolution by the irrigation system in Exercise 40. Use $\pi \approx 3.14$.

46. A tabletop is in the shape of a triangle two of whose sides are 17 inches long and the third of which is 16 inches long. The height to the 16-inch side is 15 inches. Find the area of the tabletop.

47. A box of rice is in the shape of a rectangular solid that is 6 inches long, 1 inch wide, and 8 inches high. (a) What is the volume of the box? (b) How much cardboard was used in constructing the box?

48. A can of soup is in the shape of a right circular cylinder whose diameter is 3 inches and whose height is 5 inches. (a) What is the volume of the can? (b) How much steel was used in constructing the can? Use $\pi \approx 3.14$.

Give the negative (opposite) of each number. [Section 1.5]

49. 23

50. -35

51. $|7|$

52. $|-32|$

53. $-|64|$

54. $-|-86|$

Insert the proper symbol $(=, <, or >)$ to make each statement true. [Section 1.5]

55. 5 ____ 9

56. -28 ____ 9

57. -8 ____ -10

58. $|12|$ ____ $|-12|$

59. $-|-16|$ ____ $|16|$

60. $|-32|$ ____ $-|32|$

Evaluate the following:

61. $|5 - 16|$

62. $|-6| + |-4|$

63. $|7| - |-3|$

64. $|6 - 3| - |2 - 5|$

Find the sums and/or differences. [Sections 1.6 and 1.7]

65. $19 - 13$

66. $9 - (-6)$

67. $-16 - (-9)$

68. $8 - 12 - 17$

69. $-24 - (-33) + 14$

70. $(-9 + 5) + (5 - 8)$

71. $(-53 - 37) - (32 - 27)$

72. $(-13 - 32) - (41 - 56)$

Give the name of the property illustrated by each statement. [Section 1.6]

73. $6 + (5 + 1) = (6 + 5) + 1$

74. $(3 + 8) + 5 = (8 + 3) + 5$

75. $9 + (-4 + 4) = 9 + 0$

76. $0 + (-3 + 2) = -3 + 2$

Write an expression for each phrase, and evaluate. [Sections 1.6 and 1.7]

77. The sum of -9 and 6

78. The difference of -15 and -8

79. 16 more than -5

80. -6 less than 3

81. -9 increased by 8

82. -3 decreased by -7

83. The sum of 18 and -8 added to the sum of -5 and 13

84. The difference of -5 and 9 added to the difference of -9 and -2

85. The difference of 4 and -5 subtracted from the sum of 14 and -8

86. The product of 5 and the sum of -6 and 10

Add like terms. [Section 1.7]

87. $2x - 5x$

88. $8x + 4y - 6x - 3y$

89. $-13w^3 + 7u^2 - 5 + 8w^3 - 13u^2 + 8$

90. $14x^2y - 23xy^2 - 19x^2y + 29xy^2 + 4$

Give the coefficient, variable(s), and degree of each of the following. [Section 1.8]

	Coefficient	Variable(s)	Degree
91. $9x^2$	_____	_____	_____
92. $-9x^5y^4$	_____	_____	_____

Write each expression in descending order. Identify those polynomials that are monomials, binomials, or trinomials. Give the degree of those that are polynomials. [Section 1.8]

93. $4x^3 + 6x$

94. $7y^2 - 8y + 2$

95. $5x^7$

96. $9x^4 - 7x^5 + 2x$

97. $2a^3 - 5a^2 + 14a^6 + 19$

98. $\dfrac{3x^6}{y} + 9x - 8$

Evaluate each expression for the indicated value of the variable. [Section 1.8]

99. $P(x) = 3x^2 + 2x - 8; x = 1$

100. $P(b) = 3b^4 + 5b - 5; b = 3$

Find the sums and/or differences of the given polynomials. Leave the answers in descending order. [Section 1.8]

101. $(4x + 3) + (6x - 8)$

102. $(7a + 3b) - (2a - 8b)$

103. $(u^4 - 2u^3 + u) + (4u^4 - 3u^3 - u)$

104. $(4a^2 - 7a + 2) - (2a^2 + 7a - 3)$

105. $(z^4 - 3z^3 + 6z) + (2 - 4z + 4z^3)$

106. $(3z^2 + 5z - 1) - (7z^2 - 7z + 8) + (5z^2 - 6z + 1)$

Write each statement as addition and/or subtraction and simplify. [Section 1.8]

107. Find the sum of $5x^2 - 9x + 5$ and $6x^2 - 2x - 1$.

108. Find the sum of $3x^3 + 4x^2 - 4$ and $2x^2 - 4x - 7$.

109. Find the difference of $4x^2 + 7x - 2$ and $6x^2 - 6x + 3$.

110. Subtract $3u^4 - 3u^3 + 4u^2 - 7$ from $u^4 + u^2 - 3u + 1$.

111. From the sum of $-x^3 - 4x^2 + 2x + 2$ and $2x^3 - 3x^2 - 2x + 1$, subtract $3x^3 + 6x^2 - 2x - 2$.

112. Find the perimeter of a triangle whose sides are $3x - 5$ feet, $4x + 1$ feet, and $2x - 2$ feet.

Chapter 1 Test

Write each of the following, using exponents:

1. $6 \cdot 6 \cdot 6 \cdot 6 \cdot 5 \cdot 5$

2. $4 \cdot 4 \cdot x \cdot x \cdot x \cdot x \cdot y \cdot y$

3. $3 \cdot x \cdot x \cdot x \cdot x \cdot x$

4. $(4a)(4a)(4a)$

Evaluate each of the following:

5. $36 - 2 \cdot 3^2 \div 6$

6. $5(8^2 - 16) \div 12 + 3$

7. $7 + 81 \div 3 \cdot 3 - 18$

8. $\dfrac{8^2 + 6 \cdot 4^2}{6^2 + 4 \cdot 11}$

If $x = 4$ and $y = 2$, find the value of the following:

9. $8xy \div 2y^3$

10. $\dfrac{2x^2 + y^2}{2y^2 + 1}$

Find the perimeter and area of the following:

11.

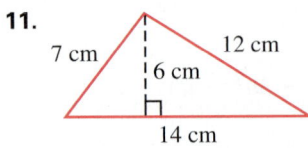

12. Express the perimeter in feet and the area in square feet.

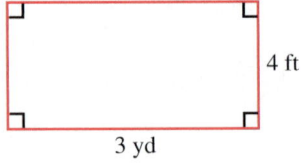

Find the area of each of the following:

13. Express the answer in square meters.

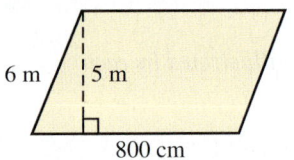

14.

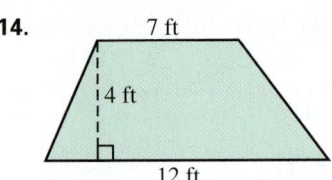

15. Find the exact and approximate circumference and area of the given circle. Use $\pi \approx 3.14$.

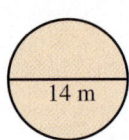

Find the volume and surface area of the following:

16.

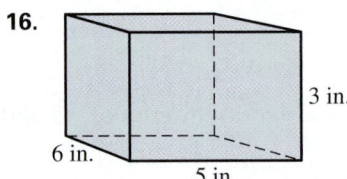

17. Use $\pi \approx 3.14$.

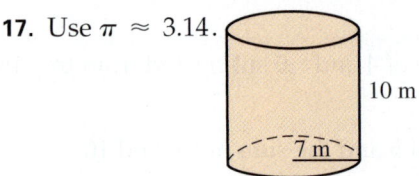

18. Magna plans to fence in a square garden plot that measures 18 feet on each side. If fencing costs $.50 per linear foot, how much will it cost to fence in the plot?

19. A rectangular wall is 15 feet long and 12 feet high. How much will it cost to paint the wall if the paint costs $1.80 per square foot?

20. A tank is in the shape of a rectangular solid with a length of 6 feet, a width of 4 feet, and a height of 3 feet. The tank is to be filled with water that weighs 62.5 pounds per cubic foot. Find the weight of the water in the tank.

Insert the proper symbol $(<, >, \text{ or } =)$ *in order to make each of the following true:*

21. -3 _____ $|-6|$

22. $-|-10|$ _____ $|10|$

Evaluate the following:

23. $|5| - |-7|$

24. $|7 - 10| - |-6 + 2|$

Find the following sums or differences:

25. $-16 + 5$

26. $-9 + (-6) + 8$

27. $(-11 + 5) - (7 - 12)$

28. $-12 - [22 - (-8 - 2)]$

Name the property illustrated by each of the following:

29. $(-5 + 7) + 4 = [7 + (-5)] + 4$

30. $8[4 + (-4)] = 8(0)$

Write an expression for each of the following and evaluate:

31. The sum of -7 and 11

32. The difference of 9 and -6 added to -12

33. The sum of -3 and 9 subtracted from the sum of 4 and -13

34. The difference of $x^2 - 6x + 2$ and $2x^2 + 5x - 3$

35. The sum of $3x + 6$ and $2x - 3$ subtracted from $4x + 5$

36. *Answer the following for the polynomial* $3x^2 + 4x^3 + 6x$:

  a. What is the numerical coefficient of the second term?

  b. What is the degree of the polynomial?

  c. Write the polynomial in descending order.

  d. Is the polynomial a *monomial, binomial, trinomial,* or *none of these*?

37. Evaluate $P(x) = 3x^2 - 2x + 8$ for $x = 2$.

Simplify the following:

38. $9xy + 6yz - 4xy - 10yz$

39. $(2x^2 - 5x + 7) + (3x^2 + 2x - 11)$

40. $(6n^3 + 7n^2 - 8) - (9n^2 - 7n - 6)$

41. Is $-x$ always a negative number? Explain.

2

Laws of Exponents, Products and Quotients of Integers and Polynomials

In this chapter, we continue our discussion of operations on integers and polynomials. In Chapter 1, we found the sums and differences of integers and polynomials. In this chapter, we find the products and quotients of integers, the products of polynomials, the quotients of monomials, and the quotients of polynomials and monomials. We will not discuss the division of polynomials by polynomials other than monomials until Chapter 6.

Section 2.1   Multiplication of Integers

OBJECTIVES *When you complete this section, you will be able to:*

A Multiply signed numbers.
B Raise signed numbers to powers.
C Identify properties of multiplication.

PREREQUISITE SKILLS *Before beginning this section, you should be able to:*

a. Multiply whole numbers (Section R.3).
b. Raise whole numbers to powers (Section 1.1).

GETTING READY FOR SECTION 2.1

Find the products of the following whole numbers:
1. $(3)(5)(4)$ **2.** $(6)(3)(4)(2)$

Raise the following whole numbers to powers:
3. 4^3 **4.** 2^4

Evaluate the following expressions for the given value of the variable(s):
5. $3xy: x = 6, y = 3$ **6.** $a^2b^3: a = 3, b = 2$

Introduction

In Sections 1.6 and 1.7, we learned to add and subtract signed numbers. In this section, we use our knowledge of how signed numbers are added to develop rules for multiplication of signed numbers. When multiplying numbers, we are finding their **product**.

Calculator Exploration Activity (Optional)

Using a calculator, complete the following:

Column A	Column B
$3(3) =$	$-3(3) =$
$3(2) =$	$-3(2) =$
$3(1) =$	$-3(1) =$
$3(0) =$	$-3(0) =$
$3(-1) =$	$-3(-1) =$
$3(-2) =$	$-3(-2) =$
$3(-3) =$	$-3(-3) =$

Based on your answers to column A, answer the following:
1. Is the product of two positive numbers positive or negative?

2. Is the product of a positive number and a negative number positive or negative?

Based on your answers to column B, answer the following:
3. Is the product of a negative number and a positive number positive or negative?
4. Is the product of two negative numbers positive or negative?
5. What is the product of any number and 0?

Conclusions based on patterns can often be misleading, so we now give a mathematical basis for the conclusions reached in questions 1–4.

OBJECTIVE **A** Multiplying signed numbers

Recall that multiplication is a shortcut for repeatedly adding the same number. For example, $3(5) = 5 + 5 + 5$. The first number, 3, tells us how many of the second number, 5, are being added. Likewise, $4(6) = 6 + 6 + 6 + 6$. In the examples that follow, we will develop rules for multiplying a positive number and a positive number, a positive number and a negative number, a negative number and a positive number, and a negative number and a negative number. The multiplication of positive numbers is already familiar to you from arithmetic.

Answers: Getting Ready: 1. 60 **2.** 144 **3.** 64 **4.** 16 **5.** 54 **6.** 72   **Calculator Exploration Activity: 1.** positive **2.** negative **3.** negative **4.** positive **5.** 0

Example 1   **Find the following products:**

a. $3 \cdot 5 = 5 + 5 + 5 = 15$ Positive 3 times positive 5 equals positive 15.

b. $4 \cdot 6 = 6 + 6 + 6 + 6 = 24$ Positive 4 times positive 6 equals positive 24.

It is apparent from the preceding example, and your prior knowledge of arithmetic, that multiplying a positive number and a positive number results in a positive number. Example 2 illustrates the product of a positive number and a negative number.

Example 2   **Find the following products:**

a. $+3(-5) = (-5) + (-5) + (-5) = -15$ Definition of multiplication. Therefore, $+3$ times -5 equals -15.

b. $+4(-6) = (-6) + (-6) + (-6) + (-6) = -24$ Definition of multiplication. Therefore, $+4$ times -6 equals -24.

The preceding example implies that a positive number times a negative number results in a negative number.

In order to discuss a negative number times a positive number, we need the **commutative property for multiplication**. We know from arithmetic that $3(4) = 4(3)$ and $6(2) = 2(6)$. These products suggest that the order in which two numbers are multiplied makes no difference. We can generalize this to the statement of the commutative property for multiplication.

> **Definition** Commutative Property for Multiplication
>
> For any two integers a and b, $a \cdot b = b \cdot a$.

Example 3   **Find the following products:**

a. $(-3)(3) =$ Apply the commutative property.
   $(3)(-3) =$ Multiply.
   -9 Therefore, -3 times $+3$ equals -9.

b. $(-6)(4) =$ Apply the commutative property.
   $(4)(-6)$ Multiply.
   -24 Therefore, -6 times $+4$ equals -24.

From Example 3, we can conclude that a negative number times a positive number results in a negative number. Unfortunately, we cannot use the definition of multiplication to develop a rule for multiplying a negative number and a negative number. For example, $(-3)(-5)$ cannot mean add -5, -3 times.

Example 4   **Look for a pattern in the following:**

$(3)(-4) = -12$
$(2)(-4) = -8$ Increase of 4
$(1)(-4) = -4$ Increase of 4
$(0)(-4) = 0$ Increase of 4

Notice that the number on the left is decreasing by 1 and the product is increasing by 4 in each case. Let's continue the pattern.

$(3)(-4) = -12$
$(2)(-4) = -8$
$(1)(-4) = -4$
$(0)(-4) = 0$
$(-1)(-4) = 4$ Increase of 4
$(-2)(-4) = 8$ Increase of 4

The preceding pattern implies that a negative number times a negative number results in a positive number.

Looking back at Examples 1 and 4, we notice that if both signs are positive or both negative, the product is positive. From Examples 2 and 3, we see that if the signs are different (one positive and one negative), the product is negative. Based on these examples, we make the following rules for multiplying signed numbers:

> ### Rule: Products of Signed Numbers
>
> The product of two numbers with the same (like) signs is positive. The product of two numbers with different (unlike) signs is negative. In symbols, this means
>
> $$(+)(+) = +, (-)(-) = +, (+)(-) = -, \text{ and } (-)(+) = -.$$

Example 5   Find the following products:

a. $4 \cdot 7 = 28$ Same signs. Therefore, the product is positive.

b. $(-5)(5) = -25$ Different signs. Therefore, the product is negative.

c. $(-6)(-7) = 42$ Same signs. Therefore, the product is positive.

d. $(5)(-8) = -40$ Different signs. Therefore, the product is negative.

PRACTICE EXERCISE 5

Find the following products:

a. $6 \cdot 3$ **b.** $(-8)(4)$ **c.** $(-8)(-3)$ **d.** $(7)(-6)$

If you need more practice, do the following Additional Practice Exercises.

Additional Practice Exercise 5   Find the following products:

a. $5 \cdot 3$ **b.** $(-6)(-4)$ **c.** $(8)(-3)$ **d.** $7(-8)$

If more than two numbers are multiplied, remember to multiply in order from left to right unless the numbers are grouped by using parentheses or brackets.

Example 6   Find the following products:

a. $4(-3)(5) =$ Multiply 4 and -3.
$(-12)(5) =$ Multiply -12 and 5.
-60 Product.

b. $(-5)(-2)(4) =$ Multiply -5 and -2.
$(10)(4) =$ Multiply 10 and 4.
40 Product.

c. $6(-4)(-5)(-3) =$ Multiply 6 and -4.
$(-24)(-5)(-3) =$ Multiply -24 and -5.
$(120)(-3) =$ Multiply 120 and -3.
-360 Product.

d. $(-2)(-3)(-5)(-1) =$ Multiply -2 and -3.
$6(-5)(-1) =$ Multiply 6 and -5.
$(-30)(-1) =$ Multiply -30 and -1.
30 Product.

By looking closely at Example 6, you may discover an easy way to determine the sign of the product in advance.

Answers: Practice Exercise 5: 1. 18 **2.** -32 **3.** 24 **4.** -42 **Additional Practice 5: a.** 15 **b.** 24 **c.** -24 **d.** -56

Be Careful

Remember that we multiply in order from left to right. A common error when multiplying signed numbers is to "distribute" the first factor over the remaining factors. For instance, in Example 6a,

$$4(-3)(5) \neq (-12)(20)$$

Rule: Products with Negative Numbers

If a product has an even number of negative signs, the product is positive. If there is an odd number of negative signs, the product is negative.

Remember, the even numbers are 0, 2, 4, 6, . . . and the odd numbers are 1, 3, 5, ...

PRACTICE EXERCISE 6

Find the following products:

a. $(6)(4)(-3)$ **b.** $(5)(-4)(-3)$ **c.** $(-3)(-4)(6)(-2)$ **d.** $(-3)(-1)(-4)(-5)$

If you need more practice, do the following Additional Practice Exercises.

Additional Practice Exercise 6   Find the following products:

a. $(4)(-3)(6)$ **b.** $(-2)(7)(-3)$ **c.** $(-3)(-2)(-5)(4)$ **d.** $(-3)(-1)(-3)(-5)$

Now that we know how to multiply signed numbers, we can raise signed numbers to powers. Recall from Section 1.1 that an exponent tells us how many times the base is used as a factor.

Calculator Exploration Activity (Optional)

Use a calculator to evaluate the expressions listed. If your calculator does not have parentheses, be very careful.

Column A	Column B	Column C*
$(-5)^2 =$	$(-5)^3 =$	$-5^2 =$
$(-3)^4 =$	$(-3)^5 =$	$-3^4 =$
$(-2)^6 =$	$(-2)^7 =$	$-2^5 =$
$(-1)^8 =$	$(-1)^9 =$	$-1^9 =$

1. Based on your answers to column A, is the result of raising a negative number to an even power positive or negative?

2. Based on your answers to column B, is the result of raising a negative number to an odd power positive or negative?

3. Based on your answers to column C, is the negative of a positive number raised to any positive integer power positive or negative?

*Some calculators without parentheses give incorrect answers to some of these.

We will now confirm these results mathematically.

OBJECTIVE Raising signed numbers to powers

In Example 7, look for a relationship between the exponent and the sign of the answer.

Example 7   **Evaluate the following:**

a. $(-3)^2 = (-3)(-3) = 9$

b. $(-5)^3 = (-5)(-5)(-5) = (25)(-5) = -125$

c. $(-1)^4 = (-1)(-1)(-1)(-1) = 1(-1)(-1) = (-1)(-1) = 1$

d. $(-2)^5 = (-2)(-2)(-2)(-2)(-2) = 4(-2)(-2)(-2)$
$= (-8)(-2)(-2) = (16)(-2) = -32$

Answers: Practice Exercise 6: a. -72 **b.** 60 **c.** -144 **d.** 60   **Additional Practice 6: a.** -72 **b.** 42 **c.** -120 **d.** 45   **Calculator Exploration Activity: 1.** positive **2.** negative **3.** negative

Since the exponent indicates how many times the base is used as a factor, the procedure for finding the sign of a product applies to finding the sign of a negative number raised to a power.

Note

Expressions of the form $-a^n$, with $a > 0$, are always negative.

Rule: Raising a Negative Number to a Power

If a negative number is raised to an even power, the answer is positive. If a negative number is raised to an odd power, the answer is negative.

Note

Another way of thinking of -3^2 is $-(3^2) = -9$.

There is a great deal of difference between $(-3)^2$ and -3^2. $(-3)^2$ is the square of -3, and -3^2 is the opposite of 3 squared. The exponent applies *only* to the symbol immediately preceding it, so the "$-$" is not part of the base unless we use parentheses. Consequently, $(-3)^2 = (-3)(-3) = 9$. But in the expression -3^2, the base of the exponent is 3 and not -3. You may think of -3^2 as $-1 \cdot 3^2$, so the meaning is $-1 \cdot 3 \cdot 3 = -3 \cdot 3 = -9$. **The only way to raise a negative number to a power is to put the negative number *inside* parentheses and the power *outside* the parentheses, as in $(-3)^2$.**

Example 8 **Evaluate the following:**

a. $(-2)^2 = (-2)(-2) = 4$

b. $-2^2 = -1 \cdot 2 \cdot 2 = -2 \cdot 2 = -4$

c. $-8^2 = -1 \cdot 8 \cdot 8 = -8 \cdot 8 = -64$

d. $(-8)^2 = (-8)(-8) = 64$

e. $(-3)^3 = (-3)(-3)(-3) = 9(-3) = -27$

f. $-3^3 = -1 \cdot 3^3 = -1 \cdot 3 \cdot 3 \cdot 3 = -3 \cdot 3 \cdot 3 = -9 \cdot 3 = -27$

NOTE Another way of thinking of parts b, c, and f are as follows: $-2^2 = -(2^2)$, $-8^2 = -(8^2)$ and $-3^3 = -(3^3)$.

PRACTICE EXERCISE 8

Evaluate the following:

a. $(-4)^2$ **b.** $(-2)^3$ **c.** $(-4)^4$

d. -2^4 **e.** $(-5)^2$ **f.** -5^2

If you need more practice, do the following Additional Practice Exercises.

Additional Practice Exercise 8 Evaluate the following:

a. $(-6)^2$ **b.** $(-4)^3$ **c.** $(-3)^4$

d. -2^5 **e.** $(-9)^2$ **f.** -4^2

In problems involving both multiplication and powers, remember to raise to powers before multiplying.

Example 9   **Find the following products:**

a. $(-2)^2(3)^3 =$ Square -2 and cube 3.

$(4)(27) =$ Multiply 4 and 27.

108 Product.

b. $(-3)^3(-1)^2 =$ Cube -3 and square -1.

$(-27)(1) =$ Multiply -27 and 1.

-27 Product.

c. $(-5)^2(-4)^3 =$ Square -5 and cube -4.

$(25)(-64) =$ Multiply 25 and -64.

-1600 Product.

PRACTICE EXERCISE 9

Find the following products:

a. $(-3)^2(2)^2$

b. $(-1)^3(-2)^3$

c. $(-4)^2(-5)^3$

If you need more practice, do the following Additional Practice Exercises.

Additional Practice Exercise 9   Find the following products:

a. $(-2)^2(4)^2$

b. $(-4)^2(-5)^2$

c. $(4)^2(-3)^3$

When evaluating expressions containing variables, remember to use parentheses when substituting a value for the variable.

Example 10   **Evaluate the following when $x = 3$ and $y = -2$:**

a. $5xy =$ Substitute 3 for x and -2 for y.

$5(3)(-2) =$ Multiply from left to right.

-30 Answer.

b. $-4x^2y^3 =$ Substitute 3 for x and -2 for y.

$-4(3)^2(-2)^3 =$ Raise to powers.

$-4(9)(-8) =$ Multiply from left to right.

288 Answer.

PRACTICE EXERCISE 10

Evaluate the following when $a = -1$ and $b = 5$:

a. $3ab$

b. $-4a^3b^3$

Additional Practice Exercise 10   Evaluate the following when $m = 2$ and $n = -3$:

a. $2mn$

b. $-4mn^2$

To indicate multiplication, we use *times*, *product*, *multiplied by*, or *of* when using *decimals*, *percents* or *fractions*, except for special cases like *twice* for two times and *thrice* for three times. For example, $4 \cdot 3$ can be written as "4 times 3" or "the product of 4 and 3" or "3 multiplied by 4."

Answers: Practice Exercise 9: a. 36 **b.** 8 **c.** -2000 **Additional Practice 9: a.** 64 **b.** 400 **c.** -432
Practice Exercise 10: a. -15 **b.** 500   **Additional Practice 10: a.** -12 **b.** -72

Example 11 **Translate the following English phrases into math and evaluate:**

a. The product of 8 and 5 increased by 4.

Solution

The product of 8 and 5 means $8 \cdot 5$. Since we need to increase the product by 4, we add 4 to $8 \cdot 5$, giving

$8 \cdot 5 + 4 =$	Multiply before adding.
$40 + 4 =$	Add.
$44.$	Answer.

NOTE $8 \cdot 5 + 4$ is the same as $(8 \cdot 5) + 4$.

b. 6 decreased by 5 times -4.

Solution

5 times -4 means $5(-4)$. Since 6 is decreased by $5(-4)$, we subtract $5(-4)$ from 6, giving

$6 - 5(-4) =$	Multiply before adding.
$6 + 20 =$	Add.
$26.$	Answer.

c. The product of 4 and -6 added to the product of -3 and -5.

Solution

The product of 4 and -6 means $4(-6)$, and the product of -3 and -5 means $(-3)(-5)$. Consequently, $4(-6)$ is added to $(-3)(-5)$, giving

$(-3)(-5) + 4(-6) =$	Multiply before adding.
$15 - 24 =$	Add.
$-9.$	Answer.

PRACTICE EXERCISE 11

Translate the following phrases into math and evaluate:

a. The product of 5 and -2 increased by 7.

b. -5 decreased by 5 times 9.

c. The product of 5 and -4 subtracted from the product of -6 and -4.

If you need more practice, do the following Additional Practice Exercises:

Additional Practice Exercise 11 Translate the following phrases into math and evaluate:

a. The product of 6 and -12 added to 7.

b. -7 decreased by 3 times 7.

c. The product of -4 and -9 added to the product of 9 and -5.

OBJECTIVE Identifying properties of multiplication

Earlier, the commutative property of multiplication was mentioned. Like addition, multiplication has a number of properties that are summarized in the following box:

> **Definition** Properties of Multiplication
>
> **Commutative**
>
> $ab = ba$ Multiplication may be performed in *any* order.
>
> **Associative**
>
> $a(bc) = (ab)c$ Multiplication may be grouped in *any* manner.
>
> *(Continued)*

Answers: **Practice Exercise 11:** **a.** $5(-2) + 7 = -3$ **b.** $-5 - 5 \cdot 9 = -50$ **c.** $-6(-4) - 5(-4) = 44$
Additional Practice 11: **a.** $7 + 6(-12) = -65$ **b.** $-7 - 3 \cdot 7 = -28$ **c.** $9(-5) + (-4)(-9) = -9$

Identity

The number 1 is the identity for multiplication, since $a \cdot 1 = 1 \cdot a = a$ for any number a.

Inverse

For any number $a \neq 0$, there exists a number $\dfrac{1}{a}$ such that $a \cdot \dfrac{1}{a} = 1$.

Multiplication by 0

$a \cdot 0 = 0 \cdot a = 0$ for all numbers a.

Example 12	**Identify the property illustrated by each of the following:**

a. $(-3)(2) = (2)(-3)$ Order is different. Commutative of multiplication.

b. $2[(-3)(-5)] = [2(-3)](-5)$ Grouping is different. Associative of multiplication.

c. $2(-3 + 5) = (-3 + 5)2$ Order is different. Commutative of multiplication.

d. $(-4)1 = -4$ Identity for multiplication.

e. $5 \cdot \dfrac{1}{5} = 1$ Inverse for multiplication.

f. $5 \cdot \dfrac{1}{5} = \dfrac{1}{5} \cdot 5$ Order is different. Commutative of multiplication.

g. $5 \cdot 0 = 0$ Multiplication by 0.

PRACTICE EXERCISE 12

Identify the property illustrated by each of the following:

a. $(5)(-6) = (-6)(5)$ **b.** $-3[2(-4)] = [(-3)2](-4)$

c. $-3(5 + 6) = (5 + 6)(-3)$ **d.** $1(-7) = -7$

e. $6 \cdot \dfrac{1}{6} = 1$ **f.** $6 \cdot \dfrac{1}{6} = \dfrac{1}{6} \cdot 6$ **g.** $8 \cdot 0 = 0$

For Extra Help

Exercise Set 2.1   MyMathLab®

Find the following products. (See Examples 5–9.)

1. $6(-4)$ **2.** $(-12)(8)$ **3.** $(-6)(4)$ **4.** $(14)(-5)$

5. $(-6)(-4)$ **6.** $(-8)(-9)$ **7.** $(-7)(3)$ **8.** $(-5)(8)$

9. $(-11)(-4)$ **10.** $(-12)(-3)$ **11.** $(-2.5)(-1.4)$ **12.** $(-6.2)(-4.3)$

Answers: Practice Exercise 12: a. commutative of multiplication **b.** associative of multiplication **c.** commutative of multiplication **d.** identity for multiplication **e.** inverse of multiplication **f.** commutative of multiplication **g.** multiplication by 0

13. $(-1.6)(5.7)$ **14.** $(-2.1)(7.6)$ **15.** $(-1)(3)(-5)$ **16.** $(-4)(-1)(6)$

17. $(4)(-3)(4)$ **18.** $(5)(-4)(6)$ **19.** $(-5)(-3)(-6)$ **20.** $(-4)(-7)(-10)$

21. $(-2)(4)(-3)(-5)$ **22.** $(-6)(4)(-3)(-1)$ **23.** $(-2)(-4)(-5)(-3)$ **24.** $(-2)(-5)(-8)(-2)$

25. -9^2 **26.** $(-9)^2$ **27.** $(-11)^2$ **28.** -11^2

29. -5^4 **30.** $(-5)^4$ **31.** $(3)^2(-2)^3$ **32.** $(-4)^2(-2)^3$

33. $(-2)^2(3)^3$ **34.** $(-4)^2(3)^2$ **35.** $(-1)^5(-5)^3$ **36.** $(-6)^2(-2)^3$

Challenge Exercises (37–44)

37. $\left(\dfrac{2}{3}\right)\left(-\dfrac{4}{5}\right)$ **38.** $\left(-\dfrac{3}{4}\right)\left(\dfrac{5}{7}\right)$ **39.** $\left(-\dfrac{8}{15}\right)\left(-\dfrac{7}{5}\right)$ **40.** $\left(-\dfrac{2}{3}\right)\left(-\dfrac{2}{5}\right)$

41. $\left(\dfrac{3}{4}\right)\left(-\dfrac{1}{5}\right)\left(-\dfrac{3}{2}\right)$ **42.** $\left(-\dfrac{1}{3}\right)\left(-\dfrac{2}{5}\right)\left(\dfrac{4}{7}\right)$ **43.** $\left(-\dfrac{1}{4}\right)\left(-\dfrac{5}{2}\right)\left(-\dfrac{9}{8}\right)$ **44.** $\left(-\dfrac{5}{2}\right)\left(-\dfrac{3}{8}\right)\left(-\dfrac{5}{18}\right)$

Calculator Exercises (45–48, Optional)

Find each product by using a calculator. The type of calculator you use determines how you will enter a negative number. On many calculators, you first enter the number and then press the $\boxed{+/-}$ *key. This key changes the sign of the number currently showing on the display. Do not attempt to use the subtraction key. On most graphing calculators, there is a special key indicated by* $\boxed{(-)}$ *that is used to indicate a negative number.*

45. $(-5.74)(8.65)$ **46.** $(98.67)(-6.45)$ **47.** $(-89.54)(-67.34)$ **48.** $(-8.546)(-6.768)$

Evaluate each of the following for $x = -2$ *and* $y = -4$. *(See Example 10.)*

49. $3x$ **50.** $4y$ **51.** $-8x$ **52.** $-2y$

53. $6x^2$ **54.** $3y^2$ **55.** $-5x^3$ **56.** $-4y^3$

57. x^2y^3 **58.** x^3y^2 **59.** x^2y^2 **60.** x^3y^3

Write an expression for each of the following and evaluate. (See Example 11.)

61. The product of 5 and -3 increased by 6

62. The product of -4 and 7 increased by 9

63. -5 decreased by the product of 4 and -7

64. -3 decreased by the product of -5 and 3

65. The product of 3 and the square of -4

66. The product of -5 and the square of 5

67. The product of -6 and -3 decreased by 8

68. The product of -3 and -5 decreased by 4

69. The product of 4 and -2 added to the product of -5 and 4

70. The product of 8 and -3 added to the product of -5 and 7

71. The product of 4 and −3 decreased by −5 times −7

72. The product of −8 and −3 decreased by −6 times −9

73. Twice the product of 3 and −2 added to the product of −8 and 4

74. Twice the product of −7 and 11 subtracted from the product of 10 and −7

75. One-half of the sum of −8 and 6

76. One-third of the difference of 7 and −5

Identify the property illustrated by each of the following. (See Example 12.)

77. $(-2)\big[(3)(-4)\big] = \big[(-2)(3)\big](-4)$

78. $\big[(-4)(9)\big](-5) = (-4)\big[(9)(-5)\big]$

79. $3(-4 + 6) = (-4 + 6)(3)$

80. $(-4)(5 - 7) = (5 - 7)(-4)$

81. $(-9) \cdot 1 = -9$

82. $1 \cdot (-6) = -6$

83. $(-4)(-7) + 2 = (-7)(-4) + 2$

84. $5 + (-5)(8) = 5 + (8)(-5)$

85. $-3 \cdot \dfrac{-1}{3} = 1$

86. $5 \cdot \dfrac{1}{5} = 1$

87. $4 \cdot \dfrac{1}{4} = \dfrac{1}{4} \cdot 4$

88. $-\dfrac{1}{6} \cdot (-6) = -6\left(-\dfrac{1}{6}\right)$

89. $0 \cdot 7 = 0$

90. $9 \cdot 0 = 0$

Complete each of the following by using the given property of multiplication. (See Example 12.)

91. $3\big[(-9)(-7)\big] =$ _____ associative property

92. $6\big[(-2)(3)\big] =$ _____ associative property

93. $3(-7) =$ _____ commutative property

94. $(-5)(8) =$ _____ commutative property

95. $4(-9 + 2) =$ _____ distributive property

96. $5(-2 + 9) =$ _____ distributive property

97. $4(-9 + 2) =$ _____ commutative property

98. $5(-2 + 9) =$ _____ commutative property

99. $(-2) \cdot 1 =$ _____ identity for multiplication

100. $(-5) \cdot 1 =$ _____ identity for multiplication

101. $8 \cdot \dfrac{1}{8} =$ _____ inverse for multiplication

102. $4 \cdot \dfrac{1}{4} =$ _____ inverse for multiplication

103. $2 \cdot 0 =$ _____ multiplication property of 0

104. $0 \cdot 5 =$ _____ multiplication property of 0

Challenge Exercises (105–108)

Evaluate each of the following:

105. $-2 \cdot 3^2 + 5 \cdot (-4)^2 \div (5)(2)$

106. $4 \cdot (-6)^2 \div (3)(2)^3$

107. $-4\big[-5 - (-6)\big]^2 + (-7)$

108. $-6 - 5\big[-3^2 - (-6)\big]^2$

Answer the following:

109. When a cold front moved through, the temperature dropped at an average rate of 4° per hour. Using a signed number, find the change in temperature after three hours.

110. On a sunny day in Florida, the temperature rose at an average rate of 3.5° per hour from 10:00 A.M. to 2:00 P.M. Using a signed number, find the change in temperature from 10:00 A.M. to 2:00 P.M.

111. On a steep mountain road, the elevation dropped 7 feet per 100 feet. Using a signed number, find the change in elevation after 1000 feet.

112. On a steep roof, there is a vertical drop of 2 feet for every horizontal change of 10 feet. Using a signed number, find the vertical change for a horizontal change of 25 feet.

113. On Black Monday, a stock dropped at an average rate of $\frac{3}{4}$ point per hour. Using a signed number, find the drop after 8 hours.

114. A feather is falling at a rate of $\frac{5}{8}$ feet per second. Using a signed number, find the number of feet that it falls in 16 seconds.

Writing Exercises (115 and 116)

115. If a product has 16 positive factors and 23 negative factors, is the product positive or negative? Why?

116. Is $(-1)^{48}$ positive or negative? Why? Is $(-1)^{97}$ positive or negative? Why?

Critical-Thinking Exercises (117–121)

117. Why is a product that contains an even number of negative signs positive?

118. Why is a product that contains an odd number of negative signs negative?

119. Compare/contrast the commutative and associative properties of multiplication.

120. Why does 0 not have a multiplicative inverse?

121. Why is $-x^2$ always nonpositive?

Group Project

122. Write an application exercise similar to Exercises 109–114. Exchange your exercise with another group and solve their exercise.

Multiplication Laws of Exponents

OBJECTIVES *When you complete this section, you will be able to:*

A Simplify expressions of the form $a^m \cdot a^n$.

B Simplify expressions of the form $(ab)^n$.

C Simplify expressions of the form $(a^m)^n$.

D Simplify expressions involving two or more of the preceding forms a through c.

PREREQUISITE SKILLS *Before beginning this section, you should be able to:*

a. Add and multiply whole numbers (Sections R.2 and R.3).

b. Add like terms (Section 1.7).

GETTING READY FOR SECTION 2.2

Add or multiply the following whole numbers:

1. $5 + 7$ **2.** $3 + 5 + 2$ **3.** $6 \cdot 3$ **4.** $3 \cdot 7$

Add the following like terms:

5. $6x^3 + 8x^3$ **6.** $14a^2b^3 - 20a^2b^3$

Introduction

Previously, we raised whole numbers and integers to powers. In this section, we will develop rules to simplify products involving exponents whose bases are the same, to raise a number to a power to another power, and to raise a product to a power. All of these rules depend on the definition of a positive integer exponent. Remember, if n is a positive integer, a^n means the product of n factors of a.

Calculator Exploration Activity 1 (Optional)

Use a calculator to evaluate each of the following columns:

Column A	Column B
1. $2^2 \cdot 2^3 =$	$2^5 =$
2. $3^3 \cdot 3^4 =$	$3^7 =$
3. $4^2 \cdot 4^4 =$	$4^6 =$

By looking at the corresponding lines in columns A and B, answer the following:

1. If two exponential expressions with the same base are multiplied, leave the _____ unchanged and _____ the exponents.

We now give a mathematical justification for the preceding observation.

OBJECTIVE Simplifying expressions of the form $a^m \cdot a^n$

By the definition of exponents,

$$(a^2)(a^3) = (a \cdot a)(a \cdot a \cdot a) = a \cdot a \cdot a \cdot a \cdot a = a^5.$$

2 factors of a + 3 factors of a = 5 factors of a

Also, $(a^4)(a^3) = (a \cdot a \cdot a \cdot a)(a \cdot a \cdot a) = a \cdot a \cdot a \cdot a \cdot a \cdot a \cdot a = a^7.$

4 factors of a + 3 factors of a = 7 factors of a

Based on the two preceding examples, we generalize to the first law of exponents for multiplication.

Answers: Getting Ready: 1. 12 **2.** 10 **3.** 18 **4.** 21 **5.** $14x^3$ **6.** $-6a^2b^3$ **Calculator Exploration Activity: 1.** base, add

> ### Rule: First Law of Exponents—Product Rule
>
> $$a^m \cdot a^n = a^{m+n}$$
>
> To multiply two expressions with the same base, leave the base unchanged and add the exponents.

The preceding property is easily extended to more than two expressions. For example, $x^3 \cdot x^4 \cdot x^2 = x^{3+4+2} = x^9$. Remember, if no exponent is given, the exponent is understood to be 1.

| **Example 1** | **Find the products. Express the answers in exponential form.** |

Be Careful

When multiplying exponential expressions with constant bases as in parts b and c, a common error is to multiply the bases as well as add the powers. The product of $3^2 \cdot 3^4$ means $(3 \cdot 3)(3 \cdot 3 \cdot 3 \cdot 3) = 3^6$, not 9^6.

a. $x^3 \cdot x^5 =$ Apply $a^m \cdot a^n = a^{m+n}$.
$x^{3+5} =$ Add the exponents.
x^8 Therefore, $x^3 \cdot x^5 = x^8$.

b. $2^4 \cdot 2^2 =$ Apply $a^m \cdot a^n = a^{m+n}$.
$2^{4+2} =$ Add the exponents. Note: The base remains 2.
2^6 Therefore, $2^4 \cdot 2^2 = 2^6$.

c. $3 \cdot 3^4 =$ Apply $a^m \cdot a^n = a^{m+n}$.
$3^{1+4} =$ Add the exponents. Remember, 3 means 3^1.
3^5 Therefore, $3 \cdot 3^4 = 3^5$.

d. $a^5 \cdot a^4 \cdot a =$ Apply $a^m \cdot a^n = a^{m+n}$.
$a^{5+4+1} =$ Add the exponents.
a^{10} Therefore, $a^5 \cdot a^4 \cdot a = a^{10}$.

PRACTICE EXERCISE 1

Find the following products:

a. $x^2 \cdot x^4$　　　　**b.** $3^2 \cdot 3^6$　　　　**c.** $5^3 \cdot 5$　　　　**d.** $b^4 \cdot b^3 \cdot b^5$

If you need more practice, do the following Additional Practice Exercises.

Additional Practice Exercise 1　Find the following products:

a. $z^4 \cdot z^3$　　　　**b.** $2^4 \cdot 2^2$　　　　**c.** $8 \cdot 8^5$　　　　**d.** $c^3 \cdot c^2 \cdot c^5$

If the terms have numerical coefficients, we use the commutative and associative properties to rearrange the order of the factors and multiply the coefficients and variable factors with the same bases separately.

| **Example 2** | **Find the following products:** |

a. $(2x^2)(3x^3) =$ Rearrange the order of the factors.
$(2 \cdot 3)(x^2 \cdot x^3) =$ Multiply 2 and 3. Apply $a^m \cdot a^n = a^{m+n}$.
$6x^{2+3} =$ Add the exponents.
$6x^5$ Therefore, $(2x^2)(3x^3) = 6x^5$.

b. $(-4x^4y^2)(6x^3y^5) =$ Rearrange the order of the factors.
$(-4 \cdot 6)(x^4 \cdot x^3)(y^2 \cdot y^5) =$ Multiply -4 and 6. Apply $a^m \cdot a^n = a^{m+n}$.
$-24x^{4+3}y^{2+5} =$ Add the exponents.
$-24x^7y^7$ Therefore, $(-4x^4y^2)(6x^3y^5) = -24x^7y^7$.

Answers: Practice Exercise 1: a. x^6 **b.** 3^8 **c.** 5^4 **d.** b^{12} **Additional Practice 1: a.** z^7 **b.** 2^6 **c.** 8^6 **d.** c^{10}

Be Careful

Do not multiply the bases!

c. $(3^2 \cdot 4^4)(3^5 \cdot 4^2) =$ Rearrange the order of the factors.

$(3^2 \cdot 3^5)(4^4 \cdot 4^2) =$ Apply $a^m \cdot a^n = a^{m+n}$.

$3^{2+5} \cdot 4^{4+2} =$ Add the exponents.

$3^7 \cdot 4^6$ Therefore, $(3^2 \cdot 4^4)(3^5 \cdot 4^2) = 3^7 \cdot 4^6$.

PRACTICE EXERCISE 2

Find the products:

a. $(4x^4)(3x^5)$ **b.** $(5y^2z^3)(-6y^3z^5)$ **c.** $(2^4 \cdot 5^3)(2^2 \cdot 5^4)$

If you need more practice, do the following Additional Practice Exercises.

Additional Practice Exercises 2   Find the following products:

a. $(5x^3)(2x^5)$ **b.** $(-4w^3z)(3w^5z^3)$ **c.** $(3^4 \cdot 6^3)(3^3 \cdot 6^4)$

Calculator Exploration Activity 2 (Optional)

Evaluate each equation by using a calculator. Be careful if your calculator does not have parentheses.

Column A	Column B
1. $(2 \cdot 3)^3 =$	$2^3 \cdot 3^3 =$
2. $(3 \cdot 5)^3 =$	$3^3 \cdot 5^3 =$
3. $(3 \cdot 4)^2 =$	$3^2 \cdot 4^2 =$

By looking at the corresponding lines of columns A and B, answer the following:

2. Raising a product to a power is the same as raising each _____ to the _____.

We now provide a mathematical justification for the preceding observation.

OBJECTIVE Simplifying expressions of the form $(ab)^n$

In an expression of the form $(ab)^3$, the base of the exponent 3 is the product ab. Consequently, $(ab)^3 = (ab)(ab)(ab)$. Using the commutative and associative properties of multiplication, $(ab)(ab)(ab) = (a \cdot a \cdot a)(b \cdot b \cdot b) = a^3b^3$. Therefore, $(ab)^3 = a^3b^3$. Likewise, $(2y)^4 = (2y)(2y)(2y)(2y) = (2 \cdot 2 \cdot 2 \cdot 2)(y \cdot y \cdot y \cdot y) = 2^4y^4 = 16y^4$. These examples suggest the following generalization:

Rule: Second Law of Exponents—Power of a Product

$(a \cdot b)^n = a^nb^n$. To raise a product to a power, raise each factor to the power.

The preceding property can be extended to include more than two factors. For example, $(2ab)^4 = (2ab)(2ab)(2ab)(2ab) = (2 \cdot 2 \cdot 2 \cdot 2)(a \cdot a \cdot a \cdot a)(b \cdot b \cdot b \cdot b) = 2^4a^4b^4 = 16a^4b^4$.

Example 3   **Rewrite the following products:**

a. $(xy)^2 = x^2y^2$

b. $(3y)^4 = 3^4y^4 = 81y^4$ *Do not forget to raise 3 to the fourth power!*

Answers: Practice Exercise 2: a. $12x^9$ **b.** $-30y^5z^8$ **c.** $2^6 \cdot 5^7$ **Additional Practice 2: a.** $10x^8$ **b.** $-12w^8z^4$ **c.** $3^7 \cdot 6^7$
Calculator Exploration Activity: 2. factor, power

c. $(-4z)^2 = (-4)^2z^2 = 16z^2$

d. $(abc)^5 = a^5b^5c^5$

e. $(-4ab)^3 = (-4)^3a^3b^3 = -64a^3b^3$

PRACTICE EXERCISE 3

Rewrite the following products:

a. $(ab)^5$ **b.** $(5x)^3$ **c.** $(-6x)^2$ **d.** $(xyz)^6$ **e.** $(3ab)^4$

If you need more practice, do the following Additional Practice Exercises.

Additional Practice Exercise 3   Rewrite the following products:

a. $(ef)^7$ **b.** $(4w)^3$ **c.** $(-5y)^4$ **d.** $(rst)^8$ **e.** $(-5st)^3$

Calculator Exploration Activity 3 (Optional)

Evaluate each of the following columns by using a calculator.

Column A	Column B
1. $(2^2)^3 =$	$2^6 =$
2. $(3^4)^2 =$	$3^8 =$
3. $(2^3)^3 =$	$2^9 =$

By comparing the corresponding lines in columns A and B, answer the following:

3. To raise a number to a power to another power, leave the _____ unchanged and _____ the exponents.

We now provide a mathematical justification for the preceding observation.

OBJECTIVE Simplifying expressions of the form $(a^m)^n$

In the expression $(a^2)^3$, the base of the exponent 3 is a^2. Consequently, $(a^2)^3 = a^2 \cdot a^2 \cdot a^2$. From the product rule, we know $a^2 \cdot a^2 \cdot a^2 = a^{2+2+2} = a^6$. Since multiplication is a shortcut for repeated additions of the same number, $2 + 2 + 2 = 3 \cdot 2$. Therefore, $(a^2)^3 = a^{(3)(2)} = a^6$. In the same manner, $(b^3)^4 = b^3 \cdot b^3 \cdot b^3 \cdot b^3 = b^{3+3+3+3} = b^{(4)(3)} = b^{12}$. These examples can be generalized into the third law of exponents.

> **Rule: Third Law of Exponents—Power to a Power**
>
> $$(a^m)^n = a^{mn}$$
>
> To raise a number to a power to another power, leave the base unchanged and multiply the powers.

Example 4   Simplify the following:

a. $(x^2)^4 =$ Apply $(a^m)^n = a^{mn}$.
$x^{(2)(4)} =$ Multiply the exponents.
x^8 Therefore, $(x^2)^4 = x^8$.

b. $(2^3)^4 =$ Apply $(a^m)^n = a^{mn}$.
$2^{(3)(4)} =$ Multiply the exponents.
2^{12} Therefore, $(2^3)^4 = 2^{12}$.

c. $(y^4)^2 =$ Apply $(a^m)^n = a^{mn}$.
$y^{(4)(2)} =$ Multiply the exponents.
y^8 Therefore, $(y^4)^2 = y^8$.

Answers: Practice Exercise 3: a. a^5b^5 **b.** $125x^3$ **c.** $36x^2$ **d.** $x^6y^6z^6$ **e.** $81a^4b^4$ **Additional Practice 3: a.** e^7f^7 **b.** $64w^3$ **c.** $625y^4$ **d.** $r^8s^8t^8$ **e.** $-125s^3t^3$ **Calculator Exploration Activity: 3.** base, multiply.

PRACTICE EXERCISE 4

Simplify the following:

a. $(a^7)^2$

b. $(4^3)^5$

c. $(z^4)^5$

If you need more practice, do the following Additional Practice Exercises.

Additional Practice Exercise 4   Simplify the following:

a. $(q^2)^5$

b. $(5^4)^4$

c. $(w^3)^6$

OBJECTIVE **D**   Simplifying expressions involving more than one law of exponents

Often it is necessary to simplify expressions that involve more than one of the laws of exponents. Be careful in distinguishing which rule applies to each situation.

| **Example 5** | **Simplify each of the following. Express answers in exponential form.** |

a. $(x^2y^3)^3 =$ Apply $(ab)^n = a^nb^n$.
$(x^2)^3(y^3)^3 =$ Apply $(a^m)^n = a^{mn}$.
x^6y^9 Answer.

b. $(2x^3y)^4 =$ Apply $(ab)^n = a^nb^n$.
$2^4(x^3)^4(y^1)^4 =$ Apply definition of exponents and $(a^m)^n = a^{mn}$.
$16x^{12}y^4$ Answer.

c. $(2xy)^2(3xy)^3 =$ Apply $(ab)^n = a^nb^n$.
$2^2 \cdot x^2 \cdot y^2 \cdot 3^3 \cdot x^3 \cdot y^3 =$ Regroup and apply the definition of exponent.
$(4 \cdot 27)(x^2 \cdot x^3)(y^2 \cdot y^3) =$ Apply $a^m \cdot a^n = a^{m+n}$.
$108x^5y^5$ Answer.

d. $(x^3y^2)^2(x^4y^3)^2 =$ Apply $(ab)^n = a^nb^n$.
$(x^3)^2(y^2)^2(x^4)^2(y^3)^2 =$ Apply $(a^m)^n = a^{mn}$.
$x^6y^4 \cdot x^8y^6 =$ Regroup.
$(x^6 \cdot x^8)(y^4 \cdot y^6) =$ Apply $a^m \cdot a^n = a^{m+n}$.
$x^{14}y^{10}$ Answer.

PRACTICE EXERCISE 5

Simplify the following:

a. $(x^4y^2)^4$

b. $(-4a^3b)^2$

c. $(3ab)^3(2ab)^4$

d. $(a^3b^4)^2(a^2b^4)^3$

If you need more practice, do the following Additional Practice Exercises.

Additional Practice Exercise 5   Simplify the following:

a. $(w^4z^3)^2$

b. $(-2x^4y^4)^4$

c. $(2cd)^2(3cd)^3$

d. $(xy^5)^3(x^3y^5)^4$

Answers: Practice Exercise 4: a. a^{14} **b.** 4^{15} **c.** z^{20} **Additional Practice 4: a.** q^{10} **b.** 5^{16} **c.** w^{18} **Practice Exercise 5: a.** $x^{16}y^8$ **b.** $16a^6b^2$
c. $432a^7b^7$ **d.** $a^{12}b^{20}$ **Additional Practice 5: a.** w^8z^6 **b.** $16x^{16}y^{16}$ **c.** $108c^5d^5$ **d.** $x^{15}y^{35}$

Sometimes, it is necessary to add like terms after applying one or more laws of exponents. Remember, when adding like terms we add the coefficients only. *We do not add the exponents.* Also remember to follow the order of operations.

| **Example 6** | **Simplify the following:** |

a. $(2x)(4x) + (6x)(2x) =$ Multiply before adding. Use $a^m \cdot a^n = a^{m+n}$.
$8x^2 + 12x^2 =$ Add like terms. Add coefficients only.
$20x^2$ Answer.

b. $(2x)^3 + (3x)(-2x^2) =$ Simplify $(2x)^3$ and multiply $(3x)(-2x^2)$.
$8x^3 - 6x^3 =$ Add like terms. Add coefficients only.
$2x^3$ Answer.

c. $(-3x^2y)(2xy) + (4xy^2)(3x^2) =$ Multiply before adding. Use $a^m \cdot a^n = a^{m+n}$.
$-6x^3y^2 + 12x^3y^2 =$ Add like terms. Add coefficients only.
$6x^3y^2$ Answer.

PRACTICE EXERCISE 6

Simplify the following:

a. $(4y)(2y) + (3y)(-5y)$ **b.** $(3r)^4 + (4r^2)(-10r^2)$ **c.** $(5a^2b^2)(-2ab^2) + (4ab^3)(3a^2b)$

If you need more practice, do the following Additional Practice Exercises.

Additional Practice Exercise 6   Simplify the following:

a. $(-2x)(4x) + (4x)(3x)$ **b.** $(4x)^2 + (5x)(-2x)$ **c.** $(6pq)(4p^2q) + (5p^2q^2)(3p)$

Study Tip 7 Preparing for Class

How well you prepare for class determines how much you will get from that class. Organize your day and set aside at least 1 hour each day to study mathematics. It is best to study in concentrated short intervals of approximately 30 minutes, with 5- to 10-minute breaks in between. The following are some things you should do:

1. Read your class notes as soon after class as possible, preferably the same day. Highlight important formulas, statements, and so on.
2. Write any definitions, rules, formulas, or important statements on your review cards. If your instructor went over these things in class, they are important. Check the chapter summary.
3. Read the textbook slowly and carefully with pencil and paper available. Mark down things that you do not understand and come back to them after you have read all of the material. Check the material that was previously covered for something similar.
4. If necessary, use other textbooks, study guides, and online help. Many texts have supplemental study guides that may be purchased in the college bookstore. This textbook is available with online tutoring and many other student resources on its corresponding MyMathLab® Web site. See the preface for details.

5. Do your homework as soon as possible after class is over—most definitely before the next class meeting. Do not skip steps. The reason for a step can often be found in something that you learned previously. Putting each step down on paper will help reinforce the principles being covered and help you remember them. Careless errors often occur when steps are skipped. Review your homework just prior to attending the next class meeting. Make a list of things that you are unsure of and ask your instructor about them.

6. Preview the text to be covered in class the next day by reading the text with pencil in hand. Mark those parts you are unsure of and make a list of questions to ask your instructor if his or her explanation does not fully answer your specific question.

7. If your instructor does not have time in class to answer all of your questions, make an appointment to see him or her during office hours.

Answers: Practice Exercise 6: a. $-7y^2$ **b.** $41r^4$ **c.** $2a^3b^4$ **Additional Practice 6: a.** $4x^2$ **b.** $6x^2$ **c.** $39p^3q^2$

Exercise Set 2.2

For Extra Help

MyMathLab®

Simplify the following by using the laws of exponents and leave answers in exponential form. (See Examples 1–5.)

1. $r^2 \cdot r^6$

2. $s^3 \cdot s^5$

3. $4^3 \cdot 4^4$

4. $5^4 \cdot 5^5$

5. $x^3 \cdot x^4 \cdot x^2$

6. $y^3 \cdot y^6 \cdot y^5$

7. $(4x^5)(5x^4)$

8. $(3y^5)(6y^6)$

9. $(-7a^4)(-3a^4)$

10. $(-6b^6)(5b^6)$

11. $(x^5y^3)(x^3y^4)$

12. $(a^3b^6)(a^7b^3)$

13. $(4a^5b^4)(2a^5b^7)$

14. $(5y^6z^2)(7y^5z^5)$

15. $(ab)^5$

16. $(xy)^7$

17. $(2y)^4$

18. $(3a)^5$

19. $(-4x)^2$

20. $(-7x)^2$

21. $-(5x)^2$

22. $-(8b)^4$

23. $-(-3c)^3$

24. $-(-5a)^3$

25. $(-3b)^3$

26. $(-4n)^4$

27. $(y^3)^6$

28. $(x^4)^5$

29. $(4^5)^3$

30. $(3^4)^2$

31. $(xy)^2(xy)^5$

32. $(ab)^5(ab)^4$

33. $(4x)^5(4x)^3$

34. $(2d)^4(2d)^5$

35. $(c^4d^3)^6$

36. $(a^3b^5)^5$

37. $(4^5a^3)^4$

38. $(3^4x^5)^3$

39. $-(4^4b^5)^5$

40. $-(2^5a^4)^3$

41. $(-6^2b^3)^4$

42. $(-5^3x^2)^4$

43. $(3a^5b^2)^3(2ab^3)^2$

44. $(2x^3y^4)^3(3xy^3)^2$

45. $(4^4b^4)^3(4^2b^4)^4$

46. $(3^2a^4)^3(3^4a^5)^3$

47. $(a^3b^2c^3)^2(a^2b^3c^4)^4$

48. $(x^2y^3z^2)^3(x^4y^4z^2)^3$

Simplify the following. (See Example 6.)

49. $(6a)(-2a) + (7a)(4a)$

50. $(5y)(4y) + (3y)(-6y)$

51. $(3x)^3 + (-4x)(6x^2)$

52. $(2z)^3 + (3z)(-5z^2)$

53. $(-2x)^6 + (5x^3)^2$

54. $(-2n)^4 + (2n^2)^2$

55. $(-6a^3)(2a) - (-4a^2)(-3a^2)$

56. $(3x^2)(-2x) - (-3x)(-5x^2)$

57. $(4m^2n^3)^2 + (-3mn^5)(6m^3n)$

58. $(2x^2y)^2 + (3x^3y)(-4xy)$

Challenge Exercises (59–68)

Simplify the following. (See Examples 1–5.)

59. $a^{5x} \cdot a^{3x}$

60. $x^{2n} \cdot x^{3n}$

61. $a^{3b-2} \cdot a^{2b-1}$

62. $x^{2m+1} \cdot x^{3m-2}$

63. $(2^{3x}a^{2y}b^z)(2^{4x}a^{3y}b^{2z})$

64. $(x^ay^b)(x^{2a}y^{3b})$

65. $(2x^ay^b)^c$

66. $(3a^mb^n)^p$

67. $(b^{2n-1})^3$

68. $(x^{n-1})^2$

Writing Exercises (69–72)

69. Are $(-2)^2$ and -2^2 equal? Explain why or why not.

70. Are $(-2)^3$ and -2^3 equal? Are their meanings the same? Explain why or why not.

71. Are 3^2 and $3 \cdot 2$ equal? Explain why or why not.

72. If we simplify $(2x^3)(3x^3)$, we multiply the coefficients and add the exponents. If we simplify $2x^3 + 3x^3$, we add the coefficients only and do not change the exponents. Why?

Section 2.3 Products of Polynomials

OBJECTIVES *When you complete this section, you will be able to:*

Ⓐ Find the product of a monomial and a polynomial.

Ⓑ Find the product of two polynomials.

PREREQUISITE SKILLS *Before beginning this section, you should be able to:*

a. Apply the distributive property (Section 1.7).

b. Apply $a^m \cdot a^n = a^{m+n}$ (Section 2.2).

c. Multiply integers (Section 2.1).

d. Add like terms (Section 1.7).

e. Find the areas of rectangles and triangles (Section 1.3).

GETTING READY FOR SECTION 2.3

1. Simplify $-3(2 - 5)$ by using the distributive property.

2. Simplify $3a^2b(-2a^4b^3)$ by using $a^m \cdot a^n = a^{m+n}$.

3. Multiply the integers: $4(-6)$.

4. Simplify $3a^2 - 5a - 5 + 2a^2 + 3a - 4$ by adding like terms.

5. Find the area of the triangle whose base is 10 cm and whose height is 7 cm.

Introduction

The laws of exponents developed in the previous section are used extensively in finding the products of polynomials. In showing that $a^m \cdot a^n = a^{m+n}$ and in subsequent examples, we were multiplying monomials by monomials. For example, $(2x^3y^2)(3x^3y^4) = 6x^6y^6$ is the product of two monomials.

OBJECTIVE Finding the product of a monomial and a polynomial

In multiplying a monomial and a polynomial, we use the distributive property, which you will recall is $a(b + c) = ab + ac$ or $(b + c)a = ba + ca$. For example, $3(2 + 4) = 3 \cdot 2 + 3 \cdot 4$. The distributive property is easily extended to include multiplication of a monomial over polynomials of more than two terms. For example, $a(b + c + d) = ab + ac + ad$ and $a(b + c + d + e) = ab + ac + ad + ae$, and so on.

Example 1 | **Find the following products:**

a. $2(x + 4) =$ Apply the distributive property.

$2 \cdot x + 2 \cdot 4 =$ Multiply 2 and 4.

$2x + 8$ These are not like terms. Therefore, this is the product.

b. $-4(x + 6) =$ Apply the distributive property.

$-4 \cdot x + (-4)6 =$ Multiply -4 and 6.

$-4x - 24$ These are not like terms. Therefore, this is the product.

c. $3(4x - 2y) =$ Apply the distributive property.

$3 \cdot 4x - 3 \cdot 2y =$ $3 \cdot 4x = 12x$ and $-3 \cdot 2y = -6y$.

$12x - 6y =$ These are not like terms. Therefore, this is the product.

Answers: Getting Ready: 1. 9 **2.** $-6a^6b^4$ **3.** -24 **4.** $5a^2 - 2a - 9$ **5.** 35 cm^2

Be Careful

You cannot add unlike terms! A common mistake is to find the product and then "add" the results by adding the coefficients and exponents. For example, $2x(x^2 + 3x) = 2x^3 + 6x^2 \neq 8x^5$.

d. $2x^2(x - 5) =$ Apply the distributive property.

$2x^2 \cdot x - 2x^2 \cdot 5 =$ $2x^2 \cdot x = 2x^3$ and $2x^2 \cdot 5 = 10x^2$.

$2x^3 - 10x^2$ These are not like terms. Therefore, this is the product.

e. $3x^2y(2xy^2 + 4x^3y^2) =$ Apply the distributive property.

$(3x^2y)(2xy^2) + (3x^2y)(4x^3y^2) =$ Apply $a^m \cdot a^n = a^{m+n}$.

$6x^3y^3 + 12x^5y^3$ These are not like terms. Therefore, this is the product.

f. $-2x^2(3x^2 - 4x + 6) =$ Apply the distributive property.

$(-2x^2)(3x^2) - (-2x^2)(4x) + (-2x^2)(6) =$ Apply $a^m \cdot a^n = a^{m+n}$.

$-6x^4 + 8x^3 - 12x^2$ These are not like terms. Therefore, this is the product.

PRACTICE EXERCISE 1

Find the following products:

a. $3(x + 5)$ **b.** $-6(a - 5)$ **c.** $4(3c - 4d)$

d. $3x^3(x - 6)$ **e.** $2ab^2(4a^2b - 3a^2b^2)$ **f.** $-2y^2(4y^3 - 3y - 5)$

If you need more practice, do the following Additional Practice Exercises.

Additional Practice Exercise 1   Find the following products:

a. $4(a + 2)$ **b.** $-4(c - 2)$ **c.** $6(2a - 3b)$

d. $a(3a - 6)$ **e.** $3yz^2(3yz - 4y^3z^2)$ **f.** $-4a^2(3a^3 - 5a^2 + 4)$

It is often necessary to combine multiples of polynomials as shown in the following example:

Example 2   Simplify the following:

a. $2(x - 4) + 3(x + 2) =$ Apply the distributive property.

$2x - 8 + 3x + 6 =$ Add like terms.

$5x - 2$ Therefore, $2(x - 4) + 3(x + 2) = 5x - 2$.

b. $2x(x - 3) - 3(x - 3) =$ Apply the distributive property.

$2x^2 - 6x - 3x + 9 =$ Add like terms.

$2x^2 - 9x + 9$ Therefore, $2x(x - 3) - 3(x - 3) = 2x^2 - 9x + 9$

c. $4(2y - 5) - 2(3y - 6) =$ Apply the distributive property.

$8y - 20 - 6y + 12 =$ Add like terms.

$2y - 8$ Therefore, $4(2y - 5) - 2(3y - 6) = 2y - 8$.

d. $3a(a^2 - 3a + 5) + 2(a^2 - 3a + 5) =$ Apply the distributive property.

$3a^3 - 9a^2 + 15a + 2a^2 - 6a + 10 =$ Add like terms.

$3a^3 - 7a^2 + 9a + 10$ Therefore, $3a(a^2 - 3a + 5) + 2(a^2 - 3a + 5) = 3a^3 - 7a^2 + 9a + 10$.

Answers: Practice Exercise 1: a. $3x + 15$ **b.** $-6a + 30$ **c.** $12c - 16d$ **d.** $3x^4 - 18x^3$ **e.** $8a^3b^3 - 6a^3b^4$
f. $-8y^5 + 6y^3 + 10y^2$ **Additional Practice 1: a.** $4a + 8$ **b.** $-4c + 8$ **c.** $12a - 18b$ **d.** $3a^2 - 6a$
e. $9y^2z^3 - 12y^4z^4$ **f.** $-12a^5 + 20a^4 - 16a^2$

PRACTICE EXERCISE 2

Simplify the following:

a. $2(r - 5) + 3(2r + 1)$

b. $3b(2b - 4) - 3(2b - 4)$

c. $-3(2r - 3) - 4(3r - 2)$

d. $2r(3r^2 - r + 5) - 3(3r^2 - r + 5)$

If you need more practice, do the following Additional Practice Exercises.

Additional Practice Exercise 2   Simplify the following:

a. $3(x - 1) + 2(x + 3)$

b. $4a(2a + 3) - 2(2a + 3)$

c. $6(2x - 1) - 3(x + 2)$

d. $3c(c^2 - 2c - 4) - 2(c^2 - 2c - 4)$

OBJECTIVE **B**   Finding the product of two polynomials

In multiplying a binomial and a polynomial of two or more terms, we need to multiply each term of the polynomial by each term of the binomial. The easiest way to do this is to use the distributive property twice. We distribute each term of the binomial over the polynomial, as illustrated by the following examples:

Example 3   **Find the following products.**

a. $(x + 2)(x + 3) =$ Rewrite by using the distributive property.

$x(x + 3) + 2(x + 3) =$ Apply the distributive property.

$x^2 + 3x + 2x + 6 =$ Combine like terms.

$x^2 + 5x + 6$ Product.

b. $(2x - 4)(3x - 2) =$ Rewrite by using the distributive property.

$2x(3x - 2) - 4(3x - 2) =$ Apply the distributive property.

$6x^2 - 4x - 12x + 8 =$ Combine like terms.

$6x^2 - 16x + 8$ Product.

c. $(2a - 3)(2a^2 - 4a + 5) =$ Rewrite by using the distributive property.

$2a(2a^2 - 4a + 5) - 3(2a^2 - 4a + 5) =$ Apply the distributive property.

$4a^3 - 8a^2 + 10a - 6a^2 + 12a - 15 =$ Combine like terms.

$4a^3 - 14a^2 + 22a - 15$ Product.

Be Careful

Notice that $(a + 5)^2 \neq a^2 + 5^2$. Do not confuse $(a + b)^2$ with $(a \cdot b)^2$.

d. $(a + 5)^2 =$ Rewrite as a product.

$(a + 5)(a + 5) =$ Rewrite by using the distributive property.

$a(a + 5) + 5(a + 5) =$ Apply the distributive property.

$a^2 + 5a + 5a + 25 =$ Combine like terms.

$a^2 + 10a + 25$ Product.

Answers: Practice Exercise 2: a. $8r - 7$ **b.** $6b^2 - 18b + 12$ **c.** $-18r + 17$ **d.** $6r^3 - 11r^2 + 13r - 15$
Additional Practice 2: a. $5x + 3$ **b.** $8a^2 + 8a - 6$ **c.** $9x - 12$ **d.** $3c^3 - 8c^2 - 8c + 8$

PRACTICE EXERCISE 3

Find the following products:

a. $(y + 4)(y + 6)$

b. $(3x - 2)(4x - 1)$

c. $(3a + 2)(a^2 - 3a + 2)$

d. $(x - 4)^2$

If you need more practice, do the following Additional Practice Exercises.

Additional Practice Exercise 3   Find the following products:

a. $(b + 4)(b + 1)$

b. $(2a - 4)(4a + 3)$

c. $(2b + 3)(b^2 + 4b - 1)$

d. $(a - 6)^2$

Optional Topic

Vertical Multiplication of Polynomials

Multiplication of polynomials can also be done vertically in much the same manner that whole numbers are multiplied. Since any two terms can be multiplied, it is not necessary to align like terms before multiplying, as was necessary with addition. When multiplying whole numbers, it is necessary to line up digits with the same place value before adding to get a final answer. *When multiplying polynomials vertically, it is necessary to align like terms before adding.* Study the following examples:

Example 4 **Find the product by multiplying vertically:**

a. Find the product of 46 and 67.

$$
\begin{array}{r}
46 \\
67 \\
\hline
322 \\
276 \\
\hline
3082
\end{array}
$$

322   Product of 7 and 46.

276   Product of 60 and 46. Align place values and add.

3082   Product.

b. Find the product of $3x + 2$ and $2x - 7$.

$$
\begin{array}{r}
3x + 2 \\
2x - 7 \\
\hline
-21x - 14 \\
6x^2 + 4x \\
\hline
6x^2 - 17x - 14
\end{array}
$$

$-21x - 14$ Product of $3x + 2$ and -7.

$6x^2 + 4x$ Product of $3x + 2$ and $2x$. Align like terms and add.

$6x^2 - 17x - 14$ Product.

c. Find the product of $2x^2 - 4x + 2$ and $3x + 4$.

$$
\begin{array}{r}
2x^2 - 4x + 2 \\
3x + 4 \\
\hline
8x^2 - 16x + 8 \\
6x^3 - 12x^2 + 6x \\
\hline
6x^3 - 4x^2 - 10x + 8
\end{array}
$$

$8x^2 - 16x + 8$ Product of 4 and $2x^2 - 4x + 2$.

$6x^3 - 12x^2 + 6x$ Product of $3x$ and $2x^2 - 4x + 2$. Align like terms and add.

$6x^3 - 4x^2 - 10x + 8$ Product.

Answers: Practice Exercise 3: **a.** $y^2 + 10y + 24$ **b.** $12x^2 - 11x + 2$ **c.** $3a^3 - 7a^2 + 4$ **d.** $x^2 - 8x + 16$
Additional Practice 3: **a.** $b^2 + 5b + 4$ **b.** $8a^2 - 10a - 12$ **c.** $2b^3 + 11b^2 + 10b - 3$ **d.** $a^2 - 12a + 36$

PRACTICE EXERCISE 4

Find the following products vertically:

a. 63
27

b. $3x - 5$
$2x + 3$

c. $3x^2 - 5x + 2$
$3x - 1$

If you need more practice, do the following Additional Practice Exercises.

Additional Practice Exercise 4   Find the following products vertically:

a. 36
73

b. $2x - 1$
$3x - 2$

c. $4x^2 - 5x + 1$
$x - 4$

If the dimensions of a geometric figure are given in terms of variables, it is possible to represent the area in terms of these same variables. Remember, the formula for the area of a rectangle is $A = LW$, and the formula for the area of a triangle is $A = \frac{bh}{2}$.

Example 5   **Write an expression for the area of each of the following, using the given dimensions:**

a. A rectangle with $L = 2x - 3$ and $W = 3x - 4$

$A = LW$ Substitute for L and W.
$A = (2x - 3)(3x - 4)$ Rewrite by using the distributive property.
$A = 2x(3x - 4) - 3(3x - 4)$ Apply the distributive property.
$A = 6x^2 - 8x - 9x + 12$ Add like terms.
$A = 6x^2 - 17x + 12$ Therefore, the area is represented as $6x^2 - 17x + 12$.

b. A triangle with $b = 4x$ and $h = 5x$

$A = \dfrac{bh}{2}$ Substitute for b and h.

$A = \dfrac{(4x)(5x)}{2}$ Multiply $4x$ and $5x$.

$A = \dfrac{20x^2}{2}$ Divide 2 into 20.

$A = 10x^2$ Therefore, the area is represented as $10x^2$.

PRACTICE EXERCISE 5

Write an expression for the area of each of the following, using the given dimensions:

a. A rectangle with $L = x - 3$ and $W = 2x + 2$

b. A triangle with $b = 5x$ and $h = 6x$

If you need more practice, do the following Additional Practice Exercises.

Additional Practice Exercise 5   Write an expression for the area of each of the following, using the given dimensions:

a. Rectangle with $L = x - 5$ and $W = x + 3$

b. Triangle with $b = 7a$ and $h = 4a$

Answers: Practice Exercise 4: a. 1701 **b.** $6x^2 - x - 15$ **c.** $9x^3 - 18x^2 + 11x - 2$ **Additional Practice 4: a.** 2628 **b.** $6x^2 - 7x + 2$ **c.** $4x^3 - 21x^2 + 21x - 4$ **Practice Exercise 5: a.** $A = 2x^2 - 4x - 6$ **b.** $A = 15x^2$ **Additional Practice 5: a.** $A = x^2 - 2x - 15$ **b.** $A = 14a^2$

Exercise Set 2.3

For Extra Help

MyMathLab®

Find the products of the following monomials:

1. $(7ax^2)(-a^2x)$

2. $(5rq^3)(-3r^4q)$

3. $(6fkb^5)(-7f)(-2f^5k)$

4. $(-5ab^2c)(2b)(-3ab^3c^2)$

5. $(-3p^3r)(4p^2qr^4)(-q^4r^3)$

6. $(5p^4r^2)(-4p^3q^2)(-q^3r^4)$

Calculator Exercises (7–10, Optional)

Find the following products by using a calculator:

7. $(-6.78x^3y^5)(7.93x^6y^8)$

8. $(8.52a^6b^2)(-9.97a^2b)$

9. $(7.68x^4y^5)(-4.5x^7y^4)$

10. $(-67.4m^4n^8)(-56.9m^2n^8)$

Find the products of the following monomials and polynomials. (See Example 1.)

11. $2(x + 3)$

12. $5(y + 6)$

13. $-4(2x - 3)$

14. $-3(4y - 2)$

15. $4(3a - 2b)$

16. $6(3x - 5y)$

17. $-5(4s - 6t)$

18. $-8(6m - 2n)$

19. $3(2x^2 - 6x + 5)$

20. $4(3a^2 + 2a - 5)$

21. $-2(6b^2 - 2b + 4)$

22. $-5(2a^2 + 5a - 7)$

23. $2b(2b^2 - 4b + 5)$

24. $4x(5x^2 - 6x - 2)$

25. $2h^3(h^2 + 8h - 5)$

26. $3y^2(y^2 + 5y - 9)$

27. $2b^4(b^2 + 7b - 6)$

28. $2b^5(b^2 + 6b - 5)$

29. $-2r^3(3r^3 - 4r^2 + 2r - 3)$

30. $-3s^2(4s^4 - 5s^2 + 6s - 5)$

31. $5x^2y(2x^2y^2 - 4xy^3)$

32. $3ab^2(6a^3b - 2a^2b^2)$

33. $-4p^2q^3(3p^4q - p^2q^2 + q)$

34. $-3u^2v^2(4u + 3u^4v^2 - u^2v^4)$

Calculator Exercises (35 and 36, Optional)

Find the following products by using a calculator:

35. $4.3x^3(5.3x^2 - 7.8x - 7.2)$

36. $-6.4y^2(-4.1y^2 - 9.3y + 9.2)$

Simplify the following. (See Example 2.)

37. $2(2x - 3) + 3(x + 4)$

38. $4(3x + 1) + 2(2x - 4)$

39. $3(6a - 4) - 2(3a + 2)$

40. $5(2b - 6) - 4(3b + 2)$

41. $4y(2y - 3) + 2(2y - 3)$

42. $3x(6x - 2) + 4(6x - 2)$

43. $3a(7a - 2b) + 2b(7a - 2b)$

44. $4c(2c - 3d) + 3d(2c - 3d)$

45. $-2x(4x - 4) - 3(2x + 5)$

46. $-6x(2x - 2) - 2(3x - 5)$

47. $3(4c^2 - 3c + 2) + 4(3c^2 + 2c - 3)$

48. $6(2m^2 + 5m - 4) + 3(3m^2 - 6m + 6)$

49. $5r(3r^2 + 2r - 5) + 2(3r^2 + 2r - 5)$

50. $8a(3a^2 - 2a + 1) + 6(3a^2 - 2a + 1)$

51. $2a(4a^2 - 3a + 1) - 3(4a^2 - 3a + 1)$

52. $3r(2r^2 + 3r - 4) - 2(2r^2 - 3r + 4)$

Challenge Exercises (53 and 54)

Simplify the following:

53. $3(a^2 + 2a - 6) + 2[3(2a^2 - 3a + 2) - 2(a^2 - 4a - 5)]$

54. $4(2b^2 - 3b + 5) - 3[2(b^2 + 5b - 6) - 3(2b^2 - 3b + 1)]$

Find the products of the following polynomials. (See Example 3.)

55. $(a + 4)(a + 5)$

56. $(q + 7)(q + 2)$

57. $(z + 5)(z - 8)$

58. $(w - 3)(w + 5)$

59. $(x + 4)(x - 4)$

60. $(x - 5)(x + 5)$

61. $(2x - y)(2x + y)$

62. $(3a + 2b)(3a - 2b)$

63. $(x + y)(x - 2y)$

64. $(r - d)(r + 3d)$

65. $(4w + 3)(5w - 8)$

66. $(2c - 1)(9c + 5)$

67. $(3z - 4)(2z - 6)$

68. $(7a - 5)(2a - 3)$

69. $(2a - b)(3a + 4b)$

70. $(4x - 3y)(2x + 5y)$

71. $(x - 1)(x^2 + x + 3)$

72. $(x - 1)(x^2 + x - 4)$

73. $(x + 3)(2x^2 - 4x + 3)$

74. $(x + 4)(3x^2 - 2x + 5)$

75. $(x + 2)(x^2 - 2x + 4)$

76. $(x - 3)(x^2 + 3x + 9)$

77. $(2x - 3)(3x^2 + 4x - 3)$

78. $(3x - 4)(3x^2 - 5x - 2)$

▦ Calculator Exercises (79 and 80, Optional)

Find the following products by using a calculator:

79. $(5.3x - 4.8)(7.1x + 3.6)$

80. $(6.2x + 3.8)(2.6x - 7.2)$

Challenge Exercises (81–88)

Find the following products. (See Examples 1–3.)

81. $(3a^2 + 2b^2)(4a^2 - 5b^2)$

82. $(4x^2 + 3y^2)(2x^2 - 3y^2)$

83. $(2a^2 + 3)(2a^2 + 4a - 5)$

84. $(3x^2 - 1)(2x^2 - 3x + 2)$

85. $(a^2 + 3a + 2)(a^2 - 4a + 3)$

86. $(x^2 - 2x + 5)(2x^2 + 3x - 2)$

87. $(x^n + 2)(x^n - 3)$

88. $(2x^a - 3)(3x^a + 4)$

Find the following products. (See Example 4.)

89. $2x + 4$
    $\underline{x + 3}$

90. $3x + 2$
    $\underline{x + 5}$

91. $3x + 5$
    $\underline{2x - 5}$

92. $4x - 3$
    $\underline{3x + 2}$

93. $a^2 - 5a + 6$
    $\underline{2a + 4}$

94. $b^2 - 4b + 7$
    $\underline{3b + 5}$

95. $2y^2 + 3y - 2$
    $\underline{4y - 2}$

96. $3t^2 - 5t + 4$
    $\underline{5t - 1}$

Recall that the formula for the area of a rectangle is $A = LW$. Write an expression for the area of each of the following rectangles with the given length and width. (See Example 5.)

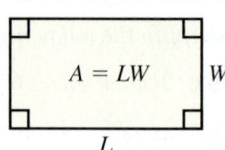

97. $L = x, W = 2x$

98. $L = x, W = 3x$

99. $L = 4x, W = x + 6$

100. $L = 5x, W = 3x - 5$

101. $L = x + 2, W = 2x - 3$

102. $L = x - 3, W = 3x + 1$

103. $L = x^2 - x + 2, W = 3x + 2$

104. $L = x^2 + 2x + 9, W = 4x - 3$

Recall that the formula for the area of a triangle is $A = \frac{bh}{2}$. Write an expression for the area of each of the following triangles. (See Example 5.)

105. $b = 2y, h = 4y$

106. $b = 4x, h = 6x$

107. $b = 6z, h = 3z$

108. $b = 4a, h = 5a$

109. The width of a living room is 8 feet less than twice the length. If y represents the length of a living room, how would you represent the amount of carpet needed to carpet the living room?

110. A rectangular pasture is $(3x + 5)$ feet long and $(3x - 2)$ feet wide. How many square feet of land are in the pasture?

111. The Martinez family wants to sod their backyard. They found the yard's length is 3 feet more than twice its width. If x represents the width of their backyard, how would you represent the amount of sod they need to sod their backyard?

112. A rectangular bathroom wall is $(2x + 8)$ feet long and $(3x - 9)$ feet high. How many square feet of tile would it take to cover it?

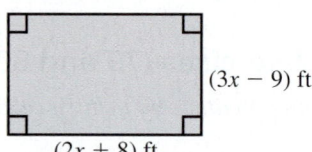

Challenge Exercises (113 and 114)

113. One brand of pile carpet sells for $12.50 per square yard. How would you represent the cost to carpet a room $(x + 3)$ yards long and x yards wide?

114. A certain type of sod sells for $.50 per square foot. How would you represent the cost to sod a rectangular yard that is $(2x + 4)$ feet long and $(x + 2)$ feet wide?

Writing Exercises (115–117)

115. Does $(a + 3)^2 = a^2 + 3^2$? Why or why not?

116. Is the product of a monomial and a trinomial always a trinomial? Explain.

117. Can the product of two binomials be a binomial? Explain.

Section 2.4 Special Products

OBJECTIVES *When you complete this section, you will be able to:*

A Multiply binomials using FOIL.
B Find products of the form $(a + b)(a - b)$.
C Square binomials.
D Optional: Recognize products of the form $(a + b)(a^2 - ab + b^2)$ and $(a - b)(a^2 + ab + b^2)$.

PREREQUISITE SKILLS *Before beginning this section, you should be able to:*

a. Apply the distributive property (Section 1.7).
b. Add like terms (Section 1.7).
c. Apply $a^m \cdot a^n = a^{m+n}$ (Section 2.2).
d. Multiply integers (Section 2.1).
e. Raise integers to powers (Section 2.1).

GETTING READY FOR SECTION 2.4

1. Simplify $x(3x + 4)$ by using the distributive property.
2. Add the like terms: $-6a + 2a$.
3. Simplify $-4v \cdot 5v$, using $a^m \cdot a^n = a^{m+n}$.
4. Multiply the integers: $-4(-9)$.
5. Raise the following integer to the indicated power: $(-3)^2$.

Introduction

The ability to multiply binomials quickly and recognize special products will be very important in Chapter 6. In the previous section, we found the products of binomials by applying the distributive property twice. We need to be able to multiply binomials more quickly and easily.

OBJECTIVE **Multiplying binomials by using FOIL**

Consider the following:

$$(x + 3)(2x + 4) = x(2x + 4) + 3(2x + 4) = 2x^2 + 4x + 6x + 12$$

The term $2x^2$ is the product of the <u>F</u>irst terms of the two binomials, x and $2x$. The term $4x$ is the product of the <u>O</u>utside terms (terms farthest apart), x and 4. To find the term $6x$, multiply the <u>I</u>nside terms (terms closest together), 3 and $2x$. The term 12 is the product of the <u>L</u>ast terms of the two binomials, 3 and 4. The underlined letters spell the word FOIL, which is an easy way to remember which terms to multiply and allows us to multiply binomials mentally. The only thing left to do is add like terms and get the final product of $2x^2 + 10x + 12$. The following diagram illustrates the use of FOIL:

First   Last   F   O   I   L

$(x + 3)(2x + 4) = 2x^2 + 4x + 6x + 12$

Inside

Outside

Now combine the like terms $4x$ and $6x$ to get the product $2x^2 + 10x + 12$.

Answers: Getting Ready: 1. $3x^2 + 4x$ **2.** $-4a$ **3.** $-20v^2$ **4.** 36 **5.** 9

Example 1 Find the following products by using FOIL:

a.

$$(x - 4)(x + 2) = x^2 + 2x - 4x - 8 = x^2 - 2x - 8$$

First   Inside   Outside   Last

b. $(2x - 5)(2x + 3) =$ Apply FOIL.
$2x \cdot 2x + 2x \cdot 3 - 5 \cdot 2x - 5 \cdot 3 =$ Multiply.
$4x^2 + 6x - 10x - 15 =$ Add like terms.
$4x^2 - 4x - 15$ Product.

c. $(4a - b)(3a - 2b) =$ Apply FOIL
$4a \cdot 3a - 4a \cdot 2b - b \cdot 3a + b \cdot 2b =$ Multiply.
$12a^2 - 8ab - 3ab + 2b^2 =$ Add like terms.
$12a^2 - 11ab + 2b^2$ Product.

d. $(x + 2y)(3a + b) =$ Apply FOIL.
$3ax + bx + 6ay + 2by$ Product, since there are no like terms.

PRACTICE EXERCISE 1

Find the following products by using FOIL:

a. $(q + 2)(q - 4)$ **b.** $(3a - 2)(5a - 3)$ **c.** $(2w + 5t)(3w - 2t)$ **d.** $(3a + 2b)(4c + 5d)$

If you need more practice, do the following Additional Practice Exercises.

Additional Practice Exercise 1   Find the following products by using FOIL:
a. $(r + 5)(r - 3)$ **b.** $(5x - 2)(3x + 4)$ **c.** $(4y + 3z)(2y + 7z)$ **d.** $(2a + 4b)(3c - 5d)$

OBJECTIVE Finding products of the form $(a + b)(a - b)$

Products of the form $(a + b)(a - b)$ are of particular interest, since the answer is a binomial instead of the usual trinomial. Notice the factors are the sum and difference of the exact same two terms. Products of this form are often referred to as **conjugate pairs**. As you will see, the sum of the outer and inner products is always zero, which leaves the difference of two squares.

Example 2 Find the following product:

$(a + b)(a - b) =$ Apply FOIL.
$a^2 - ab + ab - b^2 =$ Combine like terms.
$a^2 - b^2$ Sum of outer and inner products is 0. The product is the difference of squares.

This example leads to the following observation:

> **Rule: Products of the Form $(a + b)(a - b)$ (Conjugate Pairs)**
>
> The product of two binomials that are the sum and difference of the same terms results in a binomial that is a difference of squares. In symbols, $(a + b)(a - b) = a^2 - b^2$.

From the preceding example, we see that in the product of conjugate pairs the outside and inside products always have a sum of zero. Therefore, it is not necessary to compute them.

Answers: Practice Exercise 1: a. $q^2 - 2q - 8$ **b.** $15a^2 - 19a + 6$ **c.** $6w^2 + 11tw - 10t^2$
d. $12ac + 15ad + 8bc + 10bd$ **Additional Practice 1: a.** $r^2 + 2r - 15$ **b.** $15x^2 + 14x - 8$
c. $8y^2 + 34yz + 21z^2$ **d.** $6ac - 10ad + 12bc - 20bd$

Example 3   Find the following products:

a. $(3x - 4)(3x + 4) =$ Form of $(a + b)(a - b)$.
$9x^2 - 16$ Multiply first by first and
last by last.

b. $(3b + 4c)(3b - 4c) =$ Form of $(a + b)(a - b)$.
$9b^2 - 16c^2$ Multiply first by first and
last by last.

Note that in each case the answer is the difference of squares.

PRACTICE EXERCISE 3

Find the following products:

a. $(x + 6)(x - 6)$

b. $(2a - 6b)(2a + 6b)$

If you need more practice, do the following Additional Practice Exercises.

Additional Practice Exercise 3   Find the following products:

a. $(a + 1)(a - 1)$

b. $(2p + 5q)(2p - 5q)$

OBJECTIVE Squaring a binomial

Another special product is the square of a binomial. In the following illustration, the outside and inside products are always the same. This suggests a quick way of squaring a binomial mentally.

Example 4   Square the following binomials:

a. $(a + b)^2 =$ Apply the definition of exponent.
$(a + b)(a + b) =$ Apply FOIL.
$a^2 + ab + ab + b^2 =$ Add like terms. Note that the outside and inside products are both ab.
$a^2 + 2ab + b^2$ Product. Note that the middle term is twice the product ab.

b. $(a - b)^2 =$ Apply the definition of exponent.
$(a - b)(a - b) =$ Apply FOIL.
$a^2 - ab - ab + b^2 =$ Add like terms. Note that the outside and inside products are both $-ab$.
$a^2 - 2ab + b^2$ Product. Note that the middle term is twice the product $-ab$.

This suggests the following procedure for squaring a binomial quickly:

Procedure: Square of a Binomial

To square a binomial, follow these steps:

1. Square the first term.

2. Add two times the product of the two terms.

3. Add the square of the last term.

In symbols, $(a + b)^2 = a^2 + 2ab + b^2$ and $(a - b)^2 = a^2 - 2ab + b^2$.

Answers: Practice Exercise 3: a. $x^2 - 36$ **b.** $4a^2 - 36b^2$ **Additional Practice 3: a.** $a^2 - 1$ **b.** $4p^2 - 25q^2$

This procedure is diagrammed as follows for the square of the sum of two terms:

First term squared   Last term squared
$$\downarrow \qquad\qquad \downarrow$$
$$(a + b)^2 = a^2 + 2ab + b^2$$
$$\uparrow$$
Two times the product of the terms of the binomial

Example 5   Square the following binomials:

a. $(x + 3)^2 =$ Apply $(a + b)^2 = a^2 + 2ab + b^2$.
$x^2 + 2(x)(3) + 3^2 =$ Multiply.
$x^2 + 6x + 9$ Answer.

b. $(y - 3)^2 =$ Apply
     $(a - b)^2 = a^2 - 2ab + b^2$.
$y^2 - 2(y)(3) + (-3)^2 =$ Multiply.
$y^2 - 6y + 9$ Answer.

c. $(w - 2q)^2 =$ Apply
     $(a - b)^2 = a^2 - 2ab + b^2$.

$w^2 - 2(w)(2q) + (-2q)^2 =$ Multiply.
$w^2 - 4wq + 4q^2$ Answer.

d. $(2a + 3b)^2 =$ Apply
     $(a + b)^2 = a^2 + 2ab + b^2$.

$(2a)^2 + 2(2a)(3b) + (3b)^2 =$ Multiply and raise to
     powers.
$4a^2 + 12ab + 9b^2$ Answer.

PRACTICE EXERCISE 5

Square the following binomials:

a. $(a + 2)^2$ **b.** $(b - 5)^2$ **c.** $(c + 3d)^2$ **d.** $(3x - 2y)^2$

If you need more practice, do the following Additional Practice Exercises.

Additional Practice Exercise 5   Square the following binomials:

a. $(q + 5)^2$ **b.** $(r - 2)^2$ **c.** $(c - 2d)^2$ **d.** $(4a + 2b)^2$

OBJECTIVE **D** Recognizing products of the form $(a + b)(a^2 - ab + b^2)$ and $(a - b)(a^2 + ab + b^2)$ (Optional)

Optional Topic

Products That Result in the Sum or Difference of Cubes

Another special product is that of a binomial (two terms) multiplied by a trinomial (three terms), where the trinomial has the following properties: (1) The first term of the trinomial is the square of the first term of the binomial. (2) The second term of the trinomial is the negative of the product of the terms of the binomial. (3) The third term of the trinomial is the square of the last term of the binomial. The following illustrates the form of this product:

Square of first term   Square of last term
$$\downarrow \qquad\qquad \downarrow$$
$$(a + b)(a^2 - ab + b^2)$$
$$\uparrow$$
Negative of the product of the two terms of the binomial

A product of this form always results in the sum of two cubes if the binomial is a sum and the difference of two cubes if the binomial is a difference. The terms of the product are the cubes of the terms of the binomial.

Example 6 | **Find the following products:**

a. $(a + b)(a^2 - ab + b^2) =$ Rewrite by using distributive property.
$a(a^2 - ab + b^2) + b(a^2 - ab + b^2) =$ Apply the distributive property.
$a^3 - a^2b + ab^2 + a^2b - ab^2 + b^3 =$ Add like terms.
$a^3 + b^3$ Sum of the cubes of the terms of the binomial.

b. $(x - y)(x^2 + xy + y^2) =$ Rewrite by using distributive property.
$x(x^2 + xy + y^2) - y(x^2 + xy + y^2) =$ Apply the distributive property.
$x^3 + x^2y + xy^2 - x^2y - xy^2 - y^3 =$ Add like terms.
$x^3 - y^3$ Difference of the cubes of the terms of the binomial.

c. $(x + 2)(x^2 - 2x + 4) =$ Form of $(a + b)(a^2 - ab + b^2)$. See Example 6a.
$x^3 + 2^3 =$ Sum of the cubes of the terms of the binomial.
$x^3 + 8$ Answer.

d. $(2x - 3y)(4x^2 + 6xy + 9y^2) =$ Form of $(a - b)(a^2 + ab + b^2)$. See Example 6b.
$(2x)^3 - (3y)^3 =$ Difference of cubes. Apply $(ab)^n = a^n b^n$.
$8x^3 - 27y^3$ Answer.

PRACTICE EXERCISE 6

Find the following products:

a. $(c + d)(c^2 - cd + d^2)$ **b.** $(w - z)(w^2 + wz + z^2)$

c. $(y + 3)(y^2 - 3y + 9)$ **d.** $(3a - b)(9a^2 + 3ab + b^2)$

If you need more practice, do the following Additional Practice Exercises.

Additional Practice Exercises 6   Find the following products:

a. $(r + s)(r^2 - rs + s^2)$ **b.** $(a - b)(a^2 + ab + b^2)$

c. $(a + 4)(a^2 - 4a + 16)$ **d.** $(2a + 3b)(4a^2 - 6ab + 9b^2)$

Study Tip 8   Making Effective Use of Class Time

You need to get as much out of your time in class as possible. It is there that you have the benefit of having full access to an expert on the subject—your instructor. The following are some things you should do in order to get the most out of class time:

1. Attend all class meetings. Be on time so that you do not miss anything—otherwise you will not have anyone to explain the material to you. If you do miss a class, be sure to find out what the assignment was and complete it *before* the next class meeting.
2. Participate in class. Do not just listen to the instructor. Think along with the instructor, and anticipate the next step when he or she is doing an example. Answer questions that the instructor asks. Ask questions when you do not

understand what the teacher has said or why he or she has said it.
3. Make an effort to understand as much as possible before you leave class.
4. Sit as close to the front of the classroom as possible so you can see and hear everything. You will also find yourself more involved with classroom activities.
5. Keep a mathematics notebook that has your class notes in one section and your homework in another. When taking notes, leave extra space so you will have room to add additional comments as you review your notes.
6. If you do not understand an example, ask the instructor to explain the first step that you do not understand.

(Continued)

Answers: Practice Exercise 6: a. $c^3 + d^3$ **b.** $w^3 - z^3$ **c.** $y^3 + 27$ **d.** $27a^3 - b^3$
Additional Practice 6: a. $r^3 + s^3$ **b.** $a^3 - b^3$ **c.** $a^3 + 64$ **d.** $8a^3 + 27b^3$

7. Get help outside of class either from the instructor, a classmate, or, if available, the math learning center. Forming a study group with others in the class is an excellent idea. Use Student's Solution Manual, Math Tutor Center, and Math XL.

8. Try tape-recording the class, and use the recording to supplement your notes. Listen to the recording as you review your notes. (Don't forget to ask the instructor's permission to record the class.)

Exercise Set 2.4

For Extra Help

MyMathLab®

Find the product of the following binomials by using FOIL. (See Example 1.)

1. $(x + 3)(x + 2)$

2. $(y + 4)(y + 5)$

3. $(a + 3)(a - 4)$

4. $(b - 4)(b + 6)$

5. $(c - 5)(c - 2)$

6. $(q - 5)(q - 3)$

7. $(a + 3b)(a + 4b)$

8. $(t + 5s)(t + 2s)$

9. $(r - 5s)(r + 3s)$

10. $(x + 4y)(x - 6y)$

11. $(2a + 3b)(3a + 2b)$

12. $(3s + 5t)(2s + 3t)$

13. $(3x - 5y)(2x + 3y)$

14. $(4x + 3y)(5x - 2y)$

15. $(2x - 5y)(3x - 4y)$

16. $(3a - 5b)(2a - 7b)$

▦ Calculator Exercises (17 and 18, Optional)

Find the following products by using a calculator:

17. $(3.1x + 6.3y)(8.2x - 5.2y)$

18. $(8.3x - 3.1y)(9.5x + 4.8y)$

Find the following products of the form $(a + b)(a - b)$. (See Examples 2 and 3.)

19. $(x + 3)(x - 3)$

20. $(x + 5)(x - 5)$

21. $(p - q)(p + q)$

22. $(r - s)(r + s)$

23. $(a + 4b)(a - 4b)$

24. $(w + 3q)(w - 3q)$

25. $(6a - b)(6a + b)$

26. $(3x + y)(3x - y)$

27. $(4a + 5)(4a - 5)$

28. $(5a + 2)(5a - 2)$

29. $(2x + 5y)(2x - 5y)$

30. $(3x - 5y)(3x + 5y)$

▦ Calculator Exercises (31 and 32, Optional)

Find the following products by using a calculator:

31. $(7.2a + 6.7)(7.2a - 6.7)$

32. $(2.8x - 8.3)(2.8x + 8.3)$

Square the following binomials. (See Examples 4 and 5.)

33. $(x + 2)^2$

34. $(y + 6)^2$

35. $(a - 4)^2$

36. $(b - 7)^2$

37. $(r + s)^2$

38. $(p + q)^2$

39. $(t - s)^2$

40. $(r - s)^2$

41. $(2x + 3)^2$ **42.** $(3x + 4)^2$ **43.** $(3a - 1)^2$ **44.** $(5a - 2)^2$

45. $(2x + 5y)^2$ **46.** $(3a + 4b)^2$ **47.** $(4x - 3y)^2$ **48.** $(5x - 6y)^2$

49. $(5t^2 - w)^2$ **50.** $(b + 2z^2)^2$

▦ Calculator Exercises (51 and 52, Optional)

Square the following binomials by using a calculator:

51. $(4.7x + 3.7y)^2$ **52.** $(5.7a - 7.4b)^2$

Find the following products. (See Examples 1–5.)

53. $(2a + 5b)(3a - 2b)$ **54.** $(4x - 3y)(2x + 5y)$ **55.** $(4a - 3b)(4a + 3b)$

56. $(3s + 5t)(3s - 5t)$ **57.** $(3a + 5)^2$ **58.** $(2a - 7)^2$

59. $(2 - 3b)(4 - 3b)$ **60.** $(3 + 5c)(3 + 2c)$ **61.** $(6 + y)(6 - y)$

62. $(7 + x)(7 - x)$ **63.** $(4 + 3w)^2$ **64.** $(5 + 4c)^2$

65. $(4c + 7d)(4c - 7d)$ **66.** $(5x - 8y)(5x + 8y)$

Write an expression or equation representing each of the following and simplify when possible:

67. The square of the sum of 3 and x **68.** The square of the difference of y and 4

69. 3 less the square of x **70.** 5 less than the square of y

71. The square of the difference of x and 5 is equal to 25.   **72.** The square of the sum of x and 7 is equal to 16.

Recall that the formula for the area of a rectangle is $A = LW$. Represent the area of the following rectangles in Exercises 73–76. For example, if $L = x + 2$ and $W = x - 5$, then $A = (x + 2)(x - 5) = x^2 - 3x - 10$.

73. $L = x + 5$, $W = x - 8$ **74.** $L = y - 7$, $W = y + 9$

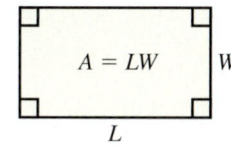

75. $L = 4a - 5$, $W = 3a + 9$ **76.** $L = 5x - 6$, $W = 2x + 5$

Recall that the area of a square is $A = s^2$, where s is the length of a side. For example, the area of a square each of whose sides is $x + 7$ is $A = (x + 7)^2 = x^2 + 14x + 49$. Represent the area of each of the squares in Exercises 77–80:

77. $s = x + 8$ **78.** $s = y - 9$

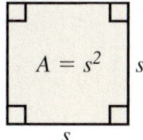

79. $s = 3a - 5b$ **80.** $s = 4y + 3z$

Optional Exercises (81–88)

Find the following products. (See Example 6.)

81. $(p + q)(p^2 - pq + q^2)$

82. $(t - s)(t^2 + ts + s^2)$

83. $(x + 3)(x^2 - 3x + 9)$

84. $(a + 5)(a^2 - 5a + 25)$

85. $(b - 4)(b^2 + 4b + 16)$

86. $(r - 2)(r^2 + 2r + 4)$

87. $(3a + 2b)(9a^2 - 6ab + 4b^2)$

88. $(2z - 4w)(4z^2 + 8wz + 16w^2)$

Writing Exercises (89–91)

89. In finding products of the form $(a + b)(a - b)$, there is no middle term. Why?

90. When squaring a binomial, why is it necessary to multiply the product of the two terms by 2 to get the middle term?

91. By assigning values to a and b, show that $(a + b)^2 \neq a^2 + b^2$. In general, show that $(a + b)^n \neq a^n + b^n$ for $n = 2, 3$, and 5.

Group Project

92. Using the formulas for the area of a rectangle $(A = LW)$ and the area of a square $(A = s^2)$, construct a geometric figure that illustrates that $(a + b)^2 = a^2 + 2ab + b^2$. (*Hint:* Begin with a square each of whose sides is of length $a + b$.)

Division of Integers and Order of Operations with Integers

OBJECTIVES *When you complete this section, you will be able to:*

A Divide integers.

B Evaluate expressions containing integers that involve combinations of addition, subtraction, multiplication, division, and raising to powers.

C Evaluate expressions containing variables when given the value(s) of the variables.

PREREQUISITE SKILLS *Before beginning this section, you should be able to:*

a. Divide whole numbers (Section R.3).

b. Add, subtract, and multiply integers (Sections 1.6, 1.7, and 2.1).

c. Raise integers to powers (Section 2.1).

d. Apply the order of operations (Section 1.1).

e. Substitute values for variables and evaluate (Section 1.1).

GETTING READY FOR SECTION 2.5

1. Divide the whole numbers: $\dfrac{56}{7}$.

Add, subtract, or multiply the following integers:

2. $6 - (-8)$

3. $-4(5)$

4. Raise the following integer to the indicated power: 3^4.

5. Simplify $-5 + 3 \cdot 24 \div 2 \cdot 4$ by using the order of operations.

Substitute the following value(s) for the variable(s) and evaluate:

6. $3a^2 - 2b$: $a = 2, b = 4$

7. $\dfrac{2m^2 + 4n}{m^3 - n^2}$: $m = 4, n = 6$

Introduction

In Section 2.1, we discussed the multiplication of integers and found that $(+)(+) = +, (+)(-) = -, (-)(+) = -$, and $(-)(-) = +$. In general, we found that the product of two numbers with the same sign is positive and the product of two numbers with opposite signs is negative. In this section, we will develop similar rules for division, but first we need to define division.

We know that $\frac{24}{3} = 8$, because $3 \cdot 8 = 24$. Likewise, $\frac{12}{4} = 3$ because $4 \cdot 3 = 12$. Division can always be checked by multiplication. Generalizing on these examples, we state the following definition of division:

Definition Division

For all numbers a, b, and c with $b \neq 0$, if $\frac{a}{b} = c$, then $b \cdot c = a$.

In this definition, $b \neq 0$. In other words, you cannot divide by 0. This is because of the definition of division. If $\frac{a}{0} = c$, then $0 \cdot c = a$, which is impossible since 0 times any real number is 0. For example, if $\frac{3}{0} = \_\_\_\_$ then $0 \cdot \_\_\_\_ = 3$, which is impossible since 0 times any number is 0. We would say that $\frac{3}{0}$ is undefined.

Answers: Getting Ready: **1.** 8 **2.** 14 **3.** −20 **4.** 81 **5.** 139 **6.** 4 **7.** 2

Calculator Exploration Activity (Optional)

Evaluate each of the following columns by using a calculator:

Column A	Column B
$\dfrac{12}{6}$ =	$\dfrac{-12}{6}$ =
$\dfrac{18}{2}$ =	$\dfrac{18}{-2}$ =
$\dfrac{-16}{-4}$ =	$\dfrac{-16}{4}$ =
$\dfrac{-21}{-3}$ =	$\dfrac{21}{-3}$ =

1. Based on your answers to column A, is the quotient of two numbers with the same sign positive or negative?

2. Based on your answers to column B, is the quotient of two numbers with different signs positive or negative?

We now give justification for the preceding observations.

OBJECTIVE Dividing integers

From our prior knowledge of arithmetic, we know that a positive number divided by a positive number is a positive number. For example, $\frac{18}{6} = 3$ and $\frac{28}{7} = 4$. Stated in symbols, $\frac{(+)}{(+)} = (+)$, because $(+)(+) = (+)$.

Using the definition of division just stated, we will develop rules for dividing a positive number by a negative number, a negative number by a positive number, and a negative number by another negative number.

Example 1 **Dividing a positive number by a negative number**

a. Find $\frac{12}{-3}$.

If $\frac{12}{-3} =$ _____, then $-3 \cdot$ _____ $= 12$. Since (-3) $\underline{(-4)} = 12$, the blank would be filled by -4.

Therefore, $\frac{12}{-3} = -4$. CHECK: $(-3)(-4) = 12$.

b. Find $\frac{32}{-4}$.

If $\frac{32}{-4} =$ _____, then $-4 \cdot$ _____ $= 32$. Since (-4) $\underline{(-8)} = 32$, the blank would be filled by -8.

Therefore, $\frac{32}{-4} = -8$. CHECK: $(-4)(-8) = 32$.

From Example 1, we can conclude that a positive number divided by a negative number results in a negative number. Stated in symbols, $\frac{(+)}{(-)} = (-)$, because $(-)(-) = (+)$.

Example 2 **Dividing a negative number by a positive number**

a. Find $\frac{-18}{3}$.

If $\frac{-18}{3} =$ _____, then $3 \cdot$ _____ $= -18$. Since (3) $\underline{(-6)} = -18$, the blank would be filled with -6.

Therefore, $\frac{-18}{3} = -6$. CHECK: $(3)(-6) = -18$.

b. Find $\frac{-12}{4}$.

If $\frac{-12}{4} =$ _____, then $4 \cdot$ _____ $= -12$. Since (4) $\underline{(-3)} = -12$, the blank would be filled with -3.

Therefore, $\frac{-12}{4} = -3$. CHECK: $(4)(-3) = -12$.

From Example 2, we can conclude that a negative number divided by a positive number results in a negative number. Written symbolically, $\frac{(-)}{(+)} = (-)$, because $(+)(-) = (-)$.

Answers: Calculator Exploration Activity: 1. positive **2.** negative

Example 3   **Dividing a negative number by a negative number**

a. Find $\frac{-8}{-2}$.

If $\frac{-8}{-2} =$ _____, then $-2 \cdot$ _____ $= -8$. Since (-2) $\underline{(4)} = -8$, the blank would be filled by 4.

Therefore, $\frac{-8}{-2} = 4$. CHECK: $(-2)(4) = -8$.

b. Find $\frac{-27}{-3}$.

If $\frac{-27}{-3} =$ _____, then $-3 \cdot$ _____ $= -27$. Since (-3) $\underline{(9)} = -27$, the blank would be filled by 9.

Therefore, $\frac{-27}{-3} = 9$. CHECK: $(-3)(9) = -27$.

From Example 3, we can conclude that a negative number divided by a negative number results in a positive number. Stated symbolically, $\frac{(-)}{(-)} = (+)$, because $(-)(+) = (-)$.

Note that the rules for division of signed numbers are exactly the same as the rules for multiplication, as summarized in the following box:

> **Rule: Division of Signed Numbers**
>
> The quotient of two numbers with the same sign is positive, and the quotient of two numbers with opposite signs is negative. In symbols. $\frac{(+)}{(+)} = (+)$, $\frac{(+)}{(-)} = (-)$, $\frac{(-)}{(+)} = (-)$, and $\frac{(-)}{(-)} = (+)$.

Example 4   **Find the following quotients:**

a. $\dfrac{10}{-2} = -5$ A positive divided by a negative is a negative.

b. $\dfrac{-14}{-7} = 2$ A negative divided by a negative is a positive.

c. $\dfrac{24}{0}$ is undefined   Division by 0 is undefined.

d. $\dfrac{-26}{13} = -2$ A negative divided by a positive is a negative.

PRACTICE EXERCISE 4

Find the following quotients:

a. $\dfrac{24}{-3}$ **b.** $\dfrac{-21}{-7}$ **c.** $\dfrac{4}{0}$ **d.** $\dfrac{-36}{4}$

If you need more practice, do the following Additional Practice Exercises.

Additional Practice Exercise 4   Find the following quotients:

a. $\dfrac{-16}{8}$ **b.** $\dfrac{-24}{-6}$ **c.** $\dfrac{15}{0}$ **d.** $\dfrac{54}{-9}$

Answers: Practice Exercise 4: a. -8 **b.** 3 **c.** undefined **d.** -9 **Additional Practice 4: a.** -2 **b.** 4 **c.** undefined **d.** -6

When translating from English to mathematics, one way of indicating division is by using the word *quotient*. When using *quotient*, look for the words *of* and *and*. If division is indicated by using a fraction, the expression following "of" is the numerator and the expression following "and" is the denominator.

Example 5   Write an expression for each of the following and simplify:

a. The quotient of 24 and -6 increased by 7

$\dfrac{24}{-6} + 7 =$ Divide before adding.

$-4 + 7 =$ Add.

3 Answer.

Solution

The quotient of 24 and -6 means $\frac{24}{-6}$. The quotient is increased by 7, which means add 7 to the quotient. Consequently, the quotient of 24 and -6 increased by 7 is expressed as

b. The quotient of -15 and -3 subtracted from the quotient of 36 and -9

$\dfrac{36}{-9} - \dfrac{-15}{-3} =$ Find each quotient.

$-4 - 5 =$ Add.

-9 Answer.

Solution

The quotient of -15 and -3 means $\frac{-15}{-3}$, and the quotient of 36 and -9 means $\frac{36}{-9}$. The first quotient is subtracted from the second, which is expressed as

PRACTICE EXERCISE 5

Write an expression for each of the following and simplify:

a. The quotient of -35 and 7, decreased by 8.

b. The quotient of 48 and -12 subtracted from the quotient of -56 and 8.

If you need more practice, do the following Additional Practice Exercises.

Additional Practice Exercise 5   Write an expression for each of the following and simplify:

a. Three less than the quotient of -48 and -12.

b. The quotient of -10 and 5 less than the quotient of 18 and -6.

OBJECTIVE Evaluating expressions containing integers that involve combinations of addition, subtraction, multiplication, division, and raising to powers

Because integers had not been discussed in Section 1.1 when we did order of operations, we were limited to operations with whole numbers only. Now that we know how to perform operations on integers, we will revisit the order of operations. We repeat the order of operations.

Answers: Practice Exercise 5: a. $\dfrac{-35}{7} - 8 = -13$ **b.** $\dfrac{-56}{8} - \dfrac{48}{-12} = -3$

Additional Practice 5: a. $\dfrac{-48}{-12} - 3 = 1$ **b.** $\dfrac{18}{-6} - \dfrac{-10}{5} = -1$

Order of Operations

If an expression contains more than one operation, operations are to be performed in the following order:

1. If parentheses or other inclusion symbols (braces or brackets) are present, begin within the innermost and work outward, using the order in steps 3–5 in doing so.

2. If a fraction bar is present, simplify above and below the fraction bar separately in the order given in steps 3–5.

3. Evaluate all indicated powers.

4. Perform all multiplication or division in the order in which they occur as you work from left to right.

5. Perform all addition or subtraction in the order in which they occur as you work from left to right.

Example 6   **Evaluate the following using the order of operations:**

a. $-5 - 6 + 8 =$ Add and subtract in order from left to right.

$-11 + 8 =$ Continue adding.

-3 Answer.

b. $8 - 15 \div (-3) =$ Divide before adding. $-15 \div (-3) = 5$.

$8 + 5 =$ Add.

13 Answer.

c. $-7 - 3(-2)^2 =$ Raise to powers first. $(-2)^2 = 4$.

$-7 - 3 \cdot 4 =$ Multiplication before subtraction.

$-7 - 12 =$ Add.

-19 Answer.

d. $6 - (12 - 18) \div 3 =$ Add numbers inside parentheses first.

$6 - (-6) \div 3 =$ Division before subtraction.

$6 - (-2) =$ $-(-2) = 2$.

$6 + 2 =$ Add.

8 Answer.

e. $-3 + 5 \cdot 6^2 \div (-9)(-2) + 6 =$ Raise to powers first.

$-3 + 5 \cdot 36 \div (-9)(-2) + 6 =$ Multiply 5 and 36.

$-3 + 180 \div (-9)(-2) + 6 =$ Divide 180 and -9.

$-3 + (-20)(-2) + 6 =$ Multiply -20 and -2.

$-3 + 40 + 6 =$ Add -3 and 40.

$37 + 6 =$ Add 37 and 6.

43 Answer.

f. $-3(5^2 - 4) \div (-7) =$ Evaluate power inside parentheses first.

$-3(25 - 4) \div (-7) =$ Add inside parentheses.

$-3(21) \div (-7) =$ Multiply before dividing in order from left to right.

$-63 \div (-7) =$ Divide.

9 Answer.

g. $|4(3 - 7)| - |2 \cdot 3^2 - 20| =$ Simplify inside the absolute value marks first and separately.
$|4(-4)| - |2 \cdot 9 - 20| =$ Keep simplifying inside the absolute value marks.
$|-16| - |18 - 20| =$ $|-16| = 16$. Keep simplifying inside the second absolute value marks.
$16 - |-2| =$ $|-2| = 2$.
$16 - 2 =$ Subtract.
14 Answer.

h. $\dfrac{4 - 6^2}{4^2 - 8} =$ Simplify the numerator and the denominator separately.

$\dfrac{4 - 36}{16 - 8} =$ Add in the numerator and the denominator.

$\dfrac{-32}{8} =$ Divide.

-4 Answer.

i. $\dfrac{|4(2^2 - 5)|}{|3 \cdot 2^2 - 2 \cdot 5|} =$ Simplify inside the absolute value marks by using the order of operations.

$\dfrac{|4(4 - 5)|}{|3 \cdot 4 - 2 \cdot 5|} =$ Keep simplifying inside the absolute value marks.

$\dfrac{|4(-1)|}{|12 - 10|} =$ Keep simplifying inside the absolute value marks.

$\dfrac{|-4|}{|2|} =$ Find the absolute values.

$\dfrac{4}{2} =$ Divide.

2

j. $\dfrac{-5(-2) - [(-4)(-5) + 6^2]}{(-2)^2(6) - 3} =$ Evaluate the power inside the brackets in the numerator and the power in the denominator first.

$\dfrac{-5(-2) - [(-4)(-5) + 36]}{(4)(6) - 3} =$ Multiply inside the brackets in the numerator and multiply in the denominator.

$\dfrac{-5(-2) - [20 + 36]}{24 - 3} =$ Add inside the brackets in the numerator and add in the denominator.

$\dfrac{-5(-2) - 56}{21} =$ Multiply in the numerator.

$\dfrac{10 - 56}{21} =$ Add in the numerator.

$\dfrac{-46}{21} = -\dfrac{46}{21}$ Answer.

PRACTICE EXERCISE 6

Evaluate the following by using the order of operations:

a. $-12 + 5 - 8$ **b.** $-9 - 18 \div (-9)$ **c.** $6 - (-3)(-4)^2$

d. $8 - (5 - 13) \div (-4)$ **e.** $6(7^2 - 13) \div (-9)$ **f.** $7 - 3 \cdot 8^2 \div (-8)(-3) + 5$

g. $|-4(-3 + 8)| - |3 \cdot 2^3 - 27|$ **h.** $\dfrac{6 - 6^2}{5^2 - 8}$

(Continued)

Answers: **Practice Exercise 6:** **a.** -15 **b.** -7 **c.** 54 **d.** 6 **e.** -24 **f.** -60 **g.** 17 **h.** $\dfrac{-30}{17} = -\dfrac{30}{17}$

i. $\dfrac{|-3(2^3 - 3)|}{|5 \cdot 3^2 - 2^3 \cdot 5|}$

j. $\dfrac{4(-2) - [(-5)(3) - 21]}{(-3)^2(-2) + 15}$

If you need more practice, do the following Additional Practice Exercises.

Additional Practice Exercise 6   Evaluate the following by using the order of operations:

a. $9 - 12 - 8$

b. $-6 - (-8) \div 2$

c. $9 - 4(-3)^3$

d. $-7 - (7 - 19) \div (-3)$

e. $-2(8^2 - 22) \div (-14)$

f. $-4 + 2 \cdot 4^3 \div (-16) - 7$

g. $|8 - (4 - 10) \div 2| - |-2^2(6 - 4^2)|$

h. $\dfrac{3^3 - 2^4}{4^2 - 9}$

i. $\dfrac{|-2^2(6 - 3^3)|}{|2^4 - 3 \cdot 2^2|}$

j. $\dfrac{4(-3) - [(-8)(3) - 4]}{(-4)^2(-1) - 9}$

OBJECTIVE **Evaluating expressions containing variables when given the value(s) of the variables**

In Sections 1.1 and 1.2, variables were introduced and we evaluated expressions containing variables. The values used for the variables were whole numbers, since operations with integers had not been discussed. We can now use integer values, as illustrated by the following examples:

Example 7   **Evaluate each of the following for $x = -2$, $y = -3$, and $z = 6$:**

Note

Remember, $5x^2 - 2y$ means $5(x^2) - 2(y)$.

a. $5x^2 - 2y =$ Substitute -2 for x and -3 for y.
$5(-2)^2 - 2(-3) =$ First raise to powers.
$5(4) - 2(-3) =$ Multiply in order from left to right.
$20 + 6 =$ Add.
26 Answer.

b. $-3xy^2z =$ Substitute for x, y, and z.
$-3(-2)(-3)^2(6) =$ First raise to powers.
$-3(-2)(9)(6) =$ Multiply in order from left to right.
$6 \cdot 9 \cdot 6 =$ Continue multiplying.
$54 \cdot 6$ Continue multiplying.
324 Answer.

c. $x - y(x^4 - 2z) =$ Substitute for x, y, and z.
$(-2) - (-3)[(-2)^4 - 2 \cdot 6] =$ Raise to powers inside the parentheses.
$(-2) - (-3)(16 - 2 \cdot 6) =$ Multiply inside the parentheses.
$(-2) - (-3)(16 - 12) =$ Add inside the parentheses.
$(-2) - (-3)(4) =$ Multiply before adding.
$(-2) - (-12) =$ $-(-12) = 12$.
$(-2) + 12 =$ Add.
10 Answer.

Answers: Practice Exercise 6: **i.** 3 **j.** $\dfrac{28}{-3} = -\dfrac{28}{3}$ **Additional Practice 6:** **a.** -11 **b.** -2 **c.** 117 **d.** -11 **e.** 6 **f.** -19 **g.** -29 **h.** $\dfrac{11}{7}$ **i.** 21

j. $\dfrac{16}{-25} = -\dfrac{16}{25}$

d. $\dfrac{2x^2 - 3y}{2z^2 - 3x^4} =$ Substitute for x, y, and z.

$\dfrac{2(-2)^2 - 3(-3)}{2(6)^2 - 3(-2)^4} =$ First raise to powers.

$\dfrac{2 \cdot 4 - 3(-3)}{2 \cdot 36 - 3 \cdot 16} =$ Multiply in order, left to right.

$\dfrac{8 + 9}{72 - 48} =$ Add.

$\dfrac{17}{24}$ Answer.

PRACTICE EXERCISE 7

Evaluate each of the following for $x = -3$, $y = 2$, and $z = -4$:

a. $3x^2 - 4z$ **b.** $-2xy^2z^2$ **c.** $-2y - x(3y^2 - 2z^2)$ **d.** $\dfrac{4y - 5x^2}{3y^3 - z^2}$

If you need more practice, do the following Additional Practice Exercises.

Additional Practice Exercise 7   Evaluate each of the following for $x = -4$, $y = 3$, and $z = -1$:

a. $4z^2 - 2x^2$ **b.** $-5x^2yz^2$ **c.** $-3y + 2x(2z^2 - 2x^2)$ **d.** $\dfrac{3y^2 - 5z}{3x^2 - 5y}$

Study Tip 9 Preparing for a Test

Proper test preparation is one of the most important factors in determining how well you do in a course. The following suggestions should help:

1. Set your goal high. Try for a score of 100% rather than just a passing grade. The higher you set your goal, the more complete your preparation will be and the higher you should score.
2. Avoid developing a mental block. Thoroughly prepare for the test. Inadequate preparation causes a loss of confidence that can produce a mental block. So be prepared.
3. Begin your test preparation early. Do not wait until the night before the test to begin studying. Begin your preparation at least a week in advance and study at least an hour a day for the exam.
4. Be organized. Make a list of specific topics to be covered on the test. Find and solve specific problems for each topic. Be sure to include all types of problems that could be contained within each topic.
5. Start at the beginning of the material to be tested and work through each section in turn. Master each section before going to the next.

6. Practice by doing the chapter review in your textbook. Do all of the exercises if you can. If you can't do an exercise, go back to the indicated section and study the examples. If you still can't do an exercise, ask your instructor.
7. In your review, answer each problem, confirm that the answer is correct (check your work!), and examine your understanding of the problem. Do not allow yourself to get stuck. If you can't do an exercise after 10 minutes, go to the next exercise.
8. Review both your notes and the text to clear up any questions that you might have. Think about the material as you review it. How does it relate to previous material? How do different parts of the material relate to other parts?
9. Be able to distinguish between the different types of problems that might be on the test.
10. Try to find or construct a practice test. The practice test might be one of your teacher's previous tests, tests from the text, or tests from a study guide.

Answers: Practice Exercise 7: a. 43 **b.** 384 **c.** -64 **d.** $-\dfrac{37}{8}$ **Additional Practice 7: a.** -28 **b.** -240 **c.** 231 **d.** $\dfrac{32}{33}$

Exercise Set 2.5

For Extra Help

MyMathLab®

Find the following quotients. (See Examples 1–4.)

1. $\dfrac{36}{-9}$

2. $\dfrac{-36}{-6}$

3. $\dfrac{-42}{7}$

4. $\dfrac{63}{9}$

5. $\dfrac{-48}{-8}$

6. $\dfrac{56}{-7}$

7. $\dfrac{6}{0}$

8. $\dfrac{-8}{0}$

9. $\dfrac{144}{-18}$

10. $\dfrac{-216}{24}$

Challenge Exercises (11–14)

11. $\dfrac{3}{4} \div \left(-\dfrac{5}{3} \right)$

12. $-\dfrac{2}{3} \div \left(-\dfrac{5}{4} \right)$

13. $-\dfrac{3}{5} \div \dfrac{5}{4}$

14. $\left(-\dfrac{7}{3} \right) \div \left(-\dfrac{9}{2} \right)$

Write an expression for each of the following and simplify. (See Example 5.)

15. The quotient of 12 and -4, increased by 8

16. The quotient of -15 and 5, increased by 9

17. -6 subtracted from the quotient of -18 and -9

18. -9 subtracted from the quotient of -24 and -4

19. 12 less the quotient of 36 and -9

20. -5 less the quotient of 45 and -5

21. The quotient of 48 and -16, subtracted from 7

22. The quotient of -28 and -7, subtracted from -8

23. The quotient of -9 and 3 added to the quotient of 14 and -7

24. The quotient of -24 and 6 added to the quotient of -32 and 8

25. The quotient of -36 and 18 subtracted from the quotient of 48 and -8

26. The quotient of -42 and 6 subtracted from the quotient of 54 and -9

Answer the following:

27. When a cold front moved through an area, the temperature dropped 28° in 4 hours. Using a signed number, find the average drop per hour.

28. Paula went on a diet and lost 12 pounds in 4 weeks. Using a signed number, find the average number of pounds per week that Paula lost.

29. On a steep mountain road the elevation dropped at the rate of 9 feet per 150 feet. Using signed numbers, find the rate at which the road dropped per foot.

30. The roof of a large building slopes so that it drops 8 feet per 120 feet. Using signed numbers, find the rate at which the roof dropped per foot.

31. A parachutist descends 54 feet in 6 seconds. Using a signed number, find the rate at which she is descending per second.

32. A football team lost 12 yards in three plays. Using a signed number, find the average loss per play.

Evaluate the following. (See Example 6.)

33. $-3 + 7 - 6$

34. $-7 + 9 - 5$

35. $12 - 5 + 15$

36. $14 - 9 + 6$

37. $8 - 4 \cdot 6$

38. $5 - 7 \cdot 3$

39. $-5 - 4(-7)$

40. $-3 - 7(-6)$

41. $6 - 12 \div 3$

42. $12 - 18 \div 3$

43. $9 - 21 \div (-7)$

44. $4 - 15 \div (-5)$

45. $10 - 2(-4)^2$

46. $14 - 4(-2)^2$

47. $8 - (8 - 16) \div 4$

48. $6 - (3 - 12) \div 3$

49. $-8 - (15 - 5) \div (-5)$

50. $-3 - (13 - 7) \div (-3)$

51. $25 + 4 \cdot 5^2 \div 25(-2) + 3$

52. $32 + 3 \cdot 8^2 \div 32(-3) + 5$

53. $-5 - 2 \cdot 6^2 \div (-36)(-4) + 8$

54. $-9 - 6 \cdot 4^2 \div (-16)(-5) + 7$

55. $4(6^2 - 4) \div (-8)$

56. $6(4^2 + 12) \div (-7)$

57. $-3^2(6 - 3^2)$

58. $-4^2(8 - 4^2)$

59. $-36 \div (4^2 - 7)$

60. $-48 \div (5^2 - 9)$

61. $|6(2 - 5)| - |4 \cdot 2^2 - 24|$

62. $|2 \cdot 3^2 - 30| - |10(4 - 6)|$

63. $|-6 - 2(14 - 6) \div 4| + |6(4 - 3^2) \div 2|$

64. $|7 - 3(8 - 2) \div 6| + |4(2 - 4^2) \div 2|$

65. $|4 - 8 \cdot 2| \cdot |3 - 4^2 \div 2|$

66. $|3 - 7 \cdot 4| \cdot |2 - 6^2 \div 3|$

67. $\dfrac{7^2 - 13}{5^2 - 7}$

68. $\dfrac{9^2 - 17}{6^2 - 4}$

69. $\dfrac{2(-3)^2 - 6}{-3(-2)^2 + 6}$

70. $\dfrac{2(-3)^2 + 6}{3(-3)^2 - 3}$

71. $\dfrac{2(4^2) - 7^2}{5^2 - 3(-3)^2}$

72. $\dfrac{3(-2)^2 - 5^2}{9^2 - 4(-5)^2}$

73. $\dfrac{|5(3 - 6^2)|}{|3 \cdot 5^2 - 2^2 \cdot 15|}$

74. $\dfrac{|4^2(8 - 4^2)|}{|2(-3)^3 - 5 \cdot 2|}$

75. $\dfrac{(-3)(-5) - [5(-4) + 7^2]}{(-3)^2(4) - 7}$

76. $\dfrac{(-2)(-6) - [6(-3) + 5^2]}{(-4)^2(3) - 21}$

▦ Calculator Exercises (77–86, Optional)

Evaluate each of the following using a calculator. If necessary, round off to the nearest hundredth.

77. $7.02 + (5.72)(9.43)$

78. $9.40 + (7.34)(6.92)$

79. $18.2 - 208.32 \div 16.8$

80. $14.9 - 325.26 \div (-23.4)$

81. $(56.2)^2 + (-82.4)^2$

82. $(-12.4)^3 - (58.9)^2$

83. $68.23 - 34.7(56.1^2 - 67.2) \div 39.1$

84. $95.35 + 62.5(9.8^2 - 68.9) \div 62.8$

85. $\dfrac{(13.1)^2 + (32.1)(45.3)^2}{(12.6)^3 - (76.3)(18.5)^2}$

86. $\dfrac{(5.7)^4 - (19.7)^2(96.3)}{(26.7)^2 + (14.6)^2(31.1)^2}$

Evaluate each of the following for $x = -2$, $y = 3$, and $z = -3$. (See Example 7.)

87. $4x^2 - 2y$

88. $5z^2 - 3x$

89. $-3y^3 + 4x$

90. $-5x^2 + 3z$

91. $5x^2 - 3y^2$

92. $2y^2 - 3x^2$

93. $3x^2y$

94. $4yz^2$

95. $-5x^2y$

96. $-6yz^2$

97. $3x^2y^2z$

98. $2xy^2z^2$

99. $3y - 2(3x - z)$

100. $3x - 4(2y - x)$

101. $-3x + z(3x - 4y)$

102. $-3z + x(4y - 3x)$

103. $-y^2 - (2x^2 - 3z)$ **104.** $-x^2 - (3z^2 - 4x)$ **105.** $2x^3y - x^2(3z^2 - 4x^2)$ **106.** $3y^3x - z^2(5x^2 - y^2)$

107. $\dfrac{3x + 3y}{3y + 2z}$ **108.** $\dfrac{3x + 3z}{4y + 3z}$ **109.** $\dfrac{x^2 - y^2}{y^2 + z^2}$ **110.** $\dfrac{y^2 + x^2}{x^2 - z^2}$

111. $\dfrac{2x^3 - 3y^2}{3x^3 - 2z^2}$ **112.** $\dfrac{4x^3 - 2y^2}{5x^3 - 3z^3}$

Write an expression for each of the following by using signed numbers, and evaluate:

113. The owner of a produce stand sold 55 pounds of tomatoes at a profit of 15 cents per pound. As the tomatoes began to age, she sold 22 pounds at a loss of 8 cents per pound. What was her net profit/loss on the sale of the tomatoes?

114. The produce stand owner in Exercise 113 also sold 20 pounds of onions at a profit of 12 cents per pound and later sold 8 pounds at a loss of 15 cents per pound. What was her net profit/loss on the sale of the onions?

115. During a recent price war among the airlines, 30 seats on one flight were sold at a profit of $40 per seat and 60 seats were sold at a loss of $25 per seat. Find the net profit/loss for this flight.

116. During the same price war as in Exercise 115, another flight sold 45 seats at a profit of $25 per seat and 50 seats at a loss of $25 per seat. Find the net profit/loss for this flight.

117. Ms. Jones recently sold some stock. She made a profit of $30 per share on 45 shares, had a loss of $12 per share on 20 shares, and made a profit of $8 per share on 36 shares. What was her net profit/loss on the sale of the stock?

118. Mr. Lopez sold 20 shares of stock at a loss of $12 per share, 32 shares at a profit of $20 per share, and 40 shares at a loss of $7 per share. What was his net profit/loss from the sale of the stock?

Writing Exercises (119–124)

119. Why is the quotient of a negative number and a positive number equal to a negative number?

120. Why is the quotient of a negative number and a negative number equal to a positive number?

121. Why is $\frac{2}{0}$ undefined? (Do not say because you cannot divide by 0!)

122. Is division commutative? Why or why not?

123. What is the error in the following: $8 - 3(2^2 + 4) = 8 - 3(4 + 4) = 8 - 3(8) = 5(8) = 40$? Rework the problem correctly.

124. What is the error in the following: $5 - 2 \cdot 4^2 = 5 - 8^2 = 5 - 64 = -59$? Rework the problem correctly.

Group Project

115. Write a problem involving the order of operations on integers similar to Exercises 113–118. Exchange your problem with another group and solve the problem that your group receives.

Section 2.6 — Quotient Rule and Integer Exponents

OBJECTIVES *When you complete this section, you will be able to:*

A Simplify exponential expressions by using the property $\dfrac{a^m}{a^n} = a^{m-n}$.

B Simplify expressions with zero exponents.

C Simplify expressions with negative integer exponents.

PREREQUISITE SKILLS *Before beginning this section, you should be able to:*

a. Subtract whole numbers (Section R.2).

b. Add integers (Section 1.6).

c. Apply $(ab)^n = a^n b^n$ and $(a^m)^n = a^{mn}$ (Section 2.2).

d. Raise integers to powers (Section 2.1).

GETTING READY FOR SECTION 2.6

Add or subtract the following integers:

1. $3 - 8$ **2.** $-2 - (-5)$

Simplify the following by using $(a^m)^n = a^{mn}$ or $(ab)^n = a^n b^n$:

3. $(4^3)^2$ **4.** $(3x)^2$

Raise the following integers to powers:

5. 5^3 **6.** -3^4

Introduction

In Section 2.2, we developed laws of exponents that involved products of expressions with exponents. For reference, these were (1) $a^m \cdot a^n = a^{m+n}$, (2) $(a^m)^n = a^{mn}$, and (3) $(ab)^n = a^n b^n$. In this section, we develop similar laws for quotients of expressions with exponents.

Calculator Exploration Activity 1 (Optional)

Evaluate each of the following columns by using a calculator:

Column A   **Column B**

1. $\dfrac{2^5}{2^2} =$ $2^3 =$

2. $\dfrac{3^7}{3^4} =$ $3^3 =$

3. $\dfrac{5^6}{5^4} =$ $5^2 =$

By looking at the corresponding lines of columns A and B, answer the following:

1. To divide two numbers with the same base, leave the _____ unchanged and _____ the exponents.

We will now give a mathematical justification for the preceding observation.

OBJECTIVE Simplifying exponential expressions by using the property $\dfrac{a^m}{a^n} = a^{m-n}$

Remember, the exponent indicates how many times the base is to be used as a factor. Another fact that we will need is that any number (other than 0) divided by itself is equal to 1. We also need the procedure for multiplying fractions. Recall that we multiply numerator times numerator and denominator times denominator. For example, $\dfrac{2}{3} \cdot \dfrac{5}{7} = \dfrac{2 \cdot 5}{3 \cdot 7} = \dfrac{10}{21}$.

| **Example 1** | **Simplify the following by using the definition of exponents. Express the answer in exponential form.** |

a. $\dfrac{2^6}{2^3} =$ Rewrite by using the definition of exponents.

$\dfrac{2 \cdot 2 \cdot 2 \cdot 2 \cdot 2 \cdot 2}{2 \cdot 2 \cdot 2} =$ Rewrite as the multiplication of fractions.

$\dfrac{2}{2} \cdot \dfrac{2}{2} \cdot \dfrac{2}{2} \cdot \dfrac{2}{1} \cdot \dfrac{2}{1} \cdot \dfrac{2}{1} =$ $\dfrac{2}{2} = 1$ and $\dfrac{2}{1} = 2$.

$1 \cdot 1 \cdot 1 \cdot 2 \cdot 2 \cdot 2 =$ Apply the identity for multiplication.

$2 \cdot 2 \cdot 2 =$ Rewrite by using the definition of exponents.

2^3 Answer.

Notice that three of the six factors of 2 in the numerator were divided by the three factors of 2 in the denominator, leaving $6 - 3 = 3$ factors of 2 in the numerator. Therefore, $\dfrac{2^6}{2^3} = 2^{6-3} = 2^3$. Notice that the base remained as 2.

b. $\dfrac{x^6}{x^2} =$ Rewrite by using the definition of exponents.

$\dfrac{x \cdot x \cdot x \cdot x \cdot x \cdot x}{x \cdot x} =$ Rewrite as the multiplication of fractions.

$\dfrac{x}{x} \cdot \dfrac{x}{x} \cdot \dfrac{x}{1} \cdot \dfrac{x}{1} \cdot \dfrac{x}{1} \cdot \dfrac{x}{1} =$ $\dfrac{x}{x} = 1$, if $x \neq 0$, and $\dfrac{x}{1} = x$.

$1 \cdot 1 \cdot x \cdot x \cdot x \cdot x =$ Apply the identity for multiplication.

$x \cdot x \cdot x \cdot x =$ Rewrite by using the definition of exponents.

x^4 Answer.

Again, notice that two of the factors of x in the numerator were divided by the two factors of x in the denominator, leaving $6 - 2 = 4$ factors of x in the numerator. Therefore, $\dfrac{x^6}{x^2} = x^{6-2} = x^4$.

Note that in each of the preceding examples, the exponent in the numerator is larger than the exponent in the denominator. Based on these examples, we generalize to the following law of exponents for division of expressions with the same base but different exponents:

> ### Rule: Fourth Law of Exponents—Quotient Rule
>
> For any two positive integers m and n, with $m > n$ and $a \neq 0$, $\dfrac{a^m}{a^n} = a^{m-n}$. To divide two numbers with the same base, take the top exponent and subtract the bottom exponent, leaving the base unchanged.

Example 2 Simplify, using the quotient rule. Express the quotients in exponential form. All variables represent nonzero quantities.

a. $\dfrac{3^5}{3^2} =$ Apply $\dfrac{a^m}{a^n} = a^{m-n}$.

$3^{5-2} =$ Subtract the exponents.

3^3 Quotient.

b. $\dfrac{5^6}{5^2} =$ Apply $\dfrac{a^m}{a^n} = a^{m-n}$.

$5^{6-2} =$ Subtract the exponents.

5^4 Quotient.

c. $\dfrac{z^9}{z^4} =$ Apply $\dfrac{a^m}{a^n} = a^{m-n}$.

$z^{9-4} =$ Subtract the exponents.

z^5 Quotient.

Be Careful

A common error is to divide the bases as well as subtract the exponents. The base does not change under this law of exponents.

For example, $\dfrac{3^5}{3^3} \neq 1^{5-3}$. By the quotient rule, $\dfrac{3^5}{3^3} = 3^{5-3} = 3^2$.

PRACTICE EXERCISE 2

Simplify, using the quotient rule. Express the answers in exponential form. All variables represent nonzero quantities.

a. $\dfrac{4^5}{4^3} =$

b. $\dfrac{6^8}{6^4} =$

c. $\dfrac{x^6}{x^3} =$

If you need more practice, do the following Additional Practice Exercises.

Additional Practice Exercise 2 Simplify, using the quotient rule. Express the answers in exponential form. All variables represent nonzero quantities.

a. $\dfrac{5^6}{5^4}$

b. $\dfrac{9^7}{9^4}$

c. $\dfrac{r^7}{r^2}$

All the exponents that we have discussed to this point have been positive integers. If the restriction that $m > n$ is removed from the quotient rule, it would be possible to have zero or negative exponents. By the quotient rule, $\frac{2^3}{2^3} = 2^{3-3} = 2^0$. What does 2^0 mean? It surely cannot mean "Use 2 as a factor 0 times." Example 3 gives meaning to 0 as an exponent.

Example 3 Simplify the following:

a. $\dfrac{2^4}{2^4} = \dfrac{2 \cdot 2 \cdot 2 \cdot 2}{2 \cdot 2 \cdot 2 \cdot 2} = \dfrac{2}{2} \cdot \dfrac{2}{2} \cdot \dfrac{2}{2} \cdot \dfrac{2}{2} = 1 \cdot 1 \cdot 1 \cdot 1 = 1.$

However, if we use the quotient rule, $\dfrac{2^4}{2^4} = 2^{4-4} = 2^0$. Since $\dfrac{2^4}{2^4} = 1$ and 2^0,

we conclude that $2^0 = 1$.

b. $\dfrac{x^3}{x^3} = \dfrac{x \cdot x \cdot x}{x \cdot x \cdot x} = \dfrac{x}{x} \cdot \dfrac{x}{x} \cdot \dfrac{x}{x} = 1 \cdot 1 \cdot 1 = 1.$

If we use the quotient rule, $\dfrac{x^3}{x^3} = x^{3-3} = x^0$. Since $\dfrac{x^3}{x^3} = 1$ and x^0,

we conclude that $x^0 = 1$ for $x \neq 0$.

Answers: Practice Exercise 2: a. 4^2 **b.** 6^4 **c.** x^3 **Additional Practice 2: a.** 5^2 **b.** 9^3 **c.** r^5

OBJECTIVE **B** Simplifying expressions with zero exponents

Based on Example 3, we make the following definition for 0 exponents:

> **Definition** 0 Exponents
>
> For all $a \neq 0$, $a^0 = 1$. Therefore, any nonzero number raised to the 0 power is equal to 1.

In using 0 as an exponent, great care must be taken in determining the base of the exponent, as illustrated in Example 4. Remember, the base is the symbol immediately preceding the exponent, unless parentheses are used, in which case the base is everything inside the parentheses.

Note

0^0 is undefined because it implies division by 0.

Example 4 | **Evaluate. Assume all variables have nonzero values.**

a. $5^0 = 1$ Definition of 0 exponents.

b. $5x^0 = 5 \cdot 1 = 5$ The base of the 0 exponent is x only, not $5x$. Remember, $x \neq 0$.

c. $(5x)^0 = 1$ The base of the 0 exponent is $5x$.

d. $-3y^0 + 7x^0 =$ The bases of the 0 exponents are y and x only. Apply the definition of 0 exponents.

     $-3 \cdot 1 + 7 \cdot 1 =$ Apply the identity for multiplication.

     $-3 + 7 =$ Add -3 and 7.

     4 Answer.

e. $(4x)^0 - 6x^0 =$ The bases of the 0 exponents are $4x$ and x. Apply the definition of 0 exponents.

     $1 - 6 \cdot 1 =$ Apply the identity for multiplication.

     $1 - 6 =$ Add 1 and -6.

     -5 Answer.

Be Careful

$(-2)^0 \neq -1$. Anything, except 0, raised to the 0 power is 1.

PRACTICE EXERCISE 4

Evaluate. Assume that x and $z \neq 0$.

a. 8^0 **b.** $8x^0$ **c.** $(8x)^0$ **d.** $4z^0 - 9z^0$ **e.** $6x^0 - (-3x)^0$

If you need more practice, do the following Additional Practice Exercises.

Additional Practice Exercise 4   Evaluate. Assume that $a, x,$ and $y \neq 0$.

a. 10^0 **b.** $10y^0$ **c.** $(10x)^0$ **d.** $3a^0 - 7a^0$ **e.** $(-5x)^0 - 8x^0$

Now we will consider negative exponents.

Answers: Practice Exercise 4: a. 1 **b.** 8 **c.** 1 **d.** -5 **e.** 5   **Additional Practice 4: a.** 1 **b.** 10 **c.** 1 **d.** -4 **e.** -7

Calculator Exploration Activity 2 (Optional)

Evaluate each of the following columns by using a calculator:

Column A	Column B
1. $2^{-3} =$	$\dfrac{1}{2^3} =$
2. $5^{-2} =$	$\dfrac{1}{5^2} =$
3. $4^{-4} =$	$\dfrac{1}{4^4} =$

By looking at the corresponding lines of columns A and B, answer the following:

2. Raising a number to a negative power is the same as _____.

We will now give a mathematical justification for the previous observation.

OBJECTIVE Simplifying expressions with negative integer exponents

If we use the quotient rule, $\frac{3^2}{3^5} = 3^{2-5} = 3^{-3}$. What does 3^{-3} mean? It cannot mean to use 3 as a factor -3 times. The following examples will give meaning to these types of expressions:

Example 5 Simplify the following:

a. $\dfrac{3^2}{3^4} = \dfrac{3 \cdot 3}{3 \cdot 3 \cdot 3 \cdot 3} = \dfrac{3}{3} \cdot \dfrac{3}{3} \cdot \dfrac{1}{3} \cdot \dfrac{1}{3} = 1 \cdot 1 \cdot \dfrac{1}{3} \cdot \dfrac{1}{3} = \dfrac{1}{3^2}.$

If we use the quotient rule, we get $\dfrac{3^2}{3^4} = 3^{2-4} = 3^{-2}.$

Since $\dfrac{3^2}{3^4} = \dfrac{1}{3^2}$ and 3^{-2}, then $3^{-2} = \dfrac{1}{3^2}.$

b. $\dfrac{x^3}{x^4} = \dfrac{x \cdot x \cdot x}{x \cdot x \cdot x \cdot x} = \dfrac{x}{x} \cdot \dfrac{x}{x} \cdot \dfrac{x}{x} \cdot \dfrac{1}{x} = 1 \cdot 1 \cdot 1 \cdot \dfrac{1}{x} = \dfrac{1}{x}.$

If we use the quotient rule, we get $\dfrac{x^3}{x^4} = x^{3-4} = x^{-1}.$

Since $\dfrac{x^3}{x^4} = \dfrac{1}{x}$ and x^{-1}, then $x^{-1} = \dfrac{1}{x}.$

From the preceding example, we generalize to the following definition for negative integer exponents:

Definition Negative Exponents

For all $x \neq 0, x^{-n} = \dfrac{1}{x^n}.$

For an alternative explanation, note that $x^n \cdot x^{-n} = x^{n+(-n)} = x^0 = 1, x \neq 0$. This means that x^n and x^{-n} are multiplicative inverses (reciprocals). Therefore, $x^{-n} = \frac{1}{x^n}$. For example, $2^1 \cdot 2^{-1} = 2^{1+(-1)} = 2^0 = 1$. This means that 2^1 and 2^{-1} are multiplicative inverses. Therefore, $2^{-1} = \frac{1}{2}$. Another way of stating it is that x^n denotes multiplication and x^{-n} denotes division for $n > 0$.

Example 6 Rewrite each expression with positive exponents only. Assume that all variables represent nonzero quantities.

a. $4^{-3} = \dfrac{1}{4^3} = \dfrac{1}{64}$ Definition of negative exponents and $4^3 = 64$.

b. $x^{-5} = \dfrac{1}{x^5}$ Definition of negative exponents.

Answers: Calculator Exploration Activity: 2. putting 1 over the base with a positive exponent

c. $3a^{-4} = 3 \cdot \dfrac{1}{a^4} = \dfrac{3}{a^4}$ The base of the exponent is a only, not $3a$.

d. $(4x)^{-2} = \dfrac{1}{(4x)^2} =$ The base of the exponent is $(4x)$.

$\dfrac{1}{4^2x^2} = \dfrac{1}{16x^2}$ Apply $(ab)^n = a^n b^n$ and $4^2 = 16$.

Be Careful

A common mistake is to confuse negative exponents with negative numbers. The value of $3^{-2} \neq -9$, and also $3^{-2} \neq (-2)(3)$. The value of $3^{-2} = \dfrac{1}{3^2} = \dfrac{1}{9}$. Another common error occurs in exercises like Example 6c, $3a^{-4}$. This is often mistakenly written as $\dfrac{1}{3a^4}$.

PRACTICE EXERCISE 6

Rewrite each expression with positive exponents only. Assume that all variables represent nonzero quantities.

a. 5^{-3} **b.** y^{-6} **c.** $6y^{-4}$ **d.** $(7z)^{-3}$

If you need more practice, do the following Additional Practice Exercises.

Additional Practice Exercise 6   Rewrite each of the following with positive exponents only:

a. 8^{-3} **b.** r^{-5} **c.** $-3a^{-5}$ **d.** $(4x)^{-2}$

We can now simplify expressions of the form $\dfrac{x^m}{x^n}$ where $m < n$.

Example 7   **Simplify each expression. Write answers in exponential form with positive exponents only. Assume all variables represent nonzero quantities.**

a. $\dfrac{3^2}{3^5} =$ Apply $\dfrac{x^m}{x^n} = x^{m-n}$.   **b.** $\dfrac{a^4}{a^9} =$ Apply $\dfrac{x^m}{x^n} = x^{m-n}$.

$3^{2-5} =$ Subtract the exponents. $a^{4-9} =$ Subtract the exponents.

$3^{-3} =$ Apply $x^{-n} = \dfrac{1}{x^n}$. $a^{-5} =$ Apply $x^{-n} = \dfrac{1}{x^n}$.

$\dfrac{1}{3^3} =$ Therefore, $\dfrac{3^2}{3^5} = \dfrac{1}{3^3}$. $\dfrac{1}{a^5}$ Therefore, $\dfrac{a^4}{a^9} = \dfrac{1}{a^5}$.

PRACTICE EXERCISE 7

Simplify each expression. Write answers in exponential form with positive exponents only. Assume all variables represent nonzero quantities.

a. $\dfrac{2^4}{2^6}$ **b.** $\dfrac{x^3}{x^{10}}$

If you need more practice, do the following Additional Practice Exercises.

(Continued)

Answers: Practice Exercise 6: a. $\dfrac{1}{5^3}$ **b.** $\dfrac{1}{y^6}$ **c.** $\dfrac{6}{y^4}$ **d.** $\dfrac{1}{(7z)^3} = \dfrac{1}{7^3z^3}$ **Additional Practice 6: a.** $\dfrac{1}{8^3}$ **b.** $\dfrac{1}{r^5}$ **c.** $\dfrac{-3}{a^5}$ **d.** $\dfrac{1}{(4x)^2} = \dfrac{1}{16x^2}$

Practice Exercise 7: a. $\dfrac{1}{2^2}$ **b.** $\dfrac{1}{x^7}$

Additional Practice Exercise 7   Simplify each expression.

a. $\dfrac{5^7}{5^{10}}$

b. $\dfrac{n^5}{n^6}$

Since 0 and negative exponents now have meaning, it is no longer necessary to restrict $m > n$ in the quotient rule. Therefore, $\frac{a^m}{a^n} = a^{m-n}$ for all integer values of m and n.

Rule: Generalized Quotient Rule

For all integer values of m and n, $\dfrac{a^m}{a^n} = a^{m-n}$ for $a \neq 0$.

Consequently, $\dfrac{x^7}{x^3} = x^{7-3} = x^4$, $\dfrac{x^6}{x^6} = x^{6-6} = x^0 = 1$, and $\dfrac{x^2}{x^6} = x^{2-6} = x^{-4} = \dfrac{1}{x^4}$.

Look for a pattern in simplifying an expression of the form $\frac{1}{x^{-n}}$.

Calculator Exploration Activity 3 (Optional)

Evaluate each of the following columns by using a calculator:

Column A	Column B
1. $\dfrac{1}{2^{-3}} =$	$2^3 =$
2. $\dfrac{1}{5^{-2}} =$	$5^2 =$
3. $\dfrac{1}{3^{-4}} =$	$3^4 =$

By looking at the corresponding lines of columns A and B, answer the following:

3. One divided by a number raised to a negative power is the same as _____
_____.

We will now give a mathematical justification for the previous observation.

Example 8

a. In the expression $(2^{-3})^{-1}$, the base of the exponent -1 is (2^{-3}). Therefore, by the definition of a negative exponent, $(2^{-3})^{-1} = \frac{1}{2^{-3}}$. However, using $(a^m)^n = a^{mn}$, which can be shown to also apply to negative exponents, $(2^{-3})^{-1} = 2^{(-3)(-1)} = 2^3$. Since $(2^{-3})^{-1} = \frac{1}{2^{-3}}$ and 2^3, then $\frac{1}{2^{-3}} = 2^3$.

b. In the expression $(x^{-4})^{-1}$, the base of the exponent -1 is (x^{-4}). Therefore, by the definition of a negative exponent, $(x^{-4})^{-1} = \frac{1}{x^{-4}}$. But, if we use $(a^m)^n = a^{mn}$, $(x^{-4})^{-1} = x^4$. Therefore, $\frac{1}{x^{-4}} = x^4$, since they both equal $(x^{-4})^{-1}$.

From the preceding examples, we make the following observation:

Rule: Observation

For all $x \neq 0$, $\dfrac{1}{x^{-n}} = x^n$.

Answers: Additional Practice 7: a. $\dfrac{1}{5^3}$ **b.** $\dfrac{1}{n}$ **Calculator Exploration Activity: 3.** the number raised to the corresponding positive power

Example 9   Write each expression with positive exponents only. Assume that all variables represent nonzero quantities.

a. $\dfrac{1}{2^{-4}} = 2^4$ Apply $\dfrac{1}{x^{-n}} = x^n$.

b. $\dfrac{1}{x^{-6}} = x^6$ Apply $\dfrac{1}{x^{-n}} = x^n$.

c. $\dfrac{2}{x^{-2}} = 2x^2$ Apply $\dfrac{1}{x^{-n}} = x^n$.

PRACTICE EXERCISE 9

Write each fraction with positive exponents only. Assume that all variables represent nonzero quantities.

a. $\dfrac{1}{5^{-7}}$ **b.** $\dfrac{1}{x^{-8}}$ **c.** $\dfrac{4}{y^{-3}}$

It can be shown that all the laws of exponents hold for zero and negative exponents. The examples which follow make that assumption. Remember, $a^m \cdot a^n = a^{m+n}$, $(ab)^n = a^n b^n$, and $(a^m)^n = a^{mn}$.

Example 10   Simplify each expression. Express answers in exponential form with positive exponents only. Assume that all variables represent nonzero quantities.

a. $3^{-3}3^5 =$ Apply $a^m a^n = a^{m+n}$.

     $3^{-3+5} =$ Add exponents.

     3^2 Answer.

b. $a^4 a^{-6} =$ Apply $a^m a^n = a^{m+n}$.

     $a^{4+(-6)} =$ Add exponents.

     $a^{-2} =$ Apply $a^{-n} = \dfrac{1}{a^n}$.

     $\dfrac{1}{a^2}$ Answer.

c. $\dfrac{x^{-4}}{x^2} =$ Apply $\dfrac{a^m}{a^n} = a^{m-n}$.

     $x^{-4-2} =$ Subtract exponents.

     $x^{-6} =$ Apply $a^{-n} = \dfrac{1}{a^n}$.

     $\dfrac{1}{x^6}$ Answer.

d. $\dfrac{a^{-3}}{a^{-5}} =$ Apply $\dfrac{a^m}{a^n} = a^{m-n}$.

     $a^{-3-(-5)} =$ $-(-5) = 5$.

     $a^{-3+5} =$ Add the exponents.

     a^2 Answer.

e. $(w^{-3})^2 =$ Apply $(a^m)^n = a^{mn}$.

     $w^{(-3)(2)} =$ Multiply the exponents.

     $w^{-6} =$ Apply $a^{-n} = \dfrac{1}{a^n}$.

     $\dfrac{1}{w^6}$ Answer.

Answers: Practice Exercise 9: a. 5^7 **b.** x^8 **c.** $4y^3$

PRACTICE EXERCISE 10

Simplify each expression. Express answers in exponential form with positive exponents only. Assume that all variables represent nonzero quantities.

a. $4^{-5}4^7$ **b.** w^3w^{-5} **c.** $\dfrac{b^{-4}}{b^2}$ **d.** $\dfrac{h^4}{h^{-3}}$ **e.** $(a^{-4})^{-2}$

If you need more practice, do the following Additional Practice Exercises.

Additional Practice Exercise 10   Simplify each expression. Express answers in exponential form with positive exponents only. Assume that all variables represent nonzero quantities.

a. $2^{-5} \cdot 2^3$ **b.** $r^{-2}r^5$ **c.** $\dfrac{z^{-3}}{z^3}$ **d.** $\dfrac{r^2}{r^{-5}}$ **e.** $(a^{-4})^2$

Exercise Set 2.6   MyMathLab®

For Extra Help

Simplify each expression. Express answers in exponential form with positive exponents only. Assume that all variables represent nonzero quantities. *(See Examples 1–10.)*

1. $\dfrac{2^6}{2^2}$ **2.** $\dfrac{3^8}{3^3}$ **3.** $\dfrac{c^5}{c}$ **4.** $\dfrac{d^7}{d}$ **5.** $\dfrac{z^9}{z^5}$

6. $\dfrac{a^8}{a^6}$ **7.** 2^{-4} **8.** 3^{-2} **9.** a^{-6} **10.** b^{-7}

11. $\dfrac{1}{5^{-3}}$ **12.** $\dfrac{1}{8^{-4}}$ **13.** $\dfrac{1}{y^{-5}}$ **14.** $\dfrac{1}{z^{-9}}$ **15.** 8^0

16. 6^0 **17.** $3a^0$ **18.** $7z^0$ **19.** $(-4s)^0$ **20.** $(-7h)^0$

21. $(3x)^0 + (2y)^0$ **22.** $(6w)^0 - (2z)^0$ **23.** $3x^0 - 4z^0$ **24.** $8q^0 - 2q^0$

25. $2x^{-3}$ **26.** $5a^{-2}$ **27.** $-7b^{-4}$ **28.** $-9c^{-6}$

29. $(3x)^4$ **30.** $(7a)^5$ **31.** $(2x)^{-3}$ **32.** $(5a)^{-2}$

33. $8^2 \cdot 8^4$ **34.** $7^3 \cdot 7^6$ **35.** $2^{-3} \cdot 2^5$ **36.** $4^5 \cdot 4^{-2}$

37. $5^{-6} \cdot 5^4$ **38.** $6^3 \cdot 6^{-7}$ **39.** $a^4 \cdot a^6$ **40.** $r^3 \cdot r^9$

41. $x^{-3}x^{-2}$ **42.** $y^{-4}y^{-5}$ **43.** $z^{-4}z^2$ **44.** q^5q^{-2}

45. $\dfrac{a^7}{a^3}$ **46.** $\dfrac{r^{10}}{r^4}$ **47.** $\dfrac{x^2}{x^5}$ **48.** $\dfrac{y^4}{y^5}$

49. $\dfrac{5^3}{5^6}$ **50.** $\dfrac{4^2}{4^6}$ **51.** $\dfrac{3^{-2}}{3^2}$ **52.** $\dfrac{2^{-4}}{2^5}$

Answers: Practice Exercise 10: a. 4^2 **b.** $\dfrac{1}{w^2}$ **c.** $\dfrac{1}{b^6}$ **d.** h^7 **e.** a^8 **Additional Practice 10: a.** $\dfrac{1}{2^2}$ **b.** r^3 **c.** $\dfrac{1}{z^6}$ **d.** r^7 **e.** $\dfrac{1}{a^8}$

53. $\dfrac{y^{-4}}{y^5}$

54. $\dfrac{x^{-4}}{x^3}$

55. $\dfrac{6^5}{6^{-2}}$

56. $\dfrac{4^4}{4^{-5}}$

57. $\dfrac{r^2}{r^{-5}}$

58. $\dfrac{x^6}{x^{-3}}$

59. $\dfrac{6^{-2}}{6^{-3}}$

60. $\dfrac{5^{-3}}{5^{-5}}$

61. $\dfrac{z^{-8}}{z^{-4}}$

62. $\dfrac{x^{-5}}{x^{-2}}$

63. $(8^3)^6$

64. $(5^2)^5$

65. $(r^5)^4$

66. $(p^4)^4$

67. $(6^3)^{-5}$

68. $(4^5)^{-4}$

69. $(y^7)^{-3}$

70. $(t^6)^{-2}$

71. $(8^{-3})^6$

72. $(6^{-1})^4$

73. $(z^3)^{-5}$

74. $(b^4)^{-2}$

75. $(9^{-5})^{-2}$

76. $(6^{-2})^{-5}$

Challenge Exercises (77–86)

Simplify each expression. Express answers with positive exponents only. Assume that all variables represent nonzero quantities. (See Example 9.)

77. $(2x^{-3}y^2)^2(-3x^3y^{-3})^3$

78. $(-4a^2b^{-3})^3(5a^{-3}b^{-3})^2$

79. $(2m^{-4}n^2)^{-2}(-2m^2n^{-4})^3$

80. $(8c^{-5}d^3)^{-1}(8c^3d^{-5})^2$

81. $\dfrac{(4a^{-4}b^2)^3}{(2a^3b^{-4})^3}$

82. $\dfrac{(6m^{-5}n^2)^2}{(3m^2n^{-6})^2}$

83. $\dfrac{x^{3a}}{x^a}$

84. $\dfrac{y^{4b}}{y^{2b}}$

85. $\dfrac{a^{3m-1}}{a^{2m-2}}$

86. $\dfrac{b^{3n+6}}{b^{2n-4}}$

Writing Exercises (87–89)

87. How does 4^{-3} differ from $(-3)(4)$?

88. Why is $3^{-3} \neq -27$?

89. Explain what is wrong with the following: $\dfrac{4^5}{4^3} = 1^{5-3} = 1^2 = 1$

Critical-Thinking Exercises (90 and 91)

90. Is it possible to raise a positive number to a power and get a negative answer? Why or why not?

91. We know that $x^0 = 1$ if $x \neq 0$. Why can $x \neq 0$?

Section 2.7 Power Rule for Quotients and Using Combined Laws of Exponents

OBJECTIVES *When you complete this section, you will be able to:*

A Simplify exponential expressions by using the property $\left(\dfrac{a}{b}\right)^n = \dfrac{a^n}{b^n}$.

B Simplify exponential expressions by using the property $\left(\dfrac{a}{b}\right)^{-n} = \left(\dfrac{b}{a}\right)^n$.

C Simplify expressions that involve integer exponents by using more than one law of exponents.

PREREQUISITE SKILLS *Before beginning this section, you should be able to:*

a. Raise integers to powers (Section 2.1).

b. Add, subtract, and multiply integers (Sections 1.6, 1.7, and 2.1).

c. Apply $a^m \cdot a^n = a^{m+n}$, $(ab)^n = a^n b^n$, and $(a^m)^n = a^{mn}$ (Section 2.2).

d. Apply $\dfrac{a^m}{a^n} = a^{m-n}$, $x^0 = 1$, and $x^{-n} = \dfrac{1}{x^n}$ (Section 2.6).

GETTING READY FOR SECTION 2.7

1. Find the following power of the integer: $(-2)^6$

Add, subtract, or multiply the following integers:

2. $-3 + 6$　　　　　　　　　　　**3.** $6(-3)$

Simplify the following, using $a^m \cdot a^n = a^{m+n}$, $(ab)^n = a^n b^n$, and $(a^m)^n = a^{mn}$:

4. $(2a^3b)^2$　　　　　　　　　　**5.** $(x^3)^2(x^4)^3$

Simplify the following, using $\dfrac{a^m}{a^n} = a^{m-n}$, $x^0 = 1$, and $x^{-n} = \dfrac{1}{x^n}$:

6. $\dfrac{x^4}{x^4}$　　　　　　　　　　　**7.** $\dfrac{a^2}{a^6}$

Introduction

One of the laws of exponents for products is $(ab)^n = a^n b^n$. This means that to raise a product to a power, you raise each of the factors to the power. There is a similar property for quotients.

Calculator Exploration Activity (Optional)

Evaluate each of the columns by using a calculator.

Column A	Column B
1. $\left(\dfrac{2}{5}\right)^2 =$	$\dfrac{2^2}{5^2} =$
2. $\left(\dfrac{3}{4}\right)^3 =$	$\dfrac{3^3}{4^3} =$
3. $\left(\dfrac{3}{2}\right)^4 =$	$\dfrac{3^4}{2^4} =$

By comparing corresponding lines in columns A and B, answer the following:

1. Raising a fraction to a power is the same as _____ _____ .

We will now give a mathematical justification for the previous observation.

OBJECTIVE **A** Simplifying exponential expressions by using the property $\left(\dfrac{a}{b}\right)^n = \dfrac{a^n}{b^n}$

Remember: To multiply fractions, you multiply numerator times numerator and divide by denominator times denominator. In general, $\dfrac{a}{b} \cdot \dfrac{c}{d} = \dfrac{a \cdot c}{b \cdot d}$.

Example 1 Simplify, using the definition of exponents. Express the answer in exponential form. Assume that all variables represent nonzero quantities.

a. $\left(\dfrac{2}{3}\right)^3 =$ Rewrite by using the definition of exponents.

$\dfrac{2}{3} \cdot \dfrac{2}{3} \cdot \dfrac{2}{3} =$ Multiply the fractions.

$\dfrac{2 \cdot 2 \cdot 2}{3 \cdot 3 \cdot 3} =$ Rewrite by using the definition of exponents.

$\dfrac{2^3}{3^3}$ Answer.

Notice that *both* the numerator and the denominator are raised to the third power.

b. $\left(\dfrac{x}{y}\right)^4 =$ Rewrite by using the definition of exponents.

$\dfrac{x}{y} \cdot \dfrac{x}{y} \cdot \dfrac{x}{y} \cdot \dfrac{x}{y} =$ Multiply the fractions. Rewrite by using the definition of exponents.

$\dfrac{x^4}{y^4}$ Answer.

Notice that *both* the numerator and the denominator are raised to the fourth power.

Based on the preceding example, we generalize to the following law for raising a quotient to a power:

> **Rule: Fifth Law of Exponents: Power Rule for Quotients**
>
> For any positive integer n and for $b \neq 0$,
> $$\left(\dfrac{a}{b}\right)^n = \dfrac{a^n}{b^n}.$$
>
> To raise a quotient to a power, raise both the numerator and the denominator to the power.

Example 2 Simplify, using the power rule for quotients. Express the answer in exponential form. Assume that all variables represent nonzero quantities.

a. $\left(\dfrac{2}{5}\right)^4 = \dfrac{2^4}{5^4}$ Apply $\left(\dfrac{a}{b}\right)^n = \dfrac{a^n}{b^n}$.

b. $\left(\dfrac{a}{b}\right)^5 = \dfrac{a^5}{b^5}$ Apply $\left(\dfrac{a}{b}\right)^n = \dfrac{a^n}{b^n}$.

c. $\left(\dfrac{3}{y}\right)^2 = \dfrac{3^2}{y^2}$ Apply $\left(\dfrac{a}{b}\right)^n = \dfrac{a^n}{b^n}$.

PRACTICE EXERCISE 2

Simplify, using the power rule for quotients. Express the answer in exponential form. Assume that all variables represent nonzero quantities.

a. $\left(\dfrac{5}{3}\right)^6$ **b.** $\left(\dfrac{r}{s}\right)^8$ **c.** $\left(\dfrac{z}{5}\right)^3$

(Continued)

Answers: Practice Exercise 2: a. $\dfrac{5^6}{3^6}$ **b.** $\dfrac{r^8}{s^8}$ **c.** $\dfrac{z^3}{5^3}$

If you need more practice, do the following Additional Practice Exercises.

Additional Practice Exercise 2   Simplify, using the power rule for exponents. Express the answer in exponential form.

a. $\left(\dfrac{3}{5}\right)^3$

b. $\left(\dfrac{a}{b}\right)^7$

c. $\left(\dfrac{m}{2}\right)^4$

Calculator Exploration Activity (Optional)

Evaluate each of the following columns by using a calculator:

Column A	Column B
1. $\left(\dfrac{2}{3}\right)^{-2}$	$\left(\dfrac{3}{2}\right)^{2}$
2. $\left(\dfrac{3}{4}\right)^{-3}$	$\left(\dfrac{4}{3}\right)^{3}$
3. $\left(\dfrac{2}{5}\right)^{-4}$	$\left(\dfrac{5}{2}\right)^{4}$

Answer the following based on your answers to the corresponding rows of column A and column B:

1. $\left(\dfrac{a}{b}\right)^{-n} = $ _____.

We will now give a mathematical justification of the preceding observation.

OBJECTIVE **B** Simplifying expressions by using the property $\left(\dfrac{a}{b}\right)^{-n} = \left(\dfrac{b}{a}\right)^{n}$

Another interesting property occurs when a quotient is raised to a negative power. We need to recall that $a^{-n} = \dfrac{1}{a^n}$ and $\dfrac{1}{a^{-n}} = a^n$

Example 3   **Simplify. Express the quotient with positive exponents only.**

a. $\left(\dfrac{a}{b}\right)^{-3} = $ Apply $\left(\dfrac{a}{b}\right)^{n} = \dfrac{a^n}{b^n}$.

$\dfrac{a^{-3}}{b^{-3}} = $ Rewrite as multiplication.

$\dfrac{a^{-3}}{1} \cdot \dfrac{1}{b^{-3}} = $ Apply $a^{-n} = \dfrac{1}{a^n}$ and $\dfrac{1}{a^{-n}} = a^n$.

$\dfrac{1}{a^3} \cdot b^3 = $ Multiply.

$\dfrac{b^3}{a^3} = $ Apply $\dfrac{a^n}{b^n} = \left(\dfrac{a}{b}\right)^{n}$.

$\left(\dfrac{b}{a}\right)^3$ Answer.

Therefore, $\left(\dfrac{a}{b}\right)^{-3} = \left(\dfrac{b}{a}\right)^3$. Notice the fraction has been inverted and the exponent has been changed to positive. Therefore, we have the following rule:

Answers: Additional Practice 2: a. $\dfrac{3^3}{5^3}$ **b.** $\dfrac{a^7}{b^7}$ **c.** $\dfrac{m^4}{2^4}$ **Calculator Exploration Activity:** $\left(\dfrac{b}{a}\right)^{n}$

> ## Rule: Quotients Raised to Negative Powers
>
> For $a \neq 0$ and $b \neq 0$, $\left(\dfrac{a}{b}\right)^{-n} = \left(\dfrac{b}{a}\right)^{n}$. To raise a fraction to a negative power, invert the fraction and change the power to positive.

Example 4 **Simplify. Express answers with positive exponents only. Assume that all variables represent nonzero quantities.**

a. $\left(\dfrac{3}{5}\right)^{-2} =$ Apply $\left(\dfrac{a}{b}\right)^{-n} = \left(\dfrac{b}{a}\right)^{n}$.

$\left(\dfrac{5}{3}\right)^{2} =$ Apply $\left(\dfrac{a}{b}\right)^{n} = \dfrac{a^n}{b^n}$.

$\dfrac{5^2}{3^2} =$ Evaluate the powers.

$\dfrac{25}{9}$ Answer.

b. $\left(\dfrac{x}{y}\right)^{-3} =$ Apply $\left(\dfrac{a}{b}\right)^{-n} = \left(\dfrac{b}{a}\right)^{n}$.

$\left(\dfrac{y}{x}\right)^{3} =$ Apply $\left(\dfrac{a}{b}\right)^{n} = \dfrac{a^n}{b^n}$.

$\dfrac{y^3}{x^3}$ Answer.

PRACTICE EXERCISE 4

Simplify. Express answers with positive exponents only. Assume that all variables represent nonzero quantities.

a. $\left(\dfrac{3}{5}\right)^{-3}$

b. $\left(\dfrac{a}{b}\right)^{-5}$

If you need more practice, do the following Additional Practice Exercises.

Additional Practice Exercise 4 Simplify. Express answers with positive exponents only. Assume that all variables represent nonzero quantities.

a. $\left(\dfrac{2}{7}\right)^{-4}$

b. $\left(\dfrac{m}{n}\right)^{-6}$

OBJECTIVE Simplifying expressions using more than one law of exponents

Often, it is necessary to use two or more of the laws of exponents to simplify an expression. For reference, a list of these laws and the properties of integer exponents follows:

> ## Rule: Laws of Exponents
>
> For any integers m and n and any real numbers a and b,
>
> **1.** $a^m a^n = a^{m+n}$ Product rule
>
> **2.** $(ab)^n = a^n b^n$ Power rule for products
>
> **3.** $(a^m)^n = a^{mn}$ Power to a power
>
> **4.** $\dfrac{a^m}{a^n} = a^{m-n}, a \neq 0$ Quotient rule

(Continued)

Answers: **Practice Exercise 4: a.** $\dfrac{5^3}{3^3}$ **b.** $\dfrac{b^5}{a^5}$ **Additional Practice 4: a.** $\dfrac{7^4}{2^4}$ **b.** $\dfrac{n^6}{m^6}$

5. $\left(\dfrac{a}{b}\right)^n = \dfrac{a^n}{b^n}, b \neq 0$ Power rule for quotients

6. $\left(\dfrac{a}{b}\right)^{-n} = \left(\dfrac{b}{a}\right)^n$ if $a, b, \neq 0$ Quotients to negative powers

7. $a^0 = 1$ if $a \neq 0$ Definition of zero exponents

8. $a^{-n} = \dfrac{1}{a^n}$ if $a \neq 0$ Definition of negative exponents

9. $\dfrac{1}{a^{-n}} = a^n$ if $a \neq 0$ Definition of negative exponents

Example 5   **Simplify. Express answers in exponential form with positive exponents only. Assume that all variables represent nonzero quantities.**

a. $\dfrac{x^2 x^4}{x^3} =$ Apply $a^m a^n = a^{m+n}$ in the numerator.

$\dfrac{x^6}{x^3} =$ Apply $\dfrac{a^m}{a^n} = a^{m-n}$.

$x^{6-3} =$ Subtract the exponents.

x^3 Answer.

b. $(3x^{-2}y^3)^{-3} =$ Apply $(ab)^n = a^n b^n$.

$(3^{-3})(x^{-2})^{-3}(y^3)^{-3} =$ Apply $a^{-n} = \dfrac{1}{a^n}$ and $(a^m)^n = a^{mn}$.

$\dfrac{1}{3^3}x^6 y^{-9} =$ Apply $a^{-n} = \dfrac{1}{a^n}$ and $3^3 = 27$.

$\dfrac{1}{27}x^6 \cdot \dfrac{1}{y^9} =$ Multiply the fractions.

$\dfrac{x^6}{27y^9}$ Answer.

c. $\dfrac{(4x^{-3})^2}{x^2 x^{-4}} =$ Apply $(a^m)^n = a^{mn}$ in the numerator and $a^m a^n = a^{m+n}$ in the denominator.

$\dfrac{4^2 x^{-6}}{x^{-2}} =$ Apply $\dfrac{a^m}{a^n} = a^{m-n}$ and $4^2 = 16$.

$16x^{-6-(-2)} =$ $-(-2) = +2$.

$16x^{-6+2} =$ Add exponents.

$16x^{-4} =$ Apply $x^{-n} = \dfrac{1}{x^n}$.

$\dfrac{16}{x^4}$ Answer.

d. $\left(\dfrac{x^4}{y^2}\right)^3 =$ Apply $\left(\dfrac{a}{b}\right)^n = \dfrac{a^n}{b^n}$.

$\dfrac{(x^4)^3}{(y^2)^3} =$ Apply $(a^m)^n = a^{mn}$.

$\dfrac{x^{12}}{y^6}$ Answer.

e. $\dfrac{(x^{-2})^3(x^3)^4}{(x^{-3})^3} =$ Apply $(a^m)^n = a^{mn}$.

$\dfrac{x^{-6}x^{12}}{x^{-9}} =$ Apply $a^m a^n = a^{m+n}$ in the numerator.

$\dfrac{x^6}{x^{-9}} =$ Apply $\dfrac{a^m}{a^n} = a^{m-n}$.

$x^{6-(-9)} =$ $-(-9) = +9$.

$x^{6+9} =$ Add exponents.

x^{15} Answer.

PRACTICE EXERCISE 5

Simplify each of the following. Express answers in exponential form with positive exponents only. Assume that all variables represent nonzero quantities.

a. $\dfrac{x^3x^5}{x^4}$

b. $(2a^{-3}b^2)^{-4}$

c. $\dfrac{(3x^4)^{-2}}{x^{-4}x^3}$

d. $\left(\dfrac{a^4}{b^2}\right)^4$

e. $\dfrac{(a^3)^{-3}(a^4)^{-2}}{(a^{-4})^2}$

If you need more practice, do the following Additional Practice Exercises.

Additional Practice Exercise 5   Simplify. Express answers in exponential form with positive exponents only. Assume that all variables represent nonzero quantities.

a. $\dfrac{x^5x^3}{x^6}$

b. $(4p^4q^{-3})^{-2}$

c. $\dfrac{(3x^{-1})^4}{x^{-3}x^{-2}}$

d. $\left(\dfrac{p^3}{q^2}\right)^5$

e. $\dfrac{(b^{-1})^4(b^{-3})^{-2}}{(b^3)^{-1}}$

Study Tip 10 Taking a Math Test

The type of test where you show all of your work and write out the answers is called an *open-ended test.* The type of test where you choose the correct answer from those presented is called a *multiple-choice test.* There are many things you can do before and during the test that can improve your score:

1. Arrive early so that you will be ready when the test is handed out.
2. As soon as you get the test, write down any formulas or definitions that you might need.
3. Read the test over from front to back. Mark the questions that you know you can answer with a check mark. Mark the ones you are unsure about with a question mark, and mark the ones you know you cannot answer with an X.
4. Answer the questions in this order:
   a. Those you know how to do, starting with the ones you think are easiest.
   b. Those you are unsure about.
   c. Those you initially thought you did not know how to do. Sometimes, doing the ones that you know reminds you of how to do others.
5. Estimate the amount of time needed for each question by dividing the number of minutes allowed for the test by the number of items on the

test. If you are spending more than this amount of time on an item, go to the next item and come back to it later if you have time.
6. Read the directions to each question carefully. Underline the key words, such as *not equal.*
7. On an open-ended test, write down the information given, what you are asked to find, and any relevant formulas, definitions, or theorems. Sometimes an estimate of the answer will give you a clue about how to solve the problem.
8. On an open-ended test, show all your work in a neat and organized manner. Box in your answers.
9. Check all your answers, if time permits.
10. If you are taking a multiple-choice test that has a penalty for wrong answers and you are able to narrow the answer to two choices, then guess. *If there is no penalty,* guess at any unanswered questions and leave nothing blank.
11. Take all of the time permitted for the test. Never worry about being the last one to leave a test. Generally, the first students to leave a test are either the ones who know everything or those who know very little.

Remember, the idea in taking a test is to show what you know, so attempt what you know first.

Exercise Set 2.7   For Extra Help   MyMathLab®

Write each of the following without exponents and evaluate. (See Examples 1–4.)

1. $\left(\dfrac{2}{3}\right)^2$

2. $\left(\dfrac{3}{4}\right)^3$

3. $\left(\dfrac{5}{2}\right)^3$

4. $\left(\dfrac{4}{3}\right)^2$

5. $\left(\dfrac{3}{2}\right)^{-2}$

6. $\left(\dfrac{2}{5}\right)^{-2}$

7. $\left(\dfrac{3}{7}\right)^{-4}$

8. $\left(\dfrac{5}{4}\right)^{-3}$

Answers: Practice Exercise 5: a. x^4 **b.** $\dfrac{a^{12}}{16b^8}$ **c.** $\dfrac{1}{9x^7}$ **d.** $\dfrac{a^{16}}{b^8}$ **e.** $\dfrac{1}{a^9}$ **Additional Practice 5: a.** x^2 **b.** $\dfrac{q^6}{16p^8}$ **c.** 81x **d.** $\dfrac{p^{15}}{q^{10}}$ **e.** b^5

Write each exercise term with positive exponents only. Assume that all variables represent nonzero quantities. (See Example 5.)

9. $\left(\dfrac{a}{b}\right)^3$

10. $\left(\dfrac{r}{s}\right)^2$

11. $\left(-\dfrac{p}{q}\right)^{-4}$

12. $\left(-\dfrac{x}{z}\right)^{-4}$

13. $\left(\dfrac{2x}{y}\right)^3$

14. $\left(\dfrac{3a}{b}\right)^4$

15. $\left(-\dfrac{p}{3q}\right)^{-3}$

16. $\left(-\dfrac{3x}{y}\right)^{-4}$

17. $(a^3b^{-5})^4$

18. $(r^{-2}s^2)^3$

19. $(x^4y^{-2})^{-6}$

20. $(z^{-2}w^3)^{-4}$

21. $(4x^2)^{-3}$

22. $(2x^3)^{-2}$

23. $(-2y^{-5})^{-2}$

24. $(-4x^{-3})^{-4}$

25. $(5a^{-2}b^{-5})^2$

26. $(2x^{-3}y^{-2})^3$

27. $(2w^3z^{-2})^{-2}$

28. $(3x^3y^{-4})^{-4}$

29. $\left(\dfrac{w^3}{z^6}\right)^3$

30. $\left(\dfrac{x^3}{y^5}\right)^3$

31. $\left(\dfrac{x^4}{z^6}\right)^{-5}$

32. $\left(\dfrac{a^5}{b^3}\right)^{-5}$

33. $\dfrac{(4w^3)(6w^4)}{(w^2)(w^2)}$

34. $\dfrac{(2x^2)(3x^7)}{(x^3)(x^3)}$

35. $\dfrac{(x^4)(3x^6)}{(-5x^8)(2x^3)}$

36. $\dfrac{(-5a^2)(3a^5)}{(a^6)(a^4)}$

37. $\dfrac{(x^4)^2}{(x^2)^3}$

38. $\dfrac{(y^3)^4}{(y^5)^2}$

39. $\dfrac{(a^{-3})^3}{(a^3)^2}$

40. $\dfrac{(z^{-2})^4}{(z^5)^2}$

41. $\dfrac{(a^{-2})^{-3}}{(a^{-4})^2}$

42. $\dfrac{(p^{-4})^{-2}}{(p^{-5})^3}$

43. $\dfrac{(2y^2)^4}{(y^4)^4}$

44. $\dfrac{(3q^3)^3}{(q^2)^6}$

45. $\dfrac{(x^2)^5}{x^2x^4}$

46. $\dfrac{(y^4)^3}{y^3y^2}$

47. $\dfrac{(3x^4)^0}{(2x^2)^2}$

48. $\dfrac{(5c^3)^2}{(8c^6)^0}$

49. $\dfrac{(7^0g^4)^3}{(2g^2)^4}$

50. $\dfrac{(4a^4)^3}{(5^0a^2)^4}$

51. $\dfrac{(x^{-2})^4}{x^3x^4}$

52. $\dfrac{(z^{-1})^5}{z^5z^2}$

53. $\dfrac{(q^2)^3(-q^4)^2}{(q^3)^5}$

54. $\dfrac{(-p^4)^2(p^5)^3}{(p^2)^7}$

55. $\dfrac{(4a^{-4})^3(2a^2)^{-3}}{(2a^{-2})^3}$

56. $\dfrac{(3b^5)^{-3}(2b^3)^{-3}}{(36b^4)^{-2}}$

57. $\dfrac{(2x^{-3})^3}{x^2x}$

58. $\dfrac{(2y^{-4})^4}{y^5y^2}$

59. $\dfrac{(-3a^{-3})^4}{(a^{-3})^5}$

60. $\dfrac{(-2b^{-2})^3}{(b^{-2})^4}$

Challenge Exercises (61–65)

Simplify. Leave answers with positive exponents only. Assume that all variables represent nonzero quantities.

61. $\dfrac{(4a^{-3}b^{-2})^{-2}(6a^3b^{-5})^3}{(2a^{-2}b^{-2})^{-4}}$

62. $\dfrac{(2x^{-3}y^2)^4(3xy^{-3})^{-2}}{(3x^{-4}y^{-2})^{-4}}$

63. $(a^mb^n)^q$

64. $\dfrac{(x^m)^n}{(x^p)^q}$

65. $\dfrac{(x^ay^b)^c}{(x^my^n)^d}$

Section 2.8 Division of Polynomials by Monomials

OBJECTIVES *When you complete this section, you will be able to:*

A Divide monomials by monomials.

B Divide polynomials of more than one term by monomials.

PREREQUISITE SKILLS *Before beginning this section, you should be able to:*

a. Divide integers (Section 2.5).

b. Apply $\dfrac{a^n}{a^m} = a^{m-n}$ and $a^{-n} = \dfrac{1}{a^n}$ (Section 2.6).

c. Subtract integers (Section 1.7).

d. Apply $(ab)^n = a^n b^n$ and $(a^m)^n = a^{mn}$ (Section 2.2).

GETTING READY FOR SECTION 2.8

1. Divide the following integers: $\dfrac{-56}{-8}$.

Simplify the following, using $\dfrac{a^m}{a^n} = a^{m-n}$ and $x^{-n} = \dfrac{1}{x^n}$:

2. $2a^{-5}$

3. $\dfrac{15x^4}{-3x^2}$

4. Subtract the following integer: $-4 - (-8)$

5. Simplify $(5m^2n^4)^2$, using $(ab)^n = a^n b^n$ and $(a^m)^n = a^{m+n}$

Introduction

Recall from Section 1.8 that a monomial is a polynomial with only one term. Examples are $4, 5x^2, 6xy, -4z^2y^3$, and $8p^3q^2r^5$. Remember, the exponent on any variable must be a whole number. This means the exponent on a variable cannot be negative, and there cannot be a variable in the denominator. Recall also that the number in front of the variable is called the *coefficient*. In order to divide a monomial by another monomial, we use the quotient rule discussed in Section 2.6. The division of polynomials by polynomials other than monomials is discussed in Section 6.7.

OBJECTIVE **A** Dividing a monomial by a monomial

Example 1 **Find the quotients of the monomials. Express answers with positive exponents only. Assume that all variables represent nonzero quantities only.**

a. $\dfrac{4x^4}{2x^3} =$ Divide the coefficients and variables separately.

$\dfrac{4}{2} \cdot \dfrac{x^4}{x^3} =$ Divide the coefficients and apply $\dfrac{a^m}{a^n} = a^{m-n}$ to the variables.

$2x$ Answer.

b. $\dfrac{-12x^5y^3}{3x^2y^2} =$ Divide the coefficients and like variables separately.

$\dfrac{-12}{3} \cdot \dfrac{x^5}{x^2} \cdot \dfrac{y^3}{y^2} =$ Divide the coefficients and apply $\dfrac{a^m}{a^n} = a^{m-n}$ to the variables.

$-4x^3y$ Answer.

Answers: Getting Ready: 1. 7 **2.** $\dfrac{2}{a^5}$ **3.** $-5x^2$ **4.** 4 **5.** $25m^4n^8$

c. $\dfrac{6x^3}{-2x^5} =$ Divide the coefficients and the variables separately.

$\dfrac{6}{-2} \cdot \dfrac{x^3}{x^5} =$ Divide the coefficients and apply $\dfrac{a^m}{a^n} = a^{m-n}$ to the variables.

$-3x^{-2} =$ Apply $x^{-n} = \dfrac{1}{x^n}$.

$-3 \cdot \dfrac{1}{x^2} =$ Multiply.

$\dfrac{-3}{x^2} = -\dfrac{3}{x^2}$ Answer.

d. $\dfrac{24a^4b^3z^6}{8a^2b^3z^8} =$ Divide the coefficients and like variables separately.

$\dfrac{24}{8} \cdot \dfrac{a^4}{a^2} \cdot \dfrac{b^3}{b^3} \cdot \dfrac{z^6}{z^8} =$ Divide the coefficients and apply $\dfrac{a^m}{a^n} = a^{m-n}$ to the variables.

$3a^2b^0z^{-2} =$ Apply $x^0 = 1$ and $x^{-n} = \dfrac{1}{x^n}$.

$3a^2 \cdot 1 \cdot \dfrac{1}{z^2} =$ Multiply.

$\dfrac{3a^2}{z^2}$ Answer.

e. $\dfrac{(4p^2q^3)^2}{8p^3q^4} =$ Apply $(ab)^n = a^n b^n$ in the numerator.

$\dfrac{(4)^2(p^2)^2(q^3)^2}{8p^3q^4} =$ Apply $(a^m)^n = a^{mn}$ in the numerator.

$\dfrac{16p^4q^6}{8p^3q^4} =$ Divide the coefficients and apply $\dfrac{a^m}{a^n} = a^{m-n}$ to the variables.

$2pq^2$ Answer.

PRACTICE EXERCISE 1

Find the quotients of the monomials. Express answers with positive exponents only. Assume that all variables represent nonzero quantities only.

a. $\dfrac{8y^6}{2y^3}$ **b.** $\dfrac{-18x^8y^7}{6x^3y^2}$ **c.** $\dfrac{16p^3}{-4p^5}$ **d.** $\dfrac{-15x^3y^5z^2}{3x^3y^3z^2}$ **e.** $\dfrac{(6x^3y^4)^2}{9x^5y^4}$

If you need more practice, do the following Additional Practice Exercises.

Additional Practice Exercise 1   Find the quotients of the monomials. Express answers with positive exponents only. Assume that all variables represent nonzero quantities.

a. $\dfrac{18a^6}{9a^4}$ **b.** $\dfrac{-24p^6q^6}{6p^4q^5}$ **c.** $\dfrac{-32r^5}{-4r^7}$ **d.** $\dfrac{-12a^4b^5c^3}{2a^3b^5c^6}$ **e.** $\dfrac{(4a^4b^4)^3}{8a^6b^8}$

OBJECTIVE Dividing polynomials of more than one term by monomials

Now we will divide polynomials by monomials. From Section R.4, we know how to add fractions with a common denominator, $\dfrac{a}{c} + \dfrac{b}{c} = \dfrac{a+b}{c}$. If we "turn this expression around," it gives us the method for dividing a polynomial by a monomial, $\dfrac{a+b}{c} = \dfrac{a}{c} + \dfrac{b}{c}$. We summarize in the following rule:

Answers: Practice Exercise 1: a. $4y^3$ **b.** $-3x^5y^5$ **c.** $-\dfrac{4}{p^2}$ **d.** $-5y^2$ **e.** $4xy^4$ **Additional Practice 1: a.** $2a^2$ **b.** $-4p^2q$ **c.** $\dfrac{8}{r^2}$ **d.** $-\dfrac{6a}{c^3}$ **e.** $8a^6b^4$

Rule: Division of a Polynomial by a Monomial

$$\frac{a + b}{c} = \frac{a}{c} + \frac{b}{c}$$

To divide a polynomial by a monomial, divide each term of the polynomial by the monomial.

Example 2 | **Find the quotient of each polynomial and monomial. Assume that all variables represent nonzero quantities only.**

a. $\dfrac{4x + 6}{2} =$ Divide each term of the polynomial by the monomial.

$\dfrac{4x}{2} + \dfrac{6}{2} =$ Divide the coefficients.

$2x + 3$ Quotient.

b. $\dfrac{6r^4 + 9r^5}{3r^2} =$ Divide each term of the polynomial by the monomial.

$\dfrac{6r^4}{3r^2} + \dfrac{9r^5}{3r^2} =$ Divide the coefficients and apply $\dfrac{a^m}{a^n} = a^{m-n}$ to the variables.

$2r^2 + 3r^3$ Quotient.

c. $\dfrac{12x^3 + 2x}{2x} =$ Divide each term of the polynomial by the monomial.

$\dfrac{12x^3}{2x} + \dfrac{2x}{2x} =$ Divide the coefficients and apply $\dfrac{a^m}{a^n} = a^{m-n}$ to the variables.

$6x^2 + 1 =$ Quotient.

d. $\dfrac{8y^4 - 12y^2}{4y^3} =$ Divide each term of the polynomial by the monomial.

$\dfrac{8y^4}{4y^3} - \dfrac{12y^2}{4y^3} =$ Divide the coefficients and apply $\dfrac{a^m}{a^n} = a^{m-n}$ to the variables.

$2y - 3y^{-1} =$ Apply $a^{-n} = \dfrac{1}{a^n}$.

$2y - \dfrac{3}{y}$ Quotient.

e. $\dfrac{10n^5 + 15n^4 - 3n^3}{5n^2} =$ Divide each term of the polynomial by the monomial.

$\dfrac{10n^5}{5n^2} + \dfrac{15n^4}{5n^2} - \dfrac{3n^3}{5n^2} =$ Divide the coefficients and apply $\dfrac{a^m}{a^n} = a^{m-n}$ to the variables.

$2n^3 + 3n^2 - \dfrac{3n}{5}$ Answer.

f. $\dfrac{12x^3y^4 - 16x^4y^3 + 3x^2y^5}{4x^3y^4} =$ Divide each term of the polynomial by the monomial.

$\dfrac{12x^3y^4}{4x^3y^4} - \dfrac{16x^4y^3}{4x^3y^4} + \dfrac{3x^2y^5}{4x^3y^4} =$ Divide the coefficients and apply $\dfrac{a^m}{a^n} = a^{m-n}$ to the variables.

$3x^0y^0 - 4xy^{-1} + \dfrac{3}{4}x^{-1}y =$ Apply $x^0 = 1$ and $x^{-n} = \dfrac{1}{x^n}$.

$3 - \dfrac{4x}{y} + \dfrac{3y}{4x}$ Answer.

PRACTICE EXERCISE 2

Find the quotient of each polynomial and monomial. Assume that all variables represent nonzero quantities only.

a. $\dfrac{6x + 12}{2}$

b. $\dfrac{10a^5 + 25a^3}{5a^2}$

c. $\dfrac{14x^2 + 7x}{7x}$

d. $\dfrac{12r^4 - 20r^2}{4r^3}$

e. $\dfrac{24p^6 - 12p^4 + 5p^7}{6p^3}$

f. $\dfrac{21p^2q^4 - 14p^5q^3 + 6p^3q^7}{7p^3q^5}$

If you need more practice, do the following Additional Practice Exercises.

Additional Practice Exercise 2 Find the quotient of each polynomial and monomial. Assume that all variables represent nonzero quantities only:

a. $\dfrac{8x - 12}{4}$

b. $\dfrac{21x^6 + 15x^3}{3x^2}$

c. $\dfrac{32a^4 + 4a}{4a}$

d. $\dfrac{15b^5 - 25b^2}{5b^4}$

e. $\dfrac{8x^6 + 16x^5 - 3x^4}{8x^3}$

f. $\dfrac{9r^4s^5 - 15r^2s^3 + 2r^3s^7}{3r^3s^4}$

Exercise Set 2.8

For Extra Help

MyMathLab®

Find the quotients of the following monomials. (See Example 1.)

1. $\dfrac{5x^2}{x}$

2. $\dfrac{6y^3}{y}$

3. $\dfrac{-3z^4}{z^2}$

4. $\dfrac{-5a^5}{a^3}$

5. $\dfrac{6x^4}{3x^2}$

6. $\dfrac{14b^5}{2b^2}$

7. $\dfrac{-15a^4}{3a^7}$

8. $\dfrac{-24w^5}{8w^9}$

9. $\dfrac{18n^6}{-3n^4}$

10. $\dfrac{28u^8}{-4u^2}$

11. $\dfrac{-30c^4}{-15c^4}$

12. $\dfrac{-32w^6}{-16w^6}$

13. $\dfrac{x^4y^5}{x^2y^3}$

14. $\dfrac{a^6b^4}{a^4b^3}$

15. $\dfrac{r^3s^2}{-r^5s^2}$

16. $\dfrac{-q^5r^2}{q^5r^5}$

17. $\dfrac{18x^5y^3}{6x^3y}$

18. $\dfrac{28y^6z^4}{7y^3z^2}$

19. $\dfrac{-32x^3y^5}{8x^6y^2}$

20. $\dfrac{-16a^2b^6}{8a^4b^3}$

21. $\dfrac{m^3n^2p^5}{m^4np^2}$

22. $\dfrac{a^5b^3c^8}{a^3b^7c^4}$

23. $\dfrac{48x^4y^6z^8}{-12x^6y^3z^{10}}$

24. $\dfrac{36a^3b^5c^2}{-18a^5b^2c}$

25. $\dfrac{-28m^2n^5p^3}{-14m^2n^3p^6}$

26. $\dfrac{-32x^4y^2z^6}{-8x^4y^6z^3}$

27. $\dfrac{(6x^3y^5)^2}{4x^4y^8}$

28. $\dfrac{(8a^4b^3)^2}{16a^6b^4}$

29. $\dfrac{-(5x^5y^2)^2}{5x^{10}y^6}$

30. $\dfrac{-(3m^3n^4)^2}{3m^{10}n^8}$

Answers: Practice Exercise 2: a. $3x + 6$ **b.** $2a^3 + 5a$ **c.** $2x + 1$ **d.** $3r - \dfrac{5}{r}$ **e.** $4p^3 - 2p + \dfrac{5p^4}{6}$ **f.** $\dfrac{3}{pq} - \dfrac{2p^2}{q^2} + \dfrac{6q^2}{7}$

Additional Practice 2: a. $2x - 3$ **b.** $7x^4 + 5x$ **c.** $8a^3 + 1$ **d.** $3b - \dfrac{5}{b^2}$ **e.** $x^3 + 2x^2 - \dfrac{3}{8}x$ **f.** $3rs - \dfrac{5}{rs} + \dfrac{2}{3}s^3$

Find the following quotients of polynomials and monomials. (See Example 2.)

31. $\dfrac{3x + 9}{3}$

32. $\dfrac{2x + 8}{2}$

33. $\dfrac{3a^2 - a}{a}$

34. $\dfrac{7x^2 - x}{x}$

35. $\dfrac{3y + 7}{y}$

36. $\dfrac{2x + 4}{x}$

37. $\dfrac{12z^2 - 6z}{6z}$

38. $\dfrac{4x^2 - 2x}{2x}$

39. $\dfrac{24y^3 + 16y}{8y^2}$

40. $\dfrac{12x^3 + 8x}{4x^2}$

41. $\dfrac{20a^5b^3 - 10a^3b^6}{5a^2b^2}$

42. $\dfrac{6x^3y^4 - 12x^4y^5}{3x^2y^2}$

43. $\dfrac{18mn^4 + 27m^4n^2}{9m^2n^3}$

44. $\dfrac{12x^2y^4 + 24x^5y^2}{6x^4y^3}$

45. $\dfrac{6x^2 + 9x - 3}{3}$

46. $\dfrac{4x^2 - 6x + 2}{2}$

47. $\dfrac{y^4 - y^3 - y}{y}$

48. $\dfrac{x^3 - x^2 - x}{x}$

49. $\dfrac{15m^4 - 5m^3 + 10m^2}{5m}$

50. $\dfrac{9x^4 - 3x^3 + 6x^2}{3x}$

51. $\dfrac{12y^4 - 16y^3 + 8y}{4y^2}$

52. $\dfrac{6x^3 - 12x^2 + 9x}{3x^2}$

53. $\dfrac{24m^3n^2 - 16m^4n^3 + 8m^2n^2}{8m^2n^2}$

54. $\dfrac{21x^2y^4 - 12x^3y^5 + 3xy^2}{3xy^2}$

55. $\dfrac{12b^4 - 3b^2}{4b^3}$

56. $\dfrac{6x^3 - 4x}{3x^2}$

57. $\dfrac{24mn^5 - 18m^4n^2}{6m^3n^4}$

58. $\dfrac{5a^6b^2 - 15a^3b^5}{5a^5b^3}$

59. $\dfrac{28r^7 - 14r^5 + 21r^3}{7r^5}$

60. $\dfrac{16x^5 + 8x^4 - 32x^3}{8x^4}$

61. $\dfrac{30r^7 - 12r^5 + 5r^4}{6r^5}$

62. $\dfrac{24a^5 + 8a^4 - 3a^3}{4a^4}$

63. $\dfrac{45r^9 - 18r^3 - 2r^2}{9r^4}$

64. $\dfrac{28b^8 - 14b^4 + 4b^3}{7b^5}$

65. $\dfrac{32p^3q^5 - 3p^6q^4 + 20p^4q^8}{4p^4q^6}$

66. $\dfrac{36m^4n^5 - 27m^3n^3 + 4m^5n^4}{9m^3n^4}$

Write an expression for each of the following, and simplify:

67. Find the quotient of $14m^4n^8$ and $7m^3n^6$.

68. Find the quotient of $18x^6y^9$ and $9x^4y^3$.

69. Divide $-32m^4n^2$ by $-16m^7n$.

70. Divide $-24a^3b^2$ by $-12a^5b$.

71. Find the quotient of $16a^7 - 24a^5$ and $8a^3$.

72. Find the quotient of $32x^5 - 36x^4$ and $4x^2$.

73. Divide $12y^5 - 16y^3 - 8$ by $4y^3$.

74. Divide $16x^4 - 32x^2 + 24$ by $8x^2$.

75. Divide the sum of $8x^3y^3$ and $-2x^3y^3$ by $3xy$.

76. Divide the sum of $6x^2y^3$ and $-2x^2y^3$ by $2xy$.

77. Divide the sum of $4y^5 - 6y^3 + y^2$ and $2y^5 - 4y^3 + 7y^2$ by $2y^2$.

78. Divide the sum of $2x^4 - 7x^3 + 3x^2$ and $4x^4 - 2x^3 - 6x^2$ by $3x^2$.

Writing Exercises (79–82)

79. Is the quotient of a polynomial with a monomial always a polynomial? Why or why not? If not, give examples.

80. Is the quotient of a monomial with a monomial always a monomial? Why or why not? If not, give examples.

81. Explain what is wrong with the following:

$$\frac{4x^2 + 5x + 6}{6} = \frac{4x^2 + 5x + 6^1}{\cancel{6}_1} = 4x^2 + 5x + 1.$$

82. Does the quotient of a trinomial with a monomial always have three terms?

Critical-Thinking Exercises (83–86)

83. The quotient of a polynomial and $3xy$ is $2x^2y - 6x^3$. What is the polynomial? Why?

84. Without using a calculator, evaluate $\dfrac{x^{1000} - x^{999}}{x^{999}}$ when $x = 1000$.

85. If the area of a rectangle is $8x^3 + 12x^2 + 4$ and the width is x^2, find the length.

86. If $\dfrac{12x^a - 18x^b + 9x^c + 6x^d}{3x} = 4x^3 - 6x^2 + 3x + 2$, find $a, b, c,$ and d.

<table>
<tr><td>**Section 2.9**</td><td>An Application of Exponents: Scientific Notation</td></tr>
</table>

OBJECTIVES *When you complete this section you will be able to:*

A Change numbers written in scientific notation to numbers in standard notation.

B Write numbers in scientific notation.

C Multiply and divide very large and/or very small numbers using scientific notation.

D Solve problems by using scientific notation.

PREREQUISITE SKILLS *Before beginning this section, you should be able to:*

a. Add and subtract integers (Sections 1.6 and 1.7).

b. Multiply and divide whole numbers and decimals (Sections R.4 and R.6).

c. Apply $a^m \cdot a^n = a^{m+n}$ (Section 2.2).

d. Apply $\dfrac{a^m}{a^n} = a^{m-n}$ (Section 2.6).

GETTING READY FOR SECTION 2.9

1. Add or subtract the following integers: $-4 + 7$

Multiply or divide the following whole numbers and decimals:

2. $(3.6)(6.8)$

3. $\dfrac{1.065}{4.26}$

4. Simplify $10^4 \cdot 10^{-6}$ by using $a^m \cdot a^n = a^{m+n}$. Leave the answers as a power of 10.

Simplify, by using $\dfrac{a^m}{a^n} = a^{m-n}$. Leave the answers as a power of 10.

5. $\dfrac{10^{-5}}{10^2}$

6. $\dfrac{10^{-3}}{10^{-6}}$

Introduction

A certain computer can perform 1,500,000,000,000 computations per second. Therefore, in a 30-day month it can perform $(1{,}500{,}000{,}000{,}000)(60)(60)(24)(30)$ computations. When this calculation was performed on one calculator, the result was 3.888E18. What does this mean?

Very large and very small numbers are used frequently in both the sciences and in everyday situations to refer to such things as the national debt, the gross national product, and the size of an atom. Frequently (as in the case in the previous paragraph), these numbers have so many digits that a calculator cannot display all of them. In such cases, we use an alternative method of writing extremely large and small numbers called **scientific notation**.

> **Definition** Scientific Notation
>
> A number is written in scientific notation if it is in the form of **$a \times 10^n$**, where $1 \le a < 10$ and n is an integer.

To say $1 \le a < 10$ means that there is one nonzero digit to the left of the decimal point. Remember that the integers are $\{\ldots -3, -2, -1, 0, 1, 2, 3, \ldots\}$. Examples of numbers written in scientific notation are

$$2.3 \times 10^4, \ 4.06 \times 10^7, \ 9.23 \times 10^{-5}, \ \text{and} \ 5.34 \times 10^{-3}.$$

Answers: Getting Ready: 1. 3 **2.** 24.48 **3.** .25 **4.** 10^{-2} **5.** 10^{-7} **6.** 10^3

Examples of numbers not written in scientific notation are

$$32 \times 10^5, 4 \times 8^{-4}, \text{ and } .78 \times 10^{-2}.$$

In performing the calculation (1,500,000,000,000)(60)(60)(24)(30), the calculator gave the answer in scientific notation with the number following the "E" representing the exponent of 10. Consequently, this computer can perform 3.888×10^{18} calculations per month. What does this number represent?

Before we write numbers in scientific notation, we need to make some observations regarding powers of 10 and how multiplying and dividing by powers of 10 affects the movement of the decimal point. Study the following table:

Power of 10	Value
10^0	1
10^1	10
10^2	100
10^3	1000
10^4	10,000

Observe that the exponent of 10 is the same as the number of zeros following the 1 in the value of the power of 10. Consequently, 10^7 equals 1 followed by seven 0s, or 10,000,000. Study the following table:

Power of 10	Value
10^{-1}	$\dfrac{1}{10} = .1$
10^{-2}	$\dfrac{1}{10^2} = \dfrac{1}{100} = .01$
10^{-3}	$\dfrac{1}{10^3} = \dfrac{1}{1000} = .001$
10^{-4}	$\dfrac{1}{10^4} = \dfrac{1}{10,000} = .0001$

Notice that the total number of decimal places is the same as the absolute value of the exponent of 10. Consequently, 10^{-6} has a total of 6 decimal places and equals .000001.

Study the following examples and look for a relationship between the exponent of 10 and the movement of the decimal point. When graphing integers on the number line, movement to the right is positive and movement to the left is negative. We will use the same idea in moving the decimal point.

Example 1 **Find the following products:**

a. 3.2×10^3 $10^3 = 1000$. Write vertically and multiply.

$$\begin{array}{r} 3.2 \\ \times\ 1000 \\ \hline 3200.0 \end{array}$$

Place three 0's behind 32 and mark off one decimal place. Notice that the decimal has been moved three places to the right, which is the same as the exponent of 10.

b. 4.67×10^5 Write vertically and multiply.

$$\begin{array}{r} 4.67 \\ \times\ 100000 \\ \hline 467000.00 \end{array}$$

Place five 0's behind 467 and mark off two decimal places. Notice that the decimal has been moved five places to the right, which is the same as the exponent of 10.

c. 2.452×10^{-4} $10^{-4} = .0001$. Write vertically and multiply.

$$\begin{array}{r} 2.452 \\ \times\ .0001 \\ \hline .0002452 \end{array}$$

$1 \times 2452 = 2452$. Mark off seven decimal places. Notice that the decimal has been moved four places to the left, which is the same as the absolute value of the exponent of 10.

d. 9.6×10^{-2} $10^{-2} = .01$. Write vertically and multiply.

$$\begin{array}{r} 9.6 \\ \times\ .01 \\ \hline .096 \end{array}$$

$1 \times 96 = 96$. Mark off three decimal places. Notice that the decimal has been moved two places to the left, which is the same as the absolute value of the exponent of 10.

Based on the results of Example 1, we make the following observation:

> **Procedure: Multiplying by Powers of 10**
>
> To find a product of the form $a \times 10^n$, move the decimal n places to the right if n is positive, or move the decimal $|n|$ places to the left if n is negative.

An easy way to remember this rule is that a positive exponent on 10 makes the product larger, so we move the decimal point to the right. A negative exponent makes the product smaller, so we move the decimal point to the left.

OBJECTIVE Changing numbers written in scientific notation to numbers in standard notation

Since all the numbers in Example 1 were given in scientific notation, the procedure for multiplying by powers of 10 is used to change a number from scientific notation into standard notation.

Example 2   **Change each of the following from scientific notation into standard notation:**

a. 6.89×10^4 Since 4 is positive, make the product larger by moving the decimal point four places to the right.

68,900   Standard notation.

b. 3.01×10^{-3} Since −3 is negative, make the product smaller by moving the decimal point three places to the left.

.00301   Standard notation.

PRACTICE EXERCISE 2

Change each of the following from scientific notation to standard notation:

a. 9.42×10^5 **b.** 1.72×10^{-5}

If you need more practice, do the following Additional Practice Exercises.

Additional Practice Exercise 2   Change each of the following from scientific notation to standard notation:

a. 7.82×10^2 **b.** 2.05×10^{-6}

OBJECTIVE Writing numbers in scientific notation

Now that we know how to multiply numbers by powers of 10 and change numbers from scientific notation to standard notation, we are ready to write numbers in scientific notation. Remember that in order for a number to be written in scientific notation, $a \times 10^n$, the "a" must have one nonzero digit to the left of the decimal. Therefore, our first task is to properly place the decimal. Our second task is to determine the exponent of 10 so that the number written in scientific notation is equal to the original number.

Answers: Practice Exercise 2: a. 942,000 **b.** .0000172   **Additional Practice 2: a.** 782 **b.** .00000205

Example 3   Write the following numbers in scientific notation:

a. 91,000

Solution

Since we need one nonzero digit to the left of the decimal, place the decimal between 9 and 1. Therefore, 91,000 written in scientific notation is of the form 9.1×10^n. Since $9.1 < 91,000$, we need to multiply 9.1 by a power of 10 that will make the product larger by moving the decimal point four places to the right. Therefore, $n = 4$. Consequently, $91,000 = 9.1 \times 10^4$.

b. .000091

Solution

Again, we place the decimal between 9 and 1. Therefore, .000091 written in scientific notation is of the form 9.1×10^n. Since $9.1 > .000091$, we need to multiply 9.1 by a power of 10 that will make the product smaller by moving the decimal point five places to the left. Therefore, $n = -5$. Consequently, $.000091 = 9.1 \times 10^{-5}$.

c. 43,600,000

Solution

Since we need one nonzero digit to the left of the decimal, place the decimal between 4 and 3. Therefore, 43,600,000 written in scientific notation is of the form 4.36×10^n. Since $4.36 < 43,600,000$, we need to multiply 4.36 by a power of 10 that will make the product larger by moving the decimal point seven places to the right. Therefore, $n = 7$. Consequently, $43,600,000 = 4.36 \times 10^7$.

d. .00361

Solution

Since we need one nonzero digit to the left of the decimal, place the decimal between 3 and 6. Therefore, .00361 written in scientific notation is of the form 3.61×10^n. Since $3.61 > .00361$, we need to multiply 3.61 by a power of 10 that will make the product smaller by moving the decimal point three places to the left. Therefore, $n = -3$. Consequently, $.00361 = 3.61 \times 10^{-3}$.

PRACTICE EXERCISE 3

Write the following numbers in scientific notation:

a. 470,000   **b.** .00000056   **c.** 5630   **d.** .000972

If you need more practice, do the following Additional Practice Exercises.

Additional Practice Exercise 3   Write the following in scientific notation:

a. 69,000,000   **b.** .00000000042   **c.** 925,000   **d.** .00000936

There is frequently a need for scientific notation in the various branches of the natural sciences, such as chemistry, physics, biology, and astronomy.

Example 4

a. The average distance between Earth and the Sun is 93,000,000 miles. Express this distance in scientific notation.

Solution

Place the decimal between 9 and 3. So, $93,000,000 = 9.3 \times 10^n$. If $9.3 \times 10^n = 93,000,000$, the decimal must be moved seven places to the right. Therefore, $n = 7$. Consequently, the average distance between Earth and the Sun is 9.3×10^7 miles.

Answers: Practice Exercise 3: a. 4.7×10^5 **b.** 5.6×10^{-7} **c.** 5.63×10^3 **d.** 9.72×10^{-4} **Additional Practice 3: a.** 6.9×10^7 **b.** 4.2×10^{-10} **c.** 9.25×10^5 **d.** 9.36×10^{-6}

Note

A joule is a unit of measure of energy. 4.184 J = 1 calorie. Hence, 4.184 J is the amount of heat required to raise the temperature of 1 gram of water 1 degree Celsius.

b. Einstein stated that the energy E equivalent to the mass m can be calculated by the equation $E = mc^2$ where c is the speed of light. According to this equation, 9.0×10^7 joules of energy is equivalent to 1.0×10^{-6} grams of mass. Write the number of grams of mass in standard form and give the fractional part of a gram that it represents.

Solution

Since the exponent of 10 is negative, move the decimal six places to the left. Therefore, 1.0×10^{-6} grams = .000001 grams. As a fraction, .000001 $= \frac{1}{1,000,000}$. Hence, according to Einstein's equation, one one-millionth of a gram of mass is converted into 90,000,000 joules of energy. This accounts for the tremendous amount of energy released during an atomic reaction.

PRACTICE EXERCISE 4

a. The age of Earth is estimated at about 4,500,000,000 years. Express the age of Earth in scientific notation.

b. The radius of an atom of a certain substance is 1.42×10^{-7} millimeters. Write the radius of this atom in standard form.

If you need more practice, do the following Additional Practice Exercises.

Additional Practice Exercise 4

a. The age of the universe is estimated to be at least 11,200,000,000 years. Express the age of the universe in scientific notation.

b. The distance between the carbon atoms in a diamond is 1.541×10^{-14} meters. Write the distance in standard form.

OBJECTIVE **Multiplying and dividing very large and/or very small numbers, using scientific notation**

In order to perform operations on numbers written in scientific notation, we need to recall two laws of exponents: $a^m \cdot a^n = a^{m+n}$ and $\frac{a^m}{a^n} = a^{m-n}$. The method that we use to multiply and divide numbers written in scientific notation is very much like the method used to multiply and divide monomials. We will refer to the "a" of $a \times 10^n$ as the coefficient and the "10^n" as the exponential factor. Consider the following:

Monomials	Scientific Notation
$(3.2x^2)(1.6x^4)$	$(3.2 \times 10^2)(1.6 \times 10^4)$
Regroup the factors so that the coefficients and variable factors are together.	Regroup the factors so that the coefficients and exponential factors are together.
$(3.2 \cdot 1.6)(x^2 \cdot x^4)$	$(3.2 \cdot 1.6) \times (10^2 \cdot 10^4)$
Multiply the coefficients and the variable factors.	Multiply the coefficients and the exponential factors.
$5.12x^6$	5.12×10^6
Therefore, $(3.2x^2)(1.6x^4) = 5.12x^6$.	Therefore, $(3.2 \times 10^2)(1.6 \times 10^4) = 5.12 \times 10^6$.
$\dfrac{2.6x^5}{1.3x^3}$	$\dfrac{2.6 \times 10^5}{1.3 \times 10^3}$
We think of this as coefficient divided by coefficient, and variable factors divided by variable factors.	We think of this as coefficient divided by coefficient, and exponential factors divided by exponential factors.
$\dfrac{2.6}{1.3} \cdot \dfrac{x^5}{x^3}$	$\dfrac{2.6}{1.3} \times \dfrac{10^5}{10^3}$

(Continued)

Answers: Practice Exercise 4: a. 4.5×10^9 years **b.** .000000142 mm   **Additional Practice 4: a.** 1.12×10^{10} years **b.** .00000000000001541

Monomials	Scientific Notation
Divide the coefficients and the variables separately. $2x^2$ Therefore, $\dfrac{2.6x^5}{1.3x^3} = 2x^2$.	Divide the coefficients and the exponentials separately. 2×10^2 Therefore, $\dfrac{2.6 \times 10^5}{1.3 \times 10^3} = 2 \times 10^2$.

When performing operations on numbers using scientific notation, the answers may not be in scientific notation because the a in $a \times 10^n$ is not between 0 and 10, as in Examples 5c and d which follow. Note how the numbers are converted into scientific notation.

Example 5   **Simplify, using scientific notation. Express answers in both scientific and standard notation.**

a. $(4.1 \times 10^3)(2.3 \times 10^{-6}) =$ Regroup the factors so that the coefficients and exponential parts are together.

$(4.1 \cdot 2.3) \times (10^3 \cdot 10^{-6}) =$ Multiply coefficients: $10^3 \cdot 10^{-6} = 10^{3+(-6)} = 10^{-3}$.

$9.43 \times 10^{-3} =$ Answer in scientific notation. Move the decimal three places to the left.

$.00943$ Answer in standard notation.

b. $\dfrac{7.44 \times 10^{-2}}{3.1 \times 10^4} =$ Divide coefficient by coefficient and exponential part by exponential part.

$\dfrac{7.44}{3.1} \times \dfrac{10^{-2}}{10^4} =$ Divide coefficients. $\dfrac{10^{-2}}{10^4} = 10^{-2-4} = 10^{-6}$.

$2.4 \times 10^{-6} =$ Answer in scientific notation. Move the decimal six places to the left.

$.0000024$ Answer in standard notation.

c. $(4100)(56,000) =$ Rewrite each number in scientific notation.

$(4.1 \times 10^3)(5.6 \times 10^4) =$ Regroup the factors so that the coefficients and exponential parts are together.

$(4.1 \cdot 5.6) \times (10^3 \cdot 10^4) =$ Multiply coefficients. $10^3 \cdot 10^4 = 10^{3+4} = 10^7$.

$22.96 \times 10^7 =$ Write 22.96 in scientific notation.

$(2.296 \times 10^1) \times 10^7 =$ Regroup using the associative property.

$2.296 \times (10^1 \times 10^7) =$ Apply $a^m \cdot a^n = a^{m+n}$.

2.296×10^8 Answer in scientific notation. Move the decimal 8 places to the right.

$229,600,000$ Answer in standard notation.

d. $\dfrac{(120,000)(.0018)}{3600} =$ Rewrite each number in scientific notation.

$\dfrac{(1.2 \times 10^5)(1.8 \times 10^{-3})}{3.6 \times 10^3} =$ Regroup in the numerator.

$\dfrac{(1.2 \cdot 1.8) \times (10^5 \cdot 10^{-3})}{3.6 \times 10^3} =$ Multiply coefficients. $10^5 \cdot 10^{-3} = 10^{5-3} = 10^2$.

$\dfrac{2.16 \times 10^2}{3.6 \times 10^3} =$ Divide coefficient by coefficient and exponential part by exponential part.

$\dfrac{2.16}{3.6} \times \dfrac{10^2}{10^3} =$ Divide. $\dfrac{10^2}{10^3} = 10^{2-3} = 10^{-1}$.

$.6 \times 10^{-1} =$ Write .6 in scientific notation.

$(6.0 \times 10^{-1}) \times 10^{-1} =$ Regroup using the associative property.

$6.0 \times (10^{-1} \times 10^{-1}) =$ Apply $a^m \cdot a^n = a^{m+n}$.

$6.0 \times 10^{-2} =$ Answer in scientific notation. Move the decimal 2 places to the left.

$.06$ Answer in standard form.

PRACTICE EXERCISE 5

Simplify each number by using scientific notation. Express answers in both scientific and standard notation:

a. $(4.4 \times 10^4)(2.3 \times 10^3)$ **b.** $\dfrac{9.3 \times 10^7}{3.0 \times 10^3}$ **c.** $(5300)(.000025)$ **d.** $\dfrac{.00345}{15000}$

If you need more practice, do the following Additional Practice Exercises.

Additional Practice Exercise 5 Simplify by using scientific notation. Express answers in both scientific and standard notation:

a. $(3.4 \times 10^{-5})(2.6 \times 10^8)$ **b.** $\dfrac{4.8 \times 10^7}{1.2 \times 10^4}$ **c.** $(.0000036)(68,000)$ **d.** $\dfrac{.0000840}{24,000}$

OBJECTIVE Solving problems by using scientific notation

Computations involving scientific notation often occur in application problems.

Example 6 **Solve, using scientific notation. Express the answers in standard notation.**

a. A certain type of computer can perform a calculation in .000000000003 of a second. How long would it take this computer to perform 5 billion (5,000,000,000) calculations?

Solution

The time required to perform 5,000,000,000 calculations is equal to

$(5,000,000,000)(.000000000003) =$ Write each number in scientific notation.
$(5 \times 10^9)(3 \times 10^{-12}) =$ Regroup the factors.
$(5 \cdot 3) \times (10^9 \cdot 10^{-12}) =$ Multiply coefficients. $10^9 \cdot 10^{-12} = 10^{9-12} = 10^{-3}$.
$15 \times 10^{-3} =$ Move the decimal three places to the right.
$.015$

> **Note**
>
> $15 \times 10^{-3} = 1.5 \times 10^{-2}$ written in scientific notation.

Therefore, it takes the computer .015 of a second to perform 5 billion computations.

b. The solubility constant of barium sulfate is 1.5×10^{-9}, and the solubility constant of silver bromide is 5×10^{-13}. How many times greater is the solubility constant of barium sulfate than that of silver bromide?

Solution

The number of times greater the solubility constant of barium sulfate is than that of silver bromide is equal to the solubility constant of barium sulfate divided by the solubility constant of silver bromide.

$\dfrac{1.5 \times 10^{-9}}{5 \times 10^{-13}}$ Divide coefficient by coefficient and exponential part by exponential part.

$\dfrac{1.5}{5} \times \dfrac{10^{-9}}{10^{-13}}$ Divide. $\dfrac{10^{-9}}{10^{-13}} = 10^{-9-(-13)} = 10^{-9+13} = 10^4$.

$.3 \times 10^4$ Move the decimal four places to the right.
3000

> **Note**
>
> $.3 \times 10^4 = 3 \times 10^3$ written in scientific notation.

Therefore, the solubility constant of barium sulfate is 3000 times greater than the solubility constant of silver bromide.

Answers: **Practice Exercise 5:** **a.** 1.012×10^8; 101,200,000 **b.** 3.1×10^4; 31,000 **c.** 1.325×10^{-1}; .1325 **d.** 2.3×10^{-7}; .00000023
Additional Practice 5: **a.** 8.84×10^3; 8840 **b.** 4.0×10^3; 4000 **c.** 2.448×10^{-1}; .2448 **d.** 3.5×10^{-9}; .0000000035

PRACTICE EXERCISE 6

Solve by using scientific notation. Express the answers in standard notation.

a. The mass of a helium atom is 6.65×10^{-24} grams. Find the mass of a sample of helium that contains 3.0×10^{26} atoms.

b. The approximate distance from Earth to the planet Saturn is 7.942×10^8 miles. How many hours would it take a spacecraft from Earth traveling at 20,000 miles per hour to reach Saturn?

If you need more practice, do the following Additional Practice Exercises.

Additional Practice Exercise 6   Solve by using scientific notation. Express the answers in standard notation:

a. The mass on one phosphorous atom is 5.0×10^{-23} grams. Find the mass of a piece of phosphorous that contains 4.1×10^{27} atoms.

b. The distance from Earth to the nearest star, Alpha Centauri, is approximately 3.8×10^{16} meters. The speed of light is approximately 3.0×10^8 meters per second. Find how long it takes light from Alpha Centauri to reach Earth.

Exercise Set 2.9

For Extra Help

MyMathLab®

Write the following in standard notation. (See Example 2.)

1. 3.5×10^4

2. 4.7×10^2

3. 9.5×10^{-3}

4. 6.3×10^{-5}

5. 4.79×10^6

6. 3.07×10^8

7. 9.24×10^{-6}

8. 2.19×10^{-2}

9. 4×10^2

10. 7×10^5

11. 1×10^{-8}

12. 9×10^{-4}

Write the following in scientific notation. (See Example 3.)

13. 7600

14. 83,000

15. .00035

16. .0026

17. 857,000

18. 13,400,000

19. .000000498

20. .000913

21. 600,000

22. 40,000,000

23. .00000001

24. .00001

Simplify by using scientific notation. Express answers in both scientific and standard notation. (See Examples 5a and 5b.)

25. $(1.2 \times 10^3)(2.5 \times 10^5)$

26. $(3.4 \times 10^4)(1.7 \times 10^2)$

27. $(4.3 \times 10^6)(3.2 \times 10^{-4})$

28. $(8.3 \times 10^7)(5.1 \times 10^{-5})$

29. $(3.12 \times 10^{-6})(4.23 \times 10^3)$

30. $(7.21 \times 10^{-8})(4.82 \times 10^4)$

31. $(6.28 \times 10^{-3})(4.21 \times 10^{-5})$

32. $(3.04 \times 10^{-2})(2.5 \times 10^{-4})$

33. $\dfrac{4.2 \times 10^7}{2.1 \times 10^4}$

34. $\dfrac{3.9 \times 10^6}{1.3 \times 10^2}$

35. $\dfrac{3.6 \times 10^{-3}}{2.4 \times 10^4}$

36. $\dfrac{4.5 \times 10^{-5}}{1.8 \times 10^2}$

37. $\dfrac{1.8 \times 10^3}{1.5 \times 10^{-3}}$

38. $\dfrac{1.065 \times 10^5}{4.26 \times 10^{-5}}$

39. $\dfrac{1.28 \times 10^{-3}}{2.56 \times 10^{-5}}$

Answers: Practice Exercise 6: a. 1995 grams **b.** 39,710 hours   **Additional Practice 6: a.** 205,000 grams **b.** approximately 126,666,667 seconds

40. $\dfrac{7.29 \times 10^{-5}}{2.43 \times 10^{-2}}$

41. $\dfrac{(1.25 \times 10^{-5})(5 \times 10^{7})}{2.5 \times 10^{-2}}$

42. $\dfrac{(3.6 \times 10^{6})(4.8 \times 10^{-4})}{1.44 \times 10^{-3}}$

43. $\dfrac{(5.4 \times 10^{-3})(7.2 \times 10^{6})}{(2.7 \times 10^{-2})(1.6 \times 10^{-1})}$

44. $\dfrac{(7.5 \times 10^{5})(1.0 \times 10^{-3})}{(1.5 \times 10^{-2})(2.5 \times 10^{8})}$

Simplify by using scientific notation. Express answers in both scientific and standard notation. (See Examples 5c and 5d.)

45. $(45,000)(2200)$

46. $(350,000)(2,600,000)$

47. $(5600)(.000045)$

48. $(7,200,000)(.00036)$

49. $(.0000025)(.000000036)$

50. $(.00000075)(.0012)$

51. $\dfrac{345,000,000}{15,000}$

52. $\dfrac{1,280,000}{1600}$

53. $\dfrac{1800}{72,000,000}$

54. $\dfrac{22,500}{112,500,000}$

55. $\dfrac{37,400}{.0017}$

56. $\dfrac{9,240,000}{.0000021}$

57. $\dfrac{.000612}{.0034}$

58. $\dfrac{.0084}{.00000056}$

59. $\dfrac{(120,000)(.0018)}{3600}$

60. $\dfrac{(24,000)(.00000028)}{560,000}$

61. $\dfrac{(8000)(.000252)}{(.00063)(400)}$

62. $\dfrac{(112)(.0015)}{(.00002)(1400)}$

Answer the following. (See Example 4.)

63. The astronomical unit is 150,000,000 kilometers. Write the astronomical unit in scientific notation.

64. As of April 22, 2010, the national debt was approximately \$12,870,000,000. Write the national debt in scientific notation.

65. The density of the element mercury is 13,600 kilograms per cubic meter. Express the density of mercury in scientific notation.

66. The coefficient of linear expansion of aluminum is .000024. Express the linear expansion coefficient of aluminum in scientific notation.

67. The wavelength of an X-ray is .0000013 meters. Express the wavelength of the X-ray in scientific notation.

68. The solubility constant of lead sulfate is .000000013. Express the solubility constant of lead sulfate in scientific notation.

69. The speed of light is approximately 2.997925×10^{8} meters per second. Express the speed of light in standard notation.

70. The half-life of a radioactive substance is the time required for one-half of the amount present to decay. The half-life of actinium is 7.04×10^{8} years. Express the half-life of actinium in standard notation.

71. The density of gold is 1.93×10^{4} kilograms per cubic meter. Express the density of gold in standard notation.

72. The density of hydrogen is 8.99×10^{-2} kilograms per cubic meter. Express the density of hydrogen in standard notation.

Solve the following. Leave answers in standard notation. (See Example 6.)

73. The mass of a hydrogen atom is 1.673×10^{-24} grams. Find the mass of the atoms in a sample of hydrogen that contains 1,000,000,000 hydrogen atoms.

74. If a computer can execute a command in .00005 second, how long will it take the computer to execute 5 million commands?

75. The distance from Earth to the nearest star, Alpha Centauri, is approximately 25,200,000,000,000 miles. A light-year is the distance that light can travel in one year and is approximately 6,000,000,000,000 miles. Find the distance to Alpha Centauri to the nearest tenth of a light-year.

76. Density is defined as $\frac{mass}{volume}$. If the mass of Earth is 5.98×10^{27} grams and the volume of Earth is 1.08×10^{27} cubic centimeters, find the density of Earth to the nearest tenth of a gram per cubic centimeter.

77. In chemistry, the number of moles (a measure of the amount of a substance) is equal to $\frac{\text{number of molecules of the substance}}{6.02 \times 10^{23}}$. Find the number of moles of a particular substance if 9.03×10^{23} molecules are present.

78. As of November 2009 the world's fastest super-computer was the Cray XT5 Jaguar located at the Oak Ridge National Laboratory, which could perform 1.759×10^{15} floating point operations per second (FLOPS). If the computer runs twenty-four hours a day, how many FLOPS can it perform in a week? Leave the answer in scientific notation.

Challenge Exercises (79–81). *(See Example 6.)*

79. The mass of a hydrogen atom is 1.673×10^{-24} grams. How many atoms are in a 10-gram sample of hydrogen?

80. In Exercise 75, the distance from Earth to Alpha Centauri was given as approximately 25,200,000,000,000 miles. Spacecraft travel at approximately 25,000 miles per hour. Approximately how many years would it take a spacecraft to reach Alpha Centauri? Is this a problem for manned space travel?

81. The mass of a helium atom is 6.65×10^{-24} grams. How many atoms are in a 4-gram sample of helium?

Writing Exercise

82. Why is scientific notation preferable when performing operations on extremely large or extremely small numbers?

Group Project

83. Interview some science instructors, engineers, astronomers, or others, and find at least five examples of ways they have used scientific notation.

Chapter 2   Summary

Concept/Procedure	Example
Rules for Multiplying Signed Numbers: [Section 2.1] • The product of two numbers with the same sign is positive: $(+)(+) = (+)$ and $(-)(-) = (+)$ • The product of two numbers with opposite signs is negative: $(+)(-) = (-)$ and $(-)(+) = (-)$	**Example 1:** Multiply $\quad (4)(-7) = -28 \quad$ Positive times a negative is negative. $\quad\quad\quad\quad (-5)(-9) = 45 \quad$ Negative times a negative is positive.
Signed Numbers to Powers: [Section 2.1] • A negative number raised to an even power is positive. • A negative number raised to an odd power is negative. • The base of $(-a)^n$ is $-a$, so $(-a)^n$ means multiply $-a$, n times. • The base of $-a^n$ is a, so $-a^n$ means $-1 \cdot a^n$.	**Example 2:** Evaluate $\quad (-6)^4 = 1296 \quad$ Negative number to an even power is positive. $\quad\quad\quad\quad (-3)^5 = -243 \quad$ Negative number to an odd power is negative. $\quad\quad\quad -5^4 = -1 \cdot 5^4 = -625 \quad$ The opposite of 5^4 is negative.
Properties of Multiplication: [Section 2.1] • Commutative: $ab = ba$ Multiplication may be performed in any order. • Associative: $a(bc) = (ab)c$ Factors may be grouped in any manner. • Identity: The number 1 is the identity for multiplication, since $a \cdot 1 = 1 \cdot a = a$. • Inverse: For any number $a \neq 0$, there exists a number $\frac{1}{a}$ such that $a \cdot \frac{1}{a} = 1$.	**Example 3:** $4 \cdot 3 = 3 \cdot 4$ $(2x)y = 2(xy)$ $7 \cdot 1 = 1 \cdot 7 = 7$ $8 \cdot \frac{1}{8} = 1$
Multiplication Laws of Exponents: [Section 2.2] • Product rule: $a^m \cdot a^n = a^{m+n}$ • Power of a product: $(ab)^n = a^n b^n$ • Power to a power: $(a^m)^n = a^{mn}$	**Example 4:** Simplify $\quad x^4 \cdot x^6 = x^{4+6} = x^{10} \quad$ Apply $a^m \cdot a^n = a^{m+n}$. $\quad\quad\quad (4a)^4 = 4^4 a^4 = 256a^4 \quad$ Apply $(ab)^n = a^n b^n$. $\quad\quad\quad (b^4)^3 = b^{4 \cdot 3} = b^{12} \quad$ Apply $(a^m)^n = a^{mn}$.

Concept/Procedure	Example
Products of Polynomials: [Section 2.3] • Monomial by monomial: Multiply coefficients and multiply variables with the same base using the product rule. • Polynomials by monomials: Use the distributive property to multiply each term of the polynomial by the monomial. If necessary, add like terms. • Polynomial by polynomial: Use the distributive property to multiply the second polynomial by each term of the first polynomial. $$(a + b)(c + d + e)$$ $$= a(c + d + e) + b(c + d + e)$$	**Example 5:** Multiply **a.** $(3x^2y^4)(8xy^5) = (3 \cdot 8)(x^2 \cdot x)(y^4 \cdot y^5) = 24x^3y^9$ **b.** $3x^3(2x^2 - 4x + 5) = 3x^3 \cdot 2x^2 - 3x^3 \cdot 4x + 3x^3 \cdot 5 =$ $6x^5 - 12x^4 + 15x^3$ **c.** $(3x + 4)(2x^2 - 3x + 4) = 3x(2x^2 - 3x + 4) +$ $4(2x^2 - 3x + 4) = 6x^3 - 9x^2 + 12x +$ $8x^2 - 12x + 16 = 6x^3 - x^2 + 16$
Special Products: [Section 2.4] • Binomial by binomial—FOIL • Sum and difference of the same terms: Multiply the first and last only. The product is the difference of squares. $$(a + b)(a - b) = a^2 - b^2$$ • Square of a binomial: Square the first term, add twice the product of the terms of the binomial, add the square of the last term. $$(a + b)^2 = a^2 + 2ab + b^2$$ $$(a - b)^2 = a^2 - 2ab + b^2$$ • Products resulting in the sum or difference of cubes: $$(a + b)(a^2 - ab + b^2) = a^3 + b^3$$ $$(a - b)(a^2 + ab + b^2) = a^3 - b^3$$	**Example 6:** Multiply **a.** $(2m - 3)(4m + 5) = 8m^2 + 10m - 12m - 15 =$ $8m^2 - 2m - 15$. Apply FOIL. **b.** $(4x + 3)(4x - 3) = (4x)^2 - 3^2 =$ $16x^2 - 9$ Apply $(a + b)(a - b) = a^2 - b^2$. **c.** $(3x + 7)^2 = (3x)^2 + 2(3x)(7) + 7^2 =$ $9x^2 + 42x + 49$ Apply $(a + b)^2 = a^2 + 2ab + b^2$. **d.** $(x + 3)(x^2 - 3x + 9) = x^3 + 3^3 = x^3 + 27$ **e.** $(a - 5)(a^2 + 5a + 25) = a^3 - 5^3 = a^3 - 125$
Definition of Division: [Section 2.5] • If $\frac{a}{b} = c$, then $bc = a$, $b \neq 0$.	**Example 7:** If $\dfrac{-18}{6} = -3$, then $6(-3) = -18$.
Rules for Dividing Signed Numbers: [Section 2.5] • The quotient of two numbers with the same sign is positive: $$\frac{(+)}{(+)} = (+) \text{ and } \frac{(-)}{(-)} = (+)$$ • The quotient of two numbers with opposite signs is negative: $$\frac{(+)}{(-)} = (-) \text{ and } \frac{(-)}{(+)} = (-)$$	**Example 8:** Divide $\dfrac{-24}{-6} = 4$ Negative divided by a negative is positive. $\dfrac{-18}{9} = -2$ Negative divided by a positive is negative.

Concept/Procedure	Example

Order of Operations on Integers: [Section 2.5]

- The order of operations on integers is the same as with whole numbers. If needed, see the Chapter 1 summary.

Example 9:

Evaluate

$15 + 4 \cdot 6^2 \div (9)(-2) - 8 =$ Raise to powers.

$15 + 4 \cdot 36 \div (9)(-2) - 8 =$ Multiply $4 \cdot 36$.

$15 + 144 \div (9)(-2) - 8 =$ Divide $144 \div 9$.

$15 + 16(-2) - 8 =$ Multiply $16(-2)$.

$15 - 32 - 8 =$ Add and subtract from left to right.

-25 Answer.

Laws of Exponents for Quotients: [Sections 2.6 and 2.7]

- Quotient rule: $\dfrac{a^m}{a^n} = a^{m-n}, a \neq 0$

- Power rule for quotients:

$$\left(\frac{a}{b}\right)^n = \frac{a^n}{b^n}, b \neq 0$$

Example 10:

Simplify

a. $\dfrac{x^4}{x^{-2}} = x^{4-(-2)} = x^{4+2} = x^6$ Apply $\dfrac{a^m}{a^n} = a^{m-n}$.

b. $\left(\dfrac{y}{2}\right)^4 = \dfrac{y^4}{2^4} = \dfrac{y^4}{16}$ Apply $\left(\dfrac{a}{b}\right)^n = \dfrac{a^n}{b^n}$.

Integer Exponents: [Sections 2.6 and 2.7]

1. $x^0 = 1$, if $x \neq 0$

2. $a^{-n} = \dfrac{1}{a^n}, a \neq 0$

3. $\dfrac{1}{a^{-n}} = a^n, a \neq 0$

4. $\left(\dfrac{a}{b}\right)^{-n} = \left(\dfrac{b}{a}\right)^n, a, b \neq 0$

Example 11:

Rewrite each expression without zero or negative exponents.

a. $3x^0 + (2y)^0 = 3 \cdot 1 + 1 = 3 + 1 = 4$ Apply $x^0 = 1$.

b. $3^{-4} = \dfrac{1}{3^4} = \dfrac{1}{81}$ Apply $x^{-n} = \dfrac{1}{x^n}$.

c. $\dfrac{1}{m^{-5}} = m^5$ Apply $\dfrac{1}{x^{-n}} = x^n$.

d. $\left(\dfrac{p}{q}\right)^{-5} = \left(\dfrac{q}{p}\right)^5 = \dfrac{q^5}{p^5}$ Apply $\left(\dfrac{a}{b}\right)^{-n} = \left(\dfrac{b}{a}\right)^n$.

Division of a Polynomial by a Monomial: [Section 2.8]

- To divide a polynomial by a monomial, divide each term of the polynomial by the monomial. In symbols,

$$\frac{a+b}{c} = \frac{a}{c} + \frac{b}{c}, c \neq 0.$$

Example 12:

Divide

$$\frac{24a^4 - 12a^2}{6a^2} = \frac{24a^4}{6a^2} - \frac{12a^2}{6a^2} = 4a^2 - 2$$ Apply $\dfrac{a+b}{c} = \dfrac{a}{c} + \dfrac{b}{c}$.

Concept/Procedure	Example

Scientific Notation: [Section 2.9]

1. A number is written in scientific notation if it is in the form $a \times 10^n$ with $1 \leq a < 10$ and n an integer.

2. To change a number from scientific notation to standard notation, move the decimal n places to the right if n is positive and $|n|$ places to the left if n is negative.

3. To write a number written in standard notation in scientific notation, place the decimal so that there is one nonzero digit to the left of the decimal. Then determine n by the number of places the decimal would have to be moved so that the number in scientific notation will be equal to the number in standard notation.

4. To multiply two numbers in scientific notation, multiply the coefficients and the exponential parts separately.

5. To divide two numbers in scientific notation, divide the coefficients and exponential parts separately.

Example 13:
Write 3.45×10^{-8} in standard notation.

Solution: $3.45 \times 10^{-8} = 0.0000000345$ Move the decimal 8 places left.

Example 14:
Evaluate by using scientific notation.
$$\frac{432{,}000}{.000000108} = \frac{4.32 \times 10^5}{1.08 \times 10^{-7}} = \frac{4.32}{1.08} \times \frac{10^5}{10^{-7}} = 4 \times 10^{5-(-7)} =$$
$$4 \times 10^{12} = 4{,}000{,}000{,}000{,}000$$

Chapter 2   Review Exercises

Find the products. [Section 2.1]

1. $(-4)(9)$

2. $(-5)(-3)$

3. $(4)(-6)(3)$

4. $(-4)(5)(-3)(-2)$

5. $(-3)^2(4)^2$

6. $(-1)^3(-4)^2$

Evaluate each term for $x = 3$ and $y = -2$. [Section 2.1]

7. $5x$

8. $-3y^2$

9. x^2y^3

10. $-4xy^3$

Write a mathematical expression for each English expression and evaluate. [Section 2.1]

11. 7 more than the product of -4 and 8

12. -6 decreased by the product of -5 and 5

13. The sum of the product of -6 and 3 and the product of 6 and -4

Give the name of the property illustrated by each of the following. [Section 2.1]

14. $(-4)(5) = (5)(-4)$

15. $(-2 \cdot 5)(6) = -2(5 \cdot 6)$

16. $(-5)(-3)(7) = (-3)(-5)(7)$

17. $-5(4 + 7) = (-5)(4) + (-5)(7)$

18. $9 \cdot \dfrac{1}{9} = 1$

19. $1 \cdot 13 = 13$

Simplify, using the multiplication laws of exponents. [Section 2.2]

20. $w^5 \cdot w^8$

21. $4^3 \cdot 4^5 \cdot 4^6$

22. $(8x^3)(-5x^7)$

23. $(6a^4b^6)(-5a^2b^6)$

24. $(xy)^7$

25. $(5d)^3$

26. $(-2ab)^4$

27. $(xy)^5(xy)^4$

28. $(4x)^3(-2x)^5$

29. $(x^5)^4$

30. $(4^3)^8$

31. $(a^4b^6)^4$

32. $(4a^3b^7)^4$

33. $(-a^4b^3)^3(a^5b^2)^5$

34. $(2m^3n^3)^3(3m^2n^5)^2$

Find the products of the following polynomials.
[Sections 2.3 and 2.4]

35. $(-3x^4y^3)(-5x^6y^2)$

36. $6y(3y^2 - 2)$

37. $-5x^2(4x^2 - 5x)$

38. $7c^3(3c^2 - 7c + 2)$

39. $-2x^2y^3(3x^4y - 8xy^5 + 4x^3)$

40. $2(3x - 4) + 3(2x + 7)$

41. $3(5x - 5) - 4(3x - 6)$

42. $-2(2a^2 + 4a - 5) + 3(3a^2 - 5a + 1)$

43. $5(b^2 + 2b - 1) - 3(2b^2 - 3b - 6)$

44. $(x + 2)(x + 5)$

45. $(4m - 5)(3m - 6)$

46. $(5m - 6n)(4m + 3n)$

47. $(x - 9)(x + 9)$

48. $(5x - 6)(5x + 6)$

49. $(2s + 5t)(2s - 5t)$

50. $(m + n)^2$

51. $(3a - 2)^2$

52. $(6a + 2b)^2$

53. $(a + 4)(a^2 - 5a + 4)$

54. $(2m - 5)(3m^2 + 2m - 4)$

55. $(4x - 2y)(3x^2 - 5xy + 2y^2)$

Write a mathematical expression for each English expression and simplify. [Section 2.3]

56. Find the sum of two times $3x + 7$ and five times $-2x + 1$.

57. Find the difference of four times $4m - 2$ and three times $2m + 5$.

Find the products (Optional Exercises 58–59). [Section 2.4]

58. $(x - 4)(x^2 + 4x + 16)$

59. $(3x + 5)(9x^2 - 15x + 25)$

60. Find the area of the rectangle with length $5a - 7$ and width $3a + 2$. [Section 2.3]

61. Find the area of the triangle with base $5x$ and height $4x$. [Section 2.3]

62. Find the area of the square each of whose sides is $3z + 8$. [Section 2.4]

Write a mathematical expression for each English expression and simplify. [Section 2.4]

63. The square of the quantity 4 less than the product of 6 and x

64. The square of the quantity 2 more than the product of 3 and z

Find the quotients. [Section 2.5]

65. $\dfrac{48}{-6}$

66. $\dfrac{-28}{-14}$

67. $\dfrac{-18}{2}$

68. $\dfrac{16}{13 - 17}$

Simplify, using the order of operations. [Section 2.5]

69. $\dfrac{23 - 35}{-6}$

70. $\dfrac{-120}{(15)(-4)}$

71. $\dfrac{(-16)(6)}{(-8)(3)}$

72. $-6 + 13 - 9$

73. $8 - 4 \cdot 5$

74. $16 + 36 \div (-4)$

75. $14 - 3(-5)^2$

76. $8 + 2(8) \div (-4)$

77. $6 - 2(9 + 4)$

78. $8 + 4 \cdot 6^2 \div (-9)(3) - 7$

79. $-4(6^2 - 2^2) \div (-8)$

80. $|6 \cdot 2^2 - 42| - |3(4 - 3^2)|$

81. $\dfrac{(-3)(-6) + 6}{(-6)(3) + 6}$

82. $\dfrac{2 \cdot 4^2 + 16}{-3 \cdot 2^2 + 4}$

83. $\dfrac{8^2 - 4^2}{-2(3^2 - 1)}$

84. $\dfrac{(-4)(5) + [6(-4) + 6^2]}{(-3)^2(4) - 28}$

85. $\dfrac{|6(3^2 - 3)|}{|3 \cdot 2^2 + 7|}$

Find the value of each expression for $x = -3$, $y = 2$, and $z = -2$. [Section 2.5]

86. $3x^2 - 4z$

87. $-4x^3y^2$

88. $3xy^2z$

89. $3x - 4(y - 3z)$

90. $-y^2 - (2x^2 - 3z^2)$

91. $\dfrac{4x - 6y}{4y - 2z}$

92. $\dfrac{2x^2 - 2z^3}{5x^2 + 3z^2}$

Simplify. Express answers in exponential form with positive exponents only. Assume that all variables represent nonzero quantities. [Section 2.6]

93. $\dfrac{2^8}{2^3}$

94. $\dfrac{m^8}{m^6}$

95. 3^{-4}

96. x^{-6}

97. $\dfrac{5^4}{5^6}$

98. $3x^0 + (5x)^0$

99. $5^{-5} \cdot 5^4$

100. $z^4 z^{-2}$

101. $x^{-4} x^{-2}$

102. $\dfrac{5^{-4}}{5^2}$

103. $\dfrac{a^{-3}}{a^{-6}}$

104. $\dfrac{7^3}{7^{-7}}$

Write each fraction without exponents and evaluate. [Section 2.7]

105. $\left(\dfrac{3}{5}\right)^2$

106. $\left(\dfrac{3}{2}\right)^{-2}$

Write each expression in simplest form with positive exponents only. [Section 2.7]

107. $(3^{-5})^2$

108. $(r^{-5})^{-4}$

109. $(x^{-4}y^4)^{-3}$

110. $(4x^{-2})^{-3}$

111. $(5a^4)^{-3}$

112. $(2x^{-4}y^6)^{-3}$

113. $\left(\dfrac{m^4}{n^3}\right)^{-2}$

114. $\dfrac{(m^{-4})^{-5}}{(m^3)^{-4}}$

115. $\dfrac{-24n^6}{-3n^3}$

116. $\dfrac{m^4 n^6}{m^6 n^2}$

117. $\dfrac{-48m^8n^5}{-8m^{10}n^5}$

118. $\dfrac{(-6x^3)(-5x^5)}{(2x^2)(-3x^4)}$

119. $\dfrac{(3m^{-4}n^3)(-8m^3n^{-6})}{(4m^5n^{-3})(-2m^{-3}n^6)}$

120. $\dfrac{(m^3)^4(m^{-6})^2}{(m^8)^{-2}(m^{-3})^{-3}}$

Find each quotient of a polynomial by a monomial. [Section 2.8]

121. $\dfrac{9x - 18}{9}$

122. $\dfrac{5y^3 + 3y}{y}$

123. $\dfrac{28a^2 - 16a}{4a}$

124. $\dfrac{36m^7n^3 + 18m^4n^7}{9m^5n^5}$

125. $\dfrac{12x^4 - 9x^3 - 21x}{3x^2}$

126. $\dfrac{8m^5n^3 - 32m^2n^7 + 24m^6n^4}{8m^6n^5}$

Write a mathematical expression for each English expression and simplify. [Section 2.8]

127. The quotient of $48m^6n^3$ and $-16m^4n^5$

128. The quotient of $24a^6 - 16a^4$ and $8a^3$

129. Divide the sum of $4x^4 - 2x^3 - 3x^2$ and $2x^4 - 7x^3 + 2x^2$ by $3x^2$.

130. The quotient of $28x^6 - 40x^3$ and $4x^4$

131. Divide the sum of $5x^2 - 7x + 5$ and $3x^2 + 3x - 9$ by $2x$.

Write each number in standard notation. [Section 2.9]

132. 4.6×10^5

133. 3.06×10^6

134. 7.03×10^{-5}

135. 7.123×10^{-3}

Write each number in scientific notation. [Section 2.9]

136. 97,000,000,000

137. 46,700

138. .00000478

139. .000307

Simplify each using scientific notation. Express answers in both scientific and standard notation. [Section 2.9]

140. $(4.5 \times 10^4)(3.1 \times 10^{-7})$

141. $(5.07 \times 10^{-5})(4.06 \times 10^{-2})$

142. $\dfrac{4.8 \times 10^4}{1.6 \times 10^6}$

143. $\dfrac{3.9 \times 10^{-5}}{1.3 \times 10^3}$

144. $(4600)(.0000012)$

145. $(.000047)(.00000034)$

146. $\dfrac{48000}{.00096}$

147. $\dfrac{.00033}{.00000022}$

Chapter 2 Test

Find the following products:

1. $(-3)(5)(-4)$

2. $(-2)^3(3)^2$

3. $(3x^4y^3)(-2xy^2)$

4. $(3x^2y)(-4x^2y)^2$

5. $2x^2(3x - 4)$

6. $p^2q(2pq^3 - 4p^2q - 2)$

7. $4(2x - 3) - 5(x - 4)$

8. $(3p + 4)(2p - 1)$

9. $(3a - 4b)(2a - b)$

10. $(3y - 1)(y^2 + 2y + 5)$

11. $(5x - 3y)(5x + 3y)$

12. $(2a - 7)^2$

Simplify. If the exercise involves exponents, express the answer with positive exponents only.

13. $\dfrac{-24 - 8}{8}$

14. $\dfrac{4 \cdot 5^2 - 25}{(-3)(-4) + 13}$

15. $(4m^3n^5)(-6m^{-6}n^2)$

16. $8x^0 - (3x)^0$

17. $\dfrac{36x^4y^3z^5}{-18x^7y^3z^2}$

18. $(3x^{-5}y^4)^{-3}$

19. $\dfrac{(4a^6b^{-3})^2}{(2a^3b^4)^3}$

20. $\dfrac{(6a^{-2}b^{-4})(-8a^{-5}b^8)}{(4a^{-3}b^4)(-6a^6b^{-6})}$

21. $\dfrac{8m^6n^3 - 32m^4n^5 + 20m^5n^2}{4m^3n^2}$

22. $\dfrac{8x^3 - 10x^2 + 6x}{2x^2}$

Evaluate each of the following:

23. $-8 - (12 - 36) \div (-4)$

24. $\dfrac{4(-2)^3 + 2}{8^2 + 2(-3)^3}$

25. Find the value of $2y - (3x^2 - 3y) \div x^3$, if $x = -2$ and $y = -4$.

Write a mathematical expression for each English expression and simplify:

26. The product of 7 and −3 subtracted from the quotient of 12 and −4

27. The product of $x + 3$ and $2x - 5$ added to $3x^2 - 7x + 5$

28. Write each of the following in scientific notation:

   a. 4,790,000,000

   b. .0000000749

29. Evaluate $(43,000,000)(.000025)$, using scientific notation. Express the answer in standard notation.

30. Evaluate the following, using scientific notation. Express the answer in scientific notation.
$$\dfrac{97,020,000}{.00231}$$

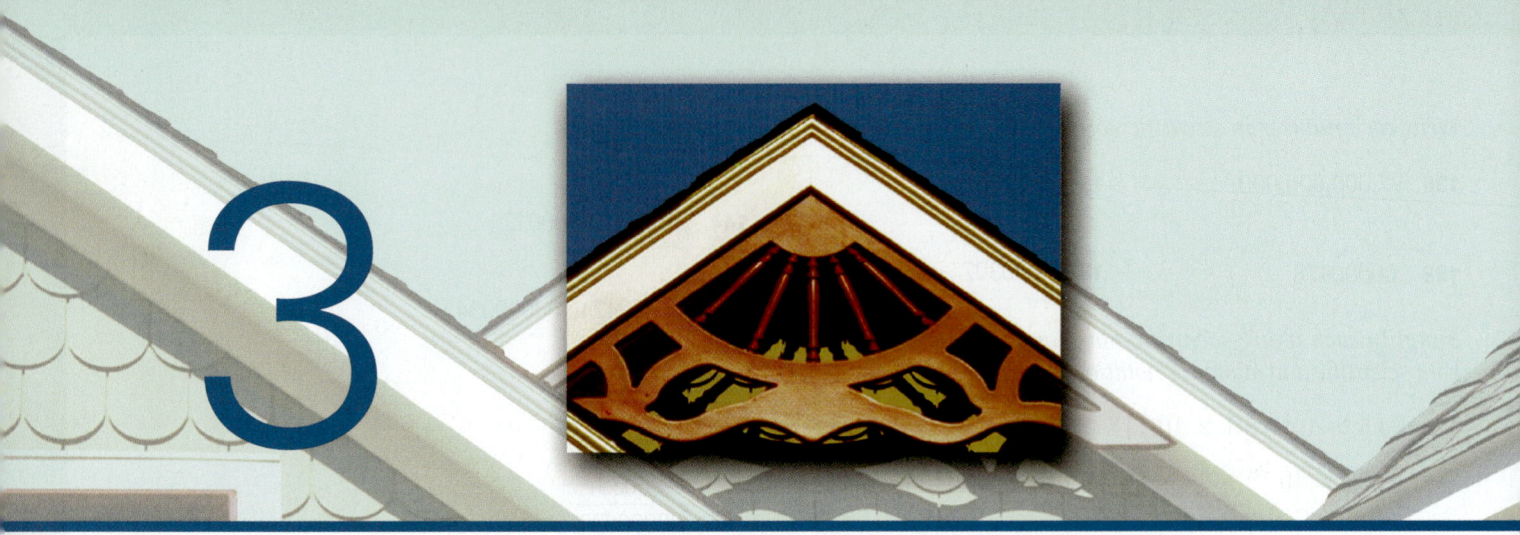

3

Linear Equations and Inequalities

— Chapter Outline —

3.1 Addition Property of Equality

3.2 Multiplication Property of Equality

3.3 Combining Properties in Solving Linear Equations

3.4 Using and Solving Formulas

3.5 General, Consecutive Integer, and Distance Application Problems

3.6 Money, Investment, and Mixture Application Problems

3.7 Geometric Application Problems

3.8 Solving Linear Inequalities

One of the primary uses of the addition, subtraction, multiplication, and division of signed numbers is in solving linear equations and inequalities. Equations and inequalities were previously mentioned in Chapter 1. In this chapter, we will expand on these concepts and use them to solve various types of real-world and not-so-real-world application problems.

<table>
<tr><td>**Section 3.1**</td><td>## Addition Property of Equality</td></tr>
</table>

OBJECTIVES *When you complete this section, you will be able to:*

A Solve linear equations, using the addition property of equality.
B Translate mathematics expressions into English and English expressions into mathematics.
C Solve various types of real-world problems.

PREREQUISITE SKILLS *Before beginning this section, you should be able to:*

a. Add and subtract integers (Sections 1.6 and 1.7).
b. Add and subtract decimals (Section R.5).
c. Apply the distributive property (Section 1.7).
d. Combine like terms (Section 1.7).

GETTING READY FOR SECTION 3.1

Add or subtract as indicated.
1. $-3 + 8$ **2.** $-4 - 5$

Add or subtract as indicated.
3. $-3.2 + 4.7$ **4.** $-3.8 - 5.4$

Perform the indicated operation.
5. $8(x - 3)$ **6.** $-2(y - 9)$

Simplify the following:
7. $3x - 5(x - 1)$ **8.** $3(3x - 2) - 4(2x - 1)$ **9.** $-5.7t + 4.6 + 2.7t + 3$

Introduction

An **equation** is a mathematical sentence indicating that two expressions are equal. For example, both $3 + 2 = 5$ and $x + 4 = 9$ are equations. A **solution** of an equation is any value of the variable that makes the equation true. For example, $x = 5$ is a solution of $x + 4 = 9$, because if we replace x with 5, we get $5 + 4 = 9$, which is a true statement. When the solutions are written as a set, we refer to it as the **solution set**. So, the solution set of $x + 4 = 9$ is $\{5\}$.

Example 1 **Determine whether the given number is a solution of the given equation.**

a. $2x - 3 = 9: x = 6$

b. $3x + 5 = -4: x = -2$

Solution

$2x - 3 = 9$	Replace x with 6 and simplify.
$2(6) - 3 = 9$	Multiply.
$12 - 3 = 9$	Subtract.
$9 = 9$	True. Therefore, $x = 6$ is a solution.

Solution

$3x + 5 = -4$	Replace x with -2 and simplify.
$3(-2) + 5 = -4$	Multiply.
$-6 + 5 = -4$	Add.
$-1 = -4$	False. Therefore, $x = -2$ is not a solution.

Answers: Getting Ready: 1. 5 **2.** -9 **3.** 1.5 **4.** -9.2 **5.** $8x - 24$ **6.** $-2y + 18$ **7.** $-2x + 5$ **8.** $x - 2$
9. $-3t + 7.6$

PRACTICE EXERCISE 1

Determine whether the given number is a solution of the given equation.

a. $-2x + 5 = 7: x = -1$　　　　　　　　　**b.** $4x - 7 = 9: x = -4$

If you need more practice, do the following Additional Practice Exercises.

Additional Practice Exercise 1　Determine whether the given number is a solution of the given equation.

a. $3x - 2 = 7: x = 4$　　　　　　　　　**b.** $5x + 6 = -4: x = -2$

Finding the solution(s) of an equation is called **solving the equation**. In this and in subsequent sections, we will discuss techniques that allow us to solve **linear equations**. A linear equation is any equation that can be put in the form of $ax + b = c$, where a, b, and c are constants and $a \neq 0$.

Suppose we begin with an equation and add the same number to both sides of the equation. Is the resulting sentence still an equation? Consider the following:

$$4 + 5 = 9$$ 　　　Add 6 to both sides.
$$4 + 5 + 6 \stackrel{?}{=} 9 + 6$$ 　　　Simplify both sides.
$$15 = 15$$ 　　　Adding 6 to both sides of the equation did not change the equality.

$$4 + 5 = 9$$ 　　　Add -8 to both sides.
$$4 + 5 - 8 \stackrel{?}{=} 9 - 8$$ 　　　Simplify both sides.
$$1 = 1$$ 　　　Adding -8 to both sides of the equation did not change the equality.

The preceding examples lead us to the following generalization:

> ### Rule: Addition Property of Equality
>
> If $a = b$, then $a + c = b + c$ for any number c. In words, if the same number is added to both sides of an equation, the result is still an equation.

OBJECTIVE Solving linear equations by using the addition property of equality

Equations that have the same solutions are called **equivalent equations**. If the addition property of equality is applied to an equation containing variables, the resulting equation is equivalent to the original equation. For example, consider the equation $x + 2 = 5$, for which $x = 3$ is a solution, since $3 + 2 = 5$. If we add 4 to both sides of the equation, we get $x + 2 + 4 = 5 + 4$, which simplifies to $x + 6 = 9$. Note that $x = 3$ is also a solution of $x + 6 = 9$, since $3 + 6 = 9$. Therefore, adding 4 to both sides of the equation $x + 2 = 5$ produced an equation with the same solution—in other words, an equivalent equation.

Solving equations by using the addition property of equality requires the use of two principles. The first is the additive inverse property, which states that every number a has an additive inverse, $-a$, such that $a + (-a) = 0$. The second property is the additive property of zero, which states that $0 + a = a + 0 = a$ for any number a.

The goal of equation solving is to get the variable on one side of the equation and a constant on the other. By doing so, we will get the simplest equivalent equation in the form of (variable) = (constant). In the following examples, we have a variable and a constant on the same side of the equation. We isolate the variable by adding the additive inverse of the constant to *both sides* of the equation and simplifying. After solving an equation, we can verify that our solution is correct by substituting that value into the original equation and simplifying.

| **Example 2** | **Solve the following:** |

a. $y - 3 = 9$ — Since 3 is the additive inverse of -3, add 3 to both sides of the equation.

$y - 3 + 3 = 9 + 3$ — Simplify both sides of the equation.

$y + 0 = 12$ — $y + 0 = y$.

$y = 12$ — Therefore, $y = 12$ is the solution and {12} is the solution set.

CHECK:

$y - 3 = 9$ — Substitute 12 for y.

$12 - 3 = 9$ — Simplify the left side of the equation.

$9 = 9$ — Therefore, 12 is the correct solution.

b. $6 = 8 + z$ — Since -8 is the additive inverse of 8, add -8 to both sides of the equation.

$-8 + 6 = -8 + 8 + z$ — Simplify both sides of the equation.

$-2 = 0 + z$ — $0 + z = z$.

$-2 = z$, or $z = -2$ — Therefore, $z = -2$ is the solution and $\{-2\}$ is the solution set.

CHECK:

$6 = 8 + z$ — Substitute -2 for z.

$6 = 8 + (-2)$ — Simplify the right side of the equation.

$6 = 6$ — Therefore, -2 is the correct solution.

PRACTICE EXERCISE 2

Solve the following:

a. $v - 8 = 7$ **b.** $r + 3 = 4$

If you need more practice, do the following Additional Practice Exercises.

Additional Practice Exercise 2   Solve the following:

a. $x - 3 = 6$ **b.** $r + 4 = -1$

It is often necessary to simplify one or both sides of an equation before applying the addition property of equality. The following steps are suggested to do so:

> **Procedure: Solving Linear Equations by Using the Addition Property of Equality**
>
> 1. If necessary, remove any inclusion symbols (parentheses, brackets, braces). This will often involve the distributive property.
>
> 2. If necessary, simplify each side of the equation by adding like terms.
>
> 3. If necessary, use the addition property of equality to get the variable on one side and the constant on the other.
>
> 4. Check the answer in the original equation.

Answers: Practice Exercise 2: a. $v = 15$ **b.** $r = 1$ **Additional Practice 2: a.** $x = 9$ **b.** $r = -5$

This procedure will be modified in Section 3.3 when we discuss the most general types of linear equations.

| **Example 3** | **Solve the following:** |

a. $t + 8 - 6 = 11 + 4$ Simplify both sides of the equation by adding like terms.

$t + 2 = 15$ Since -2 is the additive inverse of 2, add -2 to both sides of the equation.

$t + 2 + (-2) = 15 + (-2)$ Simplify both sides of the equation.

$t = 13$ Therefore, 13 is the solution.

CHECK:

$t + 8 - 6 = 11 + 4$ Substitute 13 for t.

$13 + 8 - 6 = 11 + 4$ Simplify both sides of the equation.

$15 = 15$ Therefore, $t = 13$ is the correct solution.

b. $6a + 4 - 5a - 2 = -6$ Simplify the left side of the equation.

$a + 2 = -6$ Since -2 is the additive inverse of 2, add -2 to both sides of the equation.

$a + 2 + (-2) = -6 + (-2)$ Simplify both sides of the equation.

$a = -8$ Therefore, the solution is $a = -8$.

CHECK:

The check is left as an exercise for the student.

c. $6.3u + 5.4 - 5.3u = 20.6$ Simplify the left side of the equation.

$u + 5.4 = 20.6$ Subtract 5.4 from both sides of the equation.

$u + 5.4 - 5.4 = 20.6 - 5.4$ Simplify both sides of the equation.

$u = 15.2$ Therefore, $u = 15.2$ is the solution.

CHECK:

$6.3u + 5.4 - 5.3u = 20.6$ Substitute 15.2 for u.

$6.3(15.2) + 5.4 - 5.3(15.2) = 20.6$ Multiply first.

$95.76 + 5.4 - 80.56 = 20.6$ Add.

$20.6 = 20.6$ Therefore, 15.2 is the correct solution.

d. $3x - 2(x + 4) = -5$ Apply the distributive property on the left side of the equation.

$3x - 2x - 8 = -5$ Simplify the left side.

$x - 8 = -5$ Since 8 is the additive inverse of -8, add 8 to both sides of the equation.

$x - 8 + 8 = -5 + 8$ Simplify both sides of the equation.

$x = 3$ Therefore, $x = 3$ is the solution.

CHECK:

The check is left as an exercise for the student.

e. $4 = 3(3x - 2) - 4(2x - 1)$ Apply the distributive property on the right side of the equation.

$4 = 9x - 6 - 8x + 4$ Simplify the right side.

$4 = x - 2$ Since 2 is the additive inverse of -2, add 2 to both sides of the equation.

$4 + 2 = x - 2 + 2$ Simplify both sides of the equation.

$6 = x,$ or $x = 6$ Therefore, $x = 6$ is the solution.

CHECK:

The check is left as an exercise for the student.

Note

In Examples 2a and 2b, we added -2 to both sides of the equation. Rather than adding -2, we could have subtracted 2, since subtraction is the same as the addition of the negative. Recall from Chapter 1 that $a - b = a + (-b)$ and $a + (-b) = a - b$. This is the procedure we will use in future examples.

PRACTICE EXERCISE 3

Solve, using the addition property:

a. $3 + t - 6 = 8 - 10$

b. $8b + 5 - 7b + 1 = -4$

c. $-5.7a + 4.6 + 6.7a = 16.5 - 9.2$

d. $6(x - 2) - 5x = -10$

e. $3 = 2(4x + 5) - 7(x + 2)$

If you need more practice, do the following Additional Practice Exercises.

Additional Practice Exercise 3 Solve, using the addition property:

a. $12 + s - 1 = 5$

b. $4b + 7 - 3b - 2 = -3$

c. $5.2x + 7.4 - 4.2x = 11.6$

d. $4(x + 3) - 3x = 7$

e. $6 = 5(x - 2) - 2(2x - 4)$

OBJECTIVE Translating mathematics expressions into English and English expressions into mathematics

In problem solving, we must be able to translate from English to mathematics. We will start by translating from mathematics to English and then translate from English to mathematics. The examples that follow, although they may have few applications in the real world, will serve to help you begin to think mathematically. Real-world examples will follow.

Example 4 **Translate each expression from mathematics to English. There is more than one translation.**

Mathematics	English
a. $t + 8 = 12$	Some number plus eight is twelve, or Eight added to some number is twelve, or The sum of some number and eight is twelve, or Eight more than some number is twelve, or A number increased by eight is twelve.
b. $y - 5 = 11$	Some number minus five is eleven, or Five subtracted from some number is eleven, or The difference of some number and five is eleven, or Five less than some number is eleven, or A number decreased by five equals eleven, or Some number less five is eleven.
c. $17 = u + 6$	Seventeen is some number plus six, or Seventeen equals six added to some number, or Seventeen equals a number increased by six, or Seventeen is six more than some number, or Seventeen is the sum of some number and six.
d. $4 = x - 21$	Four equals some number minus twenty-one, or Four is twenty-one subtracted from some number, or Four is some number decreased by twenty-one, or Four is twenty-one less than some number, or Four equals the difference of some number and twenty-one.

Answers: **Practice Exercise 3: a.** $t = 1$ **b.** $b = -10$ **c.** $a = 2.7$ **d.** $x = 2$ **e.** $x = 7$
Additional Practice 3: a. $s = -6$ **b.** $b = -8$ **c.** $x = 4.2$ **d.** $x = -5$ **e.** $x = 8$

PRACTICE EXERCISE 4

Translate from mathematical sentences into English sentences. There is more than one possible answer.

a. $r + 7 = 9$ **b.** $t - 4 = 17$ **c.** $15 = x + 6$ **d.** $28 = y - 19$

If you need more practice, do the following Additional Practice Exercises.

Additional Practice Exercise 4   Translate from mathematical sentences into English sentences. There is more than one possible answer.

a. $y + 3 = -9$ **b.** $r - 8 = 2$ **c.** $5 = a + 7$ **d.** $13 = m - 6$

We will now translate from English to mathematics.

Example 5   **Translate from English to mathematics and solve. Use x as the variable in each exercise.**

English	Mathematics
a. A number plus one equals four.	$x + 1 = 4$ $x + 1 - 1 = 4 - 1$ $x = 3$
b. A number minus ten is fifteen.	$x - 10 = 15$ $x - 10 + 10 = 15 + 10$ $x = 25$
c. The sum of twelve and some number is twenty.	$12 + x = 20$ $12 - 12 + x = 20 - 12$ $x = 8$
d. The difference of some number and thirteen is six.	$x - 13 = 6$ $x - 13 + 13 = 6 + 13$ $x = 19$

PRACTICE EXERCISE 5

Translate from English to mathematical sentences and solve. Use x as the variable.

a. The sum of a number and eleven is twenty-nine.

b. The difference of a number and six is fourteen.

c. A number increased by four is ten.

d. Eight subtracted from some number is twenty-three.

(Continued)

Answers: **Practice Exercise 4:** **a.** Some number plus seven is nine. **b.** Four subtracted from a number is seventeen. **c.** Fifteen is the sum of a number and six. **d.** Twenty-eight is the difference of some number and nineteen. **Additional Practice 4:** **a.** Three more than some number is negative nine. **b.** Eight less than some number is two. **c.** Five is equal to some number increased by seven. **d.** Thirteen is some number decreased by six. **Practice Exercise 5:** **a.** $x + 11 = 29, x = 18$ **b.** $x - 6 = 14, x = 20$ **c.** $x + 4 = 10, x = 6$ **d.** $x - 8 = 23, x = 31$

If you need more practice, do the following Additional Practice Exercises.

Additional Practice Exercise 5   Translate from English to mathematical sentences and solve. Use x as the variable.

a. Four more than some number is two.

b. A number decreased by three is eight.

c. Some number added to four is negative two.

d. Six less than some number is four.

The following real-world situations may be solved without using algebra if you really think about them. Later, however, the situations will be far too complicated to solve without the techniques of algebra. Consequently, even though you may be able to do these problems without algebra, use algebraic techniques as practice for the more complicated problems in upcoming sections of this chapter.

Procedure: Problem Solving with One Unknown

1. **Understand** the problem.
   a. Read and reread the problem carefully, paying special attention to key words. It is a good idea to underline key words in the problem.
   b. Identify what is given and what is to be found.
   c. Determine the unknown and how the known and unknown quantities are related to each other.
   d. When possible, make a drawing and label it with known and unknown information.

2. **Plan.** Determine the procedure you will use to solve the problem.

3. **Translate** the problem into mathematical language, which is usually an equation.
   a. Construct any necessary tables, charts, graphs, and so forth.
   b. Choose a variable to represent the unknown.
   c. Write the equation in words, using the relationship found in part 1c.
   d. Using the "word equation" in step c, write the algebraic equation.

4. **Solve** the algebraic equation.

5. **State** the solution(s) in words, using a complete English sentence(s).

6. **Check** the solution(s), using the wording of the original equation.

Example 6 | **Solve the following:**

a. John scored ten points more on his algebra test than Mike. If John scored 93, what was Mike's score?

Solution

Understand: We know John's score and the relationship between John and Mike's scores. The only unknown is Mike's score.

   Relationship: John's score is ten more than Mike's score, and John's score is 93.

Plan: Use the preceding relationship to write the equation.

Translate: The unknown is Mike's score, so **let x represent Mike's score**.

Since John scored ten points more than Mike,

| Mike's score | plus | ten points | equals | John's score |.

Using the word equation as a guide, we write the algebraic equation as follows:

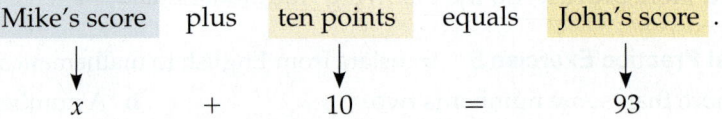

Mike's score	plus	ten points	equals	John's score .
x	$+$	10	$=$	93

Solve:

$x + 10 = 93$	Subtract 10 from both sides (add -10).
$x + 10 - 10 = 93 - 10$	Simplify both sides of the equation.
$x + 0 = 83$	$x + 0 = x$.
$x = 83$	Solution of the equation.

State: Mike's score is 83.

Check: Is John's score ten points more than Mike's score? Since 93 is 10 more than 83, the answer is yes and our solution is correct.

b. The legal limit on the number of crappie you may catch in Florida is 25 per person per day. Alice has already caught 18. How many more may she catch until she reaches her limit?

Solution

Understand: We know the total number of crappie that Alice is allowed to catch, and we know the number that she has already caught. The unknown is how many more she is allowed to catch.

   Relationship: Alice may catch a total of 25, and she has already caught 18.

Plan: Use the preceding relationship to write the equation.

Translate: The unknown is how many more crappie Alice may catch, so **let n represent the number of additional crappie she may catch.**

Since she has already caught 18 and she may catch 25,

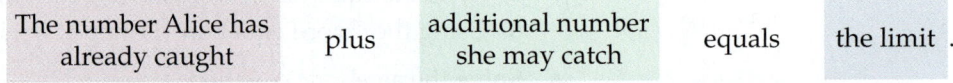

The number Alice has already caught	plus	additional number she may catch	equals	the limit .

Write an algebraic equation, using the word equation as a guide.

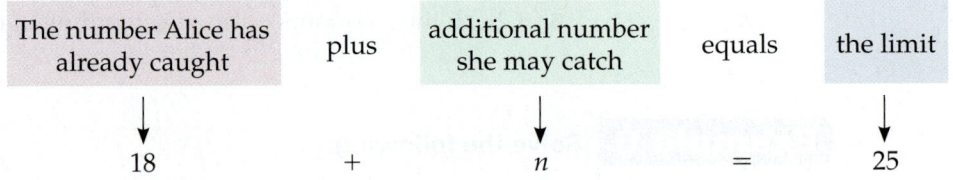

The number Alice has already caught	plus	additional number she may catch	equals	the limit .
18	$+$	n	$=$	25

Solve:

$18 + n = 25$	Subtract 18 from both sides (add -18).
$18 - 18 + n = 25 - 18$	Simplify both sides of the equation.
$0 + n = 7$	$0 + n = n$
$n = 7$	Solution of the equation.

State: Alice may catch 7 more crappie.

Check: Does the number that she has already caught (18) plus the additional number she may catch (7) equal the limit (25)? Yes. Therefore, the solution is correct.

PRACTICE EXERCISE 6

Let x represent the unknown in each exercise. Write an equation, using the given word equation as a guide, and solve.

a. The cost of a car is $2500 less than the cost of a minivan. If the cost of the car is $18,350, find the cost of the minivan.

Cost of minivan $-$ 2500 $=$ cost of car .

b. Jenny got married 6 years ago. If she is presently 28 years old, how old was she when she got married?

Age when she got married $+$ 6 $=$ present age .

If you need more practice, do the following Additional Practice Exercises.

Additional Practice Exercise 6

a. During a recent baseball game, the New York Yankees scored three runs fewer than the Chicago White Sox. If Chicago scored seven runs, how many did New York score?

b. The total cost of a computer, including taxes, is $583.25. If the taxes are $38.18, what is the cost of the computer?

Exercise Set 3.1

For Extra Help

MyMathLab®

Solve each of the following. (See Examples 2 and 3.)

1. $t + 5 = 13$

2. $z + 9 = 12$

3. $r + 8 = -2$

4. $s + 9 = -4$

5. $a - 3 = 5$

6. $b - 6 = 2$

7. $x - 6.3 = 4.2$

8. $y - 5.7 = 3.6$

9. $m + 7.3 = 6.1$

10. $v + 8.9 = 4.3$

11. $7 + w - 1 = 15$

12. $16 + u - 3 = 23$

13. $18 - 15 + x = 10$

14. $20 - 12 + y = 17$

15. $z - 5 - 4 = -8 + 2$

16. $n - 5 - 3 = -10 + 3$

17. $8r - 4 - 7r = 2 - 8$

18. $3z - 5 - 2z = -2 - 6$

19. $6.1x - 1.3 - 5.1x = 6.8$

20. $4.3x - 3.7 - 3.3x = 8.1$

21. $7 + 5 = 6r + 8 - 5r$

22. $3 + 6 = 7a + 4 - 6a$

23. $-4 - 6 = -8x - 10 + 9x$

24. $-8 - 4 = -3x - 12 + 4x$

25. $4(x + 3) - 3x = 10$

26. $3(x + 5) - 2x = 12$

27. $7x - 3(2x + 4) = -6$

28. $11x - 5(2x + 1) = -2$

29. $16x - 5(3x - 2) - 7 = 5$

30. $17x - 4(4x - 3) - 8 = 6$

31. $3(3.4x - 1.2) - 9.2x = -6.2$

32. $4(2.4x - 1.6) - 8.6x = -8.3$

33. $-2(4.7y - 5.4) + 10.4y = 3.1$

34. $-3(1.6b - 2.4) + 5.8b = -6.2$

35. $6(t + 3) - 5(t + 4) = 10$

36. $8(w - 5) - 7(w - 4) = 15$

37. $3(3x - 5) - 2(4x - 3) = -9$

38. $5(5x - 3) - 6(4x - 2) = -3$

39. $-4(5x - 4) + 7(3x - 2) - 6 = 8 - 2$

40. $-6(4x - 2) + 5(5x - 3) - 9 = 9 - 3$

Answers: Practice Exercise 6: a. $x - 2500 = 18,350$, $x = $20,850$ **b.** $x + 6 = 28$, $x = 22$ years old
Additional Practice 6: a. 4 runs **b.** $545.07

Challenge Exercises (41 and 42)

41. $3.2(1.5x - 3.4) - 3.8x + 5.32 = -6.33$ **42.** $4.6(2.5x + 1.4) - 10.5x - 7.34 = -8.21$

Translate the following mathematical sentences into English in at least two ways. (See Example 4.)

43. $x - 5 = 19$ **44.** $22 - t = 11$ **45.** $13 + y = 7$

46. $25 = u + 8$ **47.** $17 = w - 6$ **48.** $44 = v + 9$

Translate the following English sentences into mathematical sentences and solve. Use x as the variable. (See Example 5.)

49. Some number minus two is equal to nine.

50. Some number decreased by four is eight.

51. Fourteen subtracted from some number is eight.

52. Six subtracted from some number is nineteen.

53. The difference of a number and two is fifteen.

54. The difference of a number and seven is thirty-two.

55. The sum of five and some number is thirteen.

56. Five more than some number is negative two.

57. Eight less than some number is negative ten.

58. The sum of a number and negative three is negative seven.

59. The difference of some number and negative two is negative eight.

60. The difference of a number and negative four is negative three.

Let x represent the unknown in each exercise. Write an equation and solve. A typical word equation that can be expressed as an equation of the form studied in this section is given to encourage you to solve the problem by using an equation rather than arithmetic. (See Example 6.)

61. Shasta and Charlene shared the driving responsibilities on a trip. If the total trip was 550 miles and Shasta drove 275 miles, how far did Charlene drive? Let x represent the number of miles Charlene drove.

Number of miles Shasta drove + number of miles Charlene drove = total miles .

62. On a trip, Francine drove 75 miles more than Hector. If Francine drove 260 miles, how far did Hector drive? Let x represent the number of miles Hector drove.

Number of miles Hector drove + 75 = number of miles Francine drove .

63. Jenny and Bryan are going to chip in to buy their parents an anniversary gift. If the gift costs $160 and Bryan pays $75, how much does Jenny pay? Let x represent the amount Jenny pays.

Amount Jenny pays + amount Bryan pays = cost of gift .

64. Horace needs $465 to buy a new canoe. If he has already saved $312, how much more does he need? Let x represent the amount Horace needs.

Amount Horace has $+$ amount Horace needs $=$ cost of canoe .

65. In a recent football game, the Dolphins scored 8 points less than the Bills. If the Dolphins scored 24 points, how many points did the Bills score? Let x represent the number of points the Bills scored.

Number of points Bills scored $- 8 =$ number of points Dolphins scored .

66. John and Jim shared expenses on a trip from Orlando to Atlanta. John paid $30 less than Jim paid. If Jim paid $243, how much did John pay? Let x represent the amount John pays.

Amount John paid $+$ 30 $=$ amount Jim paid .

67. Connie needs a total of 800 points in order to get an A in her English I class. If she already has 632 points, how many more does she need? Let x represent the number of points Connie needs.

Number of points Connie has $+$ number of points Connie needs $=$ number of points needed for an A .

68. A mixture contains 23 more gallons of water than of alcohol. If the mixture contains 47 gallons of water, how many gallons of alcohol does it contain? Let x represent the number of gallons of alcohol.

Number of gallons of alcohol $+$ 23 $=$ number of gallons of water .

69. Pamika plans to buy a new TV. The TV costs $46 less at a wholesale store than at an electronics store. If the price of the TV is $326 at the electronics store, what is the price at the wholesale store? Let x represent the price at the wholesale store.

Price at wholesale store $+$ 46 $=$ price at electronics store .

70. The Orlando Magic scored 13 more points than the New Jersey Nets. If the Nets scored 92 points, how many did the Magic score? Let x represent the number of points the Magic scored.

Number of points Nets scored $+$ 13 $=$ number of points Magic scored .

71. Francine went on a diet and lost 23 pounds. If she weighed 118 pounds after the diet, what did she weigh before the diet? Let x represent her weight before the diet.

Weight before diet $-$ weight loss $=$ weight after diet .

72. The selling price of a computer is $1195. If the markup is $210, find the store's cost. Let x represent the store's cost.

Store's cost $+$ markup $=$ selling price .

73. A local department store received $18,000 more from the sale of men's shirts than from the sale of men's pants. If it received $40,000 for the sale of shirts, how much did it receive from the sale of men's pants? Let x represent the amount from the sale of pants.

Amount from sale of pants $+ 18,000 =$ amount from sale of shirts .

74. Mount Rainier in Washington State is 7726 feet higher than Mount Mitchell in North Carolina. If Mount Mitchell is 6684 feet high, find the height of Mount Rainier. Let x represent the height of Mount Rainier.

Height of Mount Rainier $- 7726 =$ height of Mount Mitchell .

75. The perimeter of a triangle is 18 inches. If two of the sides have lengths of 6 inches and 8 inches, what is the length of the third side? ($P = a + b + c$) Let x represent the length of the third side.

76. A board is cut into two pieces. The longer piece is 5 feet in length and is 3 feet longer than the shorter piece. Find the length of the shorter piece. Let x represent the length of the shorter piece.

Length of shorter piece $+ 3 =$ length of longer piece .

Writing Exercises (77 and 78)

77. What does it mean to say that a number "solves an equation"?

78. Why do we not need a "subtraction property of equality"?

Writing Exercise or Group Project

79. Write two application problems similar to Exercises 60–76 that require the use of the addition property of equality. Exchange with another group or person and solve.

Section 3.2 **Multiplication Property of Equality**

OBJECTIVES *When you complete this section, you will be able to:*

A Solve equations by using the multiplication property of equality.
B Solve equations that require simplifications by using the multiplication property of equality.
C Solve real-world problems.

PREREQUISITE SKILLS *Before beginning this section, you should be able to:*

a. Multiply and divide integers (Sections 2.1 and 2.5).
b. Multiply and divide decimals (Section R.6).
c. Multiply fractions (Section R.4).

GETTING READY FOR SECTION 3.2

Multiply or divide as indicated.

1. $8(-4)$ **2.** $-14(-3)$ **3.** $42 \div (-6)$
4. $-96 \div (-12)$ **5.** $(-.3)(1.6)$ **6.** $2.4 \div (-.6)$

Multiply.

7. $\dfrac{3}{2} \cdot 40$ **8.** $-5 \cdot \dfrac{1}{-5}$ **9.** $\dfrac{3}{4} \cdot \dfrac{4}{3}$

Introduction

The addition property of equality allows us to solve one type of equation. It does not, however, allow us to solve equations like $2x = 6$. If we subtracted 2 from both sides, we would get $2x - 2 = 4$, a more complicated equation than the one we began with. We need another technique to be able to solve this type of equation.

In Section 3.1, we found that we could add the same number to both sides of an equation and the equality was preserved. Would the equality be preserved if we multiply both sides of an equation by the same nonzero number?

$4 + 3 = 7$	Multiply both sides by 2.
$2(4 + 3) \stackrel{?}{=} 2(7)$	Apply the distributive property.
$2(4) + 2(3) \stackrel{?}{=} 2(7)$	Multiply before adding.
$8 + 6 \stackrel{?}{=} 14$	Add.
$14 = 14$	Therefore, the equality was preserved when both sides were multiplied by 2.
$2 \cdot 9 = 18$	Multiply both sides by $\frac{1}{6}$.
$\dfrac{1}{6} \cdot (2 \cdot 9) = \dfrac{1}{6} \cdot 18$	$2 \cdot 9 = 18$ and $\frac{1}{6} \cdot 18 = \frac{18}{6}$.
$\dfrac{18}{6} = \dfrac{18}{6}$	Divide.
$3 = 3$	Therefore, the equality was preserved when both sides were divided by 6.

Note

Multiplying by $\frac{1}{6}$ is the same as dividing by 6.

Answers: Getting Ready: 1. -32 **2.** 42 **3.** -7 **4.** 8 **5.** $-.48$ **6.** -4 **7.** 60 **8.** 1 **9.** 1

From the preceding arithmetic equations, it appears that if we multiply (or divide) both sides of an equation by any nonzero number, the equality is preserved. We formally state this as the multiplication property of equality.

> ### Rule: Multiplication Property of Equality
>
> If $a = b$ and $c \neq 0$, then $a \cdot c = b \cdot c$. In words, both sides of an equation may be multiplied (or divided) by any nonzero number, and the result will still be an equation.

OBJECTIVE ## Solving equations by using the multiplication property of equality

As with the addition property of equality, when we apply the multiplication property of equality to an equation with variables, the resulting equation is equivalent to the original.

To solve equations by using the multiplication property of equality, we need to recall two properties from Section 2.1. The first is the multiplicative inverse, also called the **reciprocal**. Two numbers are multiplicative inverses if their product is 1. The multiplicative inverse (reciprocal) of a is $\frac{1}{a}$, because $\frac{1}{a}(a) = \frac{a}{a} = 1$. The multiplicative inverse of $\frac{a}{b}$ is $\frac{b}{a}$, because $\frac{a}{b} \cdot \frac{b}{a} = \frac{ab}{ba} = 1$. For example, the reciprocal of $\frac{3}{4}$ is $\frac{4}{3}$, because $\frac{3}{4} \cdot \frac{4}{3} = \frac{3 \cdot 4}{4 \cdot 3} = \frac{12}{12} = 1$. The second property is the multiplication property of 1, which states that $1 \cdot x = x \cdot 1 = x$.

Suppose we are asked to solve the equation $3u = 15$. As in Section 3.1, the goal in solving this equation is to get the variable with a coefficient of 1 on one side of the equation and the constant on the other. This is done by multiplying both sides of the equation by the reciprocal of the coefficient of the variable, which is $\frac{1}{3}$. Note that this is the same as dividing both sides by 3.

Example 1 | **Solve the following:**

a. $3u = 15$ Since $\frac{1}{3}$ is the reciprocal of 3, multiply both sides by $\frac{1}{3}$.

$\dfrac{1}{3} \cdot 3u = \dfrac{1}{3} \cdot 15$ $\frac{1}{3} \cdot 3 = 1$ and $\frac{1}{3} \cdot 15 = \frac{15}{3} = 5$.

$1(u) = 5$ $1 \cdot u = u$.

$u = 5$ Therefore, $u = 5$ is the solution.

CHECK:

$3u = 15$ Substitute 5 for u.

$3(5) = 15$ Multiply.

$15 = 15$ Therefore, 5 is the solution.

Note: Since division by 3 is the same as multiplication by $\frac{1}{3}$, we could have done Example 2a as follows:

$3u = 15$ Divide both sides by 3.

$\dfrac{3u}{3} = \dfrac{15}{3}$ $\frac{3}{3} = 1$ and $\frac{15}{3} = 5$.

$u = 5$ Therefore, 5 is the solution.

This is the method that we will use in future examples, when appropriate.

b. $-.6v = 1.8$ Divide both sides by $-.6$.

$$\frac{-.6v}{-.6} = \frac{1.8}{-.6}$$ $\frac{-.6}{-.6} = 1$ and $\frac{1.8}{-.6} = -3$.

$1 \cdot v = -3$ $1 \cdot v = v$.

$v = -3$ Therefore, $v = -3$ is the solution.

CHECK:

$-.6v = 1.8$ Substitute -3 for v.

$-.6(-3) = 1.8$ Multiply.

$1.8 = 1.8$ Therefore, -3 is the solution.

c. $\dfrac{t}{-4} = 7$ Rewrite as a product.

$\dfrac{1}{-4}t = 7$ Since -4 is the reciprocal of $\frac{1}{-4}$, multiply both sides by -4.

$-4 \cdot \dfrac{1}{-4}t = -4 \cdot 7$ $-4 \cdot \frac{1}{-4} = 1$ and $-4 \cdot 7 = -28$.

$1 \cdot t = -28$ $1 \cdot t = t$.

$t = -28$ Therefore, -28 is the solution.

CHECK:

$\dfrac{t}{-4} = 7$ Substitute -28 for t.

$\dfrac{-28}{-4} = 7$ Divide.

$7 = 7$ Therefore, -28 is the correct solution.

d. $\dfrac{3}{2}z = 60$ Since $\frac{2}{3}$ is the reciprocal of $\frac{3}{2}$, multiply both sides by $\frac{2}{3}$.

$\dfrac{2}{3} \cdot \dfrac{3}{2}z = \dfrac{2}{3} \cdot 60$ $\frac{2}{3} \cdot \frac{3}{2} = 1$ and $\frac{2}{3} \cdot 60 = \frac{2}{3} \cdot \frac{60}{1} = \frac{120}{3} = 40$.

$1 \cdot z = 40$ $1 \cdot z = z$.

$z = 40$ Therefore, $z = 40$ is the solution.

CHECK:

$\dfrac{3}{2}z = 60$ Substitute 40 for z.

$\dfrac{3}{2}(40) = 60$ Multiply.

$\dfrac{120}{2} = 60$ Divide.

$60 = 60$ Therefore, 40 is the solution.

Note

Example 1e could have been done by dividing both sides by -1.

e. $-x = 5$ $-x = -1 \cdot x$. Multiply both sides by the reciprocal of -1, which is -1.

$-1(-x) = -1 \cdot 5$ Simplify both sides.

$x = -5$ Therefore, $x = -5$ is the solution.

CHECK:

$-x = 5$ Substitute -5 for x.

$-(-5) = 5$ Apply $-(-x) = x$.

$5 = 5$ Therefore, -5 is the correct solution.

PRACTICE EXERCISE 1

Solve the following:

a. $5v = -20$ **b.** $-.5y = 32.5$ **c.** $\dfrac{w}{2} = 6$ **d.** $-\dfrac{7}{9}u = 21$ **e.** $3 = -t$

If you need more practice, do the following Additional Practice Exercises.

Additional Practice Exercise 1   Solve the following:

a. $42 = 6n$ **b.** $.7p = -6.3$ **c.** $\dfrac{x}{-3} = 3$ **d.** $\dfrac{8}{5}s = -32$ **e.** $-x = 4$

OBJECTIVE **Solving equations that require simplifications, using the multiplication property of equality**

When applying the addition property of equality in the previous section, it was often necessary to simplify one or both sides of the equation. The same is true when the multiplication property of equality is involved.

Example 2   **Solve each equation. If the equation involves decimals, express the answer as a decimal if needed. Otherwise, when necessary, express the answer as a fraction.**

a. $8x + 4x = -48$ Simplify the left side of the equation.

$12x = -48$ Divide both sides by 12.

$\dfrac{12x}{12} = \dfrac{-48}{12}$ $\frac{12}{12} = 1$ and $\frac{-48}{12} = -4$.

$1 \cdot x = -4$ $1 \cdot x = x$.

$x = -4$ Therefore, $x = -4$ is the solution.

CHECK:

$8x + 4x = -48$ Substitute -4 for x.

$8(-4) + 4(-4) = -48$ Multiply.

$-32 - 16 = -48$ Add.

$-48 = -48$ Therefore, -4 is the solution.

b. $5y = 9 + 15$ Simplify the right side of the equation.

$5y = 24$ Divide both sides by 5.

$\dfrac{5y}{5} = \dfrac{24}{5}$ $\frac{5}{5} = 1$.

$1 \cdot y = \dfrac{24}{5}$ $1 \cdot y = y$.

$y = \dfrac{24}{5}$ Therefore, the solution is $y = \frac{24}{5}$.

CHECK:

The check is left as an exercise for the student.

Answers: **Practice Exercise 1: a.** $v = -4$ **b.** $y = -65$ **c.** $w = 12$ **d.** $u = -27$ **e.** $t = -3$
Additional Practice 1: a. $n = 7$ **b.** $p = -9$ **c.** $x = -9$ **d.** $s = -20$ **e.** $x = -4$

   c. $2.4z - 3.6z = 4.8$ Simplify the left side of the equation.

          $-1.2z = 4.8$ Divide both sides by -1.2.

          $\dfrac{-1.2z}{-1.2} = \dfrac{4.8}{-1.2}$ $\frac{-1.2}{-1.2} = 1$ and $\frac{4.8}{-1.2} = -4$.

          $1 \cdot z = -4$ $1 \cdot z = z$.

          $z = -4$ Therefore, $z = -4$ is the solution.

   CHECK:

The check is left as an exercise for the student.

PRACTICE EXERCISE 2

Solve each equation. If the equation involves decimals, express the answer as a decimal if needed. Otherwise, when necessary, express the answer as a fraction.

a. $16t - 7t = 45$ **b.** $9q = 35 - 15$ **c.** $3.3 = 5.8s - 5.5s$

If you need more practice, do the following Additional Practice Exercises.

Additional Practice Exercise 2   Solve, using the multiplication property. If the equation involves decimals, express the answer as a decimal if needed. Otherwise, express the answer as a fraction.

a. $2x + 4x = 24$ **b.** $2t = -21 + 36$ **c.** $5.7a - 2.3a = -10.2$

OBJECTIVE Solving real-world problems

As in the previous section, we will translate from mathematics to English and from English to mathematics.

Example 3   **Translate each of the expressions from mathematics to English. There may be more than one translation into English.**

Mathematics	English
a. $2y = 8$	Twice a number is eight. Two times a number equals eight. The product of two and a number is eight.
b. $\dfrac{x}{4} = 5$	The quotient of some number and four is five. Some number divided by four is five.
c. $32 = -8t$	Thirty-two is negative eight times a number. Thirty-two is the product of negative eight and a number.
d. $.6s = 3$	Six-tenths of a number is three. The product of six-tenths and a number is three. Six-tenths times some number is three.
e. $\dfrac{2}{3}x = 6$	The product of two-thirds and some number is six. Two-thirds times some number is six. Two-thirds of a number is six.

Answers: **Practice Exercise 2: a.** $t = 5$ **b.** $q = \frac{20}{9}$ **c.** $s = 11$
Additional Practice 2: a. $x = 4$ **b.** $t = \frac{15}{2}$ **c.** $a = -3$

PRACTICE EXERCISE 3

Translate from mathematical sentences to English sentences. There is more than one possible translation.

a. $9x = 45$

b. $\dfrac{v}{8} = 2$

c. $33 = -3w$

d. $.5z = -3$

e. $\dfrac{4}{5}a = 12$

If you need more practice, do the following Additional Practice Exercises.

Additional Practice Exercise 3   Translate from mathematical sentences to English sentences. There is more than one possible translation.

a. $7y = 28$

b. $\dfrac{x}{3} = -4$

c. $15 = -2x$

d. $.4t = 16$

e. $\dfrac{3}{4}y = 9$

Example 4   **Translate from English to mathematics and solve. Let x be the variable in each case.**

English	Mathematics
a. The product of four and a number is twenty.	$4x = 20$ $\dfrac{4x}{4} = \dfrac{20}{4}$ $x = 5$
b. The quotient of some number and three is negative six.	$\dfrac{x}{3} = -6$ $3 \cdot \dfrac{x}{3} = 3(-6)$ $x = -18$
c. Five and six-tenths equals negative ten times a number.	$5.6 = -10x$ $\dfrac{5.6}{-10} = \dfrac{-10x}{-10}$ $-.56 = x$
d. Eight-tenths of a number equals negative two and four-tenths.	$.8x = -2.4$ $\dfrac{.8x}{.8} = \dfrac{-2.4}{.8}$ $x = -3$
e. Three-fifths of a number is negative nine.	$\dfrac{3}{5}x = -9$ $\dfrac{5}{3} \cdot \dfrac{3}{5}x = \dfrac{5}{3}(-9)$ $1 \cdot x = -\dfrac{45}{3}$ $x = -15$

Answers: Practice Exercise 3: a. The product of nine and a number is forty-five. **b.** The quotient of a number and eight is two. **c.** Thirty-three is negative three times a number. **d.** Five-tenths of some number is negative three. **e.** Four-fifths of a number is twelve.
Additional Practice 3: a. Seven times a number is twenty-eight. **b.** The quotient of a number and three is negative four. **c.** Fifteen is negative two times some number. **d.** Four-tenths of a number is sixteen. **e.** Three-fourths of a number is nine.

PRACTICE EXERCISE 4

Translate from English to mathematical sentences and solve. Use x as the variable.

a. Five times a number is forty.

b. The quotient of some number and negative two is eight.

c. Negative four and two-tenths is negative seven times some number.

d. Seven-tenths of a number is twenty-one.

e. Two-thirds of a number is negative 8.

If you need more practice, do the following Additional Practice Exercises.

Additional Practice Exercise 4   Translate from English to mathematical sentences and solve. Use x as the variable.

a. The product of six and a number is twelve.

b. The quotient of some number and negative three is five.

c. Three and five-tenths equals seven-tenths of a number.

d. Three-tenths of a number is one and two-tenths.

e. Four-sevenths of a number is eight.

As in the previous section, most of the real-world problems presented here can be solved without the use of algebra. However, we ask that you continue to follow the format used in Section 3.1.

Sometimes, it is necessary to convert a percent into a decimal. Percent means per 100, so 16% means $\frac{16}{100}$, which is .16 written as a decimal. So, to change from percent notation to a decimal, drop the percent sign and move the decimal point two places to the left. For example, 23% = .23 and 6% = .06. Percents will be discussed in detail in Chapter 8.

Example 5

a. Jude earns $116.00 for working 8 hours. What is his hourly wage?

Solution

Understand: We know the total amount Jude earned and the number of hours that he worked. The unknown is his hourly wage.

   Relationship: Jude is paid $116.00 for working 8 hours.

Plan: We will use the preceding relationship to write the equation.

Translate: The unknown is Jude's hourly wage, so let w represent his **hourly wage**.

Write a word equation.

| Number of hours Jude worked | times | his hourly wage | equals | total earnings | . |

Write an algebraic equation, using the word equation as a guide.

| Number of hours Jude worked | times | his hourly wage | equals | total earnings | . |

$$8 \qquad \cdot \qquad w \qquad = \qquad 116$$

Solve:

$8w = 116$ Divide both sides by 8.

$\dfrac{8w}{8} = \dfrac{116}{8}$ Simplify both sides.

$w = 14.50$ Solution of the equation.

State: Jude earns $14.50 per hour.

Check: If Jude worked 8 hours at $14.50 per hour, his total earnings would be 8(14.50), which equals $116.00. Therefore, our solution is correct.

b. Sondra works as a salesperson. Her commission is 7% of her total sales. If she earned $455 one week, what were her total sales for that week?

Understand: We know Sondra's rate of commission and her earnings for one week. The unknown is her total sales for the week.

   Relationship: Sondra earns 7% commission on her sales, and she earned $455.

Plan: We use the preceding relationship to write the equation. To calculate her earnings for the week, we multiply her rate of commission and her sales for the week.

Translate: The unknown is Sondra's total sales, so let **s** represent **her sales** for the week.

Write a word equation.

Rate of commission   times   sales for the week   equals   earnings for the week .

Write an algebraic equation, using the word equation as a guide. Remember, 7% = .07.

Rate of commission   times   sales for the week   equals   earnings for the week .

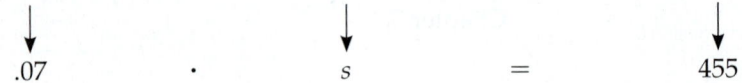

.07   ·   s   =   455

Solve:

$.07s = 455$ Divide both sides by .07.

$\dfrac{.07s}{.07} = \dfrac{455}{.07}$ Simplify both sides.

$s = 6500$ Solution of the equation.

State: Sondra's total sales for the week were $6500.

Check: Sondra's earnings are equal to $(.07)(\$6500)$, which equals $455. Therefore, our solution is correct.

c. A carpenter is installing a hardwood floor. He knows that the area of the rectangular room is 252 square feet and the length of the room is 18 feet. What is the width of the room?

Understand: We know the area and length of the room, and the unknown is the width.

   Relationship: The area of the room is 252 square feet and the length is 18 feet.

Plan: Use the preceding relationship and the formula for the area of a rectangle, $A = LW$.

Translate: We substitute into the formula and solve. We know $A = LW$, $A = 252$, and $L = 18$, which together give us

$$252 = 18W.$$

Solve:

$252 = 18W$ Divide both sides by 18.

$\dfrac{252}{18} = \dfrac{18W}{18}$ Simplify both sides.

$14 \text{ ft} = W$ Solution of the equation.

State: The width is 14 feet.

Check:

$A = LW$ Substitute for A, L, and W.

$252 = 18 \cdot 14$ Multiply 18 and 14.

$252 = 252$ Therefore, the solution is correct.

PRACTICE EXERCISE 5

Solve the following:

a. Hernando drove 230 miles in 5 hours. What is his rate in miles per hour? (rate · time = distance)

b. A bank pays 6% interest per year on savings accounts. If Sally received $1080 interest on her savings account last year, how much does she have in savings?

c. A sign company is to construct a triangular sign whose area is 72 square feet. If the base of the sign is 12 feet, what is the height? $\left(A = \dfrac{bh}{2}\right)$

Exercise Set 3.2

For Extra Help

MyMathLab®

Solve the following. (See Examples 1 and 2.)

1. $7u = 56$

2. $9v = 72$

3. $-3n = 36$

4. $-62 = 2m$

5. $42 = -14x$

6. $96 = -12y$

7. $-64 = -16p$

8. $-75 = -15q$

9. $7u = 5.6$

10. $6v = 4.8$

11. $.3x = 15$

12. $.8y = 24$

13. $-.1t = -11$

14. $-.1m = -18$

15. $\dfrac{n}{2} = 8$

16. $\dfrac{r}{5} = 3$

17. $\dfrac{q}{-3} = 6$

18. $\dfrac{w}{-4} = 6$

19. $\dfrac{8}{7}z = -56$

20. $\dfrac{8}{9}q = -72$

21. $-\dfrac{2}{3}w = -6$

22. $-\dfrac{3}{4}x = -6$

23. $-120 = \dfrac{8m}{5}$

24. $-91 = \dfrac{7n}{13}$

25. $\dfrac{2x}{3} = 6$

26. $\dfrac{3x}{5} = 9$

27. $4a = -16 + 8$

28. $6b = -36 + 24$

29. $-6y = 55 - 13$

30. $-4z = -39 + 15$

Answers: Practice Exercise 5: a. 46 miles per hour **b.** $18,000 **c.** 12 feet

31. $3x + 5x = 24$

32. $4x + 7x = 22$

33. $9x - 3x = -24$

34. $11x - 4x = -28$

35. $4x - 7x = 15 - 6$

36. $5x - 9x = 12 - 8$

37. $-2.9u + 3.6u = 6.3$

38. $-4.6z + 2.2z = 4.8$

39. $3.4x + 5.1x = 30 - 4.5$

Translate the following mathematical sentences into English sentences in at least two ways. (See Example 3.)

40. $7t = 46$

41. $4z = 62$

42. $-14 = 2h$

43. $-39 = 3k$

44. $5.6 = 7x$

45. $2.7y = 5.6$

Translate from English to mathematical sentences and solve. Use x as the variable. (See Example 4.)

46. Eight times a number is sixty-four.

47. Five times a number is thirty-five.

48. The product of six and some number is negative forty-eight.

49. The product of four and some number is negative twenty-eight.

50. Six-tenths of a number is twelve.

51. Four-tenths of a number equals two.

52. The quotient of a number and two is six.

53. The quotient of a number and four is five.

54. Some number divided by negative five is two.

55. Some number divided by negative three is eight.

56. The product of two and some number is divided by five. The result is ten. What is the number?

57. Three-sevenths of a number is nine.

58. Four-fifths of a number is twelve.

59. Two-ninths of a number is negative six.

60. Five-eighths of a number is negative ten.

Solve each of the following. (See Example 5.)

61. John is to construct a rectangular sign that is 12 feet long and has an area of 96 square feet. How wide is the sign? ($A = LW$)

62. Fred is buying carpet for his rectangular living room. He arrives at the carpet store and can remember only that the length of the room is 18 feet and the area is 216 square feet. Find the width of the room. ($A = LW$)

63. Paul is buying a triangular piece of land whose area is 1500 square yards and whose base is 100 yards. What is the height? ($A = \frac{bh}{2}$)

64. A triangular traffic sign has an area of 6 square feet and a height of 3 feet. What is the base of the sign? ($A = \frac{bh}{2}$)

65. Leslie has a piece of apple pie à la mode. The piece of pie has three times as many calories as the low-calorie ice cream served with it. If the pie has 450 calories, how many calories does the ice cream have?

66. Bill spent $14.60 for four fishing lures, all of which had the same price. How much did each cost?

67. Catalina worked 12 hours last week and earned $102. What is her hourly wage?

68. Susan is building a bookcase. How many 3.5-foot-long shelves can she cut from a board that is 14 feet long?

69. During a particularly boring speech, two-fifths of the audience left. If 46 people left, how many people were originally in the audience?

70. Two-thirds of a class passed the last exam. If 18 people passed, how many students are in the class?

71. On her morning walk, Frankie passes a tree located at a point that is one-sixth of the total distance she plans to walk from her house. If she has walked two-thirds of a mile, how far does she plan to walk?

72. After a stock split, Harry has five-fourths times as many shares of stock as he had before the split. If he has 225 shares after the split, how many shares did he have before the split?

73. During a special sale, an appliance store offers interest-free financing. If Ben buys a refrigerator that costs $1512 and pays $42.00 per month, how many months will it take him to pay off the refrigerator?

74. Teschima is renting a TV for $16.50 per week. If a new TV costs $1320, how many weeks will it take for her rental fee to equal the cost of a new TV?

75. The state sales tax in Florida is 6%. If the sales tax on a car purchased in Florida is $1050, what was the selling price of the car?

76. Katrina works as a salesperson and is paid a 6% commission on her sales. If she receives a $9000 commission on a sale, what was the amount of the sale?

77. A piece of paper is in the shape of an isosceles triangle whose base is 16 inches long. If the area of the paper is 48 square inches, what is the height of the paper? $\left(A = \frac{1}{2}bh \right)$

78. If the volume of a cone is 96 cubic meters and the area of the base is 36 square meters, what is the height of the cone? $\left(v = \frac{1}{3}Ah \right)$

Writing Exercise

79. Why do we not need a division property of equality?

Writing Exercise or Group Project

80. Write two application problems similar to Exercises 61–78 that involve the use of the multiplication property of equality. Exchange with another group or person and solve.

Section 3.3 — Combining Properties in Solving Linear Equations

OBJECTIVES *When you complete this section, you will be able to:*

Ⓐ Solve linear equations, using both the addition and multiplication properties of equality.

Ⓑ Determine whether an equation is an identity or a contradiction.

Ⓒ Translate mathematical equations into English and English sentences into mathematical equations.

Ⓓ Solve application problems.

PREREQUISITE SKILLS *Before beginning this section, you should be able to:*

a. Add and subtract integers (Sections 1.6 and 1.7).

b. Add and subtract like terms (Section 1.7).

c. Add, subtract, multiply, and divide decimals (Sections R.5 and R.6).

d. Multiply and divide integers (Sections 2.1 and 2.5).

e. Apply the distributive property (Section 1.7).

f. Solve linear equations, using the addition property of equality (Section 3.1).

g. Solve linear equations, using the multiplication property of equality (Section 3.2).

GETTING READY FOR SECTION 3.3

Add or subtract the following integers:

1. $-6 - 15$

2. $28 - 4$

Add or subtract the following like terms:

3. $4x - 6 - 3x - 15$

4. $5y + 10 - 3y$

Add, subtract, multiply, or divide the following decimals:

5. $.4 - 3.8$

6. $12(.6)$

Multiply or divide the following integers:

7. $-3(5)$

8. $\dfrac{-24}{8}$

Simplify the following, using the distributive property:

9. $7(r - 1) + 10$

10. $2(2x - 3) - 3(x + 5)$

Solve the following, using the addition property of equality:

11. $x - 3 = 7$

12. $2(x - 4) - x + 2 = 12$

Solve the following, using the multiplication property of equality:

13. $3x = -1.2$

14. $\dfrac{x}{2} = -5$

Introduction

In this section, we will combine the addition and multiplication properties of equality to solve linear equations in one variable. For example, suppose we were asked to solve $3x + 9 = -6$. In order to solve for x, we must get rid of both 3 and 9, but which do we eliminate first? The procedure for solving such equations is outlined in the following box.

Answers: Getting Ready: 1. -21 **2.** 24 **3.** $x - 21$ **4.** $2y + 10$ **5.** -3.4 **6.** 7.2 **7.** -15 **8.** -3 **9.** $7r + 3$
10. $x - 21$ **11.** $x = 10$ **12.** $x = 18$ **13.** $x = -.4$ **14.** $x = -10$

OBJECTIVE **Solving linear equations by using both the addition and multiplication properties of equality**

> **Procedure: Solving Linear Equations**
>
> 1. If necessary, simplify both sides of the equation as much as possible. This could involve using the distributive, commutative, and associative properties and combining like terms.
>
> 2. If necessary, use the addition property of equality to get all the terms with variables on one side of the equation and all the constant terms on the other.
>
> 3. If necessary, use the multiplication property of equality to eliminate any coefficient on the variable other than 1.
>
> 4. Check the answer in the original equation.

Up to this point, all the equations that we have solved have had all the variables on one side. If, after simplifying both sides of the equation, there are terms with the variables on both sides, use the addition property of equality to get the terms with the variable on one side of the equation and the constants on the other. We still use the principle of the additive inverse to accomplish this. For example, the additive inverse of $2x$ is $-2x$, since $2x + (-2x) = 0$. In solving equations like $3x + 4 = -5x + 28$, we can eliminate any one of the four terms by adding its additive inverse to both sides of the equation. In Example 1b–e, we will first eliminate a term with a variable, and then a constant. When an equation has variable terms on both sides of the equation, we will usually eliminate the variable term with the smaller coefficient in order to avoid negative coefficients in the next-to-last step, though this is not necessary.

Example 1 | **Solve the following:**

a. $3x + 9 = -6$ Subtract 9 from both sides of the equation.

$3x + 9 - 9 = -6 - 9$ Simplify both sides.

$3x = -15$ Divide both sides by 3.

$\dfrac{3x}{3} = \dfrac{-15}{3}$ Simplify both sides.

$x = -5$ Therefore, the solution is $x = -5$.

CHECK:

$3x + 9 = -6$ Substitute -5 for x.

$3(-5) + 9 = -6$ Multiply before adding.

$-15 + 9 = -6$ Add.

$-6 = -6$ Therefore, -5 is the correct solution.

b. $3x + 4 = -5x + 28$ We could eliminate any of the four terms. We will eliminate $-5x$ by adding its inverse, $5x$, to both sides of the equation.

$3x + 5x + 4 = -5x + 5x + 28$ Simplify both sides.

$8x + 4 = 28$ Subtract 4 from both sides.

$8x + 4 - 4 = 28 - 4$ Simplify both sides.

$8x = 24$ Divide both sides by 8.

$\dfrac{8x}{8} = \dfrac{24}{8}$ Simplify both sides.

$x = 3$ Therefore, $x = 3$ is the solution.

CHECK:

$3x + 4 = -5x + 28$ Substitute 3 for x.

$3(3) + 4 = -5(3) + 28$ Multiply before adding.

$9 + 4 = -15 + 28$ Add.

$13 = 13$ Therefore, 3 is the solution.

c. $12z - 3.8 = 5z + .4$ We could eliminate any of the four terms. We will eliminate $5z$ by adding its inverse, $-5z$, to both sides of the equation.

$12z - 5z - 3.8 = 5z - 5z + .4$ Simplify both sides.

$7z - 3.8 = .4$ Add 3.8 to both sides.

$7z - 3.8 + 3.8 = .4 + 3.8$ Simplify both sides.

$7z = 4.2$ Divide both sides by 7.

$\dfrac{7z}{7} = \dfrac{4.2}{7}$ Simplify both sides.

$z = .6$ Therefore, $z = .6$ is the solution.

CHECK:

$12z - 3.8 = 5z + .4$ Substitute .6 for z.

$12(.6) - 3.8 = 5(.6) + .4$ Multiply before adding.

$7.2 - 3.8 = 3.0 + .4$ Add.

$3.4 = 3.4$ Therefore, .6 is the solution.

d. $9y - 5 - 8y = 5y + 10 - 3y$ Simplify both sides of the equation by adding like terms.

$y - 5 = 2y + 10$ Subtract y from both sides.

$y - y - 5 = 2y - y + 10$ Simplify both sides.

$-5 = y + 10$ Subtract 10 from both sides.

$-5 - 10 = y + 10 - 10$ Simplify both sides.

$-15 = y$ Therefore, $y = -15$ is the solution.

The check is left as an exercise for the student.

Note

The more complicated equations involving fractions are discussed in Chapter 7, when solving equations with rational expressions is discussed.

e. $7(r - 1) + 10 = 5(2 + r)$ Apply the distributive property.

$7r - 7 + 10 = 10 + 5r$ Simplify both sides of the equation by adding like terms.

$7r + 3 = 10 + 5r$ To eliminate $5r$, subtract $5r$ from both sides.

$7r - 5r + 3 = 10 + 5r - 5r$ Simplify both sides.

$2r + 3 = 10$ Subtract 3 from both sides.

$2r + 3 - 3 = 10 - 3$ Simplify both sides.

$2r = 7$ Divide both sides by 2.

$\dfrac{2r}{2} = \dfrac{7}{2}$ Simplify both sides.

$r = \dfrac{7}{2}$ or 3.5 Therefore, $r = \frac{7}{2}$ is the solution.

The check is left as an exercise for the student.

PRACTICE EXERCISE 1

Solve each of the following:

a. $4x - 7 = 9$ **b.** $2r - 4 = 11 - 3r$ **c.** $.7 - t = 3t - 2.1$

d. $9 - 6r - 3 - 7r = 2r + 2 - 11r$ **e.** $4(r + 2) = 3(r + 4) + 6$

(Continued)

Answers: **Practice Exercise 1:** **a.** $x = 4$ **b.** $r = 3$ **c.** $t = .7$ **d.** $r = 1$ **e.** $r = 10$

Additional Practice Exercise 1 Solve the following:

a. $5x + 2 = 12$ **b.** $4x - 15 = 5 - 6x$ **c.** $2.3 - .4w = .5w - 2.2$

d. $3z + 5 + 5z = 2z - 1$ **e.** $6(p - 4) - 3p = 18 - 3p$

OBJECTIVE **Determining whether an equation is an identity or a contradiction**

In solving equations, there are two situations that require special attention. These are **identities** and **contradictions**. An *identity* is an equation that is true for all values of the variable for which the equation is defined. When solving an equation that is an identity, we arrive at a statement which is obviously true, such as $5 = 5$. Since any number solves the identity, we will denote the solutions as "all real numbers." A *contradiction* is an equation that has no solution. When solving an equation that is a contradiction, we arrive at a statement which is obviously false, such as $3 = 10$. The fact that there is no solution is usually indicated by the symbol $\varnothing$, which stands for the empty set, the set that has no elements.

Example 2 | **Determine whether each equation is an identity or a contradiction and indicate the solutions.**

a. $3(x + 2) - 4 = x + 2x + 2$ Simplify each side of the equation.

$3x + 6 - 4 = 3x + 2$ Add like terms.

$3x + 2 = 3x + 2$ Subtract $3x$ from both sides.

$3x - 3x + 2 = 3x - 3x + 2$ Simplify both sides.

$2 = 2$

This is obviously a true statement. Therefore, this is an identity and the solutions are all real numbers.

b. $4x - 3(x + 5) = 2(2x - 3) - 3(x + 5)$ Simplify both sides of the equation.

$4x - 3x - 15 = 4x - 6 - 3x - 15$ Add like terms.

$x - 15 = x - 21$ Subtract x from both sides.

$x - x - 15 = x - x - 21$ Add like terms.

$-15 = -21$

This is obviously a false statement. Therefore, this is a contradiction and there are no solutions. Equivalently, the solution set is $\varnothing$.

PRACTICE EXERCISE 2

Determine whether each equation is an identity or a contradiction and indicate the solution or solution set.

a. $2(2x + 4) - 11 = 4(x - 2) + 5$ **b.** $4(x - 1) + 2(x - 3) = 4(x - 2) + 2x$

If you need more practice, do the following Additional Practice Exercises.

Additional Practice Exercise 2 Determine whether each equation is an identity or a contradiction and indicate the solution or solution set.

a. $3(3x - 4) - x = 4(2x - 1) - 8$ **b.** $3(2x - 6) - 2(x - 4) = 5(x + 3) - x$

Answers: Additional Practice 1: a. $x = 2$ **b.** $x = 2$ **c.** $w = 5$ **d.** $z = -1$ **e.** $p = 7$
Practice Exercise 2: a. Identity, all real numbers **b.** Contradiction, no solutions, or $\varnothing$
Additional Practice 2: a. Identity, all real numbers **b.** Contradiction, no solutions, or $\varnothing$

OBJECTIVE Translating mathematical equations into English and English sentences into mathematical equations

We continue practicing translating mathematical sentences into English and English sentences into mathematics.

Example 3 Translate each of the mathematical sentences into English sentences. There may be more than one translation.

Mathematics	English
a. $2x + 3 = 5$	Three more than twice a number equals five. Twice a number plus three is five. The sum of twice a number and three is five. Twice a number increased by three is five. Three more than the product of two and a number is five.
b. $4u - 12 = 10u$	Four times a number minus twelve equals ten times the number. Four times a number decreased by twelve is ten times the number. The difference between four times a number and twelve equals ten times the number. Twelve less than four times a number is ten times that number.
c. $8(10 - t) = 96$	Eight times the difference of ten and a number is ninety-six. The product of eight and the difference of ten and some number is ninety-six.
d. $1.5 = .75v - .5v$	One and five-tenths equals seventy-five hundredths of a number decreased by five-tenths of that number. One and five-tenths is the difference of seventy-five hundredths of a number and five-tenths of that number. One and five-tenths equals seventy-five hundredths of a number minus five-tenths of that number.

PRACTICE EXERCISE 3

Translate each of the mathematical sentences into English sentences. There is more than one translation.

a. $3w - 7 = 4$ **b.** $5x + 6 = 7x$ **c.** $12(3 - m) = 48$ **d.** $x + .06x = 24$

If you need more practice, do the following Additional Practice Exercises.

Additional Practice Exercise 3 Translate each of the mathematical sentences into English sentences. There may be more than one translation.

a. $15 - 5z = 10$ **b.** $5x + 6 = 7x$ **c.** $26 = 2(a - 4)$ **d.** $.5u - 12 = .4$

Answers: Practice Exercise 3: a. The difference of three times a number and seven is four. **b.** The sum of five times a number and six equals seven times the number. **c.** The product of twelve and the difference of three and some number is forty-eight. **d.** Some number increased by six hundredths of that number is twenty-four. **Additional Practice 3: a.** Fifteen decreased by five times a number is ten. **b.** The sum of five times a number and six is seven times the number. **c.** Twenty-six is twice the difference of some number and four. **d.** Five-tenths of a number minus twelve is four-tenths.

Example 4   **Translate the English sentences into mathematical sentences and solve. Let x be the variable in each case.**

Note

The numbers are not always written in the same order in mathematics as in English. "Five subtracted from three times a number" is translated "$3x - 5$," not "$5 - 3x$."

English	Mathematics
a. Six more than four times a number equals ten.	$4x + 6 = 10$ $4x + 6 - 6 = 10 - 6$ $4x = 4$ $x = 1$
b. Five subtracted from three times a number equals four times that number.	$3x - 5 = 4x$ $3x - 3x - 5 = 4x - 3x$ $-5 = x$
c. Twice the sum of some number and five is thirty.	$2(x + 5) = 30$ $2x + 2(5) = 30$ $2x + 10 = 30$ $2x + 10 - 10 = 30 - 10$ $2x = 20$ $x = 10$
d. The total of a number, seven-tenths of the number, and negative two-tenths of the number equals six-tenths.	$x + .7x - .2x = .6$ $1.7x - .2x = .6$ $1.5x = .6$ $x = .4$

PRACTICE EXERCISE 4

Translate the English sentences into mathematical sentences and solve. Use x as the variable.

a. Eight less than three times a number is sixteen.

b. Eighteen decreased by five times a number is equal to four times the number.

c. Four times the difference of a number and five is thirty.

d. Twenty-one equals a number decreased by three-tenths of the number.

If you need more practice, do the following Additional Practice Exercises.

Additional Practice Exercise 4   Translate the English sentences into mathematical sentences and solve. Use x as the variable.

a. Five more than three times a number is negative four.

b. Six times the difference of a number and one is twenty-four.

c. The product of eight and the difference of a number and six is fourteen.

d. The sum of six-hundredths of a number and the number is seven and forty-two hundredths.

Answers: **Practice Exercise 4: a.** $3x - 8 = 16, x = 8$ **b.** $18 - 5x = 4x, x = 2$ **c.** $4(x - 5) = 30,$ $x = \dfrac{25}{2},$ or 12.5 **d.** $21 = x - .3x, x = 30.$ **Additional Practice 4: a.** $3x + 5 = -4, x = -3$ **b.** $6(x - 1) = 24, x = 5$ **c.** $8(x - 6) = 14, x = \dfrac{31}{4},$ or 7.75 **d.** $.06x + x = 7.42, x = 7$

OBJECTIVE Solving application problems

Follow the same technique as in the previous two sections to solve the following:

Example 5   **Solve the following:**

a. An appliance repairman charges a service charge of $50, plus $52 per hour. If the charges for repairing a refrigerator are $206 (excluding parts), how many hours did it take?

Solution

Understand: We know the amount of the service charge, the hourly rate, and the total charges. The unknown is the number of hours the repair took.

Relationship: The repairman charges a service charge of $50, $52 per hour, and a total of $206.

Plan: Use the preceding relationship and the fact that the total charges are equal to the service charge plus the number of hours times the charges per hour.

Translate: The unknown is the number of hours it took to repair the refrigerator, so let h represent the **number of hours**.

Write the word equation.

| Service charge | plus | charge per hour | times | number of hours | equals | total charges | . |

Write the algebraic equation.

| Service charge | plus | charge per hour | times | number of hours | equals | total charges | . |
| 50 | + | 52 | · | h | = | 206 | |

Solve:

$$50 + 52h = 206 \qquad \text{Subtract 50 from both sides.}$$
$$50 - 50 + 52h = 206 - 50 \qquad \text{Simplify both sides.}$$
$$52h = 156 \qquad \text{Divide both sides by 52.}$$
$$\frac{52h}{52} = \frac{156}{52} \qquad \text{Simplify both sides.}$$
$$h = 3 \qquad \text{Solution of equation.}$$

State: It took the service repairman 3 hours to repair the refrigerator.

Check: If the repairman works 3 hours, his hourly charges would be $3(52) = \$156$, and $\$156 + \$50 = \$206$. Therefore, the answer is correct.

b. If Pascual had $10 more than he presently has, then twice that amount would be enough to purchase a compact disc player that sells for $280. How much does he presently have?

Understand: We know the relationship between what he presently has and what he needs and that he needs a total of $280. The unknown is the amount he presently has.

Relationship: Twice $10 more than he presently has will equal $280.

Plan: Use the preceding relationship to write the equation.

Translate: The unknown is the amount of money Pascual presently has, so let a represent the **amount he presently has**. Then $a + 10$ represents the **amount he would have if he had $10 more**.

Write a word equation.

Two times $10 more than he presently has equals 280.

Write the algebraic equation.

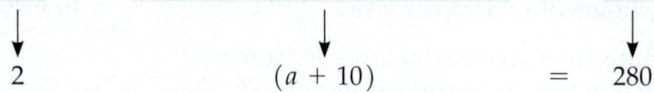

Two times $10 more than he presently has equals 280.

$$2 \qquad (a + 10) \qquad = \qquad 280$$

Solve:

$2(a + 10) = 280$	Apply the distributive property.
$2a + 20 = 280$	Subtract 20 from both sides.
$2a + 20 - 20 = 280 - 20$	Simplify both sides.
$2a = 260$	Divide both sides by 2.
$\dfrac{2a}{2} = \dfrac{260}{2}$	Simplify both sides.
$a = 130$	Solution of the equation.

State: Pascual presently has $130.

Check: If Pascual has $130, then he would have $140 if he had $10 more. Since $2(\$140) = \$280$, the answer is correct.

c. A rancher needs to fence in a rectangular corral whose perimeter is 2000 feet and whose length is 600 feet. What is the width of the corral?

Understand: The amount of fencing that he needs is the same as the perimeter of the corral.

Relationship: The perimeter is 2000 feet and the length is 600 feet.

Plan: Since the corral is rectangular, we will use the formula for the perimeter of a rectangle, which is $P = 2L + 2W$, and substitute the values from the preceding relationship.

Translate: We will substitute 2000 for P and 600 for L, which gives us $2000 = 2(600) + 2W$.

Solve:

$2000 = 2(600) + 2W$	Simplify the right side of the equation.
$2000 = 1200 + 2W$	Subtract 1200 from both sides.
$2000 - 1200 = 1200 - 1200 + 2W$	Simplify both sides.
$800 = 2W$	Divide both sides by 2.
$\dfrac{800}{2} = \dfrac{2W}{2}$	Simplify both sides.
$400 = W$	Solution of the equation.

State: The width is 400 feet.

Check:

$P = 2L + 2W$	Substitute for P, L, and W.
$2000 = 2(600) + 2(400)$	Multiply on the right.
$2000 = 1200 + 800$	Add on the right.
$2000 = 2000$	Therefore, the solution is correct.

PRACTICE EXERCISE 5

Solve the following:

a. A TV repair service charged $32 per hour, plus $85 for parts. If the bill for repairing a TV was $133, how many hours did the repair take?

b. Carla earned $550 last week for 40 hours of work. Included in the $550 was a $50 bonus. What was her hourly wage?

c. A box manufacturer receives an order for some boxes, each of which has a surface area of 292 square inches, a length of 8 inches, and a width of 6 inches. What is the height of the boxes? ($SA = 2LW + 2LH + 2WH$)

If you need more practice, do the following Additional Practice Exercises.

Additional Practice Exercise 5   Solve the following:

a. A lawn mower repairman charged $26 per hour, plus $58 for parts. If the bill for repairing a lawn mower was $123, how many hours did the repair take?

b. If Harold had $15 more than he presently has, then three times that amount would be enough to buy an iPod that costs $195. How much does he presently have?

c. A carpenter needs 54 feet of crown molding to go around the ceiling of a room whose width is 12 feet. What is the length of the room?

Exercise Set 3.3
For Extra Help
MyMathLab®

Solve the following. (See Example 1.)

1. $3x + 5 = 11$

2. $5x + 7 = 32$

3. $-7x - 6 = 8$

4. $-3x + 5 = -10$

5. $-15 = 4a - 3$

6. $-8 = 6x + 4$

7. $5x - 8 = 4x - 5$

8. $9y + 4 = 8y + 1$

9. $10 - 3u = 12 - 4u$

10. $3r + 5 = 2r + 4$

11. $7 - 9v = v - 3$

12. $4w + 5 = 15 - 6w$

13. $3n - 6 = 2n - 6$

14. $2m + 13 = 13 - 8m$

15. $9p + 3.4 = 8p + 1.4$

16. $.4q + .7 = .8 - .6q$

17. $3v + 7 - 8v = 18 - 1$

18. $6w - 4 - 8w = -14 + 46$

19. $4k + 5 = 7k - 7 + 9k$

20. $-11b - 5 + 5b = 23 - 10$

21. $3a - 4a + 9 = -12a + 43 - 6a$

22. $17z - 11z - 17 = -4z + 13 - 5z$

23. $2.3y - 1.6 - .8y = 4.4$

24. $4.7 - 5.2c - .3c = -6.3$

25. $n + 5 = 3(n + 7)$

26. $2t - 9 = 3(t - 2)$

27. $2(5r - 2) = 9r + 1$

28. $2(w - 4) = 5w - 14$

29. $6u - 17 = 4(u + 3) - 3$

30. $2v + 9 = 11 + 7(v - 1)$

31. $6(w - 2) + 8 = 4w - 16$

32. $2x + 2(3x - 4) = -23 + 5x$

33. $2y + 13(y - 1) = 13y - 8$

34. $8(z - 3) - z = 5z - 17$

Answers: **Practice Exercise 5: a.** 1.5 hours. **b.** $12.50 **c.** 7 inches   **Additional Practice 5: a.** 2.5 hours. **b.** $50 **c.** 15 feet

35. $.2(8x - 90) + .9x = .1x + 6$

36. $1.4u - .5(44 - 6u) = 66 + 2.4u$

37. $.3n + .2(4n - 1) = .4 + .5n$

38. $.3z + .7(2z + 1) = 2.9 - .5z$

39. $7(p - 3) = 3(1 - p) - 4$

40. $5(q - 2) + 20 = 5(4 - q)$

41. $15(n + 2) = 5(n + 4) - 10$

42. $8(m - 5) = 7(m - 5) - 9$

43. $5(u - 2) + 7(3 - u) = 9u$

44. $5(h + 2) - 7(h - 1) = -h$

45. $4(b - 3) + 2(b + 2) = 2(b - 2)$

46. $3(v - 8) + 2(v + 5) = 7(v - 4)$

47. $5(a - 1) - 9(a - 2) = -3(2a + 1) - 2$

48. $4(k + 4) + 13(k - 1) = -3(k - 3) + 14$

49. $9 - .2(v - 8) = .3(6v - 8)$

50. $.5(t - 6) + .4(1 - t) = -3.4$

Determine whether the following are identities or contradictions and indicate the solutions. (See Example 2.)

51. $2x + 4 = 2(x + 1) + 2$

52. $3(x + 1) - 2 = 3x - 5$

53. $3(2x - 5) - 4x = 2(x - 3) - 9$

54. $2(4x - 7) + 10 = 2(x - 5) + 2(3x + 3)$

55. $4(2x - 1) - 3(x + 5) = 5(x - 2) + 7$

56. $3(x - 7) - 4(x - 3) = 5 - (x + 3)$

Challenge Exercises (57 and 58)

Solve the following:

57. $.4(x + 2) = .3(x - 5) - .5(x - 3) + 2$

58. $1.8(x - 2) + 5(x - 5) + .8 = -.4(8x + 7)$

Translate each of the following mathematical sentences into English sentences in at least two ways. (See Example 3.)

59. $4w - 3 = 7$

60. $5x + 8 = 19$

61. $6v - 9 = 3v$

62. $16n - 8 = 6n$

63. $7y + 2.8 = 6.3$

64. $q + .05q = 45$

65. $12(y - 1) = 9$

66. $4(t + 3) = 36$

Translate the following English sentences into mathematical sentences and solve. Use x as the variable. (See Example 4.)

67. Eight less than three times a number equals thirteen.

68. One is five more than twice a number.

69. The sum of three times a number and seven is four.

70. The difference of five times a number and six is nine.

71. Zero is equal to nine times the sum of a number and three.

72. The product of seven and four more than a number is sixty-three.

73. Four times the difference of three times a number and two is sixteen.

74. The product of four and the difference of a number and five is negative twelve.

Solve the following. (See Example 5.)

75. A plumber charges $30 for a service call and $45 per hour base rate. Her charge for a repair job was $210, excluding parts. How many hours did the plumber work?

76. An appliance repairman charges $25 dollars for a service call plus his hourly wage. If a repair job cost $109 (excluding parts) and took three hours, what is his hourly wage?

77. An auto repair service charges $35 per hour plus parts. If the parts to repair a transmission cost $160 and the total bill was $335, how many hours did the job take?

78. A TV repair shop charges an hourly rate plus the cost of parts. Find the hourly rate for a job that took $\frac{3}{2}$ hours if the parts cost $65 and the bill was $113.

79. If Yo Chen had $15 more than she presently has, then with three times that amount she could buy a camera that costs $525. How much does she presently have?

80. Five times $20 less than Bryan presently has is enough to buy a new bike that costs $300. How much does Bryan presently have?

81. Frank earned $635 last week for working 40 hours. Included in the $635 was a bonus of $25. Find his hourly wage.

82. Susie earned $445 for working 28 hours. Included in the $445 was a $60 bonus. Find her hourly wage.

83. A particular copy machine requires 15 cents for the first copy and 4 cents for each additional copy. How many copies can be made for $4.95?

84. Joan uses a coupon in the paper for $25 off the cost of repairs at Joe's Garage. If her bill was $147.50, including $60 for parts and 3 hours of labor, how much did Joe charge per hour for labor?

85. Hernando earns five-fourths times his hourly wage on all hours worked in excess of 40 hours per week. Last week he worked 46 hours and earned $581.40. What is his hourly wage?

86. Sara is paid $\frac{3}{2}$ times her hourly wage for all hours worked in excess of 160 hours per month. Last month she worked 172 hours and earned $3328.60. What is her hourly wage?

87. Various long-distance phone companies have different plans. Company A has a $5.95 charge per month plus 10¢ per minute, with 30 free minutes per month. Company B charges $4.95 per month and 8¢ per minute, with no free minutes. Find the number of minutes per month so that the charges would be the same for company A as for company B.

88. A country club offers two types of memberships. Plan A charges $500 per year plus $15 per round of golf, and plan B charges $675 per year plus $10 per round of golf. How many rounds of golf would have to be played per year for the charges under plans A and B to be equal?

89. Saline solution has 1 part salt to 8 parts water. Find the number of ounces of salt and the number of ounces of water in 108 ounces of the solution.

90. Commercial bronze is 1 part tin and 9 parts copper. Find the number of pounds of tin and the number of pounds of copper in 200 pounds of commercial bronze.

91. A rectangular parking lot has a perimeter of 2000 yards and a width of 450 yards. What is the length? ($P = 2L + 2W$)

92. A rectangular patio has a perimeter of 74 feet and a length of 22 feet. What is the width? ($P = 2L + 2W$)

93. A shipping container is in the shape of a rectangular solid and has a surface area of 376 square feet. If the container is 10 feet long and 6 feet high, find the width. ($SA = 2LW + 2LH + 2WH$)

94. A storage tank is in the shape of a rectangular solid. The surface area is 236 square feet, the length is 8 feet, and the width is 6 feet. What is the height? ($SA = 2LW + 2LH + 2WH$)

Section 3.4 Using and Solving Formulas

OBJECTIVES *When you complete this section, you will be able to:*

A Solve for the unknown when given all of the values in a formula except one.

B Solve a formula for a given variable.

PREREQUISITE SKILLS *Before beginning this section, you should be able to:*

a. Solve linear equations, using the addition and multiplication properties of equality (Section 3.3).

GETTING READY FOR SECTION 3.4

Solve the following equations, using the addition and multiplication properties of equality:

1. $20 = 12 + 2w$ **2.** $60 = \dfrac{15h}{2}$ **3.** $28.26 = 6.28r$

Introduction

Formulas are equations which express relationships that apply to a whole class of problems. They are like ready-made equations without specific numbers. We used several geometric formulas in Chapter 1 and in previous sections of this chapter. In Chapter 1, we simply substituted the given values and followed the order of operations, since the formula was already solved for the quantity that we were looking for. Often, that is not the case, and we have to use the techniques for solving linear equations to find the desired value.

OBJECTIVE **Solving for the unknown when given all of the values in a formula except one**

In using a formula, we must know the values of all of the variables except the one for which we are solving. We substitute the known values for the variables and then solve for the unknown variable by using the techniques from the previous sections of this chapter.

Example 1 **Solve each of the following for the indicated variable:**

a. The formula for distance is $d = rt$. If $r = 30$ miles per hour and $d = 150$ miles, find t.

Solution

$d = rt$	Replace r with 30 and d with 150.
$150 = 30t$	Divide both sides by 30.
$5 = t$	Therefore, the time is 5 hours.

b. The equation of a line is $2x + 5y = 10$. If $x = -5$, find y.

Solution

$2x + 5y = 10$	Replace x with -5.
$2(-5) + 5y = 10$	Multiply $2(-5)$.
$-10 + 5y = 10$	Add 10 to both sides.
$5y = 20$	Divide both sides by 5.
$y = 4$	Therefore $y = 4$ when $x = -5$. Note that there are no units.

Answers: Getting Ready: 1. $w = 4$ **2.** $h = 8$ **3.** $r = 4.5$

c. The formula for finding the circumference of (distance around) a circle is $C = 2\pi r$. If $C = 28.26$ feet, find r. Use $\pi \approx 3.14$.

Solution

$$C = 2\pi r$$ — Replace C with 28.26 and π with 3.14.

$$28.26 \approx 2(3.14)r$$ — Simplify the right side of the equation.

$$28.26 \approx 6.28r$$ — Divide both sides by 6.28.

$$\frac{28.26}{6.28} \approx \frac{6.28r}{6.28}$$ — Simplify both sides.

$$4.5 \approx r$$ — Therefore, the radius is approximately 4.5 feet.

d. The formula for the area of a triangle is $A = \frac{bh}{2}$. If $A = 60$ square inches and $b = 15$ inches, find h.

Solution

$$A = \frac{bh}{2}$$ — Replace A with 60 and b with 15.

$$60 = \frac{15h}{2}$$ — Multiply both sides of the equation by $\frac{2}{15}$.

$$\frac{2}{15}(60) = \frac{2}{15} \cdot \frac{15h}{2}$$ — Simplify both sides. $\frac{2}{15}(60) = \frac{2}{15} \cdot \frac{60}{1} = \frac{120}{15} = 8$.

$$8 = 1h$$

$$\text{or } h = 8$$ — Therefore, $h = 8$ inches.

PRACTICE EXERCISE 1

Solve the following:

a. The formula for the volume of a box is $V = LWH$. If $L = 10$ feet, $H = 6$ feet, and $V = 180$ cubic feet, find W.

b. The equation of a line is $3x + 2y = 6$. If $x = 4$, find y.

c. In electricity, the formula for power is $P = IV$, where P represents power in watts, I represents current in amps, and V represents voltage. If $P = 140$ watts and $I = 10$ amps, find V.

d. The formula for rate is $r = \frac{d}{t}$, where d is the distance and t is the time. Find the distance if the rate is 45 miles per hour and the time is 4 hours.

If you need more practice, do the following Additional Practice Exercises.

Additional Practice Exercise 1   Solve the following:

a. The formula for interest is $I = prt$. If $I = \$67.50$, $p = \$1500$, and $t = .5$, find r.

b. The equation of a line is $2x - 3y = 12$. If $x = 3$, find y.

c. The formula for the area of a circle is $A = \pi r^2$. If $r = 2$ meters, find A. Use $\pi \approx 3.14$.

d. The formula for the area of a trapezoid is $A = \frac{h(B + b)}{2}$. If $A = 18$ square feet, $h = 4$ feet, and $B = 6$ feet, find b.

OBJECTIVE Solving a formula for a given variable

A formula can be solved for any variable in terms of the other variables of that formula. If we have several problems in which we are asked to find the same variable, we may want to solve the formula for that variable before substituting values for the other variables. This way, we do not have to solve several similar equations for the same variable several times.

Answers: Practice Exercise 1: a. $W = 3$ feet **b.** $y = -3$ **c.** $V = 14$ volts **d.** $d = 180$ miles
Additional Practice 1: a. $r = .09$ or 9% **b.** $y = -2$ **c.** $A = 12.56$ square meters **d.** $b = 3$ feet

Example 2   **Solve the following formulas for the indicated variable:**

a. $d = rt$, for t

b. $A = \dfrac{1}{2}bh$, for h

Solution

We treat all the other variables as if they were numbers and solve.

$d = rt$ Divide both sides of the equation by r.

$\dfrac{d}{r} = \dfrac{rt}{r}$ Simplify the right side.

$\dfrac{d}{r} = t$ Therefore, $t = \frac{d}{r}$.

$A = \dfrac{1}{2}bh$ Multiply both sides of the equation by 2.

$2A = 2\left(\frac{1}{2}\right)bh$ Simplify both sides.

$2A = bh$ Divide both sides by b.

$\dfrac{2A}{b} = \dfrac{bh}{b}$ Simplify both sides.

$\dfrac{2A}{b} = h$ Therefore, $h = \frac{2A}{b}$.

c. Solve $P = 2L + 2W$ for W.

$P = 2L + 2W$ Subtract $2L$ from both sides of the equation.

$P - 2L = 2L - 2L + 2W$ Simplify the right side.

$P - 2L = 2W$ Divide both sides by 2.

$\dfrac{P - 2L}{2} = \dfrac{2W}{2}$ Simplify the right side.

$\dfrac{P - 2L}{2} = W$ Therefore, $W = \frac{P - 2L}{2}$.

d. Solve $3x + 4y = 5$ for y.

$3x + 4y = 5$ Subtract $3x$ from both sides of the equation.

$3x - 3x + 4y = 5 - 3x$ Simplify the left side.

$4y = 5 - 3x$ Divide both sides by 4.

$\dfrac{4y}{4} = \dfrac{5 - 3x}{4}$ Simplify the left side.

$y = \dfrac{5 - 3x}{4}$ Therefore, $y = \frac{5 - 3x}{4}$.

PRACTICE EXERCISE 2

Solve each of the following formulas for the given variable:

a. $W = Fd$ for d.   **b.** $E = \dfrac{1}{2}mv^2$ for m.   **c.** $P = 2L + 2W$ for L.   **d.** $2x + 3y = 9$ for y.

If you need more practice, do the following Additional Practice Exercises.

Additional Practice Exercise 2   Solve each of the following for the variable indicated:

a. $A = LW$, for L.

b. $P = \dfrac{W}{t}$, for t. (Power)

c. $V = \pi r^2 h$, for h. (Volume of a right circular cylinder)

d. $4x + 2y = 7$, for y.

Exercise Set 3.4

For Extra Help

MyMathLab®

Solve each formula for the indicated variable with the given information. (See Example 1.)

1. $A = \dfrac{bh}{2}$ for A, with $b = 2$ miles and $h = 8$ miles

2. $A = \dfrac{bh}{2}$ for A, with $b = 6$ inches and $h = 10$ inches

---

Answers: Practice Exercise 2: a. $d = \dfrac{W}{F}$ **b.** $m = \dfrac{2E}{v^2}$ **c.** $L = \dfrac{P - 2W}{2}$ **d.** $y = \dfrac{9 - 2x}{3}$

Additional Practice 2: a. $L = \dfrac{A}{W}$ **b.** $t = \dfrac{W}{P}$ **c.** $h = \dfrac{V}{\pi r^2}$ **d.** $y = \dfrac{7 - 4x}{2}$

3. $A = \dfrac{bh}{2}$ for h, with $A = 14$ square feet and $b = 7$ feet

4. $A = \dfrac{bh}{2}$ for b, with $A = 36$ square meters and $h = 8$ meters.

5. $d = rt$ for t, with $r = 50$ kilometers per hour and $d = 75$ kilometers

6. $d = rt$ for r, with $t = 6$ hours and $d = 720$ kilometers

7. $C = 2\pi r$ for r, with $C = 25.12$ inches and $\pi = 3.14$

8. $C = 2\pi r$ for C, with $r = 5\pi$ meters and $\pi = 3.14$

9. $P = 2W + 2L$ for P, with $W = 6$ inches and $L = 15$ inches

10. $P = 2W + 2L$ for W, with $L = 7$ centimeters and $P = 24$ centimeters

In Exercises 11–20, we omit units, since many of them are unfamiliar to us. Solve each formula for the indicated variable with the given value(s). (See Example 1.)

11. $A = \dfrac{1}{2}(B + b)h$ for A, with $b = 5$, $B = 7$, and $h = 4$

12. $A = \dfrac{1}{2}(B + b)h$ for h, with $B = 6$, $b = 8$, and $A = 35$

13. $V = IR$ for V, with $I = 3$ and $R = 5$ (Ohm's law in electricity)

14. $V = IR$ for R, with $V = 120$ and $I = 3$

15. $P = IV$ for P, with $I = .5$ and $V = 120$ (Electricity: power formula)

16. $P = IV$ for I, with $P = 150$ and $V = 120$

17. $W = mg$ for W, with $m = 82$ and $g = 32$ (Physics: formula for weight)

18. $W = mg$ for m, with $W = 170$ and $g = 32$

19. $P = \dfrac{F}{A}$ for P, with $F = 180$ and $A = 15$ (Physics: pressure on an area)

20. $P = \dfrac{F}{A}$ for F, with $P = 50$ and $A = 8$

21. $3x - 4y = 12$ for y, with $x = -4$

22. $4x - 5y = 20$ for y, with $x = -5$

23. $2x + 3y = 18$ for x, with $y = 2$

24. $3x + 2y = 12$ for x, with $y = 3$.

Solve each formula for the indicated variable. (See Example 2.)

25. $K = C + 273$ for C (Conversion formula for Kelvin and Celsius temperatures)

26. $K - 273 = C$ for K

27. $P = kT$ for T (Chemistry: gas law)

28. $P = kT$ for k

29. $F = ma$ for m (Physics: law of motion)

30. $F = ma$ for a

31. $v = gt$ for t (Physics: velocity of falling object)

32. $v = gt$ for g

33. $PV = kT$ for V (Chemistry: universal gas law)

34. $PV = kT$ for T

35. $E = mc^2$ for m (Nuclear physics: conversion from mass to energy)

36. $E = mc^2$ for c^2

37. $I = prt$ for r (Finance: formula for calculating interest)

38. $V = LWH$ for W

39. $a^2 + b^2 = c^2$ for a^2 (Geometry: Pythagorean theorem for right triangles)

40. $a^2 + b^2 = c^2$ for b^2

41. $d = \frac{1}{2}gt^2$ for g (Physics: distance traveled in free fall)

42. $d = \frac{1}{2}gt^2$ for t^2

43. $D = \frac{M}{V}$ for M

44. $D = \frac{M}{V}$ for V

45. $A = P + PrT$ for T

46. $A = P + Prt$ for r

47. $s = \frac{n}{2}(a + l)$ for a

48. $s = \frac{n}{2}(a + l)$ for l

49. $v = -32T + v_0$ for T

50. $x = x_0 + vT$ for T

Solve each of the following for y:

51. $2x + y = 3$

52. $3x + y = 5$

53. $3x + 2y = 6$

54. $4x + 3y = 8$

55. $6x - 3y = 7$

56. $4x - 2y = 7$

Challenge Exercises (57–60)

Solve the given formula for the indicated variable.

57. $A = \frac{1}{2}(B + b)h$ for B

58. $A = \frac{1}{2}(B + b)h$ for b

59. $F = k\frac{q_1 q_2}{r^2}$ for q_1

60. $F = \frac{9}{5}C + 32$ for C

General, Consecutive Integer, and Distance Application Problems

OBJECTIVES *When you complete this section, you will be able to:*

A Solve real-world problems that are general in nature.

B Solve problems involving consecutive, consecutive odd, and consecutive even integers.

C Solve real-world problems involving distance, time, and rate.

PREREQUISITE SKILLS *Before beginning this section, you should be able to:*

a. Solve linear equations, using both the addition and multiplication properties of equality (Section 3.3).

GETTING READY FOR SECTION 3.5

Solve by using both the addition and multiplication properties of equality.

1. $2x + 12{,}727 = 103{,}671$

2. $175 + 0.6x = 505$

3. $n + (n + 2) + (n + 4) = 9$

4. $45t = 60(t - 2)$

Introduction

This is the first of two sections of application problems of the type traditionally found in algebra books. You will probably encounter few such problems in your everyday life. Their main purpose is to help you think mathematically and to see the power of algebra in problem solving.

In previous sections of this chapter, we solved real-world problems that involved only one unknown. The procedure was given in Section 3.1. In this section, we solve several types of application problems, many of which involve two or more unknowns. Here we give a modified problem-solving procedure for when there are more than one unknown.

Procedure: Problem Solving with More than One Unknown

1. **Understand** the problem.
   a. Read and reread the problem carefully, paying special attention to key words. It is a good idea to underline key words in the problem.
   b. Identify what is given and what is to be found.
   c. Determine the number of unknowns and how the known and unknown quantities are related to each other.
   d. When possible, make a drawing and label it with known and unknown information.

2. **Plan.** Determine the procedure you will use to solve the problem.

3. **Translate** the problem into mathematical language, which is usually an equation.
   a. Construct any necessary tables, charts, etc.
   b. Determine which unknown will be represented by the variable (usually the unknown that is acted on or the smaller quantity.)
   c. Using the appropriate relationship(s) from part 1c, represent the other unknown(s) in terms of the variable used in step 3b.
   d. Write the equation in words, using the appropriate relationship found in part 1c.
   e. Using the "word equation" in step d, write the algebraic equation.

(Continued)

Answers: Getting Ready: 1. $x = 45{,}472$ **2.** $x = 550$ **3.** $n = 1$ **4.** $t = 8$

4. **Solve** the algebraic equation.

5. **State** the solution(s) in words, using a complete English sentence(s).

6. **Check** the solution(s), using the wording of the original equation.

If there are two unknowns in a real-world problem, there must be two conditions in the problem: one that gives the relationship between the unknowns and one that gives the information necessary to write the equation. This information will most often determine which unknown to represent with a variable.

Often, there are special approaches that are used for particular types of real-world problems. In solving application problems, it is often necessary to change percents into decimals—something that you have no doubt done before. You will recall that the traditional procedure is to drop the % sign and move the decimal two places to the left. For example, 23% = .23, 6% = .06, and 182% = 1.82. A thorough discussion of percents is found in Section 8.3.

OBJECTIVE **A** Solving real-world problems that are general in nature

Example 1

a. The two largest Native American tribes are the Cherokee and the Navajo. There are 11,867 more Cherokee than Navajo, and the total number in both tribes is 550,271. Find the number in each tribe.

Solution

Understand: We are given the relationship between the number of Cherokees and Navajos, and the total number of members of both tribes. The two unknowns are the number in the Cherokee tribe and the number in the Navajo tribe.

Relationship 1: There are 11,867 more Cherokees than Navajos.

Relationship 2: The total number in both tribes is 550,271.

Plan: We will use relationship 1 to represent the number of Indians in each tribe in terms of a variable and relationship 2 to write the equation.

Translate: Using relationship 1, let x represent the **number in the Navajo tribe**. (It is usually a good idea to let the variable represent the smaller quantity.) Then $x + 11{,}867$ represents the **number in the Cherokee tribe**. (There are 11,867 more Cherokees than Navajos.)

From relationship 2, we know that

number in Cherokee tribe + number in Navajo tribe = total in both tribes .

Write an algebraic equation, using the word equation as a guide.

Number in Cherokee tribe + number in Navajo tribe = total in both tribes .

$$(x + 11{,}867) \qquad + \qquad x \qquad = \qquad 550{,}271$$

Solve:

$(x + 11{,}867) + x = 550{,}271$	Remove the parentheses.
$x + 11{,}867 + x = 550{,}271$	$x + x = 2x.$
$2x + 11{,}867 = 550{,}271$	Subtract 11,867 from both sides.
$2x + 11{,}867 - 11{,}867 = 550{,}271 - 11{,}867$	Simplify both sides.
$2x = 538{,}404$	Divide both sides by 2.
$\dfrac{2x}{2} = \dfrac{538{,}404}{2}$	Simplify both sides.
$x = 269{,}202$	Solution of the equation.

Since x represents the number in the Navajo tribe, there are 269,202 Navajos. The number of Cherokees is $x + 11,867 = 269,202 + 11,867 = 281,069$.

State: There are 269,202 Navajos and 281,069 Cherokees.

Check: Are there 11,867 more Cherokees than Navajos? Yes, since $281,069 - 269,202 = 11,867$. Is there a total of 550,271 in both tribes? Yes, since $281,069 + 269,202 = 550,271$. Therefore, our answers are correct.

b. The number of people infected with the HIV virus that causes AIDS is increasing at an alarming rate. There were an estimated 35.5 million people infected in 2010, which is 9.7 million less than twice the number infected in 1996. How many people were infected in 1996? (*Source:* www.unfpa.org/aids_clock)

Solution

Understand: We know the relationship between the number infected in 1996 and 2010, and we know the number infected in 2010. The unknown is the number infected in 1996.

Relationship: The number infected in 2010 is 9.7 million less than twice the number infected in 1996.

Plan: We will use the relationship to write an equation.

Translate: The unknown is the number infected in 1996, so let x represent the **number, in millions, of people infected with HIV in 1996**.

From relationship 1, we know that

> Twice the number in 1996 $-$ 9.7 $=$ the number in 2010 .

Write an algebraic equation, using the word equation as a guide.

> Twice the number in 1996 $-$ 9.7 $=$ the number in 2010 .

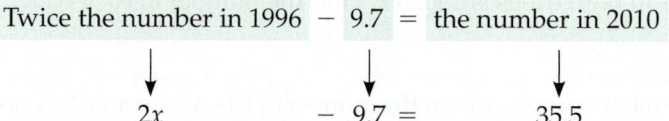

$$2x \qquad -\ 9.7 = \qquad 35.5$$

Solve:

$2x - 9.7 = 35.5$	Add 9.7 to both sides.
$2x - 9.7 + 9.7 = 35.5 + 9.7$	Simplify both sides.
$2x = 45.2$	Divide both sides by 2.
$\dfrac{2x}{2} = \dfrac{45.2}{2}$	Simplify both sides.
$x = 22.6$	Solution of the equation.

State: There were 22.6 million people infected in 1996.

Check: Since $2x - 9.7$ represents the number, in millions, infected in 2010, there were $2(22.6) - 9.7 = 45.2 - 9.7 = 35.5$ million people infected in 2010. Is 35.5 9.7 less than twice 22.6? Yes, since $2(22.6) - 9.7 = 45.2 - 9.7 = 35.5$. Therefore, our answer is correct.

c. Pablo works as a salesperson in a department store and is paid $175 per week plus 6% of his sales. Last week he earned $505. What was the amount of his sales?

Understand: We know the method of calculating how much Pablo is paid, and we know how much he was paid. The unknown is the amount of his sales.

Relationship 1: The amount Pablo is paid is $175 plus 6% of his sales.

Plan: Use the relationship to write the equation.

Translate: The unknown is the amount of his sales, so let x represent the **amount of Pablo's sales**.

Use relationship 1 to write the word equation.

> 175 $+$ 6% of his sales $=$ amount earned .

Write an algebraic equation, using the word equation as a guide. In order to represent 6% of his sales, we need to change 6% to .06. When using percents, "of" means to multiply. Consequently, since x represents the amount of his sales, 6% of his sales is .06x.

$$175 + 6\% \text{ of his sales} = \text{amount earned} .$$

$$175 + .06x = 505$$

Solve:

$175 + .06x = 505$	Subtract 175 from each side of the equation.
$175 - 175 + .06x = 505 - 175$	Simplify each side of the equation.
$.06x = 330$	Divide both sides by .06.
$\dfrac{.06x}{.06} = \dfrac{330}{.06}$	Simplify both sides of the equation.
$x = 5500$	Solution of the equation.

State: Pablo's sales were $5500.

Check: If the amount of his sales is $5500, he earns .06(5500) = $330. Then $330 + $175 = $505, which is what he earned for the week. Therefore, our solution is correct.

PRACTICE EXERCISE 1

a. One tablespoon of regular Thousand Island salad dressing has 60 calories, which is 15 less than three times the number of calories in one tablespoon of low-fat Thousand Island dressing. How many calories are in one tablespoon of low-fat Thousand Island dressing?

b. In one day, Sarah walked briskly for an hour and later did intensive aerobic exercises for another hour, burning a total of 840 calories. The number of calories she burned doing aerobic exercises was 60 fewer than two times the number she burned walking. How many calories did she burn at each activity?

c. Felicia is paid $215 per month plus 8% of her sales. Last month she earned $1655. What was the amount of her sales?

If you need more practice, do the following Additional Practice Exercises.

Additional Practice Exercise 1

a. The United States navy has 73 submarines, which is five less than three times the number of submarines in the North Korea navy. How many submarines does the North Korea navy have? (*Source:* mapsofworld.com.)

b. In 2005, the highest state gas tax per gallon was in Wisconsin, and the lowest was in Georgia. If the combined sales tax per gallon is 40.4 cents and the sales tax per gallon in Wisconsin is 4.6 cents per gallon less than five times the sales tax per gallon in Georgia, find the sales tax per gallon in each of the two states.

c. Ahmad is paid $230 per week plus 6% commission on his sales. If he earned $380 last week, how much were his sales?

Answers: Practice Exercise 1: a. 25 **b.** walking = 300, aerobics = 540 **c.** $18,000
Additional Practice 1: a. 26 submarines **b.** Georgia, 7.5 cents; Wisconsin, 32.9 cents **c.** $2500

OBJECTIVE **B**   Solving problems involving consecutive, consecutive odd, and consecutive even integers

Consecutive integers are integers that follow one another in order. For example, 1, 2, 3, 4, and so on, are consecutive integers. To get the next larger consecutive integer, we add 1 to the previous integer. For example, if we begin with 1, then $1 + 1 = 2$, $2 + 1 = 3$, $3 + 1 = 4$, and so on. Therefore, if we let $n =$ some integer, then the second integer is $n + 1$, the third is $(n + 1) + 1 = n + 2$, the fourth is $(n + 2) + 1 = n + 2 + 1 = n + 3$, and so on.

Consecutive odd integers are every other integer, starting with an odd integer. For example, 1, 3, 5, 7, and so on, are consecutive odd integers. To get the next largest consecutive odd integer, we add 2 to the previous odd integer. Therefore, if we let $n =$ the first odd integer, then the second odd integer is $n + 2$, the third odd integer is $(n + 2) + 2 = n + 4$, and the fourth odd integer is $(n + 4) + 2 = n + 6$, and so on.

Consecutive even integers are every other integer, starting with an even integer. For example, 2, 4, 6, 8, and so on, are consecutive even integers. Again, to get the next larger consecutive even integer, we add 2 to the previous even integer. Hence, if we let $n =$ the first even integer, then $n + 2$ is the second even integer, $n + 4$ is the third even integer, $n + 6$ is the fourth even integer, and so on.

Let us put all three types of integers together for comparison.

> **Note**
>
> Do not be confused because both consecutive odd and consecutive even integers are represented in the same way in the chart to the right. The difference is in whether the first integer is odd or even. If the first integer is odd, then that integer plus 2 is also odd, and likewise for even integers.

Integer	Consecutive	Consecutive Odd	Consecutive Even
First	n	n	n
Second	$n + 1$	$n + 2$	$n + 2$
Third	$n + 2$	$n + 4$	$n + 4$
Fourth	$n + 3$	$n + 6$	$n + 6$
and so on	and so on	and so on	and so on

Example 2

a. The sum of three consecutive odd integers is -9. Find the integers.

Solution

Understand: We are given the sum of three consecutive odd integers. The unknowns are the consecutive odd integers.

> Relationship 1: The integers are consecutive odd integers.
> Relationship 2: The sum of the three consecutive odd integers is -9.

Plan: Use relationship 1 to represent the consecutive odd integers with a variable and use relationship 2 to write the equation.

Translate: We represent the smallest consecutive odd integer with a variable and then represent the other consecutive odd integers in terms of this variable.

Let n represent the smallest of the three consecutive odd integers. Then

> $n + 2$ represents the second, and
> $n + 4$ represents the third.

Write a word equation. From relationship 2, we know that the sum of the three consecutive odd integers is -9. Thus,

$$\boxed{\text{1st integer}} + \boxed{\text{2nd integer}} + \boxed{\text{3rd integer}} = \boxed{-9}.$$

Write an algebraic equation, using the word equation as a guide.

$$\text{1st integer} \ + \ \text{2nd integer} \ + \ \text{3rd integer} \ = \ -9.$$

$$n \quad + \quad (n + 2) \quad + \quad (n + 4) \quad = -9.$$

Solve:

$n + (n + 2) + (n + 4) = -9$	Remove the parentheses.
$n + n + 2 + n + 4 = -9$	Add like terms.
$3n + 6 = -9$	Subtract 6 from both sides.
$3n = -15$	Divide both sides by 3.
$n = -5$	Solution of the equation.

Since n represents the smallest of the consecutive odd integers, the smallest is -5. The second is $n + 2 = -5 + 2 = -3$, and the third is $n + 4 = -5 + 4 = -1$.

State: The consecutive odd integers are -5, -3, and -1.

Check: Are -1, -3, and -5 consecutive odd integers? Yes. Is the sum of -1, -3, and -5 equal to -9? Yes. Therefore, we have the correct solution.

b. The Smiths and Jones live in apartments that are next door to each other and the apartment numbers are consecutive odd integers. If three times the larger apartment number is 61 more than twice the smaller, find the apartment numbers.

Solution

Understand: We are given that the apartment numbers are two consecutive odd integers and the relationship between the integers. The unknowns are the apartment numbers.

Relationship 1: The apartment numbers are consecutive odd numbers.

Relationship 2: Three times the larger number is 61 more than twice the smaller.

Plan: Use relationship 1 to represent the two consecutive odd integers with a variable and relationship 2 to write the equation.

Translate: Represent the smaller apartment number with a variable and then represent the larger in terms of this variable.

Since the apartment numbers are consecutive odd integers, let n represent the **smaller of the apartment numbers**. Then $n + 2$ represents **the larger**.

Write a word equation: From relationship 2 we know that 3 times the larger apartment number is 61 more than twice the smaller.

$$\text{Three times the larger} \ = \ \text{61 more than twice the smaller}$$

Write an algebraic equation using the word equation as a guide.

$$\text{Three times the larger} \ = \ \text{61 more than twice the smaller}$$

$$3(n + 2) \qquad = \qquad 2n + 61$$

Solve:

$3(n + 2) = 2n + 61$	Distribute the 3 on the left side.
$3n + 6 = 2n + 61$	Subtract $2n$ from both sides.
$3n - 2n + 6 = 2n - 2n + 61$	Simplify both sides.
$n + 6 = 61$	Subtract 6 from both sides.
$n + 6 - 6 = 61 - 6$	Simplify both sides.
$n = 55$	Solution of the equation.

Since n represents the smaller of the consecutive odd integers, the smaller is 55 and the larger is $n + 2 = 55 + 2 = 57$.

State: The apartment numbers are 55 and 57.

Check: Are 55 and 57 consecutive odd integers? Yes. Is 3 times the larger 61 more than twice the smaller? Three times the larger is $3(57) = 171$, twice the smaller is $2(55) = 110$, and 171 is 61 more than 110. So, we have the correct solutions.

PRACTICE EXERCISE 2

a. Wayne, Tonya, and Susan attend a concert. Their seat numbers are three consecutive integers whose sum is 696. Find the seat numbers.

b. Find two consecutive odd integers such that three times the larger is five less than four times the smaller.

If you need more practice, do the following Additional Practice Exercises.

Additional Practice Exercise 2

a. Jameel, Rashaad, and Vince competed in a half marathon. The order of their finishes was three consecutive even integers whose sum is 60. Find the order of their finishes.

b. Find three consecutive integers such that the sum of the two smaller integers is eight more than the largest.

OBJECTIVE **C** Solving real-world problems involving distance, time, and rate

If an object is traveling at a constant rate (speed), the distance traveled depends upon how long it has been traveling (time). This relationship is given by the equation $d = rt$, where d represents the distance traveled, r is the rate (speed), and t is the time. For example, if a truck travels at a constant rate of 50 miles per hour for 5 hours, the distance it has traveled is $d = (50)(5) = 250$ miles.

In the translation step of solving distance, rate, time problems, we recommend using a chart. Fill in the chart as follows:

1. Draw a chart and label the columns r, t, and d.
2. Fill in one column with known numerical values.
3. Fill in another column with expressions that we have represented by variables.
4. Fill in the remaining column from the first two, using the appropriate relationships between d, r, and t. In this section, $rt = d$.

Example 3

a. The distance between Atlanta and Charleston is 500 miles. A truck leaves Atlanta traveling toward Charleston at an average rate of 47 miles per hour. At the same time, a bus leaves Charleston traveling toward Atlanta at an average rate of 53 miles per hour. Assuming they are traveling on the same route, how long will it be until they meet?

Solution

Understand: We know that we have a truck and a bus traveling toward each other. We also know the initial distance between the truck and the bus and the speed of each.

Relationships: The truck and bus are 500 miles apart and traveling toward each other. The truck is traveling at an average rate of 47 miles per hour and the bus at an average rate of 53 miles per hour.

Plan: Use a chart to organize the information. Write the equation from the last column filled.

Translate:

1. Draw a chart and label the columns.

Moving Things	r	t	d
Truck			
Bus			

2. Fill in one column with known numerical values. We know the rate of the truck is 47 miles per hour and the rate of the bus is 53 miles per hour.

Moving Things	r	t	d
Truck	47		
Bus	53		

3. Assign a variable to an unknown and represent any other unknowns in terms of this variable. We are looking for the number of hours it will be until they meet. Since they left at the same time, the truck and bus will be traveling for the same number of hours. Let t represent the number of hours for each. Fill in the t column.

Moving Things	r	t	d
Truck	47	t	
Bus	53	t	

4. Fill in the remaining column, using the fact that $rt = d$. Consequently, the distance the truck travels is its rate (47) times its time (t), or $47t$. Likewise, the distance for the bus is $53t$.

Moving Things	r	t	d
Truck	47	t	47t
Bus	53	t	53t

Write the equation, using the information in the last column filled in. The last column filled in was the distance column, which means the equation must involve the distances traveled by the truck and the bus. Using the relationship from the Understand step, we draw a diagram.

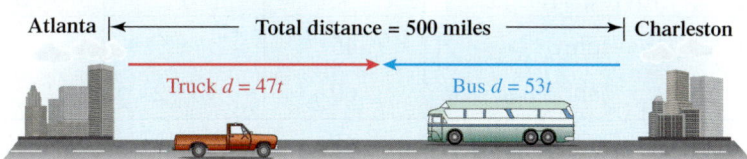

From the diagram, we can see that

Distance the truck traveled $+$ distance the bus traveled $=$ distance between the cities .

$$47t \qquad + \qquad 53t \qquad = \qquad 500$$

Solve:

$47t + 53t = 500$ Add like terms.

$100t = 500$ Divide both sides by 100.

$t = 5$ Solution of the equation.

State: The two vehicles will meet in 5 hours.

Check: The total distance traveled by the truck in 5 hours is $5(47) = 235$ miles, and the total distance the bus has traveled in 5 hours is $5(53) = 265$ miles. The total miles traveled by both in 5 hours is $235 + 265 = 500$ miles, which is the distance between the two cities. Our solution is correct.

b. John leaves home in Lake City, Florida, traveling north on I-75 at an average rate of 45 miles per hour. Two hours later, his wife Jan leaves home and takes the same route, traveling at an average rate of 60 miles per hour. How many hours will it take Jan to catch John?

Solution

Understand: We know that John and Jan are traveling in the same direction and Jan has been traveling 2 hours less than John. We also know the average rate of each. The unknown is the number of hours until Jan catches John.

Relationships: John and Jan are traveling in the same direction, with John traveling at an average rate of 45 miles per hour and Jan at 60 miles per hour. Jan will travel two hours less than John.

Plan: Use a chart and write the equation from the last column filled.

Translate:

1. Draw a chart and label the columns.

Moving Things	r	t	d
John			
Jan			

2. Fill in one column with known numerical values. We know that John's rate is 45 miles per hour and Jan's rate is 60.

Moving Things	r	t	d
John	45		
Jan	60		

3. Assign a variable to an unknown and represent any other unknowns in terms of this variable. We are looking for the number of hours it will take Jan to catch John, but John left two hours before Jan. If we let t represent John's time, then $t - 2$ will represent Jan's time. Fill in the t column.

Moving Things	r	t	d
John	45	t	
Jan	60	$t - 2$	

4. Fill in the remaining column, using the fact that $rt = d$. Consequently, the distance that John travels is his rate (45) times his time (t), or $45t$. Likewise, for Jan, $60(t - 2) = d$.

Moving Things	r	t	d
John	45	t	$45t$
Jan	60	$t - 2$	$60(t - 2)$

Equation: Write the equation, using the information in the last column filled in. The last column filled in was the distance column, which means the equation must involve the distances traveled by John and Jan. Using the relationship from the Understand step, we have the following diagram:

From the diagram we can see that

$$\text{Distance John traveled} = \text{distance Jan traveled}.$$

$$45t = 60(t-2)$$

Solve:

$$45t = 60(t-2) \qquad \text{Distribute on the right side.}$$
$$45t = 60t - 120 \qquad \text{Subtract } 60t \text{ from both sides.}$$
$$-15t = -120 \qquad \text{Divide both sides by } -15$$
$$t = 8 \qquad \text{Solution of the equation.}$$

Since t represents the number of hours that John had been traveling, John traveled for 8 hours. However, the question asked for the number of hours that Jan had been traveling. The number of hours that Jan had been traveling is $t - 2 = 8 - 2 = 6$.

State: It will take Jan 6 hours to catch John.

Check: The total distance traveled by John in 8 hours is $8(45) = 360$ miles, and the total distance Jan has traveled in 6 hours is $6(60) = 360$ miles. The total miles traveled by each is 360 miles. Thus, our solution is correct.

PRACTICE EXERCISE 3

a. A car traveling at an average rate of 44 miles per hour leaves Jacksonville, Florida, toward Jackson, Mississippi. At the same time, a bus leaves Jackson, traveling toward Jacksonville at an average rate of 36 miles per hour. Assuming they travel the same route and the distance between Jacksonville and Jackson is 600 miles, how long will it take until they meet?

b. A recreational vehicle (RV) leaves Tallahassee, Florida, heading west on I-10 for Houston, Texas, at an average speed of 50 miles per hour. Four hours later, a truck also leaves Tallahassee on I-10 for Houston at an average rate of 70 miles per hour. How many hours will it take the truck to catch up to the recreational vehicle?

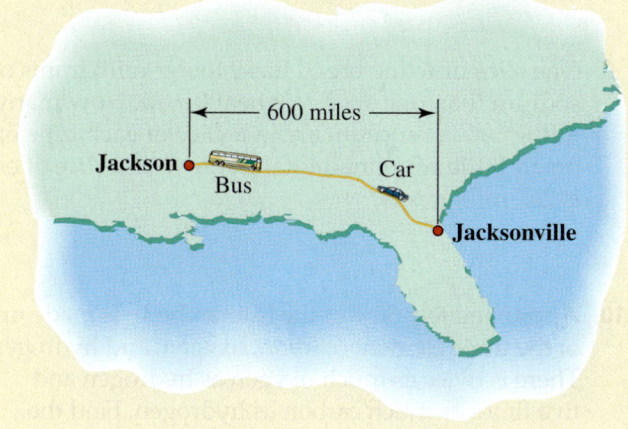

(Continued)

Answers: **Practice Exercise 3:** **a.** 7.5 hours **b.** 10 hours

If you need more practice, do the following Additional Practice Exercises.

Additional Practice Exercise 3

a. Frank leaves Dallas, traveling to Cincinnati at an average speed of 55 miles per hour. At the same time, Delores leaves Cincinnati, traveling to Dallas at an average speed of 60 miles per hour. If the distance between Dallas and Cincinnati is 920 miles, after how many hours will they meet?

b. A truck leaves Atlanta, going north on I-75 at an average rate of 56 miles per hour. Two hours later, a car leaves Atlanta, going north on I-75 at an average rate of 72 miles per hour. How many hours will it take the car to catch up to the truck?

For Extra Help

Exercise Set 3.5   MyMathLab®

Solve the following general application problems. (See Example 1.)

1. China and Iran execute more people than any other countries. In 2008, the number China executed was 12 less than five times the number Iran executed. If China executed 1718 people, find the number Iran executed. (*Data Source: National Geographic, August, 2009*)

2. The braking distance for a car traveling at 70 miles per hour is 5 feet less than twice the braking distance at 50 miles per hour. If the braking distance is 381 feet at 70 miles per hour, what is the braking distance at 50 miles per hour?

3. In a recent year, the number of deaths in the United States due to falls was approximately 13,500, which is 700 more than eight times the number of deaths due to firearms. Find the number of deaths due to firearms. (*Source: National Safety Council, Accident Facts.*)

4. Borneo is among the countries with the greatest diversity of species in the world. The number of species of birds is 27 less than three times the number of species of amphibians. If Borneo has 420 species of birds, how many species of amphibians are there? (*Data Source: National Geographic: November 2008*)

5. In 2009, the two top-selling beers were Bud Light and Coors Light. The sales for Bud Light were $726 million more than the sales for Coors Light. If the combined sales for the two beers were $2142.2 million, find the sales of each beer. (*Data Source: World Almanac and Book of Facts, 2010*)

6. A Burger King Whopper has 258 more calories than a medium order of french fries. A meal consisting of a Whopper and french fries has a total of 1002 calories. How many calories are in each?

7. One cup of canned unsweetened grapefruit juice has 5 fewer grams of carbohydrates than 1 cup of frozen diluted orange juice. If there is a combined total of 49 grams of carbohydrates in 1 cup of each, find the number of grams of carbohydrates in 1 cup of grapefruit juice and 1 cup of orange juice.

8. One slice of white bread has 9 fewer milligrams of sodium than one slice of wheat bread. How many milligrams of sodium are in a slice of each type of bread if the combined total number in a slice of each is 267 milligrams?

9. As of 2009, the U.S., France, and Japan had the most nuclear reactors for generating electricity. The U.S. had 49 more than Japan and France had 4 more than Japan. If the total number of nuclear reactors in the three counties was 218, how many did each country have? (*Data Source: World Almanac and Book of Facts, 2010*)

10. Approximately 80% of the human body is made up of the three elements carbon, oxygen, and hydrogen. There is twice as much oxygen as hydrogen and five times as much carbon as hydrogen. Find the proportion of the human body that is made up of each element.

Answers: Additional Practice 3: a. 8 hours **b.** 7 hours

11. Stress is a leading cause of illness. A rating scale assigns a score to various stressful life events, with the three most stressful being the death of a spouse, divorce, and marital separation. The score for death of a spouse is thirty-five more than the score for marital separation, and the score for divorce is eight more than the score for marital separation. If the sum of all three scores is 238, find the score of each. (*Source: Journal of Psychosomatic Research,* Vol 11.)

12. In a recent year, the three most visited national parks were the Great Smoky Mountains, the Grand Canyon, and Yosemite, with a total of 17 million visitors. There were .4 million more visitors to the Grand Canyon than Yosemite and .6 million more than twice as many visitors to the Great Smoky Mountains than Yosemite. Find the number of visitors to each park to the nearest tenth of a million. (*Source:* National Park Service.)

13. It costs $15 a day extra for after-school care at a nursery school for preschool children. If the total cost for a five-day week of nursery school including after-school care is $250, what is the charge per day for the nursery school without after-school care?

14. It costs $35 per credit hour plus a one-time lab fee of $15 to register for courses at Mountain High Community College. If Zane received a $250 student loan, what is the maximum number of credit hours that he can pay for from the loan?

15. Yolanda's job as a salesclerk pays her $25 per day plus 3% commission on all her sales. If her total wages for Monday were $70, find the amount of her sales.

16. A performer charges $1000 plus 10% of the gate receipts. She earned $16,000 for her last performance. What were the gate receipts?

17. Francoise received a raise of 4% of her present salary. If her new salary is $35,980, what is her present salary?

18. The sales tax in Orlando, Florida, is 6%. If a college cafeteria in Orlando wants the cost of the lunch special to be $5.83 including tax, what should it charge for the lunch special?

Solve the following consecutive integer problems. (See Example 2.)

19. The sum of two consecutive odd integers is 24. Find the two integers.

20. The sum of two consecutive even integers is 98. Find the integers.

21. The sum of three consecutive page numbers in a book is 75. Find the page numbers.

22. Joann, Sheila, and Francis bought tickets to a concert, and their seat numbers are consecutive integers whose sum is 711. Find their seat numbers.

23. The house numbers on the same side of Elm Street are consecutive odd integers. If the Jones, Lopez, and Wu families live in three consecutive houses and the sum of their house numbers is 375, find the house numbers.

24. The Farmers, Sings, and Smiths live in houses with numbers that are consecutive even integers whose sum is 264. Find the house numbers.

25. Johanas, Sigurd, and Hans live in the same apartment building, and their apartment numbers are consecutive integers such that the sum of the smallest and the largest is 24 more than the number in the middle. Find the apartment numbers.

26. Johanna, Murlene, and Margaret attended a play and had front row seats whose numbers were consecutive integers such that twice the smallest was 12 more than the largest. Find the seat numbers.

27. Find two consecutive odd integers such that six times the smaller is three more than five times the larger.

28. Find two consecutive integers such that four times the larger is nine less than five times the smaller.

29. Find three consecutive even integers such that the sum of the first and the second is sixteen more than the third.

30. Find three consecutive odd integers such that three times the third is seventeen more than the sum of the first and the second.

31. Find three consecutive even integers such that twice the sum of the first and the second is four more than three times the third.

32. Find three consecutive even integers such that five times the second equals twice the sum of the first and the third.

Solve the following distance, rate, and time problems. (See Example 3.)

33. The distance between Charlotte, North Carolina, and Buffalo, New York, is 700 miles. Ralph leaves Charlotte, traveling toward Buffalo at an average rate of 56 miles per hour. At the same time, Charles leaves Buffalo, traveling toward Charlotte at an average rate of 44 miles per hour. Assuming they are traveling on the same route, how long will it take until they meet?

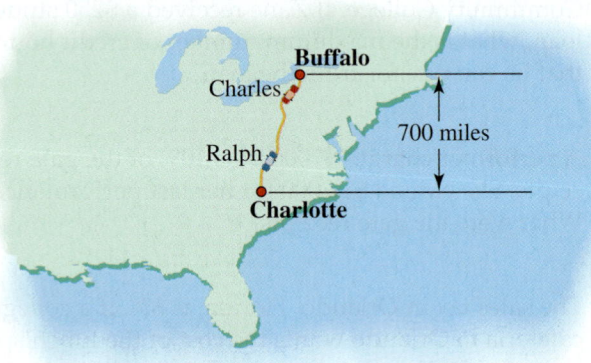

34. A commercial passenger jet leaves Miami, Florida, traveling to Reno, Nevada, at an average speed of 560 miles per hour. At the same time, a private jet leaves Reno, traveling toward Miami at an average rate of 440 miles per hour. How long will it take until they meet if the distance from Miami to Reno is 3000 miles?

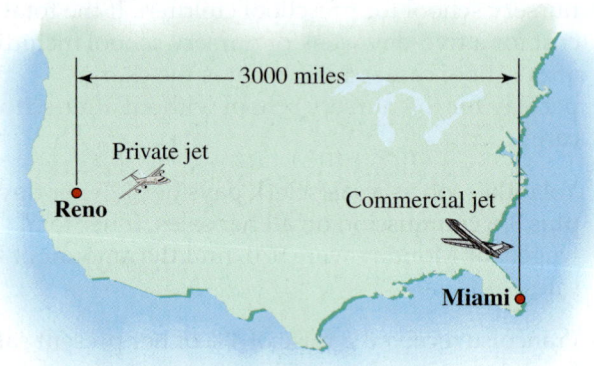

35. Frank leaves Memphis, Tennessee, traveling toward Albuquerque, New Mexico, at an average speed of 60 miles per hour. Two hours later, Marsha leaves Albuquerque, traveling toward Memphis at an average speed of 50 miles per hour. If the distance from Memphis to Albuquerque is 1000 miles, how long after Marsha leaves will it take them to meet?

36. The distance between Norfolk, Virginia, and Buffalo, New York, is 595 miles. Grandma Rosie leaves Norfolk, traveling toward Buffalo at an average speed of 35 miles per hour. Two hours later, Grandpa Julius leaves Buffalo, traveling toward Norfolk at an average speed of 40 miles per hour. How long after Grandpa Julius leaves will it take them to meet?

37. Tameka leaves Memphis, Tennessee, on I-40, traveling west at an average speed of 30 miles per hour. Three hours later, Akiva also leaves Memphis, traveling the same route at an average speed of 40 miles per hour. How long will it take Akiva to catch up to Tameka?

38. Polly leaves Savannah, Georgia, on I-95, traveling north at an average speed of 39 miles per hour. Two hours later, Eric also leaves Savannah, traveling north on I-95, but at an average speed of 52 miles per hour. How long will it take Eric to catch up to Polly?

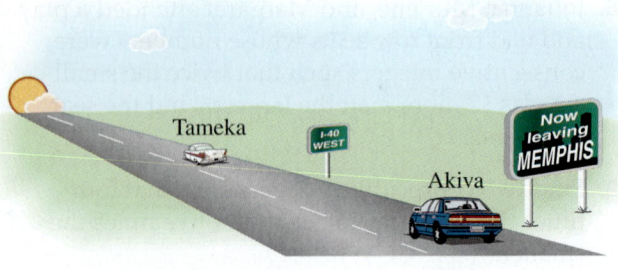

39. Alex leaves Salt Lake City, traveling west on I-80 at an average rate of 52 miles per hour. Two hours later Sondra also leaves Salt Lake City, traveling west on I-80, but at an average rate of 68 miles per hour. How long will it take Sondra to catch up to Alex?

40. Marisa leaves Denver, going south on I-25 at an average rate of 55 miles per hour. One hour later, Calvin also leaves Denver, going south on I-25, but at an average rate of 65 miles per hour. How long will it take Calvin to catch up to Marisa?

41. Pia leaves Kansas City, Missouri, traveling west on I-70 at an average speed of 42 miles per hour. At the same time, Alma leaves Kansas City, traveling east on I-70 at an average speed of 36 miles per hour. How long will it take until they are 468 miles apart?

42. A truck leaves St. Louis, Missouri, traveling north on I-55 at an average rate of 62 miles per hour. At the same time, a bus leaves St. Louis, traveling south on I-55 at an average rate of 55 miles per hour. How long will it take until they are 936 miles apart?

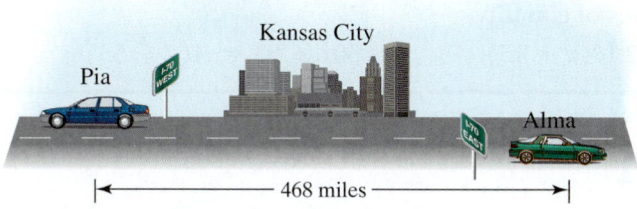

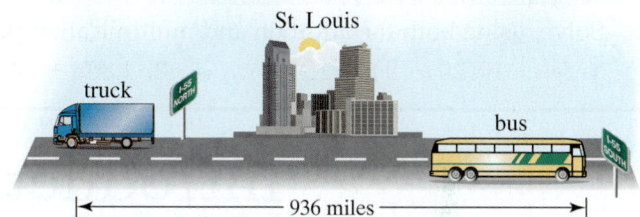

43. Harold leaves Denver, Colorado, traveling east on I-70 at an average speed of 44 miles per hour. Two hours later, Selma leaves Denver, traveling west on I-70 at an average rate of 48 miles per hour. How long after Harold leaves will it take until they are 364 miles apart?

44. A motorcycle leaves Colorado Springs, Colorado, traveling north on I-25 at an average speed of 47 miles per hour. One hour later, a car leaves Colorado Springs, traveling south on I-25 at an average speed of 54 miles per hour. How long after the motorcycle leaves will it take until they are 552 miles apart?

Challenge Exercises (45 and 46)

45. An architect wishes to build a fountain that has four pipes. If the pipes were arranged in order from the shortest to the longest, each pipe would be 3 feet longer than the previous one. If she has 54 feet of pipe, how long should each pipe be?

46. Jim and Amy planned their family so that the two children were born four years apart. If the sum of the ages of the children is 22 years, how old is each child?

Writing Exercise

47. How would you go about teaching a friend to solve application problems?

Writing Exercises or Group Projects (48 and 49)

If these exercises are done as a group project, have each group write two exercises of each type, exchange them with another group, and then solve them.

48. Write and solve an application problem involving consecutive integers.

49. Write and solve an application problem involving distance, rate, and time.

Section 3.6 # Money, Investment, and Mixture Application Problems

OBJECTIVES *When you complete this section, you will be able to:*

Ⓐ Solve real-world problems involving money.
Ⓑ Solve real-world problems involving investments.
Ⓒ Solve real-world problems involving mixtures.

PREREQUISITE SKILLS *Before beginning this section, you should be able to:*

a. Solve linear equations, using both the addition and multiplication properties of equality (Section 3.3).

GETTING READY FOR SECTION 3.6

Solve, using both the addition and multiplication properties of equality.

1. $5n + 10(20 - n) = 150$ **2.** $1.80x + 34.50 = 2.10(x + 15)$ **3.** $.30x + 33 = .4(x + 55)$

Introduction

In this section, we continue our discussion of application problems. All of the problems in this section can be solved by using a chart that is modified slightly (mostly in the headings) for each type of problem. There is an extra step involved in the problems that we did not do previously: We will need to calculate the value (in the case of money) or the amount of pure substance (in the case of mixtures) before writing an equation. Another approach to the problems in this section is given in Chapter 9, where systems of equations, the preferred method of some, are used.

A situation commonly encountered in application problems is when the total amount is known and parts of the total need to be represented by a variable.

Example 1 | **Represent each of the following:**

a. If a board is 12 feet long and a piece x feet long is cut off, represent the length of the remaining piece in terms of x.

Solution

The length of the other piece is the total length of the board minus the length of the piece cut off. Thus, the length of the remaining piece is $(12 - x)$ feet.

b. A paint contractor buys 36 gallons of paint, some of which is acrylic and some which is oil based. If he buys x gallons of acrylic, represent the number of gallons of oil-based paint in terms of x.

Solution

The number of gallons of oil-based paint is the total number of gallons purchased minus the number of gallons of acrylic paint. Therefore, the number of gallons of oil-based paint is $(36 - x)$ gallons.

PRACTICE EXERCISE 1

Represent each of the following:

a. Sharla has a piece of ribbon 65 inches long and cuts off x inches to wrap a package. How long is the remaining piece?

b. There is a total of 40 marbles in a jar, some black, the remainder white. If there are x black marbles in the jar, represent the number of white marbles in terms of x.

Answers: Getting Ready: 1. $n = 10$ **2.** $x = 10$ **3.** $x = 110$ **Practice Exercise 1: a.** $(65 - x)$ inches **b.** $(40 - x)$ white marbles

OBJECTIVE Solving real-world problems involving money

We will learn to solve two types of problems involving money. One type of problem will involve the purchase of two or more items at different prices. We will be given the total number of items purchased and the total cost, and will then be asked to find how many items of each type were purchased. The second type is similar to the first, but involves coins. In the translation step, we will use a chart as follows.

1. Draw a chart with four columns and a row for each item. Label the columns Type of Item, Price per Unit, Number of Units, and Total Value. Fill in the Type of Item column.
2. Fill in the Price per Unit column. These values will usually be given, but may involve a variable.
3. Fill in the Number of Units column. This will typically involve a variable. It is usually necessary to assign the variable to the number of units of one item and express the number of units of the other item in terms of this variable.
4. Fill in the Total Value column. The total value is the product of the price per unit and the number of units.

Example 2

a. A locksmith installed inside and outside locks in a new classroom building. The locksmith paid $1216 for 80 locks. If inside locks cost $14 each and outside locks cost $26 each, how many of each kind of lock did he buy?

Solution

Understand: We are given the total number of locks, the total cost of the locks, and the cost of each type of lock. The unknowns are the number of each type of lock.

    Relationship 1: There is a total of 80 locks.
    Relationship 2: The total cost of the locks is $1216.

Plan: Use a table. Use relationship 1 to represent the unknowns in terms of a variable and relationship 2 to write the equation.

Translate:

1. Draw a chart and label the columns. Fill in the Type of Item (in this case, locks) column.

Type of Lock	Price per Lock	Number of Locks	Total Cost
Inside			
Outside			

2. Fill in the Price per Lock column.

Type of Lock	Price per Lock	Number of Locks	Total Cost
Inside	14		
Outside	26		

3. Fill in the Number of Locks column. Since we are looking for the number of inside and outside locks, we must assign a variable to one of them. There is no particular advantage (or disadvantage) for letting our variable represent either, so let n represent the number of inside locks. From relationship 1, the number of outside locks is $80 - n$.

Type of Lock	Price per Lock	Number of Locks	Total Cost
Inside	14	n	
Outside	26	$80 - n$	

4. Fill in the Total Cost column. The total cost is the price per lock times the number of locks. Therefore, $14n$ represents the cost of the inside locks in dollars, and $26(80 - n)$ represents the cost of the outside locks in dollars.

Type of Lock	Price per Lock	Number of Locks	Total Cost
Inside	14	n	$14n$
Outside	26	$80 - n$	$26(80 - n)$

Write the word equation. Using relationship 2, we have

Cost of inside locks $+$ cost of outside locks $=$ total cost of the locks .

Write an algebraic equation, using the word equation as a guide.

Cost of inside locks $+$ cost of outside locks $=$ total cost of the locks .

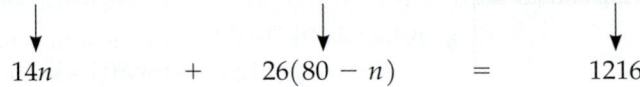

$$14n \quad + \quad 26(80 - n) \quad = \quad 1216$$

Solve:

$14n + 26(80 - n) = 1216$	Distribute 26.
$14n + 2080 - 26n = 1216$	Add like terms.
$-12n + 2080 = 1216$	Subtract 2080 from both sides.
$-12n = -864$	Divide both sides by -12.
$n = 72$	Solution of the equation.

Since n represents the number of inside locks, we have 72 inside locks. There are $80 - n = 80 - 72 = 8$ outside locks.

State: The locksmith purchased 72 inside locks and 8 outside locks.

Check: Do we have a total of 80 locks? $72 + 8 = 80$, so yes. Is the total value of the locks $1216? Since 72 inside locks are worth $72(14) = \$1008$ and 8 outside locks are worth $8(26) = \$208$, the total value of the locks is $\$1008 + \$208 = \$1216$, so our solutions are correct.

b. A collection of 23 coins is made up of nickels and dimes and is worth $1.90. How many of each type of coin is there?

Solution

This is a special case of the money problems involving coins. At first glance, this problem may seem very different from the preceding problem, but they are essentially the same. We will simply change the headings on our columns to Type of Coin, Value of Coin, Number of Coins, and Total Value.

Understand: We know the total number of coins and the value of the coins. The unknowns are the number of each type of coin.

Relationship 1: There is a total of 23 coins.
Relationship 2: The value of the coins is $1.90

Plan: Create a table. Use relationship 1 to represent the unknowns in terms of a variable and relationship 2 to write the equation.

Translate:

1. Draw the chart and label the columns. Fill in the Type of Coin column.

Type of Coin	Value of Coin	Number of Coins	Total Value
Nickel			
Dime			

2. Fill in the Value of Coin column, expressing the value in cents.

Type of Coin	Value of Coin	Number of Coins	Total Value
Nickel	5		
Dime	10		

3. Fill in the Number of Coins column. Since we are looking for the number of nickels and dimes, we need to assign a variable to one of them. Using relationship 1, if we let n represent the number of nickels, then $23 - n$ represents the number of dimes.

Type of Coin	Value of Coin	Number of Coins	Total Value
Nickel	5	n	
Dime	10	$23 - n$	

4. Fill in the Total Value column. The total value of each type of coin is the value of the coin times the number of coins. Therefore, $5n$ represents the value of the nickels in cents and $10(23 - n)$ represents the value of the dimes in cents.

Type of Coin	Value of Coin	Number of Coins	Total Value
Nickel	5	n	$5n$
Dime	10	$23 - n$	$10(23 - n)$

Write the word equation. Using relationship 2, we have

$$\text{Value of nickels} \ + \ \text{value of dimes} \ = \ \text{total value of the collection} \ .$$

Write an algebraic equation, using the word equation as a guide.

$$\text{Value of nickels} \ + \ \text{value of dimes} \ = \ \text{total value of the collection} \ .$$

$$5n \qquad + \quad 10(23 - n) \quad = \qquad 190$$

Note

Since the value of the nickels and the value of the dimes were given in cents, it was necessary to change $1.90 into 190¢.

Solve:

$5n + 10(23 - n) = 190$	Distribute 10.
$5n + 230 - 10n = 190$	Add like terms.
$-5n + 230 = 190$	Subtract 230 from both sides.
$-5n = -40$	Divide both sides by -5.
$n = 8$	Solution of the equation.

Since n represents the number of nickels, we have 8 nickels. There are $23 - n = 23 - 8 = 15$ dimes.

State: There are 8 nickels and 15 dimes.

Check: Do we have a total of 23 coins? $8 + 15 = 23$, so yes. Is the total value of the collection $1.90? Eight nickels are worth $8(.05) = \$.40$, and 15 dimes are worth $15(.10) = \$1.50$. The total value of the collection is $\$.40 + \$1.50 = \$1.90$, so our solutions are correct.

PRACTICE EXERCISE 2

a. A fishing guide buys some lures for a group that has chartered his services. He pays $6 each for some Yo-zuri lures and $5 each for some Excaliber lures. If he purchased 10 lures and paid a total of $57, how many of each did he purchase?

b. Suzie, who is 7 years old, is saving dimes and quarters to buy a new doll. Currently, she has a total of 40 coins that are worth $7.45. How many of each type of coin does she have?

If you need more practice, do the following Additional Practice Exercises.

Additional Practice Exercise 2

a. Johannes paid $188 for a total of 13 CDs and cassettes. If the CDs cost $16 each and the cassettes cost $12 each, how many of each did he buy?

b. A collection of quarters and nickels is worth $2.70. If there is a total of 22 coins, how many of each type of coin is there?

OBJECTIVE **B** Solving real-world problems involving investments

In doing investment problems, it is necessary to change percents to decimals, as in Section 3.5. Remember to drop the % sign and move the decimal two places to the left. For example, 12% = .12 and 6% = .06.

Investment problems are special types of *money problems*. Consequently, the procedure is virtually the same, except the column headings are changed to Type of Investment, Interest Rate (expressed as a decimal), Amount Invested, and Interest Earned. The amount of simple interest earned in a year is found by multiplying the annual interest rate expressed as a decimal times the amount invested. For example, the interest earned on $2000 invested at 5% interest for 1 year is (.05)(2000) = $100. Also, the total interest earned is the sum of the interests earned on each investment.

Example 3

A major corporation had a very good year, so the president received a large end-of-year bonus. She invested part of the money in bonds that pay 3% interest annually and the remainder in mutual funds that earned 4% the first year. The amount she invested in mutual funds is $150,000 more than two times the amount that she invested in bonds. If she earned $28,000 in interest the first year, how much did she invest in each?

Solution

Understand: We are given the interest rates in each account, the relationship between the amounts invested at each interest rate, and the total interest earned. The unknowns are the amounts invested at each rate.

Relationship 1: The amount invested in mutual funds is $150,000 more than two times the amount invested in bonds.

Relationship 2: The total amount of interest earned is $28,000.

Plan: Use a chart. Use relationship 1 to represent each amount in terms of a variable and use relationship 2 to write the equation.

Translate:

1. Draw and label a chart. Fill in the Type of Investment column.

Type of Investment	Interest Rate	Amount Invested	Interest Earned
Bonds			
Mutual Funds			

---

2. Fill in the Interest Rate column by converting the percents to decimals.

Type of Investment	Interest Rate	Amount Invested	Interest Earned
Bonds	.03		
Mutual Funds	.04		

3. Fill in the Amount Invested column. Using relationship 1, if we let x represent the amount invested at 3%, then $2x + 150{,}000$ is the amount invested at 4%.

Type of Investment	Interest Rate	Amount Invested	Interest Earned
Bonds	.03	x	
Mutual Funds	.04	$2x + 150{,}000$	

4. Fill in the Interest Earned column. Remember, the interest earned is the interest rate times the amount invested.

Type of Investment	Interest Rate	Amount Invested	Interest Earned
Bonds	.03	x	$.03x$
Mutual Funds	.04	$2x + 150{,}000$	$.04(2x + 150{,}000)$

Write the word equation. Using relationship 2, we have

Interest earned at 3% + interest earned at 4% = total interest earned .

Write the algebraic equation, using the word equation as a guide.

Interest earned at 3% + interest earned at 4% = total interest earned .

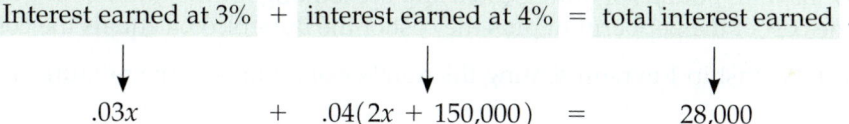

$$.03x \quad + \quad .04(2x + 150{,}000) \quad = \quad 28{,}000$$

Solve:

$.03x + .04(2x + 150{,}000) = 28{,}000$	Distribute the .04.
$.03x + .08x + 6000 = 28{,}000$	Add like terms.
$.11x + 6000 = 28{,}000$	Subtract 6000 from both sides.
$.11x = 22{,}000$	Divide both sides by .25.
$x = 200{,}000$	Solution of equation.

Since x represents the amount invested at 3%, we have found the amount invested at 3%. The amount invested at 4% is $2x + 150{,}000 = 2(200{,}000) + 150{,}000 = 400{,}000 + 150{,}000 = \$550{,}000$.

State: There was $200,000 invested at 3% and $550,000 invested at 4%.

Check: The interest earned on $200,000 at 3% is $(.03)(200{,}000) = \$6000$, and the interest earned on $550,000 at 4% is $(.04)(550{,}000) = \$22{,}000$. So the total interest on the two investments is $\$6{,}000 + \$22{,}000 = \$28{,}000$, which is what it is supposed to be. Our solutions are correct.

PRACTICE EXERCISE 3

The booster club has invested its funds in two accounts, one of which pays 5% annually. It invested $3000 less than three times that amount in an account that pays 4% annually. If the total annual interest earned from the two accounts is $1240, find the amount in each account.

(Continued)

If you need more practice, do the following Additional Practice Exercise.

Additional Practice Exercise 3

An author receives a royalty check for $60,000 and invests part of it in bonds paying 4% and the remainder in a certificate of deposit paying 5%. If the total annual interest earned from the two investments is $2650, find the amount in each investment.

OBJECTIVE Solving real-world problems involving mixtures

Mixture problems are very much like money and investment problems. We will consider two types of mixture problems: liquid and dry. We will learn to solve dry mixture problems first. To do so, we will again use a chart, but change the headings to Type of Ingredient, Price per Unit, Number of Units, and Total Value. We will also need an extra line at the bottom for the mixture. The key to solving dry mixture problems is that the sum of the values of each of the ingredients must equal the value of the mixture.

Example 4

How many ounces of peppermint candy worth $2.40 per pound must be mixed with 15 ounces of butterscotch worth $2.80 an pound to get a mixture worth $2.64 per pound?

Solution

Understand: We know the cost per pound of each type of candy, the cost per pound of the mixture, and the number of pounds of butterscotch. The unknown is the number of pounds of peppermint.

   Relationship 1: We have 15 pounds of butterscotch and are looking for the number of pounds of peppermint.
   Relationship 2: The cost of the peppermint plus the cost of the butterscotch equals the cost of the mixture.

Plan: Use a chart. Use relationship 1 in representing the number of pounds in the mixture and relationship 2 to write the equation.

Translate:

1. Draw the chart, label the columns, and fill in the types of candy.

Type of Candy	Price per Pound	Number of Pounds	Total Value
Peppermint			
Butterscotch			
Mixture			

2. Fill in the Price per Pound column.

Type of Candy	Price per Pound	Number of Pounds	Total Value
Peppermint	2.40		
Butterscotch	2.80		
Mixture	2.64		

---

3. Since we are looking for the number of pounds of peppermint candy, let x represent the number of pounds of peppermint candy. Also, we know that we have 15 pounds of butterscotch. The number of pounds of candy in the mixture is the number of pounds of peppermint plus the number of pounds of butterscotch. Consequently, the number of pounds of candy in the mixture is $x + 15$. Fill in the Number of Pounds column.

Type of Candy	Price per Pound	Number of Pounds	Total Value
Peppermint	2.40	x	
Butterscotch	2.80	15	
Mixture	2.64	$x + 15$	

4. The total value of each type of candy is the price per pound times the number of pounds. Use this fact to fill in the Total Value column.

Type of Candy	Price per Pound	Number of Pounds	Total Value
Peppermint	2.40	x	$2.40x$
Butterscotch	2.80	15	$(2.80)(15) = 42.00$
Mixture	2.64	$x + 15$	$2.64(x + 15)$

Write the word equation. Using relationship 2, we have

Value of peppermint + value of butterscotch = value of mixture .

Write the algebraic equation, using the word equation as a guide.

Value of peppermint + value of butterscotch = value of mixture .

$$2.40x \quad + \quad 42 \quad = \quad 2.64(x + 15)$$

Solve:

$2.40x + 42 = 2.64(x + 15)$	Distribute 2.64.
$2.40x + 42 = 2.64x + 39.60$	Subtract 2.60x from both sides.
$42 = .24x + 39.60$	Subtract 39.60 from both sides.
$2.4 = .24x$	Divide both sides by .24.
$10 = x$	Solution of the equation.

State: We need 10 pounds of peppermint.

Check: The value of 10 pounds of peppermint candy is $2.40(10) = \$24.00$, and from the chart we know that the value of 15 pounds of butterscotch is $42.00. Consequently, $24.00 + $42.00 = $66.00. The mixture sells for $2.64 per pound, and we have $10 + 15 = 25$ pounds. Consequently, the value of the mixture is $2.64(25) = \$66.00$, which is the same as the sum of the values of the peppermint and butterscotch. Our solution is correct.

Other variations of this type of problem include situations where the number of units and the price per unit of each of the ingredients are known and we are asked to find the price per unit of the mixture. Another type gives the price per unit of each of the ingredients, the price per unit of the mixture, and the number of units in the mixture and asks us to find the number of units of each ingredient. The approach is the same in all cases.

PRACTICE EXERCISE 4

How many pounds of coffee from the Dominican Republic that sells for $3.00 per pound should be mixed with 9 pounds of coffee from Colombia that sells for $4.50 per pound in order to get a mixture that sells for $3.90 per pound?

(Continued)

Answers: Practice Exercise 4: 6 pounds

If you need more practice, do the following Additional Practice Exercise:

Additional Practice Exercise 4

The manager of a deli received an order for a party platter consisting of a mixture of 20 pounds of ham and turkey cold cuts. Ham sells for $9.00 per pound and turkey for $5.00 per pound, and the mixture is to sell for $7.50 per pound. How many pounds of each does he use?

Liquid mixture problems are done in exactly the same way as dry mixture problems, except that we deal with the amount of pure substance in each ingredient and in the mixture instead of with values. Therefore, we will not repeat the procedure, but will illustrate with an example. The key to doing this type of problem is that the amount of pure substance in a solution (alloy, etc.) is the percent that is pure times the number of units (gallons, pounds, etc.) of the solution. For example, if we have 8 gallons of a solution that is 20% alcohol, the amount of pure alcohol in the solution is $(.20)(8) = 1.6$ gallons.

Example 5

How much of a solution that is 30% alcohol must be added to 55 liters of a solution that is 60% alcohol to obtain a solution that is 40% alcohol?

Solution

Understand: We are given the concentrations of each solution, the concentration of the mixture, and the number of liters of the 60% solution. The unknown is the number of liters of the 30% solution.

> Relationship 1: We have 55 liters of 60% solution and are looking for the number of liters of 30% solution.
> Relationship 2: The amount of pure alcohol in the mixture is the same as the amount of pure alcohol in the 30% solution plus the amount in the 60% solution.

Plan: Use a chart. Use relationship 1 in representing the total volume of the mixture and relationship 2 to write the equation.

Translate:

1. Draw a chart similar to the one for dry mixtures. Note that the headings are different. Fill in the types of solutions.

Type of Solution	Part Pure Alcohol	Volume	Amount Pure Alcohol
30%			
60%			
40% (mixture)			

2. Fill in the Part Pure Alcohol column by converting the percents to decimals.

Type of Solution	Part Pure Alcohol	Volume	Amount Pure Alcohol
30%	.30		
60%	.60		
40% (mixture)	.40		

Answers: Additional Practice 4: 12.5 pounds of ham, 7.5 pounds of turkey

3. Fill in the Volume column. Let x represent the volume of 30% alcohol. We know that we have 55 liters of 60% alcohol. Consequently, when the two are mixed, we will have $x + 55$ liters of the 40% mixture.

Type of Solution	Part Pure Alcohol	Volume	Amount Pure Alcohol
30%	.30	x	
60%	.60	55	
40% (mixture)	.40	$x + 55$	

4. Fill in the Amount Pure Alcohol column. Remember, the amount of pure alcohol in each solution equals the part pure alcohol times the volume.

Type of Solution	Part Pure Alcohol	Volume	Amount Pure Alcohol
30%	.30	x	$.30x$
60%	.60	55	$(.60)(55) = 33$
40% (mixture)	.40	$x + 55$	$.40(x + 55)$

Write the word equation. Using relationship 2, we have

Amount of pure alcohol in 30% solution $+$ amount of pure alcohol in 60% solution $=$ amount of pure alcohol in 40% mixture .

Write the algebraic equation, using the word equation as a guide.

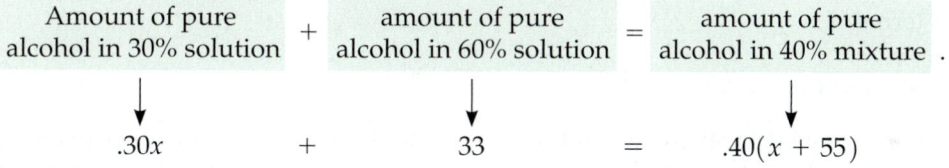

$$.30x \quad + \quad 33 \quad = \quad .40(x + 55)$$

Solve:

$.30x + 33 = .40(x + 55)$	Distribute .40 on the right side.
$.30x + 33 = .40x + 22$	Subtract .30x from both sides.
$33 = .10x + 22$	Subtract 22 from both sides.
$11 = .10x$	Divide both sides by .10
$110 = x$	Solution of the equation.

State: We need 110 ounces of 30% solution.

Check: In 110 liters of 30% solution there would be $.30(110) = 33$ liters of pure alcohol, and from the chart we know that in 55 liters of 60% solution there are 33 liters of pure alcohol. So there is a total of $33 + 33 = 66$ liters of pure alcohol in the two that are mixed. We have a total of $110 + 55 = 165$ liters of the 40% mixture. In this mixture there are $.40(165) = 66$ liters of pure alcohol, which is the same as the amount of pure alcohol in the two that are mixed to get the 40% solution. Thus, our solution is correct.

As with dry mixture problems, there are several variations of this type of problem.

PRACTICE EXERCISE 5

How much of a solution that is 20% baking soda must be added to 60 ounces of a solution that is 5% baking soda to make a solution that is 10% baking soda?

(Continued)

Answers: Practice Exercise 5: 30 ounces

If you need more practice, do the following Additional Practice Exercise:

Additional Practice Exercise 5

A chemistry lab assistant needs 150 ml of a 30% hydrochloric acid solution for an experiment. She finds that she has only 25% and 40% solutions available. How much of each solution must she mix to get the desired solution?

For Extra Help

Exercise Set 3.6 MyMathLab®

Represent each of the following. (See Example 1.)

1. Sally is taking a trip of 400 miles. If she has traveled x miles, represent the remainder of the trip in terms of x.

2. Francine walks 5 miles every morning. If she has already walked x miles, represent the distance remaining in terms of x.

3. There is a total of 24 nickels and dimes in a box. If x of these are nickels, represent the number of dimes in terms of x.

4. A seamstress has a box containing 50 buttons, some of which are brass and the remainder of which are glass. If there are x brass buttons, represent the number of glass buttons in terms of x.

5. A nursery has a total of 150 azalea and camellia plants. If there are x azaleas, represent the number of camellias in terms of x.

6. A class has 35 students. If x of the students are girls, represent the number of boys in terms of x.

Solve the following money problems. (See Example 2.)

7. An electrical contractor paid $1040 for 36 light fixtures to be installed in a hotel lobby. If the first type of fixture costs $25 each and the second type of fixture costs $32 each, how many of each kind did she buy?

8. A painting contractor paid $372 for 24 gallons of paint to paint the inside and outside of a house. If the paint for the inside costs $12 per gallon and the paint for the outside costs $18 per gallon, how many gallons of each did he buy?

9. The managers of an office building have decided to reduce the cost of air-conditioning by installing overhead fans. They paid $5225 for 105 fans, some of which measure 36 inches and the remainder of which measure 54 inches. If the 36-inch fans cost $45 each and the 54-inch fans cost $65 each, how many of each kind of fan did they buy?

10. An accounting firm purchased a supply of CDs. They paid $2020 for 60 boxes of disks. If the CD-R disks cost $22 a box and the CD-RW disks cost $57 a box, how many of each kind did it buy?

11. Winston paid $14.08 for 36 stamps. If some of the stamps cost 36¢ each and the remainder cost 44¢ each, how many of each kind did he buy?

12. Gayle paid $34.20 for 42 pens at the campus bookstore. If she bought some pens costing $.75 each and others costing $.90 each, how many of each kind did she buy?

13. A landscaping company pays $12 each for rose bushes and $9 each for azaleas. If it paid $144 for 14 plants, how many of each did it buy?

14. A building contractor has paid $2088 for 54 doors. If inside doors cost $35 each and outside doors cost $46 each, how many of each kind of door did he buy?

15. At the end of his shift, a cashier has only $5 and $20 bills in his drawer. If he has 48 bills worth a total of $900, how many of each kind of bill does he have?

16. Lamar buys $1500 worth of $20 and $50 traveler's checks. If he bought a total of 39 checks, how many $50 checks did he buy?

Answers: Additional Practice 5: 100 ml of 25%, 50 ml of 40%

17. A child goes to a bank to deposit her savings. She has $4.00 worth of nickels and dimes. If she has five more nickels than dimes, how many of each type of coin does she have?

18. If a vending machine had a total of 52 quarters and dimes whose value was $8.95, how many of each kind of coin was in the machine?

19. A parking meter will accept dimes and quarters only. At the end of the day, the meter attendants collected 55 coins whose total value was $10.75. How many coins of each type were there?

20. If a vending machine has a total of 42 nickels and dimes whose value is $3.40, how many coins of each type were in the machine?

Solve the following investment problems. (See Example 3.)

21. When he retired, Professor Gomez received a lump sum in cash for his unused sick days. He invested part of the money in bonds that paid 4% interest annually. He invested $16,000 more in certificates of deposit at 3% annual interest than he invested in bonds. The total interest earned per year from the two investments was $1180. How much money did he invest at each rate?

22. A petroleum geologist received a $50,000 bonus for discovering a new oil deposit. She invested part of the money in bonds that paid 3% interest annually and put the rest in a life insurance policy that paid 5% interest annually. How much did she put into the life insurance policy if she expects to earn $2100 total from both investments each year?

23. A professional football player received a $200,000 bonus for signing his contract. He invested part of the money in home mortgages that yield 4.5% interest annually and put the rest in a business that was yielding a 7.5% profit annually. How much did he invest in the business if he expects to earn a total of $10,200 per year from both investments?

24. A retired couple sold their home. They invested part of the money in home mortgages at 5.5% interest annually. They invested $20,000 less than this amount into certificates of deposit at 3.5% annual interest. How much money did they invest if they received $5600 per year interest from the two investments?

25. A professional organization has its cash in two investments. Part of the money is in an account paying 6% interest annually, and the remainder is in an account paying 8% interest annually. The amount in the account paying 8% interest is $7000 more than the amount in the account paying 6% interest. If the combined interest earned in 1 year from the two accounts is $3080, find the amount in each account.

26. A college support fund has its money in two investments. The first investment pays 6% interest annually and is $20,000 more than the second investment, which pays 5% interest annually. If the total interest earned from the two investments is $6700 per year, how much is in each investment?

27. A banker invests $50,000 at 4% annual interest. How much must she invest at 6% annual interest if the total interest earned per year from the two investments is $3800?

28. Frances received an inheritance of $40,000, which she invests at 7% annual interest. She already had some money invested at 6% annual interest. How much does she have invested at 6% if her combined interest from the two accounts is $4300 per year?

Solve the following mixture problems. (See Example 4.)

29. How many ounces of pipe tobacco that sells for $2.50 per ounce must be mixed with 6 ounces that sells for $3.50 per ounce if the mixture is to sell for $3.10 per ounce?

30. How many pounds of candy that sells for $2.75 per pound must be mixed with 10 pounds that sells for $3.50 per pound to get a mixture that sells for $3.25 per pound?

31. If 8 pounds of almonds that sell for $7.00 per pound are mixed with 4 pounds of peanuts that sell for $4.00 per pound, what should be the selling price of the mixture?

32. If 12 ounces of white chocolate that sells for $3.75 per ounce is mixed with 6 ounces of dark chocolate that sells for $4.50 per ounce, what should be the selling price of the mixture?

33. A specialty store owner mixes peanuts worth $3.60 per pound with cashew nuts worth $8.40 per pound. How many pounds of each type of nut is needed to make a 12-pound mixture worth $6.80 a pound?

34. A coffee company wishes to make a special blend of coffee by mixing coffee beans that cost $12.40 a pound with coffee beans that cost $16.80 per pound. How many pounds of each type of coffee bean is needed to make 40 pounds of coffee that costs $15.04 per pound?

35. How many milliliters of a 40% solution of battery acid must be added to 30 milliliters of an 8% solution to make a 20% solution of acid?

36. How many ounces of a 10% baking soda solution must be added to 40 ounces of a 2% baking soda solution to make a 5% baking soda solution?

37. How many tons of an alloy that is 5% nickel must be added to 15 tons of an alloy that is 20% nickel to make an alloy that is 10% nickel?

38. How many cups of a party mix that is 60% peanuts must be added to 5 cups of a party mix that is 30% peanuts to make a party mix that is 45% peanuts?

39. How many liters of pure hydrochloric acid must be added to 50 liters of a solution that is 10% hydrochloric acid to get a solution that is 25% hydrochloric acid?

40. How much pure water must be added to 20 liters of a solution that is 15% salt to make a solution that is 8% salt?

Solve the exercises. Types of problems are mixed. (See Examples 2–4.)

41. How many gallons of pure water must be added to 10 gallons of a solution that is 25% soap to make a solution that is 5% soap?

42. How many pounds of an alloy that is 30% tin must be added to 40 pounds of an alloy that is 15% tin to make an alloy that is 25% tin?

43. Professor Matas bought materials for a workshop for elementary school teachers. She bought number stickers for $.49 each and strings of beads for $1.19 a string. She bought 10 more number stickers than strings of beads. If she paid a total of $30.10, how many of each item did she buy?

44. Jeannette Castillo won $20,000 as part of the state lottery. After her celebration party, she put part of the money into bonds paying 7% interest. The rest, which was $10,000 less than the first part, she invested in a stock that was paying 20% in dividends. How much did she put in each investment if she earned a total of $2050 per year?

45. How much pure alcohol must be added to 50 gallons of gasohol that is 10% alcohol to make gasohol that is 20% alcohol?

46. Mr. Jacob bought two prescription medications. He paid $.83 a tablet for the first medication and $1.15 a tablet for the second medication. Altogether, there were 86 tablets for which he paid $79.70. How many of the more expensive tablets did he buy?

Challenge Exercises (47 and 48)

47. A vending machine accepts nickels, dimes, and quarters. At the end of the day, there were three more dimes than nickels and four fewer quarters than nickels. If the total value of the coins was $4.10, find the number of each type of coin.

48. A banker has three investments that pay 6%, 5%, and 8% each. She has twice as much invested at 5% as at 6% and $5000 more invested at 8% than at 6%. If the total interest earned per year from the three investments is $2800, find the amount invested at each rate.

Writing Exercises or Group Projects (49–51)

If these exercises are done as a group project, each group should write two exercises of each type, exchange them with another group, and solve them.

49. Write and solve an application problem involving money.

50. Write and solve an application problem involving mixtures.

51. Write and solve an application problem involving investments.

Section 3.7 Geometric Application Problems

OBJECTIVE *When you complete this section, you will be able to:*

A Solve real-world problems involving geometric concepts.

PREREQUISITE SKILLS *Before beginning this section, you should be able to:*

a. Solve linear equations by using both the addition and multiplication properties of equality (Section 3.3).

GETTING READY FOR SECTION 3.7

Solve, using both the addition and multiplication properties of equality.

1. $48 = 2(4x - 1) + 2x$ **2.** $21 = x + (2x - 1) + (x + 2)$ **3.** $t + 2t + (t + 12) = 180$

Introduction

Linear equations are often found in geometric applications. There are essentially two types of problems. The first type are those for which known formulas (Sections 1.2–1.4) give us a relationship among the unknowns of the problem. When solving this type of problem, either we are given a value for the variable(s) in the formula or we have to represent the unknowns in the formula in terms of variables. We then substitute into the formula and solve the resulting equation. Example 1 will be of this type. We will discuss the second type after Example 1. We do not have to look for conditions within the problem to give us the equation, since it will be some known formula. We use the same procedure as before.

OBJECTIVE **A** Solving real-world problems involving geometric concepts

Example 1

a. A silicon computer chip is rectangular in shape, and its length is one millimeter less than four times its width. If the perimeter of the chip is 48 millimeters, find the dimensions of the chip.

Solution

Understand: We are given the perimeter of a rectangle and the relationship between the length and width.

    Relationship 1: The length is one millimeter less than four times the width.
    Relationship 2: The perimeter is 48 millimeters.

Plan: Use relationship 1 to express the length and width in terms of a variable, and use relationship 2, with the formula for the perimeter of a rectangle, to write the equation.

Translate: We are looking for the length and width of the chip. Since the width is the smaller,

    Let x represent **the width**. Then, using relationship 1, $4x - 1$ represents **the length**.

Note that a word equation is not necessary. From relationship 2 our equation is the formula for the perimeter of a rectangle.

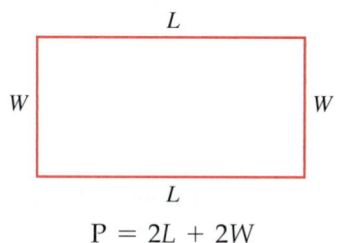

$$P = 2L + 2W$$

Substitute into the formula. We know from the problem that $P = 48$. We have also represented the width with x and the length with $4x - 1$. Therefore,

$$48 = 2(4x - 1) + 2x.$$

Solve:

$48 = 2(4x - 1) + 2x$	Distribute 2 on the right side.
$48 = 8x - 2 + 2x$	Add like terms.
$48 = 10x - 2$	Add 2 to both sides.
$50 = 10x$	Divide both sides by 10.
$5 = x$	Solution of equation.

Since x represents the width, the width is 5 millimeters. The length is $4x - 1 = 4 \cdot 5 - 1 = 20 - 1 = 19$ mm.

State: The width is 5 mm and the length is 19 mm.

Check: Is the length one less than four times the width? Yes, since $4 \cdot 5 - 1 = 19$. Is the perimeter 48 millimeters? Yes, since $2(19) + 2(5) = 38 + 10 = 48$. So, our solutions are correct.

b. The length of the longest side of a triangle is one less than twice the length of the shortest side. The length of the third side is two more than the length of the shortest side. If the perimeter of the triangle is 21 inches, find the length of the sides of the triangle.

Solution

Understand: We are given the relationships between the lengths of the sides of a triangle and the perimeter. The unknowns are the lengths of the sides of the triangle.

Relationship 1: The length of the longest side of a triangle is one less than twice the length of the shortest side, and the length of the third side is two more than the length of the shortest side.
Relationship 2: The perimeter is 21 inches.

Plan: Use relationship 1 to represent the three sides in terms of a variable, and use relationship 2 to write the equation.

Translate: Let x represent the shortest side; then from relationship 1 we know that the length of the longest side is one less than twice the length of the shortest side. So $2x - 1$ represents the length of the longest side, and also from relationship 1 we know that the length of the third side is two more than the length of the shortest side. So, $x + 2$ represents the length of the third side.

Note that again we do not need a word equation. We need the formula for the perimeter of a triangle.

$P = a + b + c$

Substitute into the formula. We know that $P = 21$ inches and the sides are represented by x, $2x - 1$, and $x + 2$. Therefore,

$$21 = x + (2x - 1) + (x + 2).$$

Solve:

$21 = x + (2x - 1) + (x + 2)$	Remove the parentheses.
$21 = x + 2x - 1 + x + 2$	Add like terms.
$21 = 4x + 1$	Subtract 1 from both sides.
$20 = 4x$	Divide both sides by 4.
$5 = x$	Solution of the equation.

Since x represents the length of the shortest side, the shortest side is 5 inches. The length of the longest side is $2x - 1 = 2 \cdot 5 - 1 = 10 - 1 = 9$ inches, and the length of the third side is $x + 2 = 5 + 2 = 7$ inches.

State: The lengths of the sides of the triangle are 5 in., 9 in., and 7 in.

Check: Is the longest side one less than twice the shortest side? Yes, since $9 = 2 \cdot 5 - 1$. Is the third side two more than the shortest side? Yes, since $7 = 5 + 2$. Is the perimeter 21 inches? Yes, since $5 + 9 + 7 = 21$. Therefore, our solutions are correct.

PRACTICE EXERCISE 1

a. A rectangular office building is four times as long as it is wide. If the perimeter is 900 feet, what are the length and width of the building?

b. The length of the longest side of a triangle is three times the length of the shortest, and the length of the third side is 4 feet more than the length of the shortest. If the perimeter is 19 feet, find the length of each of the three sides of the triangle.

If you need more practice, do the following Additional Practice Exercises.

Additional Practice Exercise 1

a. The length of a rectangular sheet of paper is 6 cm more than the width. If the perimeter of the sheet of paper is 100 cm, find the length and width.

b. A garden plot is in the shape of a right triangle in which the length of the longest side is 1 foot more than twice the shortest side, and the third side is one foot less than twice the shortest side. If the perimeter is 40 feet, find the lengths of the three sides.

The second type of geometric application problem is that for which we have to represent the unknown(s) in terms of a variable and then write an equation expressing the relationship among the unknowns, using some known geometric relationship. Example 2 will be of this type. These relationships often involve angles. An angle is usually denoted by the symbol $\angle$. The size of an angle is measured in degrees and is called the *measure* of the angle. The measure of $\angle A$ is denoted as $m\angle A$. Two angles with the same number of degrees (measure) are called **congruent** angles, and the symbol for congruent is $\cong$. These known geometric relationships may come from, but are not limited to, the following list:

1. The sum of the measures of all the angles of any triangle is $180°$.

2. Two angles are **supplementary** if the sum of their measures is $180°$. Each angle is the supplement of the other.

3. Two angles are **complementary** if the sum of their measures is $90°$. Each angle is the complement of the other.

4. Angle relationships involving parallel lines: Two or more lines are parallel if they lie in the same plane and do not intersect. Any line intersecting two or more lines is called a **transversal**. In the figure that follows, L_1 and L_2 are parallel and T is a transversal. The angle relationships are summarized as follows:

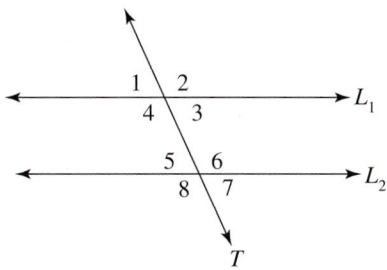

   a. Alternate interior angles are congruent. The two pairs of alternate interior angles are $\angle 4$ and $\angle 6$, and $\angle 3$ and $\angle 5$. Therefore, $m\angle 4 = m\angle 6$ and $m\angle 3 = m\angle 5$.

   b. Alternate exterior angles are congruent. The two pairs of alternate exterior angles are $\angle 1$ and $\angle 7$, and $\angle 2$ and $\angle 8$. Therefore, $m\angle 1 = m\angle 7$ and $m\angle 2 = m\angle 8$.

   c. Corresponding angles are congruent. The four pairs of corresponding angles are $\angle 1$ and $\angle 5$, $\angle 2$ and $\angle 6$, $\angle 3$ and $\angle 7$, and $\angle 4$ and $\angle 8$. Therefore, $m\angle 1 = m\angle 5$, $m\angle 2 = m\angle 6$, $m\angle 3 = m\angle 7$, and $m\angle 4 = m\angle 8$.

   d. Interior angles on the same side of the transversal are supplementary. The interior angles on the same side of the transversal are $\angle 3$ and $\angle 6$, and also $\angle 4$ and $\angle 5$. Therefore, $m\angle 3 + m\angle 6 = 180°$ and $m\angle 4 + m\angle 5 = 180°$.

Our procedure for solving will be exactly the same, except for writing the equation that in this case comes from a known geometric fact rather than a known formula.

Answers: Practice Exercise 1: a. $W = 90$ feet, $L = 360$ feet **b.** 3 feet, 9 feet, 7 feet
Additional Practice 1: a. length $= 28$ cm, width $= 22$ cm **b.** 8 feet, 15 feet, 17 feet

Example 2

a. In a triangle, the measure of the largest angle is twice the measure of the smallest angle. If the measure of the middle-sized angle is 12 more than the measure of the smallest angle, find the largest angle.

Solution

Understand: We know the relationship between the sizes of the angles of a triangle. The unknown is the size of the largest angle.

Relationship 1: The measure of the largest angle is twice the measure of the smallest angle, and the measure of the middle-sized angle is 12 more than the measure of the smallest angle.
Relationship 2: The sum of the measures of all the angles of a triangle is 180°.

Plan: Use relationship 1 to represent the three angles in terms of a variable and use relationship 2 to write the equation.

Translate: Let t represent **the measure of the smallest angle**. Then from relationship 1 we know that the largest angle is twice the smallest. So, **$2t$** represents **the measure of the largest angle**. Since the measure of the middle-sized angle is 12 more than the measure of the smallest angle, $t + 12$ represents **the measure of the middle-sized angle**.

Note that we do not need a word equation. From relationship 2, the sum of the measures of all the angles of a triangle is 180.

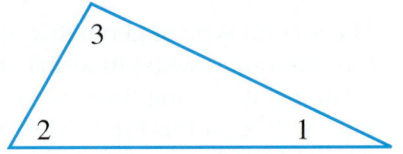

$$m\angle 1 + m\angle 2 + m\angle 3 = 180$$

Substitute into the equation. We have represented the angles by t, $2t$, and $t + 12$. So,

$$t + 2t + (t + 12) = 180.$$

Solve:

$t + 2t + (t + 12) = 180$	Remove parentheses.
$t + 2t + t + 12 = 180$	Simplify the left side.
$4t + 12 = 180$	Subtract 12 from both sides.
$4t + 12 - 12 = 180 - 12$	Simplify both sides.
$4t = 168$	Divide both sides by 4.
$\dfrac{4t}{4} = \dfrac{168}{4}$	Simplify both sides.
$t = 42$	The smallest angle is 42°.

NOTE Since the measure of an angle is the number of degrees in the angle, it is not necessary to use the degree symbol when giving the measure of an angle.

We have found t, which is the measure of the smallest angle. We are asked to find the largest angle, which is $2t = 2(42) = 84$.

State: The largest angle is 84°.

Check: The largest angle is 84°. Is this twice the smallest? The smallest angle is 42°, and $84° = 2(42°)$. So, yes. Is the middle angle 12° more than the smallest? The middle angle is $t + 12 = 42° + 12° = 54°$, and 54° is 12° more than 42°. Is the sum of the three angles 180°? Since $42° + 54° + 84° = 180°$, yes. We have the correct solution.

b. Given the accompanying figure with L_1 parallel to L_2, find the value of x and $\angle A$ and $\angle B$

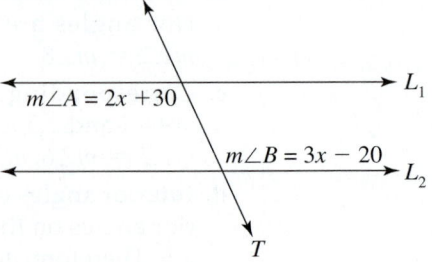

Solution

Understand: We are given two parallel lines and formulas for the measures of one pair of alternate interior angles. We are to find the value of x and the measures of $\angle A$ and $\angle B$.

Relationship: Alternate interior angles are congruent.

Plan: Use the preceding relationship to write the equation.

Translate: Since we are given a labeled figure, we do not have to assign a variable to an unknown.

Note that we do not need a word equation. Using the previous relationship, we write the equation $m\angle A = m\angle B$. So, $2x + 30 = 3x - 20$.

Solve:

$2x + 30 = 3x - 20$	Subtract $2x$ from both sides.
$30 = x - 20$	Add 20 to both sides.
$50 = x$	Therefore, $x = 50$.

$x = 50$, but we still need to find $\angle A$ and $\angle B$. $m\angle A = 2x + 30 = 2 \cdot 50 + 30 = 100 + 30 = 130$. $m\angle B = 3x - 20 = 3 \cdot 50 - 20 = 150 - 20 = 130$.

State: $\angle A$ and $\angle B$ are $130°$.

Check: Are the two angles congruent? Yes, since both have measures of 130.

c. The measure of an angle is 18 more than the measure of its complement. Find the measure of each angle.

Understand: We know the relationship between the measures of the angles and that they are complementary. The unknowns are the measures of the two angles.

> Relationship 1: The measure of an angle is 18 more than the measure of its complement.
> Relationship 2: The two angles are complementary.

Plan: Use relationship 1 to represent the angles in terms of a variable and relationship 2 to write the equation.

Translate: Let x represent the measure of the smaller angle. Then from relationship 1, $x + 18$ represents the measure of the larger angle.

Note that we do not need a word equation. From relationship 2 we know that the sum of the measures of the angles must be 90. So, $x + (x + 18) = 90$.

Solve:

$x + (x + 18) = 90$	Remove the parentheses.
$x + x + 18 = 90$	Add like terms.
$2x + 18 = 90$	Subtract 18 from both sides.
$2x = 72$	Divide both sides by 2.
$x = 36$	Solution of the equation.

Since x represents the measure of the smaller angle, the smaller angle is 36. The measure of the larger angle is $x + 18 = 36 + 18 = 54$.

State: The smaller angle is $36°$ and the larger angle is $54°$.

Check: Are the two angles complementary? Yes, since $54° + 36° = 90°$. Is the measure of one angle $18°$ more than the measure of its complement? Yes, since $54° - 18° = 36°$. Therefore, our solutions are correct.

PRACTICE EXERCISE 2

a. The measure of the smallest angle of a triangle is $18°$ less than the measure of the middle angle. If the measure of the largest angle is four times the measure of the smallest angle, find all three angles.

b. Two parallel lines are cut by a transversal, and a pair of corresponding angles have measures of $4x + 6$ and $6x - 30$. Find the value of x and each angle.

c. The measure of an angle is $30°$ less than twice the measure of its supplement. Find each of the angles.

(Continued)

Answers: Practice Exercise 2: a. $27°, 45°, 108°$ **b.** $x = 18$, each angle $= 78°$ **c.** $70°, 110°$

If you need more practice, do the following Additional Practice Exercises.

Additional Practice Exercise 2

a. In a triangle, the measure of the largest angle is 16° less than twice the measure of the smallest angle, and the measure of the middle angle is 4° more than the measure of the smallest angle. Find all three angles.

b. Two parallel lines are cut by a transversal, and two interior angles on the same side of the transversal have measures of $6x + 10$ and $4x + 30$. Find x and each angle.

c. The measure of an angle is four times the measure of its complement. Find each angle.

For Extra Help

Exercise Set 3.7   MyMathLab®

Solve the following geometric problems. (See Examples 1 and 2.)

1. The width of a rectangular lot is 45 feet shorter than the length. If the perimeter is 790 feet, how long is the lot?

2. A concert hall at Urban College is rectangular in shape and measures 50 feet longer than it is wide. If the perimeter is 420 feet, what are its length and width?

3. A playing field for soccer is rectangular in shape and is 40 yards longer than it is wide. If the perimeter is 220 yards, what are its length and width?

4. A rectangular rug on the living room floor is 4 feet longer than it is wide. If the distance around the rug is 40 feet, find the dimensions (length and width) of the rug.

5. A rectangular parking lot has a length that is 735 meters less than three times the width. If the perimeter is 2770 meters, what are the dimensions of the parking lot?

6. The length of the page proofs of a book is 5 inches less than twice the width. The perimeter of a page is 56 inches. Find the length and the width.

7. The length of the longest side of a triangle is twice the length of the shortest side, and the length of the third side is 2 inches more than the shortest side. If the perimeter is 130 inches, find the length of each of the three sides of the triangle.

8. The length of the longest side of a triangle is twice the length of the shortest side. The length of the middle-sized side is five more than the length of the shortest side. Find the lengths of the sides of the triangle if the perimeter is 33 feet.

9. Two sides of a triangular flower bed are equal in length, and the third side is 14 feet less than twice the length of the equal sides. If the perimeter of the flower bed is 94 feet, find the length of each of the three sides of the flower bed.

10. Two sides of a triangular sign are equal in length, and the length of the third side is 16 inches less than twice the length of the equal sides. If the perimeter of the sign is 152 inches, find the length of each of the three sides of the sign.

11. Two of the angles of a triangle are equal in measure and the measure of the third angle is equal to the sum of the measures of the two equal angles. Find the three angles.

12. Two of the angles of a triangle are equal in measure. The measure of the remaining angle is twice the sum of the measures of the two equal angles. Find the three angles.

13. The measure of the first angle of a triangle is three times the measure of the second angle, and the measure of the third angle is five more than the measure of the first angle. Find the three angles.

14. The measure of the smallest angle of a triangle is .3 of the measure of the largest angle, and the measure of the middle angle is .5 of the measure of the largest angle. Find all three angles.

Answers: Additional Practice 2: **a.** 48°, 52°, 80° **b.** $x = 14$, 86°, 94° **c.** 18°, 72°

15. The measure of the complement of an angle is 60° less than twice the measure of the angle. Find both angles.

16. The measure of the supplement of an angle is 10° more than four times the measure of the angle. Find both angles.

17. The measure of the supplement of an angle is 20° more than three times the measure of the angle. Find both angles.

18. The measure of the complement of an angle is 15° less than four times the measure of the angle. Find both angles.

19. Given the following figure, where L_1 and L_2 are parallel, find the value of x and $\angle A$ and $\angle B$.

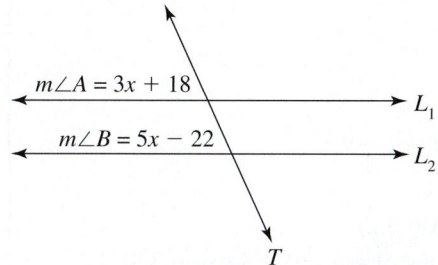

20. Given the following figure, where L_1 and L_2 are parallel, find the value of x and $\angle A$ and $\angle B$.

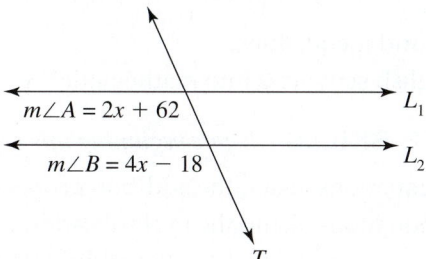

21. Given the following figure, where L_1 and L_2 are parallel, find the value of x and $\angle A$ and $\angle B$.

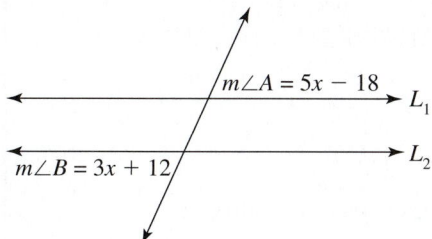

22. Given the following figure, where L_1 and L_2 are parallel, find the value of x and $\angle A$ and $\angle B$.

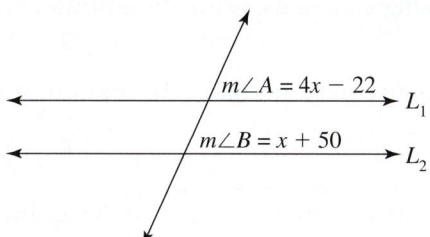

23. Given the following figure, where L_1 and L_2 are parallel, find the value of x and $\angle A$ and $\angle B$.

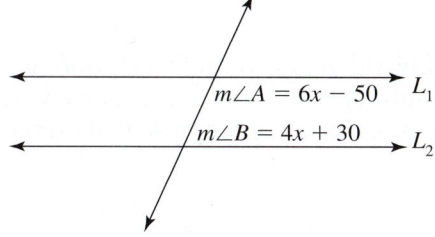

24. Given the following figure, where L_1 and L_2 are parallel, find the value of x and $\angle A$ and $\angle B$.

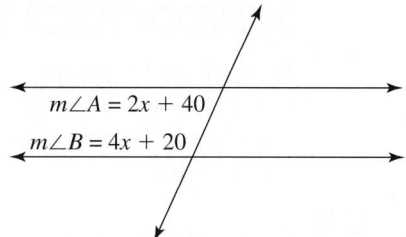

Challenge Exercise

25. This type of problem often appears on standardized tests. Find the value of x in the figure that follows:

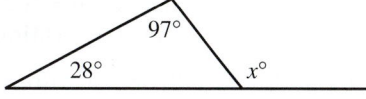

Writing Exercises or Group Projects (26 and 27)

If these exercises are done as a group project, each group should write two exercises of each type, exchange them with another group, and then solve them.

26. Write an application problem involving the perimeter of a triangle.

27. Write an application problem involving angle relationships.

OBJECTIVES *When you complete this section, you will be able to:*

A Graph inequalities on the number line.
B Solve linear inequalities, using the addition property of inequality, and graph the solutions.
C Solve linear inequalities, using the multiplication property of inequality, and graph the solutions.
D Solve linear inequalities, using both the addition and multiplication properties of inequality, and graph the solutions.
E Solve compound inequalities.
F Translate English sentences into mathematical inequalities and vice versa.

PREREQUISITE SKILLS   *Before beginning this section, you should be able to:*

  a. Solve linear equations, using the addition property of equality (Section 3.1).
  b. Solve linear equations, using the multiplication property of equality (Section 3.2).
  c. Solve linear equations, using both the addition and multiplication properties of equality (Section 3.3).

GETTING READY FOR SECTION 3.8

Solve the following equations, using the addition property of equality:
1. $x - 6 = 3$ **2.** $-2(2x - 4) + 5(x + 3) = 11$

Solve the following equations, using the multiplication property of equality:

3. $-6y = 18$ **4.** $\dfrac{x}{-2} = 1$

Solve the following equations, using both the addition and multiplication properties of equality:
5. $3x + 5 = x - 3$ **6.** $-2(y - 5) + 3y = 5(y + 6) + y$

Introduction

In Sections 3.1 through 3.3, we learned to solve linear equations. In this section, we will solve linear inequalities. Before we discuss solving linear inequalities, we will expand on the concept of inequality and how inequalities can be represented on the number line.

OBJECTIVE Graphing inequalities on the number line

The number line represents the integers and all of the numbers between the integers. The numbers between the integers are the **rational numbers** (discussed in Chapter 6) and the **irrational numbers** (discussed in Chapter 10).

Choose a number, say, 3. We put a dot at 3 on the number line. Consider some unknown number x. There are three possibilities for the position of x in relation to 3 on the number line.

<!-- number line from -6 to 6 with dot at 3, and x labels above -->

  1. x is to the right of 3, so $x > 3$; or
  2. x is right on top of 3, so $x = 3$; or
  3. x is to the left of 3, so $x < 3$.

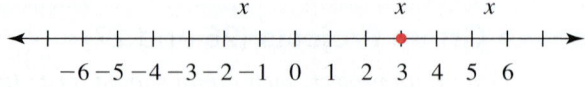

Note

Another type of notation, called *interval notation*, is frequently used when writing and graphing inequalities. When graphing, a parenthesis opening in the direction of the shading is used instead of an open dot and a bracket is used instead of a dot.

The number line has been divided into three parts, $x > 3$, $x = 3$, and $x < 3$. Expressions like $x > 3$ and $x < 3$ are called **inequalities**. Those numbers that make an inequality true are **solutions of the inequality**. One convenient method of representing the solutions of an inequality is by graphing the solutions on the number line. The expression $x \geq 3$ means $x > 3$ or $x = 3$. The graph of $x = 3$ consists of a dot on 3, and the graph of $x > 3$ consists of everything to the right of 3. So the graph of $x \geq 3$ is a dot on 3 with everything to the right shaded. An open dot is used to indicate that the point is not a part of the solution. So the graph of $x > 3$ is an open dot on 3 with everything to the right shaded.

Example 1   Graph each of the following on the number line:

a. $x > 5$

Since any number greater than 5 is to the right of 5 on the number line, we put an open dot on 5 (to show that 5 is not a solution) and shade the number line to the right of 5. The shaded region represents all values of x such that $x > 5$, and the arrow indicates that it extends to the right forever.

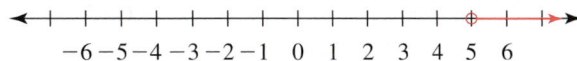

b. $x < 5$

The condition $x < 5$ is drawn in much the same way as for $x > 5$. The numbers less than 5 are to the left of 5 on the number line. We put an open dot on 5 (to show that 5 is not a solution) and then shade to the left of 5 on the number line. The shaded region extends forever, as indicated by the arrow.

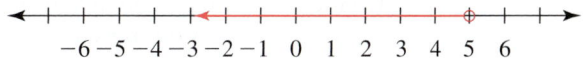

c. $x \geq -4$

The numbers greater than -4 are to the right of -4 on the number line. We put a solid dot on -4 (to indicate that -4 is a solution) and shade the number line to the right of -4. The arrow indicates that the region extends forever.

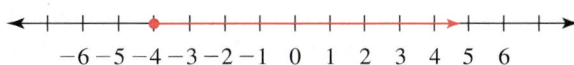

PRACTICE EXERCISE 1

Graph the following on the number line:

a. $x > 3$

b. $x < -1$

c. $x \geq 1$

(Continued)

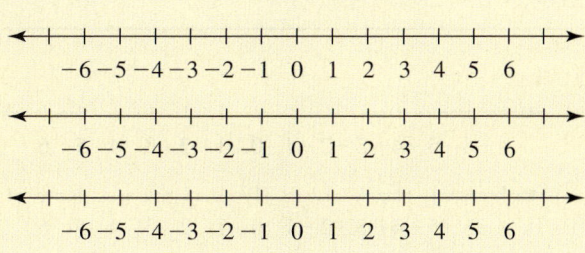
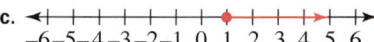

If you need more practice, do the following Additional Practice Exercises.

Additional Practice Exercise 1   Graph the following on the number line:

a. $x < -2$

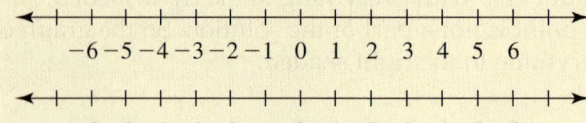

b. $x \leq 4$

c. $x \geq -3$

Inequalities of the form $a < x < b$ are called **compound inequalities**. The compound inequality $2 < x < 5$ is read, "Two is less than x and x is less than 5." On the number line, 2 is to the left of x and x is to the left of 5. Consequently, x is any number between 2 and 5. The graph of this compound inequality is shown as follows:

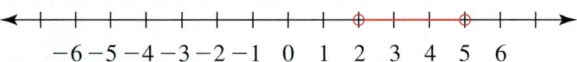

The open dots indicate that neither 2 nor 5 is a solution.

| **Example 2** | **Graph the following on the number line:** |

a. $0 \leq x < 5$

Put a solid dot on 0 (to indicate that 0 *is* a solution) and an open dot on 5 (to indicate that 5 is *not* a solution). Since 0 is to the left of x and x is to the left of 5, x can be any number between 0 and 5. Therefore, shade the region between 0 and 5.

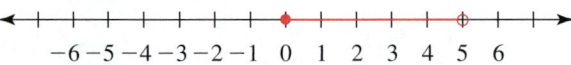

b. $-5 \leq x \leq 4$

Put a solid dot on both -5 and 4 to indicate that they are solutions. Since -5 is to the left of x and x is to the left of 4, x can be any number between -5 and 4. Therefore, shade the region between -5 and 4.

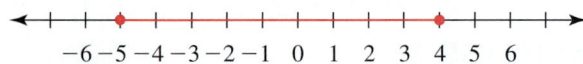

PRACTICE EXERCISE 2

Graph the following:

a. $-5 < x < 0$

b. $3 < x < 6$

(Continued)

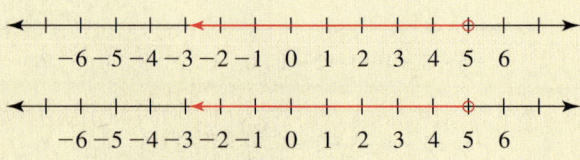

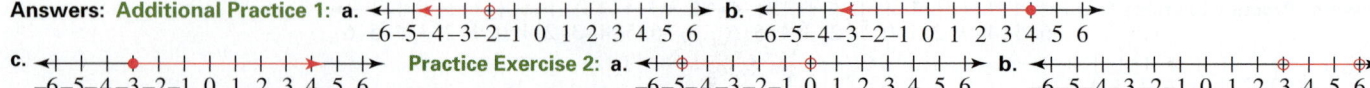

Answers: Additional Practice 1: a. [image] b. [image]
c. [image]   Practice Exercise 2: a. [image] b. [image]

If you need more practice, do the following Additional Practice Exercises.

Additional Practice Exercise 2 Graph the following:

a. $-3 < x \le 3$

$$-6\ {-5}\ {-4}\ {-3}\ {-2}\ {-1}\ \ 0\ \ 1\ \ 2\ \ 3\ \ 4\ \ 5\ \ 6$$

b. $2 < x < 5$

$$-6\ {-5}\ {-4}\ {-3}\ {-2}\ {-1}\ \ 0\ \ 1\ \ 2\ \ 3\ \ 4\ \ 5\ \ 6$$

Now we are ready to discuss solving linear inequalities. To solve linear equations, we established the addition and multiplication properties of equality. To solve linear inequalities, we will need to establish similar properties for inequalities.

Example 3

a. $3 < 8$ Add 5 to both sides of the inequality.

$3 + 5 \overset{?}{<} 8 + 5$ Simplify both sides.

$8 < 13$ This is a true statement, so adding 5 to both sides of the inequality has no effect on the order symbol.

b. $9 > 7$ Subtract 3 from both sides of the inequality (same as adding -3).

$9 - 3 \overset{?}{>} 7 - 3$ Simplify both sides.

$6 > 4$ This is a true statement, so subtracting 3 from both sides of the inequality has no effect on the order symbol.

c. $-12 < -2$ Add 8 to both sides of the inequality.

$-12 + 8 \overset{?}{<} -2 + 8$ Simplify both sides.

$-4 < 6$ This is a true statement, so adding 8 to both sides of the inequality has no effect on the order symbol.

Example 3 leads us to the following statement of the addition property of inequality:

> ### Rule: Addition Property of Inequality
>
> For all real numbers a, b, and c, if $a < b$, then $a + c < b + c$, and if $a > b$, then $a + c > b + c$. In words, any real number may be added to (or subtracted from) both sides of an inequality without affecting the order.

In other words, adding the same number to both sides of an inequality results in an equivalent inequality. The addition property of inequality is also true for the inequalities $\ge$ and $\le$.

OBJECTIVE **B** Solving linear inequalities, using the addition property of inequality, and graphing the solutions

As in solving equations with the addition property of equality, we wish to get the variable by itself on one side of the inequality and the constant on the other side. It is often easier to understand the solutions of an inequality if we graph them, so we will graph the solutions of each of the following examples:

Answers: Additional Practice 2: a.

$$-6\ {-5}\ {-4}\ {-3}\ {-2}\ {-1}\ \ 0\ \ 1\ \ 2\ \ 3\ \ 4\ \ 5\ \ 6$$

b.

$$-6\ {-5}\ {-4}\ {-3}\ {-2}\ {-1}\ \ 0\ \ 1\ \ 2\ \ 3\ \ 4\ \ 5\ \ 6$$

Example 4   **Solve the following and graph the solutions:**

a. $x - 6 < 3$ Add 6 to both sides. Adding the same quantity to both sides of an inequality does not change the order symbol.

$x - 6 + 6 < 3 + 6$ Simplify both sides.

$x < 9$ Therefore, the solutions are all numbers less than 9.

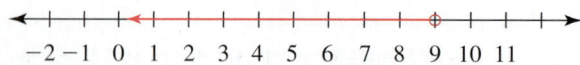

Unfortunately, there is no easy way to check the solution, because there are an infinite number of values of x that make the inequality true. However, we may choose some values of x that are less than 9 and some that are greater than 9 and substitute them into the original inequality to get an indication of whether our solution is correct.

Let $x = 8$. From $x - 6 < 3$, we have $8 - 6 < 3$, or $2 < 3$. This is a true statement, so $x = 8$ is a solution and $8 < 9$.

On the other hand, let $x = 10$. Now we have $10 - 6 < 3$, or $4 < 3$. This not true, so $x = 10$ is not a solution and $10 > 9$.

Therefore, values less than 9 seem to make the inequality true, and values greater than 9 seem to make it false. Hence, the solution is reasonable.

b. $2(y - 3) - y \geq -3$ Apply the distributive property.

$2y - 6 - y \geq -3$ Simplify the left side.

$y - 6 \geq -3$ Add 6 to both sides. Adding the same quantity to both sides of an inequality does not change the order symbol.

$y - 6 + 6 \geq -3 + 6$ Simplify both sides.

$y \geq 3$ Therefore, the solutions are all numbers greater than or equal to 3.

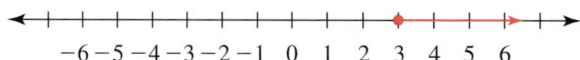

c. $-2(2x - 4) + 5(x + 3) > 11$ Apply the distributive property.

$-4x + 8 + 5x + 15 > 11$ Simplify the left side of the inequality.

$x + 23 > 11$ Subtract 23 from both sides. Subtracting the same quantity from both sides of an inequality does not change the order symbol.

$x + 23 - 23 > 11 - 23$ Simplify both sides.

$x > -12$ Therefore, the solutions are all numbers greater than -12.

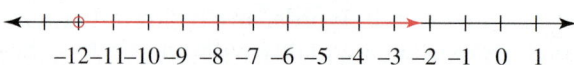

PRACTICE EXERCISE 4

Solve the following inequalities and graph the solutions:

a. $t + 3 > 1$ **b.** $3(s - 2) - 2s \leq 3$ **c.** $5(2t - 3) - 3(3t - 4) > 2$

If you need more practice, do the following Additional Practice Exercises.

Additional Practice Exercise 4   Solve the following inequalities and graph the solutions:

a. $a - 8 + 2 < 0$ **b.** $6(u + 2) - 5u < 9$ **c.** $4(4z - 2) - 5(3z - 1) \leq 4$

Answers: Practice Exercise 4: a. $t > -2$![number line](−6−5−4−3−2−1 0 1 2 3 4 5 6)   **b.** $s \leq 9$![number line](−3−2−1 0 1 2 3 4 5 6 7 8 9)

c. $t > 5$![number line](−6−5−4−3−2−1 0 1 2 3 4 5 6)   **Additional Practice 4: a.** $a < 6$![number line](−6−5−4−3−2−1 0 1 2 3 4 5 6)

b. $u < -3$![number line](−6−5−4−3−2−1 0 1 2 3 4 5 6)   **c.** $z \leq 7$![number line](−3−2−1 0 1 2 3 4 5 6 7 8 9)

OBJECTIVE **Solving linear inequalities, using the multiplication property of inequality, and graphing the solutions**

The addition property of inequality cannot be used to solve all linear inequalities, just as the addition property of equality alone will not solve all linear equations. In order to solve an inequality like $3x \leq 9$, we need to divide both sides by 3. What effect, if any, will this have on the order symbol? We investigate the effect of multiplication and division in the following examples:

Example 5

a. $4 < 9$ Multiply both sides of the inequality by 2.

$$2(4) \stackrel{?}{<} 2(9)$$ Simplify both sides.

$$8 < 18$$ This is a true statement, so multiplying both sides by 2 has no effect on the order symbol.

b. $5 < 7$ Multiply both sides of the inequality by -3.

$$(-3)5 \stackrel{?}{<} (-3)7$$ Simplify both sides.

$$-15 < -21$$ This is a false statement. In order to make it true, the $<$ symbol must be changed to a $>$ symbol.

$$-15 > -21$$

c. $6 > -1$ Multiply both sides of the inequality by 4.

$$4(6) \stackrel{?}{>} 4(-1)$$ Simplify both sides.

$$24 > -4$$ This is a true statement, so multiplying both sides by 4 has no effect on the order symbol.

d. $8 \geq -9$ Multiply both sides of the inequality by -5.

$$8(-5) \stackrel{?}{\geq} -9(-5)$$ Simplify both sides.

$$-40 \geq 45$$ This is a false statement. In order to make it true, the $\geq$ symbol must be changed to a $\leq$ symbol.

$$-40 \leq 45$$

Example 5 suggests the following: If we multiply both sides of an inequality by a positive number, as in Examples 5a and 5c, the inequality symbol remains unchanged. However, if we multiply an inequality by a negative number, as in Examples 5b and 5d, the inequality symbol changes (reverses). When the order symbol is unchanged, we say the sense of the inequality was not changed; and when the order symbol is changed, we say the sense of the inequality is reversed. These results are formally stated in the following properties:

Rule: Multiplication Properties of Inequality

Positive Factor

For all real numbers a, b, and c, if $a < b$ and $c > 0$, then $a \cdot c < b \cdot c$, and if $a > b$, then $a \cdot c > b \cdot c$. In words, if both sides of an inequality are multiplied (or divided) by a positive number, then the order symbol (sense) is unchanged.

Negative Factor

For all real numbers a, b, and c, if $a < b$ and $c < 0$, then $a \cdot c > b \cdot c$, and if $a > b$ and $c < 0$, then $a \cdot c < b \cdot c$. In words, if both sides of an inequality are multiplied (or divided) by a negative number, then the order symbol (sense) must be reversed.

The important thing to remember is that the direction of the inequality changes when we **multiply (or divide)** *both sides of the* **inequality** *by a* **negative number.** *Otherwise, solving linear inequalities is* **exactly the same as** *solving linear equations.*

Just as in the addition properties, the multiplication properties are true for the inequalities $\geq$ and $\leq$, as well as for $>$ and $<$.

| **Example 6** | **Solve and graph the solution of each of the following:** |

a. $4x \geq 20$ Divide both sides of the inequality by 4. **Dividing** by a **positive** number **does not** change the direction of the inequality symbol.

$\dfrac{4x}{4} \geq \dfrac{20}{4}$ Simplify both sides.

$x \geq 5$ Therefore, the solutions are all numbers greater than or equal to 5.

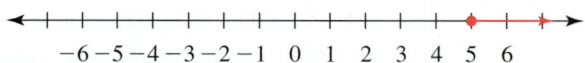

b. $-6y > 18$ Divide both sides of the inequality by -6. **Dividing** by a **negative** number **reverses** the direction of the order symbol.

$\dfrac{-6y}{-6} < \dfrac{18}{-6}$ Simplify both sides.

$y < -3$ Therefore, the solutions are all numbers less than -3.

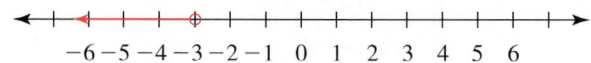

c. $\dfrac{x}{-2} \geq 1$ Multiply both sides of the inequality by -2. **Multiplying** by a **negative** number **reverses** the direction of the order symbol.

$-2 \cdot \dfrac{x}{-2} \leq -2 \cdot 1$ Simplify both sides.

$x \leq -2$ Therefore, the solutions are all numbers less than or equal to -2.

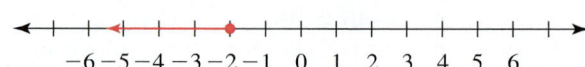

PRACTICE EXERCISE 6

Solve and graph each of the following:

a. $2z > 4$ **b.** $-5n < 35$ **c.** $\dfrac{u}{-3} > 2$

If you need more practice, do the following Additional Practice Exercises.

Additional Practice Exercise 6   Solve and graph each of the following:

a. $12 \geq 4v$ **b.** $-3r \geq 27$ **c.** $\dfrac{x}{-4} \leq -3$

Answers: Practice Exercise 6: a. $z > 2$
$$-6\ -5\ -4\ -3\ -2\ -1\ 0\ 1\ 2\ 3\ 4\ 5\ 6$$
b. $n > -7$
$$-8\ -7\ -6\ -5\ -4\ -3\ -2\ -1\ 0\ 1\ 2\ 3\ 4$$
c. $u < -6$
$$-7\ -6\ -5\ -4\ -3\ -2\ -1\ 0\ 1\ 2\ 3\ 4\ 5$$
Additional Practice 6: a. $v \leq 3$
$$-6\ -5\ -4\ -3\ -2\ -1\ 0\ 1\ 2\ 3\ 4\ 5\ 6$$
b. $r \leq -9$
$$-10\ -9\ -8\ -7\ -6\ -5\ -4\ -3\ -2\ -1\ 0\ 1\ 2$$
c. $x \geq 12$
$$1\ 2\ 3\ 4\ 5\ 6\ 7\ 8\ 9\ 10\ 11\ 12\ 13$$

OBJECTIVE Solving linear inequalities, using both the addition and multiplication properties of inequality, and graphing the solution

Up to this point, we have discussed linear inequalities that require the use of only one of the properties of inequality. The next few examples require both the addition and multiplication properties of inequality. Use the same steps as you would to solve linear equations, but remember to change the inequality symbol when multiplying or dividing a negative number.

Example 7   Solve each of the following and graph the solutions:

a. $3x + 5 > x - 3$ Subtract x from both sides of the inequality. **Subtracting** from both sides **does not** affect the order symbol.

$3x - x + 5 > x - x - 3$ Simplify both sides.

$2x + 5 > -3$ Subtract 5 from both sides. **Subtracting** from both sides **does not** affect the order symbol.

$2x + 5 - 5 > -3 - 5$ Simplify both sides.

$2x > -8$ Divide both sides by 2. **Dividing** by a **positive** number **does not** affect the order symbol.

$\dfrac{2x}{2} > \dfrac{-8}{2}$ Simplify both sides.

$x > -4$ Therefore, the solutions are all numbers greater than -4.

$$-6\ -5\ -4\ -3\ -2\ -1\quad 0\quad 1\quad 2\quad 3\quad 4\quad 5\quad 6$$

b. $16x + 8 - 4x > -16 + 10x$ Simplify both sides of the inequality.

$12x + 8 > -16 + 10x$ Subtract $12x$ from both sides. **Subtracting** from both sides **does not** affect the order symbol.

$12x - 12x + 8 > -16 + 10x - 12x$ Simplify both sides.

$8 > -16 - 2x$ Add 16 to both sides. **Adding** to both sides **does not** affect the order symbol.

$8 + 16 > -16 + 16 - 2x$ Simplify both sides.

$24 > -2x$ Divide both sides by -2. **Dividing** by a **negative** number **reverses** the order symbol.

$\dfrac{24}{-2} < \dfrac{-2x}{-2}$ Simplify both sides.

$-12 < x$ or $x > -12$ Therefore, the solutions are all numbers greater than -12.

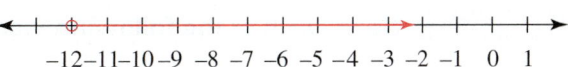
$$-12\ -11\ -10\ -9\ \ -8\ \ -7\ \ -6\ \ -5\ \ -4\ \ -3\ \ -2\ \ -1\quad 0\quad 1$$

c. $-2(y - 5) + 3y \geq 5(y + 6) + y$ Distribute -2 and 5.

$-2y + 10 + 3y \geq 5y + 30 + y$ Simplify both sides of the inequality.

$10 + y \geq 6y + 30$ Subtract $6y$ from both sides. **Subtracting** from both sides **does not** affect the order symbol.

$10 + y - 6y \geq 6y - 6y + 30$ Simplify both sides.

$10 - 5y \geq 30$ Subtract 10 from both sides. **Subtracting** from both sides **does not** affect the order symbol.

$10 - 10 - 5y \geq 30 - 10$ Simplify both sides.

$-5y \geq 20$ Divide both sides by -5. **Dividing** by a **negative** number **reverses** the order symbol.

$\dfrac{-5y}{-5} \leq \dfrac{20}{-5}$ Simplify both sides.

$y \leq -4$ Therefore, the solutions are all numbers less than or equal to -4.

$$\xleftarrow{\hspace{0.5cm}} \underset{-6\ -5\ -4\ -3\ -2\ -1\ \ 0\ \ 1\ \ 2\ \ 3\ \ 4\ \ 5\ \ 6}{|\ \ \bullet\ \ |\ \ |\ \ |\ \ |\ \ |\ \ |\ \ |\ \ |\ \ |\ \ |\ \ |} \xrightarrow{\hspace{0.5cm}}$$

PRACTICE EXERCISE 7

Solve the following and graph the solutions:

a. $5x - 3 > x - 15$

b. $4 - 3x \leq -8 + x$

c. $-4t + 7 + 16t > 10t + 12$

d. $20r - 5(6 + 2r) \leq 4r + 4(r - 7)$

If you need more practice, do the following Additional Practice Exercises.

Additional Practice Exercise 7   Solve the following:

a. $2t - 6 > 14 - 3t$

b. $6 - 2x \leq -4 + 3x$

c. $8z - 3 + 2z \leq 10z - 10 - 7z$

d. $7(-8 + u) - 2u \geq -14 + 3(2u - 15)$

OBJECTIVE E Solving compound inequalities

Compound inequalities are solved by the same techniques used for other linear inequalities. The variable is located between the inequality symbols. Our task is to eliminate the constants so that the variable will be isolated between the inequality symbols. It is important to remember that whatever you do to the middle part to isolate the variable must be done to all three parts of the inequality.

Example 8   **Solve each of the following inequalities:**

a. $-3 \leq x + 2 \leq 3$ Subtract 2 from all three expressions. **Subtracting does not** affect the order symbol.

$-3 - 2 \leq x + 2 - 2 \leq 3 - 2$ Simplify all three expressions.

$-5 \leq x \leq 1$ Therefore, the solutions are all numbers greater than or equal to -5 and less than or equal to 1.

Answers: Practice Exercise 7: **a.** $x > -3$ 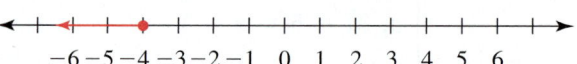 **b.** $x \geq 3$

c. $t > 2.5$ **d.** $r \leq 1$

Additional Practice 7: **a.** $t > 4$ **b.** $x \geq 2$ **c.** $z \leq -1$ **d.** $u \leq 3$

b. $1 < 6x + 7 \le 31$ Subtract 7 from all three expressions. **Subtracting does not** affect the order symbol.

$1 - 7 < 6x + 7 - 7 \le 31 - 7$ Simplify all expressions.

$-6 < 6x \le 24$ Divide each expression by 6. **Dividing** by a **positive** number **does not** affect the order symbols.

$\dfrac{-6}{6} < \dfrac{6x}{6} \le \dfrac{24}{6}$ Simplify all expressions.

$-1 < x \le 4$ Therefore, the solutions are all numbers greater than -1 and less than or equal to 4.

c. $0 \le -1.5y - 3 < 4.5$ Add 3 to all three expressions. **Adding does not** affect the order symbols.

$0 + 3 \le -1.5y - 3 + 3 < 4.5 + 3$ Simplify all expressions.

$3 \le -1.5y < 7.5$ Divide each expression by -1.5. **Dividing** by a **negative** number **reverses** the order symbols.

$\dfrac{3}{-1.5} \ge \dfrac{-1.5y}{-1.5} > \dfrac{7.5}{-1.5}$ Simplify all expressions.

$-2 \ge y > -5 \text{ or } -5 < y \le -2$ Therefore, the solutions are all numbers greater than -5 and less than or equal to -2.

PRACTICE EXERCISE 8

Solve the following inequalities:

a. $-4 \le x - 3 \le 2$ **b.** $-13 \le -4t - 5 < 15$ **c.** $3 < .2w + 3.6 < 3.6$

If you need more practice, do the following Additional Practice Exercises.

Additional Practice Exercise 8   Solve the following inequalities:

a. $3 < x + 5 < 7$ **b.** $-8 < 3v + 4 \le 19$ **c.** $-2 < .5x - 2.5 < -1$

OBJECTIVE **F**   Translating English sentences into mathematical inequalities and vice versa

Problems are often posed in terms of inequalities, as well as in terms of equations. Therefore, we need to be able to translate these relationships from mathematics to English as well.

Example 9   **Translate each expression from mathematics to English. There can be more than one translation.**

Mathematics	English
a. $4y > 12$	Four times a number is greater than twelve. The product of four and a number is more than twelve.
b. $2x - 5 \le 20$	Twice a number, decreased by five, is less than or equal to twenty. Twice a number, decreased by five, is at most twenty. The difference of twice a number and five is less than or equal to twenty. Five less than twice a number is at most 20.

(Continued)

Mathematics	English
c. $3(u + 6) < -9$	Three times the sum of a number and 6 is less than negative nine. The product of three and the sum of a number and six is less than negative nine.
d. $5(2v - 4) + 3 \geq 8$	Five times the difference of twice a number and four increased by three is greater than or equal to eight. The product of five and the difference of two times a number and four increased by three is at least eight.

PRACTICE EXERCISE 9

Translate from mathematical sentences to English sentences:

a. $8x < 4$ **b.** $15 \geq 3t + 6$ **c.** $2(x - 4) > 3$

d. $4(3x + 2) - 5 \leq 7$

Example 10   **Translate from English to mathematics and solve. There is only one mathematical translation for each English sentence. Use x as the variable.**

English	Mathematics
a. Six times a number is greater than eighteen.	$6x > 18$ $x > 3$
b. Five more than negative three times a number is less than seventeen.	$-3x + 5 < 17$ $-3x < 12$ $x > -4$
c. Seven times the difference of a number and four is at most thirty-five.	$7(x - 4) \leq 35$ $7x - 28 \leq 35$ $7x \leq 63$ $x \leq 9$
d. Three times the sum of a number and four, increased by five, is at least eleven.	$3(x + 4) + 5 \geq 11$ $3x + 12 + 5 \geq 11$ $3x + 17 \geq 11$ $3x \geq -6$ $x \geq -2$

Answers: Practice Exercise 9: *(There may be more than one English sentence for each exercise.)*
a. Eight times a number is less than four. **b.** Fifteen is greater than or equal to six more than three times a number. **c.** Two times the difference of some number and four is greater than three. **d.** The product of four and the sum of three times some number and two, decreased by five, is at most seven.

PRACTICE EXERCISE 10

Translate from English to mathematical sentences and solve. Use x as the variable.

a. The product of negative two and a number is less than eight.

b. The difference of twelve and four times a number is at most twenty.

c. Four times the difference of a number and six is at least twelve.

d. Five less than two times the sum of a number and three is at most nine.

If you need more practice, do the following Additional Practice Exercises.

Additional Practice Exercise 10   Translate from English to mathematical sentences and solve. Use x as the variable.

a. The quotient of some number and three is greater than five.

b. Six less than negative two times a number is less than eight.

c. Five times the sum of a number and two is at most thirty.

d. Seven more than three times the difference of a number and two is at least thirteen.

Exercise Set 3.8   For Extra Help   MyMathLab®

Graph each of the following. (See Examples 1 and 2.)

1. $x > -2$

$$-6 -5 -4 -3 -2 -1 \ 0 \ 1 \ 2 \ 3 \ 4 \ 5 \ 6$$

2. $x < -3$

$$-6 -5 -4 -3 -2 -1 \ 0 \ 1 \ 2 \ 3 \ 4 \ 5 \ 6$$

3. $x > 4$

$$-6 -5 -4 -3 -2 -1 \ 0 \ 1 \ 2 \ 3 \ 4 \ 5 \ 6$$

4. $x > -6$

$$-6 -5 -4 -3 -2 -1 \ 0 \ 1 \ 2 \ 3 \ 4 \ 5 \ 6$$

5. $x \geq 3$

$$-6 -5 -4 -3 -2 -1 \ 0 \ 1 \ 2 \ 3 \ 4 \ 5 \ 6$$

6. $-2 \leq x$

$$-6 -5 -4 -3 -2 -1 \ 0 \ 1 \ 2 \ 3 \ 4 \ 5 \ 6$$

7. $1 < x < 6$

$$-6 -5 -4 -3 -2 -1 \ 0 \ 1 \ 2 \ 3 \ 4 \ 5 \ 6$$

8. $-3 < x < 2$

$$-6 -5 -4 -3 -2 -1 \ 0 \ 1 \ 2 \ 3 \ 4 \ 5 \ 6$$

9. $-5 < x < 3$

$$-6 -5 -4 -3 -2 -1 \ 0 \ 1 \ 2 \ 3 \ 4 \ 5 \ 6$$

10. $-6 < x < -1$

$$-6 -5 -4 -3 -2 -1 \ 0 \ 1 \ 2 \ 3 \ 4 \ 5 \ 6$$

11. $-4 \leq x < -3$

$$-6 -5 -4 -3 -2 -1 \ 0 \ 1 \ 2 \ 3 \ 4 \ 5 \ 6$$

12. $2 < x \leq 6$

$$-6 -5 -4 -3 -2 -1 \ 0 \ 1 \ 2 \ 3 \ 4 \ 5 \ 6$$

Answers: Practice Exercise 10: a. $-2x < 8, x > -4$ **b.** $12 - 4x \leq 20, x \geq -2$ **c.** $4(x - 6) \geq 12, x \geq 9$ **d.** $2(x + 3) - 5 \leq 9, x \leq 4$

Additional Practice 10: a. $\frac{x}{3} > 5, x > 15$ **b.** $-2x - 6 < 8, x > -7$ **c.** $5(x + 2) \leq 30, x \leq 4$ **d.** $3(x - 2) + 7 \geq 13, x \geq 4$

Solve by using the addition property of inequality and graph the solutions. (See Example 4.)

13. $q + 6 < 4$

$$\longleftrightarrow\!\!+\!+\!+\!+\!+\!+\!+\!+\!+\!+\!+\!+\!+\!\longrightarrow$$
$$-6\ -5\ -4\ -3\ -2\ -1\ \ 0\ \ 1\ \ 2\ \ 3\ \ 4\ \ 5\ \ 6$$

14. $v - 8 < 5$

$$\longleftrightarrow\!\!+\!+\!+\!+\!+\!+\!+\!+\!+\!+\!+\!+\!+\!\longrightarrow$$
$$4\ \ 5\ \ 6\ \ 7\ \ 8\ \ 9\ \ 10\ 11\ 12\ 13\ 14\ 15\ 16$$

15. $6p - 5 - 5p \le -3$

$$\longleftrightarrow\!\!+\!+\!+\!+\!+\!+\!+\!+\!+\!+\!+\!+\!+\!\longrightarrow$$
$$-6\ -5\ -4\ -3\ -2\ -1\ \ 0\ \ 1\ \ 2\ \ 3\ \ 4\ \ 5\ \ 6$$

16. $4w - 2 - 3w \ge 6$

$$\longleftrightarrow\!\!+\!+\!+\!+\!+\!+\!+\!+\!+\!+\!+\!+\!+\!\longrightarrow$$
$$4\ \ 5\ \ 6\ \ 7\ \ 8\ \ 9\ \ 10\ 11\ 12\ 13\ 14\ 15\ 16$$

17. $3p - 6 - 2p < -8$

$$\longleftrightarrow\!\!+\!+\!+\!+\!+\!+\!+\!+\!+\!+\!+\!+\!+\!\longrightarrow$$
$$-6\ -5\ -4\ -3\ -2\ -1\ \ 0\ \ 1\ \ 2\ \ 3\ \ 4\ \ 5\ \ 6$$

18. $5q + 30 - 4q \le 14$

$$\longleftrightarrow\!\!+\!+\!+\!+\!+\!+\!+\!+\!+\!+\!+\!+\!+\!\longrightarrow$$
$$-21\ -20\ -19\ -18\ -17\ -16\ -15\ -14\ -13\ -12\ -11\ -10\ -9$$

19. $3(s + 2) - 2s \ge 7$

$$\longleftrightarrow\!\!+\!+\!+\!+\!+\!+\!+\!+\!+\!+\!+\!+\!+\!\longrightarrow$$
$$-6\ -5\ -4\ -3\ -2\ -1\ \ 0\ \ 1\ \ 2\ \ 3\ \ 4\ \ 5\ \ 6$$

20. $4(s - 2) - 3s \le -3$

$$\longleftrightarrow\!\!+\!+\!+\!+\!+\!+\!+\!+\!+\!+\!+\!+\!+\!\longrightarrow$$
$$-6\ -5\ -4\ -3\ -2\ -1\ \ 0\ \ 1\ \ 2\ \ 3\ \ 4\ \ 5\ \ 6$$

21. $4(2x - 5) - 7x + 16 > -2$

$$\longleftrightarrow\!\!+\!+\!+\!+\!+\!+\!+\!+\!+\!+\!+\!+\!+\!\longrightarrow$$
$$-6\ -5\ -4\ -3\ -2\ -1\ \ 0\ \ 1\ \ 2\ \ 3\ \ 4\ \ 5\ \ 6$$

22. $2(3y - 5) - 5y + 6 < 2$

$$\longleftrightarrow\!\!+\!+\!+\!+\!+\!+\!+\!+\!+\!+\!+\!+\!+\!\longrightarrow$$
$$-6\ -5\ -4\ -3\ -2\ -1\ \ 0\ \ 1\ \ 2\ \ 3\ \ 4\ \ 5\ \ 6$$

23. $8(2u - 3) - 5(3u + 4) > -3$

$$\longleftrightarrow\!\!+\!+\!+\!+\!+\!+\!+\!+\!+\!+\!+\!+\!+\!\longrightarrow$$
$$34\ 35\ 36\ 37\ 38\ 39\ 40\ 41\ 42\ 43\ 44\ 45\ 46$$

24. $7(3x - 6) - 4(5x - 6) > -10$

$$\longleftrightarrow\!\!+\!+\!+\!+\!+\!+\!+\!+\!+\!+\!+\!+\!+\!\longrightarrow$$
$$4\ \ 5\ \ 6\ \ 7\ \ 8\ \ 9\ \ 10\ 11\ 12\ 13\ 14\ 15\ 16$$

25. $-3(9t - 3) + 4(7t - 2) \le 3 - 5$

$$\longleftrightarrow\!\!+\!+\!+\!+\!+\!+\!+\!+\!+\!+\!+\!+\!+\!\longrightarrow$$
$$-6\ -5\ -4\ -3\ -2\ -1\ \ 0\ \ 1\ \ 2\ \ 3\ \ 4\ \ 5\ \ 6$$

26. $-4(8t + 5) + 3(11t + 5) \le 4 - 7$

$$\longleftrightarrow\!\!+\!+\!+\!+\!+\!+\!+\!+\!+\!+\!+\!+\!+\!\longrightarrow$$
$$-6\ -5\ -4\ -3\ -2\ -1\ \ 0\ \ 1\ \ 2\ \ 3\ \ 4\ \ 5\ \ 6$$

Solve by using the multiplication property of inequality. (See Example 6.)

27. $3y < 15$

28. $4t \ge 32$

29. $-14u > 42$

30. $-16w \le 64$

31. $-2a < -14$

32. $-6n \ge -54$

33. $\dfrac{x}{-3} < 5$

34. $\dfrac{u}{-2} \ge 4$

35. $\dfrac{r}{-1} > -2$

36. $\dfrac{b}{-6} < -1$

Solve by using the addition and multiplication properties of inequality. (See Examples 7 and 8.)

37. $2q - 16 > 8$

38. $3v - 12 < 6$

39. $-4a + 5 \le -7$

40. $-5r + 2 \ge -8$

41. $8p - 15 \ge 5p$

42. $5v + 6 \le 2v$

43. $7w + 6 > 4w - 18$

44. $11z - 6 > 4z + 8$

45. $a + .8 \ge 3a - 1.6$

46. $4b - 2.4 \ge 9b + .6$

47. $1.2 - 3t > -11t + 3.6$

48. $6.8 - 7r \ge 3r - .2$

49. $10s - 2 - 12s < 8s + 2 - 6s$

50. $7z + 16 - 4z > 10 - z + 6$

51. $12p - 9p + 12 \ge 7p - 15 + 5p$

52. $15q + 9 - 6q > 4q + 27 + 11q$

53. $7(w - 1) \ge -14$

54. $9(y - 4) \le -9$

55. $5(2m - 3) > 6m + 21$

56. $4(3n + 6) < 9n - 3$

57. $x + 6(x - 3) \le -25$

58. $2(4w - 5) - 3w < 15$

59. $7u - 10(2u + 3) \ge 22$

60. $6t - 12(2t - 5) > 6$

61. $2(1 - 4u) < -3u + 7$

62. $6(5 - 3v) \ge 9v - 24$

63. $7(n - 2) > 3(n + 4) - 26$

64. $5(m + 4) < 28 - 4(m + 2)$

65. $11z - 2(10 - 4z) \le 8(3z - 2) + 11$

66. $10s - 3(4 - 5s) > s + 6(4 + s)$

67. $3(r + .9) + .2r > 3r + 1.7$

68. $.7(t - 3) + .8 < 1.5 + .3t$

69. $-2 < x + 6 < 4$

70. $-3 < y + 7 < 3$

71. $-6 \le 2y \le 8$

72. $-4 \le 4y \le 12$

73. $-11 \le 4x - 3 < 9$

74. $-3 < 5y - 8 \le 12$

75. $3 < -2y + 5 \le 7$

76. $7 \le -3y - 2 < 16$

77. $-3.5 < 2t - 1.5 < 8.5$

78. $-1 \le 10s - 3 \le 5$

Challenge Exercises (79 and 80)

79. $6(3y - 5) + 4(2y - 7) \ge 8y - 13 + 9y$

80. $-5(3 - 2x) + 9(3x - 6) < 7(4x - 1) + 1$

Translate the following mathematical sentences into English sentences in at least two ways. (See Example 9.)

81. $5x \ge 10$

82. $9 > 2y$

83. $3s - 5 > 2$

84. $15 + 4z < 12$

85. $28 \le 4(t - 2)$

86. $31 \ge 9(2 - u)$

Translate the following English sentences into mathematical sentences and solve. Use x as the variable. (See Example 10.)

87. Six times some number is less than eighteen.

88. Negative three times a number is greater than thirty-six.

89. Nineteen decreased by four times a number is greater than eighteen.

90. Eight less than five times a number is at most twelve.

91. Three times the sum of twice a number and seven is at least negative three.

92. Sixteen is greater than or equal to twice the difference of a number and five.

93. Eighteen is less than four times a number decreased by six.

94. Twenty-two is less than or equal to five times a number plus seven.

95. Four times the difference of a number and three is at least eight.

96. Three times the sum of a number and two is at most nine.

97. Three more than two times the difference of a number and one is at most five.

98. Three less than four times the sum of a number and one is at most thirteen.

Writing Exercises (99–101)

99. List at least five situations where we use inequalities, but do not specifically state the inequality. For example, "You must be 21 to vote" means that your age must be greater than or equal to 21 years.

100. Compare solving a linear equation with solving a linear inequality. How are they similar? How do they differ?

101. How does the solution(s) of a linear equation compare with the solution(s) of a linear inequality?

Critical-Thinking Exercise

102. Recall that for any two numbers x and a, either $x < a$, $x = a$, or $x > a$. With these in mind, answer the following questions:

    a. What does "not greater than" mean?

    b. What does "not equal to" mean?

    c. What does "not less than or equal to" mean?

    d. What does "at least" mean?

Chapter 3   Summary

Concept/Procedure	Example

Definition of a Linear Equation: [Section 3.1]

- A linear equation is any equation that can be put in the form $ax + b = c$, where a, b, and c are constants and $a \neq 0$.

Addition Property of Equality: [Section 3.1]

- If $a = b$, then $a + c = b + c$ for any number c. In words, we may add any number to both sides of an equation to get an equation with the same solution(s).

Example 1:

Solve $5x - 2(2x - 3) + 4 = 12$

Solution:

$$5x - 2(2x - 3) + 4 = 12 \qquad \text{Apply the distributive property.}$$
$$5x - 4x + 6 + 4 = 12 \qquad \text{Simplify the left side.}$$
$$x + 10 = 12 \qquad \text{Subtract 10 from both sides.}$$
$$x + 10 - 10 = 12 - 10 \qquad \text{Simplify both sides.}$$
$$x = 2 \qquad \text{Solution.}$$

The check is left to the student.

Multiplication Property of Equality: [Section 3.2]

- If $a = b$ and $c \neq 0$, then $a \cdot c = b \cdot c$. In words, we may multiply both sides of an equation by a nonzero number to get another equation with the same solution(s).

Example 2:

Solve $\dfrac{3}{5}x = -12$

Solution:

$$\frac{3}{5}x = -12 \qquad \text{Multiply both sides by } \frac{5}{3}.$$
$$\frac{5}{3} \cdot \frac{3}{5}x = \frac{5}{3}(-12) \qquad \text{Simplify both sides.}$$
$$x = -\frac{60}{3} \qquad \text{Simplify the right side.}$$
$$x = -20 \qquad \text{Solution.}$$

Procedure for Solving Linear Equations: [Section 3.3]

1. If necessary, simplify both sides of the equation as much as possible. This could involve using the distributive, commutative, and associative properties and adding like terms.
2. If necessary, use the addition property of equality to get all the terms with variables on one side of the equation and all the constant terms on the other.
3. If necessary, use the multiplication property of equality to eliminate any coefficient on the variable.
4. Check the answer in the original equation.

Example 3:

Solve $5(3a - 2) = 5(a - 4) - 20$

Solution:

$$5(3a - 2) = 5(a - 4) - 20 \qquad \text{Apply the distributive property.}$$
$$15a - 10 = 5a - 20 - 20 \qquad \text{Simplify the right side.}$$
$$15a - 10 = 5a - 40 \qquad \text{Subtract } 5a \text{ from both sides.}$$
$$10a - 10 = -40 \qquad \text{Add 10 to both sides.}$$
$$10a = -30 \qquad \text{Divide both sides by 10.}$$
$$a = -3 \qquad \text{Solution.}$$

Concept/Procedure	Example

Identities and Contradictions: [Section 3.3]

- An **identity** is an equation that is true for all values of the variable for which the equation is defined. When solving an identity, we arrive at a statement that is obviously true, such as $10 = 10$. The solutions of an identity are "all real numbers."
- A **contradiction** is an equation that has no solution. When solving a contradiction, we arrive at a statement that is obviously false, such as $5 = 10$. The solution set is indicated by $\varnothing$.

Example 4:

Determine whether the following is an identity or a contradiction:

$$3(x - 7) - 5 = 4(x - 3) - (x + 3) \quad \text{Apply the distributive property.}$$
$$3x - 21 - 5 = 4x - 12 - x - 3 \quad \text{Simplify both sides.}$$
$$3x - 26 = 3x - 15 \quad \text{Subtract } 3x \text{ from both sides.}$$
$$-26 = -15 \quad \text{False. Therefore, a contradiction.}$$

Problem-Solving Procedures: [Section 3.5]

Problem-Solving Procedure:

1. **Understand** the problem.
   a. Read and reread the problem carefully, paying special attention to key words. It is a good idea to underline key words in the problem.
   b. Identify what is given and what is to be found.
   c. Determine the number of unknowns and how the known and unknown quantities are related to each other.
   d. When possible, make a drawing and label it with known and unknown information.
2. **Plan.** Determine the procedure you will use to solve the problem.
3. **Translate** the problem into mathematical language, which is usually an equation.
   a. Construct any necessary tables, charts, etc.
   b. Choose a variable to represent the unknown, and if there are two or more unknowns, represent any other unknowns in terms of this variable, using the relationship(s) found in part 1c.
   c. Write the equation in words, using the relationship(s) found in part 1c.
   d. Using the word equation in step c, write the algebraic equation.
4. **Solve** the algebraic equation.
5. **State** the solution(s) in words, using a complete English sentence(s).
6. **Check** the solution(s), using the wording of the original problem.

Example 8:

The largest ocean in the world is the Pacific Ocean, and the smallest is the Arctic Ocean. If the average depth of the Pacific Ocean is 213 meters less than four times the average depth of the Arctic Ocean and their combined average depths are 4977 meters, find the average depth of each.

Solution:

Understand:

We are given the combined average depths of the Pacific and Arctic oceans and how the average depth of the Pacific Ocean compares with the average depth of the Arctic Ocean. We are to find the average depth of each ocean.

> Relationship 1: The average depth of the Pacific Ocean is 213 meters less than four times the average depth of the Arctic Ocean.
>
> Relationship 2: The combined average depths of the two oceans are 4977 meters.

Plan: We will use relationship 1 to represent the average depths of the Pacific and Arctic oceans in terms of variables and relationship 2 to write the equation.

Translate: Using relationship 1, let x represent the average depth of the Arctic Ocean; then $4x - 213$ represents the average depth of the Pacific Ocean.

Using relationship 2, we know that

$$\text{Depth of Pacific Ocean} + \text{depth of Arctic Ocean} = 4977 \; .$$

$$(4x - 213) \qquad + \qquad x \qquad = 4977$$

Solve:

$$(4x - 213) + x = 4977 \quad \text{Remove parentheses.}$$
$$4x - 213 + x = 4977 \quad \text{Add like terms.}$$
$$5x - 213 = 4977 \quad \text{Add 213 to both sides.}$$
$$5x = 5190 \quad \text{Divide both sides by 5.}$$
$$x = 1038 \quad \text{Solution.}$$

Concept/Procedure	Example

State: Since x represents the average depth of the Arctic Ocean, we know that the average depth of the Arctic Ocean is 1038 meters. Since $4x - 213$ represents the average depth of the Pacific Ocean, the average depth is $4(1038) - 213 = 4152 - 213 = 3939$ meters.

Check: Is the average depth of the Pacific Ocean 213 m less than 4 times the average depth of the Arctic Ocean? Yes, since $3939 = 4(1038) - 213$. Is the combined depth of the two oceans 4977 m? Yes, since $3939 + 1038 = 4977$.

Distance, Rate, Time Problems: [Section 3.5]

Problem-Solving Procedure:

1. **Understand** the problem.
   a. Read and reread the problem carefully, paying special attention to key words. It is a good idea to underline key words in the problem.
   b. Identify what is given and what is to be found.
   c. Determine the number of unknowns and how the known and unknown quantities are related to each other.
   d. When possible, make a drawing and label it with known and unknown information.
2. **Plan.** Determine the procedure you will use to solve the problem.
3. **Translate** the problem into mathematical language, which is usually an equation.
   a. Construct any necessary tables, charts, etc.
   b. Choose a variable to represent the unknown, and if there are two or more unknowns, represent any other unknowns in terms of this variable, using the relationship(s) found in part 1c.
   c. Write the equation in words, using the relationship(s) found in part 1c.
   d. Using the word equation in step c, write the algebraic equation.
4. **Solve** the algebraic equation.
5. **State** the solution(s) in words, using a complete English sentence(s).
6. **Check** the solution(s), using the wording of the original problem.

Example 9:

The distance between Houston and Atlanta is approximately 800 miles. At 7:00 A.M. a car leaves Houston headed for Atlanta at an average rate of 70 miles per hour, and four hours later an RV leaves Atlanta headed for Houston at an average rate of 60 miles per hour. At what time will they meet?

Solution:

Understand:

Draw a picture of the situation.

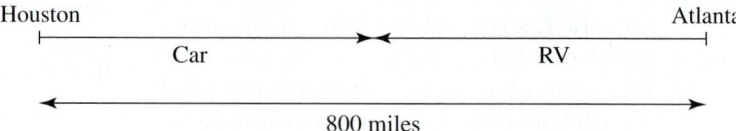

We are given the average rate of the car and of the RV, the fact that the RV travels 4 hours less than the car, and the distance between Houston and Atlanta. We are to find the time when they will meet.

Relationship 1: The RV travels for 4 hours less than the car.
Relationship 2: When the two meet, the total distance traveled by the car and the RV is 800 miles.

Plan: Use a chart. Use relationship 1 to represent the times of the car and RV in terms of a variable and use relationship 2 to write an equation. In filling in the chart, use the fact that $rt = d$.

Translate:

Category	Rate (r)	Time (t)	Distance ($rt = d$)
Car	70	x	$70x$
RV	60	$x - 4$	$60(x - 4)$

Note that if we let x represent the time the car travels until they meet, then the time for the RV is $x - 4$, since it left 4 hours later.

Concept/Procedure	Example

Example

From the foregoing diagram, we know that

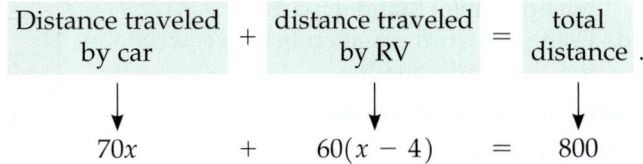

$$70x \quad + \quad 60(x-4) \quad = \quad 800$$

Solve:

$70x + 60(x-4) = 800$	Distribute 60.
$70x + 60x - 240 = 800$	Add like terms.
$130x - 240 = 800$	Add 240 to both sides.
$130x = 1040$	Divide by 130.
$x = 8$	Solution.

State: Since x represents the time the car will be traveling, the car and the RV will meet 8 hours after the car leaves. Since the car leaves at 7:00 A.M., they will meet 8 hours past 7:00, which is 3:00 P.M.

Check: In 8 hours the car will travel $8(70) = 560$ miles. Since the RV left 4 hours later, it will travel for $8 - 4 = 4$ hours. In 4 hours the RV will travel $4(60) = 240$ miles. Together they traveled $560 + 240 = 800$ miles, which is the distance between Houston and Atlanta.

Money, Investment, and Mixture Problems: [Section 3.6]

Problem-Solving Procedure:

1. **Understand** the problem.
   a. Read and reread the problem carefully, paying special attention to key words. It is a good idea to underline key words in the problem.
   b. Identify what is given and what is to be found.
   c. Determine the number of unknowns and how the known and unknown quantities are related to each other.
   d. When possible, make a drawing and label it with known and unknown information.

2. **Plan**. Determine the procedure you will use to solve the problem.

3. **Translate** the problem into mathematical language, which is usually an equation.
   a. Construct any necessary tables, charts, etc.
   b. Choose a variable to represent the unknown, and if there are two or more unknowns, represent any other unknowns in terms of this variable, using the relationship(s) found in part 1c.

Example 10:

An author of mystery books received a royalty check for $350,000 and decides to keep $50,000 and invest the remainder. He invests some in certificates of deposit (CDs) that have an annual yield of 3% and the remainder in mutual funds which have an anticipated return of 4.5%. If the total returns from the two investments are $11,625 per year, find the amount in each investment.

Solution:

Understand:

We are given the interest rate of each account, the total amount invested, and the total interest earned. We are to find the amount in each account.

There are two unknowns, which are the amount invested in CDs and the amount invested in mutual funds.

Relationship 1: There is a total of $300,000 invested ($350,000 - 50,000 = 300,000$).
Relationship 2: The total interest earned from the two investments is $11,625.

Plan: Use a chart. Use relationship 1 to represent the amounts in each investment in terms of a variable and relationship 2 to write the equation. In filling in the chart, use the fact that $rp = i$, where r is the interest rate, p is the principal, and i is the interest earned.

Concept/Procedure	Example

Concept/Procedure

   c. Write the equation in words, using the relationship(s) found in part 1c.

   d. Using the word equation in step c, write the algebraic equation.

4. Solve the algebraic equation.

5. State the solution(s) in words, using a complete English sentence(s).

6. Check the solution(s), using the wording of the original problem.

- For coin problems, change the headings to Type of Coin, Value of Coin, Number of Coins, and Total Value.

- For investment problems, change the headings to Type of Investment, Interest Rate (expressed as a decimal), Amount Invested, and Interest Earned.

- For mixture problems, change the headings to Type of Ingredient, Price per Unit, Number of Units, and Total Value. Also, add an extra row for the mixture.

Example

Translate:

Type of Investment	Interest Rate	Amount Invested	Interest Earned
Mutual Funds	.045	x	$.045x$
CDs	.03	$300,000 - x$	$.03(300,000 - x)$

Note that, according to relationship 1, if we let x represent the amount invested in mutual funds, and since there is a total of \$300,000 invested, the amount invested in CDs is $300,000 - x$.

From the chart and relationship 2, we know that

$$\begin{array}{ccc} \text{Interest earned from} & \text{interest earned} & \text{total} \\ \text{mutual funds} & + \quad \text{from CDs} & = \quad \text{interest} \end{array}.$$

$$.045x \qquad + .03(300,000 - x) = \quad 11,625$$

Solve:

$.045x + .03(300,000 - x) = 11,625$	Distribute .03.
$.045x + 9000 - .03x = 11,625$	Add like terms.
$.015x + 9000 = 11,625$	Subtract 9000 from both sides.
$.015x = 2625$	Divide by .015.
$x = 175,000$	Solution.

State: Since x represents the amount invested in mutual funds, \$175,000 is invested in mutual funds. Since $300,000 - x$ represents the amount invested in CDs, there is $300,000 - 175,000 = 125,000$ invested in CDs.

Check: The interest earned from the mutual funds is $.045(175,000) = \$7875$, and the interest earned from the CDs is $.03(125,000) = \$3750$. The total interest earned from the two investments is $7875 + 3750 = \$11,625$, which is what it is supposed to be.

Geometric Application Problems: [Section 3.7]

- Know the formulas for the perimeters, areas, and volumes of common geometric figures.

- The sum of all three angles of any triangle is $180°$.

- Two angles are *supplementary* if the sum of their measures is $180°$.

- Two angles are *complementary* if the sum of their measures is $90°$.

- If two lines are parallel, then (a) pairs of alternate interior angles are congruent, (b) pairs of alternate exterior angles are congruent, (c) pairs of corresponding angles are congruent, and (d) interior angles on the same side of the transversal are supplementary.

Example 10:

Recently a gold-colored rectangular tray was on sale on eBay. The length of the tray was 6 inches less than four times the width, and the perimeter was 48 inches. Find the dimensions of the tray.

Solution:

Understand:

We are given the relationship between the length and width, the perimeter, and that the tray is rectangular.

There are two unknowns, which are the length and the width.

    Relationship 1: The length is six less than four times the width.

    Relationship 2: The perimeter is 48 inches.

Concept/Procedure	Example
	Plan: Use relationship 1 to represent the length and width in terms of variables and use relationship 2 and the formula $P = 2L + 2W$ to write the equation. **Translate:** From relationship 1, if we let x represent the width, then $4x - 6$ represents the length. Using relationship 2 and $P = 2L + 2W$, we have the following: $P = 2L + 2W$ Substitute for p, L, and W. $48 = 2(4x - 6) + 2x$ **Solve:** $48 = 2(4x - 6) + 2x$ Distribute 2. $48 = 8x - 12 + 2x$ Add like terms. $48 = 10x - 12$ Add 12 to both sides. $60 = 10x$ Divide by 10. $6 = x$ Solution. **State:** Since x represents the width, the width is 6 inches. Since $4x - 6$ represents the length, the length is $4(6) - 6 = 24 - 6 = 18$ inches. **Check:** Is the length 6 inches less than four times the width? Yes, since $18 = 4(6) - 6$. Is the perimeter 48 inches? Yes, since $2(18) + 2(6) = 36 + 12 = 48$ inches.

Procedure for Graphing Inequalities: [Section 3.8]
 	
• Equalities are graphed by placing a dot on the number line at the appropriate point.
 	
• Inequalities are graphed by
  a. placing an open dot on the number line at the appropriate place and shading to the left for less than ($<$)or shading to the right for greater than ($>$); or
  b. placing a solid dot on the number line at the appropriate place and shading to the left for less than or equal to ($\leq$) or shading to the right for greater than or equal to ($\geq$).
	Example 5:
	
	Graph $x \geq -3$.
	
	Solution: Since the order symbol is $\geq$, put a dot at -3 and shade everything to the right.
	
	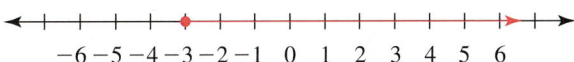

Addition Properties of Inequalities: [Section 3.8] **a.** If $a < b$, then $a + c < b + c$ for any c. **b.** If $a > b$, then $a + c > b + c$ for any c. • If we add any number to both sides of an inequality, we get an inequality with the same solutions.	Example 6: Solve $4x - 2(3x + 5) + 3x + 4 > 0$ Solution: $4x - 2(3x + 5) + 3x + 4 > 0$ Apply the distributive property. $4x - 6x - 10 + 3x + 4 > 0$ Simplify the left side. $x - 6 > 0$ Add 6 to both sides. $x - 6 + 6 > 0 + 6$ Simplify both sides. $x > 6$ Solution.

Concept/Procedure	Example
Multiplication Properties of Inequalities: **[Section 3.8]** **a.** If $a < b$ and $c > 0$, then $a \cdot c < b \cdot c$. **b.** If $a > b$ and $c > 0$, then $a \cdot c > b \cdot c$. • If we multiply both sides of the inequality by a *positive number*, we get another inequality with the same solutions. **c.** If $a < b$ and $c < 0$, then $a \cdot c > b \cdot c$. **d.** If $a > b$ and $c < 0$, then $a \cdot c < b \cdot c$. • If we multiply both sides of an inequality by a *negative number*, we must reverse the direction of the inequality symbol.	Example 7: Solve $2(3 - 4y) + 8 < -3y + 29$ Solution:

$$
\begin{array}{ll}
2(3 - 4y) + 8 < -3y + 29 & \text{Apply the distributive property.} \\
6 - 8y + 8 < -3y + 29 & \text{Simplify the left side.} \\
-8y + 14 < -3y + 29 & \text{Add } 3y \text{ to both sides.} \\
-5y + 14 < 29 & \text{Subtract 14 from both sides.} \\
-5y < 15 & \text{Divide both sides by } -5. \\
y > -3 & \text{Note that the order symbol is reversed} \\
& \text{when dividing by a negative number.}
\end{array}
$$

Chapter 3 Review Exercises

Solve, using the addition property of equality. [Section 3.1]

1. $x - 6 = -2$ **2.** $x + 3 = 4$

3. $10 + z = 7$ **4.** $8 + a - 12 = 7 - 3$

5. $5z - 28 - 4z = 12$ **6.** $4.7x + 7.2 - 3.7x = 4.6$

7. $4u + 13 = 3u + 10$ **8.** $11 + 15b = 14b + 18$

9. $5(2a - 3) - 9a = -8$

10. $-8 = 3(5z - 3) - 2(7z + 2)$

Translate each mathematical sentence into an English sentence in two ways. [Section 3.1]

11. $2x - 5 = 14$ **12.** $5 + x = -2$

Translate the English sentence into a mathematical sentence and solve. Use x as the variable. [Section 3.1]

13. Twelve more than some number is negative two.

14. A number decreased by four is six.

Solve the following: [Section 3.1]

15. John and Shelley share the cost of a TV. If the TV costs \$360 and John pays \$220, how much does Shelley pay?

16. Harold bought a suit for \$230 after a discount of \$50. What was the price of the suit before the discount?

Find the reciprocal of each. [Section 3.2]

17. $\dfrac{-2}{7}$ **18.** 4

Solve, using the multiplication property of equality. [Section 3.2]

19. $5x = -40$ **20.** $35 = -7y$

21. $\dfrac{3}{4}t = -9$ **22.** $\dfrac{x}{3} = -4$

Translate each mathematical sentence into an English sentence. [Section 3.2]

23. $14y = -28$ **24.** $\dfrac{2}{3}x = -6$

Translate each English sentence into a mathematical sentence and solve. Use x as the variable. [Section 3.2]

25. The product of four and some number is negative twelve.

26. Negative two-thirds of a number is six.

Solve the following: [Section 3.2]

27. A TV costs three times as much as a VCR. If the TV costs $468, how much does the VCR cost?

28. The area of a parallelogram is 84 square meters and the base is 14 meters. What is the height? ($A = bh$)

Solve the equations. [Section 3.3]

29. $5w - 13 = 7w + 15$ 30. $11 - 4z = 7 + 44$

Solve the following.

31. The cost of a refrigerator is three times the cost of a washing machine. If the refrigerator costs $987, find the cost of the washing machine.

32. Hans is renting a sofa for $12.75 per month. If a new sofa costs $306, how many months will it take for the rental fees to equal the cost of a new sofa?

33. A high school boasts that three-fifths of its teachers have a master's degree. If 45 teachers have a master's degree, how many teachers does the school have?

34. The value of a piece of property increased by 2% last year, which represented an increase of $3600. What was the value of the property before the 2% increase in value?

35. The perimeter of a rectangle is 42 centimeters and the length is 12 centimeters. What is the width?

36. The surface area of a rectangular solid is 236 square inches, the length is 6 inches, and the width 5 inches. What is the height?

Solve the equations. [Section 3.3]

37. $5t - 11 - 9t + 6 = 19 + 8t$

38. $4 - 5u + 12 + 16u = -6u - 18$

39. $33 - 2(7r + 9) = -9 - 6r$

40. $20 + 24s = 8 - 3(5 - 9s)$

41. $.6(8 - 5v) + .4 = v + 1.2$

42. $3.6 - 4(.7p - 15) = 2.2p - 1.4$

43. $5(q - 3) - 7(11 - 2q) = 3$

44. $8(10 - 4a) + 15 = -3(a + 7)$

Determine whether the equations are identities or contradictions. Give the solution(s) of each. [Section 3.3]

45. $2(x - 3) + 4x = 6(x - 2)$

46. $3(2x - 1) - 4x = 2(x - 4) + 5$

Translate each mathematical sentence into an English sentence. [Section 3.3]

47. $3x - 2 = 5$

48. $4(2x - 9) = 22$

Translate each English sentence into a mathematical sentence and solve. Use x as the variable. [Section 3.3]

49. The sum of four times a number and three is eleven.

50. Fifty-four is equal to six times the difference of three times a number and nine.

Solve each of the exercises. [Section 3.3]

51. The charges for repairing a transmission were $255 including parts. If the mechanic worked for 4 hours and the parts cost $75, how much does the mechanic charge per hour?

52. Rachel paid $37.10 for a blouse, including 6% for taxes. What was the cost of the blouse?

Solve the formula for the indicated variable with the given information. [Section 3.4]

53. $C = 2\pi r$, for C if $r = 3$. Use $\pi \approx 3.14$.

Solve the formula for the indicated variable. [Section 3.4]

54. $ax + b = c$, for x.

Solve each exercise. [Section 3.5]

55. In a large lecture class, there were 53 more men than women. If there were 329 students in the class, how many women were in the class?

56. The perimeter of a triangle is 16 feet, and the lengths of two of the sides are 5 feet and 4 feet. What is the length of the third side?

57. Heart disease and cancer account for approximately 56% of the deaths by disease in the United States each year. The percent of deaths due to heart disease is 9.2% more than the percent due to cancer. Find the percent of the deaths due to each disease, to the nearest tenth of a percent. (*Source:* U.S. Department of Health.)

58. There are 7 more grams of fat in 1 pound of 80% lean ground chuck than in 1 pound of 93% lean ground turkey. If there is a total of 23 grams in 1 pound of each, find the number of grams of fat in 1 pound of lean ground chuck and 1 pound of lean ground turkey.

59. The brightness of light is measured in lumens. A 60-watt bulb produces 20 more than four times the number of lumens produced by a 25-watt bulb. Combined, they produce 1095 lumens. Find the number of lumens produced by each.

60. Eggs are sized according to the weight of a dozen eggs and are classified as jumbo, extra large, large, medium, small, and pee wee. A dozen small eggs weighs 3 ounces more than a dozen pee wee, and a dozen medium eggs weighs 6 ounces more than a dozen pee wee. Find the weight of a dozen of each type of egg if the combined weight of a dozen of each type is 54 ounces.

61. At a party, there was a cheese tray that had feta, mozzarella, and provolone cheeses. One ounce of mozzarella has twice as many grams of protein as 1 ounce of feta, and 1 ounce of provolone has 3 more grams of protein than 1 ounce of feta. If the combined number of grams of protein in 1 ounce of each is 19, find the number of grams of protein in 1 ounce of each type of cheese.

62. The sum of three consecutive integers is 33. Find the integers.

63. Three consecutive odd integers are such that the sum of four times the first and the second is three times the third. Find the integers.

Solve each exercise. [Section 3.6]

64. An office purchased 50 reams of paper for which it paid $360. There were two types of paper. The lighter-weight paper cost $6.00 per ream, and the heavier paper cost $9.00 per ream. How many reams of each type of paper were purchased?

65. At the end of each day, David puts all the nickels and dimes that he has in his pockets in a jar. At the end of the week, he has a total of 32 coins whose total value is $2.20. How many of each type of coin does he have?

66. Amy matched five of the six numbers in the Florida lotto, and after the celebration party had $6500 left, which she decided to invest. She invested part of the money in municipal bonds paying 6% per year and the remainder in certificates of deposit paying 5% per year. How much did she invest in each if her total income from the two investments is $360 per year?

67. A chemist needs 500 milliliters of a solution that is 40% sulfuric acid, but all that she can find in the lab are solutions that are 25% sulfuric acid and 50% sulfuric acid. How much of each should she mix?

68. A supermarket sells a mixture of blueberries and strawberries. How many pounds of blueberries that sell for $8.00 per pound must be mixed with 12 pounds of strawberries that sell for $5.00 per pound if the mixture sells for $6.20 per pound?

Solve the exercises. [Section 3.7]

69. In a suburban area, a residential lot is typically 25 feet longer than it is wide. If the perimeter of the lot is 350 feet, what are the dimensions of the lot?

70. In a triangle, the middle angle is 45° and the largest angle is 15° more than twice the smallest angle. Find the smallest and largest angles.

71. The measure of an angle is 16° more than the measure of its complement. Find each angle.

72. Given that L_1 is parallel to L_2 in the following figure, find x and each angle:

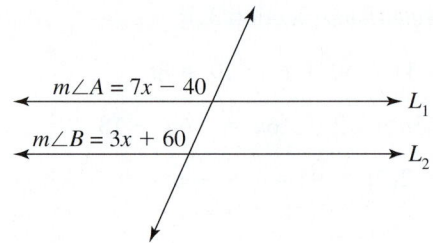

Graph each inequality on the number line. [Section 3.8]

73. $v \geq 0$

$$\xleftarrow{\hspace{0.2cm}} \underset{-6\ -5\ -4\ -3\ -2\ -1\ \ 0\ \ 1\ \ 2\ \ 3\ \ 4\ \ 5\ \ 6}{+\!\!+\!\!+\!\!+\!\!+\!\!+\!\!+\!\!+\!\!+\!\!+\!\!+\!\!+\!\!+} \xrightarrow{\hspace{0.2cm}}$$

74. $w < 4$

$$\xleftarrow{\hspace{0.2cm}} \underset{-6\ -5\ -4\ -3\ -2\ -1\ \ 0\ \ 1\ \ 2\ \ 3\ \ 4\ \ 5\ \ 6}{+\!\!+\!\!+\!\!+\!\!+\!\!+\!\!+\!\!+\!\!+\!\!+\!\!+\!\!+\!\!+} \xrightarrow{\hspace{0.2cm}}$$

75. $-1 \leq x < 5$

$$\xleftarrow{\hspace{0.2cm}} \underset{-6\ -5\ -4\ -3\ -2\ -1\ \ 0\ \ 1\ \ 2\ \ 3\ \ 4\ \ 5\ \ 6}{+\!\!+\!\!+\!\!+\!\!+\!\!+\!\!+\!\!+\!\!+\!\!+\!\!+\!\!+\!\!+} \xrightarrow{\hspace{0.2cm}}$$

76. $-3 \leq y \leq 2$

$$\xleftarrow{\hspace{0.2cm}} \underset{-6\ -5\ -4\ -3\ -2\ -1\ \ 0\ \ 1\ \ 2\ \ 3\ \ 4\ \ 5\ \ 6}{+\!\!+\!\!+\!\!+\!\!+\!\!+\!\!+\!\!+\!\!+\!\!+\!\!+\!\!+\!\!+} \xrightarrow{\hspace{0.2cm}}$$

Solve each of the inequalities by using the addition property of inequality. [Section 3.8]

77. $10p - 8 < 9p + 7$ **78.** $6 - 4q \leq -5q + 18$

79. $.6x - 4 > 12 - .4x$ **80.** $9.2 - 15y \geq -14y + 7$

81. $14z + 9 - 9z - 6z < 21$

82. $2(2n + 3) - 3(n + 3) \geq -8$

Translate the mathematical sentence into an English sentence in two ways. [Section 3.8]

83. $19 \leq t + 9$

Translate the English sentence into a mathematical sentence. [Section 3.8]

84. Eleven is greater than or equal to some number increased by six.

Solve by using the multiplication property of inequality. [Section 3.8]

85. $6t \leq -42$ **86.** $-8u < 48$

Solve by using the addition and multiplication properties of inequalities. [Section 3.8]

87. $9 + 4x > 6x + 11$

88. $33y + 6 \geq 17y + 54$

89. $4z - 32 - 5z < 6z + 13 + 2z$

90. $21 - 8r - 7 > 9r - 1 - 2r$

91. $5(2u - 2) - 6u \leq 14$

92. $7v > 8(3 - v) - 9$

93. $6(2s + 4) - 9 \geq -3(4s + 3)$

94. $3(t - 4) + 6(2t + 3) \leq 6t - 39$

95. $-3 < 2x + 5 < 7$

96. $6 \leq 3x - 9 \leq 12$

97. $-6 \leq -4x + 2 \leq 10$

98. $-5 < -3x + 4 \leq 13$

Chapter 3   Test

Solve the equations:

1. $3u + 4 = 2u - 9$

2. $30 - 5v - 19 = 8 - 6v$

3. $-7k = 42$

4. $.8m = -.54$

5. $6v - 13 = 8v + 7$

6. $5t + 7 - 9t = 4t - 33$

7. $13 - .5(y + 4) = 3.5y - 9$

8. $-3(2a - 6) + 16 = 4(a + 11) - 5a$

Solve the equations for the indicated variable:

9. $P = 2W + 2L$ for W

10. $A = \dfrac{1}{2}(B + b)h$ for B

Translate the mathematical sentences into English sentences.

11. $\dfrac{x}{3} + 4 = 6$

12. $2(s - 5) = 3s$

Translate the English sentence into a mathematical sentence.

13. Three times the difference of a number and six is one more than the number.

Solve each of the following:

14. A drywall hanger knows that the area of a wall is 128 square feet and the length of the wall is 16 feet. What is the height of the wall?

15. A trap for catching small animals is in the shape of a rectangular solid with a surface area of 1384 square inches. If the length is 26 inches and the width is 12 inches, what is the height?

16. Ericka receives $150 plus a 4% commission on all her sales. If she earned $360 last weekend, what were her sales?

17. If four pairs of shorts, each of which cost the same, total $124 and Oscar bought them using a coupon for $20 off, how much did he pay for each pair of shorts?

18. A certain wine is 12% alcohol by volume. If a bottle contains 90 milliliters of alcohol, how many milliliters of wine does the bottle contain?

19. The women's basketball team at Central College outscored their opponents by 143 points last season. If the total number of points scored by both Central and their opponents when they played each other was 4921, how many points did Central score?

20. The sum of three consecutive integers is equal to four times the second integer. Find the integers.

21. A contractor purchased a total of 25 lightbulbs, some of which were incandescent and some of which were fluorescent, at a total cost of $24. If the incandescent bulbs cost $.60 each and the fluorescent bulbs cost $1.50 each, how many of each did he buy?

22. A long-distance runner left his house running at an average rate of 8 miles per hour. Fifteen minutes (one-fourth of an hour) later, his son left his house on his bike traveling on the same route at an average rate of 12 miles per hour. How long will it take the son to catch up with his father?

23. A television is 9 centimeters longer than it is wide. If the perimeter of the tube is 142 centimeters, what are the length and width of the tube?

Solve the inequalities:

24. $12p - 3 > 1 + 11p$

25. $4.3 - 4.6q \leq 11.1 - 1.2q$

26. $5u - 10 - 3u \geq u + 7$

27. $14x - 7 + 2x > 9x + 42$

28. $10(y - 5) \leq 12y - 38$

29. $23 - 5(4z - 1) < 7(6 - 3z) + 2z$

30. $-3 < 2x + 1 < 7$

Graphing Linear Equations and Inequalities

In Chapter 3, we solved linear equations and inequalities that contained one variable. In this chapter, we will solve linear equations and inequalities that contain two variables.

In Chapter 1, we graphed whole numbers and integers on the number line. In this chapter, we will graph the solutions of equations with two variables. Graphing the solutions of an equation with two variables will require the use of something other than a number line.

In Chapter 3, we graphed the solutions of linear inequalities with one variable. In this chapter, we will graph the solutions of linear inequalities with two variables.

We end the chapter by studying some characteristics of the graphs of linear equations with two variables and by writing equations with two variables that satisfy certain given conditions.

Section 4.1 Reading Graphs and the Cartesian Coordinate System

OBJECTIVES *When you complete this section, you will be able to:*

A Read and interpret bar, pie, and line graphs.
B Verify solutions of equations with two variables.
C Write solutions of linear equations with two variables as ordered pairs.
D Determine whether a given ordered pair is a solution of a given equation.
E Find the missing member in an ordered pair for a given linear equation.
F Plot points on the rectangular coordinate system.
G Draw and interpret scatter diagrams.

PREREQUISITE SKILLS *Before beginning this section, you should be able to:*

a. Add, subtract, multiply, and divide integers. (Sections 1.6, 1.7, 2.1, 2.5)
b. Substitute values for variables and evaluate. (Section 2.5)
c. Solve linear equations. (Sections 3.1, 3.2, 3.3)

GETTING READY FOR SECTION 4.1

Add, subtract, multiply, or divide the following integers:

1. $-7 + 5$ **2.** $8 - (-3)$ **3.** $-8(7)$ **4.** $-12 \div 4$

Substitute $x = -3$ and $y = 7$ into the given expressions and evaluate.

5. $3x - 2y$ **6.** $-5x + y$

Solve the following equations:

7. $12 - 3y = 6$ **8.** $3x + 7 = 25$

Introduction

Graphs of various types can be used to convey information in a quick, concise, and accurate manner. We will concentrate on three types of graphs that frequently occur in the media—newspapers, television, magazines, etc. These are the **pie chart** (also called **circle graph**), the **bar graph**, and the **line graph.** It is important that we be able to interpret these graphs and learn what they are trying to tell us.

OBJECTIVE Reading and interpreting bar, pie, and line graphs

Pie charts, or circle graphs, are used to describe how 100% of something is divided up into parts. Consequently, the sum of all the parts must add up to 100%.

Example 1

a. The pie chart shows the percentage of service personnel in the various branches of the approximately 1,080,000 members of the U. S. armed forces worldwide in 2009. (*Source: The New York Times Almanac,* 2010)

Answers: Getting Ready: 1. -2 **2.** 11 **3.** -56 **4.** -3 **5.** -23 **6.** 22 **7.** $y = 2$ **8.** $x = 6$

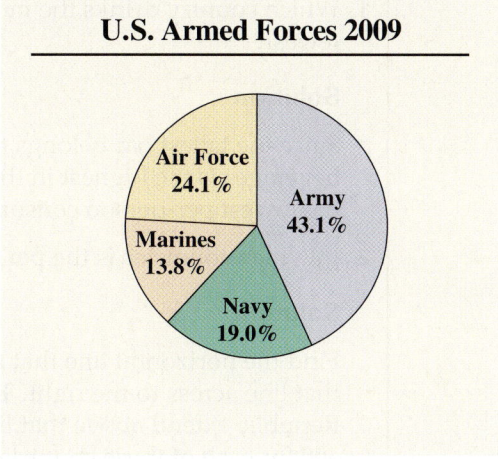

U.S. Armed Forces 2009

1. Which branch of the armed services had the largest percentage of service personnel and what is that percentage?

Solution

The largest piece of the pie belongs to the Army, so the Army has the largest percentage of service personnel with approximately 43.1%.

2. If a random sample of 1500 service personnel were chosen, approximately how many would be in the Navy?

Solution

Since 19.0% of service personnel are in the Navy, we would expect $(.19)(1500) = 285$ of the sample to be in the Navy.

3. Approximately how many service personnel are in the Marines?

Solution

Since 13.8% of service personnel are in the Marines and there are approximately 1,080,000 total, the number in the Marines is approximately $(.138)(1,080,000) = 149,040$.

A bar graph is used to show comparisons of various categories. The categories are shown along the horizontal axis, and the amount is shown along the vertical axis. Information read from a bar graph is usually approximate because of the lack of detail in the scale(s).

b. The bar graph shows the number of 12-ounce servings per person per year of carbonated beverages in 2008 for selected countries. (*Source: National Geographic, May 2010*)

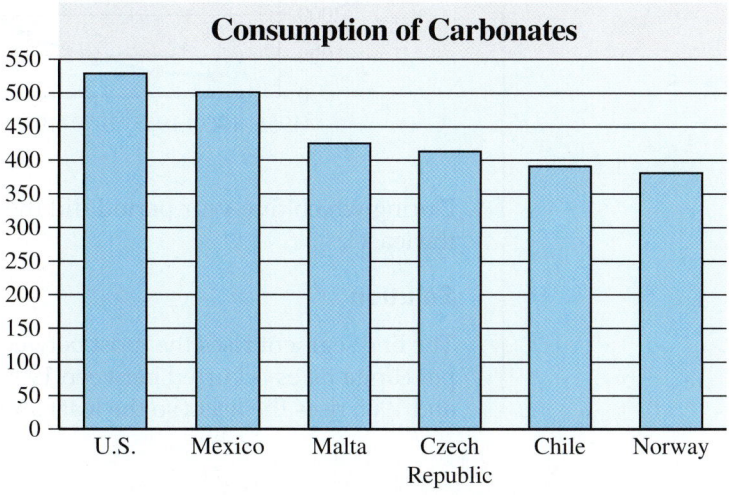

Consumption of Carbonates

1. Which country drinks the most servings of carbonated beverages per person? The fewest?

Solution

Since the tallest bar belongs to the U.S., the per-person consumption of carbonated beverages is the highest in the U.S. and since the shortest bar belongs to Norway, the lowest per-person consumption is in Norway.

2. In which countries is the per-person consumption more than 400?

Solution

Find the horizontal line that corresponds to 400 on the vertical scale and follow that line across to the right. The bars for the U.S., Mexico, Malta, and Czech Republic extend above that line, so the per-person consumption is greater than 400 for each of these countries.

3. Estimate the per-person consumption for Malta.

Solution

Locate the top of the bar for Malta and move horizontally to the left until you intersect the vertical scale and estimate the value. The value appears to be halfway between 400 and 450, so 425 is a good estimate.

4. Find the difference in the per-person consumption between Mexico and Malta and interpret the result.

Solution

The per-person consumption in Mexico is approximately 500 and Malta is 425, so the difference is $500 - 425 = 75$. This means that Mexicans consume 75 more carbonated beverages per person than people living in Malta.

c. The line graph shows The Internal Revenue Service tax per capita in five-year increments from 1960 to 2005. (*Source:* Internal Revenue Service, 2009)

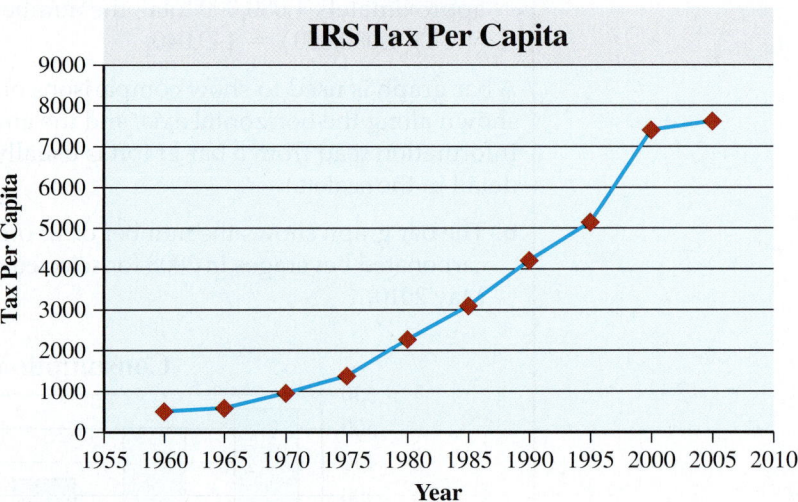

1. During which five-year period did the tax per capita increase the most? Increase the least?

Solution

The line segment rises the most between 1995 and 2000, so the greatest increase in per capita taxes occurred between 1995 and 2000. The line segment between 1960 and 1965 rises the least, so the least amount of increase was between 1960 and 1965.

2. What is the general trend in taxes per capita?

Solution

Since the line is rising from left to right, the taxes per capita are increasing. In general, the rate of increase is getting greater for each five-year period—the exception being between 2000 and 2005.

3. Estimate the taxes per capita in 1985.

Solution

Locate 1985 on the horizontal axis, go vertically until you intersect the graph, then go horizontally to the left until you intersect the vertical scale, and estimate the result. The tax per capita in 1985 is a little more than $3000 dollars per year, so $3,100 is a good estimate.

4. During what year was the tax per capita approximately $5000?

Solution

Find 5000 on the vertical scale, follow the horizontal line until it intersects the graph, then go down until you intersect the horizontal scale and estimate the result. The tax per capita was approximately $5000 in 1994.

PRACTICE EXERCISE 1

a. Exporting live animals from Southeast Asia is big business with approximately 13,360,000 being exported between 2000 and 2007. The pie graph below shows the percentage of exports from various countries. (*Source: National Geographic*, January 2010)

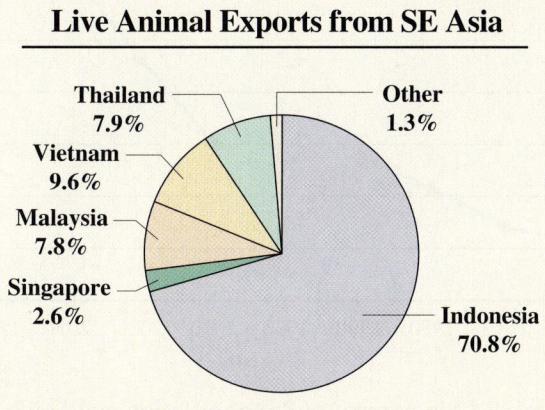

Live Animal Exports from SE Asia

- Thailand 7.9%
- Other 1.3%
- Vietnam 9.6%
- Malaysia 7.8%
- Singapore 2.6%
- Indonesia 70.8%

1. Which country exported the most live animals between 2000 and 2007? What percent of the live animals did this country export?

2. The majority of the animals are exported to the U.S. From a random sample of 500 animals exported to the U.S., how many would you expect to come from Malaysia?

3. Of the total number of live animals exported from Southeast Asia between 2000 and 2007, how many were exported from Vietnam?

(Continued)

Answers: Practice Exercise 1: a. 1. Indonesia, 70.8% **2.** 39 **3.** 1,282,560

b. Following is a bar graph showing the grade distribution on a recent test:

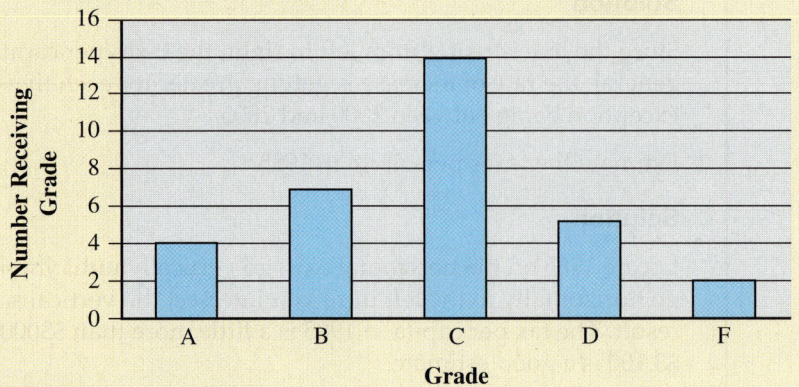

1. What grade did most students receive? The least students receive?

2. Which grades did more than six students receive?

3. How many students received a B? An F?

4. How many more students received a C than an A?

c. The population of the U.S. is aging. The line graph shows the percent and projected percent of the population that is aged 65 or older in ten-year intervals. (*Source:* U.S. Census Bureau, *Resident Population Projections by Sex and Age: 2005–2050*)

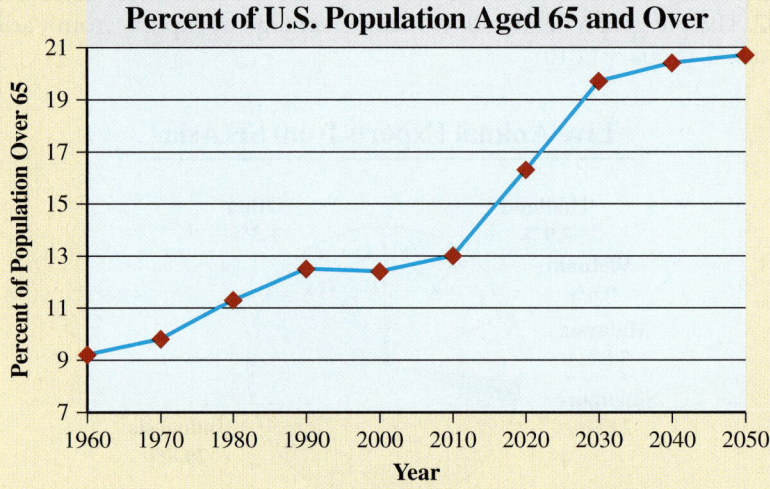

1. During which ten-year interval did the percent of the population aged 65 or older decrease?

2. What is the general trend in the percent of the population aged 65 or older?

3. Estimate the percent of the population aged 65 or older in 2010.

4. Approximately what year will the projected percent of the population aged 65 or older be 19.7%?

5. What are the implications for Social Security and Medicare?

In Chapter 3, we solved linear equations with one variable. The solution of a linear equation with one variable is usually a single number. The solution can be verified by replacing the variable with the value and simplifying. For example, the solution of

---

Answers: Practice Exercise 1: b. 1. C, F **2.** B and C **3.** 7, 2 **4.** 10 **Practice Exercise 1: c. 1.** 1990–2000 **2.** Increasing rapidly **3.** 13% **4.** around 2030 **5.** Alarming!

$3x + 7 = 19$ is $x = 4$, which we can verify by replacing x with 4 and simplifying. So, $3(4) + 7 = 12 + 7 = 19$. Therefore, the solution is correct. In this section, we study linear equations with two variables.

> **Definition** Linear Equation with Two Variables
>
> A linear equation with two variables is any equation of the form $ax + by = c$, or any equation that can be put into that form, where a and b are constants and not equal to 0 at the same time.

OBJECTIVE **B** Verifying solutions of equations with two variables

Any solution of a linear equation with two variables must have a value for each variable. Hence, any solution has two values. We check a solution of a linear equation with two variables in much the same way that we check a solution of a linear equation with one variable: We replace each variable with its value and simplify.

Example 2 | **Verify that the following are solutions of the given equations.**

a. $x + 2y = 6$; $x = 4, y = 1$ Substitute 4 for x and 1 for y.
$4 + 2 \cdot 1 = 6$ Multiply 2 and 1.
$4 + 2 = 6$ Add 4 and 2.
$6 = 6$ Therefore, the solutions are correct.

b. $x + 2y = 6$; $x = -2, y = 4$ Substitute -2 for x and 4 for y.
$-2 + 2(4) = 6$ Multiply 2 and 4.
$-2 + 8 = 6$ Add -2 and 8.
$6 = 6$ Therefore, the solutions are correct.

Notice two things from Example 2. First, the equation in part a is the same as the equation in part b, but the solutions are different. Consequently, a linear equation in two variables has more than one solution. In fact, it has an infinite number of solutions, since we can assign any value to either variable and find the corresponding value for the remaining variable. Second, it is cumbersome to write the solutions in the form given in Example 2. We would have to write, "Two of the solutions to the linear equation $x + 2y = 6$ are $x = 4$ and $y = 1$, and $x = -2$ and $y = 4$." For this reason, we introduce the notion of an **ordered pair**.

An ordered pair consists of two numbers enclosed in parentheses and separated by a comma. They are called ordered pairs because one variable is assigned to the first number of the pair and a different variable is assigned to the second. Hence, the numbers are written in a specific order determined by the variables.

OBJECTIVE **C** Writing solutions of linear equations with two variables as ordered pairs

If a set of ordered pairs is in the form (x, y), the first number in the pair is the x-value and the second is the y-value. So, in the ordered pair $(-4, 6)$, $x = -4$ and $y = 6$, and in the ordered pair $(2, -3)$, $x = 2$ and $y = -3$. In Example 2, the solution $x = 4$ and $y = 1$ is represented as the ordered pair $(4, 1)$, and the solution $x = -2$ and $y = 4$ is represented as $(-2, 4)$. This leads us to the following definition:

> **Definition** Solutions of Linear Equations of Two Variables
>
> If the solutions of $ax + by = c$ are ordered pairs in the form (x, y), the ordered pair (u, v) is a solution if the replacement of x with u and y with v results in a true statement.

OBJECTIVE **D** Determining whether a given ordered pair is a solution of a given equation

Example 3 Determine whether the given ordered pair is a solution of the given equation.

a. $3x - 4y = 9; (7, 3)$ Substitute 7 for x and 3 for y.
$3(7) - 4(3) = 9$ Multiply before adding.
$21 - 12 = 9$ Add 21 and -12.
$9 = 9$ Therefore, $(7, 3)$ is a solution.

b. $2x + 3y = 6; (-3, 0)$ Substitute -3 for x and 0 for y.
$2(-3) + 3(0) = 6$ Multiply before adding.
$-6 + 0 = 6$ Add -6 and 0.
$-6 \neq 6$ Therefore, $(-3, 0)$ is not a solution.

PRACTICE EXERCISE 3

Determine whether the given ordered pair is a solution of the given equation. Assume that the ordered pairs are in the form (x, y).

a. $2x - 4y = 10; (1, -2)$ **b.** $3x + y = 8; (-3, 1)$

If you need more practice, do the following Additional Practice Exercises.

Additional Practice Exercise 3   Determine whether the given ordered pair is a solution of the given equation. Assume the ordered pairs are in the form (x, y).

a. $x + 3y = 7; (1, 2)$ **b.** $2x - y = 8; (-4, -2)$

OBJECTIVE **E** Finding the missing member of an ordered pair for a given linear equation

Suppose we have an equation and know one member of an ordered pair. We can find the other member by substituting the known value for the variable it represents and solving for the other variable.

Example 4 **Use the given equation to find the missing member of each ordered pair. Assume that the ordered pair is in the form (x, y).**

a. $y = 2x - 6; (0, \underline{\ \ }), (\underline{\ \ }, 0), (5, \underline{\ \ }), (-3, \underline{\ \ })$

Solution

Since the ordered pairs are in the form of (x, y), the ordered pair $(0, \underline{\ \ })$ means $x = 0$ and we need to find y.

$y = 2(0) - 6$ Substitute 0 for x. $2(0) = 0$.
$y = 0 - 6$ Add 0 and -6.
$y = -6$ $0 - 6 = -6$. Therefore, the ordered pair is $(0, -6)$.

The ordered pair $(\underline{\ \ }, 0)$ means $y = 0$ and we need to find x.

$0 = 2x - 6$ Substitute 0 for y. Add 6 to both sides.
$0 + 6 = 2x - 6 + 6$ Simplify both sides.
$6 = 2x$ Divide both sides by 2.
$\dfrac{6}{2} = \dfrac{2x}{2}$ Simplify both sides.
$3 = x$ Therefore, the ordered pair is $(3, 0)$.

Answers: **Practice Exercise 3:** **a.** Yes **b.** No   **Additional Practice 3:** **a.** Yes **b.** No

In the ordered pair $(5, \_\_)$, $x = 5$ and we need to find y. Substitute 5 for x and simplify.

$y = 2(5) - 6$
$y = 10 - 6$
$y = 4$

Therefore, the ordered pair is $(5, 4)$.

In the ordered pair $(-3, \_\_)$, $x = -3$ and we need to find y. Substitute -3 for x and simplify.

$y = 2(-3) - 6$
$y = -6 - 6$
$y = -12$

Therefore, the ordered pair is $(-3, -12)$.

b. $2x - 3y = 6$; $(0, \_\_)$, $(\_\_, 0)$, $(6, \_\_)$, $(\_\_, -4)$

Solution

Since the ordered pair is in the form (x, y), the ordered pair $(0, \_\_)$ means $x = 0$ and we need to find y.

$2(0) - 3y = 6$ Substitute 0 for x. $2(0) = 0$.
$\quad 0 - 3y = 6$ $0 - 3y = -3y$.
$\quad\quad -3y = 6$ Divide both sides by -3.
$\quad\quad\quad y = -2$ Therefore, the ordered pair is $(0, -2)$.

The ordered pair $(\_\_, 0)$ means $y = 0$ and we need to find x.

$2x - 3(0) = 6$ Substitute 0 for y. $3(0) = 0$.
$\quad 2x - 0 = 6$ $2x - 0 = 2x$.
$\quad\quad 2x = 6$ Divide both sides by 2.
$\quad\quad\quad x = 3$ Therefore, the ordered pair is $(3, 0)$.

In the ordered pair $(6, \_\_)$, $x = 6$ and we need to find y. Substitute 6 for x and solve for y.

$2(6) - 3y = 6$
$12 - 3y = 6$
$\quad -3y = -6$
$\quad\quad y = 2$

Therefore, the ordered pair is $(6, 2)$.

In the ordered pair $(\_\_, -4)$, $y = -4$ and we need to find x. Substitute -4 for y and solve for x.

$2x - 3(-4) = 6$
$2x + 12 = 6$
$\quad 2x = -6$
$\quad\quad x = -3$

Therefore, the ordered pair is $(-3, -4)$.

c. $x = 2$; $(\_\_, 0)$, $(\_\_, -2)$, $(\_\_, 4)$

Solution

Notice that there is no y term in this equation. We can think of the equation as $x + 0y = 2$. For the ordered pair $(\_\_, 0)$, $y = 0$ and we need to find x.

$x + 0(0) = 2$ Substitute 0 for y. $0(0) = 0$.
$\quad x + 0 = 2$ $x + 0 = x$.
$\quad\quad x = 2$ Therefore, the ordered pair is $(2, 0)$.

In the ordered pair $(\_\_, -2)$, $y = -2$ and we need to find x. Substitute -2 for y and simplify.

$x + 0(-2) = 2$
$x + 0 = 2$
$x = 2$

Therefore, the ordered pair is $(2, -2)$.

In the ordered pair $(\_\_, 4)$, $y = 4$ and we need to find x. Substitute 4 for y and simplify.

$x + 0(4) = 2$
$x + 0 = 2$
$x = 2$

Therefore, the ordered pair is $(2, 4)$.

Note

In Example 4c, it soon becomes clear that $x = 2$ for any value of y, since $0 \cdot y = 0$ for all values of y. When one variable is missing from a linear equation, the remaining variable must always equal the given constant. For example, if $y = -2$, then all ordered pairs solving this equation must have a y-value of -2. The x-value may be anything. Some solutions are $(0, -2)$, $(-3, -2)$, and $(5, -2)$. In general, the solutions are represented by $(x, -2)$ where x can have any value.

PRACTICE EXERCISE 4

Use the given equation to find the missing member of each of the ordered pairs. Assume that the ordered pair is in the form (x, y).

a. $y = 3x + 12$; $(0, \_\_)$, $(\_\_, 0)$, $(-3, \_\_)$

b. $3x + 4y = 12$; $(0, \_\_)$, $(\_\_, 0)$, $(-4, \_\_)$

c. $y = -3$; $(0, \_\_)$, $(5, \_\_)$, $(-5, \_\_)$

If you need more practice, do the following Additional Practice Exercises.

Additional Practice Exercise 4   Use the given equation to find the missing member of each of the ordered pairs. Assume that the ordered pair is in the form (x, y).

a. $y = -3x + 9$; $(0, \_\_)$ $(\_\_, 0)$, $(4, \_\_)$

b. $x - 3y = 9$; $(0, \_\_)$, $(\_\_, 0)$, $(\_\_, -2)$

c. $x = -4$; $(\_\_, 0)$, $(\_\_, -2)$, $(\_\_, 3)$

In Chapter 3, we solved linear equations with one variable. It is possible to graph the solution of a linear equation on the number line. For example, the solution of $2x + 7 = 5$ is $x = -1$. This solution is graphed on the number line by placing a dot at -1 as follows:

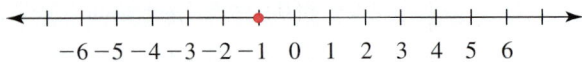

OBJECTIVE **F** Plotting points on the rectangular coordinate system

A solution of $2x + y = 8$ is $(2, 4)$. Since a solution to a linear equation in two variables is an ordered pair, we cannot graph a solution on the number line. Since there are two variables, we need two number lines—one for each variable. Suppose the ordered pairs are in the form of (x, y). We draw a horizontal number line and label it the x-line. At the 0 point of the x-line, we draw a vertical number line and put its 0 point at the point of intersection of the two lines. We call this the y-line. The x-line is called the **x-axis** and the y-line is called the **y-axis**. The point of intersection of the two axes is called the **origin**. On the x-axis, we label the points to the right of the origin with positive numbers and those to the left with negative. On the y-axis, we number the points above the origin with positive numbers and those below with negative. See the following figure:

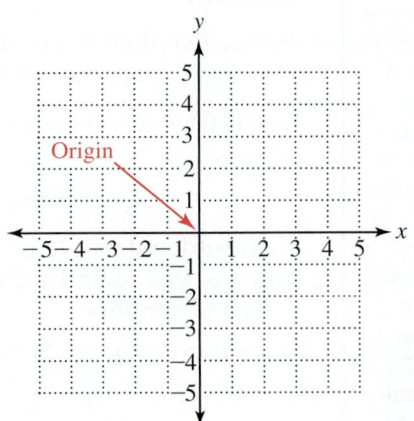

Historical Note

The invention of the rectangular coordinate system is attributed to René Descartes. One story is that the inspiration for its development came from Descartes's watching a fly crawling on a ceiling near the corner of the room. Descartes observed that he could pinpoint the location of the fly by using only two numbers that gave the perpendicular distance from each of the walls to the fly. Imagine that! A fly is responsible for one of the most important inventions in the history of mathematics!

This is known as the rectangular, or Cartesian (in honor of its inventor, René Descartes), coordinate system. The x- and y-axes divide the coordinate system into four regions called **quadrants**. The quadrants are numbered counterclockwise, with the first quadrant being the upper right quadrant. See the following figure:

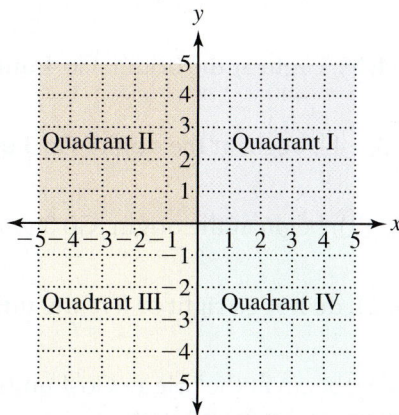

Since the x-axis is horizontal, the x-value of the ordered pair tells us how far and in which direction to go horizontally from the origin. Go to the right if x is positive and to the left if x is negative. Since the y-axis is vertical, the y-value tells us how far and in which direction to go vertically. Go up if y is positive and down if y is negative. For example, to plot the ordered pair $(4, 5)$, we begin at the origin and go 4 units to the right, and then we go 5 units up, as illustrated in the following graph:

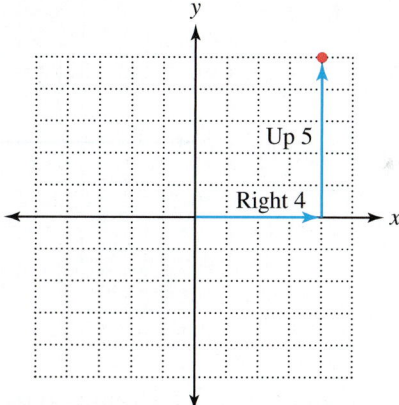

In an ordered pair, the first number is called the **abscissa** and the second number is called the **ordinate**. Together, they are called the **coordinates** of the point.

Example 5 **Plot the points represented by the following ordered pairs on the rectangular coordinate system:**

a. $(3, 5)$ **b.** $(-4, 3)$ **c.** $(-3, -4)$

d. $(3, -5)$ **e.** $(0, 2)$ **f.** $(4, 0)$

Solution

We first give a description of how the ordered pairs are plotted, and then the points are shown on the rectangular coordinate system.

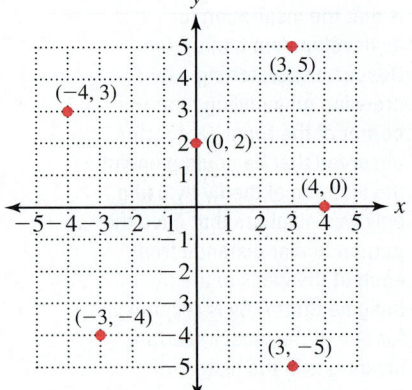

a. To plot the point $(3, 5)$, begin at the origin. Go 3 units to the right and 5 units up.

b. To plot the point $(-4, 3)$, begin at the origin. Go 4 units to the left and 3 units up.

c. To plot the point $(-3, -4)$, begin at the origin. Go 3 units to the left and 4 units down.

d. To plot the point $(3, -5)$, begin at the origin. Go 3 units to the right and 5 units down.

e. To plot the point $(0, 2)$, do not go right or left any units. Go up 2 units from the origin.

f. To plot the point $(4, 0)$, begin at the origin. Go 4 units to the right and stay there. Do not go up or down any units.

PRACTICE EXERCISE 5

Plot the points represented by the following ordered pairs on the rectangular coordinate system:

a. $(3, 4)$ b. $(-2, 5)$

c. $(-4, -1)$ d. $(4, -3)$

e. $(-6, 0)$ f. $(0, -6)$

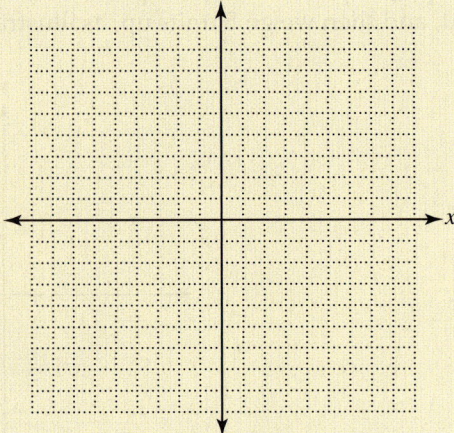

If you need more practice, do the following Additional Practice Exercises.

Additional Practice Exercise 5   Plot the points represented by the following ordered pairs:

a. $(4, 5)$ b. $(-3, 1)$ c. $(-2, -5)$ d. $(3, -6)$ e. $(5, 0)$ f. $(0, 5)$

Answers: Practice Exercise 5:

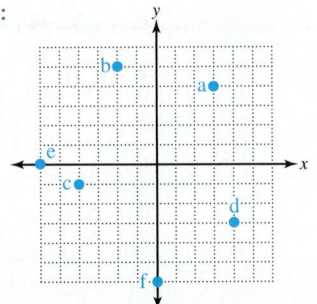

Additional Practice 5:

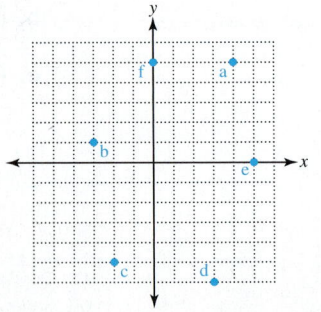

OBJECTIVE Drawing and interpreting scatter diagrams

Data can often be represented as ordered pairs and are called *paired* data. Paired data occur often in real-world situations. For example, the speed that you drive and the number of miles per gallon that you get can be written as paired data of the form (speed, miles per gallon). When paired data are graphed in the rectangular coordinate system, the resultant graph is called a **scatter diagram** and typically involves the first quadrant only. Scatter diagrams are usually used to reveal patterns in the data.

Example 6

The table shows the number of officers on active duty in the U.S. Navy from 2000 to 2009. (*Source: World Almanac and Book of Facts*, 2010)

a. Write this paired data as ordered pairs of the form (year, number of officers).

Solution

(2000, 53,698) (2001, 54,177) (2002, 55,506)
(2003, 55,852) (2004, 55,592) (2005, 54,039)
(2006, 53,209) (2007, 51,385) (2008, 52,184)
(2009, 52,233)

Year	Number of Officers
2000	53,698
2001	54,177
2002	55,506
2003	55,852
2004	55,592
2005	54,039
2006	53,209
2007	51,385
2008	52,184
2009	52,233

b. Use the paired data to draw a scatter diagram.

Solution

Since the first values are the years, we label the horizontal, or *x*-axis "years" and mark off the horizontal axis with the years. The vertical, or *y*-axis is the number of officers. The choice of the beginning value, ending value, and increment is arbitrary. Since the smallest number of officers is 51,385 and the largest number is 55,852, let's begin with 51,000 and end with 56,000 with an increment of 1000. Plot the points, and we get the following scatter diagram:

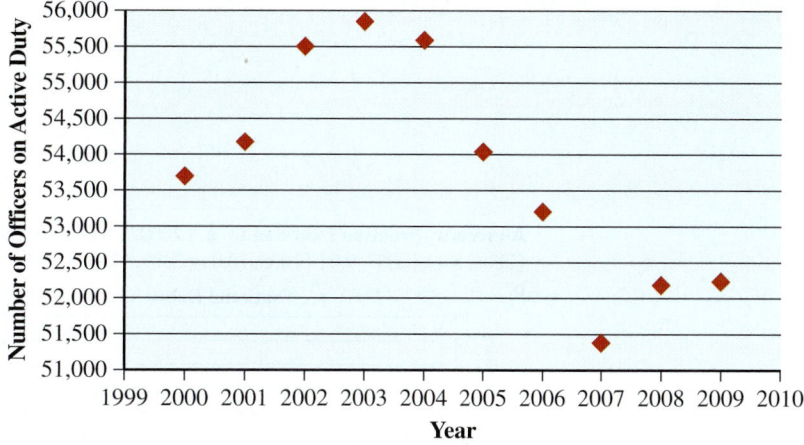

c. What trend is suggested by the scatter diagram for the years 2003–2009?

Solution

The number of officers decreased in 2003–2007 and then increased again in 2007–2009.

PRACTICE EXERCISE 6

The table shows the gross U.S. national debt rounded to the nearest tenth of a trillion dollars. (*Source: The New York Times Almanac*, 2010)

a. Write this paired data as a set of ordered pairs in the form (year, national debt).

b. Use this data to draw a scatter diagram.

c. What is the trend suggested by the scatter diagram?

Year	National Debt
2000	5.6
2001	5.8
2002	6.2
2003	6.8
2004	7.4
2005	7.9
2006	8.4
2007	9.0
2008	10.0
2009	12.9

Exercise Set 4.1

For Extra Help

MyMathLab®

For Exercises 1–6, answer the questions from the graphs. (See Example 1.)

1. The pie chart shows the percentage of sheep killed in Montana and Idaho in 2008 by various predators. (*Source: National Geographic*, March 2010)

   a. Which predator kills the smallest percentage of sheep and what is that percentage?

   b. If a rancher lost 100 sheep to predators last year, how many would he expect to have lost to dogs?

   c. If the total number of sheep lost to predators in Montana and Idaho was 46,300, how many were lost to coyotes? Round to the nearest whole number.

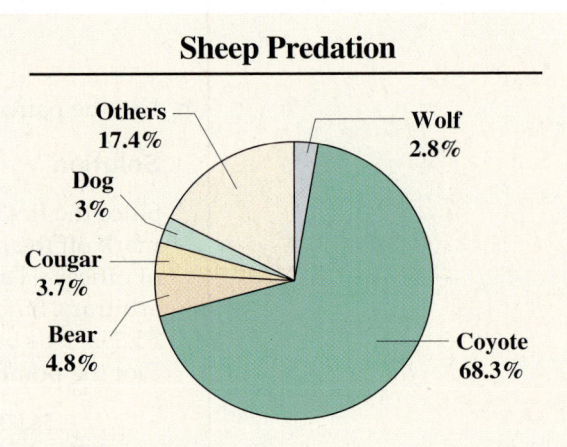

Sheep Predation

Others 17.4%
Wolf 2.8%
Dog 3%
Cougar 3.7%
Bear 4.8%
Coyote 68.3%

Answers: **Practice Exercise 6:** **a.** $(2000, 5.6)$ $(2001, 5.8)$ $(2002, 6.2)$ $(2003, 6.8)$ $(2004, 7.4)$ $(2005, 7.9)$ $(2006, 8.4)$ $(2007, 9.0)$ $(2008, 10.0)$ $(2009, 12.9)$

b.

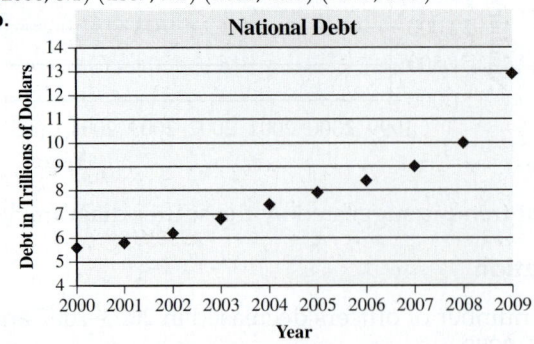

c. The national debt is increasing rapidly with a huge increase in 2009.

2. Pharmaceutical pollution of rivers and streams is becoming a serious concern. The pie graph represents the percent of four pharmaceuticals found in fish from Chicago's North Shore Channel. (*Source: National Geographic*, April 2010)

a. Which pharmaceutical is found in the greatest concentration and what is the percentage?

b. If a fish contained 45 micrograms of pharmaceuticals, how much carbamazepine would you expect to be present?

c. On a fishing trip, a fisherman catches fish that have a total of 250 micrograms of pharmaceuticals, which he and his family eat. How much diphenhydramine did they eat?

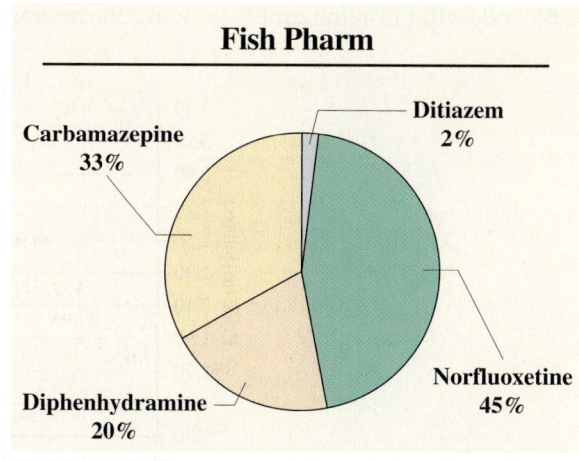

Fish Pharm

3. Following is a bar graph showing the nations that are most reliant on nuclear energy for the production of electricity and the percent of their electricity that is produced by nuclear energy:

a. Which country is most reliant on nuclear energy, and what percent of its total electricity comes from nuclear energy?

b. Which countries are more than 50% reliant on nuclear energy?

c. Estimate the percentage of electricity produced by nuclear energy for Ukraine.

d. Find the difference between the percentages for Lithuania and Sweden.

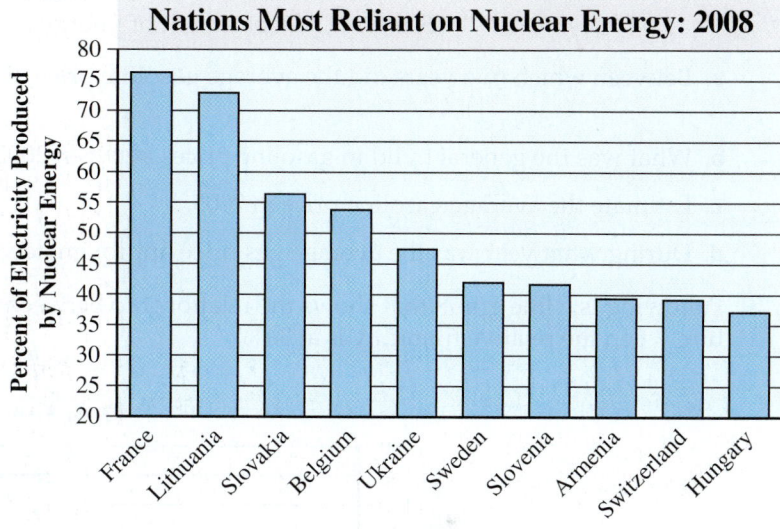

Nations Most Reliant on Nuclear Energy: 2008

(*Source: International Atomic Energy Agency, 2010.*)

4. The bar graph shows the number of barrels of oil (in thousands) imported per day from selected countries.

a. From what country does the U.S. import the most oil and approximately how many barrels per day does the U.S. import from this country?

b. From which countries does the U.S. import fewer than 600,000 barrels per day?

c. Estimate the number of barrels per day that the U.S. imports from Iraq.

d. Estimate the difference between the number of barrels imported per day from Mexico and from Ecuador.

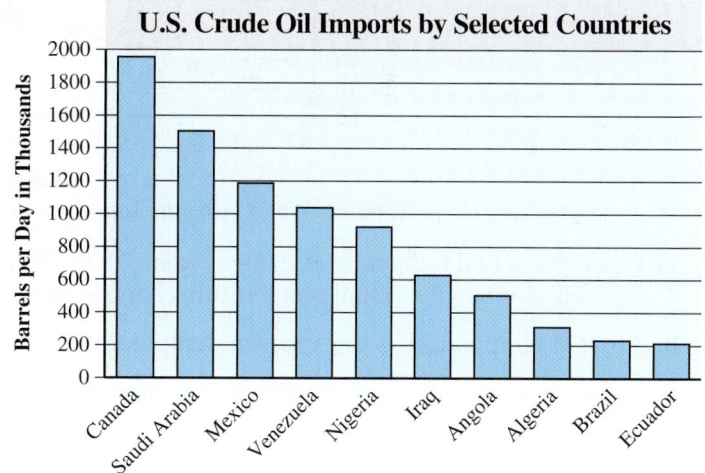

U.S. Crude Oil Imports by Selected Countries

(*Source: The World Almanac and Book of Facts, 2010.*)

5. Following is a line graph showing the average gasoline prices for regular unleaded in U.S. cities from 2000 to 2009:

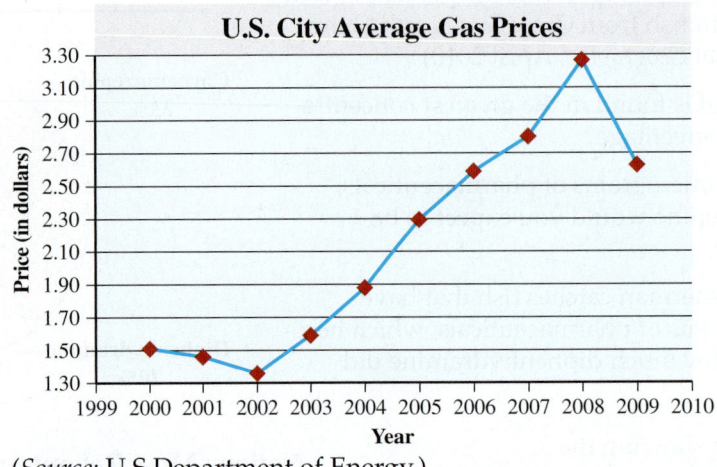

(*Source:* U.S Department of Energy.)

a. Between which two years did the average gasoline price decrease the most? Increase the most?

b. What was the general trend in gasoline prices between 2002 and 2008?

c. Estimate the average gasoline price in 2005.

d. During what year was the average gas price approximately $1.60 per gallon?

6. Following is a line graph that shows the relationship between the actual temperature and the apparent temperature when the relative humidity is at 30%:

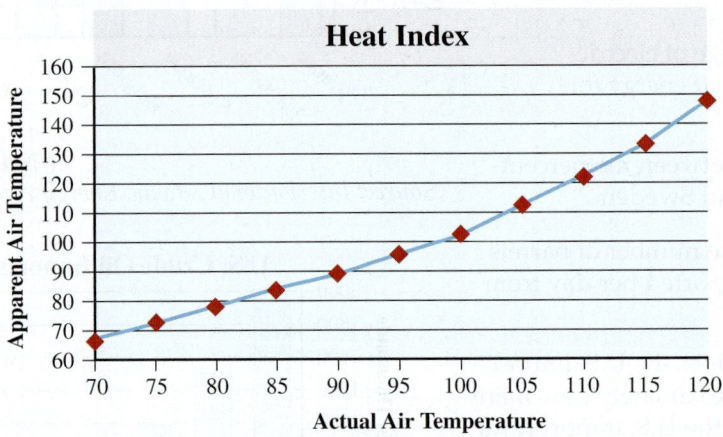

(*Source: World Almanac and Book of Facts*, 2010.)

a. When the actual temperature changes from 70° to 75°, does the apparent temperature change by the same amount as when the actual temperature changes from 110° to 115°? If no, is the change more or less?

b. At what temperature is the apparent temperature the same as the actual temperature?

c. For what temperatures is the apparent temperature greater than the actual temperature?

d. What is the approximate apparent temperature when the actual temperature is 115°?

e. What is the actual temperature when the apparent temperature is approximately 103°?

Determine whether the given ordered pair is a solution of the given equation. Assume that the ordered pairs are of the form (x, y). *(See Example 3.)*

7. $y = 3x + 2; (1, 5)$

8. $y = 2x + 4; (-1, 2)$

9. $y = -4x + 5; (-1, 1)$

10. $y = -2x - 5; (-2, 1)$

11. $2x + y = 4; (2, 0)$

12. $3x + y = 9; (0, 9)$

13. $x - 4y = 8; (0, 2)$

14. $3x - y = 12; (-4, 0)$

15. $2x + 3y = 12; (3, 2)$

16. $2x + 5y = 10; (-5, 4)$

17. $3x - 5y = 15; (-5, -6)$

18. $4x - 3y = 12; (-3, -8)$

19. $4x - 5y = -20; (0, -4)$

20. $7x - 3y = -21; (-3, 0)$

21. $2x + 3y = 6; \left(-\frac{3}{2}, 3\right)$

22. $3x + 4y = 8; \left(-\frac{4}{3}, 3\right)$

Use the given equation to find the missing member of each of the given ordered pairs. Assume that the ordered pairs are in the form (x, y). *(See Example 4.)*

23. $y = 2x - 3; (0, \_\_\_), (\_\_\_, 0), (3, \_\_\_), (\_\_\_, -7)$

24. $y = 3x - 5; (0, \_\_\_), (\_\_\_, 0), (-2, \_\_\_), (\_\_\_, 4)$

25. $y = -2x - 4; (0, \_\_\_), (\_\_\_, 0), (-3, \_\_\_), (\_\_\_, -8)$

26. $y = -3x - 5; (0, \_\_\_), (\_\_\_, 0), (-2, \_\_\_), (\_\_\_, 4)$

27. $2x - 3y = 12; (0, \_\_\_), (\_\_\_, 0), (-6, \_\_\_), (\_\_\_, -6)$

28. $3x - 2y = 18; (0, \_\_\_), (\_\_\_, 0), (4, \_\_\_), (\_\_\_, 6)$

29. $4x + 5y = 20; (0, \_\_\_), (\_\_\_, 0), (-10, \_\_\_), (\_\_\_, 8)$

30. $6x + 5y = 30; (0, \_\_\_), (\_\_\_, 0), (10, \_\_\_), (\_\_\_, 12)$

31. $2x + 3y = -6; (0, \_\_\_), (\_\_\_, 0), (9, \_\_\_), (\_\_\_, -4)$

32. $3x + 5y = -15; (0, \_\_\_), (\_\_\_, 0), (5, \_\_\_), (\_\_\_, 6)$

33. $4x - 3y = -12; (0, \_\_\_), (\_\_\_, 0), (6, \_\_\_), (\_\_\_, -12)$

34. $5x - 4y = -20; (0, \_\_\_), (\_\_\_, 0), (12, \_\_\_), (\_\_\_, -20)$

35. $x = 4; (\_\_\_, 0), (\_\_\_, 3), (\_\_\_, -2), (\_\_\_, -4)$

36. $x = -1; (\_\_\_, 2), (\_\_\_, 3), (\_\_\_, -3), (\_\_\_, -5)$

37. $y = 4; (2, \_\_\_), (3, \_\_\_), (-4, \_\_\_), (-1, \_\_\_)$

38. $y = -3; (0, \_\_\_), (2, \_\_\_), (-2, \_\_\_), (-4, \_\_\_)$

Plot the points represented by the given ordered pairs on the rectangular coordinate system. Assume that each unit represents 1. *(See Example 5.)*

39. $(3, 5), (-3, 1), (0, 5), (2, -4)$

40. $(-4, 0), (4, -2), (-3, 4), (0, 4)$

41. $(3, 5), (-2, -4), (0, 2), (4, -5)$

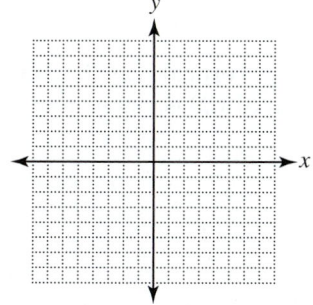

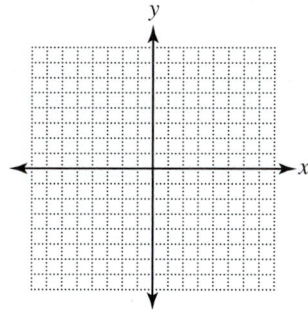

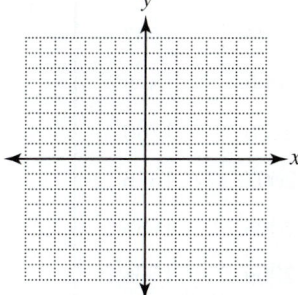

42. $(6, -2), (-4, 5), (2, -1), (3, 4)$

43. $(3, 0), (-2, 5), (3, -6), (-4, -2)$

44. $(-4, 0), (0, 3), (-1, -5), (4, -6)$

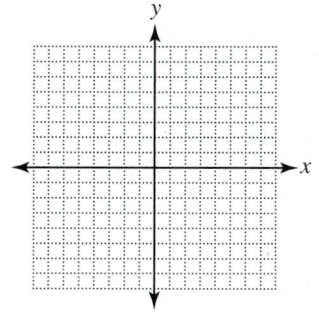

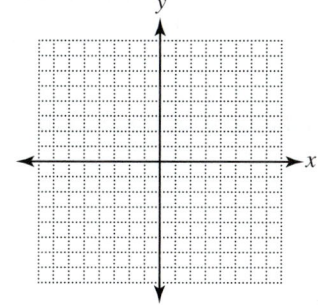

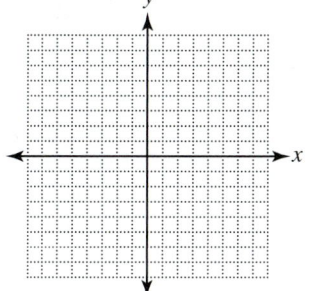

On the next grids shown, some points have been plotted. Give the coordinates of each.

45.

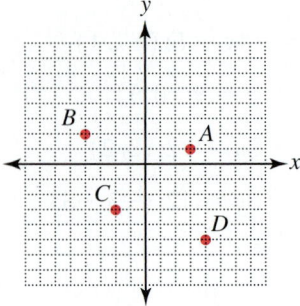

46.

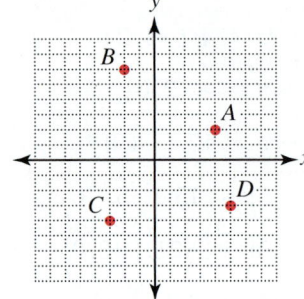

47.

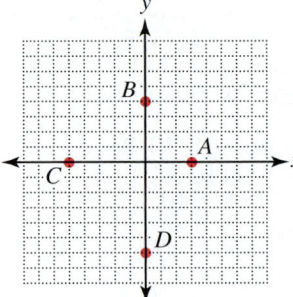

48.

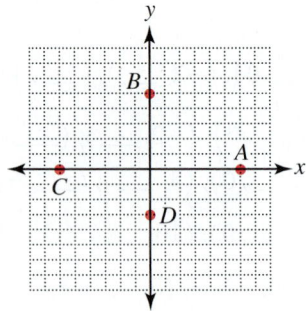

49.

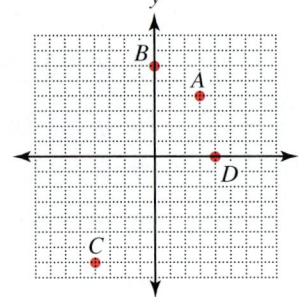

50.

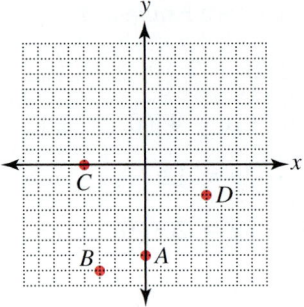

For Exercises 51 and 52, answer the questions. (See Example 6.)

51. The table shows the number of national banks in the U.S. from 2002 to 2009. (*Source:* Federal Deposit Insurance Corporation, June 30, 2009)

Year	Number
2002	2077
2003	2001
2004	1907
2005	1818
2006	1780
2007	1676
2008	1585
2009	1506

a. Write this paired data in the form (year, number of banks).

b. What does the ordered pair (2007, 1676) represent?

c. Use the ordered pairs to draw a scatter diagram. (A vertical scale beginning with 1500 and ending with 2100 with an increment of 100 is suggested.)

d. What is the general trend in the number of national banks?

52. Following is a table showing the average tuition and fee charges for public two-year colleges in the United States from 2001 to 2008. The year given in the table indicates the ending year of the school year. For example, 2006 would represent the school year 2005–2006.

Year	Tuition
2001	1333
2002	1380
2003	1483
2004	1702
2005	1849
2006	1935
2007	2018
2008	2063

a. Write this paired data in the form of (year, tuition).

b. What does the ordered pair (2004, 1702) represent?

c. Use the ordered pairs to draw a scatter diagram.

d. What has been the trend in tuition and fee costs?

Writing Exercises (53–55)

53. A linear equation with two unknowns may be written in the form $ax + by = c$. Solutions of such equations can be represented as ordered pairs of the form (x, y). A linear equation with three unknowns may be written in the form $ax + by + cz = d$. How would you represent solutions to these equations?

54. Why does a linear equation with two unknowns have more than one solution? Why does it have an infinite number of solutions?

55. If the graph of an ordered pair is in the first quadrant, both the x- and y-values of the ordered pair are positive. Describe the x- and y-values of an ordered pair whose graph is in the second quadrant. Why? Third quadrant. Why? Fourth quadrant. Why?

Section 4.2 Graphing Linear Equations with Two Variables

OBJECTIVE *When you complete this section, you will be able to:*

A Graph linear equations with two variables on the rectangular coordinate system.

PREREQUISITE SKILLS *Before beginning this section, you should be able to:*

a. Add, subtract, multiply, and divide integers. (Sections 1.6, 1.7, 2.1, 2.5)
b. Substitute values for variables and evaluate. (Section 2.5)
c. Solve linear equations. (Sections 3.1, 3.2, 3.3)
d. Plot ordered pairs on the rectangular coordinate system. (Section 4.1)

GETTING READY FOR SECTION 4.2

Add, subtract, multiply, or divide the following integers:

1. $-9 - 1$ **2.** $-10 - (-3)$

3. $-5(-3)$ **4.** $-18 \div (-9)$

Substitute the given value(s) for the variable(s) into the given expressions and evaluate:

5. $-3x + 4$: $x = 2$ **6.** $5x - 4y$: $x = 4, y = 2$

Solve the following equations:

7. $-6 - 2y = 6$ **8.** $0 = 3x - 6$

Plot the following ordered pairs on the rectangular coordinate system:

9. $(0, 2)$ **10.** $(4, -3)$ **11.** $(-4, -4)$

Introduction

With the rectangular coordinate system, it is possible to plot solutions of equations with two variables. The plot of the solutions is called the **graph** of the equation. Since there are an infinite number of solutions, it is not possible to plot them all. All the graphs in this section will be of linear equations (consequently, the graphs are straight lines) that we will graph with the procedure described next. In Sections 4.3–4.5, we will learn other methods.

Answers: Getting Ready: 1. -10 **2.** -7 **3.** 15 **4.** 2 **5.** -2 **6.** 12 **7.** $y = -6$ **8.** $x = 2$
9–11.

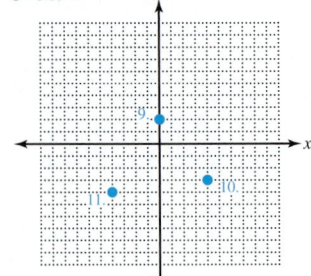

OBJECTIVE Graphing linear equations with two variables on the rectangular coordinate system

> **Procedure: Graphing Linear Equations**
>
> To graph a linear equation, we perform the following steps:
>
> **a.** Determine three ordered pairs that solve the equation by letting x have a value and finding y or letting y have a value and finding x. (Only two points are required. The third point is used as a check.)
>
> **b.** Plot the ordered pairs.
>
> **c.** Draw the line that contains the three points.

A straight line is determined by two points. However, it is a good idea to find a third point as a check, since a straight line cannot always be drawn through any three distinct points. For example, a single straight line cannot be drawn through the following three points:

Example 1 | **Draw the graph of each of the following equations:**

a. $y = 2x + 1$

Solution

We need to find some ordered pairs from the solution set. *We can let either variable have any value we choose and then solve for the other variable.* Since we have to plot the resulting ordered pairs, it is best to use values small in absolute value. This equation is solved for y, so it would be easier to assign values to x and then find y.

Let $x = 0$.	Let $x = 2$.	Let $x = -3$.
$y = 2(0) + 1$ $y = 0 + 1$ $y = 1$	$y = 2(2) + 1$ $y = 4 + 1$ $y = 5$	$y = 2(-3) + 1$ $y = -6 + 1$ $y = -5$
Therefore, the ordered pair is $(0, 1)$.	Therefore, the ordered pair is $(2, 5)$.	Therefore, the ordered pair is $(-3, -5)$.

One easy way of keeping track of the ordered pairs is to put them in a table. For example, the preceding ordered pairs could be put in a table as follows:

x	0	2	-3
y	1	5	-5

Now plot the points on the rectangular coordinate system. It appears that the three points all lie on a line, so draw a line through the three points.

Note

There are a number of things we need to be aware of at this time. First, it can be shown that the graph of any linear equation with two variables ($ax + by = c$) is a straight line. That is why they are called linear. Second, the coordinates of any point on the line solve the equation, and the graph of any ordered pair that solves the equation will be on the line. Third, the line continues in both directions indefinitely. This is usually indicated by an arrow head on each end of the line.

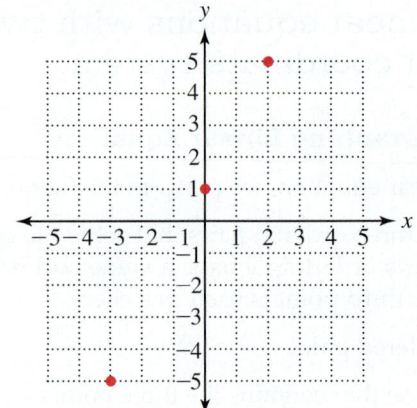

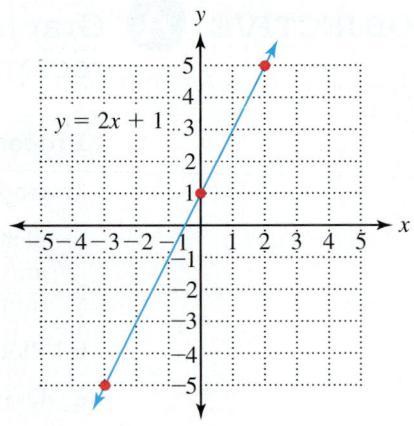

b. $3x - 2y = 6$

Solution

We find ordered pairs that solve the equation by letting either variable have a value and solving for the other variable.

Let $x = 0$.	Let $y = 0$.	Let $x = -2$.
$3(0) - 2y = 6$	$3x - 2(0) = 6$	$3(-2) - 2y = 6$
$0 - 2y = 6$	$3x - 0 = 6$	$-6 - 2y = 6$
$-2y = 6$	$3x = 6$	$-2y = 12$
$y = -3$	$x = 2$	$y = -6$
Therefore, the ordered pair is $(0, -3)$	Therefore, the ordered pair is $(2, 0)$.	Therefore, the ordered pair is $(-2, -6)$.

In a table, the ordered pairs would appear as follows:

x	0	2	-2
y	-3	0	-6

Plot the ordered pairs on the rectangular coordinate system and draw a line through them.

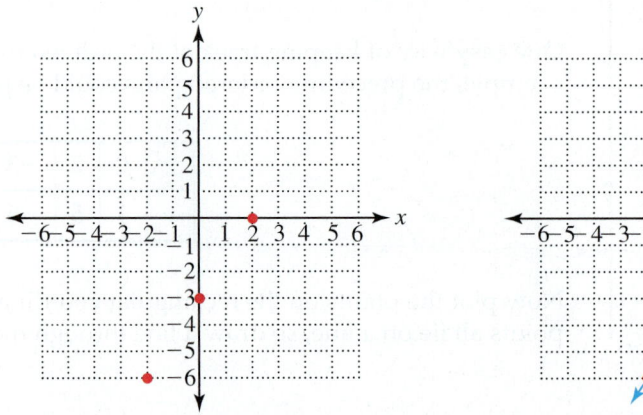

PRACTICE EXERCISE 1

Draw the graph of each of the equations. Assume that each unit on the coordinate system represents 1.

a. $y = x + 3$

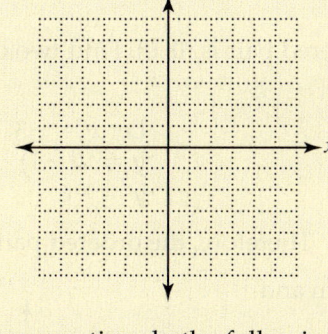

b. $2x + y = 4$

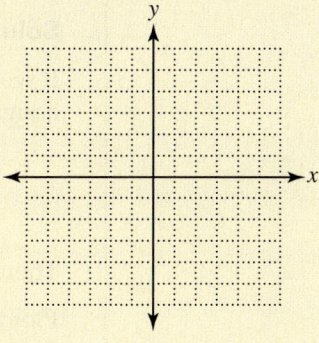

If you need more practice, do the following Additional Practice Exercises.

Additional Practice Exercise 1 Draw the graph of each of the equations. Each unit on the coordinate system represents 1.

a. $y = 3x - 2$

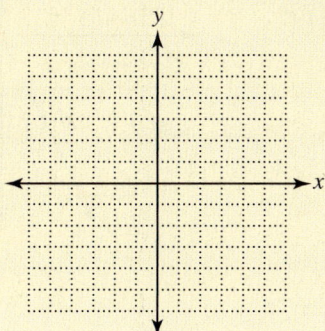

b. $4x + 2y = 8$

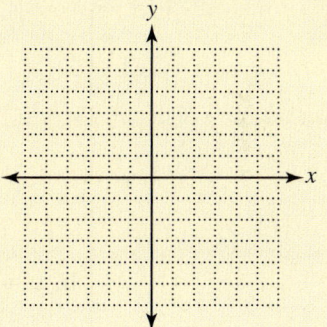

The standard form of a linear equation is $ax + by = c$, where a, b, and c are integers and $a > 0$. If one of the terms is missing, it is still a linear equation, but may require some special considerations.

If the equation is of the form $y = ax$ (or can be put into that form), then the equation is solved for y. The easiest way to find ordered pairs is to assign values to x and find y. Note that if $x = 0$, then $y = 0$, so $(0, 0)$ (which is the origin) is a point on the line. Consequently, all lines whose equations are of the form $y = ax$ pass through the origin.

Answers: Practice Exercise 1:

Additional Practice 1:

a.

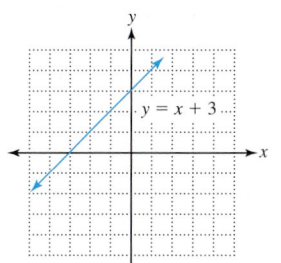

b.

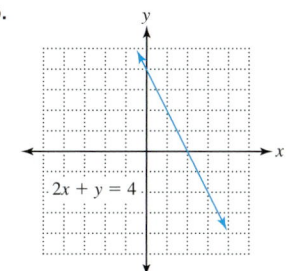

a.

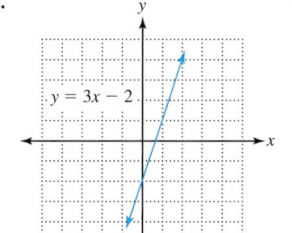

b.

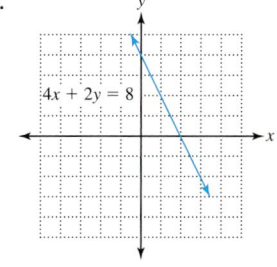

Example 2   **Graph the equations. Assume that each unit is 1.**

a. $y = 2x$

Solution

If we let $x = 0$, then $y = 0$; so one ordered pair is $(0, 0)$. Find two other points by assigning any value to x and finding y.

Let $x = 2$.
$y = 2(2)$
$y = 4$

Let $x = -3$.
$y = 2(-3)$
$y = -6$

Therefore, the ordered pair is $(2, 4)$.   Therefore, the ordered pair is $(-3, -6)$.

Plot the points on the coordinate system and draw a line through them.

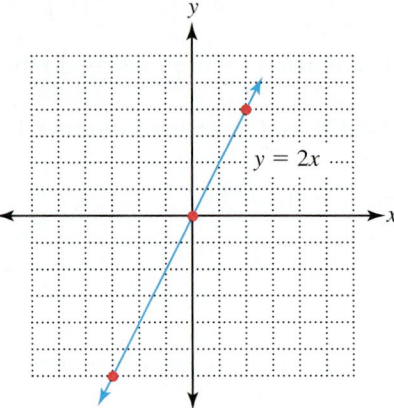

b. $y = \dfrac{1}{3}x$

Solution

If we let $x = 0$, we get $y = 0$; so one ordered pair is $(0, 0)$. Find two more points by assigning values to x and finding y. Since the denominator is 3, it would be a good idea to let x have values that are divisible by 3. Otherwise, we will get fractions that are more difficult to plot.

Let $x = 3$.

$$y = \frac{1}{3} \cdot 3 = \frac{3}{3} = 1$$

Let $x = 6$.

$$y = \frac{1}{3} \cdot 6 = \frac{6}{3} = 2$$

Therefore, the ordered pair is $(3, 1)$.   Therefore, the ordered pair is $(6, 2)$.

Plot the points on the coordinate system and draw a line through them.

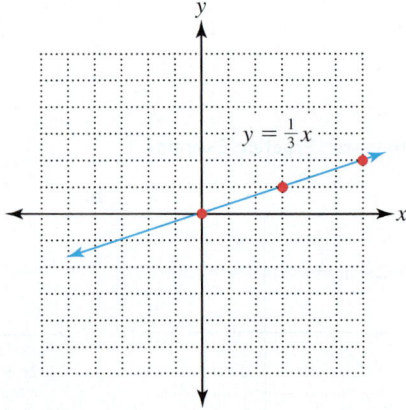

c. $y + 3x = 0$

Solution

If we solve this equation for y, then it will be in the same form as those in parts a and b.

$y + 3x = 0$ *Subtract 3x from each side of the equation.*

$y + 3x - 3x = 0 - 3x$ *Simplify each side of the equation.*

$y = -3x$ *Now generate some ordered pairs.*

If we let $x = 0$, we get $y = 0$; so one ordered pair is $(0, 0)$. Find two more ordered pairs by assigning values to x and finding y.

Let $x = 1$.　　　　　　Let $x = -2$.

$y = -3(1) = -3$　　　　$y = -3(-2) = 6$

Therefore, the ordered pair is $(1, -3)$.　　Therefore, the ordered pair is $(-2, 6)$.

Plot the points on the coordinate system and draw a line through them.

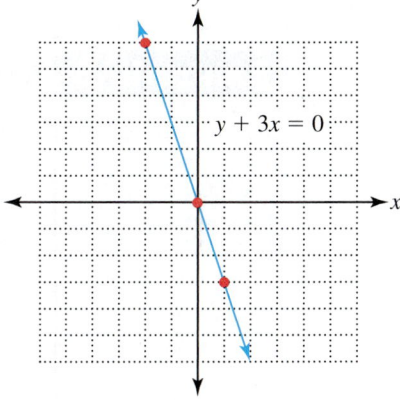

PRACTICE EXERCISE 2

Graph the equations. Assume that each unit is 1.

a. $y = 3x$　　　　　**b.** $y = \dfrac{1}{5}x$　　　　　**c.** $y + 2x = 0$

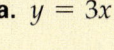 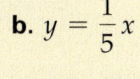 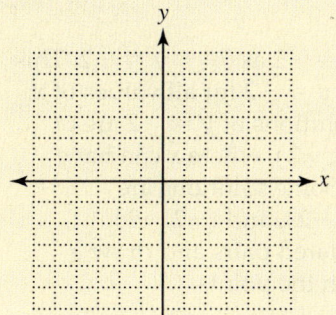 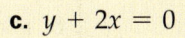 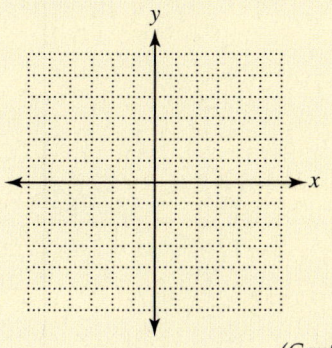

(Continued)

Answers: Practice Exercise 2: a.　　**b.**　　**c.**

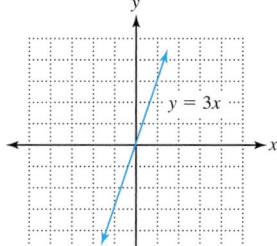 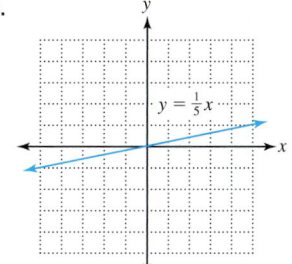 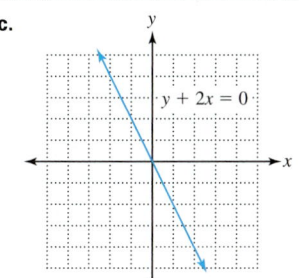

If you need more practice, do the following Additional Practice Exercises.

The other special cases of $ax + by = c$ occur if $a = 0$ or $b = 0$. If $b = 0$, the resulting equation is of the form $x = c$ and if $a = 0$, then $y = c$, where c is any constant. We can think of an equation of the form $x = c$ to be $x + 0y = c$. This means that regardless of the value for y, $x = c$. Therefore, all solutions to the equation $x = c$ are of the form (c, y), where y can have any value. Similarly, we can think of an equation of the form $y = c$ to be $0x + y = c$. This means that $y = c$ for all values of x. Therefore, all solutions of the equation $y = c$ are of the form (x, c), where x can have any value.

Example 3   **Graph the equations. Assume that each unit is 1.**

a. $x = 3$

Solution

Think of $x = 3$ as $x + 0y = 3$. Therefore, $x = 3$ for all values of y. So, all solutions of $x = 3$ are of the form $(3, y)$, where y can have any value. Thus, some solutions are $(3, 0)$, $(3, -2)$, and $(3, 4)$.

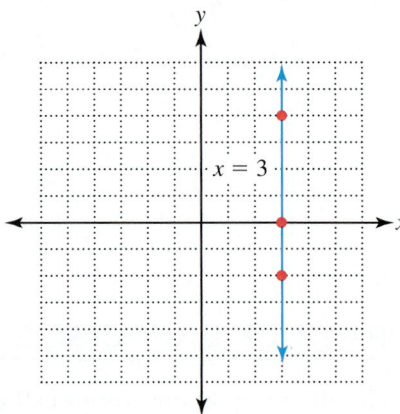

b. $y = -2$

Solution

Think of $y = -2$ as $0x + y = -2$. This means that $y = -2$ for all values of x. Thus, all solutions of $y = -2$ are of the form $(x, -2)$, where x can have any value. Some solutions are $(0, -2)$, $(3, -2)$, and $(-2, -2)$. Plot the ordered pairs and draw a line through the points.

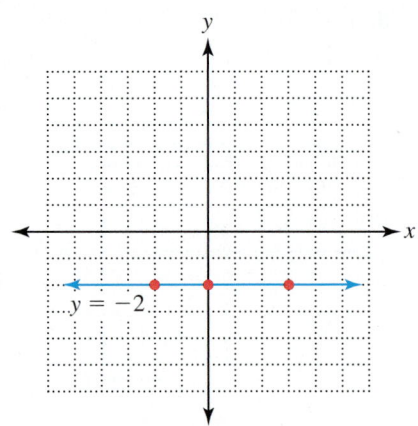

Answers: Additional Practice 2 a.

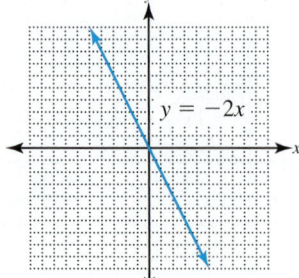

b.

c.

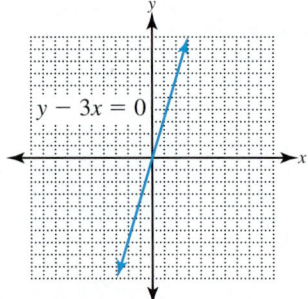

In Example 3, the graph of $x = 3$ is a vertical line and the graph of $y = -2$ is a horizontal line. The graphs of any two ordered pairs with the same x-value will always lie on the same vertical line, and the graphs of any two ordered pairs with the same y-value will always be on the same horizontal line. This leads us to the following observation:

> ### Rule: Vertical and Horizontal Lines
>
> For any constant k, the graph of an equation of the form $x = k$ is a vertical line intersecting the x-axis at $(k, 0)$.
>
> For any constant k, the graph of an equation of the form $y = k$ is a horizontal line intersecting the y-axis at $(0, k)$.

PRACTICE EXERCISE 3

Graph the equations. Assume that each unit is 1.

a. $x = -1$

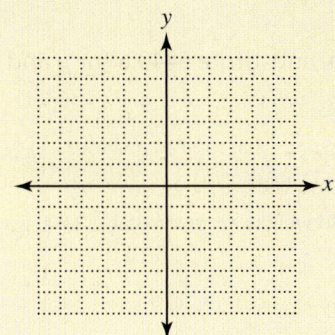

b. $y = 4$

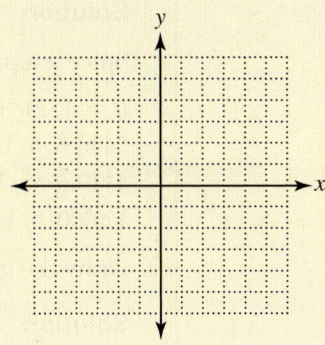

Note

There are a couple of observations we need to make regarding vertical and horizontal lines. If two or more points lie on the same vertical line, they have the same x-values. Also, if two or more points have the same x-values, they lie on the same vertical line. Likewise, if two or more points lie on the same horizontal line, they have the same y-values. Also, if two or more points have the same y-value, they lie on the same horizontal line.

Linear equations can be used to model many real-world situations.

Example 4

In administering an anesthetic to an infant or child, it is very important to have an estimate of the infant's or child's body surface area. If an infant or child weighs between 3 and 30 kilograms, the body surface area can be approximated by the

Answers: Practice Exercise 3: a.

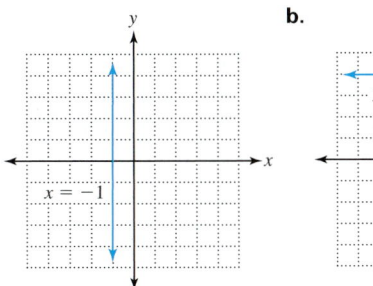

b.

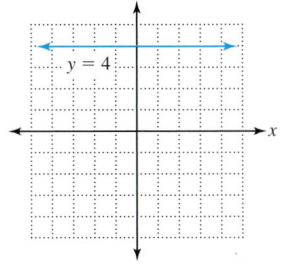

equation $y = 1321 + 0.3433x$, where y is the body surface area in square centimeters and x is the weight in grams. (*Source:* John D. Current, M.D., A Linear Equation for Estimating the Body Surface Area in Infants and Children, *Journal of Anesthesiology*.)

a. If an infant weighs 7500 grams (7.5 kg), what is its approximate surface area?

Solution

Since x represents the infant's weight, substitute 7500 for x and evaluate.

$y = 1321 + .3433x$ Substitute 7500 for x.
$y = 1321 + .3433(7500)$ Multiply.
$y = 1321 + 2574.75$ Add.
$y = 3895.75$ Therefore, the surface area is approximately 3895.75 cm^2.

b. If the surface area of a child is 5440.6 square centimeters, what is the child's approximate weight?

Solution

Since y represents the child's surface area, substitute 5440.6 for y and solve for x.

$y = 1321 + .3433x$ Substitute 5440.6 for y.
$5440.6 = 1321 + .3433x$ Subtract 1321 from both sides of the equation.
$4119.6 = .3433x$ Divide both sides by .3433.
$12,000 = x$ Therefore, the child weighs approximately 12,000 g, or 12 kg.

c. Draw the graph.

Solution

Find three ordered pairs, plot them, and connect them with a line. We found one ordered pair in part a (7500, 3895.75) and another in part b (12,000, 5440.6). Notice that this equation is true only for infants that weigh from 3 kg (3000 g) to 30 kg (30,000 g), so we should use these values to find where the graph begins and ends. (3000, 2350.9), (30,000, 11,620)

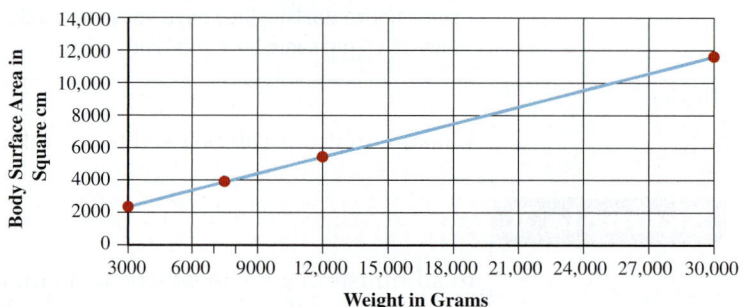

d. Use the graph to estimate the weight of a child whose body surface area is 8000 cubic centimeters.

Solution

Find the horizontal line at 8000 and follow it to the right until it intersects the graph. It looks like the point of intersection is about midway between 18,000 and 21,000, which is 19,500; so an approximate weight for the child is 19,500 g, or 19.5 kg.

PRACTICE EXERCISE 4

In calculating its budget for the coming year, the Biomedical Research and Developmental Department of the National Institutes of Health uses an index that is determined by the equation $y = 1.98 + 0.90x$, where y is the Biomedical Research and Developmental Price Index and x is the Gross Domestic Product Price Index. (*Source: National Institutes of Health.*)

a. Find the Biomedical Research and Developmental Price Index if the Gross Domestic Price Index is 2.2.

b. If the Biomedical Research and Developmental Price Index is 5.22, find the Gross Domestic Product Price Index.

c. Draw the graph. (Use values of x between 0 and 5.)

d. Use the graph to estimate the Gross Domestic Product Price Index when the Biomedical Research and Developmental Department's Price Index is 4.

Exercise Set 4.2

For Extra Help

MyMathLab®

Complete the ordered pairs and use them to draw the graph of each equation. (See Example 1.)

1. $y = 2x - 3$
$(0, \quad)(2, \quad), (-2, \quad)$

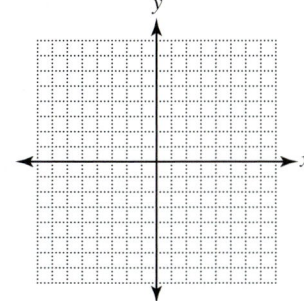

2. $y = 3x - 1$
$(0, \quad)(1, \quad)(-2, \quad)$

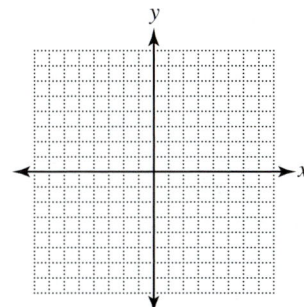

3. $x + y = 5$
$(0, \quad)(\quad, 0)(2, \quad)$

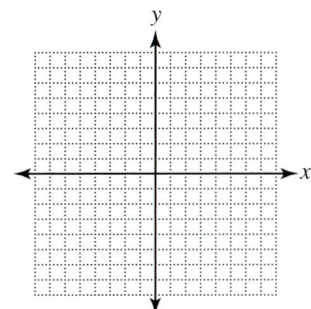

4. $x - y = 3$
$(0, \quad)(\quad, 0)(-2, \quad)$

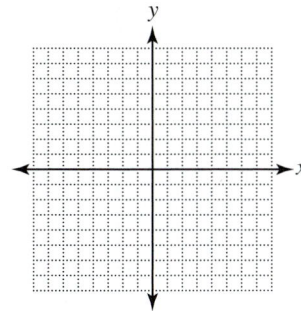

5. $2x + y = 6$
$(0, \quad)(\quad, 0)(\quad, 4)$

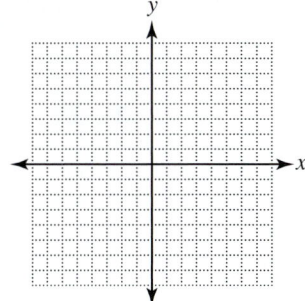

6. $x - 3y = 9$
$(0, \quad)(\quad, 0)(\quad, -2)$

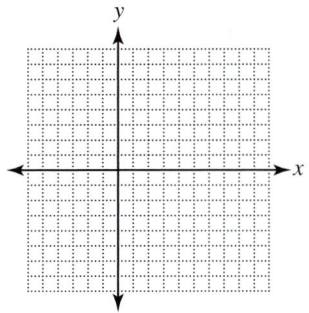

Answers: Practice Exercise 4: a. 3.96 **b.** 3.6 **c.**

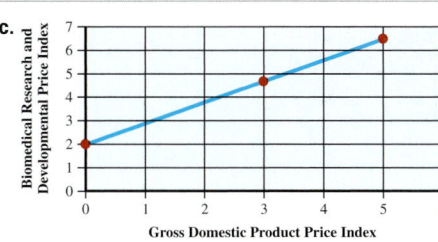

d. 2.3 (actual 2.44)

Find three ordered pairs that solve the equation and draw the graph of each. (See Examples 1–3.)

7. $y = x + 2$

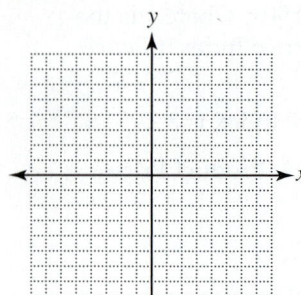

8. $y = x - 4$

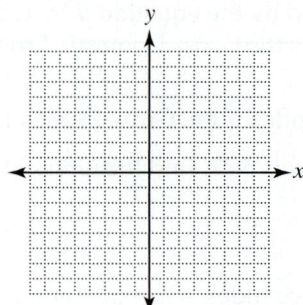

9. $y = 2x + 4$

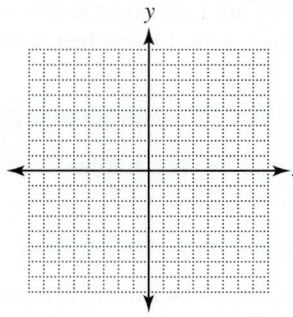

10. $y = 3x - 5$

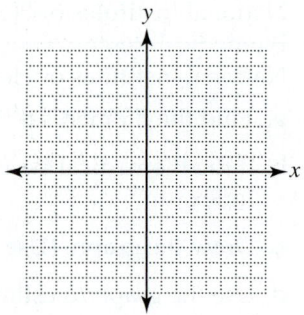

11. $y = -2x + 1$

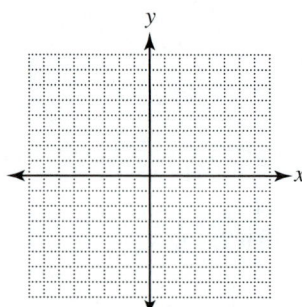

12. $y = -3x + 5$

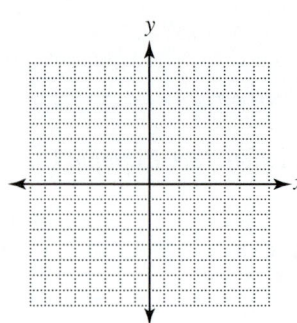

13. $x + y = 3$

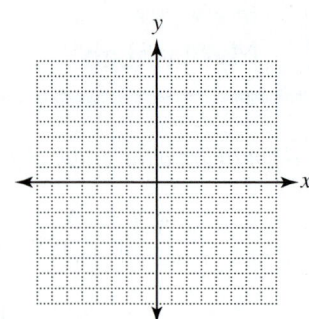

14. $x + y = 6$

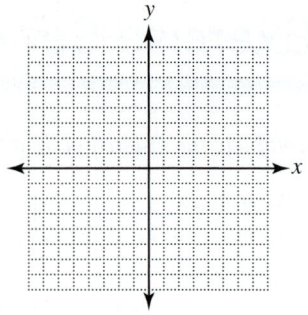

15. $x + 3y = 9$

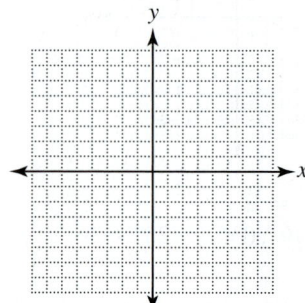

16. $x + 4y = 8$

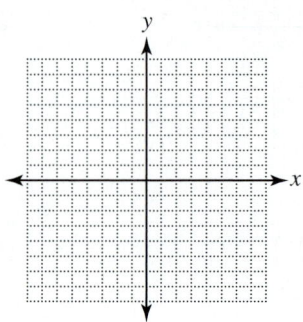

17. $x - 2y = 8$

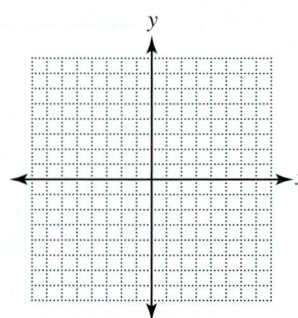

18. $x - 3y = 9$

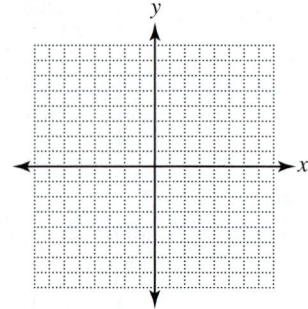

19. $4x - y = 2$

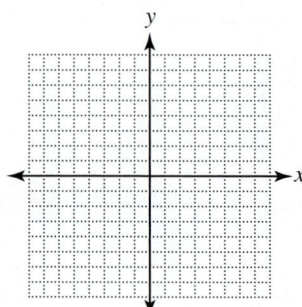

20. $x - 4y = 6$

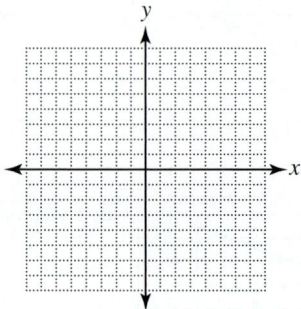

21. $2x + 3y = 6$

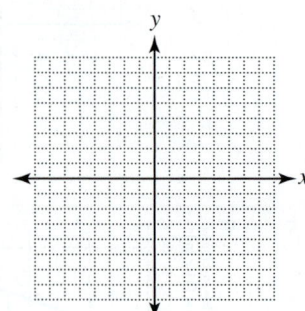

22. $3x + 2y = 6$

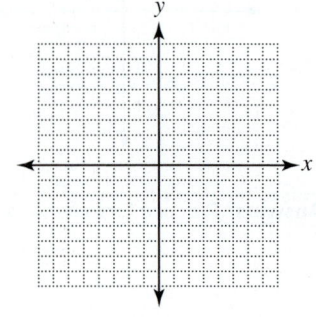

23. $2x - 5y = 10$

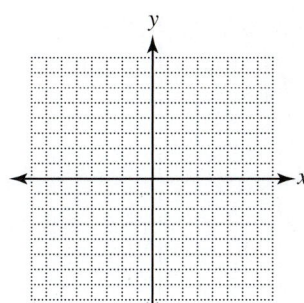

24. $5x - 2y = 10$

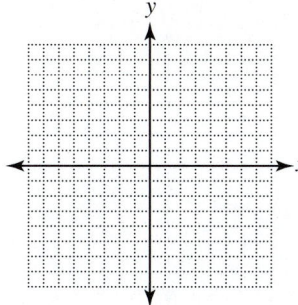

25. $2x + 6y = 12$

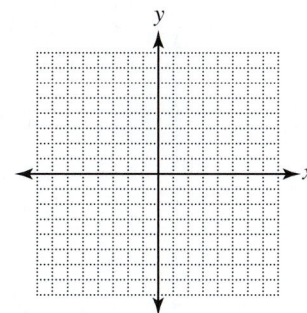

26. $6x + 2y = 12$

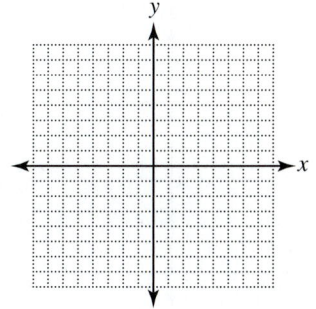

27. $3x - 4y = 12$

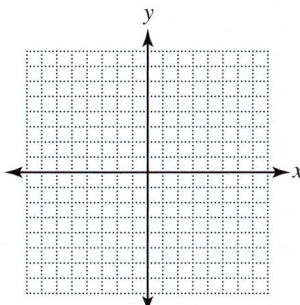

28. $5x - 3y = 15$

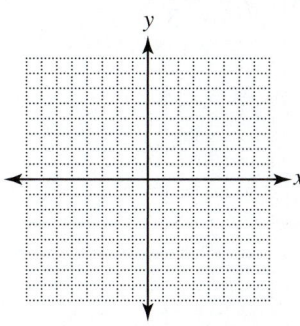

29. $2x + 5y = 5$

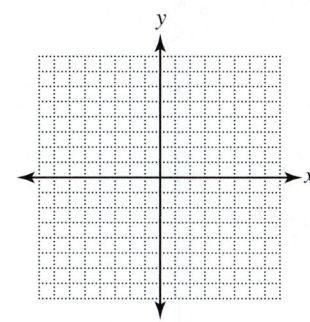

30. $2x + 7y = 7$

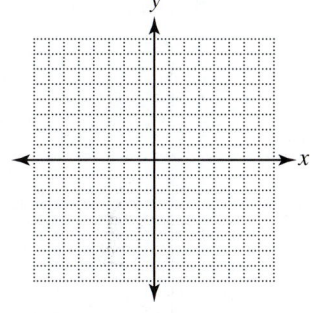

31. $3x - 5y = 10$

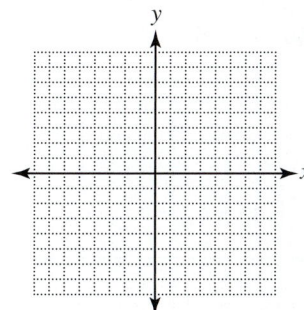

32. $4x - 3y = 8$

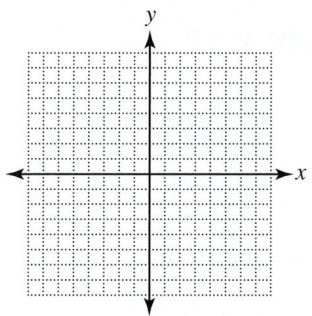

33. $4x - 3y = 10$

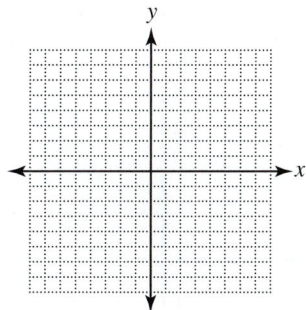

34. $3x - 4y = 14$

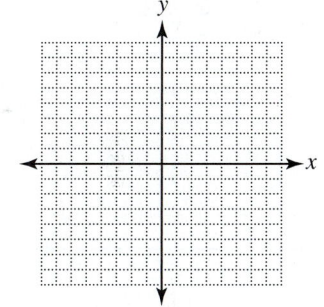

35. $y = x$

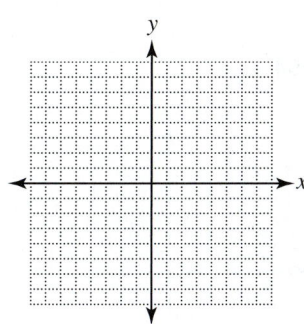

36. $y = 4x$

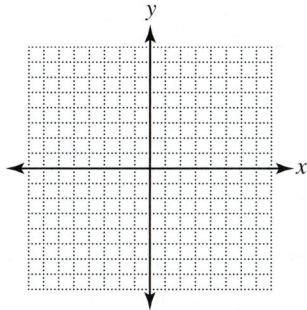

37. $y = -3x$

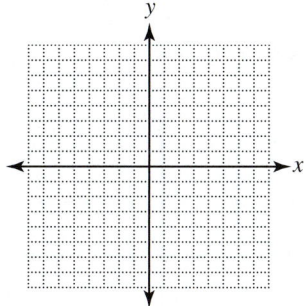

38. $y = -4x$

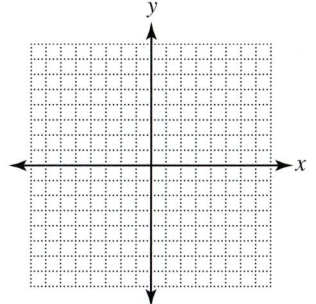

39. $2x + y = 0$

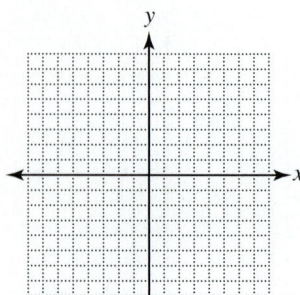

40. $3x - y = 0$

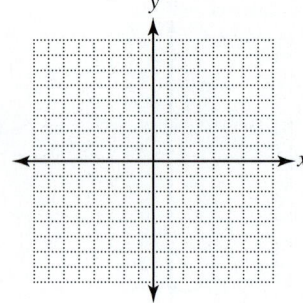

41. $-2x + y = 0$

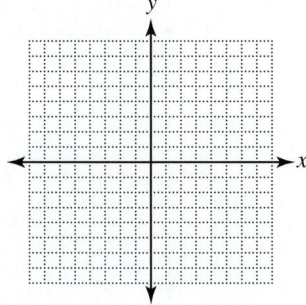

42. $-4x - y = 0$

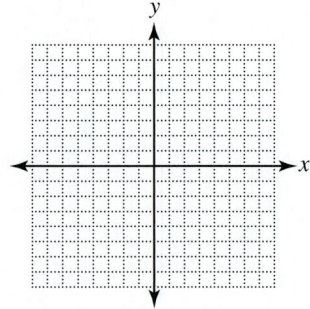

43. $y = \frac{1}{4}x$

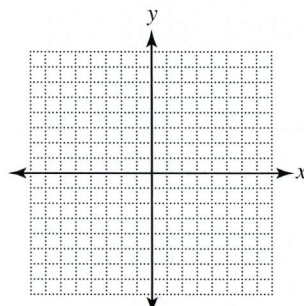

44. $y = \frac{1}{2}x$

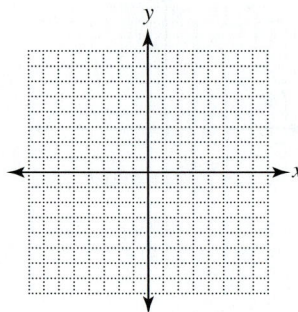

45. $y = -\frac{1}{3}x$

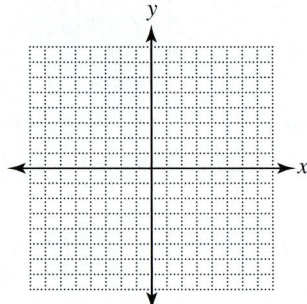

46. $y = -\frac{1}{5}x$

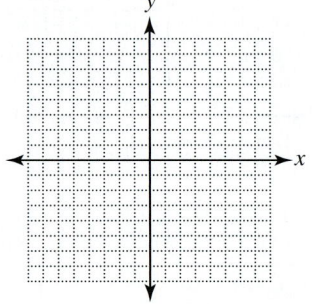

47. $y = \frac{2}{3}x$

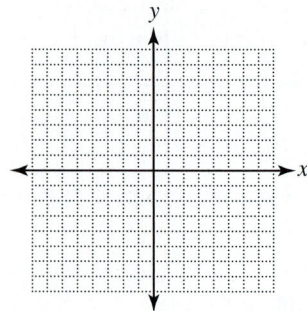

48. $y = \frac{3}{4}x$

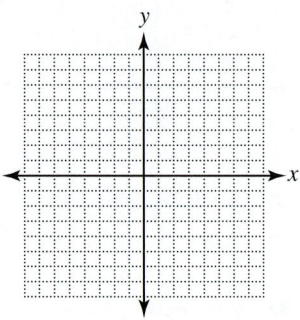

49. $x = 2$

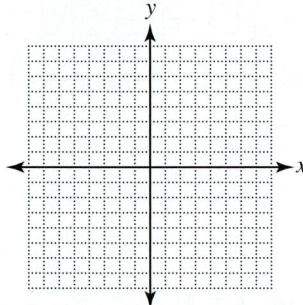

50. $x = 5$

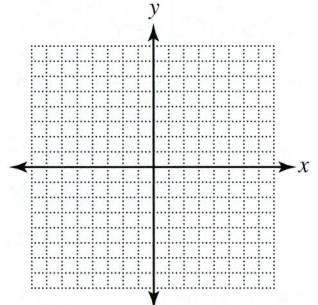

51. $y = 3$

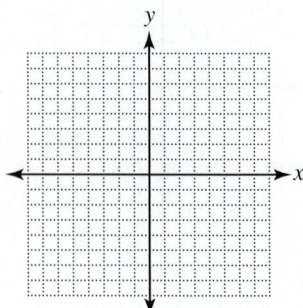

52. $y = 7$

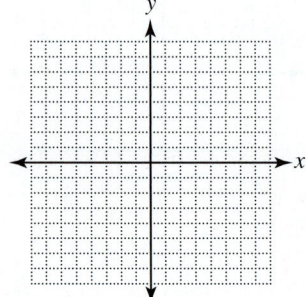

53. $x = -4$

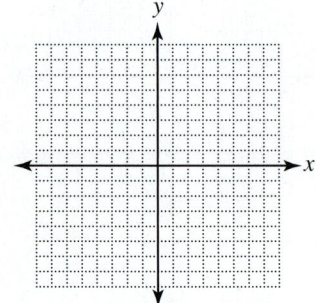

54. $x = -2$

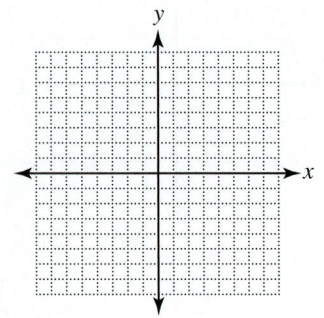

55. $y = -4$

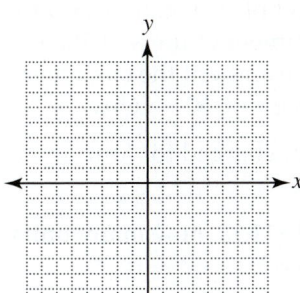

56. $y = -7$

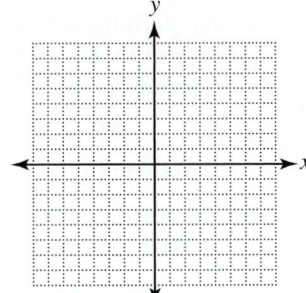

57. $y = -9$

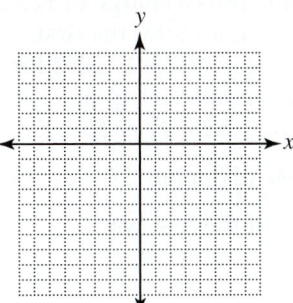

58. $x + 5 = 0$

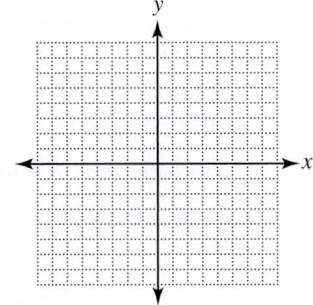

59. $x - 6 = 0$

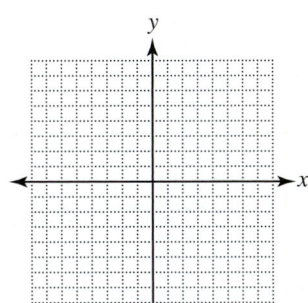

60. $y - 7 = 0$

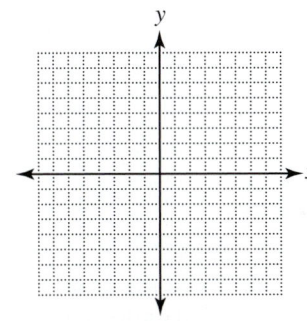

61. $y + 1 = 0$

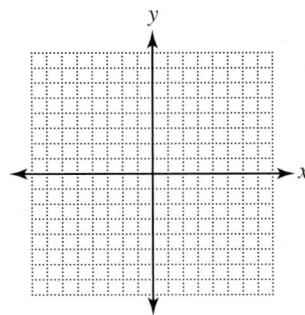

Answer the following. (See Example 4.)

62. The cost for renting a car is $25 per day plus $.15 per mile. If x represents the number of miles and y represents the cost in dollars, then $y = .15x + 25$.

   a. Find the cost of a one-day rental if the car was driven 40 miles.

   b. If the cost is $36.25 for a one-day rental, how many miles was the car driven?

   c. Draw the graph if the number of miles is less than 150.

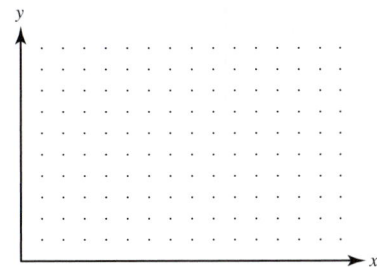

   d. Using the graph, estimate the number of miles driven if the cost of a one-day rental is $34. Verify your answer from the equation.

63. The cost to rent a small truck is $19.95 per day and $.79 per mile. If x represents the number of miles driven and y the cost in dollars for a one-day rental, then $y = .79x + 19.95$.

   a. Find the cost of a one-day rental if the truck was driven 50 miles.

   b. If the cost for a one-day rental was $87.10, find the number of miles the truck was driven.

   c. Draw the graph. (Assume that the truck was driven fewer than 100 miles.)

   d. Using the graph, estimate the number of miles the truck was driven if the one-day cost of the rental was $91.05. Verify your answer from the equation.

64. A cab company charged $4.25 for the first mile and $2.25 for each additional mile plus a charge of $2.00. If x represents the number of miles after the first mile and y is the charge, then $y = 6.25 + 2.25x$ for trips of a mile or more. For example, a trip of 4 miles would cost $y = 6.25 + 2.25(3) = \$13.00$.

a. Find the cost of a taxi ride of 8 miles.

b. Find the number of miles if the cost was $17.50.

c. Find three ordered pairs and draw the graph.

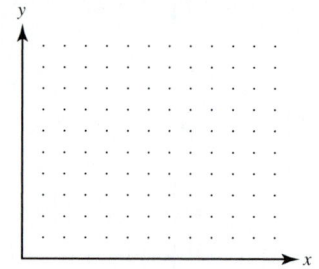

d. From the graph, estimate the number of miles if the cost was $40.00. Verify your answer by using the equation.

66. John's Lawn Service bought a new truck for $25,000, and the truck depreciates at the rate of $2500 per year. If x represents the number of years since the truck was purchased and y represents the value of the truck, then $y = 25,000 - 2500x$ for $0 \le x \le 10$.

a. Find the value of the truck 4 years after it was purchased.

b. Find the number of years after the truck was purchased if its value is $7500.

c. Find three ordered pairs and draw the graph.

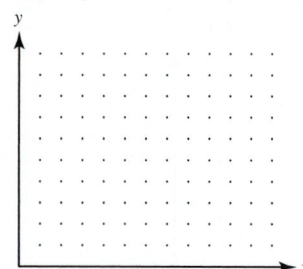

d. From the graph, estimate the number of years after the truck was purchased if the value is $10,000. Verify your answer from the equation.

65. Customers of a water system are charged $6.60 per month plus 65 cents per unit of 1000 gallons, up to 10,000 gallons. If x is the number of units of 1000 gallons of water and y is the amount the customer is charged per month, then $y = 6.60 + .65x$.

a. How much is a customer charged for four units of 1000 gallons?

b. If a customer is charged $11.15, how many units of 1000 gallons did he use?

c. Find three ordered pairs and draw the graph.

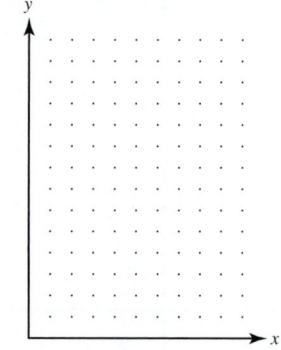

d. From the graph, estimate how many units of 1000 gallons a customer used if she was charged $12.45. Verify your result from the equation.

67. If a man and a woman have the same size feet, the woman's shoe size is 1.5 less than the man's shoe size. If x represents the man's shoe size and y the woman's, then $y = x - 1.5$.

a. If a man's shoe size is 9, what is the comparable woman's shoe size?

b. If a woman's shoe size is 7, what is the comparable man's shoe size?

c. Find three ordered pairs and draw the graph.

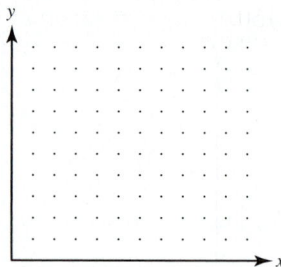

d. From the graph, estimate a woman's shoe size that corresponds to a size 10 for men. Verify your answer from the equation.

68. A salesperson at a department store works for a fixed rate plus a commission. She is paid $250 plus 8% of her sales. If x represents her sales and y represents the amount that she is paid, then $y = 250 + .08x$.

    a. Find the amount that she is paid if her sales are $800.

    b. Find the amount of her sales if she is paid $362.

    c. Find three ordered pairs and draw the graph.

    d. From the graph, estimate the amount of her sales if she is paid $330. Verify your answer from the equation.

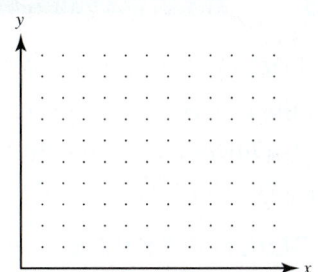

Challenge Exercises (69 and 70)

69. A rock is dropped from the top of a tall building. If y represents the velocity and x represents the time in seconds after the object is dropped, then $y = -32x$.

    a. Find the velocity at 0 seconds, 2 seconds, and 4 seconds. Write the results as ordered pairs.

    b. Draw the graph for values of x such that $0 \le x \le 5$. You will need to change the units on the y-axis.

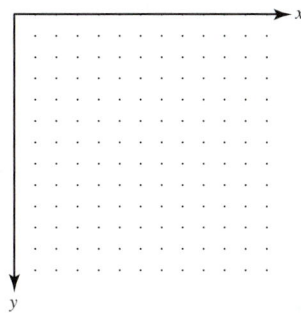

    c. Use the graph to estimate the number of seconds after the object is thrown when the velocity is -96 feet per second.

70. A stone is thrown upward from the top of a tall building, with an initial velocity of 150 feet per second. If y represents the velocity and x the number of seconds after the object is thrown, then $y = -32x + 150$.

    a. Find the velocity after 1 second and 3 seconds. Write the results as ordered pairs.

    b. Draw the graph.

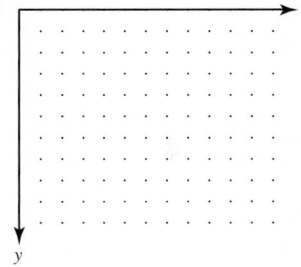

    c. Use the graph to estimate the number of seconds after the object is thrown when the velocity is -86 feet per second.

Writing Exercises (71–75)

71. What does the graph of an equation represent?

72. Why is the graph of an equation in the form $x = k$ a vertical line?

73. Why is the graph of an equation in the form $y = k$ a horizontal line?

74. Since one and only one line can be drawn through any two distinct points, why is it a good idea to always find three points when graphing a line?

75. What is the equation of the x-axis?

Section 4.3 Graphing Linear Equations by Using Intercepts

OBJECTIVES *When you complete this section, you will be able to:*

Ⓐ Find the *x*- and *y*-intercepts from the graph of a linear equation with two variables.

Ⓑ Graph a linear equation with two variables by finding the *x*- and *y*-intercepts.

Ⓒ Graph vertical and horizontal lines.

PREREQUISITE SKILLS) *Before beginning this section, you should be able to:*

a. Add, subtract, multiply, and divide integers. (Sections 1.6, 1.7, 2.1, 2.5)

b. Substitute values for variables and evaluate. (Section 2.5)

c. Solve linear equations. (Sections 3.1, 3.2, 3.3)

d. Plot ordered pairs on the rectangular coordinate system. (Section 4.1)

e. Give the coordinates of a plotted point. (Section 4.1)

GETTING READY FOR SECTION 4.3

Add, subtract, multiply, or divide the following integers:

1. $-9 + 5$ **2.** $13 - (-1)$ **3.** $-2(7)$ **4.** $-16 + 4$

Substitute $x = 0$ and $y = -4$ into the given expressions and evaluate:

5. $6x - 2y$ **6.** $-16x + 5y$

Solve the following equations:

7. $3x + 0 = 9$ **8.** $0 - 6y = 12$

Plot the following ordered pairs on the rectangular coordinate system:

9. $(0, -4)$ **10.** $(-7, 2)$

11. Give the coordinates of the points on the graph.

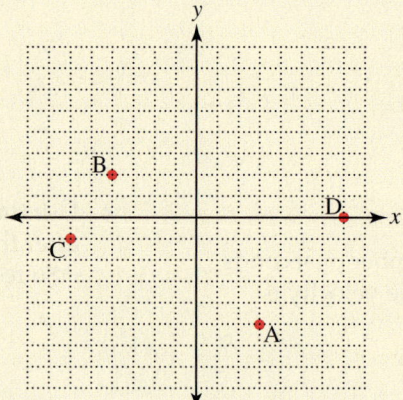

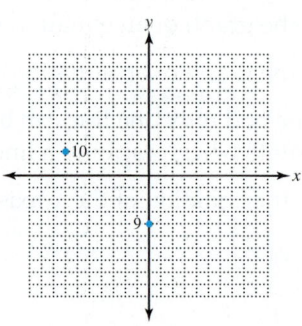

Introduction

In the previous section, we graphed linear equations by finding three ordered pairs that were solutions of the equations, plotting the ordered pairs, and drawing a line through the points. In this section, we will continue that process by finding special points called the *intercepts*.

OBJECTIVE Finding the *x*- and *y*-intercepts from the graph of a linear equation with two variables

The point where a graph intersects the *x*-axis is the **x-intercept**, and the point where the graph intersects the *y*-axis is the **y-intercept**. See the following figure:

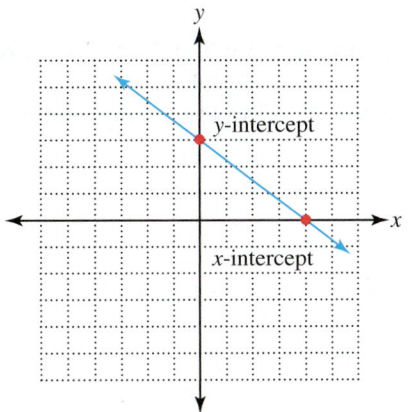

Notice that the *y*-value of the *x*-intercept is 0, because all points on the *x*-axis have a *y*-value of 0, and the *x*-value of the *y*-intercept is 0, because all points on the *y*-axis have an *x*-value of 0. Therefore, the coordinates of the *x*-intercept are of the form $(a, 0)$ and the coordinates of the *y*-intercept are of the form $(0, b)$.

Example 1 **Find the *x*- and *y*-intercepts of the given graphs. Assume that each unit is 1.**

Note

Often, the intercepts are not written as ordered pairs. Sometimes, only the value at which the line crosses the axis is given. In Example 1a, we could say that the *x*-intercept is −3 and the *y*-intercept is 6.

a.

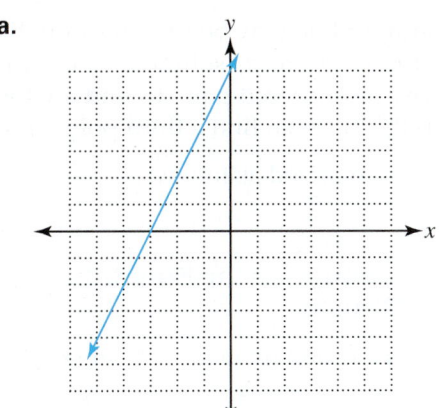
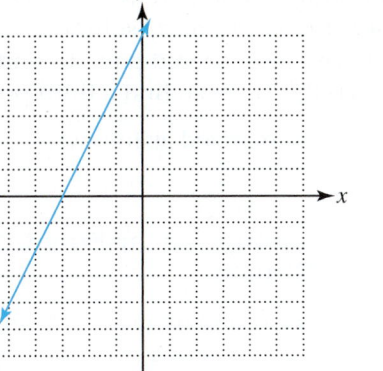

b.

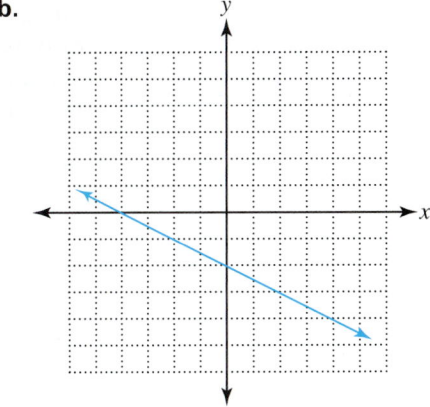

Solution

The graph crosses the *x*-axis at −3, so the *x*-intercept is $(-3, 0)$. The graph crosses the *y*-axis at 6, so the *y*-intercept is $(0, 6)$.

Solution

The graph crosses the *x*-axis at −4, so the *x*-intercept is $(-4, 0)$. The graph crosses the *y*-axis at −2, so the *y*-intercept is $(0, -2)$.

PRACTICE EXERCISE 1

Find the x- and y-intercepts of the following graphs:

a.

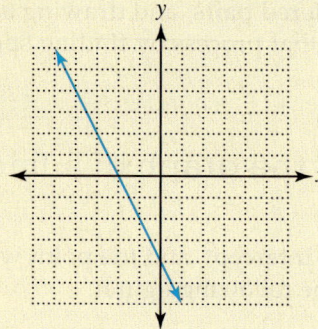

b.

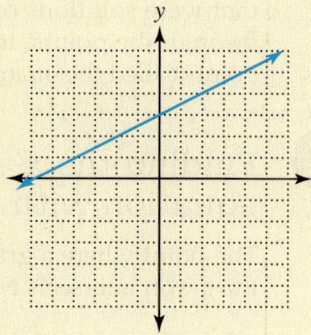

If you need more practice, do the following Additional Practice Exercises.

Additional Practice Exercise 1   Find the x- and y-intercepts of the following graphs:

a.

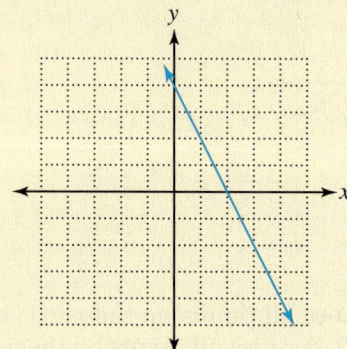

b.

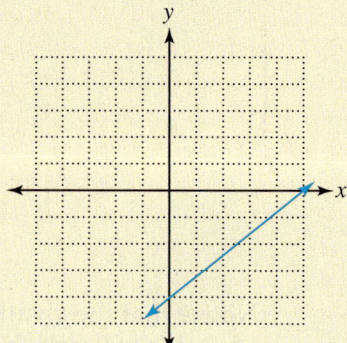

OBJECTIVE **B** ## Graphing a linear equation with two variables by finding the x- and y-intercepts

Since only one straight line may be drawn through any two distinct points, we need only two ordered pairs from the solution set to draw the graph of a linear equation. Usually, the two easiest ordered pairs to find are the x- and y-intercepts, since one of the coordinates is 0. However, find a third point as a check.

The procedure is summarized as follows:

> **Procedure: Graphing Linear Equations by Using the Intercept Method**
>
> To graph a linear equation by using the intercept method, we perform the following steps:
>
> **a.** Let $x = 0$ to find the y-intercept, and let $y = 0$ to find the x-intercept.
>
> **b.** Find a third point by letting x or y have any value, substitute into the original equation, and solve for the remaining variable.
>
> **c.** Plot the ordered pairs.
>
> **d.** Draw a line through the three points.

Answers: Practice Exercise 1: a. $(-2, 0), (0, -4)$ **b.** $(-6, 0), (0, 3)$ **Additional Practice 1: a.** $(2, 0), (0, 4)$ **b.** $(5, 0), (0, -4)$

Example 2 Draw the graph of each equation by finding the *x*- and *y*-intercepts. Find a third point as a check. Assume that each unit on the coordinate system represents one.

a. $2x + y = 6$

Solution

To find the *x*-intercept, let $y = 0$.

$$2x + 0 = 6$$
$$2x = 6$$
$$x = 3$$

Therefore, the *x*-intercept is $(3, 0)$.

To find the *y*-intercept, let $x = 0$.

$$2(0) + y = 6$$
$$0 + y = 6$$
$$y = 6$$

Therefore, the *y*-intercept is $(0, 6)$.

To find another point, let $x = -1$.

$$2(-1) + y = 6$$
$$-2 + y = 6$$
$$y = 8$$

Therefore, the ordered pair is $(-1, 8)$.

In a table, the ordered pairs appear as follows:

x	3	0	−1
y	0	6	8

Plot the points on the rectangular coordinate system and draw a line through them.

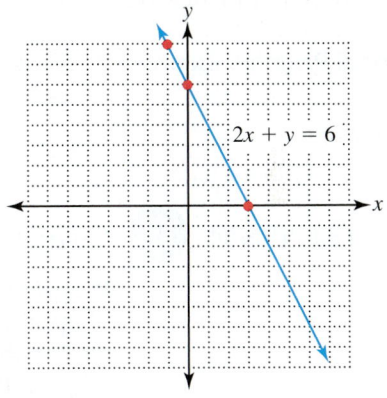

$2x + y = 6$

b. $2x + 3y = 9$

Solution

To find the *x*-intercept, let $y = 0$.

$$2x + 3(0) = 9$$
$$2x + 0 = 9$$
$$2x = 9$$
$$x = \frac{9}{2}, \text{ or } 4.5$$

Therefore, the *x*-intercept is $\left(\frac{9}{2}, 0\right)$, or $(4.5, 0)$.

To find the *y*-intercept, let $x = 0$.

$$2(0) + 3y = 9$$
$$0 + 3y = 9$$
$$3y = 9$$
$$y = 3$$

Therefore, the *y*-intercept is $(0, 3)$.

To find a third point, let $x = -3$.

$$2(-3) + 3y = 9$$
$$-6 + 3y = 9$$
$$3y = 15$$
$$y = 5$$

Therefore, the ordered pair is $(-3, 5)$.

Plot the points on the rectangular coordinate system and draw a line through them.

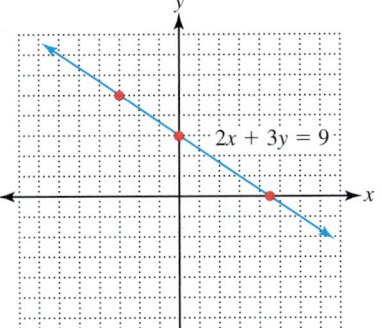

$2x + 3y = 9$

PRACTICE EXERCISE 2

Draw the graph of each equation by finding the x- and y-intercepts. Find a third point as a check. Assume that each unit on the coordinate system represents 1.

a. $x + 4y = 4$

b. $2x - 3y = 9$

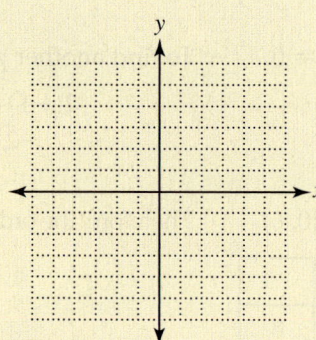
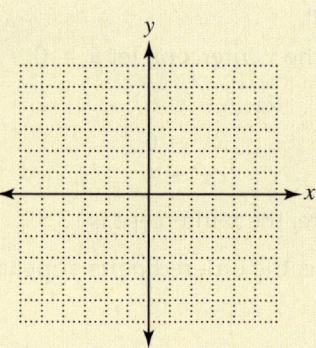

If you need more practice, do the following Additional Practice Exercises.

Additional Practice Exercise 2 Draw the graph of each equation by finding the x- and y intercepts. Find a third point as a check.

a. $3x - y = 6$

b. $3x - 2y = 9$

OBJECTIVE C Graphing vertical and horizontal lines

We graphed vertical and horizontal lines in Section 4.2, but now we have another way of approaching these graphs. Since a vertical line has an equation of the form $x = k$, we can think of a vertical line as having an x-intercept at k and no y-intercept. (Hence it is parallel to the y-axis.) Likewise, since a horizontal line has an equation of the form $y = k$, we can think of a horizontal line as having a y-intercept of k and no x-intercept. (Hence, it is parallel to the x-axis.)

Example 3 Graph each of the following:

a. $x = 4$

Solution

The graph has an x-intercept of 4 and no y-intercept. Therefore, it is a vertical line.

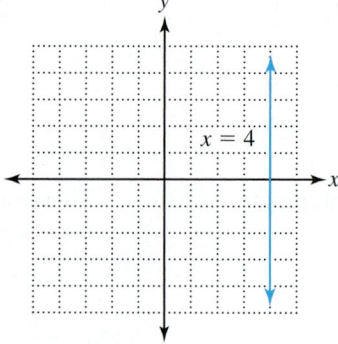

b. $y = -3$

Solution

The graph has a y-intercept of −3 and no x-intercept. Therefore, it is a horizontal line.

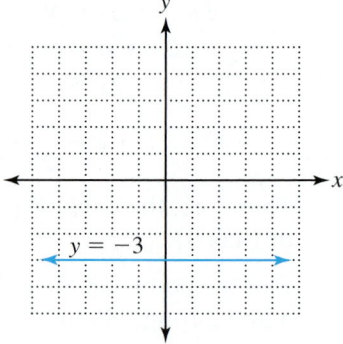

Answers: Practice Exercise 2:

a.
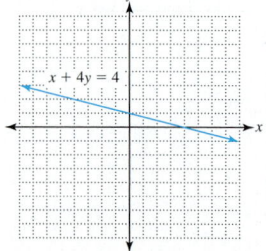

b.
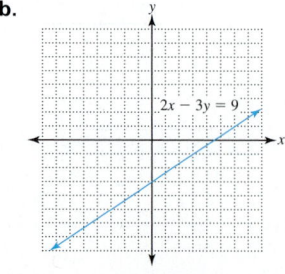

Additional Practice 2:

a.
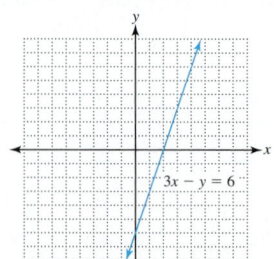

b.
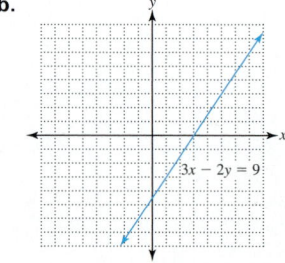

PRACTICE EXERCISE 3

Draw the graph of each of the following:

a. $x = -2$

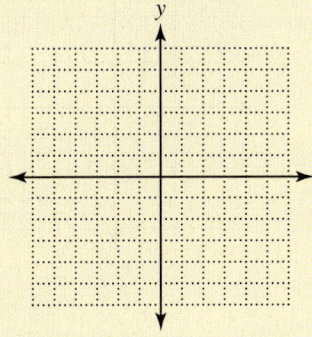

b. $y = 4$

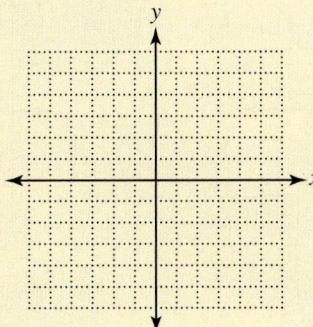

If you need more practice, do the following Additional Practice Exercises.

Additional Practice Exercise 3   Draw the graph of each of the following:

a. $x = 3$ **b.** $y = -4$

Some linear equations are graphed more easily by methods other than using the intercepts. For example, equations that are solved for y, such as $y = \frac{2}{3}x + 5$, are easier to graph by finding three ordered pairs that solve the equation or by using a method to be discussed in Section 4.5.

For Extra Help

Exercise Set 4.3   MyMathLab®

Identify the x- and y-intercepts of the following. (See Example 1.)

1.

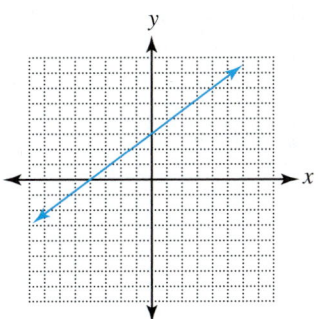

2.

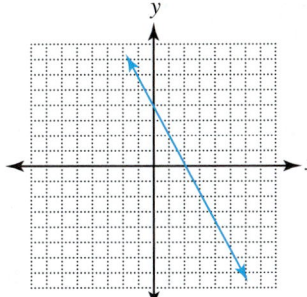

3.

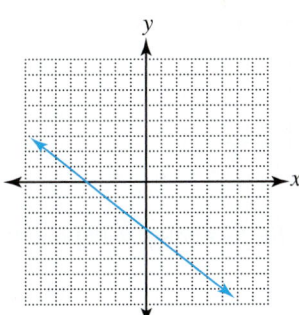

Answers: **Practice Exercise 3:**   **Additional Practice 3:**

a.

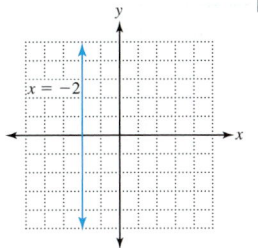

b.

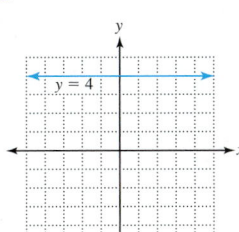

a.

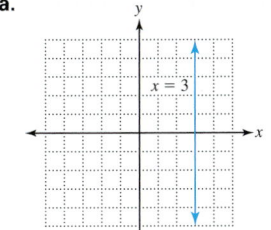

b.

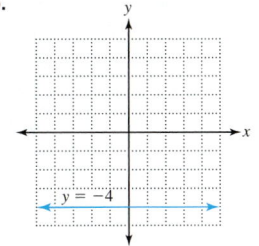

4.

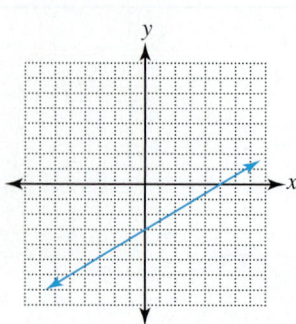

5.

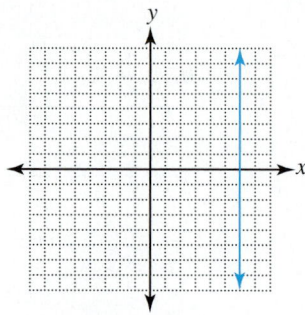

6.

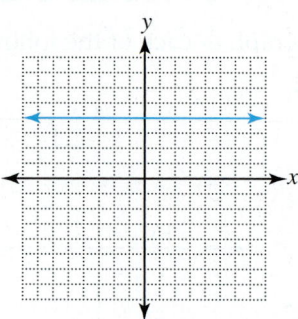

Draw the graph of the lines whose intercepts are given as follows. (See Example 2.)

7. $(3, 0), (0, 4)$

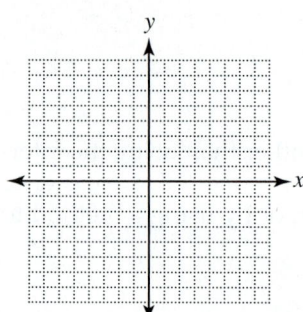

8. $(2, 0), (0, 5)$

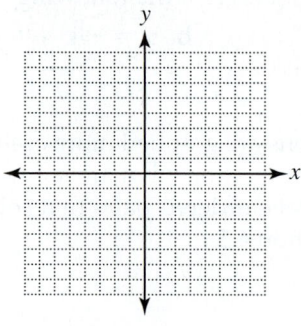

9. $(-3, 0), (0, 4)$

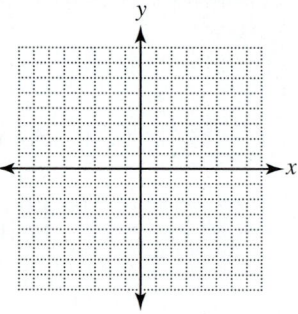

10. $(-5, 0), (0, 2)$

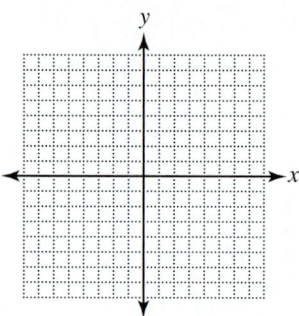

11. $(-4, 0), (0, -3)$

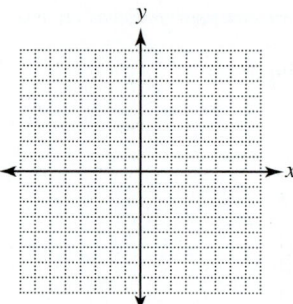

12. $(-7, 0), (0, 5)$

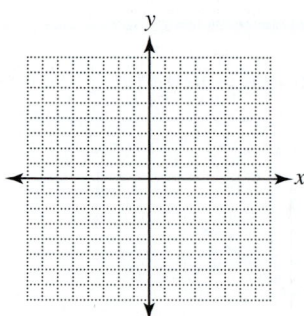

Graph each of the following, using the intercept method. (See Examples 2 and 3.)

13. $x + y = 6$

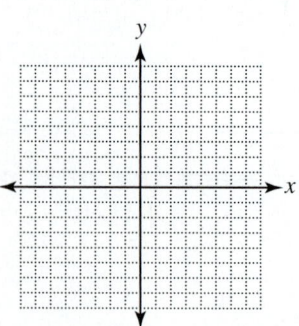

14. $x + y = 3$

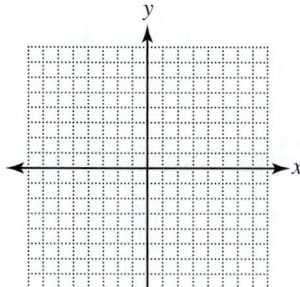

15. $x + 4y = 8$

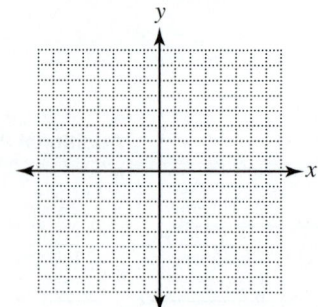

16. $x + 3y = 9$

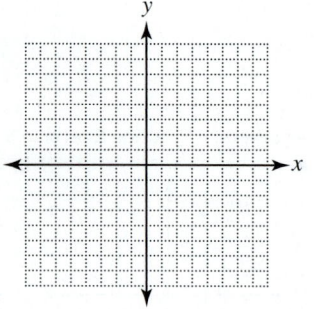

17. $x - 3y = 9$

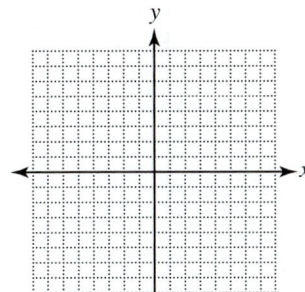

18. $x - 2y = 8$

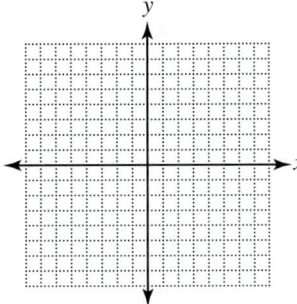

19. $x - 4y = 6$

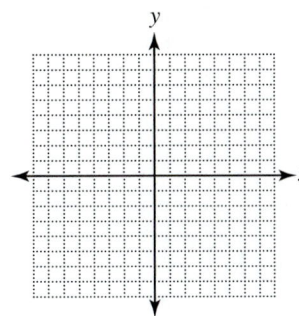

20. $4x - y = 2$

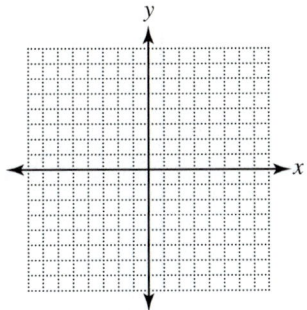

21. $3x + 2y = 6$

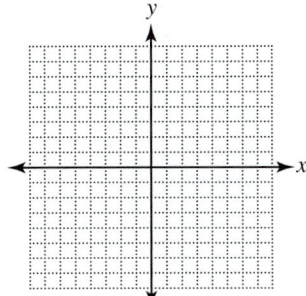

22. $2x + 3y = 6$

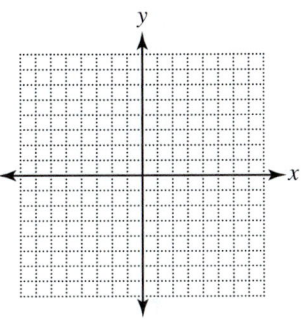

23. $5x - 2y = 10$

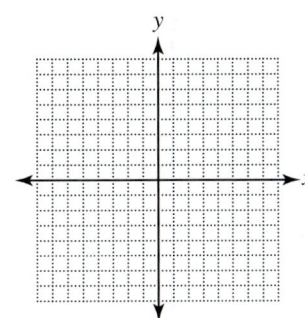

24. $2x - 5y = 10$

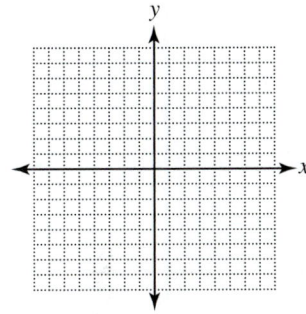

25. $6x - 2y = 12$

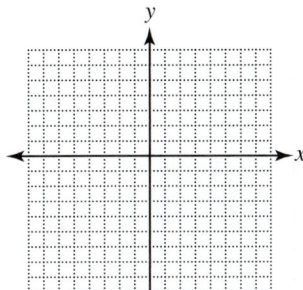

26. $2x - 6y = 12$

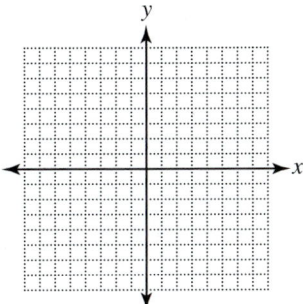

27. $5x - 3y = 15$

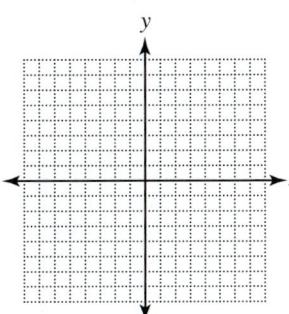

28. $3x - 4y = 12$

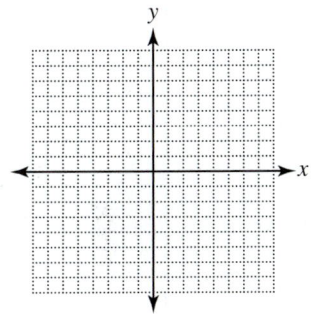

29. $2x + 7y = 7$

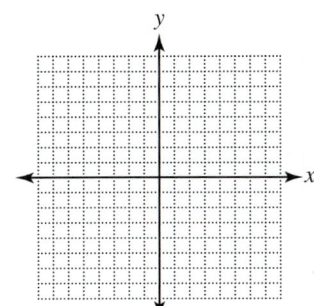

30. $2x + 5y = 5$

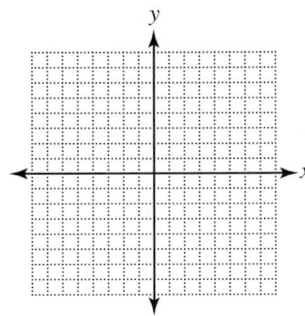

31. $4x - 3y = 8$

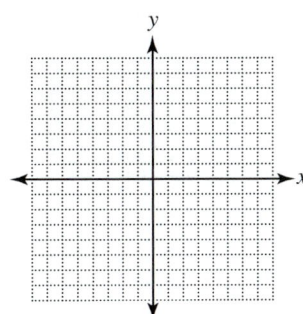

32. $3x - 5y = 10$

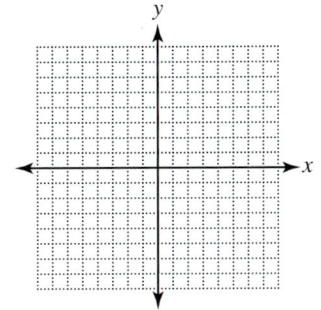

33. $3x - 4y = 14$

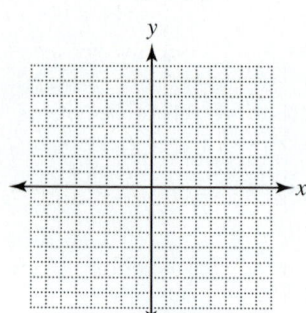

34. $4x - 3y = 10$

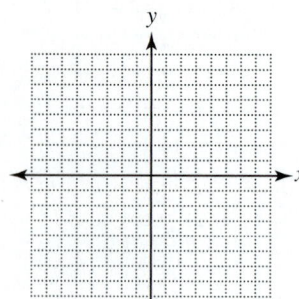

35. $x = 5$

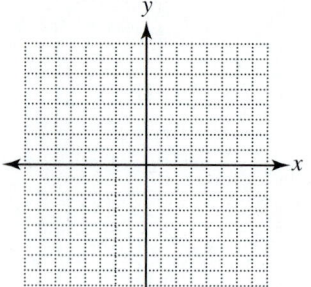

36. $x = 2$

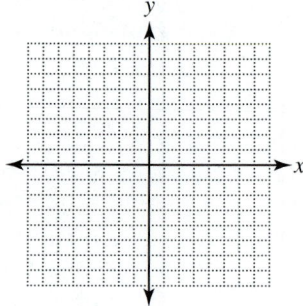

37. $y = 7$

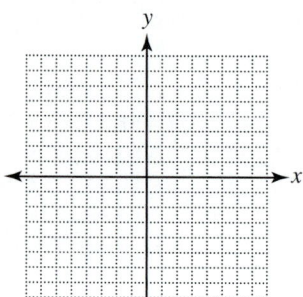

38. $y = 3$

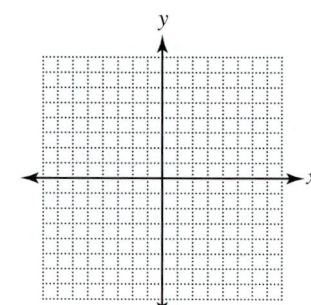

39. $x = -2$

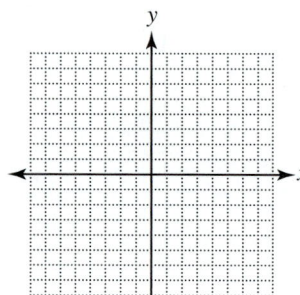

40. $x = -4$

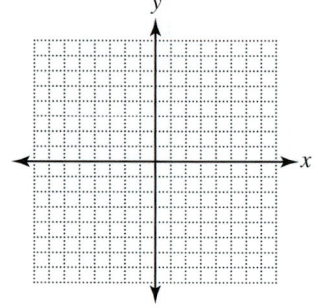

41. $y = -7$

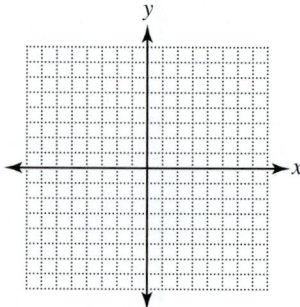

42. $y = -4$

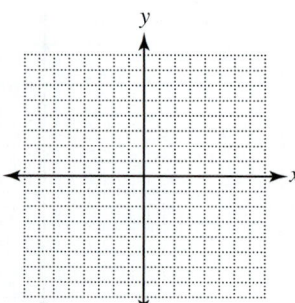

43. $x + 5 = 0$

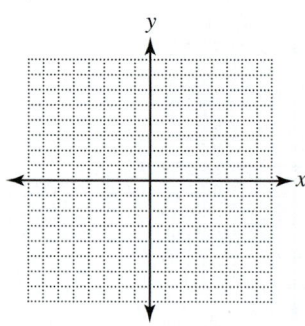

44. $y + 9 = 0$

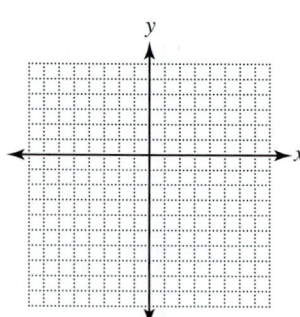

45. $y - 7 = 0$

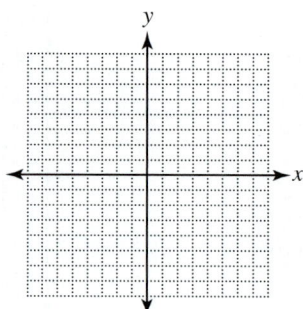

46. $x - 6 = 0$

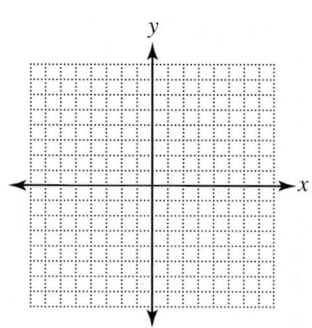

Write an equation for each of the following:

47. The Maitland Public Library was having a plant sale. If Katrina bought x plants for $4 each and y plants for $6 each, write an equation that shows the possible combinations of $4 and $6 plants if the total cost was $36.

48. If x represents the length of a rectangle and y represents the width, write an equation that shows the possible combinations of lengths and widths if the perimeter is 32 feet.

49. If there are x \$5 bills and y \$10 bills, write an equation that shows the possible combinations of \$5 and \$10 bills if the total value of the bills is \$450.

50. John receives \$6 per hour for regular time worked and \$9 per hour for overtime. If John works x hours of regular time and y hours of overtime, write an equation that shows the possible combinations of regular hours and overtime hours if John received \$420.

51. Abdul took a trip. For the first part of the trip, he averaged 50 miles per hour and for the second part he averaged 60 miles per hour. If Abdul drove x hours at 50 miles per hour and y hours at 60 miles per hour, write an equation that shows the possible combinations of hours if the trip was 500 miles long.

52. In basketball, a field goal counts for 2 points and a free throw counts for 1 point. If Jamal scored x field goals and y free throws, write an equation that shows the possible combinations of field goals and free throws if he scored 27 points.

Challenge Exercises (53–58)

53. If the graph of $ax + by = 15$ has an x-intercept of $(5, 0)$ and a y-intercept of $(0, 3)$, find a and b.

54. If the graph of $ax + by = -8$ has an x-intercept of $(4, 0)$ and a y-intercept of $(0, -2)$, find a and b.

Find the intercepts of the following:

55.

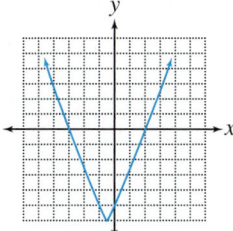

56.

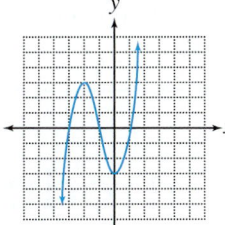

57.

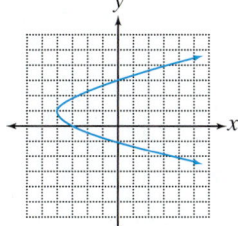

58.

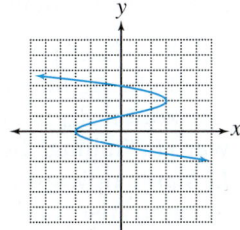

Writing Exercise

59. When graphing an equation of the form $ax + by = c$, why do we usually find the x- and y-intercepts as two of the points?

Section 4.4   Slope of a Line

OBJECTIVES *When you complete this section, you will be able to:*

A Find the slope of a line by using the slope formula.

B Find the slope of a line from its graph.

C Graph a line when given a point on the line and the slope of the line.

PREREQUISITE SKILLS *Before beginning this section, you should be able to:*

a. Add, subtract, multiply, and divide integers. (Sections 1.6, 1.7, 2.1, 2.5)

b. Substitute values for variables and evaluate. (Section 2.5)

c. Plot ordered pairs on the rectangular coordinate system. (Section 4.1)

GETTING READY FOR SECTION 4.4

Add, subtract, multiply, or divide the following integers:

1. $-4 - 4$ **2.** $5 - (-2)$ **3.** $-12(-6)$ **4.** $-18 \div (-6)$

Evaluate the following for the given values of the variables:

5. $\dfrac{a - b}{c - d}$; $a = 9, b = 3, c = 7, d = 2$ **6.** $\dfrac{a - b}{c - d}$; $a = -2, b = 3, c = 4, d = -2$

Plot the following ordered pairs on the rectangular coordinate system:

7. $(-4, 0)$ **8.** $(-6, -6)$

Introduction

Thus far, we have been graphing lines by finding at least two points on the line and then drawing a line through the points. In this section, we will learn another method.

Consider the graphs of the following lines, all of which contain the point (3, 3):

All these lines pass through the same point, yet there are substantial differences. L_1 is horizontal, L_4 is vertical, L_5 slants downward (a "falling" line) from left to right, and L_2 slants upward (a "rising" line) from left to right. Also, both L_2 and L_3 slant upward from left to right, but L_3 is steeper than L_2. One way of describing these differences is by using **slope**.

Note

You may not be familiar with the use of subscripts. In the notations L_1, L_2, L_3, and so on, the L's denote lines. The numbers are called *subscripts* and are used to denote different lines. Hence, L_1 stands for the first line, L_2 stands for the second line, and so on.

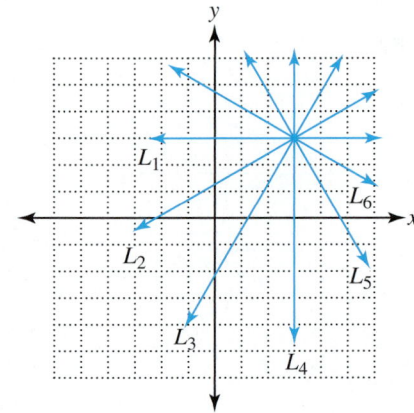

Answers: Getting Ready: 1. -8 **2.** 7 **3.** 72 **4.** 3 **5.** $\dfrac{6}{5}$ **6.** $-\dfrac{5}{6}$ **7–8.**

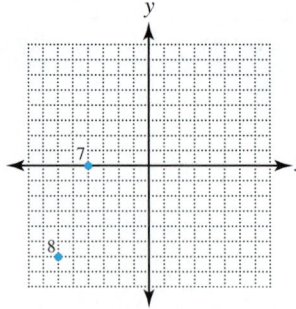

> **Definition** Slope of a Line
>
> The slope of a line is the ratio of the change in y to the change in x as you go from one point on the line to any other point on the line.

An informal way of thinking of slope is that it is the ratio of the rise in a line to the run of the line between any two points on the line, as illustrated by the accompanying figure. As such, it can be thought of as the rate of the change in y with respect to x.

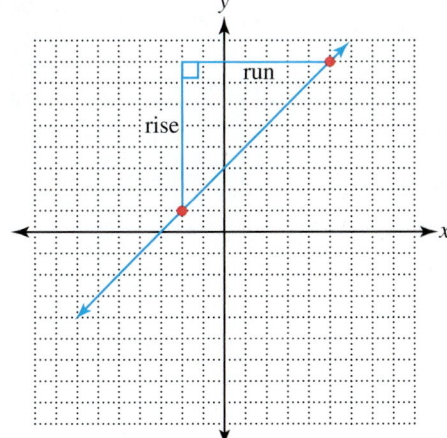

This informal way of thinking of slope is consistent with the formal definition. The rise is a vertical change and, since the y-axis is vertical, any vertical change is a change in y. Also, the run is a horizontal change, and since the x-axis is horizontal, any horizontal change is a change in x.

OBJECTIVE **Finding the slope of a line by using the slope formula**

As is often the case in mathematics, the definition gives us no indication of how to actually perform the task. Suppose we want to find the slope of the line that contains the points (2, 3) and (7, 9). Plot the points and draw the line containing these points. Then go from point (2, 3) to (7, 9) by first going vertically up from (2, 3) and then horizontally to (7, 9), as illustrated.

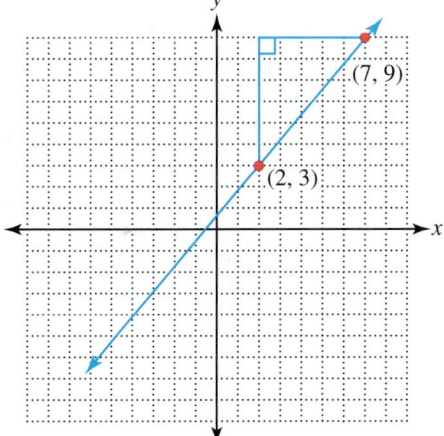

Since any two points on the same vertical line have the same x-value and any two points on the same horizontal line have the same y-value, the point of intersection of the horizontal and vertical lines is (2, 9).

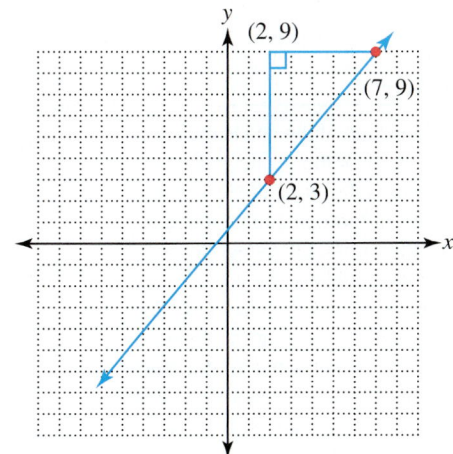

The rise is the change in y as we go from $(2, 3)$ to $(2, 9)$. The y-value changes from 3 to 9, which is a total change of 6. This change in y (rise) can be found by finding the difference of the y-values ($9 - 3 = 6$). The run is the change in x as we go from $(2, 9)$ to $(7, 9)$. The x-value changes from 2 to 7, which is a change of 5. This change in x (run) can be found by finding the difference of the x-values ($7 - 2 = 5$).

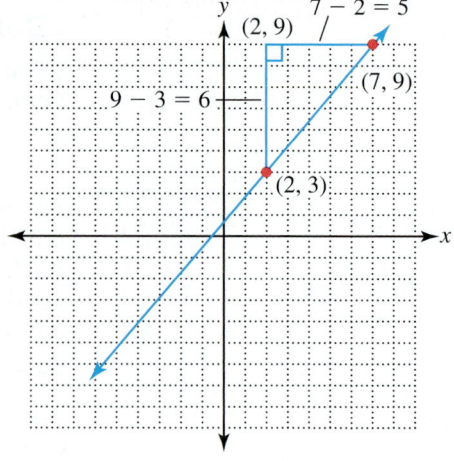

Therefore, the slope is

$$\frac{\text{rise}}{\text{run}} = \frac{\text{change in } y}{\text{change in } x} = \frac{9 - 3}{7 - 2} = \frac{6}{5}.$$

Instead of going through all of this each time we want to find the slope of a line, we use any line of the form $ax + by = c$ and any two points on the line. We denote the points as (x_1, y_1) and (x_2, y_2) and derive a formula for slope. Subscripts indicate that both points have x- and y-values, but these values may be different. The letter m is customarily used for slope. Consider the following figure:

It is not really known why the letter m is used to denote slope. One belief is that it comes from the French word "monter," which means "to climb."

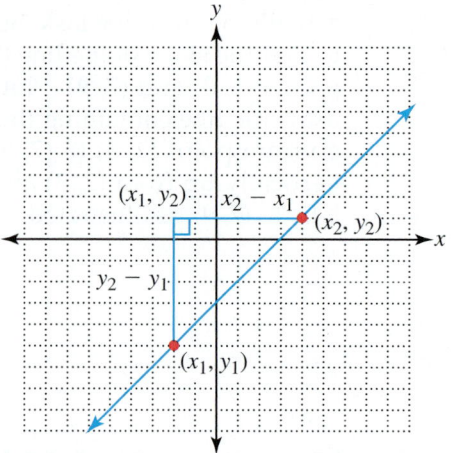

Note

The slope of the line is the difference of the y-values divided by the difference of the x-values. This difference can be found in either order. Consequently, the slope could also be $m = \frac{y_1 - y_2}{x_1 - x_2}$. However, we must be very careful. Whichever point's y-value is used first in the numerator, that point's x-value must also be used first in the denominator. Also, very simple reduction of fractions is needed in the remainder of this chapter. A thorough discussion of this topic is in Chapter 6.

Since any two points on the same vertical line have the same x-value and any two points on the same horizontal line have the same y-value, the coordinates of the point of intersection of the vertical and horizontal line segments from (x_1, y_1) to (x_2, y_2) are (x_1, y_2). See the preceding figure. As in the previous example, the change in y is the difference in the y-values. Thus, the change in y is $y_2 - y_1$. The change in x is the difference in the x-values. Thus, the change in x is $x_2 - x_1$. Therefore, the formula for the slope of a line containing the points (x_1, y_1) and (x_2, y_2) is

$$m = \frac{y_2 - y_1}{x_2 - x_1}.$$

Rule: Slope Formula

The slope of a line containing the points whose coordinates are (x_1, y_1) and (x_2, y_2) is $m = \frac{y_2 - y_1}{x_2 - x_1}$.

Example 1 | **Find the slope of the line that contains each of the following pairs of points:**

a. $(1, 3)$ and $(5, 9)$

Solution

Let $(x_1, y_1) = (1, 3)$, so $x_1 = 1$ and $y_1 = 3$; and let $(x_2, y_2) = (5, 9)$, so $x_2 = 5$ and $y_2 = 9$. Substitute these values into the slope formula.

$$m = \frac{y_2 - y_1}{x_2 - x_1} \qquad \text{Substitute for } y_2, y_1, x_2, \text{ and } x_1.$$

$$m = \frac{9 - 3}{5 - 1} \qquad \text{Simplify the numerator and the denominator.}$$

$$m = \frac{6}{4} \qquad \text{Reduce to lowest terms.}$$

$$m = \frac{3}{2} \qquad \text{Therefore, the slope is } \frac{3}{2}.$$

Alternative Solution:

If we let $(x_1, y_1) = (5, 9)$ and $(x_2, y_2) = (1, 3)$, then $m = \frac{3 - 9}{1 - 5} = \frac{-6}{-4} = \frac{3}{2}$. Notice that we get the same result.

b. $(-2, 3)$ and $(4, -2)$

Solution

Let $(x_1, y_1) = (-2, 3)$, so $x_1 = -2$ and $y_1 = 3$; and let $(x_2, y_2) = (4, -2)$, so $x_2 = 4$ and $y_2 = -2$. Substitute these values into the slope formula.

$$m = \frac{y_2 - y_1}{x_2 - x_1} \qquad \text{Substitute for } y_2, y_1, x_2, \text{ and } x_1.$$

$$m = \frac{-2 - 3}{4 - (-2)} \qquad 4 - (-2) = 4 + 2.$$

$$m = \frac{-2 - 3}{4 + 2} \qquad \text{Simplify the numerator and the denominator.}$$

$$m = \frac{-5}{6} = -\frac{5}{6} \qquad \text{Therefore, the slope is } -\frac{5}{6}.$$

c. $(3, 2)$ and $(-1, 2)$

Solution

Let $(x_1, y_1) = (3, 2)$, so $x_1 = 3$ and $y_1 = 2$; and let $(x_2, y_2) = (-1, 2)$, so $x_2 = -1$ and $y_2 = 2$. Substitute these values into the slope formula.

$$m = \frac{y_2 - y_1}{x_2 - x_1} \qquad \text{Substitute for } y_2, y_1, x_2, \text{ and } x_1.$$

$$m = \frac{2 - 2}{-1 - 3} \qquad \text{Simplify the numerator and the denominator.}$$

$$m = \frac{0}{-4} \qquad \text{Simplify.}$$

$$m = 0 \qquad \text{Therefore, the slope is 0.}$$

Note

The graph of the line containing $(3, 2)$ and $(-1, 2)$ is horizontal since the y-values are the same.

Note

The graph of the line containing (3, 6) and (3, 2) is vertical since the x-values are the same. Also, do not say there is "no slope" since no slope can be interpreted as 0 slope.

d. (3, 6) and (3, 2)

Solution

Let $(x_1, y_1) = (3, 6)$, so $x_1 = 3$ and $y_1 = 6$; and let $(x_2, y_2) = (3, 2)$, so $x_2 = 3$ and $y_2 = 2$. Substitute these values into the slope formula.

$$m = \frac{y_2 - y_1}{x_2 - x_1}$$ Substitute for y_2, y_1, x_2, and x_1.

$$m = \frac{2 - 6}{3 - 3}$$ Simplify the numerator and the denominator.

$$m = \frac{-4}{0}$$ Since division by 0 is undefined, the slope is undefined.

PRACTICE EXERCISE 1

Find the slope of the line that contains each of the following pairs of points:

a. $(2, 4), (5, 8)$ **b.** $(-2, 3), (5, -2)$ **c.** $(-3, 4), (-1, 4)$ **d.** $(3, -5), (3, -2)$

If you need more practice, do the following Additional Practice Exercises.

Additional Practice Exercise 1   Find the slope of the line that contains each of the following pairs of points:

a. $(4, 7), (6, 3)$ **b.** $(-1, 4), (4, -3)$ **c.** $(2, 2), (-4, 2)$ **d.** $(-5, -2), (-5, -4)$

OBJECTIVE Finding the slope of a line from its graph

Sometimes, it is necessary to find the slope from a graph. To do so, locate two points on the line and use their coordinates in the slope formula.

Example 2   Find the slopes of the following lines:

a.

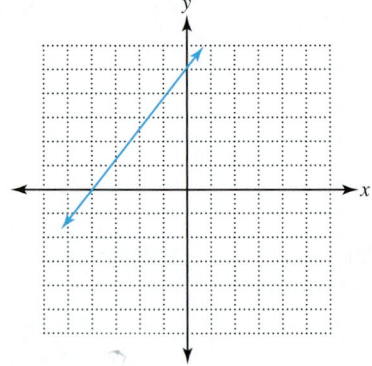

b.

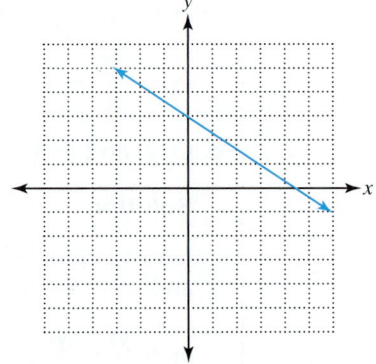

Solution

We need two points on the line. A good place to check is the intercepts. It appears that the x-intercept is $(-4, 0)$ and the y-intercept is $(0, 5)$.

$$m = \frac{y_2 - y_1}{x_2 - x_1}$$ Substitute.

$$m = \frac{5 - 0}{0 - (-4)}$$ Simplify.

$$m = \frac{5}{4}$$

Solution

We need two points. The y-intercept is $(0, 3)$, but the x-intercept does not appear to be an integer. Look for another point. The graph seems to go through $(3, 1)$.

$$m = \frac{y_2 - y_1}{x_2 - x_1}$$ Substitute.

$$m = \frac{1 - 3}{3 - 0}$$ Simplify.

$$m = -\frac{2}{3}$$

Answers: Practice Exercise 1: a. $\frac{4}{3}$ **b.** $-\frac{5}{7}$ **c.** 0 **d.** undefined   **Additional Practice 1: a.** -2 **b.** $-\frac{7}{5}$ **c.** 0 **d.** undefined

PRACTICE EXERCISE 2

Find the slopes of the following lines:

a.

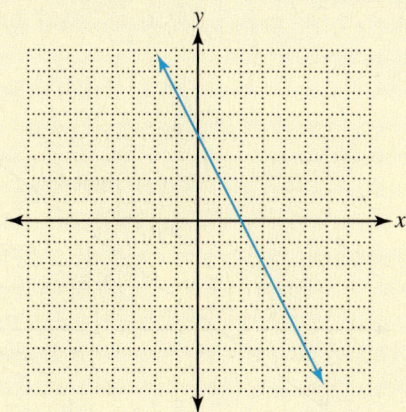

b.

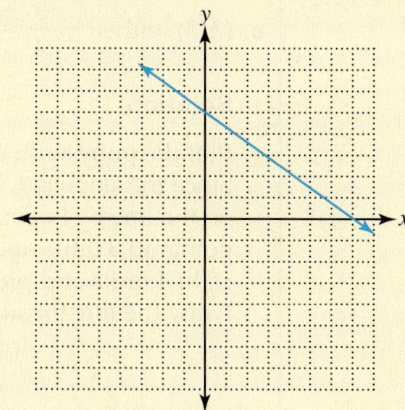

If you need more practice, do the following Additional Practice Exercises.

Additional Practice Exercise 2 Find the slope of each of the following lines:

a.

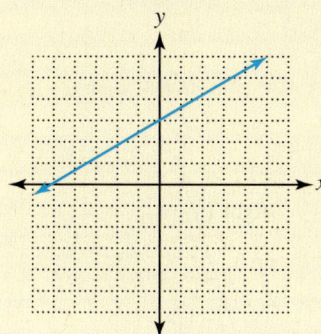

b.

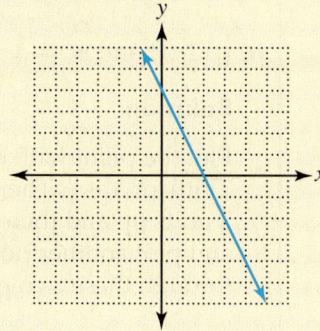

OBJECTIVE **C** Graphing a line when given a point on the line and the slope of the line

Using slope, we now have another method of graphing a line. Remember, the slope represents the change in y (vertical change) divided by the change in x (horizontal change). If we know a point on the line and the slope, we can find the graph of the line by using the following procedure:

> **Procedure: Graphing a Line, Given a Point and the Slope**
>
> 1. Plot the given point.
>
> 2. From the given point, we can find another point by going vertically (up or down) the number of units given by the numerator of the slope and horizontally (right or left) the number of units given by the denominator.
>
> 3. Draw a line through these two points.

A positive vertical change means *go up* and a negative *go down*. A positive horizontal change means *go to the right* and a negative *go to the left*.

Answers: Practice Exercise 2: a. $m = -2$ **b.** $m = -\dfrac{5}{7}$ **Additional Practice 2: a.** $\dfrac{3}{5}$ **b.** -2

Example 3 Draw the graph of each of the following lines containing the given point and with the given slope:

a. $(2, 3)$ with $m = \dfrac{3}{4}$

Solution

Plot the point with coordinates $(2, 3)$. Since the slope is $\frac{3}{4}$, locate another point on the line by beginning at $(2, 3)$ and going up 3 units and then going to the right 4 units and plot another point. Draw the line through these two points.

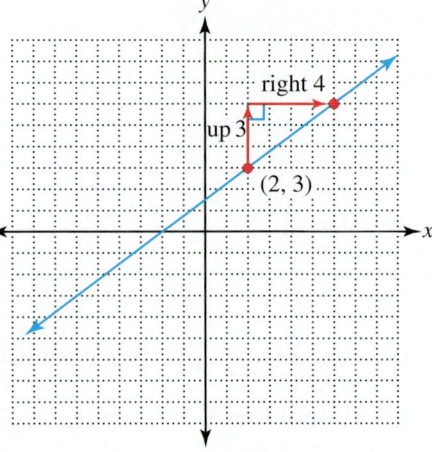

b. $(-2, 2)$ and $m = 3$

Solution

Plot the point with coordinates $(-2, 2)$. Think of 3 as $\frac{3}{1}$. Therefore, from $(-2, 2)$ go 3 units up and then go 1 unit to the right and plot another point. Draw the line through these two points.

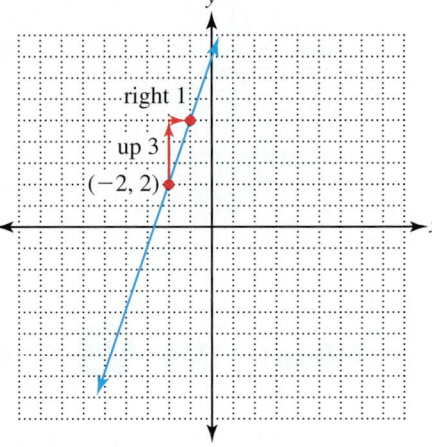

c. $(-3, -1)$ and $m = -\dfrac{2}{3}$

Solution

Plot the point with coordinates $(-3, -1)$. The slope $-\frac{2}{3}$ may be thought of as $\frac{-2}{3}$, or $\frac{2}{-3}$. We will use $\frac{-2}{3}$. Therefore, from $(-3, -1)$ go 2 units down, then go 3 units to the right and plot another point. Draw the line through these two points.

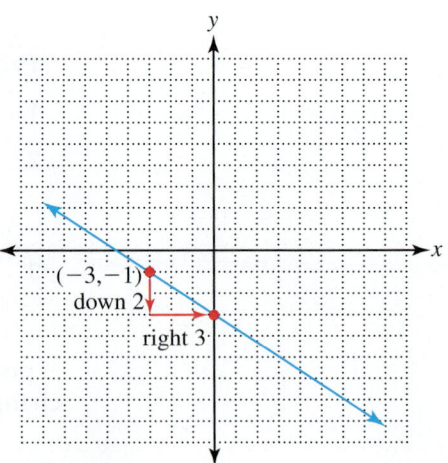

PRACTICE EXERCISE 3

Draw the graph of each of the following lines containing the given point with the given slope:

a. $(3, 4), m = \dfrac{2}{3}$

b. $(-4, 2), m = -2$

c. $(-5, 4), m = -\dfrac{5}{3}$

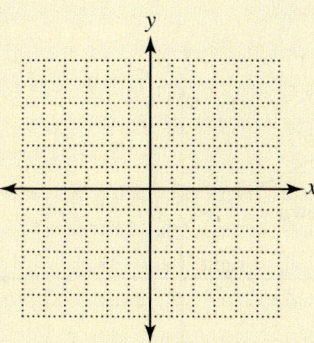

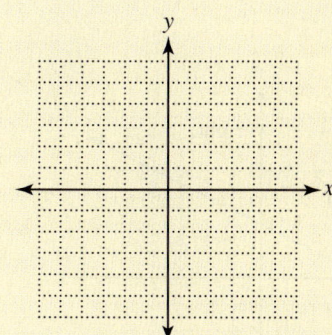

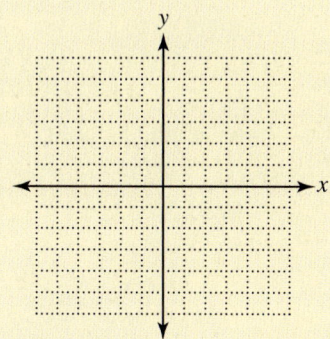

If you need more practice, do the following Additional Practice Exercises.

Additional Practice Exercise 3 Draw the graph of each of the following lines containing the given point with the given slope:

a. $(-1, 4); m = \dfrac{1}{3}$

b. $(3, -4); m = 2$

c. $(-3, 5); m = -\dfrac{2}{3}$

Answers: Practice Exercise 3: **a.**

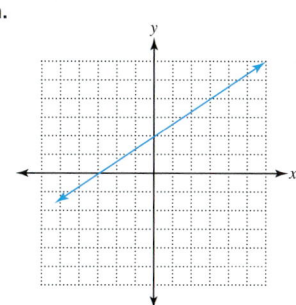

b.

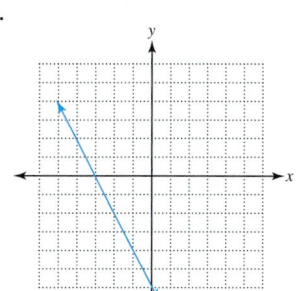

c.

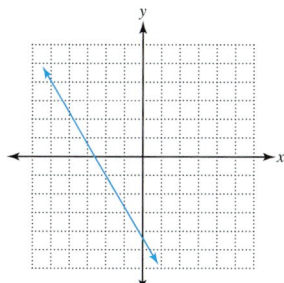

Additional Practice 3: **a.**

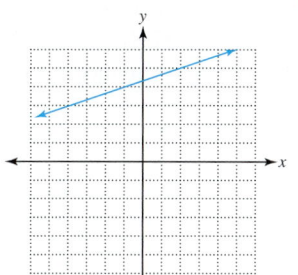

b.

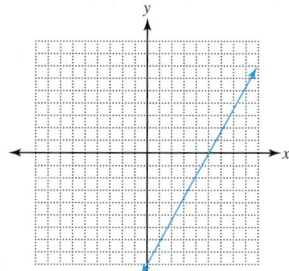

c.

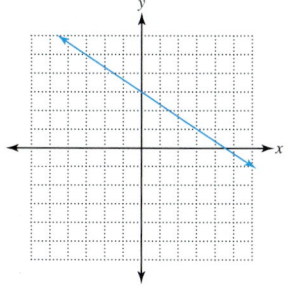

We now make two observations about the slope of a line. Since a positive slope means go up from a given point and then go to the right, a line with *positive slope* rises as we go from left to right. Since a line with a *negative slope* can mean go down from a given point and then go to the right, a line with negative slope drops as we go from left to right. See the following figures:

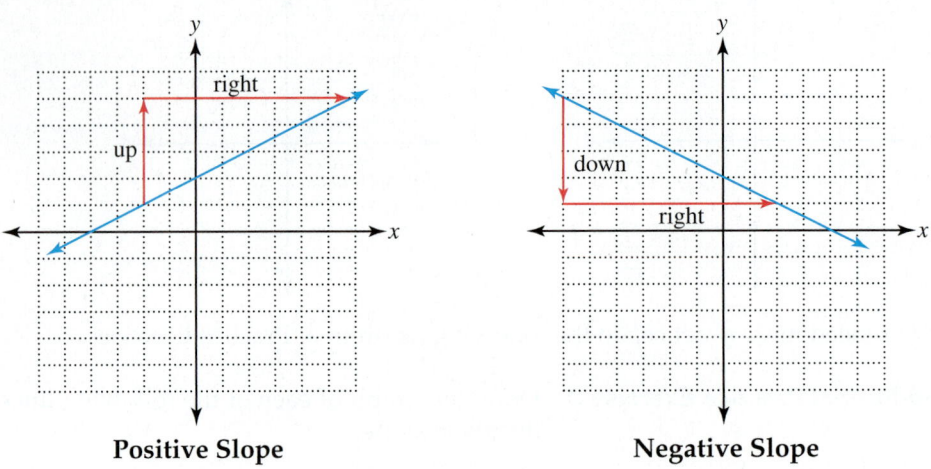

Positive Slope   **Negative Slope**

The greater the absolute value of the slope, the greater the vertical change will be in respect to the horizontal change. Therefore, the greater the absolute value of the slope, the steeper the line. See the following figures:

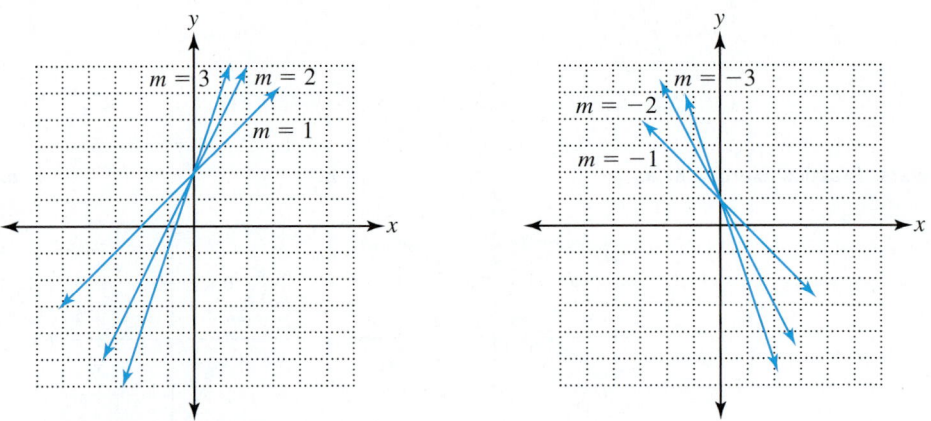

Recall that the graphs of equations of the form $y = k$ are horizontal lines and the graphs of equations of the form $x = k$ are vertical lines. To find the slopes of these lines, we will locate two points on the line and use the slope formula.

Example 4

a. Find the slope of the graph of $x = 3$.

Solution

All points on the graph of $x = 3$ have an x-value of 3, and y can have any value. So two points on the graph are $(3, 0)$ and $(3, 1)$. Substitute these values into the slope formula.

$$m = \frac{y_2 - y_1}{x_2 - x_1} \qquad \text{Substitute.}$$

$$m = \frac{1 - 0}{3 - 3} \qquad \text{Simplify.}$$

$$m = \frac{1}{0} \qquad \begin{array}{l}\textit{Division by 0 is undefined. Therefore, there is no number} \\ \textit{that represents the slope of the graph of } x = 3. \textit{ We say} \\ \textit{that the slope is undefined.}\end{array}$$

Note that there is also no y-intercept.

b. Find the slope of the graph of $y = 2$.

Solution

All points on the graph of $y = 2$ have a y-value of 2, and x can have any value. Two points on the graph are $(0, 2)$ and $(3, 2)$. Use these points in the slope formula.

$$m = \frac{y_2 - y_1}{x_2 - x_1} \qquad \text{Substitute.}$$

$$m = \frac{2 - 2}{3 - 0} \qquad \text{Simplify.}$$

$$m = \frac{0}{3} \qquad \text{Simplify.}$$

$$m = 0 \qquad \text{Therefore, the slope is 0.}$$

Note that there is no x-intercept.

Since the equation of a vertical line is in the form of $x = k$, x remains constant. Since the slope is the change in y divided by the change in x and x is a constant, there is no change in x. Consequently, the denominator is 0, which makes the slope of the vertical line undefined. Similarly, a horizontal line has an equation of the form $y = k$, which means that y is constant. Therefore, there is no change in y, so the numerator is 0. This means that the slope of the horizontal line is always 0.

From the preceding, we make the following generalizations:

Note

You can compare the slope of a line with the steepness of a road—the greater the absolute value of the slope, the steeper the road. A vertical road would be impossible to walk (undefined slope) and a horizontal road would be easy to walk (0 slope).

Rule: Slopes of Vertical and Horizontal Lines

a. The graph of any equation of the form $y = b$ is a horizontal line whose slope is 0 and whose y-intercept is b. There is no x-intercept (unless $b = 0$). If the slope of a line is 0, then it is a horizontal line.

b. The graph of any equation of the form $x = a$ is a vertical line whose slope is **undefined** and whose x-intercept is a. There is no y-intercept (unless $a = 0$). If the slope of a line is undefined, then it is a vertical line.

Example 5   **Find the slope of each of the following:**

a. $x = 4$

b. $y = -3$

Solution

The graph of $x = 4$ is a vertical line whose slope is undefined.

Solution

The graph of $y = -3$ is a horizontal line whose slope is 0.

Draw the graph of each of the following lines containing the given point and with the given slope:

c. $(3, -2)$ with $m = 0$

d. $(-5, 2)$ with undefined slope.

Solution

Since the slope is 0, this is a horizontal line through the point $(3, -2)$.

Solution

Since the slope is undefined, the graph is a vertical line through $(-5, 2)$

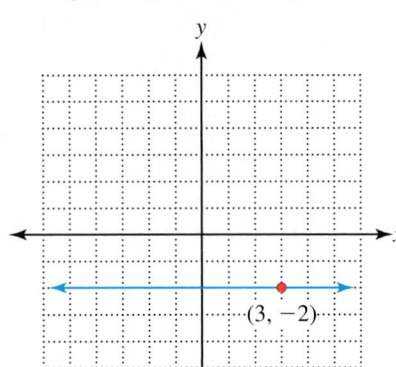

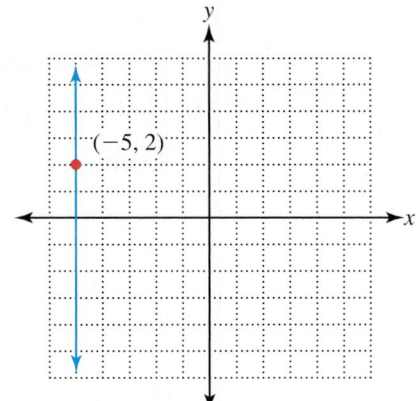

PRACTICE EXERCISE 5

Find the slope of each of the following:

a. $x = -2$

b. $y = 7$

Draw the graph of each of the following lines containing the given point and with the given slope:

c. $(-2, 4)$ with undefined slope

d. $(2, 3)$ with $m = 0$

If you need more practice, do the following Additional Practice Exercises.

Additional Practice Exercise 5   Find the slope of each of the following:

a. $y = -4$

b. $x = 5$

Draw the graph of each of the following lines containing the given point and with the given slope:

c. $(4, 5)$ with $m = 0$

d. $(-4, 2)$ with undefined slope.

Answers: Practice Exercise 5: a. undefined **b.** 0   **Additional Practice 5: a.** 0 **b.** undefined

c.   **d.**   **c.**   **d.**

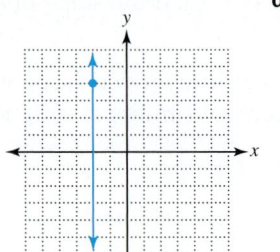

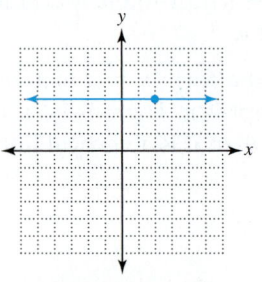

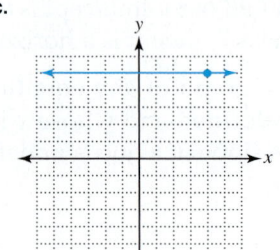

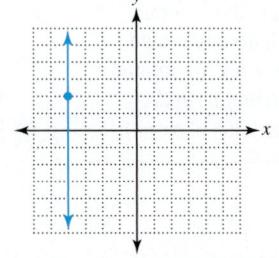

Exercise Set 4.4   MyMathLab®

Find the slope of the line that contains each of the following pairs of points. (See Example 1.)

1. $(2, 4), (4, 5)$ **2.** $(1, 5), (3, 8)$ **3.** $(3, 1), (1, 5)$

4. $(3, 2), (1, 6)$ **5.** $(-1, 4), (2, -2)$ **6.** $(-3, 2), (1, -6)$

7. $(3, -2), (-3, 2)$ **8.** $(5, -4), (1, -2)$ **9.** $(-4, -2), (-8, 2)$

10. $(-5, -2), (-1, -8)$ **11.** $(0, 3), (4, 6)$ **12.** $(6, 0), (4, -3)$

13. $(-2, -4), (6, 0)$ **14.** $(-7, -2), (-2, 0)$ **15.** $(3, 5), (-1, 5)$

16. $(4, 8), (-3, 8)$ **17.** $(2, 6), (2, 3)$ **18.** $(-3, 4), (-3, -1)$

19. $(3, 6), (-4, 6)$ **20.** $(5, -2), (5, 4)$

Find the slope of each of the following lines from the graph. (See Example 2.)

21.

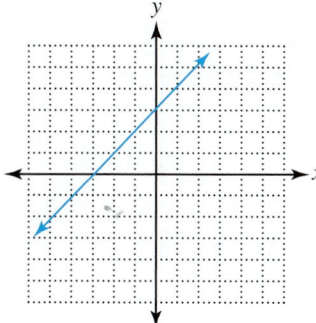

22.

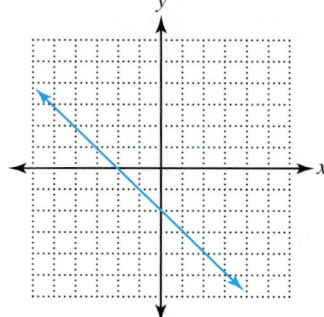

23.

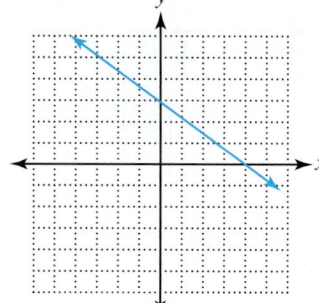

24.

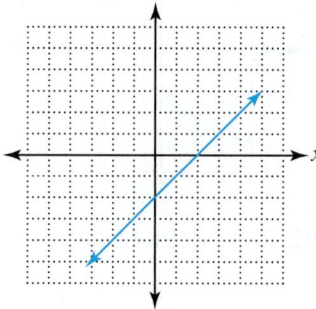

25.

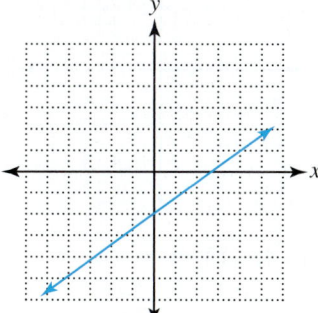

26.

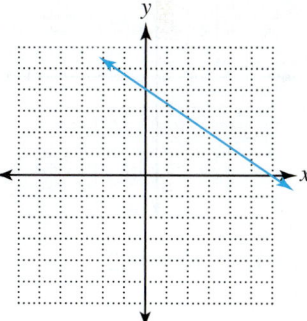

27.

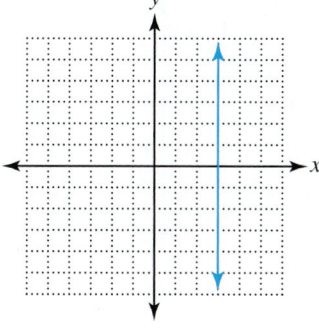

28.

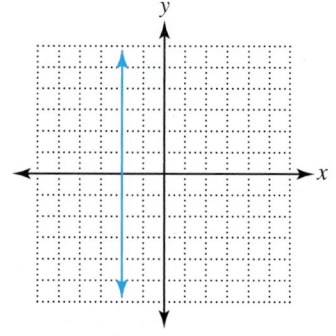

29.

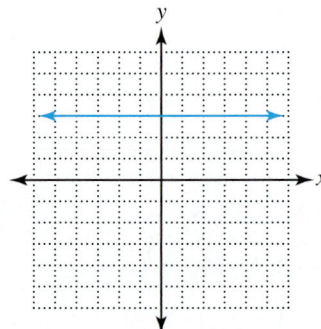

30.

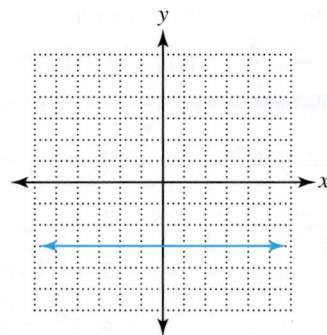

Draw the graph of each of the following lines containing the given point and with the given slope. (See Example 3.)

31. $(-3, 2), m = \dfrac{3}{4}$

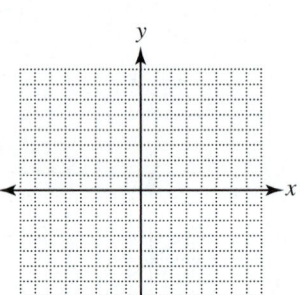

32. $(4, -3), m = \dfrac{3}{2}$

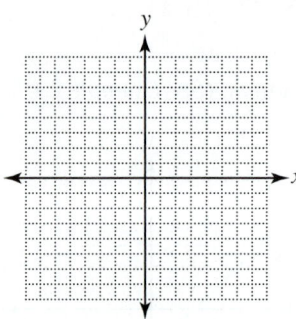

33. $(-1, -3), m = -\dfrac{5}{3}$

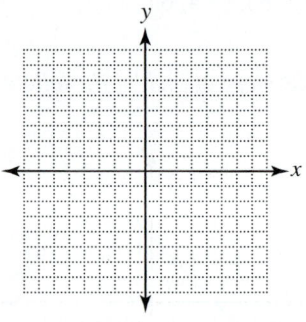

34. $(-3, -5), m = -\dfrac{4}{3}$

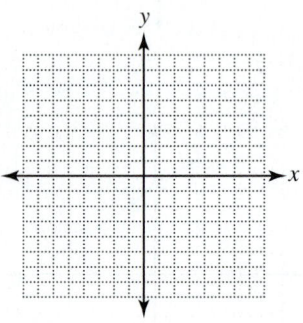

35. $(0, 2), m = 3$

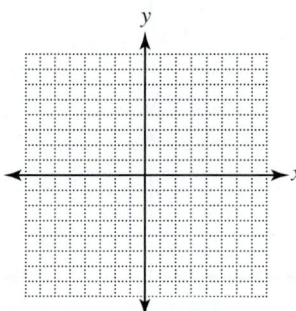

36. $(4, 0), m = 4$

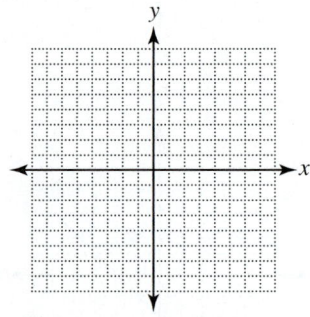

37. $(-1, 3), m = -2$

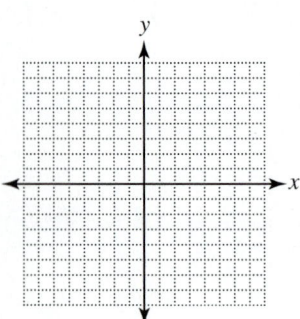

38. $(-4, -1), m = -4$

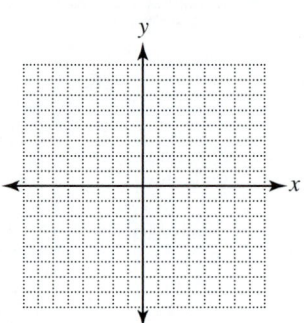

39. $(2, 5),$ undefined slope

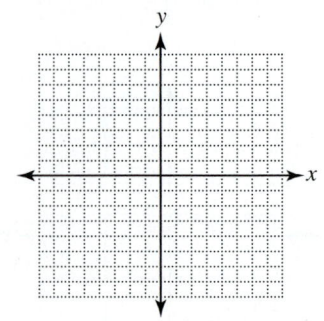

40. $(-2, 4)$, undefined slope

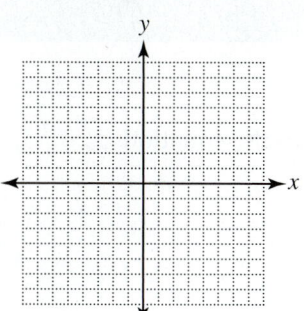

41. $(-2, 6)$, $m = 0$

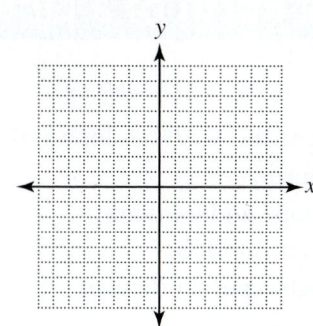

42. $(5, -2)$, $m = 0$

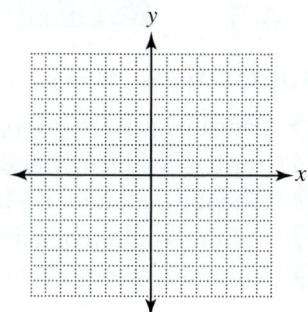

Find the slope of the graph of each of the following. (See Examples 4 and 5.)

43. $x = 5$

44. $x = 8$

45. $x = -1$

46. $x = -6$

47. $y = 4$

48. $y = 7$

49. $y = -4$

50. $y = -8$

Answer the following:

51. A mountain road rises 30 feet vertically over a horizontal distance of 500 feet. (a) What is the slope of the road? (b) When the slope of a road is written as a percent, it is called the *grade*. What is the grade?

52. The roof of a house rises 6 feet over a span of 20 feet. What is the slope of the roof? (The slope of a roof is often called the *pitch*.)

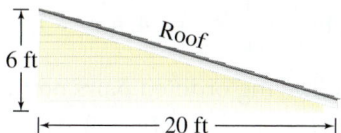

53. A carpenter is working on the outside of a house, using a ladder. If the foot of the ladder is 9 feet from the wall and the top of the ladder is at a point 24 feet from the ground, what is the slope of the ladder?

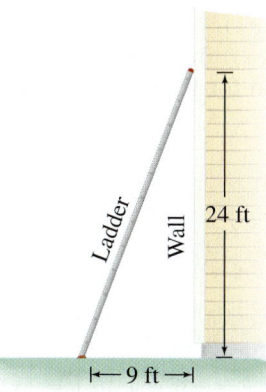

54. The top of a guy wire is attached to a utility pole at a point 30 feet above the ground, and the bottom is secured to the ground at a point 8 feet from the bottom of the pole. What is the slope of the guy wire?

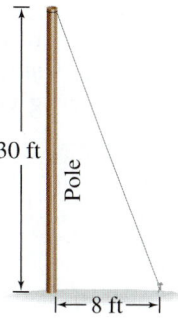

Writing Exercises (55–60)

55. How does the graph of a line whose slope is $\frac{2}{3}$ compare with the graph of a line whose slope is $-\frac{2}{3}$?

56. How does the graph of a line whose slope is $\frac{5}{2}$ compare with the graph of a line whose slope is $\frac{1}{3}$?

57. Why is the slope of a vertical line undefined?

58. Why is the slope of a horizontal line 0?

59. Why does a line with positive slope rise from left to right?

60. Why does a line with negative slope fall from left to right?

Section 4.5 Slope–Intercept Form of a Line

OBJECTIVES *When you complete this section, you will be able to:*

A Write an equation in slope–intercept form and find the slope and *y*-intercept from the equation.
B Graph a line using the slope–intercept form.
C Write the equation of a line, given the slope and *y*-intercept.
D Find the solution of a linear equation from a graph.
E Determine whether two lines are parallel or perpendicular.

PREREQUISITE SKILLS *Before beginning this section, you should be able to:*

a. Add, subtract, multiply, and divide integers. (Sections 1.6, 1.7, 2.1, 2.5)
b. Substitute values for variables and evaluate. (Section 2.5)
c. Plot ordered pairs on the rectangular coordinate system. (Section 4.1)
d. Solve an equation for one of its variables. (Section 3.4)

GETTING READY FOR SECTION 4.5

Add, subtract, multiply, or divide the following integers:
1. $-9 - 1$ **2.** $4 - (-6)$
3. $-8(-12)$ **4.** $-18 \div 2$
5. Evaluate $\dfrac{a - b}{c - d}$: $a = -4, b = -1, c = -2, d = 2$.

Plot the following ordered pairs on the rectangular coordinate system:
6. $(-4, -1)$ **7.** $(-3, 5)$ **8.** $(6, 0)$

Solve each of the following equations for *y*:
9. $3x + y = 4$ **10.** $3x - 4y = 12$

Introduction

In the last section, we found the slope of a line containing two points by using the slope formula. In this section, we learn a method for finding the slope of a line when given the equation of the line.

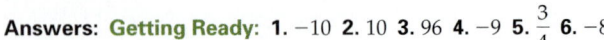

Answers: **Getting Ready: 1.** -10 **2.** 10 **3.** 96 **4.** -9 **5.** $\dfrac{3}{4}$ **6.** -8 **9.** $y = -3x + 4$ **10.** $y = \dfrac{3}{4}x - 3$

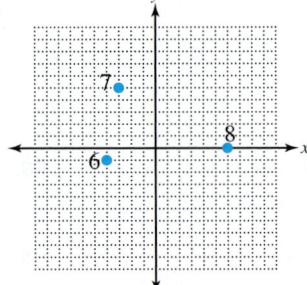

OBJECTIVE **A** Writing an equation in slope–intercept form and finding the slope and y-intercept from the equation

In Example 2a of Section 4.3, we graphed $2x + y = 6$ using the x-intercept $(3, 0)$ and y-intercept $(0, 6)$. We can find the slope of the graph of $2x + y = 6$ using these two points.

$$m = \frac{y_2 - y_1}{x_2 - x_1}$$ Substitute for $y_2, y_1, x_2,$ and x_1.

$$m = \frac{6 - 0}{0 - 3}$$ Simplify.

$$m = -2$$ Therefore, the slope is –2.

Now solve the equation $2x + y = 6$ for y.

$$2x + y = 6$$ Subtract $2x$ from both sides.

$$2x - 2x + y = -2x + 6$$ Simplify both sides.

$$y = -2x + 6$$ $2x + y = 6$ solved for y.

Notice that the slope of the graph of $2x + y = 6$ ($m = -2$) is the same as the coefficient of x when the equation is solved for y ($y = -2x + 6$). Also notice that the y-intercept, 6, is the same as the constant term when the equation is solved for y ($y = -2x + 6$). This is no accident. It can be shown that if a linear equation is solved for y, the coefficient of x is the slope and the constant term is the y-intercept.

> **Definition** Slope–Intercept Form of a Line
>
> When any linear equation is solved for y, the resulting equation is of the form $y = mx + b$, where m is the slope and b is the y-intercept of the graph of the equation. This is called the *slope–intercept form of a line*.

Example 1   Find the slope and *y*-intercept of the line represented by each of the following equations:

a. $4x + y = 5$

b. $5x - 2y = 7$

Solution

Solve the equation for y.

$$4x + y = 5$$ Subtract $4x$ from both sides of the equation.

$$4x - 4x + y = -4x + 5$$ Add like terms.

$$0 + y = -4x + 5$$ $0 + y = y$.

$$y = -4x + 5$$ Therefore, the slope is -4 and the y-intercept is 5.

Solution

Solve the equation for y.

$$5x - 2y = 7$$ Subtract $5x$ from both sides of the equation.

$$5x - 5x - 2y = -5x + 7$$ Add like terms.

$$0 - 2y = -5x + 7$$ $0 - 2y = -2y$.

$$-2y = -5x + 7$$ Divide both sides by -2.

$$\frac{-2y}{-2} = \frac{-5x + 7}{-2}$$ Simplify both sides.

$$y = \frac{5}{2}x - \frac{7}{2}$$ Therefore, the slope is $\frac{5}{2}$ and the y-intercept is $-\frac{7}{2}$.

PRACTICE EXERCISE 1

Find the slope and y-intercept of the line represented by each of the following equations:

a. $x + 2y = 6$

b. $6x - 5y = 11$

(Continued)

Answers: Practice Exercise 1: a. $m = -\frac{1}{2}$, y-intercept $= 3$ **b.** $m = \frac{6}{5}$, y-intercept $= -\frac{11}{5}$

If you need more practice, do the following Additional Practice Exercises.

OBJECTIVE Graphing a line using the slope–intercept method

The slope–intercept form of a linear equation gives us yet another method of graphing linear equations. The y-intercept is a point on the graph. By knowing a point and the slope, we can draw the graph by using the technique given in Section 4.4.

Example 2   Graph the following, using the slope and y-intercept:

a. $y = \dfrac{2}{3}x + 2$

Solution

From the equation, the slope is $\frac{2}{3}$ and the y-intercept is 2. Plot the y-intercept of 2 and from that point go up 2 and to the right 3 and plot another point. Draw a line through those two points.

b. $2x - y = 3$

Solution

First put the equation in slope–intercept form by solving for y.

$$2x - y = 3 \qquad \text{Subtract } 2x \text{ from both sides of the equation.}$$

$$-y = -2x + 3 \qquad \text{Multiply both sides by } -1.$$

$$y = 2x - 3 \qquad \text{Therefore, the slope is 2 and the } y\text{-intercept is } -3.$$

Plot the y-intercept of -3. Think of the slope, 2, as $\frac{2}{1}$. So, from the y-intercept, go up 2 and to the right 1 and plot another point. Draw a line through these two points.

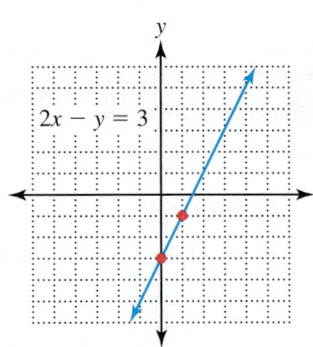

c. $3x + 4y = 8$

Solution

First put the equation into slope–intercept form by solving for y.

$$3x + 4y = 8 \qquad \text{Subtract } 3x \text{ from both sides of the equation.}$$

$$4y = -3x + 8 \qquad \text{Divide both sides by 4.}$$

$$y = \dfrac{-3x + 8}{4} \qquad \text{Simplify the right side.}$$

$$y = -\dfrac{3}{4}x + 2 \qquad \text{Therefore, the slope is } -\dfrac{3}{4} \text{ and the } y\text{-intercept is 2.}$$

Plot the y-intercept of 2. From that point, go down 3 and to the right 4 and plot another point. Draw a line through these two points.

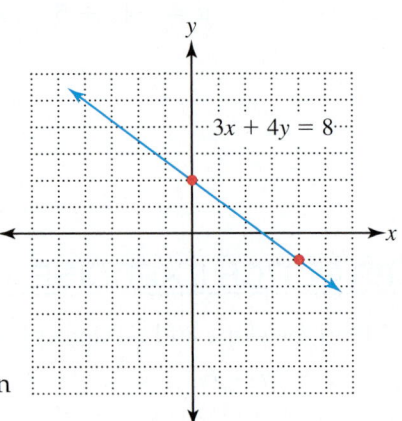

PRACTICE EXERCISE 2

Graph each of the following, using the slope–intercept method:

a. $y = \dfrac{3}{5}x - 1$

b. $x + y = 4$

c. $3x + 2y = 6$

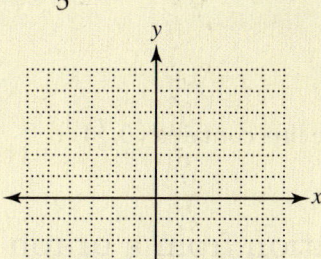
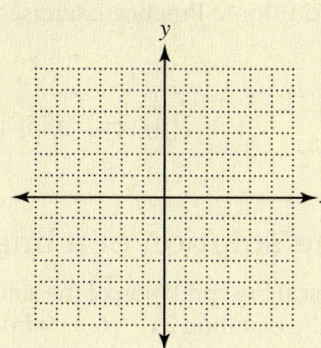
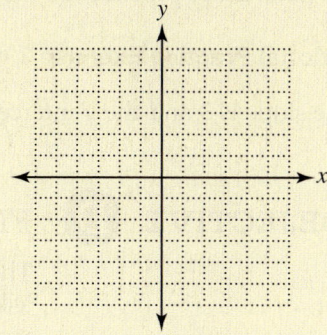

If you need more practice, do the following Additional Practice Exercises.

Additional Practice Exercise 2   Graph the following, using the slope–intercept method.

a. $y = -\dfrac{2}{3}x + 4$

b. $2x + y = 3$

c. $2x - 3y = 6$

OBJECTIVE ⒸWriting equations of lines, given the slope and *y*-intercept

If we know the slope and *y*-intercept of a line, then we can use the slope–intercept form to write the equation of the line.

Example 3   Write the equations of the given lines. Express the answers in slope–intercept form.

a. Slope of 3 and *y*-intercept of 4.

Solution

$y = mx + b$ Substitute 3 for *m* and 4 for *b*.

$y = 3x + 4$ Therefore, the equation of the line with slope of 3 and *y*-intercept of (0, 4) is $y = 3x + 4$.

b. The slope is $\frac{3}{4}$ and the line contains $\left(0, \frac{2}{3}\right)$.

Solution

The point $\left(0, \frac{2}{3}\right)$ is the *y*-intercept, so $b = \frac{2}{3}$.

$y = mx + b$ Substitute $\dfrac{3}{4}$ for *m* and $\dfrac{2}{3}$ for *b*.

$y = \dfrac{3}{4}x + \dfrac{2}{3}$ Therefore, the equation of the line with slope of $\frac{3}{4}$ and containing $\left(0, \frac{2}{3}\right)$ is $y = \frac{3}{4}x + \frac{2}{3}$.

Answers: Practice Exercise 2: a.

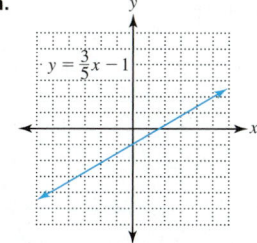

b.

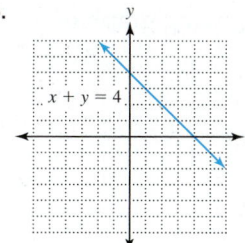

c.

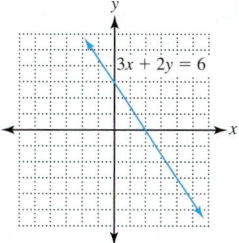

Additional Practice 2: a.

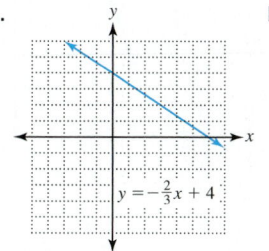

b.

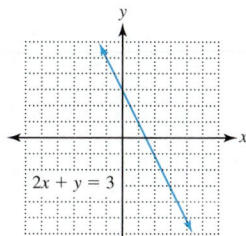

c.

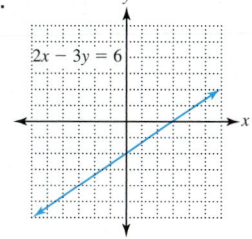

PRACTICE EXERCISE 3

Write the equations of the given lines. Express the answers in slope–intercept form.

a. The slope is $\frac{5}{4}$ and the y-intercept is -3.

b. The slope is $-\frac{3}{5}$ and the line contains $\left(0, -\frac{3}{4}\right)$.

If you need more practice, do the following Additional Practice Exercises.

Additional Practice Exercise 3

a. The slope is 3 and the y-intercept is 4.

b. The slope is $-\frac{4}{3}$ and the line contains $\left(0, \frac{7}{6}\right)$.

OBJECTIVE Finding the solution of a linear equation from a graph

There is a graphical interpretation of the solution of a linear equation. Suppose that we are asked to solve the equation $2x - 6 = 0$ from the graph of $y = 2x - 6$. If we begin with $y = 2x - 6$ and let $y = 0$, then we get the equation $0 = 2x - 6$. Recall that letting $y = 0$ means that we are finding the x-intercept of the graph. Consequently, the x-intercept of the graph of $y = 2x - 6$ is the solution of the equation $2x - 6 = 0$, as illustrated next.

Example 4

Algebraic Solution	Graphical Solution
$2x - 6 = 0$ Add 6 to both sides. $2x - 6 + 6 = 0 + 6$ Simplify both sides. $2x = 6$ Divide by 2. $\dfrac{2x}{2} = \dfrac{6}{2}$ Simplify both sides. $x = 3$ Therefore, the solution is 3.	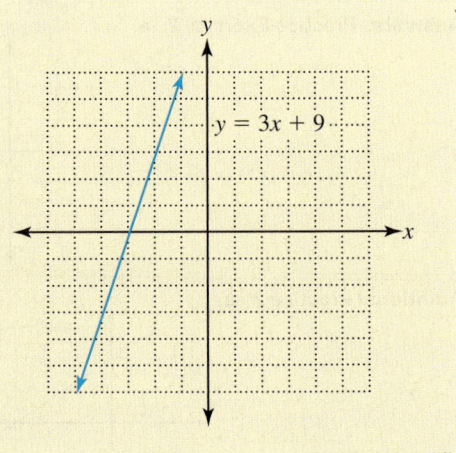

Note that the solution of the equation $(x = 3)$ is the x-intercept of the graph.

PRACTICE EXERCISE 4

Solve the equation $3x + 9 = 0$, using the following graph of $y = 3x + 9$:

(Continued)

If you need more practice, do the following Additional Practice Exercise.

Additional Practice Exercise 4

Solve the equation $-\frac{2}{3}x + 4 = 0$, using the following graph of $y = -\frac{2}{3}x + 4$:

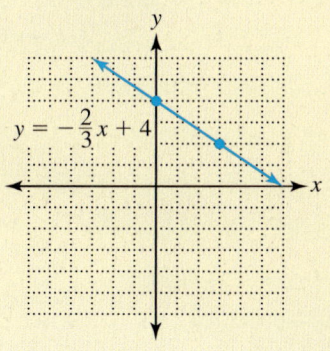

$y = -\frac{2}{3}x + 4$

OBJECTIVE **E** Determining whether two lines are parallel or perpendicular

Two lines are **parallel** if they are in the same plane and do not intersect. In the figure that follows, L_1 is the graph of a line containing $(2, 3)$ and with a slope of $\frac{1}{2}$, and L_2 is the graph of a line containing $(2, -2)$, also with a slope of $\frac{1}{2}$.

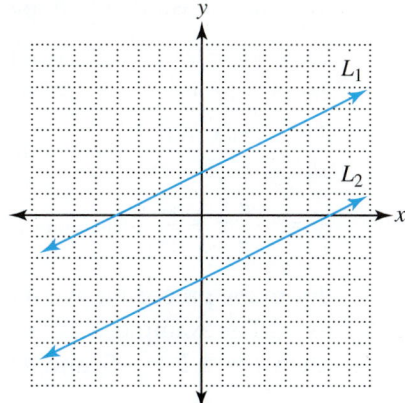

Both lines have slopes of $\frac{1}{2}$, and their graphs appear to be parallel lines. It can be shown that two lines are parallel if and only if they have equal slopes. This means that if two lines are parallel, then they have equal slopes. It also means that if two lines have equal slopes, then they are parallel.

Two lines are perpendicular if they intersect at right angles. In the figure that follows, L_1 is the graph of a line containing $(2, 3)$ and with a slope of $\frac{2}{3}$, and L_2 is the graph of a line containing $(2, 3)$ and with a slope of $-\frac{3}{2}$.

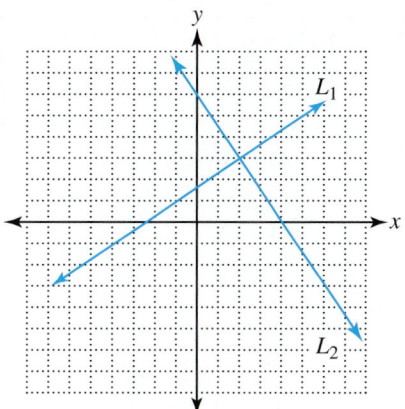

Answers: Additional Practice 4: $x = 6$

The two lines appear to be perpendicular (intersect at a 90°, or right, angle). Notice that the product of their slopes is −1. That is, $\left(\frac{2}{3}\right)\left(-\frac{3}{2}\right) = -1$. It can be shown that if two lines are perpendicular, the product of their slopes is −1. It can also be shown that if the product of the slopes of two lines is −1, the two lines are perpendicular. Consequently, if m_1 represents the slope of a line and m_2 represents the slope of any line perpendicular to the first line, then $m_1 m_2 = -1$. Solving for m_1, we get $m_1 = -\frac{1}{m_2}$. This means that m_1 is the negative of the reciprocal of m_2. Thus, if a line has a slope of 3, then any line perpendicular to it has a slope of $-\frac{1}{3}$. If a line has a slope of $-\frac{3}{4}$, then any line perpendicular to it has a slope of $\frac{4}{3}$.

> ### Definition Parallel and Perpendicular Lines
>
> Two nonvertical lines are **parallel** if and only if they have **equal slopes**.
>
> Two lines are **perpendicular** if and only if the product of their slopes is −1, provided neither is vertical nor horizontal. In other words, two nonvertical lines are perpendicular if the slopes are negatives and reciprocals. Any vertical line is perpendicular to any horizontal line, and vice versa.

Often, all we need from an equation is the slope and not the y-intercept. If we write the general equation of a linear equation $(ax + by = c)$ in slope–intercept form, we will get a formula for slope.

$ax + by = c$ Subtract ax from both sides.

$by = -ax + c$ Divide both sides by b.

$\dfrac{by}{b} = \dfrac{-ax + c}{b}$ Simplify both sides.

$y = -\dfrac{a}{b}x + \dfrac{c}{b}$ Therefore, the slope is $-\frac{a}{b}$.

From the preceding, we see that if an equation is given in $ax + by = c$ form, the slope is $-\frac{a}{b}$. For example, if $2x + 3y = 5$, then $a = 2$, $b = 3$, and $m = -\frac{2}{3}$; and if $4x - 5y = 7$, then $a = 4$, $b = -5$, and $m = -\frac{4}{-5} = \frac{4}{5}$.

Example 5   **Determine whether the lines determined by the following pairs of points are parallel, perpendicular, or neither:**

a. L_1: $(-3, 2)$ and $(1, 5)$ L_2: $(2, -1)$ and $(-2, -4)$

b. L_1: $(3, -1)$ and $(1, 3)$; L_2: $(-1, -3)$ and $(1, -2)$

Solution

In order to determine whether the lines are parallel or perpendicular, we need the slopes of the lines.

Slope of L_1	Slope of L_2
$m = \dfrac{y_2 - y_1}{x_2 - x_1}$	$m = \dfrac{y_2 - y_1}{x_2 - x_1}$
$m = \dfrac{5 - 2}{1 - (-3)}$	$m = \dfrac{-4 - (-1)}{-2 - 2}$
$m = \dfrac{5 - 2}{1 + 3}$	$m = \dfrac{-4 + 1}{-2 - 2}$
$m = \dfrac{3}{4}$	$m = \dfrac{-3}{-4}$
	$m = \dfrac{3}{4}$

Since the slopes are equal, L_1 and L_2 are parallel.

Solution

In order to determine whether the lines are parallel or perpendicular, we need the slopes of the lines.

Slope of L_1	Slope of L_2
$m = \dfrac{y_2 - y_1}{x_2 - x_1}$	$m = \dfrac{y_2 - y_1}{x_2 - x_1}$
$m = \dfrac{3 - (-1)}{1 - 3}$	$m = \dfrac{-2 - (-3)}{1 - (-1)}$
$m = \dfrac{3 + 1}{1 - 3}$	$m = \dfrac{-2 + 3}{1 + 1}$
$m = \dfrac{4}{-2} = -2$	$m = \dfrac{1}{2}$

Since $(-2)\left(\dfrac{1}{2}\right) = -1$, L_1 and L_2 are perpendicular.

c. $L_1: 2x + 5y = 10; L_2: 5x - 2y = 4$

Solution

In order to determine whether the lines are parallel or perpendicular, we need to know the slopes.

The slope of $L_1 = -\dfrac{a}{b} = -\dfrac{2}{5}$, and the slope of

$L_2 = -\dfrac{a}{b} = -\dfrac{5}{-2} = \dfrac{5}{2}$.

Since $\left(-\frac{2}{5}\right)\left(\frac{5}{2}\right) = -1$, L_1 and L_2 are perpendicular.

d. $L_1: 3x + 4y = 8; L_2: 4x + 3y = 6$

Solution

In order to determine whether the two lines are parallel or perpendicular, we need the slopes.

The slope of $L_1 = -\dfrac{a}{b} = -\dfrac{3}{4}$, and the slope of $L_2 = -\dfrac{a}{b} = -\dfrac{4}{3}$.

Since the slopes are neither equal nor is their product -1, L_1 and L_2 are neither parallel nor perpendicular.

Note

In Examples 5c and d, we could have found the slope by writing the equations in slope–intercept form instead of using $m = -\frac{a}{b}$.

PRACTICE EXERCISE 5

Determine whether the lines containing the following pairs of points are parallel, perpendicular, or neither:

a. $L_1: (4, 5)$ and $(-1, 2)$; $L_2: (-2, -1)$ and $(3, 2)$ **b.** $L_1: (5, -2)$ and $(3, 1)$; $L_2: (4, 6)$ and $(1, 4)$

Determine whether the graphs of the lines with the given equations are parallel, perpendicular, or neither.

c. $L_1: x + 3y = 6; L_2: 3x + y = 4$ **d.** $L_1: 2x - 7y = 5; L_2: 2x - 7y = -7$

If you need more practice, do the following Additional Practice Exercises.

Additional Practice Exercise 5   Determine whether the following pairs of lines are parallel, perpendicular, or neither:

a. $L_1: (1, 1)$ and $(2, 3)$; $L_2: (-3, 2)$ and $(-4, 0)$ **b.** $L_1: (-2, 2)$ and $(1, 3)$; $L_2: (2, -1)$ and $(3, 2)$

c. $L_1: 2x + y = 4; L_2: 2x + y = 6$ **d.** $L_1: 3x - 2y = 4; L_2: 2x + 3y = 4$

Exercise Set 4.5   For Extra Help   MyMathLab®

Find the slope and the y-intercept of the line represented by each of the equations. Graph each equation by using the slope–intercept method. (See Examples 1 and 2.)

1. $y = 2x + 3$ **2.** $y = x - 4$ **3.** $y = -2x + 9$ **4.** $y = -3x + 8$

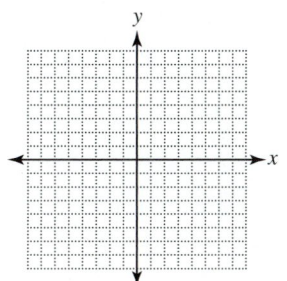 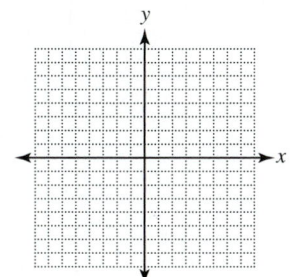 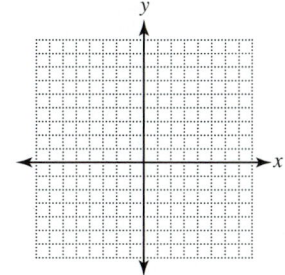

Answers: Practice Exercise 5:   a. parallel   **b.** perpendicular   **c.** neither   **d.** parallel
Additional Practice 5:   a. parallel   **b.** neither   **c.** parallel   **d.** perpendicular

5. $y = \frac{3}{4}x - 1$

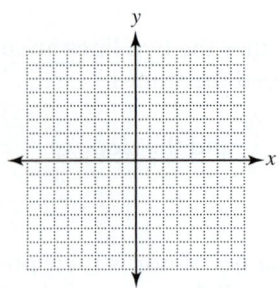

6. $y = \frac{1}{3}x + 2$

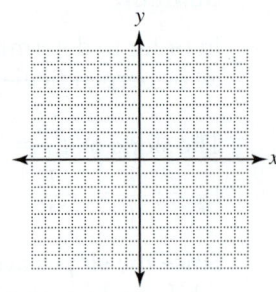

7. $y = -\frac{4}{3}x + 2$

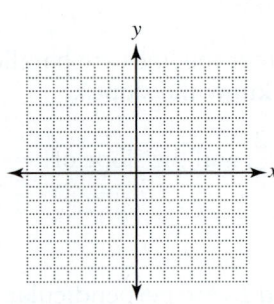

8. $y = -\frac{5}{2}x + 3$

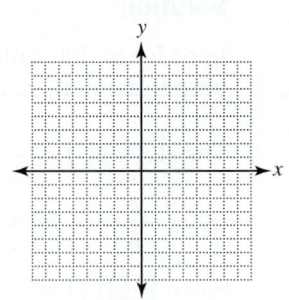

9. $3x - y = 7$

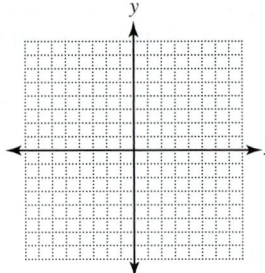

10. $2x - y = 5$

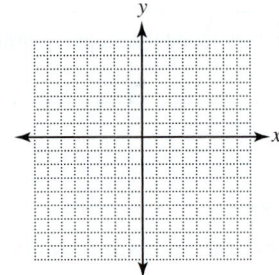

11. $2x + 3y = 9$

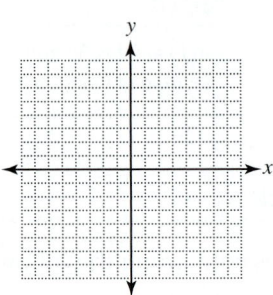

12. $3x + 2y = 12$

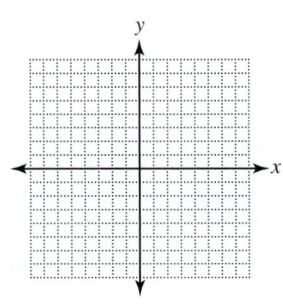

13. $4x + 5y = 20$

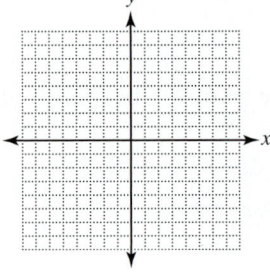

14. $3x + 5y = 15$

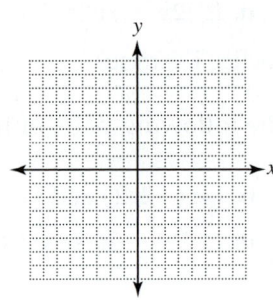

15. $5x - 3y = 15$

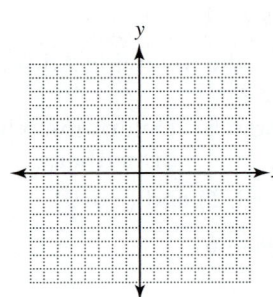

16. $2x - 5y = 10$

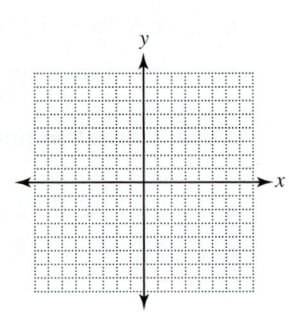

17. $2x - 5y = -10$

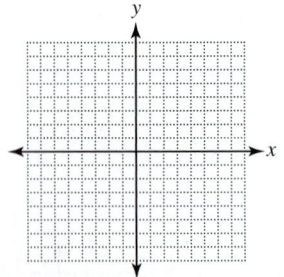

18. $5x - 3y = -15$

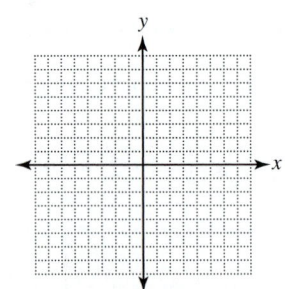

19. $2x + y = 0$

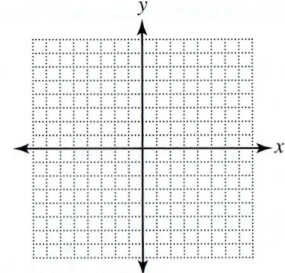

20. $3x - y = 0$

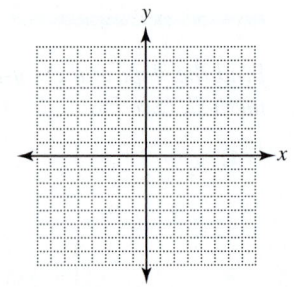

21. $3x + 2y = 0$

22. $4x + 3y = 0$

23. $3x + 2y = 5$

24. $5x - 2y = -3$

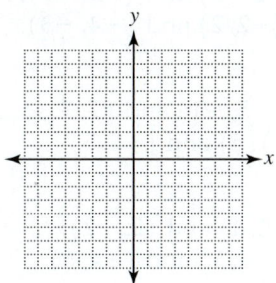

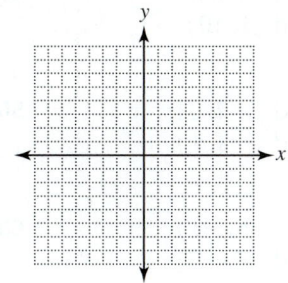

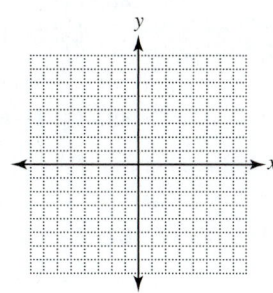

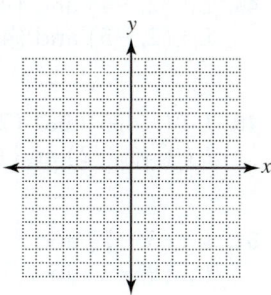

Find the slope of each of the following:

25. $x = 8$

26. $x = 5$

27. $x = -6$

28. $x = -1$

29. $y = 7$

30. $y = 4$

31. $y = -8$

32. $y = -4$

Write the equations of the given lines. Express the answers in $y = mx + b$ form. (See Example 3.)

33. Slope is 2 and y-intercept is 4.

34. Slope is -5 and y-intercept is 2.

35. Slope is $\frac{3}{5}$ and y-intercept is -1.

36. Slope is $\frac{5}{2}$ and y-intercept is -4.

37. Slope is $-\frac{2}{3}$ and y-intercept is $\frac{4}{5}$.

38. Slope is $-\frac{6}{5}$ and y-intercept is $-\frac{5}{4}$.

Solve the given equation by using the given graph. (See Example 4.)

39. Solve $2x + 8 = 0$, using the given graph of $y = 2x + 8$.

40. Solve $3x - 6 = 0$, using the given graph of $y = 3x - 6$.

41. Solve $\frac{1}{2}x - 3 = 0$, using the given graph of $y = \frac{1}{2}x - 3$.

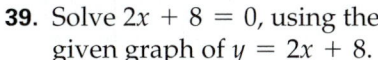

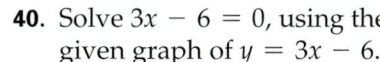

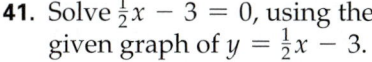

42. Solve $\frac{1}{3}x - 1 = 0$, using the given graph of $y = \frac{1}{3}x - 1$.

43. Solve $-\frac{3}{2}x + 6 = 0$, using the given graph of $y = -\frac{3}{2}x + 6$.

44. Solve $-\frac{3}{4}x - 3 = 0$, using the given graph of $y = -\frac{3}{4}x - 3$.

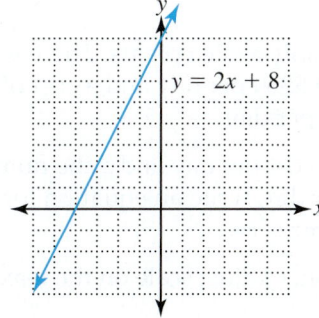

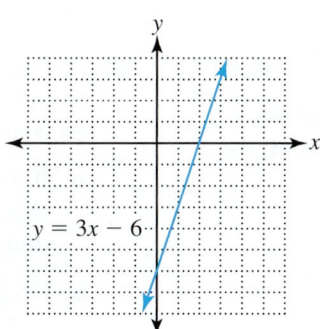

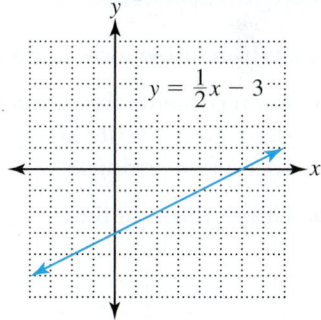

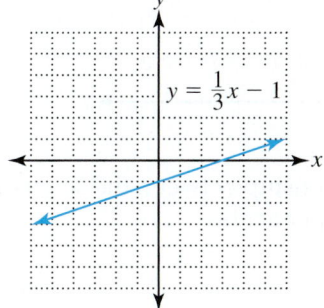

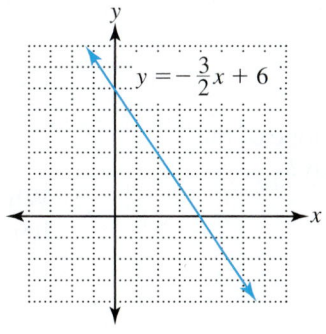

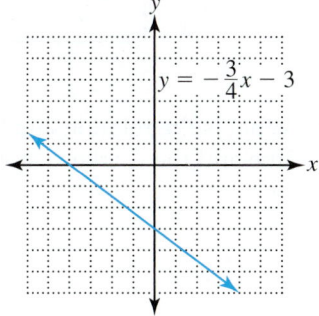

Determine whether the lines containing the given pairs of points are parallel, perpendicular, or neither. (See Example 5.)

45. L_1: $(2, -4)$ and $(4, 2)$;
L_2: $(2, -5)$ and $(4, 1)$

46. L_1: $(1, -3)$ and $(4, 6)$;
L_2: $(-1, 4)$ and $(1, 10)$

47. L_1: $(4, -3)$ and $(-1, -1)$;
L_2: $(-2, 2)$ and $(-4, -3)$

48. L_1: $(-4, 5)$ and $(2, 3)$;
L_2: $(-1, -2)$ and $(0, 1)$

49. L_1: $(-6, 2)$ and $(-2, 0)$;
L_2: $(0, -1)$ and $(-4, 1)$

50. L_1: $(-5, -1)$ and $(7, 7)$;
L_2: $(1, 5)$ and $(-2, 3)$

51. L_1: $(3, -7)$ and $(-1, -1)$;
L_2: $(8, -2)$ and $(2, 2)$

52. L_1: $(4, -5)$ and $(5, -1)$;
L_2: $(3, -4)$ and $(7, -3)$

53. L_1: $(2, 4)$ and $(2, -2)$;
L_2: $(-4, 1)$ and $(-4, 6)$

54. L_1: $(3, 5)$ and $(-2, 5)$;
L_2: $(-1, 2)$ and $(4, 2)$

55. L_1: $(-2, 3)$ and $(4, 3)$;
L_2: $(6, 1)$ and $(6, 7)$

56. L_1: $(-4, 1)$ and $(-4, 6)$;
L_2: $(-3, 5)$ and $(3, 5)$

Determine whether the graphs of the lines with the given equations are parallel, perpendicular, or neither:

57. L_1: $4x + y = 7$;
L_2: $4x + y = -3$

58. L_1: $3x - y = 5$;
L_2: $3x - y = 9$

59. L_1: $4x + 3y = 6$;
L_2: $3x - 4y = 12$

60. L_1: $5x - 3y = 11$;
L_2: $3x + 5y = 8$

61. L_1: $4x + 6y = 5$;
L_2: $6x + 4y = 9$

62. L_1: $5x + 10y = 15$;
L_2: $4x + 2y = 5$

63. L_1: $x = 4$;
L_2: $x = -1$

64. L_1: $y = 5$;
L_2: $y = -5$

65. L_1: $x = 6$;
L_2: $y = 2$

66. L_1: $y = -5$;
L_2: $x = 3$

Challenge Exercises (67–70)

67. A lawn-maintenance worker earns $6.00 per hour.

   a. If y represents the worker's earnings for working x hours, write an equation representing the worker's earnings.

   b. Graph the equation.

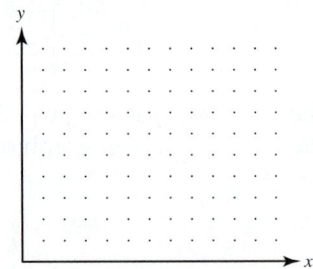

   c. What is the relationship between the slope of the line and the worker's hourly rate of pay?

68. A local nursery sells only azalea plants. There is an initial start-up cost of $360 per day and a cost of $3.00 per azalea sold per day.

   a. If y represents the cost per day and x the number of azaleas sold per day, write an equation for the daily costs of the nursery.

   b. Graph the equation. (Adjust scale on the y-axis.)

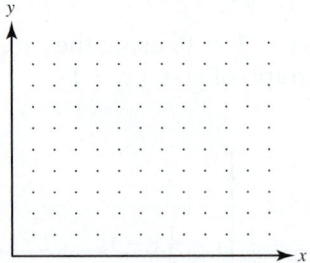

   c. What is the relationship between the slope of the line and the cost per azalea?

69. A salesperson at a paint store earns $100 per week plus a commission of $2.00 on every gallon of paint he or she sells.

   a. If y represents the salesperson's salary for selling x gallons of paint, write an equation that represents the salesperson's salary.

   b. Graph the equation. (Adjust scale on the y-axis.)

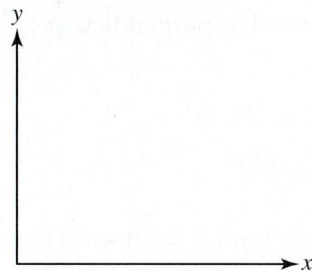

   c. What is the relationship between the slope of the line and the commission per gallon of paint sold?

70. A piece of machinery costs $3000 and depreciates at the rate of $600 per year.

   a. If y represents the value of the machinery after x years, write an equation representing the value of the machinery.

   b. Graph the equation. (Adjust scale on the y-axis.)

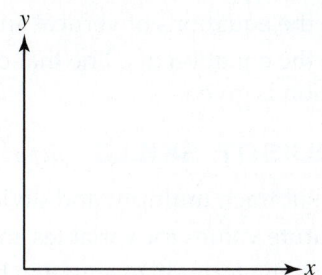

   c. What is the relationship between the slope of the line and the per-year depreciation?

Writing Exercises (71–77)

71. Give the equation of a line in $ax + by = c$ form that has **a.** positive slope, and **b.** negative slope.

72. Given the equations $y = -2x - 3$, $y = -\frac{1}{2}x - 3$, $y = -.75x - 3$, and $y = x - 3$, which equation has the steepest graph? Why?

73. The standard equation of a line is $ax + by = c$. What is the relationship between a and b if the line has a positive slope? A negative slope?

74. Why must parallel lines have equal slopes?

75. Are the lines whose equations are $y = 2x + 3$ and $y = -2x + 3$ perpendicular? Why or why not?

76. Can two perpendicular lines both have negative slopes? Why or why not?

77. How can you tell if two lines are parallel, without graphing them?

Section 4.6   Point–Slope Form of a Line

OBJECTIVES *When you complete this section, you will be able to:*

A Write the equation of a line that contains a given point and has a given slope.

B Write the equation of a line that contains two given points.

C Write the equations of vertical and horizontal lines.

D Write the equation of a line that contains a given point and is parallel or perpendicular to a line whose equation is given.

PREREQUISITE SKILLS *Before beginning this section, you should be able to:*

a. Add, subtract, multiply, and divide integers. (Sections 1.6, 1.7, 2.1, 2.5)

b. Substitute values for variables and evaluate. (Section 2.5)

c. Change the form of an equation by adding or subtracting the same term to both sides (Section 3.1) or multiplying both sides by the same quantity. (Section 3.2)

d. Apply the distributive property. (Section 1.7)

e. Find the slope of a line containing two points. (Section 4.4)

f. Know the relationships between slope and parallel or perpendicular lines. (Section 4.5)

g. Solve linear equations. (Sections 3.1, 3.2, 3.3)

GETTING READY FOR SECTION 4.6

Add, subtract, multiply, or divide the following integers:

1. $10 - (-4)$

2. $-49 \div 7$

Evaluate the following for the given value(s) of the variable(s):

3. $\dfrac{2}{3}x + 7 : x = 3$

4. $\dfrac{a - b}{c - d} : a = 2, b = 4, c = -1, d = 2$

5. Simplify $3(x - 4)$, using the distributive property.

6. Write $3(y - 4) = 3 \cdot \dfrac{2}{3}(x - 3)$ in $ax + by = c$ form.

7. Solve $3(y - 4) = 3 \cdot \dfrac{2}{3}(x - 2)$ for y.

8. Find the slope of the line that contains $(-4, 5), (-1, 7)$.

9. Find the slope of the graph of $2x - 3y = 12$.

State whether the following pairs of lines are parallel, perpendicular, or neither:

10. $2x - y = -8, -6x + 4y = 5$

11. $3x + 5y = 8, 5x - 3y = 9$

12. Find the value of y: $y = \dfrac{2}{3}(1200) + 500$.

Introduction

If we are given a point on a line and the slope of the line, we can draw a line that contains the given point and that has the given slope. Also, if we are given two points, we can draw the line containing the two points. How do we find the equations of these lines? These are the questions we will answer in this section.

Answers: Getting Ready: 1. 14 **2.** −7 **3.** 9 **4.** $\dfrac{2}{3}$ **5.** $3x - 12$ **6.** $2x - 3y = -6$ **7.** $y = \dfrac{2}{3}x + \dfrac{8}{3}$ **8.** $m = \dfrac{2}{3}$ **9.** $m = \dfrac{2}{3}$ **10.** neither
11. perpendicular **12.** 1300

OBJECTIVE **Writing the equation of a line that contains a given point and has a given slope**

Suppose we want to write the equation of a line that contains $(2, 3)$ and has a slope of 2. Let us draw the graph.

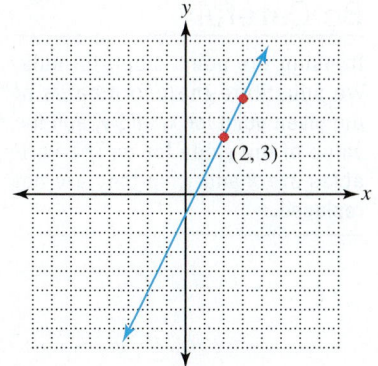

The slope of the line is the ratio of the change in y to the change in x. In this case, this ratio is $\frac{2}{1}$. This does not mean only that we go up 2 units and to the right 1 unit in order to get from one point of the line to another, but also that we could go up 4 and to the right 2. We could even go fractional parts of a unit, as long as the ratio of the change in y to the change in x is 2. That is, since $\frac{4}{2} = \frac{-10}{-5} = \frac{2}{1}$, we could use any of these ratios to indicate a slope of 2. The ratio is the same for any two points on the line.

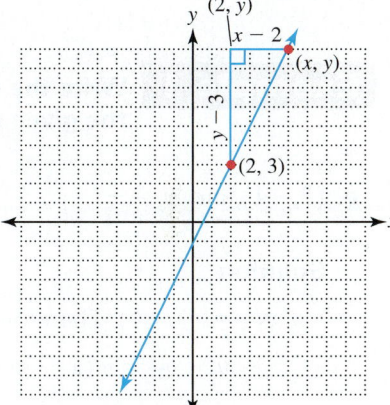

If we let (x, y) represent any other point on the line, then the slope from $(2, 3)$ to (x, y) is 2. If we go from $(2, 3)$ to (x, y) by moving vertically and then horizontally, the point of intersection of the vertical and horizontal line segments is $(2, y)$. The change in y is $y - 3$, and the change in x is $x - 2$. See the figure to the right.

To write the equation of the line, use the slope formula and let $(x_1, y_1) = (2, 3)$ and $(x_2, y_2) = (x, y)$. You will not be asked to write equations by this procedure, but we need to learn this procedure to understand where the formula comes from.

$$m = \frac{y_2 - y_1}{x_2 - x_1}$$ Substitute for m, x_1, y_1, x_2, and y_2.

$$2 = \frac{y - 3}{x - 2}$$ Multiply both sides by $x - 2$.

$$2(x - 2) = \frac{y - 3}{x - 2}(x - 2)$$ $\frac{y - 3}{x - 2}(x - 2) = \frac{y - 3}{1} \cdot \frac{1}{x - 2} \cdot \frac{x - 2}{1} =$

$(y - 3) \cdot \frac{x - 2}{x - 2} = (y - 3) \cdot 1 = y - 3.$

$$2(x - 2) = y - 3$$ Apply the distributive property.

$$2x - 4 = y - 3$$ Add 4 to both sides.

$$2x = y + 1$$ Subtract y from both sides.

$$2x - y = 1$$ Therefore, the equation of the line through $(2, 3)$ with slope of 2 is $2x - y = 1$.

Now let us use the preceding procedure to develop a formula for the equation of a line when we are given a point on the line and the slope of the line. Let (x_1, y_1) be the coordinates of the given point on the line and m be the slope of the line. Let (x, y) be any other point on the line. The slope between (x_1, y_1) and (x, y) is m. Substitute into the slope formula with $(x_2, y_2) = (x, y)$. Consequently,

$$m = \frac{y - y_1}{x - x_1}$$ Multiply both sides by $x - x_1$.

$$m(x - x_1) = \frac{y - y_1}{x - x_1}(x - x_1)$$
$$\frac{y - y_1}{x - x_1}(x - x_1) = \frac{y - y_1}{1} \cdot \frac{1}{x - x_1} \cdot \frac{x - x_1}{1} =$$

Be Careful

In using the point–slope formula, we substitute the coordinates of the given point for x_1 and y_1 and the value of the slope for m. The variables are x and y, so they are not replaced.

$$(y - y_1) \cdot \frac{x - x_1}{x - x_1} = (y - y_1) \cdot 1 = y - y_1.$$

$m(x - x_1) = y - y_1$ Rewrite as $y - y_1 = m(x - x_1)$.

$y - y_1 = m(x - x_1)$ Therefore, the equation of a line containing the point (x_1, y_1) and with slope of m is $y - y_1 = m(x - x_1)$.

We now state this more formally as the point–slope formula.

> ### Rule: Point–Slope Formula
>
> The equation of the line containing the given point (x_1, y_1) and with the given slope of m is $y - y_1 = m(x - x_1)$.

Example 1   Write the equation of each of the following lines that contain the given point and have the given slope in $ax + by = c$ form with a and b integers:

a. $(3, -4)$ and $m = \dfrac{2}{3}$

Solution

$(x_1, y_1) = (3, -4)$. Therefore, $x_1 = 3$ and $y_1 = -4$.

$y - y_1 = m(x - x_1)$ Substitute for x_1, y_1, and m.

$y - (-4) = \dfrac{2}{3}(x - 3)$ $y - (-4) = y + 4$, and multiply both sides by 3.

$3(y + 4) = 3 \cdot \dfrac{2}{3}(x - 3)$ Multiply 3 by $\dfrac{2}{3}$: $\left(3 \cdot \dfrac{2}{3} = \dfrac{3}{1} \cdot \dfrac{2}{3} = \dfrac{6}{3} = 2\right)$.

$3(y + 4) = 2(x - 3)$ Distribute on both sides.

$3y + 12 = 2x - 6$ Subtract 12 from both sides.

$3y + 12 - 12 = 2x - 6 - 12$ Simplify both sides.

$3y = 2x - 18$ Subtract $2x$ from both sides.

$3y - 2x = 2x - 2x - 18$ Write in $ax + by = c$ form.

$-2x + 3y = -18$ It is preferred that the coefficient of x be positive, so multiply both sides by -1.

$-1(-2x + 3y) = -1(-18)$ Simplify both sides.

$2x - 3y = 18$ Therefore, the equation of the line containing $(3, -4)$ with slope of $\frac{2}{3}$ is $2x - 3y = 18$.

b. $(-2, 4)$ and $m = -\dfrac{3}{4}$

Solution

$(x_1, y_1) = (-2, 4)$, so $x_1 = -2$ and $y_1 = 4$.

$y - y_1 = m(x - x_1)$ Substitute for x_1, y_1, and m.

$y - 4 = -\dfrac{3}{4}(x - (-2))$ $x - (-2) = x + 2$. Multiply both sides by 4.

$4(y - 4) = 4\left(-\dfrac{3}{4}\right)(x + 2)$ $4\left(-\dfrac{3}{4}\right) = \left(\dfrac{4}{1}\right)\left(-\dfrac{3}{4}\right) = -\dfrac{12}{4} = -3$

$4(y - 4) = -3(x + 2)$ Apply the distributive property.

$4y - 16 = -3x - 6$ Add 16 to both sides.

$4y = -3x + 10$ Add $3x$ to both sides. Write the x-term first.

$3x + 4y = 10$ Therefore, the equation of the line containing $(-2, 4)$ and with slope of $-\dfrac{3}{4}$ is $3x + 4y = 10$.

Alternative Method:

It is also possible to write the equations of a line when given a point and the slope by using the slope–intercept formula, $y = mx + b$. Since we are given a point and the slope, we know a value for x, y, and m. We substitute these values into $y = mx + b$ and solve for b. We illustrate this procedure by using Example 1a, where we were given the point $(3, -4)$. (So, $x = 3$ and $y = -4$.) We were also given that $m = \frac{2}{3}$.

$y = mx + b$ Substitute 3 for x, -4 for y, and $\frac{2}{3}$ for m.

$-4 = \frac{2}{3}(3) + b$ $\frac{2}{3}(3) = \frac{2}{3} \cdot \frac{3}{1} = \frac{6}{3} = 2$

$-4 = 2 + b$ Subtract 2 from both sides of the equation.

$-6 = b$ Substitute $\frac{2}{3}$ for m and -6 for b in $y = mx + b$.

$y = \frac{2}{3}x - 6$ Equation in slope–intercept form.

PRACTICE EXERCISE 1

Write the equation of each of the following lines that contain the given point and have the given slope in $ax + by = c$ form with a and b integers:

a. $(-1, 3)$ and $m = \frac{3}{5}$ **b.** $(4, -2)$ and $m = -\frac{3}{2}$

If you need more practice, do the following Additional Practice Exercises.

Additional Practice Exercise 1   Write the equations of the following lines in $ax + by = c$ form with a and b integers:

a. $(1, -3)$ and $m = \frac{1}{4}$ **b.** $(-3, 5)$ and $m = -3$

OBJECTIVE B Writing the equation of a line that contains two given points

The point–slope formula requires that we know the slope and a point on the line. If we know the coordinates of two points on the line, we do not know the slope, but we can find it. Consequently, we can use the point–slope formula to write the equation of a line if we know two points on the line.

Example 2   **Find the equation of the line that contains the given pairs of points. Express answers in slope–intercept form.**

a. $(2, 4)$ and $(-1, 2)$

Solution

Using the slope formula with $(x_1, y_1) = (2, 4)$ and $(x_2, y_2) = (-1, 2)$, find the slope of the line.

$m = \dfrac{y_2 - y_1}{x_2 - x_1}$ Substitute for x_1, y_1, x_2, and y_2.

$m = \dfrac{2 - 4}{-1 - 2}$ Simplify.

$m = \dfrac{-2}{-3} = \dfrac{2}{3}$ Therefore, the slope is $\frac{2}{3}$.

Answers: **Practice Exercise 1:** **a.** $3x - 5y = -18$ **b.** $3x + 2y = 8$ **Additional Practice 1:** **a.** $x - 4y = 13$ **b.** $3x + y = -4$

Since the slope between any two points on the line is the same, we may use either of the given points in the point–slope formula. Use $(2, 4)$ as (x_1, y_1).

$$y - y_1 = m(x - x_1)$$ Substitute for m, x_1, and y_1.

$$y - 4 = \frac{2}{3}(x - 2)$$ Multiply both sides by 3.

$$3(y - 4) = 3 \cdot \frac{2}{3}(x - 2)$$ Multiply 3 and $\frac{2}{3}$.

$$3(y - 4) = 2(x - 2)$$ Apply the distributive property on both sides.

$$3y - 12 = 2x - 4$$ Add 12 to both sides.

$$3y = 2x + 8$$

$$\frac{3y}{3} = \frac{2x + 8}{3}$$ Divide both sides by 3.

$$y = \frac{2}{3}x + \frac{8}{3}$$ Simplify both sides. Therefore, the equation of the line containing $(2, 4)$ and $(-1, 2)$ is $y = \frac{2}{3}x + \frac{8}{3}$.

b. $(-3, -3)$ and $(-1, 1)$

Solution

Find the slope of the line, using the slope formula with $(x_1, y_1) = (-3, -3)$ and $(x_2, y_2) = (-1, 1)$.

$$m = \frac{y_2 - y_1}{x_2 - x_1}$$ Substitute for x_1, y_1, x_2, and y_2.

$$m = \frac{1 - (-3)}{-1 - (-3)}$$ Simplify the numerator and the denominator.

$$m = \frac{1 + 3}{-1 + 3}$$ Continue simplifying.

$$m = \frac{4}{2} = 2$$ Therefore, the slope is 2.

Since the slope between any two points on the line is the same, we may use either of the given points in the point–slope formula. Use $(-1, 1)$.

$$y - y_1 = m(x - x_1)$$ Substitute for m, x_1, and y_1.

$$y - 1 = 2(x - (-1))$$ $x - (-1) = x + 1$.

$$y - 1 = 2(x + 1)$$ Distribute on the right side.

$$y - 1 = 2x + 2$$ Add 1 to both sides.

$$y = 2x + 3$$ Therefore, the equation of the line containing $(-3, -3)$ and $(-1, 1)$ is $y = 2x + 3$.

PRACTICE EXERCISE 2

Find the equation of the line that contains the given pairs of points. Express the answers in slope–intercept form.

a. $(4, 3)$ and $(-1, -7)$

b. $(-2, 5)$ and $(-4, 2)$

If you need more practice, do the following Additional Practice Exercises.

Additional Practice Exercise 2   Find the equation of the line that contains the given pairs of points. Express the answers in slope–intercept form.

a. $(5, 2)$ and $(3, 8)$

b. $(0, 3)$ and $(-2, 4)$

Answers: Practice Exercise 2: a. $y = 2x - 5$ **b.** $y = \frac{3}{2}x + 8$ **Additional Practice 2: a.** $y = -3x + 17$ **b.** $y = -\frac{1}{2}x + 3$

OBJECTIVE **Writing the equations of vertical and horizontal lines**

Recall that the graph of a linear equation with two variables of the form $x = k$ is a vertical line. Thus, all the coordinates of the points on a vertical line have the same x-value. To write the equation of a vertical line, all we need to know is the x-coordinate of one point on the line. The equation is $x = $ (the known x-coordinate). Recall also that the slope of a vertical line is undefined.

You will also recall that the graph of a linear equation with two variables of the form $y = k$ is a horizontal line. Thus, all the points on a horizontal line have the same y-coordinate. To write the equation of a horizontal line, all we need to know is the y-coordinate of one point on the line. The equation is $y = $ (the known y-coordinate). Also recall that the slope of a horizontal line is 0.

Example 3   **Write the equation of each of the following lines:**

a. containing $(3, 5)$ and vertical

Solution

Since the line is vertical, the equation is of the form $x = k$. We know that the x-coordinate of one point on the line is 3. Therefore, the equation is $x = 3$.

b. containing $(-2, 4)$, with undefined slope

Solution

Since the slope is undefined, we know the line is vertical. Therefore, the equation is of the form $x = k$. We know that the x-coordinate of one point on the line is -2. Therefore, the equation is $x = -2$.

c. containing $(3, 5)$ and horizontal

Solution

Since the line is horizontal, the equation of the line is in the form of $y = k$. We know that the y-coordinate of one point on the line is 5. Therefore, the equation is $y = 5$.

d. containing $(-3, -2)$, and $m = 0$

Solution

Since $m = 0$, we know the line is horizontal. Therefore, the equation is of the form $y = k$. We know that the y-coordinate of one point on the line is -2. Therefore, the equation of the line is $y = -2$.

PRACTICE EXERCISE 3

Write the equation of each of the following lines:
a. containing $(-3, 6)$ and vertical

b. containing $(2, 1)$, with $m = 0$

c. containing $(4, -1)$ and horizontal

d. containing $(-3, 2)$, with undefined slope

If you need more practice, do the following Additional Practice Exercises.

Additional Practice Exercise 3   Write the equation of each of the following lines:
a. vertical and containing $(2, 7)$

b. horizontal and containing $(-5, 7)$

c. containing $(-3, 2)$, with undefined slope.

d. containing $(5, -1)$, with $m = 0$

OBJECTIVE **Writing the equation of a line that contains a given point and is parallel or perpendicular to a line whose equation is given**

If we know a point on the line and the slope of the line, then we can write the equation of the line. Suppose we are given the equation of a line and the coordinates of a point not on the line. Since parallel lines have equal slopes, we can write the equation of a

---

Answers: Practice Exercise 3: **a.** $x = -3$ **b.** $y = 1$ **c.** $y = -1$ **d.** $x = -3$
Additional Practice 3: **a.** $x = 2$ **b.** $y = 7$ **c.** $x = -3$ **d.** $y = -1$

line containing the given point and parallel to the given line by using the slope of the given line. Similarly, we can write the equation of a line containing the given point and perpendicular to the given line by using the negative of the reciprocal of the slope of the given line. Recall that the slope of a line in $ax + by = c$ form is $m = -\frac{a}{b}$.

Example 4 | **Write the equations of the following lines:**

a. containing $(-2, 3)$ and parallel to the graph of $3x + 5y = 8$

Solution

To write the equation of a line, we need a point on the line and the slope of the line. We know that the point whose coordinates are $(-2, 3)$ is on the line. The slope of $3x + 5y = 8$ is $-\frac{a}{b} = -\frac{3}{5}$, and the slope of any line parallel to this line is also $-\frac{3}{5}$. Therefore, we are looking for the equation of a line containing $(-2, 3)$ and with slope of $-\frac{3}{5}$. Since we know a point and the slope, we use the point–slope formula.

$y - y_1 = m(x - x_1)$	Substitute for m, x_1, and y_1.
$y - 3 = -\dfrac{3}{5}(x - (-2))$	$x - (-2) = x + 2$, and multiply both sides by 5.
$5(y - 3) = 5\left(-\dfrac{3}{5}\right)(x + 2)$	Multiply 5 and $-\dfrac{3}{5}$.
$5(y - 3) = -3(x + 2)$	Distribute on both sides.
$5y - 15 = -3x - 6$	Add 15 to both sides.
$5y - 15 + 15 = -3x - 6 + 15$	Simplify both sides.
$5y = -3x + 9$	Add $3x$ to both sides.
$3x + 5y = -3x + 3x + 9$	Simplify the right side.
$3x + 5y = 9$	Therefore, the equation of the line containing $(-2, 3)$ and parallel to the graph of $3x + 5y = 8$ is $3x + 5y = 9$.

b. containing $(4, -3)$ and perpendicular to the graph of $2x + 5y = 7$

Solution

To write the equation of a line, we need a point on the line and the slope of the line. We know that the point whose coordinates are $(4, -3)$ is on the line. The slope of $2x + 5y = 7$ is $-\frac{a}{b} = -\frac{2}{5}$, so the slope of any line perpendicular to the given line is the negative reciprocal of $-\frac{2}{5}$, which is $\frac{5}{2}$. Hence, we are looking for the equation of a line containing $(4, -3)$ and with slope of $\frac{5}{2}$. Use the point–slope formula.

$y - y_1 = m(x - x_1)$	Substitute for m, x_1, and y_1.
$y - (-3) = \dfrac{5}{2}(x - 4)$	$y - (-3) = y + 3$, and multiply both sides by 2.
$2(y + 3) = 2\left(\dfrac{5}{2}\right)(x - 4)$	Multiply 2 and $\dfrac{5}{2}$.
$2(y + 3) = 5(x - 4)$	Distribute on both sides.
$2y + 6 = 5x - 20$	Subtract $5x$ from both sides.
$-5x + 2y + 6 = 5x - 5x - 20$	Simplify both sides.
$-5x + 2y + 6 = -20$	Subtract 6 from both sides.
$-5x + 2y + 6 - 6 = -20 - 6$	Simplify both sides.
$-5x + 2y = -26$	Multiply both sides by -1.
$-1(-5x + 2y) = -1(-26)$	Simplify both sides.
$5x - 2y = 26$	Therefore, the equation of the line containing $(4, -3)$ and perpendicular to the graph of $2x + 5y = 7$ is $5x - 2y = 26$.

PRACTICE EXERCISE 4

Write the equations of the given lines. Express answers in the form of $ax + by = c$.

a. containing $(-1, 2)$ and parallel to the graph of $3x - 6y = 8$

b. containing $(-3, -4)$ and perpendicular to $4x + 2y = 7$

If you need more practice, do the following Additional Practice Exercises.

Additional Practice Exercise 4 Write the equations of the given lines. Express answers in the form of $ax + by = c$.

a. containing $(-3, 1)$ and parallel to the graph of $3x - 4y = 8$

b. containing $(-5, -2)$ and perpendicular to the graph of $4x + 2y = 7$

Frequently, real-world situations can be modeled by using linear equations with two variables. These models frequently are meaningful for nonnegative values of x and y only.

Example 5

a. A salesperson works for a salary plus commission. If x represents the value of the merchandise sold and y represents the salary earned per month, then $(15,000, 3150)$ indicates that, for sales of \$15,000, the salary was \$3150, and $(10,000, 2150)$ means that, on \$10,000 worth of sales, the salary was \$2150.

1. Write a linear equation that gives the salary, y, in terms of the sales, x.

Solution

We are being asked to write the equation of a line that contains the points $(15,000, 3150)$ and $(10,000, 2150)$ and to express the answer in the form of $y = mx + b$. We will use the point–slope formula, so we need the slope.

$$m = \frac{y_2 - y_1}{x_2 - x_1} \qquad \text{Substitute.}$$

$$m = \frac{3150 - 2150}{15,000 - 10,000} \qquad \text{Simplify.}$$

$$m = \frac{1000}{5000} \qquad \text{Reduce.}$$

$$m = \frac{1}{5}$$

Now use either of the points and the point–slope formula. We use $(10,000, 2150)$.

$$y - y_1 = m(x - x_1) \qquad \text{Substitute.}$$

$$y - 2150 = \frac{1}{5}(x - 10,000) \qquad \text{Distribute } \frac{1}{5}.$$

$$y - 2150 = \frac{1}{5}x - 2000 \qquad \text{Add 2150 to both sides.}$$

$$y = \frac{1}{5}x + 150 \qquad \text{Therefore, the salary, } y\text{, in terms of the sales, } x\text{, is } y = \frac{1}{5}x + 150.$$

Answers: Practice Exercise 4: a. $x - 2y = -5$ **b.** $x - 2y = 5$
Additional Practice 4: a. $3x - 4y = -13$ **b.** $x - 2y = -1$

2. Use the equation found in part 1 to find the salary for sales of $18,000.

Solution

Since x represents the sales, we replace x with 18,000 and find y, the salary.

$$y = \frac{1}{5}x + 150 \qquad \text{Substitute 18,000 for } x.$$

$$y = \frac{1}{5}(18,000) + 150 \qquad \text{Multiply } \frac{1}{5} \text{ and 18,000.}$$

$$y = 3600 + 150 \qquad \text{Add.}$$

$$y = 3750 \qquad \text{Therefore, the salesperson earns \$3750 for sales of \$18,000.}$$

3. Use the equation found in part 1 to find the total value of the sales in a month when the salary was $1150.

Solution

Since y represents the salary, we will replace y with $1150 and find x, the sales.

$$y = \frac{1}{5}x + 150 \qquad \text{Substitute 1150 for } y.$$

$$1150 = \frac{1}{5}x + 150 \qquad \text{Subtract 150 from both sides.}$$

$$1000 = \frac{1}{5}x \qquad \text{Multiply both sides by 5.}$$

$$5000 = x \qquad \text{Therefore, the salesperson's sales were \$5000 in order to earn \$1150 in a month.}$$

b. A college buys a mower for $10,000. After 6 years, it has depreciated to a value of $4000. Assuming that the mower depreciates the same each year (called linear depreciation), answer the following:

1. Write a linear equation giving the value of the mower, y, in terms of the number of years after it was purchased, x. Express the answer in $y = mx + b$ form.

Solution

Since x represents the number of years after it was purchased and y represents the value of the mower, the fact that the mower cost $10,000 can be represented as the ordered pair $(0, 10,000)$, and the fact that after 6 years the mower was worth $4000 can be written as $(6, 4000)$. We need the equation of a line containing these two points and to express the answer in $y = mx + b$ form. First find the slope.

$$m = \frac{y_2 - y_1}{x_2 - x_1} \qquad \text{Substitute.}$$

$$m = \frac{4000 - 10,000}{6 - 0} \qquad \text{Simplify.}$$

$$m = \frac{-6000}{6} \qquad \text{Divide.}$$

$$m = -1000$$

Now use the slope–intercept formula and $(0, 10,000)$, since 10,000 is the y-intercept, b.

$$y = mx + b \qquad \text{Substitute } -1000 \text{ for } m \text{ and 10,000 for } b.$$

$$y = -1000x + 10,000 \qquad \text{Therefore, the value of the mower, } y, \text{ in terms of the number of years after it was purchased, } x, \text{ is } y = -1000x + 10,000.$$

2. Use the equation found in part 1 to find the value of the mower 8 years after it was purchased.

Solution

Since x represents the number of years after it was purchased, replace x with 8 and find y.

$$y = -1000x + 10,000 \qquad \text{Substitute 8 for } x.$$

$$y = -1000(8) + 10,000 \qquad \text{Multiply } -1000 \text{ and 8.}$$

$$y = -8000 + 10,000 \qquad \text{Add.}$$

$$y = 2000 \qquad \text{Therefore, 8 years after it was purchased, the mower will be worth \$2000.}$$

3. How many years after it was purchased will the mower have no value?

Solution

Since y represents the value of the mower, we are being asked to find x when $y = 0$.

$$y = -1000x + 10,000 \quad \text{Substitute 0 for } y.$$
$$0 = -1000x + 10,000 \quad \text{Add } 1000x \text{ to both sides of the equation.}$$
$$1000x = 10,000 \quad \text{Divide both sides of the equation by 1000.}$$
$$x = 10 \quad \text{Therefore, the mower will have no value after 10 years.}$$

PRACTICE EXERCISE 5

Answer the following:

a. John, an appliance salesperson, earns a salary of $100 per week plus $7.50 for each appliance sold.

  1. Write a linear equation that gives his weekly earnings, y, in terms of the number of appliances sold, x.

  2. Use the equation found in part 1 to find John's earnings during a week in which he sold 15 appliances.

  3. How many appliances would he have to sell in order to make $400 in 1 week?

b. The Jones family purchased a land lot for $20,000. After 5 years, the value of the lot had increased to $30,000. Assuming that the lot increases by the same amount each year, answer the following:

  1. Write a linear equation that gives the value of the lot, y, in terms of the number of years after it was purchased, x.

  2. Use the equation found in part 1 to find the value of the lot 9 years after it was purchased.

  3. How many years after it was purchased will the lot have a value of $50,000?

Exercise Set 4.6

For Extra Help

MyMathLab®

Write the equation of each of the following lines that contain the given point and have the given slope in $ax + by = c$ form with a and b integers. (See Examples 1 and 3.)

1. $(3, 1)$ and $m = 2$

2. $(5, 2)$ and $m = 3$

3. $(-2, 4)$ and $m = -2$

4. $(4, -2)$ and $m = -3$

5. $(-3, 5)$ and $m = \dfrac{1}{4}$

6. $(4, -5)$ and $m = \dfrac{1}{2}$

7. $(2, -4)$ and $m = -\dfrac{1}{3}$

8. $(-2, -1)$ and $m = -\dfrac{1}{5}$

9. $(5, 1)$ and $m = \dfrac{4}{3}$

10. $(6, -2)$ and $m = \dfrac{5}{2}$

11. $(-2, -5)$ and $m = -\dfrac{2}{3}$

12. $(4, -1)$ and $m = -\dfrac{5}{3}$

13. $(3, 4)$ and vertical

14. $(-2, 5)$ and vertical

15. $(2, 3)$ and horizontal

16. $(-1, -7)$ and horizontal

17. $(5, 2)$ and undefined slope

18. $(-6, 2)$ and undefined slope

19. $(5, 4)$ and $m = 0$

20. $(-6, -1)$ and $m = 0$

Answers: Practice Exercise 5:   a.   1. $y = 7.5x + 100$ **2.** $212.50 **3.** 40 appliances **b.   1.** $y = 2000x + 20,000$ **2.** $38,000 **3.** 15 years

Write the equation of the line that contains the given pairs of points. Express answers in slope–intercept form. (See Examples 2 and 3.)

21. $(4, -3)$ and $(-1, 7)$ **22.** $(1, -6)$ and $(-2, 3)$ **23.** $(1, 5)$ and $(-3, 1)$ **24.** $(4, -7)$ and $(-2, -1)$

25. $(2, -8)$ and $(-4, 1)$ **26.** $(4, -4)$ and $(-4, 2)$ **27.** $(3, 2)$ and $(-3, 6)$ **28.** $(0, 4)$ and $(3, 5)$

29. $(5, -3)$ and $(5, 1)$ **30.** $(-2, 4)$ and $(-2, -1)$ **31.** $(5, -4)$ and $(-1, -4)$ **32.** $(7, 2)$ and $(-3, 2)$

Write the equations of the following lines in ax + by = c form with a, b, and c integers. (See Example 4.)

33. containing $(3, -1)$ and parallel to the graph of $2x + 4y = 7$

34. containing $(-4, 2)$ and parallel to the graph of $3x + 9y = 10$

35. containing $(-3, -4)$ and parallel to the graph of $2x - 3y = 7$

36. containing $(-4, 6)$ and parallel to the graph of $3x + 5y = 4$

37. containing $(4, -1)$ and perpendicular to the graph of $x + 4y = 7$

38. containing $(-5, 0)$ and perpendicular to the graph of $3x - y = 6$

39. containing $(-5, 3)$ and perpendicular to the graph of $2x + 6y = 7$

40. containing $(4, -1)$ and perpendicular to the graph of $8x + 4y = 15$

41. containing $(6, -3)$ and parallel to the graph of $x = -5$

42. containing $(7, -3)$ and parallel to the graph of $y = 2$

43. containing $(4, -4)$ and perpendicular to the graph of $x = 6$

44. containing $(-2, 3)$ and perpendicular to the graph of $y = 3$

Answer the following. (See Example 5.)

45. The cost to drive a rental car depends on the number of miles driven. The ordered pair $(100, 40)$ indicates that it costs $40 to drive 100 miles, and the ordered pair $(200, 55)$ indicates that it costs $55 to drive 200 miles.

   a. Write a linear equation giving the cost, y, in terms of the number of miles, x.

   b. Using the equation found in part a, find how much it costs to drive 500 miles.

   c. How many miles would you have to drive if the cost is $62.50?

46. A salesperson works for a fixed salary plus a commission on sales. The ordered pair $(1000, 250)$ indicates earnings of $250 for $1000 in sales, and the ordered pair $(3000, 450)$ indicates earnings of $450 for $3000 in sales.

   a. Write a linear equation giving the salary, y, in terms of the value of merchandise sold, x.

   b. Use the equation in part a to find the salary for sales of $4000.

   c. Find the total sales for a week in which the salesperson earned $750.

47. The cost to ride a taxi depends on the number of miles that you ride. The ordered pair $(2, 2.80)$ means that a ride of 2 miles costs $2.80, and $(6, 6)$ means that a ride of 6 miles costs $6.

   a. Write a linear equation that gives the cost, y, in terms of the number of miles, x.

   b. Use the equation found in part a to find the cost of a taxi ride of 8 miles.

   c. Find the length of a taxi ride that costs $5.20.

48. The cost of an item depends on the number produced. The ordered pair $(2, 80)$ indicates that when two items are produced the cost per item is $80. The ordered pair $(3, 70)$ indicates that when three items are produced the cost per item is $70.

   a. Write a linear equation giving the cost per item, y, in terms of the number of items produced, x.

   b. Use the equation in part a to find the cost per item if five items are produced.

   c. Use the equation in part a to find the number of items produced if the cost per item is $40.

49. A farmer purchases a tractor for $50,000. After 5 years, the tractor has depreciated to a value of $25,000. Assuming linear depreciation, answer the following:

　a. Write a linear equation that gives the value of the tractor, y, in terms of the number of years after it was purchased, x.

　b. Use the equation found in part a to find the value of the tractor after 8 years.

　c. Use the equation found in part a to find the number of years until the tractor has no value.

51. Sally is a salesperson at a model home center, and her salary is $200 per month plus a 5% commission on her sales.

　a. Write a linear equation that gives her salary, y, in terms of her sales, x.

　b. Use the equation found in part a to find how much she will earn during a month in which her sales are $250,000.

　c. Use the equation found in part a to find her sales for a month in which she earned $8950.

50. A basketball trading card cost $1.50 new, and after 4 years has a value of $2.70. Assuming that the card increases in value the same amount each year, answer the following:

　a. Write a linear equation that gives the value of the card, y, in terms of the number of years after it was purchased, x.

　b. Use the equation found in part a to find the value of the card 10 years after it was purchased.

　c. Use the equation found in part a to find the number of years until the card is worth $9.00.

52. The cost of renting a car is $30 per day plus $.15 per mile.

　a. Write a linear equation giving the cost, y, in terms of the number of miles driven, x, for a one-day rental.

　b. Use the equation found in part a to find the cost if the car is driven 300 miles in one day.

　c. Use the equation found in part a to find how many miles the car was driven in one day if the cost was $52.50.

Section 4.7

Graphing Linear Inequalities with Two Variables

OBJECTIVE *When you complete this section, you will be able to:*

A Graph linear inequalities with two unknowns.

PREREQUISITE SKILLS *Before beginning this section, you should be able to:*

a. Add, subtract, multiply, and divide integers. (Sections 1.6, 1.7, 2.1, 2.5)
b. Substitute values for variables and evaluate. (Section 2.5)
c. Graph linear equations. (Sections 4.2, 4.3, 4.5)

GETTING READY FOR SECTION 4.7

Add, subtract, multiply, or divide the following integers:

1. $-9 - 1$ **2.** $4 - (-6)$ **3.** $-8(-12)$ **4.** $-25 \div (-5)$

Substitute $x = -4$ and $y = 3$ into the given expressions and evaluate:

5. $-2x + 5y$ **6.** $6x - 3y$

Graph each of the following linear equations:

7. $3x + 2y = 6$ **8.** $2x - y = 0$ **9.** $y = 3$ **10.** $x = 5$

Introduction

In Section 4.2, we graphed linear equations. In this section, we will graph **linear inequalities**. A linear inequality with two variables is in the form of $ax + by > c$, $ax + by \geq c$, $ax + by < c$, or $ax + by \leq c$ or can be put into one of these forms. Just as any ordered pair of numbers that makes a linear equation true is a solution of the equation, any ordered pair that makes a linear inequality true is a solution of the inequality.

Example 1   **Determine whether the given ordered pair is a solution of the given inequality.**

a. $y > 3x - 1; (2, 7)$

Solution

$y > 3x - 1$ Substitute 2 for x and 7 for y.
$7 > 3(2) - 1$ Simplify the right side of the equation.
$7 > 5$ This is true, so $(2, 7)$ is a solution.

b. $3x + 2y \geq 6; (-4, 5)$

Solution

$3(-4) + 2(5) \geq 6$ Substitute -4 for x and 5 for y.
                         Multiply before adding.
$-12 + 10 \geq 6$ Add.
$-2 \geq 6$ This is false, so $(-4, 5)$ is not a solution.

Answers: Getting Ready: 1. -10 **2.** 10 **3.** 96 **4.** 5 **5.** 23 **6.** -33

7.
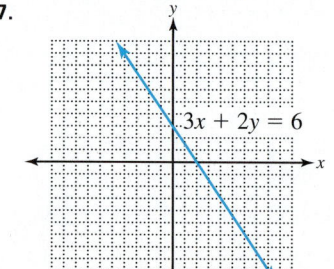
$3x + 2y = 6$

8.
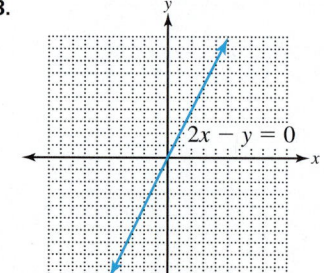
$2x - y = 0$

9.
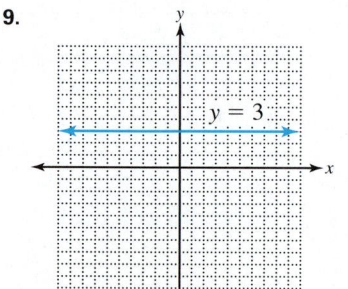
$y = 3$

10.
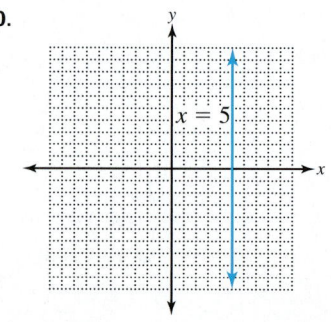
$x = 5$

In Section 3.5, we solved and graphed the solutions of linear inequalities with one unknown. For example, the solution of $2x - 3 > 5$ is $x > 4$. To graph $x > 4$, we first graph $x = 4$ with an open dot to indicate that 4 is not a part of the solution. The graph of $x > 4$ is the set of all points of the line on one side of 4 (either to the right or left of 4). To determine which side, use a test value for x, say, 5. Substitute 5 for x, and we get $5 > 4$, which is a true statement. Since 5 is to the right of 4, the graph of $x > 4$ is all the points to the right of 4. See the following graph:

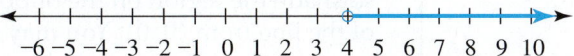

To graph $x \geq 4$, we would use a solid dot, rather than an open dot, to indicate that 4 is a part of the solution.

The graph of the solutions of a linear inequality with one variable of the form $x > a$ or $x < a$ is all the points of the number line that are on one side of $x = a$ or the other. Similarly, the graph of the solutions of a linear inequality with two variables is all the points of the rectangular coordinate system on one side of the graph of $ax + by = c$ or the other. In other words, all the points on the graph of $ax + by = c$ make the equation true. All other points in the coordinate system make either $ax + by > c$ or $ax + by < c$ true. We indicate the solutions by shading all the points on the side of the line that solve the inequality.

To graph the solutions of a linear inequality with two variables, we follow a procedure similar to graphing the solutions of a linear inequality with one variable. The procedure is outlined as follows:

OBJECTIVE Graphing linear inequalities with two unknowns

> **Procedure: Graphing Linear Inequalities with Two Variables**
>
> To graph a linear inequality with two variables:
>
> 1. Graph the equality $ax + by = c$. If the line is part of the solution ($\leq$ or $\geq$), draw a solid line. If the line is not part of the solution ($<$ or $>$), draw a dashed line.
>
> 2. Pick a test point not on the line. If the coordinates of the test point solve the inequality, then all points on the same side of the line as the test point solve the inequality. If the coordinates of the test point do not solve the inequality, then all the points on the other side of the line from the test point solve the inequality.
>
> 3. Shade the region on the side of the line that contains the solutions of the inequality.

Example 2 | **Graph the following linear inequalities with two variables:**

a. $2x + y > 8$

Solution

Draw the graph of $2x + y = 8$ with a dashed line, since the points on the line do not solve the inequality. The x-intercept is $(4, 0)$, the y-intercept is $(0, 8)$, and we use the point $(2, 4)$ as a check. Remember, to find the x-intercept, let $y = 0$, and to find the y-intercept, let $x = 0$.

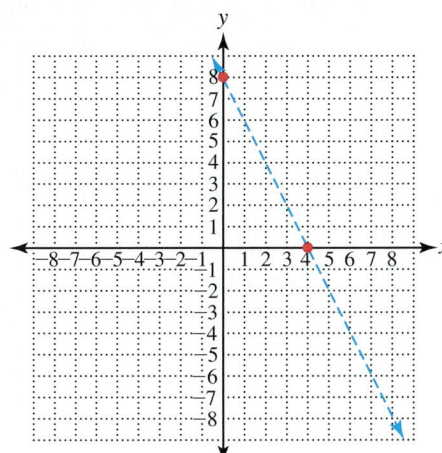

Note

You could also solve for y ($y = -2x + 8$) and graph by using the slope and y-intercept.

Pick a test point. The origin, $(0, 0)$, is often a convenient test point. Substitute $(0, 0)$ into the inequality. $2(0) + 0 > 8, 0 + 0 > 8, 0 > 8$. This is a false statement, so $(0, 0)$ is not a solution of $2x + y > 8$. Therefore, the solutions are on the opposite side of the line from $(0, 0)$, so shade the region on the opposite side of the line from $(0, 0)$. You may want to test a point on the shaded side just to be sure.

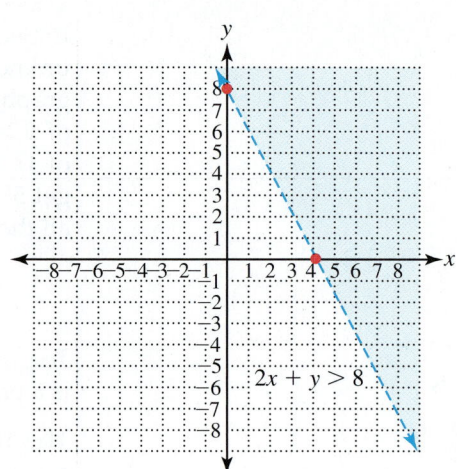

b. $3x - y \leq 6$

Solution

Draw the graph of $3x - y = 6$ as a solid line, since the points on the line satisfy the inequality less than or *equal to*. Use the x-intercept $(2, 0)$, the y-intercept $(0, -6)$, and another point as a check.

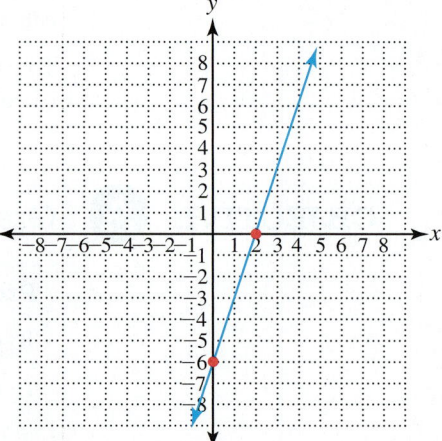

Pick a test point. Again, $(0, 0)$ is a convenient choice. Substitute $(0, 0)$ into the inequality. $3(0) - 0 \leq 6, 0 - 0 \leq 6, 0 \leq 6$. This is a true statement. Therefore, $(0, 0)$ solves the inequality. Consequently, all points on the same side of the line as $(0, 0)$ solve the inequality, so shade the region on the same side of the line as $(0, 0)$.

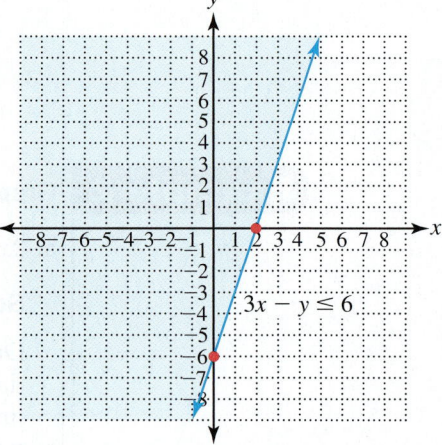

PRACTICE EXERCISE 2

a. Following is the graph of $2x + y = 6$ as a dashed line. Shade the side that solves $2x + y < 6$.

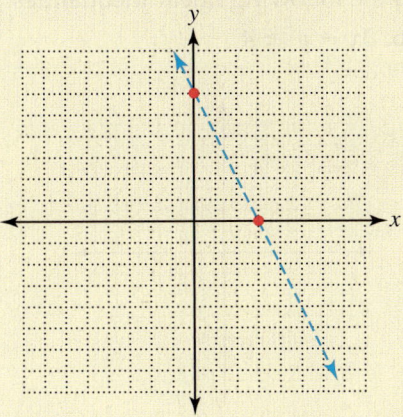

Graph the following linear inequalities with two variables:

b. $3x - 4y \geq 12$

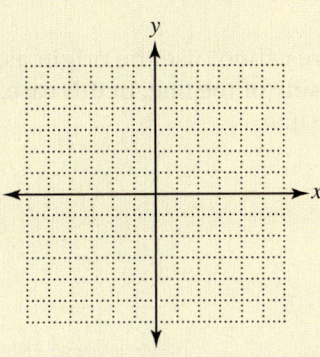

c. $5x - 2y < 10$

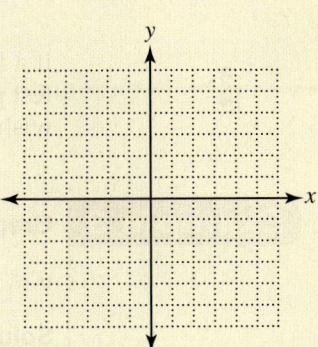

(Continued)

Answers: Practice Exercise 2: a.

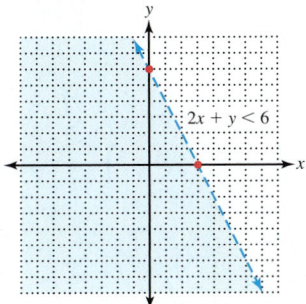

$2x + y < 6$

b.

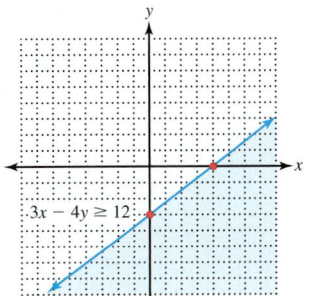

$3x - 4y \geq 12$

c.

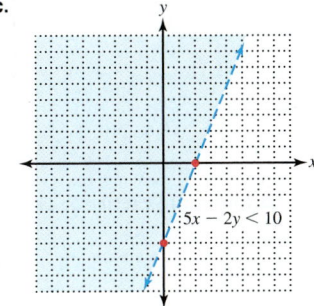

$5x - 2y < 10$

If you need more practice, do the following Additional Practice Exercises.

Additional Practice Exercise 2   Graph the following linear inequalities:

a. Following is the graph of $x + 2y = 4$ as a dashed line. Shade the side that solves $x + 2y < 4$.

b. $3x - y \geq 4$

c. $4x + y > 8$

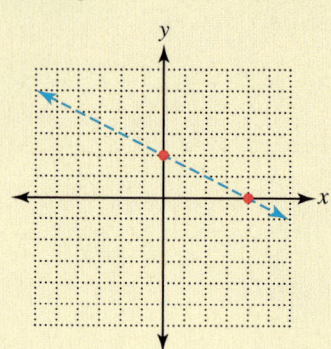

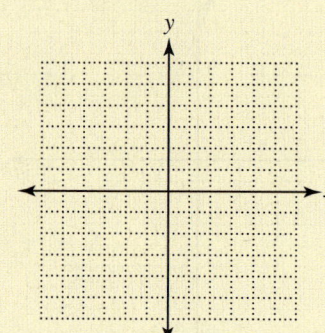

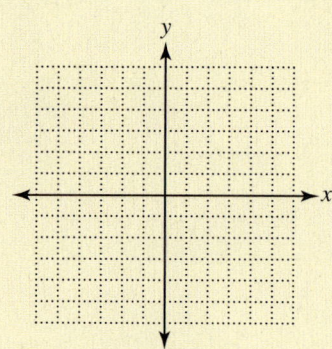

If a line passes through the origin, we cannot use the origin as a test point, since the test point must lie in a region on one side of the line. In this case, choose any point you wish as long as it is clearly not on the line.

Example 3   **Graph the following inequality:**

$y \leq 3x$

Solution

Draw the graph of $y = 3x$ as a solid line, since the points on the line satisfy the inequality less than or *equal to*. Since both the x- and y-intercepts are $(0, 0)$, we need a point other than the intercepts in order to draw the graph. Choose a value for x and solve for y. Let $x = 3$; then $y = 3(3) = 9$. Therefore, another ordered pair is $(3, 9)$.

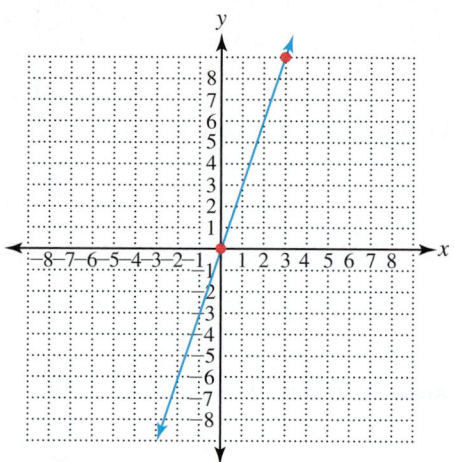

Answers: Additional Practice 2: a.

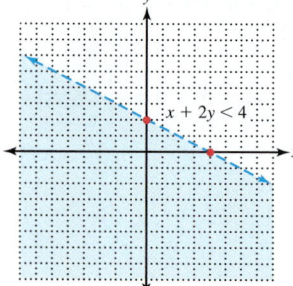

b.

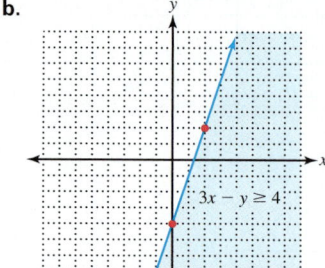

c.

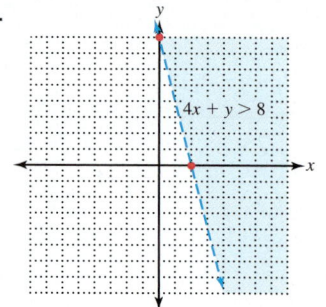

Since $(0, 0)$ lies on the line, we cannot use it as the test point. Choose any other point that clearly is not on the line, say, $(2, 0)$. Substitute $(2, 0)$ into the inequality. $0 \leq 3(2)$, $0 \leq 6$. This is true. Therefore, all the solutions are on the same side of the line as $(2, 0)$, so shade this region.

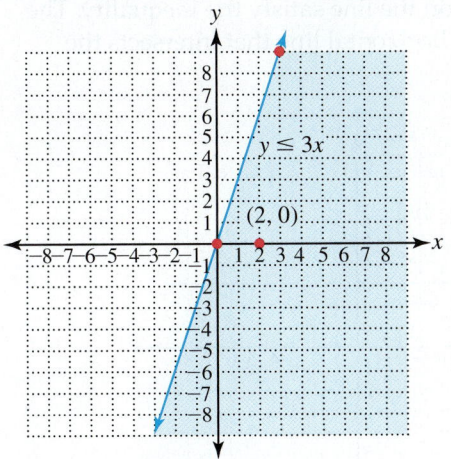

In the case of vertical and horizontal lines, it is not necessary to use a test point. The situation is very much like graphing the solutions to a linear inequality with one variable.

| **Example 4** | **Graph the following linear inequalities with two variables:** |

a. $x > -2$

Solution

Draw the graph of $x = -2$ as a dashed line, since the coordinates of the points on the line do not satisfy the inequality. The graph of $x = -2$ is a vertical line intersecting the x-axis at -2.

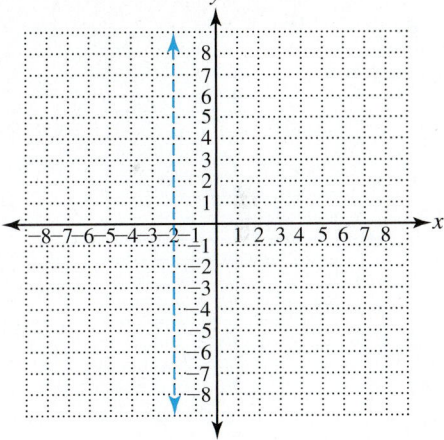

Note

You could also use a test point, if you prefer.

The solutions to $x > -2$ consist of all points whose x-values are greater than -2. Just as on the number line, these points lie to the right of the graph of $x = -2$. Therefore, shade the region to the right of $x = -2$.

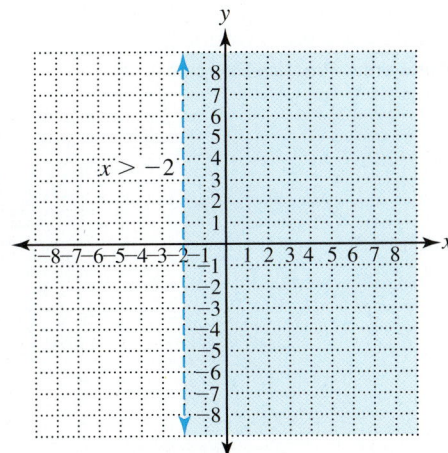

b. $y \leq 3$

Solution

Draw the graph of $y = 3$ as a solid line, since the coordinates of the points on the line satisfy the inequality. The graph of $y = 3$ is a horizontal line that intersects the y-axis at 3.

The solutions to $y \leq 3$ consist of all points whose y-values are less than 3. These points lie below the line $y = 3$. Therefore, shade the region below the line.

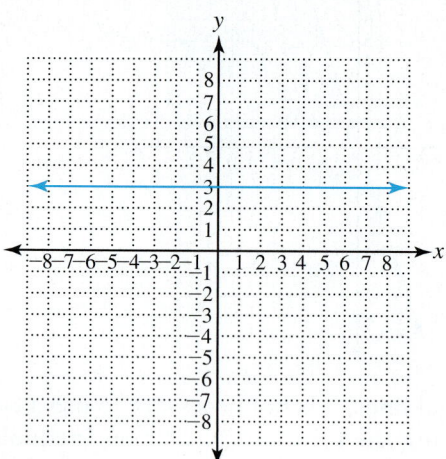

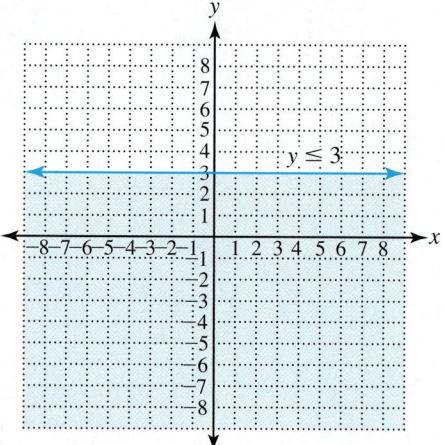

PRACTICE EXERCISE 4

Graph the given linear inequalities with two variables. Assume that each unit on the coordinate system is 1.

a. $y < -3x$

b. $x < -3$

c. $y \geq 2$

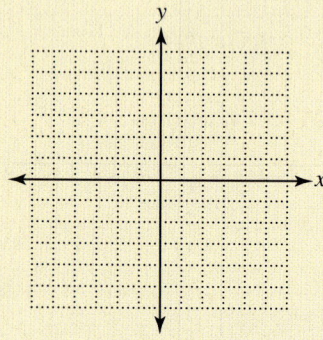

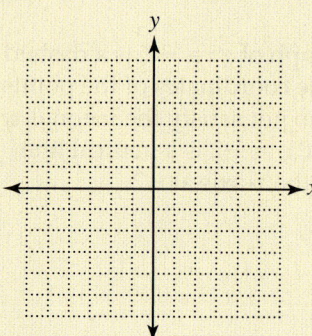

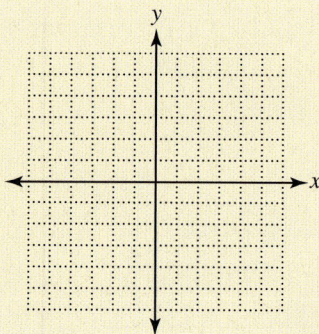

(Continued)

Answers: **Practice Exercise 4:** **a.**

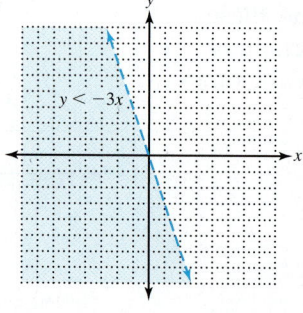

b.

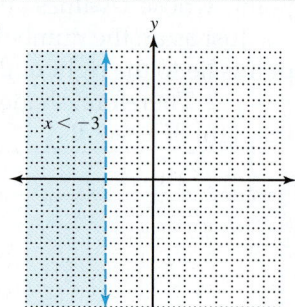

c.

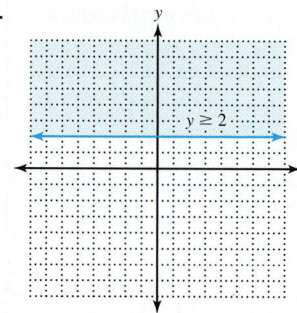

If you need more practice, do the following Additional Practice Exercises.

Additional Practice Exercise 4 Graph the given inequalities with two variables:

a. $y \geq x$ **b.** $x \geq 3$ **c.** $y < -1$

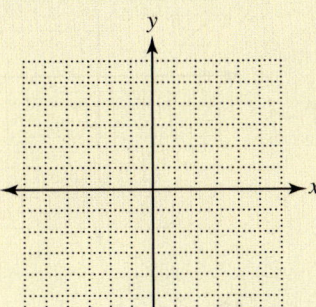 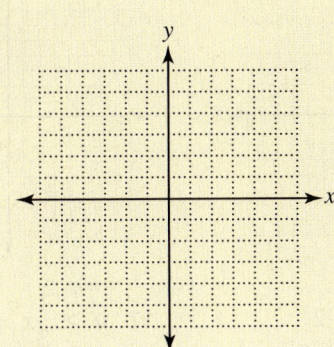 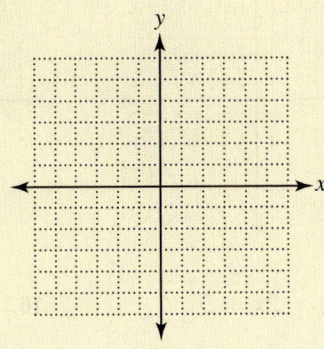

Exercise Set 4.7

For Extra Help

MyMathLab®

Determine whether the given point is a solution of the given inequality. (See Example 1.)

1. $y > 3x - 2;\ (1, 0)$ **2.** $y > -2x + 3;\ (2, 0)$ **3.** $2x + 3y \leq 6;\ (-2, 5)$

4. $3x - 4y \leq 8;\ (-2, -3)$ **5.** $3x - 2y \geq 12;\ (2, -3)$ **6.** $4x - 3y \leq 8;\ (5, 4)$

For Exercises 7–10, a dashed line has been graphed. Shade the side of the line that is the solution of the given inequality. (See Examples 2–4.)

7. $x + y > 6$ **8.** $x + y < 4$ **9.** $3x + y < -6$ **10.** $x - 4y > 8$

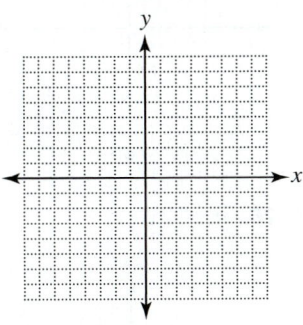 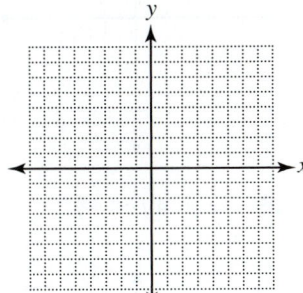 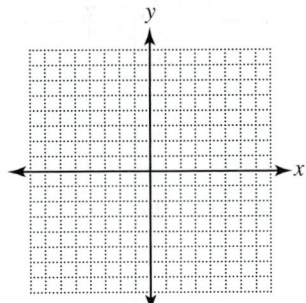 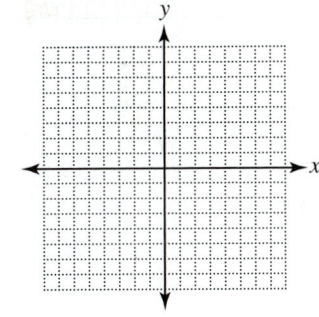

Answers: Additional Practice 4: a.

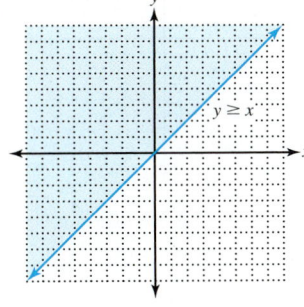 **b.** 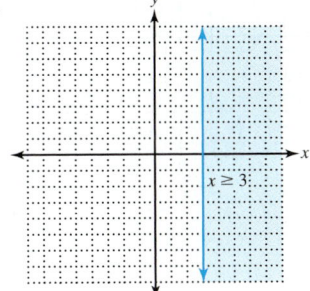 **c.**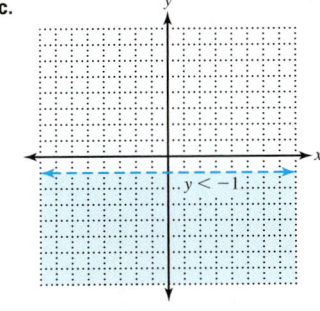

Graph the given linear inequalities with two variables.

11. $y < x + 2$

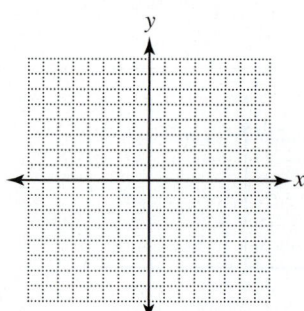

12. $y > x - 6$

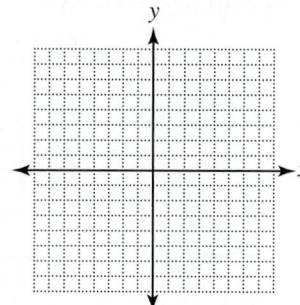

13. $y \geq 2x + 3$

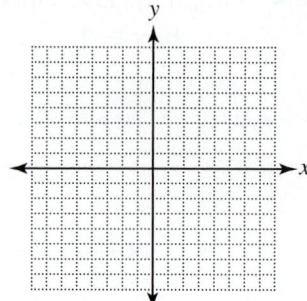

14. $y \geq 3x - 4$

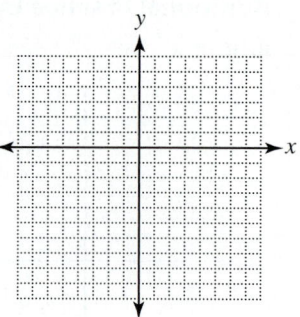

15. $y < -2x + 5$

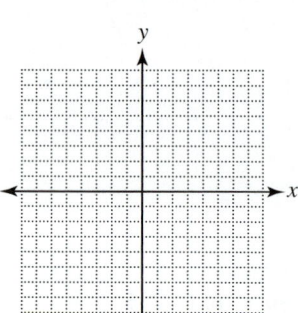

16. $y \geq -3x + 2$

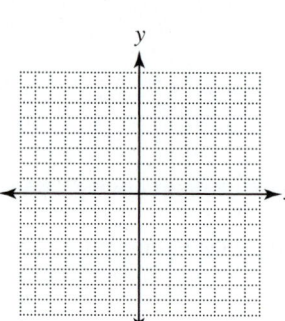

17. $y > \dfrac{1}{3}x + 2$

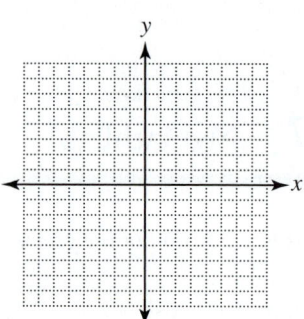

18. $y \leq \dfrac{2}{5}x - 3$

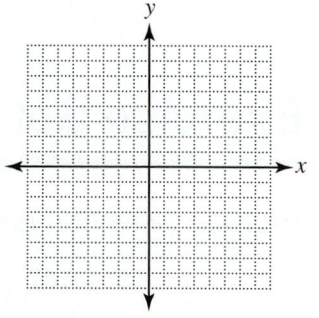

19. $2x + y > 6$

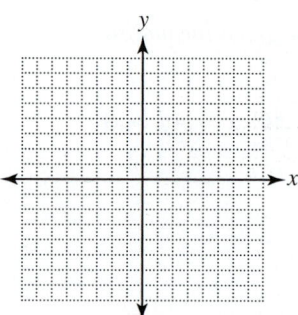

20. $3x - y \leq 9$

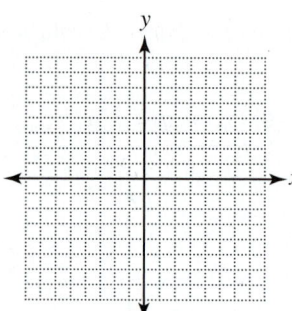

21. $x + 2y \leq 4$

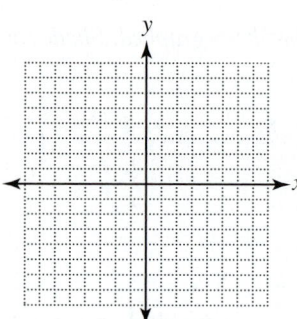

22. $x + 4y < 8$

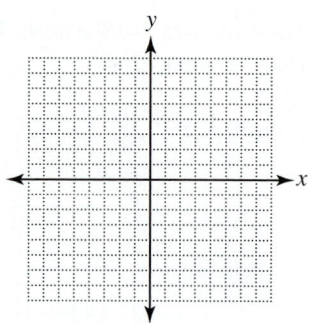

23. $2x - 3y > 6$

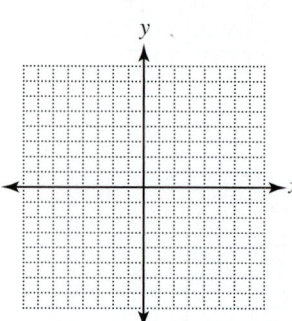

24. $5x - 2y \leq 15$

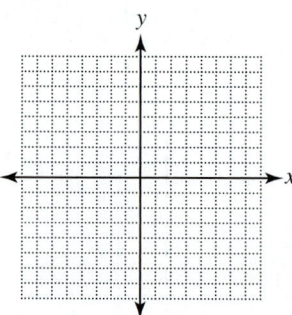

25. $3x + 4y \leq 12$

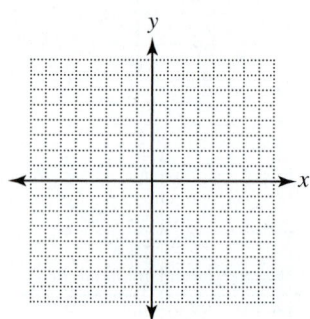

26. $y \leq 2x$

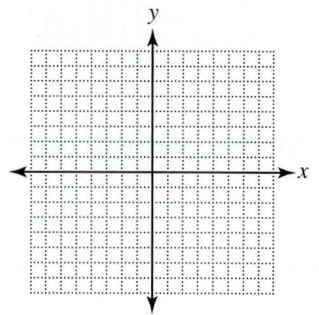

27. $y > x$

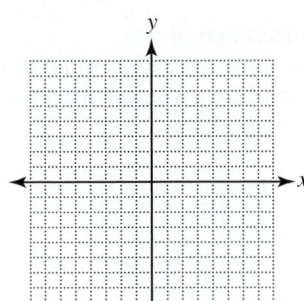

28. $y < -4x$

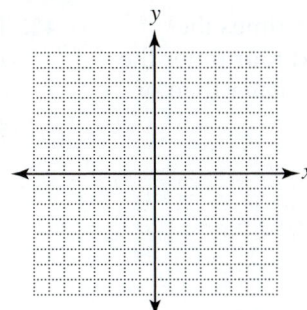

29. $y \geq -x$

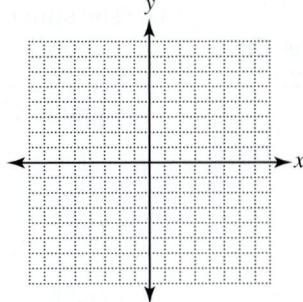

30. $y > -2x$

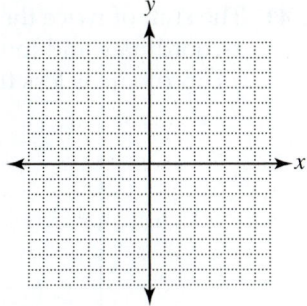

31. $y \leq -4x$

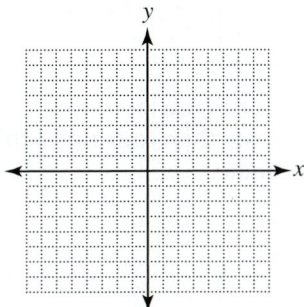

32. $x \geq 2$

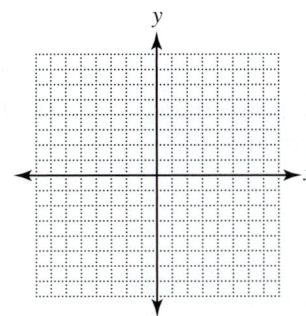

33. $x < 4$

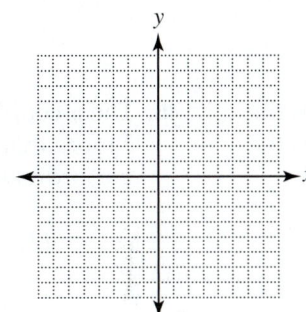

34. $y > 3$

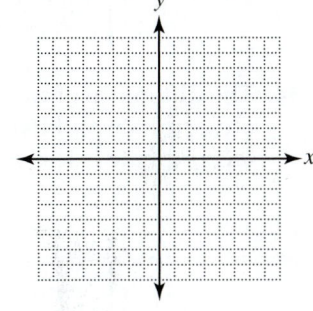

35. $y \leq 4$

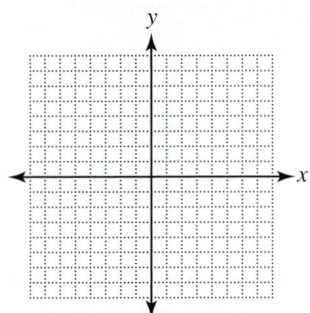

36. $x + 1 \geq 0$

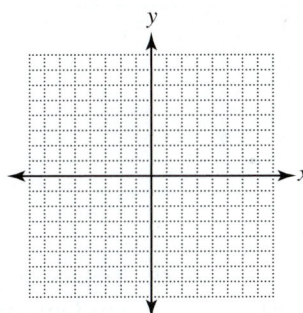

37. $x + 5 < 0$

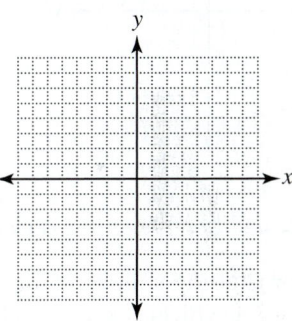

38. $y + 2 \leq 0$

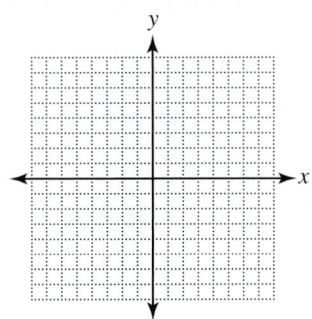

39. $y + 6 > 0$

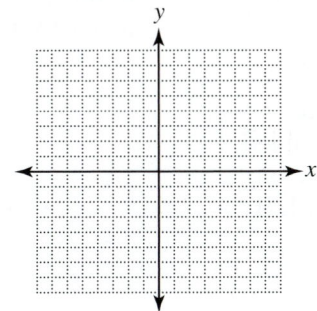

40. $y \leq 0$

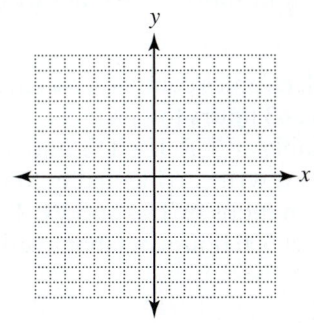

Write an inequality representing each of the following and graph:

41. The sum of twice the x-coordinate and four times the y-coordinate is less than 12.

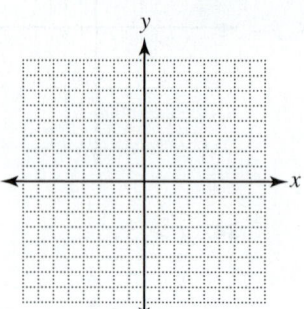

42. The sum of three times the x-coordinate and four times the y-coordinate is greater than or equal to 12.

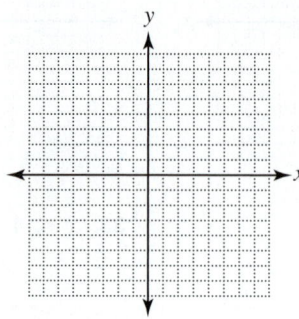

43. The difference of the x-coordinate and two times the y-coordinate is greater than or equal to 8.

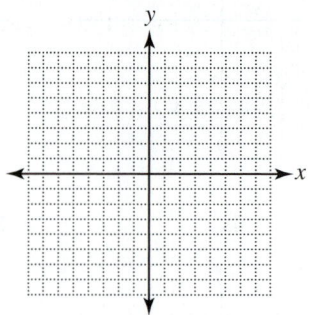

44. The difference of two times the x-coordinate and five times the y-coordinate is less than 10.

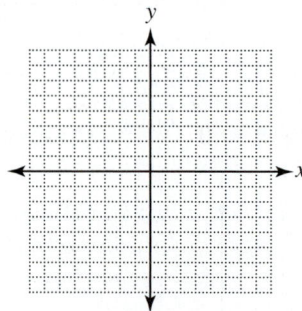

45. The y-coordinate is less than 5 more than three times the x-coordinate.

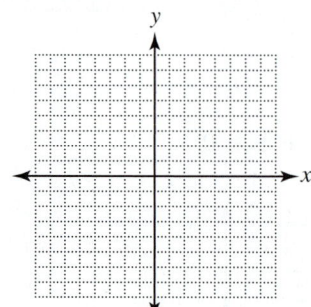

46. The y-coordinate is greater than 3 more than four times the x-coordinate.

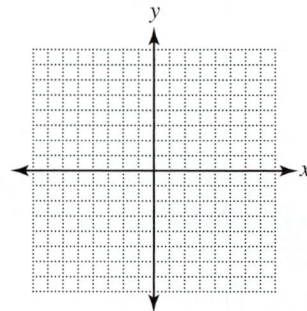

Write an inequality representing each of the following:

47. Hector is going shopping for some shirts and shorts. The shirts cost $20 each and the shorts cost $25 each. If he buys x shirts and y shorts, write an inequality that shows the possible combinations of shirts and shorts if Hector can spend no more than $150.

48. Tickets to a high school basketball game cost $3 for students and $5 for adults. If x student tickets and y adult tickets are sold, write an inequality that shows the possible combinations of student and adult tickets if the receipts from ticket sales must be at least $500.

49. An electronics company assembles TVs and DVDs. It takes 3 hours to assemble a TV and 2 hours to assemble a VCR. If x TVs and y VCRs are assembled per day, write an inequality that shows the possible combinations of TVs and VCRs if the company has at most 500 hours of labor available per day.

Challenge Exercises (50–53)

50. A nut company uses x ounces of peanuts in a regular mix and y ounces of peanuts in a special party mix. The company has at most 500 ounces of peanuts in stock.

   a. Write an inequality describing the situation.

   b. Graph the inequality.

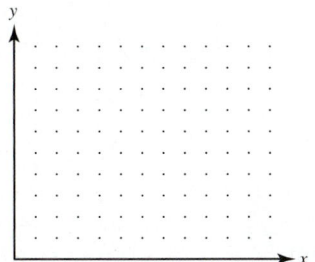

   c. Is (100, 300) a solution of this inequality?

52. A battery company sells two types of batteries. It sells battery A for $10 and battery B for $15. The company must have sales of at least $3000 per week to break even.

   a. Let x represent the number of A batteries sold per week and y represent the number of B batteries sold each week. Write an inequality describing the situation.

   b. Graph the inequality.

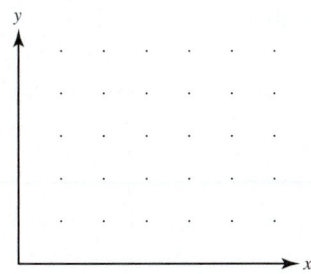

   c. Is (200, 100) a solution of the inequality?

51. A chemical company needs x liters of sulfuric acid to make chemical A and y liters to make chemical B. The company has at most 200 liters of sulfuric acid in stock.

   a. Write an inequality describing the situation.

   b. Graph the situation.

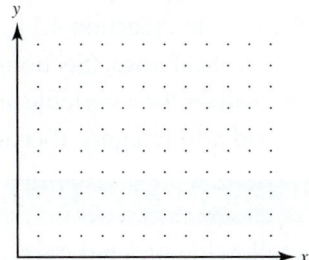

   c. Is (150, 100) a solution of the inequality?

53. A small company is in the business of manufacturing lawn chairs. It takes 2 hours of labor to make a regular chair and 3 hours of labor to make a lounge chair. There is a maximum of 84 hours per day of labor available.

   a. Let x represent the number of regular chairs produced and y represent the number of lounge chairs produced. Write an inequality describing the situation.

   b. Graph the inequality.

   c. Is (20, 20) a solution of the inequality?

Writing Exercises (54–56)

54. How is graphing a linear inequality with two variables like graphing a linear inequality with one variable? How does it differ?

55. If your test point is on one side of the boundary line and it fails to solve the inequality, why are the solutions on the other side of the line?

56. In choosing a test point, why can't you choose a point on the line?

Section 4.8   Relations and Functions

OBJECTIVES *When you complete this section you will be able to:*

A Determine the domain and range of a relation.
B Recognize functions when they are given as a set of ordered pairs, as a graph, or as an equation.
C Use functional notation.

PREREQUISITE SKILLS *Before beginning this section, you should be able to:*

a. Plot ordered pairs. (Section 4.1)
b. Recognize vertical lines. (Sections 4.2, 4.3, 4.4)
c. Substitute values for a variable and evaluate. (Section 2.5)
d. Multiply and add integers. (Sections 1.6, 2.1)

GETTING READY FOR SECTION 4.8

1. Plot the following ordered pairs on the rectangular coordinate system:
$(-4, -1), (6, -4), (6, 0), (-7, -8)$

Decide whether the graphed line is horizontal, vertical, or neither.

2. 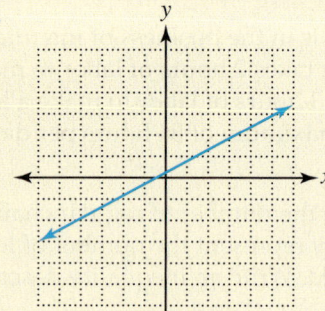   **3.** 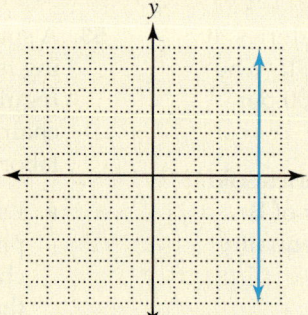   **4.**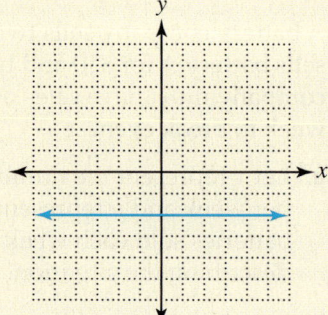

Evaluate the following expressions, using the indicated value of the variable:

5. $4x - 6, x = 2$ **6.** $x^2 - 4, x = 3$

Find the following sum or product as indicated:

7. $-7 + 4$ **8.** $8(-9)$

Introduction

In Section 1.5, a set was defined as any collection of objects. The objects that make up the set are called the **elements** of the set. In algebra, these elements are often numbers. Recall that sets are indicated by braces, { }. In this section, we will discuss sets whose elements are ordered pairs of numbers. In an ordered pair, the first number is often

Answers: Getting Ready: 1. 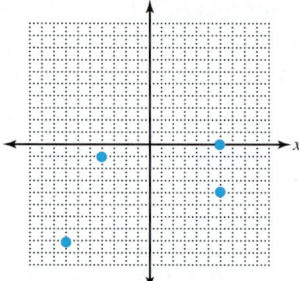   **2.** neither **3.** vertical **4.** horizontal **5.** 2 **6.** 5 **7.** −3 **8.** −72

called the *first component* and the second number the *second component*. For example, in the ordered pair $(2, -3)$, the first component is 2 and the second is -3. With this in mind, we have the following definitions:

> **Definition** Relation
>
> A **relation** is any set of ordered pairs. The set of all first components of the ordered pairs is the **domain** of the relation, and the set of all second components is called the **range**.

OBJECTIVE Determining the domain and range of a relation

If the ordered pairs are given in the form (x, y), the domain is the set of all x-values and the range is the set of all y-values.

Example 1	**Give the domain and range of each of the following relations:**

a. $\{(1, 2), (-3, 4), (6, 3), (-4, -5)\}$

Solution

The domain is the set of all first components, which is $\{1, -3, 6, -4\}$.
The range is the set of all second components, which is $\{2, 4, 3, -5\}$.

b. $\{(-3, 2), (-1, -4), (2, -4), (2, 6)\}$

Solution

The domain is the set of all first components, which is $\{-3, -1, 2\}$.
The range is the set of all second components, which is $\{2, -4, 6\}$.

Note

In Example 1b, two ordered pairs have the first component 2 and two ordered pairs have the same second component -4, but these are listed only once in the domain and range, respectively.

OBJECTIVE **B** Recognizing functions when they are given as a set of ordered pairs, as a graph, or as an equation

A **function** is a special type of relation.

> **Definition** Function
>
> A **function** is a relation in which every first component (element from the domain) is paired with exactly one second component (element from the range).

Another way of thinking of a function is that no two ordered pairs can have the same first components and different second components. More specifically, if the ordered pairs are in the form (x, y), then a function is a correspondence that assigns exactly one value of y to each value of x. As such, y is said to be a function of x. The variable x is the **independent variable**, and y is the **dependent variable**.

Since a function is a relation, the domain of the function is the set of all first components and the range is the set of all second components.

Example 2	**Determine which of the following relations are functions and give the domain and range of each:**

a. $\{(-3, 4), (-1, -3), (0, 2), (3, 5)\}$

Solution

Since no two ordered pairs have the same first component and different second components, this is a function. The domain is $\{-3, -1, 0, 3\}$, and the range is $\{4, -3, 2, 5\}$.

b. $\{(-4, 2), (-2, 4), (1, -7), (3, 4)\}$

Solution

Since no two ordered pairs have the same first component and different second components, this is a function. Notice that the ordered pairs $(-2, 4)$ and $(3, 4)$ have the same second component paired with different first components. This does not violate the definition of a function. The domain is $\{-4, -2, 1, 3\}$, and the range is $\{2, 4, -7\}$.

c. $\{(-5, 3), (3, 6), (-4, -2), (3, 1), (5, 3)\}$

Solution

This is not a function, since the first component 3 is paired with two different second components, 6 and 1. The domain is $\{-5, 3, -4, 5\}$, and the range is $\{3, 6, -2, 1\}$.

PRACTICE EXERCISE 2

Determine which of the following relations are functions and give the domain and range of each:

a. $\{(-5, -3), (-2, -1), (3, 5), (4, -6)\}$ **b.** $\{(3, -2), (-4, 6), (3, 2), (5, 7), (-3, 3)\}$

c. $\{(-5, 2), (6, -4), (-4, 5), (5, 2)\}$

If you need more practice, do the following Additional Practice Exercises.

Additional Practice Exercise 2   Determine which of the following relations are functions and give the domain and range of each:

a. $\{(-3, 1), (4, -3), (7, 4), (-1, -5)\}$ **b.** $\{(2, 5), (-3, -3), (6, 10), (2, -2)\}$

c. $\{(3, 2), (4, -2), (6, 2), (-3, 5)\}$

The relation in Example 2c was not a function. Let's take a look at the graph of this relation.

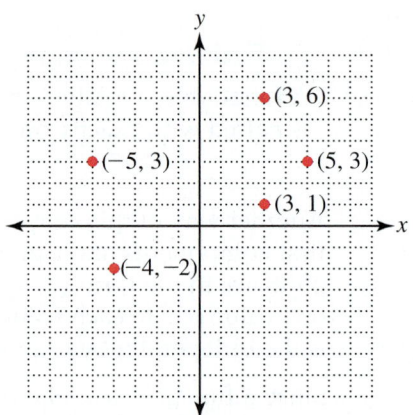

The two ordered pairs that prevented this relation from being a function were $(3, 6)$ and $(3, 1)$, because the first component, 3, was paired with two second components, 6 and 1. We know from Section 4.2 that the graph of any linear equation of the form $x = k$ is a vertical line. Consequently, if two ordered pairs have the same x-value, then

---

Answers: Practice Exercise 2: a. yes, $D = \{-5, -2, 3, 4\}, R = \{-3, -1, 5, -6\}$ **b.** no, $D = \{3, -4, 5, -3\}, R = \{-2, 6, 2, 7, 3\}$
c. yes, $D = \{-5, 6, -4, 5\}, R = \{2, -4, 5\}$ **Additional Practice 2: a.** yes, $D = \{-3, -1, 4, 7\}, R = \{-5, -3, 1, 4\}$ **b.** no, $D = \{-3, 2, 6\}$,
$R = \{-3, -2, 5, 10\}$ **c.** yes, $D = \{-3, 3, 4, 6\}, R = \{-2, 2, 5\}$

their graphs must be on the same vertical line. You will note that the graphs of $(3, 6)$ and $(3, 1)$ are on the same vertical line in the following graph:

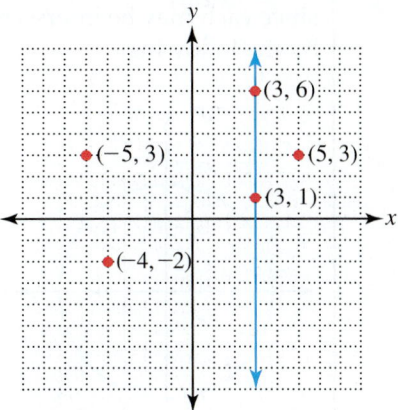

This observation leads us to the following graphical test for a function:

> ### Rule: Vertical Line Test
>
> If any vertical line intersects the graph of a relation in more than one point, then that relation is not a function.

An equivalent way of stating the preceding test is that no vertical line may intersect the graph of a function in more than one point.

Although we have not graphed anything except points and straight lines in this book, this test can be applied to any graph.

Example 3 **Determine whether each of the following is the graph of a function:**

a.

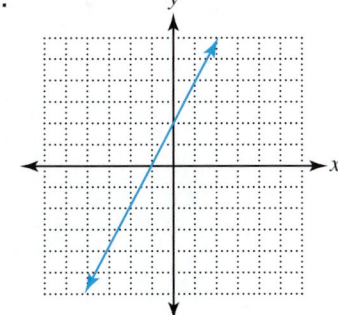

b.

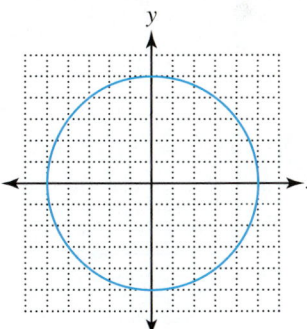

c.

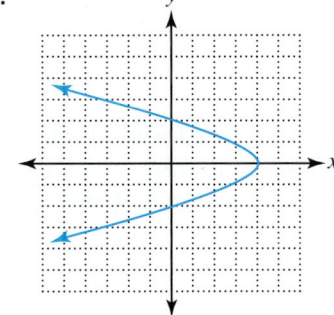

d.

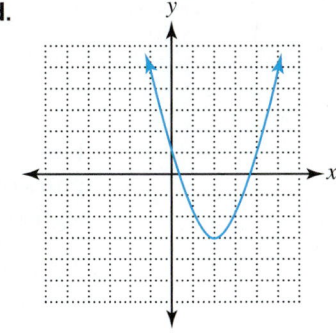

Solution

The graphs of a and d are the graphs of functions, but the graphs of b and c are not, since each may be intersected by a vertical line in more than one point, as illustrated by the following:

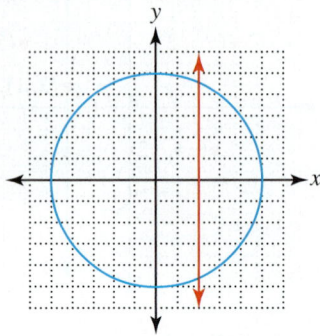 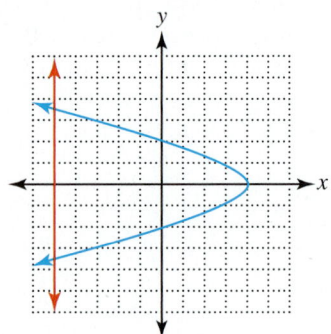

PRACTICE EXERCISE 3

Determine whether each of the following is the graph of a function:

a.

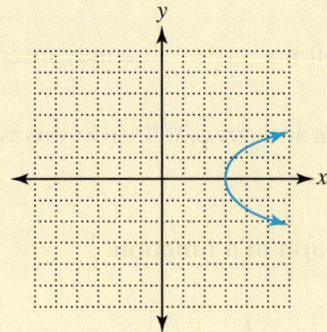

b.

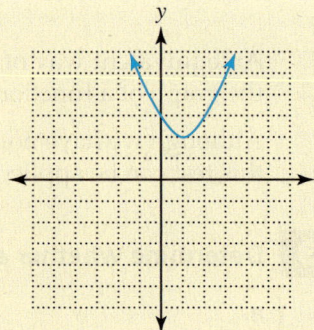

c.

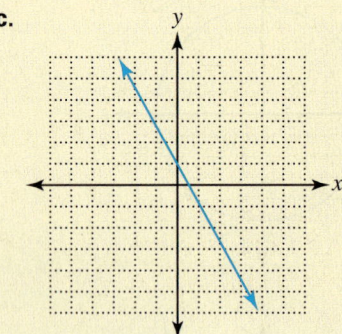

d.

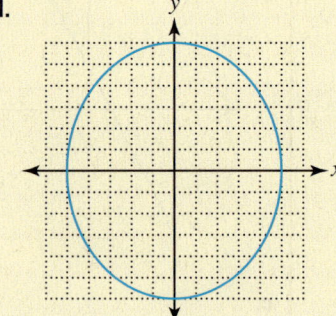

Functions are usually given in the form of equations rather than as sets of ordered pairs or graphs. If the equation defines y as a function of x, then, for each value assigned to x, the equation must give exactly one value for y.

Answers: Practice Exercise 3: a. no **b.** yes **c.** yes **d.** no

Example 4 Determine whether the following equations define *y* as a function of *x*:

a. $y = 3x + 5$

b. $y^2 = x - 3$

Solution

For each value of x, we find the corresponding value of y by multiplying the value of x by 3 and adding 5. This results in exactly one value for y. Therefore, y is a function of x. For example, if $x = 2$, then $y = 3 \cdot 2 + 5 = 6 + 5 = 11$ and only 11. This corresponds to the ordered pair $(2, 11)$, and this is the only ordered pair whose x-value is 2.

Solution

Let x have a value, say, 4. Then $y^2 = 4 - 3 = 1$. This is asking, "What number squared is equal to 1?" Since $1^2 = 1$ and $(-1)^2 = 1$, there are two numbers whose square is 1. Thus, $y = 1$ and $y = -1$. This results in the ordered pairs $(4, 1)$ and $(4, -1)$. Since there are two values for y for some values of x, this is not a function.

PRACTICE EXERCISE 4

Determine whether the following equations define y as a function of x:

a. $y = x^2$

b. $y^2 = x + 2$ (*Hint:* Let $x = -1$.)

If you need more practice, do the following Additional Practice Exercises.

Additional Practice Exercise 4   Determine whether the following equations define y as a function of x:

a. $y = 2x - 3$

b. $y^2 = x - 4$ (*Hint:* Let $x = 5$.)

OBJECTIVE **C** Using functional notation

If y is a function of x, it is convenient to name the function by a special notation called **functional notation**, where y is replaced with a symbol of the form $f(x)$ that is read "the value of f at x," or more commonly, "f of x." For example, if $y = 2x + 3$, then y is a function of x. Replacing y with $f(x)$, we have $f(x) = 2x + 3$. Function names are usually letters, so any letter may be used as a function name. So $g(x) = 2x + 3$ is the same function.

> ### Definition $f(x)$ Notation
>
> When using notation of the form $f(x)$, the following apply:
>
> **a.** f is the name of the function. (Frequently, other symbols, especially other lower-case letters, are used instead of f.)
>
> **b.** x is a value from the domain.
>
> **c.** $f(x)$ is the value of the function in the range that corresponds with the value of x from the domain. It is a y-value, so the results can be written as the ordered pair $(x, f(x))$.

Note

$f(x)$ does not mean $f \cdot (x)$. It means the value of f at x.

When finding the value of $f(x)$ for a specific value of x, we are **evaluating the function**. This is accomplished by substituting the value for x and evaluating the expression, using the order of operations. For example, if $f(x) = 2x + 3$ and we wish to find the value of the function when $x = 3$, then we are finding $f(3)$. To do this, replace x with 3 in the equation of the function and evaluate. So, $f(3) = 2 \cdot 3 + 3 = 6 + 3 = 9$. This can be written as the ordered pair $(3, 9)$.

Answers: Practice Exercise 4: **a.** yes **b.** no   **Additional Practice 4:** **a.** yes **b.** no

Actually, any letter or symbol can be used for both the name of the function and the independent variable. For example, $f(x) = 3x + 5$, $g(t) = 3t + 5$, and $h(r) = 3r + 5$ all represent the same function, since they all represent the same set of ordered pairs.

Example 5 Given $f(x) = 3x - 4$, find the following and represent each result as an ordered pair:

a. $f(2)$

Solution

To find $f(2)$, replace x with 2 and simplify.

$f(2) = 3 \cdot 2 - 4$ Multiply.
$\quad = 6 - 4$ Add.
$\quad = 2$ Therefore, $f(2) = 2$.

As an ordered pair, this is $(2, 2)$.

b. $f(-3)$

Solution

To find $f(-3)$, replace x with -3 and simplify.

$f(-3) = 3(-3) - 4$ Multiply.
$\quad = -9 - 4$ Add.
$\quad = -13$ Therefore, $f(-3) = -13$.

As an ordered pair, this is $(-3, -13)$.

c. $f(.5)$

Solution

To find $f(.5)$, replace x with .5 and simplify.

$f(.5) = 3(.5) - 4$ Multiply.
$\quad = 1.5 - 4$ Add.
$\quad = -2.5$ Therefore, $f(.5) = -2.5$.

As an ordered pair, this is $(.5, -2.5)$.

d. $f(a)$

Solution

To find $f(a)$, replace x with a.

$f(a) = 3a - 4$ Since this cannot be simplified, $f(a) = 3a - 4$.

As an ordered pair, this is $(a, 3a - 4)$.

PRACTICE EXERCISE 5

Given $g(x) = 4x - 2$, find the following and represent each result as an ordered pair:

a. $g(2)$ **b.** $g(-4)$ **c.** $g(.5)$ **d.** $g(c)$

If you need more practice, do the following Additional Practice Exercises.

Additional Practice Exercise 5 Given $h(t) = 2t + 5$, find the following and represent each result as an ordered pair:

a. $h(4)$ **b.** $h(-1)$ **c.** $h(1.3)$ **d.** $h(b)$

For Extra Help

Exercise Set 4.8 MyMathLab®

Determine which of the following relations are functions and give the domain and range of each. (See Example 2.)

1. $\{(-6, 2), (-3, -2), (5, -7), (7, 9)\}$

2. $\{(-8, -5), (-3, 6), (0, 7), (2, 3)\}$

3. $\{(-2, 3), (0, -2), (6, 4), (-5, 1), (3, 7)\}$

4. $\{(-7, 2), (-4, -3), (5, -1), (3, 6), (1, 8)\}$

5. $\{(-8, 2), (-6, 3), (-8, -3), (5, 1)\}$

6. $\{(-5, 1), (3, -2), (6, 2), (-5, 4)\}$

Answers: Practice Exercise 5: a. 6, $(2, 6)$ **b.** -18, $(-4, -18)$ **c.** 0, $(.5, 0)$ **d.** $4c - 2$, $(c, 4c - 2)$
Additional Practice 5: a. 13, $(4, 13)$ **b.** 3, $(-1, 3)$ **c.** 7.6, $(1.3, 7.6)$ **d.** $2b + 5$, $(b, 2b + 5)$

7. $\{(-7, 1), (6, -3), (-3, 2), (3, 1), (4, 7)\}$

8. $\{(1, 0), (-6, 3), (-2, -4), (5, 3), (4, 0)\}$

9. $\{(-4, 2), (-1, 3), (3, 2), (5, 1)\}$

10. $\{(7, -3), (4, 7), (-5, 2), (3, -3)\}$

11. $\{(2, 4), (5, 1), (-5, -2), (3, -1), (5, 2)\}$

12. $\{(1, 0), (0, 1), (-3, 2), (0, 6), (-5, -2)\}$

Determine whether each of the following is the graph of a function. (See Example 3.)

13.

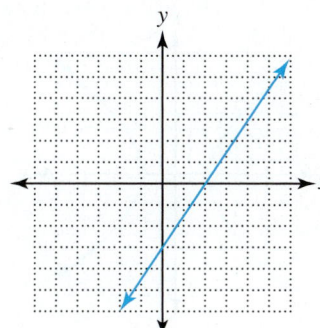

14.

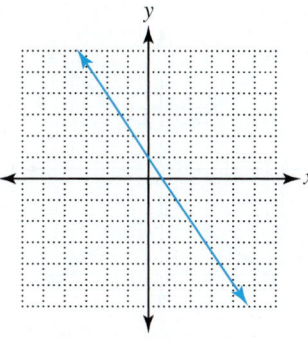

15.

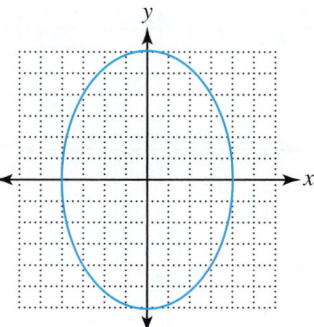

16.

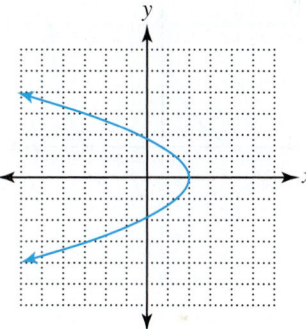

17.

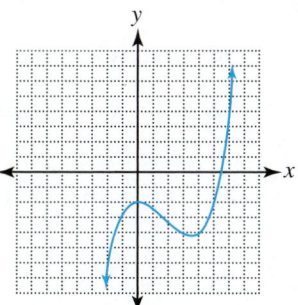

18.

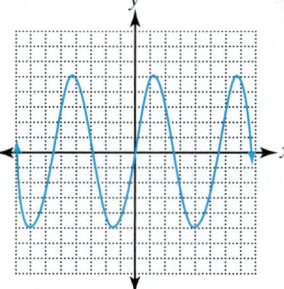

19.

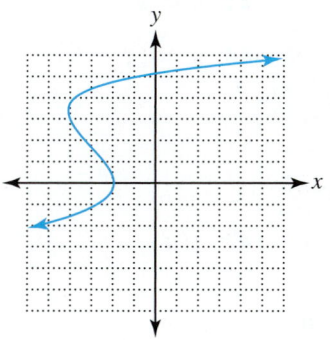

20.

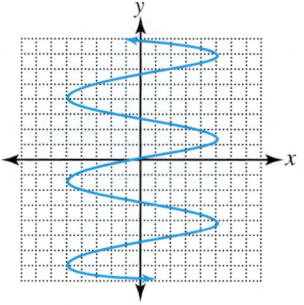

21.

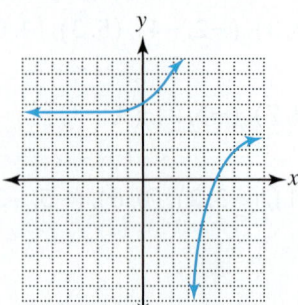

22.

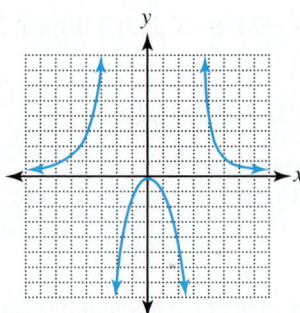

23.

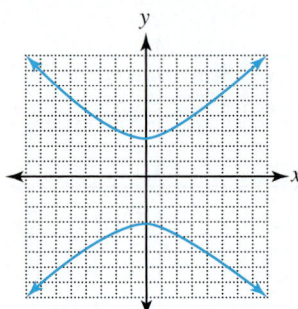

24.

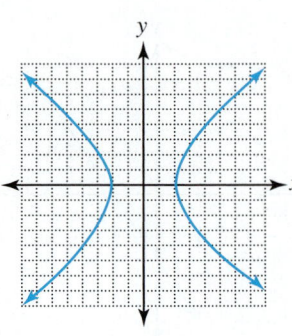

25.

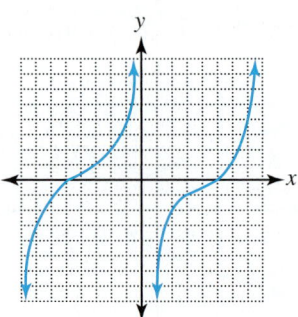

26.

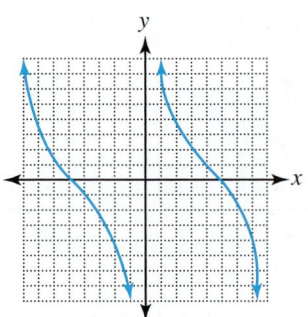

Determine whether each of the following define y as a function of x. (See Example 4.)

27. $y = -2x + 3$

28. $y = 6x - 5$

29. $y = 2x^2$

30. $y = -3x^2 + 2$

31. $y^2 = x - 3$ (*Hint:* Let $x = 4$.)

32. $y^2 = x - 2$ (*Hint:* Let $x = 3$.)

33. $y^2 = 2x + 5$ (*Hint:* Let $x = 2$.)

34. $y^2 = 4x + 1$ (*Hint:* Let $x = 2$.)

35. $|y| = 16 - x^2$ (*Hint:* Let $x = 0$.)

36. $|y| = 25 - x^2$ (*Hint:* Let $x = 0$.)

37. $y = x^3$

38. $y = x^3 - 1$

39. $y = |2x + 3|$

40. $y = |x + 7|$

41. $y < x - 6$

42. $y \geq 2x + 3$

43. $y = (x - 4)^2$

44. $y = (x + 5)^2$

Evaluate Exercises 45–63 involving functional notation. (See Example 5.)

Given $f(x) = 3x + 5$, find the following and represent each result as an ordered pair.

45. $f(0)$

46. $f(4)$

47. $f(a)$

Given $g(z) = z^2 - 2$, find the following and represent each result as an ordered pair.

48. $g(3)$

49. $g(-1)$

50. $g(a)$

Given $h(t) = 16t^2$, find the following and represent each result as an ordered pair.

51. $h(1)$

52. $h(-2)$

53. $h(a)$

Given $r(a) = a^3 + 2$, find the following and represent each result as an ordered pair.

54. $r(0)$

55. $r(-2)$

56. $r(b)$

Given $f(x) = |2x - 3|$, find the following and represent each result as an ordered pair.

57. $f(-2)$

58. $f(4)$

59. $f(z)$

60. The cost, $C(x)$, as a function of the number of items produced, x, is given by $C(x) = 20x + 150$.

   a. Find the cost of producing 10 items.

   b. Find the cost of producing 15 items.

   c. What does $C(30) = 750$ mean?

61. For a rental car, the cost, $C(x)$, as a function of the number of miles driven, x, is given by $C(x) = .12x + 22$.

   a. Find the cost of driving 100 miles.

   b. Find the cost of driving 350 miles.

   c. What does $C(250) = 52$ mean?

62. The value, $V(x)$, of a car as a function of the number of years after it was purchased, x, is given by $V(x) = -1500x + 18,000$.

   a. Find the value of the car after 3 years.

   b. Find the value of the car after 6 years.

   c. What does $V(10) = 3000$ mean?

63. The monthly salary, $S(x)$, of a salesperson as a function of her sales, x, is given by $S(x) = .10x + 175$.

   a. Find her salary for sales of $15,000.

   b. Find her salary for sales of $10,000.

   c. What does $S(20,000) = 2175$ mean?

Challenge Exercises (64–67)

Determine whether the following define y as a function of x:

64. $y = |x + 3|$

65. $|y| = 2x - 1$

66. $x = |y| + 3$

67. $|x| = y - 7$

Writing Exercises (68–70)

68. How do functions and relations differ?

69. Why does the vertical line test for a function work?

70. If $y^2 = 2x + 3$, then y is not a function of x. Why?

Chapter 4   Summary

Concept/Procedure	Example

Pie Chart: [Section 4.1]

- A pie chart is used to show how all of something is divided into parts.

Example 1: The projected population of U.S. territories in 2020 is 4,769,000. The pie chart shows the percent distribution of the population by territory. (*Source: U.S. Bureau of the Census*)

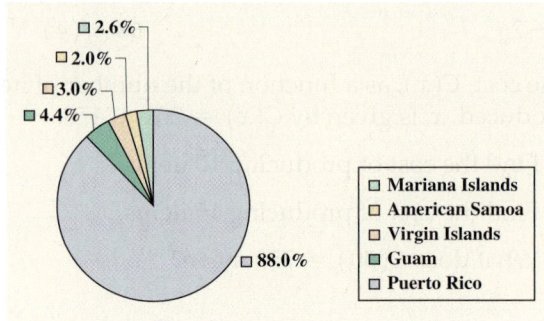

a. What will be the smallest U.S. territory in 2020 and what percentage of the population of the territories will it have?
Solution: American Samoa, 2%

b. What will be the approximate population of Guam in 2020?
Solution: 209,836

c. How many more people will be living in Puerto Rico than the Mariana Islands in 2020?
Solution: 4,072,726

Bar Graph: [Section 4.1]

- A bar graph is used to show comparisons between various categories.

Example 2: California has a massive system of canals and aqueducts. The bar graph shows the length of some of the longest.

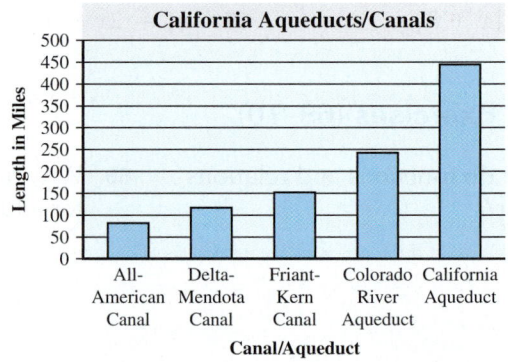

a. Which canal/aqueduct is the longest and what is its approximate length?
Solution: California Aqueduct, 445 miles

b. What is the approximate length of the Friant-Kern Canal?
Solution: 150 miles

c. How much longer is the Colorado River Aqueduct than the Delta-Mendota Canal?
Solution: 125 miles

Concept/Procedure	Example

Line Graph: [Section 4.1]

- A line graph is used to show trends over a period of time.

Example 3: Using the line graph, answer the exercises that follow.

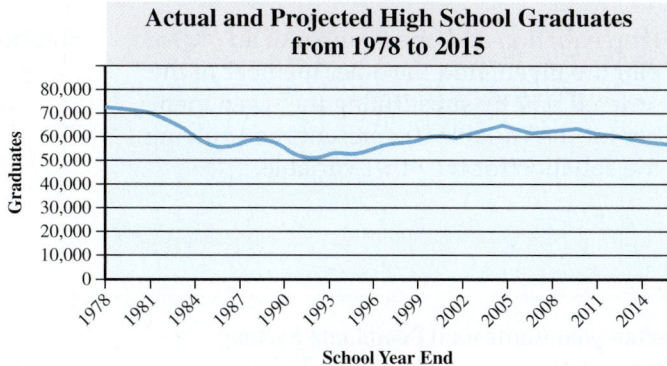

Actual and Projected High School Graduates from 1978 to 2015

School Year End

(*Source:* Minnesota Office of Higher Education)

a. What was the trend in the number of high school graduates from 1978 to 1990?
Solution: The number was decreasing.

b. In what year were there the fewest high school graduates? Approximately how many graduates were there?
Solution: 1992, 50,000

c. In what year were there approximately 70,000 graduates?
Solution: 1981

d. Will the projected number of graduates ever equal the number in 1978?
Solution: No

Linear Equations with Two Variables: [Section 4.1]

- A linear equation with two variables is any equation of the form $ax + by = c$ or any equation that can be put into that form, with a and b both not equal to zero at the same time.

Ordered Pairs: [Section 4.1]

- An ordered pair is a pair of numbers enclosed in parentheses, separated by a comma, with a variable assigned to each number. The first number is the **abscissa** and the second is the **ordinate**.

Solutions of Linear Equations with Two Variables: [Section 4.1]

- If the solutions of $ax + by = c$ are ordered pairs in the form (x, y), then the ordered pair (u, v) is a solution if the replacement of x with u and y with v into the given equation results in a true statement.

Example 4: Determine whether the ordered pair $(2, 1)$ is a solution of $4x - 2y = 10$.

Solution:

$$4x - 2y = 10 \quad \text{Replace } x \text{ with 2 and } y \text{ with 1.}$$
$$4(2) - 2(1) = 10 \quad \text{Multiply.}$$
$$8 - 2 = 10 \quad \text{Subtract.}$$
$$6 \neq 10 \quad \text{Therefore, } (2, 1) \text{ is not a solution.}$$

Concept/Procedure	Example

Finding a Missing Member of an Ordered Pair: [Section 4.1]

- If an equation and one member of an ordered pair are given, find the other member of the ordered pair by substituting the given member for the variable it represents and solving the equation for the other variable.

Example 5: if $2x - 5y = 15$, find the missing number in the ordered pair $(\ , -1)$

Solution:

$$
\begin{array}{ll}
2x - 5y = 15 & \text{Replace } y \text{ with } -1 \text{ and solve for } x. \\
2x - 5(-1) = 15 & \text{Multiply.} \\
2x + 5 = 15 & \text{Subtract 5.} \\
2x = 10 & \text{Divide by 2.} \\
x = 5 & \text{Therefore, the ordered pair is } (5, -1).
\end{array}
$$

Rectangular (Cartesian) Coordinate System: [Section 4.1]

- The rectangular coordinate system is made up of a horizontal and a vertical number line that intersect at the zero point of each. The point of intersection is called the **origin**. The horizontal number line is the *x*-**axis** and the vertical number line is the *y*-**axis**. The four regions that the *x*- and *y*-axes divide the coordinate plane into are called **quadrants**. The quadrants are numbered counterclockwise, beginning with the top right.

Points on the Rectangular Coordinate System: [Section 4.1]

- When plotting points, we assume ordered pairs to be in the form (x, y). To plot a point, begin at the origin and go to the left or right the number of units given by the first number of the ordered pair. From that point, go up or down the number of units given by the second number of the ordered pair.

Example 6: Plot the ordered pairs $(2, 5)$, $(0, 1)$, $(-3, 2)$, $(4, 0)$, and $(-3, -6)$ on the rectangular coordinate system.

Solution:

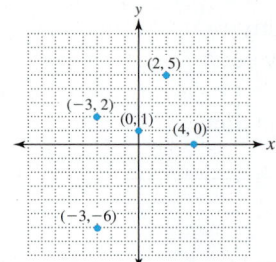

Scatter Diagram: [Section 4.1]

- A scatter diagram is a graph of paired data that usually involves only the first quadrant. Scatter diagrams can be used to show trends.

Example 7: The table shows the number of airline fatalities worldwide for selected years.

a. Write these paired data as ordered pairs of the form (year, number of fatalities).

Solution: (2000, 757), (2001, 577), (2002, 791), (2003, 466), (2004, 203), (2005, 712), (2006, 755), (2007, 587)

b. What does the ordered pair (2004, 203) mean?

Solution: In 2004, there were 203 airline fatalities.

Year	Fatalities
2000	757
2001	577
2002	791
2003	466
2004	203
2005	712
2006	755
2007	587

Concept/Procedure	Example

c. Use the ordered pairs to draw a scatter diagram.

Solution:

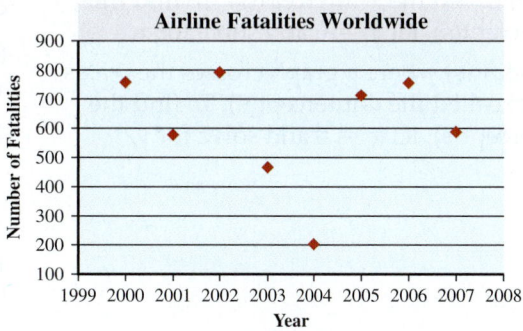

d. What trend is suggested by the diagram?

Solution: There is no apparent trend.

Graphing Linear Equations with Two Variables: [Section 4.2]

• To graph a linear equation with two variables, find at least two ordered pairs (three is recommended) that are solutions of the equation. Plot the points on the rectangular coordinate system and draw a straight line through them.

Example 8: Draw the graph of $x - 2y = -8$.

Solution: Three ordered pairs that solve the equation are $(-8, 0)$, $(0, 4)$, and $(-4, 2)$. Plot the points and draw a straight line through them.

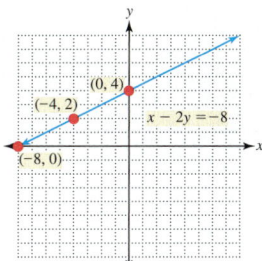

Graphing Lines That Contain the Origin: [Section 4.2]

• The graph of any equation of the form $y = ax$ contains the origin. Find at least one other point by assigning values to x and solving for y.

Example 9: Draw the graph of $y = 0.5x$.

Solution: The graph of this line contains the origin, $(0, 0)$. Find two additional points by letting x have values and solving for y. Two additional ordered pairs are $(4, 2)$ and $(-6, -3)$.

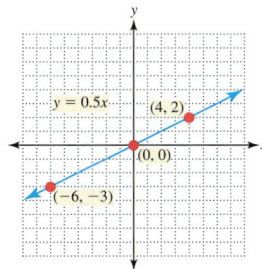

Concept/Procedure	Example

x- and y-Intercepts: [Section 4.3]

- The point(s) where a graph crosses the x-axis is (are) called the x-intercept(s). To find the x-intercept(s), let $y = 0$ and solve for x.
- The point(s) where a graph crosses the y-axis is (are) called the y-intercept(s). To find the y-intercept(s), let $x = 0$ and solve for y.

Example 10: Find the x- and y-intercepts of $4x - 5y = 20$.

Solution:

Find the y-intercept by letting $x = 0$.

$$4(0) - 5y = 20$$
$$-5y = 20$$
$$y = -4$$

Therefore, the y-intercept is $(0, -4)$.

Find the x-intercept by letting $y = 0$.

$$4x - 5(0) = 20$$
$$4x = 20$$
$$x = 5$$

Therefore, the x-intercept is $(5, 0)$.

Vertical and Horizontal Lines: [Sections 4.2, 4.3, and 4.5]

- The graph of any equation of the form $x = h$, where h is a constant, is a vertical line.
- The x-values of all points on the vertical line are h.
- The graph of any equation of the form $y = k$, where k is a constant, is a horizontal line.
- The y-values of all points on the horizontal line are k.

Example 11: Graph $x = 5$ and $y = -2$.

Solution:

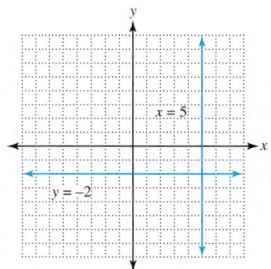

Definition of Slope: [Section 4.4]

- The **slope** of a line is the ratio of the change in y to the change in x between any two points on the line. Alternatively, the slope is the ratio of the rise over the run between any two points on the line.

Slope Formula: [Section 4.4]

- The slope of the line containing the two points (x_1, y_1) and (x_2, y_2) is $m = \frac{y_2 - y_1}{x_2 - x_1}$.

Example 12: Find the slope of the line containing $(5, -3)$ and $(-1, -2)$.

Solution: Let $(x_1, y_1) = (5, -3)$ and $(x_2, y_2) = (-1, -2)$ and use the slope formula.

$$m = \frac{y_2 - y_1}{x_2 - x_1} \qquad \text{Substitute.}$$

$$m = \frac{-2 - (-3)}{-1 - 5} \qquad \text{Simplify.}$$

$$m = \frac{-2 + 3}{-6} \qquad \text{Continue simplifying.}$$

$$m = \frac{1}{-6} \text{ or } -\frac{1}{6} \qquad \text{Therefore, the slope is } -\frac{1}{6}.$$

Concept/Procedure	Example
Graphing a Line, Given a Point and the Slope: [Section 4.4] 1. Plot the given point. 2. From the given point, we can find another point by going vertically (up or down) the number of units given by the numerator and horizontally (right or left) the number of units given by the denominator. 3. Draw a line through these two points.	Example 13: Draw the graph of the line that passes through the point whose coordinates are $(-3, -4)$ and whose slope is $\dfrac{2}{5}$. Solution: Plot the point whose coordinates are $(-3, -4)$. From that point, go up 2 and to the right 5 to find another point on the line. 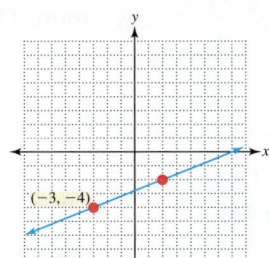
Observations about Slope: [Section 4.4] • A line with positive slope rises from left to right. • A line with negative slope falls from left to right. • The greater the absolute value of the slope is, the steeper the line will be.	
Slopes of Vertical and Horizontal Lines: [Section 4.4] • The slope of a vertical line is **undefined**. The slope of a horizontal line is 0.	
Determining Slope and y-Intercept from the Equation: [Section 4.5] • To find the slope of the graph of a line from its equation, solve the equation for y. The resulting equation is in the form $y = mx + b$, where m is the slope and b is the y-intercept.	Example 14: Find the slope and y-intercept of the graph of $4x + 3y = 9$. Solution: Write the equation in the form $y = mx + b$, solving for y. $4x + 3y = 9$ Subtract $4x$ from both sides. $3y = -4x + 9$ Divide both sides by 3. $y = \dfrac{-4x + 9}{3}$ Simplify by applying $\dfrac{a+b}{c} = \dfrac{a}{c} + \dfrac{b}{c}$. $y = -\dfrac{4}{3}x + 3$ Therefore, the slope is $-\dfrac{4}{3}$ and the y-intercept is 3.
Graphical Solutions of Equations: [Section 4.5] • The solution of an equation of the form $ax + b = 0$ is the x-intercept of the graph of $y = ax + b$.	

Concept/Procedure	Example

Parallel and Perpendicular Lines: [Section 4.5]

- Two lines are **parallel** if and only if their slopes are equal. Two lines are **perpendicular** if and only if the product of their slopes is -1.

Example 15: Determine whether the graphs of the lines whose equations are $y = 3x + 6$ and $y = -\dfrac{1}{3}x - 1$ are parallel, perpendicular, or neither.

Solution: The slope of $y = 3x + 6$ is 3, and the slope of $y = -\dfrac{1}{3}x - 1$ is $-\dfrac{1}{3}$. Since $3\left(-\dfrac{1}{3}\right) = -1$, the graphs are perpendicular.

Writing Equations of Lines: [Sections 4.5 and 4.6]

1. To write the equation of a line, given a point (x_1, y_1) on the line and the slope m of the line, use the point–slope formula, $y - y_1 = m(x - x_1)$. Substitute for m, x_1, and y_1. Never substitute for x and y.

2. To write the equation of a line that contains two given points, follow these steps: (a) Find the slope of the line. (b) Substitute into the point–slope formula, using the slope that you found and either of the two given points as (x_1, y_1).

3. To write the equation of a vertical line, set x equal to the x-value of any point on the line.

4. To write the equation of a horizontal line, set y equal to the y-value of any point on the line.

5. To write the equation of a line containing a given point and parallel to a given line, determine the slope of the line. The slope of the line you are looking for is equal to the slope of the given line. Substitute the slope and the coordinates of the given point into the point–slope formula.

6. To write the equation of a line containing a given point and perpendicular to a given line, determine the slope of the line. The slope of the line you are looking for is equal to the negative reciprocal of the slope of the given line. Substitute the slope and the coordinates of the given point into the point–slope formula.

7. To write the equation of a line, given the slope and y-intercept, substitute for the slope and the y-intercept in the slope–intercept formula, $y = mx + b$. [Section 4.5]

Example 16: Write the equation of the line that contains the point whose coordinates are $(4, -2)$ and whose slope is $-\dfrac{3}{4}$. Leave answer in standard form.

Solution: Use the point–slope formula, substitute and simplify.

$$y - y_1 = m(x - x_1) \qquad \text{Substitute.}$$

$$y - (-2) = -\frac{3}{4}(x - 4) \qquad \text{Simplify left side.}$$

$$y + 2 = -\frac{3}{4}(x - 4) \qquad \text{Multiply by 4.}$$

$$4(y + 2) = 4\left(-\frac{3}{4}\right)(x - 4) \qquad \text{Simplify both sides.}$$

$$4y + 8 = -3(x - 4) \qquad \text{Simplify the right side.}$$

$$4y + 8 = -3x + 12 \qquad \text{Add } 3x \text{ and subtract 8.}$$

$$3x + 4y = 4 \qquad \text{Therefore, the equation containing } (4, -2) \text{ and whose slope is } -\frac{3}{4} \text{ is } 3x + 4y = 4.$$

Concept/Procedure	Example

**Graphing Linear Inequalities with Two Variables:
[Section 4.7]**

- To graph a linear inequality with two variables, follow these steps:

  1. Graph the equality $ax + by = c$. If the line is part of the solution ($\leq$ or $\geq$), draw a solid line. If the line is not part of the solution ($<$ or $>$), draw a dashed line.

  2. Pick a test point not on the line. If the coordinates of the test point solve the inequality, then all points on the same side of the line as the test point solve the inequality. If the coordinates of the test point do not solve the inequality, then all the points on the other side of the line from the test point solve the inequality.

  3. Shade the region on the side of the line that contains the solutions of the inequality.

Example 17: Graph $2x + y \geq 4$.

Solution: Graph the line as a solid line and use $(0, 0)$ as a test point.

$2(0) + 0 \geq 4$.

$0 \geq 4$ is false, so the solution is the region on the other side of the line from $(0, 0)$. Shade that region.

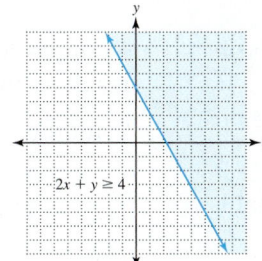

Relations, Domain, and Range: [Section 4.8]

- A **relation** is any set of ordered pairs. In algebra, relations are usually given in the form of equations, tables, or graphs. The set of all first components is the **domain**, and the set of all second components is the **range**.

Functions: [Section 4.8]

- A **function** is a relation in which each first component (element from the domain) of the ordered pair is paired with exactly one second component (element from the range). In other words, no two ordered pairs can have the same first components and different second components.

Example 18: Does the set of ordered pairs $\{(1, 3), (3, 7), (-3, 3), (-1, 7)\}$ represent a function? Why or why not? What are the domain and range?

Solution: Yes, it is a function because each first component is paired with exactly one second component.
Domain $= \{1, 3, -3, -1\}$ and range $= \{3, 7\}$.

Vertical Line Test: [Section 4.8]

- If any vertical line intersects the graph of a relation in more than one point, the graph does not represent a function.

Concept/Procedure	Example
Determining whether an Equation Represents a Function: [Section 4.8] • An equation represents a function if each value of the independent variable (often, x) results in exactly one value of the dependent variable (often, y).	Example 19: Does $x + y^2 = 6$ represent the equation of a function? Solution: No, because some values of x result in two values of y. For example, if $x = 2$, then $2 + y^2 = 6$; so $y^2 = 4$, which means that $y = 2$ or $y = -2$, since $2^2 = 4$ and $(-2)^2 = 4$.

$f(x)$ Notation: [Section 4.8]
 	
• When using $f(x)$ notation, remember these points:
  a. f is the name of the function.
  b. x is a value from the domain.
  c. $f(x)$ is the y-value from the range that is paired with x from the domain.
	Example 20: Given $f(x) = x^2 - 3x + 5$, find $f(2)$ and write the result as an ordered pair.
	
	Solution:
	$f(x) = x^2 - 3x + 5$ Replace with 2.
	$f(2) = 2^2 - 3(2) + 5$ Simplify.
	$f(2) = 4 - 6 + 5$ Continue simplifying.
	$f(2) = 3$ Therefore, $f(2) = 3$.
	
	As an ordered pair, this is (2, 3).

Chapter 4   Review Exercises

1. The pie chart shows the projected U.S. population by age groups in 2030. (*Source: U.S. Bureau of the Census*)

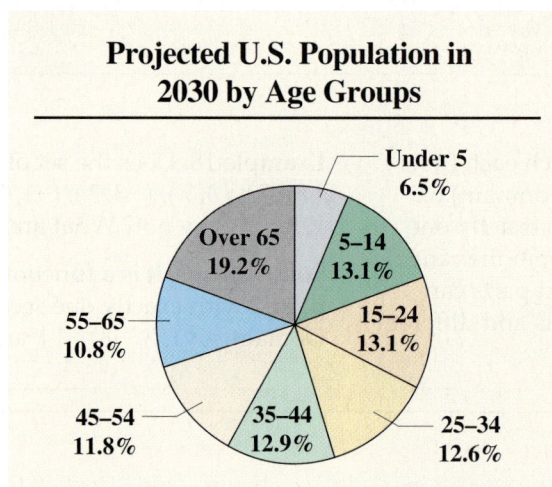

Projected U.S. Population in 2030 by Age Groups

Under 5 6.5%
5–14 13.1%
15–24 13.1%
25–34 12.6%
35–44 12.9%
45–54 11.8%
55–65 10.8%
Over 65 19.2%

a. Which age group will make up the largest percentage of the population in 2030 and what is that percentage?

b. If a random sample of 1500 Americans were chosen in 2030, how many of them would you expect to be 25 to 34 years old?

c. The projected population of the U.S. in 2030 is 373,504,000. How many Americans should be in the 15 to 24 age group?

2. Battling weight seems to be a worldwide problem. The bar graph shows the percentage of people in selected countries who have tried to lose weight. (*Source: Readers Digest,* February 2010)

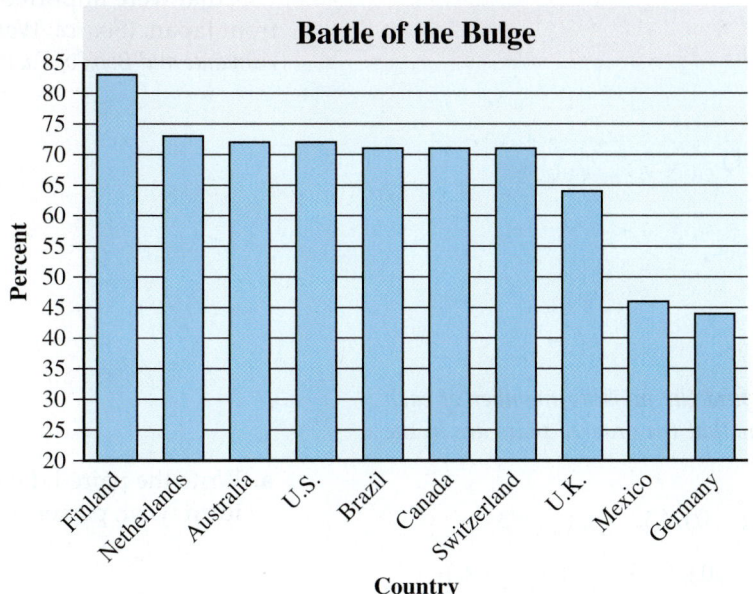

a. In which country have the highest percentage of people tried to lose weight and what is the percentage?

b. In which countries have less than 50% of the population tried to lose weight?

c. Estimate the percentage of people in the U.S. who have tried to lose weight.

d. In which country has the same percentage of people tried to lose weight as in the U.S.?

3. The line graph gives the value (in billions of dollars) for Russian Arms Transfer Agreements 2001–2008. (*Source: Congressional Research Service*)

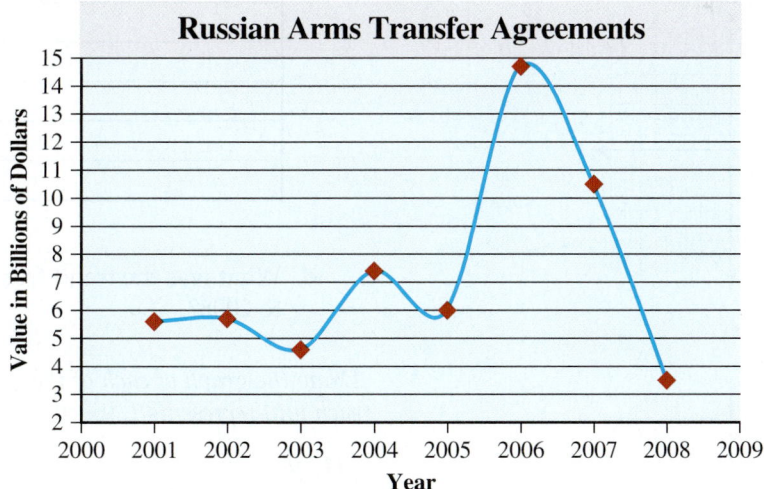

a. Between which two years was the increase in value of arms transfers the greatest and approximately how much was the increase?

b. What has been the trend in Russian arms transfers since 2006?

c. Estimate the value of Russian arms transfers in 2008.

d. In what year was the value of Russian arms transfers approximately $4.6 billion?

Determine whether the given ordered pair is a solution of the given equation. Assume that the ordered pairs are in the form of (x, y). [Section 4.1]

4. $y = 4x + 5; (-2, -3)$

5. $y = -3x + 5; (2, 1)$

6. $4x + 5y = 20; (10, -4)$

7. $7x - 2y = 14; (4, 7)$

8. $x = 3; (3, 5)$

9. $y = -5; (-5, 5)$

Use the given equation to find the missing member of each of the ordered pairs. Assume that the ordered pairs are in the form (x, y). [Section 4.1]

10. $y = -3x + 3; (0, \_), (\_, 0), (3, \_\_\_), (\_, -3)$

11. $y = 4x + 2; (0, \_), (\_, 0), (-5, \_\_\_), (\_, -6)$

12. $3x + 2y = 12; (0, \_\_), (\_\_, 0), (-4, \_\_), (\_, 3)$

13. $5x - 3y = 15; (0, \_\_), (\_, 0), (6, \_\_), (\_, 5)$

Plot the points represented by the given ordered pairs on the rectangular coordinate system. Assume that each unit represents 1. [Section 4.1]

14. $(0, 2), (-3, 4), (0, 1)$

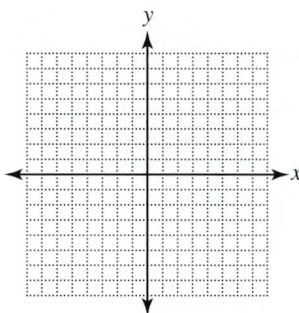

15. $(5, -2), (4, -2), (-6, 7)$

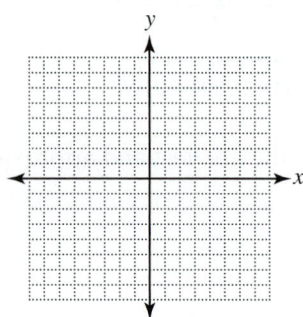

16. The table gives the percent of total new car sales in the U.S. that were imported from Japan. (*Source: World Almanac and Book of Facts*, 2010)

Year	Percent
1999	8.7
2000	9.8
2001	9.9
2002	11.5
2003	10.9
2004	10.7
2005	12.0
2006	14.8
2007	15.5
2008	16.8

a. Write the paired data as ordered pairs in the form (year, percent).

b. What does the ordered pair (2003, 10.9) represent?

c. Use the ordered pairs to draw a scatter diagram.

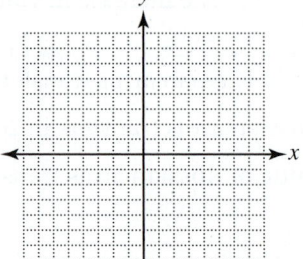

d. What was the trend from 2002 to 2004? From 2004 to 2008?

Draw the graph of each of the given equations. Assume that each unit represents 1. [Section 4.2]

17. $y = 3x - 6$ **18.** $y = -2x - 1$

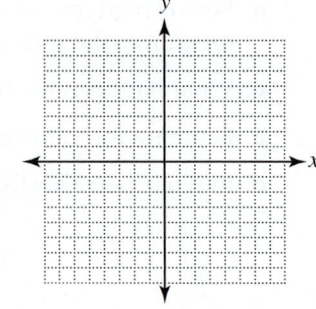

19. $3x + y = 9$

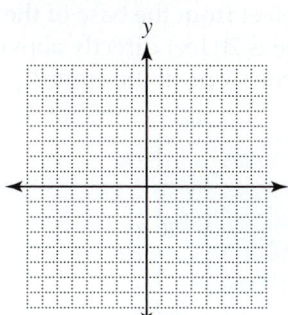

20. $4x - y = 8$

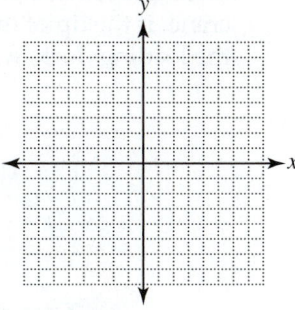

21. $2x + 3y = -6$

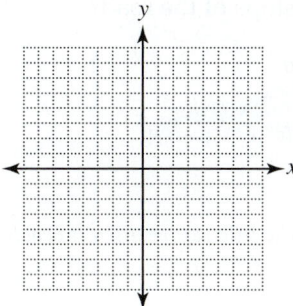

22. $2x - 6y = 12$

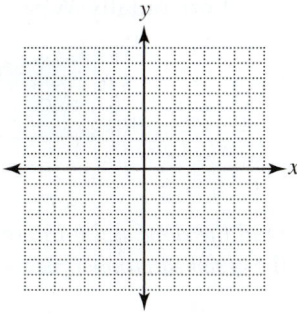

23. $y = -5x$

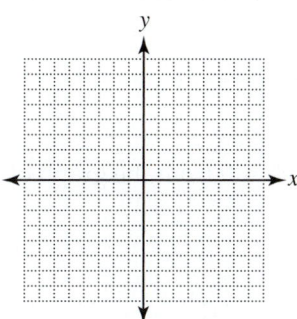

24. $y = 3x$

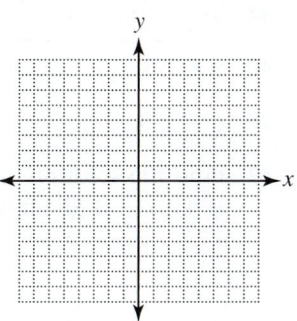

25. $x = 7$

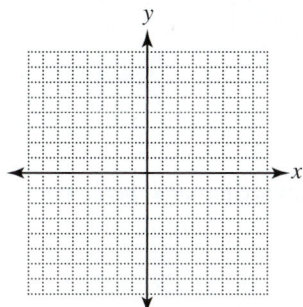

26. $y = -6$

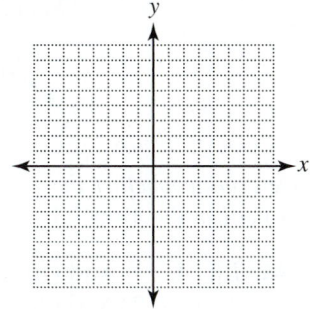

27. Francine works two part-time jobs. She receives $5 per hour when working at the supermarket and $7 per hour when working at the drugstore. If x represents the number of hours that she works at the supermarket, y represents the number of hours that she works at the drugstore, and she earns a total of $175 per week, then $5x + 7y = 175$.

a. If she worked 7 hours at the supermarket, how many hours did she work at the drugstore?

b. If she worked 15 hours at the drugstore, how many hours did she work at the supermarket?

c. Draw the graph. Remember, neither x nor y can have negative values.

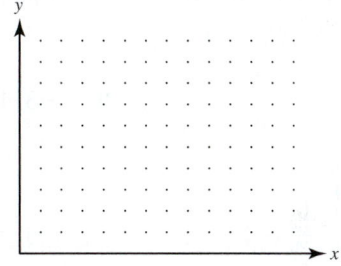

d. Using the graph, estimate the number of hours that she worked at the drugstore if she worked 21 hours at the supermarket. Verify your answer, using the equation.

28. In manufacturing, the cost of each item produced depends upon the number of items produced. Suppose that $y = -2x + 200$, where x is the number of items produced and y is the cost per item.

a. Find the cost per item if 22 items are produced.

b. Find the number of items produced if the cost per item is $110.

c. Draw the graph.

d. Using the graph, estimate the cost if 30 items are produced. Verify your answer from the equation.

Find the slope of the line that contains each of the given pairs of points. [Section 4.4]

29. $(4, 2)$ and $(5, 4)$

30. $(4, 2)$ and $(1, 6)$

31. $(4, -1)$ and $(-2, 2)$

32. $(-2, -7)$ and $(0, -2)$

33. $(4, 2)$ and $(4, -2)$

34. $(2, -5)$ and $(-4, -5)$

Draw the graph of each of the given lines containing the given point and with the given slope. Assume that each unit represents 1. [Section 4.4]

35. $(1, -4), m = \dfrac{4}{3}$

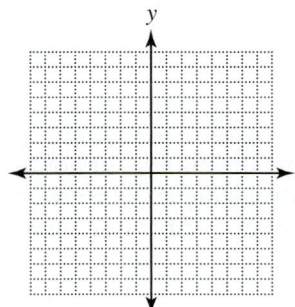

36. $(-3, 4), m = -\dfrac{2}{3}$

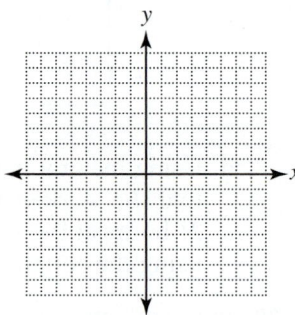

37. $(3, -1), m = 3$

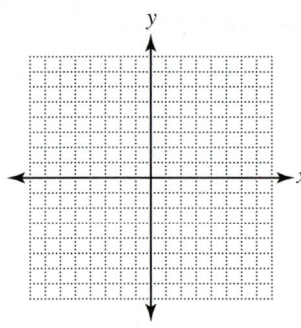

38. $(-1, -4), m = -2$

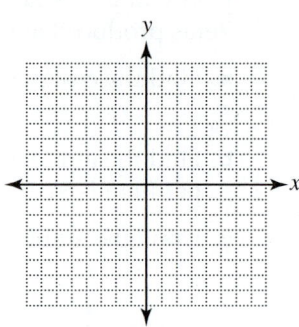

39. $(5, -2)$, undefined slope

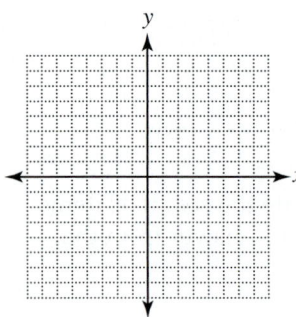

40. $(-2, 4), m = 0$

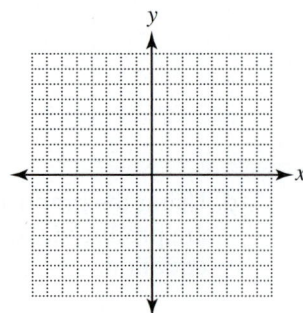

41. The operator of a crane lowers an object onto a loading dock at a point 30 feet from the base of the crane. If the tip of the crane is 20 feet directly above the loading dock, what is the slope of the crane?

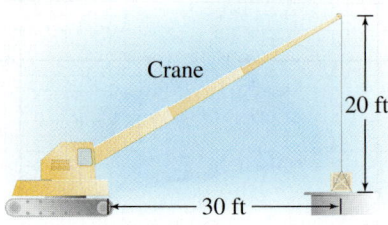

42. A road drops 8 feet vertically for every 100 feet horizontally. What is the slope of the road?

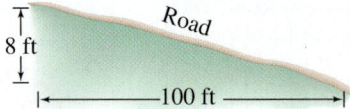

Find the slope and y-intercept of the line represented by each of the given equations. [Section 4.5]

43. $x + 2y = 8$

44. $5x - 2y = 15$

45. $x = 5$

46. $y = -3$

47. Find the solution of $\frac{2}{3}x - 4 = 0$ from the graph of $y = \frac{2}{3}x - 4$ given here.

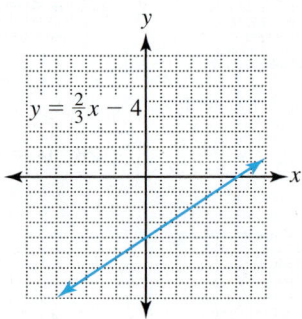

Determine whether the given lines are parallel, perpendicular, or neither. [Section 4.5]

48. L_1: $(4, -1)$ and $(10, 1)$
L_2: $(-4, 2)$ and $(2, 4)$

49. L_1: $(2, -2)$ and $(-3, -4)$
L_2: $(-1, -1)$ and $(-3, 4)$

50. L_1: $(-7, 3)$ and $(-1, -1)$
L_2: $(-2, 8)$ and $(2, 2)$

51. $L_1: 3x + 4y = 8$
$L_2: 4x - 3y = 9$

52. $L_1: 4x + 6y = 5$
$L_2: 2x + 3y = 6$

53. $L_1: 5x + 4y = 12$
$L_2: 4x + 5y = 15$

54. $L_1: x = 3$
$L_2: x = -\frac{1}{3}$

55. $L_1: y = 4$
$L_2: x = 5$

Write the equation of each of the given lines. Express the answer in slope–intercept form. [Section 4.5]

56. $m = -3$ and y-intercept is $\left(0, \frac{5}{6}\right)$

57. $m = \frac{6}{5}$ and y-intercept is $(0, 4)$

58. Horizontal line with y-intercept $(0, -3)$

Write the equation of each of the lines that contain the given point and have the given slope. Express answers in $ax + by = c$ form. [Section 4.6]

59. $(1, 3)$ and $m = 3$

60. $(-2, 5)$ and $m = -2$

61. $(4, -1)$ and $m = -\frac{5}{2}$

62. $(-5, 4)$ and $m = \frac{4}{3}$

63. $(-7, -1)$ and undefined slope

64. $(3, -5)$ and $m = 0$

65. $(6, 2)$ and horizontal

66. $(-8, 5)$ and vertical

Write the equation of the line that contains the given pairs of points. Express answers in $ax + by = c$ form. [Section 4.6]

67. $(-3, 4)$ and $(7, -1)$

68. $(5, 1)$ and $(1, -3)$

69. $(0, 6)$ and $(6, -3)$

70. $(4, 0)$ and $(5, 3)$

71. $(-3, 5)$ and $(2, 5)$

72. $(2, 6)$ and $(2, -3)$

Write the equation of each of the given lines. Express answers in $ax + by = c$ form. [Section 4.6]

73. contains $(6, -4)$ and perpendicular to the graph of $5x - 3y = 15$

74. contains $(-4, 3)$ and parallel to the graph of $9x + 3y = 11$

75. contains $(-5, 4)$ and parallel to the graph of $4x - 6y = 9$

76. contains $(-1, -1)$ and perpendicular to the graph of $4x - y = 5$

77. contains $(2, -5)$ and parallel to the graph of $x = 6$

78. contains $(4, -1)$ and parallel to the graph of $y = 3$

79. contains $(5, 3)$ and perpendicular to the graph of $x = -1$

Answer the problems. [Section 4.6]

80. The fixed cost of operating a sandwich shop is $250 per day. In addition, each sandwich sold costs the shop an average of $1.50.

 a. Write a linear equation that gives the daily operating costs, y, in terms of the number of sandwiches sold, x.

 b. Use the equation found in part a to find the cost for a day when 300 sandwiches were sold.

 c. Using the equation found in part a, find the number of sandwiches sold on a day when the costs were $587.50.

Graph the given linear inequalities with two variables. Assume that each unit represents 1. [Section 4.7]

81. $y < x + 5$

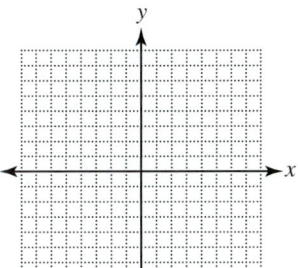

82. $y \geq \frac{5}{2}x - 3$

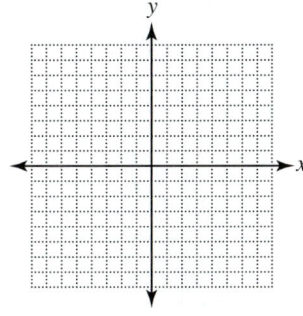

83. $2x - 5y \geq 10$

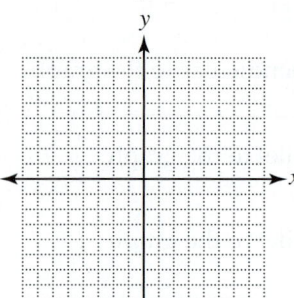

84. $y < -5x$

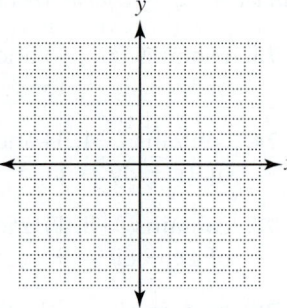

85. $x \geq -6$

86. $y \leq 7$

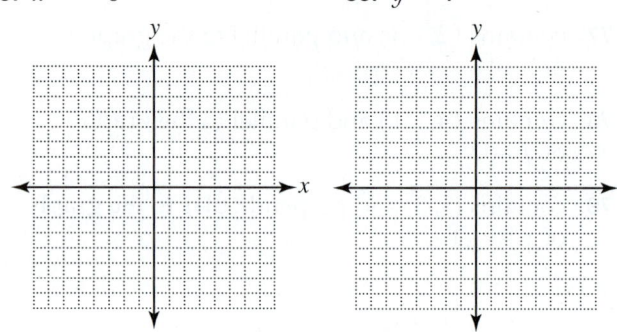

87. Write an inequality for the following: The difference of three times the y-coordinate and two times the x-coordinate is at least 8.

88. Sharon goes shopping for some slacks and blouses. The slacks cost $28 each, and the blouses cost $35 each. If x represents the number of pairs of slacks and y represents the number of blouses, write an inequality that shows the possible combinations of slacks and blouses if Sharon can spend no more than $180.

Determine whether the given relations are functions and give the domain and range of each. [Section 4.8]

89. $\{(-3, 2), (-1, 4), (0, 7), (-1, 5)\}$

90. $\{(-4, 2), (-1, 4), (3, 2), (6, 5)\}$

Determine whether each graph is the graph of a function. [Section 4.8]

91.

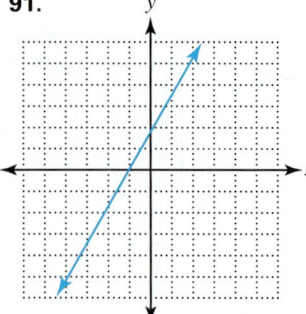

92.

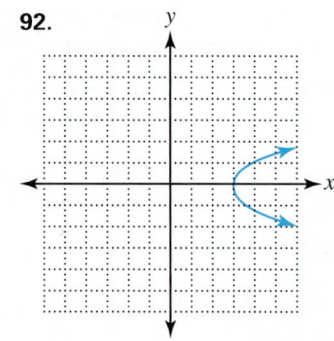

93.

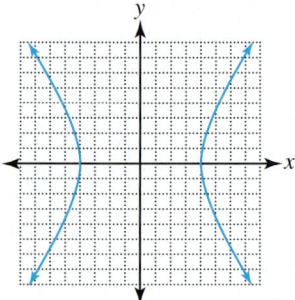

94.

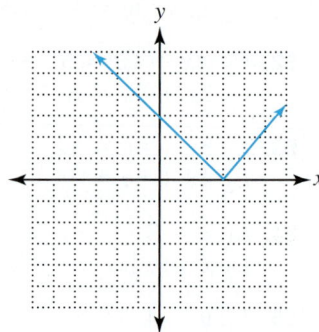

Determine whether each of the equations defines y as a function of x. [Section 4.8]

95. $y = -2x + 1$

96. $y = |2x + 6|$

97. $y^2 = x + 6$ (*Hint:* Let $x = -2$.)

98. $x^2 = y + 5$

99. Given $f(x) = 2x^2 + 3$, find each of the following and represent each result as an ordered pair:

a. $f(2)$

b. $f(-3)$

c. $f(a)$

100. Given $g(x) = x^3 - 3x^2 + 2$, find each of the following and represent each result as an ordered pair:

a. $g(-2)$

b. $g(0)$

c. $g(b)$

101. A company purchases a piece of machinery for $30,000. The value, $V(x)$, of the machinery as a function of the number of years after it was purchased, x, is given by $V(x) = -1800x + 30,000$.

a. Find the value of the machinery after 4 years.

b. Find the value of the machinery after 10 years.

c. What does $V(8) = 15,600$ mean?

Chapter 4 Test

1. Following is a bar graph showing the lengths of the Great Lakes in miles:

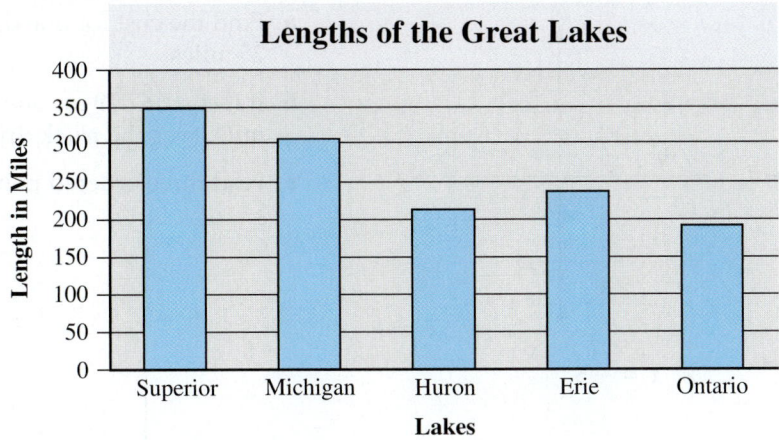

   a. Which lake is the longest?

   b. Which lake(s) has a length between 200 and 250 miles?

   c. Approximately how long is Lake Ontario?

   d. Approximately how much longer is Lake Michigan than Lake Erie?

2. If the solutions of $5x - 3y = 9$ are in the form of (x, y), is the ordered pair $(-3, -8)$ a solution?

3. Use the given equation to find the missing member of each of the given ordered pairs. Assume that the ordered pairs are in the form (x, y).

   $3x + 4y = -24; (0, \_\_\_), (\_\_\_, 0), (-4, \_\_\_), (\_\_\_, 3)$

4. Following is a scatter diagram showing the number of milk cows (in millions) on U.S. farms:

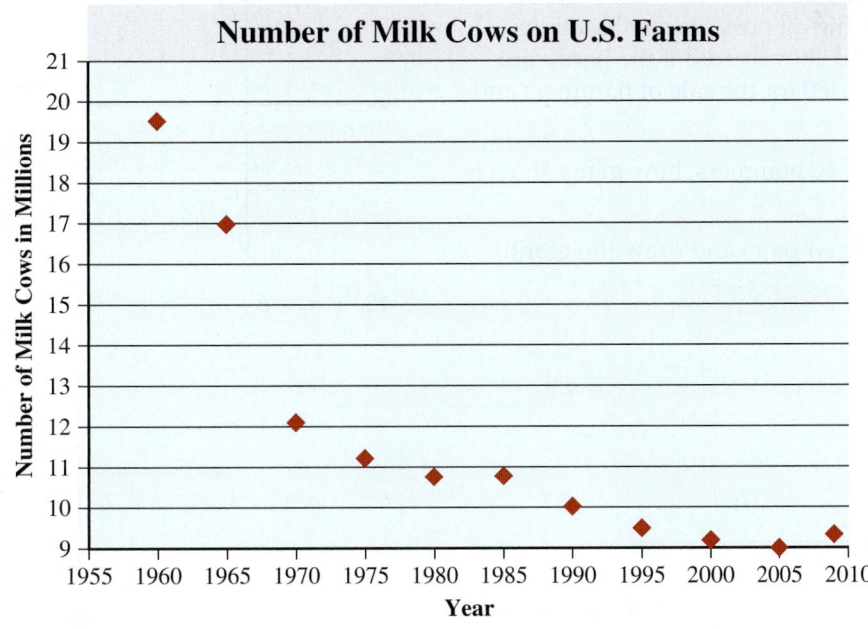

   a. What was the approximate number of cows on U.S. farms in 2005?

   b. How many more cows were on U.S. farms in 1965 than in 2005?

   c. What seems to be the trend in the number of milk cows on U.S. farms?

Graph each of the following, assuming that each unit on the coordinate system represents 1:

5. $y = 2x - 6$

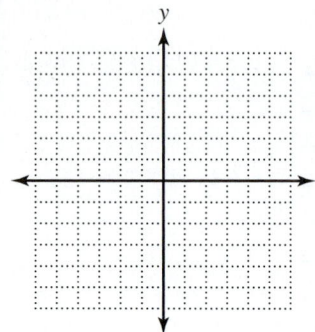

6. $3x + 5y = -15$

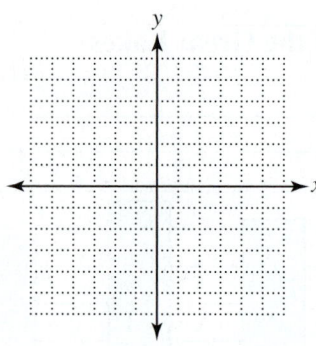

7. $3x + 4y = -6$

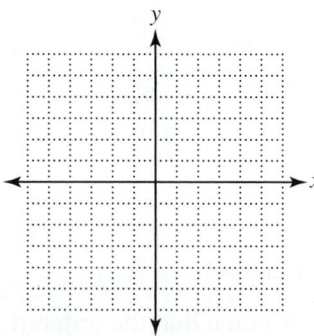

8. $x = -5$

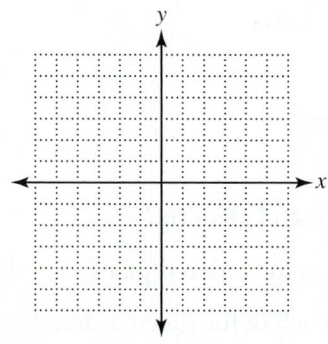

9. A hardware store sells hammers for $9 each and shovels for $20 each.

   a. If x represents the number of hammers and y represents the number of shovels, write an equation showing all possible combinations of hammers and shovels sold if the hardware store received $480 for the sale of hammers and shovels.

   b. If the store sold 40 hammers, how many shovels did it sell?

   c. Find three ordered pairs and draw the graph.

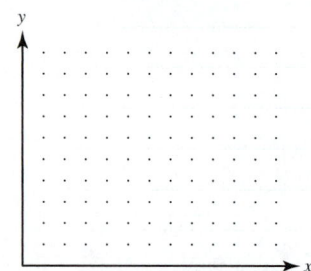

   d. From the graph, estimate the number of hammers sold if 15 shovels were sold.

10. To rent a truck for local use, the cost is $19.99 per day plus 59 cents per mile. If x represents the number of miles driven and y is the cost for one day, then $y = 19.99 + 0.59x$.

   a. Find the cost for one day if the truck was driven 25 miles.

   b. If it cost $67.19 for one-day's rental, how many miles was the truck driven?

   c. Find three ordered pairs and draw the graph.

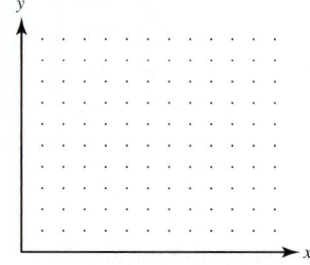

   d. From the graph, estimate the number of miles driven if the cost is $37.69. Verify your answer from the equation.

Graph each of the inequalities:

11. $y \geq -\dfrac{2}{3}x + 4$

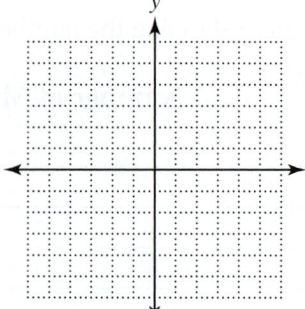

12. $4x - 5y > 20$

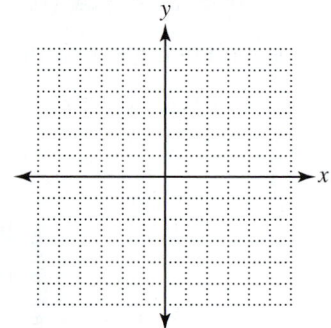

13. $y \leq -6$

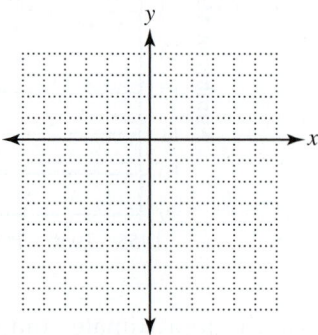

Graph the following lines.

14. containing $(-1, 4)$, with slope $= \frac{3}{5}$

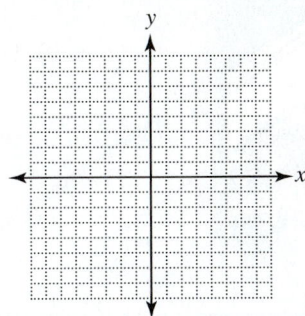

15. containing $(4, 0)$, with undefined slope

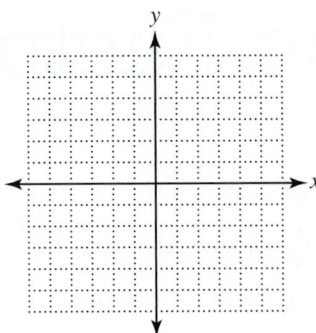

Find the slope of the lines that contain the following pairs of points:

16. $(3, -2)$ and $(-4, 5)$

17. $(4, -2)$ and $(-2, -2)$

Find the slope and y-intercept of the lines with the given equations.

18. $x - 4y = 8$

19. $5x - 2y = 3$

20. $x = -3$

21. Solve $\frac{3}{4}x + 3 = 0$ from the graph of $y = \frac{3}{4}x + 3$ given here.

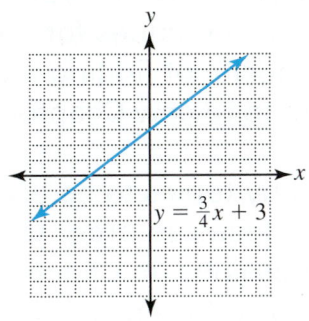
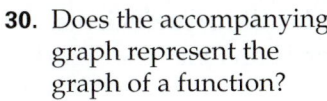

Determine whether the following pairs of lines are parallel, perpendicular, or neither parallel nor perpendicular:

22. L_1 contains $(4, -5)$ and $(-1, -3)$; L_2 contains $(-7, 2)$ and $(-3, 12)$.

23. L_1 is the graph of $4x - 7y = 5$, and L_2 is the graph of $8x - 14y = 11$.

Write the equation of each of the given lines. Express answers in $ax + by = c$ form.

24. contains $(4, -2)$, with $m = -\frac{5}{2}$

25. contains $(6, -1)$ and $(2, -3)$

26. contains $(-2, 3)$ and parallel to the graph of $x - 3y = 6$

27. contains $(5, -6)$ and is horizontal

28. How does the graph of a line whose slope is -2 compare with the graph of a line whose slope is 2?

29. Does the accompanying set of ordered pairs represent a function? Give the domain and range.
$\{(-4, 7), (8, -2), (6, -4), (8, 3)\}$

30. Does the accompanying graph represent the graph of a function?

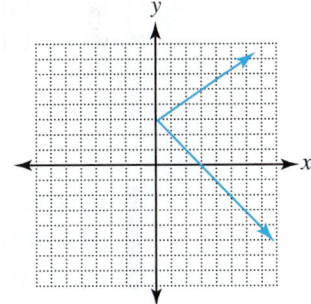

31. Does $y = 2x^2 + 1$ define y as a function of x?

32. Given $f(x) = x^2 - 3x + 2$, find each of the following and write the results as ordered pairs:

   a. $f(2)$ **b.** $f(-1)$

33. The cost, $C(x)$, of operating a small business as a function of the number of days it operates, x, is given by $C(x) = 500x + 2000$.

   a. What is the cost of operating 5 days?

   b. What does $C(10) = 7000$ mean?

5

Factors, Divisors, and Factoring

In this chapter, we continue to explore the close relationship between arithmetic and algebra. We define and find factors and divisors of natural numbers, a topic from arithmetic which we extend into algebra. In arithmetic, there are procedures for writing numbers in terms of their prime factors, but in algebra, we factor polynomials according to their type.

The prerequisite skills for this chapter are in Chapter 2, where you learned the product and quotient laws of exponents and the techniques for multiplying polynomials. Factoring a polynomial is the opposite of finding the product of polynomials. For that reason you might want to review Chapter 2 prior to beginning Chapter 5.

A good knowledge of factoring is crucial for many topics to be discussed in later chapters. In Chapter 6, we must be able to factor in order to reduce and find the products or quotients of rational expressions. In Chapter 7, you must be able to factor in order to find the sums or differences of rational expressions.

<table>
<tr><td>**Section 5.1**</td><td>**Prime Factorization and Greatest Common Factor**</td></tr>
</table>

OBJECTIVES *When you complete this section, you will be able to:*

A Recognize prime and composite numbers.

B Determine whether a number is divisible by 2, 3, or 5.

C Find the prime factorization of composite numbers.

D Find the greatest common factor for two or more natural numbers.

E Find the greatest common factor for two or more monomials that contain variables.

PREREQUISITE SKILLS *Before you begin this section, you should be able to:*

a. Multiply and divide whole numbers. (Section R.3)

b. Know the meaning of a positive exponent. (Section 1.1)

GETTING READY FOR SECTION 5.1

Find the product or quotient.

1. $105 \div 3$ **2.** $378 \div 7$

Write the following in exponential form:

3. $2 \cdot 2 \cdot 2 \cdot 3 \cdot 3 \cdot 5$ **4.** $2 \cdot 3 \cdot 3 \cdot 3 \cdot 11$

Evaluate the given expressions.

5. $3^2 \cdot 5$ **6.** $2 \cdot 4^2 \cdot 7$

Introduction

In this section, we will limit ourselves to natural numbers. We are familiar with finding the products of numbers like $5 \cdot 6 = 30$. Recall that the number 30 is called the *product* of 5 and 6 and that 5 and 6 are called *factors* of 30. The expression $5 \cdot 6$ is called a **factorization** of 30.

> **Definition** Factorization of a Number
>
> A factorization of a number c consists of writing c as the product of two or more numbers. For example, if $ab = c$, then ab is a factorization of c; a and b are factors of c.

OBJECTIVE **A** Recognizing prime and composite numbers

A number may have more than one factorization, and consequently, more than two factors. Let us consider all possible natural-number factorizations of several natural numbers by using the accompanying table. Recall that the natural numbers are $\{1, 2, 3, 4, \ldots\}$.

Answers: Getting Ready: 1. 35 **2.** 54 **3.** $2^3 \cdot 3^2 \cdot 5$ **4.** $2 \cdot 3^3 \cdot 11$ **5.** 45 **6.** 224

Number	Factorization(s)	Factors	Number of Factors
1	$1 \cdot 1$	1	1
2	$1 \cdot 2$	1, 2	2
3	$1 \cdot 3$	1, 3	2
4	$1 \cdot 4, 2 \cdot 2$	1, 2, 4	3
5	$1 \cdot 5$	1, 5	2
6	$1 \cdot 6, 2 \cdot 3$	1, 2, 3, 6	4
7	$1 \cdot 7$	1, 7	2
8	$1 \cdot 8, 2 \cdot 4$	1, 2, 4, 8	4
9	$1 \cdot 9, 3 \cdot 3$	1, 3, 9	3
10	$1 \cdot 10, 2 \cdot 5$	1, 2, 5, 10	4
and so on			

Looking through the preceding table, we observe some patterns:

1. Some numbers, like 2, 3, 5, and 7 have only two factors: 1 and the number itself.
2. Some numbers, like 4, 6, 8, 9, and 10 have more than two factors.
3. There is only one number with only one factor—the number 1.
4. All of the numbers have 1 and the number itself as factors.

These observations lead us to the following definitions:

> **Definitions** Prime Number, Composite Number, or Neither
>
> A **prime number** is a natural number with two, and only two, factors. These factors are 1 and the number itself.
>
> A **composite number** is a natural number with more than two factors. These factors are 1, the number itself, and at least one other number.
>
> The natural number 1 is neither prime nor composite.

Example 1   **Classify each of the following as prime or composite:**

a. 2   Prime, since 2 has only two factors: 1 and 2.

b. 11   Prime, since 11 has only two factors: 1 and 11.

c. 4   Composite, since 4 has three factors: 1, 2, and 4.

d. 12   Composite, since 12 has six factors: 1, 2, 3, 4, 6, and 12.

PRACTICE EXERCISE 1

Classify each of the following as prime or composite:

a. 7   **b.** 8   **c.** 10

d. 13   **e.** 41   **f.** 57

(Continued)

Answers: Practice Exercise 1: a. prime **b.** composite **c.** composite **d.** prime **e.** prime **f.** composite

If you need more practice, do the following Additional Practice Exercises.

Additional Practice Exercise 1   Classify each of the following as prime or composite:

a. 19   **b.** 27   **c.** 39

d. 51   **e.** 53   **f.** 91

The number 20 is divisible by 5, since $20 \div 5 = 4$ with 0 as the remainder. In general, the number a is **divisible** by the number b if the remainder is 0 when a is divided by b. As mentioned before, there is an important relationship between multiplication and division. By the definition of division, if $20 \div 4 = 5$, then $4 \cdot 5 = 20$. This means 4 is a factor of 20. Furthermore, if $20 \div 5 = 4$, then $5 \cdot 4 = 20$. This means 5 is also a factor of 20. From these examples, we suspect that if a number is divisible by a second number, then the second number is a factor of the first. Furthermore, a number is divisible by all its factors.

> **Definition** Divisor and Divisible
>
> If the natural number a is a **divisor** of a natural number c, then a is a nonzero factor of c. Furthermore, if the number a is a nonzero factor of the number c, then a is a divisor of c. The number c is said to be **divisible** by the number a, and a is said to divide evenly into c.

This states that nonzero factors and divisors are the same thing. So, if 18 is divisible by 3, then 3 is a factor of 18. Conversely, if 3 is a factor of 18, then 3 is a divisor of 18. Since a prime number has only itself and 1 as factors, they can be thought of as numbers that are divisible only by themselves and 1.

> **Alternative Definition** Prime Number
>
> A natural number is prime if it is divisible only by itself and 1. The number 1 is neither prime nor composite.

Example 2	**Find the divisors (factors) of each of the following:**

	Number	Answer
a.	14	Since $1 \cdot 14 = 14$ and $2 \cdot 7 = 14$, the divisors (factors) of 14 are 1, 2, 7, and 14.
b.	20	Since $1 \cdot 20 = 20$, $2 \cdot 10 = 20$, and $4 \cdot 5 = 20$, the divisors (factors) of 20 are 1, 2, 4, 5, 10, and 20.
c.	5	Since $1 \cdot 5 = 5$, the divisors (factors) of 5 are 1 and 5.
d.	48	Since $1 \cdot 48 = 48$, $2 \cdot 24 = 48$, $3 \cdot 16 = 48$, $4 \cdot 12 = 48$, and $6 \cdot 8 = 48$, the divisors (factors) of 48 are 1, 2, 3, 4, 6, 8, 12, 16, 24, and 48.

Answers: Additional Practice 1: a. prime **b.** composite **c.** composite **d.** composite **e.** prime **f.** composite

PRACTICE EXERCISE 2

Find the divisors (factors) of each of the following:

a. 9 **b.** 28 **c.** 19 **d.** 96

If you need more practice, do the following Additional Practice Exercises.

Additional Practice Exercise 2 Find the divisors (factors) of each of the following:

a. 4 **b.** 36 **c.** 29 **d.** 48

OBJECTIVE **B** Determining whether a number is divisible by 2, 3, or 5

When the numbers are somewhat larger, as in Practice Exercise 2d, it is sometimes difficult to determine whether the number is divisible by something other than 1 and itself. There are some easy ways of determining whether a number is divisible by 2, 3, or 5, as summarized in the following rules:

Rule: Rules of Divisibility by 2, 3, and 5

1. A natural number is divisible by 2 if the last digit of the number is 0, 2, 4, 6, or 8. In other words, if the number is even, then it divisible by 2.

2. A natural number is divisible by 3 if the sum of the digits of the number is divisible by 3.

3. A natural number is divisible by 5 if the last digit of the number is 0 or 5.

Example 3 Determine whether each of the following is divisible by 2, 3, 5, or none of these:

a. 20 The last digit is 0, so 20 is divisible by both 2 and 5. The sum of the digits is $2(2 + 0 = 2)$ and 2 is not divisible by 3, so 20 is not divisible by 3.

b. 39 39 is not even, so it is not divisible by 2. The last digit is not 0 or 5, so 39 is not divisible by 5. The sum of the digits is $12(3 + 9 = 12)$ and 12 is divisible by 3, so 39 is divisible by $3(39 \div 3 = 13)$.

c. 60 60 is even, so it is divisible by 2. The last digit is 0, so 60 is divisible by 5. The sum of the digits is 6 and 6 is divisible by 3, so 60 is divisible by 3.

d. 73 73 is not even, so it is not divisible by 2. The last digit is not 0 or 5, so 73 is not divisible by 5. The sum of the digits is 10 and 10 is not divisible by 3, so 73 is not divisible by 3. Does this mean 73 is prime?

PRACTICE EXERCISE 3

Determine whether each of the following is divisible by 2, 3, 5, or none of these:

a. 40 **b.** 87 **c.** 120 **d.** 91

If you need more practice, do the following Additional Practice Exercises.

Additional Practice Exercise 3 Determine whether each of the following is divisible by 2, 3, 5, or none of these.

a. 240 **b.** 51 **c.** 42 **d.** 71

Answers: Practice Exercise 2: a. 1, 3, 9 **b.** 1, 2, 4, 7, 14, 28 **c.** 1, 19 **d.** 1, 2, 3, 4, 6, 8, 12, 16, 24, 32, 48, 96 **Additional Practice 2: a.** 1, 2, 4
b. 1, 2, 3, 4, 6, 9, 12, 18, 36 **c.** 1, 29 **d.** 1, 2, 3, 4, 6, 8, 12, 16, 24, 48 **Practice Exercise 3: a.** 2 and 5 **b.** 3 **c.** 2, 3, and 5 **d.** none of these
Additional Practice 3: a. 2, 3, and 5 **b.** 3 **c.** 2 and 3 **d.** none of these

OBJECTIVE Finding the prime factorization of composite numbers

It is possible to write any composite number as the product of prime numbers. This procedure is referred to as **prime factorization**. There is one and only one prime factorization of any composite number. For example, see the chart that follows:

Number	Prime Factorization
6	$2 \cdot 3$ (Remember, $2 \cdot 3 = 3 \cdot 2$. Why?)
9	$3 \cdot 3 = 3^2$
12	$2 \cdot 6 = 2 \cdot 2 \cdot 3 = 2^2 \cdot 3$

If the numbers are small, finding the prime factorization is relatively easy. For larger numbers, the following procedure is helpful:

Procedure for Finding a Prime Factorization

1. Divide the composite number by the smallest prime number by which it is divisible.

2. Then, divide the resulting quotient by the smallest prime number by which it is divisible.

3. Continue this process until the quotient is prime.

4. The prime factorization is the product of all the divisors and the last quotient.

To assist us, the first few prime numbers are $\{2, 3, 5, 7, 11, 13, 17, 19, 23, 29, \ldots\}$. We illustrate the procedure with some examples.

Example 4   Find the prime factorization for each of the following:

a. 24   2 is the smallest prime number that divides evenly into 24.

$$\begin{array}{r} 12 \\ 2\overline{)24} \end{array}$$ 12 is also divisible by 2.

$$\begin{array}{r} 6 \\ 2\overline{)12} \\ 2\overline{)24} \end{array}$$ 6 is also divisible by 2.

$$\begin{array}{r} 3 \\ 2\overline{)6} \\ 2\overline{)12} \\ 2\overline{)24} \end{array}$$ 3 is prime.

Therefore, $24 = 2 \cdot 2 \cdot 2 \cdot 3 = 2^3 \cdot 3$, which is the product of all the divisors and the last quotient and is the prime factorization of 24.

b. 36   2 is the smallest prime number that divides evenly into 36.

$$\begin{array}{r} 18 \\ 2\overline{)36} \end{array}$$ 18 is also divisible by 2.

$$\begin{array}{r} 9 \\ 2\overline{)18} \\ 2\overline{)36} \end{array}$$ 3 is the smallest prime that divides into 9 evenly.

$$\begin{array}{r} 3 \\ 3\overline{)9} \\ 2\overline{)18} \\ 2\overline{)36} \end{array}$$ 3 is prime.

Therefore, $36 = 2 \cdot 2 \cdot 3 \cdot 3 = 2^2 \cdot 3^2$, which is the product of all the divisors and the last quotient and is the prime factorization of 36.

c. 105   3 is the smallest prime that divides evenly into 105.

$$\begin{array}{r} 35 \\ 3)\overline{105} \end{array}$$ 5 is the smallest prime that divides evenly into 35.

$$\begin{array}{r} 7 \\ 5)\overline{35} \\ 3)\overline{105} \end{array}$$ 7 is prime.

Therefore, $105 = 3 \cdot 5 \cdot 7$ is the prime factorization of 105.

d. 89   89 is not divisible by 2, 3, 5, 7, 11, 13, and so on.

We suspect that 89 might be prime, but how do we know? Do we have to try dividing 89 by every prime number until we get to 89? Fortunately, we do not. **We can stop looking for prime factors when we arrive at a prime number whose square is greater than the number we are trying to factor.** In this case, $11^2 > 89$, so there will be no prime factors greater than 11. Therefore, we conclude that 89 is prime.

> **Rule: Observation**
>
> If we try prime numbers in ascending order and a prime factor has not been found when we arrive at a prime number whose square is greater than the number we are trying to factor, then the number is prime.

The preceding observation can save us time when we attempt to factor numbers that are prime. For example, if we were attempting to factor 163, we would not try any prime factors greater than 13, since 13 is prime, $13^2 = 169$, and $169 > 163$.

Note that prime factorizations are usually written in ascending order.

PRACTICE EXERCISE 4

Find the prime factorization of each of the following:

a. 80   **b.** 84   **c.** 90   **d.** 73

If you need more practice, do the following Additional Practice Exercises.

Additional Practice Exercise 4   Find the prime factorization for each of the following:

a. 40   **b.** 54   **c.** 150   **d.** 108

OBJECTIVE Finding the greatest common factor of two or more natural numbers

Prime factorizations have many uses, one of which is finding the **greatest common factor (GCF)** of two or more natural numbers or algebraic expressions.

First, we need to know what is meant by the greatest common factor. The factors of 12 are $\{1, 2, 3, 4, 6, 12\}$, and the factors of 18 are $\{1, 2, 3, 6, 9, 18\}$. Therefore, the factors that 12 and 18 have in common are $\{1, 2, 3, 6\}$, so the largest factor that 12 and 18 have in common is 6. Therefore, 6 is the greatest common factor of 12 and 18. *Remember, a number is divisible by all its factors.* So, the greatest common factor for a group of numbers is the largest number that will divide evenly into each number of the group. For 12 and 18, 6 is the greatest common factor, so 6 is the largest number that will divide evenly into both.

Listing the factors and picking out the greatest common factor works for small numbers, but it would certainly be difficult to use this procedure for a pair of large

Answers: Practice Exercise 4: a. $2^4 \cdot 5$ **b.** $2^2 \cdot 3 \cdot 7$ **c.** $2 \cdot 3^2 \cdot 5$ **d.** prime   **Additional Practice 4: a.** $2^3 \cdot 5$ **b.** $2 \cdot 3^3$ **c.** $2 \cdot 3 \cdot 5^2$ **d.** $2^2 \cdot 3^3$

numbers like 108 and 144. We will find the greatest common factor for groups of numbers by using the procedure outlined as follows:

Procedure: Finding the Greatest Common Factor (GCF)

1. Write each number as the product of its prime factors.

2. Select those factors that are common in each prime factorization.

3. For each factor that is common, select the smallest exponent that appears in any of the prime factorizations.

4. The greatest common factor is the product of the factors found in step 3.

We demonstrate this procedure with some examples.

Example 5 **Find the greatest common factor (GCF) for each of the following by using prime factorization:**

a. 12 and 18 (We already know the answer is 6, from the introduction.)

$12 = 2^2 \cdot 3$ Prime factorization.
$18 = 2 \cdot 3^2$ Prime factorization.

The factors that are common are 2 and 3. The smallest power of 2 that appears in either factorization is 2^1 in $2 \cdot 3^2$, and the smallest power of 3 that appears in either factorization is 3^1 in $2^2 \cdot 3$. Therefore, the GCF is $2^1 \cdot 3^1 = 6$. Consequently, 6 is the largest number that will divide evenly into 12 and 18.

b. 60 and 72

$60 = 2^2 \cdot 3 \cdot 5$ Prime factorization.
$72 = 2^3 \cdot 3^2$ Prime factorization.

The factors that are common are 2 and 3 (5 is not a factor of 72). The smallest power of 2 that appears in either factorization is 2^2 in $2^2 \cdot 3 \cdot 5$, and the smallest power of 3 is 3^1 in $2^2 \cdot 3 \cdot 5$. Therefore, the GCF is $2^2 \cdot 3 = 12$. This means that 12 is the largest number that will divide into 60 and 72 evenly.

c. 6 and 35

$6 = 2 \cdot 3$ Prime factorization.
$35 = 5 \cdot 7$ Prime factorization.

There are no prime factors in common, so the greatest common factor is 1. Remember, 1 is a factor of every number. In other words, the largest number that will divide evenly into 6 and 35 is 1.

d. 270 and 315

$270 = 2 \cdot 3^3 \cdot 5$ Prime factorization.
$315 = 3^2 \cdot 5 \cdot 7$ Prime factorization.

The factors that are common are 3 and 5. (2 is not a factor of 315, and 7 is not a factor of 270.) The smallest power of 3 appearing in either factorization is 3^2 in $3^2 \cdot 5 \cdot 7$, and the smallest power of 5 is 5^1 in both prime factorizations. Therefore, the greatest common factor is $3^2 \cdot 5 = 45$. This means 45 is the largest number that will divide into 270 and 315 evenly.

e. 54, 120, and 126

$54 = 2 \cdot 3^3$ Prime factorization.

$126 = 2 \cdot 3^2 \cdot 7$ Prime factorization.

$120 = 2^3 \cdot 3 \cdot 5$ Prime factorization.

The factors that are common are 2 and 3. The smallest power of 2 that appears in any factorization is 2^1, which occurs in $2 \cdot 3^3$ and $2 \cdot 3^2 \cdot 7$. The smallest power of 3 is 3^1 in $2^3 \cdot 3 \cdot 5$. Therefore, the greatest common factor is $2 \cdot 3 = 6$. Consequently, 6 is the largest number that will divide into 54, 126, and 120 evenly.

Why does this procedure work? Since we are looking for the greatest *common* factor, we select only those factors that are *common*. Why the smallest exponent? The exponent indicates how many times the base is used as a *factor*. The number of times the *factor* is *common* is the smallest power of the *factor*.

PRACTICE EXERCISE 5

Find the greatest common factor for each of the following:

a. 24 and 30

b. 54 and 90

c. 15 and 77

d. 600 and 180

e. 56, 112, and 140

If you need more practice, do the following Additional Practice Exercises.

Additional Practice Exercise 5   Find the greatest common factor for each of the following:

a. 60 and 90

b. 60 and 140

c. 21 and 65

d. 360 and 432

e. 12, 28, and 42

OBJECTIVE **Finding the greatest common factor of two or more monomials that contain variables**

The procedure for finding the GCF of a monomial containing variables is exactly the same, except usually easier, since we do not have to factor the variables.

Example 6   Find the GCF for each of the following:

a. a^2b^3 and a^4b^2

The factors a and b are common. The smallest power of a is a^2 and the smallest power of b is b^2. Therefore, the GCF is a^2b^2. This means a^2b^2 is the monomial of greatest degree that divides evenly into a^2b^3 and a^4b^2.

b. a^2b^4 and c^3d^2

There are no factors in common; therefore, the GCF is 1.

c. $6c^2d^3$ and $8cd^5$

We need to consider the coefficients and the variables separately. The GCF of 6 and 8 is 2. The variables c and d are common, the smallest power of c is c^1, and the smallest power of d is d^3. Therefore, the GCF is $2cd^3$. So, $2cd^3$ is the monomial of greatest degree that will divide evenly into $6c^2d^3$ and $8cd^5$.

d. $12a^3b^2$ and $18b^4d$

In factored form, we have $2^2 \cdot 3a^3b^2$ and $2 \cdot 3^2b^4d$. The only factors in common are 2, 3, and b, and the smallest powers are 2, 3, and b^2, respectively. Therefore, the GCF is $6b^2$.

e. $8x^2yz^4$, $12x^4y^2z^3$, and $14x^2y^3$

In factored form, we have $2^3 \cdot x^2yz^4$, $2^2 \cdot 3x^4y^2z^3$, and $2 \cdot 7x^2y^3$. The only factors in common are 2, x and y, and the smallest powers are 2, x^2 and y^1, respectively. Therefore, the GCF is $2x^2y$.

Answers: Practice Exercise 5: a. 6 **b.** 18 **c.** 1 **d.** 60 **e.** 28   **Additional Practice 5: a.** 30 **b.** 20 **c.** 1 **d.** 72 **e.** 2

PRACTICE EXERCISE 6

Find the GCF of each of the following:

a. c^4d^2 and c^3d^5 **b.** xy^3 and c^2d **c.** $18a^3b^4$ and $24a^2b^5$

d. $15x^2y^3$ and $18x^3z^2$ **e.** $15x^4y^2z$, $9x^2z^3$, and $18xy^3z^2$

If you need more practice, do the following Additional Practice Exercises.

Additional Practice Exercise 6   Find the GCF of each of the following:

a. p^3q^2 and pq^4 **b.** $m^3n^2p^3$ and $a^4b^3c^5$ **c.** $6a^4b^2$ and $9ab^2$

d. $24m^4n^3$ and $36n^2p^3$ **e.** $10x^3yz^2$, $15y^3z^4$, and $25x^2y^4z^5$

Exercise Set 5.1

For Extra Help

MyMathLab®

Classify each of the following as prime or composite. (See Example 1.)

1. 9   **2.** 15   **3.** 17   **4.** 23

5. 28   **6.** 35   **7.** 51   **8.** 69

9. 71   **10.** 83

List all natural number factors (divisors) of each of the following. (See Example 2.)

11. 18   **12.** 12   **13.** 24   **14.** 26

15. 36   **16.** 42   **17.** 54   **18.** 72

Determine whether each of the following is divisible by 2, 3, 5, or none of these. (See Example 3.)

19. 30   **20.** 42   **21.** 57   **22.** 51   **23.** 48

24. 96   **25.** 97   **26.** 113   **27.** 145   **28.** 185

29. 732   **30.** 822   **31.** 1620   **32.** 5340

Find the prime factorization of each of the following. (See Example 4.)

33. 14   **34.** 15   **35.** 46   **36.** 38   **37.** 28

38. 20   **39.** 29   **40.** 31   **41.** 45   **42.** 63

43. 100   **44.** 225   **45.** 42   **46.** 30   **47.** 154

48. 182   **49.** 180   **50.** 126   **51.** 315   **52.** 525

53. 756   **54.** 360

Challenge Exercises (55–58)

List all the natural-number factors of the following. (See Example 2.)

55. 108   **56.** 144

Answers: Practice Exercise 6: a. c^3d^2 **b.** 1 **c.** $6a^2b^4$ **d.** $3x^2$ **e.** $3xz$ **Additional Practice 6: a.** pq^2 **b.** 1 **c.** $3ab^2$ **d.** $12n^2$ **e.** $5yz^2$

Find the prime factorization of each of the following. (See Example 4.)

57. 1512

58. 7056

Find the greatest common factor (GCF) of each of the following. (See Examples 5 and 6.)

59. 6 and 9

60. 10 and 25

61. 65 and 35

62. 77 and 21

63. 33 and 10

64. 65 and 28

65. 12 and 45

66. 18 and 75

67. 90 and 140

68. 60 and 140

69. 180 and 270

70. 210 and 280

71. 6, 13, and 21

72. 10, 31, and 55

73. 8, 16, and 24

74. 12, 30, and 42

75. 16, 32, and 56

76. 24, 48, and 60

77. 48, 96, and 120

78. 72, 108, and 180

79. mn^3 and m^2n

80. a^4b and ab^2

81. c^3d^2 and cd^3

82. pq^4 and p^3q^4

83. a^3b^5 and a^3b^2

84. m^4n^6 and m^2n^6

85. j^3k^4 and a^2b^3

86. r^3s^6 and a^3b

87. $a^3b^3c^3$ and a^2bc^2

88. $x^4y^3z^2$ and xy^4z^3

89. pq^2r^3 and p^2r^3

90. a^4b^2c and a^2c^3

91. $18pq^4$ and $30p^2q$

92. $16c^5d$ and $24cd^3$

93. $27s^3t^3$ and $36s^2t^5$

94. $12p^2q^3$ and $24p^5q^4$

95. $9a^2b^4c^2, 15a^3bc^2, 12ab^3c^2$

96. $5x^3y^6z, 10xy^2z, 20x^2y^2z$

97. $8a^4b^3c^2, 16a^6b^3c^2, 4a^2b^5c^5$

98. $12r^5s^4t^2, 24r^3s^4t^3, 28r^2s^4t^5$

99. $24p^2q^4, 33q^2r^3, 42p^3r^4$

100. $24a^4c^2, 40a^5b^2, 56b^4c^2$

Challenge Exercises (101–108)

Find the GCF for each of the following. (See Examples 5 and 6.)

101. 108 and 144

102. 240 and 384

103. $2x(x - 3), 5(x - 3)$

104. $3a(a + 2), -4(a + 2)$

105. $6a(c - 2d), 4b(c - 2d)$

106. $8x(w + 3t), 6y(w + 3t)$

107. $48a^3b^2(x + 2)^3, 64a^4b(x + 2)^2$

108. $36x^5y^7(a - 3)^4, 54x^4y^9(a - 3)^7$

Writing Exercises (109–114)

109. What is the difference between a prime and a composite number?

110. Why is 1 not considered to be a prime number?

111. How can you determine whether a number is divisible by 4? By 6?

112. Under what circumstances would a number be a common divisor of two other numbers?

113. A friend of yours is having trouble finding the GCF of two large numbers. Describe how you would help him.

114. The graphic that follows is an example of finding the GCF of 108 and 144 by a procedure called dividing out common primes. Study the example and write the procedure in words.

```
           3      4
    3 │ 9      12
    3 │ 27     36
    2 │ 54     72
    2 │ 108    144
```

Therefore, the GCF is $2 \cdot 2 \cdot 3 \cdot 3 = 2^2 \cdot 3^2 = 36$.

Section 5.2

Factoring Polynomials with Common Factors and by Grouping

OBJECTIVES *When you complete this section, you will be able to:*

Ⓐ Factor a polynomial by removing the greatest common factor.

Ⓑ Factor by grouping.

PREREQUISITE SKILLS *Before you begin this section, you should be able to:*

a. Apply the distributive property. (Section 1.7)

b. Apply $a^m \cdot a^n = a^{m+n}$. (Section 2.2)

c. Apply $\frac{a^m}{a^n} = a^{m-n}$. (Section 2.6)

d. Multiply integers. (Section 2.1)

GETTING READY FOR SECTION 5.2

Find the following products:

1. $2(2x + 3y)$ **2.** $(-3)(2x - 4y)$ **3.** $4x^2(x^2 - 5y^3)$

4. $a^2x^3(a + x)$ **5.** $xy(2x^2 - 3xy - 4x^2)$

Find the following quotients:

6. $\dfrac{8a^4b^3}{2a^3b^3}$ **7.** $\dfrac{10x^2y^3}{5x^2y}$

Find the following products:

8. $(-2)(5)$ **9.** $(-3)(-4)$

Introduction

In Section 5.1, we learned how to factor composite numbers and monomials into prime factors. This section is the first of six in which we will learn to factor other polynomials into prime factors. A prime polynomial, like a prime number, is a polynomial that can be written only as a product of itself and 1. For example, $x + 4$ is prime, but $x^2 - 4$ is not, since $x^2 - 4 = (x + 2)(x - 2)$.

When asked to "factor a polynomial," what are we actually being asked to do? We know that when two or more numbers are multiplied, each is called a factor of the product. So, factoring a polynomial means to write the polynomial as the product of two or more prime polynomials in the same way that factoring a composite number means to write the composite number as the product of two or more prime numbers. In Chapter 2, we multiplied polynomials. Factoring is "undoing the multiplication." Consider the following chart:

Composite Number	Factors	Composite Polynomial	Factors
$15 = 3 \cdot 5$	$3, 5$	$x^2 - 4 = (x + 2)(x - 2)$	$x + 2, x - 2$
$22 = 2 \cdot 11$	$2, 11$	$2x^2 + 4x = 2x(x + 2)$	$2, x, x + 2$

OBJECTIVE **A** Factoring a polynomial by removing the greatest common factor

The distributive property in the form $ab + ac = a(b + c)$ provides the technique for factoring by **removing the greatest common factor**. In the binomial $ab + ac$, a is a factor of the first term ab, and a is also a factor of the second term ac. Therefore, a is a factor common to both terms. The common factor (a in this case) is written outside the parentheses, and the polynomial inside the parentheses is determined by one of two methods:

1. We find the term that must be multiplied by the common factor in order to give the corresponding term of the polynomial being factored. In the preceding example, $ab + ac = a(\_ + \_)$. The first blank is filled by the term whose product with a is ab. This is b. The second blank is the term whose product with a is ac. This is c. Consequently, $ab + ac = a(b + c)$. This process is sometimes referred to as **factoring by inspection**.

2. Find the quotient of each of the terms of the polynomial with the common factor. Using division, divide the first term ab by the common factor a: $\frac{ab}{a} = b$. Therefore, the first term inside the parentheses is b. Divide the second term ac by the common factor a: $\frac{ac}{a} = c$. Therefore, the second term inside the parentheses is c. So, $ab + ac = a(b + c)$. This method works by the definition of division. If $\frac{x}{y} = z$, then $y \cdot z = x$. Since $\frac{ab}{a} = b$, then $a(\text{common factor}) \cdot b(\text{first term inside parentheses}) = ab(\text{first term of the polynomial})$.

Any factorization may be checked by finding the product of the factors. If the product is the same as your original polynomial, then your factorization is correct. For example, $2x + 2y = 2(x + y)$ is correct, since $2(x + y) = 2x + 2y$. However, we must be careful that the factor that we have removed is the greatest common factor. If the polynomial inside the parentheses still has a common factor, then we have not removed the greatest common factor. Try again!

When adding like terms, we were really just removing a common factor and adding the numerical coefficients that remained inside the parentheses. For example, $4x + 7x$ has a common factor of x. Rewrite $4x + 7x$ as $(4 + 7)x$ by removing the common factor x. Then add 4 and 7 and get $11x$.

In the examples and exercises that follow, you will need to remember the product rule for exponents ($a^m \cdot a^n = a^{m+n}$) and the quotient rule for exponents ($\frac{a^m}{a^n} = a^{m-n}$).

Example 1 | **Factor each of the following by removing the greatest common factor:**

a. $bx + bz$ =   *The greatest common factor of bx and bz is b. Write b outside the parentheses.*

$b(\_ + \_)$ *The first blank represents a term whose product with b is bx. The term is x. In expressions that are more complicated, we may need to divide each term of the polynomial by the common factor. If we divide the first term bx of the polynomial by the common factor b, we get the first term inside the parentheses, x. That is, $\frac{bx}{b} = x$.*

$b(x + \_)$ =   *The second blank must be a term whose product with b is bz. By inspection or division, the blank represents z. So we now have the result:*

$b(x + z)$ *Therefore, $bx + bz = b(x + z)$, where the first factor is b and the second factor is $x + z$.*

CHECK:

$b(x + z)$ = $bx + bz$ *Therefore, the factorization is correct.*

b. $4x + 2 \quad = \quad$ *The GCF of 4x and 2 is 2. Write 2 outside the parentheses.*

$2(\_ + \_) \quad = \quad$ *The first blank represents a term whose product with 2 is 4x. Since $2 \cdot 2x = 4x$, the blank represents 2x. Using division, $\frac{4x}{2} = 2x$.*

$2(2x + \_) \qquad$ *The second blank represents a term whose product with 2 is 2. Since $2 \cdot 1 = 2$, the blank represents 1. Using division, $\frac{2}{2} = 1$.*

$2(2x + 1) \qquad$ *Therefore, $4x + 2 = 2(2x + 1)$, where the first factor is 2 and the second is $2x + 1$.*

CHECK:

$2(2x + 1) = 4x + 2 \qquad$ Therefore, the factorization is correct.

c. $8x^2 + 12x^3 = \quad$ *The greatest common factor of $8x^2$ and $12x^3$ is $4x^2$. Write $4x^2$ outside the parentheses.*

$4x^2(\_ + \_) = \quad$ *The first blank is a term whose product with $4x^2$ is $8x^2$. Since $4x^2 \cdot 2 = 8x^2$, the blank represents 2. Using division, $\frac{8x^2}{4x^2} = 2$.*

$4x^2(2 + \_) = \quad$ *The second blank is a term whose product with $4x^2$ is $12x^3$. Since $4x^2 \cdot 3x = 12x^3$, the second blank represents 3x. Using division, $\frac{12x^3}{4x^2} = 3x$.*

$4x^2(2 + 3x) \qquad$ *Therefore, $8x^2 + 12x^3 = 4x^2(2 + 3x)$, where the first factor is $4x^2$ and the second is $2 + 3x$.*

CHECK:

$4x^2(2 + 3x) = 8x^2 + 12x^3 \qquad$ Therefore, the factorization is correct.

d. $12x^2y^3 - 18xy^4 \qquad$ *The GCF of $12x^2y^3$ and $18xy^4$ is $6xy^3$. Write $6xy^3$ outside the parentheses.*

$6xy^3(\_ + \_) \qquad$ *The first blank represents 2x, since $6xy^3 \cdot 2x = 12x^2y^3$, or $\dfrac{12x^2y^3}{6xy^3} = 2x$.*

$6xy^3(2x + \_) \qquad$ *The second blank represents $-3y$, since $6xy^3(-3y) = -18xy^4$, or $\dfrac{-18xy^4}{6xy^3} = -3y$.*

$6xy^3(2x - 3y) \qquad$ *Therefore, $12x^2y^3 - 18xy^4 = 6xy^3(2x - 3y)$, where the first factor is $6xy^3$ and the second is $2x - 3y$.*

CHECK:

$6xy^3(2x - 3y) = 12x^2y^3 - 18xy^4 \qquad$ Therefore, the factorization is correct.

e. $10x^2y^3 - 5x^3y - 20x^2y^2 \qquad$ *The GCF is $5x^2y$. Write $5x^2y$ outside the parentheses.*

$5x^2y(\_ + \_ + \_) \qquad$ *The first term inside the parentheses is $2y^2$, the second $-x$, and the third $-4y$. This gives us the result:*

$5x^2y(2y^2 - x - 4y) \qquad$ *Therefore, $10x^2y^3 - 5x^3y - 20x^2y^2 = 5x^2y(2y^2 - x - 4y)$.*

CHECK:

$5x^2y(2y^2 - x - 4y) = 10x^2y^3 - 5x^3y - 20x^2y^2 \qquad$ Therefore, the factorization is correct.

PRACTICE EXERCISE 1

Factor each of the following by removing the greatest common factor:

a. $pq + pr$ **b.** $6a + 9$ **c.** $8y^3 + 16y^2$

d. $10a^3b^2 - 15a^2b^2$ **e.** $9x^2 - 12x + 3$ **f.** $8a^2b^4 - 16ab^2 - 12a^2b^3$

If you need more practice, do the following Additional Practice Exercises.

Additional Practice Exercise 1   Factor each of the following by removing the greatest common factor:

a. $mn + mp$ **b.** $8p - 6$ **c.** $12p^4 + 18p^3$

d. $6c^2d^2 - 9c^3d^4$ **e.** $8a^2 - 12a + 4$ **f.** $9x^2y^4 - 15xy^3 - 18x^3y^4$

Answers: Practice Exercise 1: a. $p(q + r)$ **b.** $3(2a + 3)$ **c.** $8y^2(y + 2)$ **d.** $5a^2b^2(2a - 3)$ **e.** $3(3x^2 - 4x + 1)$ **f.** $4ab^2(2ab^2 - 4 - 3ab)$
Additional Practice 1: a. $m(n + p)$ **b.** $2(4p - 3)$ **c.** $6p^3(2p + 3)$ **d.** $3c^2d^2(2 - 3cd^2)$ **e.** $4(2a^2 - 3a + 1)$ **f.** $3xy^3(3xy - 5 - 6x^2y)$

We usually want the coefficient of the first term inside the parentheses to be positive, so it is sometimes necessary to remove a negative common factor, as illustrated by the following examples:

> **Example 2** **Factor the following by removing a negative common factor:**
>
> **a.** $-2a + 4b$ =   *Since we want the first coefficient inside the parentheses to be positive, the GCF is -2.*
>
> $-2(\_ + \_)$ =   *The first term is a, since $\dfrac{-2a}{-2} = a$, and the second term is $-2b$, since $\dfrac{4b}{-2} = -2b$.*
>
> $-2(a - 2b)$ *Therefore, $-2a + 4b = -2(a - 2b)$.*
>
> CHECK:
>
> $-2(a - 2b) = -2a + 4b$ *Therefore, the factorization is correct.*
>
> **b.** $-3x^3 + 9x^2 - 3x$ *The GCF is $-3x$. Write $-3x$ outside the parentheses.*
>
> $-3x(\_ + \_ + \_)$ *The first term is x^2, since $\dfrac{-3x^3}{-3x} = x^2$, the second term is $-3x$, since*
>
> $\dfrac{9x^2}{-3x} = -3x$, *and the third term is 1, since* $\dfrac{-3x}{-3x} = 1$.
>
> $-3x(x^2 - 3x + 1)$ *Therefore, $-3x^3 + 9x^2 - 3x = -3x(x^2 - 3x + 1)$.*
>
> CHECK:
>
> $-3x(x^2 - 3x + 1) = -3x^3 + 9x^2 - 3x$ *Therefore, the factorization is correct.*
>
> **c.** $-a + b$ *We want the first term inside the parentheses to be positive, so factor out -1. We do not write -1, just the "$-$."*
>
> $-(\_ + \_)$ *The first term is a, since $\dfrac{-a}{-1} = a$, and the second is $-b$, since $\dfrac{b}{-1} = -b$.*
>
> $-(a - b)$ *Therefore, $-a + b = -(a - b)$.*
>
> CHECK:
>
> $-(a - b) = -a + b$ *Therefore, the factorization is correct.*

PRACTICE EXERCISE 2

Factor the following by removing a negative common factor:

a. $-9x + 12y$ **b.** $-4a^4 + 2a^3 - 6a^2$ **c.** $-x + y$

If you need more practice, do the following Additional Practice Exercises.

Additional Practice Exercise 2   Factor the following by removing a negative common factor:

a. $-10p + 15q$ **b.** $-8p^3 + 4p^2 - 16p$ **c.** $-r + s$

In the previous examples and practice exercises, the common factor has been a monomial, but that is not always the case. The common factor can also be any polynomial. In the following examples, the common factor is a binomial:

Answers: Practice Exercise 2: **a.** $-3(3x - 4y)$ **b.** $-2a^2(2a^2 - a + 3)$ **c.** $-(x - y)$
Additional Practice 2: **a.** $-5(2p - 3q)$ **b.** $-4p(2p^2 - p + 4)$ **c.** $-(r - s)$

Example 3   **Factor the following by removing the common binomial factor:**

a. $a(c + d) + b(c + d) =$ *The factors of the first term are a and $(c + d)$, and the factors of the second term are b and $(c + d)$. So, the common factor is $(c + d)$. Write $c + d$ outside the parentheses. To show that the entire quantity $(c + d)$ is a factor, enclose it in parentheses.*

$(c + d)(\_ + \_)$ $=$ *The first blank is the term whose product with $(c + d)$ is $a(c + d)$, which is a.*

$(c + d)(a + \_)$ $=$ *The second blank is the term whose product with $(c + d)$ is $b(c + d)$ and is b.*

$(c + d)(a + b)$ *Therefore, $a(c + d) + b(c + d) = (c + d)(a + b)$. The check is more difficult in this case. Compare with the distributive property $x(y + z)$, where x is $(c + d)$, y is a, and z is b. Hence, $(c + d)(a + b) = (c + d)a + (c + d)b = a(c + d) + b(c + d)$ by the commutative property of multiplication.*

b. $x(x + 2) - 5(x + 2) =$ *The common factor is $(x + 2)$.*

$(x + 2)(\_ + \_)$ $=$ *The first blank is the term whose product with $(x + 2)$ is $x(x + 2)$. This is x.*

$(x + 2)(x + \_)$ $=$ *The second blank is the term whose product with $(x + 2)$ is $-5(x + 2)$. This is -5.*

$(x + 2)(x - 5)$ *Therefore, $x(x + 2) - 5(x + 2) = (x + 2)(x - 5)$.*

PRACTICE EXERCISE 3

Factor the following by removing the common binomial factor:

a. $x(y - z) + w(y - z)$

b. $y(y + 4) - 5(y + 4)$

If you need more practice, do the following Additional Practice Exercises.

Additional Practice Exercise 3   Factor the following by removing the common binomial factor:

a. $r(s + t) + p(s + t)$

b. $z(z + 5) - 6(z + 5)$

OBJECTIVE **B**   Factoring by grouping

We will now factor polynomials with four terms by removing both common monomial and binomial factors. We will group terms and remove a common monomial factor from each group. The results will have a common binomial factor that we will then remove. Study the following examples carefully:

Example 4   **Factor the following:**

a. $ab + ac + db + dc =$ *Group the first two terms together and remove the common factor, a. Group the third and fourth terms together and remove the common factor, d.*

$a(b + c) + d(b + c) =$ *We now have a common factor, $(b + c)$, which we will remove.*

$(b + c)(a + d)$ *Factorization.*

CHECK:

$(b + c)(a + d) = ab + bd + ac + cd$ *This is correct, except for the order of the terms. What property states that the order of the terms does not matter?*

Answers: Practice Exercise 3: a. $(y - z)(x + w)$ **b.** $(y + 4)(y - 5)$ **Additional Practice 3: a.** $(s + t)(r + p)$ **b.** $(z + 5)(z - 6)$

b. $xy + xz + 3y + 3z =$ Group the first two terms together and remove the common factor, x.
Group the third and fourth terms together and remove the common factor, 3.

$x(y + z) + 3(y + z) =$ Now, remove the common factor, $y + z$.

$(y + z)(x + 3)$ Factorization.

CHECK:

$(y + z)(x + 3) = xy + 3y + xz + 3z$ This is correct.

c. $4x - xy + 20 - 5y =$ Remove the common factor, x, from the first two terms and the common factor, 5, from the remaining terms.

$x(4 - y) + 5(4 - y) =$ Now, remove the common factor, $4 - y$.

$(4 - y)(x + 5)$ Factorization.

CHECK:

$(4 - y)(x + 5) = 4x + 20 - xy - 5y$ This is correct.

Note

When factoring by grouping, it is often possible to group the terms in more than one way. Example 1b could have been done as follows:

$xy + xz + 3y + 3z \quad =$ Change the order of the terms to $xy + 3y + xz + 3z$.

$xy + 3y + xz + 3z \quad =$ Group the first two terms together and remove the common factor, y.
Group the third and fourth terms together and remove the common factor, z.

$y(x + 3) + z(x + 3) =$ Now, remove the common factor, $x + 3$.

$(x + 3)(y + z)$ Factorization.

Note that the answer is the same except for the order of the factors.

PRACTICE EXERCISE 4

Factor the following:

a. $ab - ac + xb - xc$ **b.** $xy + xz + 7y + 7z$ **c.** $3y + xy + 12 + 4x$

If you need more practice, do the following Additional Practice Exercises.

Additional Practice Exercise 4   Factor the following:

a. $mn + mp + qn + qp$ **b.** $xc - xf + 5c - 5f$ **c.** $5a + ab + 15 + 3b$

The goal of this type of factoring is to have the exact same binomial inside the parentheses after we have grouped the terms and removed the common factors from each group. Consequently, it is sometimes necessary to remove a negative factor in order to make the signs of the binomial factors agree.

Answers: Practice Exercise 4: a. $(b - c)(a + x)$ **b.** $(y + z)(x + 7)$ **c.** $(3 + x)(y + 4)$
Additional Practice 4: a. $(n + p)(m + q)$ **b.** $(c - f)(x + 5)$ **c.** $(5 + b)(a + 3)$

Example 5 | **Factor the following:**

a. $ax + az - bx - bz \quad =$ Group the first two terms and remove the common factor, a.

$a(x + z) - bx - bz \quad =$ Since we want $x + z$ inside both sets of parentheses, factor $-b$ from the remaining terms. Be careful! This will change the signs of both terms inside the parentheses.

$a(x + z) - b(x + z) =$ Now remove the common factor, $x + z$.

$(x + z)(a - b)$ Factorization.

CHECK:

$(x + z)(a - b) = ax - bx + az - bz$ This is correct.

b. $3x - xr - 21 + 7r \quad =$ Remove the common factor, x, from the first two terms.

$x(3 - r) - 21 + 7r \quad =$ We want $3 - r$ inside both sets of parentheses, so remove the factor, -7, from the remaining two terms.

$x(3 - r) - 7(3 - r) =$ Now remove the common factor, $(3 - r)$.

$(3 - r)(x - 7)$ Factorization.

CHECK:

$(3 - r)(x - 7) = 3x - 21 - rx + 7r$ This is correct.

c. $3x - xy - 3 + y \quad =$ Group the first two terms and remove the common factor, x.

$x(3 - y) - 3 + y \quad =$ Notice that the two remaining terms are the same as the terms inside the parentheses, except that the signs are opposite. Factor -1 from the last two terms.

$x(3 - y) - 1(3 - y) =$ Now remove the common factor, $3 - y$.

$(3 - y)(x - 1)$ Factorization.

CHECK:

$(3 - y)(x - 1) = 3x - 3 - xy + y$ This is correct.

PRACTICE EXERCISE 5

Factor the following:

a. $xy + hy - xc - hc$ **b.** $4a - ab - 16 + 4b$ **c.** $xy - y - x + 1$

If you need more practice, do the following Additional Practice Exercises.

Additional Practice Exercise 5   Factor the following:

a. $nm + nw - xm - xw$ **b.** $4x - xr - 20 + 5r$ **c.** $xe - e - x + 1$

For Extra Help

Exercise Set 5.2   MyMathLab®

Factor the following by removing the greatest common factor. (See Example 1.)

1. $cd + cf$ **2.** $gh + gk$ **3.** $rs - rt$

4. $xy - xz$ **5.** $3x + 9$ **6.** $4x + 8$

7. $8x - 12$ **8.** $15x - 10$ **9.** $x^2 + x$

Answers: Practice Exercise 5: a. $(x + h)(y - c)$ **b.** $(4 - b)(a - 4)$ **c.** $(x - 1)(y - 1)$
Additional Practice 5: a. $(m + w)(n - x)$ **b.** $(4 - r)(x - 5)$ **c.** $(x - 1)(e - 1)$

10. $y^2 + y$

11. $r^4 - r^2$

12. $q^3 - q^2$

13. $7a^3 + 14a^2$

14. $5r^4 + 10r^2$

15. $18p^3 - 9p^5$

16. $16r^3 - 8r^4$

17. $15x^3 + 5x$

18. $12d^4 + 6d$

19. $22r^5 - 11r^2$

20. $26t^4 - 13t$

21. $12c^3 + 8c^2$

22. $14b^4 + 8b^3$

23. $18z^2 - 12z^6$

24. $15w^4 - 10w^6$

25. $x^2y^3 + x^2y^2$

26. $r^3s^4 + r^3s^3$

27. $u^4v^2 - u^3v^3$

28. $p^3q^2 - p^2q^3$

29. $c^3d^4 + cd^2$

30. $a^4b^3 + a^3b$

31. $x^2y^2 - x^3y^4$

32. $a^3b^2 - a^5b^5$

33. $18a^2b^4 + 12a^3b^2$

34. $18r^3s^2 + 27r^5s$

35. $9xy^3 - 12x^2y^2$

36. $24fg^4 - 18f^4g^3$

37. $9x^2y^4 + 3xy^3$

38. $8y^3z^2 + 4y^2z$

39. $9x^2 - 12x + 6$

40. $12x^2 - 8x + 12$

41. $15x^4 - 10x^3 - 20x^2$

42. $24y^5 - 16y^4 - 32y^2$

43. $6a^4 + 12a^2 - 24$

44. $9s^5 + 18s^3 - 27$

45. $22x^3 - 33x^2 - 11x$

46. $21a^4 - 28a^2 - 7a$

47. $16c^3d^2 - 24c^2d^4 + 36cd^2$

48. $16r^2s^4 - 8r^3s^4 + 12rs^3$

49. $14x^3y^3 - 21x^2y^4 - 7xy^2$

50. $25y^4z^4 - 30y^3z^2 - 5y^2z^2$

Factor the following by removing a negative common factor. (See Example 2.)

51. $-3m + 6n$

52. $-4a + 8b$

53. $-16c + 8d$

54. $-20p + 5q$

55. $-14x + 7$

56. $-18a + 9$

57. $-4x^2 + 8x - 16$

58. $-6x^2 + 18x - 24$

59. $-10x^3 - 15x^2 + 25x$

60. $-8a^3 - 24a^2 + 16a$

61. $-x + y$

62. $-a - b$

Factor the following by removing the common binomial factor. (See Example 3.)

63. $a(m + n) + b(m + n)$

64. $x(p + q) + y(p + q)$

65. $a(c + 2) - b(c + 2)$

66. $r(u + 5) - s(u + 5)$

67. $t(t - 3) + 6(t - 3)$

68. $y(y - 3) + 3(y - 3)$

69. $a(a - 6) - 7(a - 6)$

70. $b(b - 5) - 4(b - 5)$

71. What are the factors in the expression $(2x + 3)(3x - 5)$?

72. What are the factors in the expression $(5x - 2)(2x + 7)$?

Challenge Exercises (73–76)

Remove the greatest common factor from each of the following. (See Examples 1–3.)

73. $360x^6y^5 - 540x^8y^4$

74. $756a^9b^6 - 504a^7b^9$

75. $3(a + b)^3 + 9(a + b)^2$

76. $4(x + y)^4 - 12(x + y)^2$

Factor each of the following by grouping. (See Examples 4 and 5.)

77. $xz + xw + yz + yw$

78. $ac + ad + bc + bd$

79. $mn - 4m + 3n - 12$

80. $dc - 2d + 5c - 10$

81. $ab + 5a + 3b + 15$

82. $mn + 7n + 5m + 35$

83. $x^2 - xy + 5x - 5y$

84. $a^2 - ab + 3a - 3b$

85. $ab - 3a + b - 3$

86. $xy - 5x + y - 5$

87. $xm + xn - my - ny$

88. $cd + cf - de - ef$

89. $mn - 6n - 3m + 18$

90. $rs - 7s - 4r + 28$

91. $cd - ce - 3d + 3e$

92. $pr - pq - 7r + 7q$

93. $2x^2 - xy + 6x - 3y$

94. $3r^2 - rs + 12r - 4s$

95. $rs + 4r - s - 4$

96. $yz + 5y - z - 5$

97. $cd - 3c - d + 3$

98. $ab - 7a - b + 7$

99. $10a + 4ab + 15 + 6b$

100. $12x + 9bx + 8 + 6b$

101. $8xz - 4xw - 2yz + yw$

102. $15ac - 5ad - 3bc + bd$

103. $6ac + 4ad - 9bc - 6bd$

104. $6xz + 9xw - 8yz - 12yw$

105. $8x^2 - 2xy + 12x - 3y$

106. $6a^2 - 9ay + 10a - 15y$

Writing Exercises (107–109)

107. $3a^2b^2(6b + 4a - 8) = 18a^2b^3 + 12a^3b^2 - 24a^2b^2$. However, $3a^2b^2(6b + 4a - 8)$ is not the correct prime factorization of $18a^2b^3 + 12a^3b^2 - 24a^2b^2$. Why not?

108. How can we tell whether we have removed the greatest common factor? (That is, what will happen if the factor we remove is a factor of the polynomial, but not the greatest common factor?)

109. After factoring a polynomial, how can you determine whether your factorization is correct?

Section 5.3 Factoring General Trinomials with Leading Coefficients of One

OBJECTIVE *When you complete this section, you will be able to:*

A Factor a general trinomial whose first coefficient is one when written in descending order.

PREREQUISITE SKILLS *Before you begin this section, you should be able to:*

a. Multiply binomials by using FOIL. (Section 2.3)
b. Apply $a^m \cdot a^n = a^{m+n}$. (Section 2.2)
c. Add like terms. (Section 1.7)
d. Multiply integers. (Section 2.1)
e. Factor by removing the greatest common factor. (Section 5.2)

GETTING READY FOR SECTION 5.3

Find the following product, using FOIL:

1. $(x + 2)(x - 5)$ **2.** $(x - 3)(x - 7)$ **3.** Find the product: $ab \cdot a^2b$

Find the following sums:

4. $-3x + 5x$ **5.** $2xy - 4xy$

Find the following products:

6. $2(-5)$ **7.** $(-3)(-6)$

Factor the following:

8. $5x^2 - 20x + 15$ **9.** $2a^3c^2 - 24a^3c + 64a^3$

Introduction

When most people think of factoring, it is the factoring of trinomials that usually comes to mind.

We will consider two types of trinomials: those with positive last terms and those with negative last terms. We will first look at the products of binomials that result in trinomials whose last terms are positive.

Example 1 **Find the following products:**

a. $(x + 3)(x + 7) = x^2 + 7x + 3x + 21$ Apply FOIL.
$$= x^2 + 10x + 21$$ Add like terms.

Therefore, $x^2 + 10x + 21 = (x + 3)(x + 7)$ in factored form.

b. $(x - 3)(x - 7) = x^2 - 7x - 3x + 21$ Apply FOIL.
$$= x^2 - 10x + 21$$ Add like terms.

Therefore, $x^2 - 10x + 21 = (x - 3)(x - 7)$ in factored form.

From Example 1, we observe that the signs of the last terms of the trinomials are positive. Further, observe that the signs of the second terms of both binomial factors are the same as the sign of the second term of the trinomial.

Answers: Getting Ready: 1. $x^2 - 3x - 10$ **2.** $x^2 - 10x + 21$ **3.** a^3b^2 **4.** $2x$ **5.** $-2xy$ **6.** -10 **7.** 18 **8.** $5(x^2 - 4x + 3)$ **9.** $2a^3(c^2 - 12c + 32)$

From Example 1a:

$$x^2 + 10x + 21 = (x + 3)(x + 7)$$

Last sign positive.

$$x^2 + 10x + 21 = (x + 3)(x + 7)$$

If the last sign is positive, these three signs are the same.

From Example 1b:

$$x^2 - 10x + 21 = (x - 3)(x - 7)$$

Last sign positive.

$$x^2 - 10x + 21 = (x - 3)(x - 7)$$

If the last sign is positive, these three signs are the same.

When multiplying binomials whose signs are the same, both the inside and outside products have the same sign. Consequently, we add the absolute values of the outside and inside products in order to find the absolute value of the coefficient of the middle term of the trinomial. This will help us decide which factors of the last term to try when factoring the trinomial.

$$(x + 3)(x + 7) = x^2 + 7x + 3x + 21 = x^2 + 10x + 21$$

Same sign, so add the absolute values of 7 and 3 to get 10, which is the absolute value of the coefficient of the middle term.

$$(x - 3)(x - 7) = x^2 - 7x - 3x + 21 = x^2 - 10x + 21$$

Same sign, so add the absolute values of −7 and −3 to get 10, which is the absolute value of the coefficient of the middle term.

OBJECTIVE **A** ## Factoring a general trinomial whose first coefficient is one when written in descending order

Remember, we are doing FOIL in reverse, and if the last sign of the trinomial is positive, both binomial factors have the same sign for the second term as the sign of the middle term of the trinomial. Consider the following:

Example 2 **Factor the following:**

a. $x^2 + 4x + 3 =$ *The sign of the last term is + and the sign of the middle term is +. Therefore, both of the signs of the second terms of the binomials will be +.*

$(\_ + \_)(\_ + \_) =$ *The first blank in each binomial must be filled by terms whose product is x^2. The only reasonable factorization of x^2 is $x \cdot x$.*

$(x + \_)(x + \_) =$ *The second blank in each binomial must be filled by terms whose product is 3, but*
$(x + 1)(x + 3) =$ *which will also give the sum of the outside and inside products of 4x. The only factors of 3 are $1 \cdot 3$.*

x
$3x$
$4x$

The outside product is 3x and the inside product is x. The sum is 4x, which is what we wanted, so this is the correct choice.

$(x + 1)(x + 3)$ *Factorization.*

Note

Since multiplication is commutative,

$$(x + 3)(x + 1)$$

is also correct.

CHECK:

$$(x + 1)(x + 3) = x^2 + 3x + x + 3 = x^2 + 4x + 3$$ *This is the original trinomial. Therefore, this factorization is correct.*

b. $x^2 - 7x + 10 =$ *The sign of the last term of the trinomial is $+$ and the sign of the middle term is $-$. Therefore, both of the signs of the second terms of the binomials are $-$.*

$(\_ - \_)(\_ - \_) =$ *The first blanks of each binomial must be filled with terms whose product is x^2. The only reasonable factorization of x^2 is $x \cdot x$.*

$(x - \_)(x - \_) =$ *The second blanks must be filled by terms whose product is 10 and which give the sum of the outside and inside products of $-7x$. The possible factors of 10 are $1 \cdot 10$ and $2 \cdot 5$. Try 1 and 10.*

$(x - 1)(x - 10) =$ *The outside product is $-10x$ and the inside product is $-x$. The sum is $-11x$, which is not correct, since we need the middle term to be $-7x$. Try 2 and 5.*

$$\begin{array}{c} -x \\ -10x \\ \hline -11x \end{array}$$

$(x - 2)(x - 5)$ *The outside product is $-5x$ and the inside product is $-2x$. The sum is $-7x$, which is what we wanted, so this is the correct choice.*

$$\begin{array}{c} -2x \\ -5x \\ \hline -7x \end{array}$$

$(x - 2)(x - 5)$ Factorization.

CHECK:

$$(x - 2)(x - 5) = x^2 - 5x - 2x + 10 = x^2 - 7x + 10$$ *This is the original trinomial. Therefore, this factorization is correct.*

c. $a^2 - 6ab + 8b^2 =$ *The sign of the last term is $+$ and the sign of the middle term is $-$. Therefore, both of the signs of the second terms of the binomials are $-$.*

$(\_ - \_)(\_ - \_) =$ *The first blanks must be filled by terms whose product is a^2. The only factorization of a^2 is $a \cdot a$.*

$(a - \_)(a - \_) =$ *The second blanks must be filled with terms whose product is $8b^2$ and give a sum of the outside and inside products of $-6ab$. The choices of factors of $8b^2$ are $b \cdot 8b$ or $2b \cdot 4b$. Try $2b$ and $4b$.*

$(a - 2b)(a - 4b)$ *The outside product is $-4ab$ and the inside product is $-2ab$. The sum is $-6ab$, which is what we wanted, so this is the correct choice.*

$$\begin{array}{c} -2ab \\ -4ab \\ \hline -6ab \end{array}$$

$(a - 2b)(a - 4b)$ Factorization.

CHECK:

$$(a - 2b)(a - 4b) = a^2 - 4ab - 2ab + 8b^2 = a^2 - 6ab + 8b^2$$ *Therefore, this factorization is correct.*

PRACTICE EXERCISE 2

Factor the following:

a. $a^2 + 8a + 7$ **b.** $b^2 - 9b + 14$ **c.** $y^2 - 6yz + 5z^2$

If you need more practice, do the following Additional Practice Exercises.

Additional Practice Exercise 2   Factor the following:

a. $b^2 + 9b + 8$ **b.** $y^2 - 9yz + 14z^2$ **c.** $x^2 + 8xy + 15y^2$

Answers: Practice Exercise 2: a. $(a + 7)(a + 1)$ **b.** $(b - 2)(b - 7)$ **c.** $(y - 5z)(y - z)$ **Additional Practice 2: a.** $(b + 8)(b + 1)$
b. $(y - 2z)(y - 7z)$ **c.** $(x + 3y)(x + 5y)$

We will now look at products of binomials that result in trinomials whose last terms are negative.

<table>
<tr><td>**Example 3**</td><td>**Find the following products:**</td></tr>
</table>

a. $(x + 3)(x - 7) = x^2 - 7x + 3x - 21$ Apply FOIL.

$\qquad\qquad\qquad\quad = x^2 - 4x - 21$ Add like terms.

Therefore, $x^2 - 4x - 21 = (x + 3)(x - 7)$ in factored form.

b. $(x - 3)(x + 7) = x^2 + 7x - 3x - 21$ Apply FOIL.

$\qquad\qquad\qquad\quad = x^2 + 4x - 21$ Add like terms.

Therefore, $x^2 + 4x - 21 = (x - 3)(x + 7)$ in factored form.

Looking at Example 3, we notice the sign of the last term of the trinomial in both examples is $-$. In each case, one binomial factor has a $+$ sign and the other a $-$ sign. In other words, the signs are opposites.

From Example 3a:

$$x^2 - 4x - 21 = (x + 3)(x - 7)$$

Last sign negative.

$$x^2 - 4x - 21 = (x + 3)(x - 7)$$

If the last sign is negative, these signs are opposites.

From Example 3b:

$$x^2 + 4x - 21 = (x - 3)(x + 7)$$

Last sign negative.

$$x^2 + 4x - 21 = (x - 3)(x + 7)$$

If the last sign is negative, these signs are opposites.

Since the signs of the second terms of the binomials are opposites, the outside and inside products are opposite in sign. This means the absolute value of the coefficient of the middle term of the trinomial is found by subtracting the absolute values of the coefficients of the outside and inside products of the binomials. This will be helpful when we decide which factors of the last term to try when factoring the trinomial.

From Example 3a:

$$(x + 3)(x - 7) = x^2 - 7x + 3x - 21 = x^2 - 4x - 21$$

These are opposite in sign, so subtract the absolute values of -7 and 3 to get 4, which is the absolute value of the coefficient of the middle term of the trinomial.

From Example 3b:

$$(x - 3)(x + 7) = x^2 + 7x - 3x - 21 = x^2 + 4x - 21$$

These are opposite in sign, so subtract the absolute values of 7 and -3 to get 4, which is the absolute value of the coefficient of the middle term of the trinomial.

When factoring a trinomial whose last term is negative, we will not be concerned with the signs, initially. We will first find binomial factors that give the correct first and last terms of the trinomial and are such that the difference of the outside and inside products of the binomials gives the middle term, disregarding the sign, of the trinomial. When this is accomplished, we will decide which factor has the $+$ and which has the $-$. This will become clearer when we do some examples.

Example 4 | **Factor the following:**

a. $x^2 - x - 6$ = *The sign of the last term is $-$. Therefore, the signs of the second terms of the binomials are opposites.*

$(\_ \ \_)(\_ \ \_)$ = *The first blanks must be filled by terms whose product is x^2. The only choice is x and x. Notice that we did not put in any signs, since we will determine them last.*

$(x \ \_)(x \ \_)$ = *The second blanks must be filled by terms whose product is 6 and are such that the difference of the outside and inside products is 1. The possible factorizations of 6 are $1 \cdot 6$ and $2 \cdot 3$. Try 1 and 6.*

$(x \ 1)(x \ 6)$
$\quad x$
$\quad 6x$
$\quad 5x$

The outside product is $6x$ and the inside product is x. But $6x - x = 5x$, and the middle term of the trinomial is x, disregarding the sign. Therefore, 1 and 6 are not the correct choices. Try 2 and 3.

$(x \ 2)(x \ 3)$
$\quad 2x$
$\quad 3x$
$\quad x$

The outside product is $3x$, the inside product is $2x$, and $3x - 2x = x$, which is the middle term of the trinomial, except for the sign.

$(x + 2)(x - 3)$
$\quad 2x$
$\quad -3x$
$\quad -x$

Since the middle term of the trinomial is $-x$, we want $-3x$ and $+2x$ as the outside and inside products. Therefore, use -3 and $+2$.

$(x + 2)(x - 3)$ Factorization.

CHECK:

$(x + 2)(x - 3) = x^2 - 3x + 2x - 6 = x^2 - x - 6$ *Therefore, this factorization is correct.*

b. $x^2 + 4xy - 12y^2$ = *The last sign of the trinomial is $-$, so the binomial factors will have opposite signs.*

$(\_ \ \_)(\_ \ \_)$ = *The first blanks must be filled by terms whose product is x^2. Again, the only reasonable choice is x and x.*

$(x \ \_)(x \ \_)$ = *The second blanks must be filled with terms whose product is $12y^2$, and the difference of the outside and inside products must be $4xy$. The possible factorizations of $12y^2$ are $y \cdot 12y$, $2y \cdot 6y$, and $3y \cdot 4y$. Try y and $12y$.*

$(x \ y)(x \ 12y)$ =
$\quad xy$
$\quad 12xy$
$\quad 11xy$

The outside product is $12xy$ and the inside product is xy. However, $12xy - xy = 11xy$, and the middle term is $4xy$. Therefore, this is not the correct choice of factors of $12y^2$. Try $2y$ and $6y$.

$(x \ 2y)(x \ 6y)$
$\quad 2xy$
$\quad 6xy$
$\quad 4xy$

The outside product is $6xy$ and the inside product is $2xy$. Also, $6xy - 2xy = 4xy$, which is the correct middle term.

$(x - 2y)(x + 6y)$
$\quad -2xy$
$\quad 6xy$
$\quad 4xy$

Since the middle term is $+4xy$, we need $+6xy$ and $-2xy$ as the outside and inside products. Therefore, use $+6y$ and $-2y$.

$(x - 2y)(x + 6y)$ Factorization.

CHECK:

$(x - 2y)(x + 6y) = x^2 + 6xy - 2xy - 12y^2 = x^2 + 4xy - 12y^2$ *Therefore, this factorization is correct.*

c. $x^2 + 3x - 8 =$ *The last sign is −, so the signs of the binomials are opposites.*

$(\_ \quad \_)(\_ \quad \_) =$ *The first blanks must be filled in by terms whose product is x^2. This is, again, x and x.*

$(x \quad \_)(x \quad \_) =$ *The second blanks must be filled by terms whose product is 8, and the difference in the outside and inside product must be 3x. The possible factors of 8 are $1 \cdot 8$ and $2 \cdot 4$. Try 1 and 8.*

$(x \quad 1)(x \quad 8) =$ *The outside product is 8x and the inside product is x, but $8x - x = 7x$, which is not the correct middle term. Try 2 and 4.*

$(x \quad 2)(x \quad 4)$ *The outside product is 4x and the inside product is 2x, but $4x - 2x = 2x$, which is not the correct middle term. Since we have tried all possible combinations of the factors of 8 and none work, this trinomial is prime.*

PRACTICE EXERCISE 4

Factor the following; if the polynomial cannot be factored, write "prime":

a. $x^2 + 3x - 10$ **b.** $a^2 - 2ab - 15b^2$ **c.** $b^2 + 4b - 10$

If you need more practice, do the following Additional Practice Exercises.

Additional Practice Exercise 4   Factor the following; if the polynomial cannot be factored, write "prime":

a. $x^2 + 5x - 14$ **b.** $x^2 - 5x - 6$ **c.** $a^2 + 3a - 12$

Following is a summary of the techniques discussed in this section:

> ### Summary
>
> 1. If the last sign of the trinomial is positive, the signs of both binomial factors will be the same as the middle sign of the trinomial. Also, the absolute values of the coefficients of the outside and inside products of the binomial factors must have a sum equal to the absolute value of the coefficient of the middle term of the trinomial being factored.
>
> 2. If the last term of the trinomial is negative, the signs of the second terms of the binomial factors will be opposites. Also, the difference of the absolute values of the coefficients of the outside and inside products of the binomial factors must equal the absolute value of the coefficient of the middle term of the trinomial being factored.

We will do one of each type in the following example:

Example 5 | **Factor the following trinomials:**

a. $x^2 - 11x + 24 =$ *The sign of the last term is + and the sign of the second term is −. Therefore, the signs of both binomial factors are −.*

$(\_ - \_)(\_ - \_) =$ *The first blanks must be filled with terms whose product is x^2. As before, the only reasonable choice is x and x.*

$(x - \_)(x - \_) =$ *The second blanks must be filled with terms whose product is 24 and that give the sum of the outside and inside products as −11. The possible factors of 24 are $1 \cdot 24, 2 \cdot 12, 3 \cdot 8$, and $4 \cdot 6$. Try 3 and 8.*

Answers: Practice Exercise 4: a. $(x + 5)(x - 2)$ **b.** $(a - 5b)(a + 3b)$ **c.** prime   **Additional Practice 4: a.** $(x + 7)(x - 2)$
b. $(x - 6)(x + 1)$ **c.** prime

$$(x - 3)(x - 8) =$$

The inside product is $-3x$ and the outside product is $-8x$. The sum is $-11x$, which is the correct middle term.

$$\begin{array}{c} -3x \\ -8x \\ \hline -11x \end{array}$$

Factorization.

$$(x - 3)(x - 8)$$

CHECK:

$$(x - 3)(x - 8) = x^2 - 8x - 3x + 24 = x^2 - 11x + 24$$

This is the original trinomial. Therefore, this factorization is correct.

b. $b^2 - b - 20 =$ *The sign of the last term is $-$. Therefore, the signs of the binomials are opposites.*

$(\_ \ \_)(\_ \ \_) =$ *The first blanks must be filled by terms whose product is b^2. The only choice is b and b.*

$(b \ \_)(b \ \_)$ *The second blanks must be filled with terms whose product is 20, and the difference of the outside and inside products must be 1. The possible factors of 20 are $1 \cdot 20$, $2 \cdot 10$, and $4 \cdot 5$. Try 2 and 10.*

$$(b \ \ 2)(b \ \ 10)$$
$$\begin{array}{c} 2b \\ 10b \\ \hline 8b \end{array}$$

The outside product is $10b$ and the inside product is $2b$. However, $10b - 2b = 8b$, and the middle term of the trinomial is $-b$. Therefore, this is not the correct choice of factors of 10. Try 4 and 5.

$$(b \ \ 4)(b \ \ 5)$$
$$\begin{array}{c} 4b \\ 5b \\ \hline b \end{array}$$

The outside product is $5b$ and the inside product is $4b$. Also, $5b - 4b = b$, which is the correct middle term, except for the sign.

$$(b + 4)(b - 5)$$
$$\begin{array}{c} 4b \\ -5b \\ \hline -b \end{array}$$

To get $-b$, we need $-5b$ and $+4b$ as the outside and inside products. Therefore, use -5 and $+4$.

$$(b + 4)(b - 5)$$ *Factorization.*

CHECK:

$$(b + 4)(b - 5) = b^2 - 5b + 4b - 20 = b^2 - b - 20$$

This is the original trinomial. Therefore, this factorization is correct.

PRACTICE EXERCISE 5

Factor the following:

a. $a^2 + 13a + 40$ **b.** $r^2 - 7r - 18$ **c.** $m^2 + 2m - 24$ **d.** $x^2 + 16x + 48$

If you need more practice, do the following Additional Practice Exercises.

Additional Practice Exercise 5   Factor the following:

a. $a^2 + 11a + 30$ **b.** $x^2 - x - 42$ **c.** $x^2 + 3x - 28$ **d.** $c^2 - 12c + 32$

As in the previous sections of factoring, if a common factor is present, we must first remove it and try to factor the polynomial inside the parentheses.

Answers: Practice Exercise 5: a. $(a + 5)(a + 8)$ **b.** $(r + 2)(r - 9)$ **c.** $(m - 4)(m + 6)$
d. $(x + 4)(x + 12)$ **Additional Practice 5: a.** $(a + 5)(a + 6)$ **b.** $(x - 7)(x + 6)$ **c.** $(x - 4)(x + 7)$
d. $(c - 4)(c - 8)$

Example 6   **Factor the following completely:**

a. $3x^2 + 12x - 63 =$ *Remove the GCF, 3.*

$3(x^2 + 4x - 21) =$ *The last sign of the trinomial is −, so the signs of the binomial factors will be opposites.*

$3(\_\ \_)(\_\ \_) =$ *The first blanks must be filled with terms whose product is x^2. Again, this is x and x.*

$3(x\ \_)(x\ \_) =$ *The second blanks must be filled with terms whose product is 21, and the difference of the outside and inside products must be 4x. Try 3 and 7.*

$3(x\ \ 3)(x\ \ 7) =$ *The outside product is 7x, the inside product is 3x, and 7x − 3x = 4x, which is the desired middle term. Since we want +4x, we need +7x and −3x as the outside and inside products. Therefore, use −3 and +7.*

$3(x - 3)(x + 7)$ *Factorization.*

CHECK:

$3(x - 3)(x + 7) = 3(x^2 + 7x - 3x - 21)$ *Therefore, this factorization is correct.*
$= 3(x^2 + 4x - 21) = 3x^2 + 12x - 63$

b. $a^3b - 9a^2b^2 + 8ab^3 =$ *Remove the GCF, ab.*

$ab(a^2 - 9ab + 8b^2) =$ *The last sign of the trinomial is + and the middle sign is −. Therefore, both signs of the binomials are −.*

$ab(\_ - \_)(\_ - \_) =$ *The first blanks are filled by a and a.*

$ab(a - \_)(a - \_) =$ *The second blanks must be filled with terms whose product is $8b^2$ and that give outside and inside products whose sum is −9ab. Try b and 8b.*

$ab(a - b)(a - 8b)$ *Factorization.*

CHECK:

$ab(a - b)(a - 8b) = ab(a^2 - 8ab - ab + 8b^2)$ *Therefore, this factorization*
$= ab(a^2 - 9ab + 8b^2) = a^3b - 9a^2b^2 + 8ab^3$ *is correct.*

PRACTICE EXERCISE 6

Factor the following:

a. $2x^2 - 18x + 36$

b. $y^3z - 15y^2z^2 - 16yz^3$

c. $3x^3 + 9x^2 - 30x$

d. $6x^2z^2 + 18xyz^2 + 12y^2z^2$

If you need more practice, do the following Additional Practice Exercises.

Additional Practice Exercise 6   Factor the following:

a. $4x^2 - 40x + 36$

b. $m^3n - m^2n^2 - 12mn^3$

c. $4a^3 - 16a^2 - 48a$

d. $4a^2c^2 + 8abc^2 - 12b^2c^2$

Answers: **Practice Exercise 6:** **a.** $2(x - 3)(x - 6)$ **b.** $yz(y - 16z)(y + z)$ **c.** $3x(x - 2)(x + 5)$
d. $6z^2(x + 2y)(x + y)$ **Additional Practice 6:** **a.** $4(x - 9)(x - 1)$ **b.** $mn(m - 4n)(m + 3n)$
c. $4a(a - 6)(a + 2)$ **d.** $4c^2(a - b)(a + 3b)$

Exercise Set 5.3

For Extra Help

MyMathLab®

Factor the following, if possible; if the polynomial is not factorable, write "prime." (See Examples 2, 4, and 5.)

1. $x^2 + 3x + 2$

2. $x^2 + 6x + 5$

3. $x^2 - 4x + 3$

4. $x^2 - 8x + 7$

5. $a^2 + 5a + 6$

6. $a^2 + 7a + 10$

7. $b^2 - 6b + 8$

8. $c^2 - 12c + 27$

9. $y^2 - 4y - 5$

10. $z^2 - 2z - 3$

11. $a^2 + 2a - 35$

12. $b^2 + 4b - 21$

13. $x^2 + 10x + 16$

14. $z^2 + 11z + 18$

15. $r^2 + r - 72$

16. $s^2 + s - 20$

17. $a^2 - 12a + 27$

18. $c^2 - 13c + 40$

19. $x^2 + 4x + 8$

20. $a^2 - 6a - 8$

21. $x^2 - 12x + 35$

22. $n^2 - 14n + 33$

23. $y^2 + y - 42$

24. $x^2 + x - 72$

25. $z^2 + 15z + 36$

26. $b^2 + 11b + 24$

27. $a^2 + 4ab + 3b^2$

28. $c^2 + 12cd + 11d^2$

29. $r^2 - 8rs + 7s^2$

30. $q^2 - 6qr + 5r^2$

31. $c^2 + 7cd - 18d^2$

32. $a^2 + 10ab - 24b^2$

33. $r^2 - 7rs - 44s^2$

34. $a^2 + ab - 56b^2$

35. $y^2 + 3yz - 40z^2$

36. $m^2 + 11mn - 42n^2$

37. $a^2 - 3ab - 54b^2$

38. $x^2 - 4xy - 45y^2$

39. $r^2 - 15rs + 36s^2$

40. $a^2 - 6ab - 27b^2$

Factor the following completely. (See Example 6.)

41. $3a^2 + 15a + 18$

42. $4x^2 + 12x + 8$

43. $5z^2 - 25z + 30$

44. $6a^2 - 30a + 36$

45. $x^3 - 3x^2 - 28x$

46. $y^3 - y^2 - 20y$

47. $y^4 - 20y^3 + 75y^2$

48. $c^4 - 10c^3 + 21c^2$

49. $3a^3 - 3a^2 - 126a$

50. $4x^3 - 4x^2 - 80x$

51. $x^3y + 2x^2y - 80xy$

52. $a^3b + 10a^2b - 56ab$

53. $x^2y^2 + 9xy^2 + 20y^2$ **54.** $y^4z^3 + 13y^3z^3 + 30y^2z^3$ **55.** $4r^2s + 20rs - 56s$

56. $3a^2b + 6ab - 45b$ **57.** $3a^2b - 9ab - 84b$ **58.** $2x^2y - 10xy - 100y$

Challenge Exercises (59–69)

Factor the following:

59. $x^2 - 3x - 108$ **60.** $x^2 + 7x - 144$ **61.** $y^2 + 24y + 128$

62. $6 - x - x^2$ **63.** $24 + 2x - x^2$ **64.** $10 + 7x + x^2$

65. $18 + 9a + a^2$ **66.** $10 - 3x - x^2$ **67.** $16 - 6b - b^2$

Find all integer values of k such that each of the following is factorable over the integers:

68. $x^2 - kx + 6$ **69.** $x^2 + kx + 12$

Recall that the formula for the area of a rectangle is $A = LW$. If the following trinomials represent the areas of rectangles, find binomials that could represent each length and width.

70. $A = x^2 + 19x + 48$ **71.** $A = x^2 + 12x + 32$ **72.** $A = y^2 + 3y - 54$

The volume of a rectangular solid is $V = LWH$. If each of the following trinomials represents the volume of a rectangular solid, find monomials and/or binomials that could represent each length, width, and height:

73. $V = x^3 + 6x^2 - 40x$ **74.** $V = 4x^3 - 12x^2 - 112x$

Writing Exercises (75 and 76)

75. If the sign of the last term of a trinomial is positive, why must both signs of the second terms of the binomial factors be the same?

76. You are given $f(x) = (x + 4)(x + 2)$ and $g(x) = x^2 + 6x + 8$.

    a. Find $f(-2)$ and $g(-2)$.

    b. What is the relationship between $f(-2)$ and $g(-2)$? Why?

<table>
<tr><td>

Section 5.4

</td><td>

Factoring General Trinomials with Leading Coefficients Other Than One

</td></tr>
</table>

OBJECTIVE *When you complete this section, you will be able to:*

A Factor trinomials whose leading coefficient is a natural number other than one.

PREREQUISITE SKILLS *Before beginning this section, you should be able to:*

a. Factor trinomials with a leading coefficient of 1. (Section 5.3)
b. Factor by removing the greatest common factor. (Section 5.2)

GETTING READY FOR SECTION 5.4

Factor the following:
1. $x^2 + 2x - 35$ **2.** $x^2 - 4x + 3$ **3.** $t^3 - 5t^2$ **4.** $ay^4 + ay^3 - 3ay^2$

Introduction

In the previous section, all of the trinomials that we factored had a coefficient of 1 on the first term. The factorization of these trinomials was made easier, since all we needed to do was find the factors of the last term that had either a sum or difference that equaled the coefficient of the middle term. When a polynomial is written in descending order, the coefficient of the first term is the **leading coefficient**.

OBJECTIVE **A** Factoring trinomials whose leading coefficient is a natural number other than one

The same trial-and-error procedure used in the last section applies to trinomials whose leading coefficient is not 1. However, these are more difficult, because the location of the terms in the binomial gives different outside and inside products while using the exact same numbers. In other words, the number of possibilities is greatly increased, as the following illustrates:

Example 1 **Factor the following; if the trinomial cannot be factored, write "prime":**

a. $3x^2 + 5x + 2 =$ *The last term is + and the middle term is +, so both signs of the binomial factors are +.*

$(\_ + \_)(\_ + \_) =$ *The first blanks must be filled by terms whose product is $3x^2$. Since 3 is prime, the only choices are $3x$ and x.*

$(3x + \_)(x + \_) =$ *The second blanks must be filled by terms whose product is 2, and the sum of the outside and inside products must be $5x$. The only factors of 2 are 2 and 1. Try 1 in the first blank and 2 in the second blank.*

$(3x + 1)(x + 2)$ *The outside product is $6x$ and the inside product is x. However, $6x + x = 7x$, and we needed $5x$. Therefore,*
x
$6x$
$7x$ *this factorization is incorrect. Try 2 in the first blank and 1 in the second.*

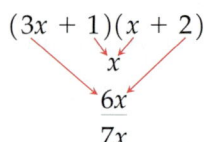

$(3x + 2)(x + 1)$ *The outside product is $3x$, the inside product is $2x$, and $3x + 2x = 5x$, which is the correct middle term.*
$2x$
$3x$
$5x$

$(3x + 2)(x + 1)$ Factorization.

CHECK:

$(3x + 2)(x + 1) = 3x^2 + 3x + 2x + 2 = 3x^2 + 5x + 2$ Therefore, this factorization is correct.

Answers: Getting Ready: 1. $(x + 7)(x - 5)$ **2.** $(x - 1)(x - 3)$ **3.** $t^2(t - 5)$ **4.** $ay^2(y^2 + y - 3)$

b. $5x^2 + 8x - 4 =$ *The last sign is −. Therefore, the signs of the binomial factors are different, so we will be looking for the difference of the outside and inside products.*

$(\_ \ \_)(\_ \ \_) =$ *The first blanks must be filled by terms whose product is $5x^2$. These are 5x and x.*

$(5x \ \_)(x \ \_) =$ *The second blanks must be filled by terms whose product is 4, and the difference of the outside and inside products must be 8x. The possible factorizations of 4 are $1 \cdot 4$ and $2 \cdot 2$. Try 1 in the first blank and 4 in the second.*

$(5x \quad 1)(x \quad 4)$ *The outside product is 20x, the inside product is x, and $20x - x = 19x$. Since we wanted 8x, this is the wrong choice. Try 4 in the first blank and 1 in the second.*

$$x$$
$$20x$$
$$\overline{19x}$$

$(5x \quad 4)(x \quad 1)$ *The outside product is 5x, the inside product is 4x, and $5x - 4x = x$. Therefore, this choice is also incorrect. This means that 1 and 4 will not work, so try 2 and 2.*

$$4x$$
$$5x$$
$$\overline{x}$$

$(5x \quad 2)(x \quad 2)$ *The outside product is 10x, the inside product is 2x, and $10x - 2x = 8x$, which is what we want.*

$$2x$$
$$10x$$
$$\overline{8x}$$

$(5x - 2)(x + 2)$ *Since the middle term is +8x, we want +10x and −2x for the outside and inside products, respectively. Hence, −2 goes in the first blank and 2 goes in the second.*

$$-2x$$
$$10x$$
$$\overline{8x}$$

$(5x - 2)(x + 2)$ Factorization.

CHECK:

$(5x - 2)(x + 2) = 5x^2 + 10x - 2x - 4 = 5x^2 + 8x - 4$ Therefore, this factorization is correct.

c. $6a^2 - 11a - 10 =$ *The last sign is −, so the binomial factors are different in sign.*

$(\_ \ \_)(\_ \ \_) =$ *The first blanks must be filled by terms whose product is $6a^2$. This could be $6a \cdot a$ or $3a \cdot 2a$. Try $3a \cdot 2a$.*

$(3a \ \_)(2a \ \_) =$ *The second blanks must be filled by terms whose product is 10 and are such that the difference of the outside and inside products is 11a. The possible factors of 10 are $1 \cdot 10$ or $2 \cdot 5$. Try 2 and 5, with 2 in the first blank and 5 in the second.*

$(3a \quad 2)(2a \quad 5)$ *This gives an outside product of 15a and an inside product of 4a. Also, $15a - 4a = 11a$, which is what we want.*

$$4a$$
$$15a$$
$$\overline{11a}$$

$(3a + 2)(2a - 5)$ *Since the middle term is −11a, we need −15a and +4a. Put + in the first factor and − in the second.*

$$4a$$
$$-15a$$
$$\overline{-11a}$$

$(3a + 2)(2a - 5)$ Factorization.

CHECK:

$(3a + 2)(2a - 5) = 6a^2 - 15a + 4a - 10$

$\qquad\qquad\qquad\quad = 6a^2 - 11a - 10$ Therefore, this factorization is correct.

Note

In any factorization, the first thing we always look for is a greatest common factor. If there is no greatest common factor other than 1 (or, once it has been removed), no factor may contain a common factor. Understanding this fact may greatly reduce the number of trial-and-error attempts.

d. $4a^2 - 11ab + 6b^2 =$

The last sign is + and the middle sign is −, so both signs of the binomial factors will be −. The first terms of the binomial factors must have a product of $4a^2$. The possibilities are $4a \cdot a$ and $2a \cdot 2a$. Let's try 2a and 2a and then show why this choice is impossible.

$(2a - \_)(2a - \_) =$

The blanks must be filled by terms whose product is $6b^2$ and are such that the sum (since the signs are the same) of the outside and inside products is 11ab. The possibilities are $6b \cdot b$ and $3b \cdot 2b$. Try 6b and b.

$(2a - 6b)(2a - b) =$

This cannot work, since the first factor has a common factor of 2. It would not help to put 6b in the second factor and b in the first, since we would then have a common factor of 2 in the second factor. Try 3b and 2b.

$(2a - 3b)(2a - 2b) =$

The second factor has a common factor of 2, so this cannot work. The same situation would occur if 2b were put in the first factor and 3b in the second. Therefore, the first terms cannot be 2a and 2a, so try 4a and a.

$(4a - \_)(a - \_) =$

Again, the possible terms for the second blanks are 6b and b or 3b and 2b. Try 6b in the first blank and b in the second.

$(4a - 6b)(a - b) =$

The first factor has a common factor of 2, so this cannot be correct. Put 6b in the second factor and b in the first.

$(4a - b)(a - 6b)$

$\underset{-ab}{\overset{}{\diagup}} \quad \underset{}{\diagdown}$

$\dfrac{-24ab}{-25ab} =$

Neither factor has a common factor, but the outside product is $-24ab$ and the inside product is $-ab$. Since $-24ab - ab = -25ab$ and the middle term is $-11ab$, this is incorrect. Try 3b in the first factor and 2b in the second.

$(4a - 3b)(a - 2b)$

$\underset{-3ab}{\overset{}{\diagup}} \quad \underset{}{\diagdown}$

$\dfrac{-8ab}{-11ab} =$

The outside product is $-8ab$, the inside product is $-3ab$, and $-8ab - 3ab = -11ab$, which is what we need. Notice that we could not have put 2b in the first factor and 3b in the second. Why?

$(4a - 3b)(a - 2b)$ *Factorization.*

CHECK:

$(4a - 3b)(a - 2b) = 4a^2 - 8ab - 3ab + 6b^2$
$\qquad\qquad\qquad\quad = 4a^2 - 11ab + 6b^2$ *Therefore, this factorization is correct.*

> **Note**
>
> After doing a few of these, you should be able to do most, if not all, of the previous trial-and-error steps mentally.

e. $3x^2 + 14x + 10 =$

The last sign is + and the middle sign is +, so both signs of the binomial factors are +. The only choice for the first terms are 3x and x.

$(3x + \_)(x + \_) =$

The choices for the second blanks are 1 and 10 or 2 and 5. Try 2 in the first blank and 5 in the second.

$(3x + 2)(x + 5)$

This gives an outside product of 15x and an inside product of 2x, and $15x + 2x = 17x$. Therefore, this is not correct. Try 5 in the first factor and 2 in the second.

$(3x + 5)(x + 2)$

This gives an outside product of 6x and an inside product of 5x, and $6x + 5x = 11x$. Therefore, this is not correct. Since 2 and 5 will not work, try 1 and 10, with 1 in the first factor and 10 in the second.

$(3x + 1)(x + 10)$

This gives an outside product of 30x and an inside product of x, and $30x + x = 31x$. Therefore, this is incorrect. Try 10 in the first factor and 1 in the second.

> **Note**
>
> If the factors were $(3x - 2)(x + 5)$, we would get the correct middle term, but the last term would be -10 instead of $+10$.

$(3x + 10)(x + 1)$

This gives an outside product of 3x and an inside product of 10x, and $3x + 10x = 13x$. Therefore, this is incorrect. We have tried all possible combinations of the factors of $3x^2$ and 10, and none of them work. Since this exhausts all the possibilities, this polynomial is prime.

Alternative Method: The *ac* Method

There is another method of factoring trinomials that is often easier than the trial-and-error method presented earlier. This is called the *ac* method.

Suppose we find the product of $(mx + n)(px + q)$:

$$(mx + n)(px + q) = \qquad \text{Apply FOIL.}$$
$$mpx^2 + mqx + npx + nq = \qquad \begin{array}{l}\text{Remove the common factor of } x \\ \text{from the middle two terms.}\end{array}$$
$$mpx^2 + (mq + np)x + nq$$

We now make some observations. If we multiply the coefficient of the x^2 term (which is mp) by the constant term (which is nq), we get $mpnq$. If we multiply the coefficients of the x terms (which are mq and np), we get $mqnp$, which, except for the order of the factors, is the same thing we got when multiplying the coefficient of the x^2 term and the constant term. This means that the coefficient of the x term must be the sum of two factors of the product of the coefficient of the x^2 term and the constant term. We summarize the procedure in the following box:

Procedure: Factoring Trinomials Using the *ac* Method

Given a trinomial in the form of $ax^2 + bx + c$, follow these steps:

1. Find the product of a and c.

2. Find two factors of ac whose sum is b.

3. Rewrite bx as the sum of two terms whose coefficients are the numbers found in step 2.

4. Factor the resulting polynomial by grouping.

The procedure is essentially the same if the trinomial is of the form $ax^2 + bxy + cy^2$. We illustrate the procedure by using the trinomials from Example 1.

Example 2   **Factor each of the following, using the ac method of factoring:**

a. $3x^2 + 5x + 2 =$
  $ac = 3 \cdot 2 = 6$ and $b = 5$, so we need two factors of 6 whose sum is 5. Six can be factored as $1 \cdot 6, (-1)(-6), 2 \cdot 3,$ or $(-2)(-3)$. Since $2 + 3 = 5$ (which is b), we rewrite $5x$ as $2x + 3x$.

$3x^2 + 2x + 3x + 2 =$
  Factor by grouping. Remove the common factor, x, from the first two terms and write $3x + 2$ as $1 \cdot (3x + 2)$.

$x(3x + 2) + 1 \cdot (3x + 2) =$ *Remove the common factor, $3x + 2$.*
$(3x + 2)(x + 1)$ *Factorization.*

b. $5x^2 + 8x - 4 =$
  $ac = (5)(-4) = -20$ and $b = 8$, so we need two factors of -20 whose sum is 8. We can factor -20 as $(-1)(20)$ or $(1)(-20), (-2)(10)$ or $(2)(-10), (-4)(5)$ or $(4)(-5)$. Since $-2 + 10 = 8$, we rewrite $8x$ as $-2x + 10x$.

$5x^2 - 2x + 10x - 4 =$ *Factor by grouping.*
$x(5x - 2) + 2(5x - 2) =$ *Remove the common factor, $5x - 2$.*
$(5x - 2)(x + 2)$ *Factorization.*

c. $6a^2 - 11a - 10 =$

$ac = 6(-10) = -60$ and $b = -11$, so we need two factors of -60 whose sum is -11. We can factor -60 as $(-1)(60)$ or $(1)(-60), (-2)(30)$ or $(2)(-30), (-3)(20)$ or $(3)(-20), (-4)(15)$ or $(4)(-15), (-5)(12)$ or $(5)(-12), (-6)(10)$ or $(6)(-10)$. Since $4 - 15 = -11$, we write $-11a$ as $4a - 15a$.

$6a^2 + 4a - 15a - 10 =$ *Factor by grouping.*

$2a(3a + 2) - 5(3a + 2) =$ *Remove the common factor, $3a + 2$.*

$(3a + 2)(2a - 5)$ *Factorization.*

d. $4a^2 + 11ab + 6b^2 =$

$ac = 4 \cdot 6 = 24$ and $b = 11$, so we need two factors of 24 whose sum is 11. We can factor 24 as $1 \cdot 24$ or $(-1)(-24), 2 \cdot 12$ or $(-2)(-12), 3 \cdot 8$ or $(-3)(-8), 4 \cdot 6$ or $(-4)(-6)$. Since $3 + 8 = 11$, we rewrite $11ab$ as $3ab + 8ab$.

$4a^2 + 3ab + 8ab + 6b^2 =$ *Factor by grouping.*

$a(4a + 3b) + 2b(4a + 3b) =$ *Remove the common factor, $4a + 3b$.*

$(4a + 3b)(a + 2b)$ *Factorization.*

e. $3x^2 + 14x + 10$

$ac = 3 \cdot 10 = 30$ and $b = 14$, so we need two factors of 30 whose sum is 14. We can factor 30 as $1 \cdot 30$ or $(-1)(-30), 2 \cdot 15$ or $(-2)(-15), 3 \cdot 10$ or $(-3)(-10), 5 \cdot 6$ or $(-5)(-6)$. None of these factors has a sum of 14, so this trinomial is prime.

Each of these methods has its advantages and disadvantages. If you use the *ac* method, you can often save yourself a lot of work by looking at the signs of the original polynomial and use that information as a guide in choosing the factors. For example, in Example 2d, both factors must be positive.

PRACTICE EXERCISE 2

Factor the following; if the trinomial will not factor, write "prime":

a. $2x^2 - 5x + 3$ **b.** $4x^2 + 17x - 15$ **c.** $6x^2 + x - 12$

d. $6c^2 - 13cd + 6d^2$ **e.** $2x^2 + 5x + 6$

If you need more practice, do the following Additional Practice Exercises.

Additional Practice Exercise 2   Factor the following; if the trinomial will not factor, write "prime":

a. $5x^2 - 9x + 4$ **b.** $7a^2 + 40a - 12$ **c.** $4b^2 + 5b - 6$

d. $6c^2 - 19cd + 15d^2$ **e.** $4x^2 - 7x - 3$

As in all cases of factoring, the first thing we should look for is a common factor, and if there is one, we should remove it.

Example 3   **Factor each expression. If the expression will not factor, write "prime."**

a. $18x^2 - 36x + 16 =$ *Remove the common factor, 2.*

$2(9x^2 - 18x + 8) =$ *Factor the trinomial inside the parentheses.*

$2(3x - 2)(3x - 4)$ *Factorization.*

CHECK:

$2(3x - 2)(3x - 4) = 2(9x^2 - 18x + 8)$

$= 18x^2 - 36x + 16$ *Therefore, this factorization is correct.*

Answers: Practice Exercise 2: a. $(2x - 3)(x - 1)$ **b.** $(4x - 3)(x + 5)$ **c.** $(3x - 4)(2x + 3)$ **d.** $(3c - 2d)(2c - 3d)$ **e.** prime
Additional Practice 2: a. $(5x - 4)(x - 1)$ **b.** $(7a - 2)(a + 6)$ **c.** $(4b - 3)(b + 2)$ **d.** $(3c - 5d)(2c - 3d)$ **e.** prime

b. $12a^3 + 27a^2 - 27a =$ Remove the common factor, $3a$.
   $3a(4a^2 + 9a - 9) =$ Factor the trinomial inside the parentheses.
   $3a(4a - 3)(a + 3)$ Factorization.

CHECK:

$3a(4a - 3)(a + 3) = 3a(4a^2 + 9a - 9)$
$\qquad\qquad\qquad\quad = 12a^3 + 27a^2 - 27a$ Therefore, this factorization is correct.

PRACTICE EXERCISE 3

Factor each expression completely. If the expression will not factor, write "prime."

a. $9b^2 + 30b - 24$ **b.** $12cd^2 - 38cd + 20c$

If you need more practice, do the following Additional Practice Exercises.

Additional Practice Exercise 3   Factor each expression completely. If the expression will not factor, write "prime."

a. $24a^2 + 52a + 24$ **b.** $16a^2b + 36ab - 10b$

For Extra Help

Exercise Set 5.4   MyMathLab®

Factor each expression completely. If the polynomial will not factor, write "prime." (See Examples 1–3.)

1. $2a^2 + 7a + 5$ **2.** $3a^2 - 22a + 7$ **3.** $2a^2 + 5a - 3$

4. $5b^2 + 3b - 2$ **5.** $7m^2 - 34m - 5$ **6.** $5y^2 - 2y - 3$

7. $3x^2 - 4x - 7$ **8.** $7x^2 - 20x - 3$ **9.** $5c^2 - 2cd + 7d^2$

10. $5c^2 - 24cd + 5d^2$ **11.** $11x^2 + 16xy + 5y^2$ **12.** $3c^2 - 10cd + 7d^2$

13. $2x^2 - 13x + 10$ **14.** $3x^2 - 11x + 15$ **15.** $5x^2 + 6x - 8$

16. $7a^2 - 9a - 10$ **17.** $4x^2 + 8x + 3$ **18.** $6x^2 + 13x + 5$

19. $10x^2 - 33x - 7$ **20.** $21x^2 + 32x - 5$ **21.** $2x^2 + 11x + 12$

22. $2b^2 + 13b - 24$ **23.** $3a^2 - 14a + 16$ **24.** $2t^2 - 13t + 18$

25. $5c^2 + 11c - 12$ **26.** $3z^2 - 13z - 30$ **27.** $3x^2 + 26x + 20$

Answers: Practice Exercise 3: a. $3(3b - 2)(b + 4)$ **b.** $2c(3d - 2)(2d - 5)$ **Additional Practice 3: a.** $4(2a + 3)(3a + 2)$
b. $2b(4a - 1)(2a + 5)$

28. $6x^2 + 7x + 5$

29. $6x^2 + 17x + 5$

30. $10r^2 + 17r + 3$

31. $8x^2 - 22x + 5$

32. $8x^2 - 18x + 7$

33. $24a^2 - 14a - 3$

34. $24c^2 - 13c - 7$

35. $16n^2 - 2n - 5$

36. $18m^2 + 9m - 5$

37. $6x^2 + 19x + 15$

38. $8x^2 + 18x + 9$

39. $12c^2 - 17c + 6$

40. $8n^2 - 22n + 15$

41. $18a^2 + 9a - 35$

42. $16d^2 - 14d - 15$

43. $8w^2 - 6w - 27$

44. $15t^2 + 7t - 30$

45. $20a^2 - 9a + 20$

46. $6a^2 + 19a - 15$

47. $18y^2 + 3y - 28$

48. $12c^2 - 8c - 15$

49. $6a^2 + 29ab + 35b^2$

50. $8a^2 + 34ab + 21b^2$

51. $18c^2 - 9cd - 20d^2$

52. $24r^2 - 25rs - 25s^2$

53. $12a^2 + 7ab - 12b^2$

54. $18x^2 + 9xy - 20y^2$

55. $12f^2 - 32fg + 21g^2$

56. $16a^2 + 48ab + 35b^2$

57. $6a^2 + 3a - 18$

58. $6r^2 - 26r - 20$

59. $4x^2y - 13xy + 3y$

60. $3a^2b^2 - 10a^2b + 8a^2$

61. $6x^3 - 21x^2 + 15x$

62. $2x^3 - 14x^2 + 20x$

63. $18x^3 - 3x^2 - 36x$

If the coefficient of the squared term of a trinomial is negative, it is easier to factor if we first remove a factor of -1, as in the following:

$$-x^2 - 3x + 54 = -1(x^2 + 3x - 54) = -1(x - 6)(x + 9)$$

Factor Exercises 64–69 by first removing a factor of -1.

64. $-x^2 + 5x + 36$

65. $-x^2 + 13x - 42$

66. $-2x^2 + 13x + 45$

67. $-3x^2 + 7x + 20$

68. $-8x^2 + 22x + 21$

69. $-12x^2 - 5x + 28$

Challenge Exercises (70–82)

Factor each expression. If the polynomial will not factor, write "prime." (See Examples 1–3.)

70. $32x^2 - 134x - 45$

71. $36x^2 + 29x - 20$

72. $24x^2 - 55x - 24$

73. $(x + 2)^2 - (x + 2) - 6$

74. $(x - 3)^2 - 2(x - 3) - 8$

75. $6 + a - 2a^2$

76. $8 + 10a + 3a^2$ **77.** $12 + b - 6b^2$ **78.** $10 + 11a - 6a^2$

79. $18 + 18n - 8n^2$ **80.** $40 - 33x - 18x^2$

Determine all integer values of k so that the following will be factorable over the integers:

81. $2x^2 + kx + 5$ **82.** $3x^2 + kx + 7$

Writing Exercises (83–86)

83. How would you explain to a classmate why $x^2 + 9x + 15$ is prime?

84. Explain why $12x^2 - x + 6$ cannot be factored correctly as $(3x + 2)(4x - 3)$.

85. Explain why $6x^2 + 21x + 15$ cannot be completely factored correctly as $(3x + 3)(2x + 5)$.

86. After factoring a trinomial, the results are $(2x - 3)(5x - 7)$. What is the trinomial? Why?

Section 5.5 — Factoring Binomials

OBJECTIVES *When you complete this section, you will be able to:*

A Factor binomials that are the difference of two squares.

B Factor the difference of squares, where one or more of the terms is a binomial.

C Factor the sum and difference of cubes. (Optional)

PREREQUISITE SKILLS *Before beginning this section, you should be able to:*

a. Multiply binomials of the form $(a + b)(a - b)$. (Section 2.4)

b. Apply $(a^m)^n = a^{mn}$ and $(ab)^n = a^n b^n$. (Section 2.2)

c. Factor by removing the greatest common factor. (Section 5.2)

d. Multiply a binomial and a trinomial. (Section 2.3)

GETTING READY FOR SECTION 5.5

Find the following product:

1. $(x + 3)(x - 3)$ **2.** $(3x + 2y)(3x - 2y)$

Simplify the following:

3. $(x^2)^2$ **4.** $(4a^2)^2$

Factor the following:

5. $12x^2 + 27y^2$

Find the following product:

6. $(2x - 3)(4x^2 + 6x + 9)$

Introduction

Recall that the product of two binomials that differ only by the signs of the second terms always results in a binomial that is the difference of two squares (Section 2.4). For example, $(a + b)(a - b) = a^2 - b^2$. Consequently, the difference of squares, $a^2 - b^2$, factors into the sum $(a + b)$ and difference $(a - b)$ of a and b, which are the expressions being squared. Thus, in factored form, $a^2 - b^2 = (a + b)(a - b)$.

OBJECTIVE **A** Factoring binomials that are the difference of two squares

> **Procedure: Factoring the Difference of Squares**
>
> A binomial that is the difference of squares factors into two binomials. One of the factors is the sum of the expressions being squared, and the other factor is the difference of the expressions being squared. Using variables, we write this as $a^2 - b^2 = (a + b)(a - b)$. Expressions of the form $a + b$ and $a - b$ are often called **conjugate pairs.**

To assist you in recognizing perfect squares, refer to Table 1 in the Appendix. Each entry in the n^2 column is the square of the corresponding entry in the n column. Consequently, each entry in the n^2 column is a perfect square. You should be able to recognize all perfect squares up to 225 from memory.

Answers: Getting Ready: 1. $x^2 - 9$ **2.** $9x^2 - 4y^2$ **3.** x^4 **4.** $16a^4$ **5.** $3(4x^2 + 9y^2)$ **6.** $8x^3 - 27$

Example 1 | **Factor the following difference of squares:**

a. $x^2 - y^2 =$ *The first term is the square of x, and the second term is the square of y. Hence, the correct factorization is the product of the sum and difference of x and y.*

$(x + y)(x - y)$ Factorization.

CHECK:

$(x + y)(x - y) = x^2 - y^2$ Therefore, the factorization is correct.

b. $a^2 - 9 =$ a^2 is the square of a and 9 is the square of 3.

$a^2 - 3^2 =$ *Therefore, the correct factorization is the product of the sum and difference of a and 3.*

$(a + 3)(a - 3)$ Factorization.

CHECK:

$(a + 3)(a - 3) = a^2 - 9$ Therefore, the factorization is correct.

c. $y^2 - 4z^2 =$ y^2 is the square of y, and $4z^2$ is the square of 2z.

$y^2 - (2z)^2 =$ *Therefore, the correct factorization is the product of the sum and difference of y and 2z.*

$(y + 2z)(y - 2z)$ Factorization.

CHECK:

$(y + 2z)(y - 2z) = y^2 - 4z^2.$ Therefore, the factorization is correct.

d. $16m^2 - 9n^2 =$ $16m^2$ is the square of 4m and $9n^2$ is the square of 3n.

$(4m)^2 - (3n)^2 =$ *Therefore, the correct factorization is the product of the sum and difference of 4m and 3n.*

$(4m + 3n)(4m - 3n)$ Factorization.

CHECK:

$(4m + 3n)(4m - 3n) = 16m^2 - 9n^2$ Therefore, the factorization is correct.

Note

Using the commutative property of multiplication, the factors may be written in either order. So, $(x - y)(x + y)$ is also correct.

Be Careful

$x^2 + y^2$ is the sum of squares and is prime. One common error is $x^2 + y^2 = (x + y)^2$. This is incorrect, since $(x + y)^2 = x^2 + 2xy + y^2$. Another common error is $x^2 + y^2 = (x + y)(x + y)$. This is the same as $(x + y)^2$. Do not confuse $(ab)^n$ with $(a + b)^n$, since $(ab)^n = a^n b^n$, but $(a + b)^n \neq a^n + b^n$.

PRACTICE EXERCISE 1

Factor the following:

a. $a^2 - b^2$ **b.** $c^2 + 16$ **c.** $x^2 - 9y^2$ **d.** $25m^2 - 4n^2$

If you need more practice, do the following Additional Practice Exercises.

Additional Practice Exercise 1   Factor the following:

a. $p^2 - q^2$ **b.** $n^2 + 25$ **c.** $9c^2 - d^2$ **d.** $4a^2 - 9b^2$

Sometimes, binomials are the differences of squares, even though they do not appear to be. Also, any factorization is not complete until each of the factors is prime. For example, $12 = 2 \cdot 6$, but the factorization is not complete, since 6 is not prime. We will also need to recall that $(a^m)^n = a^{mn}$.

Answers: **Practice Exercise 1:** **a.** $(a - b)(a + b)$ **b.** prime **c.** $(x + 3y)(x - 3y)$ **d.** $(5m + 2n)(5m - 2n)$
Additional Practice 1: **a.** $(p + q)(p - q)$ **b.** prime **c.** $(3c + d)(3c - d)$ **d.** $(2a + 3b)(2a - 3b)$

Example 2 | **Factor the following completely:**

a. $x^4 - y^4 =$ | x^4 is the square of x^2, and y^4 is the square of y^2.

$(x^2)^2 - (y^2)^2 =$ | The factorization is the product of the sum and the difference of x^2 and y^2.

$(x^2 + y^2)(x^2 - y^2)$ | $x^2 - y^2$ is still the difference of squares and must be factored again.

$(x^2 + y^2)(x + y)(x - y)$ | Complete factorization.

CHECK:

$(x^2 + y^2)(x + y)(x - y) =$
$(x^2 + y^2)(x^2 - y^2) = x^4 - y^4$ | Therefore, the factorization is correct.

b. $a^4 + b^4 = (a^2)^2 + (b^2)^2$ | This is the sum of squares and cannot be factored. It is prime.

c. $a^4 - 81 =$ | a^4 is the square of a^2, and 81 is the square of 9.

$(a^2)^2 - 9^2 =$ | The factorization is the product of the sum and the difference of a^2 and 9.

$(a^2 + 9)(a^2 - 9) =$ | $a^2 - 9$ is also the difference of squares and must be factored.

$(a^2 + 9)(a + 3)(a - 3)$ | Complete factorization.

CHECK:

$(a^2 + 9)(a + 3)(a - 3) =$
$(a^2 + 9)(a^2 - 9) = a^4 - 81$ | Therefore, the factorization is correct.

PRACTICE EXERCISE 2

Factor each expression. If the polynomial cannot be factored, write "prime."

a. $p^4 - q^4$ **b.** $r^4 + s^4$ **c.** $x^4 - 16$

If you need more practice, do the following Additional Practice Exercises.

Additional Practice Exercise 2   Factor each expression. If the polynomial cannot be factored, write "prime."

a. $m^4 - n^4$ **b.** $p^4 + q^4$ **c.** $a^4 - 1$

OBJECTIVE **B** **Factoring the difference of squares where one or more of the terms is a binomial**

In Section 5.2, we found that we could remove a common binomial factor, as in $x(x + 2) - 3(x + 2) = (x + 2)(x - 3)$. In factoring the difference of squares, it is also possible for one or more of the expressions being squared to be a binomial. Study the following examples:

Example 3 | **Factor the following:**

a. $(a + b)^2 - 4 =$ | $(a + b)^2$ is the square of $(a + b)$, and 4 is the square of 2.

$(a + b)^2 - 2^2 =$ | The factorization is the product of the sum and difference of $(a + b)$ and 2.

$[(a + b) + 2][(a + b) - 2]$ | Remove the parentheses.

$(a + b + 2)(a + b - 2)$ | Factorization.

b. $9 - (x - y)^2 =$ | 9 is the square of 3, and $(x - y)^2$ is the square of $(x - y)$.

$3^2 - (x - y)^2 =$ | The factorization is the product of the sum and difference of 3 and $(x - y)$.

$[3 + (x - y)][3 - (x - y)]$ | Remove the parentheses.

$(3 + x - y)(3 - x + y)$ | Factorization.

Answers: Practice Exercise 2: a. $(p^2 + q^2)(p + q)(p - q)$ **b.** prime **c.** $(x^2 + 4)(x + 2)(x - 2)$
Additional Practice 2: a. $(m^2 + n^2)(m + n)(m - n)$ **b.** prime **c.** $(a^2 + 1)(a + 1)(a - 1)$

PRACTICE EXERCISE 3

Factor the following:

a. $(a + b)^2 - 25$

b. $4 - (m - n)^2$

If you need more practice, do the following Additional Practice Exercises.

Additional Practice Exercise 3   Factor the following:

a. $(p - q)^2 - 9$

b. $36 - (c + d)^2$

In some cases, there may be a common factor that we must remove before factoring the difference of squares.

Example 4 | **Factor the following completely:**

a. $18x^2 - 8y^2 =$ First, remove the common factor, 2.

$2(9x^2 - 4y^2) =$ $9x^2$ is the square of $3x$, and $4y^2$ is the square of $2y$.

$2[(3x)^2 - (2y)^2] =$ Factor the difference of squares as the product of the sum and difference of $3x$ and $2y$.

$2(3x + 2y)(3x - 2y)$ Complete factorization.

b. $25x^2 - 100 =$ First, remove the common factor, 25.

$25(x^2 - 4) =$ Factor the difference of squares.

$25(x + 2)(x - 2)$ Complete factorization.

Note

In Example 4b, $25x^2 - 100 = (5x)^2 - 10^2$, so it is the difference of squares. If we had factored as $(5x + 10)(5x - 10)$, the factorization would not have been complete, since the factors are not prime. Each has a common factor of 5. Remember, always remove the GCF first.

PRACTICE EXERCISE 4

Factor the following completely:

a. $12x^2 - 27y^2$

b. $16a^2 - 64b^2$

If you need more practice, do the following Additional Practice Exercises.

Additional Practice Exercise 4   Factor the following completely:

a. $32x^2 - 2y^2$

b. $20a^2 - 45b^2$

OBJECTIVE Factoring the sum and difference of cubes (Optional)

Before factoring the sum and difference of cubes, recall that $1^3 = 1$, $2^3 = 8$, $3^3 = 27$, $4^3 = 64$, $5^3 = 125$, etc. Consequently, 1, 8, 27, 64, 125, etc., are called **perfect cubes**, because they are the cubes of other numbers. Previously, we learned that products of the form $(a + b)(a^2 - ab + b^2) = a^3 + b^3$, which is the sum of cubes

Answers: Practice Exercise 3: a. $(a + b - 5)(a + b + 5)$ **b.** $(2 + m - n)(2 - m + n)$ **Additional Practice 3: a.** $(p - q + 3)(p - q - 3)$
b. $(6 + c + d)(6 - c - d)$ **Practice Exercise 4: a.** $3(2x + 3y)(2x - 3y)$ **b.** $16(a + 2b)(a - 2b)$ **Additional Practice 4: a.** $2(4x + y)(4x - y)$
b. $5(2a + 3b)(2a - 3b)$

(Section 2.4). We also found that $(a - b)(a^2 + ab + b^2) = a^3 - b^3$, which is the difference of cubes. From these products, we can factor the sum and difference of cubes as follows:

> ### Rule: Factoring the Sum and Difference of Cubes
> $$a^3 + b^3 = (a + b)(a^2 - ab + b^2)$$
> $$a^3 - b^3 = (a - b)(a^2 + ab + b^2)$$

Let's make some observations about the factored forms. Notice that there are a binomial factor and a trinomial factor. The first term of the binomial factor is the expression whose cube is the first term of the sum or difference of cubes. The second term of the binomial is the expression whose cube is the second term of the sum or difference of cubes. The sign of the binomial is the same as the sign of the sum or difference of cubes. The first term of the trinomial is the square of the first term of the binomial. The second term of the trinomial is the negative of the product of the two terms of the binomial. The third term of the trinomial is the square of the second term of the binomial.

Square of the first term   Square of the second term

$$a^3 + b^3 = (a + b)(a^2 - ab + b^2)$$

Negative of the product of the two terms

The preceding illustration is for the sum of cubes, but the rule is exactly the same for the difference of cubes. A list of perfect cubes can be found in the n^3 column of Table 1 in the Appendix.

Example 5 Factor the following:

a. $x^3 - 27 =$

Rewrite as $x^3 - 3^3$. Therefore, this is the difference of cubes. The first term of the binomial is the expression whose cube is x^3, so it is x. The second term of the binomial is the expression whose cube is -27. Since $(-3)^3 = -27$, the second term is -3.

$(x - 3)(\_\ \_\ \_) =$

The first term of the trinomial is the square of the first term of the binomial. Therefore, the first term of the trinomial is x^2.

$(x - 3)(x^2\ \_\ \_) =$

The second term of the trinomial is the negative of the product of the terms of the binomial, which gives $-(-3x) = 3x$.

$(x - 3)(x^2 + 3x\ \_) =$

The last term of the trinomial is the square of the second term of the binomial. Since $(-3)^2 = 9$, the last term of the trinomial is 9.

$(x - 3)(x^2 + 3x + 9)$

Factorization. Check by multiplying.

b. $8x^3 + 125 =$

Rewrite as $(2x)^3 + 5^3$. Therefore, this is the sum of cubes. The first term of the binomial is the expression whose cube is $8x^3$. Therefore, the first term is $2x$. The second term is the expression whose cube is 125. Therefore, the second term is 5. The sign is $+$, since this is the sum of cubes.

$(2x + 5)(\_\ \_\ \_) =$

The first term of the trinomial is the square of the first term of the binomial. Since $(2x)^2 = 4x^2$, the first term of the trinomial is $4x^2$.

$(2x + 5)(4x^2\ \_\ \_) =$

The second term of the trinomial is the negative of the product of the terms of the binomial. Since $-(2x)(5) = -10x$, the second term of the trinomial is $-10x$.

$(2x + 5)(4x^2 - 10x\ \_) =$

The last term of the trinomial is the square of the second term of the binomial. Since $5^2 = 25$, the last term of the trinomial is 25.

$(2x + 5)(4x^2 - 10x + 25)$

Factorization. Check by multiplying.

PRACTICE EXERCISE 5

Factor each of the following:

a. $a^3 - 8$ **b.** $27x^3 + 64$

If you need more practice, do the following Additional Practice Exercises.

Additional Practice Exercise 5   Factor the following:

a. $y^3 - 1$ **b.** $8c^3 + 27$

Exercise Set 5.5   For Extra Help   MyMathLab®

Factor each difference of squares. If the polynomial cannot be factored, write "prime." (See Example 1.)

1. $m^2 - n^2$ **2.** $r^2 - s^2$ **3.** $x^2 - 4$ **4.** $p^2 - 1$

5. $r^2 - 64$ **6.** $a^2 - 81$ **7.** $m^2 + n^2$ **8.** $r^2 + s^2$

9. $c^2 - 25d^2$ **10.** $t^2 - 25s^2$ **11.** $a^2 - 100b^2$ **12.** $y^2 - 64z^2$

13. $4x^2 - 25y^2$ **14.** $9a^2 - 16b^2$ **15.** $49p^2 - 81q^2$ **16.** $25c^2 - 9d^2$

17. $9x^2 + 4y^2$ **18.** $36a^2 + 25b^2$ **19.** $121a^2 - 49b^2$ **20.** $169f^2 - 36g^2$

Factor the following completely. (See Example 2.)

21. $a^4 - b^4$ **22.** $r^4 - t^4$ **23.** $x^4 - 1$

24. $y^4 - 81$ **25.** $16x^4 - 81$ **26.** $256r^4 - 81$

Factor the following completely. (See Example 3.)

27. $(x + y)^2 - 9$ **28.** $(p + q)^2 - 25$ **29.** $36 - (x - y)^2$

30. $49 - (r - s)^2$ **31.** $4(x - y)^2 - 9$ **32.** $16(a - b)^2 - 25$

Factor the following completely. (See Example 4.)

33. $3x^2 - 75$ **34.** $5y^2 - 20$ **35.** $16x^2 - 36y^2$ **36.** $50x^2 - 32y^2$

Answers: Practice Exercise 5: a. $(a - 2)(a^2 + 2a + 4)$ **b.** $(3x + 4)(9x^2 - 12x + 16)$
Additional Practice 5: a. $(y - 1)(y^2 + y + 1)$ **b.** $(2c + 3)(4c^2 - 6c + 9)$

37. $9x^4 - 16x^2$

38. $4c^4 - 25c^2$

39. $9x^4 - 81x^2$

40. $4y^4 - 36y^2$

41. $3x^4 - 48$

42. $2y^4 - 162$

Problems 43 and 44 require the use of the formula for the area of a rectangle, which is $A = LW$. For example, if $A = x^2 - 9$, then $A = (x + 3)(x - 3)$, so L could be $x + 3$ and W could be $x - 3$. See the figure in the margin.

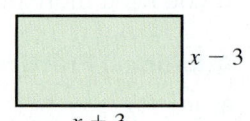

$x - 3$

$x + 3$

43. If the area of a rectangle is represented by $9x^2 - 49$, how can the length and width be represented?

44. If the area of a rectangle is represented by $16z^2 - 81$, how can the length and width be represented?

Challenge Exercises (45–56)

Factor the following completely. (See Examples 1–4.)

45. $196x^2 - 625$

46. $324a^2 - 400$

47. $256c^2 - 361d^2$

48. $441r^2 - 576s^2$

49. $338x^2 - 288$

50. $720a^2 - 125$

51. $x^8 - y^8$

52. $a^6 - b^6$

53. $81(x + 2)^2 - 16y^2$

54. $36(3x - 5)^2 - 25y^2$

55. $x^{2n} - y^{2n}$

56. $x^{4n} - 81$

Factor the following sum or difference of cubes (optional). (See Example 5.)

57. $x^3 + y^3$

58. $r^3 - s^3$

59. $c^3 + 8$

60. $d^3 + 27$

61. $n^3 - 64$

62. $r^3 - 125$

63. $m^3 + 1$

64. $b^3 - 1$

65. $8x^3 + y^3$

66. $64a^3 + b^3$

67. $27a^3 - 8$

68. $125c^3 - 64$

69. $8x^3 + 125y^3$

70. $64r^3 - 216s^3$

Writing Exercises (71–74)

71. What is wrong with this factorization? $16x^2 - 144 = (4x + 12)(4x - 12)$. What is the correct factorization?

What is wrong with these factorizations? Refactor each correctly.

72. $8x^3 + 64 = (2x + 4)(4x^2 - 8x + 16)$

73. $x^3 - 64 = (x - 4)(x^2 - 8x + 16)$

74. Given $f(x) = x^2 - 16$ and $g(x) = (x + 4)(x - 4)$,

a. Find $f(2)$ and $g(2)$.

b. What is the relationship between $f(2)$ and $g(2)$? Why?

Factoring Perfect Square Trinomials

OBJECTIVES *When you complete this section, you will be able to:*

A Recognize perfect square trinomials.

B Factor perfect square trinomials.

PREREQUISITE SKILLS *Before beginning this section, you should be able to:*

a. Square a binomial. (Section 2.4)

b. Multiply monomials. (Section 2.2)

c. Factor by removing the greatest common factor. (Section 5.2)

GETTING READY FOR SECTION 5.6

Square the following:

1. $(a - 5b)^2$

2. $(2x + 5y)^2$

Find the following product:

3. $2(3x)(4y)$

4. $2(2a)(-6)$

Factor the following:

5. $x^3y + 6x^2y + 16xy$

Introduction

Natural numbers that result from squaring integers are called **perfect squares**. The first few perfect squares are 1, 4, 9, 16, 25, 36, 49, 64, 81, 100, and so on, because $1^2 = 1, 2^2 = 4, 3^2 = 9$, and so on. For a more complete list of perfect squares, see Table 1 in the Appendix. In this section, we will be factoring **perfect square trinomials**. A perfect square trinomial results from the square of a binomial. For example, $(a + b)^2 = a^2 + 2ab + b^2$. Since $a^2 + 2ab + b^2$ is the result of squaring the binomial $a + b$, it is a perfect square trinomial.

OBJECTIVE Recognizing perfect square trinomials

How do we recognize a perfect square trinomial? The answer is in knowing the short-cut method of squaring a binomial (Section 2.4). Recall that the procedure for squaring a binomial is as follows:

1. Square the first term of the binomial.

2. Multiply the product of the two terms of the binomial by 2.

3. Square the last term of the binomial.

The procedure is diagrammed as follows for $(a + b)^2$:

First term squared   Last term squared

$$(a + b)^2 = a^2 + 2ab + b^2$$

Two times the product of the terms

For example, $(2x + 3)^2 = (2x)^2 + 2(2x)(3) + 3^2 = 4x^2 + 12x + 9$. Since $4x^2 + 12x + 9$ is the square of $2x + 3$, it is a perfect square trinomial.

Answers: Getting Ready: 1. $a^2 - 10ab + 25b^2$ **2.** $4x^2 + 20xy + 25y^2$ **3.** $24xy$ **4.** $-24a$ **5.** $xy(x^2 + 6x + 16)$

Reversing the procedure, we see that a perfect square trinomial must have the following characteristics:

> ### Rule: Characteristics of a Perfect Square Trinomial
>
> **1.** The first and last terms of the trinomial must be perfect squares.
>
> **2.** The middle term (ignore the sign for now) must be two times the product of the terms whose squares give the first and last terms of the trinomial.

From the introduction, we know that $a^2 + 2ab + b^2$ is a perfect square trinomial. The following diagram illustrates the characteristics of a perfect square trinomial, using $a^2 + 2ab + b^2$:

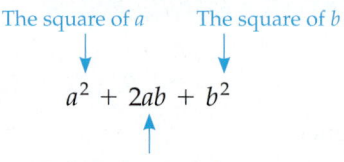

The square of a The square of b

$$a^2 + 2ab + b^2$$

Two times a times b

We illustrate with the following examples:

Example 1 | **Determine whether the following are perfect square trinomials:**

a. $x^2 - 6x + 9$ The first term is the square of x, and the last term is the square of 3.
$(x)^2 \quad 2\cdot3\cdot x \quad 3^2$ The middle term, ignoring the sign, is $2(x)(3)$. Therefore, this is a perfect square trinomial.

b. $4x^2 - 20xy + 25y^2$ The first term is the square of $2x$, and the last term is the square of $5y$.
$(2x)^2 \quad 2\cdot2x\cdot5y \quad (5y)^2$ The middle term, ignoring the sign, is $2(2x)(5y)$. Therefore, this is a perfect square trinomial.

c. $9a^2 - 12ay + 8y^2$ The last term, $8y^2$, is not the square of a term. Therefore, this is not a perfect square trinomial.

d. $9p^2 + 12p + 16$ The first term is the square of $3p$ and the last term is the square of 4. However, the middle term is not equal to $2(3p)(4)$. Therefore, this is not a perfect square trinomial.

PRACTICE EXERCISE 1

Determine whether the following are perfect square trinomials:

a. $y^2 - 8y + 16$ **b.** $16a^2 + 24ab + 9b^2$

c. $25x^2 + 20x + 32$ **d.** $9a^2 - 15a + 25$

If you need more practice, do the following Additional Practice Exercises.

Additional Practice Exercise 1   Determine whether the following are perfect square trinomials:

a. $y^2 - 16y + 64$ **b.** $9m^2 + 12mn + 4n^2$

c. $16x^2 + 24xy + 8y^2$ **d.** $25a^2 - 20a + 16$

Answers: **Practice Exercise 1:** **a.** yes **b.** yes **c.** no **d.** no   **Additional Practice 1:** **a.** yes **b.** yes **c.** no **d.** no

OBJECTIVE B Factoring perfect square trinomials

Now that we can recognize a perfect square trinomial, how do we find the binomial whose square results in the perfect square trinomial? Again, the answer lies in the procedure for squaring the binomial. In the introduction, we found that $4x^2 + 12x + 9$ is the perfect square trinomial resulting from squaring $2x + 3$. Notice that the first term of the binomial, $2x$, is the term whose square gives the first term of the trinomial, $4x^2$. The second term of the binomial, 3, is the term whose square gives the last term of the trinomial, 9. The sign between the terms of the binomial is the same as the sign of the second term of the perfect square trinomial. Based on the preceding, the procedure for factoring a perfect square trinomial is as follows:

> **Procedure: Factoring a Perfect Square Trinomial**
>
> 1. Determine whether the trinomial is a perfect square trinomial.
>
> 2. If it is, write a binomial squared. The first term of the binomial is the term whose square gives the first term of the trinomial. The second term of the binomial is the term whose square gives the last term of the trinomial.
>
> 3. The sign between the terms of the binomial is the same as the sign of the middle term of the trinomial.

From previous examples, we know that $a^2 + 2ab + b^2 = (a + b)^2$. So we will illustrate the procedure, using $a^2 + 2ab + b^2$:

$$a^2 + 2ab + b^2 = (a \quad)$$

$(a)^2$ First term is a.

$$a^2 + 2ab + b^2 = (a \quad b)$$

$(b)^2$ Second term is b.

$$a^2 + 2ab + b^2 = (a + b)$$

These two signs are the same.

Therefore, $a^2 + 2ab + b^2 = (a + b)^2$.

Example 2   **Determine whether each of the expressions is a perfect square trinomial. If it is, factor it as the square of a binomial.**

a. $x^2 - 6x + 9 =$ From Example 1a, we know that this is a perfect square trinomial.

$(x \quad _)^2 \quad = $ x^2 is the square of x, so the first term of the binomial is x.

$(x \quad 3)^2 \quad = $ 9 is the square of 3, so the second term of the binomial is 3.

$(x - 3)^2$ The sign of the middle term of the trinomial is $-$, so the sign of the binomial is $-$.

CHECK:

$(x - 3)^2 = (x)^2 - 2(x)(3) + 3^2 = x^2 - 6x + 9$

Therefore, the factorization is correct.

b. $x^2 - 10x + 25 =$ The first term is the square of x, the last term is the square of 5, and the middle term is $2(x)(5)$. Therefore, it is a perfect square trinomial.

$(x \quad _)^2 \quad = $ x^2 is the square of x, so the first term of the binomial is x.

$(x \quad 5)^2 \quad = $ 25 is the square of 5, so the second term of the binomial is 5.

$(x - 5)^2$ The sign of the middle term is $-$, so the sign of the binomial is $-$.

CHECK:

$(x - 5)^2 = (x)^2 + 2(x)(-5) + (-5)^2$
$= x^2 - 10x + 25$

Therefore, the factorization is correct.

c. $4x^2 + 20xy + 25y^2 =$ From Example 1b, we know that this is a perfect square trinomial.

$(2x\ \_\ )^2 \qquad =$ $4x^2$ is the square of $2x$, so the first term of the binomial is $2x$.

$(2x\ \ 5y)^2 \qquad =$ $25y^2$ is the square of $5y$, so the second term of the binomial is $5y$.

$(2x + 5y)^2$ The sign of the middle term is $+$, so the sign of the binomial is $+$.

CHECK:

$(2x + 5y)^2 = (2x)^2 + 2(2x)(5y) + (5y)^2$
$\qquad\qquad = 4x^2 + 20xy + 25y^2$

   Therefore, the factorization is correct.

d. $9p^2 + 24p + 16 =$ The first term is the square of $3p$, the last term is the square of 4, and the middle term is $2(3p)(4)$. Therefore, this is a perfect square trinomial.

$(3p + 4)^2$ $9p^2$ is the square of $3p$, 16 is the square of 4, and the sign of the middle term is $+$. So, $9p^2 + 24p + 16 = (3p + 4)^2$.

CHECK:

The check is left as an exercise for the student.

PRACTICE EXERCISE 2

Determine whether each of the expressions is a perfect square trinomial. If it is, factor it as the square of a binomial.

a. $y^2 - 8y + 16$ **b.** $16a^2 + 24ab + 9b^2$ **c.** $a^2 - 16a + 64$ **d.** $4x^2 + 20xy + 25y^2$

If you need more practice, do the following Additional Practice Exercises.

Additional Practice Exercise 2   Factor the following if it is a perfect square trinomial:

a. $a^2 - 4a + 4$ **b.** $25a^2 + 30ab + 9b^2$ **c.** $9a^2 - 6ab + b^2$ **d.** $9m^2 - 30mn + 25n^2$

As in the previous section, if there is a GCF, remove it and then see if the resulting polynomial can be factored further.

Example 3   **Factor the following:**

a. $3x^2 - 30x + 75 =$ First, remove the GCF, 3.

$3(x^2 - 10x + 25) =$ The resulting trinomial is a perfect square.

$3(x - 5)^2$ The first term is the square of x, the last term is the square of 5, and the sign of the second term is $-$. Check by multiplying.

b. $x^3y + 8x^2y + 16xy =$ Remove the GCF, xy.

$xy(x^2 + 8x + 16) =$ The resulting trinomial is a perfect square.

$xy(x + 4)^2$ x^2 is the square of x, 16 is the square of 4, and the sign of the second term is $+$. Check by multiplying.

PRACTICE EXERCISE 3

Factor the following:

a. $5x^2 - 60x + 180$ **b.** $x^4y^2 - 14x^3y^2 + 49x^2y^2$

If you need more practice, do the following Additional Practice Exercises.

Additional Practice Exercise 3   Factor the following:

a. $8x^2 - 16x + 8$ **b.** $x^3 - 18x^2 + 81x$

Answers: Practice Exercise 2: a. $(y - 4)^2$ **b.** $(4a + 3b)^2$ **c.** $(a - 8)^2$ **d.** $(2x + 5y)^2$ **Additional Practice 2: a.** $(a - 2)^2$ **b.** $(5a + 3b)^2$ **c.** $(3a - b)^2$ **d.** $(3m - 5n)^2$ **Practice Exercise 3: a.** $5(x - 6)^2$ **b.** $x^2y^2(x - 7)^2$ **Additional Practice 3: a.** $8(x - 1)^2$ **b.** $x(x - 9)^2$

Exercise Set 5.6

For Extra Help

MyMathLab®

Factor each expression, if the trinomial is a perfect square. If it is not a perfect square, write "not a perfect square."
(See Examples 2 and 3.)

1. $x^2 + 4x + 4$

2. $a^2 - 6a + 9$

3. $x^2 + 2x + 1$

4. $b^2 - 2b + 1$

5. $x^2 + 12x + 36$

6. $c^2 - 14c + 49$

7. $y^2 + 20y + 100$

8. $r^2 + 22r + 121$

9. $25x^2 - 10x + 1$

10. $36x^2 + 12x + 1$

11. $a^2 - 2ab + b^2$

12. $c^2 + 2cd + d^2$

13. $4a^2 + 4ab + b^2$

14. $16a^2 - 8ab + b^2$

15. $16c^2 - 24c + 9$

16. $25x^2 + 20x + 4$

17. $9x^2 + 30x + 25$

18. $4x^2 + 28x + 49$

19. $16x^2 - 8xy + y^2$

20. $49x^2 + 14xy + y^2$

21. $25x^2 + 40xy + 16y^2$

22. $36x^2 - 84xy + 49y^2$

23. $4a^2 - 14ab + 49b^2$

24. $25c^2 + 35cd + 49d^2$

25. $8x^2 - 24x + 18$

26. $27x^2 - 36x + 12$

27. $2x^2 + 28x + 98$

28. $3x^2 - 54x + 243$

29. $4x^2 + 80x + 400$

30. $5x^2 - 60x + 180$

31. $25x^4 - 30x^3 + 9x^2$

32. $4x^3y + 36x^2y + 81xy$

33. $50x^3y + 80x^2y^2 + 32xy^3$

34. $81x^4 - 36x^3y + 4x^2y^2$

Problems 35 and 36 require the use of the formula for the area of a square,
which is $A = s^2$, where s is the length of each side of the square. For example, if
$A = x^2 + 2x + 1$, then $A = (x + 1)^2$. This means that each side of the square is
$(x + 1)$. See the figure to the right.

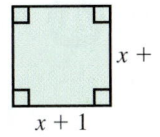

$x + 1$

$x + 1$

35. If $x^2 - 14x + 49$ represents the area of a square, what is the length of each side?

36. If $4a^2 + 36a + 81$ represents the area of a square, what is the length of each side?

Challenge Exercises (37–40)

Factor the following. (See Examples 2 and 3.)

37. $81x^2 - 90xy + 25y^2$

38. $121x^2 - 286xy + 169y^2$

39. $100a^4 + 300a^2b^2 + 225b^4$

40. $144x^4 + 192x^2y^2 + 64y^4$

Find the value of k so that each of the following will be a perfect square trinomial:

41. $x^2 + 14x + k$

42. $4x^2 + kx + 25$

43. $kx^2 + 40x + 16$

44. $9a^2 + 30ab + kb^2$

Writing Exercises (45 and 46)

45. If the middle term in a perfect square trinomial is negative, why is the middle term of the binomial whose square equals the trinomial also negative?

46. You are given $f(x) = 4x^2 + 12x + 9$ and $g(x) = (2x + 3)^2$.

    a. Find $f(1)$ and $g(1)$.

    b. What is the relationship between $f(1)$ and $g(1)$? Why?

Section 5.7 Mixed Factoring

OBJECTIVE *When you complete this section, you will be able to:*

Ⓐ Factor all types of polynomials.

PREREQUISITE SKILL *Before beginning this section, you should be able to:*

a. Factor polynomials. (Sections 5.2–5.6)

GETTING READY FOR SECTION 5.7

Factor the following completely:

1. $x^3 - 12x^2 + 36x$　　　　**2.** $16y^2 - 121$

3. $10t^2 - 14t - 12$　　　　**4.** $ax + bx - ay - by$

Introduction

In Sections 5.2–5.6, we concentrated on a particular type of factoring in each section. This made matters somewhat easier, since we knew which form to look for. When the different types of factoring are mixed, we must first recognize the form of the polynomial in order to know which procedure to use. The following checklist is recommended:

OBJECTIVE Factoring all types of polynomials

> **Procedure: Procedure for Complete Factoring**
>
> **A.** If there is a greatest common factor other than 1, remove it.
> **B.** How many terms does the polynomial have?
>　**1.** If the polynomial has two terms, follow these steps:
>　　**a.** Is it the difference of squares? If so, factor by using
>　　　$a^2 - b^2 = (a + b)(a - b)$.
>　　**b.** Is it the sum or difference of cubes? If so, factor by using
>　　　$a^3 + b^3 = (a + b)(a^2 - ab + b^2)$ or $a^3 - b^3 = (a - b)(a^2 + ab + b^2)$.
>　**2.** If the polynomial has three terms, follow these steps:
>　　**a.** If necessary, put the trinomial in descending order.
>　　**b.** Is it a perfect square trinomial? If so, factor by using
>　　　$a^2 + 2ab + b^2 = (a + b)^2$ or $a^2 - 2ab + b^2 = (a - b)^2$.
>　　**c.** If the trinomial is not a perfect square trinomial, factor by using trial and error. Check the factorization, using FOIL.
>　**3.** If the polynomial has more than three terms, try factoring by grouping.
> **C.** The factorization is not complete unless each factor is prime. Check each factor, and factor any that are not prime.

Answers: Getting Ready: **1.** $x(x - 6)^2$ **2.** $(4y + 11)(4y - 11)$ **3.** $2(5t + 3)(t - 2)$ **4.** $(a + b)(x - y)$

We will use the preceding procedure in the following examples:

Example 1 Factor each of the following completely:

a. $x^4 - 10x^3 + 25x^2$ Remove the GCF, x^2.

$x^2(x^2 - 10x + 25)$ The trinomial is a perfect square.

$x^2(x - 5)^2$ Factorization is complete.

CHECK:

$x^2(x - 5)^2 = x^2(x^2 - 10x + 25) = x^4 - 10x^3 + 25x^2$

Therefore, the factorization is correct.

b. $25x^2 - 81$ There is no common factor.

$(5x)^2 - 9^2$ The binomial is the difference of squares.

$(5x + 9)(5x - 9)$ Each factor is prime, so the factorization is complete.

CHECK:

$(5x + 9)(5x - 9) = 25x^2 - 81$

Therefore, the factorization is correct.

c. $8x^2 + 2x - 21 =$ There is no common factor. The trinomial is not a perfect square, so factor by trial and error.

$(4x + 7)(2x - 3)$ Each factor is prime, so the factorization is complete.

CHECK:

$(4x + 7)(2x - 3) = 8x^2 - 12x + 14x - 21$
$= 8x^2 + 2x - 21$

Therefore, the factorization is correct.

d. $16a^4 - 81 =$ There is no common factor.

$(4a^2)^2 - 9^2 =$ The binomial is the difference of squares.

$(4a^2 + 9)(4a^2 - 9) =$ $4a^2 + 9$ is prime, but $4a^2 - 9$ is the difference of squares.

$(4a^2 + 9)(2a + 3)(2a - 3)$ Each factor is prime, so the factorization is complete.

CHECK:

$(4a^2 + 9)(2a + 3)(2a - 3) = (4a^2 + 9)(4a^2 - 9)$
$= 16a^4 - 81$

Therefore, the factorization is correct.

e. $12bc - 9b + 20c - 15 =$ There is no common factor. There are more than three terms, so factor by grouping. Remove the factor of $3b$ from the first two terms and 5 from the remaining terms.

$3b(4c - 3) + 5(4c - 3) =$ Remove the greatest common factor, $4c - 3$.

$(4c - 3)(3b + 5)$ Each factor is prime, so the factorization is complete.

CHECK:

$(4c - 3)(3b + 5) = 12bc + 20c - 9b - 15$

Therefore, the factorization is correct.

f. $12x^2y - 22xy - 70y =$ Remove the greatest common factor, $2y$.

$2y(6x^2 - 11x - 35) =$ The trinomial factor is not a perfect square, so factor by trial and error.

$2y(2x - 7)(3x + 5)$ Each factor is prime, so the factorization is complete.

CHECK:

$2y(2x - 7)(3x + 5) = 2y(6x^2 - 11x - 35)$
$= 12x^2y - 22xy - 70y$

Therefore, the factorization is correct.

g. $6x^2 + 4x + 5$ There is no common factor. The trinomial is not a perfect square. Attempts to factor by trial and error will show that the trinomial is prime.

PRACTICE EXERCISE 1

Factor each expression completely. If the polynomial will not factor, write "prime."

a. $4x^4 + 12x^3 + 9x^2$ **b.** $81a^2 - 49$ **c.** $9x^2 + 6x - 35$

d. $16c^4 - d^4$ **e.** $6xy - 8x - 21y + 28$ **f.** $48x^2y - 20xy - 8y$

g. $4x^2 - x + 6$

(Continued)

Answers: **Practice Exercise 1:** **a.** $x^2(2x + 3)^2$ **b.** $(9a + 7)(9a - 7)$ **c.** $(3x - 5)(3x + 7)$ **d.** $(4c^2 + d^2)(2c + d)(2c - d)$ **e.** $(3y - 4)(2x - 7)$ **f.** $4y(4x + 1)(3x - 2)$ **g.** prime

If you need more practice, do the following Additional Practice Exercises.

Additional Practice Exercise 1 Factor each expression completely. If the polynomial will not factor, write "prime."

a. $8x^2 - 40x + 50$

b. $64a^2 - 1$

c. $42r^2 + 23r - 10$

d. $m^4 - 81n^4$

e. $12ab + 16a - 21b - 28$

f. $36x^2y - 15xy - 9y$

g. $6x^2 - 11x + 10$

The ability to factor quickly and accurately is one of the most important skills in algebra. This skill will be used extensively in Chapters 6 and 7.

Exercise Set 5.7

For Extra Help

MyMathLab®

Factor each of the expressions completely. If the polynomial will not factor, write "prime." (See Example 1.)

1. $16 - 8x^2$

2. $12 - 6y^2$

3. $c^2 + 12c + 27$

4. $d^2 + 13d + 40$

5. $3a^3 - 9a^2 - 54a$

6. $2x^3 + 8x^2 - 64x$

7. $x^2 - 100$

8. $a^2 - 144$

9. $c^2 - 18c - 81$

10. $r^2 - 22r - 121$

11. $ax + bx + 2a + 2b$

12. $bx - 2x + by - 2y$

13. $6a^2 - 48a + 72$

14. $5r^2 + 75r + 180$

15. $r^2 + r - 72$

16. $b^2 - b - 56$

17. $m^3n^4 + m^2n^2$

18. $a^4b^3 + a^2b$

19. $12a^4 - 75a^2$

20. $27x^4 - 48x^2$

21. $9a^2 + 30ab + 25b^2$

22. $4p^2 - 28pq + 49q^2$

23. $16r^2 + 40r - 96$

24. $30a^2 - 87a + 30$

25. $a^2b^2 - 64$

26. $p^2q^2 - 81$

27. $5n^2 + 22n + 8$

28. $3m^2 + 16m + 16$

29. $8c^2 + 4cd + d^2$

30. $10b^2 - 5bc - c^2$

31. $x(y + z) + w(y + z)$

32. $r(s - 3) - t(s - 3)$

33. $(a - b)^2 - 36$

34. $(r - s)^2 - 169$

35. $10xy - 4y - 15x + 6$

36. $10xy + 15y - 8x - 12$

37. $a^4 - 4b^2$

38. $25y^4 - 16z^2$

39. $x^2 + 4$

40. $a^2 + 25$

41. $16b^4 - 1$

42. $n^4 - 625$

Answers: Additional Practice 1: **a.** $2(2x - 5)^2$ **b.** $(8a + 1)(8a - 1)$ **c.** $(7r - 2)(6r + 5)$
d. $(m^2 + 9n^2)(m + 3n)(m - 3n)$ **e.** $(4a - 7)(3b + 4)$ **f.** $3y(3x + 1)(4x - 3)$ **g.** prime

43. $6x^2 + 31x + 5$

44. $2d^2 + d - 10$

45. $6x^2 + 9xz - 2xy - 3yz$

46. $6a^2 + 4ac - 9ab - 6bc$

47. $ab + 3a - b - 3$

48. $xy - xz - y + z$

49. $16u^3 + 46u^2v - 6uv^2$

50. $18y^3 + 24y^2z - 10yz^2$

51. $2a^2b - 5a^2b^2 + 7ab^2$

52. $x^3y - 3x^2y^2 + 5xy^3$

53. $8c^2 + 33cd + 4d^2$

54. $7r^2 - 50rs + 7s^2$

55. $9ax + 3bx - 9a - 3b$

56. $10mp - 10m + 2np - 2n$

57. $6x^2 - x - 15$

58. $8x^2 - 14x - 15$

59. $12ac - 9ad + 8bc - 6bd$

60. $8mx - 10nx - 12my + 15ny$

61. $36x^2 - 144y^2$

62. $144x^2 - 9y^2$

63. $x^2 + 2x - 120$

64. $x^2 + 18x + 80$

65. $36a^4b^3 - 39a^3b^4 - 12a^2b^5$

66. $20x^5y - 26x^4y^2 - 18x^3y^3$

Factor each of the following completely (optional):

67. $8z^3 + 125$

68. $64c^3 + 27$

69. $27c^3 - 8d^3$

70. $125t^3 - 64d^3$

71. $24c^3 - 3d^3$

72. $4t^3 + 500$

Challenge Exercises 73–81 (These do not involve the optional type of factoring):

73. $12x^2 - 23x - 24$

74. $20x^2 - 39x + 18$

75. $81a^4 - 625b^4$

76. $256r^4 - 81s^4$

77. $x^2 + 10x + 25 - y^2$

78. $a^2 - 12ab + 36b^2 - 9$

79. $x^4 - 13x^2 + 36$

80. $a^4 - 29a^2 + 100$

81. $x^2y - 4y - 3x + 6$

Writing Exercise

82. Why is it important to first remove any GCF when factoring?

Group Project

83. For each of the following techniques, write two polynomials that would be factored by using that technique:

   a. Removing the GCF.

   b. Difference of squares.

   c. Sum or difference of cubes.

   d. Perfect square trinomial.

   e. Trial and error, using FOIL.

   f. Grouping.

Solving Quadratic Equations by Factoring

OBJECTIVES *When you complete this section, you will be able to:*

A Solve quadratic equations by factoring.

B Solve application problems by using quadratic equations.

C Evaluate quadratic functions.

PREREQUISITE SKILLS *Before beginning this section, you should be able to:*

a. Factor polynomials. (Sections 5.2–5.6)

b. Solve linear equations of the form $ax + b = 0$. (Section 3.3)

c. Know the formulas for the areas of a rectangle and a triangle. (Section 1.3)

d. Know how to represent consecutive, consecutive odd, and consecutive even integers. (Section 3.6)

e. Substitute a value for a variable and evaluate by using the order of operations. (Section 2.5)

GETTING READY FOR SECTION 5.8

Factor the following completely:

1. $z^2 - 8z$ **2.** $6x^2 - 11x - 10$ **3.** $16t^2 - 80t + 96$

Solve the following equations:

1. $x + 3 = 0$ **2.** $2t + 3 = 0$

Introduction

We learned how to solve linear equations in Chapter 3 by using the addition and multiplication properties of equality. In this section, we will learn to solve a different type of equation called a **quadratic equation**. A quadratic equation is any equation of the form $ax^2 + bx + c = 0$, with $a \neq 0$, or any equation that can be put into that form. Solving quadratic equations involves the techniques of solving linear equations, combined with one additional property, the zero product property.

> **Rule: Zero Product Property**
>
> If $ab = 0$, then $a = 0$, or $b = 0$, or $a = b = 0$. In words, if the product of two factors is zero, then one or both of the factors is zero.

This property is easily extended to three or more factors. For example, if $abc = 0$, then at least one of the following is true: $a = 0$, $b = 0$, or $c = 0$. When this property is applied to the solutions of equations, it means "Find the value(s) of the variable that makes each factor equal to 0. For example, if $(x + 3)(x - 2) = 0$, then either $x + 3 = 0$ or $x - 2 = 0$, since one of the factors must be 0.

Example 1 | **Solve the following, using the zero product property:**

a. $x(3x + 2) = 0$ Set each factor equal to 0.

$x = 0$ or $3x + 2 = 0$ Solve the linear equation.

$3x = -2$

$x = -\dfrac{2}{3}$ Therefore, the solutions are 0 and $-\dfrac{2}{3}$.

Answers: Getting Ready: 1. $z(z - 8)$ **2.** $(2x - 5)(3x + 2)$ **3.** $16(t - 2)(t - 3)$ **4.** $x = -3$ **5.** $t = -\dfrac{3}{2}$

b. $(a + 3)(a - 1) = 0$ Set each factor equal to 0.

$a + 3 = 0$ or $a - 1 = 0$ Solve each linear equation.

$a = -3$ $a = 1$ Therefore, the solutions are -3 and 1.

c. $(b - 4)(2b + 5)(3b - 4) = 0$ Set each factor equal to 0.

$b - 4 = 0$ or $2b + 5 = 0$ or $3b - 4 = 0$ Solve each linear equation.

$b = 4,$ $2b = -5$ $3b = 4$

$$b = -\frac{5}{2} \qquad b = \frac{4}{3}$$ The solutions are 4, $-\frac{5}{2}$, and $\frac{4}{3}$.

PRACTICE EXERCISE 1

Solve the following:

a. $b(3b + 2) = 0$ **b.** $(n + 4)(3n - 5) = 0$ **c.** $(y - 3)(7y + 2)(5y - 3) = 0$

If you need more practice, do the following Additional Practice Exercises.

Additional Practice Exercise 1   Solve the following:

a. $w(4w - 3) = 0$ **b.** $(3x + 7)(x - 4) = 0$ **c.** $(2y - 1)(y + 3)(3y + 5) = 0$

Suppose we were to multiply the factors of Example 1b. We get

$$(a + 3)(a - 1) = 0$$
$$a^2 + 2a - 3 = 0$$

The result is a quadratic equation for which $(a + 3)(a - 1)$ is the factored form. Therefore, factoring and the zero product property give us a method for solving *some* quadratic equations. Note that this method will not work for quadratics which will not factor.

OBJECTIVE Solving quadratic equations by factoring

Procedure: Solving Quadratic Equations

1. If necessary, rewrite the equation in the form $ax^2 + bx + c = 0$ or $0 = ax^2 + bx + c$.

2. Factor $ax^2 + bx + c$.

3. Set each factor equal to 0. (zero product property)

4. Solve the resulting linear equations.

5. Check the solutions.

Answers: Practice Exercise 1: a. $b = 0, -\frac{2}{3}$ **b.** $n = -4, \frac{5}{3}$ **c.** $y = 3, -\frac{2}{7}, \frac{3}{5}$ **Additional Practice 1: a.** $w = 0, \frac{3}{4}$ **b.** $x = -\frac{7}{3}, 4$

c. $y = \frac{1}{2}, -3, -\frac{5}{3}$

Example 2   Solve the following:

a. $2x^2 + 6x = 0$ Factor the left side of the equation.

$2x(x + 3) = 0$ Set each factor equal to 0.

$2x = 0, x + 3 = 0$ Solve the linear equations.

$x = 0 \qquad x = -3$ Therefore, the solutions are 0 and -3.

CHECK:

$x = 0 \qquad x = -3$

Substitute 0 for x. $\qquad$ Substitute -3 for x.

$2(0)^2 + 6(0) = 0 \qquad 2(-3)^2 + 6(-3) = 0$

$2 \cdot 0 + 0 = 0 \qquad\quad 2(9) - 18 = 0$

$0 + 0 = 0 \qquad\qquad 18 - 18 = 0$

$0 = 0 \qquad\qquad\quad 0 = 0$

Therefore, 0 and -3 are the correct solutions.

c. $c^2 - 16 = 0$ Factor the left side of the equation.

$(c + 4)(c - 4) = 0$ Set each factor equal to 0.

$c + 4 = 0, c - 4 = 0$ Solve each linear equation.

$c = -4 \qquad c = 4$ Therefore, the solutions are 4 and -4.

CHECK:

The check is left as an exercise for the student.

e. $x(x + 10) = 24$ To put the equation in the form $ax^2 + bx + c = 0$, simplify the left side of the equation.

$x^2 + 10x = 24$ Subtract 24 from both sides.

$x^2 + 10x - 24 = 0$ Factor.

$(x + 12)(x - 2) = 0$ Set each factor equal to 0.

$x + 12 = 0, x - 2 = 0$ Solve each equation.

$x = -12 \qquad x = 2$ Therefore, the solutions are -12 and 2.

CHECK:

The check is left as an exercise for the student.

b. $x^2 - 2x - 15 = 0$ Factor the left side of the equation.

$(x - 5)(x + 3) = 0$ Set each factor equal to 0.

$x - 5 = 0, x + 3 = 0$ Solve each linear equation.

$x = 5 \qquad x = -3$ Therefore, the solutions are 5 and -3.

CHECK:

$x = 5 \qquad x = -3$

Substitute 5 for x. $\qquad$ Substitute -3 for x.

$5^2 - 2(5) - 15 = 0 \qquad (-3)^2 - 2(-3) - 15 = 0$

$25 - 10 - 15 = 0 \qquad\quad 9 + 6 - 15 = 0$

$0 = 0 \qquad\qquad\qquad 0 = 0$

Therefore, 5 and -3 are the correct solutions.

d. $6x^2 - 10 = 11x$ The equation must first be written in the form of $ax^2 + bx + c = 0$. Subtract $11x$ from both sides and write the left side in descending order.

$6x^2 - 11x - 10 = 0$ Factor the left side of the equation.

$(2x - 5)(3x + 2) = 0$ Set each factor equal to 0.

$2x - 5 = 0, 3x + 2 = 0$ Solve each linear equation.

$2x = 5 \qquad 3x = -2$

$x = \dfrac{5}{2} \qquad x = -\dfrac{2}{3}$ Therefore, the solutions are $\dfrac{5}{2}$ and $-\dfrac{2}{3}$.

CHECK:

The check is left as an exercise for the student.

PRACTICE EXERCISE 2

Solve the following:

a. $2r^2 - 6r = 0$ $\qquad$ **b.** $z^2 - 4z - 21 = 0$ $\qquad$ **c.** $m^2 - 36 = 0$

d. $8x^2 - 15 = 14x$ $\qquad$ **e.** $x(x + 3) = 40$

If you need more practice, do the following Additional Practice Exercises.

Additional Practice Exercise 2   Solve the following:

a. $4c^2 - 8c = 0$ $\qquad$ **b.** $a^2 + 9a + 14 = 0$ $\qquad$ **c.** $b^2 - 1 = 0$

d. $6x^2 - 3 = 7x$ $\qquad$ **e.** $x(x + 2) = 48$

Answers: Practice Exercise 2: a. $r = 0, 3$ **b.** $z = -3, 7$ **c.** $m = 6, -6$ **d.** $x = \dfrac{5}{2}, \dfrac{-3}{4}$ **e.** $x = 5, -8$ **Additional Practice 2: a.** $c = 0, 2$
b. $a = -7, -2$ **c.** $b = 1, -1$ **d.** $x = -\dfrac{1}{3}, \dfrac{3}{2}$ **e.** $x = 6, -8$

OBJECTIVE **B** Solving application problems, using quadratic equations

Application problems often result in quadratic equations. Care must be taken, since some solutions of the equations may not be realistic solutions to the original problem. For example, if x represents the length of the side of a rectangle, then x cannot be negative. However, the equation may have a negative solution. We will use the same technique developed in Chapter 3 to solve the following problems, but will not elaborate as much:

Example 3

a. Find two consecutive positive even integers whose product is 80.

Solution

Understand: We know the integers are consecutive positive even integers and their product is 80. The unknowns are the two consecutive positive even integers.

> Relationship 1: The two integers are consecutive positive even integers.
>
> Relationship 2: The product of the two consecutive positive even integers is 80.

Plan: Use relationship 1 to represent the integers in terms of a variable, and use relationship 2 to write the equation.

Translate: Using relationship 1:

> Let x represent the smaller of the two consecutive positive even integers. Then, $x + 2$ represents the larger of the two consecutive positive even integers.

Write the word equation, using relationship 2.

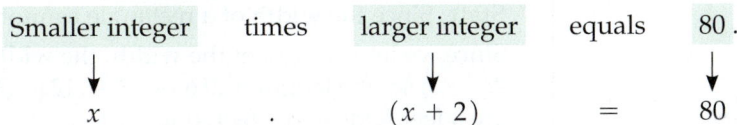

Smaller integer	times	larger integer	equals	80.
↓		↓		↓
x	$\cdot$	$(x + 2)$	$=$	80

Solve:

$x(x + 2) = 80$	Simplify the left side.
$x^2 + 2x = 80$	Subtract 80 from both sides.
$x^2 + 2x - 80 = 0$	Factor.
$(x + 10)(x - 8) = 0$	Set each factor equal to 0.
$x + 10 = 0, x - 8 = 0$	Solve each equation.
$x = -10 \qquad x = 8$	Possible solutions are -10 and 8.

State: Since we were asked for consecutive *positive* even integers, we discard -10. If 8 is the smaller of the two consecutive positive even integers, then $x + 2 = 8 + 2 = 10$ is the larger. Consequently, the integers are 8 and 10.

Check: Are 8 and 10 consecutive positive even integers? Yes. Is the product of 8 and 10 equal to 80? Yes. Then 8 and 10 are the correct solutions.

b. The length of a rectangular screen is 3 feet less than twice the width. Find the dimensions of the screen if the area is 54 square feet.

Solution

Understand: We know how the length and width are related, and we know the area. The unknowns are the length and width of the rectangle.

> Relationship 1: The length is 3 feet less than twice the width.
>
> Relationship 2: The area is 54 square feet.

Plan: Use relationship 1 to represent the length and width in terms of a variable and use relationship 2 to write the equation, using the formula for the area of a rectangle.

Translate: Using relationship 1:

Let x represent the width. Then,

$2x - 3$ represents the length. (The length is 3 less than twice the width.)

Using relationship 2 and the formula for the area of a rectangle, which is $A = LW$, we get

$A = LW$ Substitute for A, L, and W.

$54 = (2x - 3)(x)$

Solve:

$54 = (2x - 3)(x)$	Simplify the right side.
$54 = 2x^2 - 3x$	Subtract 54 from both sides.
$0 = 2x^2 - 3x - 54$	Factor.
$0 = (2x + 9)(x - 6)$	Set each factor equal to 0.
$2x + 9 = 0, x - 6 = 0$	Solve each equation.
$2x = -9, x = 6$	Solutions of the equation.

$x = -\dfrac{9}{2}$

State: Since the width of a rectangle cannot be negative, we discard $-\dfrac{9}{2}$.

Since we let x represent the width, the width is 6 feet. The length is represented by $2x - 3$; so, the length is $2(6) - 3 = 12 - 3 = 9$ feet. The dimensions of the screen are 6 feet wide and 9 feet long.

Check: Is the length, which is 9, 3 less than twice the width, which is 6? Yes. Is the area 54 square feet? Since $A = LW = 9 \cdot 6 = 54$ square feet, yes. Therefore, our solution is correct.

c. If a ball is thrown vertically upward from ground level with an initial velocity of 80 feet per second, the equation giving its height above the ground is $h = 80t - 16t^2$, where h is the height of the ball in feet and t is the number of seconds after the ball is thrown. 1. Find the number of seconds until the ball returns to the ground. 2. Find the number of seconds when the ball is 96 feet above the ground.

Solution

Understand: We know the formula for the height above the ground. In part 1, the unknown is the amount of time it takes the ball to return to the ground. In part 2, the unknown is the number of seconds until the ball is 96 feet above the ground.

Relationship: The formula for the height, h, above the ground after t seconds is
$$h = 80t - 16t^2.$$

Plan: Substitute into the formula and solve the resulting equations.

Translate:

Part 1: When the ball returns to ground level, its height above the ground is 0. So, substitute 0 for h and solve for t.

Solve:

$0 = 80t - 16t^2$ Remove the common factor of $16t$.

$0 = 16t(5 - t)$ Set each factor equal to 0.

$16t = 0, 5 - t = 0$ Solve each equation.

$t = 0, \quad 5 = t$ Solutions of the equation.

State: $t = 0$ indicates that the ball is at ground level before it is thrown. Therefore, the ball returns to ground level after 5 seconds.

Check: After 5 seconds, $h = 80(5) - 16(5^2) = 80(5) - 16(25) = 400 - 400 = 0$, which indicates that the ball is on the ground.

Translate:

Part 2: If the ball is 96 feet above the ground, $h = 96$. So, substitute 96 for h and solve for t.

Solve:

$96 = 80t - 16t^2$ Add $16t^2$ and $-80t$ to both sides.

$16t^2 - 80t + 96 = 0$ Remove the common factor of 16.

$16(t^2 - 5t + 6) = 0$ Factor $t^2 - 5t + 6$.

$16(t - 2)(t - 3) = 0$ Set $t - 2$ and $t - 3$ equal to 0. ($16 \neq 0$)

$t - 2 = 0, t - 3 = 0$ Solve each equation.

$t = 2, \quad t = 3$ Solutions of the equation.

State: Therefore, the ball is 96 feet above the ground after 2 seconds and again after 3 seconds.

Check: The check is left as an exercise for the student.

d. The formula for finding the sum of n consecutive positive integers beginning with one is $S = \dfrac{n(n + 1)}{2}$, where S is the sum and n is the number of positive integers. For example, in the sum $1 + 2 + 3 + 4 + 5 + 6 + 7 + 8 + 9 + 10, n = 10$, since we are summing the first 10 positive integers. The sum is $S = \dfrac{(10)(10 + 1)}{2} = \dfrac{(10)(11)}{2} = \dfrac{110}{2} = 55$. How many integers would it take for the sum to be 36?

Solution

Understand: We know the formula for the sum of consecutive positive integers. The unknown is the number of consecutive integers it will take for the sum to be 36.

Relationship: The formula for the sum of the first n consecutive integers is

$$S = \frac{n(n + 1)}{2}.$$

Plan: Substitute into the formula and solve.

Translate:

Since the sum is 36, substitute 36 for S and solve for n.

$36 = \dfrac{n(n + 1)}{2}$ Multiply both sides by 2.

Solve:

$$2(36) = 2 \cdot \frac{n(n+1)}{2} \quad \text{Simplify both sides.}$$

$$72 = n^2 + n \qquad\qquad \text{Subtract 72 from both sides.}$$

$$0 = n^2 + n - 72 \qquad \text{Factor.}$$

$$0 = (n+9)(n-8) \qquad \text{Set each factor equal to 0.}$$

$$n + 9 = 0, n - 8 = 0 \quad \text{Solve each equation.}$$

$$n = -9, \quad n = 8 \qquad \text{Solutions of the equation.}$$

State: Since we are looking for positive integers, the only possible answer is 8. Therefore, the sum of the first 8 positive integers is 36.

Check: Substitute 8 for n and find S. $S = \dfrac{8(8+1)}{2} = \dfrac{8(9)}{2} = \dfrac{72}{2} = 36$, which is what it is supposed to be. Therefore, our answer is correct.

PRACTICE EXERCISE 3

Solve each of the following:

a. Find two consecutive integers so that twice the square of the larger is 57 more than three times the smaller.

b. The length of a rectangular rug is 5 feet less than twice the width. If the area of the rug is 88 square feet, find the dimensions of the rug.

c. If an object is launched vertically upward from ground level with an initial velocity of 128 feet per second, the height of the object above the ground is given by $h = 128t - 16t^2$, where h is the height of the object and t is the number of seconds after it is launched. **(1)** Find how long it will take the object to return to the ground. **(2)** After how many seconds will the object be 240 feet above the ground?

d. Using the formula from Example 3d, find how many positive integers beginning with 1 it would take for the sum to be 45.

OBJECTIVE Evaluating quadratic functions

Equations of the form $ax^2 + bx + c = 0$ are called quadratic equations. Functions of the form $f(x) = ax^2 + bx + c$, accordingly, are called **quadratic functions**. Quadratic functions are evaluated in the same manner as any other functions.

Example 4 Given $f(x) = x^2 + x - 6$, find the following:

a. $f(0)$

b. $f(-3)$

Solution

To find $f(0)$, substitute 0 for x and simplify.

$$f(x) = x^2 + x - 6 \quad \text{Substitute 0 for } x.$$

$$f(0) = 0^2 + 0 - 6 \quad \text{Simplify.}$$

$$f(0) = -6$$

Therefore, $f(0) = -6$. Remember, this means that the point $(0, -6)$ lies on the graph of $y = x^2 + x - 6$.

Solution

To find $f(-3)$, substitute -3 for x and simplify.

$$f(x) = x^2 + x - 6 \qquad \text{Substitute } -3 \text{ for } x.$$

$$f(-3) = (-3)^2 + (-3) - 6 \quad \text{Raise to powers.}$$

$$f(-3) = 9 - 3 - 6 \qquad \text{Add.}$$

$$f(-3) = 0$$

Therefore, $f(-3) = 0$. This means that the point $(-3, 0)$ lies on the graph of $y = x^2 + x - 6$.

Answers: Practice Exercise 3: a. 5 and 6 **b.** $W = 8$ ft, $L = 11$ ft **c. 1.** 8 sec, **2.** 3 sec and 5 sec **d.** 9

c. $f(4)$

Solution

To find $f(4)$, substitute 4 for x and simplify.

$f(x) = x^2 + x - 6$ Substitute 4 for x.
$f(4) = 4^2 + 4 - 6$ Raise to powers.
$f(4) = 16 + 4 - 6$ Add.
$f(4) = 14$ Therefore, $f(4) = 14$, which means that the point $(4, 14)$ lies on the graph of $y = x^2 + x - 6$.

PRACTICE EXERCISE 4

Given $f(x) = x^2 + 2x - 8$, find the following:

a. $f(0)$ **b.** $f(-4)$ **c.** $f(3)$

If you need more practice, do the following Additional Practice Exercises.

Additional Practice Exercise 4 Given $f(x) = x^2 + 3x - 5$, find the following:

a. $f(0)$ **b.** $f(-2)$ **c.** $f(3)$

For Extra Help

Exercise Set 5.8   MyMathLab®

Solve the following, using the zero product property. (See Example 1.)

1. $(t - 6)(t + 4) = 0$

2. $(u + 3)(u - 8) = 0$

3. $(2v - 10)(v - 9) = 0$

4. $(w + 2)(5w + 15) = 0$

5. $r(5r - 2) = 0$

6. $s(3s - 5) = 0$

7. $6z(7z + 11) = 0$

8. $13p(8p - 15) = 0$

9. $(a - 3)(a + 2)(a - 5) = 0$

10. $(b + 4)(b - 6)(b - 1) = 0$

11. $(q + 7)(3q - 5)(4q - 1) = 0$

12. $(x - 9)(6x + 2)(13x - 5) = 0$

13. $(2y - 4)^2 = 0$

14. $(5z + 10)^2 = 0$

Solve the following. (See Example 2.)

15. $x^2 - x - 6 = 0$

16. $x^2 - x - 12 = 0$

17. $x^2 + 7x + 10 = 0$

18. $x^2 + 6x + 8 = 0$

19. $3a^2 - 11a - 4 = 0$

20. $2d^2 + 7d - 4 = 0$

21. $4m^2 - 23m + 15 = 0$

22. $2b^2 - 3b - 9 = 0$

23. $6r^2 + 11r - 10 = 0$

24. $8c^2 - 6c - 9 = 0$

25. $5s^2 - 10s = 0$

26. $21k - 7k^2 = 0$

Answers: Practice Exercise 4: a. $f(0) = -8$ **b.** $f(-4) = 0$ **c.** $f(3) = 7$ **Additional Practice 4: a.** -5 **b.** -7 **c.** 13

27. $3r^2 - 2r = 0$

28. $4h - 9h^2 = 0$

29. $v^2 - 4 = 0$

30. $u^2 - 9 = 0$

31. $3w^2 - 75 = 0$

32. $2x^2 - 72 = 0$

33. $t^2 + 12 = 7t$

34. $a^2 - 45 = 4a$

35. $3y^2 - 8 = 10y$

36. $24b^2 - 15 = -2b$

37. $c(c - 12) = -36$

38. $d(d + 10) = -25$

39. $4y(y + 5) = -25$

40. $3x(3x - 8) = -16$

41. $x^3 - 2x^2 - 24x = 0$

42. $x^3 - 3x^2 - 10x = 0$

43. $4x^3 + 18x^2 - 10x = 0$

44. $9x^3 + 21x^2 - 18x = 0$

Solve each of the following. (See Example 3.)

45. Find two consecutive positive even integers whose product is 168.

46. Find two consecutive negative odd integers whose product is 143.

47. Find two consecutive negative even integers such that the sum of their squares is 52.

48. Find two consecutive negative odd integers such that the difference of their squares is 24.

49. If the sum of an integer squared and 12 times the integer is -32, find the integer.

50. If the difference of an integer squared and twice the integer is 63, find the integer.

51. If the difference of five times the square of an integer and eight times the integer is 21, find the integer.

52. If the sum of three times the square of an integer and 13 times the integer is 30, find the integer.

53. An American flag is 4 feet longer than it is wide. Find the dimensions of the flag if the area is 32 square feet. (Recall that $A = LW$.)

54. A rectangular patio is 5 yards shorter than it is long. Find the dimensions of the patio if the area of the patio is 50 square yards. (Recall that $A = LW$.)

55. The length of the cover of a textbook is 3 inches more than its width. Find the dimensions of the cover of the book if the area of the cover is 88 square inches. (Recall that $A = LW$.)

56. The side of a rectangular box is such that the length is 7 centimeters less than twice the width. Find the dimensions of the side of the box if its area is 130 square centimeters. (Recall that $A = LW$.)

57. A search party is looking for a lost person known to be in a triangular region bounded by three highways. If one of the highways is used as the base of the region, the height to that base is 4 miles less than the base. If the area of the region is 6 square miles, find the base and height. (Recall that $A = \frac{bh}{2}$.)

58. The height of a triangular piece of plywood is 2 feet more than the base. If the area is 24 square feet, find the base and the height. (Recall that $A = \frac{bh}{2}$.)

59. A projectile is launched vertically upward from ground level with an initial velocity of 112 feet per second. The equation giving its height above the ground is $h = 112t - 16t^2$, where h is the height in feet and t is the number of seconds after it is launched. **(a)** Find the number of seconds until the projectile returns to the ground. **(b)** Find the number of seconds until the projectile is 192 feet above the ground.

60. A rock is thrown vertically upward from ground level with an initial velocity of 96 feet per second. The equation giving its height above the ground is $h = 96t - 16t^2$, where h is the height in feet and t is the number of seconds after it was thrown. **(a)** Find the number of seconds until the rock returns to the ground. **(b)** Find the number of seconds until the rock is 80 feet above the ground.

61. If a heavy object is dropped from the top of a building that is 256 feet high, the equation giving the height of the object is $h = 256 - 16t^2$, where h is the height above the ground and t is the number of seconds after the object is dropped. How long will it take the object to reach the ground?

62. If a heavy object is dropped from the top of a cliff that is 144 feet high, the equation giving the height of the object is $h = 144 - 16t^2$, where h is the height above the ground and t is the number of seconds after the object is dropped. How long will it take the object to reach the ground?

In Exercises 63–66, use the formula $S = \frac{n(n+1)}{2}$. (See Example 3d.)

63. How many positive integers, beginning with 1, will it take for the sum to be 28?

64. How many positive integers, beginning with 1, will it take for the sum to be 45?

65. Logs are stacked in a triangular-shaped pile with one log on the top row, two logs on the second row, three logs on the third row, and so on. If there is a total of 21 logs in the stack, how many logs are on the bottom row?

66. Fence posts in a lumber yard are stacked in a triangular-shaped pile with one post on the top row, two posts on the second row, three posts on the third row, and so on. If there is a total of 78 posts in the pile, find the number of posts on the bottom row.

In Exercises 67 and 68, use the formula $N = \frac{n(n-3)}{2}$, where N is the number of diagonals of a polygon whose number of vertices is n.

67. How many vertices are in a polygon with 27 diagonals?

68. How many vertices are in a polygon with 20 diagonals?

Answer the following. (See Example 4.)

69. Given $f(x) = x^2 + 2x - 8$, find the following:

    a. $f(0)$ **b.** $f(2)$ **c.** $f(-3)$

70. Given $f(x) = x^2 - 9x + 18$, find the following:

    a. $f(0)$ **b.** $f(6)$ **c.** $f(-3)$

Challenge Exercises (71–74)

Solve each of the following:

71. $x^4 - 13x^3 + 36x^2 = 0$

72. $3a^3 - 147a = 0$

73. $5t^4 + 30t^3 = 0$

74. $32x^3 - 144x^2 + 162x = 0$

Writing Exercises (75–77)

75. Compare the methods of solving linear equations and quadratic equations.

76. Suppose we are solving a quadratic equation by factoring and we arrive at $4(x - 2)(x + 5) = 0$. In order to solve this equation, we set the factors $x - 2$ and $x + 5 = 0$. Why do we not also set 4 equal to 0? What happens to the 4?

77. When solving a quadratic equation, why must one side of the equation be equal to 0?

Writing Exercises or Group Projects (78 and 79)

If these exercises are done as a group project, each group should write two exercises of each type, exchange with another group, and then solve them.

78. Write and solve an application problem involving consecutive integers.

79. Write and solve an application problem involving the area of a rectangular-shaped region.

Concept/Procedure	Example

Definitions: [Section 5.1]

- Factor: The number a is a factor of the number b if b can be written as the product of a and some other number.
- Divisible: The number b is divisible by the number a if a divides evenly into b. That is, if a is divided into b, the remainder is 0.
- Prime: A natural number is prime if it is greater than 1 and its only factors are 1 and itself. Equivalently, a natural number is prime if it is greater than 1 and divisible only by 1 and itself.
- Composite: A natural number, other than 1, is composite if it is not prime. Equivalently, a number is composite if it has at least one factor other than 1 and itself.

Finding a Prime Factorization: [Section 5.1]

1. Divide the composite number by the smallest prime number by which it is divisible.
2. Then, divide the resulting quotient by the smallest prime number by which it is divisible.
3. Continue this process until the quotient is prime.
4. The prime factorization is the product of all the divisors and the last quotient.

Example 1: Find the prime factorization of 420.

Solution: 2 is the smallest prime number that divides evenly into 420.

$$
\begin{array}{r}
7 \\
5\overline{)35} \\
3\overline{)105} \\
2\overline{)210} \\
2\overline{)420}
\end{array}
$$

Therefore, the prime factorization is $2^2 \cdot 3 \cdot 5 \cdot 7$.

Prime Factorization: [Section 5.1]

- A number is written in prime factorization form if it is written as the product of its prime factors.

Greatest Common Factor: [Section 5.1]

- The greatest common factor of two or more numbers is the largest number that is a factor of all the numbers.
- The greatest common factor of two or more monomials (other than a number) is the monomial of greatest degree that is a factor of each of the monomials.

Concept/Procedure	Example
Finding the Greatest Common Factor: [Section 5.1] 1. Write each number or monomial as the product of its prime factors. 2. Select those factors that are common in each prime factorization. 3. For each factor that is common, select the smallest exponent that occurs in any of the prime factorizations. 4. The greatest common factor is the product of each factor that is common with the smallest power appearing in any of the prime factorizations.	Example 2: Find the GCF of $24a^4b^3c^2$ and $40a^2b^4c$. Solution: The prime factorization of 24 is $2^3 \cdot 3$, and the prime factorization of 40 is $2^3 \cdot 5$. Choosing the smaller exponent on each factor that is common, the GCF is $2^3a^2b^3c$, or $8a^2b^3c$.
Removing the Greatest Common Factor: [Section 5.2] • Determine the greatest common factor of the terms of the polynomial. Use the distributive property to write the polynomial as the product of the greatest common factor and another polynomial. The terms of the polynomial within the parentheses may be determined either by inspection or by division.	Example 3: Factor $16x^2y^3 - 4x^4y^2 + 8xy^4$ by removing the greatest common factor. Solution: The GCF is $4xy^2$. $$16x^2y^3 - 4x^4y^2 + 8xy^4$$ $$= 4xy^2(4xy - x^3 + 2y^2)$$ Check by multiplying.
Factoring by Grouping: [Section 5.2] • If the polynomial has more than three terms, factor by grouping. The goal is to get the same binomial as a common factor and then remove it.	Example 4: Factor $15xz - 5xw - 3yz + yw$ by grouping. Solution: $15xz - 5xw - 3yz + yw$ Remove a common factor, $5x$, from the first two terms and $-y$ from the last two. $5x(3z - w) - y(3z - w)$ Remove the common factor of $3z - w$. $(3z - w)(5x - y)$ Factorization. Check by multiplying.
Factoring a General Trinomial: [Sections 5.3 and 5.4] 1. If the last sign of the trinomial is positive, both signs of the binomial factors will be the same as the sign of the middle term of the trinomial. The sum of the outer and inner products of the binomials will be equal to the second term of the trinomial. 2. If the last sign of the trinomial is negative, the signs of the binomial factors will be different. The difference of the outer and inner products (ignoring the signs) will equal the second term of the trinomial (ignoring the sign). Find the terms that give the correct difference, then determine the signs. 3. No binomial factor may contain a common factor.	Example 5: Factor $12x^2 - 7x - 10$, using trial and error and FOIL. Solution: The product of the first terms must be $12x^2$, the product of the last terms must be -10, and the sum of the inside and outside products must be $-7x$. $$12x^2 \qquad -10$$ $$12x^2 - 7x - 10 = (3x + 2)(4x - 5)$$ $$8x$$ $$\frac{-15x}{-7x}$$ Check by multiplying.

Concept/Procedure	Example

Factoring a Trinomial, Using the *ac* Method [Section 5.4]

- To factor a trinomial of the form $ax^2 + bx + c$ follow these steps:
  1. Find the product of a and c.
  2. Find two factors of ac whose sum is b.
  3. Rewrite bx as the sum of two terms whose coefficients are the numbers found in step 2.
  4. Factor the resulting polynomial by grouping.

Example 6: Factor $15x^2 - 19x + 6$ by using the *ac* method.

Solution:

$15x^2 - 19x + 6 =$ *$a = 15$ and $c = 6$, so $ac = 90$. Two factors of 90 whose sum is -19 are -9 and -10; so, rewrite $-19x$ as $-9x - 10x$.*

$15x^2 - 9x - 10x + 6 =$ *Factor $3x$ from the first two terms and -2 from the last two.*

$3x(5x - 3) - 2(5x - 3) =$ *Remove the GCF, $5x - 3$.*

$(5x - 3)(3x - 2)$ *Factorization. Check by multiplying.*

Difference of Squares: [Section 5.5]

- A binomial that is the difference of squares factors into the sum and difference of the expressions being squared.
$a^2 - b^2 = (a + b)(a - b)$

Example 7: Factor the $81a^2 - 49$.

Solution:

$81a^2 - 49 =$ *Rewrite as the difference of squares.*

$(9a)^2 - 7^2 =$ *Factor, using $a^2 - b^2 = (a + b)(a - b)$.*

$(9a + 7)(9a - 7)$ *Factorization. Check by multiplying.*

Factoring the Sum or Difference of Cubes: [Section 5.5: optional]

- The sum or difference of cubes factors into a binomial and a trinomial. The terms of the binomial are the expressions whose cubes give the sum or difference of cubes. The sign of the binomial is the same as the sign of the sum or difference of cubes. The first term of the trinomial is the square of the first term of the binomial. The second term of the trinomial is the negative of the product of the terms of the binomial. The third term of the trinomial is the square of the second term of the binomial.

$$a^3 + b^3 = (a + b)(a^2 - ab + b^2)$$
$$\text{and}$$
$$a^3 - b^3 = (a - b)(a^2 + ab + b^2)$$

Example 8: Factor $125x^3 + 27y^3$.

Solution:

$125x^3 + 27y^3 =$ *Rewrite as the sum of cubes.*

$(5x)^3 + (3y)^3 =$ *Factor, using $a^3 + b^3 = (a + b)(a^2 - ab + b^2)$.*

$(5x + 3y)(25x^2 - 15xy + 9y^2)$ *Factorization. Check by multiplying.*

Concept/Procedure	Example

Perfect Square Trinomial: [Section 5.6]

- A trinomial is a perfect square trinomial if the first and last terms are squares of expressions and the middle term is two times the product of the expressions being squared. The binomial whose square gives the perfect square trinomial has a first term whose square gives the first term of the trinomial, a second term whose square gives the last term of the trinomial, and the same sign as the middle term of the trinomial.

$$a^2 + 2ab + b^2 = (a + b)^2$$
and
$$a^2 - 2ab + b^2 = (a - b)^2$$

Example 9: Factor $25a^2 - 40ab + 16b^2$.

Solution:

$25a^2 - 40ab + 16b^2 =$ The first term is the square of $5a$, the last term is the square of $4b$, and the middle term is $2 \cdot 5a \cdot 4b$, so this is a perfect square trinomial.

$(5a - 4b)^2$ Factorization. Check by multiplying.

Factoring a Polynomial: [Section 5.7]

A. First, remove all common factors.

B. How many terms does the polynomial have?

  1. If the polynomial has two terms take these steps:

    a. Is it the difference of squares? If so, factor, using the form
$$a^2 - b^2 = (a + b)(a - b).$$

    b. Is it the sum or difference of cubes? If so, factor, using the form
$$a^3 + b^3 = (a + b)(a^2 - ab + b^2)$$
    or
$$a^3 - b^3 = (a - b)(a^2 + ab + b^2)$$

  2. If the polynomial has three terms take these steps:

    a. Is it a perfect square trinomial? If so, factor, using the form
$$a^2 + 2ab + b^2 = (a + b)^2$$
    or
$$a^2 - 2ab + b^2 = (a - b)^2.$$

    b. If the trinomial is not a perfect square, factor, using trial and error. Check the factorization, using FOIL.

  3. If the polynomial has more than three terms, factor by grouping.

C. The factorization is not complete unless each factor is prime. Check each factor, and if necessary, factor any that are not prime until all factors are prime.

Example 10: Factor $48x^4 - 3y^4$.

Solution:

$48x^4 - 3y^4 =$ Remove the common factor, 3.

$3(16x^4 - y^4) =$ Factor the difference of squares.

$3(4x^2 + y^2)(4x^2 - y^2) =$ Factor $4x^2 - y^2$.

$3(4x^2 + y^2)(2x + y)(2x - y)$ Factorization. Check by multiplying.

Concept/Procedure	Example

Solving Quadratic Equations: [Section 5.8]

1. The zero product property: If $ab = 0$, then $a = 0, b = 0$, or $a = b = 0$.

2. Using the zero product property to solve quadratic equations involves the following procedure:
   a. If necessary, rewrite the equation in the form
   $$ax^2 + bx + c = 0 \quad \text{or}$$
   $$0 = ax^2 + bx + c.$$
   b. Factor $ax^2 + bx + c$.
   c. Set each of the factors equal to 0.
   d. Solve the resulting equations.
   e. Check the solutions.

3. The solutions of a quadratic equation are the x-intercepts of the graph $y = ax^2 + bx + c$.

4. Solving application problems involving quadratic equations involves the following procedure:
   a. Read the problem and assign a variable to an unknown.
   b. Represent any other unknown in terms of the variable.
   c. Write the equation, using information from the problem or other known facts.
   d. Solve the equation.
   e. See if all solutions of the equation make sense when compared with the original problem.
   f. Check all solutions when compared with the original wording of the problem.

Example 11: Solve $2x^2 + 3x - 20 = 0$.

Solution:

$2x^2 + 3x - 20 = 0.$	Factor.
$(2x - 5)(x + 4) = 0$	Set each factor equal to 0.
$2x - 5 = 0, x + 4 = 0$	Solve each linear equation.
$2x = 5 \quad x = -4$	
$x = \dfrac{5}{2}$	Therefore, the solutions are $\dfrac{5}{2}$ and -4.

Check by substituting back into the original equation.

Example 12: The width of a rectangle is two less than the length, and the area is 63 square centimeters. Find the length and width.

Solution:

Understand: We know the relationship between the length and the width, and we know the area. The unknowns are the length and width.

Relationship 1: The width is two less than the length.

Relationship 2: The area is 63 square inches.

Plan: Use relationship 1 to represent the length and width in terms of a variable. Use relationship 2 and the formula for the area of a rectangle to write the equation.

Translate: Using relationship 1,
let x represent the length; then
$x - 2$ represents the width.

Use relationship 2 and the formula for the area of a rectangle to write the equation.

$A = LW$	Substitute for A, L, and W.
$63 = x(x - 2)$	

Solve:

$63 = x(x - 2)$	Distribute.
$63 = x^2 - 2x$	Subtract 63 from both sides.
$0 = x^2 - 2x - 63$	Factor.
$0 = (x - 9)(x + 7)$	Set each factor equal to 0.
$x - 9 = 0, x + 7 = 0$	Solve each equation.
$x = 9, x = -7$	Solutions of the equations.

State: Since the side of a rectangle cannot be negative, the only possible answer is 9. Since x represents the length, the length is 9 centimeters. The width is $x - 2 = 9 - 2 = 7$ centimeters.

Check: The check is left to the student.

Chapter 5   Review Exercises

Classify each of the numbers as prime or composite. [Section 5.1]

1. 87

2. 43

Determine the prime factorization of each number. [Section 5.1]

3. 60

4. 308

Find the greatest common factor of each of the pairs of numbers. [Section 5.1]

5. 126 and 210

6. 350 and 140

7. 28, 112, and 294

8. 48, 42, and 180

9. m^2n^4 and m^3n

10. a^3b^4 and b^2c^5

11. $18r^5s^3t^2$ and $24r^2s^2t^3$

12. $15a^3b^2, 10ab^3,$ and $20a^3b^2c$

Factor by removing the greatest common factor. [Section 5.2]

13. $ax + ay$

14. $6x^2 + 3$

15. $18x^2y - 24xy$

16. $r^4s^3 + r^6s^4$

17. $18x^2 - 24x + 12$

18. $8a^2b^4 - 12a^3b^5 + 2ab^3$

19. $r(s - 2) - 5(s - 2)$

20. $m(n - p) - q(n - p)$

21. Factor $-15x^3 + 5x^2 - 25x$ by removing a negative common factor.

Completely factor each polynomial by grouping. [Section 5.2]

22. $ax + bx + ay + by$

23. $rs - 8r + 3s - 24$

24. $ab - 6a - 3b + 18$

25. $xy - 7y - 4x + 28$

26. $3ab - 3a - b + 1$

27. $6x^2 + 15x - 4xy - 10y$

Completely factor each trinomial with leading coefficients of 1. [Section 5.3]

28. $x^2 + 12x + 11$

29. $x^2 - 12x + 35$

30. $x^2 + 6x - 7$

31. $m^2 - 3m - 10$

32. $x^2 - 11x + 24$

33. $r^2 + rs - 72s^2$

34. $a^2 + 25ab - 54b^2$

35. $m^2 - 19mn - 42n^2$

36. $2x^2 + 24x + 70$

37. $3y^3 + 21y^2z - 132yz^2$

Completely factor each trinomial with leading coefficient other than 1. [Section 5.4]

38. $5x^2 - 7x + 2$

39. $3z^2 - 2z - 5$

40. $5x^2 + 18x - 8$

41. $3a^2 + 26a + 16$

42. $8a^2 + 13ab - 7b^2$

43. $18a^2 + a - 5$

44. $35x^2 + 29xy + 6y^2$

45. $20a^2 + 9ab - 18b^2$

46. $20a^2 - 66a - 14$

47. $36c^3 - 51c^2d + 18cd^2$

Completely factor each difference of squares. [Section 5.5]

48. $g^2 - h^2$

49. $r^2 - 121$

50. $25x^2 - 16$

51. $49a^2 - 64b^2$

52. $6x^2 - 24$

53. $64c^2 - 16d^2$

54. $(m + n)^2 - 81$

55. $64 - (c - d)^2$

56. $r^2 + s^2$

57. $m^4 - n^4$

Completely factor each sum or difference of cubes. [Section 5.5: optional]

58. $m^3 + n^3$

59. $r^3 - 1$

60. $a^3 + 125$

61. $64x^3 + 27y^3$

Completely factor each perfect square trinomial. [Section 5.6]

62. $r^2 + 18r + 81$ **63.** $x^2 - 24x + 144$

64. $x^2 + 16x + 64$ **65.** $49a^2 - 14a + 1$

66. $49a^2 + 28ab + 4b^2$ **67.** $25c^2 + 30cd + 9d^2$

68. $18p^2 + 48p + 32$ **69.** $100a^2 - 120ab + 36b^2$

Completely factor each polynomial. The polynomials involve all types of factoring. [Section 5.7]

70. $26a^2 + 13a$ **71.** $81x^2 - y^2$

72. $2x^2 - 6x - 80$ **73.** $x^2 - 2x - 63$

74. $22a^2b - 11a^3b^2 + 33a^2b^2$ **75.** $9m^2 - 48m + 48$

76. $3xy - 27y - 4x + 36$ **77.** $4(a + b)^2 - 49$

78. $8x^2 + 28x - 56$ **79.** $49a^2 - 28ab + 4b^2$

80. $6x^2 + 21x - 45$ **81.** $6x(y - 5) + 2(y - 5)$

82. $27x^2y - 75y$ **83.** $121a^2 - 169$

84. $81x^2 - 72xy + 16y^2$ **85.** $9x^2 + 81$

86. $8x^2 + 24x - 4xy - 12y$ **87.** $r^2 + 6r - 40$

88. $x^4 + 81$ **89.** $y^4 - 625$

90. $3x^3y^2 - 12x^2y^3 - 135xy^4$

91. $8x^2 - 15x + 7$

Solve each equation. [Section 5.8]

92. $x^2 + 5x = 0$ **93.** $x^2 = 6x$

94. $4x^2 - 81 = 0$ **95.** $9x^2 = 64$

96. $x^2 - 2x - 35 = 0$ **97.** $y^2 - 4y - 21 = 0$

98. $2x^2 - 4 = 7x$ **99.** $3x^2 - 8 = -10x$

Solve each application problem. [Section 5.8]

100. The length of a rectangle is 5 less than twice the width. If the area of the rectangle is 25 square centimeters, find the dimensions of the rectangle.

101. Find two consecutive odd integers whose product is 63.

102. If 2 yards is added to the length of each side of a square, the area is increased by 28 square yards. Find the length of a side of the original square.

103. The sum of the square of an integer and three times the integer is equal to 28. Find the integer(s).

Chapter 5   Test

1. Classify each of the following as prime or composite:

    a. 111   **b.** 73

2. Find the prime factorization of 252.

Find the greatest common factor for each of the following:

    3. 72 and 96   **4.** $48x^3y^5$ and $60x^2yz^3$

Factor each polynomial. If the polynomial cannot be factored, write "prime."

    5. $6x^3 + 3x^2$ **6.** $m^2 - 18m + 81$

    7. $m^2 - 3mn - 40n^2$ **8.** $49a^2 - 9b^2$

9. $14x^3y - 28x^2y^2 + 35x^4y^3$ **10.** $9y^2 - 24y + 16$

11. $16a^2 + 25b^2$ **12.** $r(s + 5) - 7(s + 5)$

13. $(x - y)^2 - 100$ **14.** $6q^2 + 27q - 105$

15. $3a^2 - 13a - 30$ **16.** $8a^2 - 34ab + 35b^2$

17. $27x^2 - 12y^2$ **18.** $6ab - 4a - 15b + 10$

19. $5a^2 + 50ab + 125b^2$ **20.** $x^4 - 625$

21. $20a^2 - 7ab - 6b^2$

22. The area of a rectangle is represented by $x^2 - 2x - 24$. How may the length and width be represented as binomials?

Solve the following:

23. $6x^2 + 4x = 0$ **24.** $6x^2 - 19x + 10 = 0$

25. The length of a rectangle is 4 less than three times the width. Find the dimensions of the rectangle if the area is 32 square inches.

6

Multiplication and Division of Rational Numbers and Expressions

Chapter Outline

This chapter and Chapter 7 show the closest relationship between arithmetic and algebra of any chapters in the book. Polynomials are to elementary algebra as natural numbers are to arithmetic. In arithmetic, we learn to add, subtract, multiply, and divide natural numbers. Thus far, we have learned to add, subtract, and multiply polynomials. In this chapter, we will learn to divide them as well.

The next step in arithmetic is to make fractions from the natural numbers and learn to add, subtract, multiply, and divide these fractions. The corresponding step in algebra is to make rational expressions from polynomials and learn to add, subtract, multiply, and divide these rational expressions. This is what we will be doing in Chapters 6 and 7.

Section 6.1

Reducing Rational Numbers and Rational Expressions

OBJECTIVES *When you complete this section, you will be able to:*

A Graph rational numbers.

B Reduce rational numbers to lowest terms.

C Reduce rational expressions whose numerator and denominator are monomials.

PREREQUISITE SKILLS *Before beginning this chapter, you should be able to:*

a. Write a whole number in terms of its prime factorization. (Section 5.1)

b. Know the meaning of a whole-number exponent. (Section 1.1)

GETTING READY FOR SECTION 6.1

Write the following whole numbers in terms of their prime factorizations:

1. 45 **2.** 198

Write the following monomials in terms of prime factors:

3. x^2y^3 **4.** m^4n^2

Introduction

If the numerator and denominator of a fraction are both integers (and the denominator is not equal to 0), or if a number can be put into that form, the number is called a **rational number**. Examples of rational numbers are $\frac{2}{3}$, $-\frac{11}{5}$, and $8 \left(8 = \frac{8}{1} \right)$. Note that $\frac{\pi}{3}$ is not a rational number since π is not an integer.

OBJECTIVE **A** Graphing rational numbers

If the units of the number line are divided into equal parts, then we can locate a point on the number line whose distance from 0 is represented by a rational number. Remember, this procedure is called *graphing* and was first discussed in Chapter 1 when we graphed integers.

Example 1 **Graph the following rational numbers:**

a. $\dfrac{3}{4}$

b. $-\dfrac{5}{2}$

Solution

The denominator, 4, means "Divide each unit into four equal parts." Since it is positive, go to the right of 0. The numerator, 3, means "Put the dot on the third division point."

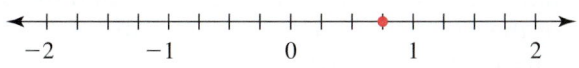

Solution

The denominator, 2, means "Divide each unit into two equal parts." The negative means "Go to the left of 0," and the numerator, 5, means "Put the dot on the fifth division point."

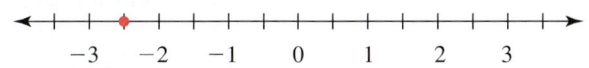

If we have several rational numbers to graph whose denominators are different, we will estimate the location of the point corresponding to each rational number.

Answers: Getting Ready: 1. $3 \cdot 3 \cdot 5$ **2.** $2 \cdot 3 \cdot 3 \cdot 11$ **3.** $x \cdot x \cdot y \cdot y \cdot y$ **4.** $m \cdot m \cdot m \cdot m \cdot n \cdot n$

Example 2 Graph the following rational numbers on the number line:

a. $\left\{-\dfrac{7}{4}, -\dfrac{4}{5}, 0, \dfrac{1}{2}, \dfrac{7}{4}, 2\right\}$

b. $\left\{-\dfrac{10}{3}, -\dfrac{5}{2}, -\dfrac{1}{4}, 1, \dfrac{5}{3}, \dfrac{13}{4}\right\}$

Solution

Since the denominators are not all the same, we will not be able to divide each unit evenly. Consequently, we will estimate the location of each point.

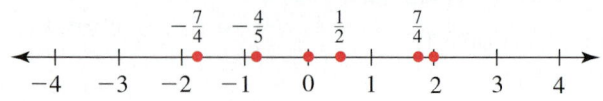

Solution

As in Example 2a, we will estimate the location of each point.

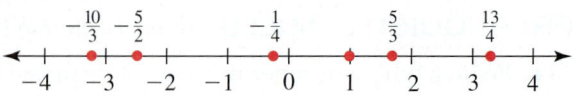

PRACTICE EXERCISE 2

Graph the following rational numbers on the number line:

a. $\left\{-\dfrac{14}{4}, -\dfrac{5}{3}, -\dfrac{1}{2}, \dfrac{3}{4}, \dfrac{11}{6}\right\}$

b. $\left\{-\dfrac{11}{3}, -\dfrac{11}{5}, -1, \dfrac{3}{8}, \dfrac{4}{3}, \dfrac{13}{4}\right\}$

In Example 1b of Section R.4, a circle is divided into four equal regions, of which two were shaded. The circle is shown here for your convenience:

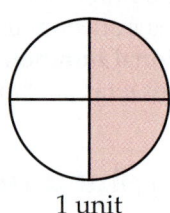

1 unit

The fraction that represents the shaded part of the circle is $\frac{2}{4}$. Notice that $\frac{2}{4}$ is also $\frac{1}{2}$ of the circle. Consequently, $\frac{2}{4}$ and $\frac{1}{2}$ represent the same part of the circle. Thus, $\frac{2}{4} = \frac{1}{2}$. Such fractions are said to be **equivalent**.

> **Definition** Equivalent Fractions
>
> Two or more fractions are **equivalent** if they represent the same quantity.

Look at the following rectangles, where each rectangle represents one unit:

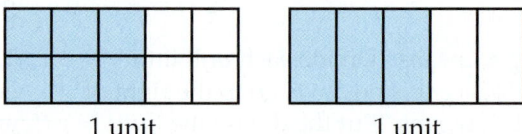

1 unit   1 unit

The fraction representing the shaded part of the first rectangle is $\frac{6}{10}$, and the fraction representing the shaded part of the second rectangle is $\frac{3}{5}$. Since the rectangles are identical in size and the exact same amount is shaded, we can see that $\frac{6}{10} = \frac{3}{5}$. In other words, $\frac{6}{10}$ and $\frac{3}{5}$ are equivalent fractions.

Answers: Practice Exercise 2:

a. 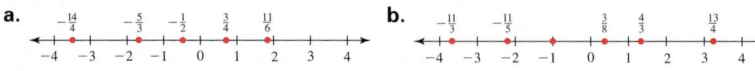   **b.**

OBJECTIVE **Reducing rational numbers to lowest terms**

In Section R.4 we reduced fractions to lowest terms by dividing the numerator and denominator by the largest number that divides evenly into both. In this section we give another procedure that is more consistent with the procedure used with algebraic fractions (see Objective C).

We can change $\frac{6}{10}$ into $\frac{3}{5}$ by the following procedure: Write the numerator and the denominator of $\frac{6}{10}$ in terms of their prime factors. Then divide the numerator and the denominator by the factors common to both. Remember that any number (except 0) divided by itself equals 1 and that the product of 1 and any number is that number. So, $\frac{6}{10} = \frac{2 \cdot 3}{2 \cdot 5} = \frac{2 \cdot 3}{2 \cdot 5} = \frac{1 \cdot 3}{1 \cdot 5} = \frac{3}{5}$. This procedure is called **reducing a fraction to lowest terms**.

> **Definition** Lowest Terms
>
> A fraction is reduced to **lowest terms** if the numerator and the denominator have no common factors other than 1.

The procedure for reducing a fraction to lowest terms was given in the previous discussion and is summarized as follows:

> **Procedure: Reducing a Fraction to Lowest Terms**
>
> 1. Write the numerator and the denominator as the product of their prime factors. (It is best not to use exponents in doing so.)
>
> 2. Divide the numerator and the denominator by the factors common to both.
>
> 3. Multiply any remaining factors.

The same procedure applies whether the fraction is proper or improper. We illustrate this procedure with some examples.

Example 3 **Reduce the following fractions to lowest terms:**

Note

Some texts refer to this procedure as "canceling." The authors of this text prefer not to use this term, since it does not adequately convey the fact that we can divide only by common factors. Using the word "canceling" could lead to serious confusion in the next section, when we reduce rational expressions, since in that process, parts of terms are often mistakenly "canceled" with parts of other terms.

a. $\dfrac{8}{12} =$ Write the numerator and the denominator as the product of prime factors.

$\dfrac{2 \cdot 2 \cdot 2}{2 \cdot 2 \cdot 3} =$ Divide the numerator and the denominator by the factors that are common.

$\dfrac{2 \cdot 2 \cdot 2}{2 \cdot 2 \cdot 3} =$ Each division leaves a quotient of 1, which is not written, since the product of 1 and any number is that number.

$\dfrac{2}{3}$ Therefore, $\frac{8}{12} = \frac{2}{3}$.

b. $\dfrac{21}{126} =$ Write the numerator and the denominator as the product of prime factors.

$\dfrac{3 \cdot 7}{2 \cdot 3 \cdot 3 \cdot 7} =$ Divide the numerator and the denominator by the factors that are common.

$\dfrac{3 \cdot 7}{2 \cdot 3 \cdot 3 \cdot 7} =$ We divide by all the factors that are in the numerator. Each time we divide, there is a quotient of 1 left. In this case, we must write the 1, since it is the only factor left in the numerator.

$\dfrac{1}{6}$ Therefore, $\frac{21}{126} = \frac{1}{6}$.

c. $\dfrac{60}{198} =$ Write the numerator and the denominator as the product of prime factors.

$\dfrac{2 \cdot 2 \cdot 3 \cdot 5}{2 \cdot 3 \cdot 3 \cdot 11} =$ Divide the numerator and the denominator by the factors that are common.

$\dfrac{2 \cdot 2 \cdot \cancel{3} \cdot 5}{2 \cdot \cancel{3} \cdot 3 \cdot 11} =$ Multiply the remaining factors.

$\dfrac{10}{33}$ Therefore, $\frac{60}{198} = \frac{10}{33}$.

d. $\dfrac{54}{45} =$ Write the numerator and the denominator as the product of prime factors.

$\dfrac{2 \cdot 3 \cdot 3 \cdot 3}{3 \cdot 3 \cdot 5} =$ Divide the numerator and the denominator by the factors that are common.

$\dfrac{2 \cdot \cancel{3} \cdot \cancel{3} \cdot 3}{\cancel{3} \cdot \cancel{3} \cdot 5} =$ Multiply the remaining factors.

$\dfrac{6}{5}$ Therefore, $\frac{54}{45} = \frac{6}{5}$.

PRACTICE EXERCISE 3

Reduce the following fractions to lowest terms:

a. $\dfrac{18}{24}$ **b.** $\dfrac{15}{60}$ **c.** $\dfrac{180}{378}$ **d.** $\dfrac{30}{12}$

If you need more practice, do the following Additional Practice Exercises.

Additional Practice Exercise 3   Reduce the following fractions to lowest terms:

a. $\dfrac{15}{40}$ **b.** $\dfrac{14}{56}$ **c.** $\dfrac{105}{315}$ **d.** $\dfrac{32}{18}$

OBJECTIVE Reducing rational expressions whose numerator and denominator are monomials

In algebra, the numerators and the denominators of fractions are polynomials. Such fractions are called **rational expressions**.

> **Definition** Rational Expression
>
> A **rational expression** is an algebraic expression of the form $\frac{P}{Q}$, where P and Q are polynomials and $Q \neq 0$.

In this section, we will limit ourselves to rational expressions whose numerators and denominators are monomials. We have already discussed this type of rational expression in Chapter 2, where we discussed the division laws of exponents. In this section, we will not be applying the laws of exponents, but will be reducing these rational expressions by the same procedure we used to reduce rational numbers. Compare the

---

Answers: Practice Exercise 3: a. $\dfrac{3}{4}$ **b.** $\dfrac{1}{4}$ **c.** $\dfrac{10}{21}$ **d.** $\dfrac{5}{2}$ **Additional Practice 3: a.** $\dfrac{3}{8}$ **b.** $\dfrac{1}{4}$ **c.** $\dfrac{1}{3}$ **d.** $\dfrac{16}{9}$

following reduction of a rational number with the reduction of a rational expression; the instructions on the right apply to both exercises:

$$\frac{45}{54} \qquad \frac{a^3b^4}{a^5b^3}$$ Write the numerator and denominator as prime factors.

$$\frac{3 \cdot 3 \cdot 5}{2 \cdot 3 \cdot 3 \cdot 3} \qquad \frac{a \cdot a \cdot a \cdot b \cdot b \cdot b \cdot b}{a \cdot a \cdot a \cdot a \cdot a \cdot b \cdot b \cdot b}$$ Divide by the common factors.

$$\frac{3 \cdot 3 \cdot 5}{2 \cdot 3 \cdot 3 \cdot 3} \qquad \frac{\cancel{a} \cdot \cancel{a} \cdot \cancel{a} \cdot \cancel{b} \cdot \cancel{b} \cdot \cancel{b} \cdot b}{\cancel{a} \cdot \cancel{a} \cdot \cancel{a} \cdot a \cdot a \cdot \cancel{b} \cdot \cancel{b} \cdot \cancel{b}}$$ Multiply the remaining factors.

$$\frac{5}{6} \qquad \frac{b}{a^2}$$ Reduced forms.

Example 4 Reduce the following rational expressions to lowest terms:

a. $\dfrac{x^2y^4}{x^5y^2} =$ Write the numerator and the denominator in terms of prime factors.

$$\frac{x \cdot x \cdot y \cdot y \cdot y \cdot y}{x \cdot x \cdot x \cdot x \cdot x \cdot y \cdot y} =$$ Divide by factors common to the numerator and the denominator.

$$\frac{\cancel{x} \cdot \cancel{x} \cdot \cancel{y} \cdot \cancel{y} \cdot y \cdot y}{\cancel{x} \cdot \cancel{x} \cdot x \cdot x \cdot x \cdot \cancel{y} \cdot \cancel{y}} =$$ Multiply the remaining factors.

$$\frac{y^2}{x^3}$$ Therefore, $\frac{x^2y^4}{x^5y^2} = \frac{y^2}{x^3}$.

b. $\dfrac{8x^2y^3}{12x^4y} =$ Write the numerator and the denominator in terms of prime factors.

$$\frac{2 \cdot 2 \cdot 2 \cdot x \cdot x \cdot y \cdot y \cdot y}{2 \cdot 2 \cdot 3 \cdot x \cdot x \cdot x \cdot x \cdot y} =$$ Divide by factors common to the numerator and the denominator.

$$\frac{2 \cdot 2 \cdot 2 \cdot \cancel{x} \cdot \cancel{x} \cdot \cancel{y} \cdot y \cdot y}{2 \cdot 2 \cdot 3 \cdot \cancel{x} \cdot \cancel{x} \cdot x \cdot x \cdot \cancel{y}} =$$ Multiply the remaining factors.

$$\frac{2y^2}{3x^2}$$ Therefore, $\frac{8x^2y^3}{12x^4y} = \frac{2y^2}{3x^2}$.

c. $\dfrac{20a^3b^2}{24ab} =$ Write the numerator and the denominator in terms of prime factors.

$$\frac{2 \cdot 2 \cdot 5 \cdot a \cdot a \cdot a \cdot b \cdot b}{2 \cdot 2 \cdot 2 \cdot 3 \cdot a \cdot b} =$$ Divide by factors common to the numerator and the denominator.

$$\frac{2 \cdot 2 \cdot 5 \cdot \cancel{a} \cdot a \cdot a \cdot \cancel{b} \cdot b}{2 \cdot 2 \cdot 2 \cdot 3 \cdot \cancel{a} \cdot \cancel{b}} =$$ Multiply the remaining factors.

$$\frac{5a^2b}{6}$$ Therefore, $\frac{20a^3b^2}{24ab} = \frac{5a^2b}{6}$.

NOTE Many times, reducing these types of rational expressions is easier than reducing arithmetic fractions because the factorization is usually easier.

PRACTICE EXERCISE 4

Reduce the following rational expressions to lowest terms:

a. $\dfrac{m^3n^6}{m^4n^2}$ **b.** $\dfrac{18x^4y^2}{21x^2y^3}$ **c.** $\dfrac{-27a^5b^2c^3}{36a^2bc^4}$

If you need more practice, do the following Additional Practice Exercises.

Additional Practice Exercise 4 Reduce the following rational expressions to lowest terms:

a. $\dfrac{a^2b^5}{a^4b}$ **b.** $\dfrac{14cd^3}{28c^3d^2}$ **c.** $\dfrac{12r^4s^4w^5}{-18r^2s^6w^3}$

The procedure of dividing the numerator and the denominator by factors common to both will be used often in the remainder of this chapter. Consequently, it is very important that you understand this procedure.

Answers: Practice Exercise 4: a. $\dfrac{n^4}{m}$ **b.** $\dfrac{6x^2}{7y}$ **c.** $-\dfrac{3a^3b}{4c}$ **Additional Practice 4: a.** $\dfrac{b^4}{a^2}$ **b.** $\dfrac{d}{2c^2}$ **c.** $-\dfrac{2r^2w^2}{3s^2}$

Exercise Set 6.1

For Extra Help

MyMathLab®

Graph the rational numbers. Estimate as needed. (See Examples 1 and 2.)

1. $\left\{-2\frac{3}{4}, -1\frac{1}{2}, -\frac{2}{3}, \frac{3}{5}, 2\right\}$

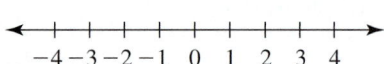

2. $\left\{-3\frac{2}{5}, -3, -2\frac{1}{2}, -\frac{2}{3}, 3\frac{3}{4}\right\}$

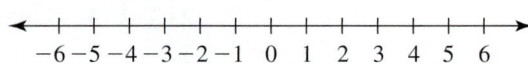

3. $\left\{-4, -2\frac{2}{5}, -\frac{3}{4}, 1\frac{4}{5}, 3\frac{5}{6}\right\}$

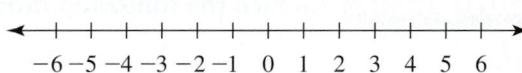

4. $\left\{-1\frac{1}{3}, -1, \frac{3}{5}, 2\frac{1}{2}, 3\frac{1}{4}\right\}$

Reduce the following fractions to lowest terms. (See Example 3.)

5. $\dfrac{3}{18}$

6. $\dfrac{4}{16}$

7. $\dfrac{12}{8}$

8. $\dfrac{18}{12}$

9. $\dfrac{40}{84}$

10. $\dfrac{16}{56}$

11. $\dfrac{36}{54}$

12. $\dfrac{32}{72}$

13. $\dfrac{90}{60}$

14. $\dfrac{100}{80}$

15. $\dfrac{72}{54}$

16. $\dfrac{84}{60}$

17. $\dfrac{72}{108}$

18. $\dfrac{120}{144}$

19. $\dfrac{120}{96}$

20. $\dfrac{72}{60}$

21. $\dfrac{210}{110}$

22. $\dfrac{300}{108}$

23. $\dfrac{90}{315}$

24. $\dfrac{180}{210}$

Reduce the following rational expressions to lowest terms. (See Example 4.)

25. $\dfrac{a^2b^3}{a^4b}$

26. $\dfrac{c^4d}{c^2d^4}$

27. $\dfrac{-r^3s^2}{r^2s^2}$

28. $\dfrac{-q^4r^3}{q^4r}$

29. $\dfrac{a^3b^4}{a^4b^6}$

30. $\dfrac{u^2v^5}{u^4v^6}$

31. $\dfrac{12x^5y^3}{3x^2y^2}$

32. $\dfrac{18x^5y^7}{6x^3y^6}$

33. $\dfrac{-16x^4y^6}{8x^3y^4}$

34. $\dfrac{-24p^6q^6}{6p^4q^5}$

35. $\dfrac{-28m^3n^5}{16m^7n^6}$

36. $\dfrac{32c^2d^6}{-24c^4d^7}$

37. $\dfrac{a^5b^3c^6}{a^3b^6c^4}$

38. $\dfrac{x^3y^6z^4}{x^2y^4z^6}$

39. $\dfrac{-32m^2n^5p^3}{-18m^2n^3p^4}$

40. $\dfrac{-27x^3y^4z^2}{-6x^3y^2z^3}$

41. $\dfrac{15r^3s^5w}{-25r^4s^6w}$

42. $\dfrac{36qr^4s^7}{-27qr^5s^9}$

43. $\dfrac{-32x^4y^3z^2}{16x^2y^2z^2}$

44. $\dfrac{42w^3x^4y^3}{-14w^2x^4y}$

Writing Exercise

45. Why is it important to reduce fractions to lowest terms?

Group Project

46. Find three examples of an everyday situation in which a fraction occurs that is not reduced to lowest terms.

Section 6.2

Further Reduction of Rational Expressions

OBJECTIVES *When you complete this section, you will be able to:*

Ⓐ Determine the value(s) of the variable(s) for which rational expressions are defined.

Ⓑ Reduce rational expressions whose numerators and denominators are not monomials.

PREREQUISITE SKILLS   *Before beginning this section, you should be able to:*

a. Know that the denominator of a fraction cannot be 0. (Section R.4)

b. Solve a linear equation for one of its variables. (Section 3.4)

c. Solve quadratic equations by factoring. (Section 5.8)

d. Reduce rational numbers to lowest terms. (Section 6.1)

e. Factor polynomials. (Sections 5.2–5.6)

f. Substitute a value for a variable and evaluate. (Section 2.5)

GETTING READY FOR SECTION 6.2

1. Evaluate $\dfrac{6}{0}$.

2. Solve $2x + 3 = 0$ for x.

3. Solve $x^2 - 2x - 24 = 0$ for x.

4. Reduce $\dfrac{45}{54}$ to lowest terms.

Factor the following polynomials:

5. $3x + 12$

6. $x^2 - 6x + 8$

7. $x^2 - 16$

8. Evaluate $\dfrac{x + 4}{x^2 - 16}$ for $x = 2$.

Introduction

In the previous section, we reduced rational expressions whose numerators and denominators were monomials. The steps we used were to factor the numerator and the denominator into prime factors and then divide by the factors that were common to both. If the numerator and the denominator are polynomials other than monomials, we use the same procedure.

OBJECTIVE Determining the value(s) of the variable(s) for which rational expressions are defined

Before we discuss reducing rational expressions, we need to find the value(s) of the variable(s) for which a rational expression is defined. Remember that if we have 0 as the denominator, the fraction is undefined. For example, if $\frac{4}{0} = x$, then $0 \cdot x = 4$, which is impossible since 0 times anything is 0. But if we have a variable in the denominator, how do we know whether the denominator is 0? A rational expression is defined for all value(s) of the variable(s) for which the denominator does not equal 0.

> **Procedure: Determining the Value(s) for Which a Rational Expression is Defined**
>
> To determine the value(s) for which a rational expression is defined, set the denominator equal to 0 and solve the resulting equation.

The variable(s) cannot equal any resulting value(s), since they make the denominator equal to 0.

Example 1 Find the value(s) of the variable(s) for which the following are defined:

a. $\dfrac{x + 2}{x + 3}$

b. $\dfrac{x - 4}{2x + 3}$

Solution

$\frac{x+2}{x+3}$ is defined for all values of x except those that make the denominator equal to 0. To find the value(s) that make the denominator equal to 0, set the denominator equal to 0 and solve the equation.

$x + 3 = 0$	Subtract 3 from both sides of the equation.
$x + 3 - 3 = 0 - 3$	Simplify both sides.
$x + 0 = -3$	Additive identity.
$x = -3$	Solution of the equation.

Since the denominator is equal to 0 for $x = -3$, $x \ne -3$. So the values of x for which the expression is defined are all numbers except -3.

Solution

$\frac{x-4}{2x+3}$ is defined for all values of x except those that make the denominator equal to 0. To find the value(s) that make the denominator equal to 0, set the denominator equal to 0 and solve the equation.

$2x + 3 = 0$	Subtract 3 from both sides of the equation.
$2x = -3$	Divide both sides by 2.
$x = -\dfrac{3}{2}$	Solution of the equation.

Since the denominator is equal to 0 for $x = -\frac{3}{2}$, $x \ne -\frac{3}{2}$. So the values of x for which the expression is defined are all numbers except $-\frac{3}{2}$.

c. $\dfrac{x - 5}{x^2 - 2x - 24}$

Solution

$\frac{x-5}{x^2-2x-24}$ is defined for all values of x except those that make the denominator equal to 0. To find these values, set the denominator equal to 0 and solve for x.

$x^2 - 2x - 24 = 0$	Factor.
$(x + 4)(x - 6) = 0$	Set each factor equal to 0.
$x + 4 = 0, x - 6 = 0$	Solve each equation.
$x = -4, \quad x = 6$	Solutions of the equation.

Since the denominator is equal to 0 when $x = -4$ or $x = 6$, $x \ne -4$ and $x \ne 6$. So the values of x for which the expression is defined are all numbers except -4 and 6.

PRACTICE EXERCISE 1

Find the values of the variable(s) for which the following are defined:

a. $\dfrac{x - 5}{x + 4}$

b. $\dfrac{4x}{3x - 2}$

c. $\dfrac{c - 1}{c^2 - 4c - 21}$

If you need more practice, do the following Additional Practice Exercises.

Additional Practice Exercise 1   Find the values of the variable(s) for which the following are defined:

a. $\dfrac{x + 6}{x + 5}$

b. $\dfrac{4a}{5a + 3}$

c. $\dfrac{m - 6}{m^2 + 6m + 8}$

We know that any number, other than 0, divided by itself is equal to 1. For example, $\frac{5}{5} = 1$ and $\frac{-4}{-4} = 1$. We used this fact in the last section to reduce rational numbers. We used an extension of this fact when reducing rational expressions with monomial

Answers: Practice Exercise 1: a. $x \ne -4$ **b.** $x \ne \frac{2}{3}$ **c.** $c \ne -3$ and $c \ne 7$ **Additional Practice 1: a.** $x \ne -5$ **b.** $a \ne -\frac{3}{5}$ **c.** $m \ne -4$ and $m \ne -2$

numerators and denominators by indicating that a variable divided by itself also has a value of 1, provided that the variable does not equal 0. For example, $\frac{x}{x} = 1$, if $x \neq 0$. We are going to extend this fact again in this section by saying that any polynomial divided by itself is equal to 1, provided that the polynomial is not equal to 0. For example, $\frac{x+3}{x+3} = 1$, if $x + 3 \neq 0$. Since $x + 3$ represents a number for any value of x, we are simply saying that any nonzero number divided by itself is 1.

OBJECTIVE Ⓑ Reducing rational expressions whose numerators and denominators are not monomials

The procedure for reducing rational expressions whose numerators and/or denominators are not monomials is exactly the same procedure as for rational numbers and rational expressions whose numerators and denominators are monomials. Compare the following:

$\dfrac{45}{54}$	$\dfrac{8x^2y^3}{12x^4y}$	$\dfrac{x^2 + 3x + 2}{x^2 - 2x - 3}$	Write the numerator and the denominator in terms of prime factors.
$\dfrac{3 \cdot 3 \cdot 5}{2 \cdot 3 \cdot 3 \cdot 3}$	$\dfrac{2 \cdot 2 \cdot 2 \cdot x \cdot x \cdot y \cdot y \cdot y}{2 \cdot 2 \cdot 3 \cdot x \cdot x \cdot x \cdot x \cdot y}$	$\dfrac{(x + 1)(x + 2)}{(x + 1)(x - 3)}$	Divide by factors common to the numerator and the denominator.
$\dfrac{3 \cdot 3 \cdot 5}{2 \cdot 3 \cdot 3 \cdot 3}$	$\dfrac{2 \cdot 2 \cdot 2 \cdot x \cdot x \cdot y \cdot y \cdot y}{2 \cdot 2 \cdot 3 \cdot x \cdot x \cdot x \cdot x \cdot y}$	$\dfrac{(x + 1)(x + 2)}{(x + 1)(x - 3)}$	Multiply the remaining factors.
$\dfrac{5}{6}$	$\dfrac{2y^2}{3x^2}$	$\dfrac{x + 2}{x - 3}$	Reduced to lowest terms.

Example 2 Reduce the rational expressions. Assume that the variables cannot equal any value for which the denominator is equal to 0.

a. $\dfrac{5(x + 1)}{8(x + 1)} =$ The numerator and the denominator are already factored, so divide the numerator and the denominator by the common factor of $x + 1$.

$\dfrac{5(x+1)}{8(x+1)} = \dfrac{x + 1}{x + 1} = 1$ if $x \neq -1$.

$\dfrac{5}{8}$ Answer if $x \neq -1$.

b. $\dfrac{4x + 16}{3x + 12} =$ Factor the numerator and the denominator.

$\dfrac{4(x + 4)}{3(x + 4)} =$ Divide the numerator and the denominator by the common factor of $x + 4$.

$\dfrac{4(x+4)}{3(x+4)} = \dfrac{x + 4}{x + 4} = 1$, if $x \neq -4$.

$\dfrac{4}{3}$ Answer if $x \neq -4$.

c. $\dfrac{x^2 - 9}{x^2 - x - 12} =$ Factor the numerator and the denominator.

$\dfrac{(x + 3)(x - 3)}{(x + 3)(x - 4)} =$ Divide the numerator and the denominator by the common factor of $x + 3$.

$\dfrac{(x+3)(x - 3)}{(x+3)(x - 4)} = \dfrac{x + 3}{x + 3} = 1$, if $x \neq -3$.

$\dfrac{x - 3}{x - 4}$ Answer if $x \neq 4$ or $x \neq -3$.

d. $\dfrac{x + 4}{x^2 - 16} =$ Factor the denominator. The numerator is prime.

$\dfrac{x + 4}{(x + 4)(x - 4)} =$ Divide the numerator and the denominator by the common factor of $x + 4$.

$\dfrac{x+4}{(x+4)(x - 4)} = \dfrac{x + 4}{x + 4} = 1$, if $x \neq -4$.

$\dfrac{1}{x - 4}$ Answer if $x \neq 4$ or -4. Remember, each time we divide by a common factor, there is a factor of 1 left. We do not write the 1 unless it is the only factor remaining in the numerator.

Be Careful

A common error is to divide by *terms* instead of *factors*. Consider the following: $\frac{6+3}{3} \neq \frac{\cancel{6}+3}{\cancel{3}} \neq 6+1=7$. If we follow the order of operations, $\frac{6+3}{3} = \frac{9}{3} = 3$. But $7 \neq 3$. We have two different answers to the same question. Since we followed the order of operations in the second case, we know that 3 is the correct answer. What is wrong with the first case? We have divided by a *term* and not a *factor*. Remember, terms are separated by $+$ and $-$ signs, and factor implies multiplication. Therefore, before you can reduce a rational expression, you must first write the numerator and the denominator in *factored* form. Then you divide by the factors that are common to both. In the following, both the incorrect and the correct methods are shown:

Incorrect

$$\frac{x^2 - 6x + 8}{x^2 + 2x - 8} = \frac{\cancel{x^2} - 6x + \cancel{8}}{\cancel{x^2} + 2x - \cancel{8}} = \frac{1 - 6x + 1}{1 + 2x - 1} = \frac{2 - 6x}{2x}$$

x^2 and 8 are *terms* common to the numerator and the denominator. They are not *factors*. **You cannot divide by terms!**

Correct

$$\frac{x^2 - 6x + 8}{x^2 + 2x - 8} = \frac{(x-4)(x-2)}{(x-2)(x+4)} = \frac{(x-4)\cancel{(x-2)}}{\cancel{(x-2)}(x+4)} = \frac{x-4}{x+4}$$

Since $x - 2$ is a factor common to both the numerator and the denominator, we can divide both the numerator and the denominator by $x - 2$.

Note

After factoring, you can divide only by an entire factor and not by a part of a factor since parts of factors are terms. Using the example above, the following is also incorrect.

$$\frac{x^2 - 6x + 8}{x^2 + 2x - 8} = \frac{(x-4)(x-2)}{(x-2)(x+4)} = \frac{(\overset{1}{\cancel{x}} - \overset{2}{\cancel{4}})(\overset{1}{\cancel{x}} - \overset{1}{\cancel{2}})}{(\underset{1}{\cancel{x}} - \underset{1}{\cancel{2}})(\underset{1}{\cancel{x}} + \underset{2}{\cancel{4}})}$$

PRACTICE EXERCISE 2

Reduce the rational expressions. Assume that the variables cannot equal any value that makes the denominator equal to 0.

a. $\dfrac{4(x+y)}{7(x+y)}$
b. $\dfrac{3x+6}{5x+10}$
c. $\dfrac{x^2-4}{x^2+9x+14}$
d. $\dfrac{x-3}{x^2-7x+12}$

* If you need more practice, do the following Additional Practice Exercises.

Additional Practice Exercise 2   Reduce the following rational expressions to lowest terms:

a. $\dfrac{2(y+6)}{5(y+6)}$
b. $\dfrac{12x+4y}{21x+7y}$
c. $\dfrac{x^2-2x-15}{x^2+7x+12}$
d. $\dfrac{x+1}{x^2+4x+3}$

Answers: Practice Exercise 2: a. $\dfrac{4}{7}$ b. $\dfrac{3}{5}$ c. $\dfrac{x-2}{x+7}$ d. $\dfrac{1}{x-4}$ **Additional Practice 2:** a. $\dfrac{2}{5}$ b. $\dfrac{4}{7}$ c. $\dfrac{x-5}{x+4}$ d. $\dfrac{1}{x+3}$

There is one instance in reducing rational expressions that requires special attention. This occurs if the numerator and the denominator are exactly the same except for the signs of the terms. For example, $\frac{a-b}{b-a}$. The numerator has $+a$ and $-b$, while the denominator has $-a$ and $+b$; so, the signs of the terms are opposites. In order to reduce this type of fraction, we need to recall a technique that was discussed earlier, where we factored out -1 when factoring by grouping. If all terms contain variables, we first write the expression in alphabetical order. If not, write the expression in descending order.

Example 3 | **Factor -1 from each of the following and rewrite the expression inside the parentheses so that the term with the positive coefficient is first:**

a. $b - a =$ Rewrite in alphabetical order.

$-a + b =$ Remove the factor of -1.

$-1(a - b)$

b. $3 - 2x =$ Rewrite in descending order.

$-2x + 3 =$ Remove the factor of -1.

$-1(2x - 3)$

PRACTICE EXERCISE 3

Factor -1 from each of the following and rewrite the expression inside the parentheses so that the term with the positive coefficient is first:

a. $4 - a$ **b.** $2 - 3y$

If you need more practice, do the following Additional Practice Exercises.

Additional Practice Exercise 3 Factor -1 from each of the following and rewrite the expression inside the parentheses so that the term with the positive coefficient is first:

a. $6 - y$ **b.** $7 - 5n$

We will now apply this technique to reducing rational expressions that contain factors in the numerator and the denominator with opposite signs.

Example 4 | **Reduce the following rational expressions to lowest terms:**

a. $\dfrac{a-b}{b-a} =$ Rewrite the denominator in alphabetical order.

$\dfrac{a-b}{-a+b} =$ Remove the common factor of -1 from the denominator and divide by the common factor $a - b$.

$\dfrac{a-b}{-1(a-b)} =$

$\dfrac{+1}{-1} =$ $\dfrac{+1}{-1} = -1$.

-1 Answer.

Be Careful

Rational expressions of the form $\frac{4 + x}{4 - x}$ are often mistaken for the previous type. In this rational expression, not all the terms in the numerator and the denominator are opposites in sign. This expression cannot be reduced, because $\frac{4 + x}{4 - x} = \frac{4 + x}{-(x - 4)}$, so there is no common factor.

b. $\dfrac{x^2 - 1}{1 - x} =$ Factor the numerator and write the denominator in descending order.

$\dfrac{(x - 1)(x + 1)}{-x + 1} =$ Remove the common factor of -1 from the denominator.

$\dfrac{(x - 1)(x + 1)}{-1(x - 1)} =$ Divide by the common factor of $x - 1$.

$\dfrac{(\cancel{x - 1})(x + 1)}{-1(\cancel{x - 1})} =$ 1 is left in both the numerator and the denominator, where the common factors were divided.

$\dfrac{1(x + 1)}{-1} =$ $\dfrac{1}{-1} = -1$.

$-1(x + 1) =$ Distribute the -1.

$-x - 1$ Answer.

PRACTICE EXERCISE 4

Reduce the following rational expressions to lowest terms:

a. $\dfrac{c - d}{d - c}$ **b.** $\dfrac{x^2 - 4}{2 - x}$

If you need more practice, do the following Additional Practice Exercises.

Additional Practice Exercise 4   Reduce the following rational expressions to lowest terms:

a. $\dfrac{r - s}{s - r}$ **b.** $\dfrac{x^2 - 9}{3 - x}$

We defined functions in Section 4.8 and discussed quadratic functions in Section 5.8. Following is the definition of another type of function:

> **Definition** Rational Function
> A function of the form $f(x) = \dfrac{p(x)}{q(x)}$, $q(x) \neq 0$, where $p(x)$ and $q(x)$ are polynomials, is called a **rational function**.

Rational functions are evaluated in the same manner as any other function.

Example 5   **Given $f(x) = \dfrac{x + 4}{x^2 - 16}$, find the following:**

a. $f(0)$

Solution

To find $f(0)$, replace x with 0 and simplify.

$f(x) = \dfrac{x + 4}{x^2 - 16}$ Replace x with 0.

$f(0) = \dfrac{0 + 4}{0^2 - 16}$ Simplify.

$f(0) = \dfrac{4}{-16}$ Reduce to lowest terms.

$f(0) = -\dfrac{1}{4}$ Therefore, $f(0) = -\dfrac{1}{4}$.

Answers: **Practice Exercise 4: a.** -1 **b.** $-x - 2$ **Additional Practice 4: a.** -1 **b.** $-x - 3$

b. $f(2)$

c. $f(4)$

Solution

To find $f(2)$, replace x with 2 and simplify.

$f(x) = \dfrac{x + 4}{x^2 - 16}$ Replace x with 2.

$f(2) = \dfrac{2 + 4}{2^2 - 16}$ Simplify by adding in the numerator and raising to a power in the denominator.

$f(2) = \dfrac{6}{4 - 16}$ Add in the denominator.

$f(2) = \dfrac{6}{-12}$ Reduce to lowest terms.

$f(2) = -\dfrac{1}{2}$ Therefore, $f(2) = -\dfrac{1}{2}$.

Solution

To find $f(4)$, replace x with 4 and simplify.

$f(x) = \dfrac{x + 4}{x^2 - 16}$ Replace x with 4.

$f(4) = \dfrac{4 + 4}{4^2 - 16}$ Simplify by adding in the numerator and raising to a power in the denominator.

$f(4) = \dfrac{8}{16 - 16}$ Add in the denominator.

$f(4) = \dfrac{8}{0}$ Since division by 0 is undefined, $f(4)$ does not exist.

$f(4)$ is undefined.

PRACTICE EXERCISE 5

Given $f(x) = \dfrac{x - 2}{x^2 + x - 6}$, find the following:

a. $f(0)$ **b.** $f(4)$ **c.** $f(-3)$

If you need more practice, do the following Additional Practice Exercises.

Additional Practice Exercise 5   Given $f(x) = \dfrac{x + 3}{x^2 + 3x - 4}$, find the following:

a. $f(-2)$ **b.** $f(2)$ **c.** $f(1)$

Rational functions play a very important role in upper-level mathematics.

Exercise Set 6.2   MyMathLab®

For Extra Help

Find the value(s) of the variable(s) for which the following are defined. (See Example 1.)

1. $\dfrac{5}{x + 5}$

2. $\dfrac{y + 3}{y + 10}$

3. $\dfrac{r + 3}{r - 8}$

4. $\dfrac{9}{a - 6}$

5. $\dfrac{x + 3}{3x + 12}$

6. $\dfrac{y - 5}{4y + 8}$

7. $\dfrac{6a}{3a - 4}$

8. $\dfrac{5b}{4b - 5}$

9. $\dfrac{5}{(x - 7)(x + 3)}$

10. $\dfrac{m}{(m + 5)(m - 3)}$

11. $\dfrac{8x}{x^2 - 9}$

12. $\dfrac{7y}{y^2 - 36}$

13. $\dfrac{x - 3}{x^2 - 3x - 10}$

14. $\dfrac{x + 2}{x^2 + x - 12}$

15. $\dfrac{r + 3}{2r^2 - 7r - 15}$

16. $\dfrac{n + 7}{3n^2 + 2n - 8}$

Answers: Practice Exercise 5: a. $\dfrac{1}{3}$ **b.** $\dfrac{1}{7}$ **c.** undefined   **Additional Practice 5: a.** $-\dfrac{1}{6}$ **b.** $\dfrac{5}{6}$ **c.** undefined

Reduce the following rational expressions to lowest terms. (See Example 2.)

17. $\dfrac{4(x + 3)}{9(x + 3)}$

18. $\dfrac{3(y - 2)}{5(y - 2)}$

19. $\dfrac{-5(2a - 7)}{6(2a - 7)}$

20. $\dfrac{-8(3b + 4)}{9(3b + 4)}$

21. $\dfrac{4(x - 10)}{6(x - 10)}$

22. $\dfrac{6(x + 5)}{8(x + 5)}$

23. $\dfrac{2x + 12}{3x + 18}$

24. $\dfrac{3x + 6}{5x + 10}$

25. $\dfrac{8x + 12y}{20x + 30y}$

26. $\dfrac{24a - 16b}{36a - 24b}$

27. $\dfrac{-3x - 15}{4x + 20}$

28. $\dfrac{-5x - 25}{6x + 30}$

29. $\dfrac{x^2 + 3x + 2}{x^2 - 2x - 3}$

30. $\dfrac{x^2 - 4x + 3}{x^2 - 5x + 6}$

31. $\dfrac{a^2 + 3a - 10}{a^2 + 8a + 15}$

32. $\dfrac{c^2 + 3c - 18}{c^2 - 5c + 6}$

33. $\dfrac{x - 6}{x^2 - 4x - 12}$

34. $\dfrac{x - 2}{x^2 + 6x - 16}$

35. $\dfrac{x^2 - 9}{x^2 - 2x - 15}$

36. $\dfrac{x^2 - 25}{x^2 - x - 30}$

37. $\dfrac{4x^2 - 9y^2}{6x^2 - xy - 15y^2}$

38. $\dfrac{9x^2 - 25y^2}{6x^2 - 5xy - 25y^2}$

39. $\dfrac{4a^2 - 4ab - 3b^2}{6a^2 - ab - 2b^2}$

40. $\dfrac{6a^2 + 13ab + 6b^2}{12a^2 - ab - 6b^2}$

Challenge Exercises (41–44)

41. $\dfrac{ac - ad + bc - bd}{ax + ay + bx + by}$

42. $\dfrac{px + py + qx + qy}{mx - nx + my - ny}$

43. $\dfrac{xy - 4x + 3y - 12}{xy - 6x + 3y - 18}$

44. $\dfrac{ab + 5a - 3b - 15}{bc - 2b + 5c - 10}$

Factor -1 from each of the following and rewrite the expression inside the parentheses so that the term with the positive coefficient is first. (See Example 3.)

45. $y - x$

46. $s - r$

47. $2d - c$

48. $3y - x$

49. $3r - 2s$

50. $3b - 4a$

Reduce the following rational expressions to lowest terms. (See Example 4.)

51. $\dfrac{m - n}{n - m}$

52. $\dfrac{q - p}{p - q}$

53. $\dfrac{2b - a}{a - 2b}$

54. $\dfrac{3m - n}{n - 3m}$

55. $\dfrac{a^2 - 25}{5 - a}$

56. $\dfrac{c^2 - 49}{7 - c}$

57. $\dfrac{x^2 - 3x - 10}{5 - x}$

58. $\dfrac{x^2 + 4x - 21}{3 - x}$

59. $\dfrac{4 - x}{x^2 - 9x + 20}$

Answer the following. (See Example 5.)

60. Given $f(x) = \dfrac{x + 7}{x^2 + 5x - 14}$, find the following:

   a. $f(0)$ **b.** $f(3)$ **c.** $f(2)$

61. Given $f(x) = \dfrac{x - 6}{x^2 - 8x + 12}$, find the following:

   a. $f(0)$ **b.** $f(-1)$ **c.** $f(2)$

Writing Exercises (62–65)

62. What is wrong with the following?

$$\frac{3x + 4}{2} = \frac{3x + \overset{2}{\cancel{4}}}{\cancel{2}} = 3x + 2$$

63. How does a term differ from a factor?

64. Are $a - b$ and $b - a$ always equal? If not, under what circumstances will $a - b$ and $b - a$ be equal?

65. Why does $\dfrac{b - a}{a - b} = -1$?

Group Project

66. You are given $f(x) = \dfrac{x - 4}{x^2 - 7x + 12}$ and $g(x) = \dfrac{1}{x - 3}$.

   a. Find $f(2)$ and $g(2)$.

   b. How are $f(2)$ and $g(2)$ related? Why?

   c. Find $f(-2)$ and $g(-2)$.

   d. How are $f(-2)$ and $g(-2)$ related? Why?

   e. Find $f(4)$ and $g(4)$.

   f. Does $f(4) = g(4)$? Why or why not?

Section
6.3
Multiplication of Rational Numbers and Expressions

OBJECTIVES *When you complete this section, you will be able to:*

Ⓐ Multiply two or more fractions.
Ⓑ Multiply mixed numbers and whole numbers with mixed numbers.
Ⓒ Multiply rational expressions with monomial numerators and denominators.

PREREQUISITE SKILLS *Before beginning this section, you should be able to:*

a. Write a whole number in terms of its prime factors. (Section 5.1)
b. Convert a mixed number into an improper fraction. (Section R.4)
c. Apply the meaning of a whole-number exponent. (Section 1.1)
d. Apply $a^m \cdot a^n = a^{m+n}$. (Section 2.2)

GETTING READY FOR SECTION 6.3

1. Write 18 in terms of prime factors.

2. Write $6\frac{3}{5}$ as an improper fraction.

3. Write $9m^2n^3$ in terms of prime factors.

4. Find the product: $5m^3n \cdot 9m^2n^3$

Introduction

In Chapter R, we introduced the procedure for multiplying fractions, but did so on a very elementary level. We needed this procedure to develop properties of exponents in Chapter 2. In Chapter 2, you were not required to multiply fractions, but only to recall the procedure. In this section, we will expand on the concepts developed in Section R.4.

In Section R.4, we used shaded regions of figures to represent a fractional part of the entire region (unit). We can represent $\frac{1}{5}$ by dividing a region into five equal parts and shading one of those parts, as the following shows:

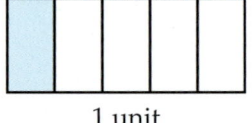

1 unit

Suppose we now wish to further shade $\frac{1}{2}$ of the $\frac{1}{5}$ that is already shaded. We divide everything into two equal parts with a horizontal line segment. The one part that was shaded has now been divided into two equal parts. So, $\frac{1}{2}$ of $\frac{1}{5}$ is one of these two equal parts that we will shade further.

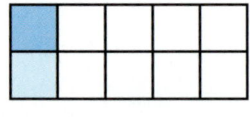

1 unit

By dividing each of the five equal parts into two equal parts, the entire region (unit) has been divided into ten equal parts. Therefore, the one part that was further shaded represents $\frac{1}{10}$ of the entire region. We would say that $\frac{1}{2}$ of $\frac{1}{5}$ is $\frac{1}{10}$. When performing operations on fractions, the word *of* is used to indicate multiplication. Consequently, $\frac{1}{2} \cdot \frac{1}{5} = \frac{1}{10}$. Note that this is the same as $\frac{1}{2} \cdot \frac{1}{5} = \frac{1 \cdot 1}{2 \cdot 5} = \frac{1}{10}$. Let us repeat this procedure with a more complicated example and see if it still works.

Answers: Getting Ready: 1. $2 \cdot 3 \cdot 3$ **2.** $\frac{33}{5}$ **3.** $3 \cdot 3 \cdot m \cdot m \cdot n \cdot n \cdot n$ **4.** $45m^5n^4$

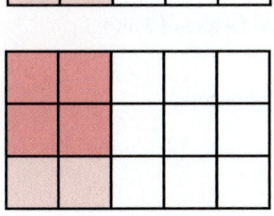

Example 1 Represent $\frac{2}{3}$ of $\frac{2}{5}$, using shaded regions, and give the result as a fractional part of the entire region.

Solution

The fraction $\frac{2}{5}$ means "Divide the region into five equal parts and shade two of them."

To find $\frac{2}{3}$ of the two parts that are shaded, we divide each of the five equal parts into three equal parts. Each of the parts that were shaded has now been divided into three equal parts. To indicate $\frac{2}{3}$ of each of these shaded parts, further shade two of these three equal parts.

You will notice that the entire figure is now divided into 15 equal parts, of which 4 have been further shaded. This represents $\frac{4}{15}$ of the entire figure. Therefore, $\frac{2}{3}$ of $\frac{2}{5}$ is $\frac{4}{15}$. As multiplication, $\frac{2}{3} \cdot \frac{2}{5} = \frac{4}{15}$. Note that this is the same as $\frac{2}{3} \cdot \frac{2}{5} = \frac{2 \cdot 2}{3 \cdot 5} = \frac{4}{15}$.

OBJECTIVE Multiplying two or more fractions

On the basis of the preceding examples, we are now ready to state the procedure for multiplication of fractions.

> **Procedure: Multiplication of Fractions**
>
> The product of two fractions is found by computing the product of the numerators and dividing by the product of the denominators. In symbols, $\frac{a}{b} \cdot \frac{c}{d} = \frac{a \cdot c}{b \cdot d}$. In other words, to find the product of two fractions multiply numerator times numerator divided by denominator times denominator.

This procedure is easily extended to three or more fractions, as follows:
$\frac{a}{b} \cdot \frac{c}{d} \cdot \frac{e}{f} = \frac{a \cdot c \cdot e}{b \cdot d \cdot f}$ and $\frac{a}{b} \cdot \frac{c}{d} \cdot \frac{e}{f} \cdot \frac{g}{h} = \frac{a \cdot c \cdot e \cdot g}{b \cdot d \cdot f \cdot h}$.

Example 2 Find the products:

a. $\frac{3}{5} \cdot \frac{2}{7} =$ Find the product of the numerators divided by the product of the denominators.

$\frac{3 \cdot 2}{5 \cdot 7} =$ Perform the multiplications.

$\frac{6}{35}$ Product.

b. $\frac{1}{4} \cdot \frac{3}{8} \cdot \frac{5}{7} =$ Find the product of the numerators divided by the product of the denominators.

$\frac{1 \cdot 3 \cdot 5}{4 \cdot 8 \cdot 7} =$ Perform the multiplications.

$\frac{15}{224}$ Product.

PRACTICE EXERCISE 2

Find the products:

a. $\frac{2}{3} \cdot \frac{4}{9}$

b. $\frac{1}{4} \cdot \frac{3}{2} \cdot \frac{5}{8}$

After finding the product of two or more fractions, it is often necessary to reduce to lowest terms. Also, you will need to recall the rules of signs for multiplication. If there is an even number of negative signs, the product is positive. If there is an odd number of negative signs, the product is negative.

Answers: Practice Exercise 2: a. $\frac{8}{27}$ **b.** $\frac{15}{64}$

Example 3   **Find the products. If necessary, reduce to lowest terms.**

a. $\dfrac{2}{5} \cdot \left(-\dfrac{3}{8}\right) =$ Find the product of the numerators divided by the product of the denominators. The product of a positive and a negative is negative.

$-\dfrac{2 \cdot 3}{5 \cdot 8} =$ Write the numerator and the denominator in terms of prime factors.

$-\dfrac{2 \cdot 3}{5 \cdot 2 \cdot 2 \cdot 2} =$ Divide by the common factor of 2.

$-\dfrac{2 \cdot 3}{5 \cdot 2 \cdot 2 \cdot 2} =$ Multiply the remaining factors.

$-\dfrac{3}{20}$ Product.

b. $\dfrac{3}{15} \cdot \dfrac{10}{9} =$ Find the product of the numerators divided by the product of the denominators.

$\dfrac{3 \cdot 10}{15 \cdot 9} =$ Write the numerator and the denominator in terms of prime factors.

$\dfrac{3 \cdot 2 \cdot 5}{3 \cdot 5 \cdot 3 \cdot 3} =$ Divide by the common factors of 3 and 5.

$\dfrac{3 \cdot 2 \cdot 5}{3 \cdot 5 \cdot 3 \cdot 3} =$ Multiply the remaining factors.

$\dfrac{2}{9}$ Product.

c. $-\dfrac{4}{5} \cdot \dfrac{15}{14} \cdot \left(-\dfrac{7}{10}\right) =$ Find the product of the numerators divided by the product of the denominators. The product of two negatives is positive.

$\dfrac{4 \cdot 15 \cdot 7}{5 \cdot 14 \cdot 10} =$ Write the numerator and the denominator in terms of prime factors.

$\dfrac{2 \cdot 2 \cdot 3 \cdot 5 \cdot 7}{5 \cdot 2 \cdot 7 \cdot 2 \cdot 5} =$ Divide by the common factors of 2, 2, 5, and 7.

$\dfrac{2 \cdot 2 \cdot 3 \cdot 5 \cdot 7}{5 \cdot 2 \cdot 7 \cdot 2 \cdot 5} =$ Multiply the remaining factors.

$\dfrac{3}{5}$ Product.

Note

You may reduce before actually multiplying. This is permitted because each numerator becomes a factor of the numerator of the product and likewise for the denominators. Therefore, we may divide by factors common to any numerator and denominator before we multiply. Example 3b could be done as follows:

$\dfrac{3}{15} \cdot \dfrac{10}{9} =$ Factor the numerators and the denominators into primes.

$\dfrac{3}{3 \cdot 5} \cdot \dfrac{2 \cdot 5}{3 \cdot 3} =$ Divide by the common factors of 3 and 5.

$\dfrac{3}{3 \cdot 5} \cdot \dfrac{2 \cdot 5}{3 \cdot 3} =$ Multiply the remaining numerators and the denominators.

$\dfrac{2}{3 \cdot 3} =$ Perform the multiplication.

$\dfrac{2}{9}$ Product.

Reducing before multiplying is more like the technique we will be using in the next section on multiplication of rational expressions. For that reason, we will use this method throughout the remainder of this section.

PRACTICE EXERCISE 3

Find the products. If necessary, reduce to lowest terms.

a. $\dfrac{3}{4} \cdot \dfrac{10}{21}$

b. $\dfrac{5}{8} \cdot \left(-\dfrac{14}{25}\right)$

c. $\dfrac{6}{7} \cdot \left(-\dfrac{21}{8}\right) \cdot \left(-\dfrac{4}{9}\right)$

If you need more practice, do the following Additional Practice Exercises.

Additional Practice Exercise 3   Find the products. If necessary, reduce to lowest terms.

a. $\dfrac{2}{9} \cdot \dfrac{15}{6}$

b. $-\dfrac{3}{8} \cdot \dfrac{4}{9}$

c. $\dfrac{10}{3} \cdot \left(-\dfrac{15}{16}\right) \cdot \dfrac{12}{5}$

OBJECTIVE **Multiplying mixed numbers and whole numbers with mixed numbers**

Since mixed numbers can be converted into improper fractions, the same technique used for multiplying fractions can be used to multiply mixed numbers. Likewise, since whole numbers may be thought of as fractions whose denominator is 1, this technique also applies to products involving fractions and whole numbers.

Example 4   **Find the products of the given numbers. Express answers reduced to lowest terms.**

a. $-2\dfrac{2}{7} \cdot 3\dfrac{1}{2} =$ Change $2\dfrac{2}{7}$ and $3\dfrac{1}{2}$ into improper fractions.

$-\dfrac{16}{7} \cdot \dfrac{7}{2} =$ Write the numerators and the denominators in terms of prime factors.

$-\dfrac{2 \cdot 2 \cdot 2 \cdot 2}{7} \cdot \dfrac{7}{2} =$ Divide by the common factors of 2 and 7.

$-\dfrac{2 \cdot 2 \cdot 2 \cdot 2}{7} \cdot \dfrac{7}{2} =$ Multiply the remaining factors.

$-\dfrac{2 \cdot 2 \cdot 2}{1} =$ Perform the multiplication.

-8 Product.

b. $12 \cdot \dfrac{5}{16} =$ $12 = \dfrac{12}{1}$

$\dfrac{12}{1} \cdot \dfrac{5}{16} =$ Write the numerators and the denominators in terms of prime factors.

$\dfrac{2 \cdot 2 \cdot 3}{1} \cdot \dfrac{5}{2 \cdot 2 \cdot 2 \cdot 2} =$ Divide by the common factors of 2 (twice).

$\dfrac{2 \cdot 2 \cdot 3}{1} \cdot \dfrac{5}{2 \cdot 2 \cdot 2 \cdot 2} =$ Multiply the remaining factors.

$\dfrac{3 \cdot 5}{1 \cdot 2 \cdot 2} =$ Perform the multiplications.

$\dfrac{15}{4}$ Product.

PRACTICE EXERCISE 4

Find the products of the given numbers. Express answers reduced to lowest terms.

a. $-3\dfrac{1}{3} \cdot 6\dfrac{3}{5}$

b. $8 \cdot \dfrac{5}{6}$

If you need more practice, do the following Additional Practice Exercises.

Additional Practice Exercise 4   Find the products of the given mixed numbers. Express answers reduced to lowest terms.

a. $\dfrac{2}{3} \cdot 4\dfrac{1}{2}$

b. $-18 \cdot \left(-1\dfrac{5}{6}\right)$

Answers: Practice Exercise 3: a. $\dfrac{5}{14}$ **b.** $-\dfrac{7}{20}$ **c.** 1 **Additional Practice 3: a.** $\dfrac{5}{9}$ **b.** $-\dfrac{1}{6}$ **c.** $-\dfrac{15}{2}$

Practice Exercise 4: a. -22 **b.** $\dfrac{20}{3}$ **Additional Practice 4: a.** 3 **b.** 33

OBJECTIVE Multiplying rational expressions with monomial numerators and denominators

If the numerators and the denominators of the rational expressions are monomials, the product is found in much the same way that we multiply rational numbers. Compare the following examples. The instructions on the right apply to both exercises. Assume that all variables have nonzero values.

Rational Number	**Rational Expression**	
$\dfrac{3}{4}\cdot\dfrac{10}{21}=$	$\dfrac{a^2}{y^2}\cdot\dfrac{y^4}{a^5}=$	Write the numerators and the denominators in terms of prime factors.
$\dfrac{3}{2\cdot 2}\cdot\dfrac{2\cdot 5}{3\cdot 7}=$	$\dfrac{a\cdot a}{y\cdot y}\cdot\dfrac{y\cdot y\cdot y\cdot y}{a\cdot a\cdot a\cdot a\cdot a}=$	Divide by the common factors.
$\dfrac{3}{2\cdot 2}\cdot\dfrac{2\cdot 5}{3\cdot 7}=$	$\dfrac{\cancel{a}\cdot\cancel{a}}{\cancel{y}\cdot\cancel{y}}\cdot\dfrac{\cancel{y}\cdot\cancel{y}\cdot y\cdot y}{\cancel{a}\cdot\cancel{a}\cdot a\cdot a\cdot a}=$	Multiply the remaining factors.
$\dfrac{5}{14}$	$\dfrac{y^2}{a^3}$	Product.

Example 5 **Find the products. Express answers reduced to lowest terms and assume that all variables have nonzero values.**

a. $\dfrac{a^2}{y^2}\cdot\dfrac{y^3}{a^3}=$ Write the numerators and the denominators in terms of prime factors.

$\dfrac{a\cdot a}{y\cdot y}\cdot\dfrac{y\cdot y\cdot y}{a\cdot a\cdot a}=$ Divide by common factors.

$\dfrac{\cancel{a}\cdot\cancel{a}}{\cancel{y}\cdot\cancel{y}}\cdot\dfrac{\cancel{y}\cdot\cancel{y}\cdot y}{\cancel{a}\cdot\cancel{a}\cdot a}=$ Multiply remaining factors.

$\dfrac{y}{a}$ Product.

b. $-\dfrac{3mn^2}{5m^3n}\cdot\dfrac{10m^2n}{9m^2n^3}=$ Write the numerators and the denominators in terms of prime factors.

$-\dfrac{3\cdot m\cdot n\cdot n}{5\cdot m\cdot m\cdot m\cdot n}\cdot\dfrac{2\cdot 5\cdot m\cdot m\cdot n}{3\cdot 3\cdot m\cdot m\cdot n\cdot n\cdot n}=$ Divide by common factors.

$-\dfrac{3\cdot\cancel{m}\cdot\cancel{n}\cdot\cancel{n}}{5\cdot\cancel{m}\cdot m\cdot m\cdot\cancel{n}}\cdot\dfrac{2\cdot 5\cdot\cancel{m}\cdot\cancel{m}\cdot\cancel{n}}{3\cdot 3\cdot m\cdot m\cdot\cancel{n}\cdot\cancel{n}\cdot n}=$ Multiply remaining factors.

$-\dfrac{2}{3m^2n}$ Product.

c. $\dfrac{3xy^2}{4a^2b^2}\cdot\dfrac{2ab}{9x^3y^3}=$ Write the numerators and the denominators in terms of prime factors.

$\dfrac{3\cdot x\cdot y\cdot y}{2\cdot 2\cdot a\cdot a\cdot b\cdot b}\cdot\dfrac{2\cdot a\cdot b}{3\cdot 3\cdot x\cdot x\cdot x\cdot y\cdot y\cdot y}=$ Divide by common factors.

$\dfrac{\cancel{3}\cdot\cancel{x}\cdot\cancel{y}\cdot\cancel{y}}{2\cdot 2\cdot\cancel{a}\cdot a\cdot\cancel{b}\cdot b}\cdot\dfrac{2\cdot\cancel{a}\cdot\cancel{b}}{3\cdot 3\cdot\cancel{x}\cdot x\cdot x\cdot\cancel{y}\cdot\cancel{y}\cdot y}=$ Multiply remaining factors. Remember, each time we divide, there is a factor of 1 left that usually is not written. Since 1 is all that is left in the numerator, it must be written.

$\dfrac{1}{6abx^2y}$ Product.

The products in Example 5 can also be found by using the laws of exponents developed in Chapter 2. Example 5b can also be done as follows:

$$-\frac{3mn^2}{5m^3n} \cdot \frac{10m^2n}{9m^2n^3} =$$ Multiply the fractions.

$$-\frac{3mn^2 \cdot 10m^2n}{5m^3n \cdot 9m^2n^3} =$$ Multiply the coefficients and apply $a^m \cdot a^n = a^{m+n}$.

$$-\frac{30m^3n^3}{45m^5n^4} =$$ Apply $\frac{a^m}{a^n} = a^{m-n}$.

$$-\frac{30}{45}m^{-2}n^{-1} =$$ Reduce the coefficient and apply $a^{-n} = \frac{1}{a^n}$.

$$-\frac{2}{3} \cdot \frac{1}{m^2} \cdot \frac{1}{n} =$$ Multiply the fractions.

$$-\frac{2}{3m^2n}$$ Product.

PRACTICE EXERCISE 5

Find the products. Express answers reduced to lowest terms and assume that all variables have nonzero values.

a. $\dfrac{x^3}{y^2} \cdot \dfrac{y^4}{x}$ **b.** $-\dfrac{8a^3b}{7a^3b} \cdot \dfrac{21ab^2}{4a^2b}$ **c.** $\dfrac{2a^2b}{15xy^3} \cdot \dfrac{5xy}{4a^3b}$

If you need more practice, do the following Additional Practice Exercises.

Additional Practice Exercise 5 Find the products. Express answers reduced to lowest terms.

a. $\dfrac{s^3}{r^2} \cdot \dfrac{r^4}{s^5}$ **b.** $\dfrac{6r^3s}{5r^2s^2} \cdot \dfrac{15rs^2}{2r^3s^2}$ **c.** $\dfrac{3rs^2}{14p^2q^3} \cdot \dfrac{7pq^2}{6r^2s^2}$

Application problems often require the use of fractions.

Example 6   Answer the following:

a. A bookcase has six shelves. If each shelf is $2\frac{2}{3}$ feet long, what is the total length of the six shelves?

b. Find the area of a triangle with a base of $3\frac{1}{5}$ inches if the altitude to that base is $3\frac{3}{4}$ inches. The formula for the area of a triangle is $A = \frac{1}{2}bh$.

Solution

Since there are six shelves, each of which is $2\frac{2}{3}$ feet long, the total length of the six shelves is the product of 6 and $2\frac{2}{3}$.

$$6 \cdot 2\frac{2}{3} =$$ Change $2\frac{2}{3}$ into an improper fraction.

$$6 \cdot \frac{8}{3} =$$ $6 = \frac{6}{1}$, and write 6 and 8 in terms of prime factors.

$$\frac{2 \cdot 3}{1} \cdot \frac{2 \cdot 2 \cdot 2}{3} =$$ Divide by the common factor of 3.

$$\frac{2 \cdot 3}{1} \cdot \frac{2 \cdot 2 \cdot 2}{3} =$$ Multiply the remaining factors.

$$16$$ Therefore, the total length of the six shelves is 16 feet.

Solution

$$A = \frac{1}{2}bh$$ Substitute into the formula.

$$A = \frac{1}{2} \cdot 3\frac{1}{5} \cdot 3\frac{3}{4}$$ Change the mixed numbers to improper fractions.

$$A = \frac{1}{2} \cdot \frac{16}{5} \cdot \frac{15}{4}$$ Write in terms of prime factors.

$$A = \frac{1}{2} \cdot \frac{2 \cdot 2 \cdot 2 \cdot 2}{5} \cdot \frac{3 \cdot 5}{2 \cdot 2}$$ Divide by the common factors.

$$A = \frac{1}{2} \cdot \frac{2 \cdot 2 \cdot 2 \cdot 2}{5} \cdot \frac{3 \cdot 5}{2 \cdot 2}$$ Multiply the remaining factors.

$$A = 6 \text{ in.}^2$$ Therefore, the area is 6 square inches.

Answers: Practice Exercise 5: a. x^2y^2 **b.** $-\dfrac{6b}{a}$ **c.** $\dfrac{1}{6ay^2}$ **Additional Practice 5: a.** $\dfrac{r^2}{s^2}$ **b.** $-\dfrac{9}{rs}$ **c.** $\dfrac{1}{4pqr}$

PRACTICE EXERCISE 6

Answer the following:

a. A beaker is $\frac{3}{4}$ full of sulfuric acid. If a student uses $\frac{2}{3}$ of the contents of the beaker in an experiment, what fractional part of a full beaker does the student use?

b. Find the area of a rectangle whose length is 12 meters and whose width is $4\frac{1}{4}$ meters. ($A = LW$)

If you need more practice, do the following Additional Practice Exercises.

Additional Practice Exercise 6   Answer the following:

a. A recipe for a cake requires $1\frac{1}{4}$ cups of flour. How much flour is needed for 10 such cakes?

b. Find the area of a triangle whose base is $2\frac{2}{3}$ inches and whose height is $2\frac{1}{4}$ inches.

Exercise Set 6.3   For Extra Help   MyMathLab®

Find the products. Express answers reduced to lowest terms and assume that all variables have nonzero values.
(See Examples 2–4.)

1. $\frac{2}{5} \cdot \frac{3}{7}$

2. $\frac{2}{7} \cdot \frac{4}{5}$

3. $\frac{5}{8} \cdot \left(-\frac{9}{11}\right)$

4. $\frac{7}{12} \cdot \left(-\frac{5}{6}\right)$

5. $\frac{21}{8} \cdot \frac{16}{3}$

6. $\frac{18}{5} \cdot \frac{10}{3}$

7. $-\frac{3}{2} \cdot \left(-\frac{8}{9}\right)$

8. $-\frac{3}{4} \cdot \left(-\frac{8}{15}\right)$

9. $\frac{5}{8} \cdot \frac{32}{15}$

10. $\frac{3}{5} \cdot \frac{10}{9}$

11. $\frac{10}{3} \cdot \frac{6}{5}$

12. $\frac{10}{3} \cdot \frac{18}{5}$

13. $-\frac{2}{3} \cdot 1\frac{2}{3}$

14. $\frac{2}{5} \cdot \left(-2\frac{2}{3}\right)$

15. $2\frac{4}{7} \cdot \frac{5}{6}$

16. $4\frac{4}{5} \cdot \frac{3}{4}$

17. $-4\frac{1}{2} \cdot \left(-\frac{8}{15}\right)$

18. $-2\frac{1}{7} \cdot \left(-\frac{21}{10}\right)$

19. $5\frac{1}{2} \cdot \left(-1\frac{5}{11}\right)$

20. $-4\frac{1}{3} \cdot 2\frac{1}{13}$

21. $2\frac{2}{3} \cdot 3\frac{3}{5}$

22. $2\frac{1}{2} \cdot 1\frac{9}{21}$

23. $-10 \cdot \frac{5}{8}$

24. $16 \cdot \left(-\frac{5}{6}\right)$

25. $-24 \cdot \left(-2\frac{5}{6}\right)$

26. $-18 \cdot \left(-3\frac{2}{9}\right)$

27. $4\frac{1}{3} \cdot 9$

28. $4\frac{3}{4} \cdot 12$

29. $-\frac{3}{7} \cdot \frac{14}{5}$

30. $\frac{5}{11} \cdot \left(-\frac{22}{7}\right)$

31. $4\frac{1}{2} \cdot 3\frac{1}{3}$

32. $6\frac{2}{3} \cdot 2\frac{2}{5}$

33. $-2\frac{1}{4} \cdot \left(-1\frac{1}{3}\right)$

34. $-2\frac{2}{5} \cdot \left(-2\frac{1}{12}\right)$

35. $15 \cdot \frac{9}{5}$

36. $28 \cdot \frac{11}{7}$

37. $\frac{7}{18} \cdot \left(-2\frac{1}{4}\right)$

38. $-\frac{8}{15} \cdot 3\frac{1}{3}$

39. $\frac{7}{13} \cdot \left(-\frac{26}{21}\right) \cdot \frac{3}{4}$

40. $\frac{6}{15} \cdot \frac{25}{8} \cdot \left(-\frac{9}{10}\right)$

41. $5\frac{3}{5} \cdot \left(-1\frac{11}{14}\right) \cdot \left(-1\frac{3}{10}\right)$

42. $-6\frac{3}{8} \cdot 2\frac{2}{15} \cdot \left(-1\frac{3}{17}\right)$

43. $3\frac{4}{7} \cdot \frac{9}{10} \cdot 14$

44. $3\frac{5}{9} \cdot \frac{15}{16} \cdot 12$

Answers: **Practice Exercise 6:** **a.** $\frac{1}{2}$ of a beaker **b.** 51 m²   **Additional Practice 6:** **a.** $\frac{25}{2}$, or $12\frac{1}{2}$ cups **b.** 3 in.²

Find the products of the following rational expressions. (See Example 5.)

45. $\dfrac{r^3}{s^3} \cdot \dfrac{s^5}{r^6}$

46. $\dfrac{a^5}{b^4} \cdot \dfrac{b^2}{a^3}$

47. $\dfrac{x^3}{y^3} \cdot \dfrac{y^5}{x}$

48. $\dfrac{a^4}{b^2} \cdot \dfrac{b^4}{a}$

49. $\dfrac{m^2}{n^2} \cdot \dfrac{n^2}{m^4}$

50. $\dfrac{c^3}{d^5} \cdot \dfrac{d^2}{c^3}$

51. $\dfrac{x^2y^2}{a^2b^3} \cdot \dfrac{a^4b}{xy^4}$

52. $\dfrac{r^4s^2}{t^3u^3} \cdot \dfrac{t^5u}{r^6s}$

53. $\dfrac{x^3y^2}{xy^3} \cdot \dfrac{x^2y^4}{x^3y^5}$

54. $\dfrac{m^2n^2}{m^3n^2} \cdot \dfrac{mn^3}{m^2n}$

55. $\dfrac{r^4s}{r^3s^2} \cdot \dfrac{r^2s^3}{r^4s^3}$

56. $\dfrac{a^2b^3}{a^3b^2} \cdot \dfrac{a^2b}{a^3b^3}$

57. $\dfrac{6x^2}{5y^3} \cdot \dfrac{10y^5}{8x}$

58. $\dfrac{12a^4}{7b} \cdot \dfrac{14b^2}{16a^2}$

59. $\dfrac{2x^2y^4}{3w^4z^2} \cdot \dfrac{15w^3z^4}{3x^5y^2}$

60. $\dfrac{21a^4b^3}{12c^2d^6} \cdot \dfrac{9c^4d^2}{14a^2b^5}$

61. $\dfrac{8x^2y^4}{12x^3y^2} \cdot \dfrac{4x^3y^3}{6xy^3}$

62. $\dfrac{10a^2b^2}{18a^2b} \cdot \dfrac{12a^3b^4}{15a^2b^4}$

63. $\dfrac{24a^2bc^3}{16ab^3c^2} \cdot \dfrac{32a^3b^2c^3}{18a^4b^3c^3}$

64. $\dfrac{30x^3y^2z}{42x^2yz^2} \cdot \dfrac{15x^2y^2z^2}{25x^4yz}$

Answer the following. (See Example 6.)

65. A fashion designer's latest creation requires $4\frac{1}{3}$ square yards of material per dress. How many square yards of material would it take to make 24 of these dresses?

66. A tailor needs $3\frac{3}{4}$ square yards of material to make a suit. How many square yards of material would be needed to make 16 suits?

67. A wiring harness for a boat trailer requires $40\frac{1}{3}$ yards of wire. How many yards of wire would the manufacturer of this trailer need for 36 trailers?

68. It takes $4\frac{5}{12}$ feet of ribbon to wrap a birthday package. How many feet of ribbon would it take to wrap 18 of these packages?

69. A chemist has a beaker that is $\frac{2}{3}$ full of hydrochloric acid. In an experiment, she uses $\frac{1}{3}$ of the acid in the beaker. What fractional part of a full beaker did she use?

70. A pharmacist fills a prescription for a blood pressure medication. He uses $\frac{3}{4}$ of a bottle that is $\frac{2}{3}$ full. What part of a full bottle does it take to fill this prescription?

71. A board is $3\frac{3}{8}$ feet long. A carpenter needs $\frac{2}{3}$ of the board to make a shelf. How long is the piece he needs?

72. A tree is $26\frac{2}{3}$ feet high. A woodpecker has drilled a hole in the tree $\frac{3}{4}$ of the way to the top. How far is the hole above the ground?

Recall from Chapter 1 that the formula for the area of a rectangle is $A = LW$, where A represents the area, L represents the length, and W represents the width. If the length of a rectangle is $2\frac{3}{4}$ inches and the width is $3\frac{2}{3}$ inches, then $A = 2\frac{3}{4} \cdot 3\frac{2}{3} = \frac{11}{4} \cdot \frac{11}{3} = \frac{121}{12}$ square inches. Find the area of each of the following rectangles:

73. $L = 2\frac{1}{2}$ centimeters and $W = 3\frac{3}{4}$ centimeters

74. $L = 3\frac{3}{5}$ meters and $W = 4\frac{2}{3}$ meters

75. $L = 12$ feet and $W = 3\frac{1}{6}$ feet

76. $L = 8\frac{2}{3}$ yards and $W = 6$ yards

Recall that the formula for the area of a triangle is $A = \frac{1}{2}bh$, where A is the area, b is the base, and h is the height. If the base of a triangle is $\frac{3}{4}$ feet and the height is $\frac{3}{5}$ feet, then $A = \frac{1}{2} \cdot \frac{3}{4} \cdot \frac{3}{5} = \frac{9}{40}$ square feet. Find the areas of the following triangles:

77. $b = \dfrac{7}{8}$ meter and $h = \dfrac{9}{10}$ meter

78. $b = \dfrac{7}{12}$ yard and $h = \dfrac{5}{6}$ yard

79. $b = 2\dfrac{1}{6}$ feet and $h = 3\dfrac{1}{4}$ feet

80. $b = 4\dfrac{3}{10}$ meters and $h = 5\dfrac{3}{5}$ meters

Recall that the formula for the volume of a rectangular box is $V = LWH$, where L is the length, W is the width, and H is the height. If the length is $2\frac{1}{4}$ inches, the width is $3\frac{1}{5}$ inches, and the height is $1\frac{3}{8}$ inches, then $V = 2\frac{1}{4} \cdot 3\frac{1}{5} \cdot 1\frac{3}{8} = \frac{9}{4} \cdot \frac{16}{5} \cdot \frac{11}{8} = \frac{99}{10}$ cubic inches. Find the volumes of the following rectangular boxes:

81. $L = 2\dfrac{3}{4}$ centimeters, $W = 3\dfrac{3}{11}$ centimeters,

and $H = 1\dfrac{2}{9}$ centimeters

82. $L = 3\dfrac{3}{5}$ inches, $W = 2\dfrac{2}{9}$ inches, and $H = 2\dfrac{1}{10}$ inches

83. $L = 3\dfrac{1}{5}$ yards, $W = 10$ yards, and $H = 2\dfrac{1}{4}$ yards

84. $L = 2\dfrac{5}{8}$ meters, $W = 16$ meters, and $H = 4\dfrac{2}{3}$ meters

Challenge Exercises (85 and 86)

Recall that the formula for finding the area of a trapezoid is $A = \frac{1}{2}h(B + b)$, where A represents the area, h represents the height, B represents the longer of the two parallel sides, and b represents the shorter of the two parallel sides. See the figure in the margin. Find the areas of the following trapezoids:

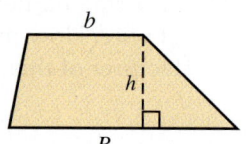

85. $h = 1\dfrac{2}{3}$ inches, $B = 8$ inches, $b = 4$ inches

86. $h = 2\dfrac{2}{5}$ centimeters, $B = 8$ centimeters,

$b = 7$ centimeters

Writing Exercise

87. In finding the product of fractions, why are we allowed to divide by factors common to the numerators and the denominators before multiplying?

Section 6.4

Further Multiplication of Rational Expressions

OBJECTIVES *When you complete this section, you will be able to:*

A Multiply rational expressions whose numerators and denominators are not monomials.

B Multiply rational expressions that contain factors of the form $a - b$ and $b - a$.

PREREQUISITE SKILLS *Before beginning this section, you should be able to:*

a. Multiply rational numbers and expressions with monomial numerators and denominators. (Section 6.3)

b. Factor polynomials. (Sections 5.2–5.6)

GETTING READY FOR SECTION 6.4

Multiply the following rational numbers and rational expressions with monomial numerators and denominators:

1. $\dfrac{3}{15} \cdot \dfrac{10}{9}$

2. $\dfrac{a^2 b^2}{a^3 b^3} \cdot \dfrac{a^2 b^4}{a^3 b}$

Factor the following polynomials:

3. $5x - 15$ **4.** $x^2 + 2x - 15$ **5.** $x^2 - 4$

Introduction

In Section 6.3, we multiplied rational numbers and rational expressions with monomial numerators and denominators, using the following procedure:

1. Factor the numerators and the denominators into prime factors.

2. Divide the numerators and the denominators by factors common to both.

3. Multiply the remaining factors in the numerators and divide this by the product of the remaining factors in the denominators.

OBJECTIVE **A** Multiplying rational expressions whose numerators and denominators are not monomials

We use the preceding technique to multiply rational expressions whose numerators and/or denominators are not integers or monomials. The only difference is the manner in which we factor the numerators and the denominators, since they are polynomials instead of integers or monomials. Compare the following, assuming that the variables cannot have values that make any denominator equal to 0:

Monomials

$$\frac{3}{15} \cdot \frac{10}{9} =$$

$$\frac{3}{3 \cdot 5} \cdot \frac{2 \cdot 5}{3 \cdot 3} =$$

$$\frac{\cancel{3}}{\cancel{3} \cdot 5} \cdot \frac{2 \cdot \cancel{5}}{3 \cdot 3} =$$

$$\frac{2}{9}$$

$$\frac{a^2 b^2}{a^3 b^3} \cdot \frac{a^2 b^4}{a^3 b} =$$

$$\frac{a \cdot a \cdot b \cdot b}{a \cdot a \cdot a \cdot b \cdot b \cdot b} \cdot \frac{a \cdot a \cdot b \cdot b \cdot b \cdot b}{a \cdot a \cdot a \cdot b} =$$

$$\frac{\cancel{a} \cdot \cancel{a} \cdot \cancel{b} \cdot \cancel{b}}{\cancel{a} \cdot \cancel{a} \cdot \cancel{a} \cdot \cancel{b} \cdot \cancel{b} \cdot \cancel{b}} \cdot \frac{\cancel{a} \cdot \cancel{a} \cdot b \cdot \cancel{b} \cdot b \cdot b}{\cancel{a} \cdot a \cdot a \cdot \cancel{b}} =$$

$$\frac{b^2}{a^2}$$

Nonmonomials

$$\frac{2x - 6}{x^2 + x - 2} \cdot \frac{x^2 + 3x - 4}{x^2 - 9} =$$

Factor the numerators and denominators.

$$\frac{2(x - 3)}{(x + 2)(x - 1)} \cdot \frac{(x + 4)(x - 1)}{(x + 3)(x - 3)} =$$

Divide by the common factors.

$$\frac{2(x \cancel{- 3})}{(x + 2)(x \cancel{- 1})} \cdot \frac{(x + 4)(x \cancel{- 1})}{(x + 3)(x \cancel{- 3})} =$$

Multiply the remaining factors.

$$\frac{2(x + 4)}{(x + 2)(x + 3)}$$

Product.

Answers: Getting Ready: 1. $\dfrac{2}{9}$ **2.** $\dfrac{b^2}{a^2}$ **3.** $5(x - 3)$ **4.** $(x + 5)(x - 3)$ **5.** $(x + 2)(x - 2)$

Example 1   Find the products. Express answers reduced to lowest terms.

a. $\dfrac{a + b}{2x^2} \cdot \dfrac{4x}{(a + b)^2} =$ Factor the numerators and the denominators.

$\dfrac{a + b}{2 \cdot x \cdot x} \cdot \dfrac{2 \cdot 2 \cdot x}{(a + b)(a + b)} =$ Divide by the common factors.

$\dfrac{\cancel{a + b}}{2 \cdot x \cdot \cancel{x}} \cdot \dfrac{2 \cdot 2 \cdot \cancel{x}}{(\cancel{a + b})(a + b)} =$ Multiply the remaining factors.

$\dfrac{2}{x(a + b)}$ Product.

b. $\dfrac{2x - 6}{3y + 12} \cdot \dfrac{2y + 8}{5x - 15} =$ Factor the numerators and the denominators.

$\dfrac{2(x - 3)}{3(y + 4)} \cdot \dfrac{2(y + 4)}{5(x - 3)} =$ Divide by the common factors.

$\dfrac{2(\cancel{x - 3})}{3(\cancel{y + 4})} \cdot \dfrac{2(\cancel{y + 4})}{5(\cancel{x - 3})} =$ Multiply the remaining factors.

$\dfrac{2 \cdot 2}{3 \cdot 5} =$ Perform the multiplications.

$\dfrac{4}{15}$ Product.

c. $\dfrac{x^2 - x - 6}{x^2 + 2x - 15} \cdot \dfrac{x^2 + 4x - 12}{x^2 - 4} =$ Factor the numerators and the denominators.

$\dfrac{(x + 2)(x - 3)}{(x + 5)(x - 3)} \cdot \dfrac{(x - 2)(x + 6)}{(x + 2)(x - 2)} =$ Divide by the common factors.

$\dfrac{(\cancel{x + 2})(\cancel{x - 3})}{(x + 5)(\cancel{x - 3})} \cdot \dfrac{(\cancel{x - 2})(x + 6)}{(\cancel{x + 2})(\cancel{x - 2})} =$ Multiply the remaining factors.

$\dfrac{x + 6}{x + 5}$ Product.

PRACTICE EXERCISE 1

Find the products. Express answers reduced to lowest terms.

a. $\dfrac{3x^2y}{(w + z)^2} \cdot \dfrac{w + z}{4xy^3}$ **b.** $\dfrac{3a + 12}{5b - 30} \cdot \dfrac{b - 6}{4a + 16}$ **c.** $\dfrac{a^2 + 6a + 8}{a^2 + 5a - 6} \cdot \dfrac{a^2 + 3a - 18}{a^2 + a - 12}$

If you need more practice, do the following Additional Practice Exercises.

Additional Practice Exercise 1   Find the products. Express answers reduced to lowest terms.

a. $\dfrac{5rs}{(r + s)^2} \cdot \dfrac{2(r + s)}{7r^2s}$ **b.** $\dfrac{4a - 12}{5b + 20} \cdot \dfrac{2b + 8}{3a - 9}$ **c.** $\dfrac{a^2 + a - 2}{a^2 - 3a - 10} \cdot \dfrac{a^2 - a - 20}{a^2 + 3a - 4}$

OBJECTIVE **B** Multiplying rational expressions that contain factors of the form $a - b$ and $b - a$

In Section 6.2, we reduced rational expressions that contained expressions of the form $\frac{a-b}{b-a}$. Recall that $\frac{a-b}{b-a} = -1$. We can also use this fact when multiplying polynomials.

Example 2 | **Find the products. Reduce answers to lowest terms.**

a. $\dfrac{4 - a}{5} \cdot \dfrac{7}{a - 4} =$ Rewrite $4 - a$ in descending order.

$\dfrac{-a + 4}{5} \cdot \dfrac{7}{a - 4} =$ Remove the common factor of -1 from $-a + 4$.

$\dfrac{-1(a - 4)}{5} \cdot \dfrac{7}{a - 4} =$ Divide by the common factors.

$\dfrac{-1(a \cancel{- 4})}{5} \cdot \dfrac{7}{a \cancel{- 4}} =$ Multiply the remaining factors.

$\dfrac{-7}{5}$ or $-\dfrac{7}{5}$ Product $\frac{-7}{5} = -\frac{7}{5}$, since a negative divided by a positive is negative.

b. $\dfrac{3 - a}{a^2 + 3a + 2} \cdot \dfrac{a^2 - 4}{a^2 + 2a - 15} =$ Factor everything that is not prime and write $3 - a$ as $-a + 3$.

$\dfrac{-a + 3}{(a + 2)(a + 1)} \cdot \dfrac{(a + 2)(a - 2)}{(a - 3)(a + 5)} =$ Factor a "-1" from $-a + 3$.

$\dfrac{-1(a - 3)}{(a + 2)(a + 1)} \cdot \dfrac{(a + 2)(a - 2)}{(a - 3)(a + 5)} =$ Divide by the common factors.

$\dfrac{-1(a \cancel{- 3})}{(a \cancel{+ 2})(a + 1)} \cdot \dfrac{(a \cancel{+ 2})(a - 2)}{(a \cancel{- 3})(a + 5)} =$ Multiply the remaining factors.

$\dfrac{-1(a - 2)}{(a + 1)(a + 5)}$ Product.

> **Note**
>
> The answer to Example 2b could be written as $-\dfrac{a - 2}{(a + 1)(a + 5)}$ or $\dfrac{2 - a}{(a + 1)(a + 5)}$. Why?

PRACTICE EXERCISE 2

Find the products. Express answers reduced to lowest terms.

a. $\dfrac{7 - x}{a} \cdot \dfrac{b}{x - 7}$

b. $\dfrac{y^2 + 3y - 10}{y^2 - 25} \cdot \dfrac{y + 4}{2 - y}$

If you need more practice, do the following Additional Practice Exercises.

Additional Practice Exercise 2   Find the products. Express answers reduced to lowest terms.

a. $\dfrac{rs}{a - 2} \cdot \dfrac{2 - a}{2a}$

b. $\dfrac{m^2 - 7m + 12}{m^2 - 2m - 8} \cdot \dfrac{m - 6}{3 - m}$

Answers: Practice Exercise 2: a. $-\dfrac{b}{a}$ **b.** $-\dfrac{y + 4}{y - 5}$ **Additional Practice 2: a.** $-\dfrac{rs}{2a}$ **b.** $-\dfrac{m - 6}{m + 2}$

For Extra Help

Exercise Set 6.4   MyMathLab®

Find the products. Express answers reduced to lowest terms. (See Examples 1 and 2.)

1. $\dfrac{(x + y)^2}{6a} \cdot \dfrac{8a^2}{x + y}$

2. $\dfrac{12x}{(a + b)^2} \cdot \dfrac{a + b}{16x^2}$

3. $\dfrac{3x^2y^4}{10(r + s)} \cdot \dfrac{5(r + s)^2}{12xy^2}$

4. $\dfrac{6(p + q)^2}{15rs} \cdot \dfrac{5r^3s^4}{18(p + q)}$

5. $\dfrac{3x + 18}{7x - 7} \cdot \dfrac{2x - 2}{5x + 30}$

6. $\dfrac{2r - 8}{3r + 6} \cdot \dfrac{5r + 10}{3r - 12}$

7. $\dfrac{8x + 12}{18x - 24} \cdot \dfrac{9x - 12}{10x + 15}$

8. $\dfrac{24x + 16}{30x - 18} \cdot \dfrac{10x - 6}{18x + 12}$

9. $\dfrac{4a - 16}{3a - 3} \cdot \dfrac{a^2 - 1}{a^2 - 16}$

10. $\dfrac{x^2 - 9}{4x + 12} \cdot \dfrac{3x - 15}{x^2 - 25}$

11. $\dfrac{4x^2 - 25}{3x - 4} \cdot \dfrac{9x^2 - 16}{2x + 5}$

12. $\dfrac{16x^2 - 9}{4x + 3} \cdot \dfrac{25x^2 - 1}{5x + 1}$

13. $\dfrac{a - 6}{7} \cdot \dfrac{9}{6 - a}$

14. $\dfrac{b - 4}{5} \cdot \dfrac{3}{4 - b}$

15. $\dfrac{3x - 4}{14} \cdot \dfrac{21}{4 - 3x}$

16. $\dfrac{5 - 2b}{18} \cdot \dfrac{12}{2b - 5}$

17. $\dfrac{4x^2y^3}{2x - 6} \cdot \dfrac{12 - 4x}{6x^3y}$

18. $\dfrac{6b^3c^3}{70 - 10b} \cdot \dfrac{5b - 35}{8b^2c^5}$

19. $\dfrac{x - 3}{x^2 + x - 2} \cdot \dfrac{x + 2}{x^2 + x - 12}$

20. $\dfrac{a - 4}{a^2 - 2a - 3} \cdot \dfrac{a + 1}{a^2 - 2a - 8}$

21. $\dfrac{x^2 - 4}{x^2 - 3x - 10} \cdot \dfrac{x^2 - 8x + 15}{x^2 - 9}$

22. $\dfrac{x^2 + 3x - 10}{x^2 - 16} \cdot \dfrac{x^2 - 9x + 20}{x^2 - 25}$

23. $\dfrac{x^2 - 2x - 15}{x^2 + x - 30} \cdot \dfrac{x^2 + 5x - 6}{x^2 + 7x + 12}$

24. $\dfrac{a^2 - 2a - 24}{a^2 + 9a + 20} \cdot \dfrac{a^2 - 3a - 10}{a^2 - 4a - 12}$

25. $\dfrac{x^2 + 2x - 15}{x^2 - 9x + 18} \cdot \dfrac{x^2 - 4x - 12}{x^2 + 3x - 10}$

26. $\dfrac{y^2 - 7y + 10}{y^2 + 4y - 12} \cdot \dfrac{y^2 + 10y + 24}{y^2 - 2y - 15}$

27. $\dfrac{6x^2 + x - 12}{2x^2 - 5x - 12} \cdot \dfrac{3x^2 - 14x + 8}{9x^2 - 18x + 8}$

28. $\dfrac{2x^2 + x - 15}{4x^2 + 11x - 3} \cdot \dfrac{4x^2 - 9x + 2}{2x^2 - 9x + 10}$

29. $\dfrac{3x^2 - 5x - 2}{4x^2 - 11x + 6} \cdot \dfrac{4x^2 + 5x - 6}{3x^2 - 8x - 3}$

30. $\dfrac{8x^2 + 10x - 3}{3x^2 + 4x - 4} \cdot \dfrac{x^2 + 6x + 8}{2x^2 + 11x + 12}$

31. $\dfrac{b^2 - 16}{b^2 + b - 12} \cdot \dfrac{b^2 + 2b - 15}{4 - b}$

32. $\dfrac{y^2 - y - 20}{y^2 - 2y - 24} \cdot \dfrac{y + 3}{5 - y}$

33. $\dfrac{2x^2 - 11x + 12}{2x^2 + 11x - 21} \cdot \dfrac{3x^2 + 20x - 7}{4 - x}$

34. $\dfrac{3x^2 - 17x + 10}{3x^2 + 7x - 6} \cdot \dfrac{2x^2 + x - 15}{5 - x}$

Challenge Exercises (35–40)

Find the products. Reduce answers to lowest terms.

35. $\dfrac{ac + 3a + 2c + 6}{ad + a + 2d + 2} \cdot \dfrac{ad - 5a + 2d - 10}{bc + 3b - 4c - 12}$

36. $\dfrac{mn + 2m + 4n + 8}{np - 3n + 2p - 6} \cdot \dfrac{pq + 5p - 3q - 15}{mq + 4m + 4q + 16}$

37. $\dfrac{ac - ad + bc - bd}{am + an - bm - bn} \cdot \dfrac{ac + ad - bc - bd}{ac + ad + bc + bd}$

38. $\dfrac{xw - xz - yw + yz}{xa - xb + ya - yb} \cdot \dfrac{xw + xz + yw + yz}{xw - xz - yw + yz}$

39. $\dfrac{a^3 - b^3}{a^2 + ab + b^2} \cdot \dfrac{2a^2 + ab - b^2}{a^2 - b^2}$

40. $\dfrac{3x^2 + 10x - 8}{x^2 - 4} \cdot \dfrac{x^3 - 8}{x^2 + 2x + 4}$

Writing Exercises (41 and 42)

41. What is wrong with the following? Rework the problem correctly.

$$\frac{a^2 + b^2}{a^2 - b^2} \cdot \frac{a^2 + 2ab - 3b^2}{a^2 + ab - 6b^2} = \frac{(a + b)(a + b)}{(a + b)(a - b)} \cdot \frac{(a - b)(a + 3b)}{(a + 3b)(a - 2b)} = \frac{(\cancel{a + b})(a + b)}{(\cancel{a + b})(a - b)} \cdot \frac{(a - b)(\cancel{a + 3b})}{(\cancel{a + 3b})(a - 2b)} = \frac{a + b}{a - 2b}$$

42. What is wrong with the following? Rework the problem correctly.

$$\frac{a^2 + b^2}{a^2 - b^2} \cdot \frac{a^2 + 2ab - 3b^2}{a^2 + ab - 6b^2} = \frac{a^2 + b^2}{a^2 - b^2} \cdot \frac{a^2 + 2ab - 3b^2}{a^2 + ab - \underset{2}{\cancel{6}}b^2} = \frac{2ab}{ab - 2}$$

Section 6.5   Division of Rational Numbers and Expressions

OBJECTIVES *When you complete this section, you will be able to:*

A Divide rational numbers.

B Solve application problems with rational numbers.

C Divide rational expressions.

PREREQUISITE SKILLS *Before beginning this section, you should be able to:*

a. Convert mixed numbers into improper fractions. (Section R.4)

b. Factor whole numbers and polynomials. (Sections 5.1–5.6)

c. Multiply rational numbers and expressions. (Sections 6.3 and 6.4)

GETTING READY FOR SECTION 6.5

1. Convert $4\frac{2}{5}$ into an improper fraction.

Factor each of the following:

2. 14

3. $x^2 - 2x$

4. $2x^2 + 7x + 6$

5. $x^2 - 36$

Multiply the following rational numbers and expressions:

6. $\dfrac{5}{8} \cdot \dfrac{4}{15}$

7. $\dfrac{r^3 s^2}{p^2 q^2} \cdot \dfrac{p^4 q}{r s^3}$

8. $\dfrac{2x^2 + 7x + 3}{x^2 - 9} \cdot \dfrac{x^2 - 3x}{2x^2 + 11x + 5}$

Introduction

In Sections 6.3 and 6.4, we multiplied rational numbers and rational expressions. In this section, we will divide rational numbers and expressions. Before we discuss division of rational numbers and expressions, we need to review the multiplicative inverse, also called the *reciprocal*. Two numbers are multiplicative inverses if their product is 1. For example, $\frac{3}{4}$ and $\frac{4}{3}$ are multiplicative inverses (reciprocals), since $\frac{3}{4} \cdot \frac{4}{3} = 1$. In general, for a and $b \neq 0$, the multiplicative inverse of $\frac{a}{b}$ is $\frac{b}{a}$, since $\frac{a}{b} \cdot \frac{b}{a} = 1$. This procedure is sometimes referred to as "inverting the fraction." To find the multiplicative inverse of a number (other than 0), we interchange the numerator and the denominator. Remember, 0 does not have a multiplicative inverse.

Suppose that a region is divided into two equal parts so that each part is $\frac{1}{2}$ of the original region, as shown in the figure:

Next, divide each of the two equal regions into two equal parts by drawing a horizontal line segment through the figure. Now further shade one of the two parts that was already shaded.

From the figure, we can see that this one part that we further shaded is $\frac{1}{4}$ of the original region. We have taken $\frac{1}{2}$ of the region and divided it by 2. This resulted in $\frac{1}{4}$ of

---

Answers: Getting Ready: 1. $\dfrac{22}{5}$ **2.** $2 \cdot 7$ **3.** $x(x - 2)$ **4.** $(2x + 3)(x + 2)$ **5.** $(x + 6)(x - 6)$ **6.** $\dfrac{1}{6}$ **7.** $\dfrac{r^2 p^2}{qs}$ **8.** $\dfrac{x}{x + 5}$

the original region. In symbols, $\frac{1}{2} \div 2 = \frac{1}{4}$. Notice that $\frac{1}{2} \cdot \frac{1}{2} = \frac{1}{4}$ also. Consequently, $\frac{1}{2} \div 2 = \frac{1}{2} \cdot \frac{1}{2}$. Without shading parts of a figure, we can see that $6 \div 3 = 2$ and $6 \cdot \frac{1}{3} = 2$, also. Thus, $6 \div 3 = 6 \cdot \frac{1}{3}$.

This suggests that dividing by a whole number, except for 0, is the same as multiplying by its multiplicative inverse (reciprocal).

To show that the same procedure applies if we are dividing by a fraction is a little more difficult. Remember, any fraction represents division, and division may be represented as a fraction. For example, $\frac{2}{3} = 2 \div 3$, and $5 \div 8 = \frac{5}{8}$.

Suppose that we have $\frac{2}{3} \div \frac{7}{5}$. Since division may be represented as a fraction, $\frac{2}{3} \div \frac{7}{5}$ may be represented as $\dfrac{\frac{2}{3}}{\frac{7}{5}}$.

In Sections 6.1 and 6.2, we reduced fractions by dividing the numerator and the denominator by factors common to both. We are now going to do something similar by multiplying the numerator and the denominator by the same number. This is permissible, since we are actually multiplying by 1. (Any number, except 0, divided by itself equals 1.) You will not be asked to actually do this in this section, but simply to understand that it can be done. This procedure will be discussed in greater detail in Chapter 7, when adding rational expressions with unlike denominators is addressed. Study the following:

$$\frac{2}{3} \div \frac{7}{5} = \qquad \text{Rewrite as a fraction.}$$

$$\dfrac{\frac{2}{3}}{\frac{7}{5}} = \qquad \text{Multiply the numerator and the denominator by the reciprocal of the denominators, } \frac{5}{7}.$$

$$\dfrac{\frac{2}{3} \cdot \frac{5}{7}}{\frac{7}{5} \cdot \frac{5}{7}} = \qquad \frac{7}{5} \cdot \frac{5}{7} = 1.$$

$$\dfrac{\frac{2}{3} \cdot \frac{5}{7}}{1} = \qquad \text{Divide by 1.}$$

$$\frac{2}{3} \cdot \frac{5}{7} = \qquad \text{Therefore, } \frac{2}{3} \div \frac{7}{5} = \frac{2}{3} \cdot \frac{5}{7}. \text{ Notice that the division by } \frac{7}{5} \text{ was changed into multiplication by } \frac{5}{7}. \text{ Multiply.}$$

$$\frac{10}{21} \qquad \text{Quotient.}$$

OBJECTIVE Dividing rational numbers

The preceding two cases suggest the following procedure for the division of rational numbers:

> **Procedure: Division of Rational Numbers**
>
> For all integers a, b, c, and d, with b, c, and $d \neq 0$, $\frac{a}{b} \div \frac{c}{d} = \frac{a}{b} \cdot \frac{d}{c}$. In words, to divide by a number, multiply by its multiplicative inverse (reciprocal).

This procedure is usually stated more loosely as "When dividing rational numbers, invert the one on the right and change the operation to multiplication." It is this form of the procedure that we will use in the subsequent examples.

We summarize the procedure for finding quotients of rational expressions as follows:

> **Procedure: Quotients of Rational Expressions**
>
> To find the quotient of two rational expressions, take these steps:
>
> 1. Invert the rational expression on the right and change the operation to multiplication.
>
> 2. Factor all numerators and denominators completely.
>
> 3. Divide by all factors common to the numerators and the denominators.
>
> 4. Find the product of all remaining factors.

Example 1   **Find the quotients. Express answers reduced to lowest terms.**

a. $\dfrac{2}{3} \div \dfrac{5}{3} =$ Invert the number on the right and change the operation to multiplication.

$\dfrac{2}{3} \cdot \dfrac{3}{5} =$ Divide by the common factor of 3.

$\dfrac{2}{\cancel{3}} \cdot \dfrac{\cancel{3}}{5} =$ Multiply the remaining factors.

$\dfrac{2}{5}$ Quotient.

b. $-\dfrac{5}{8} \div 3\dfrac{3}{4} =$ Change the mixed number to an improper fraction.

$-\dfrac{5}{8} \div \dfrac{15}{4} =$ Invert the number on the right and change the operation to multiplication.

$-\dfrac{5}{8} \cdot \dfrac{4}{15} =$ Write the numerators and the denominators in terms of prime factors.

$-\dfrac{5}{2 \cdot 2 \cdot 2} \cdot \dfrac{2 \cdot 2}{3 \cdot 5} =$ Divide by the common factors.

$-\dfrac{\cancel{5}}{2 \cdot 2 \cdot 2} \cdot \dfrac{2 \cdot 2}{3 \cdot \cancel{5}} =$ Multiply the remaining factors.

$-\dfrac{1}{6}$ Quotient.

c. $4\dfrac{2}{5} \div 11 =$ Change the mixed number to an improper fraction and write the whole number 11 as the fraction $\dfrac{11}{1}$.

$\dfrac{22}{5} \div \dfrac{11}{1} =$ Invert the number on the right and change the operation to multiplication.

$\dfrac{22}{5} \cdot \dfrac{1}{11} =$ Write the numerators and the denominators in terms of prime factors.

$\dfrac{2 \cdot 11}{5} \cdot \dfrac{1}{11} =$ Divide by the common factor of 11.

$\dfrac{2 \cdot \cancel{11}}{5} \cdot \dfrac{1}{\cancel{11}} =$ Multiply the remaining factors.

$\dfrac{2}{5}$ Quotient.

d. $-3\dfrac{1}{3} \div \left(-2\dfrac{1}{7}\right) =$ Change the mixed numbers into improper fractions.

$-\dfrac{10}{3} \div \left(-\dfrac{15}{7}\right) =$ Invert the fraction on the right and change the operation to multiplication.

$-\dfrac{10}{3} \cdot \left(-\dfrac{7}{15}\right) =$ Write the numerators and the denominators in terms of prime factors.

$-\dfrac{2 \cdot 5}{3} \cdot \left(-\dfrac{7}{3 \cdot 5}\right) =$ Divide by the common factor of 5.

$-\dfrac{2 \cdot \cancel{5}}{3} \cdot \left(-\dfrac{7}{3 \cdot \cancel{5}}\right) =$ Multiply the remaining factors.

$\dfrac{14}{9}$ Quotient.

PRACTICE EXERCISE 1

Find the quotients. Express answers reduced to lowest terms.

a. $\dfrac{3}{5} \div \dfrac{7}{5}$ **b.** $\dfrac{10}{7} \div \left(-2\dfrac{1}{2}\right)$ **c.** $3\dfrac{3}{7} \div 8$ **d.** $-5\dfrac{1}{3} \div \left(-2\dfrac{2}{5}\right)$

(Continued)

Answers: **Practice Exercise 1:** **a.** $\dfrac{3}{7}$ **b.** $-\dfrac{4}{7}$ **c.** $\dfrac{3}{7}$ **d.** $\dfrac{20}{9}$

If you need more practice, do the following Additional Practice Exercises.

Additional Practice Exercise 1 Find the quotients. Express answers reduced to lowest terms.

a. $-\dfrac{5}{8} \div \dfrac{3}{8}$ **b.** $\dfrac{8}{5} \div 2\dfrac{2}{3}$ **c.** $3\dfrac{3}{5} \div (-9)$ **d.** $-2\dfrac{1}{3} \div \left(-2\dfrac{4}{5}\right)$

OBJECTIVE Solving application problems with rational numbers

Applications with rational numbers often fall into two categories: Total ÷ Number of Parts = Size of Each Part or Total ÷ Size of Each Part = Number of Parts.

Example 2

On June 10, 2010, $3\dfrac{1}{4}$ ounces of gold cost \$4082. What was the cost of one ounce?

Solution

This is of the form Total ÷ Number of Parts = Size of Each Part, where the total is the total cost, the number of parts is the number of ounces and the size of each part is the cost per ounce. Substituting we have:

$4082 \div 3\dfrac{1}{4} = $ Cost per ounce   Change to an improper fraction.

$4082 \div \dfrac{13}{4} = $ Rewrite as multiplication.

$4082 \cdot \dfrac{4}{13} = $ Multiply.

1256 One ounce of gold cost \$1256.

PRACTICE EXERCISE 2

A piece of paper with an area of $3\dfrac{3}{4}$ square feet is cut up into pieces whose area is $\dfrac{3}{8}$ of a square foot each. How many pieces are there?

If you need more practice, do the following Additional Practice Exercise.

Additional Practice Exercise 2 A developer bought $8\dfrac{3}{4}$ acres of land for \$91,875. How much did he pay per acre?

OBJECTIVE Dividing rational expressions

Division of rational expressions is done in exactly the same manner as division of rational numbers. We invert the rational expression on the right and change the operation to multiplication.

Answers: Additional Practice 1: a. $-\dfrac{5}{3}$ **b.** $\dfrac{3}{5}$ **c.** $-\dfrac{2}{5}$ **d.** $\dfrac{5}{6}$ **Practice Exercise 2:** 10   **Additional Practice 2:** \$10,500

Example 3   **Find the quotients. Express answers reduced to lowest terms.**

a. $\dfrac{a^2b^3}{xy^2} \div \dfrac{ab^4}{x^2y^3} =$

Invert the rational expression on the right and change the operation to multiplication.

$\dfrac{a^2b^3}{xy^2} \cdot \dfrac{x^2y^3}{ab^4} =$

Write the numerators and the denominators in terms of prime factors.

$\dfrac{a \cdot a \cdot b \cdot b \cdot b}{x \cdot y \cdot y} \cdot \dfrac{x \cdot x \cdot y \cdot y \cdot y}{a \cdot b \cdot b \cdot b \cdot b} =$

Divide by the factors common to the numerator and the denominator.

$\dfrac{\cancel{a} \cdot a \cdot \cancel{b} \cdot \cancel{b} \cdot \cancel{b}}{\cancel{x} \cdot \cancel{y} \cdot \cancel{y}} \cdot \dfrac{x \cdot x \cdot \cancel{y} \cdot \cancel{y} \cdot y}{\cancel{a} \cdot \cancel{b} \cdot \cancel{b} \cdot \cancel{b} \cdot b} =$

Multiply the remaining factors.

$\dfrac{axy}{b}$

Quotient.

b. $\dfrac{4x + 8}{6} \div \dfrac{5x + 10}{9} =$

Invert the rational expression on the right and change the operation to multiplication.

$\dfrac{4x + 8}{6} \cdot \dfrac{9}{5x + 10} =$

Factor the numerators and the denominators into prime factors.

$\dfrac{2 \cdot 2(x + 2)}{2 \cdot 3} \cdot \dfrac{3 \cdot 3}{5(x + 2)} =$

Divide by the factors common to the numerator and the denominator.

$\dfrac{2 \cdot 2(\cancel{x + 2})}{\cancel{2} \cdot \cancel{3}} \cdot \dfrac{\cancel{3} \cdot 3}{5(\cancel{x + 2})} =$

Multiply the remaining factors.

$\dfrac{6}{5}$

Quotient.

c. $\dfrac{2x^2 + 7x + 3}{x^2 - 9} \div \dfrac{2x^2 + 11x + 5}{x^2 - 3x} =$

Invert the rational expression on the right and multiply.

$\dfrac{2x^2 + 7x + 3}{x^2 - 9} \cdot \dfrac{x^2 - 3x}{2x^2 + 11x + 5} =$

Factor the numerators and the denominators.

$\dfrac{(2x + 1)(x + 3)}{(x + 3)(x - 3)} \cdot \dfrac{x(x - 3)}{(2x + 1)(x + 5)} =$

Divide by common factors.

$\dfrac{(\cancel{2x + 1})(\cancel{x + 3})}{(\cancel{x + 3})(\cancel{x - 3})} \cdot \dfrac{x(\cancel{x - 3})}{(\cancel{2x + 1})(x + 5)} =$

Multiply the remaining factors.

$\dfrac{x}{x + 5}$

Quotient.

PRACTICE EXERCISE 3

Find the quotients. Express answers reduced to lowest terms.

a. $\dfrac{r^3s^2}{p^2q^2} \div \dfrac{rs^3}{p^4q}$

b. $\dfrac{8x}{2x + 6} \div \dfrac{6}{4x + 12}$

c. $\dfrac{2x^2 + 7x + 6}{x^2 - 4} \div \dfrac{2x^2 - 3x - 9}{x^2 - 2x}$

If you need more practice, do the following Additional Practice Exercises.

Additional Practice Exercise 3   Find the quotients. Express answers reduced to lowest terms.

a. $\dfrac{m^4n^2}{a^3b^2} \div \dfrac{m^2n^5}{ab^4}$

b. $\dfrac{10}{5x + 20} \div \dfrac{6}{8x + 32}$

c. $\dfrac{x^2 - x - 6}{x^2 - 16} \div \dfrac{3x^2 - 8x - 3}{x^2 + 4x}$

Answers: Practice Exercise 3: a. $\dfrac{r^2p^2}{qs}$ **b.** $\dfrac{8x}{3}$ **c.** $\dfrac{x}{x - 3}$ **Additional Practice 3: a.** $\dfrac{b^2m^2}{a^2n^3}$ **b.** $\dfrac{8}{3}$ **c.** $\dfrac{x(x + 2)}{(x - 4)(3x + 1)}$

Exercise Set 6.5 MyMathLab®

Find the quotients of fractions. Express answers reduced to lowest terms. (See Example 1.)

1. $\dfrac{1}{4} \div \dfrac{4}{7}$

2. $\dfrac{1}{5} \div \dfrac{5}{8}$

3. $\dfrac{3}{7} \div \left(-\dfrac{5}{7}\right)$

4. $-\dfrac{4}{9} \div \dfrac{7}{9}$

5. $\dfrac{2}{3} \div \dfrac{5}{9}$

6. $\dfrac{3}{4} \div \dfrac{7}{12}$

7. $-\dfrac{5}{8} \div \left(-\dfrac{11}{32}\right)$

8. $-\dfrac{8}{15} \div \left(-\dfrac{3}{5}\right)$

9. $-\dfrac{18}{25} \div \dfrac{9}{20}$

10. $\dfrac{28}{45} \div \left(-\dfrac{24}{25}\right)$

11. $\dfrac{8}{9} \div \dfrac{2}{3}$

12. $\dfrac{14}{15} \div \dfrac{7}{3}$

13. $6 \div \dfrac{2}{3}$

14. $12 \div \dfrac{3}{4}$

15. $-72 \div \dfrac{8}{3}$

16. $-144 \div \dfrac{16}{9}$

17. $24 \div \left(-\dfrac{15}{7}\right)$

18. $36 \div \left(-\dfrac{24}{5}\right)$

19. $\dfrac{10}{9} \div 5$

20. $\dfrac{4}{7} \div 2$

21. $-\dfrac{28}{9} \div (-20)$

22. $-\dfrac{30}{11} \div (-24)$

23. $2\dfrac{2}{5} \div \dfrac{4}{3}$

24. $2\dfrac{4}{7} \div \dfrac{3}{5}$

25. $-3\dfrac{1}{3} \div \dfrac{5}{9}$

26. $-2\dfrac{4}{5} \div \dfrac{7}{10}$

27. $3\dfrac{1}{2} \div 3\dfrac{3}{8}$

28. $4\dfrac{2}{3} \div 1\dfrac{2}{5}$

29. $1\dfrac{7}{8} \div \left(-4\dfrac{1}{6}\right)$

30. $2\dfrac{5}{8} \div \left(-4\dfrac{1}{12}\right)$

31. $15 \div 1\dfrac{3}{7}$

32. $24 \div 2\dfrac{5}{8}$

33. $-6\dfrac{2}{3} \div 8$

34. $2\dfrac{8}{11} \div (-12)$

35. $4\dfrac{4}{5} \div 6$

36. $3\dfrac{1}{7} \div 11$

Solve the following. (See Example 2.)

37. If $2\dfrac{3}{4}$ ounces of silver cost \$66, what is the cost of 1 ounce?

38. If a developer paid \$286,000 for $4\dfrac{2}{5}$ acres of land, how much did she pay for 1 acre?

39. How many $\dfrac{1}{4}$-acre lots can a developer get from a $3\dfrac{1}{2}$-acre tract of land?

40. How many servings of $\dfrac{2}{3}$ pound each can be gotten from a 12-pound bag of dog food?

41. A bag of cookies that is $\dfrac{2}{3}$ full is divided equally among four people. What part of a full bag did each receive?

42. It takes 10 gallons of water to fill an aquarium $\dfrac{2}{5}$ full. How many gallons of water does the aquarium hold?

Find the quotients of the given rational expressions. Express answers reduced to lowest terms. (See Example 3.)

43. $\dfrac{a}{b^4} \div \dfrac{a^3}{b^2}$

44. $\dfrac{x^3}{y^3} \div \dfrac{x^5}{y^2}$

45. $\dfrac{8a^2}{9b^3} \div \dfrac{4a^3}{3b^2}$

46. $\dfrac{4c^3}{7d^3} \div \dfrac{8c^4}{21d^2}$

47. $\dfrac{a^3b^2}{c^4d^3} \div \dfrac{ab^5}{c^2d}$

48. $\dfrac{m^4n^6}{p^2q^3} \div \dfrac{m^6n^3}{p^4q^4}$

49. $\dfrac{8x^7g^3}{15x^2g^4} \div \dfrac{3xg^2}{4x^2g^3}$

50. $\dfrac{10c^5g^5}{12c^2g^4} \div \dfrac{5cg^4}{4c^2g^3}$

51. $\dfrac{5x + 15}{8} \div \dfrac{7x + 21}{6}$

52. $\dfrac{3x - 18}{8} \div \dfrac{4x - 24}{12}$

53. $\dfrac{4a - 8}{15a} \div \dfrac{3a - 6}{5a^2}$

54. $\dfrac{6b + 24}{7b^2} \div \dfrac{7b + 28}{14b^4}$

55. $\dfrac{4p + 12q}{3p - 6q} \div \dfrac{5p + 15q}{6p - 12q}$

56. $\dfrac{5r + 10s}{3r - 12s} \div \dfrac{6r + 12s}{4r - 16s}$

57. $\dfrac{10x + 15}{18x - 6} \div \dfrac{20x + 30}{9x - 3}$

58. $\dfrac{12y + 8}{4y + 10} \div \dfrac{18y + 12}{8y + 20}$

59. $\dfrac{a - b}{6a - 9b} \div \dfrac{b - a}{4a - 6b}$

60. $\dfrac{x - y}{12x - 3y} \div \dfrac{y - x}{8x - 2y}$

61. $\dfrac{h}{h^2 + 13h + 42} \div \dfrac{4h^2 + 28h}{4h + 24}$

62. $\dfrac{b}{b^2 + 11b + 30} \div \dfrac{6b^2 + 30b}{6b + 36}$

63. $\dfrac{d^2 - d - 12}{2d^2} \div \dfrac{3d^2 + 13d + 12}{d}$

64. $\dfrac{y^2 + y - 20}{7y^2} \div \dfrac{3y^2 + 19y + 20}{y}$

65. $\dfrac{m^2 - 25}{2m - 12} \div \dfrac{3m + 15}{m^2 - 36}$

66. $\dfrac{h^2 - 36}{3h - 21} \div \dfrac{4h + 24}{h^2 - 49}$

67. $\dfrac{x + 8}{x - 8} \div \dfrac{x^2 + 16x + 64}{x^2 - 16x + 64}$

68. $\dfrac{w + 3}{w - 3} \div \dfrac{w^2 + 6w + 9}{w^2 - 6w + 9}$

69. $\dfrac{a^2 - 5a + 6}{a^2 - 9a + 18} \div \dfrac{a^2 - 6a + 8}{a^2 - 9a + 20}$

70. $\dfrac{y^2 + 3y - 40}{y^2 + 3y - 18} \div \dfrac{y^2 + 2y - 48}{y^2 + 3y - 18}$

71. $\dfrac{2x^2 - 7x - 4}{3x^2 - 14x + 8} \div \dfrac{2x^2 + 7x + 3}{3x^2 - 8x + 4}$

72. $\dfrac{8x^2 + 6x - 9}{8x^2 + 14x - 15} \div \dfrac{2x^2 + 11x + 12}{4x^2 + 19x + 12}$

Challenge Exercises (73 and 74)

Find the following quotients:

73. $\dfrac{ab + 3a + 2b + 6}{bc + 4b + 3c + 12} \div \dfrac{ac - 3a + 2c - 6}{bc + 4b - 4c - 16}$

74. $\dfrac{xy - 3x + 4y - 12}{xy + 6x - 3y - 18} \div \dfrac{xy + 5x + 4y + 20}{xy + 5x - 3y - 15}$

Writing Exercises (75 and 76)

75. When performing the division $\frac{a}{b} \div \frac{c}{d}$, why must b, c, and $d \neq 0$?

76. Write and solve an application problem involving rational numbers.

OBJECTIVE *When you complete this section, you will be able to:*

Ⓐ Divide two polynomials, using long division.

PREREQUISITE SKILLS *Before beginning this section, you should be able to:*

a. Divide whole numbers. (Section R.3)

b. Apply $a^m \cdot a^n = a^{m+n}$ and $\dfrac{a^m}{a^n} = a^{m-n}$. (Section 2.6)

c. Divide monomials. (Section 2.8)

d. Add like terms. (Section 1.7)

GETTING READY FOR SECTION 6.6

1. Divide $23\overline{)1058}$.

Simplify the following, using $a^m \cdot a^n = a^{m+n}$ and $\dfrac{a^m}{a^n} = a^{m-n}$:

2. $4x^2 \cdot 2x$

3. $\dfrac{6x^2}{2x}$

Add the following like terms:

4. $11x - 4x$

5. $x - 6x$

Introduction

In Section 6.1, we reduced fractions by factoring the numerator and the denominator and then dividing by the factors that were common to both. This is one method by which we can divide polynomials, but it has some limitations. What if the numerator and the denominator have no common factors? Can we still divide the polynomials? We can, and the method by which we perform this type of division is the topic of this section.

OBJECTIVE Ⓐ Dividing two polynomials, using long division

In Chapter R, we discussed dividing whole numbers using long division. The procedure for dividing polynomials follows almost the exact same procedure (except that it is often easier!). We will divide two whole numbers and two polynomials side by side and note the similarities in the procedure.

Divide 1058 by 23.

$$23\overline{)1058}$$

1. 23 will not go into 10, so divide 23 into 105. We guess 4. Put the 4 above the 5.

$$\begin{array}{r} 4 \\ 23\overline{)1058} \end{array}$$

Divide $2x^2 + 5x - 12$ by $x + 4$.

$$x + 4\overline{)2x^2 + 5x - 12}$$

1. Divide x into $2x^2$. There is no guessing, as in long division of whole numbers. $\frac{2x^2}{x} = 2x$. Put the $2x$ above the $2x^2$.

$$\begin{array}{r} 2x \\ x + 4\overline{)2x^2 + 5x - 12} \end{array}$$

Answers: **Getting Ready: 1.** 46 **2.** $8x^3$ **3.** $3x$ **4.** $7x$ **5.** $-5x$

2. Multiply 4 and 23 and put the product beneath 105.

$$
\begin{array}{r}
4 \\
23\overline{)1058} \\
92
\end{array}
$$

2. Multiply $2x$ and $x + 4$ and put the product beneath the like terms of $2x^2 + 5x - 12$.

$$
\begin{array}{r}
2x \\
x + 4\overline{)2x^2 + 5x - 12} \\
2x^2 + 8x
\end{array}
$$

3. Subtract 92 from 105.

$$
\begin{array}{r}
4 \\
23\overline{)1058} \\
-92 \\
\hline
13
\end{array}
$$

3. Subtract $2x^2 + 8x$ from $2x^2 + 5x$ by changing the signs of the bottom polynomial to $-2x^2 - 8x$ and adding.

$$
\begin{array}{r}
2x \\
x + 4\overline{)2x^2 + 5x - 12} \\
-2x^2 - 8x \\
\hline
-3x
\end{array}
$$

4. Bring down the 8.

$$
\begin{array}{r}
4 \\
23\overline{)1058} \\
-92 \\
\hline
138
\end{array}
$$

4. Bring down the -12.

$$
\begin{array}{r}
2x \\
x + 4\overline{)2x^2 + 5x - 12} \\
-2x^2 - 8x \\
\hline
-3x - 12
\end{array}
$$

5. Divide 23 into 138 and put the quotient over the 8.

$$
\begin{array}{r}
46 \\
23\overline{)1058} \\
-92 \\
\hline
138
\end{array}
$$

5. Divide x into $-3x$ and put the quotient (which is -3) over the $5x$.

$$
\begin{array}{r}
2x - 3 \\
x + 4\overline{)2x^2 + 5x - 12} \\
-2x^2 - 8x \\
\hline
-3x - 12
\end{array}
$$

6. Multiply 6 and 23 and put the product under 138.

$$
\begin{array}{r}
46 \\
23\overline{)1058} \\
-92 \\
\hline
138 \\
138
\end{array}
$$

6. Multiply -3 and $x + 4$ and put the product (which is $-3x - 12$) under $-3x - 12$.

$$
\begin{array}{r}
2x - 3 \\
x + 4\overline{)2x^2 + 5x - 12} \\
-2x^2 - 8x \\
\hline
-3x - 12 \\
-3x - 12
\end{array}
$$

7. Subtract 138 from 138.

$$
\begin{array}{r}
46 \\
23\overline{)1058} \\
-92 \\
\hline
138 \\
-138 \\
\hline
0
\end{array}
$$

7. Subtract $-3x - 12$ from $-3x - 12$ by changing signs of the bottom polynomial to $3x + 12$ and adding.

$$
\begin{array}{r}
2x - 3 \\
x + 4\overline{)2x^2 + 5x - 12} \\
-2x^2 - 8x \\
\hline
-3x - 12 \\
+3x + 12 \\
\hline
0
\end{array}
$$

8. Remainder of 0. Therefore, 23 divides into 1058 evenly.

CHECK:

$(23)(46) = 1058$, so 46 is the correct answer.

8. Remainder of 0. Therefore, $x + 4$ divides into $2x^2 + 5x - 12$ evenly.

CHECK:

$(x + 4)(2x - 3) = 2x^2 + 5x - 12$, so $2x - 3$ is the correct answer.

As you can see, the procedure is almost exactly the same in both cases. We further illustrate the procedure for dividing polynomials with some more examples.

Example 1 | **Find the following quotients, using long division:**

a. $\dfrac{2x^2 + x - 15}{x + 3}$

Solution

$\frac{2x^2 + x - 15}{x + 3}$ means $(2x^2 + x - 15) \div (x + 3)$, which is written as

$x + 3\overline{)2x^2 + x - 15}$, using long division.

$$\begin{array}{r} 2x \\ x + 3\overline{)2x^2 + x - 15} \end{array}$$ $\dfrac{2x^2}{x} = 2x$; put the $2x$ over $2x^2$.

$$\begin{array}{r} 2x \\ x + 3\overline{)2x^2 + x - 15} \\ 2x^2 + 6x \end{array}$$ $2x(x + 3) = 2x^2 + 6x$; put $2x^2 + 6x$ underneath the like terms of $2x^2 + x + 15$.

$$\begin{array}{r} 2x \\ x + 3\overline{)2x^2 + x - 15} \\ \underline{-2x^2 - 6x} \\ -5x \end{array}$$ Subtract $2x^2 + 6x$ by changing all the signs and adding.

$$\begin{array}{r} 2x \\ x + 3\overline{)2x^2 + x - 15} \\ \underline{-2x^2 - 6x} \\ -5x - 15 \end{array}$$ Bring down -15.

$$\begin{array}{r} 2x - 5 \\ x + 3\overline{)2x^2 + x - 15} \\ \underline{-2x^2 - 6x} \\ -5x - 15 \end{array}$$ $\dfrac{-5x}{x} = -5$; put the -5 over the $+x$.

$$\begin{array}{r} 2x - 5 \\ x + 3\overline{)2x^2 + x - 15} \\ 2x^2 - 6x \\ -5x - 15 \\ -5x - 15 \end{array}$$ $-5(x + 3) = -5x - 15$; put $-5x - 15$ under the other $-5x - 15$.

$$\begin{array}{r} 2x - 5 \\ x + 3\overline{)2x^2 + x - 15} \\ 2x^2 - 6x \\ -5x - 15 \\ \underline{+5x + 15} \\ 0 \end{array}$$ Subtract $-5x - 15$ by changing signs and adding.

The remainder is 0. Therefore, $x + 3$ divides into $2x^2 + x - 15$ evenly. Hence, $x + 3$ is a factor of $2x^2 + x - 15$.

CHECK:

$(x + 3)(2x - 5) = 2x^2 + x - 15$, so $2x - 5$ is the correct answer.

b. $\dfrac{6x^2 + 11x - 10}{3x - 2}$

Solution

Using long division, write this as $3x - 2\overline{\smash{)}6x^2 + 11x - 10}$.

$$\begin{array}{r} 2x \\ 3x - 2\overline{\smash{)}6x^2 + 11x - 10} \end{array}$$ $\dfrac{6x^2}{3x} = 2x$; put the $2x$ over $6x^2$.

$$\begin{array}{r} 2x \\ 3x - 2\overline{\smash{)}6x^2 + 11x - 10} \\ 6x^2 - 4x \end{array}$$ $2x(3x - 2) = 6x^2 - 4x$.

$$\begin{array}{r} 2x \\ 3x - 2\overline{\smash{)}6x^2 + 11x - 10} \\ -6x^2 + 4x \\ \hline 15x \end{array}$$ Subtract $6x^2 - 4x$ by changing signs and adding.

$$\begin{array}{r} 2x \\ 3x - 2\overline{\smash{)}6x^2 + 11x - 10} \\ -6x^2 + 4x \\ \hline 15x - 10 \end{array}$$ Bring down the -10.

$$\begin{array}{r} 2x + 5 \\ 3x - 2\overline{\smash{)}6x^2 + 11x - 10} \\ -6x^2 + 4x \\ \hline 15x - 10 \end{array}$$ $\dfrac{15x}{3x} = 5$; put the 5 over the $11x$.

$$\begin{array}{r} 2x + 5 \\ 3x - 2\overline{\smash{)}6x^2 + 11x - 10} \\ -6x^2 + 4x \\ \hline 15x - 10 \\ 15x - 10 \end{array}$$ $5(3x - 2) = 15x - 10$.

$$\begin{array}{r} 2x + 5 \\ 3x - 2\overline{\smash{)}6x^2 + 11x - 10} \\ -6x^2 + 4x \\ \hline 15x - 10 \\ -15x + 10 \\ \hline 0 \end{array}$$ Subtract $15x - 10$ by changing signs and adding.

The remainder is 0. Therefore, $3x - 2$ divides into $6x^2 + 11x - 10$ evenly. Hence, $3x - 2$ is a factor of $6x^2 + 11x - 10$.

CHECK:

$(3x - 2)(2x + 5) = 6x^2 + 11x - 10$, so $2x + 5$ is the correct answer.

PRACTICE EXERCISE 1

Find the following quotients, using long division:

a. $\dfrac{x^2 - 2x - 24}{x + 4}$

b. $\dfrac{8x^2 - 6x - 9}{2x - 3}$

(Continued)

Answers: **Practice Exercise 1:** **a.** $x - 6$ **b.** $4x + 3$

If you need more practice, do the following Additional Practice Exercises.

Additional Practice Exercise 1   Find the following quotients, using long division:

a. $\dfrac{3x^2 + 13x - 30}{x + 6}$

b. $\dfrac{12x^2 + 2x - 2}{3x - 1}$

Actually, all the examples we have done to this point could have been done by factoring the numerator and the denominator and dividing by common factors, as though we were reducing rational expressions. In the introductory comments, we posed the question whether we could divide two polynomials if there were no common factors. We see that we can by using long division. If the polynomial that is the divisor is not a factor of the dividend, there will be a remainder, just as in the division of whole numbers. For example, 3 is not a factor of 11, so $11 \div 3 = 3$, with a remainder of 2. Recall that in algebra the remainder is usually written as the numerator, with the divisor as the denominator. Consequently, $11 \div 3 = 3\frac{2}{3}$.

> ### Procedure: Dividing Polynomials Using Long Division
>
> 1. Write both polynomials in descending order with 0 in place of any missing terms.
>
> 2. Divide the first term of the divisor into the first term of the dividend (after both have been put into descending order and missing powers are inserted as $0 \cdot$ variable raised to the missing power).
>
> 3. Multiply the entire divisor by the first quotient (the result of the first division).
>
> 4. Align like terms and subtract the expression found in step 2.
>
> 5. Bring down the next term of the dividend.
>
> 6. Continue steps 1–5 until all terms of the dividend have been used. The remainder (if any) will have a lower degree than the divisor.

| **Example 2** | **Find the following quotients, using long division:** |

Note

In the Example 2, we will not show all the steps, only the actual division. If you get stuck, look back at Example 1.

a. $\dfrac{5x^2 + x^3 - 12 + 2x}{x + 2}$

Solution

First, we need to rewrite the numerator in descending order, then rewrite as long division.

$$\frac{5x^2 + x^3 - 12 + 2x}{x + 2} = \frac{x^3 + 5x^2 + 2x - 12}{x + 2} = x + 2 \overline{)x^3 + 5x^2 + 2x - 12}$$

Divide, as in Example 1.

$$\begin{array}{r} x^2 + 3x - 4 \\ x + 2 \overline{)x^3 + 5x^2 + 2x - 12} \\ \underline{-x^3 - 2x^2} \\ 3x^2 + 2x \\ \underline{-3x^2 - 6x} \\ -4x - 12 \\ \underline{+4x + 8} \\ -4 \end{array}$$

Answers: Additional Practice 1: a. $3x - 5$ **b.** $4x + 2$

Since x will not divide into -4, -4 is the remainder. Hence, $x + 2$ is not a factor of $x^3 + 5x^2 + 2x - 12$. Remember, the remainder is written over the divisor. Therefore,

$$\frac{x^3 + 5x^2 + 2x - 12}{x + 2} = x^2 + 3x - 4 + \frac{-4}{x + 2}$$

is the solution. Remember, in checking a division problem that has a remainder, we multiply the divisor and the quotient, then add the remainder. The result is equal to the dividend. Earlier, we stated that $11 \div 3 = 3$, with a remainder of 2. To check, we have $3 \cdot 3 + 2 = 9 + 2 = 11$. If the preceding division is correct, then $(x + 2)(x^2 + 3x - 4) + (-4) = x^3 + 5x^2 + 2x - 12$.

CHECK:

$(x + 2)(x^2 + 3x - 4) + (-4) = x(x^2 + 3x - 4) + 2(x^2 + 3x - 4) - 4 =$
$x^3 + 3x^2 - 4x + 2x^2 + 6x - 8 - 4 = x^3 + 5x^2 + 2x - 12$.

Therefore, the answer is correct.

b. $\dfrac{8x^3 - 22x + 8}{2x - 3}$

Solution

The x^2 term is missing, so insert $0x^2$ between $8x^3$ and $-22x$ and divide as before.

$$
\begin{array}{r}
4x^2 + 6x - 2 \\
2x - 3 \overline{)\, 8x^3 + 0x^2 - 22x + 8} \\
\underline{-8x^3 + 12x^2} \\
12x^2 - 22x \\
\underline{-12x^2 + 18x} \\
-4x + 8 \\
\underline{+4x - 6} \\
2
\end{array}
$$

Therefore, $\dfrac{8x^3 - 22x + 8}{2x - 3} = 4x^2 + 6x - 2 + \dfrac{2}{2x - 3}$.

CHECK:

The check is left as an exercise for the student.

PRACTICE EXERCISE 2

Use long division to find the following quotients:

a. $\dfrac{16x - 12x^2 + 3x^3 - 12}{x - 2}$

b. $\dfrac{8x^3 + 12x^2 - 6}{2x - 1}$

If you need more practice, do the following Additional Practice Exercises.

Additional Practice Exercise 2   Find the following quotients, using long division:

a. $\dfrac{-11x^2 + 3x^3 - 14x - 10}{x - 4}$

b. $\dfrac{2x^3 - 14x + 5}{2x - 4}$

Answers: Practice Exercise 2: a. $3x^2 - 6x + 4 + \dfrac{-4}{x - 2}$ **b.** $4x^2 + 8x + 4 + \dfrac{-2}{2x - 1}$ **Additional Practice 2: a.** $3x^2 + x - 10 + \dfrac{-50}{x - 4}$
b. $x^2 + 2x - 3 + \dfrac{-7}{2x - 4}$

Exercise Set 6.6

For Extra Help

MyMathLab®

Find the following quotients, using long division. (See Examples 1 and 2.)

1. $\dfrac{x^2 - 2x - 15}{x + 3}$

2. $\dfrac{x^2 + 2x - 8}{x + 4}$

3. $\dfrac{x^2 - 4x - 21}{x - 7}$

4. $\dfrac{x^2 - 2x - 24}{x - 6}$

5. $\dfrac{2x^2 - x - 10}{2x - 5}$

6. $\dfrac{3x^2 - 22x + 24}{3x - 4}$

7. $\dfrac{6x^2 + 7x - 20}{3x - 4}$

8. $\dfrac{8x^2 + 6x - 9}{4x - 3}$

9. $\dfrac{4x^2 - 25}{2x + 5}$

10. $\dfrac{9x^2 - 49}{3x + 7}$

11. $\dfrac{2x^2 + 4x - 28}{2x - 6}$

12. $\dfrac{3x^2 - 13x - 25}{3x + 5}$

13. $\dfrac{12x^2 - 5x + 6}{4x - 3}$

14. $\dfrac{12x^2 + 22x + 14}{3x + 7}$

15. $\dfrac{6x^2 + 13x - 16}{2x + 7}$

16. $\dfrac{15x^2 + 29x - 10}{5x - 2}$

17. $\dfrac{2x^3 - 14x^2 + 27x - 12}{x - 4}$

18. $\dfrac{3x^3 + 22x^2 + 33x - 10}{x + 5}$

19. $\dfrac{4x^3 - 16x + 6x^2 + 6}{2x - 1}$

20. $\dfrac{6x^3 + 25x - 10x^2 - 14}{3x - 2}$

21. $\dfrac{6x^3 + 11x^2 - 8x + 9}{2x + 5}$

22. $\dfrac{12x^3 - 17x^2 + 12x - 8}{3x - 2}$

23. $\dfrac{3x^3 - 73x - 10}{x - 5}$

24. $\dfrac{2x^3 - 35x - 12}{x + 4}$

25. $\dfrac{-2x^2 + 4x^3 + 18}{2x - 3}$

26. $\dfrac{22x^2 + 12x^3 + 12}{3x + 1}$

27. $\dfrac{16x^3 - x + 10}{4x + 3}$

28. $\dfrac{4x^3 - 42x + 23}{2x - 6}$

29. $\dfrac{x^3 - 8}{x - 2}$

30. $\dfrac{x^3 + 27}{x + 3}$

31. $\dfrac{x^4 - 16}{x - 2}$

32. $\dfrac{x^4 - 81}{x - 3}$

33. $\dfrac{x^3 + 9}{x + 3}$

34. $\dfrac{x^3 - 12}{x - 2}$

Challenge Exercises (35–38)

35. Find the value of k so that $x - 6$ will divide evenly into $2x^2 - 15x + k$.

36. Find the value of k so that $x + 4$ will divide evenly into $3x^2 + 6x + k$.

37. Find the value of k so that $x - 2$ will divide evenly into $4x^2 + kx - 6$.

38. Find the value of k so that $x + 3$ will divide evenly into $3x^2 + kx + 6$.

Writing Exercises (39 and 40)

39. Is $x + 4$ a factor of $2x^3 + 5x^2 - 11x + 4$? Why or why not?

40. Is $x - 3$ a factor of $3x^3 - 7x^2 - 8x + 6$? Why or why not?

Group Project (41 and 42)

41. Find three second-degree polynomials that are divisible by $x + 3$.

42. Find three second-degree polynomials that leave a remainder of 3 when divided by $x - 2$.

Chapter 6   Summary

Concept/Procedure	Example

Graphing Rational Numbers: [Section 6.1]

- Rational numbers are graphed on the number line by the placement of a dot at the location of the point that is the indicated distance and direction from 0. The denominator represents the number of equal parts into which the units are divided, and the numerator represents the number of equal parts from 0 to the location.

Example 1: Graph $\dfrac{7}{4}$ on the number line.

Solution: Divide each unit into fourths and put a dot on the seventh division point.

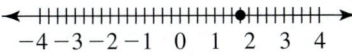

Equivalent Fractions: [Section 6.1]

- Two or more fractions are equivalent if they represent the same quantity.

Lowest Terms: [Section 6.1]

- A fraction is reduced to lowest terms if the numerator and the denominator have no common factors other than 1.

Reducing a Fraction to Lowest Terms: [Section 6.1]

a. Write the numerator and the denominator as the product of their prime factors.
b. Divide the numerator and the denominator by the common factors.

Example 2: Reduce $\dfrac{60}{84}$ to lowest terms.

Solution: Write the numerator and denominator in terms of prime factors.

$$\frac{60}{84} = \frac{2 \cdot 2 \cdot 3 \cdot 5}{2 \cdot 2 \cdot 3 \cdot 7} \qquad \text{Divide by factors common to the numerator and denominator.}$$

$$= \frac{2 \cdot 2 \cdot 3 \cdot 5}{2 \cdot 2 \cdot 3 \cdot 7} \qquad \text{Multiply the remaining factors.}$$

$$= \frac{5}{7} \qquad \text{Therefore, } \frac{60}{84} = \frac{5}{7}.$$

Rational Expressions: [Sections 6.1 and 6.2]

a. A rational expression is an algebraic expression of the form $\frac{P}{Q}$, where P and Q are polynomials and $Q \neq 0$.
b. To find the value(s) for which a rational expression is undefined, set the denominator equal to 0 and solve the resulting equation.

Example 3: Find the value(s) for which $\dfrac{2x + 4}{3x - 9}$ is undefined.

Solution: Set the denominator equal to 0 and solve.

$3x - 9 = 0$ Add 9 to both sides.

$\quad 3x = 9$ Divide both sides by 3.

$\qquad x = 3$ Therefore, $\dfrac{2x + 3}{3x - 9}$ is undefined for $x = 3$.

Concept/Procedure	Example

Reducing Rational Expressions: [Sections 6.1 and 6.2]

- A rational expression is reduced to lowest terms if the numerator and the denominator have no common factors other than 1. The procedure for reducing rational expressions is exactly the same as reducing fractions to lowest terms.

Example 4: Reduce $\dfrac{6x^2 - 5xy - 25y^2}{9x^2 - 25y^2}$ to lowest terms.

Solution: Factor the numerator and denominator.

$\dfrac{6x^2 - 5xy - 25y^2}{9x^2 - 25y^2} = \dfrac{(3x + 5y)(2x - 5y)}{(3x + 5y)(3x - 5y)}$ Divide by the common factors.

$= \dfrac{(3x + 5y)(2x - 5y)}{(3x + 5y)(3x - 5y)}$ Multiply the remaining factors.

$= \dfrac{2x - 5y}{3x - 5y}$ Therefore, $\dfrac{6x^2 - 5xy - 25y^2}{9x^2 - 25y^2} = \dfrac{2x - }{3x - }$

Multiplication of Rational Numbers: [Section 6.3]

a. To multiply two or more rational numbers, multiply numerator times numerator and denominator times denominator. In symbols, $\frac{a}{b} \cdot \frac{c}{d} = \frac{a \cdot c}{b \cdot d}$.

b. To multiply a fraction and mixed number or two or more mixed numbers, change the mixed numbers into improper fractions and multiply as with any other fractions.

c. To multiply a fraction and a whole number, think of the whole number as a fraction whose denominator is 1 and multiply as with any other fractions.

d. All products should be reduced to lowest terms by the procedure for reducing fractions to lowest terms. You may divide by factors common to the numerator and the denominator before multiplying.

Example 5: Multiply $\dfrac{15}{8} \cdot \dfrac{28}{25}$.

Solution:

$\dfrac{15}{8} \cdot \dfrac{28}{25} =$ Factor numerators and denominators.

$\dfrac{3 \cdot 5}{2 \cdot 2 \cdot 2} \cdot \dfrac{2 \cdot 2 \cdot 7}{5 \cdot 5} =$ Divide by the common factors.

$\dfrac{3 \cdot 5}{2 \cdot 2 \cdot 2} \cdot \dfrac{2 \cdot 2 \cdot 7}{5 \cdot 5} =$ Multiply the remaining factors.

$\dfrac{21}{10}$ Product.

Multiplication of Rational Expressions: [Section 6.4]

a. Rational expressions are multiplied in exactly the same manner as rational numbers. That is, multiply numerator times numerator and put the product over denominator times denominator.

b. If the product of two or more rational expressions contains factors that differ in sign only, remove a factor of -1 from one of the factors or apply the fact that $\frac{b - a}{a - b} = -1$.

Example 6: Multiply $\dfrac{3x^2 + 4x - 4}{8x^2 + 10x - 3} \cdot \dfrac{2x^2 + 11x + 12}{x^2 + 6x + 8}$.

Solution:

$\dfrac{3x^2 + 4x - 4}{8x^2 + 10x - 3} \cdot \dfrac{2x^2 + 11x + 12}{x^2 + 6x + 8} =$ Factor numerators and denominators.

$\dfrac{(3x - 2)(x + 2)}{(4x - 1)(2x + 3)} \cdot \dfrac{(2x + 3)(x + 4)}{(x + 4)(x + 2)} =$ Divide by the common factors.

$\dfrac{(3x - 2)(x + 2)}{(4x - 1)(2x + 3)} \cdot \dfrac{(2x + 3)(x + 4)}{(x + 4)(x + 2)} =$ Multiply the remaining factors.

$\dfrac{3x - 2}{4x - 1}$ Product.

Concept/Procedure	Example

Division of Rational Numbers and Expressions: [Section 6.5]

- To divide rational numbers or expressions, invert the expression on the right and change the operation to multiplication.

Example 7: Divide $\dfrac{12a^2}{10a^3} \div \dfrac{8a^4}{15a}$.

Solution:

$$\dfrac{12a^2}{10a^3} \div \dfrac{8a^4}{15a} =$$ Invert the rational expression on the right and change the operation to multiplication.

$$\dfrac{12a^2}{10a^3} \cdot \dfrac{15a}{8a^4} =$$ Factor the numerator and denominator.

$$\dfrac{2 \cdot 2 \cdot 3 \cdot a \cdot a}{2 \cdot 5 \cdot a \cdot a \cdot a} \cdot \dfrac{3 \cdot 5 \cdot a}{2 \cdot 2 \cdot 2 \cdot a \cdot a \cdot a \cdot a} =$$ Divide by the common factors.

$$\dfrac{2 \cdot 2 \cdot 3 \cdot \cancel{a} \cdot \cancel{a}}{2 \cdot 5 \cdot \cancel{a} \cdot \cancel{a} \cdot \cancel{a}} \cdot \dfrac{3 \cdot 5 \cdot \cancel{a}}{2 \cdot 2 \cdot 2 \cdot a \cdot a \cdot a \cdot a} =$$ Multiply the remaining factors.

$$\dfrac{9}{4a^4}$$ Quotient.

Division of Polynomials (Long Division): [Section 6.6]

- Long division of polynomials is done by using the same algorithm as division of whole numbers, except that like terms are aligned instead of digits with the same place value. The steps are as follows:

  1. Write both polynomials in descending order with 0 in place of any missing terms.
  2. Divide the first term of the divisor into the first term of the dividend (after both have been put into descending order and missing powers are inserted as $0 \cdot$ variable raised to the missing power).
  3. Multiply the entire divisor by the first quotient (the result of the first division).
  4. Align like terms and subtract the expression found in step 2.
  5. Bring down the next term of the dividend.
  6. Continue steps 1–5 until all terms of the dividend have been used. The remainder (if any) will have a lower degree than the divisor.

Example 8: Divide $\dfrac{15x^2 + 29x - 12}{5x - 2}$, using long division.

$$
\begin{array}{r}
3x + 7 \\
5x - 2 \overline{)15x^2 + 29x - 12} \\
\underline{-15x^2 + 6x} \\
35x - 12 \\
\underline{-35x + 14} \\
2
\end{array}
$$

Therefore, the quotient is $3x + 7 + \dfrac{2}{5x - 2}$.

Chapter 6   Review Exercises

Graph the rational numbers. Estimate as needed. [Section 6.1]

1. $\left\{ -\dfrac{5}{3}, -\dfrac{2}{5}, 0, \dfrac{3}{4}, \dfrac{9}{4} \right\}$

$$\xleftarrow{\quad\;\; | \;\;|\;\;|\;\;|\;\;|\;\;|\;\;|\;\;|\;\;|\;\;\quad}\rightarrow$$
$$-4\;-3\;-2\;-1\;\;0\;\;1\;\;2\;\;3\;\;4$$

2. $\left\{ -3\dfrac{1}{2}, -1\dfrac{1}{6}, -\dfrac{1}{3}, 1\dfrac{4}{5}, 3\dfrac{1}{2} \right\}$

$$\xleftarrow{\quad | \;|\;|\;|\;|\;|\;|\;|\;|\;|\;|\;|\;|\;\quad}\rightarrow$$
$$-6\;-5\;-4\;-3\;-2\;-1\;\;0\;\;1\;\;2\;\;3\;\;4\;\;5\;\;6$$

Reduce the fractions to lowest terms. [Section 6.1]

3. $\dfrac{36}{42}$ 4. $\dfrac{64}{72}$

5. $\dfrac{120}{124}$ 6. $\dfrac{84}{96}$

Reduce the rational expressions to lowest terms. [Section 6.1]

7. $\dfrac{x^3 y}{x^2 y^3}$ 8. $\dfrac{m^2 n^4}{m^4 n^5}$

9. $\dfrac{r^3 s^5}{r s^3}$ 10. $\dfrac{32 x^2 y^5}{8 x y^3}$

11. $\dfrac{-42 p^6 q^2}{27 p^3 q^4}$ 12. $-\dfrac{18 m^3 n^2}{54 m^5 n^5}$

Find the value(s) of the variable(s) for which the expressions are defined. [Section 6.2]

13. $\dfrac{x}{x - 9}$ 14. $\dfrac{x + 4}{2x + 5}$

15. $\dfrac{x + 3}{(x + 7)(x + 2)}$ 16. $\dfrac{3a - 4}{a^2 + 4a - 12}$

Reduce the rational expressions to lowest terms. [Section 6.2]

17. $\dfrac{7(2x + 9)}{5(2x + 9)}$ 18. $\dfrac{4a - 20}{7a - 35}$

19. $\dfrac{a^2 + 4a - 21}{a^2 + 9a + 14}$ 20. $\dfrac{2x - 3}{6x^2 - x - 12}$

21. $\dfrac{25 x^2 - 9}{10 x^2 + x - 3}$ 22. $\dfrac{2c + 3d}{8c^2 + 26cd + 21d^2}$

23. $\dfrac{3x - 4y}{4y - 3x}$ 24. $\dfrac{5 - 2x}{4x^2 - 25}$

Find the products of the rational numbers. Express answers reduced to lowest terms. [Section 6.3]

25. $\dfrac{18}{7} \cdot \dfrac{7}{12}$ 26. $\dfrac{21}{10} \cdot \left(-\dfrac{15}{14} \right)$

27. $-\dfrac{5}{16} \cdot 1\dfrac{13}{15}$ 28. $4\dfrac{4}{5} \cdot 3\dfrac{1}{3}$

29. $-32 \cdot \left(-2\dfrac{3}{8} \right)$ 30. $\dfrac{9}{16} \cdot \dfrac{28}{15} \cdot \dfrac{10}{7}$

31. A worker works for $6\frac{1}{2}$ hours and is paid \$6.80 per hour. How much did the worker earn?

32. A rope is $5\frac{1}{4}$ feet long. If $\frac{2}{3}$ of the rope is cut off, how long is the piece that was cut off?

33. Find the area of a rectangle whose length is $2\frac{1}{4}$ feet and whose width is $1\frac{1}{3}$ feet.

34. Find the area of a triangle whose base is $3\frac{3}{5}$ inches and whose height is $4\frac{2}{3}$ inches.

Find the products. Express answers reduced to lowest terms. [Section 6.4.]

35. $\dfrac{x^5}{y^3} \cdot \dfrac{y^2}{x^2}$ 36. $\dfrac{28 m^2 n^3}{15 p^2 q^6} \cdot \dfrac{25 p^4 q^3}{14 mn}$

37. $\dfrac{m - n}{32 m^3 n} \cdot \dfrac{36 m n^3}{(m - n)^2}$ 38. $\dfrac{14x - 21}{30x - 40} \cdot \dfrac{15x - 20}{8x - 12}$

39. $\dfrac{25 x^2 - 16}{16x + 24} \cdot \dfrac{12x + 18}{5x - 4}$ 40. $\dfrac{2x - 7}{12} \cdot \dfrac{10}{7 - 2x}$

41. $\dfrac{x^2 - 16}{x^2 + 6x + 8} \cdot \dfrac{x^2 - 3x - 10}{x^2 - 25}$

42. $\dfrac{8x^2 + 2x - 15}{6x^2 + x - 12} \cdot \dfrac{3x^2 - 13x + 12}{3x^2 - 7x - 6}$

43. $\dfrac{y^2 - 9}{y^2 - 3y - 18} \cdot \dfrac{y^2 - 4y - 12}{3 - y}$

Find the quotients. Express answers reduced to lowest terms.
[Section 6.5]

44. $\dfrac{5}{8} \div \dfrac{15}{24}$

45. $-\dfrac{32}{25} \div \dfrac{28}{35}$

46. $-144 \div \dfrac{9}{16}$

47. $-\dfrac{32}{19} \div (-24)$

48. $4\dfrac{1}{5} \div \left(-\dfrac{7}{14}\right)$

49. $-12 \div 3\dfrac{3}{8}$

50. If $3\frac{2}{5}$ ounces of gold cost $1224, how much does 1 ounce cost?

51. If a car used $12\frac{8}{10}$ gallons of gas on a trip of 416 miles, what was the average miles per gallon for the trip?

Find the quotients of the rational expressions. Express answers reduced to lowest terms. [Section 6.5]

52. $\dfrac{39a^2b^4}{27x^3y^2} \div \dfrac{26a^3b}{28xy^4}$

53. $\dfrac{8y - 20}{9} \div \dfrac{6y - 15}{10}$

54. $\dfrac{21a + 7b}{16a - 24b} \div \dfrac{42a + 14b}{24a - 36b}$

55. $\dfrac{x - y}{8x - 4y} \div \dfrac{y - x}{12x - 6y}$

56. $\dfrac{z^2}{z^2 - 2z - 8} \div \dfrac{9z^3 + 3z^4}{z^2 + 5z + 6}$

57. $\dfrac{4p^2 - 9}{24p + 28} \div \dfrac{10p - 15}{36p^2 - 49}$

58. $\dfrac{16p^2 - 8pq - 3q^2}{8p^2 + 22pq + 5q^2} \div \dfrac{8p^2 - 10pq + 3q^2}{10p^2 + pq - 3q^2}$

Find the quotients, using long division. [Section 6.6]

59. $\dfrac{12x^2 + 7x - 10}{3x - 2}$

60. $\dfrac{15x^2 - 29x - 16}{5x + 2}$

61. $\dfrac{6a^3 - a^2 - 9a + 7}{3a + 4}$

62. $\dfrac{16z^2 - 16z + 12}{4z - 3}$

63. $\dfrac{-27x - 2x^2 + 8x^3 + 18}{2x - 3}$

64. $\dfrac{3x^3 - 24x - 7}{x - 3}$

Chapter 6   Test

1. Graph $\left\{-\frac{5}{2}, -\frac{5}{4}, -\frac{1}{3}, 1, \frac{6}{5}\right\}$. Estimate as needed.

Find the value(s) for which the following are defined:

2. $\dfrac{4a}{3a - 12}$

3. $\dfrac{6b}{b^2 - 2b - 8}$

Simplify the following by reducing to lowest terms:

4. $\dfrac{72}{84}$

5. $-\dfrac{x^5y^2}{x^2y^6}$

6. $\dfrac{42a^3b^4}{16ab^6}$

7. $\dfrac{8x - 20}{6x - 15}$

8. $\dfrac{6x^2 + 11x - 10}{4x^2 + 4x - 15}$

9. It took $1\frac{3}{4}$ gallons of paint to paint the trim on a house. How many gallons of paint would it take to paint six such houses?

10. If $2\frac{1}{2}$ pounds of coffee cost $11.25, what is the cost of 1 pound?

Find the following products and quotients:

11. $\dfrac{18}{25} \cdot \dfrac{35}{24}$

12. $6\dfrac{2}{3} \cdot 1\dfrac{7}{8}$

13. $2\dfrac{4}{7} \div 27$

14. $\dfrac{a^3b^3}{a^5b} \cdot \dfrac{a^4b^3}{a^3b^4}$

15. $\dfrac{27a^2b^4}{14x^4y} \cdot \dfrac{35xy^3}{18a^5b^2}$

16. $\dfrac{3r - 12}{2r + 4} \div \dfrac{6r - 24}{4r + 8}$

17. $\dfrac{7 - b}{42} \cdot \dfrac{26}{b - 7}$

18. $\dfrac{a^2 + 4a - 21}{a^2 + 3a - 28} \div \dfrac{a^2 + 3a - 18}{a^2 + 8a + 12}$

19. $\dfrac{16y^2 - 25}{12y^2 - 7y - 10} \cdot \dfrac{3y^2 + 2y}{8y^2 - 2y - 15}$

20. Find $\dfrac{12x^2 - 11x - 12}{3x - 5}$, using long division.

7

Addition and Subtraction of Rational Numbers and Expressions

Chapter Outline

7.1 Addition and Subtraction of Rational Numbers and Expressions with Like Denominators

7.2 Least Common Multiple and Equivalent Rational Expressions

7.3 The Least Common Denominator of Fractions and Rational Expressions

7.4 Addition and Subtraction of Rational Numbers and Expressions with Unlike Denominators

7.5 Complex Fractions

7.6 Solving Equations Containing Rational Numbers and Expressions

7.7 Applications with Rational Expressions

Chapter 7 is a continuation of the study of rational numbers and rational expressions begun in Chapter 6. In Chapter 6, we reduced rational numbers and rational expressions to lowest terms, multiplied and divided rational numbers and expressions, and divided polynomials using long division.

We begin Chapter 7 with the addition of rational numbers and fractions with like denominators, followed by two sections that cover the skills that are necessary to be able to add and subtract rational numbers and expressions with unlike denominators. We then add and subtract rational numbers and expressions and simplify complex fractions. We end the chapter by solving equations that contain rational numbers and expressions and then use this skill to solve application problems.

Section 7.1 Addition and Subtraction of Rational Numbers and Expressions with Like Denominators

OBJECTIVES *When you complete this section, you will be able to:*

Ⓐ Add and subtract fractions with common denominators.
Ⓑ Add and subtract mixed numbers whose fractional parts have common denominators.
Ⓒ Add and subtract rational expressions with common denominators.

PREREQUISITE SKILLS *Before beginning this section, you should be able to:*

a. Add and subtract integers. (Sections 1.6 and 1.7)
b. Convert mixed numbers into improper fractions. (Section R.4)
c. Factor whole numbers and polynomials. (Sections 5.1–5.6)
d. Add and subtract polynomials. (Section 1.8)
e. Reduce rational numbers and expressions to lowest terms. (Sections 6.1 and 6.2)

GETTING READY FOR SECTION 7.1

Add or subtract the following integers:

1. $11 - (-5)$ **2.** $11 - 7 + 5$ **3.** $23 - 5 - 11$

4. Convert $2\dfrac{3}{4}$ into an improper fraction.

Factor the following whole number and polynomial:

5. 24 **6.** $x^2 - 3x - 28$

Add or subtract the following polynomials:

7. $(x^2 + 2x) + (5x - 6)$ **8.** $(x^2 + 5x) - (8x + 28)$

Reduce the following to lowest terms:

9. $\dfrac{15}{27}$ **10.** $\dfrac{2y - 6}{y - 3}$ **11.** $\dfrac{x^2 - 3x - 28}{x^2 + 7x + 12}$

Introduction

This is the first of two sections in which we will be adding fractions and rational expressions. The technique used in adding fractions depends upon the denominators. In this section, the denominators will be the same number or expression, whereas in Section 7.4, the denominators will be different.

OBJECTIVE   **Adding and subtracting fractions with common denominators**

When fractions have the same denominators, we say that the fractions have a **common denominator**.

Answers: Getting Ready: 1. 16 **2.** 9 **3.** 7 **4.** $\dfrac{11}{4}$ **5.** $2 \cdot 2 \cdot 2 \cdot 3$ **6.** $(x - 7)(x + 4)$ **7.** $x^2 + 7x - 6$
8. $x^2 - 3x - 28$ **9.** $\dfrac{5}{9}$ **10.** 2 **11.** $\dfrac{x - 7}{x + 3}$

We begin with a graphical illustration of $\frac{4}{15} + \frac{3}{15}$. The figure is divided into 15 equal regions, of which 4 are shaded. Therefore, $\frac{4}{15}$ of the unit is shaded.

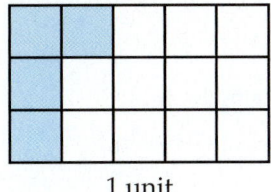

1 unit

If we shade 3 more regions, this shading represents $\frac{3}{15}$ of the unit. What part of the unit is now shaded?

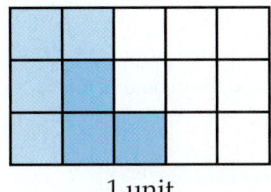

1 unit

Since 7 of the 15 regions are now shaded, $\frac{7}{15}$ of the region is shaded. Therefore, $\frac{4}{15} + \frac{3}{15} = \frac{4+3}{15} = \frac{7}{15}$. In this case, the fractions we added had common denominators. Therefore, to add fractions with common denominators, we add only the numerators and leave the denominators unchanged. This leads to the following procedure for adding fractions with common denominators:

> **Procedure: Addition of Fractions with Common Denominators**
>
> To add fractions with common denominators, add the numerators and put the sum over the common denominator. In symbols, $\frac{a}{c} + \frac{b}{c} = \frac{a+b}{c}$. Reduce if possible.

This principle is easily extended to three or more fractions: $\frac{a}{d} + \frac{b}{d} + \frac{c}{d} = \frac{a+b+c}{d}$, and so on.

Before we subtract fractions, we need to look at equivalent forms of negative fractions. By applying the rules of signs for division of signed numbers, we see that $-\frac{a}{b} = \frac{-a}{b}$. That is, the negative of a fraction is the negative of the numerator divided by the same denominator. For example, $-\frac{6}{2} = \frac{-6}{2}$, since they both equal -3. Consequently, we may change subtraction of fractions into addition as follows: The difference $\frac{a}{c} - \frac{b}{c}$ becomes the sum $\frac{a}{c} + \frac{-b}{c} = \frac{a+(-b)}{c} = \frac{a-b}{c}$. We summarize as follows:

> **Procedure: Subtraction of Fractions with Common Denominators**
>
> To subtract fractions with common denominators, put the difference of the numerators over the common denominator. In symbols, $\frac{a}{c} - \frac{b}{c} = \frac{a-b}{c}$. Reduce if possible.

Example 1   **Find the following sums or differences. Reduce answers to lowest terms.**

a. $\dfrac{2}{7} + \dfrac{4}{7} =$ Put the sum of the numerators over the common denominator.

$\dfrac{2+4}{7} =$ Add the numerators.

$\dfrac{6}{7}$ Sum.

b. $\dfrac{1}{11} + \dfrac{6}{11} + \dfrac{3}{11} =$ Put the sum of the numerators over the common denominator.

$\dfrac{1+6+3}{11} =$ Add the numerators.

$\dfrac{10}{11}$ Sum.

c. $\dfrac{11}{9} - \dfrac{-5}{9} =$ Put the difference of the numerators over the common denominator.

$\dfrac{11 - (-5)}{9} =$ $-(-5) = +5.$

$\dfrac{11 + 5}{9} =$ Add the numerators.

$\dfrac{16}{9}$ Difference.

d. $\dfrac{1}{5} - \dfrac{3}{5} =$ Put the difference of the numerators over the common denominator.

$\dfrac{1 - 3}{5} =$ $1 - 3 = -2.$

$\dfrac{-2}{5} =$ Apply equivalent form of negative fractions.

$-\dfrac{2}{5}$ Difference.

e. $\dfrac{5}{9} + \dfrac{7}{9} =$ Put the sum of the numerators over the common denominator.

$\dfrac{5 + 7}{9} =$ $5 + 7 = 12.$

$\dfrac{12}{9} =$ Factor the numerator and the denominator into prime factors.

$\dfrac{2 \cdot 2 \cdot 3}{3 \cdot 3} =$ Divide by the common factor 3.

$\dfrac{2 \cdot 2 \cdot \cancel{3}}{3 \cdot \cancel{3}} =$ Multiply the remaining factors.

$\dfrac{4}{3}$ Sum.

f. $\dfrac{11}{18} - \dfrac{7}{18} + \dfrac{5}{18} =$ Put the numerators over the common denominator.

$\dfrac{11 - 7 + 5}{18} =$ $11 - 7 + 5 = 9.$

$\dfrac{9}{18} =$ Factor the numerator and denominator into prime factors.

$\dfrac{3 \cdot 3}{2 \cdot 3 \cdot 3} =$ Divide by the common factor 3.

$\dfrac{\cancel{3} \cdot \cancel{3}}{2 \cdot \cancel{3} \cdot \cancel{3}} =$ Multiply the remaining factors.

$\dfrac{1}{2}$ Answer.

PRACTICE EXERCISE 1

Find the following sums or differences. Reduce answers to lowest terms.

a. $\dfrac{11}{15} + \dfrac{2}{15}$ **b.** $\dfrac{8}{17} + \dfrac{7}{17} + \dfrac{4}{17}$ **c.** $\dfrac{4}{9} - \dfrac{-1}{9}$

d. $\dfrac{3}{11} - \dfrac{9}{11}$ **e.** $\dfrac{11}{18} - \dfrac{7}{18}$ **f.** $\dfrac{5}{12} + \dfrac{11}{12} - \dfrac{7}{12}$

If you need more practice, do the following Additional Practice Exercises.

Additional Practice Exercise 1   Find the following sums or differences. Reduce answers to lowest terms.

a. $\dfrac{5}{17} + \dfrac{7}{17}$ **b.** $\dfrac{3}{5} + \dfrac{2}{5} + \dfrac{1}{5}$ **c.** $\dfrac{7}{11} - \dfrac{-3}{11}$

d. $\dfrac{1}{7} - \dfrac{6}{7}$ **e.** $\dfrac{9}{14} - \dfrac{3}{14}$ **f.** $\dfrac{35}{48} - \dfrac{25}{48} + \dfrac{5}{48}$

OBJECTIVE B Adding and subtracting mixed numbers whose fractional parts have common denominators

Note

In the following examples, we will assume that you remember how to reduce fractions and will omit the steps showing the prime factorizations and divisions by the common factors.

Mixed numbers may be added or subtracted in two different ways. We may combine the fractional parts and whole parts separately, or we may convert the mixed numbers to improper fractions and then combine. Converting to improper fractions avoids many of the problems associated with adding and subtracting mixed numbers and is more like the procedure used to combine algebraic fractions. For these reasons, it is the only method we will discuss, and we will express the answers as improper fractions.

Answers: Practice Exercise 1: a. $\dfrac{13}{15}$ **b.** $\dfrac{19}{17}$ **c.** $\dfrac{5}{9}$ **d.** $-\dfrac{6}{11}$ **e.** $\dfrac{2}{9}$ **f.** $\dfrac{3}{4}$ **Additional Practice 1: a.** $\dfrac{12}{17}$ **b.** $\dfrac{6}{5}$ **c.** $\dfrac{10}{11}$ **d.** $-\dfrac{5}{7}$ **e.** $\dfrac{3}{7}$ **f.** $\dfrac{5}{16}$

Example 2 | Add or subtract the following mixed numbers:

a. $2\dfrac{3}{4} + 5\dfrac{1}{4} =$ Convert to improper fractions.

$\dfrac{11}{4} + \dfrac{21}{4} =$ Put the sum of the numerators over the common denominator.

$\dfrac{11 + 21}{4} =$ $11 + 21 = 32.$

$\dfrac{32}{4} =$ $32 \div 4 = 8.$

8 Sum.

b. $4\dfrac{3}{8} - 6\dfrac{1}{8} =$ Convert to improper fractions.

$\dfrac{35}{8} - \dfrac{49}{8} =$ Put the difference of the numerators over the common denominator.

$\dfrac{35 - 49}{8} =$ $35 - 49 = -14.$

$\dfrac{-14}{8} =$ Reduce to lowest terms.

$-\dfrac{7}{4}$ Difference.

c. $6\dfrac{1}{12} - 2\dfrac{11}{12} + 3\dfrac{7}{12} =$ Convert to improper fractions.

$\dfrac{73}{12} - \dfrac{35}{12} + \dfrac{43}{12} =$ Combine fractions.

$\dfrac{73 - 35 + 43}{12} =$ $73 - 35 + 43 = 81.$

$\dfrac{81}{12} =$ Reduce to lowest terms.

$\dfrac{27}{4}$ Answer.

PRACTICE EXERCISE 2

Add or subtract the following mixed numbers:

a. $4\dfrac{1}{6} + 7\dfrac{5}{6}$ **b.** $6\dfrac{7}{8} - 9\dfrac{3}{8}$ **c.** $3\dfrac{17}{30} + 6\dfrac{9}{30} - 2\dfrac{11}{30}$

If you need more practice, do the following Additional Practice Exercises.

Additional Practice Exercise 2   Add or subtract the following mixed numbers:

a. $2\dfrac{1}{5} + 5\dfrac{3}{5}$ **b.** $3\dfrac{7}{10} - 6\dfrac{3}{10}$ **c.** $2\dfrac{11}{21} - 3\dfrac{5}{21} + 4\dfrac{1}{21}$

OBJECTIVE Adding and subtracting rational expressions with common denominators

The addition of rational expressions with common denominators is done exactly the same way as the addition of fractions. We will first do some examples in which the denominators are monomials. Compare the following:

$\dfrac{4}{13} + \dfrac{6}{13}$ $\dfrac{4}{x} + \dfrac{6}{x}$ Put the sum of the numerators over the common denominator.

$\dfrac{4 + 6}{13}$ $\dfrac{4 + 6}{x}$ Add the numerators.

$\dfrac{10}{13}$ $\dfrac{10}{x}$ Sum.

Answers: Practice Exercise 2: a. 12 **b.** $-\dfrac{5}{2}$ **c.** $\dfrac{15}{2}$ **Additional Practice 2: a.** $\dfrac{39}{5}$ **b.** $-\dfrac{13}{5}$ **c.** $\dfrac{10}{3}$

Example 3 **Find the sums of the rational expressions. Express answers reduced to lowest terms.**

a. $\dfrac{3}{a} + \dfrac{6}{a} =$ Put the sum of the numerators over the common denominator.

$\dfrac{3 + 6}{a} =$ Add the numerators.

$\dfrac{9}{a}$ Sum.

b. $\dfrac{3}{x^2} + \dfrac{y}{x^2} =$ Put the sum of the numerators over the common denominator.

$\dfrac{3 + y}{x^2}$ Sum.

c. $\dfrac{3x}{10y} - \dfrac{7x}{10y} =$ Put the difference of the numerators over the common denominator.

$\dfrac{3x - 7x}{10y} =$ $3x - 7x = -4x$.

$\dfrac{-4x}{10y} =$ Reduce to lowest terms.

$-\dfrac{2x}{5y}$ Difference.

d. $\dfrac{6x - 3y}{4x^2} - \dfrac{4x - 3y}{4x^2} =$ Put the difference of the numerators over the common denominator. Put the numerator and denominator in parentheses.

$\dfrac{(6x - 3y) - (4x - 3y)}{4x^2} =$ Remove the parentheses.

$\dfrac{6x - 3y - 4x + 3y}{4x^2} =$ Add like terms.

$\dfrac{2x}{4x^2} =$ Reduce to lowest terms.

$\dfrac{1}{2x}$ Difference.

Note

In Example 2d, both numerators contained more than one term. When the numerators were put over the common denominator, they were enclosed within parentheses to indicate that the entire numerators are being subtracted. Whenever a numerator contains more than one term, it is recommended that parentheses be used, especially if the rational expressions are being subtracted.

PRACTICE EXERCISE 3

Find the sum or difference of each of the rational expressions. Express answers reduced to lowest terms.

a. $\dfrac{6}{b} + \dfrac{4}{b}$ **b.** $\dfrac{m}{n} - \dfrac{6}{n}$ **c.** $\dfrac{3x}{16z} + \dfrac{5x}{16z}$ **d.** $\dfrac{6a^3 + 3b^2}{8a} - \dfrac{4a^3 + 3b^2}{8a}$

If you need more practice, do the following Additional Practice Exercises.

Additional Practice Exercise 3 Find the sum or difference of each of the rational expressions. Express answers reduced to lowest terms.

a. $\dfrac{8}{y} - \dfrac{3}{y}$ **b.** $\dfrac{v}{u^3} + \dfrac{w}{u^3}$ **c.** $\dfrac{5a}{14b} - \dfrac{11a}{14b}$ **d.** $\dfrac{7a^2 - 4b^3}{8b} - \dfrac{7a^2 - 6b^3}{8b}$

If the common denominator of the rational expressions is a polynomial other than a monomial, we follow exactly the same procedure as before. Compare the following:

$\dfrac{2}{7} + \dfrac{4}{7} =$ $\dfrac{2}{x} + \dfrac{4}{x} =$ $\dfrac{2}{x - 1} + \dfrac{4}{x - 1} =$ Put the sum of the numerators over the common denominator.

$\dfrac{2 + 4}{7} =$ $\dfrac{2 + 4}{x} =$ $\dfrac{2 + 4}{x - 1} =$ Add the numerators.

$\dfrac{6}{7}$ $\dfrac{6}{x}$ $\dfrac{6}{x - 1}$ Sum.

Answers: Practice Exercise 3: a. $\dfrac{10}{b}$ **b.** $\dfrac{m - 6}{n}$ **c.** $\dfrac{x}{2z}$ **d.** $\dfrac{a^2}{4}$ **Additional Practice 3: a.** $\dfrac{5}{y}$ **b.** $\dfrac{v + w}{u^3}$ **c.** $-\dfrac{3a}{7b}$ **d.** $\dfrac{b^2}{4}$

Example 4   **Find the sum or difference of each of the rational expressions. Express answers reduced to lowest terms. Assume that the denominator does not equal 0.**

a. $\dfrac{3}{a+2} + \dfrac{5}{a+2} =$ Put the sum of the numerators over the common denominator.

$\dfrac{3+5}{a+2} =$ $3 + 5 = 8$.

$\dfrac{8}{a+2}$ Sum.

b. $\dfrac{c}{c+4} - \dfrac{2}{c+4} =$ Put the difference of the numerators over the common denominator.

$\dfrac{c-2}{c+4}$ Difference.

c. $\dfrac{2y}{y-3} - \dfrac{6}{y-3} =$ Put the difference of the numerators over the common denominator.

$\dfrac{2y-6}{y-3} =$ Factor the numerator.

$\dfrac{2(y-3)}{y-3} =$ Divide by the common factor $y - 3$.

2 Difference.

d. $\dfrac{x^2}{x^2+4x-21} + \dfrac{2x-15}{x^2+4x-21} =$ Put the sum of the numerators over the common denominator.

$\dfrac{(x^2)+(2x-15)}{x^2+4x-21} =$ Remove the parentheses.

$\dfrac{x^2+2x-15}{x^2+4x-21} =$ Factor the numerator and denominator.

$\dfrac{(x+5)(x-3)}{(x+7)(x-3)} =$ Divide by the common factor $x - 3$.

$\dfrac{x+5}{x+7}$ Sum.

Be Careful

In exercises like Example 4e, where we are finding the difference of rational expressions, a common error is to forget to change all the signs of the numerator of the second fraction. To avoid making this type of error, you should write the step with the difference of the numerators in parentheses.

e. $\dfrac{x^2+5x}{x^2+7x+12} - \dfrac{8x+28}{x^2+7x+12} =$ Put the difference of the numerators over the common denominator.

$\dfrac{(x^2+5x)-(8x+28)}{x^2+7x+12} =$ Remove the parentheses.

$\dfrac{x^2+5x-8x-28}{x^2+7x+12} =$ Add like terms.

$\dfrac{x^2-3x-28}{x^2+7x+12} =$ Factor the numerator and the denominator.

$\dfrac{(x-7)(x+4)}{(x+4)(x+3)} =$ Divide by the common factor of $x + 4$.

$\dfrac{x-7}{x+3}$ Difference.

PRACTICE EXERCISE 4

Find the sum or difference of each rational expression. Express answers reduced to lowest terms.

a. $\dfrac{6}{a-5} + \dfrac{2}{a-5}$

b. $\dfrac{3a}{a+2} - \dfrac{a}{a+2}$

c. $\dfrac{x}{x-4} - \dfrac{4}{x-4}$

d. $\dfrac{x^2}{x^2+3x-18} + \dfrac{5x-6}{x^2+3x-18}$

e. $\dfrac{x^2-2x}{x^2-5x+6} - \dfrac{5x-10}{x^2-5x+6}$

(Continued)

Answers: Practice Exercise 4: a. $\dfrac{8}{a-5}$ **b.** $\dfrac{2a}{a+2}$ **c.** 1 **d.** $\dfrac{x-1}{x-3}$ **e.** $\dfrac{x-5}{x-3}$

If you need more practice, do the following Additional Practice Exercises.

Additional Practice Exercise 4   Find the sum or difference of each rational expression. Express answers reduced to lowest terms.

a. $\dfrac{12}{2x + 1} - \dfrac{5}{2x + 1}$

b. $\dfrac{y}{3y + 4} + \dfrac{3y}{3y + 4}$

c. $\dfrac{6x}{2x + 5} + \dfrac{15}{2x + 5}$

d. $\dfrac{2x^2}{x^2 - 9} + \dfrac{7x + 3}{x^2 - 9}$

e. $\dfrac{x^2}{x^2 - x - 6} - \dfrac{2x + 8}{x^2 - x - 6}$

There is one type of problem that requires special attention, so we will discuss it separately. Earlier in this section, we discussed equivalent forms of negative fractions, and we indicated that $-\frac{a}{b} = \frac{-a}{b}$. It is also true that $\frac{a}{-b} = \frac{-a}{b}$, since both are equal to $-\frac{a}{b}$. For example, $\frac{4}{-7} = \frac{-4}{7}$ and $\frac{5}{-(a - b)} = \frac{-5}{a - b}$. In general, a fraction has three signs, any two of which may be changed without changing the value of the fraction. For example, $+\frac{+2}{+3} = +\frac{-2}{-3} = -\frac{-2}{+3} = -\frac{+2}{-3}$.

Example 5   **Find the sums or differences. Express the answers reduced to lowest terms.**

a. $\dfrac{5}{x - 2} + \dfrac{3}{2 - x} =$ Rewrite $2 - x$ as $-x + 2$.

$\dfrac{5}{x - 2} + \dfrac{3}{-x + 2} =$ Factor out a negative in the denominator.

$\dfrac{5}{x - 2} + \dfrac{3}{-(x - 2)} =$ Rewrite $\dfrac{3}{-(x - 2)}$ as $\dfrac{-3}{x - 2}$.

$\dfrac{5}{x - 2} + \dfrac{-3}{x - 2} =$ Find the sum.

$\dfrac{5 + (-3)}{x - 2} =$ $5 + (-3) = 2$.

$\dfrac{2}{x - 2}$ Sum.

b. $\dfrac{a + b}{a - b} - \dfrac{2a - 3b}{b - a} =$ Rewrite $b - a$ as $-(a - b)$.

$\dfrac{a + b}{a - b} - \dfrac{2a - 3b}{-(a - b)} =$ Rewrite $\dfrac{2a - 3b}{-(a - b)}$ as $\dfrac{-(2a - 3b)}{a - b}$.

$\dfrac{a + b}{a - b} - \dfrac{-(2a - 3b)}{a - b} =$ Remove the parentheses.

$\dfrac{a + b}{a - b} - \dfrac{-2a + 3b}{a - b} =$ Find the difference.

$\dfrac{(a + b) - (-2a + 3b)}{a - b} =$ Remove the parentheses.

$\dfrac{a + b + 2a - 3b}{a - b} =$ Add like terms.

$\dfrac{3a - 2b}{a - b}$ Difference.

PRACTICE EXERCISE 5

Find the sums or differences. Express answers reduced to lowest terms.

a. $\dfrac{8}{y - 3} + \dfrac{5}{3 - y}$

b. $\dfrac{3x + 2y}{x - y} - \dfrac{x - 5y}{y - x}$

If you need more practice, do the following Additional Practice Exercises.

Additional Practice Exercise 5   Find the sums or differences. Express answers reduced to lowest terms.

a. $\dfrac{-5}{a - 4} + \dfrac{2}{4 - a}$

b. $\dfrac{2b - 3c}{b - c} - \dfrac{b - 2c}{c - b}$

Answers: Additional Practice 4: a. $\dfrac{7}{2x + 1}$ **b.** $\dfrac{4y}{3y + 4}$ **c.** 3   **d.** $\dfrac{2x + 1}{x - 3}$ **e.** $\dfrac{x - 4}{x - 3}$ **Practice Exercise 5: a.** $\dfrac{3}{y - 3}$ **b.** $\dfrac{4x - 3y}{x - y}$

Additional Practice 5: a. $\dfrac{-7}{a - 4}$ **b.** $\dfrac{3b - 5c}{b - c}$

Exercise Set 7.1

For Extra Help
MyMathLab®

Find the following sums or differences. Express the answers reduced to lowest terms. (See Example 1.)

1. $\dfrac{2}{7} + \dfrac{3}{7}$

2. $\dfrac{4}{25} + \dfrac{9}{25}$

3. $\dfrac{7}{18} + \dfrac{5}{18}$

4. $\dfrac{3}{8} + \dfrac{1}{8}$

5. $\dfrac{3}{16} + \dfrac{1}{16} + \dfrac{5}{16}$

6. $\dfrac{6}{29} + \dfrac{12}{29} + \dfrac{3}{29}$

7. $\dfrac{3}{4} - \dfrac{2}{4}$

8. $\dfrac{8}{9} - \dfrac{7}{9}$

9. $\dfrac{10}{17} - \dfrac{-5}{17}$

10. $\dfrac{9}{22} - \dfrac{-9}{22}$

11. $\dfrac{7}{11} - \dfrac{9}{11}$

12. $\dfrac{5}{12} - \dfrac{10}{12}$

13. $\dfrac{5}{14} + \dfrac{9}{14}$

14. $\dfrac{13}{24} + \dfrac{11}{24}$

15. $\dfrac{9}{10} - \dfrac{3}{10}$

16. $\dfrac{17}{25} - \dfrac{12}{25}$

17. $\dfrac{8}{15} + \dfrac{3}{15} - \dfrac{6}{15}$

18. $\dfrac{6}{21} - \dfrac{8}{21} + \dfrac{16}{21}$

19. $\dfrac{17}{18} - \dfrac{6}{18} - \dfrac{5}{18}$

20. $-\dfrac{9}{32} - \dfrac{-4}{32} + \dfrac{29}{32}$

Add or subtract the following mixed numbers. (See Example 2.)

21. $8\dfrac{2}{7} + 4\dfrac{3}{7}$

22. $3\dfrac{2}{9} + 6\dfrac{5}{9}$

23. $10\dfrac{7}{9} - 5\dfrac{4}{9}$

24. $7\dfrac{3}{14} - 12\dfrac{10}{14}$

25. $8\dfrac{10}{27} + 9\dfrac{16}{27} - 11\dfrac{5}{27}$

26. $13\dfrac{24}{33} - 9\dfrac{10}{33} + 4\dfrac{8}{33}$

Find the sum or difference of each of the rational expressions. Express your answers reduced to lowest terms. (See Examples 3 and 4.)

27. $\dfrac{4}{b} + \dfrac{6}{b}$

28. $\dfrac{2}{t} + \dfrac{6}{t}$

29. $\dfrac{u}{v} - \dfrac{5}{v}$

30. $\dfrac{8}{w} - \dfrac{z}{w}$

31. $\dfrac{11m}{21u} + \dfrac{4m}{21u}$

32. $\dfrac{13p}{30q} + \dfrac{7p}{30q}$

33. $\dfrac{7}{2y} + \dfrac{3}{2y} - \dfrac{5}{2y}$

34. $\dfrac{5}{2x} + \dfrac{7}{2x} - \dfrac{3}{2x}$

35. $\dfrac{6x}{7y^2} - \dfrac{2x}{7y^2} - \dfrac{5x}{7y^2}$

36. $\dfrac{9a}{11y^2} - \dfrac{6a}{11y^2} - \dfrac{4a}{11y^2}$

37. $\dfrac{9x^2 + 3y^2}{3x^3} + \dfrac{3x^2 - 3y^2}{3x^3}$

38. $\dfrac{12m^3 + 6n^3}{4m^4} + \dfrac{4m^3 - 6n^3}{4m^4}$

39. $\dfrac{4a^3 - 2b^2}{6b} - \dfrac{4a^3 - 4b^2}{6b}$

40. $\dfrac{2c^4 - 3d^3}{6d^2} - \dfrac{2c^4 - 6d^3}{6d^2}$

41. $\dfrac{5a + 7}{3} - \dfrac{2a + 1}{3}$

42. $\dfrac{7y + 4}{5} - \dfrac{2y - 6}{5}$

43. $\dfrac{6}{n + 5} + \dfrac{3}{n + 5}$

44. $\dfrac{4}{2 - m} + \dfrac{7}{2 - m}$

45. $\dfrac{9}{t - 8} - \dfrac{7}{t - 8}$

46. $\dfrac{8}{4 - z} - \dfrac{12}{4 - z}$

47. $\dfrac{5c}{c + 2} - \dfrac{2c}{c + 2}$

48. $\dfrac{5z}{6 - b} - \dfrac{7z}{6 - b}$

49. $\dfrac{3a}{a + 2} + \dfrac{5a}{a + 2} - \dfrac{2a}{a + 2}$

50. $\dfrac{8x}{x + 3} - \dfrac{2x}{x + 3} - \dfrac{4x}{x + 3}$

51. $\dfrac{4x - 3}{3x - 5} + \dfrac{2x - 7}{3x - 5}$

52. $\dfrac{5a + 6}{4a + 5} + \dfrac{7a + 9}{4a + 5}$

53. $\dfrac{12a - 6}{2a - 3} - \dfrac{2a + 9}{2a - 3}$

54. $\dfrac{14a - 2}{3a + 2} - \dfrac{2a - 10}{3a + 2}$

55. $\dfrac{x^2}{x^2 - x - 12} + \dfrac{5x + 6}{x^2 - x - 12}$

56. $\dfrac{y^2}{y^2 - 5y + 4} + \dfrac{y - 20}{y^2 - 5y + 4}$

57. $\dfrac{t^2 - 4t}{t^2 + 10t + 21} - \dfrac{-6t + 3}{t^2 + 10t + 21}$

58. $\dfrac{v^2 + 6}{v^2 - 3v - 10} - \dfrac{2v + 14}{v^2 - 3v - 10}$

59. $\dfrac{u^2}{u^2 - 9} + \dfrac{2u - 15}{u^2 - 9}$

60. $\dfrac{w^2 + 3w}{w^2 + 4w + 4} - \dfrac{5w + 8}{w^2 + 4w + 4}$

61. $\dfrac{x^2 - 2x}{x^2 - 12x + 36} + \dfrac{-5x + 6}{x^2 - 12x + 36}$

62. $\dfrac{q^2 - 3}{q^2 - 49} - \dfrac{5q + 11}{q^2 - 49}$

63. $\dfrac{2x^2 + x - 3}{x^2 + 6x + 5} + \dfrac{x^2 - 2x + 4}{x^2 + 6x + 5} - \dfrac{2x^2 + x + 4}{x^2 + 6x + 5}$

64. $\dfrac{3x^2 - 2x + 3}{x^2 + x - 12} - \dfrac{x^2 + x - 2}{x^2 + x - 12} - \dfrac{x^2 + 4x - 7}{x^2 + x - 12}$

Find the sums or differences. Express the answers reduced to lowest terms. (See Example 5.)

65. $\dfrac{3}{7} + \dfrac{4}{-7}$

66. $\dfrac{13}{15} + \dfrac{8}{-15}$

67. $\dfrac{-9}{17} - \dfrac{8}{-17}$

68. $\dfrac{19}{22} - \dfrac{2}{-22}$

69. $\dfrac{3}{t - 4} + \dfrac{11}{4 - t}$

70. $\dfrac{-5}{7 - t} + \dfrac{8}{t - 7}$

71. $\dfrac{v}{u - v} - \dfrac{3v}{v - u}$

72. $\dfrac{2x}{x - y} + \dfrac{-3y}{y - x}$

73. $\dfrac{2m - n}{m - n} - \dfrac{m - 3}{n - m}$

74. $\dfrac{p - 5q}{p - q} - \dfrac{4p - 6q}{q - p}$

75. $\dfrac{2a - 3b}{3a - b} + \dfrac{3a + 2b}{b - 3a}$

76. $\dfrac{4x + 3y}{2x - 5y} + \dfrac{2x - y}{5y - 2x}$

Solve the following:

77. A triangular flower bed has sides whose lengths are $3\frac{3}{8}$ feet, $2\frac{2}{8}$ feet, and $4\frac{5}{8}$ feet. How much border material will it take to enclose the bed?

78. Fred is wrapping a triangular package whose sides are $8\frac{6}{16}$ inches, $7\frac{3}{16}$ inches, and $5\frac{7}{16}$ inches. How long would a piece of ribbon need to be in order to wrap around the package?

79. A rectangular picture is $6\frac{3}{4}$ inches long and $4\frac{1}{4}$ inches wide. What is the perimeter of the picture?

80. A rectangular traffic sign is $18\frac{5}{8}$ inches long and $12\frac{3}{8}$ inches wide. What is the perimeter of the sign?

Find the perimeters of the following triangles whose sides have the given lengths:

81. $\dfrac{5}{24a}$ feet, $\dfrac{7}{24a}$ feet, $\dfrac{11}{24a}$ feet

82. $\dfrac{7}{28x}$ yard, $\dfrac{9}{28x}$ yard, $\dfrac{8}{28x}$ yard

83. $\dfrac{5a}{2a + 3b}$ decimeters, $\dfrac{3b}{2a + 3b}$ decimeters, $\dfrac{2a + 4b}{2a + 3b}$ decimeters

84. $\dfrac{3m}{2m + n}$ feet, $\dfrac{m + 2n}{2m + n}$ feet, $\dfrac{2m - n}{2m + n}$ feet

85. $\dfrac{2x^2 - 3x}{x^2 - 2x - 15}$ centimeters, $\dfrac{-x^2 + x - 4}{x^2 - 2x - 15}$ centimeters, $\dfrac{x - 8}{x^2 - 2x - 15}$ centimeters

86. $\dfrac{3x^2 + 2}{x^2 - 3x - 10}$ inches, $\dfrac{-x^2 + 4x + 1}{x^2 - 3x - 10}$ inches, $\dfrac{-x^2 + 2x + 5}{x^2 - 3x - 10}$ inches

Find the perimeters of the following rectangles, given their lengths and widths:

87. $L = \dfrac{6}{x}$ meters, $W = \dfrac{4}{x}$ meters

88. $L = \dfrac{8}{3y}$ meters, $W = \dfrac{5}{3y}$ meters

89. $L = \dfrac{a + 2b}{2a - 3b}$ feet, $W = \dfrac{2a - b}{2a - 3b}$ feet

90. $L = \dfrac{2x + 4y}{3x - 5y}$ centimeters, $W = \dfrac{x + 5y}{3x - 5y}$ centimeters

91. $L = \dfrac{4x^2 + 8x + 3}{8x^2 + 2x - 3}$ meters, $W = \dfrac{4x^2 + 10x + 6}{8x^2 + 2x - 3}$ meters

92. $L = \dfrac{x^2 + 4x + 3}{2x^2 - 5x - 12}$ inches, $W = \dfrac{x^2 + 3x + 3}{2x^2 - 5x - 12}$ inches

Challenge Exercises (93–95)

Find the following sums or differences:

93. $\dfrac{x^2 + 2x}{x^3 + 2x^2 - 15x} - \dfrac{15}{x^3 + 2x^2 - 15x}$

94. $\dfrac{y^2 + 9y}{y^3 - 36y} + \dfrac{18}{y^3 - 36y}$

95. $\dfrac{t^3 - 4t^2}{t^3 - 2t^2 + t} - \dfrac{-3t}{t^3 - 2t^2 + t}$

Writing Exercises (96–100)

96. When adding or subtracting rational numbers or rational expressions that have the same denominator, why do we add or subtract the numerators, but we do not add or subtract the denominators?

97. In this section, we began with a graphical illustration for the addition of rational numbers. Make your own illustration for the addition of fractions and explain how it works.

98. Show a graphical illustration for the subtraction of fractions and explain how it works.

99. Explain what is wrong with the following:

$$\dfrac{3x - 1}{x + 6} - \dfrac{2x - 3}{x + 6} = \dfrac{3x - 1 - 2x - 3}{x + 6} = \dfrac{x - 4}{x + 6}$$

What *is* the correct answer?

100. Recall that $\frac{6}{x} = 6x^{-1}$ when negative exponents are used. Explain how $\frac{6}{x} + \frac{4}{x}$ can be added as like terms by rewriting with negative exponents.

Section 7.2
Least Common Multiple and Equivalent Rational Expressions

OBJECTIVES *When you complete this section, you will be able to:*

A Find the least common multiple of two or more integers.

B Find the least common multiple of two or more polynomials.

C Change a fraction into an equivalent fraction with a specified denominator.

D Change a rational expression into an equivalent rational expression with a specified denominator.

PREREQUISITE SKILLS *Before beginning this section, you should be able to:*

a. Write a whole number in terms of its prime factors. (Section 5.1)

b. Factor polynomials. (Sections 5.2–5.6)

c. Divide whole numbers. (Section R.3)

d. Apply $\dfrac{a^m}{a^n} = a^{m-n}$ to divide monomials. (Sections 2.6 and 2.8)

e. Apply $a^m \cdot a^n = a^{m+n}$ to multiply monomials. (Sections 2.2 and 2.3)

f. Multiply rational numbers and rational expressions. (Sections 6.3 and 6.4)

GETTING READY FOR SECTION 7.2

1. Write 90 in terms of its prime factors.

Factor the following:

2. $4r^2 - 9$

3. $6r^2 - 11r + 3$

Divide the following whole numbers and monomials:

4. $\dfrac{108}{12}$

5. $\dfrac{28y^4}{7y^2}$

Multiply the following monomials:

6. $3x^2 \cdot 5x$

7. $6xy^3 \cdot 5x^2y$

Multiply the following rational numbers and expressions. Do not reduce the product to lowest terms.

8. $\dfrac{9}{9} \cdot \dfrac{7}{12}$

9. $\dfrac{4y^2}{4y^2} \cdot \dfrac{12x^2}{7y^2}$

10. $\dfrac{x+3}{x+3} \cdot \dfrac{7}{x(x-2)}$

Introduction

Previously, we discussed the greatest common factor of two or more integers. One interpretation of the greatest common factor (GCF) is that it is the largest integer that will divide evenly into all the integers for which it is the greatest common factor. In this section, we discuss the **least common multiple** (LCM) of two or more integers. One interpretation of the least common multiple is that it is the smallest positive integer that is divisible by all the integers for which it is the least common multiple. Note that this is the reverse of a GCF.

Before discussing the technique for finding the least common multiple, let us discuss the idea of multiples. To find the multiples of a number, multiply the number by 1, 2, 3, 4, and so on. For example, the multiples of 3 are $3 \cdot 1 = 3, 3 \cdot 2 = 6, 3 \cdot 3 = 9$, $3 \cdot 4 = 12$, and so on. Note that the multiples of a number have that number as a factor.

Answers: Getting Ready: 1. $2 \cdot 3^2 \cdot 5$ **2.** $(2r+3)(2r-3)$ **3.** $(3r-1)(2r-3)$ **4.** 9 **5.** $4y^2$ **6.** $15x^3$ **7.** $30x^3y^4$ **8.** $\dfrac{63}{108}$ **9.** $\dfrac{48x^2y^2}{28y^4}$

10. $\dfrac{7(x+3)}{x(x-2)(x+3)}$

The multiples of 3 are $\{3, 6, 9, 12, 15, 18, 21, 24, \ldots\}$. The multiples of 4 are $\{4, 8, 12, 16, 20, 24, \ldots\}$. Looking at the multiples of 3 and 4, we see that 12 and 24 are multiples of both. Consequently, we call 12 and 24 common multiples of 3 and 4. Other common multiples are 36, 48, 60, and so on. Of all the common multiples of 3 and 4, the smallest is 12. Therefore, 12 is the least common multiple of 3 and 4. Notice that each multiple of 3 is divisible by 3 and each multiple of 4 is divisible by 4. Consequently, the common multiples of 3 and 4 are divisible by both 3 and 4. Numbers that are divisible by 3 and 4 have 3 and 4 as factors.

> ### Definition Least Common Multiple (LCM)
>
> The integer a is the least common multiple of the integers b and c if a is the smallest positive integer that is a multiple of both b and c. Equivalently, a is the least common multiple of b and c if a is the smallest positive integer that is divisible by both b and c.

The preceding definition is easily extended to more than two integers.

OBJECTIVE **Finding the least common multiple of two or more integers**

The technique we use for finding the LCM depends upon finding the prime factorizations of the numbers for which we wish to find the LCM. Prime factorization was discussed in Section 5.1. You may wish to review this technique before continuing.

> ### Procedure: Finding the Least Common Multiple (LCM)
>
> Procedure for finding the LCM of two or more numbers is as follows:
>
> **1.** Write each number as the product of its prime factors.
>
> **2.** Select every factor appearing in any prime factorization. If a factor appears in more than one of the prime factorizations, select the factor with the largest exponent so that the LCM will have the original numbers as factors.
>
> **3.** The LCM is the product of the factors selected in step 2.

This technique for finding the LCM works because any factor of a number is also a factor of any multiple of that number. By taking the largest exponent of every factor that appears in any prime factorization, we are assuring ourselves that their product (the LCM) contains every factor of each number and only those numbers that are factors.

Example 1   Find the LCM of the following:

a. 3 and 7

Solution

Since 3 and 7 are both prime numbers, the different factors that appear are 3 and 7. The largest exponent on each is 1. Consequently, the LCM is their product: $3 \cdot 7 = 21$. Note that 21 is divisible by both 3 and 7.

b. 108 and 144

Solution

$108 = 2^2 \cdot 3^3$ Prime factorization.
$144 = 2^4 \cdot 3^2$ Prime factorization.

The different factors that appear in either factorization are 2 and 3. The largest exponent on 2 in either factorization is 4 in 2^4, and the largest exponent on 3 in either factorization is 3 in 3^3. Consequently, the LCM is $2^4 \cdot 3^3 = 16 \cdot 27 = 432$. Note that 432 is divisible by both 108 and 144.

c. 90 and 189

Solution

$90 = 2 \cdot 3^2 \cdot 5$ Prime factorization.
$189 = 3^3 \cdot 7$ Prime factorization.

The different factors that appear in either factorization are 2, 3, 5, and 7. The largest exponent on 2 is 1 in 2, the largest exponent of 3 is 3 in 3^3, the largest exponent of 5 is 1 in 5, and the largest exponent of 7 is 1 in 7. Consequently, the LCM is $2 \cdot 3^3 \cdot 5 \cdot 7 = 2 \cdot 27 \cdot 5 \cdot 7 = 54 \cdot 5 \cdot 7 = 270 \cdot 7 = 1890$. Verify that 1890 is divisible by both 90 and 189.

d. 50, 45, and 42

Solution

$50 = 2 \cdot 5^2$ Prime factorization.
$45 = 3^2 \cdot 5$ Prime factorization.
$42 = 2 \cdot 3 \cdot 7$ Prime factorization.

The different factors that appear in the factorizations are 2, 3, 5, and 7. The largest exponent of 2 is 1 in 2, the largest exponent of 3 is 2 in 3^2, the largest exponent of 5 is 2 in 5^2, and the largest exponent of 7 is 1 in 7. Consequently, the LCM is $2 \cdot 3^2 \cdot 5^2 \cdot 7 = 2 \cdot 9 \cdot 25 \cdot 7 = 18 \cdot 25 \cdot 7 = 450 \cdot 7 = 3150$. Verify that 3150 is divisible by 50, 45, and 42.

PRACTICE EXERCISE 1

Find the LCM of the following:

a. 7 and 11   **b.** 48 and 72   **c.** 60 and 126   **d.** 28, 30, and 33

If you need more practice, do the following Additional Practice Exercises.

Additional Practice Exercise 1   Find the LCM of the following:

a. 3 and 5   **b.** 225 and 375   **c.** 90 and 168   **d.** 20, 63, and 70

OBJECTIVE B Finding the least common multiple of two or more polynomials

The procedure for finding the LCM of two or more monomials involving variables is exactly the same as for finding the LCM from prime factorizations. Many times it is actually easier, since we do not have to find the prime factorizations for the variable factors.

If the monomial expressions have numerical coefficients, we think of the exercise in two parts. We find the prime factorizations of the coefficients and use the same method as in Example 1.

Example 2   **Find the LCM of the following:**

a. $a^2 b^3$ and $a^4 b$

Solution

The different factors that appear are a and b. The largest exponent of a is 4 in a^4, and the largest exponent of b is 3 in b^3. Consequently, the LCM is $a^4 b^3$. Note that $\frac{a^4 b^3}{a^2 b^3} = a^2$ and $\frac{a^4 b^3}{a^4 b} = b^2$. Consequently, $a^4 b^3$ is divisible by both $a^2 b^3$ and $a^4 b$.

Answers: Practice Exercise 1: a. $7 \cdot 11 = 77$ **b.** $2^4 \cdot 3^2 = 144$ **c.** $2^2 \cdot 3^2 \cdot 5 \cdot 7 = 1260$ **d.** $2^2 \cdot 3 \cdot 5 \cdot 7 \cdot 11 = 4620$
Additional Practice 1: a. $3 \cdot 5 = 15$ **b.** $3^2 \cdot 5^3 = 1125$ **c.** $2^3 \cdot 3^2 \cdot 5 \cdot 7 = 2520$ **d.** $2^2 \cdot 3^2 \cdot 5 \cdot 7 = 1260$

b. x^3y^2z and x^2y^4w

Solution

The different factors that appear are x, y, z, and w. The largest exponent on x is 3 in x^3, the largest exponent of y is 4 in y^4, the largest exponent of z is 1 in z, and the largest exponent of w is 1 in w. Consequently, the LCM is x^3y^4zw. Note that x^3y^4zw is divisible by both x^3y^2z and x^2y^4w.

c. $12r^3s^5$ and $18rs^2$

Solution

We find the prime factorizations of the coefficients and use the same method as in Example 1.

$12r^3s^5 = 2^2 \cdot 3r^3s^5$ Prime factorization of 12.
$18rs^2 = 2 \cdot 3^2rs^2$ Prime factorization of 18.

The different factors that appear are 2, 3, r, and s. The largest exponent on 2 is 2 in 2^2, the largest exponent on 3 is 2 in 3^2, the largest exponent on r is 3 in r^3, and the largest exponent on s is 5 in s^5. Therefore, the LCM is $2^2 \cdot 3^2r^3s^5 = 36r^3s^5$. Note that $36r^3s^5$ is divisible by $12r^3s^5$ and $18rs^2$.

d. $6a^5b^2c^2$, $9a^2bc^2$, and $10a^2b^3c^2$

Solution

We find the prime factorizations of the coefficients and use the same method as in Examples 1 and 2.

$6a^5b^2c^2 = 2 \cdot 3a^5b^2c^2$ Prime factorization of 6.
$9a^2bc^2 = 3^2a^2bc^2$ Prime factorization of 9.
$10a^2b^3c^2 = 2 \cdot 5a^2b^3c^2$ Prime factorization of 10.

The different factors that appear are 2, 3, 5, a, b, and c. The largest exponent on 2 is 1 in 2, the largest exponent on 3 is 2 in 3^2, the largest exponent on 5 is 1 in 5, the largest exponent on a is 5 in a^5, the largest exponent on b is 3 in b^3, and the largest exponent on c is 2 in c^2. Therefore, the LCM is $2 \cdot 3^2 \cdot 5a^5b^3c^2 = 90a^5b^3c^2$. Note that $90a^5b^3c^2$ is divisible by $6a^5b^2c^2$, $9a^2bc^2$, and $10a^2b^3c^2$.

> **Note**
>
> You can find the LCM of the coefficients without using prime factorizations.

PRACTICE EXERCISE 2

Find the LCM of the following:

a. x^2y^5 and x^4y

b. r^4st^3 and $r^2s^3u^2$

c. $15a^2b^4$ and $12a^3b^4$

d. $8x^2yz^3$, $12x^4y^2z$, and $15x^4y^2z^2$

If you need more practice, do the following Additional Practice Exercises.

Additional Practice Exercise 2   Find the LCM of the following:

a. a^3b^2 and a^2b^4

b. $m^2n^4o^2$ and $m^5n^3o^4$

c. $14r^6s^2$ and $20r^2s^2$

d. $4m^3n^3$, $6m^2n^4$, and $15m^3n^4$

In order to find the LCM of two or more polynomials that are not monomials, the procedure is still the same. However, we usually leave the LCM in factored form rather than multiply it out.

Answers: Practice Exercise 2: a. x^4y^5 **b.** $r^4s^3t^3u^2$ **c.** $60a^3b^4$ **d.** $120x^4y^2z^3$ **Additional Practice 2: a.** a^3b^4 **b.** $m^5n^4o^4$ **c.** $140r^6s^2$ **d.** $60m^3n^4$

Example 3   **Find the LCM of the following:**

a. $2x + 4$ and $3x + 6$

Solution $2x + 4 = 2(x + 2)$
$3x + 6 = 3(x + 2)$

The different factors that appear are 2, 3, and $(x + 2)$ and the largest exponent of each factor is 1. Therefore, the LCM is $2 \cdot 3(x + 2) = 6(x + 2)$. Note that $6(x + 2)$ is divisible by $2x + 4$ and $3x + 6$.

b. $y^2 + 5y$ and $2y + 10$

Solution $y^2 + 5y = y(y + 5)$
$2y + 10 = 2(y + 5)$

The different factors that appear are y, 2, and $(y + 5)$ and the largest exponent on each factor is 1. Therefore, the LCM is $2 \cdot y(y + 5) = 2y(y + 5)$. Note that $2y(y + 5)$ is divisible by both $y^2 + 5y$ and $2y + 10$.

c. $z + 4$ and $z^2 + 6z + 8$

Solution $z + 4$ $z + 4$ is a prime polynomial.
$z^2 + 6z + 8 = (z + 4)(z + 2)$ Prime factorization.

The different factors that appear are $z + 4$ and $z + 2$. The largest exponent on each is 1. Therefore, the LCM is $(z + 4)(z + 2)$. Note that $(z + 4)(z + 2)$ is divisible by $z + 4$ and $z^2 + 6z + 8$.

d. $x^2 + 2x - 15$ and $x^2 - 5x + 6$

Solution $x^2 + 2x - 15 = (x - 3)(x + 5)$ Prime factorization.
$x^2 - 5x + 6 = (x - 3)(x - 2)$ Prime factorization.

The different factors that appear are $x - 3$, $x + 5$, and $x - 2$. The largest exponent on each is 1. Therefore, the LCM is $(x - 3)(x + 5)(x - 2)$. Note that $(x - 3)(x + 5)(x - 2)$ is divisible by $x^2 + 2x - 15$ and $x^2 - 5x + 6$.

e. $x + 3$ and $x - 2$

Solution

Since $x + 3$ and $x - 2$ are both prime polynomials, the LCM is their product $(x + 3)(x - 2)$. Note that $(x + 3)(x - 2)$ is divisible by both $x + 3$ and $x - 2$.

f. $y - 6$ and $y^2 + 5y$

Solution $y - 6$ $y - 6$ is a prime polynomial.
$y^2 + 5y = y(y + 5)$ Prime factorization.

The different factors that appear in either prime factorization are y, $y - 6$, and $y + 5$ and the largest exponent of each is 1. Therefore, the LCM is $y(y - 6)(y + 5)$. Note that $y(y - 6)(y + 5)$ is divisible by $y - 6$ and $y^2 + 5y$.

g. $x^2 + 5x + 4$ and $x^2 + 2x + 1$

Solution $x^2 + 5x + 4 = (x + 1)(x + 4)$ Prime factorization.
$x^2 + 2x + 1 = (x + 1)^2$ Prime factorization.

The different factors that appear are $x + 1$ and $x + 4$. The largest exponent on $x + 1$ is 2 in $(x + 1)^2$, and the largest exponent on $x + 4$ is 1 in $x + 4$. Therefore, the LCM is $(x + 1)^2(x + 4)$. Note that $(x + 1)^2(x + 4)$ is divisible by $x^2 + 5x + 4$ and $x^2 + 2x + 1$.

h. $4r^2 - 9$, $6r^2 + 7r - 3$, and $6r^2 - 11r + 3$

Solution $4r^2 - 9 = (2r + 3)(2r - 3)$ Prime factorization.
$6r^2 + 7r - 3 = (2r + 3)(3r - 1)$ Prime factorization.
$6r^2 - 11r + 3 = (3r - 1)(2r - 3)$ Prime factorization.

The different factors that appear are $2r + 3$, $2r - 3$, and $3r - 1$. Since the largest exponent on each is 1, the LCM is $(2r + 3)(2r - 3)(3r - 1)$. Note that $(2r + 3)(2r - 3)(3r - 1)$ is divisible by $4r^2 - 9$, $6r^2 + 7r - 3$, and $6r^2 - 11r + 3$.

PRACTICE EXERCISE 3

Find the LCM of the following:

a. $3x - 12$ and $5x - 20$

b. $x^2 - 3x$ and $4x - 12$

c. $b - 4$ and $b^2 - 2b - 8$

d. $y^2 - 7y + 10$ and $y^2 + 3y - 10$

e. $a + 4$ and $a - 3$

f. $b + 3$ and $b^2 + 3b$

g. $x^2 + 4x + 4$ and $x^2 - 3x - 10$

h. $4x^2 - 25$, $8x^2 + 14x - 15$, and $8x^2 - 26x + 15$

If you need more practice, do the following Additional Practice Exercises.

Additional Practice Exercise 3 Find the LCM of the following:

a. $2x - 10$ and $7x - 35$

b. $m^2 + 6m$ and $3m + 18$

c. $m - 3$ and $m^2 - 7m + 12$

d. $c^2 + 3c - 18$ and $c^2 + c - 12$

e. $b - 5$ and $b + 2$

f. $n - 4$ and $n^2 - 4n$

g. $4x^2 - 12x + 9$ and $8x^2 - 18x + 9$

h. $9x^2 - 4$, $6x^2 + x - 2$, and $6x^2 - 7x + 2$

In Section 6.1, we introduced the idea of equivalent fractions. Look at the following rectangles where each rectangle represents one unit:

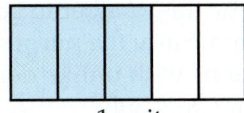
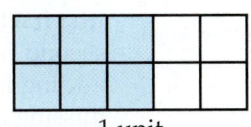

1 unit   1 unit

The fraction representing the shaded part of the first rectangle is $\frac{3}{5}$. Divide the unit horizontally into halves in order to get the second rectangle. The fraction representing the shaded part of the second rectangle is $\frac{6}{10}$. Since the shaded area is the same before and after the division, only the name representing the part of the unit that is shaded has changed. Hence, we see that $\frac{6}{10} = \frac{3}{5}$. In other words, $\frac{6}{10}$ and $\frac{3}{5}$ are equivalent fractions.

From the example, we have $\frac{6}{10} = \frac{3}{5}$. Notice that the product of the numerator of the first fraction with the denominator of the second is equal to the product of the denominator of the first fraction and the numerator of the second. That is, $6 \cdot 5 = 10 \cdot 3$. This observation leads to a more formal definition of equivalent fractions than the one given earlier in Section 6.1.

Definition Equivalent Fractions

Two or more fractions are equivalent if they represent the same quantity. Equivalently, $\frac{a}{b} = \frac{c}{d}$ if $a \cdot d = b \cdot c$.

The preceding procedure is often referred to as **cross multiplication**, because the lines of multiplication "cross" each other:

$$\frac{a}{b} \diagdown\!\!\!\!\diagup \frac{c}{d}$$

$$a \cdot d = b \cdot c$$

In reducing fractions to lowest terms, we divide the numerators and the denominators by all factors common to both to get an equivalent fraction reduced to lowest terms. For example, $\frac{6}{10} = \frac{2 \cdot 3}{2 \cdot 5} = \frac{3}{5}$. In this section, we are going to do the opposite and generate equivalent fractions by multiplying the numerator and the denominator

Answers: Practice Exercise 3: a. $15(x - 4)$ **b.** $4x(x - 3)$ **c.** $(b - 4)(b + 2)$ **d.** $(y - 2)(y - 5)(y + 5)$ **e.** $(a + 4)(a - 3)$ **f.** $b(b + 3)$
g. $(x + 2)^2(x - 5)$ **h.** $(2x + 5)(2x - 5)(4x - 3)$ **Additional Practice 3: a.** $14(x - 5)$ **b.** $3m(m + 6)$ **c.** $(m - 3)(m - 4)$
d. $(c - 3)(c + 6)(c + 4)$ **e.** $(b - 5)(b + 2)$ **f.** $n(n - 4)$ **g.** $(2x - 3)^2(4x - 3)$ **h.** $(3x + 2)(3x - 2)(2x - 1)$

of a fraction by the same number. This is permissible because any nonzero number divided by itself is 1, and any number multiplied by 1 is equivalent to the original number. Since we are multiplying the numerator and the denominator by the same number, we are multiplying the fraction by 1. This is called the **fundamental property of fractions**.

<table>
<tr><td>

Note

If we cross multiply $\frac{a}{b} = \frac{ac}{bc}$, we get $abc = bac$. By what properties of multiplication does $abc = bac$?

</td><td>

Rule: Fundamental Property of Fractions

If $\frac{a}{b}$ is a fraction and c is any number except 0, then $\frac{a}{b} = \frac{a}{b} \cdot \frac{c}{c} = \frac{ac}{bc}$. In words, we can multiply both the numerator and the denominator of a fraction by any number, except 0, and get a fraction equivalent to the original fraction.

</td></tr>
</table>

OBJECTIVE Changing a fraction into an equivalent fraction with a specified denominator

In the examples that follow, we will use the fundamental property of fractions to create equivalent fractions with a given denominator. We must decide by what we need to multiply the denominator of the first fraction in order to get the denominator of the second fraction. Then we will multiply the numerator and the denominator of the first fraction by this number to get an equivalent fraction with the denominator of the second fraction. Since the purpose of these exercises is to aid you in a later section in adding fractions with unlike denominators, we will limit our discussion to finding missing numerators only.

Example 4   **Find the missing numerator so that the fractions are equal.**

a. $\dfrac{1}{3} = \dfrac{?}{6}$ Since $2 \cdot 3 = 6$, multiply $\dfrac{1}{3}$ by $\dfrac{2}{2}$.

$\dfrac{2}{2} \cdot \dfrac{1}{3} = \dfrac{2}{6}$ Therefore, $\dfrac{1}{3} = \dfrac{2}{6}$.

CHECK: $1 \cdot 6 = 3 \cdot 2$, so the answer is correct.

b. $\dfrac{3}{4} = \dfrac{?}{12}$ Since $3 \cdot 4 = 12$, multiply $\dfrac{3}{4}$ by $\dfrac{3}{3}$.

$\dfrac{3}{3} \cdot \dfrac{3}{4} = \dfrac{9}{12}$ Therefore, $\dfrac{3}{4} = \dfrac{9}{12}$.

CHECK: $3 \cdot 12 = 4 \cdot 9$, so the answer is correct.

Note

The number by which you must multiply the first fraction to get an equivalent fraction with the given denominator can be found by dividing the denominator of the first fraction into the denominator of the new fraction. In Example 4b, $12 \div 4 = 3$. Therefore, we multiply $\frac{3}{4}$ by $\frac{3}{3}$.

c. $\dfrac{7}{12} = \dfrac{?}{108}$ It may not be obvious what we have to multiply 12 by in order to get 108. Since $108 \div 12 = 9$, multiply $\dfrac{7}{12}$ by $\dfrac{9}{9}$.

$\dfrac{9}{9} \cdot \dfrac{7}{12} = \dfrac{63}{108}$ Therefore, $\dfrac{7}{12} = \dfrac{63}{108}$.

CHECK: Does $7 \cdot 108 = 12 \cdot 63$?

d. $\dfrac{11}{8} = \dfrac{?}{96}$ It may not be obvious what we have to multiply 8 by in order to get 96. Since $96 \div 8 = 12$, multiply $\dfrac{11}{8}$ by $\dfrac{12}{12}$.

$\dfrac{12}{12} \cdot \dfrac{11}{8} = \dfrac{132}{96}$ Therefore, $\dfrac{11}{8} = \dfrac{132}{96}$.

CHECK: Does $11 \cdot 96 = 8 \cdot 132$?

PRACTICE EXERCISE 4

Find the missing numerator so that the two fractions are equal.

a. $\dfrac{1}{5} = \dfrac{?}{30}$
b. $\dfrac{2}{3} = \dfrac{?}{18}$
c. $\dfrac{5}{16} = \dfrac{?}{144}$
d. $\dfrac{13}{9} = \dfrac{?}{126}$

If you need more practice, do the following Additional Practice Exercises.

Additional Practice Exercise 4   Find the missing numerator so that the two fractions are equal.

a. $\dfrac{1}{6} = \dfrac{?}{24}$
b. $\dfrac{3}{7} = \dfrac{?}{35}$
c. $\dfrac{5}{6} = \dfrac{?}{96}$
d. $\dfrac{17}{8} = \dfrac{?}{120}$

OBJECTIVE **Changing a rational expression into an equivalent rational expression with a specified denominator**

The procedure is exactly the same if we have rational expressions instead of rational numbers. In the exercises that follow, assume that all variables in the denominator have nonzero values.

Example 5   **Find the missing numerator so that the rational expressions are equal. Assume that all variables in the denominator have nonzero values.**

a. $\dfrac{x}{y} = \dfrac{?}{yz}$

Since $y \cdot z = yz$, multiply $\dfrac{x}{y}$ by $\dfrac{z}{z}$.

$\dfrac{z}{z} \cdot \dfrac{x}{y} = \dfrac{xz}{yz}$

Therefore, $\dfrac{x}{y} = \dfrac{xz}{yz}$.

CHECK:

$xyz = yxz$, so the answer is correct.

b. $\dfrac{3}{5x} = \dfrac{?}{15x^3}$

Since $3x^2 \cdot 5x = 15x^3$, multiply $\dfrac{3}{5x}$ by $\dfrac{3x^2}{3x^2}$.

$\dfrac{3x^2}{3x^2} \cdot \dfrac{3}{5x} = \dfrac{9x^2}{15x^3}$

Therefore, $\dfrac{3}{5x} = \dfrac{9x^2}{15x^3}$.

CHECK:

$3 \cdot 15x^3 = 5x \cdot 9x^2$, since both equal $45x^3$. Therefore, the answer is correct.

c. $\dfrac{12x^2}{7y^2} = \dfrac{?}{28y^4}$

Since $4y^2 \cdot 7y^2 = 28y^4$, multiply $\dfrac{12x^2}{7y^2}$ by $\dfrac{4y^2}{4y^2}$.

$\dfrac{4y^2}{4y^2} \cdot \dfrac{12x^2}{7y^2} = \dfrac{48x^2y^2}{28y^4}$

Therefore, $\dfrac{12x^2}{7y^2} = \dfrac{48x^2y^2}{28y^4}$.

CHECK:

Does $12x^2 \cdot 28y^4 = 7y^2 \cdot 48x^2y^2$?

d. $\dfrac{8}{5x^2y} = \dfrac{?}{30x^3y^4}$

Since $30x^3y^4 \div 5x^2y = 6xy^3$, we need to multiply by $6xy^3$.

$\dfrac{6xy^3}{6xy^3} \cdot \dfrac{8}{5x^2y} = \dfrac{48xy^3}{30x^3y^4}$

Therefore, $\dfrac{8}{5x^2y} = \dfrac{48xy^3}{30x^3y^4}$.

CHECK:

Does $8 \cdot 30x^3y^4 = 5x^2y \cdot 48xy^3$?

Note

In Example 5c, it may not be obvious that we need to multiply $7y^2$ by $4y^2$ in order to get $28y^4$. We can use the same procedure we used in Example 4. Since $28y^4 \div 7y^2 = 4y^2$, we need to multiply by $4y^2$.

Answers: Practice Exercise 4: **a.** 6 **b.** 12 **c.** 45 **d.** 182   **Additional Practice 4:** **a.** 4 **b.** 15 **c.** 80 **d.** 255

PRACTICE EXERCISE 5

Find the missing numerator so that the rational expressions are equal. Assume that all variables in the denominator have nonzero values.

a. $\dfrac{a}{b} = \dfrac{?}{bd}$ **b.** $\dfrac{5}{3a} = \dfrac{?}{18a^4}$ **c.** $\dfrac{8a^3}{5b^2} = \dfrac{?}{30b^5}$ **d.** $\dfrac{6}{11a^2b^3} = \dfrac{?}{55a^3b^3c}$

If you need more practice, do the following Additional Practice Exercises.

Additional Practice Exercise 5   Find the missing numerator so that the rational expressions are equal.

a. $\dfrac{m}{n} = \dfrac{?}{np}$ **b.** $\dfrac{3}{8x} = \dfrac{?}{48x^5}$ **c.** $\dfrac{7y^2}{4z^3} = \dfrac{?}{24z^6}$ **d.** $\dfrac{4}{13x^3y^2} = \dfrac{?}{52x^5y^4z}$

If the denominators of the rational expressions are polynomials other than monomials, we will need to factor those denominators that are not prime. By doing so, we can determine the expression we need to multiply the first rational expression by so that we can convert it into a rational expression with the same denominator as the second.

| **Example 6** | **Find the missing numerator so that the rational expressions will be equal.** |

a. $\dfrac{5a}{2a + 8} = \dfrac{?}{6a + 24}$ Factor the denominators.

$\dfrac{5a}{2(a + 4)} = \dfrac{?}{6(a + 4)}$ Since $3 \cdot 2(a + 4) = 6(a + 4)$, multiply $\dfrac{5a}{2(a + 4)}$ by $\dfrac{3}{3}$.

$\dfrac{3}{3} \cdot \dfrac{5a}{2(a + 4)} = \dfrac{15a}{6(a + 4)}$ Therefore, $\dfrac{5a}{2a + 8} = \dfrac{15a}{6a + 24}$.

Note

The checks can be done in the usual manner, but may often be very difficult. For that reason, equivalent rational expressions are usually not checked.

b. $\dfrac{7}{x^2 - 2x} = \dfrac{?}{x(x - 2)(x + 3)}$ Factor $x^2 - 2x$.

$\dfrac{7}{x(x - 2)} = \dfrac{?}{x(x - 2)(x + 3)}$ The factor missing is $x + 3$, so we multiply $\dfrac{7}{x(x - 2)}$ by $\dfrac{x + 3}{x + 3}$.

$\dfrac{x + 3}{x + 3} \cdot \dfrac{7}{x(x - 2)} = \dfrac{7(x + 3)}{x(x - 2)(x + 3)}$ Multiply.

$= \dfrac{7x + 21}{x(x - 2)(x + 3)}$ Therefore, $\dfrac{7}{x^2 - 2x} = \dfrac{7x + 21}{x(x - 2)(x + 3)}$.

Note

It is customary to simplify the numerators, but not the denominators. The reason will be clear in the next section.

c. $\dfrac{7x}{x^2 + 2x - 8} = \dfrac{?}{(x - 5)(x - 2)(x + 4)}$ Factor $x^2 + 2x - 8$.

$\dfrac{7x}{(x - 2)(x + 4)} = \dfrac{?}{(x - 5)(x - 2)(x + 4)}$ Since the denominator of the first fraction is missing the factor $x - 5$, we need to multiply by $\dfrac{x - 5}{x - 5}$.

$\dfrac{x - 5}{x - 5} \cdot \dfrac{7x}{x^2 + 2x - 8} = \dfrac{7x(x - 5)}{(x - 5)(x - 2)(x + 4)}$ Multiply.

$= \dfrac{7x^2 - 35x}{(x - 5)(x - 2)(x + 4)}$ Therefore, $\dfrac{7x}{x^2 + 2x - 8} = \dfrac{7x^2 - 35x}{(x - 5)(x - 2)(x + 4)}$.

Answers: Practice Exercise 5: a. ad **b.** $30a^3$ **c.** $48a^3b^3$ **d.** $30ac$ **Additional Practice 5: a.** mp **b.** $18x^4$
c. $42y^2z^3$ **d.** $16x^2y^2z$

PRACTICE EXERCISE 6

Find the missing numerator so that the rational expressions are equal.

a. $\dfrac{7x}{3x - 6} = \dfrac{?}{9x - 18}$

b. $\dfrac{2a}{b^2 - 3b} = \dfrac{?}{b(b - 3)(b + 2)}$

c. $\dfrac{3x}{x^2 - 8x + 15} = \dfrac{?}{(x + 4)(x - 3)(x - 5)}$

If you need more practice, do the following Additional Practice Exercises.

Additional Practice Exercise 6   Find the missing numerator so that the rational expressions are equal.

a. $\dfrac{9}{4a + 8} = \dfrac{?}{12a + 24}$

b. $\dfrac{4z}{m^2 + 5m} = \dfrac{?}{m(m + 4)(m + 5)}$

c. $\dfrac{6x}{x^2 + x - 12} = \dfrac{?}{(x + 6)(x + 4)(x - 3)}$

Exercise Set 7.2

For Extra Help

MyMathLab®

Find the LCM of the following whole numbers. (See Example 1.)

1. 3 and 5

2. 2 and 7

3. 9 and 11

4. 13 and 15

5. 25 and 30

6. 42 and 60

7. 36 and 60

8. 35 and 75

9. 70 and 154

10. 30 and 105

11. 28, 63, and 42

12. 36, 60, and 72

Find the LCM of the following monomials. (See Example 2.)

13. xy and x^3y^3

14. r^3s^5 and rs

15. u^6v and u^2v^4

16. t^3w^8 and t^5w

17. a^2b and $a^3b^6c^4$

18. m^8np^2 and m^9p^2

19. x^2y^6z and $x^4y^4z^3$

20. v^7st^6 and v^8s^3t

21. $3u^2v$ and $9u^3v^5$

22. $5w^4x^6$ and $15w^8x$

23. $24y^3z^4$ and $36y^5z^2$

24. $32r^2w^7$ and $48r^3w^6$

25. $15m^5np^9$ and $45m^3p^7$

26. $18a^3b^3$ and $54ab^4c^3$

27. $10r^4s^2t^2$ and $12r^4s^8t^3$

28. $10x^5yz^7$ and $18x^2y^6z^8$

29. $9u^3v$, $36u^2v^2$, and $27u^4v^6$

30. $4w^3x^2$, $36w^6x$, and $20w^5x^5$

31. $21r^4s^9t^5$, $14r^4s^3t$, and $28r^2s^8t^6$

32. $33a^2b^5c^2$, $22a^2b^8c^3$, and $44ab^5c^4$

Find the LCM of the following polynomials. (See Example 3.)

33. $5a + 25$ and $3a + 15$

34. $4b - 16$ and $7b - 28$

35. $r^2 - 7r$ and $5r - 35$

Answers: Practice Exercise 6: a. $21x$ **b.** $2ab + 4a$ **c.** $3x^2 + 12x$
Additional Practice 6: a. 27 **b.** $4mz + 16z$ **c.** $6x^2 + 36x$

36. $x^2 + 4x$ and $6x + 24$

37. $v - 5$ and $v^2 - 2v - 15$

38. $r + 9$ and $r^2 + 8r - 9$

39. $x^2 - x - 12$ and $x^2 + 8x + 15$

40. $y^2 - 8y + 7$ and $y^2 + y - 2$

41. $y + 4$ and $y + 1$

42. $t - 5$ and $t - 2$

43. $w + 3$ and $w^2 - 6w$

44. $u - 7$ and $u^2 + 5u$

45. $z^2 - 4z - 5$ and $z^2 + 2z + 1$

46. $b^2 - 6b + 9$ and $b^2 - 8b + 15$

47. $t^2 - 9, t^2 - 2t - 15,$ and $t^2 - 7t + 12$

48. $9y^2 - 16, 3y^2 + 7y + 4, 3y^2 - 2y - 8$

Challenge Exercises (49–52)

Find the LCM of the following:

49. 66, 132, and 33

50. $105x^2yz^3, 75x^4y^5z,$ and $195x^3y^6z^7$

51. $3y^2 - 75$ and $6y^2 - 60y + 150$

52. $24a^2 + 24ab - 90b^2, 24a^2 - 4ab - 48b^2$

Find each missing numerator so that the fractions are equal. (See Example 4.)

53. $\dfrac{3}{4} = \dfrac{?}{28}$

54. $\dfrac{6}{7} = \dfrac{?}{35}$

55. $\dfrac{5}{8} = \dfrac{?}{56}$

56. $\dfrac{1}{9} = \dfrac{?}{99}$

57. $1 = \dfrac{?}{5}$

58. $2 = \dfrac{?}{4}$

59. $\dfrac{15}{2} = \dfrac{?}{8}$

60. $\dfrac{11}{3} = \dfrac{?}{21}$

61. $\dfrac{9}{5} = \dfrac{?}{135}$

62. $\dfrac{17}{9} = \dfrac{?}{117}$

Find each missing numerator so that the rational expressions are equal. (See Example 5.)

63. $\dfrac{c}{e} = \dfrac{?}{ef}$

64. $\dfrac{a}{b} = \dfrac{?}{cb}$

65. $\dfrac{4}{5x} = \dfrac{?}{30x}$

66. $\dfrac{3}{7y^2} = \dfrac{?}{42y^2}$

67. $\dfrac{5t}{8z} = \dfrac{?}{56z^3}$

68. $\dfrac{8u}{9v} = \dfrac{?}{63v^5}$

69. $\dfrac{9r}{11s^2} = \dfrac{?}{121s^5}$

70. $\dfrac{4m}{13n^3} = \dfrac{?}{169n^7}$

71. $\dfrac{9p}{4q^4} = \dfrac{?}{28q^6}$

72. $\dfrac{17w}{5x^5} = \dfrac{?}{40x^8}$

73. $\dfrac{12y^2}{7z^4} = \dfrac{?}{63z^9}$

74. $\dfrac{15a^3}{8b^2} = \dfrac{?}{72b^5}$

75. $\dfrac{9t}{14x^2y} = \dfrac{?}{42x^5y^3}$

76. $\dfrac{5r^2}{16ps^4} = \dfrac{?}{64p^4s^7}$

Find each missing numerator so that the rational expressions are equal. (See Example 6.)

77. $\dfrac{y}{y + 3} = \dfrac{?}{5y + 15}$

78. $\dfrac{z}{z - 2} = \dfrac{?}{4z - 8}$

79. $\dfrac{3t}{2t - 10} = \dfrac{?}{6t - 30}$

80. $\dfrac{5a}{3a + 12} = \dfrac{?}{12a + 48}$

81. $\dfrac{4c}{5c + 10} = \dfrac{?}{20c + 40}$

82. $\dfrac{6h}{7h - 14} = \dfrac{?}{21h - 42}$

83. $\dfrac{11}{r - 1} = \dfrac{?}{r^2 + 2r - 3}$

84. $\dfrac{2u}{4 + u} = \dfrac{?}{12 - u - u^2}$

85. $\dfrac{5v}{v^2 - 5v} = \dfrac{?}{v(v - 5)(v + 2)}$

86. $\dfrac{7x}{x^2 + 8x} = \dfrac{?}{x(x - 1)(x + 8)}$

87. $\dfrac{y - 2}{y^2 - 10y + 21} = \dfrac{?}{(y + 4)(y - 7)(y - 3)}$

88. $\dfrac{u + 5}{u^2 + 6u - 16} = \dfrac{?}{(u + 2)(u - 2)(u + 8)}$

89. $\dfrac{w - 1}{w^2 - 16} = \dfrac{?}{(w + 4)(w - 4)(w + 3)}$

90. $\dfrac{z + 8}{z^2 - 25} = \dfrac{?}{(z + 5)(z - 3)(z - 5)}$

91. $\dfrac{n - 3}{n^2 + 6n + 9} = \dfrac{?}{(n + 3)(n - 9)(n + 3)}$

92. $\dfrac{m + 5}{m^2 - 8m + 16} = \dfrac{?}{(m - 4)(m + 8)(m - 4)}$

Challenge Exercises (93 and 94)

Find each missing numerator so that the rational expressions are equal.

93. $\dfrac{t}{x^2 + 4x + 4} = \dfrac{?}{(x + 2)^3}$

94. $\dfrac{4z}{y^2 - 6y + 9} = \dfrac{?}{y^3 - 6y^2 + 9y}$

Writing Exercises (95–102)

95. In this section, we have developed the LCM for both arithmetic and algebra. How do you think we will use the LCM in arithmetic? How do you think we will use the LCM in algebra?

96. Why is the LCM divisible by each of the terms for which it is the LCM?

97. In choosing the factors for the LCM, why do we select only factors that appear in the prime factorizations of the terms whose LCM we are finding? What would happen if we selected a factor not found in the prime factorizations?

98. In selecting the factors for the LCM, why is the largest exponent on each factor in the prime factorizations selected?

99. In selecting the factors for the LCM, why must each different factor that appears in any prime factorization be selected?

100. In creating a fraction that is equivalent to a fraction with a given denominator, why is it necessary to multiply both the numerator and the denominator by the *same* expression?

101. Explain how you would find the missing denominator in $\dfrac{6}{5xy^2} = \dfrac{42x^3y^2}{?}$.

102. Explain how you would show that $\frac{3}{4}$ is equal to $\frac{6}{8}$ by using shading in a diagram.

Group Project

103. In Example 1d we found the LCM of 50, 42, and 45 to be $2 \cdot 3^2 \cdot 5^2 \cdot 7$ using prime factorization. Study the method below and explain why it works. The product of the numbers in blue is the LCM.

$$
\begin{array}{r}
\ \ 5\quad 7\quad 3 \\ \hline
3)\ \ 5\quad 21\quad 9 \\ \hline
5)25\quad 21\quad 45 \\ \hline
2)50\quad 42\quad 45
\end{array}
$$

Section 7.3 The Least Common Denominator of Fractions and Rational Expressions

OBJECTIVES *When you complete this section, you will be able to:*

A Find the least common denominator for two or more fractions and convert each fraction into an equivalent fraction with the least common denominator as its denominator.

B Find the least common denominator for two or more rational expressions and convert each rational expression into an equivalent rational expression with the least common denominator as its denominator.

PREREQUISITE SKILLS *Before beginning this section, you should be able to:*

a. Find the least common multiple. (Section 7.2)
b. Multiply monomials by using $a^m \cdot a^n = a^{m+n}$. (Sections 2.2 and 2.3)
c. Factor polynomials. (Sections 5.2–5.6)
d. Multiply rational numbers and expressions. (Sections 6.3 and 6.4)

GETTING READY FOR SECTION 7.3

Find the least common multiple for each of the following:

1. 12 and 18

2. $10x^2y^3$ and $14x^2y^2$

3. $x^2 + 3x - 4$ and $x^2 + 6x + 8$

4. Multiply $5y \cdot 10x^2y^3$.

5. Factor $b^2 - 2b - 35$.

Multiply the rational numbers and expressions. Do not reduce the product.

6. $\dfrac{4}{4} \cdot \dfrac{4}{15}$

7. $\dfrac{5y}{5y} \cdot \dfrac{3}{14x^2y^2}$

8. $\dfrac{x-1}{x-1} \cdot \dfrac{x-3}{(x+4)(x+2)}$

Introduction

In Section 7.2, we discussed how to find the least common multiple (LCM) of two or more natural numbers or polynomials. In this section, we will use this concept to find the **least common denominator** of two or more fractions. The least common denominator (LCD) of two or more fractions is the least common multiple of the denominators. You may need to review this procedure before continuing. After finding the LCD, we will write fractions as equivalent fractions with the LCD as their denominators. Finding the least common denominator and writing fractions with the least common denominator as their denominators will be very important in the next section.

Answers: Getting Ready: 1. 36 **2.** $70x^2y^3$ **3.** $(x+2)(x+4)(x-1)$ **4.** $50x^2y^4$ **5.** $(b-7)(b+5)$
6. $\dfrac{16}{60}$ **7.** $\dfrac{15y}{70x^2y^3}$ **8.** $\dfrac{x^2 - 4x + 3}{(x+2)(x+4)(x-1)}$

OBJECTIVE **A** Finding the least common denominator of fractions and writing fractions with the least common denominator

Example 1 | **Find the least common denominator. Then write each fraction as an equivalent fraction with the LCD as its denominator.**

a. $\dfrac{1}{3}$ and $\dfrac{1}{5}$

Solution

Since 3 and 5 are both prime, their LCM is $3 \cdot 5 = 15$. Therefore, the LCD for $\frac{1}{3}$ and $\frac{1}{5}$ is 15. To write $\frac{1}{3}$ as a fraction with a denominator of 15, we multiply by $\frac{5}{5}$, and to write $\frac{1}{5}$ as a fraction with a denominator of 15, we multiply by $\frac{3}{3}$.

$$\frac{1}{3} = \frac{5}{5} \cdot \frac{1}{3} = \frac{5}{15} \quad \text{and} \quad \frac{1}{5} = \frac{3}{3} \cdot \frac{1}{5} = \frac{3}{15}$$

b. $\dfrac{3}{10}$ and $\dfrac{5}{21}$

Solution $\quad\quad 10 = 2 \cdot 5$ Prime factorization of the denominators.

$\quad\quad\quad\quad\quad\quad 21 = 3 \cdot 7$

By choosing the highest power of each factor, the LCM of 10 and 21 is $2 \cdot 3 \cdot 5 \cdot 7 = 210$. So the LCD of $\frac{3}{10}$ and $\frac{5}{21}$ is 210. Notice that the LCD is the same as $10 \cdot 21$. This is because 10 and 21 have no factors in common. To write $\frac{3}{10}$ as a fraction whose denominator is 210, we multiply by $\frac{21}{21}$; and to write $\frac{5}{21}$ as a fraction whose denominator is 210, we multiply by $\frac{10}{10}$.

$$\frac{3}{10} = \frac{21}{21} \cdot \frac{3}{10} = \frac{63}{210}$$

$$\frac{5}{21} = \frac{10}{10} \cdot \frac{5}{21} = \frac{50}{210}$$

c. $\dfrac{5}{6}$ and $\dfrac{5}{12}$

Solution $\quad\quad 6 = 2 \cdot 3$ Prime factorization of the denominators.

$\quad\quad\quad\quad\quad\quad 12 = 2^2 \cdot 3$

By choosing the highest power of each factor, the LCM is $2^2 \cdot 3 = 12$. So the LCD is also 12. Note that 12 is a multiple of 6. If one denominator is a multiple of the other, the larger number is the LCD. To write $\frac{5}{6}$ as a fraction whose denominator is 12, we multiply by $\frac{2}{2}$. We do not need to multiply $\frac{5}{12}$ by anything, since it already has 12 as its denominator.

$$\frac{5}{6} = \frac{2}{2} \cdot \frac{5}{6} = \frac{10}{12} \quad \text{and} \quad \frac{5}{12} = \frac{5}{12}$$

d. $\dfrac{7}{12}$ and $\dfrac{11}{18}$

Solution $\quad\quad 12 = 2^2 \cdot 3$ Prime factorization of the denominators.

$\quad\quad\quad\quad\quad\quad 18 = 2 \cdot 3^2$

The LCM of 12 and 18 is $2^2 \cdot 3^2 = 36$. So the LCD of $\frac{7}{12}$ and $\frac{11}{18}$ is also 36. To write $\frac{7}{12}$ as a fraction whose denominator is 36, we multiply by $\frac{3}{3}$; and to write $\frac{11}{18}$ as a fraction whose denominator is 36, we multiply by $\frac{2}{2}$.

$$\frac{7}{12} = \frac{3}{3} \cdot \frac{7}{12} = \frac{21}{36}$$

$$\frac{11}{18} = \frac{2}{2} \cdot \frac{11}{18} = \frac{22}{36}$$

e. $\dfrac{4}{15}, \dfrac{5}{12},$ and $\dfrac{7}{10}$

Solution $15 = 3 \cdot 5$ Prime factorization of the denominators.
$12 = 2^2 \cdot 3$
$10 = 2 \cdot 5$

The LCM of 15, 12, and 10 is $2^2 \cdot 3 \cdot 5 = 60$. So the LCD of $\frac{4}{15}, \frac{5}{12},$ and $\frac{7}{10}$ is 60. To write $\frac{4}{15}$ as a fraction whose denominator is 60, we multiply by $\frac{4}{4}$. To write $\frac{5}{12}$ as a fraction whose denominator is 60, we multiply by $\frac{5}{5}$. To write $\frac{7}{10}$ as a fraction whose denominator is 60, we multiply by $\frac{6}{6}$.

$$\frac{4}{15} = \frac{4}{4} \cdot \frac{4}{15} = \frac{16}{60}$$
$$\frac{5}{12} = \frac{5}{5} \cdot \frac{5}{12} = \frac{25}{60}$$
$$\frac{7}{10} = \frac{6}{6} \cdot \frac{7}{10} = \frac{42}{60}$$

PRACTICE EXERCISE 1

Find the least common denominator. Then write each fraction as an equivalent fraction with the LCD as its denominator.

a. $\dfrac{1}{5}$ and $\dfrac{1}{7}$ **b.** $\dfrac{5}{14}$ and $\dfrac{8}{15}$ **c.** $\dfrac{2}{9}$ and $\dfrac{7}{18}$ **d.** $\dfrac{9}{20}$ and $\dfrac{7}{50}$ **e.** $\dfrac{3}{4}, \dfrac{5}{6},$ and $\dfrac{3}{8}$

If you need more practice, do the following Additional Practice Exercises.

Additional Practice Exercise 1 Find the least common denominator. Then write each fraction as an equivalent fraction with the LCD as its denominator.

a. $\dfrac{1}{3}$ and $\dfrac{1}{7}$ **b.** $\dfrac{1}{22}$ and $\dfrac{1}{6}$ **c.** $\dfrac{2}{3}$ and $\dfrac{4}{15}$ **d.** $\dfrac{7}{36}$ and $\dfrac{11}{24}$ **e.** $\dfrac{1}{6}, \dfrac{4}{9},$ and $\dfrac{3}{10}$

OBJECTIVE **B** **Finding the least common denominator of rational expressions and writing each with the least common denominator**

Writing rational expressions with the least common denominator follows the same procedure as writing rational numbers with the least common denominator. Compare each of the following, for which we write each pair of fractions with a common denominator:

$\dfrac{2}{5}$ and $\dfrac{3}{4}$ $\dfrac{2}{x}$ and $\dfrac{3}{y}$ Find the LCD and determine by what you need to multiply each fraction in order to convert it into a fraction with the LCD as its denominator.

$\dfrac{2}{5} \cdot \dfrac{4}{4} = \dfrac{8}{20}$ $\dfrac{2}{x} \cdot \dfrac{y}{y} = \dfrac{2y}{xy}$ Fractions written with the LCD as their denominator.

$\dfrac{3}{4} \cdot \dfrac{5}{5} = \dfrac{15}{20}$ $\dfrac{3}{y} \cdot \dfrac{x}{x} = \dfrac{3x}{xy}$

Answers: Practice Exercise 1: a. LCD $= 35, \dfrac{7}{35}, \dfrac{5}{35}$ **b.** LCD $= 210, \dfrac{75}{210}, \dfrac{112}{210}$ **c.** LCD $= 18, \dfrac{4}{18}, \dfrac{7}{18}$ **d.** LCD $= 100, \dfrac{45}{100}, \dfrac{14}{100}$ **e.** LCD $= 24, \dfrac{18}{24}, \dfrac{20}{24}, \dfrac{9}{24}$

Additional Practice 1: a. LCD $= 21, \dfrac{7}{21}, \dfrac{3}{21}$ **b.** LCD $= 66, \dfrac{3}{66}, \dfrac{11}{66}$ **c.** LCD $= 15, \dfrac{10}{15}, \dfrac{4}{15}$ **d.** LCD $= 72, \dfrac{14}{72}, \dfrac{33}{72}$ **e.** LCD $= 90, \dfrac{15}{90}, \dfrac{40}{90}, \dfrac{27}{90}$

Example 2   Find the least common denominator. Then write each rational expression as an equal rational expression with the LCD as its denominator.

a. $\dfrac{2}{a}$ and $\dfrac{3}{b}$

b. $\dfrac{b}{a^2d}$ and $\dfrac{c}{ad^2}$

Solution

Since the denominators are prime, the LCM is their product ab. So the LCD is ab. To write $\frac{2}{a}$ with a denominator of ab, we multiply by $\frac{b}{b}$; and to write $\frac{3}{b}$ with a denominator of ab, we multiply by $\frac{a}{a}$.

$$\frac{2}{a} = \frac{b}{b} \cdot \frac{2}{a} = \frac{2b}{ab}$$

$$\frac{3}{b} = \frac{a}{a} \cdot \frac{3}{b} = \frac{3a}{ab}$$

Solution

The LCM of a^2d and ad^2 is a^2d^2. Therefore, the LCD is also a^2d^2. To write $\frac{b}{a^2d}$ with a denominator of a^2d^2, we multiply by $\frac{d}{d}$; and to write $\frac{c}{ad^2}$ with a denominator of a^2d^2, we multiply by $\frac{a}{a}$.

$$\frac{b}{a^2d} = \frac{b}{a^2d} \cdot \frac{d}{d} = \frac{bd}{a^2d^2}$$

$$\frac{c}{ad^2} = \frac{c}{ad^2} \cdot \frac{a}{a} = \frac{ac}{a^2d^2}$$

c. $\dfrac{5}{10x^2y^3}$ and $\dfrac{3}{14x^2y^2}$

Solution

$10x^2y^3 = 2 \cdot 5x^2y^3$ Prime factorization of the denominators.
$14x^2y^2 = 2 \cdot 7x^2y^2$ Prime factorization of the denominators.

Therefore, the LCM of $10x^2y^3$ and $14x^2y^2$ is $2 \cdot 5 \cdot 7x^2y^3 = 70x^2y^3$. Consequently, the LCD is $70x^2y^3$. To write $\frac{5}{10x^2y^3}$ as a fraction with a denominator of $70x^2y^3$, we multiply by $\frac{7}{7}$; and to write $\frac{3}{14x^2y^2}$ as a fraction with a denominator of $70x^2y^3$, we multiply by $\frac{5y}{5y}$.

$$\frac{5}{10x^2y^3} = \frac{7}{7} \cdot \frac{5}{10x^2y^3} = \frac{35}{70x^2y^3}$$

$$\frac{3}{14x^2y^2} = \frac{5y}{5y} \cdot \frac{3}{14x^2y^2} = \frac{15y}{70x^2y^3}$$

PRACTICE EXERCISE 2

Find the least common denominator. Then write each rational expression as an equivalent rational expression with the LCD as its denominator.

a. $\dfrac{7}{m}$ and $\dfrac{3}{n}$

b. $\dfrac{m}{c^3d}$ and $\dfrac{n}{c^2d^2}$

c. $\dfrac{8}{15m^3n^3}$ and $\dfrac{7}{18m^2n}$

If you need more practice, do the following Additional Practice Exercises.

Additional Practice Exercise 2   Find the least common denominator. Then write each rational expression as an equivalent rational expression with the LCD as its denominator.

a. $\dfrac{5}{x}$ and $\dfrac{6}{y}$

b. $\dfrac{5}{r^4s^2}$ and $\dfrac{2}{r^3s^3}$

c. $\dfrac{3t}{21m^2n^5}$ and $\dfrac{8s}{12mn^2}$

Answers: Practice Exercise 2: a. LCD $= mn$, $\dfrac{7n}{mn}$, $\dfrac{3m}{mn}$ **b.** LCD $= c^3d^2$, $\dfrac{dm}{c^3d^2}$, $\dfrac{cn}{c^3d^2}$ **c.** LCD $= 90m^3n^3$, $\dfrac{48}{90m^3n^3}$, $\dfrac{35mn^2}{90m^3n^3}$

Additional Practice 2: a. LCD $= xy$, $\dfrac{5y}{xy}$, $\dfrac{6x}{xy}$ **b.** LCD $= r^4s^3$, $\dfrac{5s}{r^4s^3}$, $\dfrac{2r}{r^4s^3}$ **c.** LCD $= 84m^2n^5$, $\dfrac{12t}{84m^2n^5}$, $\dfrac{56mn^3s}{84m^2n^5}$

If the denominators are polynomials other than monomials, we still use the same procedure. It is a little more complicated, since the polynomials must first be factored, if possible, and the factors will now be binomials instead of monomials. Compare each of the following, for which we write each pair of fractions or rational expressions with a common denominator:

$$\frac{2}{5} \text{ and } \frac{3}{4} \qquad \frac{2}{x} \text{ and } \frac{3}{y} \qquad \frac{2}{x-2} \text{ and } \frac{3}{x+3}$$

Find the LCD and determine by what you need to multiply each fraction in order to convert it into a fraction with the LCD as its denominator.

$$\frac{2}{5} \cdot \frac{4}{4} = \frac{8}{20} \qquad \frac{2}{x} \cdot \frac{y}{y} = \frac{2y}{xy} \qquad \frac{2}{x-2} \cdot \frac{x+3}{x+3} = \frac{2x+6}{(x-2)(x+3)}$$

$$\frac{3}{4} \cdot \frac{5}{5} = \frac{15}{20} \qquad \frac{3}{y} \cdot \frac{x}{x} = \frac{3x}{xy} \qquad \frac{3}{x+3} \cdot \frac{x-2}{x-2} = \frac{3x-6}{(x-2)(x+3)}$$

Fractions written with the LCD as their denominator.

Example 3 **Find the least common denominator. Then write each rational expression as an equivalent rational expression with the LCD as its denominator.**

a. $\dfrac{4}{x-3}$ and $\dfrac{7}{x+4}$

Solution

Since $x - 3$ and $x + 4$ are prime, their LCM is $(x - 3)(x + 4)$. Therefore, the LCD is $(x - 3)(x + 4)$. To write $\frac{4}{x-3}$ as a fraction with a denominator of $(x - 3)(x + 4)$, we multiply it by $\frac{x+4}{x+4}$. To write $\frac{7}{x+4}$ as a fraction with a denominator of $(x - 3)(x + 4)$, we multiply it by $\frac{x-3}{x-3}$.

> **Note**
>
> It is customary to find the product of the numerators but not the denominators. You will see the reason for this in the next section.

$$\frac{4}{x-3} = \frac{x+4}{x+4} \cdot \frac{4}{x-3} = \frac{4(x+4)}{(x+4)(x-3)} = \frac{4x+16}{(x+4)(x-3)}$$

$$\frac{7}{x+4} = \frac{x-3}{x-3} \cdot \frac{7}{x+4} = \frac{7(x-3)}{(x-3)(x+4)} = \frac{7x-21}{(x-3)(x+4)}$$

b. $\dfrac{4}{2x+6}$ and $\dfrac{3x}{6x+18}$

Solution

$$2x + 6 = 2(x + 3)$$ Prime factorization of the denominators.
$$6x + 18 = 2 \cdot 3(x + 3)$$

Therefore, the LCM of $2x + 6$ and $6x + 18$ is $2 \cdot 3(x + 3) = 6(x + 3)$. So the LCD is $6(x + 3)$. To write $\frac{4}{2(x+3)}$ as a fraction with a denominator of $6(x + 3)$, we multiply it by $\frac{3}{3}$. The rational expression $\frac{3x}{6x+18} = \frac{3x}{6(x+3)}$, so it already has the LCD as its denominator.

$$\frac{4}{2x+6} = \frac{4}{2(x+3)} = \frac{3}{3} \cdot \frac{4}{2(x+3)} = \frac{12}{6(x+3)}$$

c. $\dfrac{x+2}{x^2+3x-4}$ and $\dfrac{x-3}{x^2+6x+8}$

Solution

$$x^2 + 3x - 4 = (x + 4)(x - 1)$$ Prime factorization of the denominators.
$$x^2 + 6x + 8 = (x + 4)(x + 2)$$

Therefore, the LCM of $x^2 + 3x - 4$ and $x^2 + 6x + 8$ is $(x + 4)(x - 1)(x + 2)$.
So the LCD is $(x + 4)(x - 1)(x + 2)$. Since $\frac{x + 2}{x^2 + 3x - 4} = \frac{x + 2}{(x + 4)(x - 1)}$, we need to
multiply it by $\frac{x + 2}{x + 2}$. Since $\frac{x - 3}{x^2 + 6x + 8} = \frac{x - 3}{(x + 4)(x + 2)}$, we need to multiply it by $\frac{x - 1}{x - 1}$.

$$\frac{x + 2}{x^2 + 3x - 4} = \frac{x + 2}{(x + 4)(x - 1)} = \frac{x + 2}{x + 2} \cdot \frac{x + 2}{(x + 4)(x - 1)}$$

$$= \frac{(x + 2)(x + 2)}{(x + 2)(x + 4)(x - 1)}$$

$$= \frac{x^2 + 4x + 4}{(x + 2)(x + 4)(x - 1)}$$

$$\frac{x - 3}{x^2 + 6x + 8} = \frac{x - 3}{(x + 4)(x + 2)} = \frac{x - 1}{x - 1} \cdot \frac{x - 3}{(x + 4)(x + 2)}$$

$$= \frac{(x - 1)(x - 3)}{(x - 1)(x + 4)(x + 2)}$$

$$= \frac{x^2 - 4x + 3}{(x - 1)(x + 4)(x + 2)}$$

PRACTICE EXERCISE 3

Find the least common denominator. Then write each rational expression as an equivalent rational expression with the LCD as its denominator.

a. $\dfrac{a}{a - 6}$ and $\dfrac{a}{a + 4}$

b. $\dfrac{3y}{4y - 20}$ and $\dfrac{6y}{6y - 30}$

c. $\dfrac{b - 4}{b^2 - 2b - 35}$ and $\dfrac{b + 3}{b^2 + 7b + 10}$

If you need more practice, do the following Additional Practice Exercises.

Additional Practice Exercise 3   Find the least common denominator. Then write each rational expression as an equivalent rational expression with the LCD.

a. $\dfrac{x}{x + 1}$ and $\dfrac{y}{x - 2}$

b. $\dfrac{x + 2}{6x - 12}$ and $\dfrac{x - 3}{8x - 16}$

c. $\dfrac{y + 9}{y^2 + 2y - 24}$ and $\dfrac{y - 3}{y^2 - 16}$

Exercise Set 7.3

For Extra Help

MyMathLab®

Find the least common denominator. Then write each fraction as an equivalent fraction with the LCD as its denominator. (See Example 1.)

1. $\dfrac{1}{2}$ and $\dfrac{1}{7}$

2. $\dfrac{1}{11}$ and $\dfrac{1}{13}$

3. $\dfrac{2}{3}$ and $\dfrac{3}{5}$

4. $\dfrac{4}{17}$ and $\dfrac{8}{19}$

5. $\dfrac{5}{6}$ and $\dfrac{4}{5}$

6. $\dfrac{10}{11}$ and $\dfrac{7}{10}$

Answers: Practice Exercise 3: a. LCD $= (a - 6)(a + 4)$, $\dfrac{a^2 + 4a}{(a - 6)(a + 4)}$, $\dfrac{a^2 - 6a}{(a - 6)(a + 4)}$ **b.** LCD $= 12(y - 5)$, $\dfrac{9y}{12(y - 5)}$, $\dfrac{12y}{12(y - 5)}$

c. LCD $= (b + 5)(b - 7)(b + 2)$, $\dfrac{b^2 - 2b - 8}{(b + 5)(b - 7)(b + 2)}$, $\dfrac{b^2 - 4b - 21}{(b + 5)(b - 7)(b + 2)}$

Additional Practice 3: a. LCD $= (x + 1)(x - 2)$, $\dfrac{x^2 - 2x}{(x + 1)(x - 2)}$, $\dfrac{xy + y}{(x + 1)(x - 2)}$ **b.** LCD $= 24(x - 2)$, $\dfrac{4x + 8}{24(x - 2)}$, $\dfrac{3x - 9}{24(x - 2)}$

c. LCD $= (y - 4)(y + 6)(y + 4)$, $\dfrac{y^2 + 13y + 36}{(y - 4)(y + 6)(y + 4)}$, $\dfrac{y^2 + 3y - 18}{(y - 4)(y + 6)(y + 4)}$

7. $\dfrac{2}{5}$ and $\dfrac{7}{15}$

8. $\dfrac{3}{7}$ and $\dfrac{3}{28}$

9. $\dfrac{3}{8}$ and $\dfrac{4}{9}$

10. $\dfrac{5}{16}$ and $\dfrac{11}{15}$

11. $\dfrac{1}{10}$ and $\dfrac{1}{15}$

12. $\dfrac{5}{14}$ and $\dfrac{7}{21}$

13. $\dfrac{11}{40}$ and $\dfrac{13}{30}$

14. $\dfrac{1}{18}$ and $\dfrac{1}{24}$

15. $\dfrac{9}{16}$ and $\dfrac{11}{21}$

16. $\dfrac{41}{75}$ and $\dfrac{61}{100}$

17. $\dfrac{1}{6}, \dfrac{3}{8}$, and $\dfrac{9}{10}$

18. $\dfrac{4}{9}, \dfrac{5}{12}$, and $\dfrac{7}{18}$

19. $\dfrac{11}{12}, \dfrac{5}{18}$, and $\dfrac{17}{24}$

20. $\dfrac{4}{7}, \dfrac{3}{14}$, and $\dfrac{2}{21}$

Find the least common denominator. Then write each rational expression as an equivalent expression with the LCD as its denominator. (See Example 2.)

21. $\dfrac{2}{n}$ and $\dfrac{3}{m}$

22. $\dfrac{5}{r}$ and $\dfrac{6}{t}$

23. $\dfrac{a}{b}$ and $\dfrac{c}{d}$

24. $\dfrac{x}{y}$ and $\dfrac{u}{v}$

25. $\dfrac{6}{ab}$ and $\dfrac{5}{bc}$

26. $\dfrac{8}{xy}$ and $\dfrac{10}{yz}$

27. $\dfrac{r}{stu}$ and $\dfrac{n}{mtu}$

28. $\dfrac{a}{pqr}$ and $\dfrac{b}{rtq}$

29. $\dfrac{x}{wv^2}$ and $\dfrac{y}{sv}$

30. $\dfrac{m}{p^2q}$ and $\dfrac{n}{pz}$

31. $\dfrac{2t}{a^2b^5}$ and $\dfrac{3z}{a^3b^3}$

32. $\dfrac{6u}{h^4k^2}$ and $\dfrac{4v}{h^3k^7}$

33. $\dfrac{5a}{x^2yz^4}$ and $\dfrac{7d}{xy^5z^3}$

34. $\dfrac{8p}{r^7s^3t^2}$ and $\dfrac{9q}{r^2s^2t^4}$

35. $\dfrac{7}{6u^2v^4}$ and $\dfrac{3}{8uv^3}$

36. $\dfrac{2}{9a^4b^3}$ and $\dfrac{5}{12ab^2}$

37. $\dfrac{4}{15p^6q^8}$ and $\dfrac{9}{20p^4q^5}$

38. $\dfrac{11}{12r^6t^3}$ and $\dfrac{13}{18r^4t^5}$

39. $\dfrac{r}{7x^2y}$ and $\dfrac{t}{14xy^2}$

40. $\dfrac{p}{9mn^3}$ and $\dfrac{q}{27m^3n}$

41. $\dfrac{3a}{10w^6z^7}$ and $\dfrac{6b}{25w^2z^6}$

42. $\dfrac{5r}{24ab^6}$ and $\dfrac{3s}{16a^2b^9}$

Find the least common denominator. Then write each rational expression as an equivalent expression with the LCD as its denominator. (See Example 3.)

43. $\dfrac{3}{u+8}$ and $\dfrac{9}{u-10}$

44. $\dfrac{5}{v-4}$ and $\dfrac{6}{v-2}$

45. $\dfrac{x}{x+7}$ and $\dfrac{x}{x-5}$

46. $\dfrac{y}{y-3}$ and $\dfrac{y}{y-4}$

47. $\dfrac{t}{3t-9}$ and $\dfrac{5t}{4t-12}$

48. $\dfrac{3z}{5z-10}$ and $\dfrac{z}{7z-14}$

49. $\dfrac{6y}{9y+18}$ and $\dfrac{8y}{15y+30}$

50. $\dfrac{11z}{8x-32}$ and $\dfrac{17x}{12x-48}$

51. $\dfrac{3a}{a^2-1}$ and $\dfrac{a}{a^2+3a-4}$

52. $\dfrac{4b}{b^2-9}$ and $\dfrac{2b}{b^2+b-6}$

53. $\dfrac{v-5}{v^2-2v-3}$ and $\dfrac{v+3}{v^2-5v+6}$

54. $\dfrac{w+6}{w^2+6w+5}$ and $\dfrac{w-2}{w^2+w-20}$

55. $\dfrac{m+1}{m^2+8m+16}$ and $\dfrac{m-4}{m^2+5m+4}$

56. $\dfrac{n-3}{n^2-10n+25}$ and $\dfrac{n-1}{n^2-2n-15}$

57. $\dfrac{x+5}{x^3+x^2-6x}$ and $\dfrac{x-3}{x^4+7x^3+12x^2}$

58. $\dfrac{y-2}{y^5-16y^3}$ and $\dfrac{y-1}{y^3+5y^2+4y}$

Writing Exercises (59 and 60)

59. How is finding the least common denominator of rational numbers like finding the least common denominator of rational expressions? How is it different?

60. What is the relationship between the prime factors and the LCD of an expression?

Section 7.4   Addition and Subtraction of Rational Numbers and Expressions with Unlike Denominators

OBJECTIVES *When you complete this section, you will be able to:*

A Add fractions and mixed numbers with unlike denominators.

B Add rational expressions with unlike denominators.

PREREQUISITE SKILLS *Before beginning this section, you should be able to:*

a. Convert mixed numbers to improper fractions. (Section R.4)

b. Add fractions and rational expressions with common denominators. (Section 7.1)

c. Factor polynomials. (Sections 5.2–5.6)

d. Find the least common denominator and rewrite rational numbers and expressions as equivalent expressions with the least common denominator. (Section 7.2)

e. Reduce rational numbers and expressions to lowest terms. (Sections 6.1 and 6.2)

GETTING READY FOR SECTION 7.4

1. Write $5\dfrac{7}{12}$ as an improper fraction.

Add the following fractions and rational expressions:

2. $\dfrac{25}{60} + \dfrac{14}{60}$

3. $\dfrac{6x + 12}{12x} + \dfrac{6x - 4}{12x}$

4. $\dfrac{3x + 6}{12(x - 2)} - \dfrac{2x + 8}{12(x - 2)}$

Factor the following:

5. $3x^2 - 13x - 10$

6. $c^2 - 10c + 25$

Rewrite each of the following as equivalent fractions with a common denominator:

7. $\dfrac{2}{3}$ and $\dfrac{3}{4}$

8. $\dfrac{7}{6a}$ and $\dfrac{5}{8a}$

9. $\dfrac{x + 2}{4x - 8}$ and $\dfrac{x + 4}{6x - 12}$

Introduction

In Section 7.1, we combined fractions and rational expressions with common denominators. If the denominators are not common, we have to perform an intermediate step before we can combine the expressions.

OBJECTIVE **A** Adding fractions and mixed numbers with unlike denominators

Suppose we want to find $\frac{1}{2} + \frac{1}{3}$ by using shaded regions. We first shade $\frac{1}{2}$ the region.

1 unit

Answers: Getting Ready: 1. $\dfrac{67}{12}$ **2.** $\dfrac{13}{20}$ **3.** $\dfrac{3x + 2}{3x}$ **4.** $\dfrac{1}{12}$ **5.** $(3x + 2)(x - 5)$ **6.** $(c - 5)^2$ **7.** $\dfrac{8}{12}, \dfrac{9}{12}$ **8.** $\dfrac{28}{24a}, \dfrac{15}{24a}$ **9.** $\dfrac{3x + 6}{12(x - 2)}, \dfrac{2x + 8}{12(x - 2)}$

Then we shade $\frac{1}{3}$ of the region.

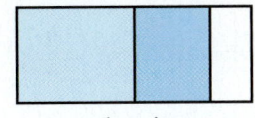

1 unit

What part of the total region is shaded? At this moment, we cannot answer that question. If we go back and divide the region into six equal parts, we shade $\frac{3}{6}$ for $\frac{1}{2}$ and $\frac{2}{6}$ for $\frac{1}{3}$.

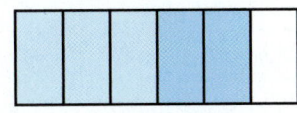

1 unit

Therefore, $\frac{1}{2} + \frac{1}{3} = \frac{3}{6} + \frac{2}{6} = \frac{5}{6}$. Note that 6 is the least common denominator of 2 and 3. What we have done is change $\frac{1}{2}$ and $\frac{1}{3}$ into equivalent fractions with the LCD as their denominators, and then add as in Section 7.1.

> ### Procedure: Adding Fractions with Unlike Denominators
>
> To add fractions with unlike denominators, change each fraction into an equivalent fraction with the least common denominator as its denominator. Then add the fractions, using the techniques of Section 7.1.

Since subtraction of fractions may be expressed in terms of addition, the same procedure applies to subtraction as well. If whole numbers or mixed numbers are involved, first convert them to improper fractions.

Example 1   Find each sum or difference. Express answers reduced to lowest terms.

a. $\dfrac{2}{3} + \dfrac{3}{4} =$ The LCD of 3 and 4 is 12. Therefore, multiply $\frac{2}{3}$ by $\frac{4}{4}$ and multiply $\frac{3}{4}$ by $\frac{3}{3}$.

$\dfrac{4}{4} \cdot \dfrac{2}{3} + \dfrac{3}{3} \cdot \dfrac{3}{4} =$ Multiply the fractions.

$\dfrac{8}{12} + \dfrac{9}{12} =$ Add the fractions.

$\dfrac{17}{12}$ Sum.

b. $3 - \dfrac{3}{5} =$ Think of 3 as $\frac{3}{1}$. The LCD of 1 and 5 is 5. Therefore, multiply 3 by $\frac{5}{5}$. Since $\frac{3}{5}$ already has the LCD, we do not multiply it by anything.

$\dfrac{5}{5} \cdot \dfrac{3}{1} - \dfrac{3}{5} =$ Multiply the fractions.

$\dfrac{15}{5} - \dfrac{3}{5} =$ Subtract the fractions.

$\dfrac{12}{5}$ Difference.

c. $\dfrac{5}{12} + \dfrac{7}{30} =$ The LCD is 60. Therefore, multiply $\frac{5}{12}$ by $\frac{5}{5}$ and multiply $\frac{7}{30}$ by $\frac{2}{2}$.

$\dfrac{5}{5} \cdot \dfrac{5}{12} + \dfrac{2}{2} \cdot \dfrac{7}{30} =$ Multiply the fractions.

$\dfrac{25}{60} + \dfrac{14}{60} =$ Add the fractions.

$\dfrac{39}{60} =$ Reduce to lowest terms.

$\dfrac{13}{20}$ Sum.

d. $\dfrac{3}{10} + \dfrac{7}{15} - \dfrac{9}{25} =$ The LCD is 150. Therefore, multiply $\frac{3}{10}$ by $\frac{15}{15}$, multiply $\frac{7}{15}$ by $\frac{10}{10}$, and multiply $\frac{9}{25}$ by $\frac{6}{6}$.

$\dfrac{15}{15} \cdot \dfrac{3}{10} + \dfrac{10}{10} \cdot \dfrac{7}{15} - \dfrac{6}{6} \cdot \dfrac{9}{25} =$ Multiply the fractions.

$\dfrac{45}{150} + \dfrac{70}{150} - \dfrac{54}{150} =$ Add and subtract the fractions.

$\dfrac{61}{150}$ Answer.

e. $3\dfrac{3}{8} - 1\dfrac{1}{2} =$ Change to improper fractions.

$\dfrac{27}{8} - \dfrac{3}{2} =$ The LCD is 8. Therefore, multiply $\dfrac{3}{2}$ by $\dfrac{4}{4}$.

$\dfrac{27}{8} - \dfrac{4}{4} \cdot \dfrac{3}{2} =$ Multiply the fractions.

$\dfrac{27}{8} - \dfrac{12}{8} =$ Subtract the fractions.

$\dfrac{15}{8}$ Difference.

f. $4 - 5\dfrac{7}{12} + 2\dfrac{3}{8} =$ Convert to improper fractions.

$\dfrac{4}{1} - \dfrac{67}{12} + \dfrac{19}{8} =$ The LCD is 24. Therefore, multiply $\dfrac{4}{1}$ by $\dfrac{24}{24}$, multiply $\dfrac{67}{12}$ by $\dfrac{2}{2}$, and multiply $\dfrac{19}{8}$ by $\dfrac{3}{3}$.

$\dfrac{24}{24} \cdot \dfrac{4}{1} - \dfrac{2}{2} \cdot \dfrac{67}{12} + \dfrac{3}{3} \cdot \dfrac{19}{8} =$ Multiply the fractions.

$\dfrac{96}{24} - \dfrac{134}{24} + \dfrac{57}{24} =$ Add and subtract the fractions.

$\dfrac{19}{24}$ Answer.

PRACTICE EXERCISE 1

Find each sum or difference. Express answers reduced to lowest terms.

a. $\dfrac{3}{5} + \dfrac{5}{6}$ **b.** $\dfrac{5}{9} + 4$ **c.** $\dfrac{9}{14} - \dfrac{1}{4}$

d. $4\dfrac{5}{6} - 6\dfrac{1}{8}$ **e.** $\dfrac{2}{3} + \dfrac{5}{8} - \dfrac{7}{9}$ **f.** $\dfrac{7}{10} - 6 + 5\dfrac{7}{12}$

If you need more practice, do the following Additional Practice Exercises.

Additional Practice Exercise 1   Find each sum or difference. Express answers reduced to lowest terms.

a. $\dfrac{1}{6} + \dfrac{5}{8}$ **b.** $6 - \dfrac{3}{4}$ **c.** $\dfrac{5}{6} - \dfrac{4}{9}$

d. $4\dfrac{1}{3} - 6\dfrac{3}{8}$ **e.** $\dfrac{2}{3} + \dfrac{3}{5} - \dfrac{7}{12}$ **f.** $7 - 4\dfrac{5}{12} - \dfrac{7}{10}$

OBJECTIVE Ⓑ Adding rational expressions with unlike denominators

In the following examples with rational expressions, the denominators are monomials that have variable factors. You will notice that the procedure is the same as with ordinary fractions. Compare the following:

$\dfrac{2}{3} + \dfrac{3}{4} =$ $\dfrac{2}{a} + \dfrac{3}{b} =$ Rewrite each with the LCD.

$\dfrac{2}{3} \cdot \dfrac{4}{4} + \dfrac{3}{4} \cdot \dfrac{3}{3} =$ $\dfrac{2}{a} \cdot \dfrac{b}{b} + \dfrac{3}{b} \cdot \dfrac{a}{a} =$ Simplify each expression.

$\dfrac{8}{12} + \dfrac{9}{12} =$ $\dfrac{2b}{ab} + \dfrac{3a}{ab} =$ Add the fractions.

$\dfrac{17}{12}$ $\dfrac{2b + 3a}{ab}$ Sum.

Answers: Practice Exercise 1   **a.** $\dfrac{43}{30}$ **b.** $\dfrac{41}{9}$ **c.** $\dfrac{11}{28}$ **d.** $-\dfrac{31}{24}$ **e.** $\dfrac{37}{72}$ **f.** $\dfrac{17}{60}$ **Additional Practice 1:**   **a.** $\dfrac{19}{24}$ **b.** $\dfrac{21}{4}$ **c.** $\dfrac{7}{18}$ **d.** $-\dfrac{49}{24}$ **e.** $\dfrac{41}{60}$ **f.** $\dfrac{113}{60}$

Example 2   **Find each sum or difference. Express the answers reduced to lowest terms.**

a. $\dfrac{4}{x} + \dfrac{7}{y} =$

The LCD is xy. Therefore, multiply $\dfrac{4}{x}$ by $\dfrac{y}{y}$ and multiply $\dfrac{7}{y}$ by $\dfrac{x}{x}$.

$\dfrac{4}{x} \cdot \dfrac{y}{y} + \dfrac{7}{y} \cdot \dfrac{x}{x} =$ Multiply the rational expressions.

$\dfrac{4y}{xy} + \dfrac{7x}{xy} =$ Add the rational expressions.

$\dfrac{4y + 7x}{xy}$ Sum.

b. $\dfrac{3}{4a} - \dfrac{5}{3a} =$

The LCD is $12a$. Therefore, multiply $\dfrac{3}{4a}$ by $\dfrac{3}{3}$ and multiply $\dfrac{5}{3a}$ by $\dfrac{4}{4}$.

$\dfrac{3}{3} \cdot \dfrac{3}{4a} - \dfrac{4}{4} \cdot \dfrac{5}{3a} =$ Multiply the rational expressions.

$\dfrac{9}{12a} - \dfrac{20}{12a} =$ Subtract the rational expressions.

$-\dfrac{11}{12a}$ Difference.

c. $\dfrac{2x + 4}{4x} + \dfrac{3x - 2}{6x} =$

The LCD is $12x$. Therefore, multiply $\dfrac{2x + 4}{4x}$ by $\dfrac{3}{3}$ and multiply $\dfrac{3x - 2}{6x}$ by $\dfrac{2}{2}$.

$\dfrac{3}{3} \cdot \dfrac{2x + 4}{4x} + \dfrac{2}{2} \cdot \dfrac{3x - 2}{6x} =$ Multiply the rational expressions.

$\dfrac{6x + 12}{12x} + \dfrac{6x - 4}{12x} =$ Add the rational expressions.

$\dfrac{(6x + 12) + (6x - 4)}{12x} =$ Remove the parentheses.

$\dfrac{6x + 12 + 6x - 4}{12x} =$ Add like terms.

$\dfrac{12x + 8}{12x} =$ Factor.

$\dfrac{4(3x + 2)}{4 \cdot 3x} =$ Divide by the common factor 4.

$\dfrac{3x + 2}{3x}$ Sum.

d. $\dfrac{a - 3}{3a^2} - \dfrac{2a - 3}{2a} =$

The LCD is $6a^2$. Therefore, multiply $\dfrac{a - 3}{3a^2}$ by $\dfrac{2}{2}$ and multiply $\dfrac{2a - 3}{2a}$ by $\dfrac{3a}{3a}$.

$\dfrac{2}{2} \cdot \dfrac{a - 3}{3a^2} - \dfrac{3a}{3a} \cdot \dfrac{2a - 3}{2a} =$ Multiply the rational expressions.

$\dfrac{2a - 6}{6a^2} - \dfrac{6a^2 - 9a}{6a^2} =$ Subtract the rational expressions.

$\dfrac{(2a - 6) - (6a^2 - 9a)}{6a^2} =$ Remove the parentheses.

$\dfrac{2a - 6 - 6a^2 + 9a}{6a^2} =$ Add like terms.

$\dfrac{-6a^2 + 11a - 6}{6a^2}$ Difference.

PRACTICE EXERCISE 2

Find each sum or difference. Express answers reduced to lowest terms.

a. $\dfrac{4}{x} + \dfrac{6}{y}$ **b.** $\dfrac{7}{6a} - \dfrac{5}{8a}$ **c.** $\dfrac{b - 5}{5b} + \dfrac{b + 6}{6b}$ **d.** $\dfrac{3x + 7}{8xy^2} - \dfrac{3x - 4}{6x^2y}$

If you need more practice, do the following Additional Practice Exercises.

Additional Practice Exercise 2   Find each sum or difference. Express answers reduced to lowest terms.

a. $\dfrac{9}{r} - \dfrac{4}{s}$ **b.** $\dfrac{2}{5d} + \dfrac{3}{10d}$ **c.** $\dfrac{2x - 3}{5x} + \dfrac{3x + 5}{3x}$ **d.** $\dfrac{a + 4}{3a^2b} - \dfrac{2a - 3}{6ab^2}$

Answers: **Practice Exercise 2: a.** $\dfrac{4y + 6x}{xy}$ **b.** $\dfrac{13}{24a}$ **c.** $\dfrac{11}{30}$ **d.** $\dfrac{9x^2 + 21x - 12xy + 16y}{24x^2y^2}$

Additional Practice 2: a. $\dfrac{9s - 4r}{rs}$ **b.** $\dfrac{7}{10d}$ **c.** $\dfrac{21x + 16}{15x}$ **d.** $\dfrac{2ab + 8b - 2a^2 + 3a}{6a^2b^2}$

If the denominators are not integers or monomials, we follow exactly the same pattern. Compare the following:

$\dfrac{2}{3} + \dfrac{3}{4} =$	$\dfrac{2}{a} + \dfrac{3}{b} =$	$\dfrac{2}{a+b} + \dfrac{3}{a-b} =$	Rewrite each with the LCD.
$\dfrac{2}{3}\cdot\dfrac{4}{4} + \dfrac{3}{4}\cdot\dfrac{3}{3} =$	$\dfrac{2}{a}\cdot\dfrac{b}{b} + \dfrac{3}{b}\cdot\dfrac{a}{a} =$	$\dfrac{2}{a+b}\cdot\dfrac{a-b}{a-b} + \dfrac{3}{a-b}\cdot\dfrac{a+b}{a+b} =$	Perform the multiplications.
$\dfrac{8}{12} + \dfrac{9}{12} =$	$\dfrac{2b}{ab} + \dfrac{3a}{ab} =$	$\dfrac{2a-2b}{(a+b)(a-b)} + \dfrac{3a+3b}{(a+b)(a-b)} =$	Add the fractions.
$\dfrac{8+9}{12} =$	$\dfrac{2b+3a}{ab} =$	$\dfrac{2a-2b+3a+3b}{(a+b)(a-b)} =$	Add the numerators if possible.
$\dfrac{17}{12}$	$\dfrac{2b+3a}{ab}$	$\dfrac{5a+b}{(a+b)(a-b)}$	Sum.

If the denominators are binomials or trinomials, it is often necessary to factor the denominators to find the LCD. If possible, factor the numerator also to see if any of the rational expressions will reduce before carrying out any operations. Sometimes we have to factor the numerator of the sum or difference and then reduce the rational expression. For this reason, we usually leave the LCD in factored form.

Example 3   **Find each sum or difference. Express answers reduced to lowest terms.**

a. $\dfrac{5}{x-2} + \dfrac{3}{x+4} =$ The LCD is $(x-2)(x+4)$.

Therefore, multiply $\dfrac{5}{x-2}$ by $\dfrac{x+4}{x+4}$

and multiply $\dfrac{3}{x+4}$ by $\dfrac{x-2}{x-2}$.

$\dfrac{x+4}{x+4}\cdot\dfrac{5}{x-2} + \dfrac{x-2}{x-2}\cdot\dfrac{3}{x+4} =$ Multiply the rational expressions.

$\dfrac{5x+20}{(x+4)(x-2)} + \dfrac{3x-6}{(x+4)(x-2)} =$ Add the rational expressions.

$\dfrac{(5x+20)+(3x-6)}{(x+4)(x-2)} =$ Remove the parentheses.

$\dfrac{5x+20+3x-6}{(x+4)(x-2)} =$ Add like terms.

$\dfrac{8x+14}{(x+4)(x-2)}$ Sum.

Note

We could factor the numerator and get $\dfrac{2(4x+7)}{(x+4)(x-2)}$, but the expression will not reduce. Consequently, we leave the answer as is.

b. $\dfrac{x+2}{4x-8} - \dfrac{x+4}{6x-12} =$ Factor the denominators.

$\dfrac{x+2}{4(x-2)} - \dfrac{x+4}{6(x-2)} =$ The LCD is $12(x-2)$. Therefore, multiply $\dfrac{x+2}{4(x-2)}$ by $\dfrac{3}{3}$ and $\dfrac{x+4}{6(x-2)}$ by $\dfrac{2}{2}$.

$\dfrac{3}{3}\cdot\dfrac{x+2}{4(x-2)} - \dfrac{2}{2}\cdot\dfrac{x+4}{6(x-2)} =$ Multiply the rational expressions.

$\dfrac{3x+6}{12(x-2)} - \dfrac{2x+8}{12(x-2)} =$ Subtract the rational expressions.

$\dfrac{(3x+6)-(2x+8)}{12(x-2)} =$ Remove the parentheses.

$\dfrac{3x+6-2x-8}{12(x-2)} =$ Add like terms.

$\dfrac{x-2}{12(x-2)} =$ Divide by the common factor $x-2$.

$\dfrac{1}{12}$ Difference.

c. $\dfrac{4c - 1}{c^2 - 10c + 25} - \dfrac{4}{c - 5} =$ Factor the denominators.

$\dfrac{4c - 1}{(c - 5)^2} - \dfrac{4}{c - 5} =$ The LCD is $(c - 5)^2$. Therefore, multiply $\dfrac{4}{c - 5}$ by $\dfrac{c - 5}{c - 5}$.

$\dfrac{4c - 1}{(c - 5)^2} - \dfrac{c - 5}{c - 5} \cdot \dfrac{4}{c - 5} =$ Multiply the rational expressions.

$\dfrac{4c - 1}{(c - 5)^2} - \dfrac{4c - 20}{(c - 5)^2} =$ Subtract the rational expressions.

$\dfrac{(4c - 1) - (4c - 20)}{(c - 5)^2} =$ Remove the parentheses.

$\dfrac{4c - 1 - 4c + 20}{(c - 5)^2} =$ Add like terms.

$\dfrac{19}{(c - 5)^2}$ Difference.

d. $\dfrac{x}{x^2 - 6x + 8} + \dfrac{4}{x^2 - 10x + 24} =$ Factor the denominators.

$\dfrac{x}{(x - 2)(x - 4)} + \dfrac{4}{(x - 4)(x - 6)} =$ The LCD is $(x - 2)(x - 4)(x - 6)$ so multiply $\dfrac{x}{(x - 2)(x - 4)}$ by $\dfrac{x - 6}{x - 6}$ and $\dfrac{4}{(x - 4)(x - 6)}$ by $\dfrac{x - 2}{x - 2}$.

$\dfrac{x - 6}{x - 6} \cdot \dfrac{x}{(x - 2)(x - 4)} + \dfrac{x - 2}{x - 2} \cdot \dfrac{4}{(x - 4)(x - 6)} =$ Multiply.

$\dfrac{x^2 - 6x}{(x - 6)(x - 2)(x - 4)} + \dfrac{4x - 8}{(x - 2)(x - 4)(x - 6)} =$ Put the numerators over the common denominator.

$\dfrac{(x^2 - 6x) + (4x - 8)}{(x - 6)(x - 2)(x - 4)} =$ Remove the parentheses.

$\dfrac{x^2 - 6x + 4x - 8}{(x - 6)(x - 2)(x - 4)} =$ Add like terms in the numerator.

$\dfrac{x^2 - 2x - 8}{(x - 6)(x - 2)(x - 4)} =$ Factor the numerator.

$\dfrac{(x - 4)(x + 2)}{(x - 6)(x - 2)(x - 4)} =$ Divide by the common factor $x - 4$.

$\dfrac{(\cancel{x - 4})(x + 2)}{(x - 6)(x - 2)(\cancel{x - 4})} =$ Multiply the remaining factors.

$\dfrac{x + 2}{(x - 6)(x - 2)}$ Sum.

PRACTICE EXERCISE 3

Find each sum or difference. Express answers reduced to lowest terms.

a. $\dfrac{2x}{x + 3} + \dfrac{4}{x - 5}$

b. $\dfrac{x + 6}{4x + 16} - \dfrac{x + 7}{6x + 24}$

c. $\dfrac{2x + 3}{x^2 + 6x + 9} - \dfrac{x + 4}{x + 3}$

d. $\dfrac{x}{x^2 + 3x + 2} - \dfrac{2}{x^2 + 5x + 6}$

(Continued)

If you need more practice, do the following Additional Practice Exercises.

Additional Practice Exercise 3 Find each sum or difference. Express answers reduced to lowest terms.

a. $\dfrac{7}{2x+1} + \dfrac{3}{3x-2}$

b. $\dfrac{x-2}{6x-30} - \dfrac{x-1}{8x-40}$

c. $\dfrac{2y}{y^2-4y+4} - \dfrac{4y+3}{y-2}$

d. $\dfrac{x-4}{x^2-x-20} + \dfrac{2x-1}{x^2+7x+12}$

Exercise Set 7.4

For Extra Help

MyMathLab®

Find each sum or difference. Express the answers as fractions reduced to lowest terms. (See Example 1.)

1. $\dfrac{4}{7} + \dfrac{1}{3}$

2. $\dfrac{2}{3} + \dfrac{1}{5}$

3. $\dfrac{3}{10} + \dfrac{2}{5}$

4. $\dfrac{5}{8} + \dfrac{3}{4}$

5. $\dfrac{7}{6} - \dfrac{2}{3}$

6. $\dfrac{5}{18} - \dfrac{1}{2}$

7. $\dfrac{5}{8} - \dfrac{5}{6}$

8. $\dfrac{4}{9} - \dfrac{9}{11}$

9. $2 + \dfrac{5}{9}$

10. $4 + \dfrac{5}{3}$

11. $\dfrac{13}{4} - 6$

12. $\dfrac{12}{7} - 2$

13. $\dfrac{1}{4} + \dfrac{5}{6}$

14. $\dfrac{1}{9} + \dfrac{4}{15}$

15. $\dfrac{8}{9} - \dfrac{11}{12}$

16. $\dfrac{14}{15} - \dfrac{7}{10}$

17. $\dfrac{7}{12} + \dfrac{1}{9} - \dfrac{5}{6}$

18. $\dfrac{7}{10} + \dfrac{4}{15} - \dfrac{9}{30}$

19. $\dfrac{7}{12} - \dfrac{11}{18} + \dfrac{1}{3}$

20. $\dfrac{7}{12} - \dfrac{5}{8} + \dfrac{1}{6}$

21. $9\dfrac{7}{8} - 3\dfrac{1}{4}$

22. $10\dfrac{7}{12} - 1\dfrac{1}{6}$

23. $7\dfrac{5}{8} + 4\dfrac{2}{12}$

24. $23\dfrac{1}{4} + 34\dfrac{2}{3}$

25. $14\dfrac{1}{9} - 8\dfrac{5}{12}$

26. $9\dfrac{9}{10} - 7\dfrac{6}{15}$

27. $9\dfrac{3}{8} - 7 + \dfrac{11}{16}$

28. $\dfrac{5}{6} + 4 - 3\dfrac{5}{9}$

29. $\dfrac{3}{4} - 4 + 2\dfrac{3}{10}$

30. $-5 + 9\dfrac{7}{18} + \dfrac{11}{24}$

Find the each sum or difference. Express the answers as rational expressions reduced to lowest terms. (See Examples 2 and 3.)

31. $\dfrac{2}{a} + \dfrac{a}{b}$

32. $\dfrac{6}{r} + \dfrac{7}{t}$

33. $\dfrac{5}{8u} - \dfrac{7}{12u}$

34. $\dfrac{6}{15v} - \dfrac{4}{9v}$

35. $\dfrac{3z}{10x} + \dfrac{5z}{4x}$

36. $\dfrac{10w}{21y} + \dfrac{7w}{18y}$

Answers: Additional Practice 3: a. $\dfrac{27x-11}{(2x+1)(3x-2)}$ **b.** $\dfrac{1}{24}$ **c.** $\dfrac{-4y^2+7y+6}{(y-2)^2}$ **d.** $\dfrac{3x^2-12x-7}{(x-5)(x+4)(x+3)}$

37. $\dfrac{m+8}{2m} - \dfrac{m-7}{3m}$

38. $\dfrac{n-9}{7n} - \dfrac{n-2}{4n}$

39. $\dfrac{3p+1}{10p^3q^2} + \dfrac{9p-2}{6p^2q}$

40. $\dfrac{5t-8}{16s^5t^2} + \dfrac{2t-7}{12s^2t^3}$

41. $\dfrac{4}{k} - \dfrac{6}{k+2}$

42. $\dfrac{5}{j-5} - \dfrac{9}{j}$

43. $\dfrac{4}{w-3} + \dfrac{5}{w+7}$

44. $\dfrac{6}{c+4} + \dfrac{11}{c-1}$

45. $\dfrac{3a}{a-3} - \dfrac{4}{a+4}$

46. $\dfrac{2r}{r+6} - \dfrac{5}{r+4}$

47. $\dfrac{t+2}{t+4} + \dfrac{t-1}{t+6}$

48. $\dfrac{x-1}{x+3} + \dfrac{x+2}{x-5}$

49. $\dfrac{2a-3}{a+4} - \dfrac{a+1}{2a+3}$

50. $\dfrac{3n-1}{2n-3} - \dfrac{n-3}{n+2}$

51. $\dfrac{x-4}{3x+9} - \dfrac{x+5}{6x+18}$

52. $\dfrac{y-3}{2y-8} - \dfrac{y-7}{6y-24}$

53. $\dfrac{2t}{t^2-8t+16} + \dfrac{6}{t-4}$

54. $\dfrac{4}{s^2+14s+49} + \dfrac{5s}{s+7}$

55. $\dfrac{2a}{a^2-1} + \dfrac{-1}{a+1}$

56. $\dfrac{2a}{a^2-4} + \dfrac{-1}{a-2}$

57. $\dfrac{18x}{9x^2-4} - \dfrac{3}{3x+2}$

58. $\dfrac{8b}{4b^2-9} - \dfrac{2}{2b-3}$

59. $\dfrac{x+1}{x^2-x-6} + \dfrac{x-5}{x^2+6x+8}$

60. $\dfrac{y+3}{y^2-3y+2} + \dfrac{y+2}{y^2+y-2}$

61. $\dfrac{z-3}{z^2+6z+9} + \dfrac{z+1}{z^2+z-6}$

62. $\dfrac{7w-2}{w^2+2w-24} + \dfrac{9w-8}{3w^2-16w+16}$

63. $\dfrac{3}{x^2-x-2} + \dfrac{2}{x^2+4x+3}$

64. $\dfrac{6}{x^2+4x+3} + \dfrac{-3}{x^2+3x+2}$

65. $\dfrac{x}{x^2+5x+6} + \dfrac{-3}{x^2+7x+12}$

66. $\dfrac{x}{x^2+3x+2} + \dfrac{-2}{x^2+5x+6}$

67. $\dfrac{3x+4}{x^2-4} + \dfrac{1}{x^2+2x}$

68. $\dfrac{x-1}{x^2-9} - \dfrac{1}{x^2-3x}$

69. $\dfrac{x+1}{x^2-x-6} + \dfrac{2x+8}{x^2+6x+8}$

70. $\dfrac{y+3}{y^2-3y+2} + \dfrac{2y+4}{y^2+y-2}$

Answer the following:

71. Margot walked $2\frac{1}{3}$ miles on Monday, $3\frac{3}{4}$ miles on Tuesday, $1\frac{2}{3}$ miles on Wednesday, $2\frac{3}{8}$ miles on Thursday, $4\frac{1}{4}$ miles on Friday, and $2\frac{5}{8}$ miles on Saturday. She did not walk on Sunday because of rain. What were her total miles walked for the week?

72. John jogged $4\frac{1}{2}$ miles on Monday, $2\frac{3}{4}$ miles on Tuesday, $3\frac{2}{3}$ miles on Wednesday, $4\frac{1}{4}$ miles on Thursday, $5\frac{1}{3}$ miles on Friday, $4\frac{2}{3}$ miles on Saturday, and $3\frac{1}{4}$ miles on Sunday. What were his total miles jogged for the week?

73. A room is $16\frac{3}{4}$ feet long and $12\frac{5}{6}$ feet wide. A carpenter is putting molding around the ceiling. If the molding costs $0.12 per foot, find the cost of the molding.

74. A triangular garden has sides that are $16\frac{5}{8}$ feet, $18\frac{2}{3}$ feet, and $14\frac{5}{6}$ feet in length. How much will it cost to fence in the garden if it costs $1.60 per foot?

Find the perimeter of each of the following triangles, the lengths of whose sides are given:

75. $\dfrac{1}{a}$ feet, $\dfrac{1}{a^2}$ feet, $\dfrac{2}{a}$ feet

76. $\dfrac{3}{b^2}$ meters, $\dfrac{4}{b}$ meters, $\dfrac{1}{b^2}$ meters

77. $\dfrac{2}{x}$ inches, $\dfrac{3}{y}$ inches, $\dfrac{4}{z}$ inches

78. $\dfrac{3}{a}$ yards, $\dfrac{2}{b}$ yards, $\dfrac{5}{c}$ yards

79. $\dfrac{4}{a+b}$ feet, $\dfrac{3}{c+d}$ feet, $\dfrac{6}{a+b}$ feet

80. $\dfrac{5}{x+y}$ centimeters, $\dfrac{7}{w+z}$ centimeters, $\dfrac{4}{w+z}$ centimeters

Find the perimeters of the following rectangles, whose lengths and widths are given:

81. $L = \dfrac{3}{x}$ feet, $W = \dfrac{4}{y}$ feet

82. $L = \dfrac{5}{a}$ yards, $W = \dfrac{2}{b}$ yards

83. $L = \dfrac{x}{x+y}$ centimeters, $W = \dfrac{y}{2x-y}$ centimeters

84. $L = \dfrac{3a}{2a+b}$ millimeters, $W = \dfrac{2b}{a-3b}$ millimeters

85. $L = \dfrac{a+5}{a^2-2a-24}$ inches, $W = \dfrac{a-2}{a^2+6a+8}$ inches

86. $L = \dfrac{x-4}{x^2+2x-15}$ centimeters, $W = \dfrac{x+2}{x^2-5x+6}$ centimeters

Writing Exercises (87–89)

87. Why is it necessary to rewrite each rational number or rational expression with the LCD as its denominator before adding or subtracting?

88. How are adding and subtracting rational numbers like adding and subtracting whole numbers? How are they different?

89. How are adding and subtracting rational expressions like adding and subtracting rational numbers? How are they different?

Complex Fractions

OBJECTIVES *When you complete this section, you will be able to:*

Ⓐ Simplify complex fractions whose numerators and/or denominators are rational numbers.

Ⓑ Simplify complex fractions whose numerators and/or denominators are rational expressions.

PREREQUISITE SKILLS *Before beginning this section, you should be able to:*

a. Divide rational numbers and expressions. (Section 6.5)

b. Reduce rational numbers and expressions to lowest terms. (Sections 6.1 and 6.2)

c. Factor polynomials. (Sections 5.2–5.6)

d. Find the least common denominator. (Section 7.3)

e. Apply the distributive property. (Section 1.7)

f. Multiply rational numbers and expressions. (Sections 6.3 and 6.4)

GETTING READY FOR SECTION 7.5

Divide the following rational numbers and expressions:

1. $\dfrac{2}{3} \div \dfrac{5}{6}$

2. $\dfrac{a^2b}{c^2} \div \dfrac{ab^3}{c^3}$

Reduce the following to lowest terms:

3. $\dfrac{24a^2cd^3}{6ac^4d}$

4. $\dfrac{x^2 - 9}{x^2 + 3x}$

5. Factor $x^2 + 7x + 12$.

6. Find the LCD for $1, \dfrac{7}{x}, \dfrac{15}{x^2}$.

Simplify the following, using the distributive property:

7. $x^2\left(1 + \dfrac{3}{x}\right)$

8. $(x - 6)\left(x + \dfrac{9}{x - 6}\right)$

Multiply the following rational numbers and expressions:

9. $\dfrac{17}{12} \cdot \dfrac{8}{11}$

10. $\dfrac{x^3y^2}{z} \cdot \dfrac{z^3}{x^2y^4}$

Introduction

One interpretation of a fraction is that it represents division. For example, $\frac{a}{b} = a \div b$. This interpretation gives us one method of simplifying **complex fractions**. A complex fraction is any fraction whose numerator and/or denominator contains a fraction. Examples of complex fractions are

$$\frac{\frac{1}{2}}{3}, \quad \frac{\frac{3}{4}}{\frac{5}{8}}, \quad \frac{\frac{2}{3} + \frac{3}{4}}{\frac{4}{5} + \frac{2}{7}}, \quad \frac{\frac{1}{x}}{\frac{x}{y}}, \quad \text{and} \quad \frac{\frac{3}{x} + x}{\frac{x^2 - 6}{x}}.$$

Answers: Getting Ready: **1.** $\dfrac{4}{5}$ **2.** $\dfrac{ac}{b^2}$ **3.** $\dfrac{4ad^2}{c^3}$ **4.** $\dfrac{x - 3}{x}$ **5.** $(x + 4)(x + 3)$ **6.** x^2 **7.** $x^2 + 3x$ **8.** $x^2 - 6x + 9$ **9.** $\dfrac{34}{33}$ **10.** $\dfrac{xz^2}{y^2}$

OBJECTIVE A Simplifying complex fractions whose numerators and/or denominators are rational numbers

To simplify a complex fraction whose numerator and/or denominator contains a rational number means to write it as an ordinary fraction in the form of $\frac{a}{b}$, where a and b are integers and $b \neq 0$. We will first use the division interpretation of a fraction and then show another, sometimes easier, technique. In simplifying complex fractions with the division interpretation of a fraction, we use the following procedure:

Procedure: Simplifying Complex Fractions by Division

1. If necessary, rewrite the numerator and the denominator as single fractions by performing whatever operations are indicated in either.

2. Rewrite the complex fraction as the division of the numerator by the denominator.

3. Perform the division by inverting the fraction on the right and multiplying.

4. If necessary, reduce to lowest terms.

Example 1   Simplify the following complex fractions by using division:

a. $\dfrac{\frac{2}{3}}{\frac{5}{6}} =$ Rewrite as division, using $\frac{a}{b} = a \div b$.

$\dfrac{2}{3} \div \dfrac{5}{6} =$ Invert $\dfrac{5}{6}$ and change division to multiplication.

$\dfrac{2}{3} \cdot \dfrac{6}{5} =$ Write 6 in terms of prime factors.

$\dfrac{2}{3} \cdot \dfrac{2 \cdot 3}{5} =$ Divide by the common factor of 3.

$\dfrac{2}{3} \cdot \dfrac{2 \cdot \cancel{3}}{5}$ Multiply the results.

$\dfrac{4}{5}$ Quotient.

b. $\dfrac{\frac{2}{3} + \frac{3}{4}}{\frac{3}{8} + 1} =$ The LCD in the numerator is 12, and the LCD of the denominator is 8. Change the fractions in the numerator into equivalent fractions whose denominators are 12, and change the fractions in the denominator into equivalent fractions whose denominators are 8.

$\dfrac{\frac{4}{4} \cdot \frac{2}{3} + \frac{3}{3} \cdot \frac{3}{4}}{\frac{3}{8} + \frac{8}{8} \cdot \frac{1}{1}} =$ Multiply the fractions.

$\dfrac{\frac{8}{12} + \frac{9}{12}}{\frac{3}{8} + \frac{8}{8}} =$ Add the fractions in the numerator and the denominator.

$\dfrac{\frac{17}{12}}{\frac{11}{8}} =$ Rewrite as division.

$\dfrac{17}{12} \div \dfrac{11}{8} =$ Rewrite as multiplication.

$\dfrac{17}{12} \cdot \dfrac{8}{11} =$ Write 12 and 8 in prime factors.

$\dfrac{17}{2 \cdot 2 \cdot 3} \cdot \dfrac{2 \cdot 2 \cdot 2}{11}$ Divide by the common factors.

$\dfrac{17}{2 \cdot 2 \cdot 3} \cdot \dfrac{2 \cdot 2 \cdot 2}{11}$ Multiply the remaining fractions.

$\dfrac{34}{33}$ Quotient.

Note

If the numerator and/or the denominator is the sum or difference of two or more fractions, we must write the numerator and the denominator as a single fraction before we can simplify.

The method used in Example 1b is quite involved and has a lot of places where errors may occur. Consequently, many people prefer a second method in which we multiply the numerator and the denominator of the complex fraction by the LCD of all the fractions that appear in the numerator and/or the denominator. This multiplies the original fraction by 1, so does not change its value. The reason we multiply by the LCD of all the fractions is that each denominator divides evenly into the LCD. Therefore, we eliminate all fractions from the numerator and the denominator of the complex fraction. This results in an ordinary fraction. The procedure is summarized as follows:

> **Procedure: Simplifying Complex Fractions by the LCD**
>
> 1. Find the LCD of all fractions that occur in the numerator and/or the denominator.
> 2. Multiply both the numerator and the denominator by the LCD.
> 3. Simplify the results.

We demonstrate this method by reworking the exercises from Example 1 plus one other case.

Example 2 Simplify the following complex fractions, using multiplication:

a.
$$\dfrac{\dfrac{2}{3}}{\dfrac{5}{6}} =$$

The LCD for $\dfrac{2}{3}$ and $\dfrac{5}{6}$ is 6. Therefore, multiply the numerator and the denominator by 6.

$$\dfrac{6\left(\dfrac{2}{3}\right)}{6\left(\dfrac{5}{6}\right)} =$$

Perform the multiplications. Note that we multiplied by 1.

$$\dfrac{4}{5}$$

Simplified form.

b.
$$\dfrac{\dfrac{2}{3} + \dfrac{3}{4}}{\dfrac{3}{8} + 1} =$$

The LCD is 24. Therefore, multiply the numerator and the denominator by 24.

$$\dfrac{24\left(\dfrac{2}{3} + \dfrac{3}{4}\right)}{24\left(\dfrac{3}{8} + 1\right)} =$$

Apply the distributive property.

$$\dfrac{24\left(\dfrac{2}{3}\right) + 24\left(\dfrac{3}{4}\right)}{24\left(\dfrac{3}{8}\right) + 24 \cdot 1} =$$

Perform the multiplications.

$$\dfrac{16 + 18}{9 + 24} =$$

Perform the additions.

$$\dfrac{34}{33}$$

Simplified form.

c.
$$\dfrac{6}{\dfrac{8}{3}} =$$

Since $6 = \dfrac{6}{1}$, the LCD of 6 and $\dfrac{8}{3}$ is 3. Therefore, multiply the numerator and the denominator by 3.

$$\dfrac{3\left(\dfrac{6}{1}\right)}{3\left(\dfrac{8}{3}\right)} =$$

Perform the multiplications.

$$\dfrac{18}{8} =$$

Reduce to lowest terms.

$$\dfrac{9}{4}$$

Simplified form.

PRACTICE EXERCISE 2

Simplify the following complex fractions by both methods:

a. $\dfrac{\dfrac{3}{8}}{\dfrac{5}{6}}$

b. $\dfrac{\dfrac{4}{3} - \dfrac{3}{8}}{1 + \dfrac{7}{6}}$

c. $\dfrac{\dfrac{8}{9}}{12}$

If you need more practice, do the following Additional Practice Exercises.

Additional Practice Exercise 2 Simplify the following complex fractions by both methods:

a. $\dfrac{\dfrac{4}{15}}{\dfrac{2}{5}}$

b. $\dfrac{3 + \dfrac{2}{3}}{\dfrac{1}{4} + \dfrac{5}{12}}$

c. $\dfrac{8}{\dfrac{4}{7}}$

Answers: Practice Exercise 2: a. $\dfrac{9}{20}$ **b.** $\dfrac{23}{52}$ **c.** $\dfrac{2}{27}$ **Additional Practice 2: a.** $\dfrac{2}{3}$ **b.** $\dfrac{11}{2}$ **c.** 14

OBJECTIVE **B** Simplifying complex fractions whose numerators and/or denominators are rational expressions

Since the second method, in which we use the LCD, is always at least as easy as the first, in which we use the division interpretation, the second method is the only method we will use for complex fractions whose numerators and/or denominators contain rational expressions.

Example 3 | **Simplify the following complex fractions:**

a. $\dfrac{\dfrac{a}{b}}{\dfrac{c}{d}} =$ The LCD for $\dfrac{a}{b}$ and $\dfrac{c}{d}$ is bd. Therefore, multiply the numerator and the denominator by bd.

$\dfrac{bd \cdot \dfrac{a}{b}}{bd \cdot \dfrac{c}{d}} =$ Perform the multiplications.

$\dfrac{ad}{bc}$ Simplified form.

b. $\dfrac{\dfrac{a^2b}{c^2}}{\dfrac{ab^3}{c^3}} =$ The LCD for $\dfrac{a^2b}{c^2}$ and $\dfrac{ab^3}{c^3}$ is c^3. Therefore, multiply the numerator and the denominator by c^3.

$\dfrac{c^3\left(\dfrac{a^2b}{c^2}\right)}{c^3\left(\dfrac{ab^3}{c^3}\right)} =$ Perform the multiplications.

$\dfrac{a^2bc}{ab^3} =$ Reduce to lowest terms.

$\dfrac{ac}{b^2}$ Simplified form.

c. $\dfrac{1 - \dfrac{9}{x^2}}{1 + \dfrac{3}{x}} =$ The least common denominator for all the fractions in the numerator and the denominator is x^2. Therefore, multiply the numerator and the denominator by x^2.

$\dfrac{x^2\left(1 - \dfrac{9}{x^2}\right)}{x^2\left(1 + \dfrac{3}{x}\right)} =$ Apply the distributive property.

$\dfrac{x^2 \cdot 1 - x^2\left(\dfrac{9}{x^2}\right)}{x^2 \cdot 1 + x^2\left(\dfrac{3}{x}\right)} =$ Perform the multiplications.

$\dfrac{x^2 - 9}{x^2 + 3x} =$ Factor the numerator and the denominator.

$\dfrac{(x + 3)(x - 3)}{x(x + 3)} =$ Divide by the common factor $(x + 3)$.

$\dfrac{x - 3}{x}$ Simplified form.

d. $\dfrac{1 - \dfrac{2}{x} - \dfrac{15}{x^2}}{1 + \dfrac{7}{x} + \dfrac{12}{x^2}} =$ The LCD of all the fractions in the numerator and the denominator is x^2. Therefore, multiply the numerator and the denominator by x^2.

$\dfrac{x^2\left(1 - \dfrac{2}{x} - \dfrac{15}{x^2}\right)}{x^2\left(1 + \dfrac{7}{x} + \dfrac{12}{x^2}\right)} =$ Apply the distributive property.

$\dfrac{x^2 \cdot 1 - x^2\left(\dfrac{2}{x}\right) - x^2\left(\dfrac{15}{x^2}\right)}{x^2 \cdot 1 + x^2\left(\dfrac{7}{x}\right) + x^2\left(\dfrac{12}{x^2}\right)} =$ Perform the multiplications.

$\dfrac{x^2 - 2x - 15}{x^2 + 7x + 12} =$ Factor the numerator and the denominator.

$\dfrac{(x - 5)(x + 3)}{(x + 4)(x + 3)} =$ Divide by the common factor $(x + 3)$.

$\dfrac{x - 5}{x + 4}$ Simplified form.

e. $\dfrac{x + \dfrac{9}{x-6}}{1 + \dfrac{3}{x-6}} =$

The LCD of all the fractions in the numerator and the denominator is $x - 6$. Therefore, multiply the numerator and the denominator by $x - 6$.

$\dfrac{(x-6)\left(x + \dfrac{9}{x-6}\right)}{(x-6)\left(1 + \dfrac{3}{x-6}\right)} =$

Apply the distributive property.

$\dfrac{(x-6)x + (x-6)\left(\dfrac{9}{x-6}\right)}{(x-6)\cdot 1 + (x-6)\left(\dfrac{3}{x-6}\right)} =$

Perform the multiplications.

$\dfrac{x^2 - 6x + 9}{x - 6 + 3} =$

Factor the numerator and combine like terms in the denominator.

$\dfrac{(x-3)^2}{x-3} =$

Divide by the common factor $x - 3$.

$x - 3$

Answer.

PRACTICE EXERCISE 3

Simplify the following complex fractions:

a. $\dfrac{\dfrac{m}{n}}{\dfrac{p}{q}}$
  b. $\dfrac{\dfrac{x^3 y^2}{z}}{\dfrac{x^2 y^4}{z^3}}$
  c. $\dfrac{1 - \dfrac{4}{x^2}}{1 + \dfrac{2}{x}}$
  d. $\dfrac{1 - \dfrac{4}{x} - \dfrac{12}{x^2}}{1 + \dfrac{6}{x} + \dfrac{8}{x^2}}$
  e. $\dfrac{x + \dfrac{16}{x-8}}{1 + \dfrac{4}{x-8}}$

If you need more practice, do the following Additional Practice Exercises.

Additional Practice Exercise 3 Simplify the following complex fractions:

a. $\dfrac{\dfrac{r}{s}}{\dfrac{t}{u}}$
  b. $\dfrac{\dfrac{4cd^3}{3a}}{\dfrac{2c^4 d}{6a^2}}$
  c. $\dfrac{1 - \dfrac{25}{a^2}}{1 - \dfrac{5}{a}}$
  d. $\dfrac{1 - \dfrac{3}{x} - \dfrac{28}{x^2}}{1 - \dfrac{1}{x} - \dfrac{20}{x^2}}$
  e. $\dfrac{4x + \dfrac{9}{x-3}}{2 + \dfrac{3}{x-3}}$

Exercise Set 7.5

For Extra Help

MyMathLab®

Simplify the following complex fractions. (See Examples 1–3.)

1. $\dfrac{\dfrac{2}{3}}{\dfrac{5}{6}}$
  2. $\dfrac{\dfrac{3}{8}}{\dfrac{5}{4}}$
  3. $\dfrac{\dfrac{7}{5}}{\dfrac{9}{10}}$
  4. $\dfrac{\dfrac{3}{7}}{\dfrac{11}{14}}$

Answers: Practice Exercise 3: a. $\dfrac{mq}{pn}$ **b.** $\dfrac{xz^2}{y^2}$ **c.** $\dfrac{x-2}{x}$ **d.** $\dfrac{x-6}{x+4}$ **e.** $x - 4$ **Additional Practice 3: a.** $\dfrac{ru}{st}$ **b.** $\dfrac{4ad^2}{c^3}$ **c.** $\dfrac{a+5}{a}$ **d.** $\dfrac{x-7}{x-5}$ **e.** $2x - 3$

5. $\dfrac{\dfrac{3}{4}}{\dfrac{1}{6}}$

6. $\dfrac{\dfrac{5}{9}}{\dfrac{4}{15}}$

7. $\dfrac{\dfrac{9}{21}}{\dfrac{3}{14}}$

8. $\dfrac{\dfrac{15}{12}}{\dfrac{5}{8}}$

9. $\dfrac{\dfrac{1}{8}}{2}$

10. $\dfrac{\dfrac{15}{8}}{20}$

11. $\dfrac{\dfrac{6}{7}}{10}$

12. $\dfrac{\dfrac{3}{1}}{5}$

13. $\dfrac{\dfrac{14}{7}}{2}$

14. $\dfrac{\dfrac{9}{12}}{5}$

15. $\dfrac{\dfrac{4}{9} - \dfrac{1}{3}}{\dfrac{7}{12} - \dfrac{5}{18}}$

16. $\dfrac{\dfrac{3}{4} - \dfrac{1}{2}}{\dfrac{5}{8} - \dfrac{7}{20}}$

17. $\dfrac{\dfrac{6}{5} - \dfrac{7}{6}}{\dfrac{1}{12} + \dfrac{2}{15}}$

18. $\dfrac{\dfrac{11}{14} + \dfrac{3}{2}}{\dfrac{13}{21} - \dfrac{1}{7}}$

19. $\dfrac{8 - \dfrac{9}{4}}{\dfrac{7}{8} + \dfrac{9}{10}}$

20. $\dfrac{\dfrac{7}{9} + \dfrac{5}{12}}{3 - \dfrac{7}{3}}$

21. $\dfrac{\dfrac{7}{6} - \dfrac{1}{3}}{1 - \dfrac{3}{2}}$

22. $\dfrac{6 - \dfrac{2}{9}}{\dfrac{8}{9} + \dfrac{3}{8}}$

23. $\dfrac{5 + \dfrac{7}{2}}{1 + \dfrac{4}{11}}$

24. $\dfrac{9 - \dfrac{5}{4}}{2 + \dfrac{7}{6}}$

25. $\dfrac{\dfrac{7}{9} + 9}{3 - \dfrac{4}{3}}$

26. $\dfrac{\dfrac{12}{5} - 6}{8 + \dfrac{7}{10}}$

27. $\dfrac{\dfrac{m}{8}}{\dfrac{n}{12}}$

28. $\dfrac{\dfrac{p}{15}}{\dfrac{q}{10}}$

29. $\dfrac{\dfrac{5}{d}}{\dfrac{3}{f}}$

30. $\dfrac{\dfrac{7}{g}}{\dfrac{2}{h}}$

31. $\dfrac{\dfrac{x}{y}}{\dfrac{r}{t}}$

32. $\dfrac{\dfrac{u}{v}}{\dfrac{w}{z}}$

33. $\dfrac{\dfrac{g}{h}}{4}$

34. $\dfrac{\dfrac{6}{r}}{s}$

35. $\dfrac{\dfrac{a}{b}}{c}$

36. $\dfrac{\dfrac{m}{n}}{p}$

37. $\dfrac{\dfrac{a}{x^2}}{\dfrac{b}{x}}$

38. $\dfrac{\dfrac{p}{y}}{\dfrac{q}{y^3}}$

39. $\dfrac{\dfrac{u}{v^3}}{\dfrac{w}{v^4}}$

40. $\dfrac{\dfrac{r}{s^5}}{\dfrac{t}{s^2}}$

41. $\dfrac{\dfrac{ac}{b^4}}{\dfrac{ac}{b^3}}$

42. $\dfrac{\dfrac{rs}{t^2}}{\dfrac{rs}{t^3}}$

43. $\dfrac{\dfrac{u^7 v^2}{w^5}}{\dfrac{u^3 v^4}{w}}$

44. $\dfrac{\dfrac{m^6 p^3}{n^7}}{\dfrac{m^5 p^6}{n^4}}$

45. $\dfrac{\dfrac{2}{z}}{1 - \dfrac{z}{2}}$

46. $\dfrac{\dfrac{s}{3}}{4 - \dfrac{1}{s}}$

47. $\dfrac{1 + \dfrac{x}{2}}{1 - \dfrac{x}{2}}$

48. $\dfrac{y - \dfrac{y}{3}}{5 + \dfrac{y}{3}}$

49. $\dfrac{\dfrac{1}{t} - 1}{\dfrac{1}{t^2} - 1}$

50. $\dfrac{\dfrac{4}{x^2} - 1}{\dfrac{2}{x} - 1}$

51. $\dfrac{1 + \dfrac{1}{v} - \dfrac{2}{v^2}}{1 - \dfrac{4}{v} + \dfrac{3}{v^2}}$

52. $\dfrac{2 + \dfrac{13}{w} + \dfrac{6}{w^2}}{1 + \dfrac{11}{w} + \dfrac{30}{w^2}}$

53. $\dfrac{\dfrac{1}{u} - \dfrac{2}{u^2}}{1 + \dfrac{4}{u} - \dfrac{12}{u^2}}$

54. $\dfrac{\dfrac{4}{p} + \dfrac{4}{p^2}}{3 - \dfrac{2}{p} - \dfrac{5}{p^2}}$

55. $\dfrac{\dfrac{2}{x} - \dfrac{7}{x^2} - \dfrac{30}{x^3}}{2 + \dfrac{11}{x} + \dfrac{15}{x^2}}$

56. $\dfrac{1 - \dfrac{3}{y} - \dfrac{4}{y^2}}{\dfrac{3}{y} - \dfrac{13}{y^2} + \dfrac{4}{y^3}}$

57. $\dfrac{5 + \dfrac{4}{r + 8}}{r + 8}$

58. $\dfrac{\dfrac{7}{s - 15} - 6}{s - 15}$

59. $\dfrac{t + \dfrac{12}{t - 7}}{2 + \dfrac{8}{t - 7}}$

60. $\dfrac{q + \dfrac{3}{q + 4}}{3 - \dfrac{3}{q + 4}}$

61. $\dfrac{\dfrac{1}{a - 1} - \dfrac{1}{a + 1}}{\dfrac{1}{a^2 - 1}}$

62. $\dfrac{\dfrac{16}{n^2 - 16}}{\dfrac{2}{n + 4} - \dfrac{2}{n - 4}}$

63. $\dfrac{1 - \dfrac{1}{2p + 1}}{\dfrac{1}{2p^2 - 5p - 3} - \dfrac{1}{p - 3}}$

64. $\dfrac{1 - \dfrac{1}{3a + 1}}{\dfrac{1}{3a^2 - 5a - 2} - \dfrac{1}{a - 2}}$

65. $\dfrac{\dfrac{x^2 - 2x - 8}{x^2 - 16}}{\dfrac{x^2 + 2x}{x^2 + 3x - 4}}$

66. $\dfrac{\dfrac{r^2 + r - 2}{r^2 + 4r}}{\dfrac{r^2 - 4}{r^2 + 2r - 8}}$

67. $\dfrac{\dfrac{2a}{a + 6} + \dfrac{1}{a}}{9a - \dfrac{3}{a + 6}}$

68. $\dfrac{\dfrac{8}{b} - \dfrac{7b}{b - 1}}{\dfrac{4}{b - 1} + \dfrac{21}{b}}$

Writing Exercises (69–71)

69. In simplifying complex fractions, why do we multiply the numerator and the denominator by the LCD of all the denominators that occur in the numerator and the denominator?

70. Consider the complex fraction $\dfrac{\frac{2}{3} + \frac{5}{6}}{\frac{1}{4} - \frac{1}{3}}$. The LCD of the numerator is 6, and the LCD of the denominator is 12. Why don't we multiply the numerator by 6 and the denominator by 12? That is, what is wrong with $\dfrac{6\left(\frac{2}{3} + \frac{5}{6}\right)}{12\left(\frac{1}{4} - \frac{1}{3}\right)}$?

71. Explain why $\dfrac{a^{-2} + b^{-2}}{a^{-1} + b^{-1}} \neq \dfrac{a + b}{a^2 + b^2}$.

Section 7.6   Solving Equations Containing Rational Numbers and Expressions

OBJECTIVES *When you complete this section, you will be able to:*

A Solve equations that contain rational numbers.

B Solve equations that contain rational expressions.

PREREQUISITE SKILLS *Before beginning this section, you should be able to:*

a. Find the least common denominator of rational numbers and expressions. (Section 7.3)

b. Apply the distributive property. (Section 1.7)

c. Solve linear equations. (Sections 3.1–3.4)

d. Solve quadratic equations. (Section 5.8)

GETTING READY FOR SECTION 7.6

Find the LCD for each of the following:

1. $\dfrac{2}{3}, \dfrac{1}{6}, \dfrac{8}{9}$

2. $\dfrac{3}{2a + 2}, \dfrac{4}{3a + 3}, \dfrac{5}{6}$

3. $\dfrac{4}{x^2 - x - 6}, \dfrac{2}{x^2 + 3x + 2}, \dfrac{4}{x^2 - 2x - 3}$

Simplify the following, using the distributive property:

4. $15\left(\dfrac{2x - 3}{5} - \dfrac{3x + 2}{3}\right)$

5. $6(a + 1)\left(\dfrac{3}{2(a + 1)} - \dfrac{4}{3(a + 1)}\right)$

Solve the following linear equations:

6. $20 - 3x = 2x$

7. $3y - 12 = 6y - 24$

8. Solve $x^2 - 3x = 4$.

Introduction

The basis for solving equations that contain fractions or rational expressions is the multiplication property of equality introduced in Section 3.2. In words, it states that we may multiply both sides of an equation by any nonzero number and the result is an equation with the same solution(s) as the original equation. In symbols, if $a = b$ and $c \neq 0$, then $ac = bc$.

OBJECTIVE Solving equations that contain rational numbers

When solving equations that contain fractions or rational expressions, we multiply both sides of the equation by the LCD of the fractions or rational expressions. Since all the denominators divide evenly into the LCD, multiplying both sides of the equation by the LCD eliminates all fractions. The resulting equation is then solved by the same methods used in Chapter 3 and Section 5.8. We illustrate with some examples.

Example 1   Solve the following equations:

a. $\dfrac{x}{3} - \dfrac{x}{5} = 2$

Multiply both sides of the equation by the LCD of 15.

$15\left(\dfrac{x}{3} - \dfrac{x}{5}\right) = 15 \cdot 2$

Distribute on the left; multiply on the right.

$15\left(\dfrac{x}{3}\right) - 15\left(\dfrac{x}{5}\right) = 30$

Perform the multiplications.

$5x - 3x = 30$

Add like terms.

$2x = 30$

Divide both sides by 2.

$\dfrac{2x}{2} = \dfrac{30}{2}$

Perform the divisions.

$x = 15$

Solution.

CHECK:

Substitute 15 for x.

$\dfrac{15}{3} - \dfrac{15}{5} = 2$

$5 - 3 = 2$

$2 = 2$ Therefore, the solution is correct.

b. $\dfrac{2}{3}a - \dfrac{1}{6}a = \dfrac{8}{9}$

Multiply both sides of the equation by the LCD of 18.

$18\left(\dfrac{2}{3}a - \dfrac{1}{6}a\right) = 18\left(\dfrac{8}{9}\right)$

Distribute on the left; multiply on the right.

$18\left(\dfrac{2}{3}a\right) - 18\left(\dfrac{1}{6}a\right) = 16$

Perform the multiplications.

$12a - 3a = 16$

Combine like terms.

$9a = 16$

Divide both sides by 9.

$a = \dfrac{16}{9}$

Answer.

CHECK:

Substitute $\dfrac{16}{9}$ for a.

$\dfrac{2}{3} \cdot \dfrac{16}{9} - \dfrac{1}{6} \cdot \dfrac{16}{9} = \dfrac{8}{9}$

Perform the multiplications.

$\dfrac{32}{27} - \dfrac{16}{54} = \dfrac{8}{9}$

Reduce $\dfrac{16}{54}$ to lowest terms.

$\dfrac{32}{27} - \dfrac{8}{27} = \dfrac{8}{9}$

Subtract the fractions on the left side.

$\dfrac{24}{27} = \dfrac{8}{9}$

Reduce the left side to lowest terms.

$\dfrac{8}{9} = \dfrac{8}{9}$

Therefore, the answer is correct.

c. $\dfrac{y + 2}{4} = y + 5$

Multiply both sides of the equation by the LCD of 4.

$4\left(\dfrac{y + 2}{4}\right) = 4(y + 5)$

Perform the multiplications.

$y + 2 = 4y + 20$

Subtract $4y$ from both sides.

$y + 2 - 4y = 4y + 20 - 4y$

Combine like terms.

$-3y + 2 = 20$

Subtract 2 from both sides.

$-3y + 2 - 2 = 20 - 2$

Perform the subtractions.

$-3y = 18$

Divide both sides by -3.

$\dfrac{-3y}{-3} = \dfrac{18}{-3}$

Perform the divisions.

$y = -6$

Solution.

CHECK:

Substitute -6 for y.

$\dfrac{-6 + 2}{4} = -6 + 5$ Simplify both sides.

$\dfrac{-4}{4} = -1$

$-1 = -1$ Therefore, the answer is correct.

d. $\dfrac{2x - 3}{5} - \dfrac{3x + 2}{3} = -\dfrac{2}{3}$

Multiply both sides of the equation by the LCD of 15.

$15\left(\dfrac{2x - 3}{5} - \dfrac{3x + 2}{3}\right) = 15\left(-\dfrac{2}{3}\right)$

Distribute on the left and multiply on the right.

$15\left(\dfrac{2x - 3}{5}\right) - 15\left(\dfrac{3x + 2}{3}\right) = -10$

Multiply on the left.

$3(2x - 3) - 5(3x + 2) = -10$

Distribute on the left.

$6x - 9 - 15x - 10 = -10$

Combine like terms.

$-9x - 19 = -10$

Add 19 to both sides.

$-9x - 19 + 19 = -10 + 19$

Perform the additions.

$-9x = 9$

Divide both sides by -9.

$\dfrac{-9x}{-9} = \dfrac{9}{-9}$

Perform the divisions.

$x = -1$

Solution.

CHECK:

The check is left as an exercise for the student.

PRACTICE EXERCISE 1

Solve the following equations and check your answer:

a. $\dfrac{x}{3} - \dfrac{x}{7} = 4$

b. $\dfrac{3}{5}x + \dfrac{3}{10}x = \dfrac{4}{5}$

c. $\dfrac{2x - 5}{5} = x + 2$

d. $\dfrac{2z - 6}{5} - \dfrac{2z + 3}{2} = \dfrac{3}{10}$

If you need more practice, do the following Additional Practice Exercises.

Additional Practice Exercise 1 Solve the following equations:

a. $\dfrac{a}{9} - \dfrac{a}{4} = \dfrac{5}{6}$

b. $\dfrac{1}{4}x = \dfrac{7}{10}x + \dfrac{1}{2}$

c. $\dfrac{3x + 2}{4} = 2x - 2$

d. $\dfrac{3r + 1}{4} = \dfrac{13}{12} + \dfrac{2r + 5}{6}$

OBJECTIVE B Solving equations that contain rational expressions

If the denominator(s) contains variables, the procedure is essentially the same, with one exception. Since division by 0 is impossible, we must eliminate any value of the variable that makes any denominator equal to zero. It is possible to solve an equation correctly and get a number that does not solve the equation because it makes one or more of the denominators equal to 0. Such numbers are called **apparent**, or **extraneous**, solutions. For this reason, it is mandatory that all answers to equations that contain variables in the denominator be checked in order to determine if they are solutions. It is helpful to identify potential extraneous solutions before solving the equation.

Example 2   Solve the following equations:

Note

$x \neq 0$, so $x = 0$ is a potential extraneous solution since 0 makes the denominator of $\dfrac{5}{x}$ equal to 0.

a. $\dfrac{5}{x} - \dfrac{3}{4} = \dfrac{1}{2}$ Multiply both sides of the equation by the LCD, $4x$.

$4x\left(\dfrac{5}{x} - \dfrac{3}{4}\right) = 4x\left(\dfrac{1}{2}\right)$ Distribute on the left; multiply on the right.

$4x\left(\dfrac{5}{x}\right) - 4x\left(\dfrac{3}{4}\right) = 2x$ Perform the multiplications.

$20 - 3x = 2x$ Add $3x$ to both sides.

$20 = 5x$ Divide both sides by 5.

$4 = x$ Possible solution.

CHECK:

Substitute 4 for x.

$\dfrac{5}{4} - \dfrac{3}{4} = \dfrac{1}{2}$

$\dfrac{2}{4} = \dfrac{1}{2}$

$\dfrac{1}{2} = \dfrac{1}{2}$ Therefore, 4 is the solution.

Answers: Practice Exercise 1: a. $x = 21$ **b.** $x = \dfrac{8}{9}$ **c.** $x = -5$ **d.** $z = -5$ **Additional Practice 1: a.** $a = -6$ **b.** $x = -\dfrac{10}{9}$ **c.** $x = 2$ **d.** $r = 4$

Note

$y \neq 4$, so $y = 4$ is a potential extraneous solution since 4 makes the denominators of $\dfrac{3y}{y-4}$ and $\dfrac{12}{y-4}$ equal to 0.

b. $\dfrac{3y}{y-4} - \dfrac{12}{y-4} = 6$ Multiply both sides of the equation by the LCD, $y - 4$.

$(y-4)\left(\dfrac{3y}{y-4} - \dfrac{12}{y-4}\right) = (y-4)(6)$ Distribute on the left; multiply on the right.

$(y-4)\left(\dfrac{3y}{y-4}\right) - (y-4)\left(\dfrac{12}{y-4}\right) = 6y - 24$ Multiply on the left.

$3y - 12 = 6y - 24$ Subtract $3y$ from both sides.

$-12 = 3y - 24$ Add 24 to both sides.

$12 = 3y$ Divide both sides by 3.

$4 = y$ Extraneous solution; see Note.

Since $y \neq 4$ and 4 is the only possible solution, the solution set is the empty set, $\varnothing$, discussed in Section 3.3.

c. $\dfrac{x}{x+4} = \dfrac{1}{x-2}$

Rather than multiply both sides of the equation by the LCD, a quicker way is to cross multiply. This procedure was discussed in Section 7.2, where it was shown that if $\frac{a}{b} = \frac{c}{d}$, then $ad = bc$. It can be used only when the equation is one fraction equal to another fraction.

Note

$x = -4$ and $x = 2$ are potential extraneous solutions.

$\dfrac{x}{x+4} = \dfrac{1}{x-2}$ Cross multiply.

$x(x-2) = (x+4)(1)$ Apply the distributive property.

$x^2 - 2x = x + 4$ Subtract x from both sides of the equation.

$x^2 - 3x = 4$ Subtract 4 from both sides.

$x^2 - 3x - 4 = 0$ Factor.

$(x-4)(x+1) = 0$ Set each factor equal to 0.

$x - 4 = 0 \qquad x + 1 = 0$ Solve each equation.

$x = 4 \qquad\quad x = -1$ Therefore, 4 and -1 are possible solutions.

CHECK:

$x = 4$ $\qquad\qquad x = -1$

Replace x with 4. $\qquad$ Replace x with -1.

$\dfrac{4}{4+4} = \dfrac{1}{4-2}$ $\qquad$ $\dfrac{-1}{-1+4} = \dfrac{1}{-1-2}$

$\dfrac{4}{8} = \dfrac{1}{2}$ $\qquad\qquad$ $\dfrac{-1}{3} = \dfrac{1}{-3}$

$\dfrac{1}{2} = \dfrac{1}{2}$ $\qquad\qquad$ $-\dfrac{1}{3} = -\dfrac{1}{3}$ Therefore, both answers are solutions.

Note

$a = -1$ is a potential extraneous solution.

d. $\dfrac{3}{2a+2} - \dfrac{4}{3a+3} = \dfrac{5}{6}$ Factor the denominators.

$\dfrac{3}{2(a+1)} - \dfrac{4}{3(a+1)} = \dfrac{5}{6}$ Multiply both sides of the equation by the LCD of $6(a+1)$.

$6(a+1)\left(\dfrac{3}{2(a+1)} - \dfrac{4}{3(a+1)}\right) = 6(a+1)\left(\dfrac{5}{6}\right)$ Distribute on left; multiply on right.

$6(a+1)\left(\dfrac{3}{2(a+1)}\right) - 6(a+1)\left(\dfrac{4}{3(a+1)}\right) = 5(a+1)$ Multiply.

$3 \cdot 3 - 2 \cdot 4 = 5a + 5$ Simplify the left side.

$9 - 8 = 5a + 5$ Continue simplifying.

$1 = 5a + 5$ Subtract 5 from both sides.

$-4 = 5a$ Divide both sides by 5.

$-\dfrac{4}{5} = a$ Possible solution.

CHECK:

As you can imagine, the check to this example is very difficult and is omitted. However, if an equation containing rational expressions is solved correctly, the extraneous solutions always make a denominator equal to 0, as in Example 2b. Often, it is advisable to see if the possible solution(s) make a denominator equal to 0 rather than doing a complete check. If so, eliminate that value(s). Since $-\frac{4}{5}$ does not make any denominator equal to 0, it is at least a possible solution if not a correct one.

e. $\dfrac{x}{x + 4} - \dfrac{1}{x + 2} = \dfrac{8}{x^2 + 6x + 8}$

Factor the denominator $x^2 + 6x + 8$.

Note

$x = -4$ and $x = -2$ are potential extraneous solutions.

$\dfrac{x}{x + 4} - \dfrac{1}{x + 2} = \dfrac{8}{(x + 2)(x + 4)}$

Multiply both sides by the LCD $(x + 4)(x + 2)$.

$(x + 4)(x + 2)\left(\dfrac{x}{x + 4} - \dfrac{1}{x + 2}\right) = (x + 4)(x + 2)\left(\dfrac{8}{(x + 2)(x + 4)}\right)$

Distribute on the left and multiply on the right.

$(x + 4)(x + 2)\left(\dfrac{x}{x + 4}\right) - (x + 4)(x + 2)\left(\dfrac{1}{x + 2}\right) = 8$

Multiply.

$(x + 2) \cdot x - (x + 4) \cdot 1 = 8.$

Multiply.

$x^2 + 2x - x - 4 = 8$

Add like terms on the left side.

$x^2 + x - 4 = 8$

Subtract 8 from both sides.

$x^2 + x - 12 = 0$

Factor.

$(x + 4)(x - 3) = 0$

Use the zero-factor theorem.

$x + 4 = 0, \qquad x - 3 = 0$

Solve both equations.

$x = -4, \qquad\quad x = 3$

From the note, we see that $x = -4$ is an extraneous solution, so $x = 3$ is the only possible solution. The check is left to the student.

PRACTICE EXERCISE 2

Solve the following equations:

a. $\dfrac{5}{3z} - \dfrac{1}{2z} = \dfrac{7}{6}$

b. $\dfrac{x}{x + 5} = 4 - \dfrac{5}{x + 5}$

c. $\dfrac{x}{2x + 1} = \dfrac{-4}{x - 3}$

d. $\dfrac{y}{3y + 6} + \dfrac{3}{4y + 8} = \dfrac{3}{4}$

e. $\dfrac{x}{x - 2} - \dfrac{4}{x - 1} = \dfrac{2}{x^2 - 3x + 2}$

If you need more practice, do the following Additional Practice Exercises.

Additional Practice Exercise 2 Solve the following equations:

a. $\dfrac{3}{x} - \dfrac{3}{2} = \dfrac{6}{x}$

b. $\dfrac{b}{b - 6} - 5 = \dfrac{6}{b - 6}$

c. $\dfrac{x}{x + 4} = \dfrac{4}{x - 2}$

d. $\dfrac{6}{5m - 10} - \dfrac{4}{2m - 4} = \dfrac{1}{5}$

e. $\dfrac{w}{w + 5} - \dfrac{2}{w - 3} = -\dfrac{16}{w^2 + 2w - 15}$

Answers: Practice Exercise 2: a. $z = 1$ **b.** $\varnothing$ **c.** $x = -4$ or $x = -1$ **d.** $y = -\dfrac{9}{5}$ **e.** $x = 3$ (2 is extraneous)

Additional Practice 2: a. $x = -2$ **b.** $\varnothing$ **c.** $x = 8$ or $x = -2$ **d.** $m = -2$ **e.** $w = 2$ (3 is extraneous)

Exercise Set 7.6

For Extra Help
MyMathLab®

Solve the following equations. (See Example 1.)

1. $\dfrac{x}{3} - \dfrac{1}{4} = \dfrac{5}{12}$

2. $\dfrac{y}{6} + \dfrac{5}{2} = 2$

3. $\dfrac{t}{10} + \dfrac{3}{5} = 2$

4. $\dfrac{v}{8} + \dfrac{3}{4} = 1$

5. $\dfrac{1}{3}u - \dfrac{1}{9}u = \dfrac{4}{3}$

6. $\dfrac{1}{4}w + \dfrac{1}{8}w = \dfrac{9}{4}$

7. $\dfrac{2m}{3} - \dfrac{4m}{5} = 1$

8. $\dfrac{3p}{8} + \dfrac{p}{4} = 5$

9. $\dfrac{3}{4}x - \dfrac{1}{3}x = \dfrac{5}{6}$

10. $\dfrac{5}{8}q - \dfrac{5}{12}q = \dfrac{5}{4}$

11. $\dfrac{1}{4}t + \dfrac{1}{3}t + \dfrac{5}{6} = \dfrac{15}{4}$

12. $\dfrac{2}{3}v + \dfrac{1}{6}v - \dfrac{1}{4} = \dfrac{7}{12}$

13. $\dfrac{3x}{4} - \dfrac{2}{5} = \dfrac{3x}{10} + \dfrac{1}{2}$

14. $\dfrac{y}{3} - \dfrac{1}{4} = \dfrac{5y}{12} + \dfrac{1}{6}$

15. $\dfrac{y-3}{4} = y - 1$

16. $t + 5 = \dfrac{t+8}{7}$

17. $\dfrac{3x+6}{2} = x + 2$

18. $2a - 3 = \dfrac{5a+6}{4}$

19. $\dfrac{4v+7}{15} = \dfrac{v+1}{5} + \dfrac{1}{3}$

20. $\dfrac{2y-1}{5} - 1 = \dfrac{y-2}{2}$

21. $\dfrac{2q+1}{4} - \dfrac{3q-5}{16} = \dfrac{7}{8}$

22. $\dfrac{p-1}{2} - \dfrac{p+2}{6} = \dfrac{7}{6}$

Solve the following equations. (See Example 2.)

23. $\dfrac{6}{m} + \dfrac{1}{2} = \dfrac{3}{4}$

24. $\dfrac{5}{8} - \dfrac{4}{n} = \dfrac{7}{12}$

25. $\dfrac{5}{r} + \dfrac{3}{r} = 12$

26. $\dfrac{6}{t} - \dfrac{2}{t} = 10$

27. $\dfrac{4}{3u} - \dfrac{1}{2u} = \dfrac{5}{6}$

28. $\dfrac{2}{v} - \dfrac{4}{3v} = \dfrac{2}{15}$

29. $\dfrac{5}{4w} - \dfrac{3}{5w} = \dfrac{-13}{40}$

30. $\dfrac{7}{2z} + \dfrac{11}{6z} = \dfrac{16}{9}$

31. $\dfrac{7}{x-3} = 8 - \dfrac{1}{x-3}$

32. $\dfrac{5}{y+2} + 9 = -\dfrac{4}{y+2}$

33. $\dfrac{t}{t+4} = 3 - \dfrac{12}{t+4}$

34. $\dfrac{8}{z-5} - 2 = \dfrac{z}{z-5}$

35. $\dfrac{4a}{a+6} = \dfrac{10}{a+6} + 2$

36. $\dfrac{5b}{b-8} - 4 = \dfrac{3}{b-8}$

37. $\dfrac{c}{10-c} = \dfrac{1}{c+2}$

38. $\dfrac{x}{2-3x} = \dfrac{1}{3x+2}$

39. $\dfrac{3y}{y-1} = \dfrac{-2}{y+1}$

40. $\dfrac{2n}{n+1} = \dfrac{-5}{n+3}$

41. $\dfrac{m}{m-1} = \dfrac{3m}{4m-3}$

42. $\dfrac{5p}{p+4} = \dfrac{p}{p-1}$

43. $\dfrac{2y}{3y-6} + \dfrac{3}{4y-8} = \dfrac{1}{4}$

44. $\dfrac{2y}{3y+6} - \dfrac{3}{4y+8} = \dfrac{1}{4}$

45. $\dfrac{a^2}{a+2} - 3 = \dfrac{a+6}{a+2} - 4$

46. $\dfrac{3b+4}{b-4} = \dfrac{b^2}{b-4} + 3$

47. $\dfrac{x}{x-2} - \dfrac{4}{x-1} = \dfrac{2}{x^2-3x+2}$

48. $\dfrac{6}{x+2} - \dfrac{4}{x-2} = \dfrac{3x-22}{x^2-4}$

49. $\dfrac{6}{t^2+t-12} = \dfrac{4}{t^2-t-6} + \dfrac{1}{t^2+6t+8}$

50. $\dfrac{7}{v^2-6v+5} - \dfrac{2}{v^2-4v-5} = \dfrac{3}{v^2-1}$

51. $\dfrac{5}{w^2+2w-8} + \dfrac{3}{w^2+w-6} = \dfrac{4}{w^2+7w+12}$

52. $\dfrac{8}{a^2+4a-12} - \dfrac{5}{a^2+a-6} = \dfrac{2}{a^2+9a+18}$

53. $\dfrac{4x}{x^2+4x} - \dfrac{2x}{x^2+2x} = \dfrac{3x-2}{x^2+6x+8}$

54. $\dfrac{9v}{v^2-4v} - \dfrac{3v}{2v^2+v} = \dfrac{v-7}{2v^2-7v-4}$

Challenge Exercises (55–57)

Solve the following equations:

55. $\dfrac{1}{b^2-b-2} = \dfrac{b}{b^2+2b-8} - \dfrac{1}{b^2+5b+4}$

56. $\dfrac{p+1}{p^2+8p+15} - \dfrac{2}{p^2+7p+10} = \dfrac{1}{p^2+5p+6}$

57. $\dfrac{3}{x^2+2x-24} + \dfrac{x-5}{x^2-16} = \dfrac{x}{x^2+10x+24}$

Writing Exercises (58–60)

58. In solving equations that involve rational numbers or rational expressions, why do we multiply both sides of the equation by the LCD?

59. Explain the difference between solving an equation involving rational expressions and adding rational expressions. For example, how is solving $\dfrac{y}{3y+5} + \dfrac{3}{4y+8} = \dfrac{3}{4}$ different from adding $\dfrac{y}{3y+6} + \dfrac{3}{4y+8} + \dfrac{3}{4}$?

60. In Example 2c, we solved $\dfrac{x}{x+4} = \dfrac{1}{x-2}$ by cross multiplying. Show that this is the same as multiplying both sides by the LCD.

Section 7.7 Applications with Rational Expressions

OBJECTIVES *When you complete this section you will be able to:*

- **A** Solve application problems involving number properties.
- **B** Solve application problems involving work.
- **C** Solve application problems involving distance, rate, and time.

PREREQUISITE SKILLS *Before beginning this section, you should be able to:*

- **a.** Solve equations containing rational numbers and expressions. (Section 7.6)
- **b.** Represent consecutive, consecutive even, and consecutive odd integers. (Section 3.5)

GETTING READY FOR SECTION 7.7

Solve the following equations:

1. $\dfrac{1}{14} + \dfrac{1}{x} = \dfrac{1}{8}$ **2.** $\dfrac{450}{x} = \dfrac{1200}{x} - \dfrac{3}{2}$ **3.** $\dfrac{20}{x-5} = \dfrac{30}{x+5}$ **4.** $\dfrac{1}{x} + \dfrac{3}{x+2} = \dfrac{13}{24}$

Represent the following, using x as the variable:

5. two consecutive integers **6.** three consecutive odd integers

Introduction

Application problems were first discussed in detail in Section 3.5. You may wish to review the procedure and suggested approach before doing this section. Our purpose in considering application problems is to help you to think more mathematically, to better understand the structure of mathematics, to help you translate English into mathematics, and to show the power of mathematics in problem solving. Another purpose of these problems is to serve as a beginning for the more realistic problems that some of you will encounter in more advanced courses.

One of the difficulties in doing applications problems is knowing where to start. All applications problems are not done the same way. There are approaches that are unique to specific types of problems. For this reason, we will discuss number problems, work problems, and distance, rate, time problems separately. We will emphasize the approach needed to do each type of problem. The first type we will discuss is problems involving numbers.

OBJECTIVE **A** Solving application problems involving number properties

Number problems are usually fairly straightforward. You do not have to know special formulas or other facts. All the information needed is contained in the problem. However, there are a few things you need to remember.

- **a.** When working with consecutive integer problems, we let x represent the smallest of the integers, $x + 1$ represents the next largest, $x + 2$ represents the next largest, and so on.
- **b.** When working with consecutive even or consecutive odd integers, we let x represent the smallest even or odd integer, $x + 2$ represents the next largest, $x + 4$ represents the next largest, and so on.
- **c.** The reciprocal of a number is found by putting 1 over the number. For example, the reciprocal of x is $\frac{1}{x}$, $x \neq 0$.
- **d.** In a fraction, the numerator is the number on top and the denominator is the number on the bottom.

We use the same procedure introduced in Chapter 3.

Answers: **Getting Ready: 1.** $x = \dfrac{56}{3}$ **2.** $x = 500$ **3.** $x = 25$ **4.** $x = -\dfrac{8}{13}, 6$ **5.** $x, x + 1$ **6.** $x, x + 2, x + 4$

Example 1

a. The denominator of a fraction is 3 more than the numerator. If 4 is added to the numerator and subtracted from the denominator, the resulting fraction is $\frac{9}{4}$. Find the original fraction.

Solution

Understand: We know how the numerator and denominator are related and the results when 4 is added and subtracted, respectively. The unknown is the original fraction.

> Relationship 1: The denominator is 3 more than the numerator.
> Relationship 2: When 4 is added to the numerator and subtracted from the denominator, the resulting fraction is equal to $\frac{9}{4}$.

Plan: Use relationship 1 to represent the numerator and denominator in terms of a variable, and use relationship 2 to write the equation.

Translate: Using relationship 1,

> Let x represent the numerator. Then, $x + 3$ represents the denominator. (The denominator is 3 more than the numerator.)

Therefore, the original fraction is $\frac{x}{x+3}$. Since 4 is added to the numerator, the numerator of the new fraction is $x + 4$. Since 4 is also subtracted from the denominator, the denominator of the new fraction is $x + 3 - 4$. Consequently, the new fraction is $\frac{x+4}{x+3-4}$. Since the new fraction is equal to $\frac{9}{4}$, the equation is:
$$\frac{x+4}{x+3-4} = \frac{9}{4}$$

Solve:

$\dfrac{x+4}{x+3-4} = \dfrac{9}{4}$	Simplify the denominator on the left.
$\dfrac{x+4}{x-1} = \dfrac{9}{4}$	Cross multiply.
$4(x+4) = (x-1)9$	Apply the distributive property.
$4x + 16 = 9x - 9$	Subtract $9x$ from both sides.
$-5x + 16 = -9$	Subtract 16 from both sides.
$-5x = -25$	Divide both sides by -5.
$x = 5$	Solution of the equation.

State: Since x represents the numerator, the numerator is 5. The denominator is $x + 3 = 5 + 3 = 8$. Since the numerator is 5 and the denominator is 8, the original fraction is $\frac{5}{8}$.

Check: If 4 is added to the numerator of $\frac{5}{8}$, the numerator becomes 9. If 4 is subtracted from the denominator, the denominator becomes 4. The new fraction is $\frac{9}{4}$, which is what it is supposed to be. Therefore, our solution is correct.

b. One number is 4 less than the other. If $\frac{1}{2}$ of the larger number is 3 more than $\frac{1}{3}$ of the smaller, find the numbers.

Solution

Understand: We know how the two numbers are related. The unknowns are the two numbers.

> Relationship 1: One number is 4 less than the other.
> Relationship 2: One-half of the larger number is 3 more than $\frac{1}{3}$ of the smaller.

Plan: Use relationship 1 to represent the numbers in terms of a variable and relationship 2 to write the equation.

Translate: Using relationship 1,

Let x represent the larger number. Then,
$x - 4$ represents the smaller number. (The smaller number is 4 less than the larger.)

Using relationship 2, we represent $\frac{1}{2}$ of the larger number by $\frac{1}{2}x$ and $\frac{1}{3}$ of the smaller number by $\frac{1}{3}(x - 4)$. Since $\frac{1}{2}$ of the larger number is 3 more than $\frac{1}{3}$ of the smaller, the equation is $\frac{1}{2}x = \frac{1}{3}(x - 4) + 3$.

Solve:

$$\frac{1}{2}x = \frac{1}{3}(x - 4) + 3 \qquad \text{\color{teal}Multiply both sides by 6.}$$

$$6\left(\frac{1}{2}x\right) = 6\left(\frac{1}{3}(x - 4) + 3\right) \qquad \text{\color{teal}Multiply on left, distribute on right.}$$

$$3x = 6 \cdot \frac{1}{3}(x - 4) + 6(3) \qquad \text{\color{teal}Multiply on the right side.}$$

$$3x = 2(x - 4) + 18 \qquad \text{\color{teal}Distribute.}$$

$$3x = 2x - 8 + 18 \qquad \text{\color{teal}Combine like terms.}$$

$$3x = 2x + 10 \qquad \text{\color{teal}Subtract $2x$ from both sides.}$$

$$x = 10 \qquad \text{\color{teal}Solution of the equation.}$$

State: Since x represents the larger number, the larger number is 10 and the smaller number is $x - 4 = 10 - 4 = 6$.

Check: The check is left as an exercise for the student.

Note

Since you should now be familiar with the techniques in equation solving, not as much detail will be given in the explanations to the right of each step. Also, some of the easier steps will be omitted.

c. For two consecutive even integers, the reciprocal of the smaller is added to three times the reciprocal of the larger. If the resulting sum is $\frac{13}{24}$, find the two consecutive even integers.

Solution

Understand: We are given that there are two consecutive even integers and a relationship between their reciprocals. The unknowns are the two consecutive even integers.

Relationship 1: There are two consecutive even integers.
Relationship 2: If the reciprocal of the smaller is added to three times the reciprocal of the larger, the resulting sum is $\frac{13}{24}$.

Plan: Use relationship 1 to represent the consecutive even integers in terms of a variable, and use relationship 2 to write the equation.

Note

Recall that the reciprocal of a number is found by putting 1 over the number.

Translate: Using relationship 1,

Let x represent the smaller of the two consecutive even integers. Then, $x + 2$ represents the larger of the two consecutive even integers. The reciprocal of the smaller is $\frac{1}{x}$, and the reciprocal of the larger is $\frac{1}{x + 2}$. Three times the reciprocal of the larger is $3\left(\frac{1}{x + 2}\right) = \frac{3}{x + 2}$. Since sum means add and the sum of the two consecutive even integers is $\frac{13}{24}$, from relationship 2 the equation is
$\frac{1}{x} + \frac{3}{x + 2} = \frac{13}{24}$.

Solve:

$$\frac{1}{x} + \frac{3}{x+2} = \frac{13}{24}$$

Multiply both sides by the LCD $24x(x+2)$.

$$24x(x+2)\left(\frac{1}{x} + \frac{3}{x+2}\right) = 24x(x+2)\left(\frac{13}{24}\right)$$

Simplify both sides.

$$24x(x+2)\left(\frac{1}{x}\right) + 24x(x+2)\left(\frac{3}{x+2}\right) = 13x(x+2)$$

Perform the multiplications.

$$24(x+2) + 24x(3) = 13x^2 + 26x$$

Simplify the left side.

$$24x + 48 + 72x = 13x^2 + 26x$$

Continue simplifying.

$$96x + 48 = 13x^2 + 26x$$

Subtract $96x$ from both sides.

$$48 = 13x^2 - 70x$$

Subtract 48 from both sides.

$$0 = 13x^2 - 70x - 48$$

Factor the right side.

$$0 = (13x + 8)(x - 6)$$

Set each factor equal to 0.

$$13x + 8 = 0, \text{ or } x - 6 = 0$$

Solve each equation.

$$13x = -8 \qquad x = 6$$

$$x = -\frac{8}{13}$$

Translate: The only acceptable answer is 6, since $-\frac{8}{13}$ is not an integer. If $x = 6$, then $x + 2 = 8$, so the two consecutive even integers are 6 and 8.

Check: The check is left as an exercise for the student.

PRACTICE EXERCISE 1

a. The denominator of a fraction is 2 more than the numerator. If 3 is added to the numerator and subtracted from the denominator, the result is equal to $\frac{9}{5}$. Find the original fraction.

b. Two numbers differ by 2. If $\frac{1}{4}$ of the larger number is 4 less than $\frac{1}{5}$ of the smaller number, find the numbers.

c. The sum of a fraction and twice its reciprocal is $3\frac{14}{15}$. Find the fraction. (*Hint:* Let $x =$ the fraction.)

If you need more practice, do the following Additional Practice Exercises.

Additional Practice Exercise 1

a. The denominator of a fraction is 2 less than the numerator. If 2 is added to the numerator and subtracted from the denominator, the result is 3. Find the original fraction.

b. Two numbers differ by 4. If $\frac{1}{2}$ of the larger is 1 more than $\frac{2}{3}$ of the smaller, find the numbers.

c. For two consecutive integers, the sum of 2 times the reciprocal of the smaller and 3 times the reciprocal of the larger is $\frac{17}{12}$. Find the integers.

OBJECTIVE Solving application problems involving work

The next type of problem to be discussed is the work problem. In this type of problem, a task can be performed in a certain amount of time by one individual or thing, and the same task can be performed by another individual or thing in a different amount of time. We are then asked how long it would take for the two working together to perform the same task. There are many variations to this problem. The key to solving this problem is to represent the part of the task that can be done in one unit of time.

Answers: Practice Exercise 1: a. $\frac{6}{8}$ **b.** -88 and -90 **c.** $\frac{10}{3}$ or $\frac{3}{5}$ **Additional Practice 1: a.** $\frac{7}{5}$ **b.** 10 and 6 **c.** 3 and 4

For example, if it takes 5 hours to do a job, then $\frac{1}{5}$ is the part of the job that can be done in 1 hour. If it takes x hours to do a job, then $\frac{1}{x}$ is the part of the job that can be done in 1 hour. We use the same procedure as before.

Example 2

a. If Mike can paint his room in 10 hours and Susan can paint the same room in 7 hours, how long would it take them to paint the room working together?

Solution

Understand: We are given how long it takes Mike and Susan each to paint the room working alone. The unknown is the amount of time it takes them to paint the room working together.

Relationship 1: It takes Mike 10 hours working alone and Susan 7 hours working alone.

Relationship 2: (The part Mike can paint in one hour.) + (the part Susan can paint in one hour) = (the part that they can paint together in one hour).

Plan: Represent the part of the room that each can paint in one hour. Represent by a variable the part they can paint in one hour working together. Write the equation, using relationship 2.

Translate: We use a chart.

	Number of Hours	Part Done in 1 Hour
Mike	10	$\dfrac{1}{10}$
Susan	7	$\dfrac{1}{7}$
Together	x	$\dfrac{1}{x}$

From relationship 2 we have

$$\underset{\frac{1}{10}}{\boxed{\text{part Mike paints in 1 hr}}} + \underset{\frac{1}{7}}{\boxed{\text{part Susan paints in 1 hr}}} = \underset{\frac{1}{x}}{\boxed{\text{part they paint together in 1 hr}}}$$

Solve:

$$\frac{1}{10} + \frac{1}{7} = \frac{1}{x} \qquad \text{Multiply both sides by } 70x.$$

$$70x\left(\frac{1}{10} + \frac{1}{7}\right) = 70x\left(\frac{1}{x}\right) \qquad \text{Distribute on the left; multiply on the right.}$$

$$70x\left(\frac{1}{10}\right) + 70x\left(\frac{1}{7}\right) = 70 \qquad \text{Perform the multiplications on the left.}$$

$$7x + 10x = 70 \qquad \text{Combine like terms.}$$

$$17x = 70 \qquad \text{Divide both sides by 17.}$$

$$x = \frac{70}{17} \text{ hours} \qquad \text{Solution of the equation.}$$

State: Therefore, it takes $\frac{70}{17}$ hours working together.

Check: Does the answer seem reasonable? If both worked at the same rate as Mike, it would take half as long as Mike took, which would be 5 hours. If both worked at the same rate as Susan, it would take 3.5 hours. The amount of time working together should be between 3.5 and 5 hours. Is $\frac{70}{17}$ between 3.5 and 5?

b. Sal and Melissa work in an electronics factory. Together they can install the picture tubes in 150 televisions in 8 hours. If Sal can install the picture tubes in 150 televisions in 14 hours, how long would it take Melissa to install 150 picture tubes working alone?

Solution

Understand: We are given the number of hours that it would take Sal working alone, and Sal and Melissa working together to install 150 picture tubes. The unknown is how long it would take Melissa working alone.

Relationship 1: It takes Sal 14 hours and Sal and Melissa 8 hours working together to install 150 tubes.

Relationship 2: (The part Sal does in one hour) + (the part Melissa does in one hour) = (the part they do working together in one hour).

Plan: Represent the part Sal can do in one hour and the part they can do together in one hour. Represent the part Melissa can do in one hour, using a variable. Use relationship 2 to write the equation.

Translate: We will again use a chart.

	Number of Hours	Part Done in 1 Hour
Melissa	x	$\frac{1}{x}$
Sal	14	$\frac{1}{14}$
Together	8	$\frac{1}{8}$

From relationship 2 we have

$$\begin{array}{ccc} \text{part Sal does} & + & \text{part Melissa does} & = & \text{part they do} \\ \text{in 1 hr} & & \text{in 1 hr} & & \text{together in 1 hr} \end{array}$$

$$\frac{1}{14} + \frac{1}{x} = \frac{1}{8}$$

Solve:

$$\frac{1}{14} + \frac{1}{x} = \frac{1}{8} \qquad \text{Multiply both sides by } 56x.$$

$$56x\left(\frac{1}{14} + \frac{1}{x}\right) = 56x\left(\frac{1}{8}\right) \qquad \text{Simplify both sides.}$$

$$56x\left(\frac{1}{14}\right) + 56x\left(\frac{1}{x}\right) = 7x \qquad \text{Continue simplifying.}$$

$$4x + 56 = 7x \qquad \text{Subtract } 4x \text{ from both sides.}$$

$$56 = 3x \qquad \text{Divide both sides by 3.}$$

$$\frac{56}{3} = x$$

State: Since x represents the number of hours for her working alone, it takes Melissa $18\frac{2}{3}$ hours working alone.

Check: Is this answer reasonable?

PRACTICE EXERCISE 2

Solve the following:

a. If one pipe can fill a tank in 12 hours and a second pipe can fill a tank in 8 hours, how long would it take to fill the tank with both pipes open?

b. John can clean the pool in 6 hours. If he and Johanna work together, they can clean the pool in 4 hours. How long would it take Johanna to clean the pool working alone?

If you need more practice, do the following Additional Practice Exercises.

Additional Practice Exercise 2

a. A vat can be drained by one pipe in 8 hours and by a larger pipe in 6 hours. How long would it take to drain the vat with both pipes open?

b. Jose can bail the hay in a field in 12 hours working alone. If Frank helps, using a smaller hay bailer, it takes the two of them 8 hours working together. How long would it take Frank working alone?

OBJECTIVE **Solving application problems involving distance, time, and rate**

The distance traveled depends upon the rate (speed) and the time traveled according to the formula $d = rt$. For example, if an object moves at the rate of 40 miles per hour for 5 hours, then $d = 40 \cdot 5 = 200$ miles. Equivalent forms of this formula are $r = \frac{d}{t}$ (divide both sides by t) and $t = \frac{d}{r}$ (divide both sides by r). For example, if an object travels 300 miles in 6 hours, then $r = \frac{300}{6} = 50$ mph. If an object travels 240 miles at a rate of 60 miles per hour, then $t = \frac{240}{60} = 4$ hours.

We use the same procedure used in Section 3.5.

Example 3   **Solve the following:**

a. Tom and John Truong live in cities that are 80 miles apart. Both leave home at 8:00 A.M., traveling toward each other riding bicycles. Tom's speed is two-thirds of John's. Find John's speed if they meet at 10:00 A.M.

Solution

Understand: We are given the distance between the cities, the time they leave, and the relationship between their speeds. The unknown is John's speed.

Relationship 1: Tom's speed is two-thirds of John's speed.
Relationship 2: The distance Tom travels plus the distance John travels is equal to the distance between the cities.

Plan: Use a chart as we did in Section 3.5. Use relationship 1 to represent their speeds in terms of a variable and use relationship 2 to write the equation.

Translate: Consider the table.

	d	r	t
Tom			
John			

Since each rider left at 8:00 A.M. and they met at 10:00 A.M., we know that both times are two hours. Put 2 in the time column for each.

	d	r	t
Tom			2
John			2

We also know that Tom's speed is two-thirds of John's. Let John's speed be x; then Tom's speed is $\frac{2}{3}x$. Put these in the table.

	d	r	t
Tom		$\frac{2}{3}x$	2
John		x	2

The rate and time columns have been filled. Since $d = rt$, fill in the d column by finding the products of the expressions in the r and t columns. Therefore, Tom's distance is $\frac{2}{3}x(2) = \frac{4}{3}x$, and John's distance is $2x$.

	d	r	t
Tom	$\frac{4}{3}x$	$\frac{2}{3}x$	2
John	$2x$	x	2

The last column filled in was the distance column, so we write our equation by using some known fact(s) about distance. Sometimes it is helpful to draw a diagram.

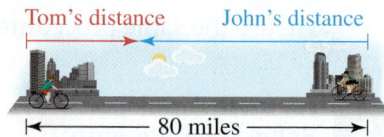

From the diagram and relationship 2, we write the work equation.

Tom's distance	plus	John's distance	equals	distance between the cities
$\downarrow$		$\downarrow$		$\downarrow$
$\frac{4}{3}x$	$+$	$2x$	$=$	80

Solve:

$\dfrac{4}{3}x + 2x = 80$ Multiply both sides by 3.

$3\left(\dfrac{4}{3}x + 2x\right) = 3 \cdot 80$ Simplify each side.

$3\left(\dfrac{4}{3}x\right) + 3 \cdot 2x = 240$ Continue simplifying.

$4x + 6x = 240$ Add like terms.

$10x = 240$ Divide both sides by 10.

$x = 24$

State: Since x represents John's rate, John's rate is 24 mph.

Check: If John's rate is 24 mph, he will travel 48 miles in two hours. Since Tom's rate is two-thirds of John's, he travels at 16 mph. In two hours, Tom will travel 32 miles. Together they have traveled $48 + 32 = 80$ miles, which is the distance between the cities. Therefore, our solution is correct.

b. An airplane flew 450 miles from city A to city B in $1\frac{1}{2}$ hours less than it took to fly 1200 miles from city C to city D. If the airplane flew at the same rate on both trips, find the rate of the airplane.

Solution

Understand: We are given that the plane flew at the same rate on both trips and the relationship between the times for the two trips. The unknown is the rate of the plane.

Relationship 1: The plane flew at the same rate on both trips.
Relationship 2: The airplane flew 450 miles from city A to city B in $1\frac{1}{2}$ hours less than it took to fly 1200 miles from city C to city D.

Plan: Use a chart as in Section 3.5. Use relationship 1 to represent the rate in terms of a variable. Use relationship 2 to write the equation.

Translate: We know the distance from city A to city B is 450 miles and the distance from city C to city D is 1200 miles. We are looking for the rate of the plane. Since it was the same on both trips, let x represent the rate of the plane. Enter these values in the table.

	d	r	t
A to B	450	x	
C to D	1200	x	

The d and r columns are filled in. Since $t = \frac{d}{r}$, fill in the t column by dividing the expressions in the d column by the expressions in the r column. Hence, the time from A to B $= \frac{450}{x}$ and the time from C to D $= \frac{1200}{x}$. Enter these values in the table.

	d	r	t
A to B	450	x	$\frac{450}{x}$
C to D	1200	x	$\frac{1200}{x}$

The last column filled is the time column. Therefore, our equation involves time. We know from relationship 2 that the trip from A to B took $1\frac{1}{2}$ hours less time than the trip from C to D. So, our equation is of the form

times from A to B $=$ time from C to D $- 1\frac{1}{2}$. Therefore, the equation is

$$\frac{450}{x} = \frac{1200}{x} - \frac{3}{2}.$$

Solve:

$$\frac{450}{x} = \frac{1200}{x} - \frac{3}{2} \qquad \text{Multiply both sides by } 2x.$$

$$2x\left(\frac{450}{x}\right) = 2x\left(\frac{1200}{x} - \frac{3}{2}\right) \qquad \text{Simplify both sides.}$$

$$2 \cdot 450 = 2x\left(\frac{1200}{x}\right) - 2x\left(\frac{3}{2}\right) \qquad \text{Continue simplifying.}$$

$$900 = 2(1200) - 3x \qquad \text{Continue simplifying.}$$

$$900 = 2400 - 3x \qquad \text{Subtract 2400 from both sides.}$$

$$-1500 = -3x \qquad \text{Divide both sides by } -3.$$

$$500 = x \qquad \text{Solution of the equation.}$$

State: Therefore, the rate of the plane is 500 mph.

Check: If the plane flies at a rate of 500 mph, the trip of 450 miles would take $\frac{450}{500} = \frac{9}{10}$ of an hour and the trip of 1200 miles would take $\frac{1200}{500} = \frac{12}{5}$ of an hour. The difference in times is $\frac{12}{5} - \frac{9}{10} = \frac{24}{10} - \frac{9}{10} = \frac{15}{10} = \frac{3}{2}$ hours. Therefore, our solution is correct.

c. A river has a current of 5 mph. A boat can make a trip of 20 miles upstream in the same time it can make a trip of 30 miles downstream. Find the rate of the boat in still water.

Solution

Understand: We are given the rate of the current and the relationship between the times for a trip upstream and a trip downstream. The unknown is the rate of the boat in still water.

> Relationship: The boat can make a 20-mile trip upstream in the same time it takes to make a 30-mile trip downstream.

Plan: Use a chart. Represent the rate of the boat in still water with a variable and use the relationship to write the equation.

Translate: We know the distance upstream is 20 miles and the distance downstream is 30 miles. Let x represent the speed of the boat in still water. Since the rate of the current is 5 mph, the current will be slowing the rate of the boat upstream by 5 mph. Therefore, the rate of the boat upstream is $x - 5$. When the boat is going downstream, the current is assisting the boat by 5 mph. Therefore, the rate of the boat downstream is $x + 5$. Enter these values in the table.

	d	r	t
Upstream	20	$x - 5$	
Downstream	30	$x + 5$	

Now we fill in the column for t. Since $t = \frac{d}{r}$, divide the expressions in the d column by the expressions in the r column. Consequently, the time going upstream is $\frac{20}{x - 5}$ and the time going downstream is $\frac{30}{x + 5}$. Enter these values in the table.

	d	r	t
Upstream	20	$x - 5$	$\frac{20}{x - 5}$
Downstream	30	$x + 5$	$\frac{30}{x + 5}$

The last column filled was the t column. Therefore, our equation involves time. Using the given relationship, we have the following work equation:

$$\text{time upstream} = \text{time downstream}$$

$$\frac{20}{x - 5} = \frac{30}{x + 5}$$

Solve:

$$\frac{20}{x - 5} = \frac{30}{x + 5} \qquad \text{Cross multiply.}$$

$20(x + 5) = 30(x - 5)$ Simplify both sides.

$20x + 100 = 30x - 150$ Subtract 20x from both sides.

$100 = 10x - 150$ Add 150 to both sides.

$250 = 10x$ Divide both sides by 10.

$25 = x$

State: Therefore, the boat travels at 25 mph in still water.

Check: If the boat travels at the rate of 25 mph in still water, then its speed against the current is 20 mph. In order to go 20 miles upstream, it would take 1 hour. The speed of the boat with the current is 30 mph. In order to go 30 miles downstream, it would also take 1 hour. The time going upstream equals the time going downstream. Therefore, our solution is correct.

PRACTICE EXERCISE 3

Solve the following:

a. Ellie and Frank live in cities that are 140 miles apart. Both leave home at 7 A.M., traveling by bicycle toward each other. What is Frank's rate if they meet at 11 A.M. and Ellie's rate is three-fourths of Frank's?

b. A plane flew 1500 miles against the wind in the same amount of time it took it to fly 1800 miles with the wind. If the speed of the wind was 50 mph, find the speed of the plane in still air.

c. A plane flew 600 miles in $1\frac{3}{4}$ hours less time than it flew 1125 miles. If the rate is the same for both trips, find the rate of the plane.

If you need more practice, do the following Additional Practice Exercises.

Additional Practice Exercise 3

a. Manuel and Maria live in cities that are 315 miles apart. Both leave home at 9:00 A.M., traveling toward each other. Manuel's speed is $\frac{3}{4}$ of Maria's speed. Find Maria's speed if they meet at 12:00 noon.

b. A cargo ship traveled 100 knots in $2\frac{1}{2}$ hours less time than it took to travel 150 knots. If the speed of the ship remained constant, find the speed of the ship.

c. The current in a river is 4 mph. A boat can go 48 miles upstream in the same time that it takes to go 64 miles downstream. Find the rate of the boat in still water.

For Extra Help

Exercise Set 7.7 MyMathLab®

Solve the following by using the same method as in Example 1:

1. The numerator and the denominator of a fraction are the same. If 2 is added to the numerator, the new fraction is equivalent to $\frac{14}{10}$. Find the original numerator.

2. The numerator and the denominator of a fraction are the same. If 4 is subtracted from the denominator, the new fraction is equivalent to $\frac{39}{27}$. Find the original numerator.

3. The denominator of a fraction is 3 less than the numerator. If 2 is added to both the numerator and the denominator, the new fraction is equivalent to $\frac{12}{9}$. Find the original fraction.

4. The numerator of a fraction is 5 more than the denominator. If 3 is subtracted from both the numerator and the denominator, the result is 6. Find the original fraction.

5. One integer is 3 more than another integer. If $\frac{1}{4}$ of the larger is equal to $\frac{1}{3}$ of the smaller, find the two integers.

6. One integer is 12 less than another integer. If $\frac{1}{7}$ of the larger is equal to $\frac{1}{5}$ of the smaller, find the two integers.

Answers: Practice Exercise 3: a. 20 mph **b.** 550 mph **c.** 300 mph   **Additional Practice 3: a.** 60 mph **b.** 20 knots per hour **c.** 28 mph

7. Two numbers differ by 6. If $\frac{2}{3}$ of the larger is 5 more than $\frac{1}{2}$ of the smaller, find the two numbers.

8. Two numbers differ by 8. If $\frac{3}{2}$ of the smaller is 6 less than $\frac{3}{4}$ of the larger, find the two numbers.

9. The sum of the reciprocals of two consecutive even integers is $\frac{5}{12}$. Find the two consecutive even integers.

10. Three times the reciprocal of the sum of two consecutive odd integers is $\frac{3}{20}$. Find the two consecutive odd integers.

11. A car costs $\frac{5}{6}$ as much as a van. If together they cost $33,000, how much does the van cost?

12. An lightbulb uses $\frac{1}{18}$ as much power as a toaster. If together they use 1140 watts of electricity, how many watts does the toaster use?

Solve the following by using the same method as in Example 2. Convert fractions to mixed numbers.

13. Jason can wash and wax his car in 4 hours. His younger sister can wash and wax the same car in 6 hours. Working together, how fast can they wash and wax the car?

14. Jon can clean the house in 7 hours, and Linda can clean the house in 5 hours. How long would it take if they worked together?

15. Grandma and Grandpa have volunteered to make quilts for HIV-positive babies. If Grandma can make a quilt in 25 days and Grandpa can make a quilt in 35 days, in how many days can they make a quilt working together?

16. It takes Alice 90 minutes to put a futon frame together, and it takes Maya 60 minutes to put the same type of frame together. If they worked together, how long would it take to put a frame together?

17. Working together, it takes two roofers 4 hours to put a new roof on a portable classroom. If the first roofer can do the job by himself in 6 hours, how many hours would it take for the second roofer to do the job by himself?

18. Working together, Rita and Tiffany can cut and trim a lawn in 2 hours. If it takes Rita 5 hours to do the lawn by herself, how long would it take Tiffany to do the lawn by herself?

19. With both the cold water and the hot water faucets open, it takes 9 minutes to fill a bathtub. The cold water faucet alone takes 15 minutes to fill the tub. How long would it take to fill the tub with the hot water faucet alone?

20. The cargo hold of a ship has two loading pipes. Used together, the two pipes can fill the cargo hold in 6 hours. If the larger pipe alone can fill the hold in 8 hours, how many hours would it take the smaller pipe to fill the hold by itself?

Solve the following by using the same method as in Example 3.

21. Two buses leave Kansas City traveling in opposite directions, one going east and one going west. The eastbound bus travels at 40 mph, and the westbound bus travels at 65 mph. If both buses leave at 10:00 A.M., at what time will they be 525 miles apart?

22. Sailing in opposite directions, an aircraft carrier and a destroyer leave their base in Hawaii at 5:00 A.M. If the destroyer sails at 30 mph and the aircraft carrier sails at 20 mph, at what time will the two ships be 300 miles apart?

23. Rapid City, South Dakota, is 360 miles from Sioux Falls. At 6:00 A.M., a freight train leaves Rapid City for Sioux Falls, and at the same time a passenger train leaves Sioux Falls for Rapid City. The two trains meet at 9:00 A.M. If the freight train travels $\frac{3}{5}$ as fast as the passenger train, how fast does the passenger train travel?

24. Houston, Texas, and Calgary, Alberta, are about 2100 miles apart. At 2:00 P.M., an airplane leaves Houston for Calgary flying at 250 mph. At the same time, an airplane leaves Calgary for Houston flying at 450 mph. How long will it take before the two airplanes meet?

25. An automobile makes a trip of 270 miles in $2\frac{1}{2}$ hours less than it can travel 420 miles at the same speed. What is the speed of the automobile?

26. A bus can travel 250 miles in $3\frac{1}{2}$ hours less than it can travel 425 miles at the same speed. Find the speed of the bus.

27. An airliner flies against the wind from Washington, D.C., to San Francisco in $5\frac{1}{2}$ hours. It flies back to Washington, D.C., with the wind in 5 hours. If the average speed of the wind is 21 mph, what is the speed of the airliner in still air?

28. A river has a current of 3 mph. A boat goes 40 miles upstream in the same amount of time as it goes 50 miles downstream. Find the speed of the boat in still water.

Solve the following. (See Examples 1–3.)

29. The denominator of a fraction is twice the numerator. If 3 is added to the numerator and 3 is subtracted from the denominator, the new fraction is $\frac{11}{13}$. Find the original fraction.

30. The same number is added to the numerator and the denominator of the fraction $\frac{4}{5}$. If the new fraction is equivalent to $\frac{33}{36}$, find the number added.

31. The pitcher of a baseball team has $\frac{2}{3}$ as many hits as the catcher, and together they have 85 hits. How many hits does each player have?

32. Two numbers differ by 6. If $\frac{3}{4}$ of the smaller number is $\frac{1}{4}$ less than $\frac{2}{3}$ of the larger, find the two numbers.

33. Hank can load a moving van in 12 hours, and Andy can load a moving van in 15 hours. Working together, how long would it take to load the moving van?

34. In a bicycle race, the first rider crossed the finish line $2\frac{1}{4}$ hours before the last rider. If the winner had an average speed of 26 mph, and the last rider had an average speed of 16 mph, how long did it take the winner to finish the race?

Challenge Exercises (35–37)

35. Find a positive number that is equal to 1 less than 12 times its reciprocal.

36. With both hot and cold water faucets on, it takes 5 minutes to fill a sink. However, the stopper is broken, so water is running out of the sink at the same time that water is running into the sink. If the leak can empty the sink in 15 minutes, how long will it take the sink to fill $\frac{1}{2}$ full with both faucets on?

37. Dr. Gilbert spent a total of 4 hours on a trip in the country. He rode out to the country at 15 mph. His bicycle broke down, and he had to walk back at 3 mph. How far out in the country did he get?

Writing Exercises or Group Activities (38–40)

If these exercises are done as a group activity, each group should write two exercises of each of the following types, exchange with another group, and then solve them:

38. Write and solve a number application problem.

39. Write and solve a work application problem.

40. Write and solve a distance, time, and rate application problem.

Chapter 7 Summary

Concept/Procedure	Example

Adding (Subtracting) Rational Numbers or Rational Expressions with Common Denominators: [Section 7.1]

- To add (subtract) rational numbers or rational expressions with common denominators, add the numerators and put the sum over the common denominator. In symbols, $\frac{a}{c} + \frac{b}{c} = \frac{a+b}{c}$. If necessary, reduce the sum to lowest terms. If situations of the form $\frac{a}{c} + \frac{b}{-c}$ occur, rewrite as $\frac{a}{c} + \frac{-b}{c}$.

Example 1:

Find the difference: $\dfrac{2x + 3y}{3x - y} - \dfrac{x - 5y}{3x - y}$.

Solution:

$\dfrac{2x + 3y}{3x - y} - \dfrac{x - 5y}{3x - y} =$ Put the difference of the numerators over the common denominator.

$\dfrac{(2x + 3y) - (x - 5y)}{3x - y} =$ Remove parentheses.

$\dfrac{2x + 3y - x + 5y}{3x - y} =$ Add like terms.

$\dfrac{x + 8y}{3x - y}$ Difference.

Least Common Multiple: [Section 7.2]

- The integer a is the **least common multiple** of the integers b and c if a is the smallest positive integer that is a multiple of both b and c. Equivalently, a is the least common multiple of b and c if a is the smallest positive integer that is divisible by both b and c.

Finding the Least Common Multiple (LCM): [Section 7.2]

1. Write each number or expression as the product of its prime factors.
2. Select every factor that appears in any prime factorization. If a factor appears in more than one of the prime factorizations, select the factor with the largest exponent.
3. The LCM is the product of the factors selected in step 2.

Example 2:

Find the LCM of $14a^4b^2$ and $20a^2b^6$.

Solution:

$14a^4b^2 = 2 \cdot 7a^4b^2$ Prime factorization of 14.

$20a^2b^6 = 2^2 \cdot 5a^2b^6$ Prime factorization of 20.

The different factors that appear are 2, 7, 5, a, and b. The largest exponent of 2 is 2, of 7 is 1, of 5 is 1, of a is 4, and of b is 6. Therefore, the LCM is $2^2 \cdot 5 \cdot 7a^4b^6 = 140a^4b^6$.

Equivalent Fractions: [Section 7.2]

- Two or more fractions are equivalent if they represent the same quantity. Equivalently, $\frac{a}{b} = \frac{c}{d}$ if $a \cdot d = b \cdot c$.

Fundamental Property of Fractions: [Section 7.2]

- If $\frac{a}{b}$ is a fraction and c is any number except 0, then $\frac{a}{b} = \frac{a}{b} \cdot \frac{c}{c} = \frac{ac}{bc}$. In words, we can multiply both the numerator and the denominator of a fraction by any number except 0 and get a fraction equivalent to the original fraction.

Concept/Procedure	Example

Finding the Missing Numerator to Get Equivalent Fractions: [Section 7.2]

- Determine the number or expression that you must multiply the denominator of the given fraction by in order to get the denominator of the new fraction. Multiply both the numerator and the denominator of the given fraction by this number to get an equivalent fraction with the given denominator.

Example 3:

Find the missing numerator so that $\dfrac{x + 2}{x - 6} = \dfrac{?}{2x^2 - 9x - 18}$.

Solution:

$\dfrac{x + 2}{x - 6} = \dfrac{?}{2x^2 - 9x - 18}$ Factor $2x^2 - 9x - 18$.

$\dfrac{x + 2}{x - 6} = \dfrac{?}{(x - 6)(2x + 3)}$ The missing factor is $2x + 3$, so multiply $\dfrac{x + 2}{x - 6}$ by $\dfrac{2x + 3}{2x + 3}$.

$\dfrac{x + 2}{x - 6} \cdot \dfrac{2x + 3}{2x + 3} = \dfrac{2x^2 + 7x + 6}{(x - 6)(2x + 3)}$ Therefore, $\dfrac{x + 2}{x - 6} = \dfrac{2x^2 + 7x + 6}{2x^2 - 9x - 18}$.

Least Common Denominator (LCD): [Section 7.3]

- The least common denominator of two or more rational numbers or rational expressions is the least common multiple of the denominators of the fractions.

Adding (Subtracting) Rational Numbers or Rational Expressions with Unlike Denominators: [Section 7.4]

- To add (subtract) rational numbers or rational expressions with unlike denominators, change each rational number or rational expression into an equivalent rational number or rational expression with the least common denominator as its denominator. Then add (subtract) the fractions, using the method of adding (subtracting) rational numbers or rational expressions with common denominators.

Example 4:

Subtract: $\dfrac{3x - 2}{4x - 12} - \dfrac{x - 4}{6x - 18}$.

Solution:

$\dfrac{3x - 2}{4x - 12} - \dfrac{x - 4}{6x - 18} =$ Factor the denominators.

$\dfrac{3x - 2}{4(x - 3)} - \dfrac{x - 4}{6(x - 3)} =$ The LCD is $12(x - 3)$, so multiply $\dfrac{3x - 2}{4(x - 3)}$ by $\dfrac{3}{3}$ and multiply $\dfrac{x - 4}{6(x - 3)}$ by $\dfrac{2}{2}$.

$\dfrac{3}{3} \cdot \dfrac{3x - 2}{4(x - 3)} - \dfrac{2}{2} \cdot \dfrac{x - 4}{6(x - 3)} =$ Multiply.

$\dfrac{9x - 6}{12(x - 3)} - \dfrac{2x - 8}{12(x - 3)} =$ Subtract the rational expressions.

$\dfrac{(9x - 6) - (2x - 8)}{12(x - 3)} =$ Remove parentheses.

$\dfrac{9x - 6 - 2x + 8}{12(x - 3)} =$ Add like terms.

$\dfrac{7x + 2}{12(x - 3)}$ Difference.

Concept/Procedure	Example

Simplifying Complex Fractions Using Division: [Section 7.5]

1. If necessary, rewrite the numerator and/or the denominator as single fractions by performing whatever operations are indicated in either.
2. Rewrite the complex fraction as the division of the numerator by the denominator.
3. Perform the division by inverting the fraction on the right and multiplying.
4. If necessary, reduce to lowest terms.

Example 5:

Simplify $\dfrac{\dfrac{3}{8}}{\dfrac{9}{20}}$ by the division method.

Solution:

$\dfrac{\dfrac{3}{8}}{\dfrac{9}{20}} = \dfrac{3}{8} \div \dfrac{9}{20}$ Rewrite as division.

$= \dfrac{3}{8} \cdot \dfrac{20}{9}$ Rewrite as multiplication.

$= \dfrac{3}{2 \cdot 2 \cdot 2} \cdot \dfrac{2 \cdot 2 \cdot 5}{3 \cdot 3}$ Rewrite as prime factors.

$= \dfrac{\cancel{3}}{2 \cdot 2 \cdot 2} \cdot \dfrac{2 \cdot 2 \cdot 5}{\cancel{3} \cdot 3}$ Divide by common factors.

$= \dfrac{5}{6}$ Multiply the remaining factors.

   Simplified form.

Simplifying Complex Fractions Using the LCD: [Section 7.5]

1. Find the LCD of all fractions that occur in the numerator and/or the denominator.
2. Multiply both the numerator and the denominator by the LCD.
3. Simplify the results.

Example 6:

Simplify $\dfrac{4 - \dfrac{9}{a^2}}{2 - \dfrac{3}{a}}$ by the LCD.

Solution:

$\dfrac{4 - \dfrac{9}{a^2}}{2 - \dfrac{3}{a}} =$ The LCD is a^2, so multiply the numerator and denominator by a^2.

$\dfrac{a^2\left(4 - \dfrac{9}{a^2}\right)}{a^2\left(2 - \dfrac{3}{a}\right)} =$ Multiply.

$\dfrac{4a^2 - 9}{2a^2 - 3a} =$ Factor.

$\dfrac{(2a - 3)(2a + 3)}{a(2a - 3)} =$ Divide by common factor of $2a - 3$.

$\dfrac{2a + 3}{a}$ Simplified form.

Concept/Procedure	Example

Solving Equations Involving Rational Numbers or Rational Expressions: [Section 7.6]

1. Determine the value(s) of the variable, if any, that would make any denominator equal to 0.
2. Multiply both sides of the equation by the LCD of all the rational numbers or rational expressions that appear in the equation.
3. Solve the resulting equation.
4. Check for extraneous solutions.

Example 7:

Solve $\dfrac{2x}{3x + 6} - \dfrac{1}{4} = \dfrac{3}{4x + 8}$.

Solution: Since the denominator cannot equal 0, $x \neq -2$.

$$\frac{2x}{3x + 6} - \frac{1}{4} = \frac{3}{4x + 8} \qquad \text{Factor the denominators.}$$

$$\frac{2x}{3(x + 2)} - \frac{1}{4} = \frac{3}{4(x + 2)} \qquad \begin{array}{l}\text{Multiply both sides by}\\ \text{the LCD } 12(x + 2).\end{array}$$

$$12(x + 2)\left(\frac{2x}{3(x + 2)} - \frac{1}{4}\right) = 12(x + 2)\left(\frac{3}{4(x + 2)}\right) \qquad \begin{array}{l}\text{Distribute on left,}\\ \text{multiply on right.}\end{array}$$

$$12(x + 2) \cdot \frac{2x}{3(x + 2)} - 12(x + 2) \cdot \frac{1}{4} = 3 \cdot 3 \qquad \text{Multiply.}$$

$$4 \cdot 2x - 3(x + 2) = 9 \qquad \text{Continue multiplying.}$$

$$8x - 3x - 6 = 9 \qquad \text{Add like terms.}$$

$$5x - 6 = 9 \qquad \text{Add 6.}$$

$$5x = 15 \qquad \text{Divide by 5}$$

$$x = 3. \qquad \text{Solution.}$$

Check: Be sure $x = 3$ does not make any denominator equal to 0.

Applications with Rational Expressions: [Section 7.7]

Example 8:

One pipe can fill a tank in 6 hours and a larger pipe can fill the same tank in 4 hours. How long would it take to fill the tank with the use of both pipes?

Understand: We are given the number of hours for each pipe to fill the tank alone. The unknown is the number of hours it will take to fill the tank with the use of both pipes.

> Relationship 1: It takes the smaller pipe 6 hours to fill the tank and the larger pipe 4 hours.
>
> Relationship 2: (The part the smaller pipe can fill in 1 hour) + (the part the larger pipe can fill in 1 hour) = (the part of the tank they can fill in one hour together).

Plan: Represent the part each pipe can fill in one hour. Represent the part they can fill together in one hour. Write the equation, using relationship 2.

Translate: Use a chart.

	Number of Hours	Part Done in 1 Hour
Smaller Pipe	6	$\dfrac{1}{6}$
Larger Pipe	4	$\dfrac{1}{4}$
Both Pipes	x	$\dfrac{1}{x}$

Concept/Procedure	Example

Example

From relationship 2, we have

(part smaller pipe fills in 1 hr) + (part larger pipe fills in 1 hr) = (part filled in 1 hr together).

$$\frac{1}{6} + \frac{1}{4} = \frac{1}{x}$$

Solve:

$$\frac{1}{6} + \frac{1}{4} = \frac{1}{x} \qquad \text{Multiply both sides by } 12x.$$

$$12x\left(\frac{1}{6} + \frac{1}{4}\right) = 12x\left(\frac{1}{x}\right) \qquad \text{Distribute on the left, multiply on the right.}$$

$$2x + 3x = 12 \qquad \text{Combine like terms.}$$

$$5x = 12 \qquad \text{Divide both sides by 5.}$$

$$x = \frac{12}{5} \qquad \text{Solution of equation.}$$

State: It will take $\frac{12}{5}$ hours to fill the tank with the use of both pipes.

Check: Does this answer seem reasonable?

Chapter 7   Review Exercises

Find the sums or differences. Express the answers reduced to the lowest terms: [Section 7.1]

1. $\dfrac{5}{7} - \dfrac{3}{7}$

2. $\dfrac{3}{8} - \dfrac{-5}{8}$

3. $\dfrac{4}{13} + \dfrac{6}{13} + \dfrac{2}{13}$

4. $\dfrac{9}{11} + \dfrac{5}{11} + \dfrac{7}{-11}$

5. $9\dfrac{7}{10} + 6\dfrac{1}{10}$

6. $19\dfrac{11}{12} - 13\dfrac{4}{12}$

7. $11\dfrac{7}{9} - 4\dfrac{2}{9} - \dfrac{5}{9}$

8. $15\dfrac{15}{17} - 8\dfrac{11}{17} + 3\dfrac{4}{17}$

9. $\dfrac{9r}{14x} + \dfrac{2p}{14x}$

10. $\dfrac{4}{y+2} - \dfrac{3}{y+2}$

11. $\dfrac{x^2 - 3x - 8}{x^2 + 7x + 10} + \dfrac{-2x - 6}{x^2 + 7x + 10}$

12. $\dfrac{x^2 + 2x - 5}{x^2 - 3x - 18} - \dfrac{3x + 7}{x^2 - 3x - 18}$

13. $\dfrac{2p^2 - p + 2}{p^2 - 16} + \dfrac{p^2 + 4p - 8}{p^2 - 16}$

14. $\dfrac{x - 3y}{x - y} - \dfrac{5x - 2y}{y - x}$

Answer the following: [Section 7.1]

15. A stock's share price began the day at $32\frac{3}{8}$ and rose $3\frac{5}{8}$ points. What was the closing price of the stock?

16. A rectangular tablecloth is $66\frac{9}{16}$ inches long and $44\frac{5}{16}$ inches wide. A border that costs $.02 per inch is sewn around the tablecloth. Find the cost of the border to the nearest cent.

17. Find the perimeter of a triangle the length of whose sides are $3\frac{2}{9}$ meters, $4\frac{5}{9}$ meters, and $6\frac{1}{9}$ meters.

18. Find the perimeter of a rectangle whose length is $\frac{7}{16}$ kilometer and whose width is $\frac{3}{16}$ kilometer.

Find the LCM of the following: [Section 7.2]

19. 6 and 7

20. 25 and 40

21. 16, 24, 32

22. 18, 27, 36

23. a^2y and ay^4

24. $12t^5v^3$ and $15t^2v^7$

25. $14x^2w^3$ and $21x^5w^4$

26. $z - 4$ and $z + 3$

27. $a + 2$ and $a^2 + 4a + 4$

28. $y^2 + 8y + 16, y + 4, y^2 + y - 12$

Find the missing numerator so that the fractions will be equal: [Section 7.2]

29. $\dfrac{6}{5} = \dfrac{?}{20}$

30. $3 = \dfrac{?}{15}$

31. $\dfrac{4r}{7p^3} = \dfrac{?}{35p^5}$

32. $\dfrac{8x^3}{13tz^4} = \dfrac{?}{52t^3z^7}$

33. $\dfrac{4a}{a + 5} = \dfrac{?}{2a + 10}$

34. $\dfrac{7}{b - 3} = \dfrac{?}{b^2 - 9}$

35. $\dfrac{p - 8}{p^2 - 10p + 25} = \dfrac{?}{(p - 5)(p - 3)(p - 5)}$

36. $\dfrac{2q}{y^2 - 5y} = \dfrac{?}{y(y - 5)(y + 3)}$

Find the least common denominator of each pair of fractions. Then write each fraction as an equivalent fraction with the LCD as its denominator: [Section 7.3]

37. $\dfrac{4}{7}$ and $\dfrac{5}{2}$

38. $\dfrac{5}{6}$ and $\dfrac{7}{18}$

39. $\dfrac{8}{9}$ and $\dfrac{11}{12}$

40. $\dfrac{6}{15}, \dfrac{3}{5},$ and $\dfrac{13}{20}$

41. $\dfrac{4}{t^2v^5}$ and $\dfrac{6}{tv^3}$

42. $\dfrac{8b}{9p^3q^4}$ and $\dfrac{11c}{12p^2q^5}$

43. $\dfrac{7}{t - 4}$ and $\dfrac{9}{t + 2}$

44. $\dfrac{y}{y - 11}$ and $\dfrac{y}{y - 3}$

45. $\dfrac{4u}{8u + 12}$ and $\dfrac{9u}{14u + 21}$

46. $\dfrac{2w}{w^2 - 16}$ and $\dfrac{6w}{w^2 + 5w + 4}$

47. $\dfrac{3a}{a^2 + 6a + 9}$ and $\dfrac{10a}{a^2 - 2a - 15}$

48. $\dfrac{m + 1}{m^2 - 2m - 8}$ and $\dfrac{m - 1}{m^2 - 3m - 4}$

Find the sums or differences. Express the answers reduced to lowest terms: [Section 7.4]

49. $\dfrac{5}{12} + \dfrac{7}{16}$

50. $\dfrac{3}{4} - \dfrac{2}{5}$

51. $\dfrac{17}{5} - 4$

52. $\dfrac{3}{14} + \dfrac{5}{7} - \dfrac{4}{21}$

53. $7\dfrac{5}{6} - 5\dfrac{1}{3}$

54. $\dfrac{11}{2} - 5 + 3\dfrac{5}{8}$

55. $\dfrac{7}{m} + \dfrac{b}{n}$

56. $\dfrac{10}{9x} - \dfrac{2}{3x}$

57. $\dfrac{y - 3}{4y} + \dfrac{y + 2}{5y}$

58. $\dfrac{2t - 3}{18t^4u^2} + \dfrac{5t + 1}{12t^3u^5}$

59. $\dfrac{8}{w - 3} - \dfrac{-3}{w}$

60. $\dfrac{6r}{r - 5} + \dfrac{2}{5 - r}$

61. $\dfrac{v + 4}{4v + 8} - \dfrac{v - 2}{2v - 6}$

62. $\dfrac{16x}{16x^2 - 25} - \dfrac{2}{4x + 5}$

63. $\dfrac{x + 5}{x^2 - 9} - \dfrac{1}{x^2 + 3x}$

64. $\dfrac{u^2 - 3u}{u^2 - 6u + 9} + \dfrac{8}{5u - 15}$

Answer the following: [Section 7.4]

65. A rectangular computer screen is $8\frac{5}{8}$ inches wide and $11\frac{1}{4}$ inches high. What is the perimeter of the screen?

66. A carpenter is installing molding around a triangular opening, the lengths of whose sides are $8\frac{3}{4}$ inches, $7\frac{5}{6}$ inches, and $8\frac{3}{8}$ inches. If the molding costs \$.04 per inch, find the cost of the molding to the nearest cent.

Simplify the following complex fractions using both the division and the LCD methods: [Section 7.5]

67. $\dfrac{\dfrac{4}{5}}{\dfrac{3}{10}}$

68. $\dfrac{\dfrac{5}{18}}{\dfrac{11}{30}}$

69. $\dfrac{\dfrac{8}{9}}{16}$

70. $\dfrac{12}{\dfrac{24}{13}}$

71. $\dfrac{\dfrac{4}{3} - \dfrac{8}{9}}{\dfrac{5}{6} - \dfrac{4}{12}}$

72. $\dfrac{6 - \dfrac{10}{7}}{\dfrac{14}{3} - 2}$

73. $\dfrac{\dfrac{x}{9}}{\dfrac{y}{27}}$

74. $\dfrac{\dfrac{p}{r}}{s}$

75. $\dfrac{\dfrac{4x}{t^2}}{\dfrac{20x}{t^3}}$

76. $\dfrac{\dfrac{u^4 v^2}{w}}{\dfrac{uv^3}{w^2}}$

77. $\dfrac{\dfrac{x^6 y^3}{t^4}}{\dfrac{x^2 y^2}{t}}$

78. $\dfrac{2w - \dfrac{w}{4}}{6 - \dfrac{w}{4}}$

79. $\dfrac{1 - \dfrac{1}{y} - \dfrac{12}{y^2}}{1 - \dfrac{6}{y} + \dfrac{8}{y^2}}$

80. $\dfrac{x + \dfrac{25}{x + 10}}{1 - \dfrac{5}{x + 10}}$

Solve the following equations: [Section 7.6]

81. $\dfrac{y}{4} - \dfrac{y}{5} = \dfrac{3}{20}$

82. $\dfrac{6p}{7} - \dfrac{5p}{14} = 1$

83. $\dfrac{7}{2}t - \dfrac{4}{9}t = \dfrac{2}{3}$

84. $\dfrac{4w}{5} + \dfrac{1}{6} = \dfrac{7w}{3} - \dfrac{13}{10}$

85. $\dfrac{2v - 5}{4} = 3 - v$

86. $\dfrac{z + 6}{7} - \dfrac{3}{14} = \dfrac{5 - 4z}{21}$

87. $\dfrac{5}{4t} - \dfrac{3}{8t} = \dfrac{1}{2}$

88. $\dfrac{12}{m - 2} = 9 + \dfrac{m}{m - 2}$

89. $\dfrac{1}{q + 4} = \dfrac{q}{3q + 2}$

90. $\dfrac{x^2 + 5x - 2}{x + 2} - 3 = \dfrac{x^2 + 6x}{x + 2} - x$

Solve the following: [Section 7.7]

91. The numerator of a fraction is 3 less than the denominator. If 5 is added to the numerator and 5 is subtracted from the denominator, the result is $\frac{3}{2}$. Find the original fraction.

92. The difference of an integer and 10 times its reciprocal is 3. Find the integer.

93. George can write a chapter for a mathematics textbook in 30 days. Leon can write a chapter in 45 days. How long would it take them to write a chapter together?

94. The Gulf Stream is an ocean current off the eastern coast of the United States. A Coast Guard ship can sail 200 miles against the current in the same time that it takes it to sail 260 miles with the current. If the current is 3 mph, how fast can the cutter sail in still water?

95. It is 510 road miles from Denver, Colorado, to Salt Lake City, Utah. A truck leaves Denver for Salt Lake City at 4:00 A.M. At the same time, an automobile leaves Salt Lake City for Denver. If the truck travels at 45 mph and the auto travels at 55 mph, what time will it be when the auto and truck meet? (Give answer to the nearest minute.)

Chapter 7   Test

Find the LCM of the following:

1. 12 and 16

2. $9x^2y^4$ and $15xy^3$

Find the missing numerator so that the fractions are equal.

3. $\dfrac{3t}{4u^3} = \dfrac{?}{20u^5w}$

4. $\dfrac{6p}{q^2 + 4q} = \dfrac{?}{q(q^2 - 16)}$

Find the least common denominator. Then write each fraction as an equivalent fraction with the LCD as its denominator.

5. $\dfrac{2}{9}, \dfrac{7}{18},$ and $\dfrac{4}{27}$

6. $\dfrac{7a}{a^2 - 3a - 18}$ and $\dfrac{b}{a + 3}$

Find the sums or differences. Reduce the answers to the lowest terms.

7. $\dfrac{9}{7} - \dfrac{5}{7} + \dfrac{1}{7}$

8. $\dfrac{r^3}{r^2 + 16} + \dfrac{r - r^3}{r^2 + 16}$

9. $\dfrac{5}{6} - \dfrac{7}{12}$

10. $\dfrac{2}{5} - \dfrac{3}{10} + \dfrac{4}{15}$

11. $\dfrac{17}{4} - 6 + 4\dfrac{5}{6}$

12. $\dfrac{v + 9}{3v} - \dfrac{v - 6}{9v^2}$

13. $\dfrac{5w}{3w - 12} + \dfrac{7w - 1}{5w - 20}$

14. $\dfrac{x}{x^2 + 5x + 6} - \dfrac{3}{x^2 + 7x + 12}$

Find the perimeters of each of the following triangles the lengths of whose sides are given:

15. $2\dfrac{3}{5}$ inches, $3\dfrac{7}{10}$ inches, $4\dfrac{2}{5}$ inches

16. $\dfrac{8}{2a + 3}$ meters, $\dfrac{5}{a - 2}$ meters, $\dfrac{6}{2a + 3}$ meters

Find the perimeters of the following rectangles the lengths of whose sides are given:

17. $L = \dfrac{6}{x}$ feet, $W = \dfrac{2}{y}$ feet

18. $L = \dfrac{4m}{3m - n}$ yards, $W = \dfrac{3n}{2m + 3n}$ yards

19. A window is $69\dfrac{1}{8}$ inches wide and $47\dfrac{3}{5}$ inches high. A decorative molding that costs $.02 per inch is put around the window. Find the cost of the molding to the nearest cent.

Simplify the following complex fractions:

20. $\dfrac{\dfrac{1}{8}}{\dfrac{5}{12}}$

21. $\dfrac{\dfrac{7}{12} + \dfrac{2}{3}}{\dfrac{9}{16} - \dfrac{6}{24}}$

22. $\dfrac{\dfrac{m^2}{3z}}{\dfrac{m}{4z}}$

23. $\dfrac{8 - \dfrac{2u}{3}}{5u + \dfrac{u}{3}}$

Solve the following equations:

24. $\dfrac{5v}{6} - \dfrac{1}{4} = 4 - \dfrac{7v}{12}$

25. $\dfrac{4x - 2}{8} + \dfrac{3}{4} = \dfrac{2x - 5}{2}$

26. $\dfrac{11}{t - 3} + 6 = \dfrac{7t}{t - 3}$

27. $\dfrac{n}{n + 1} + \dfrac{n + 1}{n + 2} = \dfrac{6n + 5}{n^2 + 3n + 2}$

Solve each of the following:

28. Two numbers differ by 4. If $\frac{3}{4}$ of the smaller number is increased by $\frac{1}{2}$ of the larger number, the sum is 2. Find the numbers.

29. Working together, it takes Elena and Eduardo 10 days to do an architectural project. If Elena can do the project by herself in 15 days, how long would it take Eduardo to do the project working by himself?

30. A cargo plane and an airliner take off at the same time from the same airport. They fly in opposite directions. The cargo plane flies at 420 mph, and the airliner flies at 530 mph. How long is it before the two airplanes are 1900 miles apart?

Ratios, Percents, and Applications

Chapter Outline

Ratios have been used to describe the relationships between numbers for thousands of years. Ratios occur in the fundamental mathematics of arithmetic, in geometry to compare lengths of parts of geometric figures, and in higher mathematics such as algebra, trigonometry, and calculus. Many applied problems that seem difficult at first glance easily yield to a solution using ratios.

Another example of a ratio is percentage. The idea of percent, or per hundred, came about because of the need for business to calculate profit and loss and for governments to levy taxes. As people of the Middle Ages became more interested in trade, money became available for lending (at interest) and governments found a new source of revenue (taxing business). Thus, there was a need for a method to handle both transactions. The symbol for percent, %, is relatively new, since it was gradually developed from older symbols over the last century.

OBJECTIVES *When you complete this section, you will be able to:*

A Recognize the ratio of two or more numbers.
B Simplify ratios to lowest terms.
C Write ratios so that each part is a natural number.
D Recognize the rate of two measurements.
E Recognize and use unit pricing.

PREREQUISITE SKILLS *Before beginning this section, you should be able to:*

a. Write a whole number in terms of its prime factors. (Section 5.1)
b. Reduce fractions to lowest terms. (Section 6.1)
c. Multiply rational numbers. (Section 6.3)
d. Simplify complex fractions (optional). (Section 7.5)

GETTING READY FOR SECTION 8.1

1. Write 63 as a product of prime factors.

2. Reduce $\dfrac{25}{45}$ to lowest terms.

3. Multiply $\dfrac{2}{5} \cdot \dfrac{3}{5}$.

4. Simplify $\dfrac{\frac{2}{5}}{\frac{3}{5}}$.

Introduction

In this section, we will explore a simple, but important concept in applied mathematics. It is useful in solving problems that otherwise could be complicated. It is defined as follows:

OBJECTIVE Recognizing the ratio of two or more numbers

> **Definition** Ratio
>
> A **ratio** is a comparison of two or more quantities by division. If *a* and *b* are quantities, then *a* : *b* is a ratio of *a* to *b* and can be written as $\frac{a}{b}$. If *a*, *b*, and *c* are quantities, then *a* : *b* : *c* is a ratio of *a* to *b* to *c*.

The quantities in a ratio may be either numbers or measurements. If two quantities have the same unit of measurement, then the ratio is a comparison of numbers. For example, 7 meters compared to 10 meters becomes the ratio 7 to 10. A ratio in the form of 7 to 10 is a **comparison**.

There may be more than two quantities in a ratio. Suppose that we have the results of a test. If there are 5 A's, 8 B's, 12 C's, 4 D's, and 2 F's on the test, then the ratio of A's to B's to C's to D's to F's is 5 : 8 : 12 : 4 : 2. Ratios may be written in several ways, but they all mean the same thing. If there are 5 men and 8 women in the math lab, we say that the ratio of men to women is 5 to 8, or 5 : 8, or $\frac{5}{8}$, or five men to eight women. We summarize this in a table.

Ways to Write Ratios

Comparison	Colon	Fraction	Words
5 to 8	5 : 8	$\frac{5}{8}$	five to eight
12 to 5	12 : 5	$\frac{12}{5}$	twelve to five
3 to 4 to 7	3 : 4 : 7	—	three to four to seven

Note

Ratios with more than two quantities cannot be written in fractional notation.

OBJECTIVE B Simplifying ratios to lowest terms

> **Definition** Ratios in Lowest Terms
>
> As with rational numbers, ratios are in lowest terms when the parts of a ratio are integers that contain no common factors other than 1.

Ratios are usually simplified or reduced to lowest terms just as fractions are reduced to lowest terms. Recall from Chapter 6 that we write the numerator and the denominator of the fractions in terms of prime factors and divide by the factors common to both. We do the same with ratios.

> **Procedure: Processes for Simplifying Ratios**
>
> a. If the ratio is in fractional form, we divide the numerator and the denominator by the factors common to both. For example,
>
> $$\frac{4}{6} = \frac{2 \cdot 2}{2 \cdot 3} = \frac{2}{3}, \text{ or 2 to 3, or 2 : 3.}$$
>
> b. If the ratio is in another form such as $a : b$ or a to b, we divide each part by the factors common to both. For example 4 : 6 becomes $2 \cdot 2 : 2 \cdot 3$. Dividing each part by the common factor of 2, we have $\frac{2 \cdot 2}{2} : \frac{2 \cdot 3}{2}$ or 2 : 3.

Let us look at a few examples.

Example 1   **Simplify the following ratios:**

a. 15 : 10 =   Write each number in terms of prime factors.

   $3 \cdot 5 : 2 \cdot 5 =$ Divide both parts by the common factor of 5.

   $\frac{3 \cdot \cancel{5}}{\cancel{5}} : \frac{2 \cdot \cancel{5}}{\cancel{5}} =$ Simplify.

   3 : 2   Hence, 15 : 10 is the same as 3 : 2.

b. $\frac{21}{27} =$ Write the numerator and the denominator in terms of prime factors.

   $\frac{3 \cdot 7}{3 \cdot 3 \cdot 3} =$ Divide by the common factor of 3.

   $\frac{\cancel{3} \cdot 7}{\cancel{3} \cdot 3 \cdot 3} =$ Multiply the remaining factors.

   $\frac{7}{9}$ Hence, $\frac{21}{27} = \frac{7}{9}$.

c. 63 to 9 =   Write each number in terms of prime factors.

$3 \cdot 3 \cdot 7$ to $3 \cdot 3$ =   Divide both parts by the common factors of $3 \cdot 3$.

$\dfrac{\cancel{3} \cdot \cancel{3} \cdot 7}{\cancel{3} \cdot \cancel{3}}$ to $\dfrac{\cancel{3} \cdot \cancel{3}}{\cancel{3} \cdot \cancel{3}}$ =   Simplify.

7 to 1   Hence, 63 to 9 is the same as 7 to 1.

PRACTICE EXERCISE 1

Simplify the following ratios:

a. $16 : 24$

b. $\dfrac{25}{45}$

c. 72 to 8

Often, we need to write English sentences as ratios so that we may perform the necessary mathematical operations.

Example 2 | **Write each of the following as ratios in lowest terms:**

a. At a large university, there are 30,000 students. There are 16,000 women and 14,000 men. What is the ratio of women to men?

Solution

We compare the number of women students to the number of men students. As a comparison, the ratio is of the following form:

"number of women" to "number of men"   Substitute. (This could be written in any of the three forms.)

16,000 to 14,000 =   Simplify. Divide by the common factor of 1000.

$\dfrac{16,000}{1000}$ to $\dfrac{14,000}{1000}$ =   Simplify.

16 to 14 =   Factor into prime factors.

$2 \cdot 2 \cdot 2 \cdot 2$ to $2 \cdot 7$ =   Divide by the common factor of 2.

$\dfrac{2 \cdot 2 \cdot 2 \cdot 2}{2}$ to $\dfrac{2 \cdot 7}{2}$ =   Simplify.

8 to 7   Hence, the ratio of women to men is 8 to 7.

b. A recipe for homemade cereal calls for 6 cups of rolled oats, 1 cup of sunflower seeds, 2 cups of chopped walnuts, and 4 cups of dried fruit. Write the ratio of the ingredients in the order given.

Solution

Since they are all measured in cups, we have $6 : 1 : 2 : 4$.

PRACTICE EXERCISE 2

Write each of the following as ratios in lowest terms:

a. A citrus grove produced 18,000 boxes of oranges and 12,000 boxes of grapefruit. What is the ratio of the number of boxes of oranges to the number of boxes of grapefruit?

b. A recipe for fruit punch calls for 96 ounces of water, 16 ounces of tea, 12 ounces of frozen lemonade, 64 ounces of cranberry juice, 32 ounces of apple juice, and 16 ounces of orange juice. Write the ratio of the ingredients in the order given.

Answers: **Practice Exercise 1: a.** $2 : 3$ **b.** $\dfrac{5}{9}$ **c.** 9 to 1 **Practice Exercise 2: a.** $3 : 2$ **b.** $24 : 4 : 3 : 16 : 8 : 4$

OBJECTIVE **C** Writing ratios so that each part is a natural number

Unless a special case is called for, a ratio should be simplified so that the parts of the ratio are natural numbers.

Example 3 Write each of the following as a ratio of natural numbers:

a. In a 1-liter beaker, we have a solution that is $\frac{2}{5}$ alcohol and $\frac{3}{5}$ water. What is the ratio of alcohol to water?

Solution

Compare the part that is alcohol to the part that is water. Both of the measurements are in terms of liters, so the units are the same and we write

$\frac{2}{5}$ alcohol to $\frac{3}{5}$ water, or

$\frac{2}{5} : \frac{3}{5} =$ Multiply both parts of the ratio by 5.

$5\left(\frac{2}{5}\right) : 5\left(\frac{3}{5}\right) =$

$2 : 3$ Hence, the ratio of alcohol to water is 2 : 3.

This means that for every 2 parts of alcohol there are 3 parts of water.

Note

This could also be done by using complex fractions, as follows:

$\dfrac{\frac{2}{5}}{\frac{3}{5}} =$ Multiply numerator and denominator by 5.

$\dfrac{\frac{2}{5} \cdot 5}{\frac{3}{5} \cdot 5} =$ Multiply.

$\dfrac{2}{3}$ Answer.

b. Recently, a research study showed that 2.4 out of every 100 women giving birth at a certain hospital were HIV positive. Write the ratio of HIV-positive women giving birth to the total number of women giving birth.

Solution

The number of HIV-positive women to the total was 2.4 : 100. The number 2.4 is not a natural number. We multiple both parts by 10 to make 2.4 a natural number.

2.4 : 100   Multiple both parts by 10.
2.4(10) : 100(10)   Complete multiplication.
24 : 1000   Reduce to lowest terms.
3 : 125   This means that 3 out of every 125 women tested were HIV positive.

PRACTICE EXERCISE 3

Write each of the following as a ratio of natural numbers:

a. Homemade candy is composed of $\frac{1}{6}$ cup of chocolate, $\frac{2}{6}$ cup of nuts, and $\frac{3}{6}$ cup of caramel. What is the ratio of chocolate to caramel?

b. At a community college, 56.6 out of 100 faculty members are male. What is the ratio of men to total faculty?

Chemists and other people who are concerned about pollution of the environment by toxic chemicals often use ratios that have very large denominators. Our goal is to have natural numbers in both parts of the ratio.

Example 4 | **Solve the following:**

a. Fish with less than 5 parts of mercury per 10,000,000 parts of flesh are considered safe to eat. Is a fish with .3 parts of mercury per 100,000 parts of flesh safe to eat?

Solution

We need to know if .3 parts per 100,000 is less than 5 parts per 10,000,000. Therefore, we must have the same number in the second parts of the ratios so that we can compare the first parts.

.3 : 100,000	Multiply both parts by 100 to get 10,000,000 in the second part.
.3(100) : 100,000(100)	Perform multiplication.
30 : 10,000,000	Since the second part is 10,000,000, we can compare this ratio to 5 : 10,000,000.

Is 30 : 10,000,000 less than 5 : 10,000,000?
No!
Conclusion: Do not eat the fish.

PRACTICE EXERCISE 4

If the level of insecticide to total fluid is over 15 parts per 1,000,000, the fluid cannot be processed for release into streams and rivers. Can fluid that has a level of .12 parts insecticide per 1000 parts of fluid be processed?

OBJECTIVE **D** **Recognizing the rate of two measurements**

Ratios that have different units of measurement are called **rates**. Rates are usually stated in lowest terms and use the expression *per* or *for every* between the units of the numerator and denominator. Since rates are a special type of ratio, all of the properties that we have discussed for ratios are true for rates as well. Rates are different from ratios in that they are often expressed in a form where the second part is equal to 1.

Answers: **Practice Exercise 3:** **a.** 1 : 3 **b.** 283 : 500   **Practice Exercise 4:** 120 : 1,000,000; no

Example 5 Express each rate with the second part equal to 1.

a. Professor Smart drove her automobile 240 miles in 6 hours. At what average rate did she drive in miles per hour?

Solution

$$\frac{240 \text{ miles}}{6 \text{ hours}} =$$ Ratio of miles to hours. Reduce to lowest terms, but leave the denominator as 1.

$$\frac{40 \text{ miles}}{1 \text{ hour}} =$$ This is read, "40 miles in 1 hour."

$$\frac{40 \text{ miles}}{\text{hour}} =$$ This is read, "40 miles per hour."

b. There are 160 students in five sections of Integrated Algebra and Arithmetic. What is the average number of students per section?

Solution

$$\frac{160 \text{ students}}{5 \text{ sections}} =$$ Ratio of students to sections. Reduce to lowest terms, but leave the denominator as 1.

$$\frac{32 \text{ students}}{1 \text{ section}} =$$ This is read, "32 students in 1 section."

$$\frac{32 \text{ students}}{\text{section}} =$$ This is read, "32 students per section."

PRACTICE EXERCISE 5

Express each rate with the second part equal to 1.

a. If 10 children at a daycare center must share 160 pieces of a construction toy, how many pieces are there per child?

b. A family of five people just won $200,000 in the lottery. What are the average winnings per person?

OBJECTIVE Recognizing and using unit pricing

A rate that expresses the cost compared to one unit is called a **unit price**. Sometimes, expressing the second part of a ratio or rate as 1 may result in the first part of the ratio or rate being a fraction or decimal.

Example 6 Express each of the following as a unit price:

a. Jose Valdez paid $360 to take 12 credits at Grand Valley University. How much did he pay per credit?

Solution

$$\frac{\$360}{12 \text{ credits}}$$ Ratio of cost to credits. Divide 12 into 360 and leave the denominator as 1.

$$\frac{\$30}{1 \text{ credit}}$$ This is read, "$30 per 1 credit."

$$\frac{\$30}{\text{credit}}$$ This is read "$30 per credit."

b. A 10-pound bag of potatoes sells for $4.49. What is the unit price rounded to the nearest cent?

Solution

$$\frac{\$4.49}{10 \text{ pounds}} =$$ Ratio of cost to the number of pounds. Change $4.49 to 449 cents.

$$\frac{449 \text{ cents}}{10 \text{ pounds}}$$ Divide 10 into 449 and write the denominator as 1.

$$\frac{44.9 \text{ cents}}{1 \text{ pound}} =$$ Round to the nearest cent.

$$\frac{45 \text{ cents}}{\text{pound}}$$ Therefore, the potatoes cost 45 cents per pound.

PRACTICE EXERCISE 6

Express each of the following as a unit price to the nearest cent:

a. If 6 ounces of yogurt cost $1.29, what is the price per ounce?

b. If 24 hamburger rolls cost $2.39, what is the price per hamburger roll?

Answers: Practice Exercise 5: a. 16 pieces per child **b.** $40,000 per person **Practice Exercise 6: a.** 22 cents **b.** 10 cents

Exercise Set 8.1

For Extra Help

MyMathLab®

Simplify the following ratios. (See Example 1.)

1. 8 to 12

2. 10 to 25

3. 21 : 35

4. 12 : 28

5. 30 to 12

6. 48 to 18

7. 27 : 9

8. 30 : 5

9. $\dfrac{15}{35}$

10. $\dfrac{14}{21}$

11. $\dfrac{48}{36}$

12. $\dfrac{72}{42}$

13. $\dfrac{22}{55}$

14. $\dfrac{30}{84}$

15. $\dfrac{52}{28}$

16. $\dfrac{65}{39}$

Write each of the following as a ratio in simplest form with each part of the ratio a natural number. (See Examples 2–4.)

17. One year, an auto dealer sold 500 cars and 245 trucks. What was the ratio of cars to trucks?

18. Last month, the college bookstore sold 624 T-shirts and 360 sweatshirts. What was the ratio of T-shirts to sweatshirts?

19. In a county referendum, 50,000 people voted for higher taxes and 65,000 voted against higher taxes. What was the ratio of people who voted for the higher taxes to those who voted against higher taxes?

20. On a freight train, there are 72 flatcars and 42 box-cars. What is the ratio of boxcars to flatcars?

21. Gasohol is $\frac{1}{5}$ alcohol and $\frac{4}{5}$ gasoline. What is the ratio of gasoline to alcohol?

22. At a certain temperature, an antifreeze mixture should be $\frac{4}{10}$ antifreeze and $\frac{6}{10}$ water. What is the ratio of antifreeze to water?

23. A recipe calls for 3 cups of water and $\frac{3}{4}$ cup of grits. What is the ratio of water to grits? What is the ratio of grits to water?

24. A biscuit recipe calls for $2\frac{1}{4}$ cups of baking mix and $\frac{2}{3}$ cup of milk. What is the ratio of mix to milk? What is the ratio of milk to mix?

25. In a baking mix, there are 140 milligrams of sodium in $\frac{1}{2}$ ounce of mix. What is the ratio of sodium to mix?

26. In a popular cereal, there is $\frac{1}{2}$ gram of soluble fiber and 1.0 gram of insoluble fiber. What is the ratio of soluble to insoluble fiber?

27. At an urban community college, 20.2 out of 100 faculty members have doctorate degrees. What is the ratio of doctorate-holding faculty to total faculty?

28. A taxpayer pays .007 of a dollar for every dollar of assessed value of his house for school taxes. What is the ratio of tax to value of the house?

29. Water that has over a level of 1 part of a carcinogen (cancer-producing agent) per 1 billion parts of water should not be consumed. If a sample contains 0.006 parts of carcinogen per million parts of water, is it safe to drink?

30. An acceptable level of emission of lead from an incinerator is 2.788 grams per second. What is the ratio of grams to seconds?

Express each rate with the second part equal to 1. (See Example 5.)

31. Dr. Small drove 350 miles in 7 hours. At what rate did he drive his car?

32. Hank Diaz drove 320 miles and used 8 gallons of gasoline. What is the number of miles per gallon?

33. Four students pooled their money and bought a case of sodas. If there are 24 cans of soda in a case, what is the number of cans per student?

34. A study shows that 36,000 people rode in 20,000 cars. What is the average number of people per car?

Express each of the following as a unit price. If necessary, round answers to the nearest cent. (See Example 6.)

35. Six bananas cost $0.24. What is the cost per banana?

36. An 8-pound bag of ice costs $1.92. What is the cost per pound?

37. Eight sticks of cheese cost $3.04. What is the cost per stick?

38. A box containing 20 ounces of cereal costs $2.95. What is the cost per ounce?

39. A prescription medicine costs $11.19 for 30 capsules. What is the cost per capsule?

40. A bottle with 50 aspirin tablets costs $2.49. What is the cost per tablet?

41. A hotel charges $250 for a room with four people. What is the cost per person?

42. A family bought a season pass that permits two adults and three children to sit in the end zone at the local college football games. If the pass cost $140, what is the average cost per person?

Many food items come in various sizes and, generally speaking, the larger the size, the better the price. Determine the unit price (to the nearest tenth of a cent) for each of the different sizes and indicate how much less the larger size is per unit. All prices were found at Publix Supermarket in Longwood, Florida, on July 5, 2010. (See Example 6.)

43. A 50-ounce bottle of Woolite sells for $8.79 and a 75-ounce bottle sells for $10.79.

44. An 18-ounce box of Quaker Oats sells for $2.69 and a 42-ounce box for $4.55.

45. An 11.4-ounce box of Kellogg's Special K Blueberry cereal sells for $3.99 and a 15.5 ounce box for 4.99

46. A 12.8-ounce box of General Food's Cinnamon Toast Crunch sells for $3.49 and a 24.9-ounce box for $4.99.

47. An $8\frac{3}{4}$-ounce can of Del Monte Whole Kernel Corn sells for $.83, and a $15\frac{1}{4}$-ounce can sells for $1.29.

48. A 8.75-ounce can of Publix cream-style corn sells for $0.67 and a 14.75-ounce can sells for $0.77.

Challenge Exercises (49 and 50)

49. A can of mixed nuts contained 250 peanuts, 100 cashews, 50 Brazil nuts, 75 almonds, 50 filberts, and 100 pecans. What is the ratio of peanuts to cashews, to Brazil nuts, to almonds, to filberts, to pecans?

50. If $\frac{1}{4}$ cup of cereal weighs 28.35 grams, what is the ratio of grams per cup?

Writing Exercise

51. Name and give an example of at least five situations that can be described by a ratio.

Section 8.2   Proportions

OBJECTIVES *When you complete this section, you will be able to:*

A Verify proportions.

B Solve proportions.

C Solve application problems by using proportions.

PREREQUISITE SKILLS *Before beginning this section, you should be able to:*

a. Multiply whole numbers. (Section R.3)

b. Solve linear equations. (Sections 3.1–3.3)

c. Solve quadratic equations. (Section 5.8)

d. Multiply and divide decimals. (Section R.6)

e. Multiply rational numbers. (Section 6.3)

f. Solve a literal equation for one of its variables. (Section 3.8)

GETTING READY FOR SECTION 8.2

1. Find the product: $8 \cdot 36$

Solve the following linear equations:

2. $4x = 60$

3. $10x + 10 = 12x + 4$

4. Solve the quadratic equation $x^2 - x - 6 = 6$.

5. Find the product: $4.2(8)$

6. Find the quotient: $\dfrac{105}{2.5}$

7. Find the product: $12 \cdot \dfrac{5}{3}$

8. Solve $T = PV$ for P.

Introduction

In the last section, we looked at the general idea of ratio. In this section, we emphasize the idea of a ratio as a fraction. If ratios are equal, we make the following definition:

> **Definition** Proportion
>
> A **proportion** is an equation that states that two or more ratios are equal. In symbols, $\frac{a}{b} = \frac{c}{d}$ is a proportion.

OBJECTIVE Verifying proportions

We have worked with proportions before in Section 7.2 when we studied equivalent, or equal, fractions. Determining whether two ratios form a proportion is the same as determining whether two fractions are equal. We cross multiply, and if the cross products are the same, the two fractions are equal and we have a proportion.

Answers: Getting Ready: **1.** 288 **2.** $x = 15$ **3.** $x = 3$ **4.** $x = -3, 4$ **5.** 33.6 **6.** 42 **7.** 20 **8.** $P = \dfrac{T}{V}$

> **Rule: Verifying Proportions**
>
> If $\frac{a}{b} = \frac{c}{d}$, then $a \cdot d = b \cdot c$ if $b, d \neq 0$. It is also true that if $a \cdot d = b \cdot c$, then $\frac{a}{b} = \frac{c}{d}$.

Consequently, we use cross multiplication to determine whether two ratios can form a proportion.

Example 1 **Determine if the following fractions can form a proportion:**

a. $\frac{4}{5}$ and $\frac{16}{20}$

$\frac{4}{5} \diagdown \frac{16}{20}$ Cross multiply.

$4(20) \overset{?}{=} 5(16)$ Multiply.

$80 = 80$ The ratios are equal and can form a proportion. So, $\frac{4}{5} = \frac{16}{20}$.

b. $\frac{2}{3}$ and $\frac{9}{12}$

$\frac{2}{3} \diagdown \frac{9}{12}$ Cross multiply.

$2(12) \overset{?}{=} 3(9)$ Multiply.

$24 \neq 27$ The ratios are not equal and cannot form a proportion. So, $\frac{2}{3} \neq \frac{9}{12}$.

PRACTICE EXERCISE 1

Determine whether the following fractions can form a proportion:

a. $\frac{8}{9}$ and $\frac{30}{36}$ **b.** $\frac{16}{28}$ and $\frac{4}{7}$

If you need more practice, do the following Additional Practice Exercises.

Additional Practice Exercise 1   Determine whether the following fractions can form a proportion:

a. $\frac{3}{4}$ and $\frac{15}{30}$ **b.** $\frac{18}{21}$ and $\frac{12}{14}$

OBJECTIVE **B** Solving proportions

If one of the quantities of a proportion is unknown, we may solve the proportion for that quantity by cross multiplying and solving the resulting equation.

> **Procedure: Solving Proportions**
>
> When solving a proportion, perform the following steps:
>
> **1.** If desired, reduce all fractions to lowest terms.
>
> **2.** Cross multiply the proportion.
>
> **3.** Solve the resulting equation.

Answers: Practice Exercise 1: a. no **b.** yes   **Additional Practice 1: a.** no **b.** yes

Example 2 | Solve the following proportions for x:

a. $\dfrac{x}{20} = \dfrac{9}{12}$ Reduce $\dfrac{9}{12}$ to lowest terms.

$\dfrac{x}{20} = \dfrac{3}{4}$ Cross multiply.

$4x = 20(3)$ $20(3) = 60$.

$4x = 60$ Divide both sides of the equation by 4.

$x = 15$ Solution.

b. $\dfrac{4}{x} = \dfrac{7}{9}$ Cross multiply.

$4(9) = 7x$ Multiply 4 and 9.

$36 = 7x$ Divide both sides by 7.

$\dfrac{36}{7} = x$ Solution.

c. $\dfrac{2x + 2}{4} = \dfrac{3x + 1}{5}$ Cross multiply.

$5(2x + 2) = 4(3x + 1)$ Simplify both sides of the equation.

$10x + 10 = 12x + 4$ Subtract $10x$ from both sides of the equation.

$10x - 10x + 10 = 12x - 10x + 4$ Simplify both sides.

$10 = 2x + 4$ Subtract 4 from both sides of the equation.

$10 - 4 = 2x + 4 - 4$ Simplify both sides of the equation.

$6 = 2x$ Divide both sides of the equation by 2.

$3 = x$ Solution.

d. $\dfrac{x + 2}{3} = \dfrac{2}{x - 3}$ Cross multiply.

$(x + 2)(x - 3) = 3 \cdot 2$ Multiply.

$x^2 - x - 6 = 6$ Subtract 6 from both sides of the equation.

$x^2 - x - 12 = 0$ Factor.

$(x - 4)(x + 3) = 0$ Set each factor equal to 0.

$x - 4 = 0, x + 3 = 0$ Solve each linear equation.

$x = 4, x = -3$ Therefore, 4 and -3 are the solutions.

NOTE Proportions are checked in the same manner that any equation is checked.

PRACTICE EXERCISE 2

Solve the following proportions:

a. $\dfrac{12}{16} = \dfrac{y}{20}$ **b.** $\dfrac{y}{3} = \dfrac{7}{11}$ **c.** $\dfrac{4}{3} = \dfrac{2x + 2}{2x - 1}$ **d.** $\dfrac{x - 4}{2} = \dfrac{8}{x + 2}$

If you need more practice, do the following Additional Practice Exercises.

Additional Practice Exercise 2   Solve the following proportions:

a. $\dfrac{6}{y} = \dfrac{12}{8}$ **b.** $\dfrac{3}{2} = \dfrac{8}{v}$ **c.** $\dfrac{3}{2x + 1} = \dfrac{5}{4x - 1}$ **d.** $\dfrac{4}{x + 5} = \dfrac{x - 3}{5}$

Proportions may have fractions, decimals, or only variables. They are solved in exactly the same manner as with whole numbers.

Example 3 | Solve the following proportions:

a. $\dfrac{\frac{1}{2}}{5} = \dfrac{x}{40}$ Cross multiply.

$\dfrac{1}{2}(40) = 5x$ Multiply $\dfrac{1}{2}$ and 40.

$20 = 5x$ Divide both sides of the equation by 5.

$4 = x$ Solution.

b. $\dfrac{y}{4.2} = \dfrac{8}{3}$ Cross multiply.

$3y = 4.2(8)$ Multiply 4.2 and 8.

$3y = 33.6$ Divide both sides of the equation by 3.

$y = 11.2$ Solution.

Answers: Practice Exercise 2: a. $y = 15$ **b.** $y = \dfrac{21}{11}$ **c.** $x = 5$ **d.** $x = 6, -4$ **Additional Practice 2: a.** $y = 4$ **b.** $v = \dfrac{16}{3}$ **c.** $x = 4$ **d.** $x = -7, 5$

Solve each of the following for the indicated variable:

c. $\dfrac{C}{D} = \pi; C$

d. $\dfrac{T}{P} = V; P$

Solution

$\dfrac{C}{D} = \pi$ Think of π as $\dfrac{\pi}{1}$.

$\dfrac{C}{D} = \dfrac{\pi}{1}$ Cross multiply.

$C = D \cdot \pi$ Solution.

Solution

$\dfrac{T}{P} = V$ Think of V as $\dfrac{V}{1}$.

$\dfrac{T}{P} = \dfrac{V}{1}$ Cross multiply.

$T = P \cdot V$ Divide both sides of the equation by V.

$\dfrac{T}{V} = P$ Solution.

PRACTICE EXERCISE 3

Solve each of the following proportions:

a. $\dfrac{2}{\frac{3}{4}} = \dfrac{8}{x}$

b. $\dfrac{7}{2.5} = \dfrac{z}{15}$

Solve for the indicated variable:

c. $D = \dfrac{W}{V}; W$

d. $i = \dfrac{V}{R}; R$

If you need more practice, do the following Additional Practice Exercises.

Additional Practice Exercise 3   Solve the following proportions:

a. $\dfrac{y}{12} = \dfrac{\frac{5}{3}}{10}$

b. $\dfrac{3.2}{r} = \dfrac{4}{5}$

Solve for the indicated variable:

c. $\dfrac{A}{L} = W; A$

d. $\dfrac{i}{p} = r; p$

OBJECTIVE **C** Solving application problems by using proportions

There are many applications of proportions. There is more than one correct way of setting up a proportion from an application problem. For example, all of the following are equivalent proportions: $\frac{a}{b} = \frac{c}{d}, \frac{a}{c} = \frac{b}{d}, \frac{d}{b} = \frac{c}{a}, \frac{b}{a} = \frac{d}{c}$. Note that if we cross multiply each of these, we get $ad = bc$.

Answers: Practice Exercise 3: a. $x = 3$ **b.** $z = 42$ **c.** $W = DV$ **d.** $R = \dfrac{V}{i}$

Additional Practice 3: a. $y = 2$ **b.** $r = 4$ **c.** $A = LW$ **d.** $p = \dfrac{i}{r}$

Example 4 **Solve each of the following by using a proportion:**

a. An automobile travels 224 miles on 8 gallons of gasoline. How far can it travel on 15 gallons?

Solution

We let x equal the number of miles the car can travel on 15 gallons. The key to the solution is the assumption that the rate of consumption in the first ratio will be the same for the second ratio. One method of solution is to set up the first ratio by comparing miles to gallons. Then we set up the second ratio by keeping the same pattern of comparing miles to gallons. In words,

$$\frac{\text{First number of miles}}{\text{First number of gallons}} = \frac{\text{Second number of miles}}{\text{Second number of gallons}} \qquad \text{Substitute.}$$

$$\frac{224 \text{ miles}}{8 \text{ gallons}} = \frac{x \text{ miles}}{15 \text{ gallons}} \qquad \text{Drop units.}$$

$$\frac{224}{8} = \frac{x}{15} \qquad \text{Reduce } \frac{224}{8}.$$

$$\frac{28}{1} = \frac{x}{15} \qquad \text{Cross multiply.}$$

$$28(15) = 1(x) \qquad \text{Multiply.}$$

$$420 = x \qquad \text{Therefore, the car can travel 420 miles on 15 gallons of gasoline.}$$

> **Note**
>
> Other proportions that could also be used are $\frac{15}{8} = \frac{x}{224}$, $\frac{224}{x} = \frac{8}{15}$, and $\frac{8}{224} = \frac{15}{x}$. Notice that all of these give the same expression when we cross multiply.

b. The state wildlife department wants to estimate the number of deer in a game preserve. In September, it captures a sample of 34 deer, tags them, and releases them. Later, it captures a sample of 45 deer and finds that 17 are tagged. Estimate the number of deer in the preserve.

Solution

Let y be the number of deer in the preserve. We assume that the ratio of tagged deer to total deer in the sample is the same as the ratio of tagged deer to total deer in the preserve. We know there are 34 tagged deer in the preserve. In words,

$$\frac{\text{Number of tagged deer in sample}}{\text{Number of deer in sample}} = \frac{\text{Number of tagged deer in preserve}}{\text{Number of deer in preserve}}$$

$$\frac{17 \text{ tagged}}{45 \text{ total}} = \frac{34 \text{ tagged}}{y \text{ total}} \qquad \text{Drop labels.}$$

$$\frac{17}{45} = \frac{34}{y} \qquad \text{Cannot simplify. Cross multiply.}$$

$$17y = 45(34) \qquad \text{Multiply 45 and 34.}$$

$$17y = 1530 \qquad \text{Divide both sides by 17.}$$

$$y = 90 \qquad \text{Therefore, there are about 90 deer in the preserve.}$$

> **Note**
>
> Other proportions that could have been used are $\frac{y}{45} = \frac{34}{17}$, $\frac{17}{34} = \frac{45}{y}$, and $\frac{45}{17} = \frac{y}{34}$. Notice that all of them give the same expression when we cross multiply.

c. The distance between two cities is 150 miles. The scaled distance between the cities on a map is $1\frac{1}{2}$ inches. If the distance between two other cities is 900 miles, what is the scaled distance between them on the map?

Solution

Let d represent the scaled distance between the two cities on the map. The ratio of miles between the cities to inches on the map will remain the same for all distances. Therefore, we set up the ratio of miles between cities to inches on the map as follows:

In words,

$$\frac{\text{First number of miles}}{\text{First number of inches}} = \frac{\text{Second number of miles}}{\text{Second number of inches}}$$ Substitute.

$$\frac{150 \text{ miles}}{1\frac{1}{2} \text{ inches}} = \frac{900 \text{ miles}}{d \text{ inches}}$$ Drop units.

$$\frac{150}{1\frac{1}{2}} = \frac{900}{d}$$ Convert mixed number to improper fraction or decimal.

$$\frac{150}{\frac{3}{2}} = \frac{900}{d}$$ Cross multiply.

$$150d = \frac{3}{2}(900)$$ Multiply $\frac{3}{2}$ and 900.

$$150d = 1350$$ Divide both sides by 150.

$$d = 9$$ Therefore, the scaled distance between the cities on the map is 9 inches.

Note

Other proportions that could also be used are $\frac{d}{1\frac{1}{2}} = \frac{900}{150}$, $\frac{150}{900} = \frac{1\frac{1}{2}}{d}$ and $\frac{1\frac{1}{2}}{150} = \frac{d}{900}$. Notice that all of these give the same expression when we cross multiply.

PRACTICE EXERCISE 4

Solve each of the following:

a. A chef knows that he can serve eight people with 3 pounds of fish fillets. How many pounds of fish fillets does he need to serve 32 people?

b. A manufacturer tested 400 randomly chosen cell phones as they came off the production line. If she found 15 defective cell phones, how many defectives should she expect to find in 2000 cell phones?

c. On a map, $1\frac{1}{4}$ inches represents 100 miles on the surface of Earth. What is the distance on Earth between two locations that are 15 inches apart on the map?

If you need more practice, do the following Additional Practice Exercises.

Additional Practice Exercise 4

a. A recipe for ambrosia for 6 people calls for (among other things) 2 cups of orange slices. How many cups of orange slices are needed to make enough ambrosia for 21 people?

b. A wildlife manager wants to estimate the number of wild hogs in his hunting lease. He captures 40 hogs, tags them, and releases them. Later he captures 84 hogs, of which 21 are tagged. Find the number of wild hogs in his hunting lease.

c. On a map, $2\frac{1}{4}$ inches represents 450 miles. What is the distance between two points that are 4 inches apart on the map?

Answers: Practice Exercise 4: a. 12 pounds **b.** 75 cell phones **c.** 1200 miles   **Additional Practice 4: a.** 7 cups **b.** 160 hogs **c.** 800 miles

Exercise Set 8.2   For Extra Help   MyMathLab®

Determine whether the following fractions can form a proportion. (See Example 1.)

1. $\dfrac{4}{5}$ and $\dfrac{12}{15}$

2. $\dfrac{20}{24}$ and $\dfrac{25}{30}$

3. $\dfrac{2}{3}$ and $\dfrac{8}{12}$

4. $\dfrac{5}{6}$ and $\dfrac{4}{3}$

5. $\dfrac{4}{9}$ and $\dfrac{20}{45}$

6. $\dfrac{5}{11}$ and $\dfrac{7}{9}$

7. $\dfrac{7}{12}$ and $\dfrac{17\frac{1}{2}}{30}$

8. $\dfrac{3}{8}$ and $\dfrac{15}{40}$

Solve the following proportions. (See Examples 2 and 3.)

9. $\dfrac{4}{5} = \dfrac{12}{x}$

10. $\dfrac{6}{5} = \dfrac{24}{y}$

11. $\dfrac{4}{6} = \dfrac{t}{18}$

12. $\dfrac{9}{5} = \dfrac{y}{15}$

13. $\dfrac{y}{8} = \dfrac{4}{7}$

14. $\dfrac{22}{33} = \dfrac{n}{44}$

15. $\dfrac{13}{6} = \dfrac{t}{42}$

16. $\dfrac{3}{7} = \dfrac{9}{y}$

17. $\dfrac{28}{16} = \dfrac{49}{p}$

18. $\dfrac{56}{16} = \dfrac{q}{10}$

19. $\dfrac{9}{w} = \dfrac{22.5}{1.5}$

20. $\dfrac{y}{1.8} = \dfrac{.6}{3.6}$

21. $\dfrac{y}{14} = \dfrac{\frac{3}{7}}{2}$

22. $\dfrac{\frac{6}{5}}{8} = \dfrac{r}{20}$

23. $\dfrac{3\frac{1}{2}}{x} = \dfrac{14}{12}$

24. $\dfrac{9}{m} = \dfrac{6}{4\frac{2}{3}}$

25. $\dfrac{.6}{t} = \dfrac{5}{8}$

26. $\dfrac{.7}{5} = \dfrac{n}{3}$

27. $\dfrac{14}{3.5} = \dfrac{w}{5}$

28. $\dfrac{v}{8} = \dfrac{13}{2.6}$

29. $\dfrac{2}{3} = \dfrac{8}{x+8}$

30. $\dfrac{3}{4} = \dfrac{9}{x+5}$

31. $\dfrac{x+5}{6} = \dfrac{7}{3}$

32. $\dfrac{x+4}{20} = \dfrac{3}{4}$

33. $\dfrac{3}{4} = \dfrac{x+1}{x+3}$

34. $\dfrac{2}{3} = \dfrac{x+1}{x+6}$

35. $\dfrac{3}{2x} = \dfrac{5}{3x+2}$

36. $\dfrac{4}{3x-3} = \dfrac{7}{4x+1}$

37. $\dfrac{4x-4}{8} = \dfrac{2x+1}{5}$

38. $\dfrac{7x-1}{8} = \dfrac{2x+4}{3}$

39. $\dfrac{x-3}{2} = \dfrac{5}{x+6}$

40. $\dfrac{x+5}{7} = \dfrac{1}{x-1}$

41. $\dfrac{5}{2x+3} = \dfrac{x-5}{3}$

42. $\dfrac{2}{3x+2} = \dfrac{x-3}{17}$

Solve for the indicated variable. (See Examples 3c and 3d.)

43. $P = \dfrac{W}{t}$; W

44. $\dfrac{q}{t} = i$; t

45. $\dfrac{PV}{T} = K$; P

46. $P = \dfrac{F}{LW}$; F

Solve each of the following. (See Example 4.)

47. A truck travels 440 miles in 11 hours. How far will the truck travel in 7 hours?

48. An airplane uses 350 pounds of fuel to fly 1400 miles. How much fuel will it use to fly 3500 miles?

49. In 8 months, Joel can read 5 novels. How many novels can he read in 12 months?

50. A machine can produce 160 toys in 3 hours. How long will it take to produce 400 toys?

51. A baseball player got 6 hits in 14 innings of play. How many innings must he play to get 33 hits?

52. Martha can do 32 mathematics exercises in 15 minutes. How long will it take her to do 96 exercises?

53. Two rivers that are 270 miles apart are drawn as 3 inches apart on a map. How many inches apart on the map will two locations be that are 675 miles apart on Earth?

54. Two cities that are 600 miles apart on Earth's surfaces are 6 inches apart on a map. How many inches apart on the map are two cities that are 75 miles apart on Earth's surface?

55. The state game commission wants to estimate how many trout there are in a lake. It releases 145 tagged trout into the lake. Six months later, it returns and captures 160 trout, 58 of which are tagged. Estimate the number of trout in the lake.

56. The wildlife commission wants to determine how many alligators are in a swamp. Conservation officers capture, tag, and release 33 alligators. One year later, they capture 60 alligators and find that 12 are tagged. Approximately how many alligators are in the swamp?

57. A model airplane is built at a scale of 1 : 40. If a wing on the model is six inches long, how long is the wing of the airplane in feet?

58. A model car is built at a scale of 1 : 15. If the bumper of the model car is 4 inches long, how long is the bumper of the car in feet?

59. If a tile store will install 23 square feet of tile for $92, how much would it charge to install 62 square feet?

60. If a flooring company can install 50 square feet of hardwood flooring for $1200, how much would it charge to install 80 square feet?

Challenge Exercises (61–63)

61. One of the professors at Urban University can walk 3 miles in 39 minutes. Assuming that he walks at the same pace, how long will it take him to walk 5 miles?

62. A physician can see 40 patients in 7 hours. How many can he see in 35 hours (one week)? At the rate of $60 per patient, how much can he make in 50 weeks (one year)?

63. A 44-pound bag of fertilizer will cover 5000 square feet of lawn. How many pounds of fertilizer are needed to cover 17,500 square feet of lawn? If the fertilizer costs $23.95 a bag, how much will it cost to fertilize the 17,500 square feet? Assume that it is not possible to buy a fractional part of a bag.

Writing Exercises (64–66)

64. What is the underlying assumption on which all of the preceding proportion exercises are based?

65. Name three types of situations where proportions are used.

66. Explain why $\frac{a}{b} = \frac{c}{d}$ and $\frac{a}{c} = \frac{b}{d}$ are essentially the same proportion.

Writing Exercises or Group Projects (67 and 68)

67. Interview science teachers or students, nursing teachers or students, and other professionals and make a list of five situations where proportions are used in their fields/professions.

68. There is a special ratio called the golden ratio that often appears in art, architecture, mathematics, and many other fields. Go to the library, research the topic, and write a one-page report on the golden ratio.

Section 8.3   Percent

OBJECTIVES *When you complete this section, you will be able to:*

- **A** Convert percent notation to fractions.
- **B** Convert percent notation to decimals.
- **C** Convert decimals to percent notation.
- **D** Convert fractions to percent notation.
- **E** Add and subtract in percent notation.

PREREQUISITE SKILLS *Before beginning this section, you should be able to:*

- **a.** Reduce fractions. (Section 6.1)
- **b.** Convert a mixed number into an improper fraction. (Section R.4)
- **c.** Multiply rational numbers. (Section 6.3)
- **d.** Divide whole numbers and decimals. (Sections R.3 and R.6)
- **e.** Add whole numbers and decimals. (Sections R.2 and R.5)

GETTING READY FOR SECTION 8.3

1. Reduce $\dfrac{8}{10,000}$ to lowest terms.

2. Change $5\dfrac{4}{7}$ into an improper fraction.

Find the following products:

3. $\dfrac{17}{2} \cdot \dfrac{1}{100}$

4. $12.6(.01)$

5. Divide $\dfrac{400}{9}$ and round the answer to the nearest hundredth.

Add or subtract the following:

6. $9.3 - 6$

7. $5 + 9\dfrac{1}{2}$

Introduction

Throughout recorded history, business people and governments have looked for a method that they could use to calculate profits, taxes, and so on. The most common method is the use of the fraction $\frac{1}{100}$. Over the years, the number of hundredths has become known as "percent." The word *percent* comes from the Latin phrase *per centum*, or per hundred, and has a special symbol, %.

> **Definition** Percent Notation
>
> **Percent notation** is a special way of writing a fraction with a denominator of 100. For any number n, $n\% = \frac{n}{100}$.

Answers: Getting Ready: 1. $\dfrac{1}{1250}$ **2.** $\dfrac{39}{7}$ **3.** $\dfrac{17}{200}$ **4.** $.126$ **5.** 44.44 **6.** 3.3 **7.** $14\dfrac{1}{2}$

OBJECTIVE A Converting percent notation to fractions

This definition means that all we have to do to change a number in percent notation into a fraction is write it as a fraction with a denominator of 100 and simplify. For example, $17\% = \frac{17}{100}$, and $25\% = \frac{25}{100} = \frac{1}{4}$. However, when converting something like $5\frac{4}{7}\%$ or $.06\%$ to a fraction, the arithmetic becomes very messy. For example,

$$5\frac{4}{7}\% = \frac{5\frac{4}{7}}{100} = \frac{\frac{39}{7}}{100} = \frac{39}{7} \div 100 = \frac{39}{7} \cdot \frac{1}{100} = \frac{39}{700}.$$

An alternative method that leads to nicer arithmetic involves thinking of $x\%$ as $x \cdot 1\%$. By definition, $1\% = \frac{1}{100}$. Consequently, $x\% = x \cdot 1\% = x \cdot \frac{1}{100}$. Since we are converting a percent to a fraction, we replace the % sign with the fraction $\frac{1}{100}$.

Note

Rather than thinking of $x\%$ as $x \cdot 1\%$ and then replacing 1% with $\frac{1}{100}$, it is a common practice to replace the % sign with $\frac{1}{100}$, which amounts to the same thing.

Procedure: Converting from Percent to a Fraction

To convert from percent notation to a fraction:

a. Think of $x\%$ as $x \cdot 1\%$.

b. Replace 1% with $\frac{1}{100}$.

c. Multiply.

In using the preceding technique, we also need to recall how to change decimals into fractions.

Example 1 Convert each of the following percents to fractions:

a. $75\% =$ Think of 75% as $75 \cdot 1\%$ and replace 1% with $\frac{1}{100}$.

$75 \cdot \dfrac{1}{100} =$ Multiply. Remember, think of 75 as $\frac{75}{1}$.

$\dfrac{75}{100} =$ Write numerator and denominator in prime factorizations and divide by common factors.

$\dfrac{3 \cdot 5 \cdot 5}{2 \cdot 2 \cdot 5 \cdot 5} =$ Multiply the remaining factors.

$\dfrac{3}{4}$ Hence, $75\% = \frac{3}{4}$.

b. $.08\% =$ Think of .08% as $.08 \cdot 1\%$ and replace 1% with $\frac{1}{100}$.

$.08 \cdot \dfrac{1}{100} =$ Change .08 into the fraction $\frac{8}{100}$.

$\dfrac{8}{100} \cdot \dfrac{1}{100} =$ Write in terms of prime factors.

$\dfrac{2 \cdot 2 \cdot 2}{2 \cdot 2 \cdot 5 \cdot 5} \cdot \dfrac{1}{2 \cdot 2 \cdot 5 \cdot 5} =$ Divide by common factors.

$\dfrac{2 \cdot 2 \cdot 2}{2 \cdot 2 \cdot 5 \cdot 5} \cdot \dfrac{1}{2 \cdot 2 \cdot 5 \cdot 5} =$ Multiply the remaining factors.

$\dfrac{1}{1250}$ Hence, $.08\% = \frac{1}{1250}$.

c. 8.5% Think of 8.5% as $8.5 \cdot 1\%$ and replace 1% with $\frac{1}{100}$.

$8.5 \cdot \dfrac{1}{100}$ Change 8.5 into $8\frac{5}{10}$.

$8\dfrac{5}{10} \cdot \dfrac{1}{100}$ Reduce $\frac{5}{10}$ to lowest terms.

$8\dfrac{1}{2} \cdot \dfrac{1}{100}$ Change $8\frac{1}{2}$ into an improper fraction.

$\dfrac{17}{2} \cdot \dfrac{1}{100}$ Multiply.

$\dfrac{17}{200}$ Hence, $8.5\% = \frac{17}{200}$.

d. $5\frac{4}{7}\% =$ Think of $5\frac{4}{7}\%$ as $5\frac{4}{7} \cdot 1\%$ and change 1% to $\frac{1}{100}$.

$5\dfrac{4}{7} \cdot \dfrac{1}{100} =$ Rewrite $5\frac{4}{7}$ as an improper fraction.

$\dfrac{39}{7} \cdot \dfrac{1}{100} =$ Multiply.

$\dfrac{39}{700}$ Hence, $5\frac{4}{7}\% = \frac{39}{700}$.

PRACTICE EXERCISE 1

Convert each of the following percents to fractions:

a. 90%   **b.** .09%   **c.** 6.8%   **d.** $8\frac{6}{7}$%

If you need more practice, do the following Additional Practice Exercises.

Additional Practice Exercise 1   Convert each of the following percents to fractions:

a. 62%   **b.** .75%   **c.** 2.3%   **d.** $5\frac{7}{12}$%

Before converting percents to decimals, we need to recall from Chapter R the procedure for converting fractions to decimals. Remember that we divide the denominator into the numerator and, if necessary, round the answer to a specified number of decimal places. For example, we write $\frac{5}{8}$ as a decimal as follows:

$8\overline{)5}$ Write the decimal point behind the 5 and bring the decimal point
              up to the quotient. Write zeroes after the 5 as needed.

$$
\begin{array}{r}
.625 \\
8\overline{)5.000} \\
-48 \\
\hline
20 \\
-16 \\
\hline
40 \\
-40 \\
\hline
0
\end{array}
$$
    Therefore, $\dfrac{5}{8} = .625.$

If we were to write $\frac{5}{6}$ as a decimal, the division would never terminate, so it is necessary to round the answer to a specified number of decimal places. If we wanted to write $\frac{5}{6}$ as a decimal rounded to the nearest hundredth, we would proceed as follows:

Note

Some conversions to repeating decimals occur so often in applications that they should be memorized to help computations. Examples of these are $.\overline{3} = \frac{1}{3}$ and $.\overline{6} = \frac{2}{3}$, $.\overline{1} = \frac{1}{9}$, $.\overline{2} = \frac{2}{9}$, $.\overline{3} = \frac{3}{9}$, $.\overline{4} = \frac{4}{9}$, and so on.

$6\overline{)5}$ Write the decimal point behind the 5 and bring the decimal point
              up to the quotient. Write zeroes after the 5 as needed.

$$
\begin{array}{r}
.833 \\
6\overline{)5.000} \\
-48 \\
\hline
20 \\
-18 \\
\hline
20 \\
-18 \\
\hline
2
\end{array}
$$
    Since we want to round to the nearest hundredth,
    we need to see the thousandths digit. Round .833
    to the nearest hundredth. (See Section R.6.)

    Therefore, $\dfrac{5}{6} \approx .83$ to the nearest hundredth.

OBJECTIVE   Converting percent notation to decimals

We can also use the definition of percent to change a percent into a decimal. However, the arithmetic is much nicer if we use a method similar to the one used to convert percents to fractions. We use the fact that $1\% = \frac{1}{100} = .01$. So, when changing percent to decimals, we think of $x\%$ as being $x \cdot 1\%$ and replace 1% with $.01$. Since we are converting a percent to a decimal, we replace the % sign with the decimal $.01$.

Answers: Practice Exercise 1: a. $\dfrac{9}{10}$ **b.** $\dfrac{9}{10{,}000}$ **c.** $\dfrac{17}{250}$ **d.** $\dfrac{31}{350}$ **Additional Practice 1: a.** $\dfrac{31}{50}$ **b.** $\dfrac{3}{400}$ **c.** $\dfrac{23}{1000}$ **d.** $\dfrac{67}{1200}$

> ### Procedure: Converting from Percent to Decimals
>
> To convert from percent notation to a decimal:
>
> **a.** Think of $x\%$ as $x \cdot 1\%$.
>
> **b.** Replace 1% with .01.
>
> **c.** Multiply.

Example 2 | **Convert the given percents to decimals. If necessary, round answers to the nearest hundredths.**

a. 67%   Think of 67% as 67 · 1% and replace 1% with .01.

67 · .01   Multiply.

.67   Hence, 67% = .67.

b. 180% =   Think of 180% as 180 · 1% and replace 1% with .01.

180 · .01 =   Multiply.

1.8   Hence, 180% = 1.8.

c. 12.6% =   Think of 12.6% as 12.6 · 1% and replace 1% with .01.

12.6 · .01 =   Multiply.

.126   Hence, 12.6% = .126.

d. $4\frac{5}{7}\% =$ Think of $4\frac{5}{7}\%$ as $4\frac{5}{7} \cdot 1\%$ and replace 1% with .01.

$4\frac{5}{7} \cdot .01 =$ Change $\frac{5}{7}$ into a decimal.

$(4.714285\ldots)(.01) =$ Multiply.

.04714285 =   Round to the nearest hundredth.

.05   Hence, $4\frac{5}{7}\% = .05$ to the nearest hundredth.

PRACTICE EXERCISE 2

Convert the given percents to decimals. If necessary, round to the nearest thousandth.

a. 35%   **b.** 9%   **c.** 55.7%   **d.** $11\frac{8}{17}\%$

If you need more practice, do the following Additional Practice Exercises.

Additional Practice Exercise 2   Convert the given percents to decimals:

a. 47%   **b.** 8.4%   **c.** 44.3%   **d.** $\frac{3}{4}\%$

OBJECTIVE Converting decimals to percent notation

If we were to write 100% as a decimal, we would have 100% = 100 · .01 = 1. Thus 100% = 1. We also know that 1 is the identity for multiplication, because the product of any number and 1 is identical to the number itself. Consequently, to convert a decimal to a percent, we will write the decimal as the product of itself and 1 and then change 1 to 100% and multiply. We summarize as follows:

Answers: **Practice Exercise 2:** **a.** .35 **b.** .09 **c.** .557 **d.** .115   **Additional Practice 2:** **a.** .47 **b.** .084 **c.** .443 **d.** .008

Procedure: Converting Decimals to Percents

To convert a decimal into percent notation:

a. Write the decimal as the product of itself and 1.

b. Change 1 to 100%.

c. Multiply and, if necessary, round to the specified number of places.

Example 3   Convert the following decimals to percent notation:

a. .17 =   Write .17 as .17 · 1.
.17 · 1 =   Change 1 into 100%.
.17 · 100% =   Multiply.
17%   Hence, .17 = 17%.

b. .008 =   Write .008 as .008 · 1.
.008 · 1 =   Change 1 into 100%.
.008 · 100% =   Multiply.
.8%   Hence, .008 = .8%.

c. 2.54 =   Write 2.54 as 2.54 · 1.
2.54 · 1 =   Change 1 into 100%.
2.54 · 100% =   Multiply.
254%   Hence, 2.54 = 254%.

d. .333 . . . =   Think of .333 . . . as (.333 . . .)(1).
(.333 . . .)(1) =   Change 1 into 100%.
(.333 . . .)(100%) =   Multiply.
33.333 . . . % =   Replace .333 . . . with fractional equivalent, $\frac{1}{3}$.

$33\frac{1}{3}\%$ Hence, .333 . . . = $33\frac{1}{3}\%$.

e. 8   Think of 8 as 8 · 1.
8 · 1   Change 1 into 100%.
8 · 100%   Multiply.
800%   Hence, 8 = 800%.

PRACTICE EXERCISE 3

Convert the following decimals to percent notation:
a. .87   **b.** .006   **c.** 2.53   **d.** 5.333 . . .   **e.** 9

If you need more practice, do the following Additional Practice Exercises.

Additional Practice Exercise 3   Convert the following decimals to percent notation:
a. .95   **b.** .055   **c.** 1.64   **d.** 4.666 . . .   **e.** 4

OBJECTIVE   Converting fractions to percent notation

To convert a fraction to a percent, we use the same procedure used to convert a decimal to a percent. If we change 100% to a fraction, we would have $100\% = 100 \cdot \frac{1}{100} = \frac{100}{100}$. Therefore, 100% as a fraction is 1. Consequently, we will write the fraction as the product of itself and 1, and then change 1 to 100% and multiply.

Answers: Practice Exercise 3: a. 87% **b.** .6% **c.** 253% **d.** $533\frac{1}{3}\%$ **e.** 900%

Additional Practice 3: a. 95% **b.** 5.5% **c.** 164% **d.** $466\frac{2}{3}\%$ **e.** 400%

> **Procedure: Converting Fractions to Percents:**
>
> To convert a fraction to percent notation:
>
> **a.** Write the fraction as the product of itself and 1.
>
> **b.** Change 1 to 100%.
>
> **c.** Multiply and, if necessary, round to a specified number of decimal places.

Example 4 Convert the following fractions to percent notation:

a. $\dfrac{5}{8} =$ Write $\dfrac{5}{8}$ as $\dfrac{5}{8} \cdot 1$.

$\dfrac{5}{8} \cdot 1 =$ Replace 1 with 100%.

$\dfrac{5}{8} \cdot 100\% =$ Multiply.

$\dfrac{500}{8}\% =$ Divide.

$62\dfrac{1}{2}$ or 62.5%   Therefore, $\dfrac{5}{8} = 62\dfrac{1}{2}\%$ or 62.5%.

b. $\dfrac{2}{3} =$ Write $\dfrac{2}{3}$ as $\dfrac{2}{3} \cdot 1$.

$\dfrac{2}{3} \cdot 1 =$ Replace 1 with 100%.

$\dfrac{2}{3} \cdot 100\% =$ Multiply.

$\dfrac{200}{3}\% =$ Divide.

$66\dfrac{2}{3}$ or 66.66 . . . %   Hence, $\dfrac{2}{3} = 66\dfrac{2}{3}\%$.

c. $\dfrac{4}{9} =$ Think of $\dfrac{4}{9}$ as $\dfrac{4}{9} \cdot 1$.

$\dfrac{4}{9} \cdot 1 =$ Replace 1 with 100%.

$\dfrac{4}{9} \cdot 100\% =$ Multiply.

$\dfrac{400}{9}\% =$ Divide.

$44\dfrac{4}{9}$ or 44.4444 . . . %   Hence, $\dfrac{4}{9} = 44\dfrac{4}{9}\%$.

d. $\dfrac{3}{7} =$ Think of $\dfrac{3}{7}$ as $\dfrac{3}{7} \cdot 1$.

$\dfrac{3}{7} \cdot 1 =$ Replace 1 with 100%.

$\dfrac{3}{7} \cdot 100\% =$ Multiply.

$\dfrac{300}{7}\% =$ Divide.

$42\dfrac{6}{7}$ or 42.86%   Hence, $\dfrac{3}{7} = 42\dfrac{6}{7}\%$.

PRACTICE EXERCISE 4

Convert the given fractions to percent notation. If necessary, round to the nearest hundredth of a percent.

a. $\dfrac{7}{8}$ **b.** $\dfrac{1}{6}$ **c.** $\dfrac{2}{11}$ **d.** $\dfrac{5}{13}$

If you need more practice, do the following Additional Practice Exercises.

Additional Practice Exercise 4 Convert the given fractions to percent notation. If necessary, round to the nearest hundredth of a percent.

a. $\dfrac{3}{8}$ **b.** $\dfrac{5}{9}$ **c.** $\dfrac{1}{8}$ **d.** $\dfrac{5}{6}$

Answers: Practice Exercise 4: a. 87.5% **b.** 16.67% **c.** 18.18% **d.** 38.46% **Additional Practice 4: a.** 37.5% **b.** 55.56% **c.** 12.5% **d.** 83.33%

OBJECTIVE Adding and subtracting in percent notation

Since percent represents hundredths, adding and subtracting percents is the same as adding and subtracting fractions whose common denominator is 100. In doing so, we add the numerators and put the sum over the common denominator of 100. Since the amount of percent is the numerator, we simply add and subtract the percents as with any other rational numbers. For example, $5\% + 23\% = \frac{5}{100} + \frac{23}{100} = \frac{5+23}{100} = \frac{28}{100} = 28\%$. So, $5\% + 23\% = (5 + 23)\% = 28\%$.

Example 5   **Perform each indicated operation:**

a. $9.3\% - 6\%$

Solution

$9.3\% - 6\% = 3.3\%$ Subtract percents as you would subtract rational numbers.

b. $5\% + 9\frac{1}{2}\%$

Solution

$5\% + 9\frac{1}{2}\% = 14\frac{1}{2}\%$ Add percents as you would add rational numbers.

c. Arthur Gomez paid 14% of his salary in federal income tax last year. Since he lives in New York City, he also paid a 2% income tax to the city. If he made $45,000 last year, what percent of his salary did he pay in income taxes?

Solution

Ignore the $45,000.   It has nothing to do with the question asked.

$14\% + 2\% = 16\%$ Since both the percents are calculated on the same thing, his salary, we can add them. Hence, he spent 16% on income taxes.

d. Jenny Song pays 25% of her monthly income for housing and 18% for food. How much greater is the percent of her income spent on housing than on food?

Solution

Percent spent on housing minus the percent spent on food.   Since both percents are calculated on the same thing, the income, we can subtract them.

$25\% - 18\% = 7\%$ Hence, she spent 7% more on housing.

PRACTICE EXERCISE 5

Perform each indicated operation:

a. $25\% + 7\frac{3}{4}\%$

b. $48.2\% - 14\%$

Solve the following:

c. The price tag on the nurse's uniform read "25% off." However, the clerk told Francesca that since she worked at the county hospital, the manager would give her another 10% off the original price. What percent did Francesca get off the uniform?

d. When Anne went to pay her bill at the restaurant, she found that a 7% sales tax and a 1.5% restaurant tax had been added to the bill. What percent more did she pay for the sales tax than for the restaurant tax?

(Continued)

Answers: Practice Exercise 5: a. $32\frac{3}{4}\%$ **b.** 34.2% **c.** 35% **d.** 5.5%

If you need more practice, do the following Additional Practice Exercises.

Additional Practice Exercise 5 Perform each indicated operation.

a. $12\frac{3}{4}\% - 7\%$

b. $16.2\% - 5.6\%$

c. In a particular county there is a sales tax of 6.5% plus a resort tax of 3.2% on hotel rooms. What is the total tax on a hotel room?

d. A rental car company is giving a 15% discount to everyone. AAA (triple-A) members get an additional 10% discount. What is the total discount that a AAA member would receive?

Being able to convert between percents, decimals, and fractions is a very important everyday skill.

Exercise Set 8.3

For Extra Help

MyMathLab®

Convert each of the following percents to fractions. (See Example 1.)

1. 16%

2. 5%

3. .09%

4. .2%

5. 13.6%

6. 15.6%

7. 325%

8. 550%

9. $5\frac{1}{3}\%$

10. $3\frac{1}{8}\%$

11. $12\frac{1}{7}\%$

12. $15\frac{10}{11}\%$

Convert the given percents to decimals. If necessary, round to the nearest thousandths. (See Example 2.)

13. 42%

14. 15%

15. .6%

16. .9%

17. 37.5%

18. 62.5%

19. 120%

20. 230%

21. 8.08%

22. 4.02%

23. $12\frac{7}{9}\%$

24. $11\frac{5}{9}\%$

Convert the following decimals to percent notation. (See Example 3.)

25. .2

26. .4

27. .38

28. .19

29. .008

30. .006

31. 3.21

32. 5.67

33. .555 . . .

34. .3888 . . .

35. 5

36. 8

37. 10

38. 25

Convert the following fractions to percent notation, using the method that works best for each fraction. (See Example 4.)

39. $\frac{2}{5}$

40. $\frac{7}{10}$

41. $\frac{9}{20}$

42. $\frac{3}{4}$

Answers: Additional Practice 5: a. $5\frac{3}{4}\%$ **b.** 10.6% **c.** 9.7% **d.** 25%

43. $\dfrac{19}{25}$ **44.** $\dfrac{13}{50}$ **45.** $\dfrac{9}{5}$ **46.** $\dfrac{13}{5}$

47. $\dfrac{3}{8}$ **48.** $\dfrac{5}{16}$ **49.** $\dfrac{4}{3}$ **50.** $\dfrac{7}{6}$

51. $\dfrac{11}{9}$ **52.** $\dfrac{5}{9}$ **53.** $\dfrac{7}{13}$ **54.** $\dfrac{4}{13}$

Perform each indicated operation. (See Example 5.)

55. $25\% + 14\%$ **56.** $15\% + 8\%$ **57.** $37.5\% - 16.6\%$

58. $8\dfrac{3}{4}\% - 5\dfrac{1}{3}\%$ **59.** $6.23\% + 8\%$ **60.** $19.61\% + 12\%$

61. $11\% - 4.6\%$ **62.** $76\% - 18.2\%$

Solve the following. (See Example 5.)

63. Greg Mendes noticed a 7% sales tax and a 2.5% room tax on his hotel bill. What percent of his hotel bill was tax?

64. The owner of a piece of downtown business property pays a 7% real estate tax on the business. On her home she pays a 2% real estate tax. How much greater is the percentage of taxes that she pays on her business than on her home?

65. Albert Young is a writer. In his home country, he would have to pay 48% of his income in taxes. In his adopted country, he must pay only 21% of his income in taxes. What percent does he save in taxes by living in his adopted country?

66. Janet Sanchez received two bonuses last year. She received 10% of her salary for meeting a sales quota and 5% for signing a big client for the firm. What percent of her salary did she receive in bonuses?

Challenge Exercises (67–70)

Convert each fraction to percent notation. Round to the nearest one hundredth of a percent.

67. $\dfrac{3}{17}$

Convert each decimal to percent notation. Round to the nearest tenth of a percent.

68. $.8333\ldots$

Convert the percent to a fraction.

69. $9\dfrac{4}{11}\%$

Convert the percent to a decimal.

70. $18\dfrac{2}{3}\%$

Writing Exercises (71–74)

71. Why do we have percent notation?

72. Can you multiply in percent notation? Why or why not? Give an example.

73. Can you divide in percent notation? Why or why not? Give an example.

74. Why is $9\% + 3 \neq 12\%$?

Section 8.4 — Applications of Percent

OBJECTIVES *When you complete this section, you will be able to:*

A Translate mathematical expressions involving percent into English.
B Solve simple percent problems.
C Solve simple percent problems with proportions.
D Solve application problems involving percent.

PREREQUISITE SKILLS *Before beginning this section, you should be able to:*

a. Convert percents into fractions and percents into decimals. (Section 8.3)
b. Multiply and divide decimals. (Section R.6)
c. Solve linear equations, using the multiplication property of equality. (Section 3.2)
d. Solve a proportion. (Section 8.2)

GETTING READY FOR SECTION 8.4

1. Convert 90% into a fraction.   **2.** Convert 4% into a decimal.

Multiply or divide the following decimals:

3. $.121(32{,}000)$

4. $\dfrac{250}{.05}$

Solve the following equations:

5. $.9y = 63$ **6.** $1.5p = 123$

Solve the following proportions:

7. $\dfrac{3}{4} = \dfrac{A}{50}$ **8.** $\dfrac{P}{100} = \dfrac{6}{5}$

Introduction

The applications of percent are numerous. Percent is used in business, education, economics, medicine, and many other areas. In this section, we will begin with simple applications of percent and continue to the more complicated applications in the next section.

Let us begin by familiarizing ourselves with expressions involved in a **simple percent situation**. Simple percent situations are represented by equations that relate percent, the number of which we are taking the percent, and the result of taking the percent. These situations can be represented as _____% of _____ is _____, where we are given two of these values and are asked to find the third. The first blank represents the percent, the second blank represents the base, and the third blank represents the amount, or part. Consequently, this method can be thought of as percent · base = amount.

OBJECTIVE Translating mathematical expressions involving percent into English

In translating from mathematics to English, we use the following translations: multiplication symbol " · " translates as "of," equal sign "=" translates as "is," and the variable translates as "what number," "some number," or "a number."

Answers: Getting Ready: 1. $\dfrac{9}{10}$ **2.** .04 **3.** 3872 **4.** 5000 **5.** $y = 70$ **6.** $p = 82$ **7.** $A = 37.5$ **8.** $P = 120$

Example 1   **Translate each percent equation into an English sentence.**

a. $75\% \cdot 80 = x$

b. $150\% \cdot x = 18$

Solution

Solution

75% of 80 is what number?   The symbol "·" becomes "of,"
                               "=" becomes "is," and "x"
                               becomes "what number."

150% of what number is 18?   The symbol "·" becomes "of,"
                                "x" becomes "what number,"
                                and "=" becomes "is."

c. $x\% \cdot 40 = 14$

Solution

What percent of 40 is 14?   "x%" becomes "what
                              percent," "·" becomes "of,"
                              and "=" becomes "is."

PRACTICE EXERCISE 1

Translate each of the following percent equations into an English sentence:

a. $80\% \cdot 60 = x$ **b.** $75\% \cdot y = 30$ **c.** $t\% \cdot 500 = 60$

If you need more practice, do the following Additional Practice Exercises.

Additional Practice Exercise 1   Translate each percent equation into an English sentence.

a. $x = 85\% \cdot 400$ **b.** $150 = 250\% \cdot y$ **c.** $t\% \cdot 75 = 45$

OBJECTIVE Solving simple percent problems

Since we cannot multiply or divide in percent notation, we will change the percents into decimals, because the arithmetic is usually easier than with fractions. Consequently, "_____ % of _____ is _____" becomes "_____ · _____ = _____," where the first blank is a decimal. Recall that % means .01 or $\frac{1}{100}$.

Example 2   **Translate each expression into an equation and solve. Use *y* as the variable.**

a. 4% of 1500 is what number?   "Of" means multiply, "is" means equals, and "what
                                       number" is the variable.

$4\% \cdot 1500 = y$ Write 4% as a decimal.

$.04(1500) = y$ Multiply.

$60 = y$ Therefore, 4% of 1500 is 60.

Note

Instead of changing 4% into a decimal, we could have changed it into a fraction and done Example 2a as follows:

$4\% \cdot 1500 = y$ Write 4% as a fraction.

$\dfrac{4}{100} \cdot 1500 = y$ Multiply.

$\dfrac{6000}{100} = y$ Divide.

$60 = y$ Therefore, 4% of 1500 is 60.

Answers: Practice Exercise 1: a. 80% of 60 is what number? **b.** 75% of what number is 30? **c.** What % of 500 is 60?
Additional Practice 1: a. What number is 85% of 400? **b.** 150 is 250% of what number? **c.** What % of 75 is 45?

b. 90% of what number is 63?   "Of" means multiply, "what number" is the variable, and "is" means equals.

$$90\% \cdot y = 63$$ Change 90% to a decimal.

$$.90y = 63$$ Divide both sides of the equation by .90.

$$\frac{.90y}{.90} = \frac{63}{.90}$$ Divide.

$$y = 70$$ Hence, 90% of 70 is 63.

Note

Just as in Example 2a, we could have changed 90% into a fraction instead of a decimal.

c. What percent of 40 is 8?   "What percent" is the variable, "of" means multiply, and "is" means equals.

$$y\% \cdot 40 = 8$$ Since we cannot convert $y\%$ to a decimal, think of $y\%$ as $y \cdot 1\%$ and change 1% to .01.

$$y(.01)(40) = 8$$ Multiply .01 and 40.

$$.40\,y = 8$$ Divide both sides of the equation by .40.

$$\frac{.40y}{.40} = \frac{8}{.40}$$ Simplify both sides.

$$y = 20$$ Hence, 20% of 40 is 8.

PRACTICE EXERCISE 2

Translate each expression into an equation and solve.

a. 10% of 46 is what number?   **b.** 180% of what number is 45?

c. What percent of 500 is 110?

If you need more practice, do the following Additional Practice Exercises.

Additional Practice Exercise 2   Translate each expression into an equation and solve.

a. What number is 20% of 150?   **b.** 45 is 90% of what number?

c. 54 is what percent of 300?

OBJECTIVE Solving simple percent problems with proportions

Alternative Method: Using proportions to solve percent problems.

Using proportions to solve percent problems may be the preferred method. Remember that percent is the number of hundredths. Any percent may be expressed as a ratio with the second part or the denominator as 100. That is, 35% may be written as a ratio 35 : 100, or $\frac{35}{100}$. The 35 is the **number of percent**. The number of which we are taking the percent is called the **base** and follows the word *of* in the English sentence because we are taking a percentage of the base. The result of taking the percent is called the **amount**, or **part**.

Answers: Practice Exercise 2: a. 4.6 **b.** 25 **c.** 22%   **Additional Practice 2: a.** 30 **b.** 50 **c.** 18%

Rule: Percent Proportion

If P represents the number of percent, A represents the amount, and B represents the base, the relationship between these quantities is given by the following proportion:

$$\frac{P}{100} = \frac{A}{B}$$

Consider the following:
35% of 20 = 7 becomes the proportion $\frac{35}{100} = \frac{7}{20}$. This is read "35 is to 100 as 7 is to 20."

Procedure: Solving Simple Percent Situations Using Proportions

The procedure for solving a percent situation with the percent proportion is as follows:

a. Identify the number of percent, the amount, and the base of the second ratio.

b. Substitute the appropriate values into the percent proportion.

c. Solve the proportion for the indicated variable.

Example 3   **Solve the following using proportions:**

a. 75% of 50 is what number?

Solution

The number of percent is 75, the base is 50, because it follows "of" and we are looking for the amount.

$\dfrac{P}{100} = \dfrac{A}{B}$ Substitute 75 for P and 50 for B.

$\dfrac{75}{100} = \dfrac{A}{50}$ Reduce $\dfrac{75}{100}$ to lowest terms.

$\dfrac{3}{4} = \dfrac{A}{50}$ Cross multiply.

$3(50) = 4A$ $3(50) = 150$. Divide both sides of the equation by 4.

$\dfrac{150}{4} = \dfrac{4A}{4}$

$37.5 = A$ Therefore, 75% of 50 is 37.5.

Note

It is not necessary to reduce $\frac{75}{100}$ to lowest terms, especially if we are using a calculator. Reducing just makes the numbers smaller and easier to work with.

b. 50% of what number is 90?

Solution

The number of percent is 50, the amount is 90, and we are looking for the base.

$$\frac{P}{100} = \frac{A}{B}$$ Substitute 50 for P and 90 for A.

$$\frac{50}{100} = \frac{90}{B}$$ Reduce $\frac{50}{100}$ to lowest terms.

$$\frac{1}{2} = \frac{90}{B}$$ Cross multiply.

$$1(B) = 2(90)$$ Multiply.

$$B = 180$$ Therefore, 50% of 180 is 90.

c. What percent of 60 is 72?

Solution

The amount is 72, the base of the second ratio is 60 because it follows "of" and we are looking for the percent.

$$\frac{P}{100} = \frac{A}{B}$$ Substitute 72 for A and 60 for B.

$$\frac{P}{100} = \frac{72}{60}$$ Reduce $\frac{72}{60}$ to lowest terms.

$$\frac{P}{100} = \frac{6}{5}$$ Cross multiply.

$$5P = 100(6)$$ $100(6) = 600$. Divide both sides of the equation by 5.

$$\frac{5P}{5} = \frac{600}{5}$$ Divide.

$$P = 120$$ Hence, 120% of 60 is 72.

PRACTICE EXERCISE 3

Solve the following equations by using proportions:

a. 45% of 80 is what number?   **b.** 15 is 2.5% of what number?   **c.** 43.4 is what percent of 70?

If you need more practice, do the following Additional Practice Exercises.

Additional Practice Exercise 3   Solve the following equations by using proportions:

a. What number is 3.5% of 400?   **b.** 24% of what number is 12?   **c.** 48 is what percent of 30?

OBJECTIVE Solving application problems involving percent

Application problems involving percent are solved in much the same manner as other word problems. We solve them by rewriting them in the form of _____ % of _____ is _____, translating into an equation or proportion, and solving. We summarize as follows:

Answers: Practice Exercise 3: a. 36 **b.** 600 **c.** 62%   **Additional Practice 3: a.** 14 **b.** 50 **c.** 160%

Procedure: Solving Percent Problems

1. Rewrite the problem in the form of _____ % of _____ is _____.
2. Translate the statement in step 1 into an equation or proportion.
3. Solve the equation or proportion.
4. Check your answer.

Example 4   **Solve the following and round answers to the nearest percent or hundredth where appropriate:**

a. Tanya correctly answered 123 questions out of 150 questions on an arithmetic final exam. What percent of the questions did she answer correctly?

Solution

Rewrite the problem in the form of _____ % of _____ is _____. The question is, "What percent *of the questions* (150) did she answer correctly (123)?"

This translates as:

What percent of 150 is 123?   Rewrite as an equation. Let p represent the percent.

$p\% \cdot 150 = 123$ Since we cannot change $p\%$ into a decimal, think of $p\%$ as $p \cdot 1\%$ and change 1% to .01.

$p(.01) \cdot 150 = 123$ Multiply .01 and 150.

$1.5p = 123$ Divide both sides by 1.5.

$p = 82$ Therefore, she answered 82% of the questions correctly.

CHECK:

Is 82% of 150 equal to 123? $(82\%)(150) = (.82)(150) = 123$. Yes. Therefore, our answer is correct.

Note

This example could have been done by using the proportion $\frac{y}{100} = \frac{123}{150}$.

b. This term, Harold spent $250 on books. If this amount represents 5% of his student loan, how much was the loan?

Solution

Rewrite the problem in the form of _____ % of _____ is _____. The problem states that the amount Harold spent for books represents 5% *of his loan*, and we are asked to find the amount of the loan. Consequently, this translates as follows:

5% of what is $250?   Let y represent the amount of the loan and write an equation.

$5\% \cdot y = 250$ Change 5% to a decimal.

$.05y = 250$ Divide both sides by .05.

$\dfrac{.05y}{.05} = \dfrac{250}{.05}$ Divide.

$y = \$5000$ Therefore, Harold's student loan is $5000.

CHECK:

Is 5% of $5000 equal to $250? $5\%(5000) = .05(\$5000) = \$250$. Yes. So, our solution is correct.

Note

This example could have been done by using the proportion $\frac{5}{100} = \frac{250}{y}$.

c. A study showed that 12.1% of the women giving birth at a certain hospital had taken cocaine within 72 hours of giving birth. If 32,000 women gave birth at this hospital last year, how many of them had taken cocaine within 72 hours of giving birth?

Solution

Rewrite the problem in the form of _____ % of _____ is _____. The problem states that 12.1% *of the women* who gave birth had taken cocaine and that 32,000 women had given birth. This translates as follows:

12.1% of 32,000 is what number?   Let z represent the number of women who had taken cocaine and write an equation.

$12.1\% \cdot 32{,}000 = z$ Convert 12.1% to a decimal.

$.121(32{,}000) = z$ Multiply.

$z = 3872$ Therefore, 3872 mothers had taken cocaine within 72 hours of giving birth.

CHECK:

Is 12.1% of 32,000 = 3872? Yes, $.121(32{,}000) = 3872$.

Note

This example could have been done by using the proportion $\frac{12.1}{100} = \frac{z}{32{,}000}$.

PRACTICE EXERCISE 4

Solve and round answers to the nearest percent or hundredth where appropriate.

a. Out of every 5000 adults in the United States, 4000 do not exercise properly. What percent of the adults do not exercise properly?

b. A score of 70% on a test corresponds to 84 correctly answered test items. How many items were on the test?

c. A family must pay 22% of its income for income tax. How much is the tax if the family earned $65,000 that year?

If you need more practice, do the following Additional Practice Exercises.

Additional Practice Exercise 4   Solve and round answer to the nearest percent or hundredth.

a. Thirty-five out of 700 lightbulbs failed after 1000 hours of operation. What percent of the bulbs failed?

b. During a recent downsizing, 8% of the employees of a company were laid off. If 160 people were laid off, how many employees did the company have before the downsizing?

c. At a local college, 15% of the employees are administrators or staff for the administration. If 320 people are employed by the college, how many are administrators or staff for the administrators?

Answers: Practice Exercise 4: a. 80% **b.** 120 items **c.** $14,300   **Additional Practice 4: a.** 5% **b.** 2000 employees **c.** 48 people

Exercise Set 8.4

For Extra Help

MyMathLab®

Translate the following percent equations into English sentences. (See Example 1.)

1. $x = 25\% \cdot 36$

2. $50\% \cdot 32 = x$

3. $40\% \cdot 60 = v$

4. $v = 90\% \cdot 150$

5. $20\% \cdot y = 8$

6. $25 = 35\% \cdot y$

7. $52 = 65\% \cdot w$

8. $32\% \cdot w = 16$

9. $t\% \cdot 30 = 21$

10. $32 = t\% \cdot 20$

11. $4.8 = z\% \cdot 80$

12. $z\% \cdot 90 = 7.2$

Translate each expression into an equation and solve. (See Example 2.)

13. 5% of 200 is what number?

14. What number is 15% of 80?

15. What number is 80% of 35?

16. 40% of 60 is what number?

17. 60 is 75% of what number?

18. 50% of what number is 45?

19. 9% of what number is 54?

20. 18 is 37.5% of what number?

21. 24 is what percent of 80?

22. What percent of 90 is 36?

23. What percent of 96 is 12?

24. 9 is what percent of 24?

25. $12\frac{1}{2}\%$ of 40 is what number?

26. $8\frac{3}{4}\%$ of 160 is what number?

27. $6\frac{2}{3}\%$ of what number is 3?

28. $16\frac{2}{3}\%$ of what number is 6?

Solve the following using proportions. (See Example 3.)

29. What number is 55% of 160?

30. 92% of 650 is what number?

31. 5% of 35 is what number?

32. What number is 18.5% of 40?

33. 3.6 is 5% of what number?

34. 12% of what number is 17.4?

35. 100 is 125% of what number?

36. 140% of what number is 350?

37. What percent of 25 is 1?

38. 1 is what percent of 20?

39. 141 is what percent of 600?

40. What percent of 800 is 356?

41. $18\frac{1}{3}\%$ of 120 is what number?

42. $4\frac{2}{7}\%$ of 210 is what number?

43. $4\frac{4}{9}\%$ of what number is 8?

44. $3\frac{1}{8}\%$ of what number is 5?

Solve each of the following and round the answers to the nearest percent or nearest hundredth where appropriate. (See Example 4.)

45. A recent survey indicated that 405 out of 500 women did not consider physical attraction an essential requirement for a spouse. This represents what percent of the women surveyed?

46. Nicky has read 198 pages of a book that has 660 pages. What percent of the book has she read?

47. Research shows that 63% of all reported traffic accidents involve men. If 1500 traffic accidents are reported, how many involve men?

48. In a study, 34% of adults age 20 and over were obese. If 600 adults were studied, how many were obese?

49. By volume a wine is 14% alcohol. If a bottle contains 750 milliliters of wine, how many milliliters are alcohol?

50. A solution of salt and water is called a saline solution. One saline solution is 8% salt. How much salt is in 50 milliliters of solution?

51. In mixing salad dressing, we use 3 ounces of water and 2 ounces of olive oil. What percent of the salad dressing is olive oil?

52. If we heat metals to liquid form and combine them, we get a mixture called an alloy. If 18 pounds of copper is mixed with 12 pounds of tin, what percent of the alloy is copper?

53. A family paid $2210 in real estate taxes last year. If this is a tax rate of 2.6%, how much is their house worth?

54. Out of 800 greeting cards sold at a card store, 85.5% were bought by women. How many cards were bought by women?

55. The body of a male human is 63% water by weight. If a man's body contains 120 pounds of water, how much does he weigh to the nearest pound?

56. At the Lonesome Pine Ranch, 2475 head of cattle became infected with the hoof-and-mouth disease. If this represents 45% of the herd, how many head of cattle were in the herd?

Challenge Exercises (57–60)

Solve the following;

57. $33\frac{1}{3}$% of 120 is what number?

58. What number is $6\frac{1}{2}$% of 900?

59. Vinegar is 5% acetic acid by volume. How much vinegar would 5 milliliters of acetic acid make?

60. Muriatic acid is a 2% solution of hydrochloric acid. How many gallons of muriatic acid can be made from 1 gallon of hydrochloric acid?

Writing Exercise

61. Write down five household items that are mixtures and list their ingredients by percent.

<table>
<tr><td>

Section 8.5
</td><td>

Further Applications of Percent
</td></tr>
</table>

OBJECTIVES *When you complete this section, you will be able to:*

A Compute sales tax.
B Compute discount.
C Compute commission.
D Compute simple interest.
E Compute compound interest.
F Calculate percent increase and percent decrease.

PREREQUISITE SKILLS *Before beginning this section, you should be able to:*

a. Solve linear equations by the multiplication property of equality. (Section 3.2)
b. Change percent into fractions and percent into decimals. (Section 8.3)
c. Multiply and divide decimals. (Section R.6)
d. Solve a proportion. (Section 8.2)
e. Multiply rational numbers. (Section 6.3)
f. Convert a fraction into percent. (Section 8.3)

GETTING READY FOR SECTION 8.5

1. Solve $.06P = 15$ for P.

2. Change 35% into a decimal.

3. Change 16% into a fraction.

4. Multiply: $.045(400)$

5. Divide: $\dfrac{15}{.06}$

6. Solve: $\dfrac{6}{100} = \dfrac{15}{p}$

7. Multiply: $1250(.11)\dfrac{1}{4}$

8. Change $\dfrac{1}{5}$ into a percent.

Introduction

In the last section, we used the simple percent situation to solve percent problems. In this section, we will apply the simple percent situation to the specific situations of computing sales tax, discounts, and commission. We will also use percent to compute interest and percent increase and decrease.

Sales tax is the amount of money that we pay to the government when we make some types of purchases. Sales tax is calculated as a fractional part, usually expressed as a percent, of the amount we pay for the goods purchased. We find the amount of the sales tax by multiplying the purchase price by a percent called the **sales tax rate.**

OBJECTIVE Computation of sales tax

Calculating sales tax is a special type of simple percent situation and thus takes the form of _____ % of _____ is _____. In this case, it becomes:

Note

Computation of sales tax can also be done using the proportion

$$\frac{\text{Sales tax rate}}{100} = \frac{\text{Sales tax}}{\text{Purchase price}}.$$

> **Rule: Calculating Sales Tax**
>
> Sales tax rate · purchase price = sales tax.

Answers: Getting Ready: 1. $P = 250$ **2.** $.35$ **3.** $\dfrac{4}{25}$ **4.** 18 **5.** 250 **6.** $p = 250$ **7.** 34.375 **8.** 20%

The total cost of an item is the purchase amount plus the sales tax.

Total cost = purchase amount + sales tax.

Example 1 | Solve the following involving sales tax:

a. Find the sales tax on a purchase of $400 if the sales tax rate is 4.5%

Solution

We think of this in the form of _____ % of _____ is _____, or *sales tax rate · purchase price = sales tax.* We know the sales tax rate is 4.5% and the purchase price is $400. Consequently,

4.5% of $400 is sales tax	Write as an equation, using T as the taxes.
$4.5\% \cdot 400 = T$	Convert 4.5% to a decimal.
$.045 \cdot 400 = T$	Multiply.
$18.00 = T$	Hence, the sales tax on $400 at 4.5% is $18.00.

Note

This example could have been done using the proportion $\frac{4.5}{100} = \frac{T}{400}$.

b. Rebecca purchased a refrigerator for $600 and paid $30 in sales tax. What is the tax rate?

Solution

We think of this in the form of _____ % of _____ is _____, or *sales tax rate · purchase price = sales tax.* We know the purchase price is $600 and the sales tax is $30. Consequently,

Sales tax rate · 600 is 30	Write as an equation, using r as the sales tax rate.
$r\% \cdot 600 = 30$	Think of $r\%$ as $r \cdot 1\%$ and change 1% to .01.
$r(.01)(600) = 30$	Multiply .01 by 600.
$6r = 30$	Divide both sides by 6.
$r = 5$	Therefore, the sales tax rate is 5%.

Note

This example could have been done using the proportion $\frac{r}{100} = \frac{30}{600}$.

c. The sales tax in a certain state is 6%. If Rebecca paid $15 in sales tax, what was the amount of purchase? What was the total cost?

Solution

We think of this in the form of _____ % of _____ is _____, or *sales tax rate · purchase price = the sales tax.* We know the sales tax rate is 6% and the sales tax is $15. Consequently,

6% of purchase price is $15	Write as an equation, using P as the purchase price.
$6\% \cdot P = 15$	Convert 6% to a decimal.
$.06 \cdot P = 15$	Divide both sides of the equation by .06.
$\dfrac{.06 \cdot P}{.06} = \dfrac{15}{.06}$	Divide.
$P = 250$	Hence, the purchase amount is $250.

To find the total cost, recall the relationship

Total cost = purchase amount + sales tax
Total cost = $250 + $15   Substitute values and add.
Total cost = $265   Thus, the cost is $265.

Note

This example could have been done using the proportion $\frac{6}{100} = \frac{15}{P}$.

PRACTICE EXERCISE 1

Solve the following involving sales tax:

a. Find the sales tax on an item that has a purchase price of $120 if the sales tax rate is 6.5%.

b. John purchased a bow for $250 and paid $17.50 in sales tax. What was the sales tax rate?

c. The sales tax rate in a certain city is 1.5%. If Juan Carlos paid $6.75 in sales tax, what was the amount of purchase? What was the total cost?

(Continued)

Answers: **Practice Exercise 1: a.** $7.80 **b.** 7% **c.** $450, $456.75

If you need more practice, do the following Additional Practice Exercises.

Additional Practice Exercise 1   Solve the following involving sales tax:

a. John purchased a stereo for $450 and paid a local city sales tax of 3.5%. Find the sales tax amount.

b. Find the purchase price if the sales tax is $16 and the sales tax rate is 8%.

c. Julia bought an item for $70. If the sales tax rate is 7%, what is the amount of the sales tax? What is the total cost?

OBJECTIVE **Computation of discount**

A **discount** is the amount of money that is taken off (subtracted from) the original price of an item. The amount of the discount is calculated as a percent of the price marked on the goods purchased. This percent is called the **discount rate**.

This is another special type of simple percent situation. If we think of it in terms of _____ % of _____ is _____, it becomes the following relationship:

> **Rule: Calculating Discounts**
>
> Discount rate · original price = the amount of the discount.

Note

Computation of discount problems can also be done using the proportion

$$\frac{\text{Discount rate}}{100} = \frac{\text{Amount of discount}}{\text{Original price}}.$$

The **sale price** (S) of an item is the original price minus the discount:

$$\text{Sale price} = \text{original price} - \text{discount}.$$
$$S = O - D$$

Example 2   **Solve the following involving discounts:**

a. Find the amount of discount of a bedroom suite priced at $1400 with a 25% discount.

Solution

We think of this in the form of _____ % of _____ is _____, or *discount rate · original price = amount of the discount*. We know the discount rate is 25% and the original price is $1400. Consequently,

25% of 1400 is the discount	Write as an equation, using D as the discount.
$25\% \cdot 1400 = D$	Convert 25% to a decimal.
$.25 \cdot 1400 = D$	Multiply.
$350 = D$	Therefore, the discount is $350.

Note

This example could have been done using the proportion $\frac{25}{100} = \frac{D}{1400}$.

Answers: Additional Practice 1: a. $15.75 **b.** $200 **c.** $4.90, $74.90

b. Find the original price of a TV if the discount is $157.50 and the discount rate is 35%.

Solution

We think of this in the form of _____% of _____ is _____, or *discount rate · original price = amount of the discount*. We know the amount of the discount is $157.50 and the discount rate is 35%. Consequently,

35% of original price is 157.50	Write as an equation, using O as the original price.
$35\% \cdot O = 157.50$	Convert 35% to a decimal.
$.35 \cdot O = 157.50$	Divide both sides of the equation by .35.
$\dfrac{.35 \cdot O}{.35} = \dfrac{157.50}{.35}$	Divide.
$O = 450$	Therefore, the original price is $450.

> **Note**
>
> This example could have been done using the proportion $\dfrac{35}{100} = \dfrac{157.50}{O}$.

c. Mark paid $280 for a suit that was discounted $120. What was the original price of the suit? What was the discount rate?

Solution

First, we solve for the original price. Recall the sale price relationship:

Sale price = original price − discount.

$S = O - D$	Substitute values for variables.
$280 = O - 120$	Add $120 to both sides of the equation.
$280 + 120 = O - 120 + 120$	Add.
$400 = O$	Hence, the original price is $400.

Now that we know the original price, we can solve for the discount rate.

We think of this in the form of _____% of _____ is _____, or *discount rate · original price = amount of the discount*. We know the original price is $400 and the amount of the discount is $120. Consequently,

The discount rate of 400 is 120	Write as an equation, using r as the discount rate.
$r\% \cdot 400 = 120$	Think of $r\%$ as $r \cdot 1\%$ and change 1% to .01.
$r(.01)(400) = 120$	Multiply .01 and 400.
$4r = 120$	Divide both sides of the equation by 4.
$\dfrac{4r}{4} = \dfrac{120}{4}$	Divide.
$r = 30$	Therefore, the rate of discount is 30%.

> **Note**
>
> This example could have been done using the proportion $\dfrac{r}{100} = \dfrac{120}{400}$.

PRACTICE EXERCISE 2

Solve the following involving discounts:

a. Find the amount of the discount if an item has an original price of $150 and the discount rate is 20%.

b. Jin Ho bought a bed whose original price was $460 with a discount of $69. What was the discount rate?

c. Monica paid $150 for a suit that was discounted $100. What was the original price of the suit? What was the discount rate?

If you need more practice, do the following Additional Practice Exercises.

Additional Practice Exercise 2   Solve the following involving discounts:

a. Find the original price if the discount rate is 35% and the amount of the discount is $77.

b. Pamika bought a blouse whose original price was $65 with a discount of 25%. Find the sales price.

c. A sale sign said "45% off." If an item is originally $120, how much is the sale price?

Answers: Practice Exercise 2: a. $30 **b.** 15% **c.** $250, 40%   **Additional Practice 2: a.** $220 **b.** $48.75 **c.** $66

OBJECTIVE Computation of commission

A **commission** is the amount of money that is paid to a person for selling an item. Commission is calculated as a percent of the price paid for the goods purchased. This price is called the **sales**. We find the amount of the commission by multiplying the sales price by a percent called the **commission rate**.

This, too, is an application of the simple percent situation. In this case, _____ % of _____ is _____ becomes the following relationship:

> **Rule: Calculating Commission**
>
> Commission rate · sales = amount of commission.

Note

Calculating commissions can also be done using the proportion

$$\frac{\text{Commission rate}}{100} = \frac{\text{Amount of the commission}}{\text{Total sales}}.$$

Example 3 **Solve the following involving commissions:**

a. Find the amount of the commission on sales of $560 if the commission rate is 16%.

Solution

We think of this in the form of _____ % of _____ is _____, or *commission rate · sales = amount of the commission*. We know the commission rate is 16% and the sales are $560. Consequently,

16% of $560 is commission	Write as an equation, using C as the commission.
$16\% \cdot 560 = C$	Convert 16% to a decimal.
$.16(560) = C$	Multiply.
$\$89.60 = C$	Therefore, the commission is $89.60.

Note

This example could have been done using the proportion $\frac{16}{100} = \frac{C}{560}$.

b. Karla sells jewelry and is paid a 30% commission on all her sales. What was the selling price on a bracelet for which she received a commission of $78?

Solution

We think of this in the form of _____ % of _____ is _____, or *commission rate · sales = amount of the commission*. We know the commission rate is 30% and her commission is $78. Consequently,

30% of sales is $78.	Write as an equation, using S for sales.
$30\% \cdot S = 78$	Convert 30% to a decimal.
$.3S = 78$	Divide both sides of the equation by .3.
$S = 260$	Therefore, the selling price, or sales, is $260.

Note

This example could have been done using the proportion $\frac{30}{100} = \frac{78}{S}$.

c. Sam received $287.50 commission on a refrigerator selling for $1150. What was the commission rate?

Solution

We think of this in the form of _____ % of _____ is _____, or *commission rate · sales = amount of the commission*. We know the sales are $1150 and his commission is $287.50. Consequently,

Commission rate of 1150 is 287.50	Write as an equation, using r for rate.
$r\% \cdot 1150 = 287.50$	Write $r\%$ as $r \cdot 1\%$ and write 1% as .01.
$r(.01)(1150) = 287.50$	Multiply .01 and 1150.
$11.5r = 287.50$	Divide both sides of the equation by 11.5.
$\dfrac{11.5r}{11.5} = \dfrac{287.50}{11.5}$	Divide.
$25 = r$	Therefore, the commission rate is 25%.

PRACTICE EXERCISE 3

Solve the following involving commissions:

a. Find the amount of the commission for an item that sold for $175 if the commission rate is 12%.

b. Karl works at a department store and receives a 5% commission on all sales. If Karl received a commission of $11.50 for the sale of a suit, what was the selling price of the suit?

c. Marla received a commission of $160 for selling an $800 stove. What was her commission rate?

If you need more practice, do the following Additional Practice Exercises.

Additional Practice Exercise 3 Solve the following involving commissions:

a. Find the amount of the commission on a sale of $124 if the commission rate is 8.5%.

b. Sue received a commission of $135 on a computer monitor that sold for $540. What was her rate of commission?

c. Dion received a commission of $420 for selling a refrigerator. If she has a commission rate of 35%, what was the sales amount?

OBJECTIVE Computation of simple interest

Interest is money that you must pay to someone or to an institution to use their money, or money paid to you by someone who uses your money. The usual way to use someone's money is to get a loan, and the usual way for someone to use your money is for you to make a deposit at a bank or savings institution. The amount of money borrowed or deposited is called the **principal**.

Interest is calculated as a part or percentage of the money being loaned or deposited. This amount is found by multiplying the principal by a percent called the **annual rate of interest**, then multiplying the resulting product by the length of time (usually, in years) the loan is active. Thus, we have the relationship

$$\text{Interest} = (\text{principal})(\text{annual interest rate})(\text{time}).$$

Hence, we have the formula for calculating interest.

Note

In applying this formula, r must be expressed as a decimal.

> ### Rule: Interest Calculation Formula
>
> If I = interest, p = principal, r = annual interest rate, and t = time in years, then interest is calculated by the following formula:
>
> $$I = p \cdot r \cdot t$$

Example 4 | **Solve each of the following for the unknown value:**

a. $p = \$800, r = 18\%, t = 1$ year, $I = ?$

Solution

$I = p \cdot r \cdot t$	Recall the interest calculation formula. Substitute values for variables.
$I = 800 \cdot 18\% \cdot 1$	Convert 18% to a decimal.
$I = (800)(.18) \cdot 1$	Multiply.
$I = 144$	Therefore, the interest on $800 at 18% interest rate for 1 year is $144.

Answers: Practice Exercise 3: a. $21 **b.** $230 **c.** 20% **Additional Practice 3: a.** $10.54 **b.** 25% **c.** $1200

b. $I = \$22.26 \; p = \$742, t = \frac{1}{2}$ year, $r = ?$

Solution

$I = p \cdot r \cdot t$	Recall the interest calculation formula. Substitute values for variables.
$\$22.26 = 742 \cdot r \cdot \dfrac{1}{2}$	Divide 2 into 742.
$22.26 = 371 \cdot r$	Divide both sides of the equation by 371.
$\dfrac{22.26}{371} = \dfrac{371 \cdot r}{371}$	Divide.
$.06 = r$	Convert to percent notation.
$6\% = r$	Therefore, the interest on $742 for $\frac{1}{2}$ year at 6% interest rate is $22.26.

c. Rosa Lee borrowed some money at 12% for 3 years. If she paid $1800 in interest, how much did she borrow?

Solution

$I = p \cdot r \cdot t$	Recall the interest calculation formula. Substitute values for variables.
$\$1800 = p \cdot 12\% \cdot 3$	Convert 12% to a decimal.
$1800 = p \cdot .12 \cdot 3$	Multiply.
$1800 = .36p$	Divide both sides of the equation by .36.
$\dfrac{1800}{.36} = \dfrac{.36p}{.36}$	Divide.
$5000 = p$	Therefore, she paid $1800 interest on $5000 over the 3 years.

PRACTICE EXERCISE 4

Solve each of the following for the unknown value:

a. $p = \$15,000, r = 8\%, t = 2$ years, $I = ?$ **b.** $I = \$833.40, r = 12\%, t = \frac{3}{4}$ years, $p = ?$

c. Edward deposited $2500 at a 16% interest rate. If he was paid $1600 interest, how many years was the money on deposit?

If you need more practice, do the following Additional Practice Exercises.

Additional Practice Exercise 4   Solve each of the following for the unknown value:

a. $p = \$15,000, r = 4\%, t = 1$ year, $I = ?$ **b.** $I = \$40.50, p = \$900, t = \frac{1}{4}$ year, $r = ?$

c. Lilian borrowed $2500 for $\frac{1}{2}$ year at an 8% interest rate. How much interest did she pay?

From Example 4, we see that the length of time used in interest calculations is always measured in years. Any other expression of time must be converted to years before a calculation can be made. In order to make calculations easier, the banking world has set up its own system of time by making a banking month equal to 30 days. See the following table:

Banking Time

30 banking days = 1 banking month
12 banking months = 1 banking year
360 banking days = 1 banking year

We may calculate interest for a certain number of days or months as parts of a year.

Answers: Practice Exercise 4: a. $2400 **b.** $9260 **c.** 4 years   **Additional Practice 4: a.** $600 **b.** 18% **c.** $100

Example 5   **Calculate the interest for each of the following sets of information:**

a. $r = 11\%, p = \$1250, t = 90$ days

Solution

First, we need to change 90 days into a part of a year. 90 days $= \frac{90}{360}$ years $= \frac{1}{4}$ year.

$I = p \cdot r \cdot t$ Substitute values for variables. Express 11% as .11.

$I = 1250(.11)\dfrac{1}{4}$ Multiply.

$I = \dfrac{137.5}{4}$ Divide.

$I = 34.375$ Round to nearest cent.

$I = 34.38$ Hence, the interest on \$1250 at 11% for 90 days is \$34.38 to the nearest cent.

b. $r = 7.5\%, p = \$5500, t = 4$ months

Solution

First, we need to change 4 months into a part of a year:
4 months $= \frac{4}{12}$ year $= \frac{1}{3}$ year.

$I = p \cdot r \cdot t$ Substitute values for variables and express 7.5% as .075.

$I = 5500(.075)\dfrac{1}{3}$ Multiply.

$I = \dfrac{412.5}{3}$ Divide.

$I = 137.5$ Hence, the interest on \$5500 at 7.5% for 4 months is \$137.50.

c. Felicia took out a book loan for \$460 at a 6% interest rate. If the loan is to be repaid in 10 months, how much interest will she have to pay?

Solution

First, we need to change 10 months into years: 10 months $= \frac{10}{12}$ years $= \frac{5}{6}$ year.

$I = p \cdot r \cdot t$ Substitute values for variables and express 6% as .06.

$I = 460(.06)\dfrac{5}{6}$ Multiply.

$I = \dfrac{138}{6}$ Divide.

$I = 23$ Hence, Felicia will pay \$23.00 interest on a loan of \$460 at an interest rate of 6% for 10 months.

PRACTICE EXERCISE 5

Calculate the interest for each of the following sets of information:

a. $p = \$900, r = 18\%, t = 30$ days   **b.** $p = \$2400, r = 5\%, t = 8$ months

c. Andre put \$3200 in an 18-month certificate of deposit that paid a 9% interest rate. How much interest did he earn?

If you need more practice, do the following Additional Practice Exercises.

Additional Practice Exercise 5   Calculate the interest for each of the following sets of information:

a. $p = \$400, r = 12\%, t = 6$ months   **b.** $p = \$6000, r = 3\%, t = 9$ months

c. Monica got a 60-day loan for \$300 at 19.2%. How much interest did she pay?

Answers: **Practice Exercise 5: a.** \$13.50 **b.** \$80 **c.** \$432   **Additional Practice 5: a.** \$24 **b.** \$135 **c.** \$9.60

OBJECTIVE Ⓔ Computation of compound interest

Until now, we have been discussing **simple interest**. Simple interest is interest calculated only on the principal. **Compound interest** is interest computed on previously earned interest in addition to principal. Compound interest is calculated at various lengths of time expressed as parts of a year. Some lengths of time are used so often that they have been given names. See the following chart:

Common Lengths of Time

Part of a Year	Phrase	Number of Times a Year Interest Is Computed
1	annually	1
$\frac{1}{2}$	semiannually	2
$\frac{1}{4}$	quarterly	4
$\frac{1}{12}$	monthly	12
$\frac{1}{360}$	daily	360

The **balance** of an account is the amount of money in the account at a particular time. To **post** an amount means to credit the account with the money deposited by adding this amount to the balance.

$$\text{Balance} = \text{principal} + \text{interest}.$$

Example 6 **Calculate the balance after 1 year for an account that pays 8% interest compounded semiannually on a beginning balance of $700.00.**

Solution

Interest must be calculated two times, once for the first half of the year and once for the second half of the year. The principal for the second half of the year will be the original principal plus the interest earned during the first half of the year. We may consider the posting dates as January 1, 2012; July 1, 2012; and January 1, 2013.

Balance	$700.00	_____	_____
Date	Jan. 1, 2012	July 1, 2012	Jan. 1, 2013

The first calculation is simple interest.

$I = p \cdot r \cdot t$ Substitute values for variables.

$I = 700 \cdot 8\% \cdot \dfrac{1}{2}$ Convert 8% to a decimal or fraction.

$I = 700 \cdot .08 \cdot \dfrac{1}{2}$ Multiply.

$I = \dfrac{56}{2}$ Divide.

$I = 28$ The interest for the first half year is $28.00.

We post this interest by adding it to the original principal. Therefore, the balance on July 1 is $700 + $28 = $728.

Balance	$700.00	$728.00	_____
Date	Jan. 1, 2012	July 1, 2012	Jan. 1, 2013

The second calculation will include the interest from the first calculation and hence will be compound interest. Therefore, the principal for the second half of the year is $728.

$$I = p \cdot r \cdot t$$ Substitute values for variables.

$$I = 728 \cdot 8\% \cdot \frac{1}{2}$$ Convert 8% to a decimal.

$$I = 728 \cdot .08 \cdot \frac{1}{2}$$ Multiply.

$$I = \frac{58.24}{2}$$ Divide.

$$I = 29.12$$ The interest for the second half year is $29.12.

We post this interest by adding it to the July 1, 2012, balance to get the January 1, 2013, balance. Therefore, the balance on January 1 is $728 + $29.12 = $757.12.

Balance	$700.00	$728.00	$757.12
Date	Jan. 1, 2012	July 1, 2012	Jan. 1, 2013

Hence, the final balance is $757.12, which includes $28 + $29.12 = $57.12 interest.

Note

In the business world, there is a formula that is used for computing compound interest. See Challenge Exercises 51 and 52.

PRACTICE EXERCISE 6

Calculate the balance after 1 year for an account that pays 6.5% interest compounded semiannually on a beginning balance of $1300.00.

If you need more practice, do the following Additional Practice Exercise.

Additional Practice Exercise 6

Calculate the balance after 3 years for an account that pays 10% interest compounded annually on a beginning balance of $884.00.

OBJECTIVE   Calculation of percent increase and percent decrease

Another situation involving percent is where there is an increase or decrease in a quantity. We start with a value called the **base**, or original amount, and increase or decrease the value. The amount of the increase or decrease may be a number or a percent of the base. To find the percent increase or decrease, divide the amount of the increase or decrease by the *original amount* and convert to a percent.

Answers: Practice Exercise 6: $1385.87   **Additional Practice 6:** $1176.60

> **Procedure: Finding the Percent Increase or Decrease**
>
> To find the percent increase or decrease, find $\dfrac{\text{amount of increase or decrease}}{\text{original amount}}$ and convert to a percent.

Example 7 **The price of a lawn mower changed from $2100 to $2163. What is the percent increase in price?**

Solution

We know the original price of the mower. To find the amount of increase in price subtract the old price from the new price which gives us $2163 - 2100 = 63$. Consequently, we find that

$$\frac{\text{Amount of increase}}{\text{Original amount}} = \qquad \text{Substitute for the amount of increase and the original amount.}$$

$$\frac{63}{2100} = \qquad \text{Reduce } \frac{63}{2100} \text{ to lowest terms.}$$

$$\frac{3}{100} = \qquad \text{Change } \frac{3}{100} \text{ to a percent.}$$

$$3\%$$

Therefore, the lawn mower increased 3% in price.

CHECK:

Is the original price ($2100) plus 3% equal to the new price ($2163)? Yes, since 3% of $2100 = (.03)($2100) = $63 and $2100 + $63 = $2163.

PRACTICE EXERCISE 7

In 2010, a computer cost $3100. In 2011, the same computer cost $1612. What was the percent decrease in price?

If you need more practice, do the following Additional Practice Exercise.

Additional Practice Exercise 7

The price of gasoline rose from $3.60 per gallon to $3.84 per gallon. What was the percent increase to the nearest tenth of a percent?

Exercise Set 8.5

For Extra Help

MyMathLab®

Solve the following involving sales tax. (See Example 1.)

1. Find the tax on an item that sells for $90 if the tax rate is 7%.

2. Find the tax on an item that sells for $120 if the tax rate is 6%.

3. If the tax on a microwave oven that sells for $460 is $20.70, find the tax rate.

4. If the tax on a DVD player that sells for $310 is $17.05, find the tax rate.

Answers: Practice Exercise 7: 48% **Additional Practice 7:** 6.7%

5. The sales tax on a TV set is $15.30. If the sales tax rate is 4.5%, what is the purchase amount?

6. The sales tax on a microwave oven is $11.70. If the sales tax rate is 6.5%, what is the purchase amount?

7. The sales tax rate in a particular state is 7.5%. If you purchase a washing machine for $550, what is the total cost?

8. Olga buys a refrigerator that costs $1350 and pays 5.5% sales tax. What is her total cost?

Solve the following involving discounts. (See Example 2.)

9. If an item has an original price of $180 and the discount rate is 15%, find the discount.

10. If an item has an original price of $240 and the discount rate is 25%, find the discount.

11. A blouse has an original price of $55 and is discounted by $11. What is the rate of the discount?

12. A cleaning kit has an original price of $75 and is discounted by $9. Find the rate of the discount.

13. Victor paid $105 for a coat that was discounted $45. What was the original price of the coat? What was the discount rate?

14. Karen paid $165 for an outfit that was discounted $55. What was the original price of the outfit? What was the discount rate?

Solve the following involving commissions. (See Example 3.)

15. If the commission rate is 15%, find the commission on sales of $300.

16. If the rate of commission is 30%, find the commission on sales of $750.

17. John works in a department store. There was a shirt that had been on the rack for a long time. John's boss told him that he would give him 10% commission if he could sell it. If John's commission was $12.50, what was the selling price of the shirt?

18. Sing Le worked for 25% commission. If she received a commission of $64.50 on the sale of a washer/dryer combination, what was the selling price?

19. Sharon received $20 commission for selling a wooden worktable that sold for $200. What was the commission rate?

20. Tim received $196 commission for selling a $560 vacuum cleaner. What was the commission rate?

Given the interest formula $I = p \cdot r \cdot t$, solve each exercise for the unknown value. If necessary, round to the nearest cent, percent, or year. (See Example 4.)

21. $r = 6\%, p = \$400, t = 2$ years, $I =$

22. $r = 4\%, p = \$800, t = 3$ years, $I =$

23. $r = 8.5\%, I = \$221, t = \frac{3}{4}, p =$

24. $r = 9.5\%, I = \$23.75, t = \frac{1}{2}, p =$

25. $r = 7\%, I = \$109.20, p = \$520, t =$

26. $r = 3\%, I = \$240, p = \$4000, t =$

27. $p = \$1200, I = \$480, t = 5$ years, $r =$

28. $p = \$1500, I = \$180, t = 3$ years, $r =$

29. James borrowed $3000 at an interest rate of 5% for 4 years. How much interest did he pay?

30. Irene deposited $2500 in her bank at an interest rate of 6% for 2 years. How much interest did she receive?

31. Blanca received $306 in interest on money deposited for 5 years at 9% interest. How much did she deposit?

32. Miki received $427.20 in interest on money deposited for 8 years at 6% interest. How much did she deposit?

Calculate the simple interest to the nearest cent for each of the following sets of information. (See Example 5.)

33. $p = \$1000, r = 17\%, t = 60$ days

34. $p = \$1800, r = 16\%, t = 120$ days

35. $p = \$4800, r = 19.2\%, t = 5$ months

36. $p = \$3600, r = 18.6\%, t = 9$ months

37. Janice borrowed $1600 for 8 months at an 18% interest rate. How much interest did she have to pay?

38. Diego deposited $720 in the student depository for 7 months. If he received 6% interest on his deposit, how much interest did he earn?

Calculate the compound interest to the nearest cent. (See Example 6.)

39. Calculate the balance after 1 year for an account that pays 8% compounded semiannually on a beginning balance of $3500.

40. Calculate the balance after 1 year for an account that pays 6% compounded semiannually on a beginning balance of $1400.

Solve the exercises, using percent increase and percent decrease. If necessary, round to the nearest tenth. (See Example 7.)

41. Christopher just went for his yearly physical examination. Last year he measured 50 inches tall, and this year he measured 54 inches tall. What is the percent of increase in his height?

42. In Siberia, houses sometimes sink into the frozen ground because the heat from the house melts the ice. If 20% of a 12-foot-tall house sinks, how high is the remaining portion of the house above ground?

43. Through exercise and diet, Willie lost 15 pounds. When he began this weight reduction program, he weighed 180 pounds. What was the percent of weight loss?

44. An airline pilot requests a change in altitude from 28,000 feet to 35,000 feet to avoid bad weather. What was the percent of change?

45. On a very bad day, the value of a mutual fund fell from $14.25 to $9.12. What was the percent decrease in value?

46. If the total sales of a company increased from $4,500,000 to $4,590,000, what was the percent increase in sales?

Challenge Exercises (47–52)

47. Calculate the balance after 1 year for an account that pays 5% compounded quarterly on a beginning balance of $6500.

48. In a certain county, the county sales tax is 1.5% and the hotel room tax is 1%. If the state sales tax rate is 5.5%, how much tax must a tourist pay on a hotel room that rents for $120 a day for 5 days?

49. A sewing machine that is marked at $849.99 is on sale for $499.99. What is the percent of discount?

50. Marie received $5400 for selling a house. If the commission rate is 6%, how much did the house sell for?

The formula used for computing compound interest is $A = P(1 + \frac{r}{n})^{nt}$, where A is the accumulated amount, P is the principal (amount invested or borrowed), r is the interest rate written as a decimal, n is the number of times per year that the interest is compounded, and t is the number of years. For example, if $1000 is invested at 6% interest compounded quarterly (four times per year) for 10 years, the accumulated amount would be $A = 1000(1 + \frac{.06}{4})^{(4)(10)} = 1000(1 + .015)^{40} = 1000(1.015)^{40} = \$1814.02$. We used a calculator to perform the last computation.

51. Find the accumulated amount if $5000 is invested at 5% interest compounded semiannually (twice per year) for 12 years.

52. Find the accumulated amount if $8000 is invested at 9% compounded monthly (12 times per year) for 15 years.

Writing Exercise

53. Name at least five business applications of percent not mentioned in the text.

Concept/Procedure	Example
Ratio: [Section 8.1] • A ratio is a comparison of two or more quantities by division. If a and b are quantities, then $a : b$, or $\frac{a}{b}$, is a ratio of a to b. If a, b, c are quantities, then $a : b : c$ is a ratio of a to b to c.	
Ratios in Lowest Terms: [Section 8.1] • As with rational numbers, ratios are in lowest terms when the parts of a ratio are integers that contain no common factors other than 1.	

Processes for Simplifying Ratios: [Section 8.1]
 	
• To put a ratio into lowest terms, we do one of the following:
  a. If the ratio is in fractional form, factor the numerator and the denominator into prime factors and divide the numerator and denominator by the factors common to both.
  b. If the ratio is in the form $a : b$ or a to b, factor both sides into prime factors and divide both sides by the factors common to both.
	Example 1:
	Write the ratio $20 : 16$ in lowest terms.
	
	Solution:
	$20 : 16$ Factor both sides into prime factors.
	$2 \cdot 2 \cdot 5 : 2 \cdot 2 \cdot 2 \cdot 2$ Divide both parts by the common factors of $2 \cdot 2$.
	$\dfrac{2 \cdot 2 \cdot 5}{2 \cdot 2} : \dfrac{2 \cdot 2 \cdot 2 \cdot 2}{2 \cdot 2}$ Simplify.
	$5 : 4$ So, $20 : 16$ is the same as $5 : 4$.

Rates: [Section 8.1] • **Rates** are ratios with different units of measurement.	**Example 2:** There are 208 students in 8 sections of Beginning Algebra. What is the average number of students per section? Solution: $\dfrac{208 \text{ students}}{8 \text{ sections}} =$ Ratio of students to sections. Reduce and leave the denominator 1. $\dfrac{26 \text{ students}}{1 \text{ section}} =$ This is read "26 students in 1 section." $\dfrac{26 \text{ students}}{\text{section}}$ This is read "26 students per section."
Unit Price: [Section 8.1] • A **unit price** is a rate that expresses the cost compared to one unit.	
Proportion: [Section 8.2] • A proportion is an equation that states that two or more ratios are equal. In symbols, $\frac{a}{b} = \frac{c}{d}$ is a proportion. The fractions $\frac{a}{b}$ and $\frac{c}{d}$ form a proportion if $a \cdot d = b \cdot c$.	

Concept/Procedure	Example

Procedure for Solving Proportions: [Section 8.2]

- When solving a proportion, we perform the following steps:
  1. Simplify all fractional ratios.
  2. Cross multiply the fractional ratios.
  3. Solve the resulting equation.

Example 3:

Solve the proportion $\dfrac{5}{3} = \dfrac{4x - 1}{2x + 1}$.

Solution:

$\dfrac{5}{3} = \dfrac{4x - 1}{2x + 1}$ Cross multiply.

$5(2x + 1) = 3(4x - 1)$ Distribute.

$10x + 5 = 12x - 3$ Subtract $10x$ from both sides.

$5 = 2x - 3$ Add 3 to both sides.

$8 = 2x$ Divide both sides by 2.

$4 = x$ Solution. (4 is not an extraneous solution)

Percent Notation: [Section 8.3]

- **Percent notation** is a special way of writing a fraction with a denominator of 100. That is, if n is some number, then $\frac{n}{100} = n\%$.

Conversions: [Section 8.3]

- Percent to a Fraction
  a. Think of $x\%$ as $x \cdot 1\%$.
  b. Replace 1% with $\frac{1}{100}$.
  c. Multiply.

Example 4:

Convert 37% to a fraction.

Solution:

$37\% = 37 \cdot 1\%$ Replace 1% with $\dfrac{1}{100}$.

$= 37 \cdot \dfrac{1}{100}$ Multiply.

$= \dfrac{37}{100}$ Therefore, $37\% = \dfrac{37}{100}$.

- Percent to a Decimal
  a. Think of $x\%$ as $x \cdot 1\%$.
  b. Replace 1% with .01.
  c. Multiply.

Example 5:

Convert 2.5% to a decimal.

Solution:

$2.5\% = 2.5 \cdot 1\%$ Replace 1% with .01.

$= 2.5(.01)$ Multiply.

$= .025$ Therefore, $2.5\% = .025$.

- Decimal to Percent
  a. Write the decimal as the product of itself and 1.
  b. Change 1 to 100%.
  c. Multiply.

Example 6:

Convert .193 into a percent.

Solution:

$.193 = .193(1)$ Change 1 to 100%.

$= .193(100\%)$ Multiply.

$= 19.3\%$ Therefore, $.193 = 19.3\%$.

Concept/Procedure	Example

• Fraction to Percent

  a. Write the fraction as the product of itself and 1.

  b. Change 1 to 100%.

  c. Multiply, and if necessary, round to a specified number of decimal places.

Example 7:

Convert $\dfrac{7}{20}$ into a percent.

Solution:

$$\dfrac{7}{20} = \dfrac{7}{20} \cdot 1 \qquad \text{Change 1 to 100\%.}$$

$$= \dfrac{7}{20} \cdot 100\% \qquad \text{Multiply.}$$

$$= \dfrac{700\%}{20} \qquad \text{Divide.}$$

$$= 35\% \qquad \text{Therefore, } \dfrac{7}{20} = 35\%.$$

Simple Percent Situation: [Section 8.4]

• Simple percent situations are represented by equations that relate percent, the number of which we are taking the percent, and the result of taking the percent. Simple percent situations can be represented as _____ % of _____ is _____. The first blank is called the *percent*, the second the *base*, and the third the *amount*.

Example 8:

45 is 30% of what number?

Solution:

45 is 30% of what number?   "Is" means equals, "of" means multiply, and "what number" is the variable.

$45 = 30\% \cdot x$ Convert 30% to a decimal.

$45 = .30x$ Divide both sides by .30.

$150 = x$ Solution.

Percent Proportion: [Section 8.4]

• If P represents the number of percent, A represents the amount, and B represents the base, the relationship between these quantities is given by the following proportion:

$$\dfrac{P}{100} = \dfrac{A}{B}$$

Procedure for Using Proportions to Solve Simple Percent Situations: [Section 8.4]

• The procedure for solving a percent situation with the percent proportion, $\dfrac{P}{100} = \dfrac{A}{B}$

  a. Identify the number of percent, the amount, and the base of the second ratio.

  b. Substitute the appropriate values into the simple percent proportion.

  c. Solve the proportion for the indicated variable.

Example 9:

What percent of 48 is 72? Solve by using a proportion.

Solution:

The base is 48, the amount is 72, and we are looking for the percent.

$$\dfrac{P}{100} = \dfrac{A}{B} \qquad \text{Substitute for } A \text{ and } B.$$

$$\dfrac{P}{100} = \dfrac{72}{48} \qquad \text{Reduce } \dfrac{72}{48}.$$

$$\dfrac{P}{100} = \dfrac{3}{2} \qquad \text{Cross multiply.}$$

$$2P = 300 \qquad \text{Divide both sides by 2.}$$

$$P = 150 \qquad \text{Therefore, 72 is 150\% of 48.}$$

Concept/Procedure	Example

Procedure for Solving Percent Problems:
[Section 8.4]

1. Rewrite the problem in the form of
   _____% of _____ is _____.
2. Translate the statement in step 1 into an equation or proportion.
3. Solve the equation or proportion.
4. Check your answer.

Example 10:

A brine solution is 14% salt and the remainder is water. How much salt is in 500 liters of the solution?

Solution:

Rewrite the question in the form if _____% of _____ is _____. We are told that 14% *of the solution* (500 liters) is salt. This translates as follows:

14% of 500 is what number? — Rewrite as an equation, letting x represent the amount of salt.

$14\% \cdot 500 = x$ — Convert 14% to a decimal.

$.14 \cdot 500 = x$ — Multiply.

$70 = x$ — Therefore, there are 70 liters of salt.

Sales Tax: [Section 8.5]

- **Sales tax** is the amount of money that we pay to the government when we make certain types of purchases.

Sales Tax Rate: [Section 8.5]

- The **sales tax rate** is a portion of the amount paid for the goods purchased. The sales tax rate is expressed as a percent.

Sales Tax Calculation: [Section 8.5]

- Calculating sales tax is a special type of simple percent situation and thus takes the form of _____ % of _____ is _____. In this case, it becomes *sales tax rate · purchase price = the sales tax*.

Example 11:

The sales tax on an HDTV was $84.00. Find the purchase price of the HDTV if the sales tax rate is 7%.

Solution:

We know the sales tax rate is 7% = .07 and the sales tax is $84.00. Use *sales tax rate · purchase price = the sales tax* and substitute.

$.07 \cdot p = 84$ — Divide both sides by .07.

$\dfrac{.07p}{.07} = \dfrac{84}{.07}$ — Simplify.

$p = 1200$ — The HDTV cost $1200.

Total Cost = Purchase Amount + Sales Tax:
[Section 8.5]

- That is, $C = A + T$.

Discount: [Section 8.5]

- A **discount** is the amount of money that is taken off (subtracted from) the original price of an item.

Concept/Procedure	Example
Discount Rate: [Section 8.5] • The **discount rate** is a portion of the original price of the item. It is expressed in terms of percent.	
Discount Calculation: [Section 8.5] • Calculating discount is a special type of simple percent situation and takes the form of _____ % of _____ is _____ , which becomes *discount rate · original price = the amount of the discount.*	Example 12: A pair of hiking shoes originally priced at $210 was discounted by $25.20. Find the discount rate. Solution: We know the original price was $210 and the discount was $25.20. Use *discount rate · original price = amount of discount* and substitute. $r\% \cdot 210 = 25.20$ Think of $r\%$ and $r \cdot 1\%$ and change 1% to 0.1. $r(.01)(210) = 25.20$ Multiply .01 and 210. $2.1r = 25.20$ Divide both sides by 2.1. $r = 12$ Therefore, the rate of discount is 12%.
Sale Price = Original Price − Discount: [Section 8.5] • That is, $S = O - D$.	
Commission: [Section 8.5] • **Commission** is the amount of money that is paid to a person for selling an item.	
Commission Rate: [Section 8.5] • The **commission rate** is a portion of the amount of sales. It is expressed in terms of percent.	
Commission Calculation: [Section 8.5] • Calculating commission is a special type of simple percent situation. It takes the form of _____% of _____ is _____ and becomes *commission rate · sales = amount of commission.*	Example 13: A house sold for $640,000. The real estate office charged a commission of $6\frac{1}{2}\%$. How much was the commission? Solution: We know the house sold for $640,000 and the commission rate was $6\frac{1}{2}\%$. Use *commission rate · sales = amount of commission* and substitue. As a decimal, $6\frac{1}{2}\% = .065$. $.065 \cdot 640,000 = c$ Multiply. $41,600 = c$ The commission was $41,600.

Concept/Procedure	Example
Interest: [Section 8.5] • Interest is money that you must pay to someone or to an institution to use their money, or money paid to you by someone who uses your money.	
Principal: [Section 8.5] • **Principal** is the amount of money borrowed or deposited into an account.	
Annual Rate of Interest: [Section 8.5] • The **annual rate of interest** is a portion of the principal expressed as a percent.	
Interest Calculation Formula: [Section 8.5] • If I = interest, p = principal, r = interest rate as a decimal, and t = time in years, then interest is calculated by the following formula: $$I = p \cdot r \cdot t$$	Example 14: Find the simple interest paid on a loan of $5000 at 8% interest for 4 years. Solution: $I = p \cdot r \cdot t$ Substitute. $I = (5000)(.08)(4)$ Multiply. $I = 1600$ The interest is $1600.
Balance: [Section 8.5] • The **balance** of an account is the amount of money in the account at a particular time. $$\text{Balance} = \text{principal} + \text{interest}.$$	
Percent Increase or Decrease: [Section 8.5] $\dfrac{\text{Amount of increase or decrease}}{\text{original amount}}$ and convert to a percent.	Example 15: In one year, Kasey grew from a height of 25 inches to 27 inches. What was the percent increase in her height? Solution: $\text{Percent increase} = \dfrac{\text{amount of increase}}{\text{original amount}}$ Substitute. $\text{Percent increase} = \dfrac{2}{25}$ Convert to a percent. $\text{Percent increase} = \dfrac{2}{25} \cdot 1$ Change 1 to 100%. $\text{Percent increase} = \dfrac{2}{25} \cdot 100\%$ Multiply. $\text{Percent increase} = 8\%$ Therefore, Kasey had an increase of 8% in height.

Chapter 8 Review Exercises

Simplify the following ratios: [Section 8.1]

1. 12 : 20

2. 35 to 30

3. $\dfrac{27}{15}$

4. $\dfrac{18}{24}$

Write each of the following as ratios in lowest terms: [Section 8.1]

5. In a college parking lot, there were 625 compact cars and 350 middle-sized cars. What was the ratio of compact cars to middle-sized cars?

6. A party mix calls for 4 cups of pretzels, 2 cups of peanuts, and 2 cups of raisins. What is the ratio of the ingredients?

Write each of the following as a ratio of whole numbers: [Section 8.1]

7. A recipe for homemade coconut candy calls for $\frac{4}{3}$ cups of fresh coconut, $\frac{2}{3}$ cup of sugar, and $\frac{1}{3}$ cup of margarine. What is the ratio of the ingredients?

8. There are 2.27 kilograms in 5 pounds. What is the ratio of kilograms to pounds?

Express each ratio with the second part equal to 1: [Section 8.1]

9. A group of 15 students orders five large pizzas that are each cut into 12 pieces. How many pieces should each student get?

10. There are five families in a food co-operative. If they buy 100 pounds of rice, how much does each family receive?

Express each of the following as a unit price. If necessary, round to the nearest cent: [Section 8.1]

11. If 8 ounces of tomato sauce cost $.65, what is the price per ounce?

12. If 75 square feet of aluminum foil cost $2.99, what is the price per square foot?

Determine whether the following fractions can form a proportion: [Section 8.2]

13. $\dfrac{12}{15}$ and $\dfrac{20}{25}$

14. $\dfrac{.8}{.24}$ and $\dfrac{6}{18}$

Solve the following proportions: [Section 8.2]

15. $\dfrac{2}{x} = \dfrac{4}{5}$

16. $\dfrac{8}{5} = \dfrac{y}{60}$

17. $\dfrac{48}{15} = \dfrac{32}{t}$

18. $\dfrac{u}{14} = \dfrac{0}{4}$

19. $\dfrac{y}{2} = \dfrac{8.4}{3}$

20. $\dfrac{11}{3} = \dfrac{x}{.6}$

21. $\dfrac{\frac{2}{3}}{5} = \dfrac{v}{15}$

22. $\dfrac{6}{\frac{4}{5}} = \dfrac{35}{v}$

23. $\dfrac{9}{4\frac{1}{2}} = \dfrac{10}{t}$

24. $\dfrac{w}{15} = \dfrac{3\frac{2}{5}}{17}$

25. $\dfrac{3}{4} = \dfrac{9}{x+5}$

26. $\dfrac{4}{5} = \dfrac{2}{x-1}$

27. $\dfrac{y+5}{2} = \dfrac{y}{-3}$

28. $\dfrac{3(5t-1)}{7} = \dfrac{4t+2}{2}$

29. $\dfrac{x+2}{2} = \dfrac{9}{x-5}$

30. $\dfrac{3}{2x-3} = \dfrac{x+2}{3}$

Solve each of the following: [Section 8.2]

31. A riding mower uses 2 quarts of gasoline to mow $\frac{2}{3}$ acre of grass. How many acres can be mowed with 9 quarts of gasoline?

32. A machine can fill 3600 cans in 25 minutes. How many cans can be filled in 70 minutes?

33. On a map, two airports are $4\frac{1}{2}$ inches apart. If 90 miles is represented as $\frac{3}{4}$ inch apart on the map, how many miles apart are the two airports?

34. A naturalist group is interested in how many birds are nesting in a conservation area. They capture 210 birds, tag them, and release them. Later, 288 birds are captured and 32 of them are tagged. How many birds are in the conservation area?

Convert the given fractions to percent notation. If necessary, round to the nearest tenth of a percent: [Section 8.3]

35. $\dfrac{1}{4}$

36. $\dfrac{7}{8}$

37. $\dfrac{5}{2}$

38. 4

39. $\dfrac{3}{7}$

40. $\dfrac{5}{6}$

Convert the following decimals to fractions: [Section 8.3]

41. .4

42. 1.09

43. .082

44. .333 . . .

Convert the following decimals to percent notation: [Section 8.3]

45. .025

46. 6.57

47. .3

48. .0461

Perform each indicated operation. [Section 8.3]

49. 88% − 27.5%

50. $32\dfrac{1}{3}\% + 18\%$

Convert each of the following percents to fractions: [Section 8.3]

51. 8%

52. 12.5%

53. $9\dfrac{1}{4}\%$

54. .04%

Convert the following percents to decimals: [Section 8.3]

55. 18%

56. $6\dfrac{3}{4}\%$

57. 2.6%

58. .005%

59. 327%

60. 9.7%

Solve the following by simple percent sentences: [Section 8.4]

61. 24.5 is 70% of what number?

62. 48 is what percent of 60?

63. 60 is what percent of 40?

64. 26 is 65% of what number?

Solve each of the following and round the answers to the nearest percent or nearest hundredth where appropriate: [Section 8.4]

65. A college football quarterback completed 16 out of 25 passes attempted. What percent of his passes were completed?

66. At one point during the Persian Gulf conflict, the United States lost four aircraft, which was said to be a 1% loss rate. How many U.S. aircraft were involved in this war?

67. The federal government spends $77 million a year investigating ways to prevent breast cancer and $648 million for investigating ways to prevent heart disease. What percent of the amount spent on heart disease is the amount spent on breast cancer?

68. On the basis of calories consumed, the average adult eats a diet that is 42% fat. If a person consumes 2000 calories of food per day, how many calories are from fat?

Solve the following exercises involving sales tax: [Section 8.5]

69. Find the tax rate if the tax on a refrigerator that sold for $980 was $63.70.

70. The sales tax on an automobile purchase is $750. If the sales tax rate is 6%, what was the total cost of the automobile, including tax?

Solve the following exercises involving discounts: [Section 8.5]

71. Find the original price of a dining room suite that was discounted by $345 if the discount rate was 30%.

72. Mark paid $664 for a suit that was discounted 20%. What was the original price of the suit?

Solve the following exercises involving commissions: [Section 8.5]

73. If Hans sold a used car for $5800 and received a commission of $696, what was his rate of commission?

74. Jim earned a 2.5% commission for selling a $198,000 house. How much did he earn for selling the house?

Given the interest formula I = p · r · t, solve for the missing value: [Section 8.5]

75. $r = 8\%, I = \$120, p = ?, t = \frac{1}{2}$ year

76. Sophia paid $1280 to borrow $4000 for 4 years. What was the annual interest rate?

77. Patrick borrowed $560 for 90 days at 6% interest. How much interest did he pay?

78. Calculate the balance after 1 year for an account that pays 7% interest compounded quarterly on a beginning balance of $2500.

Solve the exercise. If necessary, round to the nearest tenth: [Section 8.5]

79. The federal government expected an 85% increase on its costs to control pollution from 2000 to the year 2010. If the 2000 costs were $100 billion, what was the expected cost in the year 2010?

Chapter 8 Test

Simplify the ratio.

1. $8 : 6 : 10$

Write the ratio in lowest terms.

2. At a certain college, 840 faculty members teach 84,000 students. What is the ratio of faculty to students?

Write as a ratio of whole numbers.

3. A recipe for pancakes calls for 2 cups of pancake mix, $\frac{1}{4}$ cup of cholesterol-free egg product, and $1\frac{1}{4}$ cups of skim milk. What is the ratio of the ingredients?

Express the ratio with the second part equal to 1.

4. A grant of $5,600,000 was given to seven centers for educational research. How much money was given to each center?

Express as a unit price. If necessary, round to the nearest cent.

5. If $\frac{1}{2}$ gallon (64 ounces) of ice cream costs $2.88, what is the price per ounce?

Solve the following proportions:

6. $\dfrac{15}{27} = \dfrac{y}{18}$

7. $\dfrac{3}{4} = \dfrac{2x + 1}{4x - 4}$

8. $\dfrac{x + 2}{-2} = \dfrac{4}{x + 8}$

Solve each of the following:

9. If a washing machine can wash three loads of clothes in 75 minutes, how many minutes will it take to wash seven loads of clothes?

10. In Europe, a forester wants to know how many wild hogs there are in a forest. He captures 12 hogs, tags them, and releases them. A few months later, he captures 18 hogs and finds that 4 are tagged. Estimate the number of hogs in the forest.

Convert the fractions to percents. If necessary, round to the nearest tenth of a percent.

11. $\dfrac{8}{5}$

12. $\dfrac{4}{7}$

Convert the decimals to percents. If necessary, round to the nearest tenth of a percent.

13. .308

14. 7.666 . . .

Convert the percents to fractions.

15. 37.5%

16. $9\dfrac{1}{4}\%$

Solve each of the exercises. Round the answers to the nearest percent or nearest hundredth where appropriate.

17. 969 is 102% of what number?

18. If 350 jobs are 8% of the jobs in a company, how many jobs are in the company?

19. A major airline decided to cancel 97 of its 291 flights. What percent of the flights did it cancel?

737

Solve each of the following:

20. Fred borrowed $2400 at an 8.5% interest rate for a period of 90 days. How much interest did he pay?

21. Helene paid $25.50 in sales tax on a dishwasher that costs $425. What was the sales tax rate?

22. An item marked at $350 was discounted 15%. How much was the purchase price?

23. A book sales representative earns 5% commission on all books sold. If he sold 20,000 books at $35 each, what was his commission?

24. A corporation announced a 15% cut in management salaries. If an executive was making $85,000 a year, what was his new salary?

25. Find the balance after 1 year in a savings account that pays 6% interest compounded semiannually if the beginning balance was $12,000.

26. Francine purchased a boat for $18,500. If the sales tax rate is 7%, what was the total cost of the boat?

Systems of Linear Equations

Chapter Outline

In this chapter, we continue our study of linear equations with two variables. We know from Chapter 4 that a linear equation with two variables has an infinite number of solutions that are written as ordered pairs. In this chapter, we will be finding the single ordered pair (if there is one) that solves two linear equations with two variables. We will learn three techniques for finding this point, if it exists. These techniques also provide us with another technique for solving application problems with two unknowns.

Section 9.1   Defining Linear Systems and Solving by Graphing

OBJECTIVES *When you complete this section, you will be able to:*

A Determine whether a given ordered pair is the solution of a linear system with two variables.

B Solve a linear system with two variables by graphing.

C Determine whether a linear system with two variables is consistent, inconsistent, or dependent.

PREREQUISITE SKILLS *Before beginning this section, you should be able to:*

a. Determine if an ordered pair is a solution of an equation. (Section 4.1)

b. Recognize that the coordinates of any point on the graph of a line is a solution of the equation. (Section 4.2)

c. Determine whether two lines are parallel. (Section 4.5)

GETTING READY FOR SECTION 9.1

Determine whether the given ordered pair is a solution to the given equation.

1. $x + 3y = 1; (-2, 1)$ 　　　　　　 **2.** $3x - 2y = 12; (4, -3)$

Given a value of x or y, determine the other coordinate that will make the ordered pair lie on the line represented by the given equation.

3. $x - 4y = 8; x = 0$ 　　　　　　 **4.** $5x - 2y = 3; y = 1$

Are the following pairs of lines parallel?

5. $2x + 3y = 12$
   $2x + 3y = -2$

6. $3x - 4y = 8$
   $x - 2y = 6$

Introduction

As mentioned in the introduction, in Chapter 4 we found that a linear equation with two variables has an infinite number of solutions, each of which is an ordered pair. In this section, we will be looking for one ordered pair that is the solution of two linear equations. For example, some solutions of $x + y = 4$ are $(0, 4), (4, 0), (1, 3), (2, 2),$ $(5, -1), (-1, 5),$ and $(-2, 6)$. Some solutions of $x - y = 6$ are $(0, -6), (6, 0), (1, -5),$ $(3, -3), (5, -1), (-2, -8),$ and $(-3, -9)$. By examining the ordered pairs that are solutions of each equation, we see that the ordered pair $(5, -1)$ is a solution of both equations. The equations $x + y = 4$ and $x - y = 6$ are an example of a **system of linear equations**. Therefore, $(5, -1)$ is a solution of this system. Systems of equations are usually written with one equation beneath the other, as in the following:

$$\begin{cases} x + y = 4 \\ x - y = 6 \end{cases}$$

> **Definition** System of Linear Equations
>
> A **system of linear equations** is two or more linear equations with the same variables.

We observed that the ordered pair $(5, -1)$ solved each of the equations $x + y = 4$ and $x - y = 6$. Any ordered pair that solves all the equations in a system is a **solution of the system**.

> **Definition** Solution(s) of a System of Linear Equations
>
> The solution(s) of a system of linear equations with two variables consist(s) of all ordered pairs that solve all the equations of the system.

Answers: **Getting Ready:** **1.** Yes **2.** No **3.** $y = -2$ **4.** $x = 1$ **5.** Yes **6.** No

OBJECTIVE **Determining whether a given ordered pair is the solution of a linear system with two variables**

To determine whether an ordered pair is a solution of a system of linear equations, we must show that the ordered pair solves each equation of the system. This means that if we replace x in each equation with the first element of the ordered pair and replace y with the second element of the ordered pair, *both* equations will result in true statements when simplified.

Example 1 | **Determine whether the given ordered pair is a solution of the system of linear equations. The ordered pair is in the form (x, y).**

a. $(3, -1)$
$$\begin{cases} 2x + y = 5 \\ 3x - y = 10 \end{cases}$$

Solution

To determine whether the given ordered pair is a solution of the system, we must determine whether it solves each equation in the system. Therefore, we substitute 3 for x and -1 for y in each equation and simplify.

$$2x + y = 5 \qquad\qquad 3x - y = 10$$
$$2(3) + (-1) = 5 \qquad 3(3) - (-1) = 10$$
$$6 - 1 = 5 \qquad\qquad 9 + 1 = 10$$
$$5 = 5 \qquad\qquad 10 = 10$$

Since $(3, -1)$ solves both equations of the system, $(3, -1)$ is a solution of the system.

b. $(2, -3)$
$$\begin{cases} 4x + 3y = -1 \\ 2x - 5y = -11 \end{cases}$$

Solution

Substitute 2 for x and -3 for y in each equation and simplify.

$$4x + 3y = -1 \qquad\qquad 2x - 5y = -11$$
$$4(2) + 3(-3) = -1 \qquad 2(2) - 5(-3) = -11$$
$$8 - 9 = -1 \qquad\qquad 4 + 15 = -11$$
$$-1 = -1 \qquad\qquad 19 \neq -11$$

Since $(2, -3)$ does not solve both of the equations in the system, $(2, -3)$ is not a solution of the system.

PRACTICE EXERCISE 1

Determine whether the given ordered pair is a solution of the system of linear equations. The ordered pair is in the form (x, y).

a. $(4, 1)$
$$\begin{cases} 3x + y = 13 \\ 2x + y = 9 \end{cases}$$

b. $(-3, 2)$
$$\begin{cases} 2x + 3y = -12 \\ 4x - 5y = 2 \end{cases}$$

(Continued)

Answers: Practice Exercise 1: a. yes **b.** no

If you need more practice, do the following Additional Practice Exercises.

Additional Practice Exercise 1   Determine whether the given ordered pair is a solution of the system of linear equations. The ordered pair is in the form (x, y).

a. $(2, -4)$
$$\begin{cases} x - 2y = -10 \\ x + 4y = -14 \end{cases}$$

b. $(2, -1)$
$$\begin{cases} 5x + 3y = 7 \\ 4x - y = 9 \end{cases}$$

OBJECTIVE **B**   Solving a linear system with two variables by graphing

In the previous examples and practice exercises, the ordered pair was given and we were asked to determine whether the ordered pair was a solution of the system. Now we will find the ordered pair that is the solution of a system of linear equations. Recall from Chapter 4 that if a point lies on the graph of a line, then the coordinates of the point must solve the equation of the line. Consequently, if the graphs of two lines intersect at a point, then the coordinates of the point of intersection must solve both equations. This leads us to the following method of solving a system of linear equations with two variables:

Note

We must draw the graphs carefully and accurately in order to find the point of intersection.

> **Procedure: Solving a System of Linear Equations by Graphing**
>
> To solve a system of linear equations with two variables, graph each equation of the system. If the graphs of the equations intersect, the coordinates of the point(s) of intersection are the solutions of the system.

Example 2   **Find the solution(s) of the following systems of linear equations by graphing:**

a. $\begin{cases} x + y = 6 \\ x - y = -2 \end{cases}$

Solution

To find the solution(s) of the system, we graph both equations on the same coordinate system and find the point(s) of intersection of the graphs. Remember, a quick way to graph the equations is by using the intercepts. To find the x-intercept, we let $y = 0$; and to find the y-intercept, we let $x = 0$.

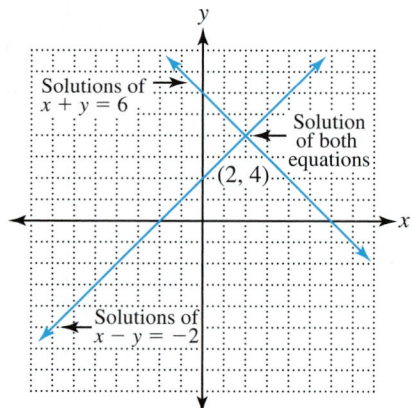

From the graph, we see that the two lines seem to intersect at the point whose coordinates are $(2, 4)$. Therefore, $(2, 4)$ is a possible solution of the system.

CHECK:

If $(2, 4)$ is the solution of the system, then $(2, 4)$ must solve both equations. Substitute 2 for x and 4 for y into both equations and simplify.

$$\begin{array}{ll} x + y = 6 & x - y = -2 \\ 2 + 4 = 6 & 2 - 4 = -2 \\ 6 = 6 & -2 = -2 \end{array}$$

Since $(2, 4)$ solves both equations, $(2, 4)$ is the solution of the system.

Answers: Additional Practice 1: a. no **b.** yes

b. $\begin{cases} y = -2x + 3 \\ y = 3x - 7 \end{cases}$

Solution

To find the solution(s) of the system, we graph both equations on the same coordinate system and find the point(s) of intersection of the graphs. Note that both equations are written in slope-intercept form.

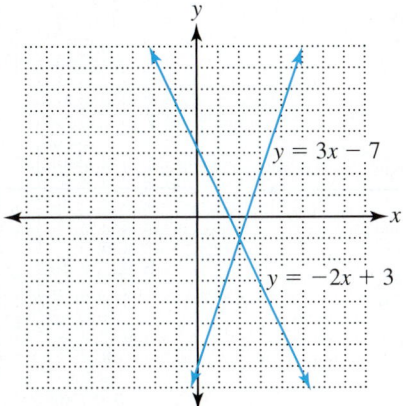

From the graph, we see that the two lines seem to intersect at the point whose coordinates are $(2, -1)$. Therefore, $(2, -1)$ is a possible solution of the system.

CHECK:

If $(2, -1)$ is the solution of the system, then $(2, -1)$ must solve both equations in the system. Substitute 2 for x and -1 for y into both equations and simplify.

$$\begin{array}{ll} y = -2x + 3 & \qquad y = 3x - 7 \\ -1 = -2(2) + 3 & \qquad -1 = 3(2) - 7 \\ -1 = -4 + 3 & \qquad -1 = 6 - 7 \\ -1 = -1 & \qquad -1 = -1 \end{array}$$

Since $(2, -1)$ solves both equations, $(2, -1)$ is the solution of the system.

c. $\begin{cases} x + 2y = 4 \\ x = -2 \end{cases}$

Solution

We need to graph the equations on the same coordinate system and find the point(s) of intersection. Note that $x = -2$ is the equation of a vertical line.

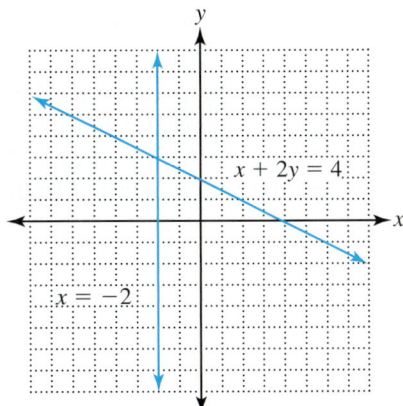

From the graph, we see that the lines seem to intersect at the point whose coordinates are $(-2, 3)$. Therefore, $(-2, 3)$ is a possible solution of the system. To verify that $(-2, 3)$ is the solution, substitute -2 for x and 3 for y and simplify as in Examples 2a and 2b.

PRACTICE EXERCISE 2

Find the solution(s) of the given systems of linear equations with two variables by graphing. Assume that each unit on the coordinate system represents 1.

a. $\begin{cases} x + y = 1 \\ x - y = 5 \end{cases}$ **b.** $\begin{cases} y = -3x - 1 \\ y = 2x - 6 \end{cases}$ **c.** $\begin{cases} 2x - 3y = 12 \\ x = 3 \end{cases}$

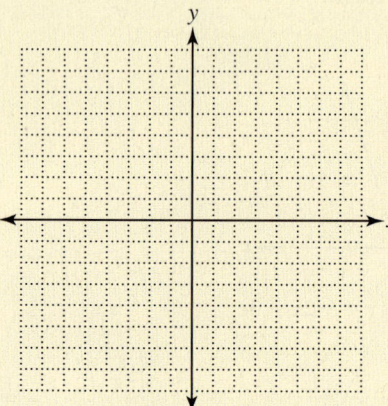   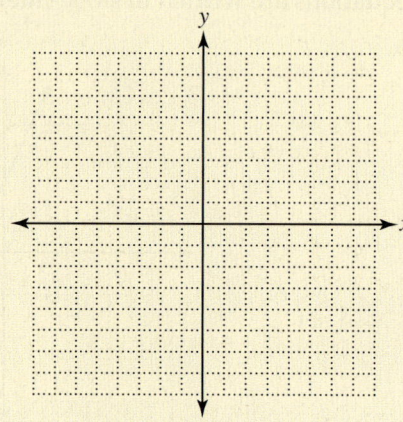   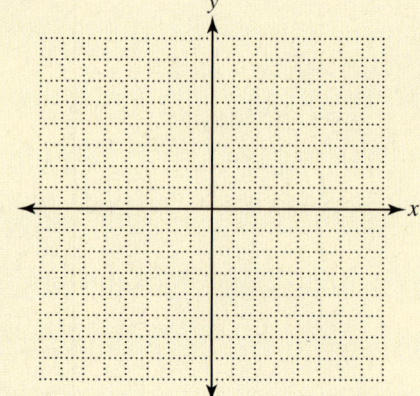

If you need more practice, do the following Additional Practice Exercises.

Additional Practice Exercise 2 Find the solution(s) of the given linear systems with two variables by graphing. Assume that each unit on the coordinate system represents 1.

a. $\begin{cases} x - y = -6 \\ x + 2y = 0 \end{cases}$ **b.** $\begin{cases} 3x - 2y = -12 \\ x - 2y = -8 \end{cases}$ **c.** $\begin{cases} y = 2x - 5 \\ y = 3 \end{cases}$

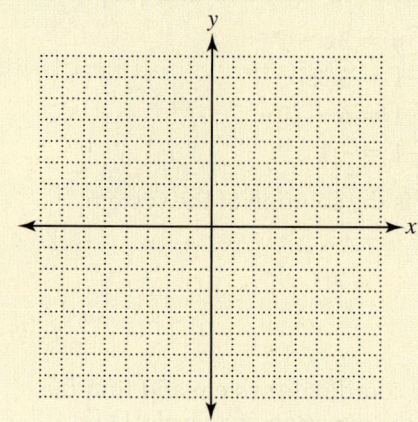   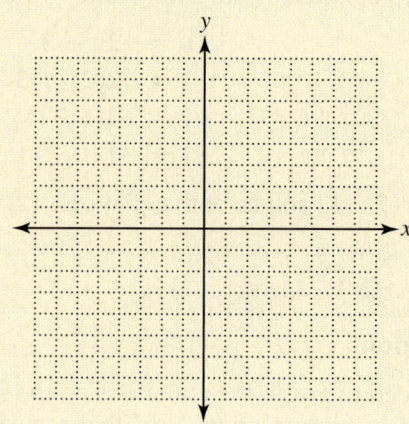   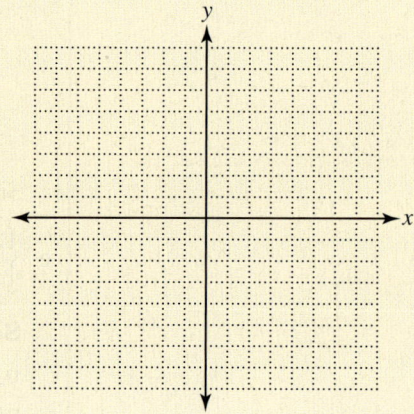

One of the problems with solving systems of equations graphically is that the two lines do not always intersect at a point whose coordinates are integers. In this case, it may not be possible to determine the exact solution of the system by the graphing method, as illustrated in the following example:

Answers: Practice Exercise 2: a. $(3, -2)$ **b.** $(1, -4)$ **c.** $(3, -2)$ **Additional Practice 2: a.** $(-4, 2)$ **b.** $(-2, 3)$ **c.** $(4, 3)$

Example 3 **Find the solution(s) of the given linear system with two variables by graphing. Assume that each unit on the coordinate system represents 1.**

a. $\begin{cases} 6x + 6y = 5 \\ 2x - 3y = -10 \end{cases}$

Solution

To find the solution(s) of the system, we graph both equations on the same coordinate system and find the point(s) of intersection of the graphs.

From the graph, we see that the x-value of the point of intersection is between -1 and -2 and the y-value is between 2 and 3. The only way to find the exact coordinates of the point of intersection by the graphical method is to approximate the coordinates and then substitute them into both equations. This can be very time consuming, and there is no guarantee that the exact solution will ever be found. For that reason, in this section we will limit ourselves to systems whose graphs intersect at points whose coordinates are integers. We will learn two other methods that are more efficient for solving systems of linear equations in Sections 9.2 and 9.3.

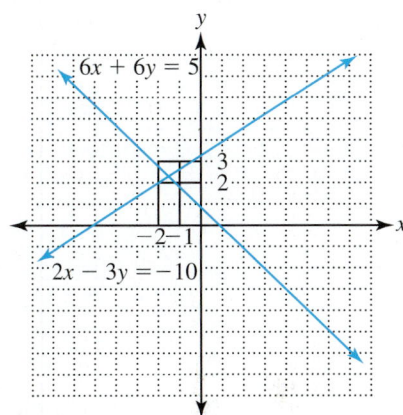

OBJECTIVE **C** Determining whether a linear system with two variables is consistent, inconsistent, or dependent

If a system of two linear equations with two variables is graphed on one coordinate system, there are three things that can happen.

1. The lines can intersect in exactly one point, as in Examples 2 and 3. If this occurs, the system is said to be **consistent** and the system has one solution, which is the coordinates of the point of intersection.

2. The lines can be parallel. If this occurs, the system is said to be **inconsistent** and the system has no solution, since parallel lines do not intersect.

3. The lines can coincide. If this occurs, the system is said to be **dependent** and there are an infinite number of solutions, since the lines intersect at an infinite number of points. Any ordered pair that solves one equation of a dependent system also solves the other. In a dependent system, one equation is a constant multiple of the other.

Examples of inconsistent and dependent systems follow:

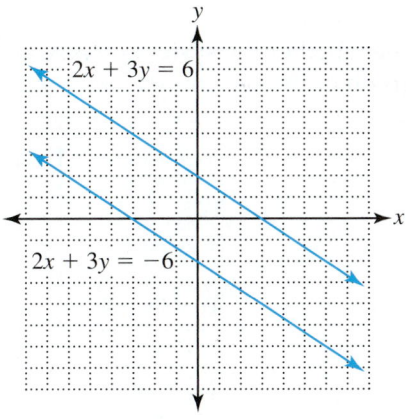

Inconsistent system:
No solution

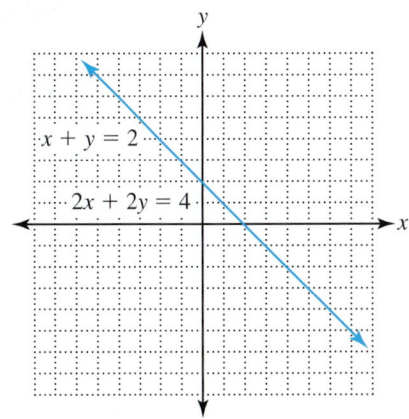

Dependent system:
Infinite number of solutions

Example 4 | Determine whether the given systems are consistent, inconsistent, or dependent. If the system is consistent, find the solution of the system.

a. $\begin{cases} 2x + y = 4 \\ 4x + 2y = -7 \end{cases}$

Solution

To determine whether the system is consistent, inconsistent, or dependent, we can draw the graph of the system on one coordinate system.

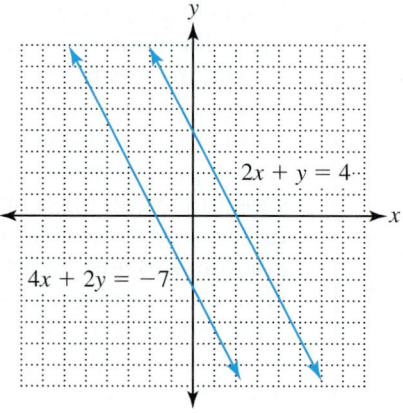

The two lines appear to be parallel. To confirm that they are, we need to find the slope of each line. Remember that parallel lines have equal slopes. To find the slope of each line, write each equation in slope-intercept form $(y = mx + b)$, where m is the slope.

$$2x + y = 4 \qquad\qquad\qquad 4x + 2y = -7$$
$$2x - 2x + y = -2x + 4 \qquad 4x - 4x + 2y = -4x - 7$$
$$y = -2x + 4 \qquad\qquad\qquad 2y = -4x - 7$$

Therefore, the slope is -2.
$$y = -2x - \frac{7}{2}$$

Therefore, the slope is -2.

Since both lines have the same slope, the lines are parallel. Since the lines have different y-intercepts, they are not the same line. Consequently, the system is inconsistent and has no solutions, which is denoted by $\varnothing$.

b. $\begin{cases} 4x + 6y = 4 \\ 6x + 9y = 6 \end{cases}$

Solution

To determine whether the system is consistent, inconsistent, or dependent, we can draw the graph of the system on one coordinate system.

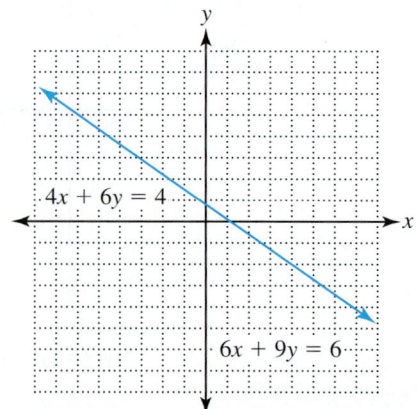

Note

It is possible to determine whether a system is inconsistent without graphing by writing each equation in slope-intercept form. The lines of an inconsistent system are parallel, so they must have the same slope. In order to be distinct lines, rather than the same line, they must have different y-intercepts.

The graphs of the two lines appear to be the same line. To confirm that they are, write each equation in slope-intercept form.

$$4x + 6y = 4$$
$$4x - 4x + 6y = -4x + 4$$
$$6y = -4x + 4$$
$$\frac{6y}{6} = \frac{-4x + 4}{6}$$
$$y = -\frac{4}{6}x + \frac{4}{6}$$
$$y = -\frac{2}{3}x + \frac{2}{3}$$

Therefore, the slope is $-\frac{2}{3}$ and the y-intercept is $\frac{2}{3}$.

$$6x + 9y = 6$$
$$6x - 6x + 9y = -6x + 6$$
$$9y = -6x + 6$$
$$\frac{9y}{9} = \frac{-6x + 6}{9}$$
$$y = -\frac{6}{9}x + \frac{6}{9}$$
$$y = -\frac{2}{3}x + \frac{2}{3}$$

Therefore, the slope is $-\frac{2}{3}$ and the y-intercept is $\frac{2}{3}$.

> **Note**
>
> It is possible to determine whether a system is dependent by writing both equations in slope-intercept form. If both lines have the same slope and the same y-intercept, they are the same line. Therefore, the system is dependent.

Since both lines have the same slope and the same y-intercept, they are the same line. Consequently, the system is dependent. The solutions of a dependent system are usually written as the set of ordered pairs that solve either of the equations. The solutions of this system could be written as $\{(x, y): 4x + 6y = 4\}$. This is read "the set of all ordered pairs (x, y) such that $4x + 6y = 4$."

c. $\begin{cases} 2x - y = 5 \\ 3x + 4y = 24 \end{cases}$

Solution

To determine whether the system is consistent, inconsistent, or dependent, we draw the graph of the system on one coordinate system.

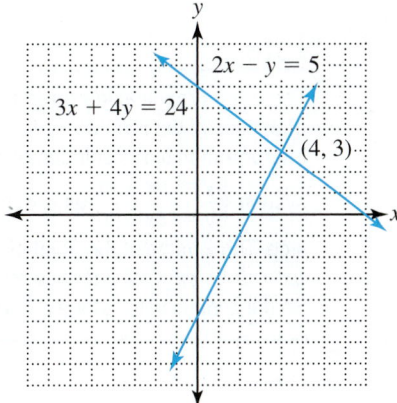

> **Note**
>
> It is possible to determine whether a system is consistent without graphing by writing the equations of the system in slope-intercept form. Since any two lines in the same plane that are not parallel must intersect, the system is consistent if the slopes of the lines are not the same.

Since the lines intersect at a single point, the system is consistent. The coordinates of the point of intersection are $(4, 3)$. Therefore, $(4, 3)$ is a possible solution of the system.

CHECK:

The check is left as an exercise for the student.

PRACTICE EXERCISE 4

Determine whether the given systems are consistent, inconsistent, or dependent. If the system is consistent, find the solution of the system.

a. $\begin{cases} 3x + 2y = 12 \\ 6x + 4y = -12 \end{cases}$

b. $\begin{cases} x + 2y = 6 \\ 2x = 12 - 4y \end{cases}$

c. $\begin{cases} 5x + 3y = 30 \\ x + y = 8 \end{cases}$

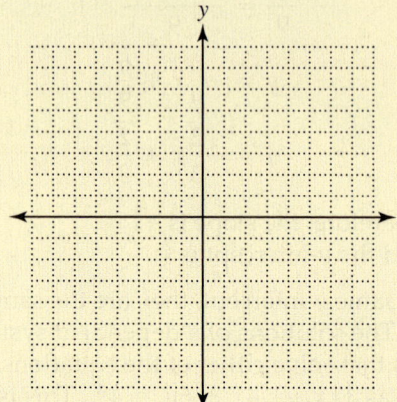   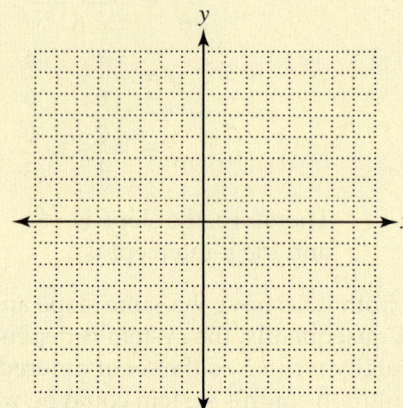   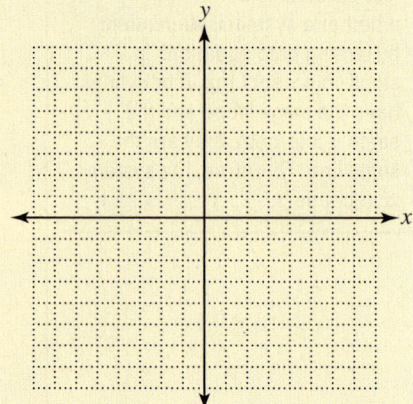

If you need more practice, do the following Additional Practice Exercises.

Additional Practice Exercise 4   Determine whether the given systems are consistent, inconsistent, or dependent. If the system is consistent, find the solution of the system.

a. $\begin{cases} x + 2y = 6 \\ 2x = -4y - 7 \end{cases}$

b. $\begin{cases} 4x + y = 7 \\ 2y = -8x + 14 \end{cases}$

c. $\begin{cases} x + 2y = 3 \\ y = -2x + 12 \end{cases}$

Note

There is a graphical interpretation of the solution of a linear equation. In Chapter 3, we solved linear equations of the form $ax + b = cx + d$ or simplified to that form. We can give a graphical interpretation of the solution of such equations by using systems of equations. We create the system $\begin{cases} y = ax + b \\ y = cx + d \end{cases}$ where $ax + b$ is the left side of the equation and $cx + d$ is the right side. The x-value of the point of intersection of these two graphs is the solution of the equation $ax + b = cx + d$, because the coordinates of the point of intersection solve both equations. So the x-value will make $ax + b = cx + d$.

For Extra Help

Exercise Set 9.1   MyMathLab®

Determine whether the given ordered pair is a solution of the system of linear equations. The ordered pair is in the form (x, y). (See Example 1.)

1. $(3, 4)$
$\begin{cases} x + 3y = 15 \\ x + 2y = 11 \end{cases}$

2. $(-4, 1)$
$\begin{cases} 2x + y = -7 \\ 3x + y = -11 \end{cases}$

3. $(-2, -3)$
$\begin{cases} 4x + 2y = 5 \\ x + 3y = 1 \end{cases}$

4. $(1, 6)$
$\begin{cases} x - y = -1 \\ 2x + y = 7 \end{cases}$

Answers: Practice Exercise 4: a. inconsistent **b.** dependent **c.** consistent; $(3, 5)$
Additional Practice 4: a. inconsistent **b.** dependent **c.** consistent; $(7, -2)$

5. $(-1, 2)$
$$\begin{cases} x - 3y = -7 \\ 6x + 2y = -2 \end{cases}$$

6. $(3, -1)$
$$\begin{cases} 2x - y = 7 \\ x + 2y = 1 \end{cases}$$

7. $(-2, 0)$
$$\begin{cases} 2x + 3y = 4 \\ x + 2y = 4 \end{cases}$$

8. $(0, 3)$
$$\begin{cases} 5x + 3y = 9 \\ -4x + 7y = 17 \end{cases}$$

The graphs shown here are of systems of equations. (a) *Give the type of system (consistent, inconsistent, or dependent).* (b) *Give the number of solutions, if any.* (c) *If there is only one solution, give the solution. (See Examples 2 and 4.)*

9. $\begin{cases} 3x + 2y = 1 \\ -2x + 3y = -18 \end{cases}$

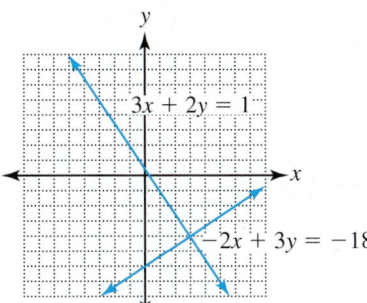

10. $\begin{cases} x + 2y = -5 \\ -3x + y = 1 \end{cases}$

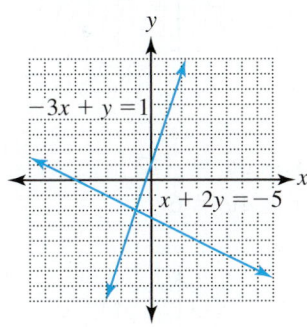

11. $\begin{cases} 2x + y = 4 \\ 4x + 2y = 12 \end{cases}$

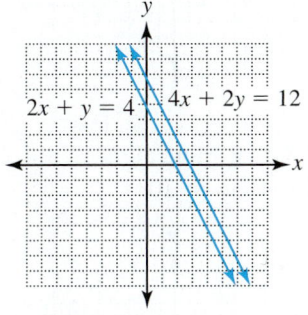

12. $\begin{cases} 3x - y = 6 \\ 6x - 2y = 2 \end{cases}$

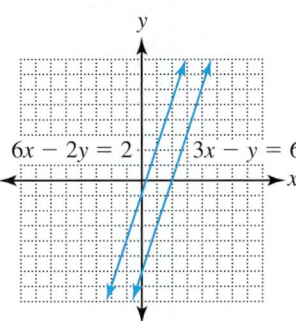

13. $\begin{cases} x + y = 4 \\ 3x + 3y = 12 \end{cases}$

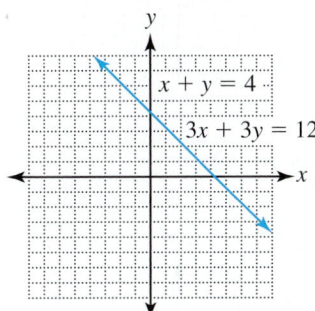

14. $\begin{cases} 2x - y = 4 \\ 6x - 3y = 12 \end{cases}$

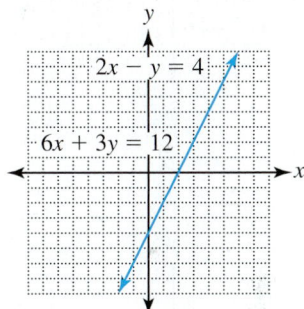

Find the solution(s) of the given systems of linear equations by graphing. Assume that each unit on the coordinate system represents 1. (See Examples 2 and 3.)

15. $\begin{cases} x + y = 1 \\ x - y = -1 \end{cases}$

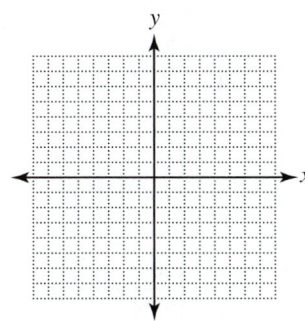

16. $\begin{cases} x + y = -1 \\ x - y = 3 \end{cases}$

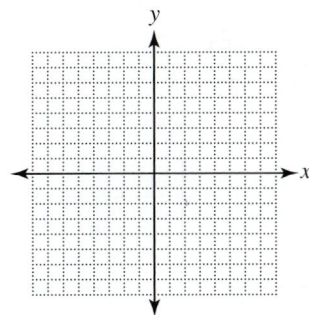

17. $\begin{cases} 3x - 2y = -12 \\ 2x + y = -1 \end{cases}$

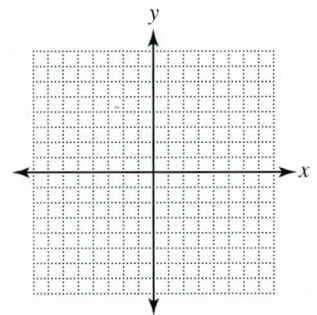

18. $\begin{cases} x - 2y = 2 \\ x + 4y = 8 \end{cases}$

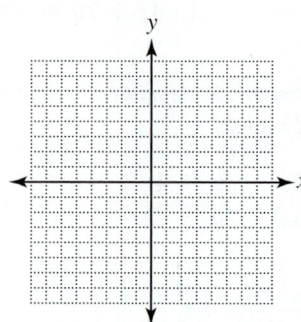

19. $\begin{cases} 2x - y = -4 \\ 2x - 3y = 0 \end{cases}$

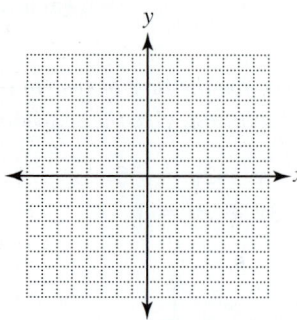

20. $\begin{cases} x - 4y = 8 \\ x + 4y = 0 \end{cases}$

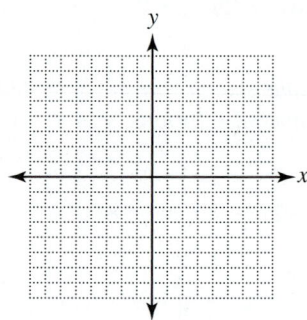

21. $\begin{cases} x - y = 4 \\ x + 5y = 10 \end{cases}$

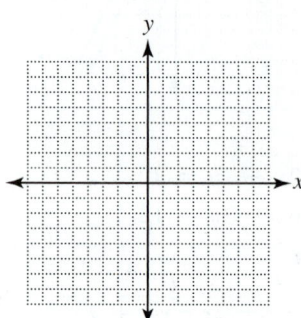

22. $\begin{cases} x - 2y = -12 \\ x + 4y = 6 \end{cases}$

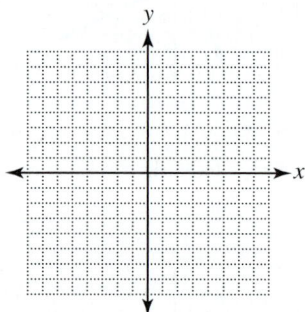

23. $\begin{cases} y = -x - 1 \\ y = -2x + 2 \end{cases}$

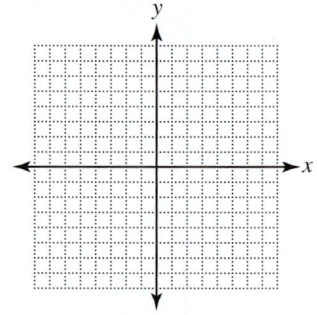

24. $\begin{cases} y = -x + 2 \\ y = 2x - 1 \end{cases}$

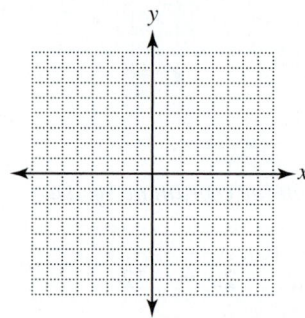

25. $\begin{cases} y = -2x + 9 \\ y = 2x - 3 \end{cases}$

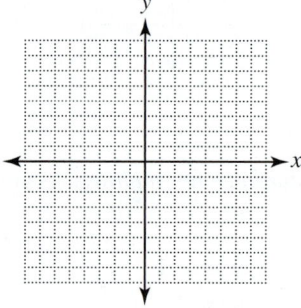

26. $\begin{cases} y = 3x + 6 \\ y = \frac{1}{3}x - 2 \end{cases}$

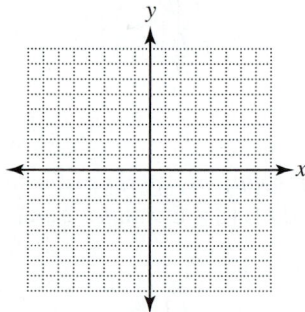

27. $\begin{cases} y = -\frac{3}{2}x - 7 \\ y = \frac{1}{2}x - 3 \end{cases}$

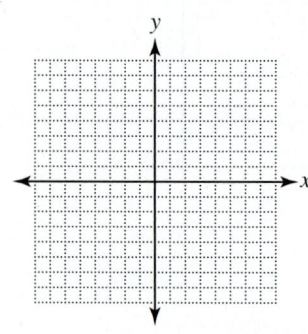

28. $\begin{cases} y = -\frac{2}{3}x - 3 \\ y = \frac{4}{3}x + 3 \end{cases}$

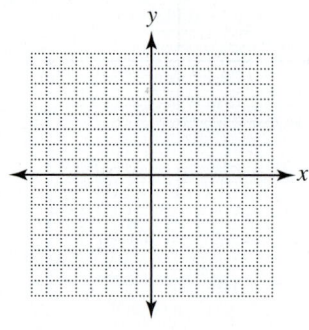

29. $\begin{cases} x - y = -7 \\ x = -3 \end{cases}$

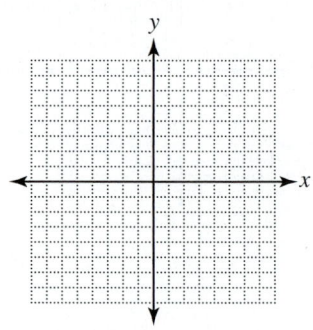

30. $\begin{cases} y = 5 \\ x + y = 4 \end{cases}$

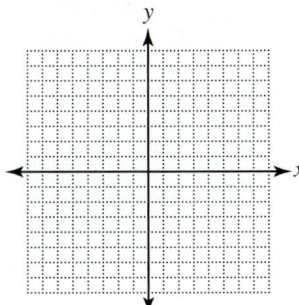

31. $\begin{cases} y = 3 \\ 2x - y = 3 \end{cases}$

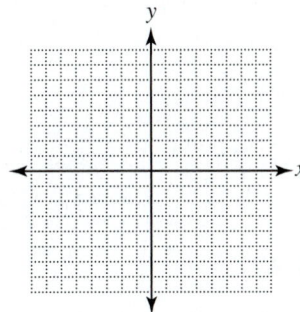

32. $\begin{cases} x = -5 \\ x - 2y = 3 \end{cases}$

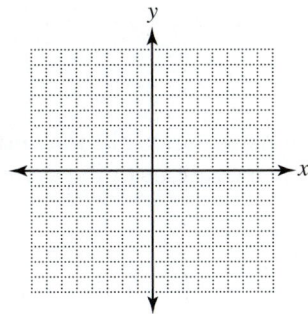

33. $\begin{cases} 2x + 3y = 0 \\ 3x - 2y = 0 \end{cases}$

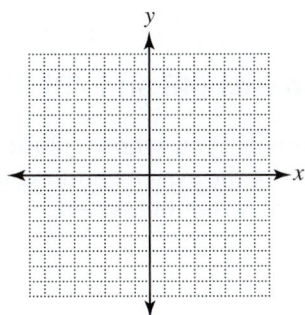

34. $\begin{cases} 4x - 2y = 4 \\ 2x + 3y = -6 \end{cases}$

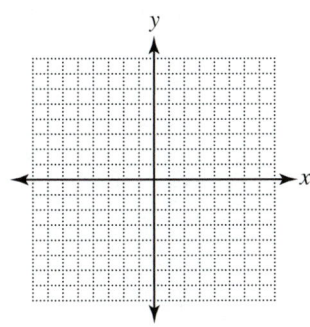

Determine whether the systems given are consistent, inconsistent, or dependent. If the system is consistent, find the solution of the system. Grids are not provided as with Exercises 15–34. (See Example 4.)

35. $\begin{cases} x - 3y = -12 \\ 4x - 12y = -36 \end{cases}$

36. $\begin{cases} 3x + 2y = 6 \\ 6x + 4y = 24 \end{cases}$

37. $\begin{cases} 2x + 3y = 6 \\ 4x + 6y = 12 \end{cases}$

38. $\begin{cases} 3x + y = 9 \\ 6x + 2y = 18 \end{cases}$

39. $\begin{cases} 3x + y = -6 \\ 9x + 3y = 9 \end{cases}$

40. $\begin{cases} -2x - y = 4 \\ 6x + 3y = 18 \end{cases}$

41. $\begin{cases} x + y = -5 \\ 6x - y = 12 \end{cases}$

42. $\begin{cases} 3x - 2y = 12 \\ x + y = -1 \end{cases}$

Challenge Exercises (43 and 44)

Determine whether the given ordered pair is a solution of the system of linear equations. The ordered pair is in the form (x, y). (See Example 1.)

43. $\left(\dfrac{1}{4}, 2\right)$

$$\begin{cases} 4x + y = 3 \\ 8x + 3y = 8 \end{cases}$$

44. $(-24, 12)$

$$\begin{cases} 2x - 3y = -84 \\ x + 5y = 36 \end{cases}$$

Writing Exercises (45–47)

45. If the slopes of two lines are not equal, what type of system is composed of these lines? Why?

46. If the slopes of two lines are equal, what type of system(s) is (are) composed of these lines? Why?

47. Why are the coordinates of the point of intersection of two lines the solution of the system?

Critical-Thinking Exercises (48 and 49)

48. Find a system of linear equations whose solution is (2, 3), draw the graph of the system, and explain how you found the equations of the system.

49. Find a system of equations with no solution and describe how you found the equations of the system.

<table>
<tr><td>Section
9.2</td><td>Solving Systems of Linear Equations
by Using Elimination by Addition</td></tr>
</table>

OBJECTIVES *When you complete this section, you will be able to:*

Ⓐ Use the addition method to solve consistent systems of linear equations with two variables by eliminating a variable.

Ⓑ Recognize inconsistent and dependent systems of equations when using the addition method.

PREREQUISITE SKILLS *Before beginning this section, you should be able to:*

a. Add, subtract, multiply, and divide signed numbers. (Sections 1.6, 1.7, 2.1, 2.5)

b. Combine like terms. (Section 1.7)

c. Determine whether an ordered pair is a solution of an equation. (Section 4.1)

d. Multiply a whole number and a fraction. (Section R.4)

e. Rewrite an equation in an equivalent form by using the addition property of equality. (Section 3.1)

f. Solve simple linear equations. (Sections 3.1, 3.2, 3.3)

g. Recognize consistent, dependent, and inconsistent systems. (Section 9.1)

h. Use the distributive property. (Section 1.7)

GETTING READY FOR SECTION 9.2

Add, subtract, multiply, or divide as indicated.

1. $8 + (-6)$ **2.** $10 - (-5)$

3. $(-4)(7)$ **4.** $6 \div (-2)$

Combine like terms.

5. $3y + 8 - 5y$ **6.** $-6w - 4 + 9w + 7$

7. Is $(1, -3)$ a solution of $x + y = -2$?   **8.** Multiply: $\left(-\dfrac{1}{3}\right)(15)$

9. Rewrite $y = -2x + 6$ in the $ax + by = c$ form.   **10.** If $y = -\dfrac{2}{3}$ and $x + 3y = -1$, find x.

11. Simplify $2(3x - 4y)$, using the distributive property.

Introduction

We found in Section 9.1 that one of the difficulties with solving systems of equations by the graphing method was that the lines did not always intersect at a point whose coordinates were integers. In this section, we will develop another method of solving systems of linear equations with two variables without graphing the system. It can be shown that if two lines intersect at a point, then the sum of the equations of the lines, or any multiples of these equations, is the equation of another line that intersects the two given lines at the same point of intersection.

Answers: Getting Ready: 1. 2 **2.** 15 **3.** -28 **4.** -3 **5.** $8 - 2y$ **6.** $3w + 3$ **7.** Yes **8.** -5 **9.** $2x + y = 6$
10. $x = 1$ **11.** $6x - 8y$

In Chapter 3, we introduced the addition property of equality, which stated if $a = b$, then $a + c = b + c$. We will need an extension of this property, which is given as follows:

> ### Rule: Addition of Equals
>
> If $a = b$ and $c = d$, then $a + c = b + d$. In words, equals added to equals results in equals.

We repeat here Example 2a from Section 9.1.

Example 1   **Find the solution(s) of the following systems of linear equations by graphing:**

$$\begin{cases} x + y = 6 \\ x - y = -2 \end{cases}$$

Solution

From the graph, we see that the two lines seem to intersect at the point whose coordinates are $(2, 4)$. Therefore, $(2, 4)$ is a possible solution of the system.

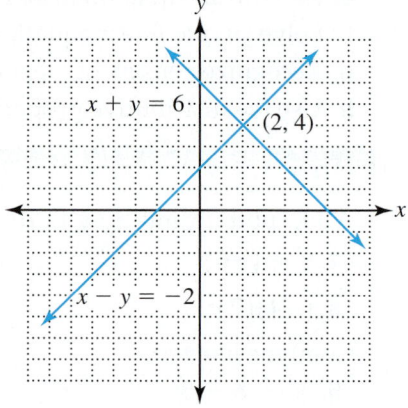

Notice that if we added the two equations, the y's would drop out.

$$\begin{array}{ll} x + y = 6 & \\ \underline{x - y = -2} & \text{Add the equations.} \\ 2x \quad\; = 4 & \text{Divide both sides of the equation by 2.} \\ \;\;\; x = 2 & \end{array}$$

This is the x-value of the point of intersection of the two lines, because it represents the equation of a vertical line that passes through the point of intersection. To find the y-value, substitute 2 for x in either of the original equations.

Use $x + y = 6$.

$$\begin{array}{ll} x + y = 6 & \text{Substitute 2 for } x. \\ 2 + y = 6 & \text{Subtract 2 from both sides of the equation.} \\ \quad\;\; y = 4 & \end{array}$$

This is the y-value of the point of intersection. Therefore, we have solved the system without graphing it.

OBJECTIVE **Using the addition method to solve consistent systems of linear equations with two variables by eliminating a variable**

The foregoing leads us to the following procedure for solving systems of equations:

> ### Procedure: Solving Systems of Linear Equations by Using Elimination by Addition
>
> 1. If necessary, rewrite each equation in the form $ax + by = c$, with a, b, and c integers.
>
> 2. Choose the variable to be eliminated.

(Continued)

3. If necessary, multiply one or both of the equations of the system by the appropriate constant(s) so that the coefficients of the variable to be eliminated are additive inverses (or opposites).

4. Add the equations.

5. If necessary, solve the equation resulting from step 4 for the remaining variable.

6. Substitute the value found for the variable in step 5 into either of the equations of the original system, and solve for the other variable.

7. Check the solution in both equations of the system.

Recall that two numbers are additive inverses if their sum is zero. We illustrate this procedure with some examples.

Example 2 **Solve the following systems of equations using elimination by addition:**

a. $\begin{cases} x + y = 3 & \text{(Equation 1)} \\ x - y = 9 & \text{(Equation 2)} \end{cases}$

Solution

It is easier to eliminate y than to eliminate x, because we can eliminate y simply by adding the two equations.

$$\begin{array}{ll} x + y = 3 & \\ \underline{x - y = 9} & \text{Add the equations.} \\ 2x \quad\;\; = 12 & \text{Divide both sides of the equation by 2.} \\ \qquad x = 6 & \text{Substitute } x = 6 \text{ in either equation to find } y. \\ x + y = 3 & \text{Use } x + y = 3. \\ 6 + y = 3 & \text{Substitute 6 for } x. \text{ Subtract 6 from both sides of the equation.} \\ \qquad y = -3 & \text{Therefore, the solution of the system is } (6, -3). \end{array}$$

CHECK:

Substitute 6 for x and -3 for y in both equations.

$$\begin{array}{ll} x + y = 3 & x - y = 9 \\ 6 + (-3) = 3 & 6 - (-3) = 9 \\ 3 = 3 & 6 + 3 = 9 \\ & 9 = 9 \end{array}$$

Since $(6, -3)$ solves both equations, it is the correct solution.

b. $\begin{cases} 2x + 3y = 1 & \text{(Equation 1)} \\ x + 4y = -7 & \text{(Equation 2)} \end{cases}$

Solution

Notice that neither variable will be eliminated if we add the equations just as they are. In order for a variable to be eliminated, the coefficients of the variable must be additive inverses. It appears easier to make the coefficients of x additive inverses, since we would have to multiply only one equation by a constant. If we multiply Equation 2 by -2, the x-term becomes $-2x$ which is the additive inverse of $2x$ in Equation 1.

$$\begin{array}{lll} 2x + 3y = 1 & \xrightarrow{\;\;\text{Leave unchanged.}\;\;} & 2x + 3y = 1 \quad \text{(Equation 1)} \\ x + 4y = -7 & \xrightarrow{\;\;\text{Multiply by } -2.\;\;} & -2x - 8y = 14 \quad \text{(Equation 2 rewritten)} \end{array}$$

Now we add the equations to eliminate x.

$$2x + 3y = 1$$
$$\underline{-2x - 8y = 14} \qquad \text{Add the equations.}$$
$$-5y = 15 \qquad \text{Divide both sides of the equation by } -5.$$
$$y = -3 \qquad \text{Substitute } -3 \text{ for } y \text{ in either equation to find } x.$$
$$\text{Use } x + 4y = -7.$$

$$x + 4y = -7 \qquad \text{Substitute } -3 \text{ for } y.$$
$$x + 4(-3) = -7 \qquad \text{Multiply } 4 \text{ and } -3.$$
$$x - 12 = -7 \qquad \text{Add 12 to both sides of the equation.}$$
$$x = 5 \qquad \text{Therefore, the solution is } (5, -3).$$

CHECK:

Substitute 5 for x and -3 for y in both equations.

$$2x + 3y = 1 \qquad\qquad x + 4y = -7$$
$$2(5) + 3(-3) = 1 \qquad 5 + 4(-3) = -7$$
$$10 - 9 = 1 \qquad\qquad 5 - 12 = -7$$
$$1 = 1 \qquad\qquad\qquad -7 = -7$$

Since $(5, -3)$ solves both equations, it is the correct solution.

c. $\begin{cases} 3x - 2y = 7 & \text{(Equation 1)} \\ 4x - 3y = 10 & \text{(Equation 2)} \end{cases}$

Solution

If we add the equations as they are, neither variable would be eliminated. Also, we cannot multiply just one equation by an integer to make the coefficients of either variable additive inverses. In order to make the coefficients of one of the variables additive inverses, find the smallest number that each coefficient of that variable will divide into and change each coefficient into that number, except with opposite signs. For example, the coefficients of x are 3 and 4 and the smallest number that we can divide both into is 12. So, if we decide to eliminate x, we would change one coefficient into 12 and the other into -12. The coefficients of y are 2 and 3, and the smallest number that we can divide 2 and 3 into is 6. So if we decide to eliminate y, we would need to change one coefficient into 6 and the other into -6. Since 6 is smaller than 12 (it's as good a reason as any!), we choose to eliminate y by making one coefficient 6 and the other -6. Consequently, we multiply Equation 1 by 3 and Equation 2 by -2.

$$3x - 2y = 7 \xrightarrow{\text{Multiply by 3.}} 9x - 6y = 21$$
$$4x - 3y = 10 \xrightarrow{\text{Multiply by } -2.} -8x + 6y = -20$$

NOTE The y-terms are now additive inverses.

Now we add the rewritten equations to eliminate y.

$$9x - 6y = 21$$
$$\underline{-8x + 6y = -20} \qquad \text{Add the equations.}$$
$$x = 1 \qquad \text{Substitute 1 for } x \text{ in either equation and solve for } y.$$
$$\text{Use } 3x - 2y = 7.$$

$$3x - 2y = 7 \qquad \text{Substitute 1 for } x.$$
$$3(1) - 2y = 7 \qquad \text{Multiply 3 and 1.}$$
$$3 - 2y = 7 \qquad \text{Subtract 3 from both sides of the equation.}$$
$$-2y = 4 \qquad \text{Divide both sides of the equation by } -2.$$
$$y = -2 \qquad \text{Therefore, the solution is } (1, -2).$$

CHECK:

The check is left as an exercise for the student.

Note

Example 1d exhibits the advantage of the addition method when the solutions are fractions instead of integers.

d. $\begin{cases} x + 3y = -1 & \text{(Equation 1)} \\ 3x + 6y = -1 & \text{(Equation 2)} \end{cases}$

Solution

Let us eliminate x by multiplying Equation 1 by -3 and adding the equations.

$$x + 3y = -1 \xrightarrow[\text{Leave unchanged.}]{\text{Multiply by } -3.} -3x - 9y = 3$$

$$3x + 6y = -1 \xrightarrow{\hspace{2cm}} 3x + 6y = -1$$

NOTE The x-terms are now additive inverses.

Now we add the equations to eliminate x.

$$\begin{array}{ll} -3x - 9y = 3 & \\ \underline{3x + 6y = -1} & \text{Add the equations.} \\ -3y = 2 & \text{Divide both sides of the equation by } -3. \\ y = -\dfrac{2}{3} & \text{Substitute } -\dfrac{2}{3} \text{ for } y \text{ in either equation and solve for } x. \\ & \text{Use } x + 3y = -1. \end{array}$$

$$\begin{array}{ll} x + 3y = -1 & \text{Substitute } -\dfrac{2}{3} \text{ for } y. \\ x + 3\left(-\dfrac{2}{3}\right) = -1 & 3\left(-\dfrac{2}{3}\right) = \dfrac{3}{1}\left(-\dfrac{2}{3}\right) = -\dfrac{6}{3} = -2. \\ x - 2 = -1 & \text{Add 2 to both sides of the equation.} \\ x = 1 & \text{Therefore, } \left(1, -\dfrac{2}{3}\right) \text{ is the solution.} \end{array}$$

CHECK:

Substitute 1 for x and $-\frac{2}{3}$ for y in both equations.

$$\begin{array}{cc} x + 3y = -1 & 3x + 6y = -1 \\ 1 + 3\left(-\dfrac{2}{3}\right) = -1 & 3(1) + 6\left(-\dfrac{2}{3}\right) = -1 \\ 1 - 2 = -1 & 3 - 4 = -1 \\ -1 = -1 & -1 = -1 \end{array}$$

Since $\left(1, -\frac{2}{3}\right)$ solves both equations, $\left(1, -\frac{2}{3}\right)$ is the correct solution.

PRACTICE EXERCISE 2

Solve the following systems of equations by using elimination by addition:

a. $\begin{cases} x - y = 5 \\ x + y = 1 \end{cases}$ **b.** $\begin{cases} 3x + 4y = 6 \\ x - 3y = -11 \end{cases}$ **c.** $\begin{cases} 4x - 3y = -23 \\ 5x - 4y = -30 \end{cases}$ **d.** $\begin{cases} 12x + 18y = 17 \\ 6x + 10y = 9 \end{cases}$

If you need more practice, do the following Additional Practice Exercises.

Additional Practice Exercise 2   Solve the following systems of equations:

a. $\begin{cases} x + 2y = 10 \\ x - 2y = -6 \end{cases}$ **b.** $\begin{cases} x + 3y = -3 \\ 5x + 2y = 24 \end{cases}$ **c.** $\begin{cases} 4x + 5y = 2 \\ 3x - 2y = -10 \end{cases}$ **d.** $\begin{cases} 4x - 4y = -1 \\ 12x - 8y = 3 \end{cases}$

When one of the variables has a fractional value and this value is substituted back into one of the equations to find the other variable, it is often necessary to combine fractions, which is discussed in Section 7.4. Sometimes it is easier to repeat the elimination procedure to find the remaining variable rather than to substitute a fraction and solve the resulting equation with fractions.

Answers: Practice Exercise 2: a. $(3, -2)$ **b.** $(-2, 3)$ **c.** $(-2, 5)$ **d.** $\left(\frac{2}{3}, \frac{1}{2}\right)$ **Additional Practice 2: a.** $(2, 4)$ **b.** $(6, -3)$ **c.** $(-2, 2)$ **d.** $\left(\frac{5}{4}, \frac{3}{2}\right)$

Example 3 | **Solve the following system of linear equations:**

$$\begin{cases} 3x - 4y = 15 & \text{(Equation 1)} \\ 4x + 2y = 7 & \text{(Equation 2)} \end{cases}$$

Solution

Since we can eliminate y by multiplying Equation 2 by 2 and then adding, we choose to eliminate y.

$$3x - 4y = 15 \xrightarrow{\text{Leave unchanged.}} 3x - 4y = 15$$
$$4x + 2y = 7 \xrightarrow{\text{Multiply by 2.}} 8x + 4y = 14$$

Now we can add the equations to eliminate y.

$$\begin{array}{ll} 3x - 4y = 15 & \\ \underline{8x + 4y = 14} & \text{Add the equations.} \\ 11x \quad = 29 & \text{Divide both sides of the equation by 11.} \\ x = \dfrac{29}{11} & \text{Therefore, the } x\text{-value of the solutions is } \frac{29}{11}. \end{array}$$

We could find y by substituting for x in either equation and solving for y, but that does not look like it would be a whole lot of fun. Instead, let us repeat the elimination procedure and eliminate x this time. Since 12 is the smallest number we can divide 3 and 4 into, make one coefficient 12 and the other -12.

$$3x - 4y = 15 \xrightarrow{\text{Multiply by 4.}} 12x - 16y = 60$$
$$4x + 2y = 7 \xrightarrow{\text{Multiply by } -3.} -12x - 6y = -21$$

Now we add the equations to eliminate x.

$$\begin{array}{ll} 12x - 16y = 60 & \\ \underline{-12x - 6y = -21} & \text{Add the equations.} \\ -22y = 39 & \text{Divide both sides of the equation by } -22. \\ y = -\dfrac{39}{22} & \text{Therefore, the } y\text{-value of the solution is } -\frac{39}{22}. \end{array}$$

Therefore, the solution of the system is $\left(\frac{29}{11}, -\frac{39}{22}\right)$.

CHECK:

The check is left as an exercise for the student.

PRACTICE EXERCISE 3

Solve the following systems of equations:

a. $\begin{cases} 3x + 2y = 9 \\ 2x - 5y = 12 \end{cases}$
b. $\begin{cases} 4x + y = 6 \\ 3x + 2y = 9 \end{cases}$

If you need more practice, do the following Additional Practice Exercises.

Additional Practice Exercise 3 Solve the following systems of equations:

a. $\begin{cases} 5x + 3y = 15 \\ 3x - 2y = 6 \end{cases}$
b. $\begin{cases} 2x + 4y = 3 \\ 4x - 6y = 5 \end{cases}$

Answers: Practice Exercise 3: a. $\left(\frac{69}{19}, -\frac{18}{19}\right)$ **b.** $\left(\frac{3}{5}, \frac{18}{5}\right)$ **Additional Practice 3: a.** $\left(\frac{48}{19}, \frac{15}{19}\right)$ **b.** $\left(\frac{19}{14}, \frac{1}{14}\right)$

Before the elimination by addition method can be used, both equations should be written in the form of $ax + by = c$, where a, b, and c are integers. If necessary, rewrite the equations of the system in this form before attempting to solve the system.

| **Example 4** | **Solve the following systems of equations by using elimination by addition:** |

$$\begin{cases} x = 3y + 22 \\ 5y = -2x - 22 \end{cases}$$

Solution

We first have to rewrite each equation in $ax + by = c$ form.

$$x = 3y + 22 \qquad\qquad 5y = -2x - 22$$
$$x - 3y = 3y - 3y + 22 \qquad 2x + 5y = -2x + 2x - 22$$
$$x - 3y = 22 \qquad\qquad 2x + 5y = -22$$

Now solve the system $x - 3y = 22$ and $2x + 5y = -22$ by eliminating x.

$$x - 3y = 22 \xrightarrow[\text{Leave unchanged.}]{\text{Multiply by } -2.} -2x + 6y = -44$$
$$2x + 5y = -22 \xrightarrow{} 2x + 5y = -22$$

Now we add the equations to eliminate x.

$$\begin{array}{r} -2x + 6y = -44 \\ \underline{2x + 5y = -22} \\ 11y = -66 \\ y = -6 \end{array}$$

$\qquad\qquad\qquad\qquad$ Add the equations.

$\qquad\qquad\qquad\qquad$ Divide both sides of the equation by 11.

$\qquad\qquad\qquad\qquad$ Substitute $y = -6$ in either of the original equations and find x. Use $x = 3y + 22$.

$$\begin{aligned} x &= 3y + 22 && \text{Substitute } -6 \text{ for } y. \\ x &= 3(-6) + 22 && \text{Multiply 3 and } -6. \\ x &= -18 + 22 && \text{Add } -18 \text{ and 22.} \\ x &= 4 && \text{Therefore, the solution of the system is } (4, -6). \end{aligned}$$

CHECK:

Substitute 4 for x and -6 for y in both of the original equations. The check is left as an exercise for the student.

PRACTICE EXERCISE 4

$$\begin{cases} 5y = 5 - 3x \\ x = -2y + 1 \end{cases}$$

If you need more practice, do the following Additional Practice Exercise.

Additional Practice Exercise 4

$$\begin{cases} y = 2x + 2 \\ x = -3y + 13 \end{cases}$$

Solving equations containing fractions was discussed in great detail in Section 7.6. We can eliminate the fractions from an equation by multiplying both sides of the equation by the least common denominator. For example, given the equation $\frac{1}{5}x + \frac{1}{2}y = \frac{1}{5}$, we see that the denominators are 5 and 2 and the least common denominator is 10, so we

Answers: Practice Exercise 4: $(5, -2)$ **Additional Practice 4:** $(1, 4)$

multiply both sides of the equation by 10, as shown. For a review of multiplying fractions and whole numbers, see Section R.4.

$$10\left(\frac{1}{5}x + \frac{1}{2}y\right) = 10\left(\frac{1}{5}\right) \quad \text{Distribute on the left side. } 10\left(\frac{1}{5}\right) = \frac{10}{1}\cdot\frac{1}{5} = \frac{10}{5} = 2.$$

$$10\cdot\frac{1}{5}x + 10\cdot\frac{1}{2}y = 2 \quad 10\cdot\frac{1}{5} = \frac{10}{1}\cdot\frac{1}{5} = \frac{10}{5} = 2 \text{ and } 10\cdot\frac{1}{2} = \frac{10}{1}\cdot\frac{1}{2} = \frac{10}{2} = 5.$$

$$2x + 5y = 2 \quad \text{Equivalent equation without the fractions.}$$

We use this technique in the following example:

Example 5 **Solve the following systems of equations:**

$$\begin{cases} \dfrac{1}{5}x + \dfrac{1}{2}y = \dfrac{1}{5} \\ \dfrac{1}{2}x + \dfrac{1}{3}y = -\dfrac{4}{3} \end{cases}$$

Solution

The coefficients of x and y should be integers, since it is difficult to work with fractions.

$$\frac{1}{5}x + \frac{1}{2}y = \frac{1}{5} \quad \text{The least common denominator is 10, so multiply both sides by 10.}$$

$$\frac{1}{2}x + \frac{1}{3}y = -\frac{4}{3} \quad \text{The least common denominator is 6, so multiply both sides by 6.}$$

$$10\left(\frac{1}{5}x + \frac{1}{2}y\right) = 10\left(\frac{1}{5}\right) \quad \text{Simplify both sides of the equation.}$$

$$6\left(\frac{1}{2}x + \frac{1}{3}y\right) = 6\left(-\frac{4}{3}\right) \quad \text{Simplify both sides of the equation.}$$

$$10\cdot\frac{1}{5}x + 10\cdot\frac{1}{2}y = 2$$

$$6\cdot\frac{1}{2}x + 6\cdot\frac{1}{3}y = -8 \quad \text{Simplify the left side of the equations.}$$

$$2x + 5y = 2$$
$$3x + 2y = -8 \quad \text{The system without the fractions.}$$

$$2x + 5y = 2 \xrightarrow[\text{Multiply by 2.}]{\text{Multiply by }-3.} -6x - 15y = -6$$
$$3x + 2y = -8 \xrightarrow{\hspace{2cm}} 6x + 4y = -16$$

Now we add the equations to eliminate x.

$$-6x - 15y = -6$$
$$\underline{6x + 4y = -16} \quad \text{Add the equations.}$$
$$-11y = -22 \quad \text{Divide both sides of the equation by }-11.$$
$$y = 2 \quad \text{Substitute } y = 2 \text{ into either of the original equations to find } x.$$

$$\text{Use } \frac{1}{5}x + \frac{1}{2}y = \frac{1}{5}.$$

$$\frac{1}{5}x + \frac{1}{2}y = \frac{1}{5} \quad \text{Substitute 2 for } y.$$

$$\frac{1}{5}x + \frac{1}{2}(2) = \frac{1}{5} \quad \text{Multiply } \frac{1}{2} \text{ and 2.}$$

$$\frac{1}{5}x + 1 = \frac{1}{5} \quad \text{Subtract 1 from both sides of the equation.}$$

$$\frac{1}{5}x + 1 - 1 = \frac{1}{5} - 1 \quad \text{Add. Write 1 as } \frac{5}{5}.$$

$$\frac{1}{5}x = \frac{1}{5} - \frac{5}{5} \qquad \text{Add.}$$

$$\frac{1}{5}x = -\frac{4}{5} \qquad \text{Multiply both sides of the equation by 5.}$$

$$5\left(\frac{1}{5}x\right) = 5\left(-\frac{4}{5}\right) \qquad \text{Simplify both sides of the equation.}$$

$$x = -4 \qquad \text{Therefore, the solution of the system is } (-4, 2).$$

CHECK:

The check is left as an exercise for the student.

PRACTICE EXERCISE 5

Solve the following system of equations.

$$\begin{cases} \frac{1}{4}x - \frac{1}{2}y = -\frac{1}{4} \\ \frac{1}{3}x + \frac{1}{6}y = \frac{4}{3} \end{cases}$$

If you need more practice, do the following Additional Practice Exercise.

Additional Practice Exercise 5   Solve the following system of equations:

$$\begin{cases} \frac{1}{3}x + \frac{1}{5}y = \frac{1}{3} \\ x + \frac{1}{2}y = \frac{1}{2} \end{cases}$$

OBJECTIVE   Ⓑ Recognizing inconsistent and dependent systems of equations when using the addition method

Remember, if a system is inconsistent, it has no solutions. If a system is dependent, it has an infinite number of solutions. The following examples show how to determine whether a system is inconsistent or dependent without graphing:

Example 6   **Solve the following systems of equations:**

a. $\begin{cases} 4x - 2y = 7 \\ 2x - y = 4 \end{cases}$

Solution

Let us eliminate y. Make one coefficient -2 and the other 2.

$$4x - 2y = 7 \xrightarrow[\text{Multiply by } -2.]{\text{Leave unchanged.}} 4x - 2y = 7$$
$$2x - y = 4 \xrightarrow{\hspace{2cm}} -4x + 2y = -8$$

Now add the equations.

$$\begin{array}{r} 4x - 2y = 7 \\ \underline{-4x + 2y = -8} \qquad \text{Add the equations.} \\ 0 = -1 \end{array}$$

This is a false statement, which means that there is no solution, which is denoted by the symbol $\varnothing$. The lines are parallel and the system is inconsistent.

Answers: **Practice Exercise 5:** $(3, 2)$ **Additional Practice 5:** $(-2, 5)$

b. $\begin{cases} 4x + 2y = 6 \\ 6x + 3y = 9 \end{cases}$

Solution

Since the coefficients of y are smaller than the coefficients of x, we will eliminate y. Since 2 and 3 both divide into 6, we will make one coefficient 6 and the other -6.

$$4x + 2y = 6 \xrightarrow{\text{Multiply by 3.}} 12x + 6y = 18$$
$$6x + 3y = 9 \xrightarrow{\text{Multiply by }-2.} -12x - 6y = -18$$

Now add the equations.

$$\begin{array}{r} 12x + 6y = 18 \\ \underline{-12x - 6y = -18} \quad \text{Add the equations.} \\ 0 = 0 \end{array}$$

This is a true statement for all values of x and y that solve either equation. Hence, there are an infinite number of solutions, which means the system is dependent. We indicate the solutions as $\{(x, y): 4x + 2y = 6\}$ or $\{(x, y): 6x + 3y = 9\}$.

PRACTICE EXERCISE 6

Solve the given systems of equations. Indicate whether the system is consistent, inconsistent, or dependent.

a. $\begin{cases} 6x + 2y = 6 \\ y = -3x + 3 \end{cases}$
b. $\begin{cases} 12x + 6y = 9 \\ 4x + 2y = 5 \end{cases}$

If you need more practice, do the following Additional Practice Exercises.

Additional Practice Exercise 6 Solve the given systems of equations. Indicate whether the system is consistent, inconsistent, or dependent.

a. $\begin{cases} 6x + 3y = 5 \\ 4x + 2y = 7 \end{cases}$
b. $\begin{cases} x = -2y + 6 \\ \dfrac{1}{2}x + y = 3 \end{cases}$

Exercise Set 9.2

For Extra Help

MyMathLab®

Solve the following systems of equations by using elimination by addition, and verify the solution from the graph:

1. $\begin{cases} x - y = 2 \\ x + y = 2 \end{cases}$
2. $\begin{cases} y + x = 2 \\ y - x = 8 \end{cases}$
3. $\begin{cases} x + y = 3 \\ 3x - y = 1 \end{cases}$
4. $\begin{cases} x + 2y = 4 \\ -x + 3y = 6 \end{cases}$

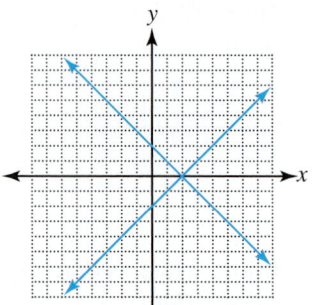
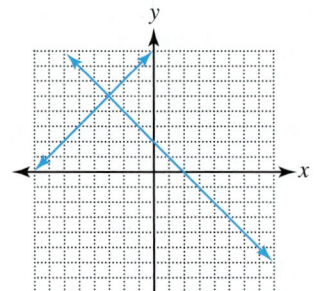
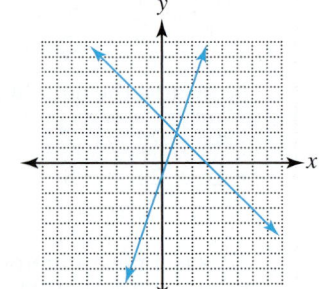
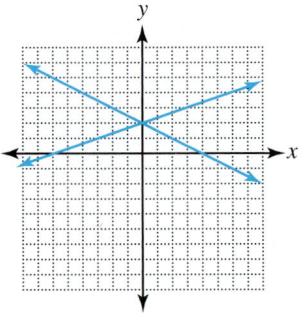

Answers: Practice Exercise 6: a. dependent **b.** inconsistent **Additional Practice 6: a.** inconsistent **b.** dependent

Solve the following systems of equations by using elimination by addition. (See Examples 2–5.)

5. $\begin{cases} x + 3y = -9 \\ -x + 2y = -11 \end{cases}$

6. $\begin{cases} -x + 3y = 13 \\ x - 2y = -10 \end{cases}$

7. $\begin{cases} 5x + y = -17 \\ 4x - y = -10 \end{cases}$

8. $\begin{cases} 3x - y = -3 \\ 2x + y = -7 \end{cases}$

9. $\begin{cases} 5x + y = -1 \\ x - 3y = -13 \end{cases}$

10. $\begin{cases} 3x - 5y = -17 \\ 4x + y = -15 \end{cases}$

11. $\begin{cases} x - 2y = 3 \\ -2x + 5y = -6 \end{cases}$

12. $\begin{cases} 4x + 5y = 24 \\ -x + 3y = -6 \end{cases}$

13. $\begin{cases} 7x - 4y = 4 \\ 5x + y = 26 \end{cases}$

14. $\begin{cases} 5x - 2y = 0 \\ -x + 4y = 0 \end{cases}$

15. $\begin{cases} 2x + 3y = 4 \\ -6x + 5y = 16 \end{cases}$

16. $\begin{cases} 3x + 2y = 10 \\ 5x - 6y = -2 \end{cases}$

17. $\begin{cases} 3x + 2y = 6 \\ 5x + 4y = 8 \end{cases}$

18. $\begin{cases} 3x + 2y = 1 \\ 6x + 5y = 7 \end{cases}$

19. $\begin{cases} 12x - 2y = -54 \\ 13x + 4y = -40 \end{cases}$

20. $\begin{cases} 10x - 3y = 14 \\ -3x + 9y = 12 \end{cases}$

21. $\begin{cases} 4x + 5y = -4 \\ 3x + 8y = -20 \end{cases}$

22. $\begin{cases} 3x + 7y = 8 \\ 2x - 4y = 14 \end{cases}$

23. $\begin{cases} 3x - 4y = -14 \\ 5x + 7y = 45 \end{cases}$

24. $\begin{cases} -5x + 4y = -31 \\ 13x - 6y = 85 \end{cases}$

25. $\begin{cases} 5x + 2y = 13 \\ 2x - 5y = 11 \end{cases}$

26. $\begin{cases} 4x - 9y = -30 \\ 5x + 4y = -7 \end{cases}$

27. $\begin{cases} 13x - 22y = 70 \\ 26x + 33y = -14 \end{cases}$

28. $\begin{cases} 15x + 6y = -33 \\ 5x - 8y = -31 \end{cases}$

29. $\begin{cases} 5x + 2y = -19 \\ 8x - 5y = -14 \end{cases}$

30. $\begin{cases} 3x - 4y = 24 \\ 5x + 3y = 11 \end{cases}$

31. $\begin{cases} 4x + y = -2 \\ 8x - y = 11 \end{cases}$

32. $\begin{cases} 3x - y = 6 \\ 9x + 2y = -2 \end{cases}$

33. $\begin{cases} 3x + 2y = 8 \\ x + 4y = 1 \end{cases}$

34. $\begin{cases} 14x + 7y = 6 \\ 7x + 6y = 8 \end{cases}$

35. $\begin{cases} 9x + 6y = 2 \\ 6x + 9y = -2 \end{cases}$

36. $\begin{cases} 8x - 12y = 9 \\ 4x + 8y = 1 \end{cases}$

37. $\begin{cases} 2x - 7y = 3 \\ 5x + 3y = 4 \end{cases}$

38. $\begin{cases} 3x - 4y = 6 \\ 5x + 3y = -7 \end{cases}$

39. $\begin{cases} x + 2y = -5 \\ 4x = 3y + 24 \end{cases}$

40. $\begin{cases} 5x = -2y - 1 \\ 3x - y = 6 \end{cases}$

41. $\begin{cases} x + 4y + 16 = 0 \\ 3y = 9 - 6x \end{cases}$

42. $\begin{cases} 4x = 8 - 2y \\ 3x = 6y + 21 \end{cases}$

43. $\begin{cases} \dfrac{1}{4}x - \dfrac{1}{2}y = \dfrac{3}{4} \\ \dfrac{1}{2}x + \dfrac{1}{3}y = \dfrac{1}{6} \end{cases}$

44. $\begin{cases} \dfrac{3}{4}x - \dfrac{2}{7}y = -5 \\ \dfrac{5}{8}x + \dfrac{1}{4}y = -\dfrac{3}{4} \end{cases}$

45. $\begin{cases} \dfrac{3}{2}x + \dfrac{1}{4}y = 1 \\ \dfrac{2}{5}x + \dfrac{1}{3}y = \dfrac{4}{3} \end{cases}$

46. $\begin{cases} \dfrac{1}{2}x + \dfrac{1}{3}y = \dfrac{2}{3} \\ \dfrac{2}{3}x + \dfrac{1}{2}y = 1 \end{cases}$

47. $\begin{cases} .4x + .3y = 4.2 \\ 1.2x + .6y = 12 \end{cases}$

48. $\begin{cases} 2.4x + 3y = 7.2 \\ .8x + 2.4y = 8 \end{cases}$

49. $\begin{cases} .3x - 1.8y = 2.7 \\ 3.6x + 4y = 6.8 \end{cases}$

50. $\begin{cases} 6.7x + 3.2y = 2.6 \\ .08x + .5y = 2.34 \end{cases}$

Solve the given systems of equations. Indicate whether the system is consistent, inconsistent, or dependent. (See Example 6.)

51. $\begin{cases} 4x - 2y = 3 \\ -8x + 4y = -6 \end{cases}$

52. $\begin{cases} 5x + y = 4 \\ 15x + 3y = 12 \end{cases}$

53. $\begin{cases} -2x + 5y = 8 \\ x + 4y = 22 \end{cases}$

54. $\begin{cases} 3x + y = 1 \\ -4x - 2y = -6 \end{cases}$

55. $\begin{cases} 2x - 5y = 4 \\ -8x + 20y = -20 \end{cases}$

56. $\begin{cases} x - 2y = 6 \\ -3x + 6y = 9 \end{cases}$

Challenge Exercises (57–60)

Solve the following systems of equations:

57. $\begin{cases} 6x + 9y = -3 \\ 5x + 37y = -32 \end{cases}$

58. $\begin{cases} 5x - 7y = -73 \\ -12x + 6y = 132 \end{cases}$

59. $\begin{cases} 3x - 8y = 11 \\ 12x + 10y = 23 \end{cases}$

60. $\begin{cases} 85 = 24x + 24y \\ 55 = 18x + 16y \end{cases}$

Writing Exercises (61 and 62)

61. If the result of applying the elimination by addition method is the equation $-2 \neq 0$, what does this mean?

62. If the result of applying the elimination by addition method is the equation $4 = 4$, what does this mean?

Solving Systems of Linear Equations by Using Substitution

OBJECTIVE *When you complete this section, you will be able to:*

A Solve systems of linear equations with two variables by using substitution.

PREREQUISITE SKILLS *Before beginning this section, you should be able to:*

a. Solve an equation for one of its variables. (Section 3.4)

b. Solve simple linear equations. (Sections 3.1, 3.2, 3.3)

c. Determine whether an ordered pair is a solution of an equation. (Section 4.1)

d. Add, subtract, multiply, and divide signed numbers. (Sections 1.6, 1.7, 2.1, 2.5)

e. Multiply whole numbers and fractions. (Section R.4)

f. Recognize consistent, dependent, and inconsistent systems. (Section 9.1)

GETTING READY FOR SECTION 9.3

1. Solve $2x + y = 6$ for y.

2. If $x = 8$ and $3x + y = 20$, find y.

3. Is $(0, 4)$ a solution of $4x - 3y = 12$?

Add, subtract, multiply, or divide as indicated.

4. $5 + (-8)$ **5.** $4 - (-6)$ **6.** $(-2)(-9)$ **7.** $18 \div (-6)$

8. Multiply. $-15\left(\dfrac{4}{5}\right)$

9. Is the system of equations $\begin{cases} 3x - 2y = 4 \\ -6x + 4y = -8 \end{cases}$ consistent, dependent, or inconsistent?

Introduction

In this section, we will learn another method for solving linear systems with two variables. This method is based on the principle of substitution, which allows us to replace a quantity with any other quantity that is equal to it.

OBJECTIVE Solving systems of linear equations with two variables by using substitution

To use substitution to solve a system of linear equations with two variables, we will solve one equation for one of its variables, if necessary. This expression is then substituted for that variable in the other equation, because at the point of intersection these two expressions are equal. This results in an equation with one variable that we can solve by previously learned techniques. We will illustrate with an example and then formalize the technique.

Answers: Getting Ready: 1. $y = -2x + 6$ **2.** $y = -4$ **3.** no **4.** -3 **5.** 10 **6.** 18 **7.** -3 **8.** -12
9. dependent

Example 1   Solve the following system of equations by using substitution:

$$\begin{cases} 2x + y = 7 \\ 3x + 2y = 12 \end{cases}$$

Solution

Solve $2x + y = 7$ for y, since it is the easier variable to solve for in either equation. Substitute the resulting expression for y in the other equation, $3x + 2y = 12$.

$2x + y = 7$	Subtract $2x$ from both sides of the equation.
$2x - 2x + y = 7 - 2x$	Add like terms.
$y = 7 - 2x$	Substitute $7 - 2x$ for y in $3x + 2y = 12$.
$3x + 2(7 - 2x) = 12$	Distribute 2.
$3x + 14 - 4x = 12$	Add like terms.
$-x + 14 = 12$	Subtract 14 from both sides of the equation.
$-x = -2$	Multiply both sides of the equation by -1.
$-1(-x) = -1(-2)$	Perform the multiplications.
$x = 2$	Substitute 2 for x in either of the original equations and solve for y. Use $2x + y = 7$.
$2x + y = 7$	Substitute 2 for x and solve for y.
$2(2) + y = 7$	Perform the multiplication.
$4 + y = 7$	Subtract 4 from both sides of the equation.
$y = 3$	Therefore, $(2, 3)$ is the solution of the system.

CHECK:

Substitute 2 for x and 3 for y in both of the original equations.

$2x + y = 7$	$3x + 2y = 12$
$2(2) + 3 = 7$	$3(2) + 2(3) = 12$
$4 + 3 = 7$	$6 + 6 = 12$
$7 = 7$	$12 = 12$

Since $(2, 3)$ solves both equations, it is the correct solution.

We summarize the procedure in the following box:

Note

Recall that this means the graphs of $2x + y = 7$ and $3x + 2y = 12$ intersect at the point whose coordinates are $(2, 3)$.

Procedure: Solving Systems of Linear Equations by Using Substitution

1. If necessary, solve one of the equations for one of its variables.

2. Substitute the expression found in step 1 for that variable in the other equation of the system.

3. Solve the resulting equation.

4. Substitute the value found in step 3 into either of the original equations of the system and solve for the other variable.

5. Check the solution of the system in both of the original equations.

The substitution method is most practical if the coefficient of one of the variables is 1 or -1, since solving for that variable will avoid fractions. If neither of the variables has a coefficient of 1 in either equation, then solving for either variable may result in fractions, which are not easy to work with.

The substitution method gives us another way of solving systems of linear equations. This technique will be useful in later courses when the systems are not linear.

| **Example 2** | **Solve the following systems of equations by using substitution:** |

a. $\begin{cases} 4x + 3y = 10 \\ y = -3x \end{cases}$

Solution

Since $y = -3x$ is already solved for y, we do not need to solve one of the equations for a variable.

$$y = -3x \qquad \text{Substitute } -3x \text{ for } y \text{ in } 4x + 3y = 10.$$
$$4x + 3(-3x) = 10 \qquad \text{Multiply 3 and } -3x.$$
$$4x - 9x = 10 \qquad \text{Add like terms.}$$
$$-5x = 10 \qquad \text{Divide both sides of the equation by } -5.$$
$$x = -2 \qquad \text{Substitute } -2 \text{ for } x \text{ in either of the original equations and solve for } y. \text{ Use } y = -3x.$$

$$y = -3x \qquad \text{Substitute } -2 \text{ for } x.$$
$$y = -3(-2) \qquad \text{Multiply.}$$
$$y = 6 \qquad \text{Therefore, the solution to the system is } (-2, 6).$$

CHECK:

Since the check is done in the same manner as solving by the graphical and elimination by addition methods, the check will be omitted.

b. $\begin{cases} x + y = -2 \\ 2x + 7y = -29 \end{cases}$

Solution

We must solve one of the equations for one of its variables. If possible, we would like to avoid fractions. Since the coefficients of both x and y are 1 in the equation $x + y = -2$, solving for either x or y would avoid fractions.

$$x + y = -2 \qquad \text{Solve for } x. \text{ Subtract } y \text{ from both sides of the equation.}$$
$$x + y - y = -2 - y \qquad \text{Add like terms.}$$
$$x = -2 - y \qquad \text{Substitute } -2 - y \text{ for } x \text{ in } 2x + 7y = -29.$$

$$2(-2 - y) + 7y = -29 \qquad \text{Distribute 2.}$$
$$-4 - 2y + 7y = -29 \qquad \text{Add like terms.}$$
$$-4 + 5y = -29 \qquad \text{Add 4 to both sides of the equation.}$$
$$5y = -25 \qquad \text{Divide both sides of the equation by 5.}$$
$$y = -5 \qquad \text{Substitute } -5 \text{ for } y \text{ in either of the original equations and solve for } x. \text{ Use } x + y = -2.$$

$$x + y = -2 \qquad \text{Substitute } -5 \text{ for } y.$$
$$x + (-5) = -2 \qquad \text{Add 5 to both sides of the equation.}$$
$$x = 3 \qquad \text{Therefore, the solution is } (3, -5).$$

c. $\begin{cases} 2x + 5y = -4 \\ 3x + y = 20 \end{cases}$

Solution

The only variable with a coefficient of 1 is y in $3x + y = 20$. Therefore, we will solve $3x + y = 20$ for y.

$$3x + y = 20 \qquad \text{Subtract } 3x \text{ from both sides of the equation.}$$
$$3x - 3x + y = 20 - 3x \qquad \text{Add like terms.}$$
$$y = 20 - 3x \qquad \text{Substitute } 20 - 3x \text{ for } y \text{ in } 2x + 5y = -4.$$

$$2x + 5(20 - 3x) = -4 \qquad \text{Distribute 5.}$$
$$2x + 100 - 15x = -4 \qquad \text{Add like terms.}$$
$$-13x + 100 = -4 \qquad \text{Subtract 100 from both sides of the equation.}$$
$$-13x = -104 \qquad \text{Divide both sides of the equation by } -13.$$
$$x = 8 \qquad \text{Substitute 8 for } x \text{ in either of the original equations and solve for } y. \text{ Use } 3x + y = 20.$$

$$3x + y = 20 \qquad \text{Substitute 8 for } x.$$
$$3(8) + y = 20 \qquad \text{Multiply 3 and 8.}$$
$$24 + y = 20 \qquad \text{Subtract 24 from both sides of the equation.}$$
$$y = -4 \qquad \text{Therefore, the solution is } (8, -4).$$

PRACTICE EXERCISE 2

Solve the following systems of equations, using substitution:

a. $\begin{cases} 2x + 3y = 8 \\ y = -2x \end{cases}$
b. $\begin{cases} 3x + 4y = 13 \\ x + y = 5 \end{cases}$
c. $\begin{cases} x - 3y = -5 \\ 4x + 5y = -3 \end{cases}$

If you need more practice, do the following Additional Practice Exercises.

Additional Practice Exercise 2 Solve the following systems of equations, using substitution:

a. $\begin{cases} x + 4y = 12 \\ x = 2y \end{cases}$
b. $\begin{cases} x - 3y = -26 \\ x + y = 2 \end{cases}$
c. $\begin{cases} 2x + 3y = 1 \\ x + 4y = -7 \end{cases}$

If neither variable has a coefficient of 1 in either equation, the substitution method is made more difficult by the introduction of fractions. Examine both equations carefully and try to determine which variable would be best to solve for. As before, substitute the resulting expression into the other equation for that variable. Actually, solving linear equations of this type is usually easier by the addition method of Section 9.2. It is included here because substitution is often the only method that can be used to solve nonlinear systems of equations found in more advanced algebra courses.

Example 3 **Solve the following systems of equations by using substitution:**

a. $\begin{cases} 3x + 2y = 4 \\ 2x + 3y = 1 \end{cases}$

Solution

Neither variable has a coefficient of 1 in either equation. There seems to be no particular advantage or disadvantage in solving either equation for either variable. We have to make a choice, so let us solve $3x + 2y = 4$ for y.

$$3x + 2y = 4 \qquad \text{Subtract } 3x \text{ from both sides of the equation.}$$
$$2y = 4 - 3x \qquad \text{Divide both sides of the equation by 2.}$$
$$\frac{2y}{2} = \frac{4 - 3x}{2} \qquad \text{Apply } \frac{a + b}{c} = \frac{a}{c} + \frac{b}{c}.$$
$$y = 2 - \frac{3}{2}x \qquad \text{Substitute } 2 - \frac{3}{2}x \text{ for } y \text{ in } 2x + 3y = 1.$$

Answers: Practice Exercise 2: a. $(-2, 4)$ **b.** $(7, -2)$ **c.** $(-2, 1)$ **Additional Practice 2: a.** $(4, 2)$
b. $(-5, 7)$ **c.** $(5, -3)$

$$2x + 3\left(2 - \frac{3}{2}x\right) = 1 \qquad \text{Distribute 3.}$$

$$2x + 6 - \frac{9}{2}x = 1 \qquad \text{Multiply both sides of the equation by 2 to eliminate fractions.}$$

$$2\left(2x + 6 - \frac{9}{2}x\right) = 2(1) \qquad \text{Distribute on the left. Multiply on the right.}$$

$$4x + 12 - 9x = 2 \qquad \text{Add like terms.}$$

$$-5x + 12 = 2 \qquad \text{Subtract 12 from both sides of the equation.}$$

$$-5x = -10 \qquad \text{Divide both sides of the equation by } -5.$$

$$x = 2 \qquad \text{Substitute 2 for } x \text{ in either of the original equations and solve for } y. \text{ Use } 3x + 2y = 4.$$

$$3x + 2y = 4 \qquad \text{Substitute 2 for } x \text{ and solve for } y.$$

$$3(2) + 2y = 4 \qquad \text{Multiply 3 and 2.}$$

$$6 + 2y = 4 \qquad \text{Subtract 6 from both sides of the equation.}$$

$$2y = -2 \qquad \text{Divide both sides of the equation by 2.}$$

$$y = -1 \qquad \text{Therefore, the solution is } (2, -1).$$

b. $\begin{cases} 3x + 2y = 2 \\ 4x - 3y = 14 \end{cases}$

Solution

Neither equation has a variable with a coefficient of 1, so we will solve $3x + 2y = 2$ for y.

$$3x + 2y = 2 \qquad \text{Subtract } 3x \text{ from both sides of the equation.}$$

$$2y = 2 - 3x \qquad \text{Divide both sides by 2.}$$

$$\frac{2y}{2} = \frac{2 - 3x}{2} \qquad \text{Apply } \frac{a + b}{c} = \frac{a}{c} + \frac{b}{c} \text{ to the right side.}$$

$$y = 1 - \frac{3}{2}x \qquad \text{Substitute } 1 - \frac{3}{2}x \text{ for } y \text{ in } 4x - 3y = 14.$$

$$4x - 3\left(1 - \frac{3}{2}x\right) = 14 \qquad \text{Distribute } -3.$$

$$4x - 3 + \frac{9}{2}x = 14 \qquad \text{Multiply both sides of the equation by 2 to eliminate fractions.}$$

$$2\left(4x - 3 + \frac{9}{2}x\right) = 2 \cdot 14 \qquad \text{Distribute on the left and multiply on the right.}$$

$$8x - 6 + 9x = 28 \qquad \text{Add like terms.}$$

$$17x - 6 = 28 \qquad \text{Add 6 to both sides of the equation.}$$

$$17x = 34 \qquad \text{Divide both sides of the equation by 17.}$$

$$x = 2 \qquad \text{Substitute 2 for } x \text{ in either of the original equations. Use } 3x + 2y = 2.$$

$$3x + 2y = 2 \qquad \text{Substitute 2 for } x.$$

$$3(2) + 2y = 2 \qquad \text{Multiply.}$$

$$6 + 2y = 2 \qquad \text{Subtract 6 from both sides of the equation.}$$

$$2y = -4 \qquad \text{Divide both sides by 2.}$$

$$y = -2 \qquad \text{Therefore, the solution is } (2, -2).$$

PRACTICE EXERCISE 3

Solve the following systems by using the substitution method:

a. $\begin{cases} 2x + 5y = 19 \\ 4x - 3y = -27 \end{cases}$

b. $\begin{cases} 3x + 5y = 8 \\ 3x - 10y = -10 \end{cases}$

If you need more practice, do the following Additional Practice Exercises.

Additional Practice Exercise 3   Solve the following systems by using the substitution method:

a. $\begin{cases} 7x - 3y = 34 \\ 6x + 5y = 14 \end{cases}$

b. $\begin{cases} 12x + 3y = 8 \\ 6x + 6y = 1 \end{cases}$

We recognize inconsistent and dependent systems the same way when using the substitution method as when using the addition method. That is, if we arrive at a false statement, we know the system is inconsistent. If we arrive at a true statement, we know the system is dependent.

Example 4   If possible, solve the given systems of equations. If the system is not consistent, indicate whether it is inconsistent or dependent.

a. $\begin{cases} 2x - y = 7 \\ 4x - 2y = -3 \end{cases}$

Solution   Since the coefficient of y in $2x - y = 7$ is -1, solve $2x - y = 7$ for y.

$2x - y = 7$	Subtract $2x$ from both sides of the equation.
$2x - 2x - y = 7 - 2x$	Add like terms.
$-y = 7 - 2x$	Multiply both sides of the equation by -1.
$-1(-y) = -1(7 - 2x)$	Multiply on the left. Distribute on the right.
$y = -7 + 2x$	Substitute $-7 + 2x$ for y in $4x - 2y = -3$.
$4x - 2(-7 + 2x) = -3$	Distribute -2.
$4x + 14 - 4x = -3$	Add like terms.
$14 = -3$	

Since this is a false statement, the system is inconsistent and has no solution. Therefore, the solution set is $\varnothing$.

b. $\begin{cases} 3x - y = 4 \\ 2y - 6x = -8 \end{cases}$

Solution   Since the coefficient of y in $3x - y = 4$ is -1, solve $3x - y = 4$ for y.

$3x - y = 4$	Subtract $3x$ from both sides of the equation.
$3x - 3x - y = 4 - 3x$	Add like terms.
$-y = 4 - 3x$	Multiply both sides of the equation by -1.
$-1(-y) = -1(4 - 3x)$	Multiply on the left. Distribute on the right.
$y = -4 + 3x$	Substitute $-4 + 3x$ for y in $2y - 6x = -8$.
$2(-4 + 3x) - 6x = -8$	Distribute 2.
$-8 + 6x - 6x = -8$	Add like terms.
$-8 = -8$	

Since this is a true statement, the system is dependent and all solutions are common. Therefore, the solutions may be written as $\{(x, y) : 3x - y = 4\}$.

Answers: Practice Exercise 3: a. $(-3, 5)$ **b.** $\left(\dfrac{2}{3}, \dfrac{6}{5}\right)$ **Additional Practice 3: a.** $(4, -2)$ **b.** $\left(\dfrac{5}{6}, -\dfrac{2}{3}\right)$

Note

Recall from Section 9.1 that it is possible to determine whether a system of linear equations is consistent, inconsistent, or dependent without attempting to solve the system. Rewrite the equations in slope-intercept form. If the slopes are different, the system is consistent; if the slopes are the same and the y-intercepts are different, the system is inconsistent and if the slopes and y-intercepts are both the same, the system is dependent.

PRACTICE EXERCISE 4

If possible, solve the given systems of equations. If the system is not consistent, indicate whether it is inconsistent or dependent.

a. $\begin{cases} x + 3y = 7 \\ 6y + 2x = -9 \end{cases}$

b. $\begin{cases} 2x - y = 6 \\ -4x + 2y = -12 \end{cases}$

If you need more practice, do the following Additional Practice Exercises.

Additional Practice Exercise 4   If possible, solve the following systems of equations. If the system is not consistent, indicate whether it is inconsistent or dependent.

a. $\begin{cases} 2x - y = 3 \\ -4x + 2y = -5 \end{cases}$

b. $\begin{cases} x - 4y = 6 \\ -3x + 12y = -18 \end{cases}$

Exercise Set 9.3   For Extra Help   MyMathLab®

Solve the given systems of equations by using substitution. Verify your solution by using the graph. (See Example 1.)

1. $\begin{cases} x = 2 \\ 4x + 5y = 3 \end{cases}$

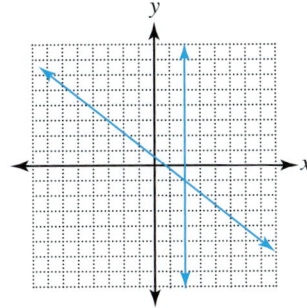

2. $\begin{cases} 3x + y = 5 \\ y = 2 \end{cases}$

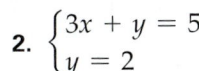
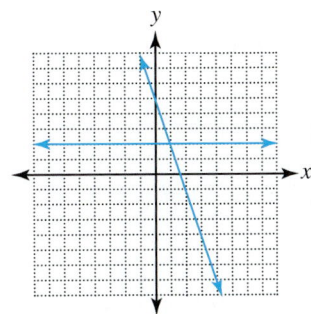

3. $\begin{cases} 3x - y = -2 \\ 5x = y \end{cases}$

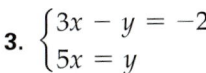
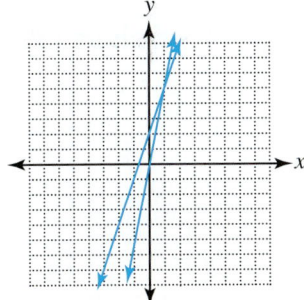

4. $\begin{cases} 2x - y = 9 \\ x = 2y \end{cases}$

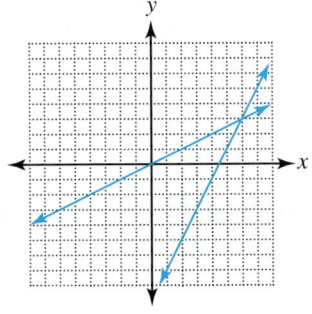

5. $\begin{cases} 4x + y = 5 \\ 3x = -2y \end{cases}$

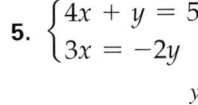
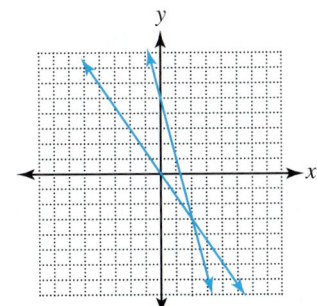

6. $\begin{cases} 3y = -4x \\ 2x + y = -2 \end{cases}$

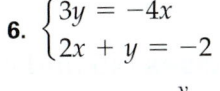
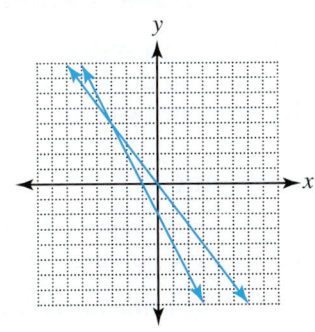

Answers: Practice Exercise 4: a. inconsistent **b.** dependent   **Additional Practice 4: a.** inconsistent **b.** dependent

Solve the following system of equations by the substitution method. (See Examples 1–3.)

7. $\begin{cases} 3x + 2y = -3 \\ y = -2x - 1 \end{cases}$

8. $\begin{cases} 2x - 3y = 14 \\ y = 3x - 14 \end{cases}$

9. $\begin{cases} x = -3y + 10 \\ 5x + 2y = -2 \end{cases}$

10. $\begin{cases} x = 3y - 4 \\ 2x + 5y = 3 \end{cases}$

11. $\begin{cases} x + y = -5 \\ x - 2y = -2 \end{cases}$

12. $\begin{cases} x + 3y = -11 \\ x + y = -7 \end{cases}$

13. $\begin{cases} 2x - 3y = -6 \\ x - y = -1 \end{cases}$

14. $\begin{cases} 2x + 5y = -7 \\ x - y = -7 \end{cases}$

15. $\begin{cases} x + y = 6 \\ 2x - y = 3 \end{cases}$

16. $\begin{cases} x + y = 10 \\ 4x - 7y = -15 \end{cases}$

17. $\begin{cases} x - 2y = -9 \\ 2x + 3y = -4 \end{cases}$

18. $\begin{cases} x - 3y = 10 \\ 4x + 5y = 6 \end{cases}$

19. $\begin{cases} 2x - y = -4 \\ x - 3y = 3 \end{cases}$

20. $\begin{cases} x - 3y = -19 \\ 2x - y = -13 \end{cases}$

21. $\begin{cases} -x + 2y = -12 \\ 2x - 3y = 20 \end{cases}$

22. $\begin{cases} -x + 2y = 9 \\ 4x + 5y = 3 \end{cases}$

23. $\begin{cases} 2x - y = -6 \\ 4x - 3y = -16 \end{cases}$

24. $\begin{cases} 3x - y = -3 \\ 4x - 5y = 7 \end{cases}$

25. $\begin{cases} 3x + y = 2 \\ 6x + 5y = 7 \end{cases}$

26. $\begin{cases} x + 4y = 1 \\ 3x + 8y = 0 \end{cases}$

27. $\begin{cases} 8x - 3y = 10 \\ 4x + 3y = 14 \end{cases}$

28. $\begin{cases} 2x - 5y = 18 \\ 3x + 2y = 8 \end{cases}$

29. $\begin{cases} 5x - 6y = -3 \\ 3x - 5y = 1 \end{cases}$

30. $\begin{cases} 2x + 3y = -1 \\ 2x + 5y = 5 \end{cases}$

If possible, solve the given systems of equations. If the system is not consistent, indicate whether it is inconsistent or dependent. (See Example 4.)

31. $\begin{cases} x - 2y = 6 \\ -2x + 4y = 12 \end{cases}$

32. $\begin{cases} x - 3y = -4 \\ -5x + 15y = 6 \end{cases}$

33. $\begin{cases} 8x + 5y = 11 \\ 2x + y = 5 \end{cases}$

34. $\begin{cases} 2x + 5y = 8 \\ 2x - y = -16 \end{cases}$

35. $\begin{cases} 3x + y = 4 \\ 9x + 3y = 12 \end{cases}$

36. $\begin{cases} x + 2y = 1 \\ 4x + 8y = 4 \end{cases}$

37. $\begin{cases} 3x - 2y = 6 \\ -6x + 4y = 15 \end{cases}$

38. $\begin{cases} -5x + 3y = 15 \\ 15x - 9y = 30 \end{cases}$

Challenge Exercises (39–44) *(See Examples 2 and 3.)*

39. $\begin{cases} \dfrac{1}{2}x + y = 1 \\ \dfrac{1}{4}x + y = 0 \end{cases}$

40. $\begin{cases} \dfrac{2}{3}x + y = 1 \\ \dfrac{5}{6}x + \dfrac{2}{5}y = -3 \end{cases}$

41. $\begin{cases} 2.1x + .5y = 16.6 \\ 4x - 5y = -16 \end{cases}$

42. $\begin{cases} 6x - 7y = 14 \\ 4.8x + 3.2y = -6.4 \end{cases}$

43. $\begin{cases} \dfrac{1}{2}x + \dfrac{2}{9}y = 2 \\ .8x + .03y = 3.2 \end{cases}$

44. $\begin{cases} \dfrac{3}{4}x - \dfrac{2}{5}y = 6 \\ 15x - 8y = 120 \end{cases}$

Writing Exercises (45 and 46)

45. When solving a system of linear equations by substitution, how do you determine which equation to solve for which variable?

46. If the result of solving a system of linear equations by substitution is a statement like $8 = 8$, which type of system is it?

Group Project

If one or more of the equations in a system is not linear, the system is called nonlinear and can sometimes be solved using substitution. Solving these systems often requires solving a quadratic equation. (Section 5.8.)

Solve the following nonlinear systems using substitution:

47. $\begin{cases} y = 2x^2 \\ 2x + y = 4 \end{cases}$

48. $\begin{cases} x^2 + 2y = 1 \\ 2x + y = 2 \end{cases}$

Solving Application Problems by Using Systems of Equations

OBJECTIVES *When you complete this section, you will be able to solve the following types of application problems, using systems of linear equations:*

A Number
B Money
C Distance, rate, and time
D Percent
E Mixture

PREREQUISITE SKILLS *Before beginning this section, you should be able to:*

a. Solve systems of linear equations. (Sections 9.2, 9.3)
b. Multiply whole numbers and decimals. (Section R.6)
c. Change percents into decimals. (Section 8.3)

GETTING READY FOR SECTION 9.4

Solve the following systems of equations:

1. $\begin{cases} x + y = 2 \\ 2x + 3y = 7 \end{cases}$

2. $\begin{cases} x + y = 23 \\ 5x + 10y = 190 \end{cases}$

Multiply as indicated.

3. $100(0.06)$

4. $10(3.5)$

Change the following percents to decimals:

5. 6%

6. 40%

7. 12%

Introduction

In previous chapters, we did application problems by using only one variable. One of the most difficult things in doing application problems that have more than one unknown is representing all the unknowns in terms of the one variable. Using systems of equations, we can assign a different variable to each unknown and eliminate this problem. Some of the examples and exercises in this section are the same as in Chapter 3, except that the approach is different. Therefore, when you complete this section you will have a choice of methods to use when solving some types of application problems.

An important thing to remember is that there must be the same number of equations as there are unknowns. Since the systems we have been solving involve two variables, the application problems will have two unknowns and will require two equations.

Since there are two unknowns, it is necessary to repeat the Solve and State steps of the problem-solving procedure.

Answers: **Getting Ready:** **1.** $(-1, 3)$ **2.** $(8, 15)$ **3.** 6 **4.** 35 **5.** .06 **6.** .4 **7.** .12

OBJECTIVE Solving number problems by using systems of linear equations

Example 1

The top-grossing North American Concert tour of all time was the 2005 Rolling Stones tour. This tour grossed $23.1 million more than the second top-grossing tour, which was the U2 tour of 2005. If together they grossed $300.9 million, find the amount that each tour grossed. (*Source: The World Almanac and Book of Facts 2010*)

Solution

Understand: We are given the relationship between the amounts grossed by each tour and the combined amount grossed by both tours. The unknowns are the amounts grossed by each tour.

> Relationship 1: The Stones tour grossed $23.1 million more than the U2 tour.
> Relationship 2: The combined amount grossed by both tours was $300.9 million.

Plan: Use two unknowns—one for the amount grossed by the Stones and another for the amount grossed by U2. Use relationship 1 to write one equation and relationship 2 to write another. Solve the system.

Translate:

Let x represent the amount grossed by the 2005 Rolling Stones tour, and
Let y represent the amount grossed by the 2005 U2 tour. Then,

$$x = y + 23.1 \quad \text{The Stones tour grossed 23.1 million more than the U2 tour.}$$
$$x + y = 300.9 \quad \text{Together they grossed 300.9 million.}$$

Solve: Since $x = y + 23.1$ is already solved for x, we will use the substitution method by replacing x in $x + y = 300.1$ with $y + 23.1$.

$$x = y + 23.1 \quad \text{Substitute } y + 23.1 \text{ for } x \text{ in } x + y = 300.9.$$
$$y + 23.1 + y = 300.9 \quad \text{Add like terms.}$$
$$2y + 23.1 = 300.9 \quad \text{Subtract 23.1 from both sides of the equation.}$$
$$2y = 277.8 \quad \text{Divide both sides of the equation by 2.}$$
$$y = 138.9 \quad \text{Solution.}$$

State: Therefore, the U2 tour grossed $138.9 million.

Solve: To find the amount grossed by the Stones tour, substitute 138.9 for y into either original equation. $x = y + 23.1$ would be easier.

$$x = y + 23.1 \quad \text{Substitute 138.9 for } y.$$
$$x = 138.9 + 23.1 \quad \text{Add.}$$
$$x = 162 \quad \text{Solution.}$$

State: Therefore, the Stones tour grossed $162 million.

Check: Did the Stones tour gross 23.1 more than the U2 tour? Yes, since $162 - 138.9 = 23.1$. Did they gross 300.9 million together? Yes, since $162 + 138.9 = 300.9$. Therefore, our solutions are correct.

PRACTICE EXERCISE 1

As of February 6, 2011, *The Phantom of the Opera* was the longest-running Broadway play in history, with 2094 more performances than the second-longest-running play, which was *CATS*. Find the number of performances of each if together they had a total of 17,064 performances. (***Source:*** *wikipedia*)

If you need more practice, do the following Additional Practice Exercise.

Additional Practice Exercise 1

The top-grossing movie in 2008 was *The Dark Knight*, which grossed $212.6 million more than second place, *Iron Man*. Find the amount that each grossed if together they grossed $849.2 million. (***Source:*** *World Almanac and Book of Facts 2010*)

OBJECTIVE **Solving money problems by using systems of linear equations**

Example 2

A paint contractor paid $372 for 24 gallons of paint to paint the inside and outside of a house. If the paint for the inside costs $12 per gallon and the paint for the outside costs $18 per gallon, how many gallons of each did he buy?

Solution

Understand: We are given the total number of gallons of paint, the cost per gallon for each type of paint, and the total amount paid for the paint. The unknowns are the number of gallons of each type of paint.

> Relationship 1: He bought a total of 24 gallons of paint.
>
> Relationship 2: The total amount paid was $372.

Plan: Use two unknowns—one for the number of gallons of inside paint and the other for the number of gallons of outside paint. Use relationship 1 to write one equation and relationship 2 to write the other. Organize the information in a chart.

Translate:

Let x represent the number of gallons for the inside, and

Let y represent the number of gallons for the outside.

Type of Paint	Price per Gallon	Number of Gallons	Total Cost
Inside	12	x	$12x$
Outside	18	y	$18y$
Total		24	372

From the table, we see that $x + y = 24$. (The number of gallons of inside paint plus the number of gallons of outside paint equals the total number of gallons of paint purchased, by relationship 1.) The table also shows that $12x + 18y = 372$. (The cost of the inside paint plus the cost of the outside paint equals the total cost of the paint, by relationship 2.)

Answers: **Practice Exercise 1:** *The Phantom of the Opera* $= 9579$; *Cats* $= 7485$ **Additional Practice 1:** *The Dark Knight* $= $530.9 millions$; *Iron Man* $= $318.3 million$

Solve: Let's eliminate x.

$$\begin{array}{l} x + y = 24 \xrightarrow[\text{Leave unchanged.}]{\text{Multiply by } -12} -12x - 12y = -288 \\ 12x + 18y = 372 \xrightarrow{} 12x + 18y = 372 \end{array}$$

Now add the equations to eliminate x.

$$\begin{array}{rl} -12x - 12y = -288 & \\ \underline{12x + 18y = 372} & \text{Add the equations.} \\ 6y = 84 & \text{Divide both sides by 6.} \\ y = 14 & \text{Solution.} \end{array}$$

State: Therefore, he bought 14 gallons of paint for the outside.

Solve: Substitute 14 for y in either of the original equations and solve for x. Use $x + y = 24$.

$$\begin{array}{rl} x + y = 24 & \text{Substitute 14 for } y. \\ x + 14 = 24 & \text{Subtract 14 from both sides.} \\ x = 10 & \text{Solution.} \end{array}$$

State: Therefore, he bought 10 gallons of paint for the inside.

Check: Did he buy a total of 24 gallons? Yes, since $14 + 10 = 24$. Does 10 gallons of inside paint at $12 per gallon and 14 gallons of outside paint at $18 per gallon cost $372? $10(12) + 14(18) = 120 + 252 = 372$. Yes, so our solutions are correct.

PRACTICE EXERCISE 2

A landscaping company pays $7 each for rose bushes and $5 each for azaleas. If it paid $82 for 14 plants, how many of each did it buy?

If you need more practice, do the following Additional Practice Exercise.

Additional Practice Exercise 2

A mechanic paid $2 each for spark plugs and $6 each for some wrenches. If he paid a total of $84 for 30 items, how many of each did he buy?

Example 3

A collection of 23 coins, made up of nickels and dimes, is worth $1.90. How many of each type of coin is there?

Solution

Understand: We are given the types of coins, the total number of coins, and the value of the coins. The unknowns are the number of each type of coin.

Relationship 1: There is a total of 23 coins.

Relationship 2: The total value of the coins is $1.90.

Plan: Use a table. Use two unknowns, one for the number of nickels and one for the number of dimes. Use relationship 1 to write one equation and relationship 2 to write another. Solve the system.

Translate:

Let x represent the number of nickels, and

Let y represent the number of dimes.

Answers: Practice Exercise 2: 6 rose bushes, 8 azalea bushes **Additional Practice 2:** 24 spark plugs and 6 wrenches

Type of Coin	Value of Coin	Number of Coins	Total Value
Nickel	5	x	$5x$
Dime	10	y	$10y$
Total		23	190

From the table, we see that $x + y = 23$. (The number of nickels + the number of dimes $= 23$, from relationship 1.) We also see that $5x + 10y = 190$. (Value of the nickels + value of the dimes $= 190$, from relationship 2.)

Solve: Let's eliminate x.

$$x + y = 23 \xrightarrow[\text{Leave unchanged.}]{\text{Multiply by } -5} -5x - 5y = -115$$
$$5x + 10y = 190 \xrightarrow{\phantom{\text{Leave unchanged.}}} 5x + 10y = 190$$

Now add the equations to eliminate x.

$$\begin{array}{rl} -5x - 5y = -115 & \\ \underline{5x + 10y = 190} & \text{Add the equations.} \\ 5y = 75 & \text{Divide both sides by 5.} \\ y = 15 & \end{array}$$

State: Therefore, there are 15 dimes.

Solve: Substitute 15 for y in either of the original equations and solve for x. Use $x + y = 23$.

$$\begin{array}{rl} x + y = 23 & \text{Substitute 15 for } y. \\ x + 15 = 23 & \text{Subtract 25 from both sides.} \\ x = 8 & \text{Solution.} \end{array}$$

State: Therefore, there are 8 nickels.

Check: Is there a total of 23 coins? Yes, since $15 + 8 = 23$. Is the value of 8 nickels and 15 dimes equal to 1.90? $8(5) + 15(10) = 40 + 150 = 190¢ = 1.90$. Yes, so our solutions are correct.

PRACTICE EXERCISE 3

A cashier has a total of 30 nickels and quarters in the cash register. If the coins are worth $3.90, how many of each does he have?

If you need more practice, do the following Additional Practice Exercise.

Additional Practice Exercise 3

Amanda is saving her money to buy a CD. She notices that among the coins there is a total of 69 dimes and quarters that are worth $10.35. How many of each does she have?

OBJECTIVE Solving distance, rate, and time problems by using systems of linear equations

Answers: Practice Exercise 3: 18 nickels, 12 quarters   **Additional Practice 3:** 46 dimes and 23 quarters

Example 4

In 4 hours, a boat can go 152 kilometers downstream or 104 kilometers upstream. Find the speed of the boat in still water and the speed of the current.

Solution

Understand: We are given the distance the boat can go upstream and downstream in 4 hours. The unknowns are the speed of the boat and the speed of the current.

  Relationship 1: The distance downstream is 152 kilometers.

  Relationship 2: The distance upstream is 104 kilometers.

Plan: Use a chart and two unknowns—the rate of the boat and the rate of the water. Use relationship 1 and the formula $d = rt$ to write one equation and relationship 2 and the formula $d = rt$ to write another. Solve the system.

Translate:

Let x represent the speed of the boat in still water, and

Let y represent the speed of the current.

	d	r	t
Downstream	152	$x + y$	4
Upstream	104	$x - y$	4

Since the rate of the boat is x and the rate of the current is y, the rate downstream is $x + y$ and the rate upstream is $x - y$.

Since $d = rt$,

$152 = (x + y)4$ (distance downstream) = (rate downstream)(time downstream), and

$104 = (x - y)4$ (distance upstream) = (rate upstream)(time upstream)

Solve:

$152 = 4x + 4y$

$\underline{104 = 4x - 4y}$ Add the equations.

$256 = 8x$ Divide both sides by 8.

$32 = x$ Solution.

State: Therefore, the rate of the boat in still water is 32 kilometers per hour.

Solve: Substitute 32 for x in either original equation and solve for y.
Use $152 = (x + y)4$.

$152 = (x + y)4$ Substitute 32 for x.

$152 = (32 + y)4$ Distribute on the right side.

$152 = 128 + 4y$ Subtract 128 from both sides.

$24 = 4y$ Divide both sides by 4.

$6 = y$ Solution.

State: Therefore, the speed of the current is 6 kilometers per hour.

Check: Will the boat go 152 kilometers downstream in 4 hours? The rate downstream is the rate of the boat plus the rate of the current, which is $32 + 6 = 38$ kilometers per hour. In 4 hours the distance downstream is $4(38) = 152$ kilometers. Will the boat go 104 kilometers upstream in 4 hours? The rate upstream is the rate of the boat minus the rate of the current, which is $32 - 6 = 26$ kilometers per hour. In 4 hours the distance upstream is $4(26) = 104$ kilometers. Therefore, our solutions are correct.

Note

Both $152 = (x + y)4$ and $104 = (x - y)4$ can be simplified by dividing both sides of each equation by 4 resulting in the system

$$\begin{cases} 38 = x + y \\ 26 = x - y \end{cases}$$

PRACTICE EXERCISE 4

In 3 hours, a plane can go 1170 miles with the wind and 930 miles against the wind. Find the speed of the plane in still air and the speed of the wind.

If you need additional practice, do the following Additional Practice Exercise.

Additional Practice Exercise 4

In 5 hours, a ship can travel 100 knots with the current of a river and 60 knots against the current. Find the speed of the ship in still water and the speed of the current.

OBJECTIVE **Solving percent problems by using systems of linear equations**

The next example involves changing a percent to a decimal. Remember, to change a percent to a decimal, drop the percent sign and move the decimal point two places to the left. For example, 23% = .23 and 6% = .06.

Example 5

Roxanne received an inheritance of $10,000. She invested part of it at 8% and the remainder at 11%. How much did she invest at each rate if the total interest from both investments was $1010 per year?

Solution

Understand: We are given the total amount invested, the interest rate of each investment, and the total amount of interest earned. The unknowns are the amount invested at each interest rate.

    Relationship 1: The total amount invested is $10,000.

    Relationship 2: The total interest earned is $1010.

Plan: Use a chart and two unknowns—one for the amount invested at 8% and another for the amount invested at 11%. Use relationship 1 to write one equation and relationship 2 to write another. Solve the system.

Translate:

Let x represent the amount invested at 8%, and

Let y represent the amount invested at 11%.

Type of Investment	Interest Rate	Amount Invested	Interest Earned
8%	.08	x	$.08x$
11%	.11	y	$.11y$
Total		10,000	1010

Using relationship 1, we see that $x + y = 10,000$. (The total amount invested was $10,000.) Using relationship 2, we see that $.08x + .11y = 1010$. (The interest earned at 8% + the interest earned at 11% = $1010.)

Solve: Eliminate the decimals in the second equation.

$$x + y = 10,000 \xrightarrow{\text{Leave unchanged.}} x + y = 10,000$$
$$.08x + .11y = 1010 \xrightarrow{\text{Multiply by 100.}} 8x + 11y = 101,000$$

Solve the System.

$$x + y = 10,000 \xrightarrow[\text{Leave unchanged.}]{\text{Multiply by } -8} -8x - 8y = -80,000$$
$$8x + 11y = 101,000 \xrightarrow{\hspace{3cm}} 8x + 11y = 101,000$$

Now add the equations to eliminate x.

$$
\begin{aligned}
-8x - 8y &= -80,000 \\
\underline{8x + 11y} &= \underline{101,000} \quad \text{Add the equations.} \\
3y &= 21,000 \quad \text{Divide both sides by 3.} \\
y &= 7000 \quad \text{Solution.}
\end{aligned}
$$

State: Therefore, there was $7000 invested at 11%.

Solve: Substitute 7000 for y in either of the original equations. Use $x + y = 10,000$.

$$
\begin{aligned}
x + y &= 10,000 \quad \text{Substitute 7000 for } y. \\
x + 7000 &= 10,000 \quad \text{Subtract 7000 from both sides.} \\
x &= 3000 \quad \text{Solution.}
\end{aligned}
$$

State: Therefore, there was $3000 invested at 8%.

Check: Is there a total of $10,000 invested? Yes, since $7000 + $3000 = $10,000. Is the total interest earned $1010? The interest earned on $3000 at 8% = .08(3000) = $240. The interest earned on $7000 at 11% is .11(7000) = $770. Since $240 + $770 = $1010, our solutions are correct.

PRACTICE EXERCISE 5

Jerome received $50,000 from the sale of some property. He invested part of it in a certificate of deposit at 5% and the remainder in bonds at 7%. Find the amount invested in each if the total interest received from the two investments is $3100 per year.

If you need more practice, do the following Additional Practice Exercise.

Additional Practice Exercise 5

Franz inherited $80,000. He invested part in municipal bonds at 4% and the remainder in an annuity paying 6%. Find the amount invested in each if the total annual interest received from the two investments is $3900.

OBJECTIVE Solving mixture problems by using systems of linear equations

Example 6

A candy store owner mixes some candy worth $2.50 per pound with some worth $3.50 per pound. How many pounds of each would he need if he wants 10 pounds of the mixture worth $2.90 per pound?

Solution

Understand: We are given the price of each type of candy, the price of the mixture, and the number of pounds of the mixture. The unknowns are the number of pounds of each type of candy.

> Relationship 1: There are 10 pounds of the mixture.
>
> Relationship 2: The value of the mixture is the same as the value of the individual candies in the mix.

Answers: Practice Exercise 5: $20,000 at 5%, $30,000 at 7%   **Additional Practice 5:** $45,000 in municipal bonds, $35,000 in the annuity

Plan: Use a chart as in Chapter 3. Use two unknowns, one for the number of pounds of $2.50 candy and another for the $3.50 candy. Use relationship 1 to write one equation and relationship 2 to write another. Solve the system.

Translate:

Let x represent the number of pounds of candy worth $2.50 per pound, and

Let y represent the number of pounds of candy worth $3.50 per pound.

Fill in the chart.

Type of Candy	Price per Pound	Number of Pounds	Total Value
$2.50 per pound	2.50	x	$2.50x$
$3.50 per pound	3.50	y	$3.50y$
Mixture	2.90	10	$2.90(10) = 29$

From relationship 1 we see that that $x + y = 10$. (The number of pounds of $2.50 candy + the number of pounds of $3.50 candy = the number of pounds in the mixture.) From relationship 2 we see that $2.50x + 3.50y = 29$. (The value of the $2.50 per pound candy) + (the value of the $3.50 per pound candy) = (the value of the mixture).

Sometimes a diagram of this type of problem, referred to as a mixture problem, is helpful.

$2.50 per lb$x$ lb	+	$3.50 per lb$y$ lb	=	$2.90 per lb10 lb
value of x lb	+	value of y lb	=	value of mixture
$2.50x$	+	$3.50y$	=	$2.90(10)$

Solve: Eliminate the decimals from the second equation.

$$x + y = 10 \xrightarrow{\text{Leave unchanged.}} x + y = 10$$
$$2.5x + 3.5y = 29 \xrightarrow{\text{Multiply both sides by 10.}} 25x + 35y = 290$$

Note

$25x + 35y = 290$ can be simplified to $5x + 7y = 58$ by dividing both sides of the equation by 5.

Solve the system using the substitution method. So, solve $x + y = 10$ for y.

$x + y = 10$	Subtract x from both sides.
$x - x + y = 10 - x$	Add like terms.
$y = 10 - x$	Substitute $10 - x$ for y in $25x + 35y = 290$.
$25x + 35(10 - x) = 290$	Distribute 35.
$25x + 350 - 35x = 290$	Add like terms.
$-10x + 350 = 290$	Subtract 350 from both sides.
$-10x = -60$	Divide both sides by -10.
$x = 6$	Solution.

State: Therefore, he would need 6 pounds of candy worth $2.50 per pound.

Solve: Substitute 6 for x in either of the original equations and find y. Use $x + y = 10$.

$6 + y = 10$	Subtract 6 from both sides.
$y = 4$	Solution.

State: Therefore, he needs 4 pounds of candy worth $3.50 per pound.

Check: Does the number of pounds of $2.50 per pound candy + the number of pounds of $3.50 per pound candy = the number of pounds of candy in the mixture? Yes, since 6 + 4 = 10. Does the value of the $2.50 per pound candy + the value of the $3.50 per pound candy = the value of the mixture? The value of the $2.50 per pound candy is 6($2.50) = $15.00. The value of the $3.50 per pound candy is 4($3.50) = $14.00. The value of the mixture is 10($2.90) = $29. Since $15.00 + $14.00 = $29.00, the answer is yes. Therefore, our solutions are correct.

PRACTICE EXERCISE 6

The Smokehouse Restaurant specializes in a barbecue that is a mixture of pork and beef. It needs 20 pounds of this mixture for a picnic it is catering. If pork barbecue sells for $3.50 per pound and beef barbecue sells for $5.00 per pound, how many pounds of each will be needed if the mixture sells for $4.10 per pound?

If you need more practice, do the following Additional Practice Exercise.

Additional Practice Exercise 6

A rancher has found that a mixture of fast-maturing and slow-maturing grass makes the best pasture for his livestock. Seeds for the fast-maturing grass sell for $2.50 per pound, and seeds for the slow-maturing grass sell for $3.00 per pound. If he purchases 500 pounds of the mixture that sells for $2.80 per pound, how many pounds of each type of grass seeds did he buy?

Example 7

How much of a solution that is 40% alcohol must be added to a solution that is 60% alcohol to obtain 50 liters of a solution that is 55% alcohol?

Solution

Understand: We are given the concentrations of the individual solutions and the mixture. We are also given the volume of the mixture. The unknowns are the amount of 40% solution and the amount of 60% solution.

Relationship 1: The total volume of the mixture is 50 liters.

Relationship 2: The amount of alcohol in the mixture is the same as the sum of the amounts in the individual solutions.

Plan: Use a chart and two unknowns, one for the amount of 40% solution and one for the amount of 60% solution. Use relationship 1 to write one equation and relationship 2 to write the other. Solve the system.

Translate:

Let x represent the number of liters of 40% alcohol, and

Let y represent the number of liters of 60% alcohol.

Fill in the chart.

Type of Solution	Part Pure Alcohol	Volume	Amount Pure Alcohol
40%	.40	x	$.40x$
60%	.60	y	$.60y$
55% (mixture)	.55	50	$.55(50) = 27.5$

Using relationship 1, we see that $x + y = 50$. (The number of liters of 40% alcohol + the number of liters of 60% alcohol = the number of liters in the mixture.) Using relationship 2, we have $.40x + .60y = 27.5$. (The amount of pure alcohol in the 40% solution + the amount of pure alcohol in the 60% solution = the amount of pure alcohol in the mixture.)

Answers: Practice Exercise 6: 12 lb pork, 8 lb beef   **Additional Practice 6:** 200 pounds at $2.50 and 300 pounds at $3.00

Solve:

$$x + y = 50 \qquad \text{Solve for } x.$$
$$.40x + .60y = 27.5 \qquad \text{Leave unchanged.}$$

$$x = 50 - y \qquad \text{Substitute for } x \text{ in } .40x + .60y = 27.5.$$
$$.40(50 - y) + .60y = 27.5 \qquad \text{Distribute .40.}$$
$$20 - .40y + .60y = 27.5 \qquad \text{Add like terms.}$$
$$20 + .20y = 27.5 \qquad \text{Subtract 20 from both sides.}$$
$$.20y = 7.5 \qquad \text{Divide both sides by .20.}$$
$$y = 37.5 \qquad \text{Solution.}$$

State: Therefore, 37.5 liters of the 60% solution are needed.

Solve: To find x, substitute 37.5 for y in the equation $x + y = 50$.

$$x + 37.5 = 50 \qquad \text{Subtract 37.5 from both sides.}$$
$$x = 12.5 \qquad \text{Solution.}$$

State: Therefore, 12.5 liters of the 40% solution are needed.

Check: The check is left as an exercise for the student.

PRACTICE EXERCISE 7

How many pounds of an alloy that is 30% tin are to be mixed with an alloy that is 70% tin in order to get 100 pounds of an alloy that is 60% tin?

If you need more practice, do the following Additional Practice Exercise.

Additional Practice Exercise 7

How many liters of a brine solution that is 8% salt must be mixed with a brine solution that is 12% salt to get 80 liters of a brine solution that is 9% salt?

Exercise Set 9.4

For Extra Help

MyMathLab®

For Exercises 1–8, solve the number problems using a system of linear equations. (See Example 1.)

1. From 1961 to 6/16/10, the United States has had 14 more than three times the number of individuals who have flown in space as Russia. If there has been a total of 438 from both countries, how many individuals have flown in space from each country? (*Source: wikipedia.org*)

2. From 1957 to 2009, Russia had 1379 less than three times the number of successful rocket launches as the United States. If together they have had 4905 successful launches, how many has each country had? (*Source: World Almanac and Book of Facts 2010*)

3. The percentage of expenditures for advertising on television is 7.2% less than three times the percentage spent for advertising in newspapers. If together 63.6% of all advertising expenditures is for television or newspaper, find the percentage spent on each medium. (*Source: The New York Times Almanac, 2010*)

4. The U.S. budget deficit in 2009 was $6.8 billion more than four times the deficit in 2008. If the sum of the deficits for the two years was $2299.8 billion, find the deficit for each years. (*Source: The New York Times Almanac, 2010*)

Answers: Practice Exercise 7: 25 lb **Additional Practice 7:** 20 liters of 12% and 60 liters of 8%

5. The tallest man-made structure in the world is Burj Khalifa in Dubai, United Arab Emirates. Burj Khalifa is 199 meters higher than the KVLY-TV Mast in Blanchard, North Dakota, which is the tallest structure in the United States. If the sum of the heights of the two structures is 1457 meters, find the height of each. (***Source:*** *wikipedia.org*)

6. The highest waterfall in the world is Angel Falls in Venezuela and the second highest is Yosemite Falls in the United States. If the sum of their heights is 5657 feet and the difference of their heights is 767 feet, find the height of each. (***Source:*** *Webster's New World Book of Facts 2010*)

7. The largest asteroid is Ceres, and the second largest is Pallas. The diameter of Ceres is 146 miles less than twice the diameter of Pallas. If the difference of their diameters is 219 miles, find the diameter of each. (***Source:*** *Webster's New World Book of Facts 2010*)

8. The largest moon of Neptune is Triton, and the next largest is Nereid. The diameter of Triton is 201 miles more than seven times the diameter of Nereid. If the difference in their diameters is 1467 miles, find the diameter of each. (***Source:*** *Webster's New World Book of Facts 2010*)

For Exercises 9–16, solve the money problems using a system of linear equations. (See Examples 2 and 3.)

9. An electrical contractor paid $1040 for 36 light fixtures to be installed in a hotel lobby. If the first type of fixture costs $25 each and the second type of fixture costs $32 each, how many of each kind did she buy?

10. A locksmith installed inside and outside locks in a new classroom building. The locksmith paid $1216 for 80 locks. If inside locks cost $14 each and outside locks cost $26 each, how many of each kind of lock did he buy?

11. The managers of an office building plan to reduce the cost of air-conditioning by installing overhead fans. They paid $5225 for 105 fans, some of which measure 36 inches and the remainder of which measure 54 inches. If the 36-inch fans cost $45 each and the 54-inch fans cost $65 each, how many of each kind of fan did they buy?

12. An accounting firm purchased a supply of computer disks. They paid $725 for 60 boxes of disks. If the CD-ROM disks cost $8 a box and the DVD disks cost $15 a box, how many of each kind were purchased?

13. A vending machine accepts only dimes and quarters. If the person who services the machine collects 52 coins with a total value of $8.95, how many of each kind of coin was in the machine?

14. At the end of his shift, a cashier has only $5 and $20 bills in his drawer. If he has 48 bills worth a total of $900, how many of each kind of bill does he have?

15. Winston paid $7.46 for 36 stamps. If he bought eight fewer 25-cent stamps than 18-cent stamps, how many of each kind did he buy?

16. Gayle paid $17.40 for 42 pens at the campus bookstore. If she bought six fewer pens costing $0.35 each than pens costing $0.50 each, how many of each kind did she buy?

For Exercises 17–22, solve the distance, rate, time problems using a system of linear equations. (See Example 4.)

17. In 7 hours, a Coast Guard cutter can sail 126 miles with an ocean current, or it can sail 84 miles against the current. Find the speed of the cutter in still water and find the speed of the current.

18. In 5 hours, a cruise ship in Alaska's Inside Passage can travel 75 miles with the current or 45 miles against the current. Find the speed of the ship in still water and the speed of the current.

19. A ship can travel the 80 miles from Venice, Louisiana to New Orleans against the current of the Mississippi river in 10 hours and return with the current in 5 hours. Find the speed of the ship in still water and the speed of the current.

20. An airplane flies 3000 miles coast to coast in 6 hours with a tailwind pushing it. When it flies back against the wind, it takes $7\frac{1}{2}$ hours. What is the speed of the airplane in still air and what is the speed of the wind?

21. A freight train and a passenger train leave Memphis, Tennessee, at the same time, traveling in opposite directions. The freight train is traveling 15 miles per hour slower than the passenger train. Find the rate of each train if the distance between them is 525 miles after 5 hours.

22. At 8:00 A.M., two airplanes leave cities that are 2250 miles apart, flying toward each other. The rate of one airplane is 150 miles per hour less than twice the rate of the other. If the planes meet at 11:00 A.M., how fast is each plane flying?

For Exercises 23–26, solve the percent problems using a system of linear equations. (See Example 5.)

23. Rochelle received a $20,000 bonus for selling over $1,000,000 worth of life insurance. She put part of the money into a savings account paying 6% interest and the rest of the money into municipal bonds paying 8.5% interest. How much did she invest at each rate if the total interest from both investments was $1650 per year?

24. The manager of a supermarket received a $5000 bonus for exceeding the sales quota for her store. She put part of the money into a savings account paying 5% interest and the rest into a certificate of deposit paying 9% interest. How much did she invest at each rate if the total interest from both investments was $370 per year?

25. A college-development fund has raised $250,000. Part of the money was invested in home mortgages with an expected return of 12% per year. The rest was invested in common stock with an expected return of 8% per year. If the fund earns $26,000 in one year, how much was put into each type of investment?

26. A college invests its operating funds in two types of U.S. government securities. The amount invested is $750,000, which is split between U.S. government guaranteed mortgages at 12% interest and U.S. government treasury bills at 9% interest. If a total of $79,500 is earned per year, how much is invested in each type of security?

For Exercises 27–30, solve the mixture problems using a system of linear equations. (See Examples 6 and 7.)

27. A specialty store owner mixes peanuts worth $1.80 per pound with cashew nuts worth $5.40 per pound. How many pounds of each type of nut is needed to make a 12-pound mixture worth $3.00 a pound?

28. A coffee company makes a blend of coffee by mixing beans that cost $3.20 per pound with beans that cost $6.20 per pound. How many pounds of each type of bean are needed to make 40 pounds of coffee that costs $3.95 per pound?

29. A 12% solution of salt is mixed with a 4% solution. How many liters of each are needed to obtain 30 liters of an 8% solution of salt?

30. A 70% solution of alcohol is mixed with a 30% solution. How many gallons of each are needed to obtain 25 gallons of a 60% solution of alcohol?

For Exercises 31–48, solve the application problems using a system of linear equations. (See Examples 1–7.)

31. The sum of two numbers is 25. If the difference of these numbers is 5, find the two numbers.

32. One-half of a number is equal to three times another number. If the difference of the two numbers is 20, find the numbers.

33. The owners of an exterminator business have replaced their fleet of 27 cars and trucks by purchasing new cars for $12,000 each and new trucks for $9500 each. If they paid a total of $294,000, how many of each type of vehicle did they buy?

34. A building contractor has paid $2088 for 54 doors. If inside doors cost $35 each and outside doors cost $46 each, how many of each kind of door did he buy?

35. The metro rail system accepts only quarters and silver dollars. If at the end of the day 900 coins worth $375 are collected, how many of each type of coin were in the machines?

36. Mr. and Mrs. Wong had a birthday party at the movie theater for their 8-year-old daughter. They paid $89 for 22 people to attend the movie. If a child's ticket is $3.50 and an adult ticket is $5.50, how many children and how many adults attended the party?

37. In 4 hours, a speedboat can go 180 kilometers downstream or 140 kilometers upstream. Find the speed of the boat in still water and the speed of the current.

38. Two ships leave port at 5:00 A.M., both traveling in the same direction. At noon, the ships are 56 miles apart and the sum of the distances they have traveled is 224 miles. How fast is each ship traveling?

39. A book salesperson received $3000 for signing six authors last year. She invested part of the money in a mutual fund paying 18% interest and the remainder in a certificate of deposit paying 9% interest. If she earns $414 per year interest from her investments, how much did she put in each investment?

40. An author received a $160,000 royalty check for his new novel. He invests part of the money in bonds at 8% and part in stocks at a 6% expected return. If at the end of the year he expects to earn $11,200, how much did he put in each investment?

41. Pure rice (100%) and a mixture that is 50% rice and 50% beans are mixed. How many cups of each are needed to get 20 cups of a mixture that is 70% rice and 30% beans?

42. A solution that is 15% baking soda is mixed with a solution that is 5% baking soda. How many liters of each is needed to obtain 12 liters of a 10% solution of baking soda?

43. The length of a rectangular sheet of plywood is 4 feet more than the width. If the perimeter is 32 feet, find the length and width.

44. The length of a rectangular picture is 2 inches less than twice the width. If the perimeter is 26 inches, find the length and width of the picture.

45. One side of a triangular sign is 8 centimeters. The length of the smaller of the remaining two sides is 2 centimeters less than the length of the longer side. If the perimeter is 20 centimeters, find the lengths of the remaining two sides.

46. The parallel sides of a trapezoid are 6 meters and 9 meters in length. The length of the longer of the two nonparallel sides is 1 less than twice the length of the shorter side. If the perimeter is 23 meters, find the lengths of the nonparallel sides.

47. Recall from Section 3.8 that two angles are supplementary if the sum of their measures is 180. If two angles are supplementary and the measure of the larger angle is 30 more than the measure of the smaller angle, find the measure of each angle.

48. Recall from Section 3.8 that two angles are complementary if the sum of their measures is 90. If two angles are complementary and the measure of the smaller angle is 14 less than the measure of the larger angle, find the measure of each angle.

Writing Exercise

49. Which do you think is easier? Solving application problems by using a system of two variables or solving application problems by using one variable, as we did in Chapter 3? Why?

Writing Exercises or Group Project (50–52)

If these are done as a group project, each group should write two exercises of each of the following types, exchange with another group, and then solve them:

50. Write and solve a number problem, using a system of linear equations.

51. Write and solve a rate, time, and distance problem, using a system of linear equations.

52. Write and solve a mixture problem, using a system of linear equations.

OBJECTIVE *When you complete this section, you will be able to:*

A Graphically represent the solutions of a system of linear inequalities.

PREREQUISITE SKILL *Before beginning this section, you should be able to:*

a. Graph linear inequalities. (Section 4.7)

GETTING READY FOR SECTION 9.5

Graph the following inequalities:

1. $x - 2y < 5$ **2.** $3x + 2y \leq 6$ **3.** $y \leq 2$

Introduction

In Section 4.7, we learned to graph linear inequalities with two variables. The procedure is given in the following box:

> **Procedure: Graphing Linear Inequalities with Two Variables**
>
> **1.** Graph the equality $ax + by = c$. If the line is part of the solution ($\leq$ or $\geq$), draw a solid line. If the line is not part of the solution ($<$ or $>$), draw a dashed line.
>
> **2.** Pick a test point not on the line. If the coordinates of the test point satisfy the inequality, then all points on the same side of the line as the test point solve the inequality. If the coordinates of the test point do not satisfy the inequality, then all points on the other side of the line from the test point solve the inequality.
>
> **3.** Shade the region on the side of the line that contains the solutions of the inequality.

Example 2b from Section 4.7 is repeated here to remind you of this procedure.

Answers: Getting Ready: 1.   **2.**   **3.**

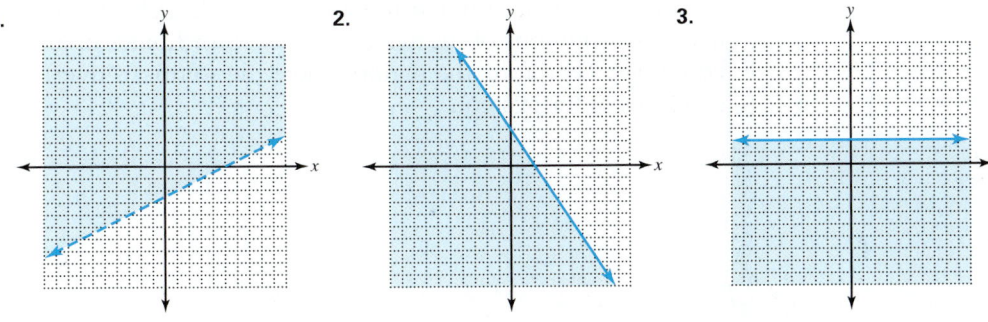

Example 1   Graph $3x - y \le 6$.

Solution

Draw the graph of $3x - y = 6$ as a solid line, since the points on the line solve the inequality less than or *equal to*. Use the x-intercept $(2, 0)$ and the y-intercept $(0, -6)$.

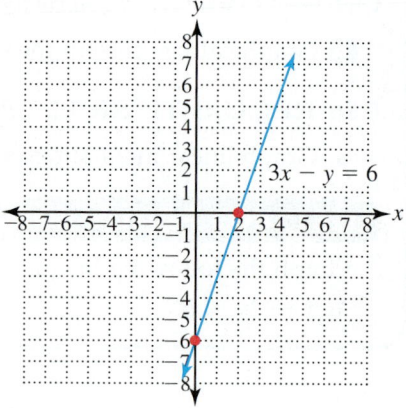

Pick a test point. The origin, $(0, 0)$, is a convenient choice. Substitute $(0, 0)$ into the inequality. $3(0) - 0 \le 6, 0 - 0 \le 6, 0 \le 6$. This is a true statement. Therefore, $(0, 0)$ solves the inequality. Consequently, the coordinates of all points on the same side of the line as $(0, 0)$ solve the inequality. Therefore, shade the region on the same side of the line as $(0, 0)$.

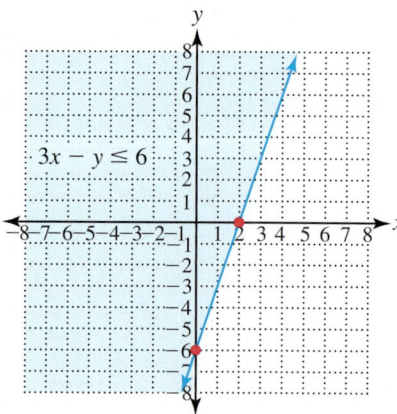

OBJECTIVE **Graphically representing the solutions of a system of linear inequalities**

A system of linear inequalities with two variables is much like a system of linear equations with two variables.

> **Definition** System of Linear Inequalities
>
> A system of linear inequalities is two or more linear inequalities with the same variables.

To solve a system of linear inequalities, we graph each inequality on the same set of axes. The solutions of the system are the coordinates of all points that solve all the inequalities. Therefore, the region where the solutions overlap (intersect) represents the solutions of the system.

Example 2   **Graph the solutions of the following systems of linear inequalities:**

a. $\begin{cases} x + y > 6 \\ x - y < -2 \end{cases}$

Solution

The solutions of the system are represented by the region where the solutions of $x + y > 6$ and $x - y < -2$ overlap. Therefore, we draw the graphs of $x + y > 6$ and $x - y < -2$ on the same set of axes. First draw the graph of $x + y > 6$.

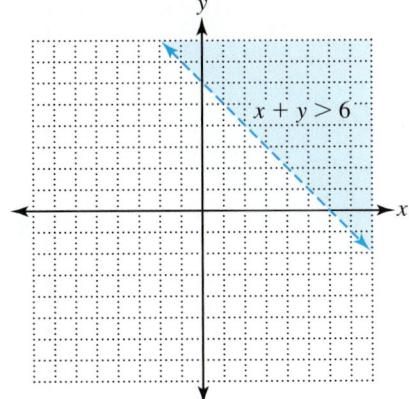

Now draw the graph of $x - y < -2$ on the same set of axes as the graph of $x + y > 6$.

The heavily shaded region where the solutions of $x + y > 6$ and $x - y < -2$ overlap represents the points whose coordinates are the solutions of the system. Since the lines are dashed, the coordinates of the points on the lines are not solutions of the system.

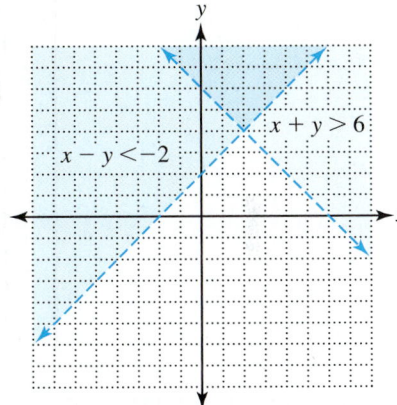

b. $\begin{cases} 2x + 3y \leq 12 \\ 2x + y \leq 8 \end{cases}$

Solution

The solutions of the system are represented by the region where the solutions of $2x + 3y \leq 12$ and $2x + y \leq 8$ overlap. Therefore, draw the graphs of $2x + 3y \leq 12$ and $2x + y \leq 8$ on the same set of axes.

The heavily shaded region where the solutions of $2x + 3y \leq 12$ and $2x + y \leq 8$ overlap (the intersection of the two sets) represents the points whose coordinates are the solutions of the system. Since the lines are solid, the coordinates of the points on the lines (within the shaded region) are also solutions of the system.

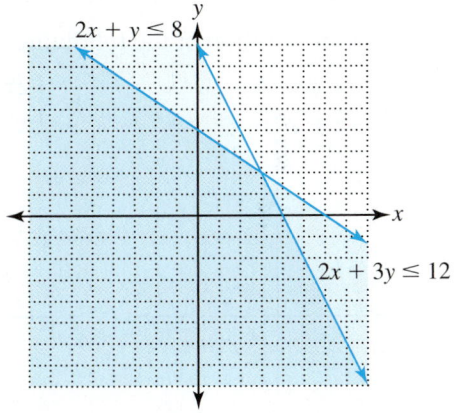

c. $\begin{cases} x + 4y > 8 \\ x \geq 5 \end{cases}$

Solution

The solutions of the system are represented by the region where the solutions of $x + 4y > 8$ and $x \geq 5$ overlap. Therefore, draw the graphs of $x + 4y > 8$ and $x \geq 5$ on the same set of axes.

The heavily shaded region where the solutions of $x + 4y > 8$ and $x \geq 5$ overlap represents the points whose coordinates are the solutions of the system. Within the shaded region, the coordinates of the points on the line $x + 4y = 8$ are not solutions of the system, but the coordinates of the points on the line $x = 5$ are.

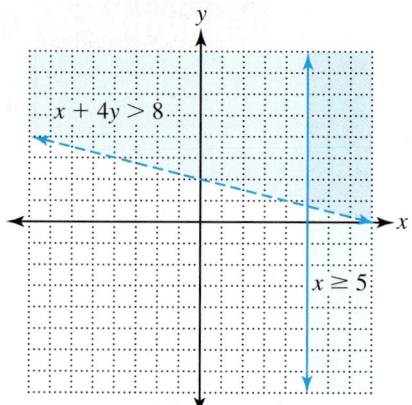

d. $\begin{cases} x \leq -4 \\ y > 3 \end{cases}$

Solution

The solutions of the system are represented by the region where the solutions of $x \leq -4$ and $y > 3$ overlap. Therefore, draw the graphs of $x \leq -4$ and $y > 3$ on the same set of axes.

The heavily shaded region, where the solutions of $x \leq -4$ and $y > 3$ overlap, represents the points whose coordinates are the solutions of the system. Within the shaded region, the coordinates of the points on the line $x = -4$ are solutions of the system, but the coordinates of the points on the line $y = 3$ are not.

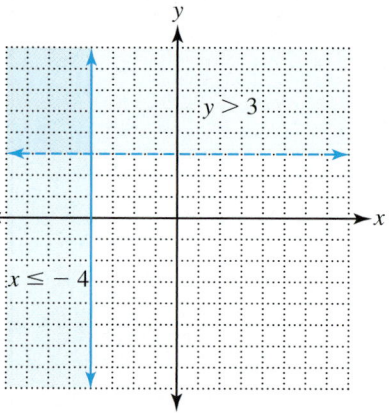

PRACTICE EXERCISE 2

Graph the solution of the following systems of linear inequalities:

a. $\begin{cases} x + y > 4 \\ x - y < 6 \end{cases}$ **b.** $\begin{cases} 2x - y \leq 6 \\ x + 2y \geq 8 \end{cases}$ **c.** $\begin{cases} 3x - 2y \leq 6 \\ y > 5 \end{cases}$ **d.** $\begin{cases} x \geq 4 \\ y < -3 \end{cases}$

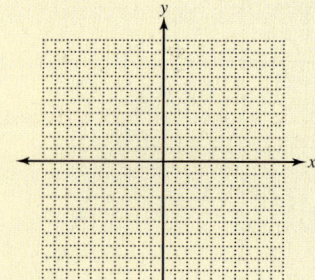 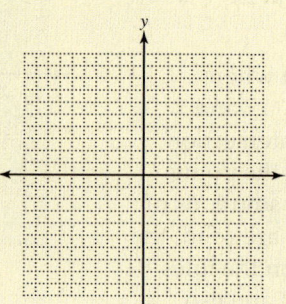 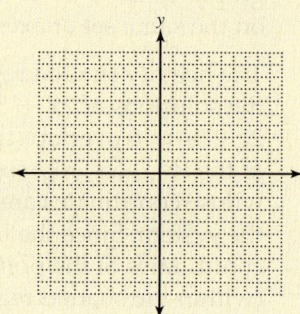 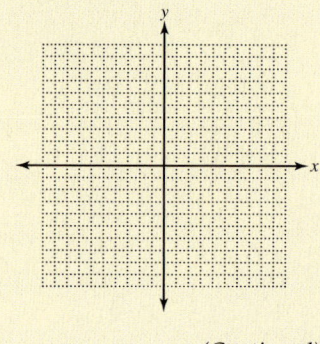

(Continued)

Answers: Practice Exercise 2:

a.

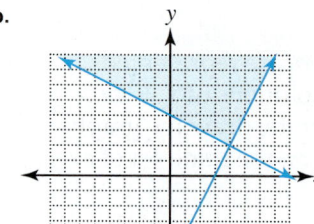

b.

c.

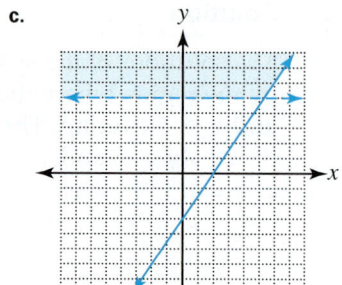

d.

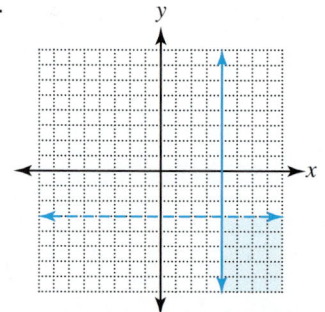

If you need more practice, do the following Additional Practice Exercises.

Additional Practice Exercise 2 Graph the solutions of the following systems of linear inequalities:

a. $\begin{cases} x + y < 3 \\ x - y > 5 \end{cases}$

b. $\begin{cases} x + 3y \le 6 \\ x - 2y \ge 2 \end{cases}$

c. $\begin{cases} 3x - 2y \ge 6 \\ x < 6 \end{cases}$

d. $\begin{cases} x > 4 \\ y < -3 \end{cases}$

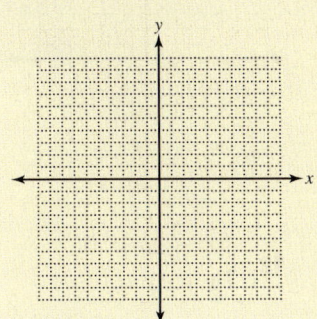
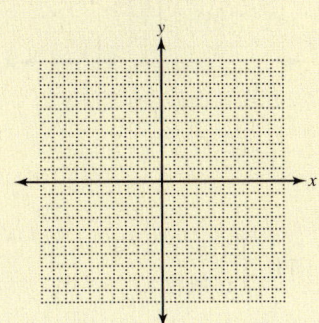
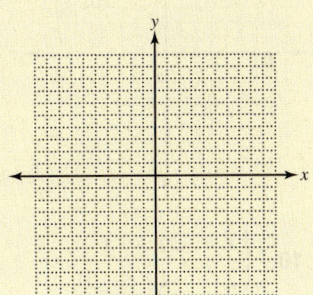
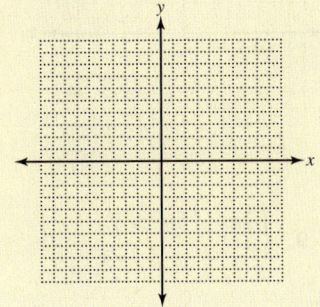

Exercise Set 9.5

For Extra Help

MyMathLab®

Graph the solutions of the following systems of linear inequalities. (See Examples 1 and 2.)

1. $\begin{cases} x - y > 2 \\ x + y > 4 \end{cases}$

2. $\begin{cases} x - y < -5 \\ 2x - y < -7 \end{cases}$

3. $\begin{cases} x + y < -2 \\ x - y > 6 \end{cases}$

4. $\begin{cases} y - x > 5 \\ x + y < 3 \end{cases}$

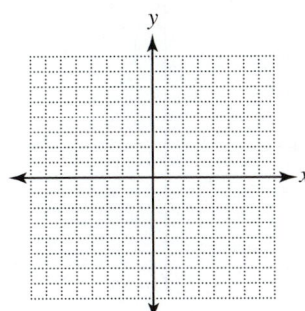
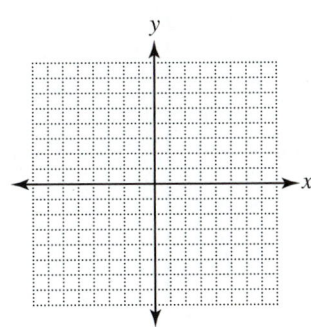
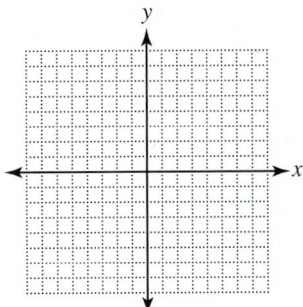
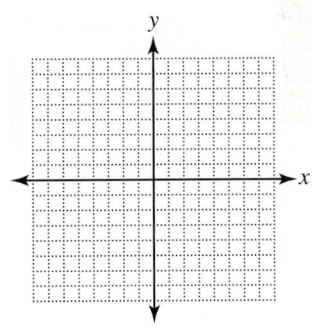

Answers: Additional Practice 2:

a.

b.

c.

d.

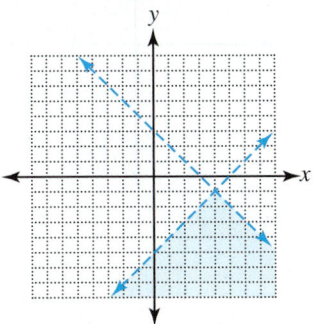
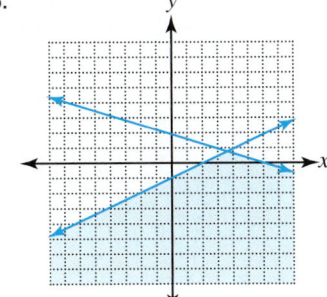
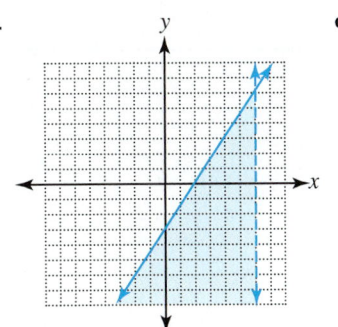
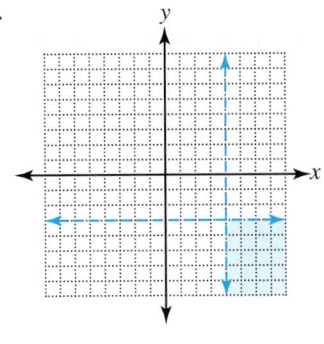

5. $\begin{cases} 2x + y \geq -1 \\ x - y \leq 5 \end{cases}$

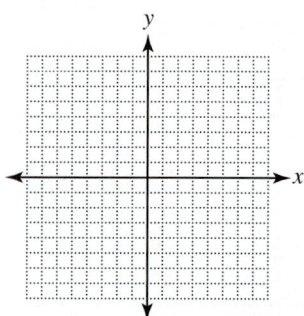

6. $\begin{cases} x + 3y \leq 11 \\ x - 2y \leq 1 \end{cases}$

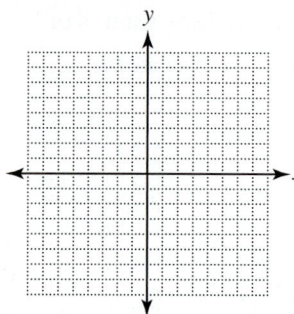

7. $\begin{cases} 2x + 3y < 1 \\ x - 4y \geq 3 \end{cases}$

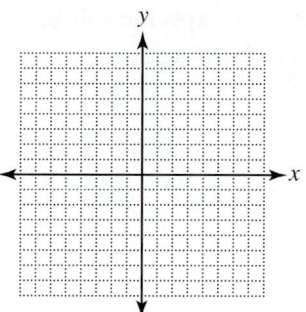

8. $\begin{cases} 4x - y > -14 \\ 5x + 2y \leq 2 \end{cases}$

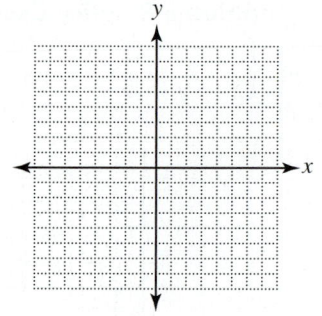

9. $\begin{cases} 2x + 5y \leq 7 \\ 3x + y > -9 \end{cases}$

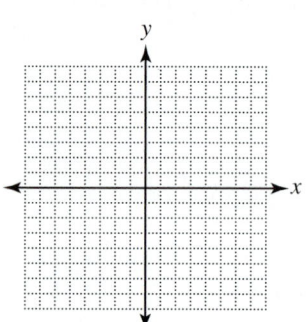

10. $\begin{cases} x - 2y > 3 \\ 2x - 4y \leq 20 \end{cases}$

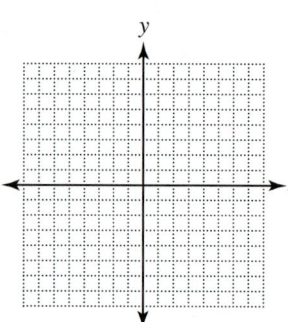

11. $\begin{cases} x + y \geq 4 \\ y \geq 2 \end{cases}$

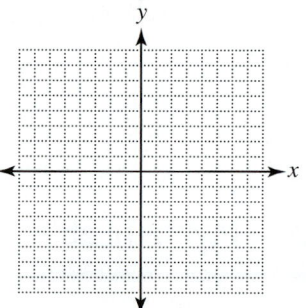

12. $\begin{cases} x - 2y > 0 \\ x < 0 \end{cases}$

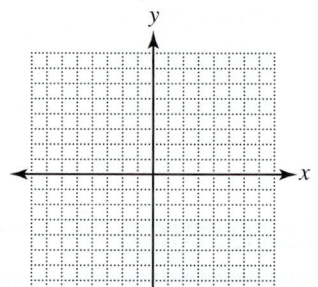

13. $\begin{cases} 2x + y \geq 0 \\ x < 3 \end{cases}$

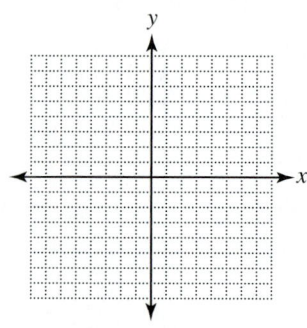

14. $\begin{cases} 5x - 6y > -12 \\ y > 2 \end{cases}$

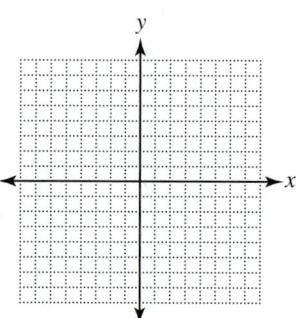

15. $\begin{cases} 5x + 3y < -4 \\ 2y \geq 4 \end{cases}$

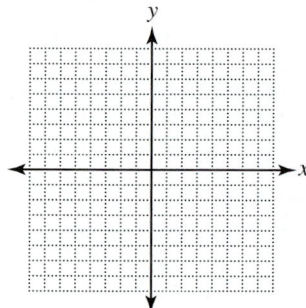

16. $\begin{cases} 3x - y \geq -13 \\ 4x < 12 \end{cases}$

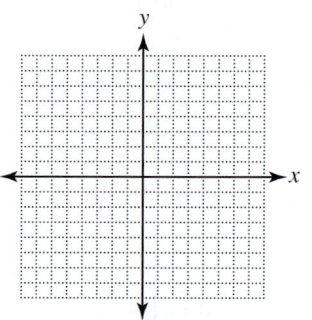

17. $\begin{cases} x \geq -2 \\ y < 4 \end{cases}$

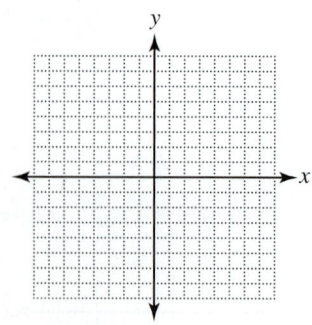

18. $\begin{cases} y \geq -4 \\ x < 3 \end{cases}$

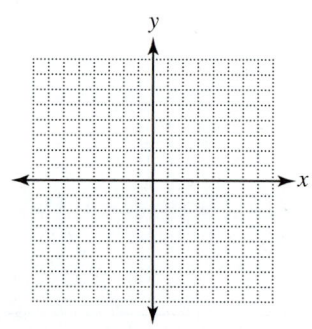

19. $\begin{cases} x < 1 \\ y \geq 0 \end{cases}$

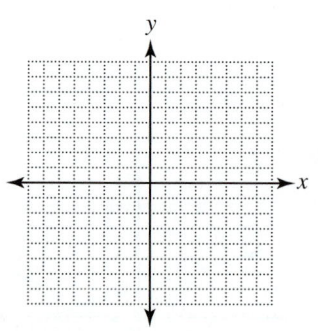

20. $\begin{cases} x \geq -2 \\ y \geq 1 \end{cases}$

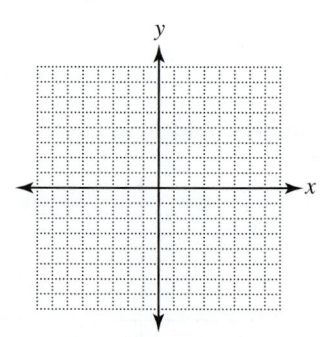

Challenge Exercises (21 and 22)

21. $\begin{cases} 2x + y \geq 1 \\ x - y > -1 \\ x > 2 \end{cases}$

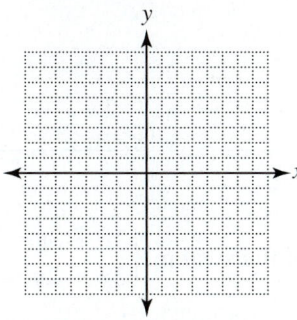

22. $\begin{cases} x + y \geq 1 \\ y - x \geq -5 \\ y > -2 \end{cases}$

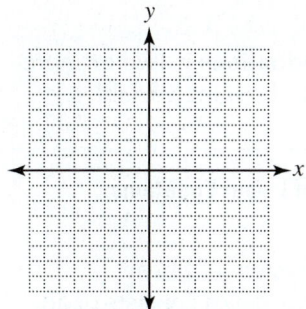

Writing Exercise

23. Describe the region that is the solution of the following systems of inequalities:

a. $x > 0$ and $y > 0$ **b.** $x > 0$ and $y < 0$

Concept/Procedure	Example
Definition of a System of Linear Equations: [Section 9.1] • A **system of linear equations** is two or more linear equations with the same variables.	

Solution(s) of a System of Linear Equations: [Section 9.1]

• The solution(s) of a system of linear equations with two variables consists of all ordered pairs that solve both equations of the system.

Example 1:

Verify that $(4, -2)$ is a solution of the system $\begin{cases} 2x + 3y = 2 \\ 3x - 2y = 16 \end{cases}$.

Solution: Substitute $x = 4$ and $y = -2$ into each equation and simplify.

$$
\begin{array}{ll}
2x + 3y = 2 & 3x - 2y = 16 \\
2(4) + 3(-2) = 2 & 3(4) - 2(-2) = 16 \\
8 - 6 = 2 & 12 + 4 = 16 \\
2 = 2 & 16 = 16
\end{array}
$$

Since $(4, -2)$ solves both equations, it is a solution to the system.

Solving a System of Linear Equations by Graphing: [Section 9.1]

• To graphically solve a system of linear equations with two variables, graph each equation of the system. The coordinates of the point(s) of intersection are the solutions of the system.

Example 2:

Solve the system $\begin{cases} 2x + 3y = 12 \\ 2x - y = 4 \end{cases}$ graphically.

Solution: Graph each line. The point at which they cross is the solution of the system.

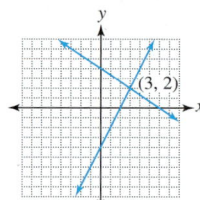

The lines intersect at the point whose coordinates are (3, 2). Therefore, (3, 2) is the solution of the system.

The check is left to the reader.

Types of Systems of Equations: [Section 9.1]

• A system is **consistent** if the graphs of the equations of the system intersect at a single point. The coordinates of the point of intersection are the solution(s) of the system.

• A system is **inconsistent** if the graphs of the equations of the system are parallel lines. Therefore, the system has no solution. This is usually indicated by the symbol $\varnothing$.

• A system is **dependent** if the graphs of the equations of the system are the same line. There are an infinite number of solutions that are represented by $\{(x, y): ax + by = c\}$, where $ax + by = c$ is either of the equations.

Example 3:

Determine whether the system $\begin{cases} 2x + 3y = 6 \\ y = -\dfrac{2}{3}x + 6 \end{cases}$ is consistent, inconsistent, or dependent, and determine the number of solutions.

Solution: Graph the lines and determine whether they intersect in one point, are parallel, or are the same line.

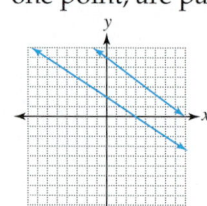

The lines are parallel; therefore, the system is inconsistent and there are no solutions. The solution set is $\varnothing$.

Concept/Procedure	Example
Solving Equations of the Form $ax + b = cx + d$: [Section 9.1] • To solve an equation of the form $ax + b = cx + d$, form the system $y = ax + b$ and $y = cx + d$. Graph the system and the x-value of the point where the lines intersect is the solution of the equation.	

Solving Linear Systems, Using Elimination by Addition: [Section 9.2]

1. If necessary, rewrite each equation in the form $ax + by = c$, with a, b, and c integers.
2. Choose the variable to be eliminated.
3. If necessary, multiply one or both of the equations of the system by the appropriate constant(s) so that the coefficients of the chosen variable will be additive inverses.
4. Add the equations.
5. If necessary, solve the equation resulting from step 4 for the remaining variable.
6. Substitute the value found for the variable in step 5 into either of the equations of the original system and solve for the other variable.
7. Check the solution in both equations.

Example 4:

Solve the system $\begin{cases} 5x + 4y = -10 \\ 3x + 5y = 7 \end{cases}$, using the elimination by addition method.

Solution: We will eliminate x by multiplying the first equation by 3 and the second equation by -5.

$5x + 4y = -10$ — Multiply by 3. → $15x + 12y = -30$
$3x + 5y = 7$ — Multiply by -5. → $-15x - 25y = -35$

Add the equations to eliminate x.

$$\begin{array}{l} 15x + 12y = -30 \\ \underline{-15x - 25y = -35} \\ \qquad\quad -13y = -65 \end{array}$$ Divide both sides by -13.

$y = 5$ Substitute 5 for y in $3x + 5y = 7$.

$3x + 5(5) = 7$ Simplify.

$3x + 25 = 7$ Subtract 25 from both sides.

$3x = -18$ Divide both sides by 3.

$x = -6$ Therefore, the solution of the system is $(-6, 5)$.

The check is left to the reader.

Recognizing Types of Systems When Using Addition: [Section 9.2]

• If the system is consistent, the solution of the system is found.
• If the system is inconsistent, a false statement will result when the equations are added.
• If the system is dependent, a true statement will result when the equations are added.

Example 5:

Determine whether the system $\begin{cases} 6x + 9y = 15 \\ 4x + 6y = 10 \end{cases}$ is consistent, inconsistent, or dependent by using the elimination by addition method.

Solution: Multiply the first equation by 2 and the second by -3 to eliminate x.

$6x + 9y = 15$ — Multiply by 2. → $12x + 18y = 30$
$4x + 6y = 10$ — Multiply by -3. → $-12x - 18y = -30$

Add the equations.

$$\begin{array}{l} 12x + 18y = 30 \\ \underline{-12x - 18y = -30} \\ \qquad\qquad 0 = 0 \end{array}$$

Since this is a true statement, the system is dependent.

Concept/Procedure	Example

Solving Systems of Linear Equations, Using Substitution: [Section 9.3]

1. If necessary, solve one of the equations for a chosen variable.

2. Substitute the expression found in step 1 for the chosen variable in the other equation.

3. Solve the resulting equation.

4. Substitute the value found in step 3 into either of the original equations and solve for the remaining variable.

5. Check the solution of the system in both of the original equations.

Example 6:

Solve the system $\begin{cases} 3x + 2y = 2 \\ 2x - y = -8 \end{cases}$, using the substitution method.

Solution: Since the coefficient of y is -1 in the equation $2x - y = -8$, solve for y.

$2x - y = -8$	Subtract $2x$ from both sides.
$-y = -2x - 8$	Divide both sides by -1.
$y = 2x + 8$	Substitute $2x + 8$ for y in $3x + 2y = 2$.
$3x + 2(2x + 8) = 2$	Distribute 2.
$3x + 4x + 16 = 2$	Add $3x$ and $4x$.
$7x + 16 = 2$	Subtract 16 from both sides.
$7x = -14$	Divide both sides by 7.
$x = -2$	Substitute -2 for x in $3x + 2y = 2$.
$3(-2) + 2y = 2$	Multiply.
$-6 + 2y = 2$	Add 6 to both sides.
$2y = 8$	Divide both sides by 2.
$y = 4$	Therefore, the solution of the system is $(-2, 4)$.

The check is left to the reader.

Recognizing Types of Systems When Using Substitution: [Section 9.3]

• This is the same as when using the addition method.

Solving Applications Problems, Using Systems of Equations: [Section 9.4]

We use the same procedure as in Chapter 3 with the following modifications:

1. Represent each of the unknowns in the problem with a different variable.

2. From the information given in the problem, write two equations.

3. Solve the system of equations, using either the addition or the substitution method.

4. Check the solutions against the wording of the original problem.

Example 7:

A minor league baseball player receives a $50,000 signing bonus. He invests a portion of the bonus in bonds that yield 5% annual interest and the remainder in mutual funds that yield 7% annual returns. How much does he have in each investment if the annual return on both investments is $3060?

Solution:

Understand: We know the total amount invested and the total annual returns on the two investments. The unknowns are the amounts invested in bonds and mutual funds.

　　Relationship 1: The total amount invested is $50,000.

　　Relationship 2: The total annual returns are $3060.

Plan: Use a chart and two variables, one for the amount invested in bonds and another for the amount invested in mutual funds. Use relationship 1 to write one equation and relationship 2 to write the other. Solve the system.

Translate:

Let x represent the amount invested in bonds.
Let y represent the amount invested in mutual funds.

Concept/Procedure	Example

Example

Fill in the chart.

Type of Investment	Interest Rate	Amount Invested	Amount of Returns
Bonds	.05	x	$.05x$
Mutual Funds	.07	y	$.07y$
Total		50,000	$3060

Using relationship 1, we get $x + y = 50{,}000$; and using relationship 2, we get $.05x + .07y = 3060$.

Solve:

$x + y = 50{,}000$ Leave unchanged.
$.05x + .07y = 3060$ Multiply by 100 to eliminate the decimals.
$x + y = 50{,}000$ Multiply by -5 to eliminate x.
$5x + 7y = 306{,}000$ Leave unchanged.

$$
\begin{aligned}
-5x - 5y &= -250{,}000 \\
\underline{5x + 7y} &= \underline{306{,}000} \qquad \text{Add the equations.} \\
2y &= 56{,}000 \qquad \text{Divide both sides by 2.} \\
y &= 28{,}000 \qquad \text{Solution.}
\end{aligned}
$$

State: Therefore, he invested $28,000 in mutual funds.

Solve: We need to solve for x, so we substitute 28,000 for y in $x + y = 50{,}000$.

$x + 28{,}000 = 50{,}000$ Subtract 28,000 from both sides.
$x = 22{,}000$ Solution.

State: Therefore, he invested $22,000 in bonds.

Check: The check is left to the reader.

Systems of Linear Inequalities: [Section 9.5]

- To graphically represent the solutions of a system of linear inequalities, graph all the inequalities of the system on the same coordinate axes. The region where the individual solutions overlap represents all points whose coordinates solve the system.

Example 8:

Solve the system $\begin{cases} x - 2y < 2 \\ x + 2y \leq 8 \end{cases}$.

Solution: Graph each inequality, and the region where the solutions overlap is the solution of the system.

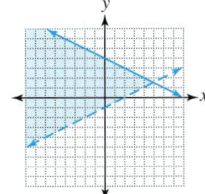

Chapter 9   Review Exercises

Determine whether the given ordered pair is a solution of the system of linear equations. The ordered pair is in the form (x, y). [Section 9.1]

1. $(5, -5)$

$$\begin{cases} 7x - 3y = 36 \\ 4x + 2y = 2 \end{cases}$$

2. $\left(\dfrac{2}{3}, -2\right)$

$$\begin{cases} 3x + 4y = -6 \\ 3x - y = 4 \end{cases}$$

3. $(2, 1)$

$$\begin{cases} \dfrac{1}{2}x + 7y = 8 \\ \dfrac{1}{4}x + 3y = \dfrac{7}{2} \end{cases}$$

4. $(-3, 4)$

$$\begin{cases} .3x + .7y = -1.9 \\ x - .5y = 5 \end{cases}$$

Find the solution(s) of the given systems of linear equations by graphing. Assume that each unit on the coordinate system represents 1. [Section 9.1]

5. $\begin{cases} x = -1 \\ 4x + y = -2 \end{cases}$

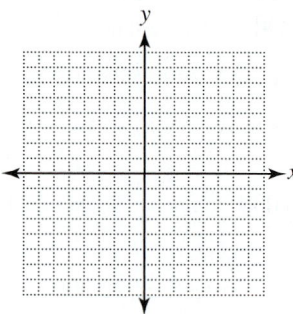

6. $\begin{cases} x + y = 1 \\ 2x - 3y = 12 \end{cases}$

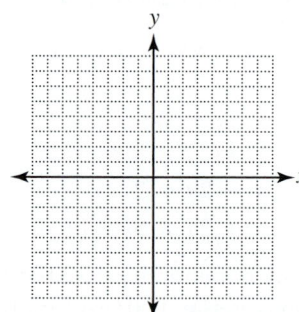

7. $\begin{cases} x - y = 6 \\ 2x + 3y = -8 \end{cases}$

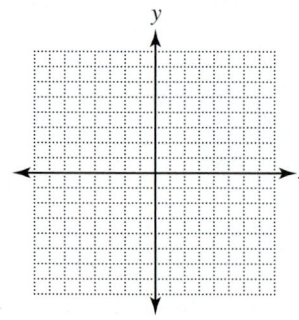

8. $\begin{cases} 2x - 3y = 14 \\ 2x + y = -2 \end{cases}$

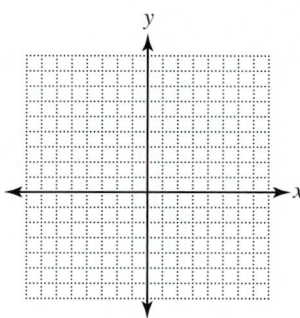

Determine whether each system is consistent, inconsistent, or dependent. If the system is consistent, find the solution of the system. Use the method of your choice. [Sections 9.1–9.3]

9. $\begin{cases} 3x - 7y = 4 \\ 12x - 28y = 16 \end{cases}$

10. $\begin{cases} 3x - 5y = 42 \\ x + y = -2 \end{cases}$

11. $\begin{cases} 4x + 6y = 8 \\ 6x + 9y = 16 \end{cases}$

12. $\begin{cases} -6x + 24y = -9 \\ 2x - 8y = 3 \end{cases}$

Solve the systems of equations by the elimination by addition method. If the system is not consistent, indicate whether it is inconsistent or dependent. [Section 9.2]

13. $\begin{cases} x - y = 12 \\ x + y = 14 \end{cases}$

14. $\begin{cases} 2x - y = -2 \\ 2x - 3y = -30 \end{cases}$

15. $\begin{cases} 4x + 2y = -4 \\ 5x - 3y = -38 \end{cases}$

16. $\begin{cases} 3x - 7y = -10 \\ 8x - 3y = -11 \end{cases}$

17. $\begin{cases} x - 8y = -5 \\ 3x + 12y = -6 \end{cases}$

18. $\begin{cases} 6x + 6y = 7 \\ 9x - 10y = 20 \end{cases}$

19. $\begin{cases} 2x - 3y = 9 \\ 5x - 4y = 5 \end{cases}$

20. $\begin{cases} x = 1 - 7y \\ 2y = 3x + 20 \end{cases}$

21. $\begin{cases} \dfrac{1}{2}x - \dfrac{3}{4}y = -8 \\ \dfrac{1}{4}x - \dfrac{3}{8}y = 2 \end{cases}$

22. $\begin{cases} .5x + .3y = -1.1 \\ .4x - .7y = -3.7 \end{cases}$

23. $\begin{cases} 3x - 4y = 2 \\ -15x + 20y = -10 \end{cases}$ **24.** $\begin{cases} 6x - 3y = 9 \\ -4x + 2y = -8 \end{cases}$

Solve the given systems of equations by the substitution method. If the system is not consistent, indicate whether it is inconsistent or dependent. [Section 9.3]

25. $\begin{cases} x = -2y \\ 6x - 5y = -17 \end{cases}$ **26.** $\begin{cases} x - y = -2 \\ x - 2y = -7 \end{cases}$

27. $\begin{cases} 2x - y = 2 \\ 8x - 7y = -10 \end{cases}$ **28.** $\begin{cases} 9x + 4y = 18 \\ 2x - y = -13 \end{cases}$

29. $\begin{cases} 4x - 3y = 7 \\ 8x - 3y = -13 \end{cases}$ **30.** $\begin{cases} 5x - y = 3 \\ 10x - 2y = 5 \end{cases}$

Solve the systems of linear equations by the most appropriate method. If the system is not consistent, indicate whether it is inconsistent or dependent. [Section 9.1–9.3]

31. $\begin{cases} x + y = -5 \\ 3x - y = -7 \end{cases}$ **32.** $\begin{cases} 2x = 1 + y \\ 5x - 4y = -8 \end{cases}$

33. $\begin{cases} 6x - 15y = 9 \\ -2x + 5y = -3 \end{cases}$ **34.** $\begin{cases} 5x - 2y = 0 \\ 3x + 3y = -21 \end{cases}$

35. $\begin{cases} y = -4x \\ 11x + 5y = 9 \end{cases}$ **36.** $\begin{cases} 8x - 11y = 7 \\ 16x - 22y = 10 \end{cases}$

37. $\begin{cases} 4x + y = 4 \\ x - .5y = -.5 \end{cases}$ **38.** $\begin{cases} 8x - 5y = 4 \\ \dfrac{1}{2}x - y = -\dfrac{1}{40} \end{cases}$

Solve each of the following by using a system of linear equations: [Section 9.4]

39. The men's world record in the 200-meter dash is 2.15 seconds less than the women's world record. If the sum of the two world record times is 40.53 seconds, find each time. (*Source: Webster's New World Book of Facts 2010.*)

40. A university changes to a new phone system. It pays a total of $12,495 for 425 basic phone sets and 17 executive phone sets. If the cost of a basic phone and an executive phone together is $63, how much did each type of phone set cost?

41. A 9-year-old boy finds that his piggy bank has $5.25 in nickels and dimes. If there are 65 coins in the piggy bank, how many nickels and how many dimes does he have?

42. In 4 hours, a plane can travel 1160 miles with the wind or 1000 miles against the wind. Find the speed of the plane in still air and the speed of the wind.

43. A professional athlete received a $50,000 bonus for signing a contract to skate in an ice show. She invested part of the money in a certificate of deposit that pays 12% interest and part of the money in tax-free bonds paying an average of 7.5% interest. If she earns $4200 interest, how much did she put in each investment?

44. Pure antifreeze (100%) and a solution that is 40% antifreeze are mixed. How many quarts of each are needed to obtain 8 quarts of 70% antifreeze solution?

Graph the solutions of the following systems of linear inequalities: [Section 9.5]

45. $\begin{cases} x + y > 5 \\ x - y < -1 \end{cases}$

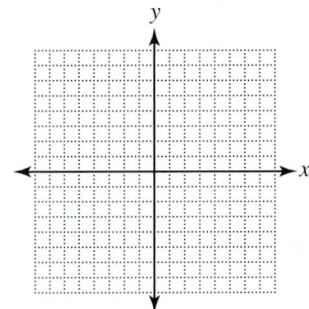

46. $\begin{cases} x + y \geq 2 \\ x + 2y \leq 6 \end{cases}$

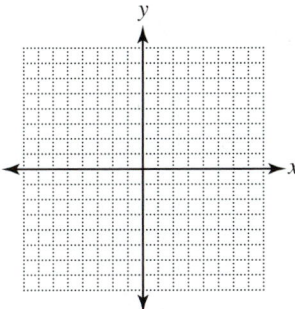

47. $\begin{cases} 2x - y \le 3 \\ 2x > 6 \end{cases}$

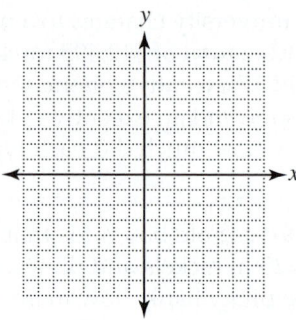

49. $\begin{cases} 2x + 3y \le 4 \\ y + x > 0 \end{cases}$

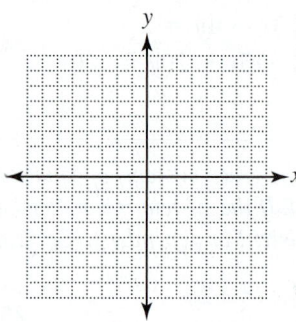

48. $\begin{cases} 3x < 15 \\ y \le -4 \end{cases}$

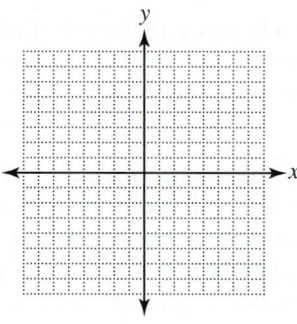

50. $\begin{cases} y - x \le 0 \\ x \ge 1 \end{cases}$

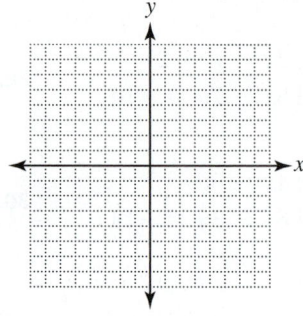

Chapter 9 Test

Determine whether the given ordered pair is a solution of the system of linear equations. The ordered pair is in the form (x, y).

1. $(-3, -1)$
$$\begin{cases} 2x + 5y = 1 \\ \dfrac{1}{3}x - y = 2 \end{cases}$$

Find the solution(s) of the given systems of linear equations by graphing. Assume that each unit on the coordinate system represents 1.

2. $\begin{cases} x - y = 8 \\ 2x - y = 12 \end{cases}$

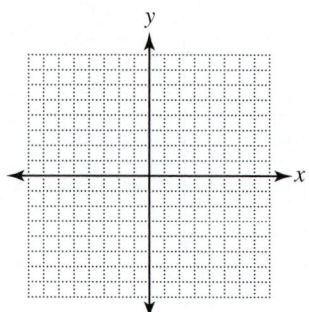

3. $\begin{cases} 4x - 3y = -18 \\ 2x - 3y = -12 \end{cases}$

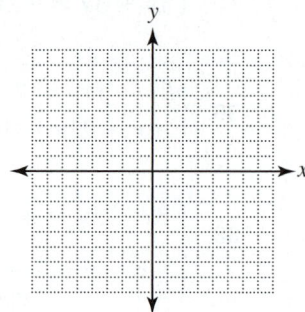

Determine whether the system is consistent, inconsistent, or dependent. If the system is consistent, find the solution of the system.

4. $\begin{cases} 4x - 6y = 12 \\ 2x - 3y = 6 \end{cases}$

Solve the given systems of equations, using the elimination by addition method. If the system is not consistent, indicate whether it is inconsistent or dependent.

5. $\begin{cases} 2x + 3y = 24 \\ 5x + 4y = 46 \end{cases}$

6. $\begin{cases} .9x - .2y = 3.1 \\ 11x - 8y = -1 \end{cases}$

7. $\begin{cases} \dfrac{2}{3}x + \dfrac{5}{8}y = 3 \\ \dfrac{5}{6}x - \dfrac{1}{8}y = -\dfrac{7}{2} \end{cases}$

Solve each system of equations by the substitution method. If the system is not consistent, indicate whether it is inconsistent or dependent.

8. $\begin{cases} x + 2y = -6 \\ 3x + 4y = -10 \end{cases}$

9. $\begin{cases} 4x + y = 17 \\ 3x + 2y = 14 \end{cases}$

10. $\begin{cases} 3x + 4y = 14 \\ 2x - 3y = -19 \end{cases}$

Solve the systems of linear equations by the most appropriate method. If the system is not consistent, indicate whether it is inconsistent or dependent.

11. $\begin{cases} 3x = y - 5 \\ 4x = 3y + 5 \end{cases}$

12. $\begin{cases} 10x + 5y = -2 \\ 5x + 2y = 0 \end{cases}$

13. $\begin{cases} y - x = 1 \\ .8x - .7y = -.4 \end{cases}$

14. $\begin{cases} x = \dfrac{1}{2}y - 6 \\ \dfrac{2x}{7} - 3y = 4 \end{cases}$

Solve each of the following by using a system of linear equations:

15. As of 2009, Panama had the world's largest merchant marine fleet by flag of registry. Panama has 1159 less than four times the number of ships of second-place Liberia. Together, Panama and Liberia have 10,371 ships. Find the number of ships registered to each country.

16. The machines at a coin laundry take only dimes and quarters. At the end of the day, one washer has 90 coins with a total value of $13.50. How many of each kind of coin were in the washer?

17. In 2 hours a canoeist can row 16 kilometers with the current or 8 kilometers against the current. Find the speed of the canoeist in still water and the speed of the current.

18. Christopher won $16,000 in a video game contest. His parents put part of the money into a mutual fund paying an expected return of 22% and part of the money into tax-free bonds paying 7% interest. If he earns $2920 in interest in a year, how much was put into each investment?

Graph the solutions of the following systems of linear inequalities:

19. $\begin{cases} x - y \geq 5 \\ 2x + 3y > -3 \end{cases}$

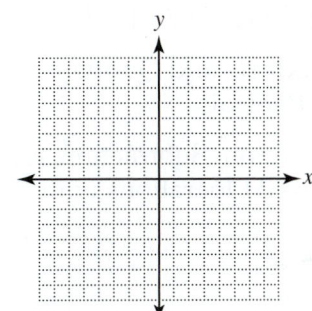

20. $\begin{cases} y - x < 3 \\ x \geq -2 \end{cases}$

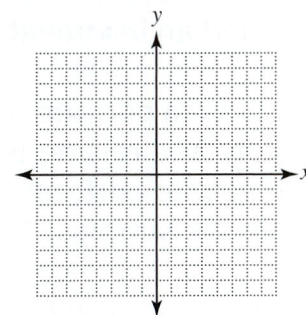

10

Roots and Radicals

In Section 1.1, we learned to raise numbers to powers. In this chapter, we will learn the inverse procedure called *finding a root*. The result of finding a root is not always a rational number. Thus, we will introduce a new type of number called an **irrational number** that will be used in solving equations in Chapter 11.

We know how to perform the operations of addition, subtraction, multiplication, division, and raising to powers on rational numbers. In this chapter, we will learn how to perform these same operations on radicals. In addition, we will define the simplest form for a radical and learn to write radicals in this form.

We end the chapter with a very important relationship from geometry involving the sides of a right triangle. Finding one side of a right triangle when given the other two sides often results in an irrational number.

<table>
<tr><td>**Section 10.1**</td><td>Defining and Finding Roots</td></tr>
</table>

OBJECTIVES *When you complete this section, you will be able to:*

A Find square roots and higher order roots of numbers.
B Determine whether a root is rational or irrational.
C Find decimal approximations of irrational square and cube roots, using a calculator or a table.
D Find roots of variables raised to powers.

PREREQUISITE SKILLS *Before beginning this section, you should be able to:*

  a. Raise a number to a power. (Section 1.1)
  b. Apply $(a^m)^n = a^{mn}$. (Section 2.2)

GETTING READY FOR SECTION 10.1

Evaluate the following powers of numbers:

1. $(-3)^2$ **2.** 2^5 **3.** $(-4)^3$ **4.** 8^3

Simplify the following, using $(a^m)^n = a^{mn}$:

5. $(y^3)^2$ **6.** $(w^3)^4$

Introduction

The operations of addition and subtraction are inverse operations, since one undoes the other. Begin with a number, say, 3. Add any number to 3. From this sum subtract the same number you previously added. The result will be 3. Multiplication and division (except by 0) are also inverse operations.

OBJECTIVE **Finding square roots and higher order roots of numbers**

In Section 1.1, we learned to raise numbers to positive integral powers. In this section, we will learn the inverse operation, which, with some restrictions, is finding the roots of numbers.

The inverse of squaring a number is finding the square root. A square root of a number is a number that we must square to get the given number. Since $4^2 = 16$, a square root of 16 is 4. Another square root of 16 is -4, since $(-4)^2 = 16$.

The inverse of cubing a number is finding the cube root. We know that $2^3 = 8$. Therefore, the cube root of 8 is 2. Note that 2 is the only real number whose cube is 8, so 2 is the only real cube root of 8.

The inverse of raising a number to the fourth power is finding the fourth root. We know that $3^4 = 81$. So, a fourth root of 81 is 3. Since $(-3)^4 = 81$, -3 is also a fourth root of 81.

On the basis of these examples, we make the following definitions:

> **Definition** Roots of a Number
> If the number **b** is a square root of the number **a**, then $b^2 = a$.
> If the number **b** is a cube root of the number **a**, then $b^3 = a$.
> If the number **b** is a fourth root of the number **a**, then $b^4 = a$.
> If the number **b** is an *n*th root of the number **a**, then $b^n = a$.

Answers: Getting Ready: 1. 9 **2.** 32 **3.** -64 **4.** 512 **5.** y^6 **6.** w^{12}

In words, the square root of a given number is a number whose square equals the given number. The cube root of a given number is a number whose cube equals the given number. The fourth root of a given number is a number whose fourth power equals the given number. In general, the **nth root** of a given number is a number whose nth power equals the given number.

To indicate the root of a number, we use the symbol $\sqrt{}$, called a **radical sign**. In the expression $\sqrt[n]{a}$, read "the nth root of a," n is called the **root index** and indicates which root we are to find. If no root index is given, we assume it to be 2, which means find the square root. The number a is called the **radicand**. The entire expression is called a **radical**, and any expression containing a radical is called a **radical expression**.

Using radicals, we restate the definition previously given:

> ## Definition Roots of Numbers Using Radicals
>
> If $\sqrt{a} = b$, then $b^2 = a$. That is, $\sqrt{a}$ is the number whose square is a.
> If $\sqrt[3]{a} = b$, then $b^3 = a$. That is, $\sqrt[3]{a}$ is the number whose cube is a.
> If $\sqrt[4]{a} = b$, then $b^4 = a$. That is, $\sqrt[4]{a}$ is the number whose fourth power is a.
> In general, if $\sqrt[n]{a} = b$, then $b^n = a$. That is, $\sqrt[n]{a}$ is the number whose nth power is a.

Based on the preceding definition, if $\sqrt{9} = 3$, then $3^2 = 9$. Also, since $5^2 = 25$, $\sqrt{25} = 5$. If $\sqrt[3]{27} = 3$ then $3^3 = 27$. Since $4^3 = 64$, then $\sqrt[3]{64} = 4$. Similar statements can be made for other powers and roots.

Using radicals, $\sqrt{4}$ is read "the square root of 4," $\sqrt[3]{8}$ is read "the cube root of 8," $\sqrt[4]{16}$ is read "the fourth root of 16," and $\sqrt[5]{32}$ is read "the fifth root of 32."

Previously, we noted that 16 has two square roots, 4 and -4. This is usually written as ± 4. Also, 81 has two fourth roots, ± 3. Consequently, without some agreement, this can lead to confusion as to what the symbols represent. If the root index is even, then $\sqrt[n]{a}$ denotes the nonnegative root only and is called the **principal root**. We write the negative even root of a number as $-\sqrt[n]{a}$. We write both the positive and negative even roots as $\pm\sqrt[n]{a}$. Consider the following symbols and their meanings for n even:

Symbol	Meaning
$\sqrt{16}$	4. (The positive square root of 16.)
$-\sqrt{16}$	-4. (The negative square root of 16.)
$\pm\sqrt{16}$	± 4. (The positive and negative square roots of 16.)
$\sqrt[4]{81}$	3. (The positive fourth root of 81.)
$-\sqrt[4]{81}$	-3. (The negative fourth root of 81.)
$\pm\sqrt[4]{81}$	± 3. (The positive and negative fourth roots of 81.)
$\sqrt[n]{a}$	The positive nth root of a.
$-\sqrt[n]{a}$	The negative nth root of a.
$\pm\sqrt[n]{a}$	The positive and negative nth roots of a.

It should also be noted that if the root index is even, the radicand must be nonnegative. For example, $\sqrt{-4}$ does not exist as a real number, since there is no real number whose square is -4. Also, $\sqrt[4]{-81}$ does not exist as a real number, since there is no real number whose fourth power is -81. These types of numbers are called **imaginary numbers** and are discussed in Chapter 11.

If the root index is odd, there is no problem with finding roots of any real number. For example, $\sqrt[3]{27} = 3$ and only 3, since 3 is the only real number whose cube is 27. Also,

$\sqrt[3]{-27} = -3$ and only -3, since -3 is the only real number whose cube is -27. We summarize as follows:

> **Rule: Existence of Roots**
>
> **a.** If n is even and $a \geq 0$, then $\sqrt[n]{a}$ exists as a real number.
>
> **b.** If n is even and $a < 0$, then $\sqrt[n]{a}$ does not exist as a real number.
>
> **c.** If n is odd, then $\sqrt[n]{a}$ exists as a real number for all values of a.

Example 1 | **Find the roots if they exist. If the root does not exist, write "This root does not exist as a real number."**

a. $\sqrt{9} = 3$, since 3 is positive and $3^2 = 9$.

b. $-\sqrt{9} = -3$, since -3 is negative and $(-3)^2 = 9$.

c. $\sqrt{-9}$ does not exist as a real number, since there is no real number whose square is -9.

d. $\pm\sqrt{9} = \pm 3$, since we want both square roots and $(\pm 3)^2 = 9$.

e. $\sqrt[3]{8} = 2$, since $2^3 = 8$.

f. $\sqrt[3]{-8} = -2$, since $(-2)^3 = -8$.

g. $\sqrt[4]{16} = 2$ since 2 is positive and $2^4 = 16$.

PRACTICE EXERCISE 1

Find the roots if they exist. If the root does not exist, write "This root does not exist as a real number."

a. $\sqrt{36}$ **b.** $-\sqrt{36}$ **c.** $\sqrt{-36}$ **d.** $\pm\sqrt{36}$

e. $\sqrt[3]{64}$ **f.** $-\sqrt[3]{64}$ **g.** $\sqrt[3]{-64}$ **h.** $-\sqrt[4]{81}$

If you need more practice, do the following Additional Practice Exercises.

Additional Practice Exercise 1   Find the roots if they exist. If the root does not exist, write "This root does not exist as a real number."

a. $\sqrt{64}$ **b.** $-\sqrt{64}$ **c.** $\sqrt{-64}$ **d.** $\pm\sqrt{64}$

e. $\sqrt[5]{32}$ **f.** $-\sqrt[5]{32}$ **g.** $\sqrt[5]{-32}$ **h.** $-\sqrt[3]{125}$

OBJECTIVE **B** ## Determining whether a root is rational or irrational

Only certain numbers have roots that are rational numbers. Remember, a rational number is any number that can be written as the ratio of two integers. As a decimal, a rational number either terminates or repeats.

The numbers 1, 4, 9, 16, 25, 36, . . . , which have rational square roots, are called **perfect squares**. The numbers $\pm 1, \pm 8, \pm 27, \pm 64, . . .$, which have rational cube roots, are called **perfect cubes**. Similarly, we have numbers that have rational fourth roots and fifth roots, and so on. Table 1 in the Appendix gives the squares and cubes of the

---

Answers: Practice Exercise 1: a. 6 **b.** -6 **c.** This root does not exist as a real number. **d.** ± 6 **e.** 4 **f.** -4 **g.** -4 **h.** -3
Additional Practice 1: a. 8 **b.** -8 **c.** This root does not exist as a real number. **d.** ± 8 **e.** 2 **f.** -2 **g.** -2 **h.** -5

integers from 1 to 100. Therefore, Table 1 also gives the perfect square integers from 1 to 10,000 and the perfect cube integers from 1 to 1,000,000.

If a number is not a perfect square, then it is impossible to represent the square root exactly as a terminating or repeating decimal. For example, we cannot represent $\sqrt{3}$ exactly as a decimal, since there is no decimal whose square is 3. However, we may approximate $\sqrt{3}$ as closely as we please by using a sufficient number of decimal places. To the nearest tenth, $\sqrt{3} = 1.7$ and $(1.7)^2 = 2.89$. To the nearest hundredth, $\sqrt{3} = 1.73$ and $(1.73)^2 = 2.9929$. This is closer to 3 than $(1.7)^2$, so 1.73 is a better approximation of $\sqrt{3}$ than 1.7. To the nearest thousandth, $\sqrt{3} = 1.732$ and $(1.732)^2 = 2.999824$. This is closer to 3 than $(1.73)^2$, so 1.732 is a better approximation of $\sqrt{3}$ than 1.73.

Numbers like $\sqrt{3}$ are called **irrational** numbers. Irrational numbers cannot be represented by terminating or repeating decimals, since terminating and repeating decimals are rational numbers. Hence, irrational numbers are nonrepeating, nonterminating decimals. In general, any root of a number that is not rational is irrational. We may use a calculator or tables to find decimal approximations of irrational roots. Table 1 in the Appendix gives decimal approximations of square and cube roots of numbers from 1 to 100 and square roots of 10 times any number from 1 to 100. Calculators afford a useful tool for approximating irrational roots.

OBJECTIVE Finding decimal approximations of irrational square and cube roots, using a calculator or a table

The headings at the top of Appendix Table 1 appear as follows:

n	n^2	n^3	$\sqrt{n}$	$\sqrt[3]{n}$	$\sqrt{10n}$

Under the n column are the integers 1–100. The other columns are self-explanatory. For example, if we go down to 8 in the n column, we find the following row:

n	n^2	n^3	$\sqrt{n}$	$\sqrt[3]{n}$	$\sqrt{10n}$
8	64	512	2.828	2.000	8.944

From this row, we see that $8^2 = 64$, $8^3 = 512$, $\sqrt{8} \approx 2.828$, $\sqrt[3]{8} = 2.000$, and $\sqrt{80} = 8.944$ to the nearest thousandths.

We may also use the table in another way. Since 64 is in the n^2 column, $8^2 = 64$, which means $\sqrt{64} = 8$. Likewise, since 512 is in the n^3 column, $8^3 = 512$, which means $\sqrt[3]{512} = 8$. In general, any number in the n column is the square root of the corresponding number in the n^2 column and the cube root of any number in the n^3 column.

Example 2

Identify each of the numbers as rational or irrational. If the number is rational, find the root exactly. If the number is irrational, approximate the number to the nearest thousandth by using a calculator or Appendix Table 1.

a. $\sqrt{12}$ Irrational. $\sqrt{12} \approx 3.464$.

b. $\sqrt[3]{30}$ Irrational. $\sqrt[3]{30} \approx 3.107$.

c. $\sqrt{81}$ Rational. $\sqrt{81} = 9$.

d. $\sqrt{5184}$ Rational. $5184 = 72^2$, so $\sqrt{5184} = 72$.

e. $\sqrt{9.61}$ Rational. Since $961 = 31^2$, $9.61 = 3.1^2$, so $\sqrt{9.61} = 3.1$.

f. $\sqrt[3]{753,571}$ Rational. $753,571 = 91^3$, so $\sqrt[3]{753,571} = 91$.

g. $\sqrt[3]{68}$ Irrational. $\sqrt[3]{68} \approx 4.082$.

h. $\sqrt{670}$ Irrational. $\sqrt{670} \approx 25.884$. (Look under $\sqrt{10n}$ column, where $n = 67$.)

PRACTICE EXERCISE 2

Identify each of the numbers as rational or irrational. If the number is rational, find the root exactly. If the number is irrational, approximate the number to the nearest thousandth by using a calculator or Table 1.

a. $\sqrt{19}$ **b.** $\sqrt[3]{63}$ **c.** $\sqrt{121}$ **d.** $\sqrt{3364}$

e. $\sqrt{12.25}$ **f.** $\sqrt[3]{262,144}$ **g.** $\sqrt[3]{86}$ **h.** $\sqrt{440}$

If you need more practice, do the following Additional Practice Exercises.

Additional Practice Exercise 2   Identify each of the numbers as rational or irrational. If the number is rational, find the root exactly. If the number is irrational, approximate the number to the nearest thousandth by using a calculator or Table 1.

a. $\sqrt{24}$ **b.** $\sqrt[3]{49}$ **c.** $\sqrt{49}$ **d.** $\sqrt{1156}$

e. $\sqrt{8.41}$ **f.** $\sqrt[3]{2744}$ **g.** $\sqrt[3]{21}$ **h.** $\sqrt{450}$

For convenience in finding powers and roots, we include the following table of powers of numbers that occur often in this chapter:

<table>
<tr><th>n</th><th>n^2</th><th>n^3</th><th>n^4</th><th>n^5</th><th>n^6</th></tr>
<tr><td>2</td><td>4</td><td>8</td><td>16</td><td>32</td><td>64</td></tr>
<tr><td>3</td><td>9</td><td>27</td><td>81</td><td>243</td><td>729</td></tr>
<tr><td>4</td><td>16</td><td>64</td><td>256</td><td></td><td></td></tr>
<tr><td>5</td><td>25</td><td>125</td><td>625</td><td></td><td></td></tr>
<tr><td>6</td><td>36</td><td>216</td><td>1296</td><td></td><td></td></tr>
<tr><td>7</td><td>49</td><td>343</td><td></td><td></td><td></td></tr>
</table>

> **Note**
>
> Remember, a more complete list of perfect squares and perfect cubes can be found in Table 1 in the Appendix.

This table can be used to find specific roots of some numbers. For example, 243 is found under the n^5 column. Read to the left, and we find the number 3 in the "n" column, which means that $3^5 = 243$. Consequently, $\sqrt[5]{243} = 3$. Likewise, 1296 appears in the n^4 column for $n = 6$, which means that $6^4 = 1296$. Consequently, $\sqrt[4]{1296} = 6$.

OBJECTIVE   Finding roots of variables raised to powers

The definition of a root can also be used to find roots with variable radicands. Since even roots of negative radicands do not exist, all variables are assumed to represent nonnegative numbers. Recall that $(a^m)^n = a^{mn}$.

Example 3   **Find the roots. Assume that all variables represent nonnegative numbers.**

a. $\sqrt{x^2} = (\quad)$ By definition of square root, $(\quad)^2 = x^2$. Since $(x)^2 = x^2$, the $(\quad)$ represents x.
$\sqrt{x^2} = x$ Therefore, $\sqrt{x^2} = x$. Remember, x must be nonnegative.

> **Note**
>
> If we do not assume that $x \geq 0$, then $\sqrt{x^2} \neq x$, since the radical always denotes the nonnegative square root. For example, $\sqrt{(-2)^2} \neq -2$, since $\sqrt{(-2)^2} = \sqrt{4} = 2$. Consequently, $\sqrt{x^2} = |x|$, since $|x|$ is also nonnegative. This situation is usually addressed in intermediate or college algebra and is why we will assume, when finding even roots, that all variables have nonnegative values.

Answers: Practice Exercise 2: a. irrational, 4.359 **b.** irrational, 3.979 **c.** rational, 11 **d.** rational, 58 **e.** rational, 3.5 **f.** rational, 64
g. irrational, 4.414 **h.** irrational, 20.976 **Additional Practice 2: a.** irrational, 4.899 **b.** irrational, 3.659 **c.** rational, 7 **d.** rational, 34
e. rational, 2.9 **f.** rational, 14 **g.** irrational, 2.759 **h.** irrational, 21.213

b. $\sqrt{a^4} = (\quad)$ By the definition of a square root, $(\quad)^2 = a^4$. Since $(a^2)^2 = a^4$, the $(\quad)$ represents a^2.

   $\sqrt{a^4} = a^2$ Therefore, $\sqrt{a^4} = a^2$.

c. $\sqrt[3]{y^6} = (\quad)$ By the definition of a cube root, $(\quad)^3 = y^6$. Since $(y^2)^3 = y^6$, the $(\quad)$ represents y^2.

   $\sqrt[3]{y^6} = y^2$ Therefore, $\sqrt[3]{y^6} = y^2$.

Note

You may have noticed that to find the root of a variable radicand, the exponent on the variable must be divisible by the root index. In fact, the exponent on the root is the exponent on the radicand divided by the index. For example, to find $\sqrt[4]{y^8}$, divide 8 by 4. Since $8 \div 4 = 2$, the exponent on the root is 2. Therefore, $\sqrt[4]{y^8} = y^2$. This works because of the definition of multiplication. If $\sqrt[4]{y^8} = y^n$, then $(y^n)^4 = y^8$, which gives $y^{4n} = y^8$. Consequently, $4n = 8$, or $n = \frac{8\,(the\ exponent)}{4\,(the\ index)}$.

PRACTICE EXERCISE 3

Find the roots. Assume that all expressions have nonnegative values.

a. $\sqrt{z^2}$ **b.** $\sqrt{c^6}$ **c.** $\sqrt[4]{w^{12}}$

If you need more practice, do the following Additional Practice Exercises.

Additional Practice Exercise 3   Find the roots. Assume that all variables have nonnegative values.

a. $\sqrt{r^4}$ **b.** $\sqrt{p^8}$ **c.** $\sqrt[5]{x^{15}}$

Often, the radicands of radical expressions are not monomials. However, the same procedures apply.

Example 4   **Find the roots. Assume that all expressions have nonnegative values.**

a. $\sqrt{(a-2)^2} =$ We need the expression whose square is $(a-2)^2$, which is $a-2$.

   $a - 2$ Therefore, $\sqrt{(a-2)^2} = a - 2$.

b. $\sqrt{(2x-3y)^2} =$ We need the expression whose square is $(2x-3y)^2$, which is $2x-3y$.

   $2x - 3y$ Therefore, $\sqrt{(2x-3y)^2} = 2x - 3y$.

c. $\sqrt{x^2 + 6x + 9} =$ Rewrite $x^2 + 6x + 9$ as $(x+3)^2$.

   $\sqrt{(x+3)^2} =$ We need the expression whose square is $(x+3)^2$, which is $x+3$.

   $x + 3$ Therefore, $\sqrt{x^2 + 6x + 9} = x + 3$.

PRACTICE EXERCISE 4

Find the roots. Assume that all expressions have nonnegative values.

a. $\sqrt{(x+4)^2}$ **b.** $\sqrt{(4x-5y)^2}$ **c.** $\sqrt{a^2 + 4a + 4}$

If you need more practice, do the following Additional Practice Exercises.

Additional Practice Exercise 4   Find the roots. Assume that all expressions have nonnegative values.

a. $\sqrt{(x-3)^2}$ **b.** $\sqrt{(3a+5b)^2}$ **c.** $\sqrt{x^2 - 10x + 25}$

Answers: Practice Exercise 3: a. z **b.** c^3 **c.** w^3 **Additional Practice 3: a.** r^2 **b.** p^4 **c.** x^3
Practice Exercise 4: a. $x+4$ **b.** $4x-5y$ **c.** $a+2$ **Additional Practice 4: a.** $x-3$ **b.** $3a+5b$ **c.** $x-5$

For Extra Help

Exercise Set 10.1 MyMathLab®

Find the roots if they exist. If the root does not exist, write, "This root does not exist as a real number." (See Example 1.)

1. $\sqrt{25}$

2. $\sqrt{-25}$

3. $-\sqrt{25}$

4. $\pm\sqrt{25}$

5. $-\sqrt{100}$

6. $\sqrt{100}$

7. $\sqrt{-100}$

8. $\pm\sqrt{100}$

9. $\sqrt{625}$

10. $-\sqrt{289}$

11. $\sqrt{169}$

12. $\pm\sqrt{256}$

13. $\sqrt[3]{27}$

14. $-\sqrt[3]{27}$

15. $\sqrt[3]{-27}$

16. $-\sqrt[3]{216}$

17. $\sqrt[3]{-216}$

18. $\sqrt[3]{216}$

19. $-\sqrt[3]{-216}$

20. $\sqrt[4]{625}$

21. $\sqrt[4]{-625}$

22. $\sqrt[5]{243}$

23. $-\sqrt[5]{243}$

24. $\sqrt[5]{-243}$

Identify each of the numbers as rational or irrational. If the number is rational, find the root exactly. If the number is irrational, approximate the number to the nearest thousandth by using a calculator or Table 1. (See Example 2.)

25. $\sqrt{24}$

26. $\sqrt{32}$

27. $\sqrt{144}$

28. $\sqrt{400}$

29. $\sqrt[3]{50}$

30. $\sqrt[3]{21}$

31. $\sqrt[3]{64}$

32. $\sqrt[3]{27}$

33. $\sqrt{1.69}$

34. $\sqrt{.64}$

35. $\sqrt{810}$

36. $\sqrt{640}$

37. $\sqrt[3]{.125}$

38. $\sqrt[3]{.216}$

39. $\sqrt[3]{100}$

40. $\sqrt[3]{80}$

41. $\sqrt{2116}$

42. $\sqrt{3844}$

43. $\sqrt[3]{9261}$

44. $\sqrt[3]{24,389}$

Find the roots. Assume that all variables represent nonnegative numbers. (See Example 3.)

45. $\sqrt{b^2}$

46. $\sqrt{c^2}$

47. $\sqrt{n^4}$

48. $\sqrt{m^4}$

49. $\sqrt{r^8}$

50. $\sqrt{a^{10}}$

51. $\sqrt[3]{n^3}$

52. $\sqrt[3]{m^3}$

53. $\sqrt[3]{r^6}$

54. $\sqrt[3]{a^{12}}$

55. $\sqrt[3]{c^9}$

56. $\sqrt[4]{d^8}$

57. $\sqrt[5]{s^{10}}$

58. $\sqrt[5]{y^{15}}$

59. $\sqrt[6]{a^6}$

60. $\sqrt[6]{d^{12}}$

Find the roots. Assume that all expressions have nonnegative values. (See Example 4.)

61. $\sqrt{(x-5)^2}$

62. $\sqrt{(r+1)^2}$

63. $\sqrt{(2a+b)^2}$

64. $\sqrt{(3x-2y)^2}$

65. $\sqrt{x^2+10x+25}$

66. $\sqrt{a^2-8a+16}$

67. $\sqrt{4a^2-12ab+9b^2}$

68. $\sqrt{9a^2-30ab+25b^2}$

Writing Exercises (69 and 70)

69. Why is $\sqrt{-16}$ not a real number?

70. In order for $\sqrt[n]{x}$ to exist as a real number when n is even, x must be nonnegative. However, if n is odd, then $\sqrt[n]{x}$ exists for all values of x whether positive or negative. Explain why this is true.

Group Project

71. Interview some science instructors, engineering instructors, and so on, and make a list of five situations where radical expressions are used in areas other than mathematics.

Section 10.2   Simplifying Radicals

OBJECTIVES *When you complete this section, you will be able to:*

A Simplify square roots by using the product rule.

B Use the alternative method to simplify radicals.

C Simplify higher order roots by using the product rule.

PREREQUISITE SKILLS *Before beginning this section, you should be able to:*

a. Multiply and divide whole numbers. (Section R.3)

b. Write a whole number in terms of prime factors. (Section 5.1)

c. Find the square root of a perfect square. (Section 10.1)

d. Find the cube root of a perfect cube. (Section 10.1)

e. Apply $a^m \cdot a^n = a^{m+n}$. (Section 2.2)

GETTING READY FOR SECTION 10.2

Multiply or divide the following whole numbers:

1. $36 \cdot 2$

2. $27 \cdot 4$

3. $\dfrac{108}{36}$

4. $\dfrac{80}{16}$

5. Write 72 in terms of its prime factors.

Find the following square roots:

6. $\sqrt{16}$

7. $\sqrt{81}$

Find the following cube roots:

8. $\sqrt[3]{27}$

9. $\sqrt[3]{125}$

Simplify the following:

10. $x^9 \cdot x^2$

11. $c^4 d^2 \cdot cd$

Introduction

Fractions are not considered to be in their simplest form unless they are reduced to their lowest terms. Likewise, radical expressions are not considered to be in their simplest form unless they satisfy certain conditions. In order to simplify radicals, we need to develop some properties of radicals.

Calculator Exploration Activity (Optional)

Use your calculator to evaluate columns A and B and round your answer to the nearest ten thousandth. Compare the results obtained in each line of the columns.

Column A	Column B
a. $\sqrt{2} \cdot \sqrt{3} =$ _____	$\sqrt{6} =$ _____
b. $\sqrt{3} \cdot \sqrt{6} =$ _____	$\sqrt{18} =$ _____
c. $\sqrt{5} \cdot \sqrt{6} =$ _____	$\sqrt{30} =$ _____
d. $\sqrt{20} \cdot \sqrt{30} =$ _____	$\sqrt{600} =$ _____

1. Based on your answers to the corresponding lines in columns A and B, $\sqrt{a} \cdot \sqrt{b} =$ _____.

Answers: Getting Ready: 1. 72 **2.** 108 **3.** 3 **4.** 5 **5.** $2 \cdot 2 \cdot 2 \cdot 3 \cdot 3$ **6.** 4 **7.** 9 **8.** 3 **9.** 5 **10.** x^{11} **11.** $c^5 d^3$ **Calculator Exploration Activity: 1.** $\sqrt{ab}$

Consider the following: If we find each root and then multiply, we have $\sqrt{9} \cdot \sqrt{16} = 3 \cdot 4 = 12$. If we multiply the radicands and then find the root, we have $\sqrt{9} \cdot \sqrt{16} = \sqrt{9 \cdot 16} = \sqrt{144} = 12$. Note that we arrived at the same answer by either method. This suggests the following rule for multiplying radicals:

Rule: Products of Radicals

If $\sqrt[n]{a}$ and $\sqrt[n]{b}$ are both real numbers, then $\sqrt[n]{a} \cdot \sqrt[n]{b} = \sqrt[n]{a \cdot b}$.

OBJECTIVE Simplifying square roots by using the product rule

In this section, we will be using this rule in the form $\sqrt[n]{ab} = \sqrt[n]{a} \cdot \sqrt[n]{b}$. For example, $\sqrt{4 \cdot 3} = \sqrt{4} \cdot \sqrt{3} = 2\sqrt{3}$.

A radical is in **simplified form** if the radicand contains no factor that can be written to a power greater than or equal to the index. In other words, the radicand of a square root cannot contain a factor that is a perfect square, the radicand of a cube root cannot contain a factor that is a perfect cube, and so on. Consequently, the first step in simplifying a *square* root is to write the radicand as the product of the largest possible perfect square and a number that has no perfect-square factors.

Recall from Section 5.1 that any number is divisible by all of its factors. Consequently, if a radicand has a perfect square factor, it must be divisible by an integer that is a perfect square. If so, by the definition of division, the number of times that the perfect square divides into the radicand is the other factor of the radicand. For example, 9 is a perfect square and $45 \div 9 = 5$. Therefore, $9 \cdot 5 = 45$. Remember, the integers that are perfect squares are 1, 4, 9, 16, 25, 36, ... Since 1 is a factor of every number, we ignore 1 when finding perfect-square factors of a radicand.

Example 1 **Write each of the following as the product of the largest possible perfect square and another number that has no perfect-square factors:**

a. 27

Solution

See if 27 is divisible by a perfect square integer. Is 27 divisible by 4? No. Is 27 divisible by 9? Yes, $27 \div 9 = 3$. Since 3 has no perfect square factors, 9 is the largest perfect square factor of 27. Therefore, $27 = 9 \cdot 3$.

b. 150

Solution

See if 150 is divisible by a perfect square integer. Is 150 divisible by 4? No. Is 150 divisible by 9? No. Is 150 divisible by 16? No. Is 150 divisible by 25? Yes, $150 \div 25 = 6$. Since 6 has no perfect square factors, 25 is the largest perfect square factor of 150. Therefore, $150 = 25 \cdot 6$.

c. 80

Solution

See if 80 is divisible by a perfect square integer. Is 80 divisible by 4? Yes, $80 \div 4 = 20$. Does 20 have a perfect square factor? Yes, $20 \div 4 = 5$. Consequently, 4 is not the largest perfect square factor of 80. Keep going. Is 80 divisible by 9? No. Is 80 divisible by 16? Yes, $80 \div 16 = 5$. Since 5 has no perfect square factors, 16 is the largest perfect square factor of 80. Therefore, $80 = 16 \cdot 5$.

We can determine the greatest perfect square factor of a number by using a procedure similar to the one we used to write a number in terms of its prime factors. The procedure is as follows:

Procedure: Determining the Greatest Perfect-Square Factor

1. Divide the number by any perfect-square factor.

2. If possible, divide the resulting quotient by any perfect-square factor.

3. Continue dividing the quotients by perfect-square factors until you arrive at a quotient with no perfect-square factors.

4. The greatest perfect-square factor is the product of all the perfect-square divisors, and the last quotient is the other factor.

Using the procedure from Example 1c would result in the following solution:

4 is a perfect-square factor of 80, so divide 4 into 80.

$$\begin{array}{r} 5 \\ 4\overline{)20} \\ 4\overline{)80} \end{array}$$

5 has no perfect-square factors.

20 has a perfect-square factor of 4, so divide 4 into 20.

Therefore, $80 = 4 \cdot 4 \cdot 5 = 16 \cdot 5$, where 16 is the greatest perfect-square factor of 80.

Using the same procedure on 72, we have the following:

4 is a perfect-square factor of 72, so divide 4 into 72.

$$\begin{array}{r} 2 \\ 9\overline{)18} \\ 4\overline{)72} \end{array}$$

2 has no perfect-square factors.

9 is a perfect-square factor of 18, so divide 9 into 18.

Therefore, $72 = 4 \cdot 9 \cdot 2 = 36 \cdot 2$, where 36 is the greatest perfect-square factor of 72.

PRACTICE EXERCISE 1

Write each of the following as the product of the largest possible perfect square and another number that has no perfect-square factors:

a. 40   **b.** 125   **c.** 96

If you need more practice, do the following Additional Practice Exercises.

Additional Practice Exercise 1   Write each of the following as the product of the largest possible perfect square and another number that has no perfect-square factors.

a. 54   **b.** 24   **c.** 108

The procedure for simplifying *square roots* is outlined as follows:

Procedure: Simplifying Square Roots

1. Write the radicand as the product of the largest possible perfect square and a number that has no perfect-square factors.

2. Apply the product rule in the form $\sqrt{a \cdot b} = \sqrt{a} \cdot \sqrt{b}$, where a is a perfect-square factor.

3. Find the square root of the perfect-square radicand.

Answers: Practice Exercise 1: a. $4 \cdot 10$ **b.** $25 \cdot 5$ **c.** $16 \cdot 6$ **Additional Practice 1: a.** $9 \cdot 6$ **b.** $4 \cdot 6$ **c.** $36 \cdot 3$

Example 2 | **Simplify the following radical expressions:**

a. $\sqrt{18} =$ The largest perfect-square factor of 18 is 9. Rewrite 18 as $9 \cdot 2$.

$\sqrt{9 \cdot 2} =$ Apply $\sqrt{ab} = \sqrt{a} \cdot \sqrt{b}$.

$\sqrt{9} \cdot \sqrt{2} =$ $\sqrt{9} = 3$.

$3\sqrt{2}$ Therefore, $\sqrt{18} = 3\sqrt{2}$.

b. $\sqrt{72} =$ The largest perfect-square factor of 72 is 36. Rewrite 72 as $36 \cdot 2$.

$\sqrt{36 \cdot 2} =$ Apply $\sqrt{ab} = \sqrt{a} \cdot \sqrt{b}$.

$\sqrt{36} \cdot \sqrt{2} =$ $\sqrt{36} = 6$.

$6\sqrt{2}$ Therefore, $\sqrt{72} = 6\sqrt{2}$.

Note

Even though 4 is a perfect-square factor of 72, we should not write $\sqrt{72} = \sqrt{4 \cdot 18}$, because 18 has 9 as a perfect-square factor. However, we could proceed as follows:

$$\sqrt{72} = \sqrt{4 \cdot 18} = \sqrt{4} \cdot \sqrt{18} = 2\sqrt{18} = 2\sqrt{9 \cdot 2}$$
$$= 2\sqrt{9} \cdot \sqrt{2} = 2 \cdot 3\sqrt{2} = 6\sqrt{2}$$

and arrive at the same answer, but with more steps. That is why it is preferable to find the largest perfect-square factor.

c. $2\sqrt{80} =$ The largest perfect-square factor of 80 is 16. Rewrite 80 as $16 \cdot 5$.

$2\sqrt{16 \cdot 5} =$ Apply $\sqrt{ab} = \sqrt{a} \cdot \sqrt{b}$.

$2\sqrt{16} \cdot \sqrt{5} =$ $\sqrt{16} = 4$.

$2 \cdot 4\sqrt{5} =$ Multiply 2 and 4.

$8\sqrt{5}$ Therefore, $2\sqrt{80} = 8\sqrt{5}$.

d. $5\sqrt{108} =$ The largest perfect-square of 108 is 36. Rewrite 108 as $36 \cdot 3$.

$5\sqrt{36 \cdot 3} =$ Apply $\sqrt{ab} = \sqrt{a} \cdot \sqrt{b}$.

$5\sqrt{36} \cdot \sqrt{3} =$ $\sqrt{36} = 6$.

$5 \cdot 6\sqrt{3} =$ Multiply 5 and 6.

$30\sqrt{3}$ Therefore, $5\sqrt{108} = 30\sqrt{3}$.

OBJECTIVE Using the alternative method to simplify radicals

Optional Topic

Alternative Method

Sometimes, finding the greatest perfect-square factor of the radicand, especially if the radicand is a large number, can be troublesome. Another method, using the prime factorization of the radicand, is often useful under these circumstances. A perfect-square integer can be thought of as an integer that can be expressed as the product of two identical factors. For example, $49 = 7 \cdot 7$ and $121 = 11 \cdot 11$. The square root of a perfect square is one of these two identical factors.

For example, $\sqrt{49} = \sqrt{7 \cdot 7} = 7$, which is one of the two identical factors of 49. We can use this idea to simplify a radical by writing the radicand as the product of its prime factors without using exponents. Each factor that appears twice will have a square root equal to one of the two factors. The remaining factors stay underneath the radical sign. We illustrate with Examples 2b and 2c.

b. $\sqrt{72} =$ Write 72 as the product of its prime factors.

$\sqrt{2 \cdot 2 \cdot 2 \cdot 3 \cdot 3} =$ $\sqrt{2 \cdot 2} = 2$ and $\sqrt{3 \cdot 3} = 3$.

$2 \cdot 3\sqrt{2} =$ $2 \cdot 3 = 6$.

$6\sqrt{2}$ Therefore, $\sqrt{72} = 6\sqrt{2}$.

c. $2\sqrt{80} =$ Write 80 as the product of its prime factors.

$2\sqrt{2 \cdot 2 \cdot 2 \cdot 2 \cdot 5} =$ $\sqrt{2 \cdot 2} = 2$.

$2 \cdot 2 \cdot 2\sqrt{5} =$ $2 \cdot 2 \cdot 2 = 8$.

$8\sqrt{5}$ Therefore, $2\sqrt{80} = 8\sqrt{5}$.

PRACTICE EXERCISE 2

Simplify the following square roots:

a. $\sqrt{8}$ **b.** $\sqrt{50}$ **c.** $2\sqrt{54}$ **d.** $5\sqrt{108}$

If you need more practice, do the following Additional Practice Exercises.

Additional Practice Exercise 2   Simplify the following square roots:

a. $\sqrt{20}$ **b.** $\sqrt{162}$ **c.** $3\sqrt{45}$ **d.** $5\sqrt{44}$

We can apply either procedure for simplifying radicals whose radicands contain variables. Remember, to find the root of a variable expression, the exponent on the variable must be divisible by the root index. For example, $\sqrt{x^8} = x^{\frac{8}{2}} = x^4$. In general, $\sqrt{x^n} = x^{n/2}$. Consequently, variable factors that are perfect squares must have even exponents. Therefore, to find the square root, we rewrite the radicand as the product of the variable with the largest possible even exponent and the variable to the first power.

Example 3   **Simplify the roots. Assume that all variables have nonnegative values.**

a. $\sqrt{x^3} =$ The largest perfect-square factor of x^3 is x^2. Rewrite x^3 as $x^2 \cdot x$.

$\sqrt{x^2 \cdot x} =$ Apply $\sqrt{ab} = \sqrt{a} \cdot \sqrt{b}$.

$\sqrt{x^2} \cdot \sqrt{x} =$ $\sqrt{x^2} = x$.

$x\sqrt{x}$ Therefore, $\sqrt{x^3} = x\sqrt{x}$.

b. $a^2\sqrt{a^7} =$ The largest perfect-square factor of a^7 is a^6. Rewrite a^7 as $a^6 \cdot a$.

$a^2\sqrt{a^6 \cdot a} =$ Apply $\sqrt{ab} = \sqrt{a} \cdot \sqrt{b}$.

$a^2\sqrt{a^6} \cdot \sqrt{a} =$ $\sqrt{a^6} = a^3$.

$a^2 \cdot a^3\sqrt{a} =$ Multiply $a^2 \cdot a^3$.

$a^5\sqrt{a}$ Therefore, $a^2\sqrt{a^7} = a^5\sqrt{a}$.

c. $\sqrt{c^5 d^3} =$ The largest perfect-square factor of c^5 is c^4, and the largest perfect-square factor of d^3 is d^2. Rewrite $c^5 d^3$ as $c^4 d^2 \cdot cd$.

$\sqrt{c^4 d^2 \cdot cd} =$ Apply $\sqrt{ab} = \sqrt{a} \cdot \sqrt{b}$.

$\sqrt{c^4 d^2} \cdot \sqrt{cd} =$ Apply $\sqrt{ab} = \sqrt{a} \cdot \sqrt{b}$ to $\sqrt{c^4 d^2}$.

$\sqrt{c^4} \cdot \sqrt{d^2} \cdot \sqrt{cd} =$ $\sqrt{c^4} = c^2$ and $\sqrt{d^2} = d$.

$c^2 d\sqrt{cd}$ Therefore, $\sqrt{c^5 d^3} = c^2 d\sqrt{cd}$.

Answers: Practice Exercise 2: a. $2\sqrt{2}$ **b.** $5\sqrt{2}$ **c.** $6\sqrt{6}$ **d.** $30\sqrt{3}$ **Additional Practice 2: a.** $2\sqrt{5}$ **b.** $9\sqrt{2}$ **c.** $9\sqrt{5}$ **d.** $10\sqrt{11}$

Note

If we used the alternative method of simplifying square roots, Example 3c would appear as follows:

$$\sqrt{c^5 d^3} =$$ Write $c^5 d^3$ as the product of its prime factors.

$$\sqrt{c \cdot c \cdot c \cdot c \cdot c \cdot d \cdot d \cdot d} =$$ $\sqrt{c \cdot c} = c$ and $\sqrt{d \cdot d} = d$.

$$c \cdot c \cdot d \sqrt{c \cdot d} =$$ $c \cdot c = c^2$.

$$c^2 d \sqrt{cd}$$ Therefore, $\sqrt{c^5 d^3} = c^2 d \sqrt{cd}$.

d. $3n \sqrt{28 n^9} =$ The largest perfect-square factor of 28 is 4, and the largest perfect-square factor of n^9 is n^8. Rewrite $28 n^9$ as $4 n^8 \cdot 7n$.

$$3n \sqrt{4 n^8 \cdot 7n} =$$ Apply $\sqrt{ab} = \sqrt{a} \cdot \sqrt{b}$.

$$3n \sqrt{4 n^8} \cdot \sqrt{7n} =$$ Apply $\sqrt{ab} = \sqrt{a} \cdot \sqrt{b}$ to $\sqrt{4 n^8}$.

$$3n \sqrt{4} \cdot \sqrt{n^8} \cdot \sqrt{7n} =$$ $\sqrt{4} = 2$ and $\sqrt{n^8} = n^4$.

$$3n \cdot 2 n^4 \sqrt{7n} =$$ Multiply $3n \cdot 2 n^4$.

$$6 n^5 \sqrt{7n}$$ Therefore, $3n \sqrt{28 n^9} = 6 n^5 \sqrt{7n}$.

Note

We can simplify the roots of numbers, using the same method as for variables, by prime factorization:
$$\sqrt{72} = \sqrt{2^3 \cdot 3^2} = \sqrt{2^2 \cdot 2 \cdot 3^2} = 2 \cdot 3 \sqrt{2} = 6 \sqrt{2}.$$

PRACTICE EXERCISE 3

Simplify the following square roots:

a. $\sqrt{a^5}$ **b.** $c^2 \sqrt{c^{11}}$ **c.** $\sqrt{r^7 s^3}$ **d.** $4b \sqrt{27 b^7}$

If you need more practice, do the following Additional Practice Exercises.

Additional Practice Exercise 3   Simplify the following square roots:

a. $\sqrt{n^9}$ **b.** $y^3 \sqrt{y^7}$ **c.** $\sqrt{m^5 n^{11}}$ **d.** $7r \sqrt{18 r^{13}}$

OBJECTIVE Simplifying higher order roots by using the product rule

To simplify cube roots, we rewrite the radicand as the product of the largest perfect-cube factor and some number that has no perfect-cube factors. Recall that the integers that are perfect cubes are 1, 8, 27, 64, 125, ... To determine whether the radicand has a perfect-cube factor, see whether the radicand is divisible by any integer that is a perfect-cube. For example, 32 has the perfect-cube factor 8, since $32 \div 8 = 4$. Consequently, $32 = 8 \cdot 4$.

Remember: In finding cube roots of variables, the exponent on the radicand must be divisible by the index 3. For example, $\sqrt[3]{x^9} = x^{9/3} = x^3$. In general, $\sqrt[3]{x^n} = x^{n/3}$. So, variable factors must be written as the product of the variable to the largest possible power divisible by 3 and the variable to a power smaller than 3.

Answers: Practice Exercise 3: a. $a^2 \sqrt{a}$ **b.** $c^7 \sqrt{c}$ **c.** $r^3 s \sqrt{rs}$ **d.** $12 b^4 \sqrt{3b}$
Additional Practice 3: a. $n^4 \sqrt{n}$ **b.** $y^6 \sqrt{y}$ **c.** $m^2 n^5 \sqrt{mn}$ **d.** $21 r^7 \sqrt{2r}$

Example 4	Simplify the following radical expressions:

a. $\sqrt[3]{24} =$ The largest perfect-cube factor of 24 is 8. Rewrite 24 as $8 \cdot 3$.

$\sqrt[3]{8 \cdot 3} =$ Apply the product rule.

$\sqrt[3]{8} \cdot \sqrt[3]{3} =$ $\sqrt[3]{8} = 2$.

$2\sqrt[3]{3}$ Therefore, $\sqrt[3]{24} = 2\sqrt[3]{3}$.

b. $5\sqrt[3]{108} =$ The largest perfect-cube factor of 108 is 27. Rewrite 108 as $27 \cdot 4$.

$5\sqrt[3]{27 \cdot 4} =$ Apply the product rule.

$5\sqrt[3]{27} \cdot \sqrt[3]{4} =$ $\sqrt[3]{27} = 3$.

$5 \cdot 3\sqrt[3]{4} =$ Multiply 5 and 3.

$15\sqrt[3]{4}$ Therefore, $5\sqrt[3]{108} = 15\sqrt[3]{4}$.

c. $\sqrt[3]{x^5} =$ The largest perfect-cube factor of x^5 is x^3. Rewrite x^5 as $x^3 \cdot x^2$.

$\sqrt[3]{x^3 \cdot x^2} =$ Apply the product rule.

$\sqrt[3]{x^3} \cdot \sqrt[3]{x^2} =$ $\sqrt[3]{x^3} = x$.

$x\sqrt[3]{x^2}$ Therefore, $\sqrt[3]{x^5} = x\sqrt[3]{x^2}$.

d. $3x\sqrt[3]{40x^{10}} =$ The largest perfect-cube factor of 40 is 8, and the largest perfect-cube factor of x^{10} is x^9. Rewrite $40x^{10}$ as $8x^9 \cdot 5x$.

$3x\sqrt[3]{8x^9 \cdot 5x} =$ Apply the product rule.

$3x\sqrt[3]{8x^9} \cdot \sqrt[3]{5x} =$ Apply the product rule to $\sqrt[3]{8x^9}$.

$3x\sqrt[3]{8} \cdot \sqrt[3]{x^9} \cdot \sqrt[3]{5x} =$ $\sqrt[3]{8} = 2$ and $\sqrt[3]{x^9} = x^3$.

$3x \cdot 2x^3\sqrt[3]{5x} =$ Multiply.

$6x^4\sqrt[3]{5x}$ Therefore, $3x\sqrt[3]{40x^{10}} = 6x^4\sqrt[3]{5x}$.

Optional Topic

Alternative Method

The alternative method of simplifying radicands is easily extended to roots other than square roots. For example, a perfect cube can be thought of as a number that can be written as the product of three identical factors, such as $8 = 2 \cdot 2 \cdot 2$.

Therefore, the cube root of a perfect cube is one of its three identical factors. For example, $\sqrt[3]{8} = \sqrt[3]{2 \cdot 2 \cdot 2} = 2$, which is one of the three identical factors of 8. By the alternative method, Example 4d would appear as follows:

$3x\sqrt[3]{40x^{10}} =$ Write $40x^{10}$ as the product of its prime factors.

$3x\sqrt[3]{2 \cdot 2 \cdot 2 \cdot 5 \cdot x \cdot x \cdot x \cdot x \cdot x \cdot x \cdot x \cdot x \cdot x \cdot x} =$ $\sqrt[3]{2 \cdot 2 \cdot 2} = 2$ and $\sqrt[3]{x \cdot x \cdot x} = x$.

$3x \cdot 2 \cdot x \cdot x \cdot x\sqrt[3]{5x}$ $x \cdot x \cdot x = x^3$.

$6x^4\sqrt[3]{5x} =$ Therefore, $3x^3\sqrt{40x^{10}} = 6x^4 \sqrt{5x}$.

PRACTICE EXERCISE 4

Simplify the following radical expressions:

a. $\sqrt[3]{56}$ **b.** $6\sqrt[3]{128}$ **c.** $\sqrt[3]{a^8}$ **d.** $2c^2\sqrt[3]{81c^{13}}$

If you need more practice, do the following Additional Practice Exercises.

Additional Practice Exercise 4   Simplify the following roots:

a. $\sqrt[3]{72}$ **b.** $4\sqrt[3]{48}$ **c.** $\sqrt[3]{r^4}$ **d.** $3n^3\sqrt[3]{250n^{14}}$

Answers: Practice Exercise 4: a. $2\sqrt[3]{7}$ **b.** $24\sqrt[3]{2}$ **c.** $a^2\sqrt[3]{a^2}$ **d.** $6c^6\sqrt[3]{3c}$ **Additional Practice 4: a.** $2\sqrt[3]{9}$ **b.** $8\sqrt[3]{6}$ **c.** $r\sqrt[3]{r}$ **d.** $15n^7\sqrt[3]{2n^2}$

Exercise Set 10.2 MyMathLab®

Simplify the following square roots. (See Examples 2 and 3.)

1. $\sqrt{12}$ **2.** $\sqrt{28}$ **3.** $\sqrt{98}$ **4.** $\sqrt{48}$

5. $\sqrt{128}$ **6.** $\sqrt{180}$ **7.** $\sqrt{320}$ **8.** $\sqrt{405}$

9. $5\sqrt{54}$ **10.** $3\sqrt{63}$ **11.** $6\sqrt{80}$ **12.** $4\sqrt{50}$

13. $5\sqrt{112}$ **14.** $3\sqrt{245}$ **15.** $7\sqrt{243}$ **16.** $6\sqrt{242}$

17. $\sqrt{a^3}$ **18.** $\sqrt{d^5}$ **19.** $\sqrt{b^{13}}$ **20.** $\sqrt{m^{23}}$

21. $\sqrt{x^2 y^4}$ **22.** $\sqrt{a^6 b^2}$ **23.** $\sqrt{x^6 y^8 z^{10}}$ **24.** $\sqrt{p^4 q^8 r^8}$

25. $\sqrt{x^5 y^8}$ **26.** $\sqrt{a^{10} b^{15}}$ **27.** $\sqrt{m^{17} n^{19}}$ **28.** $\sqrt{c^7 d^{13}}$

29. $c\sqrt{c^3 d^6}$ **30.** $x\sqrt{x^7 y^4}$ **31.** $rs^2\sqrt{r^9 s^3}$ **32.** $a^2 b^3 \sqrt{a^{11} b^3}$

33. $\sqrt{20x^6}$ **34.** $\sqrt{45y^4}$ **35.** $3\sqrt{72x^5}$ **36.** $6\sqrt{75d^3}$

37. $5xy^2\sqrt{150x^7 y^{10}}$ **38.** $4a^2 b\sqrt{216a^6 b^9}$ **39.** $8a^3 c\sqrt{99a^{13} c^4}$ **40.** $11r^3 s^2 \sqrt{405r^{11} s^7}$

Simplify the following higher order roots. (See Example 4.)

41. $\sqrt[3]{32}$ **42.** $\sqrt[3]{54}$ **43.** $4\sqrt[3]{40}$ **44.** $7\sqrt[3]{81}$

45. $9\sqrt[3]{56}$ **46.** $7\sqrt[3]{320}$ **47.** $\sqrt[3]{x^7}$ **48.** $\sqrt[3]{b^{11}}$

49. $\sqrt[3]{x^6 y^5}$ **50.** $\sqrt[3]{m^{13} n^9}$ **51.** $\sqrt[3]{128z^8}$ **52.** $\sqrt[3]{48h^{14}}$

53. $2\sqrt[3]{24}$ **54.** $4\sqrt[3]{250}$ **55.** $4a\sqrt[3]{162a^7}$ **56.** $-3p^2 \sqrt[3]{500p^{11}}$

The amount of time it takes an object to fall s feet is given by the formula $t = \dfrac{\sqrt{s}}{4}$ where t is in seconds. Find the number of seconds for an object of fall the following distances.

57. 144 ft.   **58.** 196 feet

Ships use the formula $s = 1.17\sqrt{h}$ to estimate the distance to the horizon where s is in nautical miles and h is the distance (in feet) that your eye is above the water. Estimate the distance to the horizon for a ship's captain whose eyes are the following distances above the water. (Source: BoatSafe.com)

59. 81 feet.   **60.** 64 feet.

If the sides of a triangle are of length a, b, and c, the area can be found using the formula $A = \sqrt{s(s - a)(s - b)(s - c)}$ where $s = \dfrac{a + b + c}{2}$. Find the areas of the following triangles.

61. $a = 3$ in., $b = 4$ in., $c = 5$ in.   **62.** $a = 6$ cm, $b = 8$ cm, $c = 10$ cm.

63. $a = 4$ ft., $b = 5$ ft., $c = 7$ ft.   **64.** $a = 9$ mm, $b = 7$ mm, $c = 8$ mm

Challenge Exercises (65–72)

Simplify the following:

65. $\sqrt[4]{162}$

66. $\sqrt[4]{x^9}$

67. $\sqrt[4]{a^6 b^7}$

68. $\sqrt[5]{64}$

69. $\sqrt[5]{972}$

70. $\sqrt[5]{x^{12}}$

71. $\sqrt[5]{a^9 b^{16}}$

72. $\sqrt[6]{x^9 y^{16}}$

Writing Exercises (73 and 74)

73. When simplifying square root radicals, we wrote the radicand as the product of a perfect-square and another number that had no perfect-square factors. When simplifying cube roots, we wrote the radicand as the product of a perfect cube and another number with no perfect-cube factors. Using this same line of reasoning, how would you rewrite the radicand of a fourth root before simplifying? A fifth root?

74. To find the root of a variable expression, the exponent must be divisible by the root index. If the root index is 4, what exponents would you want on the factor of the radicand whose fourth root you will find? Fifth root?

Critical-Thinking Exercises (75 and 76)

75. Write a procedure for finding the largest perfect-cube factor similar to that for finding the largest perfect-square factor, following Example 1c.

76. Use $a = 6$ and $b = 8$ to answer the following:
  a. Does $\sqrt{a^2 + b^2} = \sqrt{a^2} + \sqrt{b^2} = a + b$?
  b. Does $\sqrt{(a + b)^2} = a + b$?

Section 10.3   Products and Quotients of Radicals

OBJECTIVES *When you complete this section, you will be able to:*

A Find the products of radicals and simplify the results.

B Find the quotients of radicals and simplify the results.

C Simplify radical expressions by using both the product and quotient rules.

PREREQUISITE SKILLS   *Before beginning this section, you should be able to:*

a. Multiply and divide whole numbers. (Section R.3)

b. Find the square root of a perfect square. (Section 10.1)

c. Simplify a radical expression. (Section 10.2)

d. Multiply rational numbers. (Section 6.3)

GETTING READY FOR SECTION 10.3

Find the following products or quotients:

1. $3 \cdot 15$ **2.** $5a^2 \cdot 10a^5$ **3.** $\dfrac{756}{7}$ **4.** $\dfrac{81x^4y^3}{3xy^2}$

Find the following square roots:

5. $\sqrt{64}$ **6.** $\sqrt{x^6y^4}$

Simplify the following radical expressions:

7. $\sqrt{50a^7}$ **8.** $\sqrt{108a^4b^3}$

Introduction

We know how to perform the operations of addition, subtraction, multiplication, and division on rational numbers. But how do we perform these operations on radical expressions that are irrational numbers? In this section, we will multiply and divide radicals. In Section 10.4, we will learn how to add and subtract radical expressions.

OBJECTIVE Finding the products of radicals and simplifying the results

We learned how to multiply radicals in Section 10.2. For reference, the product rule for radicals is repeated here.

> **Rule: Products of Radicals**
>
> If $\sqrt[n]{a}$ and $\sqrt[n]{b}$ are both real numbers, then $\sqrt[n]{a} \cdot \sqrt[n]{b} = \sqrt[n]{a \cdot b}$.

In Section 10.2, we used this rule in the form $\sqrt[n]{ab} = \sqrt[n]{a} \cdot \sqrt[n]{b}$. In this section, we will be using this rule in the form $\sqrt[n]{a} \cdot \sqrt[n]{b} = \sqrt[n]{ab}$.

Answers: Getting Ready: 1. 45 **2.** $50a^7$ **3.** 108 **4.** $27x^3y$ **5.** 8 **6.** x^3y^2 **7.** $5a^3\sqrt{2a}$ **8.** $6a^2b\sqrt{3b}$

Example 1   Find the products of radicals. Assume that all variables have nonnegative values.

a. $\sqrt{3} \cdot \sqrt{5} =$ Apply $\sqrt{a} \cdot \sqrt{b} = \sqrt{ab}$.
$\sqrt{3 \cdot 5} =$ Multiply 3 and 5.
$\sqrt{15}$ Therefore, $\sqrt{3} \cdot \sqrt{5} = \sqrt{15}$.

b. $\sqrt{2} \cdot \sqrt{8} =$ Apply $\sqrt{a} \cdot \sqrt{b} = \sqrt{ab}$.
$\sqrt{2 \cdot 8} =$ Multiply 2 and 8.
$\sqrt{16} =$ Find the $\sqrt{16}$.
4   Therefore, $\sqrt{2} \cdot \sqrt{8} = 4$.

c. $\sqrt{7} \cdot \sqrt{x} =$ Apply $\sqrt{a} \cdot \sqrt{b} = \sqrt{ab}$.
$\sqrt{7x}$ Therefore, $\sqrt{7} \cdot \sqrt{x} = \sqrt{7x}$.

d. $\sqrt{x} \cdot \sqrt{x} =$ Apply $\sqrt{a} \cdot \sqrt{b} = \sqrt{ab}$.
$\sqrt{x \cdot x} =$ Multiply x and x.
$\sqrt{x^2} =$ Find the $\sqrt{x^2}$.
x Therefore, $\sqrt{x} \cdot \sqrt{x} = x$.

PRACTICE EXERCISE 1

Find the products of radicals. Assume that all variables have nonnegative values.

a. $\sqrt{3} \cdot \sqrt{7}$ **b.** $\sqrt{3} \cdot \sqrt{27}$ **c.** $\sqrt{5} \cdot \sqrt{a}$ **d.** $\sqrt{r} \cdot \sqrt{r}$

If you need more practice, do the following Additional Practice Exercises.

Additional Practice Exercise 1   Find the products of radicals. Assume that all variables have nonnegative values.
a. $\sqrt{6} \cdot \sqrt{5}$ **b.** $\sqrt{2} \cdot \sqrt{32}$ **c.** $\sqrt{7} \cdot \sqrt{p}$ **d.** $\sqrt{q} \cdot \sqrt{q}$

It is often necessary to simplify the products of radicals. If the radical expressions have coefficients, multiply the coefficients and then multiply the radicals in much the same manner monomials are multiplied. See the following:

Monomials	**Radicals**	
$2x \cdot 3x^2$	$2\sqrt{3} \cdot 3\sqrt{5}$	Regroup.
$(2 \cdot 3)(x \cdot x^2)$	$(2 \cdot 3)(\sqrt{3} \cdot \sqrt{5})$	Multiply.
$6x^3$	$6\sqrt{15}$	Product.

Example 2   Find the products. Express answers in simplified form. Assume that all variables represent nonnegative values.

a. $\sqrt{2} \cdot \sqrt{6} =$ Apply the product rule in the form $\sqrt{a} \cdot \sqrt{b} = \sqrt{ab}$.
$\sqrt{2 \cdot 6} =$ Multiply.
$\sqrt{12} =$ 12 has a perfect-square factor of 4. Rewrite 12 as $4 \cdot 3$.
$\sqrt{4 \cdot 3} =$ Apply the product rule in the form $\sqrt{ab} = \sqrt{a} \cdot \sqrt{b}$.
$\sqrt{4} \cdot \sqrt{3} =$ $\sqrt{4} = 2$.
$2\sqrt{3}$ Therefore, $\sqrt{2} \cdot \sqrt{6} = 2\sqrt{3}$.

b. $5\sqrt{3} \cdot 3\sqrt{15} =$ Regroup and multiply 5 and 3, and $\sqrt{3}$ and $\sqrt{15}$.
$(5 \cdot 3)(\sqrt{3} \cdot \sqrt{15}) =$ Multiply.
$15\sqrt{45} =$ 45 has a perfect-square factor of 9. Rewrite 45 as $9 \cdot 5$.
$15\sqrt{9 \cdot 5} =$ Apply the product rule in the form $\sqrt{ab} = \sqrt{a} \cdot \sqrt{b}$.
$15\sqrt{9} \cdot \sqrt{5} =$ $\sqrt{9} = 3$.
$15 \cdot 3\sqrt{5} =$ Multiply 15 and 3.
$45\sqrt{5}$ Therefore, $5\sqrt{3} \cdot 3\sqrt{15} = 45\sqrt{5}$.

Answers: Practice Exercise 1: a. $\sqrt{21}$ **b.** 9 **c.** $\sqrt{5a}$ **d.** r **Additional Practice 1: a.** $\sqrt{30}$ **b.** 8 **c.** $\sqrt{7p}$ **d.** q

c. $\sqrt{x^3} \cdot \sqrt{x^5} =$ Apply the product rule in the form $\sqrt{a} \cdot \sqrt{b} = \sqrt{ab}$.

$\sqrt{x^3 \cdot x^5} =$ $x^3 \cdot x^5 = x^{3+5} = x^8$.

$\sqrt{x^8} =$ Find the root.

x^4 Therefore, $\sqrt{x^3} \cdot \sqrt{x^5} = x^4$.

d. $2\sqrt{5a^2} \cdot 3\sqrt{10a^5} =$ Regroup and multiply 2 and 3, and $\sqrt{5a^2}$ and $\sqrt{10a^5}$.

$(2 \cdot 3)\left(\sqrt{5a^2} \cdot \sqrt{10a^5}\right) =$ Apply the product rule in the form $\sqrt{a} \cdot \sqrt{b} = \sqrt{ab}$.

$2 \cdot 3\sqrt{5a^2 \cdot 10a^5} =$ Multiply.

$6\sqrt{50a^7} =$ The largest perfect-square factor of 50 is 25, and the largest perfect-square factor of a^7 is a^6. Rewrite $50a^7$ as $25a^6 \cdot 2a$.

$6\sqrt{25a^6 \cdot 2a} =$ Apply the product rule in the form $\sqrt{ab} = \sqrt{a} \cdot \sqrt{b}$.

$6\sqrt{25a^6} \cdot \sqrt{2a} =$ $\sqrt{25a^6} = 5a^3$.

$6 \cdot 5a^3\sqrt{2a} =$ Multiply 6 and 5.

$30a^3\sqrt{2a}$ Therefore, $2\sqrt{5a^2} \cdot 3\sqrt{10a^5} = 30a^3\sqrt{2a}$.

PRACTICE EXERCISE 2

Find the products. Leave answers in simplified form. Assume that all variables have nonnegative values.

a. $\sqrt{3} \cdot \sqrt{21}$ **b.** $4\sqrt{6} \cdot 7\sqrt{15}$ **c.** $\sqrt{a^5} \cdot \sqrt{a^5}$ **d.** $4\sqrt{10c} \cdot 2\sqrt{6c^4}$

If you need more practice, do the following Additional Practice Exercises.

Additional Practice Exercise 2   Find the products of radicals. Express answers in simplified form. Assume that all variables have nonnegative values.

a. $\sqrt{6} \cdot \sqrt{15}$ **b.** $4\sqrt{14} \cdot 2\sqrt{6}$ **c.** $\sqrt{a} \cdot \sqrt{a^3}$ **d.** $5\sqrt{10m^2} \cdot \sqrt{15m^5}$

The product rule for radicals can be used to develop a very important property of radicals. In Example 1d, we found that the product $\sqrt{x} \cdot \sqrt{x} = x$. By the definition of an exponent, $\sqrt{x} \cdot \sqrt{x} = (\sqrt{x})^2$. Consequently, $(\sqrt{x})^2 = x$. Also, $\sqrt[3]{x} \cdot \sqrt[3]{x} \cdot \sqrt[3]{x} = \sqrt[3]{x^3} = x$. But $\sqrt[3]{x} \cdot \sqrt[3]{x} \cdot \sqrt[3]{x} = (\sqrt[3]{x})^3$ also. Consequently, $(\sqrt[3]{x})^3 = x$. This leads to the following generalization:

> ### Rule: Powers of Roots
> If $\sqrt[n]{a}$ is a real number, then $(\sqrt[n]{a})^n = a$.

Example 3   **Find the following, assuming that all variables have nonnegative values:**

a. $(\sqrt{2})^2 = 2$ **b.** $(\sqrt[3]{a})^3 = a$ **c.** $(\sqrt[4]{x})^4 = x$

d. $(\sqrt[5]{9})^5 = 9$ **e.** $(\sqrt[3]{3x^2y})^3 = 3x^2y$

PRACTICE EXERCISE 3

Find the following, assuming that all variables have nonnegative values:

a. $(\sqrt{5})^2$ **b.** $(\sqrt[3]{r})^3$ **c.** $(\sqrt[4]{y})^4$

d. $(\sqrt[5]{16})^5$ **e.** $(\sqrt[4]{6ab})^4$

Answers: Practice Exercise 2: a. $3\sqrt{7}$ **b.** $84\sqrt{10}$ **c.** a^5 **d.** $16c^2\sqrt{15c}$ **Additional Practice 2: a.** $3\sqrt{10}$ **b.** $16\sqrt{21}$ **c.** a^2 **d.** $25m^3\sqrt{6m}$
Practice Exercise 3: a. 5 **b.** r **c.** y **d.** 16 **e.** $6ab$

OBJECTIVE Ⓑ **Finding the quotients of radicals and simplifying the results**

Look for a pattern for quotients of radicals in the following Calculator Exploration Activity:

Calculator Exploration Activity (Optional)

Use your calculator to evaluate columns A and B, and round your answer to the nearest ten thousandth. Compare the results obtained in each line.

Column A   **Column B**

a. $\dfrac{\sqrt{10}}{\sqrt{2}} = $ _____ $\sqrt{5} = $ _____

b. $\dfrac{\sqrt{18}}{\sqrt{6}} = $ _____ $\sqrt{3} = $ _____

c. $\dfrac{\sqrt{48}}{\sqrt{8}} = $ _____ $\sqrt{6} = $ _____

d. $\dfrac{\sqrt{42}}{\sqrt{6}} = $ _____ $\sqrt{7} = $ _____

1. Based on your answers to the corresponding lines of columns A and B,
$\dfrac{\sqrt{a}}{\sqrt{b}} = $ _____.

A rule similar to the product rule can be established for quotients. Consider the following: Simplify $\dfrac{\sqrt{36}}{\sqrt{4}}$ by first evaluating the numerator and the denominator.

$\dfrac{\sqrt{36}}{\sqrt{4}} = \dfrac{6}{2} = 3$. If we divide the radicands, we get $\dfrac{\sqrt{36}}{\sqrt{4}} = \sqrt{\dfrac{36}{4}} = \sqrt{9} = 3$. Note

that we arrived at the same answer. Consequently, $\dfrac{\sqrt{36}}{\sqrt{4}} = \sqrt{\dfrac{36}{4}}$. This suggests the

following rule for quotients of radicals:

Rule: Quotients of Radicals

If $\sqrt[n]{a}$ and $\sqrt[n]{b}$ both exist and $b \neq 0$, then $\dfrac{\sqrt[n]{a}}{\sqrt[n]{b}} = \sqrt[n]{\dfrac{a}{b}}$.

Like the product rule, this rule may be applied in either order as $\dfrac{\sqrt[n]{a}}{\sqrt[n]{b}} = \sqrt[n]{\dfrac{a}{b}}$ or as

$\sqrt[n]{\dfrac{a}{b}} = \dfrac{\sqrt[n]{a}}{\sqrt[n]{b}}$.

Calculator Exploration Activity: 1. $\sqrt{\dfrac{a}{b}}$

Example 4   Find the following quotients, assuming that all variables have positive values:

a. $\dfrac{\sqrt{28}}{\sqrt{7}} =$ Apply $\dfrac{\sqrt{a}}{\sqrt{b}} = \sqrt{\dfrac{a}{b}}$.

$\sqrt{\dfrac{28}{7}} =$ Divide.

$\sqrt{4} =$ Simplify.

2 Therefore, $\dfrac{\sqrt{28}}{\sqrt{7}} = 2$.

b. $\sqrt{\dfrac{4}{9}} =$ Apply $\sqrt{\dfrac{a}{b}} = \dfrac{\sqrt{a}}{\sqrt{b}}$.

$\dfrac{\sqrt{4}}{\sqrt{9}} =$ Simplify the numerator and the denominator.

$\dfrac{2}{3}$ Therefore, $\sqrt{\dfrac{4}{9}} = \dfrac{2}{3}$.

c. $\dfrac{\sqrt{32x^5}}{\sqrt{2x}} =$ Apply $\dfrac{\sqrt{a}}{\sqrt{b}} = \sqrt{\dfrac{a}{b}}$.

$\sqrt{\dfrac{32x^5}{2x}} =$ Divide.

$\sqrt{16x^4} =$ Simplify.

$4x^2$ Therefore, $\dfrac{\sqrt{32x^5}}{\sqrt{2x}} = 4x^2$.

d. $\dfrac{\sqrt{81x^4y^3}}{\sqrt{3xy^2}} =$ Apply $\dfrac{\sqrt{a}}{\sqrt{b}} = \sqrt{\dfrac{a}{b}}$.

$\sqrt{\dfrac{81x^4y^3}{3xy^2}} =$ Divide.

$\sqrt{27x^3y} =$ Rewrite $27x^3y$ as $9x^2 \cdot 3xy$.

$\sqrt{9x^2 \cdot 3xy} =$ Apply $\sqrt{ab} = \sqrt{a} \cdot \sqrt{b}$.

$\sqrt{9x^2} \cdot \sqrt{3xy} =$ $\sqrt{9x^2} = 3x$.

$3x\sqrt{3xy}$ Therefore, $\dfrac{\sqrt{81x^4y^3}}{\sqrt{3xy^2}} = 3x\sqrt{3xy}$.

e. $\dfrac{8\sqrt{756a^8b^5}}{2\sqrt{7a^4b^2}} =$ Divide coefficient into coefficient and radical into radical.

$\dfrac{8}{2} \cdot \dfrac{\sqrt{756a^8b^3}}{\sqrt{7a^4b^2}} =$ Divide coefficients and apply $\dfrac{\sqrt{a}}{\sqrt{b}} = \sqrt{\dfrac{a}{b}}$.

$4\sqrt{\dfrac{756a^8b^5}{7a^4b^2}} =$ Divide under the radical sign.

$4\sqrt{108a^4b^3} =$ Rewrite $108\,a^4b^3$ as $36a^4b^2 \cdot 3b$.

$4\sqrt{36a^4b^2 \cdot 3b} =$ Apply $\sqrt{ab} = \sqrt{a} \cdot \sqrt{b}$.

$4\sqrt{36a^4b^2} \cdot \sqrt{3b} =$ Find the square roots.

$4 \cdot 6a^2b\sqrt{3b} =$ Multiply.

$24a^2b\sqrt{3b}$ Answer.

Therefore, $\dfrac{8\sqrt{756a^8b^5}}{2\sqrt{7a^4b^2}} = 24a^2b\sqrt{3b}$.

PRACTICE EXERCISE 4

Find the following quotients, assuming that all variables have positive values:

a. $\dfrac{\sqrt{75}}{\sqrt{3}}$

b. $\sqrt{\dfrac{16}{25}}$

c. $\dfrac{\sqrt{48x^3}}{\sqrt{12x}}$

d. $\dfrac{\sqrt{16a^5b^4}}{\sqrt{2a^2b^2}}$

e. $\dfrac{12\sqrt{480a^7b^6}}{3\sqrt{6a^3b^3}}$

(Continued)

Answers: Practice Exercise 4: a. 5 **b.** $\dfrac{4}{5}$ **c.** $2x$ **d.** $2ab\sqrt{2a}$ **e.** $16a^2b\sqrt{5b}$

If you need more practice, do the following Additional Practice Exercises.

Additional Practice Exercise 4 Find the following quotients, assuming that all variables have positive values:

a. $\dfrac{\sqrt{45}}{\sqrt{5}}$

b. $\sqrt{\dfrac{64}{25}}$

c. $\dfrac{\sqrt{54x^7}}{\sqrt{6x^3}}$

d. $\dfrac{\sqrt{32m^4n^3}}{\sqrt{4m^2n^2}}$

e. $\dfrac{14\sqrt{315x^{11}y^8}}{2\sqrt{5x^6y^5}}$

OBJECTIVE Simplifying radical expressions by using both the product and quotient rules

We can combine the product and quotient rules to simplify radical expressions involving both multiplication and division.

Example 5 **Simplify the following:**

a. $\sqrt{\dfrac{2}{3}} \cdot \sqrt{\dfrac{10}{3}} =$ Apply the product rule.

$\sqrt{\dfrac{2}{3} \cdot \dfrac{10}{3}} =$ Multiply the fractions.

$\sqrt{\dfrac{20}{9}} =$ Apply the quotient rule.

$\dfrac{\sqrt{20}}{\sqrt{9}} =$ Rewrite 20 as $4 \cdot 5$ and $\sqrt{9} = 3$.

$\dfrac{\sqrt{4 \cdot 5}}{3} =$ Apply the product rule.

$\dfrac{\sqrt{4} \cdot \sqrt{5}}{3} =$ $\sqrt{4} = 2$.

$\dfrac{2\sqrt{5}}{3}$ Therefore, $\sqrt{\dfrac{2}{3}} \cdot \sqrt{\dfrac{10}{3}} = \dfrac{2\sqrt{5}}{3}$.

Note

We could also have applied the quotient rule first, and then the product rule, and done Example 5a as follows:

$$\sqrt{\frac{2}{3}} \cdot \sqrt{\frac{10}{3}} = \frac{\sqrt{2}}{\sqrt{3}} \cdot \frac{\sqrt{10}}{\sqrt{3}} = \frac{\sqrt{2 \cdot 10}}{\sqrt{3 \cdot 3}} = \frac{\sqrt{20}}{\sqrt{9}} = \frac{\sqrt{4 \cdot 5}}{3} = \frac{\sqrt{4} \cdot \sqrt{5}}{3} = \frac{2\sqrt{5}}{3}.$$

Answers: Additional Practice 4: a. 3 **b.** $\dfrac{8}{5}$ **c.** $3x^2$ **d.** $2m\sqrt{2n}$ **e.** $21x^2y\sqrt{7xy}$

b. $\sqrt{\dfrac{x}{5}} \cdot \sqrt{\dfrac{x^3}{125}} =$ Apply the product rule.

$\sqrt{\dfrac{x}{5} \cdot \dfrac{x^3}{125}} =$ Multiply the rational expressions.

$\sqrt{\dfrac{x^4}{625}} =$ Apply the quotient rule.

$\dfrac{\sqrt{x^4}}{\sqrt{625}} =$ $\sqrt{x^4} = x^2$ and $\sqrt{625} = 25$.

$\dfrac{x^2}{25}$ Therefore, $\sqrt{\dfrac{x}{5}} \cdot \sqrt{\dfrac{x^3}{125}} = \dfrac{x^2}{25}$.

PRACTICE EXERCISE 5

Simplify the following:

a. $\sqrt{\dfrac{3}{5}} \cdot \sqrt{\dfrac{6}{5}}$

b. $\sqrt{\dfrac{2}{a}} \cdot \sqrt{\dfrac{8}{a}}$

If you need more practice, do the following Additional Practice Exercises.

Additional Practice Exercise 5 Simplify the following:

a. $\sqrt{\dfrac{2}{7}} \cdot \sqrt{\dfrac{6}{7}}$

b. $\sqrt{\dfrac{3}{x}} \cdot \sqrt{\dfrac{27}{x}}$

Exercise Set 10.3 MyMathLab®

For Extra Help

Find the products or powers of roots. Express all answers in simplified form. Assume that all variables have nonnegative values. (See Examples 1–3.)

1. $\sqrt{6} \cdot \sqrt{5}$

2. $\sqrt{6} \cdot \sqrt{7}$

3. $\sqrt{7} \cdot \sqrt{5}$

4. $\sqrt{3} \cdot \sqrt{13}$

5. $\sqrt{2} \cdot \sqrt{18}$

6. $\sqrt{2} \cdot \sqrt{50}$

7. $\sqrt{3} \cdot \sqrt{75}$

8. $\sqrt{27} \cdot \sqrt{3}$

9. $\sqrt{5} \cdot \sqrt{x}$

10. $\sqrt{11} \cdot \sqrt{c}$

11. $\sqrt{15} \cdot \sqrt{y}$

12. $\sqrt{13} \cdot \sqrt{k}$

13. $\sqrt{x} \cdot \sqrt{y}$

14. $\sqrt{a} \cdot \sqrt{b}$

15. $\sqrt{k} \cdot \sqrt{k}$

16. $\sqrt{m} \cdot \sqrt{m}$

17. $\sqrt{x} \cdot \sqrt{x^3}$

18. $\sqrt{s^3} \cdot \sqrt{s}$

19. $\left(\sqrt{w}\right)^2$

20. $\left(\sqrt{z}\right)^2$

21. $\left(\sqrt{x^3}\right)^2$

22. $\left(\sqrt{b^5}\right)^2$

23. $\left(\sqrt[3]{w}\right)^3$

24. $\left(\sqrt[4]{x}\right)^4$

25. $\left(\sqrt[7]{2ab}\right)^7$

26. $\left(\sqrt[5]{4x^2}\right)^5$

27. $\sqrt{2} \cdot \sqrt{10}$

28. $\sqrt{3} \cdot \sqrt{6}$

29. $\sqrt{6} \cdot \sqrt{8}$

30. $\sqrt{12} \cdot \sqrt{6}$

31. $3\sqrt{5} \cdot 4\sqrt{15}$

32. $2\sqrt{7} \cdot 3\sqrt{14}$

33. $4\sqrt{6} \cdot 2\sqrt{15}$

34. $5\sqrt{10} \cdot 3\sqrt{14}$

35. $2\sqrt{6} \cdot 5\sqrt{21}$

36. $4\sqrt{10} \cdot \sqrt{20}$

37. $\sqrt{a^3} \cdot \sqrt{a}$

38. $\sqrt{r^3} \cdot \sqrt{r^5}$

39. $\sqrt{y^3} \cdot \sqrt{y^2}$

40. $\sqrt{n^5} \cdot \sqrt{n^2}$

Answers: Practice Exercise 5: a. $\dfrac{3\sqrt{2}}{5}$ **b.** $\dfrac{4}{a}$ **Additional Practice 5: a.** $\dfrac{2\sqrt{3}}{7}$ **b.** $\dfrac{9}{x}$

41. $\sqrt{m^6n^5} \cdot \sqrt{m^3n}$

42. $\sqrt{x^2y^3} \cdot \sqrt{x^4y^3}$

43. $x\sqrt{x^5y^2} \cdot y\sqrt{xy^3}$

44. $c\sqrt{c^5d^3} \cdot c\sqrt{c^5d^2}$

45. $3\sqrt{12m^3} \cdot 5\sqrt{5m^5}$

46. $2\sqrt{5a} \cdot 3\sqrt{10a^3}$

47. $4\sqrt{6c^3} \cdot 3\sqrt{10c^6}$

48. $6\sqrt{15c^2} \cdot 2\sqrt{10c^5}$

Find the quotients, assuming that all variables have positive values. Express answers in simplest form. (See Example 4.)

49. $\sqrt{\dfrac{1}{4}}$

50. $\sqrt{\dfrac{9}{25}}$

51. $\sqrt{\dfrac{49}{64}}$

52. $\sqrt{\dfrac{81}{64}}$

53. $\dfrac{\sqrt{32}}{\sqrt{8}}$

54. $\dfrac{\sqrt{125}}{\sqrt{5}}$

55. $\dfrac{\sqrt{98}}{\sqrt{2}}$

56. $\dfrac{\sqrt{27}}{\sqrt{3}}$

57. $\dfrac{6\sqrt{48}}{2\sqrt{6}}$

58. $\dfrac{8\sqrt{54}}{4\sqrt{3}}$

59. $\dfrac{9\sqrt{160}}{3\sqrt{8}}$

60. $\dfrac{10\sqrt{280}}{2\sqrt{10}}$

61. $\dfrac{\sqrt{x^3}}{\sqrt{x}}$

62. $\dfrac{\sqrt{a^5}}{\sqrt{a^3}}$

63. $\dfrac{\sqrt{c^6d^5}}{\sqrt{cd^3}}$

64. $\dfrac{\sqrt{m^6n^5}}{\sqrt{m^3n^3}}$

65. $\dfrac{\sqrt{18x^3}}{\sqrt{2x}}$

66. $\dfrac{\sqrt{20r^5}}{\sqrt{5r^3}}$

67. $\dfrac{8\sqrt{45a^5}}{2\sqrt{5a}}$

68. $\dfrac{6\sqrt{48n^7}}{3\sqrt{3n^3}}$

69. $\dfrac{12\sqrt{72c^9}}{4\sqrt{6c^2}}$

70. $\dfrac{15\sqrt{48a^7}}{5\sqrt{2a^2}}$

71. $\dfrac{36\sqrt{96x^6y^{11}}}{4\sqrt{3x^2y^4}}$

72. $\dfrac{54\sqrt{240r^9s^{10}}}{9\sqrt{5r^6s^4}}$

Simplify the following. (See Example 5.)

73. $\sqrt{\dfrac{3}{7}} \cdot \sqrt{\dfrac{8}{7}}$

74. $\sqrt{\dfrac{8}{5}} \cdot \sqrt{\dfrac{6}{5}}$

75. $\sqrt{\dfrac{a^3}{2}} \cdot \sqrt{\dfrac{a^5}{2}}$

76. $\sqrt{\dfrac{c^2}{6}} \cdot \sqrt{\dfrac{c^5}{6}}$

77. $\sqrt{\dfrac{3x^5}{2}} \cdot \sqrt{\dfrac{15x^5}{8}}$

78. $\sqrt{\dfrac{5y}{3}} \cdot \sqrt{\dfrac{10}{27}}$

Writing Exercise

79. Compare simplifying $(3x)(2x)$ with simplifying $(3\sqrt{2})(2\sqrt{2})$.

Section 10.4

Addition, Subtraction, and Mixed Operations with Radicals

OBJECTIVES *When you complete this section, you will be able to:*

Ⓐ Add and subtract like radicals.

Ⓑ Simplify radicals and then add and/or subtract like radicals.

Ⓒ Simplify radical expressions that contain mixed operations.

Ⓓ Find products of radical expressions involving sums or differences.

PREREQUISITE SKILLS   *Before beginning this section, you should be able to:*

a. Add like terms. (Section 1.7)

b. Simplify a radical expression. (Section 10.2)

c. Multiply and divide radical expressions. (Section 10.3)

d. Use the distributive property. (Section 1.7)

e. Apply FOIL. (Section 2.4)

f. Square a binomial. (Section 2.4)

g. Multiply expressions of the form $(a + b)(a - b)$. (Section 2.4)

h. Reduce an expression to lowest terms. (Section 6.1)

GETTING READY FOR SECTION 10.4

Add the following like terms:

1. $4x - 8x$ **2.** $3a^2b - 7a^2b$ **3.** Simplify $3\sqrt{80}$.

Multiply or divide the following radical expressions:

4. $\sqrt{3} \cdot \sqrt{15}$ **5.** $\dfrac{\sqrt{140}}{\sqrt{5}}$

6. Simplify $2a(3 + 5a)$, using the distributive property.

7. Multiply $(3x + y)(2x - 5y)$, using FOIL.   **8.** Simplify $(5x + 2y)^2$.

9. Multiply $(x - 3y)(x + 3y)$.   **10.** Reduce $\dfrac{6 + 12x}{6}$ to lowest terms.

Introduction

We have found that $\sqrt[n]{a} \cdot \sqrt[n]{b} = \sqrt[n]{ab}$ and $\dfrac{\sqrt[n]{a}}{\sqrt[n]{b}} = \sqrt[n]{\dfrac{a}{b}}$, provided that $\sqrt[n]{a}$ and $\sqrt[n]{b}$ are real numbers. Does $\sqrt[n]{a} + \sqrt[n]{b} = \sqrt[n]{a + b}$?

Answers: Getting Ready: 1. $-4x$ **2.** $-4a^2b$ **3.** $12\sqrt{5}$ **4.** $3\sqrt{5}$ **5.** $2\sqrt{7}$ **6.** $6a + 10a^2$
7. $6x^2 - 13xy - 5y^2$ **8.** $25x^2 + 20xy + 4y^2$ **9.** $x^2 - 9y^2$ **10.** $1 + 2x$

Calculator Exploration Activity I (Optional)

Use your calculator to evaluate columns A and B, and round your answers to the nearest ten thousandth. Compare the results of the corresponding lines.

Column A	Column B
a. $\sqrt{2} + \sqrt{3} =$ _____	$\sqrt{5} =$ _____
b. $\sqrt{5} + \sqrt{12} =$ _____	$\sqrt{17} =$ _____

c. $\sqrt{23} + \sqrt{19} =$ _____ $\sqrt{42} =$ _____
d. $\sqrt{31} + \sqrt{43} =$ _____ $\sqrt{74} =$ _____

1. Based on your answers to the corresponding lines in columns A and B, does $\sqrt{a} + \sqrt{b} = \sqrt{a + b}$?

Calculator Exploration Activity II (Optional)

Use your calculator to evaluate columns A and B. Round your answers to the nearest ten thousandth. Compare the results of the corresponding lines.

Column A	Column B
a. $2\sqrt{2} + 3\sqrt{2} =$ _____	$5\sqrt{2} =$ _____
b. $4\sqrt{6} + 5\sqrt{6} =$ _____	$9\sqrt{6} =$ _____

c. $-4\sqrt{3} + 7\sqrt{3} =$ _____ $3\sqrt{3} =$ _____
d. $-3\sqrt{5} - 6\sqrt{5} =$ _____ $-9\sqrt{5} =$ _____

2. By looking at the results of the corresponding lines, state a rule for adding radical expressions with the same root indices and the same radicands. Following is justification for this rule:

OBJECTIVE **A** Adding and subtracting like radicals

Addition and subtraction of radicals is done in much the same manner as addition and subtraction of variables. Remember that like terms have the same variables and the same exponents on the variables. Similarly, **like radical expressions** have the same radicand and the same root indices. The distributive property is used to add like terms and will be used to add like radicals. Study the following comparisons of addition of terms and addition of radical expressions:

a. $3x + 5x$ $3\sqrt{2} + 5\sqrt{2}$

These are like terms, so we can apply the distributive property and add.

These are like radical expressions, so we can apply the distributive property and add.

$$3x + 5x = (3 + 5)x = 8x$$ $$3\sqrt{2} + 5\sqrt{2} = (3 + 5)\sqrt{2} = 8\sqrt{2}$$

b. $3x + 5y$ $3\sqrt{2} + 5\sqrt{3}$

These are not like terms, because the variables are different. Consequently, we cannot add these terms.

These are not like radical expressions, because the radicands are different. Consequently, we cannot add these radicals.

c. $3x^2 + 5x^3$ $3\sqrt{2} + 5\sqrt[3]{2}$

These are not like terms, because the exponents on the variables are different. Consequently, we cannot add these terms.

These are not like radical expressions, because the root indices are different. Consequently, we cannot add these radicals.

Answers: Calculation Exploration Activity I: 1. no   **Calculator Exploration Activity II: 2.** Add the coefficients and express the radical portions unchanged.

Note

Since $a - b = a + (-b)$, we do not need a separate rule for differences.

Instead of applying the distributive property each time we combine like terms, we observe that you add the coefficients and leave the variable portion unchanged. Similarly, we have the following rule for adding like radical expressions:

Rule: Adding Like Radical Expressions

To add like radical expressions, add the coefficients and leave the radical portion unchanged.

Example 1 | **Find the following sums of radicals:**

Be Careful

A very common error is to add radicands when finding the sums of radicals. The following illustrates that this cannot be done:

$$\sqrt{9} + \sqrt{16} \neq \sqrt{9 + 16}$$
$$3 + 4 \neq \sqrt{25}$$
$$7 \neq 5$$

Never add radicands!

a. $7\sqrt{5} + 2\sqrt{5} =$ Add the coefficients and leave the radicals unchanged.
$9\sqrt{5}$ Therefore, $7\sqrt{5} + 2\sqrt{5} = 9\sqrt{5}$.

b. $4\sqrt{x} - 8\sqrt{x} =$ Add the coefficients and leave the radicals unchanged.
$-4\sqrt{x}$ Therefore, $4\sqrt{x} - 8\sqrt{x} = -4\sqrt{x}$.

c. $6\sqrt[3]{4} + \sqrt[3]{4} =$ Add the coefficients and leave the radicals unchanged.
$7\sqrt[3]{4}$ Therefore, $6\sqrt[3]{4} + \sqrt[3]{4} = 7\sqrt[3]{4}$.

d. $6x\sqrt[4]{3} - 2x\sqrt[4]{3} =$ Add the coefficients and leave the radicals unchanged.
$4x\sqrt[4]{3}$ Therefore, $6x\sqrt[4]{3} - 2x\sqrt[4]{3} = 4x\sqrt[4]{3}$.

e. $3a^2\sqrt{2a} - 7a^2\sqrt{2a} =$ Add the coefficients and leave the radicals unchanged.
$-4a^2\sqrt{2a}$ Therefore, $3a^2\sqrt{2a} - 7a^2\sqrt{2a} = -4a^2\sqrt{2a}$.

PRACTICE EXERCISE 1

Find the following sums of radicals:

a. $5\sqrt{6} + 2\sqrt{6}$ **b.** $3\sqrt{c} - 9\sqrt{c}$ **c.** $5\sqrt[4]{5} + \sqrt[4]{5}$

d. $7y\sqrt[3]{7} - 12y\sqrt[3]{7}$ **e.** $7y^3\sqrt{6y} + 8y^3\sqrt{6y}$

If you need more practice, do the following Additional Practice Exercises.

Additional Practice Exercise 1   Find the following sums of radicals:

a. $4\sqrt{5} + 6\sqrt{5}$ **b.** $3\sqrt{y} - 7\sqrt{y}$ **c.** $9\sqrt[3]{3x} - \sqrt[3]{3x}$

d. $z\sqrt{5} + 8z\sqrt{5}$ **e.** $5a^2\sqrt{7a} + 3a^2\sqrt{7a}$

OBJECTIVE **B** Simplifying radicals and then adding and/or subtracting like radicals

It is often necessary to simplify radicands before they can be added.

Example 2 | **Find the following sums, assuming that all variables represent nonnegative quantities:**

a. $\sqrt{12} + \sqrt{75} =$ Rewrite 12 as $4 \cdot 3$ and 75 as $25 \cdot 3$.
$\sqrt{4 \cdot 3} + \sqrt{25 \cdot 3} =$ Apply the product rule.
$\sqrt{4} \cdot \sqrt{3} + \sqrt{25} \cdot \sqrt{3} =$ Find $\sqrt{4}$ and $\sqrt{25}$.
$2\sqrt{3} + 5\sqrt{3} =$ Add like radicals.
$7\sqrt{3}$ Therefore, $\sqrt{12} + \sqrt{75} = 7\sqrt{3}$.

Answers: Practice Exercise 1: a. $7\sqrt{6}$ **b.** $-6\sqrt{c}$ **c.** $6\sqrt[4]{5}$ **d.** $-5y\sqrt[3]{7}$ **e.** $15y^3\sqrt{6y}$
Additional Practice 1: a. $10\sqrt{5}$ **b.** $-4\sqrt{y}$ **c.** $8\sqrt[3]{3x}$ **d.** $9z\sqrt{5}$ **e.** $8a^2\sqrt{7a}$

b. $3\sqrt{80} - 2\sqrt{245} =$ Rewrite 80 as $16 \cdot 5$ and 245 as $49 \cdot 5$.

$3\sqrt{16 \cdot 5} - 2\sqrt{49 \cdot 5} =$ Apply the product rule.

$3\sqrt{16} \cdot \sqrt{5} - 2\sqrt{49} \cdot \sqrt{5} =$ Find $\sqrt{16}$ and $\sqrt{49}$.

$3 \cdot 4\sqrt{5} - 2 \cdot 7\sqrt{5} =$ Multiply.

$12\sqrt{5} - 14\sqrt{5} =$ Add like radicals.

$-2\sqrt{5}$ Therefore, $3\sqrt{80} - 2\sqrt{245} = -2\sqrt{5}$.

Note

You can verify the answers to Examples 2a and 2b by using a calculator.

c. $\sqrt{48x^3} + \sqrt{12x^3} =$ Rewrite $48x^3$ as $16x^2 \cdot 3x$ and $12x^3$ as $4x^2 \cdot 3x$.

$\sqrt{16x^2 \cdot 3x} + \sqrt{4x^2 \cdot 3x} =$ Apply the product rule.

$\sqrt{16x^2} \cdot \sqrt{3x} + \sqrt{4x^2} \cdot \sqrt{3x} =$ Find $\sqrt{16x^2}$ and $\sqrt{4x^2}$.

$4x\sqrt{3x} + 2x\sqrt{3x} =$ Add like radicals.

$6x\sqrt{3x}$ Therefore, $\sqrt{48x^3} + \sqrt{12x^3} = 6x\sqrt{3x}$.

PRACTICE EXERCISE 2

Find the following sums, assuming that all variables represent nonnegative quantities:

a. $\sqrt{45} + \sqrt{20}$ **b.** $4\sqrt{24} - 6\sqrt{54}$ **c.** $\sqrt{50x^5} - \sqrt{18x^5}$

If you need more practice, do the following Additional Practice Exercises.

Additional Practice Exercise 2   Find the following sums of radicals, assuming that all variables represent nonnegative quantities:

a. $\sqrt{72} + \sqrt{32}$ **b.** $5\sqrt{48} - 2\sqrt{75}$ **c.** $\sqrt{28a^5} - \sqrt{63a^5}$

OBJECTIVE Simplifying radical expressions that contain mixed operations

Now that we know how to perform the operations of addition, subtraction, multiplication, and division with radicals, it is possible for us to simplify radical expressions that contain more than one operation. The order of operations applies to radical expressions just as it does to any other numbers. All answers should be left in simplest form, as described in the following box:

Procedure: Simplest Form of Radicals

A radical expression is in its simplest form if

1. All rational roots have been found.

2. No factors that can be removed are left in a radicand. For example, no square root may have a radicand with a perfect square factor.

3. All possible products, quotients, sums, and differences have been found.

4. There are no radicals in the denominator of a fraction. (This will be discussed in Section 10.5.)

Answers: **Practice Exercise 2:** **a.** $5\sqrt{5}$ **b.** $-10\sqrt{6}$ **c.** $2x^2\sqrt{2x}$ **Additional Practice 2:** **a.** $10\sqrt{2}$
b. $10\sqrt{3}$ **c.** $-a^2\sqrt{7a}$

Example 3	**Perform the following operations:**

a. $\sqrt{2}\cdot\sqrt{10} + \sqrt{3}\cdot\sqrt{15} =$ Apply the product rule.

$\sqrt{2\cdot10} + \sqrt{3\cdot15} =$ Find the products.

$\sqrt{20} + \sqrt{45} =$ Rewrite 20 as $4\cdot5$ and 45 as $9\cdot5$.

$\sqrt{4\cdot5} + \sqrt{9\cdot5} =$ Apply the product rule.

$\sqrt{4}\cdot\sqrt{5} + \sqrt{9}\cdot\sqrt{5} =$ Find $\sqrt{9}$ and $\sqrt{4}$.

$2\sqrt{5} + 3\sqrt{5} =$ Add like radicals.

$5\sqrt{5}$ Therefore, $\sqrt{2}\cdot\sqrt{10} + \sqrt{3}\cdot\sqrt{15} = 5\sqrt{5}$.

b. $\dfrac{\sqrt{54}}{\sqrt{3}} + \sqrt{32} =$ Apply the quotient rule and rewrite 32 as $16\cdot2$.

$\sqrt{\dfrac{54}{3}} + \sqrt{16\cdot2} =$ Divide 54 by 3. Apply the product rule to $\sqrt{16\cdot2}$.

$\sqrt{18} + \sqrt{16}\cdot\sqrt{2} =$ Rewrite 18 as $9\cdot2$ and find $\sqrt{16}$.

$\sqrt{9\cdot2} + 4\sqrt{2} =$ Apply the product rule to $\sqrt{9\cdot2}$.

$\sqrt{9}\cdot\sqrt{2} + 4\sqrt{2} =$ Find $\sqrt{9}$.

$3\sqrt{2} + 4\sqrt{2} =$ Add like radicals.

$7\sqrt{2}$ Therefore, $\dfrac{\sqrt{54}}{\sqrt{3}} + \sqrt{32} = 7\sqrt{2}$.

PRACTICE EXERCISE 3

Perform the following operations:

a. $2\sqrt{3}\cdot\sqrt{6} + 4\sqrt{7}\cdot\sqrt{14}$

b. $\sqrt{63} + \dfrac{\sqrt{140}}{\sqrt{5}}$

If you need more practice, do the following Additional Practice Exercises.

Additional Practice Exercise 3   Perform the following operations:

a. $\sqrt{5}\cdot\sqrt{15} + 2\sqrt{7}\cdot\sqrt{21}$

b. $\dfrac{\sqrt{72}}{\sqrt{6}} + \sqrt{75}$

OBJECTIVE Finding products of radical expressions involving sums or differences

Products involving sums and differences of radicals are found in much the same way as products of polynomials. For example, we apply the distributive property to find $x(x + 4) = x\cdot x + x\cdot4 = x^2 + 4x$. Likewise, we would apply the distributive property to find $\sqrt{3}(\sqrt{2} + 3) = \sqrt{3}\cdot\sqrt{2} + \sqrt{3}\cdot3 = \sqrt{6} + 3\sqrt{3}$. We apply FOIL to find $(x + 2)(x - 3) = x^2 - 3x + 2x - 6 = x^2 - x - 6$. Likewise, we would apply FOIL to find $(\sqrt{3} + 2)(\sqrt{3} - 3) = \sqrt{3}\cdot\sqrt{3} - 3\sqrt{3} + 2\sqrt{3} - 6 = \sqrt{9} - \sqrt{3} - 6 = 3 - \sqrt{3} - 6 = -\sqrt{3} - 3$. We apply the rule for squaring a binomial to find $(a + b)^2 = a^2 + 2\cdot a\cdot b + b^2$. Likewise, we would use the same rule to find $(\sqrt{3} + 2)^2 = (\sqrt{3})^2 + 2\cdot2\cdot\sqrt{3} + 2^2 = 3 + 4\sqrt{3} + 4 = 7 + 4\sqrt{3}$.

Answers: Practice Exercise 3: a. $34\sqrt{2}$ **b.** $5\sqrt{7}$ **Additional Practice 3: a.** $19\sqrt{3}$ **b.** $7\sqrt{3}$

Example 4   **Find the following products, assuming that all variables have nonnegative values:**

a. $\sqrt{3}(\sqrt{3} + \sqrt{15}) =$ Apply the distributive property.

$\sqrt{3} \cdot \sqrt{3} + \sqrt{3} \cdot \sqrt{15} =$ Apply the product rule.

$\sqrt{3 \cdot 3} + \sqrt{3 \cdot 15} =$ Find the products.

$\sqrt{9} + \sqrt{45} =$ Find $\sqrt{9}$ and simplify $\sqrt{45}$.

$3 + 3\sqrt{5}$ Therefore, $\sqrt{3}(\sqrt{3} + \sqrt{15}) = 3 + 3\sqrt{5}$.

b. $2\sqrt{6}(3 + 5\sqrt{5}) =$ Apply the distributive property.

$2\sqrt{6} \cdot 3 + 2\sqrt{6} \cdot 5\sqrt{5} =$ Apply the product rule.

$2 \cdot 3\sqrt{6} + 2 \cdot 5\sqrt{6 \cdot 5} =$ Find the products.

$6\sqrt{6} + 10\sqrt{30}$ Therefore, $2\sqrt{6}(3 + 5\sqrt{5}) = 6\sqrt{6} + 10\sqrt{30}$.

c. $(2 + \sqrt{3})(\sqrt{5} - \sqrt{6}) =$ Multiply, using FOIL.

$2\sqrt{5} - 2\sqrt{6} + \sqrt{3} \cdot \sqrt{5} - \sqrt{3} \cdot \sqrt{6} =$ Apply the product rule.

$2\sqrt{5} - 2\sqrt{6} + \sqrt{15} - \sqrt{18} =$ Simplify $\sqrt{18}$.

$2\sqrt{5} - 2\sqrt{6} + \sqrt{15} - 3\sqrt{2}$ Therefore, $(2 + \sqrt{3})(\sqrt{5} - \sqrt{6}) = 2\sqrt{5} - 2\sqrt{6} + \sqrt{15} - 3\sqrt{2}$.

d. $(3\sqrt{x} + \sqrt{y})(2\sqrt{x} - 5\sqrt{y}) =$ Multiply, using FOIL.

$3\sqrt{x} \cdot 2\sqrt{x} - 3\sqrt{x} \cdot 5\sqrt{y} + \sqrt{y} \cdot 2\sqrt{x} - \sqrt{y} \cdot 5\sqrt{y} =$ Apply the product rule.

$6x - 15\sqrt{xy} + 2\sqrt{xy} - 5y =$ Add like radicals.

$6x - 13\sqrt{xy} - 5y$ Therefore, $(3\sqrt{x} + \sqrt{y})(2\sqrt{x} - 5\sqrt{y}) = 6x - 13\sqrt{xy} - 5y$.

e. $(5 + \sqrt{3})^2 =$ Apply $(a + b)^2 = a^2 + 2ab + b^2$.

$5^2 + 2 \cdot 5\sqrt{3} + (\sqrt{3})^2 =$ Simplify.

$25 + 10\sqrt{3} + 3 =$ Add 25 and 3.

$28 + 10\sqrt{3}$ Therefore, $(5 + \sqrt{3})^2 = 28 + 10\sqrt{3}$.

PRACTICE EXERCISE 4

Find the following products, assuming that all variables represent nonnegative quantities.

a. $\sqrt{7}(\sqrt{3} + \sqrt{7})$ **b.** $2\sqrt{11}(3 - 3\sqrt{6})$ **c.** $(4 - \sqrt{3})(\sqrt{2} - \sqrt{11})$

d. $(2\sqrt{a} + 3\sqrt{b})(\sqrt{a} - 3\sqrt{b})$ **e.** $(3 + \sqrt{5})^2$

If you need more practice, do the following Additional Practice Exercises.

Additional Practice Exercise 4   Find the following products:

a. $\sqrt{5}(\sqrt{6} - \sqrt{2})$ **b.** $7\sqrt{13}(3\sqrt{2} - 4)$ **c.** $(\sqrt{3} - 5)(\sqrt{7} - \sqrt{5})$

d. $(3\sqrt{n} + 2\sqrt{m})(5\sqrt{n} - 3\sqrt{m})$ **e.** $(4 - \sqrt{3})^2$

One form of products of sums and differences of radicals is of particular interest. Recall that the product of binomials of the form $(a + b)(a - b) = a^2 + ab - ab - b^2 = a^2 - b^2$. For example, $(x + 3)(x - 3) = x^2 - 3^2 = x^2 - 9$. If square roots are involved, expressions of the form $a + b$ and $a - b$ are called **conjugates**. For example, $3 + \sqrt{2}$ and $3 - \sqrt{2}$ are conjugates, and their product is $(3 + \sqrt{2})(3 - \sqrt{2}) = 3^2 - (\sqrt{2})^2 = 9 - 2 = 7$. (Remember that $(\sqrt{x})^2 = x$.)

Example 5 | **Find the following products, using the fact that $(a + b)(a - b) = a^2 - b^2$:**

a. $(2 + \sqrt{3})(2 - \sqrt{3}) =$ Apply $(a + b)(a - b) = a^2 - b^2$.

$2^2 - (\sqrt{3})^2 =$ $2^2 = 4$ and $(\sqrt{3})^2 = 3$.

$4 - 3 =$ Add 4 and -3.

1 Therefore, $(2 + \sqrt{3})(2 - \sqrt{3}) = 1$.

b. $(\sqrt{5} - 3\sqrt{3})(\sqrt{5} + 3\sqrt{3}) =$ Apply $(a + b)(a - b) = a^2 - b^2$.

$(\sqrt{5})^2 - (3\sqrt{3})^2 =$ $(\sqrt{5})^2 = 5$ and $(3\sqrt{3})^2 = 3^2(\sqrt{3})^2 = 9 \cdot 3$.

$5 - 9 \cdot 3 =$ $9 \cdot 3 = 27$.

$5 - 27 =$ Add 5 and -27.

-22 Therefore, $(\sqrt{5} - 3\sqrt{3})(\sqrt{5} + 3\sqrt{3}) = -22$.

> **Note**
>
> The product of conjugates always results in a rational number. This will be useful in the next section on rationalizing the denominator.

PRACTICE EXERCISE 5

Find the following products:

a. $(4 + \sqrt{10})(4 - \sqrt{10})$

b. $(3\sqrt{5} + 2\sqrt{6})(3\sqrt{5} - 2\sqrt{6})$

If you need more practice, do the following Additional Practice Exercises.

Additional Practice Exercise 5   Find the following products:

a. $(\sqrt{7} - 3)(\sqrt{7} + 3)$

b. $(5\sqrt{7} - \sqrt{3})(5\sqrt{7} + \sqrt{3})$

In Section 11.3, it will be necessary to simplify and reduce expressions of the form $\dfrac{a + b\sqrt{c}}{d}$ when solving quadratic equations. It will be helpful if we practice this procedure at this time. Remember, when reducing fractions, you must divide the numerator and the denominator by the greatest common factor of both the numerator and the denominator.

Example 6 | **Simplify the following:**

a. $\dfrac{4 + 2\sqrt{5}}{6} =$ Remove the common factor of 2 from the numerator.

$\dfrac{2(2 + \sqrt{5})}{6} =$ Divide the numerator and the denominator by 2.

$\dfrac{2 + \sqrt{5}}{3}$ Therefore, $\dfrac{4 + 2\sqrt{5}}{6} = \dfrac{2 + \sqrt{5}}{3}$.

b. $\dfrac{6 + 4\sqrt{18}}{6} =$ Simplify $\sqrt{18}$.

$\dfrac{6 + 4 \cdot 3\sqrt{2}}{6} =$ Multiply 4 and 3.

$\dfrac{6 + 12\sqrt{2}}{6} =$ Remove the common factor of 6 from the numerator.

$\dfrac{6(1 + 2\sqrt{2})}{6} =$ Divide the numerator and the denominator by 6.

$1 + 2\sqrt{2}$ Therefore, $\dfrac{6 + 4\sqrt{18}}{6} = 1 + 2\sqrt{2}$.

Answers: Practice Exercise 5: a. 6 **b.** 21   **Additional Practice 5: a.** -2 **b.** 172

PRACTICE EXERCISE 6

Simplify the following:

a. $\dfrac{8 - 4\sqrt{2}}{2}$

b. $\dfrac{10 + 3\sqrt{75}}{5}$

If you need more practice, do the following Additional Practice Exercises.

Additional Practice Exercise 6 Simplify the following:

a. $\dfrac{12 + 6\sqrt{3}}{3}$

b. $\dfrac{4 - 3\sqrt{8}}{4}$

Exercise Set 10.4 MyMathLab®

For Extra Help

Find the sums of radicals. Assume that all variables have nonnegative values. (See Example 1.)

1. $2\sqrt{7} + 5\sqrt{7}$

2. $3\sqrt{3} + 8\sqrt{3}$

3. $9\sqrt{6} - 15\sqrt{6}$

4. $2\sqrt{10} - 13\sqrt{10}$

5. $7\sqrt{a} + 2\sqrt{a}$

6. $5\sqrt{y} + 7\sqrt{y}$

7. $-5\sqrt{n} + 2\sqrt{n}$

8. $-8\sqrt{m} + 4\sqrt{m}$

9. $6x\sqrt[3]{9} - 3x\sqrt[3]{9}$

10. $4y\sqrt[3]{3} - y\sqrt[3]{3}$

11. $6x^2\sqrt[4]{5x} + 12x^2\sqrt[4]{5x}$

12. $3y^3\sqrt[4]{8y} - 9y^3\sqrt[4]{8y}$

Simplify the radicals and then find the sums. Assume that all variables have nonnegative values. (See Example 2.)

13. $\sqrt{27} + \sqrt{12}$

14. $\sqrt{8} + \sqrt{32}$

15. $\sqrt{48} - \sqrt{75}$

16. $\sqrt{20} - \sqrt{80}$

17. $\sqrt{80y} - \sqrt{125y}$

18. $\sqrt{27x} - \sqrt{75x}$

19. $\sqrt{24n^2} + \sqrt{96n^2}$

20. $\sqrt{48x^2} + \sqrt{75x^2}$

21. $\sqrt{18y^5} + \sqrt{72y^5}$

22. $\sqrt{48a^7} + \sqrt{108a^7}$

23. $12\sqrt{5} + \sqrt{45}$

24. $14\sqrt{6} - \sqrt{24}$

25. $\sqrt{80} - 4\sqrt{45}$

26. $\sqrt{20} - 2\sqrt{180}$

27. $3\sqrt{96} - 2\sqrt{54}$

28. $5\sqrt{63} + 2\sqrt{28}$

29. $6\sqrt{48a^3} - 2\sqrt{75a^3}$

30. $4\sqrt{98y^5} - 7\sqrt{128y^5}$

31. $\sqrt{150} - \sqrt{54} + \sqrt{24}$

32. $\sqrt{20} + \sqrt{125} - \sqrt{80}$

33. $2\sqrt{8} - 3\sqrt{48} + 2\sqrt{98} - \sqrt{75}$

34. $3\sqrt{216} - \sqrt{147} - 4\sqrt{96} - \sqrt{108}$

35. $4\sqrt{72x^2y} - 2x\sqrt{128y} + 5\sqrt{32x^2y}$

36. $3\sqrt{24a^3b^2} - 2a\sqrt{150ab^2} + 4ab\sqrt{96a}$

Simplify the following. (See Example 3.)

37. $\sqrt{3} \cdot \sqrt{15} + \sqrt{8} \cdot \sqrt{10}$

38. $\sqrt{6} \cdot \sqrt{8} + \sqrt{5} \cdot \sqrt{15}$

39. $3\sqrt{3} \cdot \sqrt{18} - 4\sqrt{18} \cdot \sqrt{12}$

40. $6\sqrt{8} \cdot \sqrt{10} - 5\sqrt{12} \cdot \sqrt{15}$

Answers: Practice Exercise 6: a. $4 - 2\sqrt{2}$ **b.** $2 + 3\sqrt{3}$ **Additional Practice 6: a.** $4 + 2\sqrt{3}$ **b.** $\dfrac{2 - 3\sqrt{2}}{2}$

41. $\dfrac{\sqrt{40}}{\sqrt{5}} + \sqrt{50}$

42. $\dfrac{\sqrt{60}}{\sqrt{5}} + \sqrt{48}$

43. $\dfrac{\sqrt{540}}{\sqrt{3}} - 4\sqrt{125}$

44. $\dfrac{\sqrt{288}}{\sqrt{6}} - 6\sqrt{108}$

Simplify the products. Assume that all variables have nonnegative values. (See Example 4.)

45. $\sqrt{2}(3 + \sqrt{2})$

46. $\sqrt{5}(4 - \sqrt{5})$

47. $\sqrt{3}(\sqrt{3} - \sqrt{15})$

48. $\sqrt{6}(\sqrt{6} + \sqrt{2})$

49. $\sqrt{5}(\sqrt{3} + 2\sqrt{15})$

50. $\sqrt{7}(\sqrt{5} - 3\sqrt{14})$

51. $4\sqrt{3}(2\sqrt{3} - 4\sqrt{6})$

52. $6\sqrt{2}(3\sqrt{2} + 2\sqrt{10})$

53. $(3 + \sqrt{5})(4 - \sqrt{2})$

54. $(3 + \sqrt{7})(7 - \sqrt{3})$

55. $(3 + \sqrt{x})(2 + \sqrt{x})$

56. $(5 + \sqrt{a})(2 + \sqrt{a})$

57. $(4 - 2\sqrt{6})(2 - 5\sqrt{6})$

58. $(9 - 2\sqrt{2})(1 - 4\sqrt{2})$

59. $(2 + 3\sqrt{3})(3 + 5\sqrt{2})$

60. $(7 - 3\sqrt{5})(2 - 2\sqrt{10})$

61. $(\sqrt{2} + \sqrt{3})(\sqrt{3} + \sqrt{5})$

62. $(\sqrt{5} + \sqrt{2})(\sqrt{2} + \sqrt{7})$

63. $(\sqrt{x} + \sqrt{y})(\sqrt{x} - 2\sqrt{y})$

64. $(2\sqrt{a} + \sqrt{b})(\sqrt{a} + \sqrt{b})$

65. $(\sqrt{3} + 2\sqrt{2})(\sqrt{3} - 4\sqrt{2})$

66. $(\sqrt{5} - 4\sqrt{6})(\sqrt{5} + 2\sqrt{6})$

67. $(5\sqrt{2} - 3\sqrt{5})(2\sqrt{2} + 6\sqrt{5})$

68. $(2\sqrt{2} + 3\sqrt{3})(3\sqrt{2} - 4\sqrt{3})$

69. $(4\sqrt{2} + 2\sqrt{5})(3\sqrt{7} - 3\sqrt{3})$

70. $(8\sqrt{2} - 2\sqrt{3})(2\sqrt{5} + 3\sqrt{10})$

71. $(2\sqrt{a} + 3\sqrt{b})(4\sqrt{a} - \sqrt{b})$

72. $(\sqrt{m} - 4\sqrt{n})(2\sqrt{m} - 3\sqrt{n})$

73. $(4 + \sqrt{6})^2$

74. $(1 - \sqrt{2})^2$

75. $(6 + \sqrt{3})^2$

76. $(5 + \sqrt{2})^2$

77. $(2 + 2\sqrt{3})^2$

78. $(3 + 2\sqrt{5})^2$

79. $(6 - \sqrt{x})^2$

80. $(3 + \sqrt{y})^2$

81. $(3 - \sqrt{x - 2})^2$

82. $(4 + \sqrt{x - 4})^2$

Find the products of the given conjugates. Assume that all variables have nonnegative values. (See Example 5.)

83. $(2 + \sqrt{3})(2 - \sqrt{3})$

84. $(3 + \sqrt{5})(3 - \sqrt{5})$

85. $(4 + 2\sqrt{3})(4 - 2\sqrt{3})$

86. $(5 + 3\sqrt{3})(5 - 3\sqrt{3})$

87. $(3 + \sqrt{x})(3 - \sqrt{x})$

88. $(5 + \sqrt{y})(5 - \sqrt{y})$

89. $(\sqrt{2} + 4)(\sqrt{2} - 4)$

90. $(\sqrt{7} - 6)(\sqrt{7} + 6)$

91. $(5\sqrt{2} + 5)(5\sqrt{2} - 5)$

92. $(6\sqrt{3} - 2)(6\sqrt{3} + 2)$

93. $\left(\sqrt{3} + \sqrt{2}\right)\left(\sqrt{3} - \sqrt{2}\right)$

94. $\left(\sqrt{5} - \sqrt{3}\right)\left(\sqrt{5} + \sqrt{3}\right)$

95. $\left(\sqrt{x} + \sqrt{y}\right)\left(\sqrt{x} - \sqrt{y}\right)$

96. $\left(\sqrt{a} - \sqrt{b}\right)\left(\sqrt{a} + \sqrt{b}\right)$

97. $\left(3\sqrt{7} + \sqrt{13}\right)\left(3\sqrt{7} - \sqrt{13}\right)$

98. $\left(4\sqrt{5} - 3\sqrt{2}\right)\left(4\sqrt{5} + 3\sqrt{2}\right)$

99. $\left(3\sqrt{5} - 2\sqrt{6}\right)\left(3\sqrt{5} + 2\sqrt{6}\right)$

100. $\left(5\sqrt{2} - 2\sqrt{7}\right)\left(5\sqrt{2} + 2\sqrt{7}\right)$

Simplify the following. (See Example 6.)

101. $\dfrac{6 + 3\sqrt{2}}{3}$

102. $\dfrac{8 + 4\sqrt{5}}{4}$

103. $\dfrac{10 + 20\sqrt{6}}{5}$

104. $\dfrac{12 - 18\sqrt{2}}{3}$

105. $\dfrac{3 + 2\sqrt{45}}{3}$

106. $\dfrac{6 - 3\sqrt{20}}{6}$

107. $\dfrac{9 - 2\sqrt{54}}{3}$

108. $\dfrac{10 - 3\sqrt{50}}{5}$

109. $\dfrac{4 + 6\sqrt{3}}{8}$

110. $\dfrac{6 - 15\sqrt{5}}{9}$

111. $\dfrac{8 - 4\sqrt{6}}{6}$

112. $\dfrac{12 + 6\sqrt{7}}{9}$

Challenge Exercises (113–120)

Perform the following operations on radical expressions whose indices are greater than 2:

113. $\sqrt[3]{16} + \sqrt[3]{54}$

114. $\sqrt[3]{24} + \sqrt[3]{81}$

115. $\sqrt[3]{250x^4y^5} + \sqrt[3]{128x^4y^5}$

116. $\sqrt[3]{108a^6b^4} + \sqrt[3]{32a^6b^4}$

117. $\sqrt[4]{48} + \sqrt[4]{243}$

118. $\sqrt[4]{64} - \sqrt[4]{324}$

119. $\left(\sqrt[3]{2} + \sqrt[3]{4}\right)\left(2\sqrt[3]{2} - 3\sqrt[3]{4}\right)$

120. $\left(\sqrt[3]{9} - 2\sqrt[3]{3}\right)\left(2\sqrt[3]{9} + 3\sqrt[3]{3}\right)$

Writing Exercises (121–124)

121. Compare finding the product $(2x + 3y)(3x - 4y)$ with finding the product $\left(2\sqrt{2} + 3\sqrt{5}\right)$ $\left(3\sqrt{2} - 4\sqrt{5}\right)$. How are they alike and how are they different?

122. Compare adding $3x + 2x$ with adding $3\sqrt{2} + 2\sqrt{2}$.

123. Why is the product of conjugates always a rational number instead of an irrational number?

124. Why is $\sqrt{a} + \sqrt{b} \neq \sqrt{a + b}$? Use examples if necessary.

Section 10.5 — Rationalizing the Denominator

OBJECTIVES *When you complete this section, you will be able to:*

A Rationalize the denominator when the denominator contains a single radical that is a square root or a cube root.

B Rationalize the denominator when the radicand is a fraction.

C Rationalize denominators of the form $a \pm b$, where a and/or b are square roots.

PREREQUISITE SKILLS *Before beginning this section, you should be able to:*

a. Multiply radical expressions. (Section 10.3)

b. Find square and cube roots. (Section 10.1)

c. Write a whole number in terms of its prime factors. (Section 5.1)

GETTING READY FOR SECTION 10.5

Multiply the following radical expressions:

1. $\sqrt{2x} \cdot \sqrt{8x^3}$ **2.** $\sqrt[3]{9x^2} \cdot \sqrt[3]{3x}$ **3.** $3(4 - \sqrt{5})$ **4.** $(4 + \sqrt{2})(4 - \sqrt{2})$

5. $(1 + \sqrt{y})(\sqrt{x} + \sqrt{y})$

Find the following roots:

6. $\sqrt{16x^4}$ **7.** $\sqrt[3]{27x^3}$

8. Write $9a^3$ in terms of prime factors.   **9.** Simplify $\dfrac{10 + 6\sqrt{2}}{14}$.

Introduction

In Section 10.2, we discussed certain conditions that must be satisfied in order for a radical to be in the simplest form. We did not discuss the condition that the denominator of a fraction cannot contain a radical that is an irrational number. The process of making the denominator a rational number is called **rationalizing the denominator.** One of the reasons for rationalizing the denominator is to make it easier to find decimal approximations if a calculator is not available.

Consider the following two methods of approximating $\dfrac{2}{\sqrt{3}}$:

Method 1

$\dfrac{2}{\sqrt{3}}$ Replace $\sqrt{3}$ with 1.732.

$\dfrac{2}{1.732}$ Divide.

1.155 Therefore, $\dfrac{2}{\sqrt{3}} \approx 1.155$.

Method 2

$\dfrac{2}{\sqrt{3}}$ Multiply by $\dfrac{\sqrt{3}}{\sqrt{3}}$.

$\dfrac{2}{\sqrt{3}} \cdot \dfrac{\sqrt{3}}{\sqrt{3}}$ Simplify.

$\dfrac{2\sqrt{3}}{\sqrt{9}}$ Replace $\sqrt{3}$ with 1.732 and find $\sqrt{9} = 3$.

$\dfrac{2(1.732)}{3}$ Simplify.

1.155 Therefore, $\dfrac{2}{\sqrt{3}} = 1.155$

Answers: **Getting Ready:** **1.** $4x^2$ **2.** $3x$ **3.** $12 - 3\sqrt{5}$ **4.** 14 **5.** $\sqrt{x} + \sqrt{y} + \sqrt{xy} + y$ **6.** $4x^2$ **7.** $3x$ **8.** $3 \cdot 3 \cdot a \cdot a \cdot a$ **9.** $\dfrac{5 + 3\sqrt{2}}{7}$

Method 2 is a little longer, but the arithmetic is much easier, since it is easier to multiply by 2 and divide by 3 than to divide by 1.732. With the use of calculators, ease in approximating radical expressions is no longer an important reason for rationalizing the denominator. In more advanced math courses, there are other reasons for rationalizing the denominator (and the numerator) that make it an important skill to possess.

OBJECTIVE **Rationalizing the denominator when the denominator contains a single radical that is a square root or a cube root**

The technique used in rationalizing the denominator is to multiply the fraction by an appropriate radical expression divided by itself (hence, its value is 1) so that the denominator of the fraction becomes a radical expression whose value is rational. This means that if the denominator of the original fraction is a square root, we will multiply this fraction by a square root divided by itself (thus, its value is 1) so that the radicand in the denominator is a perfect square. If the radicand of the original fraction contains variables, we must multiply the fraction by a radical expression divided by itself that will make the exponent(s) of the variable(s) in the radicand of the denominator divisible by 2. Cube roots will be discussed later in this section.

Example 1   **Rationalize the denominators of the following fractions, assuming that all variables have positive values:**

a. $\dfrac{3}{\sqrt{5}} =$ Since $\sqrt{5} \cdot \sqrt{5} = \sqrt{25} = 5$, multiply by $\dfrac{\sqrt{5}}{\sqrt{5}}$, which equals 1.

$\dfrac{3}{\sqrt{5}} \cdot \dfrac{\sqrt{5}}{\sqrt{5}} =$ Multiply.

$\dfrac{3\sqrt{5}}{\sqrt{25}} =$ $\sqrt{25} = 5$.

$\dfrac{3\sqrt{5}}{5}$ Therefore, $\dfrac{3}{\sqrt{5}} = \dfrac{3\sqrt{5}}{5}$.

b. $\dfrac{8}{\sqrt{6}} =$ Since $\sqrt{6} \cdot \sqrt{6} = \sqrt{36} = 6$, multiply by $\dfrac{\sqrt{6}}{\sqrt{6}}$, which equals 1.

$\dfrac{8}{\sqrt{6}} \cdot \dfrac{\sqrt{6}}{\sqrt{6}} =$ Multiply.

$\dfrac{8\sqrt{6}}{\sqrt{36}} =$ $\sqrt{36} = 6$.

$\dfrac{8\sqrt{6}}{6} =$ Reduce to lowers terms.

$\dfrac{4\sqrt{6}}{3}$ Therefore, $\dfrac{8}{\sqrt{6}} = \dfrac{4\sqrt{6}}{3}$.

c. $\dfrac{\sqrt{3}}{\sqrt{8}} =$ Since $\sqrt{8} \cdot \sqrt{2} = \sqrt{16} = 4$, multiply by $\dfrac{\sqrt{2}}{\sqrt{2}}$.

$\dfrac{\sqrt{3}}{\sqrt{8}} \cdot \dfrac{\sqrt{2}}{\sqrt{2}} =$ Multiply.

$\dfrac{\sqrt{6}}{\sqrt{16}} =$ $\sqrt{16} = 4$.

$\dfrac{\sqrt{6}}{4}$ Therefore, $\dfrac{\sqrt{3}}{\sqrt{8}} = \dfrac{\sqrt{6}}{4}$.

Note

One method of determining the radicand of the radical expression by which you multiply the original fraction is to find the smallest perfect square that is divisible by the radicand of the original fraction. For example, in Example 1c the smallest perfect square that is divisible by 8 is 16 and $16 \div 8 = 2$. Therefore, the radicand of the expression that we multiply by is 2. As another example, given $\dfrac{1}{\sqrt{12}}$, the smallest perfect square divisible by 12 is 36, and $36 \div 12 = 3$. Therefore, we would multiply $\dfrac{1}{\sqrt{12}}$ by $\dfrac{\sqrt{3}}{\sqrt{3}}$.

d. $\dfrac{3}{\sqrt{x}} =$ Since $\sqrt{x} \cdot \sqrt{x} = \sqrt{x^2} = x$, multiply by $\dfrac{\sqrt{x}}{\sqrt{x}}$.

$\dfrac{3}{\sqrt{x}} \cdot \dfrac{\sqrt{x}}{\sqrt{x}} =$ Multiply.

$\dfrac{3\sqrt{x}}{\sqrt{x^2}} =$ $\sqrt{x^2} = x$, Remember, $x \geq 0$.

$\dfrac{3\sqrt{x}}{x}$ Therefore, $\dfrac{3}{\sqrt{x}} = \dfrac{3\sqrt{x}}{x}$.

Optional Topic

Alternative Method

In Section 10.2, we gave an alternative method of simplifying radicals in which we wrote the radicand in terms of prime factors. Each time a factor appeared two times, the square root was one of the two equal factors. For example, $\sqrt{2 \cdot 2} = 2$, $\sqrt{2 \cdot 2 \cdot 2 \cdot 2} = 2 \cdot 2 = 4$, and $\sqrt{x \cdot x} = x$. We illustrate this technique by reworking Examples 1a and 1c.

a. $\dfrac{3}{\sqrt{5}} =$ Since $\sqrt{5 \cdot 5} = 5$, we need another factor of 5 in the radicand of the denominator. So, multiply by $\dfrac{\sqrt{5}}{\sqrt{5}}$.

$\dfrac{3}{\sqrt{5}} \cdot \dfrac{\sqrt{5}}{\sqrt{5}} =$ Multiply.

$\dfrac{3\sqrt{5}}{\sqrt{5 \cdot 5}} =$ $\sqrt{5 \cdot 5} = 5$.

$\dfrac{3\sqrt{5}}{5}$ Therefore, $\dfrac{3}{\sqrt{5}} = \dfrac{3\sqrt{5}}{5}$.

b. $\dfrac{\sqrt{3}}{\sqrt{8}} =$ Write 8 in terms of prime factors.

$\dfrac{\sqrt{3}}{\sqrt{2 \cdot 2 \cdot 2}} =$ Since we need an even number of factors of 2 in the radicand of the denominator, multiply by $\dfrac{\sqrt{2}}{\sqrt{2}}$.

$\dfrac{\sqrt{3}}{\sqrt{2 \cdot 2 \cdot 2}} \cdot \dfrac{\sqrt{2}}{\sqrt{2}} =$ Multiply.

$\dfrac{\sqrt{6}}{\sqrt{2 \cdot 2 \cdot 2 \cdot 2}} =$ $\sqrt{2 \cdot 2 \cdot 2 \cdot 2} = 2 \cdot 2 = 4$.

$\dfrac{\sqrt{6}}{4}$ Therefore, $\dfrac{\sqrt{3}}{\sqrt{8}} = \dfrac{\sqrt{6}}{4}$.

PRACTICE EXERCISE 1

Rationalize the denominators of the fractions. Assume that all variables have positive values.

a. $\dfrac{5}{\sqrt{2}}$ **b.** $\dfrac{8}{\sqrt{10}}$ **c.** $\dfrac{\sqrt{5}}{\sqrt{12}}$ **d.** $\dfrac{b}{\sqrt{a}}$

If you need more practice, do the following Additional Practice Exercises.

Additional Practice Exercise 1   Rationalize the denominators. Assume that all variables have positive values.

a. $\dfrac{7}{\sqrt{5}}$ **b.** $\dfrac{10}{\sqrt{18}}$ **c.** $\dfrac{\sqrt{3}}{\sqrt{7}}$ **d.** $\dfrac{c}{\sqrt{d}}$

Answers: Practice Exercise 1: 1. $\dfrac{5\sqrt{2}}{2}$ **2.** $\dfrac{4\sqrt{10}}{5}$ **3.** $\dfrac{\sqrt{15}}{6}$ **4.** $\dfrac{b\sqrt{a}}{a}$ **Additional Practice 1: a.** $\dfrac{7\sqrt{5}}{5}$ **b.** $\dfrac{5\sqrt{2}}{3}$ **c.** $\dfrac{\sqrt{21}}{7}$ **d.** $\dfrac{c\sqrt{d}}{d}$

OBJECTIVE **B** Rationalizing the denominator when the radicand is a fraction

If the radicand is a fraction, apply the quotient rule for radicals. This results in a fraction with a radical in the denominator that is rationalized using the technique of Example 1.

Example 2	**Simplify the following, assuming that all variables have positive values:**

a. $\sqrt{\dfrac{7}{2}} =$ Rewrite, using the quotient rule.

$\dfrac{\sqrt{7}}{\sqrt{2}} =$ Since $\sqrt{2}\cdot\sqrt{2} = \sqrt{4} = 2$, multiply by $\dfrac{\sqrt{2}}{\sqrt{2}}$.

$\dfrac{\sqrt{7}}{\sqrt{2}}\cdot\dfrac{\sqrt{2}}{\sqrt{2}} =$ Multiply.

$\dfrac{\sqrt{14}}{\sqrt{4}} =$ $\sqrt{4} = 2$.

$\dfrac{\sqrt{14}}{2}$ Therefore, $\sqrt{\dfrac{7}{2}} = \dfrac{\sqrt{14}}{2}$.

b. $\sqrt{\dfrac{a}{b^3}} =$ Rewrite, using the quotient rule.

$\dfrac{\sqrt{a}}{\sqrt{b^3}} =$ Since $\sqrt{b^3}\cdot\sqrt{b} = \sqrt{b^4} = b^2$, multiply by $\dfrac{\sqrt{b}}{\sqrt{b}}$.

$\dfrac{\sqrt{a}}{\sqrt{b^3}}\cdot\dfrac{\sqrt{b}}{\sqrt{b}} =$ Multiply.

$\dfrac{\sqrt{ab}}{\sqrt{b^4}} =$ $\sqrt{b^4} = b^2$.

$\dfrac{\sqrt{ab}}{b^2}$ Therefore, $\sqrt{\dfrac{a}{b^3}} = \dfrac{\sqrt{ab}}{b^2}$.

c. $\sqrt{\dfrac{2}{3x}} =$ Rewrite, using the quotient rule.

$\dfrac{\sqrt{2}}{\sqrt{3x}} =$ Since $\sqrt{3x}\cdot\sqrt{3x} = \sqrt{9x^2} = 3x$, multiply by $\dfrac{\sqrt{3x}}{\sqrt{3x}}$.

$\dfrac{\sqrt{2}}{\sqrt{3x}}\cdot\dfrac{\sqrt{3x}}{\sqrt{3x}} =$ Multiply.

$\dfrac{\sqrt{6x}}{\sqrt{9x^2}} =$ $\sqrt{9x^2} = 3x$.

$\dfrac{\sqrt{6x}}{3x}$ Therefore, $\sqrt{\dfrac{2}{3x}} = \dfrac{\sqrt{6x}}{3x}$.

PRACTICE EXERCISE 2

Simplify the following, assuming that all variables have positive values:

a. $\sqrt{\dfrac{5}{3}}$ 　　　**b.** $\sqrt{\dfrac{c}{d^3}}$ 　　　**c.** $\sqrt{\dfrac{5}{2y}}$

If you need more practice, do the Additional Practice Exercises.

Additional Practice Exercise 2 Simplify the following, assuming that all variables have positive values:

a. $\sqrt{\dfrac{3}{5}}$ 　　　**b.** $\sqrt{\dfrac{a}{b^5}}$ 　　　**c.** $\sqrt{\dfrac{7}{3r}}$

If the denominator contains a cube root instead of a square root, the technique is essentially the same. We multiply the original fraction by the cube root of a quantity divided by itself so that the resulting radicand in the denominator is a perfect cube.

Example 3 Rationalize the denominators of the following:

a.
$$\frac{3}{\sqrt[3]{2}} =$$ 　　Since $\sqrt[3]{2} \cdot \sqrt[3]{4} = \sqrt[3]{8} = 2$, multiply by $\dfrac{\sqrt[3]{4}}{\sqrt[3]{4}}$.

$$\frac{3}{\sqrt[3]{2}} \cdot \frac{\sqrt[3]{4}}{\sqrt[3]{4}} =$$ 　　Multiply.

$$\frac{3\sqrt[3]{4}}{\sqrt[3]{8}} =$$ 　　$\sqrt[3]{8} = 2$.

$$\frac{3\sqrt[3]{4}}{2}$$ 　　Therefore, $\dfrac{3}{\sqrt[3]{2}} = \dfrac{3\sqrt[3]{4}}{2}$.

b.
$$\frac{\sqrt[3]{a}}{\sqrt[3]{b}} =$$ 　　Since $\sqrt[3]{b} \cdot \sqrt[3]{b^2} = \sqrt[3]{b^3} = b$, multiply by $\dfrac{\sqrt[3]{b^2}}{\sqrt[3]{b^2}}$.

$$\frac{\sqrt[3]{a}}{\sqrt[3]{b}} \cdot \frac{\sqrt[3]{b^2}}{\sqrt[3]{b^2}} =$$ 　　Multiply.

$$\frac{\sqrt[3]{ab^2}}{\sqrt[3]{b^3}} =$$ 　　$\sqrt[3]{b^3} = b$.

$$\frac{\sqrt[3]{ab^2}}{b}$$ 　　Therefore, $\dfrac{\sqrt[3]{a}}{\sqrt[3]{b}} = \dfrac{\sqrt[3]{ab^2}}{b}$.

c.
$$\sqrt[3]{\frac{5}{9a^2}} =$$ 　　Rewrite $\sqrt[3]{\dfrac{5}{9a^2}}$ as $\dfrac{\sqrt[3]{5}}{\sqrt[3]{9a^2}}$.

$$\frac{\sqrt[3]{5}}{\sqrt[3]{9a^2}} =$$ 　　Since $\sqrt[3]{9a^2} \cdot \sqrt[3]{3a} = \sqrt[3]{27a^3} = 3a$, multiply by $\dfrac{\sqrt[3]{3a}}{\sqrt[3]{3a}}$.

$$\frac{\sqrt[3]{5}}{\sqrt[3]{9a^2}} \cdot \frac{\sqrt[3]{3a}}{\sqrt[3]{3a}} =$$ 　　Multiply.

$$\frac{\sqrt[3]{15a}}{\sqrt[3]{27a^3}} =$$ 　　$\sqrt[3]{27a^3} = 3a$.

$$\frac{\sqrt[3]{15a}}{3a}$$ 　　Therefore, $\sqrt[3]{\dfrac{5}{9a^2}} = \dfrac{\sqrt[3]{15a}}{3a}$.

Note

We can also determine the radicand of the radical expression by which we multiply the original fraction by finding the smallest perfect cube divisible by the radicand. For example, in Example 3a the smallest perfect cube that is divisible by 2 is 8, and $8 \div 2 = 4$. Therefore, the radicand of the radical expression that we multiply by is 4.

Optional Topic

Alternative Method

The method of prime factorization also works very well for radicands containing cube roots. Each time a factor appears three times, the cube root is one of the three equal factors. For example, $\sqrt[3]{a \cdot a \cdot a} = a$, and $\sqrt[3]{2 \cdot 2 \cdot 2 \cdot y \cdot y \cdot y \cdot y \cdot y} = 2y\sqrt[3]{2y^2}$. We illustrate this technique by reworking Examples 3a and 3c.

Answers: Practice Exercise 2: a. $\dfrac{\sqrt{15}}{3}$ **b.** $\dfrac{\sqrt{cd}}{d^2}$ **c.** $\dfrac{\sqrt{10y}}{2y}$ **Additional Practice 2: a.** $\dfrac{\sqrt{15}}{5}$ **b.** $\dfrac{\sqrt{ab}}{b^3}$ **c.** $\dfrac{\sqrt{21r}}{3r}$

a. $\dfrac{3}{\sqrt[3]{2}} =$ Since $\sqrt[3]{2 \cdot 2 \cdot 2} = 2$, we need two more factors of 2 in the radicand of the denominator. So, multiply by $\dfrac{\sqrt[3]{2 \cdot 2}}{\sqrt[3]{2 \cdot 2}}$.

$\dfrac{3}{\sqrt[3]{2}} \cdot \dfrac{\sqrt[3]{2 \cdot 2}}{\sqrt[3]{2 \cdot 2}} =$ Multiply.

$\dfrac{3\sqrt[3]{2 \cdot 2}}{\sqrt[3]{2 \cdot 2 \cdot 2}} =$ $\sqrt[3]{2 \cdot 2 \cdot 2} = 2$.

$\dfrac{3\sqrt[3]{4}}{2}$ Therefore, $\dfrac{3}{\sqrt[3]{2}} = \dfrac{3\sqrt[3]{4}}{2}$.

c. $\sqrt[3]{\dfrac{5}{9a^2}} =$ Rewrite $\sqrt[3]{\dfrac{5}{9a^2}}$ as $\dfrac{\sqrt[3]{5}}{\sqrt[3]{9a^2}}$.

$\dfrac{\sqrt[3]{5}}{\sqrt[3]{9a^2}} =$ Write $9a^2$ in terms of prime factors.

$\dfrac{\sqrt[3]{5}}{\sqrt[3]{3 \cdot 3 \cdot a \cdot a}} =$ Since $\sqrt[3]{3 \cdot 3 \cdot 3} = 3$ and $\sqrt[3]{a \cdot a \cdot a} = a$, we need another factor of 3 and a in the radicand of the denominator. So, multiply by $\dfrac{\sqrt[3]{3a}}{\sqrt[3]{3a}}$.

$\dfrac{\sqrt[3]{5}}{\sqrt[3]{3 \cdot 3 \cdot a \cdot a}} \cdot \dfrac{\sqrt[3]{3a}}{\sqrt[3]{3a}} =$ Multiply.

$\dfrac{\sqrt[3]{15a}}{\sqrt[3]{3 \cdot 3 \cdot 3 \cdot a \cdot a \cdot a}} =$ $\sqrt[3]{3 \cdot 3 \cdot 3 \cdot a \cdot a \cdot a} = 3a$.

$\dfrac{\sqrt[3]{15a}}{3a}$ Therefore, $\sqrt[3]{\dfrac{5}{9a^2}} = \dfrac{\sqrt[3]{15a}}{3a}$.

PRACTICE EXERCISE 3

Rationalize the denominators of the following:

a. $\dfrac{6}{\sqrt[3]{3}}$ **b.** $\sqrt[3]{\dfrac{m}{n}}$ **c.** $\dfrac{4}{\sqrt[3]{4y^2}}$

If you need more practice, do the following Additional Practice Exercise.

Additional Practice Exercise 3   Rationalize the denominators of the following:

a. $\dfrac{5}{\sqrt[3]{9}}$ **b.** $\sqrt[3]{\dfrac{3}{x^2}}$ **c.** $\dfrac{8}{\sqrt[3]{25c}}$

OBJECTIVE **Rationalizing denominators of the form $a \pm b$, where a and/or b are square roots**

Recall from Section 10.4 that conjugates are of the form $a + b$ and $a - b$, and their product always results in a rational number. Consequently, if the denominator contains an expression of the form $a + b$, where a and/or b are square roots, multiply the original fraction by the conjugate of the denominator divided by itself. For example, if the denominator is $2 + \sqrt{3}$, multiply by $\dfrac{2 - \sqrt{3}}{2 - \sqrt{3}}$.

Answers: Practice Exercise 3: **a.** $2\sqrt[3]{9}$ **b.** $\dfrac{\sqrt[3]{mn^2}}{n}$ **c.** $\dfrac{2\sqrt[3]{2y}}{y}$

Additional Practice 3: **a.** $\dfrac{5\sqrt[3]{3}}{3}$ **b.** $\dfrac{\sqrt[3]{3x}}{x}$ **c.** $\dfrac{8\sqrt[3]{5c^2}}{5c}$

Example 4 Rationalize the denominators. Assume that all variables have nonnegative values.

a. $\dfrac{3}{4+\sqrt{5}} =$

Since the conjugate of $4+\sqrt{5}$ is $4-\sqrt{5}$, multiply by $\dfrac{4-\sqrt{5}}{4-\sqrt{5}}$.

$\dfrac{3}{4+\sqrt{5}} \cdot \dfrac{4-\sqrt{5}}{4-\sqrt{5}} =$ Multiply the fractions.

$\dfrac{3(4-\sqrt{5})}{(4+\sqrt{5})(4-\sqrt{5})} =$ Perform the multiplications.

$\dfrac{12-3\sqrt{5}}{4^2-(\sqrt{5})^2} =$ Simplify the denominator.

$\dfrac{12-3\sqrt{5}}{16-5} =$ Add 16 and -5.

$\dfrac{12-3\sqrt{5}}{11}$

Therefore, $\dfrac{3}{4+\sqrt{5}} = \dfrac{12-3\sqrt{5}}{11}$.

b. $\dfrac{2+\sqrt{2}}{4-\sqrt{2}} =$

Since the conjugate of $4-\sqrt{2}$ is $4+\sqrt{2}$, multiply by $\dfrac{4+\sqrt{2}}{4+\sqrt{2}}$.

$\dfrac{2+\sqrt{2}}{4-\sqrt{2}} \cdot \dfrac{4+\sqrt{2}}{4+\sqrt{2}} =$ Multiply the fractions.

$\dfrac{(2+\sqrt{2})(4+\sqrt{2})}{(4-\sqrt{2})(4+\sqrt{2})} =$ Perform the multiplications.

$\dfrac{8+2\sqrt{2}+4\sqrt{2}+\sqrt{4}}{4^2-(\sqrt{2})^2} =$ Simplify.

$\dfrac{8+6\sqrt{2}+2}{16-2} =$ Continue simplifying.

$\dfrac{10+6\sqrt{2}}{14} =$ Factor the numerator.

$\dfrac{2(5+3\sqrt{2})}{14} =$ Reduce to lowest terms.

$\dfrac{5+3\sqrt{2}}{7}$

Therefore, $\dfrac{2+\sqrt{2}}{4-\sqrt{2}} = \dfrac{5+3\sqrt{2}}{7}$.

c. $\dfrac{1+\sqrt{y}}{\sqrt{x}-\sqrt{y}} =$

Since the conjugate of $\sqrt{x}-\sqrt{y}$ is $\sqrt{x}+\sqrt{y}$, multiply by $\dfrac{\sqrt{x}+\sqrt{y}}{\sqrt{x}+\sqrt{y}}$.

$\dfrac{1+\sqrt{y}}{\sqrt{x}-\sqrt{y}} \cdot \dfrac{\sqrt{x}+\sqrt{y}}{\sqrt{x}+\sqrt{y}} =$ Multiply the fractions.

$\dfrac{(1+\sqrt{y})(\sqrt{x}+\sqrt{y})}{(\sqrt{x}-\sqrt{y})(\sqrt{x}+\sqrt{y})} =$ Perform the multiplications.

$\dfrac{\sqrt{x}+\sqrt{y}+\sqrt{xy}+\sqrt{y^2}}{(\sqrt{x})^2-(\sqrt{y^2})} =$ Simplify.

$\dfrac{\sqrt{x}+\sqrt{y}+\sqrt{xy}+y}{x-y}$

Therefore, $\dfrac{1+\sqrt{y}}{\sqrt{x}-\sqrt{y}} = \dfrac{\sqrt{x}+\sqrt{y}+\sqrt{xy}+y}{x-y}$.

PRACTICE EXERCISE 4

Rationalize the denominators. Assume that all variables have nonnegative values.

a. $\dfrac{7}{6+\sqrt{6}}$

b. $\dfrac{4-\sqrt{2}}{3-\sqrt{2}}$

c. $\dfrac{\sqrt{3}+\sqrt{5}}{\sqrt{3}-\sqrt{5}}$

If you need more practice, do the following Additional Practice Exercises.

Additional Practice Exercise 4 Rationalize the denominators of the following:

a. $\dfrac{7}{3+\sqrt{7}}$

b. $\dfrac{3+\sqrt{3}}{5-\sqrt{3}}$

c. $\dfrac{\sqrt{5}+\sqrt{2}}{\sqrt{5}-\sqrt{2}}$

Answers: Practice Exercise 4: a. $\dfrac{42-7\sqrt{6}}{30}$ **b.** $\dfrac{10+\sqrt{2}}{7}$ **c.** $-4-\sqrt{15}$ **Additional Practice 4: a.** $\dfrac{21-7\sqrt{7}}{2}$ **b.** $\dfrac{9+4\sqrt{3}}{11}$ **c.** $\dfrac{7+2\sqrt{10}}{3}$

For Extra Help

Exercise Set 10.5 MyMathLab®

Rationalize the denominators. Assume that all variables have positive values. (See Example 1.)

1. $\dfrac{3}{\sqrt{7}}$

2. $\dfrac{2}{\sqrt{3}}$

3. $\dfrac{5}{\sqrt{5}}$

4. $\dfrac{6}{\sqrt{6}}$

5. $\dfrac{6}{\sqrt{2}}$

6. $\dfrac{12}{\sqrt{3}}$

7. $\dfrac{20}{\sqrt{10}}$

8. $\dfrac{30}{\sqrt{15}}$

9. $\dfrac{4}{\sqrt{x}}$

10. $\dfrac{6}{\sqrt{y}}$

11. $\dfrac{m}{\sqrt{n}}$

12. $\dfrac{r}{\sqrt{s}}$

13. $\dfrac{\sqrt{5}}{\sqrt{2}}$

14. $\dfrac{\sqrt{7}}{\sqrt{5}}$

15. $\dfrac{\sqrt{3}}{\sqrt{8}}$

16. $\dfrac{\sqrt{11}}{\sqrt{18}}$

17. $\dfrac{\sqrt{5}}{\sqrt{32}}$

18. $\dfrac{\sqrt{7}}{\sqrt{27}}$

19. $\dfrac{\sqrt{a}}{\sqrt{b}}$

20. $\dfrac{\sqrt{c}}{\sqrt{d}}$

Simplify. Assume that all variables have positive values. (See Example 2.)

21. $\sqrt{\dfrac{5}{3}}$

22. $\sqrt{\dfrac{3}{2}}$

23. $\sqrt{\dfrac{7}{12}}$

24. $\sqrt{\dfrac{7}{8}}$

25. $\sqrt{\dfrac{p}{q}}$

26. $\sqrt{\dfrac{u}{v}}$

27. $\sqrt{\dfrac{7}{x^5}}$

28. $\sqrt{\dfrac{3}{y^3}}$

29. $\sqrt{\dfrac{7}{5y}}$

30. $\sqrt{\dfrac{3}{2x}}$

31. $\sqrt{\dfrac{9}{7a}}$

32. $\sqrt{\dfrac{4}{3b}}$

Rationalize the denominators. Assume that all variables have positive values. (See Example 3.)

33. $\dfrac{2}{\sqrt[3]{9}}$

34. $\dfrac{5}{\sqrt[3]{4}}$

35. $\dfrac{10}{\sqrt[3]{25}}$

36. $\dfrac{8}{\sqrt[3]{2}}$

37. $\dfrac{\sqrt[3]{r}}{\sqrt[3]{s}}$

38. $\dfrac{\sqrt[3]{x}}{\sqrt[3]{y}}$

39. $\dfrac{\sqrt[3]{x}}{\sqrt[3]{y^2}}$

40. $\dfrac{\sqrt[3]{a^2}}{\sqrt[3]{b^2}}$

41. $\sqrt[3]{\dfrac{5}{4b}}$

42. $\sqrt[3]{\dfrac{2}{3a}}$

43. $\sqrt[3]{\dfrac{7}{2d^2}}$

44. $\sqrt[3]{\dfrac{5}{9x^2}}$

Rationalize the denominators of the following. (See Example 4.)

45. $\dfrac{-1}{2 + \sqrt{5}}$

46. $\dfrac{23}{5 + \sqrt{2}}$

47. $\dfrac{7}{3 - \sqrt{3}}$

48. $\dfrac{4}{4 - \sqrt{6}}$

49. $\dfrac{a}{7 + \sqrt{a}}$

50. $\dfrac{x}{5 - \sqrt{x}}$

51. $\dfrac{4 + \sqrt{3}}{3 - \sqrt{3}}$

52. $\dfrac{3 + \sqrt{2}}{2 + \sqrt{2}}$

53. $\dfrac{\sqrt{5} - 6}{\sqrt{5} + 4}$

54. $\dfrac{\sqrt{6} - 5}{\sqrt{6} + 3}$

55. $\dfrac{5 + \sqrt{3}}{4 - \sqrt{5}}$

56. $\dfrac{3 + \sqrt{2}}{3 + \sqrt{3}}$

57. $\dfrac{7 + \sqrt{a}}{2 - \sqrt{a}}$ **58.** $\dfrac{2 - \sqrt{b}}{3 + \sqrt{b}}$ **59.** $\dfrac{\sqrt{5} - \sqrt{3}}{\sqrt{5} + \sqrt{3}}$

60. $\dfrac{\sqrt{3} + \sqrt{2}}{\sqrt{3} - \sqrt{2}}$ **61.** $\dfrac{\sqrt{x} + \sqrt{y}}{\sqrt{x} - \sqrt{y}}$ **62.** $\dfrac{\sqrt{a} + \sqrt{b}}{\sqrt{a} - \sqrt{b}}$

63. $\dfrac{5 - \sqrt{6}}{\sqrt{2}}$ **64.** $\dfrac{4 + \sqrt{2}}{\sqrt{3}}$

Challenge Exercises (65–72)

Rationalize the denominators of the following:

65. $\dfrac{2\sqrt{5} + 3\sqrt{6}}{3\sqrt{2} - 3\sqrt{7}}$ **66.** $\dfrac{4 - 2\sqrt{3}}{3\sqrt{2} - 2\sqrt{5}}$ **67.** $\dfrac{3}{\sqrt[4]{9}}$

68. $\dfrac{2}{\sqrt[4]{2}}$ **69.** $\dfrac{y^2}{\sqrt[4]{y^3}}$ **70.** $\dfrac{x^2}{\sqrt[4]{x^2}}$

71. $\dfrac{6}{\sqrt[5]{8}}$ **72.** $\dfrac{2}{\sqrt[5]{2}}$

Writing Exercises (73–75)

73. In your own words, describe the procedures used in rationalizing the denominator.

74. What is wrong with the following: $\dfrac{9}{16} = \dfrac{3}{4}$?

75. If we were to rationalize the denominator of $\dfrac{1}{\sqrt[3]{a} + \sqrt[3]{b}}$, we would not multiply by $\dfrac{\sqrt[3]{a} - \sqrt[3]{b}}{\sqrt[3]{a} - \sqrt[3]{b}}$. Why not?

OBJECTIVES *When you complete this section, you will be able to:*

A Solve equations that contain one radical (square root).
B Solve equations that have two radicals (square roots).

PREREQUISITE SKILLS *Before beginning this section, you should be able to:*

a. Apply $\left(\sqrt{x}\right)^2 = x$. (Section 10.3)
b. Square a binomial containing radicals. (Section 2.4)
c. Solve linear equations. (Sections 3.1–3.3)
d. Solve quadratic equations. (Section 5.8)
e. Find the square root of a number. (Section 10.1)

GETTING READY FOR SECTION 10.6

1. Simplify $\left(\sqrt{2x+3}\right)^2$.   **2.** Simplify $\left(3 - \sqrt{x-2}\right)^2$.

For exercises 3 and 4, solve the equations:

3. $4x - 3 = 2x + 7$ **4.** $3x + 10 = x^2 + 4x + 4$ **5.** Find $\sqrt{\dfrac{25}{4}}$.

Introduction

Solving equations which contain radicals that are square roots depends upon two concepts. The first, presented in Section 10.3, is $\left(\sqrt{x}\right)^2 = x$. For example, $\left(\sqrt{3}\right)^2 = 3$, $\left(\sqrt{x+2}\right)^2 = x + 2$, and so on. The second, presented in Section 2.4, is $(a + b)^2 = a^2 + 2ab + b^2$. For example, $(x + 5)^2 = x^2 + 2 \cdot 5x + 5^2 = x^2 + 10x + 25$, and $\left(2 + \sqrt{x}\right)^2 = 2^2 + 2 \cdot 2\sqrt{x} + \left(\sqrt{x}\right)^2 = 4 + 4\sqrt{x} + x$.

OBJECTIVE **Solving equations that contain one radical (square roots)**

The key to solving equations that contain one radical is to isolate the radical on one side of the equation. After this is accomplished, square both sides of the equation. However, there may be a problem in doing so. When both sides of an equation are raised to a power, *the resulting equation may have solutions that do not solve the original equation.* Consider the equation $x = 4$.

$x = 4$	The solution is obviously 4. Square both sides.
$x^2 = 4^2$	$4^2 = 16$.
$x^2 = 16$	Subtract 16 from both sides.
$x^2 - 16 = 0$	Factor $x^2 - 16$.
$(x + 4)(x - 4) = 0$	Set each factor equal to 0.
$x + 4 = 0, x - 4 = 0$	Solve each equation.
$x = -4, x = 4$	Therefore, the solutions are 4 and -4.

You will note that the equation that resulted from squaring both sides of the original equation not only has the solution of the original equation (4), but also has another solution (-4). This leads to the following observation:

Answers: Getting Ready: 1. $2x + 3$ **2.** $x + 7 - 6\sqrt{x-2}$ **3.** $x = 5$ **4.** $x = -3, 2$ **5.** $\dfrac{5}{2}$

> **Observation**
>
> If both sides of an equation are raised to a power, the resulting equation contains all the solutions of the original equation and perhaps some solutions that are not solutions of the original equation. Any solution of the resulting equation that does not solve the original equation is called an **extraneous** solution and is discarded.

This means that when we solve an equation by raising both sides to a power, we will get all the solutions of the original equation, but we may get some answers that will not solve the original equation (extraneous solutions). Consequently, we must check all solutions of the resulting equation in the original equation. This difficulty results because the symbol $\sqrt{}$ denotes the positive square root only. Following is the procedure for solving equations that contain one square root:

> **Procedure: Solving Equations Containing One Square Root**
>
> If an equation contains one square root, it can be solved by
>
> 1. Isolating the square root on one side of the equation.
> 2. Squaring both sides of the equation.
> 3. Solving the resulting equation.
> 4. Checking all solutions in the original equation.

Example 1 **Solve the following equations:**

a.

$\sqrt{x} = 4$ Since $(\sqrt{x})^2 = x$, square both sides of the equation.

$(\sqrt{x})^2 = 4^2$ Simplify both sides.

$x = 16$ Therefore, 16 is a possible solution.

CHECK:

$\sqrt{x} = 4$ Substitute 16 for x.

$\sqrt{16} = 4$ Remember, the radical sign denotes the positive root only.

$4 = 4$ Therefore, 16 is the correct solution.

b.

$\sqrt{2x + 3} - 5 = 0$ Isolate the radical by adding 5 to both sides of the equation.

$\sqrt{2x + 3} - 5 + 5 = 0 + 5$ Simplify both sides.

$\sqrt{2x + 3} = 5$ Square both sides.

$(\sqrt{2x + 3})^2 = 5^2$ Simplify both sides.

$2x + 3 = 25$ Subtract 3 from both sides.

$2x = 22$ Divide both sides by 2.

$x = 11$ Therefore, 11 is a possible solution.

CHECK:

$\sqrt{2x + 3} - 5 = 0$ Substitute 11 for x.

$\sqrt{2(11) + 3} - 5 = 0$ Simplify.

$\sqrt{22 + 3} - 5 = 0$ Continue simplifying.

$\sqrt{25} - 5 = 0$ $\sqrt{25} = 5$.

$5 - 5 = 0$ Simplify.

$0 = 0$ Therefore, 11 is the correct solution.

c. $\sqrt{3x - 3} + 6 = 0$ Isolate the radical by subtracting 6 from both sides of the equation.

$\sqrt{3x - 3} = -6$ Square both sides.

$(\sqrt{3x - 3})^2 = (-6)^2$ Simplify both sides.

$3x - 3 = 36$ Add 3 to both sides.

$3x = 39$ Divide both sides by 3.

$x = 13$ Therefore, 13 is a possible solution.

CHECK:

$\sqrt{3x - 3} + 6 = 0$ Substitute 13 for x.

$\sqrt{3(13) - 3} + 6 = 0$ Simplify.

$\sqrt{39 - 3} + 6 = 0$ Continue simplifying.

$\sqrt{36} + 6 = 0$ $\sqrt{36} = 6$.

$6 + 6 = 0$ Simplify.

$12 \neq 0$ Therefore, 13 is not a solution.

> ### Note
>
> We know there will be no solution from the step $\sqrt{3x - 3} = -6$, since we know the $\sqrt{}$ symbol represents the positive root only. So the left side of the equation is positive and the right is negative, which is impossible.

Since 13 is the only possible solution, $\sqrt{3x - 3} + 6 = 0$ has no solution. Remember, this is usually denoted by $\varnothing$, which is called the *empty set*, or *null set*. This means that 13 is an extraneous solution.

d. $\sqrt{3x + 10} = x + 2$ Square both sides of the equation.

$(\sqrt{3x + 10})^2 = (x + 2)^2$ Simplify both sides.

$3x + 10 = x^2 + 2 \cdot 2x + 4$ Simplify the right side.

$3x + 10 = x^2 + 4x + 4$ Subtract $3x$ and 10 from both sides.

$3x - 3x + 10 - 10 = x^2 + 4x - 3x + 4 - 10$ Simplify both sides.

$0 = x^2 + x - 6$ Factor the right side.

$0 = (x + 3)(x - 2)$ Set each factor equal to 0.

$x + 3 = 0, x - 2 = 0$ Solve each equation.

$x = -3, x = 2$ Therefore, the possible solutions are -3 and 2.

CHECK:

$x = -3$

$\sqrt{3x + 10} = x + 2$ Substitute -3 for x.

$\sqrt{3(-3) + 10} = (-3) + 2$ Simplify.

$\sqrt{-9 + 10} = -1$ Continue simplifying.

$\sqrt{1} = -1$ $\sqrt{1} = 1$.

$1 \neq -1$ Therefore, -3 is not a solution.

$x = 2$

$\sqrt{3x + 10} = x + 2$ Substitute 2 for x.

$\sqrt{3(2) + 10} = (2) + 2$ Simplify.

$\sqrt{6 + 10} = 4$ Continue simplifying.

$\sqrt{16} = 4$ $\sqrt{16} = 4$.

$4 = 4$ Therefore, 2 is a solution.

Since -3 did not check, 2 is the only solution of the equation and -3 is an extraneous solution. Also, in the check, we could tell that -3 was not going to work when we arrived at the line $\sqrt{-9 + 10} = -1$, because the left side of the equation is positive (the radical denotes the positive root only) and the right is negative. Consequently, they cannot be equal.

PRACTICE EXERCISE 1

Solve the following equations:

a. $\sqrt{a} - 6 = 0$ **b.** $\sqrt{4x - 8} - 4 = 0$ **c.** $\sqrt{2x - 5} + 7 = 0$ **d.** $\sqrt{x - 1} = x - 3$

If you need more practice, do the following Additional Practice Exercises.

Additional Practice Exercise 1 Solve the following equations:

a. $\sqrt{b} + 5 = 0$ **b.** $\sqrt{3x + 6} = 6$ **c.** $\sqrt{4 - 2x} + 8 = 0$ **d.** $\sqrt{3x + 3} = 2x - 1$

OBJECTIVE Solving equations that have two radicals (square roots)

If the equation contains more than one radical, the procedure for solving is more complicated and is outlined as follows:

> **Procedure: Solving Equations Containing Two Square Roots**
>
> If an equation contains two square roots, it can be solved by
>
> **1.** Isolating one of the radicals on one side of the equation.
>
> **2.** Squaring both sides of the equation and, if necessary, simplifying each side.
>
> **3.** If the equation still contains a radical, isolate the radical and square both sides of the equation again and then solve the resulting equation.
>
> **4.** Checking all solutions in the original equation.

Example 2 **Solve the following equations:**

a.

$$\sqrt{4x - 3} - \sqrt{2x + 7} = 0$$ Isolate $\sqrt{4x - 3}$ by adding $\sqrt{2x + 7}$ to both sides of the equation.

$$\sqrt{4x - 3} - \sqrt{2x + 7} + \sqrt{2x + 7} = 0 + \sqrt{2x + 7}$$ Simplify.

$$\sqrt{4x - 3} = \sqrt{2x + 7}$$ Square both sides.

$$\left(\sqrt{4x - 3}\right)^2 = \left(\sqrt{2x + 7}\right)^2$$ Simplify.

$$4x - 3 = 2x + 7$$ Subtract $2x$ from both sides.

$$2x - 3 = 7$$ Add 3 to both sides.

$$2x = 10$$ Divide both sides by 2.

$$x = 5$$ Therefore, 5 is a possible solution.

CHECK:

$$\sqrt{4x - 3} - \sqrt{2x + 7} = 0$$ Substitute 5 for x.

$$\sqrt{4(5) - 3} - \sqrt{2(5) + 7} = 0$$ Simplify.

$$\sqrt{20 - 3} - \sqrt{10 + 7} = 0$$ Continue simplifying.

$$\sqrt{17} - \sqrt{17} = 0$$ Add like radicals.

$$0 = 0$$ Therefore, 5 is the solution.

Answers: Practice Exercise 1: a. $a = 36$ **b.** $x = 6$ **c.** $\varnothing$ **d.** $x = 5$ **Additional Practice 1: a.** $\varnothing$
b. $x = 10$ **c.** $\varnothing$ **d.** $x = 2$

b. $\sqrt{x+4} = 3 - \sqrt{x-2}$ Since one radical is isolated, square both sides of the equation.

$$(\sqrt{x+4})^2 = (3 - \sqrt{x-2})^2$$ Simplify both sides.

$$x + 4 = 3^2 - 2 \cdot 3\sqrt{x-2} + (\sqrt{x-2})^2$$ Simplify the right side.

$$x + 4 = 9 - 6\sqrt{x-2} + x - 2$$ Continue simplifying the right side.

$$x + 4 = x + 7 - 6\sqrt{x-2}$$ Subtract x from both sides.

$$4 = 7 - 6\sqrt{x-2}$$ Subtract 7 from both sides.

$$-3 = -6\sqrt{x-2}$$ Divide both sides by -3.

$$1 = 2\sqrt{x-2}$$ Square both sides.

$$1^2 = (2\sqrt{x-2})^2$$ Simplify both sides.

$$1 = 4(x-2)$$ Simplify the right side.

$$1 = 4x - 8$$ Add 8 to both sides.

$$9 = 4x$$ Divide both sides by 4.

$$\frac{9}{4} = x$$ Therefore, $\frac{9}{4}$ is a possible solution.

Be Careful

When squaring $2\sqrt{x-2}$, don't forget to square 2.

CHECK:

$$\sqrt{x+4} = 3 - \sqrt{x-2}$$ Substitute $\frac{9}{4}$ for x.

$$\sqrt{\frac{9}{4} + 4} = 3 - \sqrt{\frac{9}{4} - 2}$$ Get common denominators.

$$\sqrt{\frac{9}{4} + \frac{16}{4}} = 3 - \sqrt{\frac{9}{4} - \frac{8}{4}}$$ Add the fractions.

$$\sqrt{\frac{25}{4}} = 3 - \sqrt{\frac{1}{4}}$$ Find the square roots.

$$\frac{5}{2} = 3 - \frac{1}{2}$$ Common denominator on the right.

$$\frac{5}{2} = \frac{6}{2} - \frac{1}{2}$$ Add the fractions.

$$\frac{5}{2} = \frac{5}{2}$$ Therefore, $\frac{9}{4}$ is the correct solution.

PRACTICE EXERCISE 2

Solve the following equations:

a. $\sqrt{3x+2} - \sqrt{x+4} = 0$ **b.** $\sqrt{12+x} = 6 - \sqrt{x}$

If you need more practice, do the following Additional Practice Exercises.

Additional Practice Exercise 2   Solve the following equations:

a. $\sqrt{5x+2} + \sqrt{x-6} = 0$ **b.** $\sqrt{x+4} - \sqrt{x-4} = 4$

Radical equations are frequently used in real-world situations.

Answers: Practice Exercise 2: a. $x = 1$ **b.** $x = 4$ **Additional Practice 2: a.** $\varnothing$ **b.** $\varnothing$

Example 3

a. The distance a car will skid when the brakes are applied is approximated by the formula $s = k\sqrt{d}$, where s is the speed of the car in miles per hour, d is the distance the car will skid in feet, and k is a constant determined by road conditions, weight of the car, type of tires, and so on. If $k = 4.21$, find the following:

1. If the length of the skid mark is 196 feet, find the speed of the car.

Solution

$s = k\sqrt{d}$ Substitute 4.21 for k and 196 for d.

$s = 4.21\sqrt{196}$ $\sqrt{196} = 14$.

$s = 4.21(14)$ Multiply.

$s = 58.94$ Therefore, the speed of the car was about 59 miles per hour.

2. If the car is traveling at a speed of 63.15 miles per hour, how far will it take it to skid to a stop?

Solution

$s = k\sqrt{d}$ Substitute 63.15 for s and 4.21 for k.

$63.15 = 4.21\sqrt{d}$ Divide both sides by 4.21.

$15 = \sqrt{d}$ Square both sides of the equation.

$225 = d$ Therefore, it would take the car 225 feet to stop.

b. Under certain conditions, the approximate distance to the horizon is given by $s = \sqrt{1.50h}$, where s is the distance in miles and h is the elevation of the viewer in feet. (*Source:* wikipedia.org)

1. How far can a person see to the horizon if his or her elevation is 1536 feet?

Solution

$s = \sqrt{1.50h}$ Substitute 1536 for h.

$s = \sqrt{(1.50)(1536)}$ Multiply.

$s = \sqrt{2304}$ $\sqrt{2304} = 48$.

$s = 48$ Therefore, one would be able to see 48 miles to the horizon.

2. What would the elevation of the viewer have to be in order to see 36 miles to the horizon?

Solution

$s = \sqrt{1.50h}$ Substitute 36 for s.

$36 = \sqrt{1.50h}$ Square both sides.

$1296 = 1.50h$ Divide both sides by 1.50.

$864 = h$ Therefore, the elevation would have to be 864 feet.

PRACTICE EXERCISE 3

Answer the following:

a. The number of amperes in an electrical system is calculated by the formula

$$\text{amperes} = \sqrt{\frac{\text{watts}}{\text{ohms}}}.$$

1. How many amperes are used if there are 8000 watts and 20 ohms?

2. How many watts of power are used by an appliance that uses 8 amperes when the resistance is 35 ohms?

b. The amount of time it takes an object to fall s feet is given by the formula $t = \dfrac{\sqrt{s}}{4}$, where t is the time in seconds and s is distance in feet.

1. How long will it take an object to reach the ground if it is dropped from a height of 256 feet?

2. If an object is dropped from the top of a tall building and it takes 6 seconds for it to hit the ground, how tall is the building?

Answers: Practice Exercise 3: a1. 20 amperes **2.** 2240 watts **b1.** 4 sec. **2.** 576 ft

For Extra Help

Exercise Set 10.6 MyMathLab®

Solve the following equations. (See Examples 1 and 2.)

1. $\sqrt{a} = 2$ **2.** $\sqrt{c} = 9$ **3.** $\sqrt{z} - 7 = 0$ **4.** $\sqrt{n} - 5 = 0$

5. $\sqrt{a} + 4 = 0$ **6.** $\sqrt{b} + 2 = 0$ **7.** $2\sqrt{x} = 8$ **8.** $3\sqrt{y} = 9$

9. $\sqrt{y} - 4 = 4$ **10.** $\sqrt{y} - 4 = 2$ **11.** $\sqrt{x} + 3 = 8$ **12.** $\sqrt{x} + 2 = 6$

13. $\sqrt{5h - 21} = 3$ **14.** $\sqrt{2d + 39} = 7$ **15.** $\sqrt{7r - 26} - 4 = 0$ **16.** $\sqrt{7d - 24} - 2 = 0$

17. $\sqrt{8r - 15} = r$ **18.** $\sqrt{6h - 8} = h$ **19.** $\sqrt{2m + 24} - m = 0$ **20.** $\sqrt{4d + 32} - d = 0$

21. $\sqrt{w - 3} = w - 5$ **22.** $\sqrt{2t + 9} = t - 3$ **23.** $\sqrt{4k + 25} = k - 5$ **24.** $\sqrt{4f + 1} = f - 1$

25. $\sqrt{w - 4} - w + 6 = 0$ **26.** $\sqrt{x - 2} - x + 4 = 0$ **27.** $\sqrt{y - 4} - y + 4 = 0$

28. $\sqrt{n - 3} - n + 3 = 0$ **29.** $\sqrt{4f - 20} = \sqrt{f - 2}$ **30.** $\sqrt{10n - 3} = \sqrt{4n + 3}$

31. $\sqrt{5n - 21} = \sqrt{n - 1}$ **32.** $\sqrt{6f - 4} = \sqrt{4f + 8}$ **33.** $\sqrt{3a - 7} - \sqrt{a - 1} = 0$

34. $\sqrt{6r - 26} - \sqrt{r - 1} = 0$ **35.** $\sqrt{4 - x} = 2 - \sqrt{x}$ **36.** $\sqrt{16 - x} = 4 - \sqrt{x}$

37. $\sqrt{a - 4} = 4 - \sqrt{a + 4}$ **38.** $\sqrt{b + 8} = 6 - \sqrt{b - 4}$ **39.** $\sqrt{c + 10} + \sqrt{c + 2} = 2$

40. $\sqrt{n + 7} + \sqrt{n - 4} = 1$ **41.** $\sqrt{x + 5} + \sqrt{x - 3} = 4$ **42.** $\sqrt{m - 4} + \sqrt{m + 5} = 3$

Answer the following. (See Example 3.)

43. Using the formula $s = k\sqrt{d}$ from Example 3a with $k = 3.7$, find the following:

  a. How fast a car was traveling if the length of the skid mark is 324 feet.

  b. The length of the skid mark for a car traveling 74 miles per hour.

44. Using the formula $s = \sqrt{1.50\,h}$ from Example 3b, find the following:

  a. How far can a person see to the horizon from a height of 216 feet?

  b. How high would a person have to be in order to see a distance of 9 miles?

45. Using the formula $\text{amperes} = \sqrt{\dfrac{\text{watts}}{\text{ohms}}}$ given in Practice Exercise 3a, find the following:

  a. The number of amperes if the number of watts is 4000 and the number of ohms is 10.

  b. The number of watts used by an electrical system that draws 15 amperes when the resistance is 120 ohms.

46. Using the formula $t = \dfrac{\sqrt{s}}{4}$ given in Practice Exercise 3b, find the following:

  a. How long it takes an object to hit the ground if it is dropped from a height of 3600 feet.

  b. The height of a cliff if it takes 16 seconds for a rock dropped from the top of the cliff to reach the ground below.

47. The equation for the length of time it takes a pendulum to make one swing is $t = 2\pi\sqrt{\dfrac{L}{32}}$, where t is the time in seconds and L is the length of the pendulum in feet. Find the following:

  a. How long it takes a pendulum that is 2 feet long to make one swing.

  b. The length of the pendulum, to the nearest one hundredth of a foot, if the pendulum makes one swing every 1.5 seconds.

48. If air resistance is neglected, the velocity (v) of an object after it has fallen a distance (d) is given by $v = 8\sqrt{d}$, where v is the velocity in feet per second and d is the distance in feet. Find the following:

  a. The velocity of an object that has fallen 289 feet.

  b. How far an object has fallen if it hits the ground with a velocity of 120 feet per second.

Challenge Exercises (49–58)

Solve the following equations:

49. $\sqrt{2x - 3} - \sqrt{2x - 11} = 2$

50. $\sqrt{2x + 20} + \sqrt{2x + 5} = 3$

51. $\sqrt[3]{x} = 2$

52. $\sqrt[3]{x} = -3$

53. $\sqrt[3]{x + 3} = -2$

54. $\sqrt[3]{x - 5} = 1$

55. $\sqrt[4]{x} = 2$

56. $\sqrt[4]{x + 2} = 1$

57. $\sqrt[4]{2x - 1} = 3$

58. $\sqrt[4]{2x + 4} = 2$

Writing Exercise

59. What is wrong with the following solution of the equation?

$$\sqrt{3x - 9} = 5 + \sqrt{2x + 6}$$
$$\left(\sqrt{3x - 9}\right)^2 = \left(5 + \sqrt{2x + 6}\right)^2$$
$$3x - 9 = 25 + 2x + 6$$
$$3x - 9 = 2x + 31$$
$$x - 9 = 31$$
$$x = 40$$

Section 10.7

Pythagorean Theorem

OBJECTIVES *When you complete this section, you will be able to:*

A Find the unknown side of a right triangle.

B Find the length of a diagonal of a rectangle or square.

C Find the length of the missing side of a rectangle when given the length of a diagonal and the length of one side.

D Solve application problems involving right triangles.

PREREQUISITE SKILLS *Before beginning this section, you should be able:*

a. Square a radical expression. (Section 10.3)

b. Apply the addition property of equality. (Section 3.1)

c. Simplify a radical expression. (Section 10.2)

GETTING READY FOR SECTION 10.7

Simplify the following:

1. $\left(2\sqrt{3}\right)^2$ **2.** $\left(4\sqrt{5}\right)^2$ **3.** Solve $49 + b^2 = 64$ for b^2.   **4.** Simplify $\sqrt{68}$.

Historical Note

There is some doubt as to whether Pythagoras himself actually discovered the relationship between the sides of a right triangle or whether it was discovered by one of his followers or perhaps by the Egyptians or Babylonians.

Introduction

One situation where radicals are used in the real world is in finding an unknown side of a right triangle. The procedure for finding an unknown side of a right triangle is attributed to the Greek mathematician and philosopher Pythagoras and for that reason is called the Pythagorean theorem.

A triangle, one of whose angles is a right (90°) angle, is called a **right triangle**. The sides that form the right angle are called the **legs** of the right triangle and the side opposite the right angle is called the **hypotenuse**. See the figure that follows:

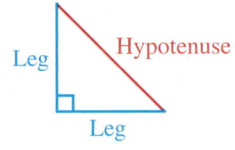

OBJECTIVE **A** Finding the unknown side of a right triangle

Conclusions drawn from observations are never accepted as fact until they are proven mathematically. Many proofs of the Pythagorean theorem exist, but none are included in this text. Once a fact has been proven, it is called a theorem.

The Pythagorean theorem probably arose from observing that if squares were constructed on each of the sides of a right triangle, the sum of the areas of the squares on the legs was equal to the area of the square on the hypotenuse. See the following figure, where the lengths of the legs are 3 and 4 and the length of the hypotenuse is 5:

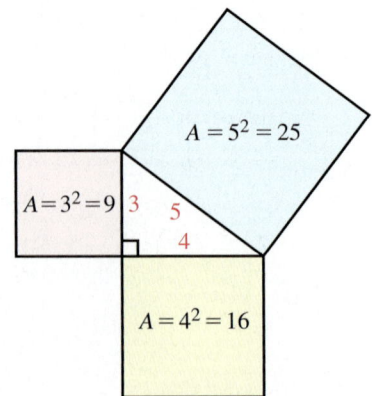

Answers: Getting Ready: 1. 12 **2.** 80 **3.** $b^2 = 15$ **4.** $2\sqrt{17}$

From the figure, we see that $3^2 + 4^2 = 5^2$, because $9 + 16 = 25$. Thus, the sum of the areas of the squares on the legs is equal to the area of the square on the hypotenuse. The Pythagorean theorem is no longer stated in terms of areas of squares, but is stated as follows:

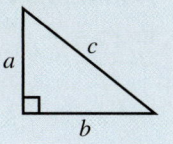

Rule: Pythagorean Theorem

If the legs of a right triangle have lengths of a and b and the hypotenuse has length of c, then $a^2 + b^2 = c^2$.

In words, the sum of the squares of the lengths of the legs of a right triangle is equal to the square of the length of the hypotenuse.

Before finding an unknown side of a right triangle, we need one other concept, which will be discussed in detail in Section 11.1. Consider an equation like $x^2 = 16$. This asks the question, "The square of what number is 16?" This is the definition of square root. Consequently, the numbers whose squares equal 16 are the square roots of 16, which are 4 and -4. Algebraically, we would write this as follows: If $x^2 = 16$, then $\sqrt{x^2} = \pm\sqrt{16}$; so $x = 4$ or $x = -4$.

This leads to the following rule:

Rule: Solutions of Equations of the form $x^2 = a$

An equation of the form $x^2 = a$, with $a > 0$, has solutions of $x = \pm\sqrt{a}$. In words, if $x^2 = a$, then x is the positive and negative square roots of a.

Example 1   **Solve the following equations:**

a. $x^2 = 36$

Solution

$x^2 = 36$ Take the positive and negative square roots of 36.
$x = \pm\sqrt{36}$ $\sqrt{36} = 6$.
$x = \pm 6$ Therefore, $x = 6$ and $x = -6$.

b. $x^2 = 3$

Solution

$x^2 = 3$ Take the positive and negative square roots of 3.
$x = \pm\sqrt{3}$ Therefore, $x = \sqrt{3}$ and $-\sqrt{3}$.

c. $x^2 = 24$

Solution

$x^2 = 24$ Take the positive and negative square roots of 24.
$x = \pm\sqrt{24}$ Simplify $\sqrt{24}$.
$x = \pm 2\sqrt{6}$ Therefore, $x = 2\sqrt{6}$ and $-2\sqrt{6}$.

PRACTICE EXERCISE 1

Solve the following equations:

a. $x^2 = 81$ **b.** $x^2 = 7$ **c.** $x^2 = 27$

If you need more practice, do the following Additional Practice Exercises.

Additional Practice Exercise 1   Solve:

a. $x^2 = 4$ **b.** $x^2 = 6$ **c.** $x^2 = 32$

Since the lengths of sides of triangles are positive, we will be limited to taking the *positive square roots only*. Since a and b represent the legs, we may substitute the lengths of a leg for either a or b, interchangeably.

Example 2   **Find the length of the unknown side of the following right triangles:**

a.

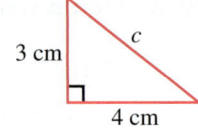

Solution

The lengths of the legs are 3 centimeters and 4 centimeters. We need to find the length of the hypotenuse. Use the Pythagorean theorem.

$a^2 + b^2 = c^2$ Substitute for a and b.

$3^2 + 4^2 = c^2$ Square 3 and 4.

$9 + 16 = c^2$ Add 9 and 16.

$25 = c^2$ Take the positive square root of 25.

$\sqrt{25} = c$ Find $\sqrt{25}$.

$5 = c$ Therefore, the length of the hypotenuse is 5 centimeters.

b.

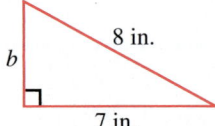

Solution

We know the lengths of the hypotenuse and one of the legs. Use the Pythagorean theorem to find the length of the other leg.

$a^2 + b^2 = c^2$ Substitute for a and c.

$7^2 + b^2 = 8^2$ Square 7 and 8.

$49 + b^2 = 64$ Subtract 49 from both sides of the equation.

$b^2 = 15$ Take the positive square root of 15.

$b = \sqrt{15}$ Therefore, the length of the other leg is $\sqrt{15}$ inches.

Answers: Practice Exercise 1:   **a.** $x = 9, -9$ **b.** $x = \sqrt{7}, -\sqrt{7}$ **c.** $x = 3\sqrt{3}, -3\sqrt{3}$
Additional Practice 1:   **a.** $x = 2, -2$ **b.** $x = \sqrt{6}, -\sqrt{6}$ **c.** $x = 4\sqrt{2}, -4\sqrt{2}$

c.

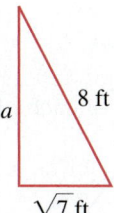

Solution

We know the lengths of the hypotenuse and one leg. Use the Pythagorean theorem to find the length of the other leg.

$$a^2 + b^2 = c^2 \qquad \text{Substitute for } b \text{ and } c.$$
$$a^2 + \left(\sqrt{7}\right)^2 = 8^2 \qquad \text{Square } \sqrt{7} \text{ and } 8.$$
$$a^2 + 7 = 64 \qquad \text{Subtract 7 from both sides of the equation.}$$
$$a^2 = 57 \qquad \text{Take the positive square root of 57.}$$
$$a = \sqrt{57} \qquad \text{Therefore, the length of the other leg is } \sqrt{57} \text{ feet.}$$

d. Find the length of the other leg of a right triangle, one of whose legs is $2\sqrt{3}$ feet and whose hypotenuse is $4\sqrt{5}$ feet.

Solution

Use the Pythagorean theorem. Substitute for a or b, and for c.

$$a^2 + b^2 = c^2 \qquad \text{Substitute for } b \text{ and } c.$$
$$a^2 + \left(2\sqrt{3}\right)^2 = \left(4\sqrt{5}\right)^2 \qquad \text{Apply } (ab)^n = a^n b^n.$$
$$a^2 + 2^2 \cdot \left(\sqrt{3}\right)^2 = 4^2 \cdot \left(\sqrt{5}\right)^2 \qquad \text{Square.}$$
$$a^2 + 4 \cdot 3 = 16 \cdot 5 \qquad \text{Multiply.}$$
$$a^2 + 12 = 80 \qquad \text{Subtract 12 from both sides of the equation.}$$
$$a^2 = 68 \qquad \text{Take the positive square root of 68.}$$
$$a = \sqrt{68} \qquad \text{Simplify } \sqrt{68}.$$
$$a = 2\sqrt{17} \qquad \text{Therefore, the unknown leg is } 2\sqrt{17} \text{ feet.}$$

PRACTICE EXERCISE 2

Find the length of the unknown side of each of the following right triangles:

a.

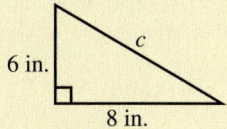

b.

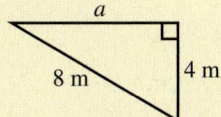

c.

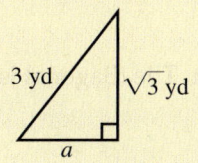

d. Find the length of the other leg of a right triangle, one of whose legs is $4\sqrt{3}$ inches and whose hypotenuse is $4\sqrt{6}$ inches.

(Continued)

Answers: Practice Exercise 2: a. $c = 10$ in. **b.** $a = 4\sqrt{3}$ m **c.** $a = \sqrt{6}$ yd **d.** $4\sqrt{3}$ in.

If you need more practice, do the following Additional Practice Exercises.

Additional Practice Exercise 2   Find the length of the unknown side of each of the following right triangles:

a.

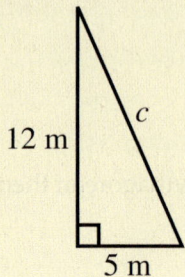

12 m

5 m

b.

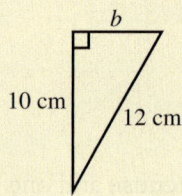

b

10 cm   12 cm

c.

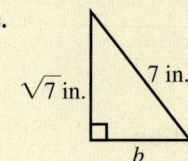

$\sqrt{7}$ in.   7 in.

b

d. The hypotenuse of a right triangle is $4\sqrt{3}$ inches, and one leg is $3\sqrt{2}$ inches. Find the length of the other leg.

OBJECTIVE **B** Finding the length of a diagonal of a rectangle or square

OBJECTIVE **C** Finding the length of the missing side of a rectangle when given the length of a diagonal and the length of one side

A diagonal of a square or a rectangle divides the square or rectangle into two right triangles. See the following figure:

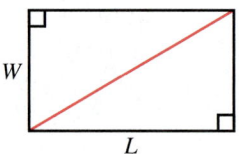

W

L

The sides of the rectangle become the legs of the right triangle, and the diagonal of the rectangle becomes the hypotenuse. Consequently, if we know the lengths of the sides of the rectangle, we can find the length of the diagonal. Also, if we know the length of the diagonal and the length of one of the sides of the rectangle, we can find the length of the other side.

Example 3

a. Find the length of the diagonal of the following rectangle:

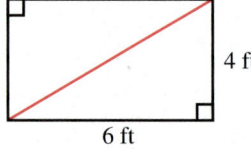

4 ft

6 ft

Solution

The lengths of the legs of the right triangle are 6 feet and 4 feet. The diagonal is the hypotenuse of a right triangle.

$$a^2 + b^2 = c^2 \quad \text{Substitute for } a \text{ and } b.$$
$$6^2 + 4^2 = c^2 \quad \text{Square 6 and 4.}$$
$$36 + 16 = c^2 \quad \text{Add 36 and 16.}$$
$$52 = c^2 \quad \text{Take the positive square root of 52.}$$
$$\sqrt{52} = c \quad \text{Simplify } \sqrt{52}.$$
$$2\sqrt{13} = c \quad \text{Therefore, the diagonal of the rectangle is } 2\sqrt{13} \text{ feet.}$$

Answers: Additional Practice 2: a. $c = 13$ m **b.** $b = 2\sqrt{11}$ cm **c.** $b = \sqrt{42}$ in. **d.** $\sqrt{30}$ in.

b. The length of a diagonal of a rectangle is $\sqrt{26}$ meters, and the width of the rectangle is $2\sqrt{2}$ meters. Find the length.

Solution

Draw and label a figure.

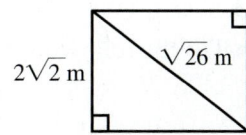

The hypotenuse is $\sqrt{26}$ meters long, and one leg is $2\sqrt{2}$ meters long.

$a^2 + b^2 = c^2$	Substitute for a (or b) and c.
$(2\sqrt{2})^2 + b^2 = (\sqrt{26})^2$	Square $2\sqrt{2}$ and $\sqrt{26}$.
$2^2(\sqrt{2})^2 + b^2 = 26$	Square 2 and $\sqrt{2}$.
$4 \cdot 2 + b^2 = 26$	Multiply 4 and 2.
$8 + b^2 = 26$	Subtract 8 from both sides of the equation.
$b^2 = 18$	Take the positive square root of 18.
$b = \sqrt{18}$	Simplify $\sqrt{18}$.
$b = \sqrt{18} = 3\sqrt{2}$	Therefore, the length is $3\sqrt{2}$ meters.

c. The diagonal of a square is $4\sqrt{2}$ inches long. Find the length of the sides of the square.

Solution

Draw and label a figure. Remember, all four sides of a square are equal in length. Therefore, let each side have a length of x.

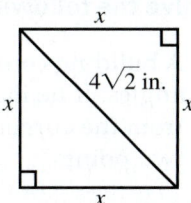

The legs are each x inches long, and the hypotenuse is $4\sqrt{2}$ inches long.

$a^2 + b^2 = c^2$	Substitute x for a and b, and $4\sqrt{2}$ for c.
$x^2 + x^2 = (4\sqrt{2})^2$	Add x^2 and x^2, and square $4\sqrt{2}$.
$2x^2 = 4^2 \cdot (\sqrt{2})^2$	Square 4 and $\sqrt{2}$.
$2x^2 = 16 \cdot 2$	Multiply 16 and 2.
$2x^2 = 32$	Divide both sides of the equation by 2.
$x^2 = 16$	Take the positive square root 16.
$x = \sqrt{16} = 4$	Therefore, each side of the square is 4 inches long.

PRACTICE EXERCISE 3

a. Find the length of the diagonal of the following rectangle:

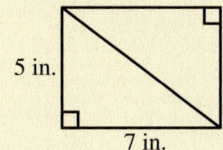

(Continued)

Answers: **Practice Exercise 3:** **a.** $\sqrt{74}$ in.

b. The diagonal of a rectangle is $2\sqrt{11}$ centimeters, and the length is $2\sqrt{7}$ centimeters. Find the width.

c. The diagonal of a square is $3\sqrt{2}$ yards. Find the length of the sides of the square.

If you need more practice, do the following Additional Practice Exercises.

Additional Practice Exercise 3

a. Find the length of the diagonal of the following rectangle:

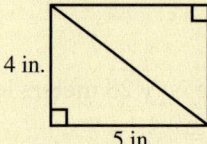

4 in.

5 in.

b. The diagonal of a rectangle is $\sqrt{30}$ feet, and the width is $\sqrt{14}$ feet. Find the length.

c. The diagonal of a square is $4\sqrt{3}$ meters. Find the length of each side of the square.

OBJECTIVE **D** Solving application problems involving right triangles

Applications of right triangles are found in many real-world situations. Knowledge of how to find the unknown side of a right triangle is important in solving many types of problems.

| **Example 4** | **Solve the following:** |

a. A building contractor wants to be sure that the walls of a building meet at right angles. If he marks a point 6 feet from the corner of one wall and a point 8 feet from the corner of the other wall, what should the distance be between these two points?

Solution

If the walls meet at right angles, the distances measured along each wall will be the legs of a right triangle and the distance between the points will be the hypotenuse. See the following figure:

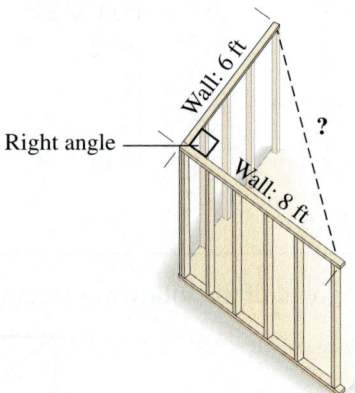

Right angle

Wall: 6 ft

Wall: 8 ft

?

Use the Pythagorean theorem to find the distance between the points.

$a^2 + b^2 = c^2$ Substitute for a and b.

$6^2 + 8^2 = c^2$ Square 6 and 8.

$36 + 64 = c^2$ Add 36 and 64.

$100 = c^2$ Take the positive square root of 100.

$10 = c$ If the walls meet at right angles, there will be 10 feet between the two points.

b. A house is 48 feet wide. The roof is supported by rafters of equal length that are 7 feet above the level of the walls at the center of the house. Find the length of each rafter. See the following figure:

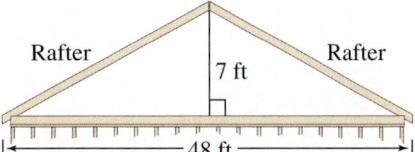

Solution

The length of each rafter is the length of the hypotenuse of a right triangle. Since the rafters meet at the center of the house, the point below where the rafters meet is 24 feet from each wall. See the following figure:

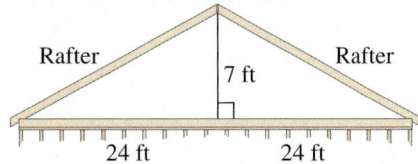

Thus, we have two right triangles whose legs are 24 feet and 7 feet. Use the Pythagorean theorem to find the hypotenuse.

$a^2 + b^2 = c^2$ Substitute for a and b.

$24^2 + 7^2 = c^2$ Square 24 and 7.

$576 + 49 = c^2$ Add 576 and 49.

$625 = c^2$ Take the positive square root of 625.

$25 = c$ Therefore, each rafter is 25 feet long.

PRACTICE EXERCISE 4

Solve the following:

a. Television screens are measured diagonally. For example, if a television is advertised as having a 17-inch screen, it means the screen measures 17 inches diagonally. If a television is advertised as having a 25-inch rectangular screen and the screen is 20 inches long, how wide is the screen?

b. The range for a particular brand of marine radio is 20 miles. Bill and Don are in separate boats that are equipped with these radios. Their boats are traveling on courses that are at right angles to each other. If Bill is 8 miles from port and Don is 15 miles from the same port, can they contact each other by radio? Why or why not?

Answers: Practice Exercise 4: a. 15 in. **b.** Yes, because the distance between Don and Bill is 17 miles and the range of the radio is 20 miles.

For Extra Help
Exercise Set 10.7 MyMathLab®

Find the value of x in each of the following right triangles. (See Example 2.)

1.
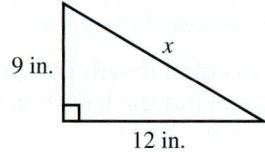
9 in.
12 in.
x

2.
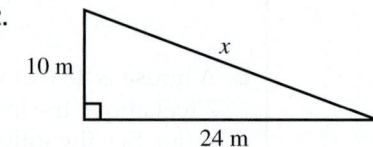
10 m
24 m
x

3.
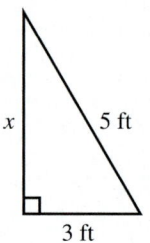
x
5 ft
3 ft

4.
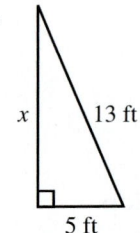
x
13 ft
5 ft

5.
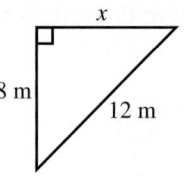
x
8 m
12 m

6.
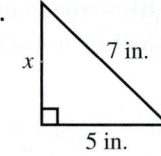
x
7 in.
5 in.

7.
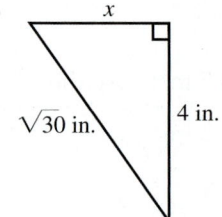
x
$\sqrt{30}$ in.
4 in.

8.
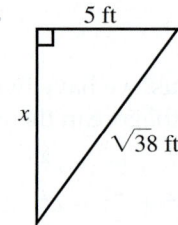
5 ft
x
$\sqrt{38}$ ft

9.
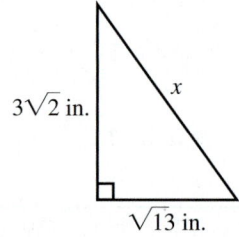
$3\sqrt{2}$ in.
x
$\sqrt{13}$ in.

10.
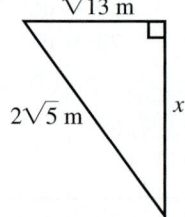
$\sqrt{13}$ m
$2\sqrt{5}$ m
x

Find the length of the unknown side of each of the following right triangles. (See Example 2.)

11. The lengths of the legs are 6 feet and 9 feet.

12. The lengths of the legs are 8 inches and 10 inches.

13. One leg is 4 meters long, and the hypotenuse is 6 meters long.

14. One leg is 10 centimeters long, and the hypotenuse is 12 centimeters long.

15. The lengths of the legs are $\sqrt{38}$ inches and $\sqrt{14}$ inches.

16. The lengths of the legs are $\sqrt{10}$ yards and $\sqrt{22}$ yards.

17. One leg is $2\sqrt{6}$ centimeters long, and the hypotenuse is $4\sqrt{2}$ centimeters long.

18. One leg is $2\sqrt{7}$ inches long, and the hypotenuse is $4\sqrt{3}$ inches long.

Find the lengths of the diagonals of the following rectangles or squares. (See Example 3.)

19.
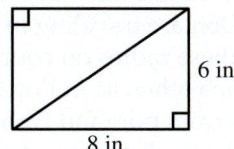
6 in.
8 in.

20.
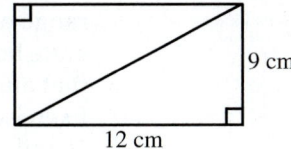
9 cm
12 cm

21.
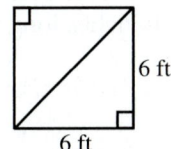
6 ft
6 ft

22.
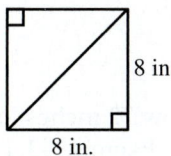
8 in.
8 in.

23.
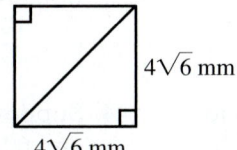
$4\sqrt{6}$ mm
$4\sqrt{6}$ mm

24.
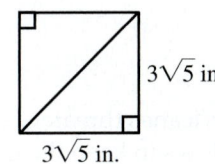
$3\sqrt{5}$ in.
$3\sqrt{5}$ in.

Find the length of the unknown side(s) of each of the following rectangles. (See Example 3.)

25.
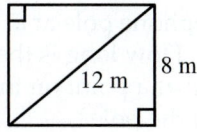
8 m
12 m

26.
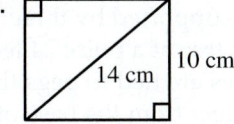
10 cm
14 cm

27.
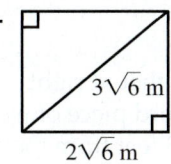
$3\sqrt{6}$ m
$2\sqrt{6}$ m

28.
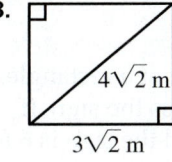
$4\sqrt{2}$ m
$3\sqrt{2}$ m

Solve the following. (See Example 3.)

29. The length of a rectangle is 5 centimeters, and the width is 8 centimeters. Find the length of a diagonal.

30. The length of a rectangle is 9 meters, and the width is 5 meters. Find the length of a diagonal.

31. The length of a diagonal of a rectangle is 12 inches, and the length is $4\sqrt{5}$ inches. Find the width.

32. The length of a diagonal of a rectangle is 14 yards, and the length is $4\sqrt{6}$ yards. Find the width.

33. The length of each side of a square is 7 feet. Find the length of a diagonal.

34. The length of each side of a square is 5 feet. Find the length of a diagonal.

35. The length of each side of a square is $3\sqrt{3}$ centimeters. Find the length of a diagonal.

36. The length of each side of a square is $2\sqrt{7}$ feet. Find the length of a diagonal.

37. The length of a diagonal of a square is 8 inches. Find the length of each side of the square.

38. The length of a diagonal of a square is 10 yards. Find the length of each side of the square.

Solve the following. (See Example 4.)

39. In Example 4a, a building contractor wanted to be sure that the walls of a building met at right angles. Suppose that he marks a point on one wall 8 feet from the corner and a point 15 feet from the corner on the other wall. If the walls are perpendicular, how far is it between the two points?

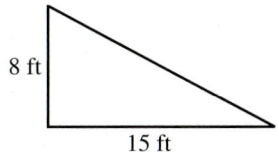
8 ft
15 ft

40. In Example 4b, we found the length of some rafters. How long would the rafters be for a house 48 feet wide if the rafters meet at the center of the house at a point 10 feet above the level of the walls?

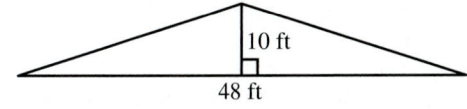
10 ft
48 ft

41. Practice Exercise 4a dealt with the measurement of television screens. If a television has a 13-inch screen that is 10 inches long, how wide is it?

42. Practice Exercise 4b dealt with a particular brand of marine radio. Suppose that a marine radio has a range of 25 miles. Bill and Don are traveling in separate boats equipped with these radios on courses that are at right angles to each other, as in Practice Exercise 9. If Bill is 12 miles from port and Don is 16 miles from the same port, can they contact each other by radio? Why or why not?

43. When hurricanes threaten, a common practice is to tape windows to keep the glass from shattering. If a rectangular window is 40 inches long and 30 inches wide, how long is a piece of tape that goes diagonally across the window?

44. Suppose that a plate glass window 72 inches long and 60 inches wide is taped as in Exercise 43. How long is a piece of tape that goes diagonally across this window?

45. A newly transplanted tree is supported by three ropes that are attached to the tree at a point 24 feet above the ground. These ropes are tied to pegs that are driven into the ground 7 feet from the base of the tree. What is the total length of the three pieces of rope?

46. A guy wire is attached to a telephone pole at a point 30 feet above the ground. How long is the wire if the other end is attached at a point on the ground 10 feet from the base of the pole?

47. An advertising sign is in the shape of a rectangle. A stripe is painted diagonally across the sign. If the length of the stripe is 8 feet and the sign is 6 feet long, find the width of the sign.

48. Two pieces of wood are nailed together at right angles. To give added support, a third piece of wood is nailed to each of these at points 3 feet from the corner. Find the length of this piece of wood.

Writing Exercise

49. Write and solve an application problem involving the Pythagorean theorem.

Group Project

50. Make a list of five everyday situations where right triangles have a practical application.

Chapter 10   Summary

Concept/Procedure	Example

Roots of a Number: [Section 10.1]

- A number b is an **nth** root of a number a if $b^n = a$. Written symbolically, $\sqrt[n]{a} = b$ if $b^n = a$. If n is even, then a must be nonnegative. If n is odd, then a may have any real-number value.

Radical: [Section 10.1]

- A **radical** is an algebraic expression of the form $\sqrt[n]{a}$. The symbol $\sqrt{}$ is the radical sign, n is the **root index**, and a is the **radicand**.

Perfect nth Powers: [Section 10.1]

- The perfect squares are $\{1, 4, 9, 16, 25, \ldots\}$.
- The perfect cubes are $\{\pm 1, \pm 8, \pm 27, \pm 64, \pm 125, \ldots\}$.
- The perfect fourth powers are $\{1, 16, 81, 256, \ldots\}$.
- And so on.
- Numbers that are perfect nth powers have rational nth roots.

Example 1:

Find the following roots:

a. $\sqrt{36}$ b. $\sqrt[3]{-64}$ c. $\sqrt[4]{16}$

Solution:

a. $\sqrt{36} = 6$, because $6^2 = 36$.

b. $\sqrt[3]{-64} = -4$, because $(-4)^3 = -64$.

c. $\sqrt[4]{16} = 2$, because $2^4 = 16$.

Irrational Numbers: [Section 10.1]

- An **irrational** number is any real number that is not rational. Written as a decimal, irrational numbers neither terminate nor repeat. Examples are $\sqrt{2}$, $\sqrt[3]{5}$, and $\sqrt[5]{8}$.

Products of Radicals: [Section 10.2]

- If $\sqrt[n]{a}$ and $\sqrt[n]{b}$ both exist, then $\sqrt[n]{a} \cdot \sqrt[n]{b} = \sqrt[n]{a \cdot b}$. This rule is also used in the form of $\sqrt[n]{ab} = \sqrt[n]{a} \cdot \sqrt[n]{b}$.

Example 2:

Simplify $\sqrt{192x^5}$.

Solution:

Rewrite $192x^5$ as the product of a perfect square and a number with no perfect-square factors.

$$\sqrt{192x^5} = \sqrt{64x^4 \cdot 3x} \qquad \text{Apply } \sqrt{ab} = \sqrt{a} \cdot \sqrt{b}.$$
$$= \sqrt{64x^4} \cdot \sqrt{3x} \qquad \sqrt{64x^4} = 8x^2.$$
$$= 8x^2\sqrt{3x} \qquad \text{Therefore, } \sqrt{192x^5} = 8x^2\sqrt{3x}.$$

Example 3:

Find the product of $2\sqrt{14} \cdot 5\sqrt{10}$.

Solution:

$$2\sqrt{14} \cdot 5\sqrt{10} = \qquad \text{Apply } \sqrt{a} \cdot \sqrt{b} = \sqrt{ab}.$$
$$10\sqrt{140} = \qquad \text{Write 140 as } 4 \cdot 35.$$
$$10\sqrt{4 \cdot 35} = \qquad \text{Apply } \sqrt{ab} = \sqrt{a} \cdot \sqrt{b}.$$
$$10\sqrt{4} \cdot \sqrt{35} = \qquad \sqrt{4} = 2.$$
$$10 \cdot 2\sqrt{35} = \qquad \text{Multiply.}$$
$$20\sqrt{35} \qquad \text{Therefore, } 2\sqrt{14} \cdot 5\sqrt{10} = 20\sqrt{35}.$$

Concept/Procedure	Example

Radicals to Powers: [Section 10.2]

- In general, if $\sqrt[n]{a}$ exists, then $\left(\sqrt[n]{a}\right)^n = a$.

Quotients of Radicals: [Section 10.3]

- If $\sqrt[n]{a}$ and $\sqrt[n]{b}$ both exist and $b \neq 0$, then $\dfrac{\sqrt[n]{a}}{\sqrt[n]{b}} = \sqrt[n]{\dfrac{a}{b}}$. This rule is also used in the form $\sqrt[n]{\dfrac{a}{b}} = \dfrac{\sqrt[n]{a}}{\sqrt[n]{b}}$.

Example 4:

Simplify $\dfrac{\sqrt{54}}{\sqrt{6}}$.

Solution: Apply $\dfrac{\sqrt{a}}{\sqrt{b}} = \sqrt{\dfrac{a}{b}}$.

$\dfrac{\sqrt{54}}{\sqrt{6}} = \sqrt{\dfrac{54}{6}}$ $\dfrac{54}{6} = 9$.

$= \sqrt{9}$ $\sqrt{9} = 3$.

$= 3$ Therefore, $\dfrac{\sqrt{54}}{\sqrt{6}} = 3$.

Like Radical Expressions: [Section 10.4]

- **Like radical expressions** must have the same root index and the same radicand.

Adding Like Radical Expressions: [Section 10.4]

- To add like radicals, add the coefficients and leave the radical portion unchanged. Never add the radicands. It is often necessary to simplify the radical expressions before they can be added.

Example 5:

Simplify $\sqrt{63a^5} + \sqrt{28a^5}$.

Solution:

Write $63a^5$ as $9a^4 \cdot 7a$ and $28a^5$ as $4a^4 \cdot 7a$.

$\sqrt{63a^5} + \sqrt{28a^5} = \sqrt{9a^4 \cdot 7a} + \sqrt{4a^4 \cdot 7a}$ Apply $\sqrt{ab} = \sqrt{a} \cdot \sqrt{}$

$= \sqrt{9a^4} \cdot \sqrt{7a} + \sqrt{4a^4} \cdot \sqrt{7a}$ Simplify square roots.

$= 3a^2\sqrt{7a} + 2a^2\sqrt{7a}$ Add like radicals.

$= 5a^2\sqrt{7a}$ Sum.

Simplest Form of Radicals: [Sections 10.4 and 10.5]

1. Find all rational roots.

2. Remove all possible factors from the radicand. For example, no square root may have a radicand with a perfect-square factor.

3. Find all possible products, quotients, sums, and differences.

4. Rationalize all the denominators. This also means no radicand may contain a fraction.

Concept/Procedure	Example

Rationalizing the Denominator: [Section 10.5]

1. If the denominator is in the form of $\sqrt{a}$, multiply the numerator and the denominator of the original fraction by the same appropriate radical so that the denominator of the original fraction becomes a radical expression whose value is rational.

2. If the denominator is of the form $a + b$, where a and/or b are square roots, multiply the numerator and the denominator of the fraction by the conjugate of the denominator. The conjugate is found by changing the middle sign of the expression $a + b$, where a and/or b are square roots.

Example 6:

Rationalize the denominator of $\dfrac{3}{\sqrt{8x}}$.

Solution:

Since $\sqrt{8x} \cdot \sqrt{2x} = \sqrt{16x^2} = 4x$, multiply by $\dfrac{\sqrt{2x}}{\sqrt{2x}}$.

$\dfrac{3}{\sqrt{8x}} \cdot \dfrac{\sqrt{2x}}{\sqrt{2x}} =$ Multiply.

$\dfrac{3\sqrt{2x}}{\sqrt{16x^2}} =$ $\sqrt{16x^2} = 4x$.

$\dfrac{3\sqrt{2x}}{4x}$ Therefore, $\dfrac{3}{\sqrt{8x}} = \dfrac{3\sqrt{2x}}{4x}$.

Example 7:

Rationalize the denominator of $\dfrac{3}{4 - \sqrt{5}}$.

Solution:

Multiply both numerator and denominator by the conjugate of the denominator, $4 + \sqrt{5}$.

$\dfrac{3}{4 - \sqrt{5}} \cdot \dfrac{4 + \sqrt{5}}{4 + \sqrt{5}} =$ Multiply.

$\dfrac{12 + 3\sqrt{5}}{16 - 5} =$ $16 - 5 = 11$.

$\dfrac{12 + 3\sqrt{5}}{11}$ Therefore, $\dfrac{3}{4 - \sqrt{5}} = \dfrac{12 + 3\sqrt{5}}{11}$.

Solutions of Equations Containing Radicals: [Section 10.6]

- If both sides of an equation are raised to a power, the resulting equation contains all the solutions of the original equation and perhaps some solutions that do not solve the original equation.

Extraneous Solutions: [Section 10.6]

- Any solution of an equation resulting from raising both sides of an equation to a power that is not a solution of the original equation is called an **extraneous** solution and is discarded.

Concept/Procedure	Example

Solving Equations Containing Square Roots: [Section 10.6]

1. If the equation contains only one square root, isolate the radical on one side of the equation. Then square both sides of the equation and solve the resulting equation. Check all solutions in the original equation.

2. If the equation contains two square roots, isolate one radical on one side of the equation. Then square both sides of the equation and simplify the resulting equation. Isolate the remaining radical on one side of the equation and square both sides again. Solve the resulting equation and check all solutions in the original equation.

Example 8:

Solve $\sqrt{3x + 3} - 2x = -1$.

Solution:

Isolate the radical and square both sides.

$\sqrt{3x + 3} - 2x = -1$	Add $2x$ to both sides.
$\sqrt{3x + 3} = 2x - 1$	Square both sides.
$3x + 3 = 4x^2 - 4x + 1$	Subtract $3x$ and 3 from both sides.
$0 = 4x^2 - 7x - 2$	Factor.
$0 = (4x + 1)(x - 2)$	Set each factor equal to 0.
$4x + 1 = 0, x - 2 = 0$	Solve each equation.
$4x = -1 \quad x = 2$	
$x = -\dfrac{1}{4}$	

A check will verify that $x = 2$ is the only solution, since $x = -\dfrac{1}{4}$ is extraneous.

Pythagorean Theorem: [Section 10.7]

- It the lengths of the legs of a right triangle are a and b and the length of the hypotenuse is c, then $a^2 + b^2 = c^2$.

Example 9:

The length of one leg of a right triangle is 5 m, and the hypotenuse is $\sqrt{37}$ m. Find the other leg.

Solution:

Use the Pythagorean theorem.

$a^2 + b^2 = c^2$	Substitute for a and c.
$5^2 + b^2 = (\sqrt{37})^2$	Simplify.
$25 + b^2 = 37$	Subtract 25.
$b^2 = 12$	Take the positive square root of each side.
$\sqrt{b^2} = \sqrt{12}$	Simplify.
$b = 2\sqrt{3}$	Therefore, the other leg is $2\sqrt{3}$ m.

Chapter 10   Review Exercises

Find the roots if they exist. If the root does not exist, write, "This root does not exist as a real number." Assume that all variables have nonnegative values. [Sections 10.1 and 10.2]

1. $\sqrt{16}$

2. $\sqrt{-16}$

3. $-\sqrt{16}$

4. $\sqrt[3]{125}$

5. $\sqrt[3]{-125}$

6. $-\sqrt[3]{125}$

7. $\sqrt[4]{81}$

8. $\sqrt[4]{-81}$

9. $-\sqrt[4]{81}$

10. $\sqrt{b^4}$

11. $\sqrt{n^5}$

12. $\sqrt[3]{m^5}$

13. $\sqrt[5]{z^8}$

14. $\sqrt{m^5 n^6}$

15. $\sqrt[3]{a^4 b^6}$

Identify each of the numbers as rational or irrational. If the number is rational, find the root exactly. If the number is irrational, approximate the number to the nearest thousandth by using a calculator or Appendix Table 1. [Section 10.1]

16. $\sqrt{32}$

17. $\sqrt{169}$

18. $\sqrt[3]{63}$

19. $\sqrt[3]{1331}$

20. $\sqrt{1.44}$

21. $\sqrt[3]{.027}$

Simplify the following radicals: [Section 10.2]

22. $\sqrt{63}$

23. $5\sqrt{108}$

24. $\sqrt[3]{54}$

25. $6\sqrt[3]{32}$

26. $\sqrt{162x^7}$

27. $\sqrt{200a^8}$

28. $3\sqrt{40z^5y^3}$

29. $\sqrt[3]{128a^4}$

30. $5\sqrt[3]{108n^3m^7}$

Find the products or powers of roots. Express answers in simplified form. Assume that all variables have nonnegative values. [Section 10.3]

31. $\left(\sqrt{r}\right)^2$

32. $\left(\sqrt{x^3}\right)^2$

33. $\sqrt{5}\cdot\sqrt{7}$

34. $\sqrt{m}\cdot\sqrt{n}$

35. $\sqrt{7}\cdot\sqrt{7}$

36. $\sqrt{5}\cdot\sqrt{35}$

37. $4\sqrt{6}\cdot 2\sqrt{24}$

38. $7\sqrt{18}\cdot 3\sqrt{6}$

39. $\sqrt{c^3}\cdot\sqrt{c}$

40. $\sqrt{s^5}\cdot\sqrt{s^3}$

41. $\sqrt{v}\cdot\sqrt{v^4}$

42. $2\sqrt{30n^3}\cdot 3\sqrt{10n^3}$

Find the quotients. Express all answers in simplified form. Assume that all variables have positive values. [Section 10.3]

43. $\sqrt{\dfrac{81x^2}{4y^4}}$

44. $\sqrt{\dfrac{13}{25}}$

45. $\dfrac{6\sqrt{72}}{3\sqrt{8}}$

46. $\dfrac{\sqrt{x^7}}{\sqrt{x^3}}$

47. $\dfrac{\sqrt{c^5d^4}}{\sqrt{c^2d^2}}$

48. $\dfrac{6a^3\sqrt{140n^5}}{2a\sqrt{5n^2}}$

Simplify. Express answers in simplified form. Assume that all variables have positive values. [Section 10.3]

49. $\sqrt{\dfrac{3}{2}}\cdot\sqrt{\dfrac{5}{8}}$

50. $\sqrt{\dfrac{n^2}{m^3}}\cdot\sqrt{\dfrac{n^5}{m}}$

51. $\sqrt{\dfrac{7a}{18b}}\cdot\sqrt{\dfrac{3a}{2b}}$

Find the sums. Express answers in simplified form. Assume that all variables have nonnegative values. [Section 10.4]

52. $7\sqrt{5} - 9\sqrt{5}$

53. $4a\sqrt[3]{x} - 6a\sqrt[3]{x}$

54. $\sqrt{48} - \sqrt{3}$

55. $4\sqrt{27} - 5\sqrt{12}$

56. $2a\sqrt{20} - 5a\sqrt{45}$

57. $4\sqrt{63a^3} + 6\sqrt{64a^3}$

Simplify the following: [Section 10.4]

58. $\sqrt{6}\cdot\sqrt{2} + \sqrt{6}\cdot\sqrt{8}$

59. $\dfrac{\sqrt{300}}{\sqrt{2}} - \sqrt{96}$

60. $\sqrt{5}\left(3 - \sqrt{30}\right)$

61. $\sqrt{3}\left(\sqrt{3} + 2\sqrt{6}\right)$

62. $\left(4 + \sqrt{5}\right)\left(5 - \sqrt{5}\right)$

63. $\left(2\sqrt{3} + \sqrt{5}\right)\left(\sqrt{3} - 4\sqrt{5}\right)$

64. $\left(3\sqrt{3} + \sqrt{6}\right)\left(3\sqrt{3} - \sqrt{6}\right)$

65. $\left(2 - \sqrt{7}\right)^2$

Reduce the following to lowest terms: [Section 10.4]

66. $\dfrac{6 - 2\sqrt{2}}{12}$

67. $\dfrac{8 + 4\sqrt{12}}{4}$

Rationalize the denominators. Express answers in simplest form. Assume that all variables have positive values. [Section 10.5]

68. $\dfrac{6}{\sqrt{7}}$

69. $\dfrac{8}{\sqrt{18}}$

70. $\sqrt{\dfrac{5}{3}}$

71. $\dfrac{a}{\sqrt{b}}$

72. $\dfrac{5}{\sqrt{5n}}$

73. $\dfrac{4}{\sqrt[3]{9}}$

74. $\dfrac{4a}{\sqrt[3]{2a}}$

75. $\dfrac{4}{2 - \sqrt{5}}$

76. $\dfrac{6}{\sqrt{3} - \sqrt{5}}$

77. $\dfrac{\sqrt{12}}{3 - \sqrt{6}}$

78. $\dfrac{2 + \sqrt{5}}{2 - \sqrt{5}}$

Solve the following equations: [Section 10.6]

79. $\sqrt{a} = 9$ **80.** $3\sqrt{n} - 18 = 0$

81. $\sqrt{x + 4} = 5$ **82.** $2\sqrt{2x + 4} + 6 = 2$

83. $4\sqrt{s} = \sqrt{15s - 5}$ **84.** $\sqrt{4x - 12} = x - 3$

85. $\sqrt{n + 2} + \sqrt{n - 6} = 4$

Find the value of x in each of the following: [Section 10.7]

86.

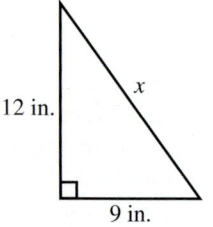

87.

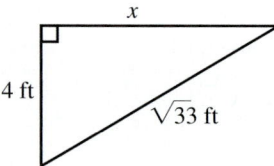

88.

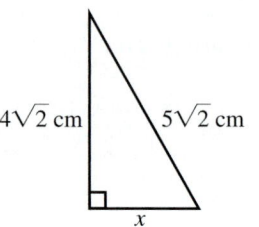

89.

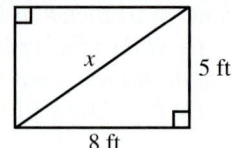

Solve the following: [Section 10.7]

90. The lengths of the legs of a right triangle are 9 inches and 10 inches. Find the length of the hypotenuse.

91. The length of the hypotenuse of a right triangle is 12 meters, and the length of one leg is 4 meters. Find the length of the other leg.

92. A rectangle has a diagonal length of 9 yards and a width of $3\sqrt{2}$ yards. Find the length.

93. The length of a diagonal of a square is $3\sqrt{10}$ feet. Find the length of each side of the square.

94. A television antenna is supported by two guy wires. The wires are attached to the antenna at a point 15 feet above the ground. The other ends of the wires are anchored to the ground at a point 5 feet from the base of the antenna. Find the length of each of the wires.

95. A tree fell during a storm, with the broken portion still resting on the stump, forming a right triangle. The point where the tree broke is 15 feet above the ground. The top of the tree is on the ground at a point 36 feet from the base of the tree. How tall was the tree before it broke?

Chapter 10   Test

Find the roots if they exist. If the root does not exist, write, "The root does not exist as a real number."

1. $\sqrt{81}$ **2.** $\sqrt[3]{-64}$

3. $\sqrt{x^8}$ **4.** $\sqrt[3]{c^9 d^6}$

Simplify the following:

5. $\sqrt{96}$ **6.** $3\sqrt{98x^5}$

7. $4\sqrt[3]{40}$ **8.** $3\sqrt[3]{64y^7}$

Find the products and/or quotients. Express answers in simplified form.

9. $2\sqrt{35} \cdot 3\sqrt{5}$ **10.** $3\sqrt{15a^3} \cdot 4\sqrt{10a^4}$

11. $\dfrac{\sqrt{96}}{\sqrt{6}}$ **12.** $\dfrac{8\sqrt{56c^7}}{\sqrt{7c^2}}$

13. $\sqrt{\dfrac{21c^3}{5}} \cdot \sqrt{\dfrac{3c}{5}}$

Find the following sums:

14. $6\sqrt{6} - 12\sqrt{6}$

15. $3\sqrt{80} + 5\sqrt{20}$

16. $\sqrt{180a^3} + 4\sqrt{45a^3}$

Simplify. Express answers in simplified form. Assume that all variables have nonnegative values.

17. $\sqrt{3}\left(\sqrt{6} + 4\sqrt{18}\right)$

18. $\left(\sqrt{5} + 2\sqrt{6}\right)\left(3\sqrt{5} - \sqrt{6}\right)$

19. $\left(4 + 2\sqrt{7}\right)\left(4 - 2\sqrt{7}\right)$

20. $\left(5 - \sqrt{x}\right)^2$

Rationalize the denominators of the following:

21. $\dfrac{8}{\sqrt{12}}$

22. $\dfrac{8}{\sqrt[3]{16}}$

23. $\sqrt{\dfrac{m}{n}}$

24. $\dfrac{12}{7 - \sqrt{5}}$

25. $\dfrac{5 + \sqrt{6}}{2 - \sqrt{6}}$

Solve the equations. If there is no solution, write $\varnothing$.

26. $2\sqrt{b} - 10 = 0$

27. $\sqrt{7z + 58} - 11 = 0$

28. $\sqrt{x + 27} - \sqrt{x + 6} = 3$

29. Find the value of x in the following right triangle:

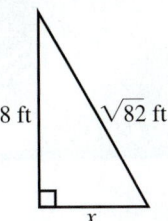

8 ft $\sqrt{82}$ ft

x

30. Two cars approach the same intersection at right angles to each other. If one car is 45 feet from the intersection and the other is 60 feet from the intersection, find the distance between the two cars.

11

Solving Quadratic Equations

Chapter Outline

11.1 Solving Incomplete Quadratic Equations

11.2 Solving Quadratic Equations by Completing the Square

11.3 Solving Quadratic Equations by the Quadratic Formula

11.4 Quadratic Equations with Complex Solutions

11.5 Applications Involving Quadratic Equations

Quadratic equations were introduced in Section 5.8. Recall that a quadratic equation is any equation that can be put in the form of $ax^2 + bx + c = 0$, with $a \neq 0$. We solved quadratic equations by factoring, and the solutions were rational numbers.

However, if asked to solve the equation $x^2 + x - 5 = 0$ by factoring, we could not, since $x^2 + x - 5$ cannot be factored. In this chapter, we will learn two methods of solving this type of equation. One method is called *completing the square* and involves creating a perfect-square trinomial, which is the square of a binomial. The second method is using the *quadratic formula*. The solutions of quadratic equations are often irrational numbers, which were discussed in Chapter 10.

Quadratic equations also allow us to solve a wide variety of application problems. We end the chapter with a section on application problems that can be solved only by the use of quadratic equations.

Section 11.1 — Solving Incomplete Quadratic Equations

OBJECTIVES *When you complete this section, you will be able to:*

A Solve quadratic equations of the form $ax^2 + bx = 0$.

B Solve quadratic equations of the form $x^2 = k$ by extraction of roots.

C Solve quadratic equations of the form $(ax + b)^2 = c$ by extraction of roots.

D Solve application problems involving incomplete quadratic equations.

PREREQUISITE SKILLS *Before beginning this section, you should be able to:*

a. Factor by removing the greatest common factor. (Section 5.2)

b. Solve linear equations. (Sections 3.1–3.3)

c. Find the square roots of perfect squares. (Section 10.1)

d. Simplify radicals. (Section 10.2)

e. Rationalize a denominator. (Section 10.5)

f. Square a radical expression. (Section 10.3)

GETTING READY FOR SECTION 11.1

1. Factor $4x^2 - 8x$.

2. Solve $4x + 3 = -3$.

3. Find $\sqrt{49}$.

4. Simplify $\sqrt{27}$.

5. Rationalize the denominator of $\dfrac{\sqrt{7}}{\sqrt{3}}$.

6. Simplify $(-3\sqrt{3})^2$.

Introduction

In Chapter 5, we solved quadratic equations by factoring. Recall that after the equation was written in $ax^2 + bx + c = 0$ form, we factored $ax^2 + bx + c$ and set each factor equal to 0. We then solved the resulting linear equations. For example, solve $x^2 + 2x - 15 = 0$.

$x^2 + 2x - 15 = 0$ Factor $x^2 + 2x - 15$.

$(x + 5)(x - 3) = 0$ Set each factor equal to 0.

$x + 5 = 0, x - 3 = 0$ Solve each equation.

$x = -5, x = 3$ Therefore, the solutions are -5 and 3.

This method, however, cannot be used to solve all quadratic equations, since not all quadratic polynomials are factorable. For example, $x^2 + x + 3 = 0$ cannot be solved by factoring, since $x^2 + x + 3$ cannot be factored. In this section, we will learn a technique that will be used in Section 11.2 to develop a procedure that will allow us to solve any quadratic equation.

OBJECTIVE Solving quadratic equations of the form $ax^2 + bx = 0$

A quadratic equation in the form of $ax^2 + bx + c = 0$, in which either b or $c = 0$, is called an **incomplete quadratic equation**. If $c = 0$, the equation is of the form $ax^2 + bx = 0$. This type can be solved by factoring and was discussed in Section 5.8. We will do a couple of examples to refresh your memory.

Answers: Getting Ready: 1. $4x(x - 2)$ **2.** $x = -\dfrac{3}{2}$ **3.** 7 **4.** $3\sqrt{3}$ **5.** $\dfrac{\sqrt{21}}{3}$ **6.** 27

Example 1   Solve the following equations:

a. $2x^2 + 3x = 0$ — Remove the common factor x.

$x(2x + 3) = 0$ — Set each factor equal to 0.

$x = 0$, or $2x + 3 = 0$ — $x = 0$ is one solution. Solve $2x + 3 = 0$.

$2x = -3$

$x = -\dfrac{3}{2}$ — Therefore, the solutions are 0 and $-\dfrac{3}{2}$.

CHECK:

$x = 0$ $x = -\dfrac{3}{2}$

Substitute 0 for x.   Substitute $-\frac{3}{2}$ for x.

$2x^2 + 3x = 0$ $2x^2 + 3x = 0$

$2(0)^2 + 3(0) = 0$ $2\left(-\dfrac{3}{2}\right)^2 + 3\left(-\dfrac{3}{2}\right) = 0$

$2(0) + 0 = 0$ $2\left(\dfrac{9}{4}\right) - \dfrac{9}{2} = 0$

$0 + 0 = 0$ $\dfrac{9}{2} - \dfrac{9}{2} = 0$

$0 = 0$ $0 = 0$

Therefore, $x = 0$ and $x = -\frac{3}{2}$ are the correct solutions.

b. $4x^2 = 8x$ — Subtract $8x$ from both sides of the equation.

$4x^2 - 8x = 0$ — Remove the common factor of $4x$.

$4x(x - 2) = 0$ — Set each factor equal to 0.

$4x = 0$, or $x - 2 = 0$ — Solve each equation.

$x = 0$ $x = 2$ — Therefore, the solutions are 0 and 2.

CHECK:

$x = 0$ $x = 2$

Substitute 0 for x.   Substitute 2 for x.

$4x^2 = 8x$ $4x^2 = 8x$

$4(0)^2 = 8(0)$ $4(2)^2 = 8(2)$

$4(0) = 0$ $4(4) = 16$

$0 = 0$ $16 = 16$

Therefore, $x = 0$ and $x = 2$ are the correct solutions.

PRACTICE EXERCISE 1

Solve the following equations:

a. $3x^2 + 9x = 0$

b. $5x^2 = 2x$

If you need more practice, do the following Additional Practice Exercises.

Additional Practice Exercise 1   Solve the following equations:

a. $4x^2 - 12x = 0$

b. $3x^2 = 7x$

OBJECTIVE B Solving quadratic equations of the form $x^2 = k$ by extraction of roots

Incomplete quadratic equations with $b = 0$ have the form $ax^2 + c = 0$. We solved some special cases of this type of equation in Section 5.8 by factoring. For example, $x^2 - 9 = 0$ can be solved by factoring. However, $x^2 - 3 = 0$ cannot be solved by factoring. The technique we will use in this section to solve equations of the form $ax^2 + c = 0$ is known as **extraction of roots** and was used in Section 10.7 on the Pythagorean theorem. We will now expand on this method.

Answers: Practice Exercise 1: a. $x = 0, -3$ **b.** $x = 0, \dfrac{2}{5}$ **Additional Practice 1: a.** $x = 0, 3$ **b.** $x = 0, \dfrac{7}{3}$

We begin with the equation $x^2 - 9 = 0$. This equation can be solved by factoring as follows:

$$x^2 - 9 = 0$$
$$(x + 3)(x - 3) = 0$$
$$x + 3 = 0, x - 3 = 0$$
$$x = -3, \quad x = 3$$

Therefore, the solutions are -3 and 3.

However, $x^2 - 9 = 0$ can be written as $x^2 = 9$. This equation is asking the question, "The square of what number is equal to 9?" This is precisely the definition of square root. Thus, we can solve the equation $x^2 = 9$ by extracting the two square roots of 9, which are 3 and -3. This procedure is summarized as follows:

Note

If k is a negative number, then $\sqrt{k}$ is not a real number. Therefore, the equation has no real-number solutions. As mentioned in Section 10.1, the square roots of negative numbers result in **imaginary numbers** and will be discussed in Section 11.4.

Procedure: Solving Quadratic Equations by Extraction of Roots

To solve quadratic equations of the form $ax^2 + c = 0$, follow these steps:

1. Solve the equation for x^2. This results in an equation of the form $x^2 = k$, where k is a constant.

2. The solutions of $x^2 = k$ are both the positive and negative square roots of k, written as $x = \pm\sqrt{k}$ (read "x equals the positive and negative square roots of k"). That is, $x = \sqrt{k}$ or $x = -\sqrt{k}$.

3. Write the answers in simplified form.

Example 2

Solve the equations. If the solutions are not real numbers, write "no real-number solution."

a. $x^2 = 16$ Apply if $x^2 = k$, then $x = \pm\sqrt{k}$.

$x = \pm\sqrt{16}$ $\pm\sqrt{16} = \pm 4$

$x = \pm 4$ Therefore, the solutions are 4 and -4.

CHECK: $x = 4$ $x = -4$

Substitute 4 for x.   Substitute -4 for x.

$x^2 = 16$ $x^2 = 16$

$4^2 = 16$ $(-4)^2 = 16$

$16 = 16$ $16 = 16$

Therefore, ± 4 are the correct solutions.

b. $x^2 - 27 = 0$ Add 27 to both sides of the equation.

$x^2 = 27$ Apply if $x^2 = k$, then $x = \pm\sqrt{k}$.

$x = \pm\sqrt{27}$ Simplify the $\sqrt{27}$.

$x = \pm 3\sqrt{3}$ Therefore, the solutions are $3\sqrt{3}$ and $-3\sqrt{3}$.

CHECK: $x = 3\sqrt{3}$ $x = -3\sqrt{3}$

Substitute $3\sqrt{3}$ for x.   Substitute $-3\sqrt{3}$ for x.

$x^2 - 27 = 0$ $x^2 - 27 = 0$

$(3\sqrt{3})^2 - 27 = 0$ $(-3\sqrt{3})^2 - 27 = 0$

$3^2(\sqrt{3})^2 - 27 = 0$ $(-3)^2(\sqrt{3})^2 - 27 = 0$

$9 \cdot 3 - 27 = 0$ $9 \cdot 3 - 27 = 0$

$27 - 27 = 0$ $27 - 27 = 0$

$0 = 0$ $0 = 0$

Therefore, $\pm 3\sqrt{3}$ are the correct solutions.

c. $2x^2 - 6 = 0$ Add 6 to both sides of the equation.

$2x^2 = 6$ Divide both sides of the equation by 2.

$x^2 = 3$ Apply if $x^2 = k$, then $x = \pm\sqrt{k}$.

$x = \pm\sqrt{3}$ Therefore, the solutions are $\sqrt{3}$ and $-\sqrt{3}$.

CHECK: $x = \sqrt{3}$ $x = -\sqrt{3}$.

Substitute $\sqrt{3}$ for x.   Substitute $-\sqrt{3}$ for x.

$2x^2 - 6 = 0$ $2x^2 - 6 = 0$

$2(\sqrt{3})^2 - 6 = 0$ $2(-\sqrt{3})^2 - 6 = 0$

$2(3) - 6 = 0$ $2(3) - 6 = 0$

$6 - 6 = 0$ $6 - 6 = 0$

$0 = 0$ $0 = 0$

Therefore, $\pm\sqrt{3}$ are the correct solutions.

d. $3x^2 - 7 = 0$ Add 7 to both sides of the equation.

$3x^2 = 7$ Divide both sides by 3.

$x^2 = \dfrac{7}{3}$ Apply if $x^2 = k$, then $x = \pm\sqrt{k}$.

$x = \pm\sqrt{\dfrac{7}{3}}$ Apply the quotient rule for radicals.

$x = \pm\dfrac{\sqrt{7}}{\sqrt{3}}$ Rationalize the denominator.

$x = \pm\dfrac{\sqrt{7}}{\sqrt{3}}\cdot\dfrac{\sqrt{3}}{\sqrt{3}}$ Apply the product rule for radicals.

$x = \pm\dfrac{\sqrt{21}}{\sqrt{9}}$ $\sqrt{9} = 3$.

$x = \pm\dfrac{\sqrt{21}}{3}$ Therefore, the solutions are $\dfrac{\sqrt{21}}{3}$ and $-\dfrac{\sqrt{21}}{3}$.

CHECK:

The check is left as an exercise for the student.

e. $x^2 + 4 = 0$ Subtract 4 from both sides of the equation.

$x^2 = -4$ Apply if $x^2 = k$, then $x = \pm\sqrt{k}$.

$x = \pm\sqrt{-4}$ Since $\sqrt{-4}$ does not exist as a real number, this equation has no real-number solutions.

$\varnothing$

Be Careful

Do not confuse solving equations like $x^2 = 9$ with finding $\sqrt{9}$. Remember, the symbol $\sqrt{}$ means find the positive square root only. Hence, $\sqrt{9} = 3$ only. There is no radical sign in $x^2 = 9$, so we are not limited to just the positive square root. We are asked to find the number(s) whose square is 9. *Consequently, we find both the positive and negative square roots of 9 which are* ± 3.

PRACTICE EXERCISE 2

Solve the equations. If the solutions are not real numbers, write "no real-number solutions."

a. $x^2 = 49$ **b.** $x^2 - 8 = 0$

c. $4x^2 - 24 = 0$ **d.** $5x^2 - 2 = 0$

e. $x^2 + 16 = 0$

If you need more practice, do the Additional Practice Exercises. If the solutions are not real numbers, write "no real-number solutions."

Additional Practice Exercise 2   Solve the equations:

a. $x^2 = 81$ **b.** $x^2 - 18 = 0$

c. $6x^2 - 42 = 0$ **d.** $7x^2 - 5 = 0$

e. $2x^2 + 18 = 0$

OBJECTIVE Solving quadratic equations of the form $(ax + b)^2 = c$ by extraction of roots

The method of extraction of roots can be extended to solve equations like $(2x + 3)^2 = 9$. We illustrate this with some examples.

Answers: Practice Exercise 2: **a.** $x = \pm 7$ **b.** $x = \pm 2\sqrt{2}$ **c.** $x = \pm\sqrt{6}$ **d.** $x = \pm\dfrac{\sqrt{10}}{5}$ **e.** no real-number solutions

Additional Practice 2: **a.** $x = \pm 9$ **b.** $x = \pm 3\sqrt{2}$ **c.** $x = \pm\sqrt{7}$ **d.** $x = \pm\dfrac{\sqrt{35}}{7}$ **e.** no real-number solutions

Example 3   **Solve the equations. If the solutions are not real numbers, write "no real-number solutions."**

a. $(x + 3)^2 = 16$

Apply if $x^2 = k$, then $x = \pm \sqrt{k}$.

$x + 3 = \pm 4$ — Rewrite as two equations.

$x + 3 = 4, x + 3 = -4$ — Solve each equation.

$x = 1, \qquad x = -7$ — Therefore, the solutions are 1 and −7.

CHECK: $x = 1$ $\qquad$ $x = -7$

Substitute 1 for x. $\qquad$ Substitute −7 for x.

$(x + 3)^2 = 16$ $\qquad$ $(x + 3)^2 = 16$

$(1 + 3)^2 = 16$ $\qquad$ $(-7 + 3)^2 = 16$

$4^2 = 16$ $\qquad\quad$ $(-4)^2 = 16$

$16 = 16$ $\qquad\quad$ $16 = 16$

Therefore, 1 and −7 are the correct solutions.

b. $(x + 3)^2 = 5$

Apply if $x^2 = k$, then $x = \pm \sqrt{k}$.

$x + 3 = \pm \sqrt{5}$ — Rewrite as two equations.

$x + 3 = \sqrt{5}, x + 3 = -\sqrt{5}$ — Solve each equation.

$x = -3 + \sqrt{5}, \qquad x = -3 - \sqrt{5}$ — Therefore, the solutions are $-3 + \sqrt{5}$ and $-3 - \sqrt{5}$.

Note

The solutions of Example 3b are usually written as $-3 \pm \sqrt{5}$.

CHECK: $x = -3 + \sqrt{5}$ $\qquad$ $x = -3 - \sqrt{5}$

Substitute $-3 + \sqrt{5}$ for x. $\qquad$ Substitute $-3 - \sqrt{5}$ for x.

$(x + 3)^2 = 5$ $\qquad$ $(x + 3)^2 = 5$

$(-3 + \sqrt{5} + 3)^2 = 5$ $\qquad$ $(-3 - \sqrt{5} + 3)^2 = 5$

$(\sqrt{5})^2 = 5$ $\qquad$ $(-\sqrt{5})^2 = 5$

$5 = 5$ $\qquad\qquad$ $5 = 5$

Therefore, $-3 + \sqrt{5}$ and $-3 - \sqrt{5}$ are the correct solutions.

c. $(2x + 4)^2 = 25$

Apply if $x^2 = k$, then $x = \pm \sqrt{k}$.

$2x + 4 = \pm 5$ — Rewrite as two equations.

$2x + 4 = 5, 2x + 4 = -5$ — Solve each equation.

$2x = 1, \quad 2x = -9$

$x = \dfrac{1}{2}, \qquad x = -\dfrac{9}{2}$ — Therefore, the solutions are $\dfrac{1}{2}$ and $-\dfrac{9}{2}$.

CHECK: The check is left as an exercise for the student.

d. $(3x - 1)^2 = 7$

Apply if $x^2 = k$, then $x = \pm \sqrt{k}$.

$3x - 1 = \pm \sqrt{7}$ — Rewrite as two equations.

$3x - 1 = \sqrt{7}, 3x - 1 = -\sqrt{7}$ — Solve each equation.

$3x = 1 + \sqrt{7}, 3x = 1 - \sqrt{7}$

$x = \dfrac{1 + \sqrt{7}}{3}, \qquad x = \dfrac{1 - \sqrt{7}}{3}$ — Therefore, the solutions are $\dfrac{1 + \sqrt{7}}{3}$ and $\dfrac{1 - \sqrt{7}}{3}$.

Note

The solutions to Example 3d are often written as $\dfrac{1 \pm \sqrt{7}}{3}$.

CHECK: $x = \dfrac{1 + \sqrt{7}}{3}$ $\qquad$ $x = \dfrac{1 - \sqrt{7}}{3}$

Substitute $\dfrac{1 + \sqrt{7}}{3}$ for x. $\qquad$ Substitute $\dfrac{1 - \sqrt{7}}{3}$ for x.

$(3x - 1)^2 = 7$ $\qquad$ $(3x - 1)^2 = 7$

$\left[3\left(\dfrac{1 + \sqrt{7}}{3}\right) - 1 \right]^2 = 7$ $\qquad$ $\left[3\left(\dfrac{1 - \sqrt{7}}{3}\right) - 1 \right]^2 = 7$

$(1 + \sqrt{7} - 1)^2 = 7$ $\qquad$ $(1 - \sqrt{7} - 1)^2 = 7$

$(\sqrt{7})^2 = 7$ $\qquad$ $(-\sqrt{7})^2 = 7$

$7 = 7$ $\qquad\qquad$ $7 = 7$

Therefore, $\dfrac{1 \pm \sqrt{7}}{3}$ are the correct solutions.

e. $(2a - 3)^2 + 12 = 0$ Subtract 12 from both sides of the equation.
$(2a - 3)^2 = -12$ Apply if $x^2 = k$, then $x = \pm\sqrt{k}$.
$2a - 3 = \pm\sqrt{-12}$ Since $\sqrt{-12}$ does not exist as a real number, this equation has no real-number solutions.

PRACTICE EXERCISE 3

Solve the equations. If there are no real solutions, write "no real-number solutions."

a. $(x + 5)^2 = 36$

b. $(x - 4)^2 = 8$

c. $(3x - 2)^2 = 4$

d. $(5x - 3)^2 = 10$

e. $(3x - 1)^2 + 16 = 0$

If you need more practice, do the following Additional Practice Exercises.

Additional Practice Exercise 3 Solve the following equations:

a. $(x - 6)^2 = 64$

b. $(x + 4)^2 = 2$

c. $(4x + 3)^2 = 9$

d. $(5x - 1)^2 = 11$

e. $(x + 5)^2 + 25 = 0$

OBJECTIVE **D** Solving application problems involving incomplete quadratic equations

Incomplete quadratic equations are also used in the solution of application problems.

Example 4

a. The square of 3 less than a number is 36. Find the numbers.

b. Find the length of each side of a square whose area is 49 square inches.

Solution

Let $x = $ the number. Then, $x - 3$ is 3 less than the number. Therefore, the equation is

$(x - 3)^2 = 36$ Apply if $x^2 = k$, then $x = \pm\sqrt{k}$.
$x - 3 = \pm 6$ Rewrite as two equations.
$x - 3 = 6, x - 3 = -6$ Solve each equation.
$x = 9, \quad x = -3$ Therefore, the numbers are 9 and -3.

CHECK:

3 less than 9 is $9 - 3 = 6$, and $6^2 = 36$. Therefore, 9 is a solution. 3 less than -3 is $-3 - 3 = -6$, and $(-6)^2 = 36$. Therefore, -3 is also a solution.

Solution

The formula for the area of a square is $A = s^2$, where s is the length of a side of the square.

$A = s^2$ Substitute 49 for A.
$49 = s^2$ Apply if $x^2 = k$, then $x = \pm\sqrt{k}$.
$\pm 7 = s$ Since the length of a side of a square must be positive, the only solution is 7 inches.
$7 = s$

Answers: Practice Exercise 3: a. $x = 1, -11$ **b.** $x = 4 \pm 2\sqrt{2}$ **c.** $x = 0, \dfrac{4}{3}$ **d.** $x = \dfrac{3 \pm \sqrt{10}}{5}$ **e.** no real-number solutions

Additional Practice 3: a. $x = 14, -2$ **b.** $x = -4 \pm \sqrt{2}$ **c.** $x = 0, -\dfrac{3}{2}$ **d.** $x = \dfrac{1 \pm \sqrt{11}}{5}$ **e.** no real-number solutions

PRACTICE EXERCISE 4

a. The square of the sum of a number and 4 is 64. Find the numbers.

b. Find the length of each side of a square whose area is 169 square centimeters.

If you need more practice, do the following Additional Practice Exercises.

Additional Practice Exercise 4

a. The square of the difference of a number and 5 is 81. Find the numbers.

b. Find the length of each side of a square whose area is 196 square inches.

For Extra Help

Exercise Set 11.1 MyMathLab®

Solve the equations. If there are no real solutions, write "no real-number solutions." (See Examples 1–3.)

1. $4x^2 + 8x = 0$ **2.** $2x^2 + 10x = 0$ **3.** $7a^2 = -21a$ **4.** $5r^2 = -45r$

5. $9z^2 = 6z$ **6.** $12t^2 = 30t$ **7.** $x(x + 2) = 4x$ **8.** $n(n - 4) = 2n$

9. $r^2 = 25$ **10.** $y^2 = 100$ **11.** $a^2 = -36$ **12.** $p^2 = -64$

13. $x^2 - 121 = 0$ **14.** $x^2 - 144 = 0$ **15.** $x^2 - 7 = 0$ **16.** $x^2 - 11 = 0$

17. $x^2 + 5 = 0$ **18.** $t^2 + 7 = 0$ **19.** $x^2 - 48 = 0$ **20.** $x^2 - 32 = 0$

21. $w^2 - 28 = 0$ **22.** $m^2 - 45 = 0$ **23.** $2x^2 - 8 = 0$ **24.** $3x^2 - 27 = 0$

25. $4x^2 - 20 = 0$ **26.** $5x^2 - 30 = 0$ **27.** $2m^2 - 40 = 0$ **28.** $3r^2 - 54 = 0$

29. $5q^2 - 40 = 0$ **30.** $7b^2 - 56 = 0$ **31.** $7x^2 + 28 = 0$ **32.** $3a^2 + 30 = 0$

33. $2x^2 - 7 = 0$ **34.** $3x^2 - 5 = 0$ **35.** $5a^2 - 11 = 0$ **36.** $7r^2 - 5 = 0$

37. $3z^2 - 4 = 0$ **38.** $6a^2 - 25 = 0$ **39.** $(x + 4)^2 = 25$ **40.** $(a - 2)^2 = 100$

41. $(x + 5)^2 = 44$ **42.** $(n - 7)^2 = 54$ **43.** $(3c + 4)^2 = 49$ **44.** $(2a - 5)^2 = 64$

45. $(2r - 3)^2 = 121$ **46.** $(3s - 6)^2 = 81$ **47.** $(x - 3)^2 = 7$ **48.** $(x + 4)^2 = 3$

49. $(m + 5)^2 = 11$ **50.** $(n - 9)^2 = 13$ **51.** $(2a + 4)^2 = 2$ **52.** $(3s - 2)^2 = 5$

53. $(s - 2)^2 = 98$ **54.** $(n - 5)^2 = 75$ **55.** $(3x + 2)^2 = 52$ **56.** $(2a - 6)^2 = 72$

Answers: Practice Exercise 4: a. $4, -12$ **b.** 13 cm   **Additional Practice 4: a.** $x = 14, -4$ **b.** 14 inches

Solve the application problems. Remember from Section 5.8 that if the solution of an application problem requires a quadratic equation, not all solutions of the equation will always satisfy the conditions of the problem. Always check the solution(s) against the wording of the problem. (See Example 4.)

57. The square of 3 more than a number is 25. Find the number.

58. The square of 2 less than three times a number is 16. Find the number.

Exercises 59 through 62 require the use of the formula for the area of a square. The formula for the area of a square is $A = s^2$, where s represents the length of each side.

59. Find the length of each side of a square if the area is 64 square inches.

60. Find the length of each side of a square if the area is 121 square meters.

61. If the length of each side of a square is doubled and then increased by 4, the area would be 100 square centimeters. Find the length of each side of the square.

62. If the length of each side of a square is tripled and then decreased by 3, the area would be 36 square feet. Find the length of each side of the square.

Challenge Exercises (63–66)

Solve the equations. If there are no real-number solutions, write "no real-number solutions."

63. $3(x + 5)^2 = 7$

64. $2(a - 3)^2 = 5$

65. $5(2x + 3)^2 - 11 = 0$

66. $7(4x - 1)^2 - 15 = 0$

Writing Exercises (67 and 68)

67. Compare and contrast the meanings of $\sqrt{49}$ and $x^2 = 49$. How are they similar and how are they different?

68. Explain in detail why the equation $(x + 2)^2 = -9$ has no real-number solutions.

Section 11.2 Solving Quadratic Equations by Completing the Square

OBJECTIVES *When you complete this section, you will be able to:*

A Determine the number to be added to a polynomial of the form $x^2 + bx$ in order to form a perfect-square trinomial.

B Solve quadratic equations by completing the square.

PREREQUISITE SKILLS *Before beginning this section, you should be able to:*

a. Square a binomial. (Section 2.4)

b. Square a rational number. (Section 2.1)

c. Factor a perfect-square trinomial. (Section 5.6)

d. Multiply rational numbers. (Section 6.3)

e. Solve quadratic equations by extraction of roots. (Section 11.1)

f. Find the square roots of perfect squares. (Section 10.1)

g. Simplify radicals. (Section 10.2)

h. Perform operations on radical expressions. (Section 10.4)

i. Rationalize a denominator. (Section 10.5)

GETTING READY FOR SECTION 11.2

Simplify the following:

1. Simplify $(2 + \sqrt{2})^2$.

2. $\left(a + \dfrac{7}{2}\right)^2$.

3. Simplify $\left(\dfrac{7}{6}\right)^2$.

4. Factor $a^2 + 5a + \dfrac{25}{4}$.

5. Simplify $\dfrac{1}{2} \cdot \dfrac{5}{4}$.

For Exercises 6 and 7, solve the equations:

6. $\left(a + \dfrac{5}{2}\right)^2 = \dfrac{9}{4}$.

7. $(x - 2)^2 = 2$.

8. Find $\sqrt{\dfrac{9}{4}}$.

9. Simplify $\sqrt{45}$.

10. Add $-\dfrac{7}{2} + \dfrac{3\sqrt{2}}{2}$.

11. Combine by putting over a common denominator: $-1 - \dfrac{\sqrt{6}}{3}$.

12. Rationalize the denominator: $\sqrt{\dfrac{2}{3}}$.

Introduction

Recall from Section 5.6 that a perfect-square trinomial is a trinomial that is the square of a binomial. For example, $(x + 3)^2 = (x)^2 + 2(x)(3) + 3^2 = x^2 + 6x + 9$. Since $x^2 + 6x + 9 = (x + 3)^2$, it is a perfect-square trinomial. By examining $x^2 + 6x + 9 = (x + 3)^2$ and the procedure for squaring a binomial, we make the following observations about a binomial whose square gives a perfect-square trinomial:

1. The first term of the binomial is the square root of the first term of the trinomial.

2. The second term of the binomial is the square root of the last term of the trinomial.

3. The sign of the binomial is the same as the sign of the middle term of the trinomial.

Answers: Getting Ready: 1. $6 + 4\sqrt{2}$ **2.** $a^2 + 7a + \dfrac{49}{4}$ **3.** $\dfrac{49}{36}$ **4.** $\left(a + \dfrac{5}{2}\right)^2$ **5.** $\dfrac{5}{8}$ **6.** $a = -1, -4$ **7.** $x = 2 \pm \sqrt{2}$ **8.** $\dfrac{3}{2}$ **9.** $3\sqrt{5}$ **10.** $\dfrac{-7 + 3\sqrt{2}}{2}$

11. $\dfrac{-3 - \sqrt{6}}{3}$ **12.** $\dfrac{\sqrt{6}}{3}$

Using the preceding example, we have

$$x^2 + 6x + 9$$

$$(\sqrt{x^2} + \sqrt{9})^2.$$

That is, $x^2 + 6x + 9 = (\sqrt{x^2} + \sqrt{9})^2 = (x + 3)^2$.

OBJECTIVE Determining the number to be added to a polynomial of the form $x^2 + bx$ in order to form a perfect-square trinomial

Suppose we know the first two terms of a trinomial whose squared term has a coefficient of 1, and we want to find the constant term necessary for the trinomial to be a perfect-square trinomial. For example, what constant term is necessary for $x^2 + 6x +$ _____ to be a perfect-square trinomial, and what is the binomial whose square gives this perfect-square trinomial?

If we represent the last term of the preceding trinomial with a^2, we have $x^2 + 6x + a^2$. We need to find a. The first term of the binomial whose square gives $x^2 + 6x + a^2$ is $\sqrt{x^2} = x$. The second term of the binomial is $\sqrt{a^2} = a$. Since the sign of the second term of the trinomial is $+$, we have

$$x^2 + 6x + a^2 = (x + a)^2.$$

Expand $(x + a)^2$, and we have

$$x^2 + 6x + a^2 = x^2 + 2ax + a^2.$$

Since two polynomials are equal if and only if their corresponding terms are equal, we conclude that

$$6 = 2a$$
$$\frac{6}{2} = a$$
$$3 = a.$$

Since the last term of the perfect-square trinomial is a^2 and $a = 3$, the number needed in order for $x^2 + 6x +$ _____ to be a perfect-square trinomial is $3^2 = 9$. Hence, $x^2 + 6x + 9$ is a perfect-square trinomial and $x^2 + 6x + 9 = (x + 3)^2$.

The procedure for finding the number that will make a polynomial of the form $x^2 + bx$ a perfect-square trinomial is called **completing the square**. We do not want to go through all these steps each time we need to complete the square. The key to the procedure lies in the steps that we repeat as follows:

$$6 = 2a$$
$$\frac{6}{2} = a$$
$$3 = a$$

We then squared 3 to get the last term of the perfect-square trinomial. Since 6 is the coefficient of the first-degree term of the trinomial, we found the last term of the perfect-square trinomial by taking half the coefficient of the first-degree term and squaring it. We generalize this discussion to the following procedure for completing the square:

Procedure: Completing the Square

To find the number needed to make a polynomial of the form $x^2 + bx$ a perfect-square trinomial, take $\frac{1}{2}$ of b and square it. Consequently, add $\left(\dfrac{b}{2}\right)^2 = \dfrac{b^2}{4}$. Since the square of a negative number is positive, we can ignore the sign of b.

Example 1 Find the number needed to make each of the polynomials a perfect-square trinomial. Then write each trinomial as the square of a binomial.

a. $x^2 + 8x + \underline{\quad} =$ Take $\dfrac{1}{2}$ of 8 and square it. $\dfrac{1}{2}(8) = 4$, and $4^2 = 16$.

$x^2 + 8x + 16 =$ Therefore, the number needed is 16.

$(x + 4)^2$ Write as the square of a binomial.

b. $a^2 - 10a + \underline{\quad} =$ Take $\dfrac{1}{2}$ of 10 and square it. $\dfrac{1}{2}(10) = 5$, and $5^2 = 25$.

$a^2 - 10a + 25 =$ Therefore, the number needed is 25.

$(a - 5)^2$ Write as the square of a binomial.

c. $b^2 + 3b + \underline{\quad} =$ Take $\dfrac{1}{2}$ of 3 and square it. $\dfrac{1}{2}(3) = \dfrac{3}{2}$, and $\left(\dfrac{3}{2}\right)^2 = \dfrac{9}{4}$.

$b^2 + 3b + \dfrac{9}{4} =$ Therefore, the number needed is $\dfrac{9}{4}$.

$\left(b + \dfrac{3}{2}\right)^2$ Write as the square of a binomial.

d. $r^2 - \dfrac{7}{3}r + \underline{\quad} =$ Take $\dfrac{1}{2}$ of $\dfrac{7}{3}$ and square it. $\dfrac{1}{2}\left(\dfrac{7}{3}\right) = \dfrac{7}{6}$, and $\left(\dfrac{7}{6}\right)^2 = \dfrac{49}{36}$.

$r^2 - \dfrac{7}{3}r + \dfrac{49}{36} =$ Therefore, the number needed is $\dfrac{49}{36}$.

$\left(r - \dfrac{7}{6}\right)^2$ Write as the square of a binomial.

PRACTICE EXERCISE 1

Find the number needed to make each polynomial a perfect-square trinomial. Then write each trinomial as the square of a binomial.

a. $x^2 + 2x + \underline{\quad}$ **b.** $n^2 - 14n + \underline{\quad}$ **c.** $m^2 + 7m + \underline{\quad}$ **d.** $y^2 - \dfrac{5}{2}y + \underline{\quad}$

If you need more practice, do the following Additional Practice Exercises.

Additional Practice Exercise 1 Find the number needed to make each polynomial a perfect-square trinomial. Then write each trinomial as the square of a binomial.

a. $x^2 + 12x + \underline{\quad}$ **b.** $r^2 - 4r + \underline{\quad}$ **c.** $c^2 + 9c + \underline{\quad}$ **d.** $r^2 - \dfrac{3}{5}r + \underline{\quad}$

OBJECTIVE Solving quadratic equations by completing the square

By combining the techniques of completing the square and the extracting of roots, we can solve any quadratic equation with the procedure outlined in the following box:

> **Procedure: Solving Quadratic Equations by Completing the Square**
>
> To solve a quadratic equation by completing the square, follow these steps:
>
> **1.** If the coefficient of the x^2 term is not 1, divide both sides of the equation by the coefficient of x^2.
>
> **2.** Write the equation in the form $x^2 + bx = c$.
>
> **3.** Add the constant needed to make $x^2 + bx$ a perfect square to both sides of the equation.
>
> **4.** Write the perfect-square trinomial as the square of a binomial.
>
> **5.** Solve by extraction of roots.
>
> **6.** Check all solutions in the original equation.

Answers: Practice Exercise 1: a. 1; $(x + 1)^2$ **b.** 49; $(n - 7)^2$ **c.** $\dfrac{49}{4}$; $\left(m + \dfrac{7}{2}\right)^2$ **d.** $\dfrac{25}{16}$; $\left(y - \dfrac{5}{4}\right)^2$

Additional Practice 1: a. 36; $(x + 6)^2$ **b.** 4; $(r - 2)^2$ **c.** $\dfrac{81}{4}$; $\left(c + \dfrac{9}{2}\right)^2$ **d.** $\dfrac{9}{100}$; $\left(r - \dfrac{3}{10}\right)^2$

Although the equations in Example 2 could be solved by factoring, for illustrative purposes we will solve them by completing the square.

Example 2 **Solve the following equations by completing the square:**

a. $x^2 - 2x - 8 = 0$ Add 8 to both sides of the equation.

$x^2 - 2x = 8$ Find the number that must be added to $x^2 - 2x$ to make it a perfect-square trinomial, and add it to both sides of the equation.

$x^2 - 2x + \underline{\hspace{1cm}} = 8 + \underline{\hspace{1cm}}$ Take $\frac{1}{2}$ of 2 and square it. $\frac{1}{2}(2) = 1$, and $1^2 = 1$.

Therefore, add 1 to both sides of the equation.

$x^2 - 2x + 1 = 8 + 1$ Write $x^2 - 2x + 1$ as the square of a binomial.

$(x - 1)^2 = 9$ Solve, using extraction of roots.

$x - 1 = \pm 3$ Write as two equations.

$x - 1 = 3, x - 1 = -3$ Solve each equation.

$x = 4, \quad\quad x = -2$ Therefore, the solutions are 4 and -2.

CHECK: $x = 4$ $x = -2$

Substitute 4 for x.   Substitute -2 for x.

$x^2 - 2x - 8 = 0$ $x^2 - 2x - 8 = 0$

$4^2 - 2(4) - 8 = 0$ $(-2)^2 - 2(-2) - 8 = 0$

$16 - 8 - 8 = 0$ $4 + 4 - 8 = 0$

$0 = 0$ $0 = 0$

Therefore, 4 and -2 are the correct solutions.

b. $a^2 + 5a + 4 = 0$ Subtract 4 from both sides of the equation.

$a^2 + 5a = -4$ Find the number that must be added to $a^2 + 5a$ to make it a perfect-square trinomial, and add it to both sides of the equation.

$a^2 + 5a + \underline{\hspace{1cm}} = -4 + \underline{\hspace{1cm}}$ Take $\frac{1}{2}$ of 5 and square it. $\frac{1}{2}(5) = \frac{5}{2}$, and $\left(\frac{5}{2}\right)^2 = \frac{25}{4}$.

Therefore, add $\frac{25}{4}$ to both sides of the equation.

$a^2 + 5a + \dfrac{25}{4} = -4 + \dfrac{25}{4}$ Write the left side of the equation as the square of a binomial and write the right side with the LCD of 4.

$\left(a + \dfrac{5}{2}\right)^2 = -\dfrac{16}{4} + \dfrac{25}{4}$ Simplify the right side.

$\left(a + \dfrac{5}{2}\right)^2 = \dfrac{9}{4}$ Solve by using extraction of roots.

$a + \dfrac{5}{2} = \pm \dfrac{3}{2}$ Write as two equations.

$a + \dfrac{5}{2} = \dfrac{3}{2}, a + \dfrac{5}{2} = -\dfrac{3}{2}$ Solve each equation.

$a = \dfrac{3}{2} - \dfrac{5}{2}, a = -\dfrac{3}{2} - \dfrac{5}{2}$

$a = -\dfrac{2}{2} = -1, \quad a = -\dfrac{8}{2} = -4$ Therefore, the solutions are -1 and -4.

CHECK: To check, substitute -1 and -4 in the original equation. This is left as an exercise for the student.

PRACTICE EXERCISE 2

Solve the following equations by completing the square:

a. $x^2 - 8x + 12 = 0$

b. $r^2 + 3r - 18 = 0$

If you need more practice, do the following Additional Practice Exercises.

Additional Practice Exercise 2   Solve the following equations by completing the square:

a. $x^2 - 6x + 8 = 0$

b. $b^2 + 3b + 2 = 0$

As previously indicated, the equations in Example 2 could have been solved by factoring. The real advantage to solving equations by completing the square is in solving equations that cannot be solved by factoring.

Example 3   **Solve the equations by completing the square. If the equation has no real solutions, write "no real-number solutions."**

a. $x^2 - 4x + 2 = 0$
    Subtract 2 from both sides of the equation.

$x^2 - 4x = -2$
    Find the number that must be added to $x^2 - 4x$ to make it a perfect-square trinomial and add it to both sides of the equation.

$x^2 - 4x + \underline{\qquad} = -2 + \underline{\qquad}$
    Take $\frac{1}{2}$ of 4 and square it. $\frac{1}{2}(4) = 2$, and $2^2 = 4$. Therefore, add 4 to both sides.

$x^2 - 4x + 4 = -2 + 4$
    Write the left side as the square of a binomial.

$(x - 2)^2 = 2$
    Solve, using extraction of roots.

$x - 2 = \pm\sqrt{2}$
    Write as two equations.

$x - 2 = \sqrt{2}, x - 2 = -\sqrt{2}$
    Solve each equation.

$x = 2 + \sqrt{2}, \quad x = 2 - \sqrt{2}$
    Therefore, the solutions are $2 \pm \sqrt{2}$.

CHECK:

$x = 2 + \sqrt{2}$	$x = 2 - \sqrt{2}$
Substitute $2 + \sqrt{2}$ for x.	Substitute $2 - \sqrt{2}$ for x.
$x^2 - 4x + 2 = 0$	$x^2 - 4x + 2 = 0$
$(2 + \sqrt{2})^2 - 4(2 + \sqrt{2}) + 2 = 0$	$(2 - \sqrt{2})^2 - 4(2 - \sqrt{2}) + 2 = 0$
$4 + 4\sqrt{2} + 2 - 8 - 4\sqrt{2} + 2 = 0$	$4 - 4\sqrt{2} + 2 - 8 + 4\sqrt{2} + 2 = 0$
$0 = 0$	$0 = 0$

Therefore, the correct solutions are $2 \pm \sqrt{2}$.

b. $a^2 = -7a - 1$
    Write the equation in $a^2 + ba = c$ form by adding $7a$ to both sides of the equation.

$a^2 + 7a = -1$
    Find the number that must be added to $a^2 + 7a$ to make it a perfect-square trinomial, and add it to both sides of the equation.

$a^2 + 7a + \underline{\qquad} = -1 + \underline{\qquad}$
    Take $\frac{1}{2}$ of 7 and square it. $\frac{1}{2}(7) = \frac{7}{2}$, and $\left(\frac{7}{2}\right)^2 = \frac{49}{4}$. Therefore, add $\frac{49}{4}$ to both sides.

$a^2 + 7a + \frac{49}{4} = -1 + \frac{49}{4}$
    Write the left side as the square of a binomial, and write the right side with the LCD of 4.

Note

Radical expressions like $a + \sqrt{b}$ and $a - \sqrt{b}$ are called **conjugate pairs**. Irrational solutions to quadratic equations *always* occur as conjugate pairs.

Answers: Practice Exercise 2: a. $x = 2, 6$ **b.** $r = 3, -6$ **Additional Practice 2: a.** $x = 2, 4$
b. $b = -2, -1$

$$\left(a + \frac{7}{2}\right)^2 = -\frac{4}{4} + \frac{49}{4}$$ Simplify the right side.

$$\left(a + \frac{7}{2}\right)^2 = \frac{45}{4}$$ Solve, using extraction of roots.

$$a + \frac{7}{2} = \pm\sqrt{\frac{45}{4}}$$ Apply the quotient rule for radicals, and write 45 as $9 \cdot 5$.

$$a + \frac{7}{2} = \pm\frac{\sqrt{9 \cdot 5}}{\sqrt{4}}$$ Apply the product rule to $\sqrt{9 \cdot 5}$ and $\sqrt{4} = 2$.

$$a + \frac{7}{2} = \pm\frac{\sqrt{9} \cdot \sqrt{5}}{2}$$ $\sqrt{9} = 3$.

$$a + \frac{7}{2} = \pm\frac{3\sqrt{5}}{2}$$ Write as two equations.

$$a + \frac{7}{2} = \frac{3\sqrt{5}}{2}, a + \frac{7}{2} = -\frac{3\sqrt{5}}{2}$$ Solve each equation.

$$a = -\frac{7}{2} + \frac{3\sqrt{5}}{2}, a = -\frac{7}{2} - \frac{3\sqrt{5}}{2}$$ Add the fractions.

$$a = \frac{-7 + 3\sqrt{5}}{2}, a = \frac{-7 - 3\sqrt{5}}{2}$$ Therefore, the solutions are $x = \frac{-7 \pm 3\sqrt{5}}{2}$.

CHECK:

The check is complicated and is left as Challenge Exercise 55.

c. $x^2 + 8x + 20 = 0$ Subtract 20 from both sides of the equation.

$$x^2 + 8x + \underline{\hspace{1cm}} = -20 + \underline{\hspace{1cm}}$$ Take $\frac{1}{2}$ of 8 and square it. $\frac{1}{2}(8) = 4$, and $4^2 = 16$. Therefore, add 16 to both sides.

$$x^2 + 8x + 16 = -20 + 16$$ Write the left side as the square of a binomial. Simplify the right side.

$$(x + 4)^2 = -4$$ Solve, using extraction of roots.

$$x + 4 = \pm\sqrt{-4}$$ Since $\sqrt{-4}$ is not a real number, this equation has no real-number solutions.

$$\emptyset$$

PRACTICE EXERCISE 3

Solve the equations by completing the square. If there is no real solution, write "no real-number solutions."

a. $x^2 - 4x - 6 = 0$ **b.** $b^2 - 5 = -5b$

c. $x^2 - 10x + 34 = 0$

If you need more practice, do the following Additional Practice Exercises.

Additional Practice Exercise 3 Solve the equations by completing the square: If there is no real solution, write "no real-number solutions."

a. $r^2 = -6r + 4$ **b.** $x^2 - 4 = -5x$

c. $a^2 - 6a + 13 = 0$

Answers: **Practice Exercise 3: a.** $x = 2 \pm \sqrt{10}$ **b.** $b = \dfrac{-5 \pm 3\sqrt{5}}{2}$ **c.** no real-number solutions

Additional Practice 3: **a.** $-3 \pm \sqrt{13}$ **b.** $x = \dfrac{-5 \pm \sqrt{41}}{2}$ **c.** no real-number solutions

The procedure for completing the square works only if the coefficient of the second-degree term is 1. If the coefficient of the second-degree term is not 1, we must divide both sides of the equation by that coefficient in order to make it 1. We illustrate with some examples.

Example 4 | **Solve the following equations by completing the square:**

a. $2x^2 - 12 = 5x$ Rewrite in the form $ax^2 + bx = c$.

$2x^2 - 5x = 12$ Divide both sides of the equation by 2 to get x^2.

$\dfrac{2x^2 - 5x}{2} = \dfrac{12}{2}$ Simplify both sides.

$x^2 - \dfrac{5}{2}x = 6$ Find the number that must be added to $x^2 - \dfrac{5}{2}x$ to make it a perfect-square trinomial, and add it to both sides of the equation.

$x^2 - \dfrac{5}{2}x + \underline{\hphantom{xxx}} = 6 + \underline{\hphantom{xxx}}$ Take $\dfrac{1}{2}$ of $\dfrac{5}{2}$ and square it. $\dfrac{1}{2} \cdot \dfrac{5}{2} = \dfrac{5}{4}$, and $\left(\dfrac{5}{4}\right)^2 = \dfrac{25}{16}$.

Add $\dfrac{25}{16}$ to both sides.

$x^2 - \dfrac{5}{2}x + \dfrac{25}{16} = 6 + \dfrac{25}{16}$ Write the left side as the square of a binomial, and write the right side with the LCD of 16.

$\left(x - \dfrac{5}{4}\right)^2 = \dfrac{96}{16} + \dfrac{25}{16}$ Simplify the right side.

$\left(x - \dfrac{5}{4}\right)^2 = \dfrac{121}{16}$ Solve, using extraction of roots.

$x - \dfrac{5}{4} = \pm\dfrac{11}{4}$ Write as two equations.

$x - \dfrac{5}{4} = \dfrac{11}{4}, x - \dfrac{5}{4} = -\dfrac{11}{4}$ Solve each equation.

$x = \dfrac{11}{4} + \dfrac{5}{4}, x = -\dfrac{11}{4} + \dfrac{5}{4}$ Add the fractions.

$x = \dfrac{16}{4} = 4, x = -\dfrac{6}{4} = -\dfrac{3}{2}$ Therefore, the solutions are 4 and $-\dfrac{3}{2}$.

CHECK: The check is left as an exercise for the student.

b. $3x^2 + 6x + 1 = 0$ Subtract 1 from both sides of the equation.

$3x^2 + 6x = -1$ Divide both sides by 3 to get x^2.

$\dfrac{3x^2 + 6x}{3} = \dfrac{-1}{3}$ Simplify the left side.

$x^2 + 2x = -\dfrac{1}{3}$ Find the number that must be added to $x^2 + 2x$ to make it a perfect-square trinomial and add it to both sides.

$x^2 + 2x + \underline{\hphantom{xxx}} = -\dfrac{1}{3} + \underline{\hphantom{xxx}}$ Take $\dfrac{1}{2}$ of 2 and square it. $\dfrac{1}{2}(2) = 1$, and $1^2 = 1$.

Add 1 to both sides.

$x^2 + 2x + 1 = -\dfrac{1}{3} + 1$ Write the left side as a binomial squared, and write the right side with the LCD of 3.

$(x + 1)^2 = -\dfrac{1}{3} + \dfrac{3}{3}$ Simplify the right side.

$(x + 1)^2 = \dfrac{2}{3}$ Solve, using extraction of roots.

$x + 1 = \pm\sqrt{\dfrac{2}{3}}$ Apply the quotient rule for radicals.

$$x + 1 = \pm \frac{\sqrt{2}}{\sqrt{3}}$$ Rationalize the denominator.

$$x + 1 = \pm \frac{\sqrt{2}}{\sqrt{3}} \cdot \frac{\sqrt{3}}{\sqrt{3}}$$ Simplify the right side.

$$x + 1 = \pm \frac{\sqrt{6}}{3}$$ Rewrite as two equations.

$$x + 1 = \frac{\sqrt{6}}{3}, x + 1 = -\frac{\sqrt{6}}{3}$$ Solve each equation.

$$x = -1 + \frac{\sqrt{6}}{3}, x = -1 - \frac{\sqrt{6}}{3}$$ Write the right side of each equation with the LCD of 3.

$$x = -\frac{3}{3} + \frac{\sqrt{6}}{3}, x = -\frac{3}{3} - \frac{\sqrt{6}}{3}$$ Add the fractions.

$$x = \frac{-3 + \sqrt{6}}{3}, x = \frac{-3 - \sqrt{6}}{3}$$ Therefore, the solutions are $\frac{-3 \pm \sqrt{6}}{3}$.

CHECK: The check is complicated and is left as Challenge Exercise 56.

PRACTICE EXERCISE 4

Solve the following equations by completing the square:

a. $3x^2 + 2 = 5x$ **b.** $2x^2 + 6x = -3$

If you need more practice, do the following Additional Practice Exercises.

Additional Practice Exercise 4   Solve the following equations by completing the square:

a. $-3x + 5 = 2x^2$ **b.** $3x^2 + 12x = 5$

As you have no doubt noticed, solving quadratic equations by completing the square can be very tedious and sometimes difficult. In the next section, we will use completing the square to develop another method that is usually easier.

For Extra Help

Exercise Set 11.2   MyMathLab®

Find the number needed to make each polynomial a perfect-square trinomial. Then, write each trinomial as the square of a binomial. (See Example 1.)

1. $a^2 + 6a +$ _____

2. $r^2 + 14r +$ _____

3. $x^2 - 18x +$ _____

4. $a^2 - 20a +$ _____

5. $n^2 + 5n +$ _____

6. $m^2 + 11m +$ _____

7. $b^2 - 7b +$ _____

8. $r^2 - 9r +$ _____

9. $x^2 + \frac{5}{2}x +$ _____

10. $c^2 + \frac{5}{3}c +$ _____

11. $a^2 - \frac{3}{7}a +$ _____

12. $n^2 - \frac{3}{5}n +$ _____

Answers: Practice Exercise 4: a. $x = 1, \frac{2}{3}$ **b.** $x = \frac{-3 \pm \sqrt{3}}{2}$ **Additional Practice 4: a.** $x = 1, -\frac{5}{2}$ **b.** $x = \frac{-6 \pm \sqrt{51}}{3}$

Solve each equation by completing the square. Most of Exercises 13 through 32 have rational solutions or no real solutions. If the equation has no real solutions, write "no real-number solutions." (See Example 2.)

13. $r^2 - 8r + 15 = 0$

14. $x^2 - 6x + 5 = 0$

15. $x^2 = -8x - 12$

16. $x^2 = 6x - 8$

17. $z^2 - 4z = 12$

18. $b^2 - 2b = 24$

19. $x^2 - 10 = -3x$

20. $a^2 - 36 = -9a$

21. $n^2 + 9n = -20$

22. $n^2 - 5n = 14$

23. $x^2 - \dfrac{7}{2}x + 3 = 0$

24. $x^2 - \dfrac{13}{2}x + 10 = 0$

25. $x^2 - \dfrac{14}{3}x = 8$

26. $y^2 - \dfrac{24}{5}y = 1$

27. $2x^2 = 3x + 20$

28. $2y^2 = 15y - 18$

29. $3x^2 + 6 = 11x$

30. $3x^2 - 24 = 14x$

31. $a^2 - 4a + 6 = 0$

32. $n^2 + 6n + 10 = 0$

Solve each equation by completing the square. Most of Exercises 33 through 52 have irrational solutions or no real solutions. If the equation has no real solutions, write "no real-number solutions." (See Examples 3 and 4.)

33. $x^2 + 4x - 1 = 0$

34. $a^2 + 6a - 4 = 0$

35. $r^2 - 8r = -5$

36. $z^2 - 10z = -6$

37. $a^2 - 2 = 14a$

38. $t^2 - 38 = 6t$

39. $x^2 + 8x + 3 = 0$

40. $y^2 + 6y + 2 = 0$

41. $x^2 - 10x = 5$

42. $z^2 - 2z = 6$

43. $t^2 - 3 = -5t$

44. $z^2 - 1 = -7z$

45. $2x^2 = 4x + 3$

46. $2a^2 = 5a + 4$

47. $2x^2 = -6x + 5$

48. $4x^2 = 12x + 1$

49. $3x^2 + 6x = 1$

50. $3s^2 - 6s = 1$

51. $2x^2 + 3x + 5 = 0$

52. $3x^2 - 5x + 8 = 0$

Challenge Exercises (53–56)

Solve the following by completing the square:

53. $x^2 + \dfrac{1}{3}x - \dfrac{2}{9} = 0$

54. $x^2 + \dfrac{5}{2}x + \dfrac{21}{16} = 0$

55. Check Example 3b.

56. Check Example 4b.

Writing Exercises (57 and 58)

57. Why do some quadratic equations have no real-number solutions?

58. What is the advantage to solving quadratic equations by completing the square?

Critical-Thinking Exercise

59. Write a quadratic equation that has no real solutions and explain how you arrived at that equation.

OBJECTIVE   *When you complete this section, you will be able to:*

Ⓐ Solve quadratic equations by the quadratic formula.

PREREQUISITE SKILLS   *Before beginning this section you should be able to:*

   a. Solve quadratic equations by completing the square.
   b. Substitute values for variables and evaluate. (Sections 1.1 and 2.5)
   c. Find the square roots of perfect squares. (Section 10.1)
   d. Simplify radicals. (Section 10.2)
   e. Reduce an expression to lowest terms. (Section 6.1)

GETTING READY FOR SECTION 11.3

1. Solve $x^2 + 6x - 4 = 0$ by completing the square.

   Given $\dfrac{-b \pm \sqrt{b^2 - 4ac}}{2a}$, evaluate for the following values of a, b, and c:

2. $a = 1, b = -2, c = -8$

3. $a = 2, b = -7, c = 1$

4. Simplify $\sqrt{52}$.

6. Reduce $\dfrac{-6 + 2\sqrt{13}}{2}$ to lowest terms.

Introduction

Thus far, we have learned two methods of solving quadratic equations: factoring and completing the square. The disadvantage of factoring is that not all quadratic equations can be solved by factoring. The disadvantage of completing the square is that it is complicated, and there are many places where errors can be made. In this section, we will learn a third method of solving quadratic equations that also allows us to solve *any* quadratic equation. This method involves using the **quadratic formula**, which we will now develop.

Let us begin with the general form of a quadratic equation and solve for x by completing the square. This will result in a formula that gives the solutions of the equation in terms of its coefficients.

$ax^2 + bx + c = 0$ Subtract c from both sides of the equation.

$ax^2 + bx = -c$ Divide both sides by a to get x^2.

$\dfrac{ax^2 + bx}{a} = -\dfrac{c}{a}$ Simplify the left side.

$x^2 + \dfrac{b}{a}x = -\dfrac{c}{a}$ Find the expression needed to make the left side a perfect-square trinomial, and add it to both sides.

$x^2 + \dfrac{b}{a}x + \underline{} = \underline{} -\dfrac{c}{a}$ $\dfrac{1}{2} \cdot \dfrac{b}{a} = \dfrac{b}{2a}$, and $\left(\dfrac{b}{2a}\right)^2 = \dfrac{b^2}{4a^2}$. Therefore, add $\dfrac{b^2}{4a^2}$ to both sides.

$x^2 + \dfrac{b}{a}x + \dfrac{b^2}{4a^2} = \dfrac{b^2}{4a^2} - \dfrac{c}{a}$ Write the left side as the square of a binomial, and write the right side with the LCD of $4a^2$.

$\left(x + \dfrac{b}{2a}\right)^2 = \dfrac{b^2}{4a^2} - \dfrac{c}{a} \cdot \dfrac{4a}{4a}$ Simplify the right side.

Answers: Getting Ready: 1. $-3 \pm \sqrt{13}$ **2.** $-2, 4$ **3.** $\dfrac{7 \pm \sqrt{41}}{4}$ **4.** $2\sqrt{13}$ **5.** $-3 + \sqrt{13}$

$$\left(x + \frac{b}{2a}\right)^2 = \frac{b^2}{4a^2} - \frac{4ac}{4a^2}$$ Add the fractions on the right side.

$$\left(x + \frac{b}{2a}\right)^2 = \frac{b^2 - 4ac}{4a^2}$$ Solve, using extraction of roots.

$$x + \frac{b}{2a} = \pm\sqrt{\frac{b^2 - 4ac}{4a^2}}$$ Apply the quotient rule for radicals.

$$x + \frac{b}{2a} = \pm\frac{\sqrt{b^2 - 4ac}}{\sqrt{4a^2}}$$ $\sqrt{4a^2} = 2a$ if $a > 0$.

$$x + \frac{b}{2a} = \pm\frac{\sqrt{b^2 - 4ac}}{2a}$$ Subtract $\frac{b}{2a}$ from both sides.

$$x = -\frac{b}{2a} \pm \frac{\sqrt{b^2 - 4ac}}{2a}$$ Add the fractions.

$$x = \frac{-b \pm \sqrt{b^2 - 4ac}}{2a}$$ Therefore, the solutions of $ax^2 + bx + c = 0$ are $\frac{-b + \sqrt{b^2 - 4ac}}{2a}$ and $\frac{-b - \sqrt{b^2 - 4ac}}{2a}$.

Note

The letters a, b, and c in the quadratic formula come from the equation written in the form $ax^2 + bx + c = 0$. This means that the solutions of a quadratic equation are determined by its coefficients.

Rule: The Quadratic Formula

The solutions of any equation written in the form $ax^2 + bx + c = 0$ are

$$x = \frac{-b \pm \sqrt{b^2 - 4ac}}{2a}.$$

The first task necessary in using the quadratic formula is to find the values of a, b, and c. This is done by rewriting the given equation, if necessary, in the form of $ax^2 + bx + c = 0$ and comparing the coefficients.

Example 1   Find the values of a, b, and c in each of the following:

a. $3x^2 + 2x - 5 = 0$ Compare with $ax^2 + bx + c = 0$.

$a = 3, b = 2, c = -5$

b. $x^2 - 3x = 5$ Subtract 5 from both sides of the equation.

$x^2 - 3x - 5 = 0$ Compare with $ax^2 + bx + c = 0$.

$a = 1, b = -3, c = -5$

c. $x^2 = x - 7$ Subtract x from and add 7 to both sides of the equation.

$x^2 - x + 7 = 0$ Compare with $ax^2 + bx + c = 0$.

$a = 1, b = -1, c = 7$

d. $3 = x^2 + 7x$ Subtract 3 from both sides of the equation.

$0 = x^2 + 7x - 3$ Rewrite in the form $ax^2 + bx + c = 0$.

$x^2 + 7x - 3 = 0$ Compare with $ax^2 + bx + c = 0$.

$a = 1, b = 7, c = -3$

PRACTICE EXERCISE 1

Find the values of a, b, and c in each of the following:

a. $4x^2 + 5x - 6 = 0$ **b.** $x^2 - x = 9$ **c.** $6x + 2 = -4x^2$ **d.** $x^2 + 5 = -4x$

OBJECTIVE   Solving quadratic equations by the quadratic formula

To solve quadratic equations by the quadratic formula, we use the following procedure:

Answers: Practice Exercise 1: a. $a = 4, b = 5, c = -6$ **b.** $a = 1, b = -1, c = -9$ **c.** $a = 4, b = 6, c = 2$ **d.** $a = 1, b = 4, c = 5$

> **Procedure: Solving Quadratic Equations with the Quadratic Formula**
>
> 1. If necessary, rewrite the equation in the form $ax^2 + bx + c = 0$.
> 2. Find a, b, and c.
> 3. Write the quadratic formula.
> 4. Substitute the values for a, b, and c into the formula.
> 5. Simplify the results.
> 6. Check the solution(s) in the original equation.

In Example 2, the equations could be solved by factoring, but for illustrative purposes we solve them by the quadratic formula.

Example 2 | Solve the following equations by using the quadratic formula:

a. $x^2 - 2x - 8 = 0$ — Find a, b, and c.

$a = 1, b = -2, c = -8$ — Write the quadratic formula.

$$x = \frac{-b \pm \sqrt{b^2 - 4ac}}{2a}$$ — Substitute for a, b, and c.

$$x = \frac{-(-2) \pm \sqrt{(-2)^2 - 4(1)(-8)}}{2(1)}$$ — Simplify. Watch the signs!

$$x = \frac{2 \pm \sqrt{4 + 32}}{2}$$ — Add 4 and 32.

$$x = \frac{2 \pm \sqrt{36}}{2}$$ — $\sqrt{36} = 6$.

$$x = \frac{2 \pm 6}{2}$$ — Write as two separate fractions.

$$x = \frac{2 + 6}{2}, x = \frac{2 - 6}{2}$$ — Simplify each fraction.

$$x = \frac{8}{2}, \quad x = \frac{-4}{2}$$ — Continue simplifying.

$$x = 4, \quad x = -2$$ — Therefore, the solutions are 4 and −2.

b. $6x^2 + x = 12$ — Rewrite in the form $ax^2 + bx + c = 0$.

$6x^2 + x - 12 = 0$ — Find a, b, and c.

$a = 6, b = 1, c = -12$ — Write the quadratic formula.

$$x = \frac{-b \pm \sqrt{b^2 - 4ac}}{2a}$$ — Substitute for a, b, and c.

$$x = \frac{-1 \pm \sqrt{1^2 - 4(6)(-12)}}{2(6)}$$ — Simplify.

$$x = \frac{-1 \pm \sqrt{1 + 288}}{12}$$ — Continue simplifying.

$$x = \frac{-1 \pm \sqrt{289}}{12}$$ — $\sqrt{289} = 17$.

$$x = \frac{-1 \pm 17}{12}$$ — Rewrite as two separate fractions.

$$x = \frac{-1 + 17}{12}, x = \frac{-1 - 17}{12}$$ — Simplify each fraction.

$$x = \frac{16}{12}, \quad x = \frac{-18}{12}$$ — Reduce each to lowest terms.

$$x = \frac{4}{3}, \quad x = -\frac{3}{2}$$ — Therefore, the solutions are $\frac{4}{3}$ and $-\frac{3}{2}$.

Note

These equations are checked as in Section 11.2. Consequently, the checks are omitted here.

PRACTICE EXERCISE 2

Solve the following equations by using the quadratic formula:

a. $x^2 + 7x + 10 = 0$

b. $6x^2 - x - 2 = 0$

(Continued)

If you need more practice, do the following Additional Practice Exercises.

The real advantage to solving quadratic equations by the quadratic formula is in solving equations that do not factor. Although this type of equation can be solved by completing the square, the quadratic formula is usually easier.

Example 3   **Solve the following by using the quadratic formula. If the equation has no real solutions, write "no real-number solutions."**

a. $x^2 + 3x - 1 = 0$ Find a, b, and c.

$a = 1, b = 3, c = -1$ Write the quadratic formula.

$x = \dfrac{-b \pm \sqrt{b^2 - 4ac}}{2a}$ Substitute for a, b, and c.

$x = \dfrac{-3 \pm \sqrt{3^2 - 4(1)(-1)}}{2(1)}$ Simplify.

$x = \dfrac{-3 \pm \sqrt{9 + 4}}{2}$ Continue simplifying.

$x = \dfrac{-3 \pm \sqrt{13}}{2}$ Therefore, the solutions are $\dfrac{-3 + \sqrt{13}}{2}$ and $\dfrac{-3 - \sqrt{13}}{2}$.

b. $x(x + 6) = 4$ Distribute x.

$x^2 + 6x = 4$ Subtract 4 from both sides of the equation to put in $ax^2 + bx + c = 0$ form.

$x^2 + 6x - 4 = 0$ Find a, b, and c.

$a = 1, b = 6, c = -4$ Write the quadratic formula.

$x = \dfrac{-b \pm \sqrt{b^2 - 4ac}}{2a}$ Substitute for a, b, and c.

$x = \dfrac{-6 \pm \sqrt{6^2 - 4(1)(-4)}}{2(1)}$ Simplify.

$x = \dfrac{-6 \pm \sqrt{36 + 16}}{2}$ Continue simplifying.

$x = \dfrac{-6 \pm \sqrt{52}}{2}$ Simplify $\sqrt{52}$.

$x = \dfrac{-6 \pm 2\sqrt{13}}{2}$ Factor 2 from the numerator.

$x = \dfrac{2(-3 \pm \sqrt{13})}{2}$ Divide by the common factor of 2.

$x = -3 \pm \sqrt{13}$ Therefore, the solutions are $-3 + \sqrt{13}$ and $-3 - \sqrt{13}$.

c. $2x^2 = 7x - 1$ Subtract $7x$ from and add 1 to both sides of the equation to put it in $ax^2 + bx + c = 0$ form.

$2x^2 - 7x + 1 = 0$ Find a, b, and c.

$a = 2, b = -7, c = 1$ Write the quadratic formula.

$x = \dfrac{-b \pm \sqrt{b^2 - 4ac}}{2a}$ Substitute for a, b, and c.

$x = \dfrac{-(-7) \pm \sqrt{(-7)^2 - 4(2)(1)}}{2(2)}$ Simplify.

$x = \dfrac{7 \pm \sqrt{49 - 8}}{4}$ Continue simplifying.

$x = \dfrac{7 \pm \sqrt{41}}{4}$ Therefore, the solutions are $\dfrac{7 + \sqrt{41}}{4}$ and $\dfrac{7 - \sqrt{41}}{4}$.

d. $2x^2 + 6 = -2x$ Add $2x$ to both sides of the equation and write in $ax^2 + bx + c = 0$ form.

$2x^2 + 2x + 6 = 0$ Find a, b, and c.

$a = 2, b = 2, c = 6$ Write the quadratic formula.

$x = \dfrac{-b \pm \sqrt{b^2 - 4ac}}{2a}$ Substitute for a, b, and c.

$x = \dfrac{-2 \pm \sqrt{2^2 - 4(2)(6)}}{2(2)}$ Simplify.

$x = \dfrac{-2 \pm \sqrt{4 - 48}}{4}$ Continue simplifying.

$x = \dfrac{-2 \pm \sqrt{-44}}{4}$ Since $\sqrt{-44}$ does not exist as a real number, this equation has no real-number solutions.

Answers: Additional Practice 2: a. $x = 3, -4$ **b.** $x = \dfrac{1}{4}, -\dfrac{3}{2}$

PRACTICE EXERCISE 3

Solve the equations by using the quadratic formula. If the equation has no real-number solutions, write "no real-number solutions."

a. $x^2 + 6x = -6$ **b.** $x^2 = 4x + 23$ **c.** $3x^2 - 9x + 3 = 0$ **d.** $x(3x + 2) = -6$

If you need more practice, do the following Additional Practice Exercises.

Additional Practice Exercise 3   Solve the equations by using the quadratic formula. If the equation has no real-number solutions, write "no real-number solutions."

a. $x^2 - 4x = -1$ **b.** $x(x + 6) = 3$ **c.** $2x^2 - 6x - 1 = 0$ **d.** $2x^2 + 5 = -2x$

Given a choice of which method to use in solving a quadratic equation, we recommend the following procedure:

> ### Procedure: Solving Quadratic Equations
>
> 1. Try to solve by factoring.
>
> 2. If $b = 0$, use extraction of roots.
>
> 3. If the equation cannot be solved by factoring, or if it is difficult to factor, then solve it by using the quadratic formula.

You might ask why we learned to solve quadratic equations by completing the square if we do not use it. Remember, we needed to know how to solve quadratic equations by completing the square in order to derive the quadratic formula. Though completing the square is rarely used in solving quadratic equations, this procedure has many important applications, including graphing of second-degree equations.

Exercise Set 11.3   For Extra Help   MyMathLab®

Solve answers using the quadratic formula. Most of the solutions to Exercises 1 through 20 are rational numbers. If the equation has no real solutions, write "no real-number solutions." (See Example 2.)

1. $x^2 - 6x + 5 = 0$ **2.** $r^2 - 8r + 15 = 0$ **3.** $x^2 = 6x - 8$ **4.** $x^2 = -8x - 12$

5. $b^2 - 2b = 24$ **6.** $z^2 - 4z = 12$ **7.** $a^2 + 6 = -7a$ **8.** $x^2 - 10 = -3x$

9. $n^2 - 5n = 14$ **10.** $n^2 + 9n = -20$ **11.** $2y^2 - 15y + 18 = 0$ **12.** $2x^2 - 3x - 20 = 0$

Answers: Practice Exercise 3: a. $-3 \pm \sqrt{3}$ **b.** $2 \pm 3\sqrt{3}$ **c.** $\dfrac{3 \pm \sqrt{5}}{2}$ **d.** no real-number solutions

Additional Practice 3: a. $2 \pm \sqrt{3}$ **b.** $-3 \pm 2\sqrt{3}$ **c.** $\dfrac{3 \pm \sqrt{11}}{2}$ **d.** no real-number solutions

13. $3x^2 - 14x = 24$ **14.** $3x^2 - 11x = -6$ **15.** $n^2 = -6n - 10$ **16.** $a^2 = -4a - 6$

17. $8x^2 - 14x - 15 = 0$ **18.** $9x^2 - 6x - 8 = 0$ **19.** $15a^2 = 29a + 2$ **20.** $6x^2 = 25x - 25$

Solve the quadratic equations by using the quadratic formula. Most of Exercises 21 through 48 have irrational solutions. If the equation has no real solutions write "no real-number solutions." (See Example 3.)

21. $x^2 + 3x - 1 = 0$ **22.** $x^2 + 5x - 2 = 0$ **23.** $a^2 = -a + 3$ **24.** $b^2 - 5b = -3$

25. $2c^2 - 3c = 4$ **26.** $2n^2 - 5n + 1 = 0$ **27.** $5r^2 - r - 3 = 0$ **28.** $4y^2 = 5y + 2$

29. $3q^2 + 3 = -2q$ **30.** $2m^2 + 2 = -2m$ **31.** $x^2 + 7x + 1 = 0$ **32.** $2x^2 + 9x = 1$

33. $2a^2 = 9a + 3$ **34.** $2a^2 = 5a + 4$ **35.** $x^2 - 2x = 2$ **36.** $a^2 + 4a = -2$

37. $x^2 - 10x = -20$ **38.** $x^2 + 12x = -34$ **39.** $b^2 = 8b - 8$ **40.** $w^2 + 8w = 11$

41. $2x^2 - 4x - 5 = 0$ **42.** $2x^2 - 6x = -3$ **43.** $4r^2 + 10r + 5 = 0$ **44.** $4x^2 - 12x + 7 = 0$

45. $3x^2 = -2x + 4$ **46.** $3x^2 = -8x + 2$ **47.** $5z^2 + 2z + 10 = 0$ **48.** $6a^2 - 4a + 2 = 0$

Solve the following application problems:

49. Find two consecutive odd integers the sum of whose squares is 34.

50. Find two consecutive even integers the sum of whose squares is 52.

Exercises 51 through 54 require the use of the formula for the area of a rectangle. The formula for the area of a rectangle is $A = LW$.

51. The length of a rectangle is 6 inches more than the width. If the area is 55 square inches, find the length and width.

52. The width of a rectangle is 3 meters less than the length. If the area is 70 square meters, find the length and width.

53. The length of a rectangle is 1 yard less than twice the width. If the area is 28 square yards, find the length and width.

54. The length of a rectangle is 2 centimeters more than three times the width. If the area is 33 square centimeters, find the length and width.

A function of the form $f(x) = ax^2 + bx + c$ *is a quadratic function.*

55. If $f(x) = x^2 + 2x - 3$, find the following:
  a. $f(0)$ **b.** $f(2)$ **c.** $f(-3)$

56. If $f(x) = x^2 - 4x + 5$, find the following:
  a. $f(0)$ **b.** $f(3)$ **c.** $f(-1)$

57. If $f(x) = 3x^2 - 2x + 1$, find the following:
  a. $f(0)$ **b.** $f(1)$ **c.** $f(-2)$

58. If $f(x) = 2x^2 + 4x - 3$, find the following:
  a. $f(0)$ **b.** $f(2)$ **c.** $f(-2)$

Challenge Exercises (59–62)

Solve the following quadratic equations by using the quadratic formula:

59. $x^2 - \dfrac{7}{2}x + 3 = 0$

60. $x^2 - \dfrac{13}{2}x + 10 = 0$

61. $x^2 - \dfrac{7}{3}x - 2 = 0$

62. $y^2 - \dfrac{24}{5}y - 1 = 0$

Writing Exercises (63–65)

63. What are the advantages of using the quadratic formula instead of completing the square?

64. In the quadratic formula, $x = \dfrac{-b \pm \sqrt{b^2 - 4ac}}{2a}$, the expression $b^2 - 4ac$ is called the *discriminant*. For what values (types of numbers) of the discriminant will a quadratic equation have

  a. No real solutions? Why?

  b. Two irrational solutions? Why?

  c. Two rational solutions? Why?

  d. Exactly one solution? Why?

65. Given a quadratic equation in the form $ax^2 + bx + c = 0$, if b is an even integer, then $\sqrt{b^2 - 4ac}$ can always be simplified. Why?

<table>
<tr><td>

Section 11.4

</td><td>

Quadratic Equations with Complex Solutions

</td></tr>
</table>

OBJECTIVES *When you complete this section, you will be able to:*

Ⓐ Write complex numbers in terms of i.

Ⓑ Add and subtract complex numbers.

Ⓒ Multiply and divide complex numbers.

Ⓓ Solve quadratic equations that have complex solutions.

PREREQUISITE SKILLS

a. Simplify radicals. (Section 10.2)

b. Add like terms. (Section 1.7)

c. Apply the distributive property. (Section 1.7)

d. Multiply binomials. (Section 2.4)

e. Apply $\frac{a+b}{c} = \frac{a}{c} + \frac{b}{c}$. (Section 2.8)

f. Solve quadratic equations by extraction of roots. (Section 11.1)

g. Solve quadratic equations by the quadratic formula. (Section 11.3)

GETTING READY FOR SECTION 11.4

1. Simplify $\sqrt{18}$.

2. Simplify $3 - 6i - 2 + 4i$ by adding like terms.

3. Simplify $2i(4 - 3i)$ by using the distributive property.

Multiply the following binomials:

4. $(2 + 3i)(3 - 4i)$

5. $(3 + 2i)(3 - 2i)$

6. Rewrite $\dfrac{3 - 11i}{13}$, using $\dfrac{a+b}{c} = \dfrac{a}{c} + \dfrac{b}{c}$.

7. Solve $(x - 3)^2 = 16$ by using extraction of roots.

8. Solve $x^2 - 2x - 4 = 0$ by using the quadratic formula.

Historical Note

When the idea of finding square roots of negative numbers was first introduced, members of the established mathematics community said that this type of number existed only in the imagination of those finding them. Hence, they were given the name "imaginary numbers," which has stuck; but these numbers are just as real as any other numbers.

Introduction

In Sections 11.1 through 11.3, we sometimes encountered quadratic equations whose solutions involved square roots of negative numbers. In those sections, we indicated that these equations had no real-number solutions. In this section, we will learn how to deal with these types of equations by introducing a new type of number known as a **complex number**.

OBJECTIVE Writing complex numbers in terms of i.

In Chapter 10, we learned that a radical like $\sqrt{-9}$ does not exist as a real number because there is no real number whose square is -9. We now extend our number system to include such numbers by defining $\sqrt{-1} = i$, where i is called the **imaginary unit**. So, by definition of a square root, i is the number whose square is -1. Thus, $i^2 = -1$. It is important to remember that i is a number—not a variable. We summarize next.

Answers: Getting Ready: 1. $3\sqrt{2}$ **2.** $1 - 2i$ **3.** $8i - 6i^2$ **4.** $6 + i - 12i^2$ **5.** $9 - 4i^2$ **6.** $\dfrac{3}{13} - \dfrac{11}{13}i$ **7.** $x = -1, 7$ **8.** $1 \pm \sqrt{5}$

> **Definition** The Imaginary Unit
>
> The imaginary unit, i, is the number whose square is -1. In symbols, $\sqrt{-1} = i$ and $i^2 = -1$.

To write the square roots of numbers in terms of i, we use the property of radicals which states that $\sqrt[n]{ab} = \sqrt[n]{a} \cdot \sqrt[n]{b}$.

Example 1   Write each of the following as the product of a real number and i:

a. $\sqrt{-9} =$ Write -9 as $-1 \cdot 9$.

$\sqrt{-1 \cdot 9} =$ Apply $\sqrt[n]{ab} = \sqrt[n]{a} \cdot \sqrt[n]{b}$.

$\sqrt{-1} \cdot \sqrt{9} =$ Replace $\sqrt{-1}$ with i and $\sqrt{9} = 3$.

$i \cdot 3 = 3i$ Answer. Notice the answer is written in the form of bi.

b. $\sqrt{-5} =$ Write -5 as $-1 \cdot 5$.

$\sqrt{-1 \cdot 5} =$ Apply $\sqrt[n]{ab} = \sqrt[n]{a} \cdot \sqrt[n]{b}$.

$\sqrt{-1} \cdot \sqrt{5} =$ Replace $\sqrt{-1}$ with i.

$i\sqrt{5}$ Answer.

c. $\sqrt{-18} =$ Write -18 as $-1 \cdot 18$.

$\sqrt{-1 \cdot 18} =$ Apply $\sqrt[n]{ab} = \sqrt[n]{a} \cdot \sqrt[n]{b}$.

$\sqrt{-1} \cdot \sqrt{18} =$ Write $\sqrt{-1}$ as i and write $\sqrt{18}$ as $\sqrt{9 \cdot 2}$.

$i\sqrt{9 \cdot 2} =$ Simplify $\sqrt{9 \cdot 2}$.

$i \cdot 3\sqrt{2} =$ Rewrite as $3i\sqrt{2}$.

$3i\sqrt{2}$ Answer.

PRACTICE EXERCISE 1

Write each of the following as the product of a real number and i:

a. $\sqrt{-16}$ **b.** $\sqrt{-7}$ **c.** $\sqrt{-28}$

If you need more practice, do the following Additional Practice Exercises.

Additional Practice Exercise 1   Write each of the following as the product of a real number and i:

a. $\sqrt{-49}$ **b.** $\sqrt{-11}$ **c.** $\sqrt{-45}$

Numbers like $2i$, $i\sqrt{3}$, and $4i\sqrt{7}$ are called **imaginary numbers**. If we combine the real numbers and the imaginary numbers into numbers of the form $a + bi$, where a and b are real, we have the **complex number system**.

> **Definition** Complex Number System
>
> A complex number is any number that can be written in the form of $a + bi$, where a and b are real numbers and $i = \sqrt{-1}$. Any number written in the form of $0 + bi = bi$ is an imaginary number.

When a complex number is written in the form $a + bi$, it is said to be in **standard form**. When a complex number is written in standard form and $b = 0$, it becomes $a + 0i = a + 0 = a$, which is a real number. Hence, a is called the **real part** of a complex number. When a complex number is written in standard form and $a = 0$, it becomes $0 + bi = bi$, which is an imaginary number. Hence, bi is called the **imaginary part**.

Answers: Practice Exercise 1: a. $4i$ **b.** $i\sqrt{7}$ **c.** $2i\sqrt{7}$ **Additional Practice 1: a.** $7i$ **b.** $i\sqrt{11}$ **c.** $3i\sqrt{5}$

<table>
<tr><td>**Example 2**</td><td>Write each of the following as a complex number in standard form $a + bi$:</td></tr>
</table>

a. 5

Solution

To be in standard form, 5 must be written in the form $a + bi$. Consequently,
$5 = 5 + 0 \cdot i$.

b. $4i$

Solution

To be in standard form, $4i$ must be written in the form $a + bi$. Consequently,
$4i = 0 + 4i$.

c. $\sqrt{45}$

Solution

To be in standard form, $\sqrt{45}$ must be written in the form $a + bi$. First we
need to simplify $\sqrt{45}$. $\sqrt{45} = \sqrt{9 \cdot 5} = \sqrt{9} \cdot \sqrt{5} = 3\sqrt{5}$. Consequently,
$\sqrt{45} = 3\sqrt{5} + 0 \cdot i$.

d. $\sqrt{-32}$

Solution

To be in standard form, $\sqrt{-32}$ must be written in the form $a + bi$. First we need
to simplify $\sqrt{-32}$.
$\sqrt{-32} = \sqrt{-1 \cdot 16 \cdot 2} = \sqrt{-1} \cdot \sqrt{16} \cdot \sqrt{2} = i \cdot 4\sqrt{2} = 4i\sqrt{2}$. Consequently,
$\sqrt{-32} = 0 + 4i\sqrt{2}$.

e. $4 + \sqrt{-9}$

Solution

To be in standard form, $\sqrt{-9}$ must be written in the form $a + bi$. First we need to
simplify $\sqrt{-9}$. $\sqrt{-9} = 3i$. Therefore, $4 + \sqrt{-9} = 4 + 3i$.

Note

Technically, $0 + 4i\sqrt{2}$ is not in
$a + bi$ form, since the i is not at
the very end. When b contains a
radical, we usually put i before
the radical, because when it
is placed after the radical, it is
sometimes thought to be under
the radical sign.

PRACTICE EXERCISE 2

Write each of the following as a complex number in standard form $a + bi$:

a. 12   **b.** $3i$ **c.** $\sqrt{20}$ **d.** $\sqrt{-8}$ **e.** $6 + \sqrt{-25}$

If you need more practice, do the following Additional Practice Exercises.

Additional Practice Exercise 2   Write each of the following as a complex number in standard form $a + bi$:

a. 15   **b.** $5i$ **c.** $\sqrt{44}$ **d.** $\sqrt{-54}$ **e.** $-3 + \sqrt{-36}$

OBJECTIVE **B** Adding and subtracting complex numbers

We add and subtract complex numbers in exactly the same manner that we add and
subtract polynomials.

Answers: Practice Exercise 2: a. $12 + 0i$ **b.** $0 + 3i$ **c.** $2\sqrt{5} + 0i$ **d.** $0 + 2i\sqrt{2}$ **e.** $6 + 5i$
Additional Practice 2: a. $15 + 0i$ **b.** $0 + 5i$ **c.** $2\sqrt{11} + 0i$ **d.** $0 + 3i\sqrt{6}$ **e.** $-3 + 6i$

Example 3   Find the sums and differences of the complex numbers. Express answers in standard form $a + bi$.

a. $(2 + 3i) + (-5 - 4i) =$ Remove the parentheses.

$2 + 3i - 5 - 4i =$ Add the real parts and add the imaginary parts.

$-3 - i$ Sum.

b. $(3 - 6i) - (2 - 4i) =$ Remove the parentheses.

$3 - 6i - 2 + 4i =$ Add the real parts and add the imaginary parts.

$1 - 2i$ Difference.

c. $6i + (-7 - 3i) =$ Remove the parentheses.

$6i - 7 - 3i =$ Add the imaginary parts.

$-7 + 3i$ Answer.

PRACTICE EXERCISE 3

Find the sums and differences of the complex numbers. Express answers in standard form $a + bi$.

a. $(5 + 4i) + (-7 + 3i)$ **b.** $(-4 + 2i) - (3 - 6i)$ **c.** $8i + (3 - 3i)$

If you need more practice, do the following Additional Practice Exercises.

Additional Practice Exercise 3   Find the sums and differences of the complex numbers. Express answers in standard form $a + bi$.

a. $(7 - 3i) + (-4 - 5i)$ **b.** $(6 - 2i) - (-4 - 5i)$ **c.** $9i + (-5 - 4i)$

OBJECTIVE C Multiplying and dividing complex numbers

We multiply complex numbers in the same way that we multiply polynomials. We use the distributive property and FOIL in doing so. However, we must remember that $i^2 = -1$.

Example 4   Find the products of the complex numbers. Express answers in standard form $a + bi$.

a. $2i(4 - 3i) =$ Apply the distributive property.

$2i \cdot 4 - 2i \cdot 3i =$ Multiply and treat i just as though it were a variable.

$8i - 6i^2 =$ Replace i^2 with -1.

$8i - 6(-1) =$ $-6(-1) = 6$.

$8i + 6 =$ Rewrite in standard form.

$6 + 8i$ Answer.

b. $(2 + 3i)(3 - 4i) =$ Apply FOIL.

$2 \cdot 3 + 2 \cdot (-4i) + 3i \cdot 3 + 3i(-4i) =$ Multiply.

$6 - 8i + 9i - 12i^2 =$ Add $-8i$ and $9i$, and replace i^2 with -1.

$6 + i - 12(-1) =$ $-12(-1) = 12$.

$6 + i + 12 =$ Add 6 and 12.

$18 + i$ Product.

c. $(3 + 2i)(3 - 2i) =$ Apply FOIL.

$9 - 6i + 6i - 4i^2 =$ Add $-6i$ and $6i$, and replace i^2 with -1.

$9 - 4(-1) =$ $-4(-1) = 4$.

$9 + 4 =$ Add.

$13 =$ Product.

$13 + 0i$ Product in standard form.

Answers: Practice Exercise 3: a. $-2 + 7i$ **b.** $-7 + 8i$ **c.** $3 + 5i$ **Additional Practice 3: a.** $3 - 8i$
b. $10 + 3i$ **c.** $-5 + 5i$

PRACTICE EXERCISE 4

Find the products of the complex numbers. Express answers in standard form $a + bi$.

a. $3i(4 + 3i)$ **b.** $(2 + 5i)(3 - 2i)$ **c.** $(2 + 4i)(2 - 4i)$

If you need more practice, do the following Additional Practice Exercises.

Additional Practice Exercise 4   Find the products of the complex numbers. Express answers in standard form $a + bi$.

a. $5i(3 - 2i)$ **b.** $(3 - 2i)(4 - i)$ **c.** $(3 + i)(3 - i)$

In Chapter 10, we called expressions like $2 + \sqrt{3}$ and $2 - \sqrt{3}$ *conjugates*. Since $i = \sqrt{-1}$ and numbers of the form $a + bi$ are complex numbers, we call complex numbers of the form $a + bi$ and $a - bi$ **complex conjugates**. For example, $2 + 6i$ and $2 - 6i$ are complex conjugates. In this section, the numbers in Example 4c, Practice Exercise 4c, and Additional Practice Exercise 4c are all complex conjugates. Notice that in each case the product is a real number. The product of complex conjugates is always a real number, because $(a + bi)(a - bi) = a^2 - b^2 i^2 = a^2 - b^2(-1) = a^2 + b^2$, which is a real number. (There is no i term.) This means that we can divide complex numbers by the same technique that we used to rationalize denominators—multiply the numerator and denominator by the complex conjugate of the denominator.

Example 5   Find the quotients of the complex numbers. Express the answers in standard form.

a. $\dfrac{2 + 3i}{4 - 2i} =$

Multiply the numerator and the denominator by $4 + 2i$, which is the conjugate of the denominator.

$\dfrac{2 + 3i}{4 - 2i} \cdot \dfrac{4 + 2i}{4 + 2i} =$ Multiply.

$\dfrac{8 + 16i + 6i^2}{16 - 4i^2} =$ Replace i^2 with -1.

$\dfrac{8 + 16i + 6(-1)}{16 - 4(-1)} =$ Multiply.

$\dfrac{8 + 16i - 6}{16 + 4} =$ Add.

$\dfrac{2 + 16i}{20} =$ Write in standard form by applying $\dfrac{a + b}{c} = \dfrac{a}{c} + \dfrac{b}{c}$.

$\dfrac{2}{20} + \dfrac{16}{20}i =$ Reduce to lowest terms.

$\dfrac{1}{10} + \dfrac{4}{5}i$ Quotient.

b. $\dfrac{3 - i}{2 + 3i} =$

Multiply the numerator and the denominator by $2 - 3i$, which is the conjugate of the denominator.

$\dfrac{3 - i}{2 + 3i} \cdot \dfrac{2 - 3i}{2 - 3i} =$ Multiply.

$\dfrac{6 - 11i + 3i^2}{4 - 9i^2} =$ Replace i^2 with -1.

$\dfrac{6 - 11i + 3(-1)}{4 - 9(-1)} =$ Multiply.

$\dfrac{6 - 11i - 3}{4 + 9} =$ Add.

$\dfrac{3 - 11i}{13} =$ Write in standard form by applying $\dfrac{a + b}{c} = \dfrac{a}{c} + \dfrac{b}{c}$.

$\dfrac{3}{13} - \dfrac{11}{13}i$ Quotient.

Answers: **Practice Exercise 4:** **a.** $-9 + 12i$ **b.** $16 + 11i$ **c.** 20   **Additional Practice 4:** **a.** $10 + 15i$ **b.** $10 - 11i$ **c.** 10.

PRACTICE EXERCISE 5

Find the quotients of the complex numbers. Express the answers in standard form.

a. $\dfrac{5 - 2i}{1 + 2i}$

b. $\dfrac{4 + 2i}{2 - 2i}$

If you need more practice, do the following Additional Practice Exercises.

Additional Practice Exercise 5   Find the quotients of the complex numbers. Express the answers in standard form.

a. $\dfrac{1 + i}{2 + i}$

b. $\dfrac{-3 + 2i}{3 - 3i}$

OBJECTIVE **Solving quadratic equations that have complex solutions**

In Sections 11.1 through 11.3, we encountered quadratic equations whose solutions contained the square roots of negative numbers. At that time, we indicated that there were no real-number solutions to these equations. Now we can find these solutions as complex numbers. Remember that the quadratic formula is

$$x = \frac{-b \pm \sqrt{b^2 - 4ac}}{2a}.$$

Example 6   **Find the solutions of the quadratic equations. Leave answers in standard form.**

a. $(x - 3)^2 = -9$

Take the positive and negative square roots of both sides.

$\sqrt{(x - 3)^2} = \pm\sqrt{-9}$ Simplify both sides.

$x - 3 = \pm 3i$ Add 3 to both sides.

$x - 3 + 3 = 3 \pm 3i$ Simplify the left side of the equation.

$x = 3 \pm 3i$ Therefore, the solutions are $x = 3 + 3i$ and $x = 3 - 3i$.

b. $x^2 - 4x + 13 = 0$ Identify a, b, and c.

$a = 1, b = -4, c = 13$ Write the quadratic formula.

$x = \dfrac{-b \pm \sqrt{b^2 - 4ac}}{2a}$ Substitute into the quadratic formula.

$x = \dfrac{-(-4) \pm \sqrt{(-4)^2 - 4(1)(13)}}{2(1)}$ Simplify.

$x = \dfrac{4 \pm \sqrt{16 - 52}}{2}$ Simplify under the radical sign.

$x = \dfrac{4 \pm \sqrt{-36}}{2}$ $\sqrt{-36} = 6i$.

$x = \dfrac{4 \pm 6i}{2}$ Apply $\dfrac{a + b}{c} = \dfrac{a}{c} + \dfrac{b}{c}$.

$x = \dfrac{4}{2} \pm \dfrac{6i}{2}$ Simplify.

$x = 2 \pm 3i$ Therefore, $x = 2 + 3i$ and $x = 2 - 3i$ are the solutions.

Answers: Practice Exercise 5: a. $\dfrac{1}{5} - \dfrac{12}{5}i$ **b.** $\dfrac{1}{2} + \dfrac{3}{2}i$ **Additional Practice 5: a.** $\dfrac{3}{5} + \dfrac{1}{5}i$ **b.** $-\dfrac{5}{6} - \dfrac{1}{6}i$

c. $x^2 - 2x + 4 = 0$ Write the quadratic formula.

$x = \dfrac{-b \pm \sqrt{b^2 - 4ac}}{2a}$ Substitute for a, b, and c.

$x = \dfrac{-(-2) \pm \sqrt{(-2)^2 - 4(1)(4)}}{2(1)}$ Simplify.

$x = \dfrac{2 \pm \sqrt{4 - 16}}{2}$ Simplify under the radical sign.

$x = \dfrac{2 \pm \sqrt{-12}}{2}$ $\sqrt{-12} = \sqrt{-1 \cdot 4 \cdot 3} = 2i\sqrt{3}$.

$x = \dfrac{2 \pm 2i\sqrt{3}}{2}$ Factor 2 from the numerator.

$x = \dfrac{2(1 \pm i\sqrt{3})}{2}$ Divide by the common factor of 2.

$x = 1 \pm i\sqrt{3}$ Therefore, $x = 1 + i\sqrt{3}$ and $x = 1 - i\sqrt{3}$.

d. $x^2 + x + 3 = 0$ Write the quadratic formula.

$x = \dfrac{-b \pm \sqrt{b^2 - 4ac}}{2a}$ Substitute for a, b, and c.

$x = \dfrac{-1 \pm \sqrt{1^2 - 4(1)(3)}}{2(1)}$ Simplify.

$x = \dfrac{-1 \pm \sqrt{1 - 12}}{2}$ Simplify under the radical sign.

$x = \dfrac{-1 \pm \sqrt{-11}}{2}$ $\sqrt{-11} = \sqrt{-1(11)} = i\sqrt{11}$.

$x = \dfrac{-1 \pm i\sqrt{11}}{2}$ Therefore, $x = \dfrac{-1 + i\sqrt{11}}{2}$ and $x = \dfrac{-1 - i\sqrt{11}}{2}$.

PRACTICE EXERCISE 6

Find the solutions of the quadratic equations. Express answers in standard form.

a. $(x - 5)^2 = -4$
b. $x^2 - 2x + 5 = 0$
c. $x^2 + 2x + 6 = 0$
d. $x^2 + 3x + 5 = 0$

If you need more practice, do the following Additional Practice Exercises.

Additional Practice Exercise 6   Find the solutions of the quadratic equations. Express answers in standard form.

a. $(x + 2)^2 = -100$
b. $x^2 - 4x + 20 = 0$
c. $x^2 - 4x + 6 = 0$
d. $x^2 - x + 5 = 0$

Exercise Set 11.4

For Extra Help

MyMathLab®

Write each of the following as the product of a real number and i. (See Example 1.)

1. $\sqrt{-4}$ **2.** $\sqrt{-36}$ **3.** $\sqrt{-64}$ **4.** $\sqrt{-100}$

5. $\sqrt{-6}$ **6.** $\sqrt{-3}$ **7.** $\sqrt{-15}$ **8.** $\sqrt{-21}$

9. $\sqrt{-24}$ **10.** $\sqrt{-63}$ **11.** $\sqrt{-75}$ **12.** $\sqrt{-32}$

Write each of the following as a complex number in standard form $a + bi$. (See Example 2.)

13. 6 **14.** 8 **15.** -7 **16.** -9

17. $6i$ **18.** $8i$ **19.** $-4i$ **20.** $-11i$

Answers: Practice Exercise 6: a. $5 + 2i, 5 - 2i$ **b.** $1 + 2i, 1 - 2i$ **c.** $-1 + i\sqrt{5}, -1 - i\sqrt{5}$ **d.** $\dfrac{-3 + i\sqrt{11}}{2}, \dfrac{-3 - i\sqrt{11}}{2}$

Additional Practice 6: a. $-2 + 10i, -2 - 10i$ **b.** $2 + 4i, 2 - 4i$ **c.** $2 + i\sqrt{2}, 2 - i\sqrt{2}$ **d.** $\dfrac{1 + i\sqrt{19}}{2}, \dfrac{1 - i\sqrt{19}}{2}$

21. $\sqrt{27}$ **22.** $\sqrt{54}$ **23.** $\sqrt{80}$ **24.** $\sqrt{75}$

25. $\sqrt{-18}$ **26.** $\sqrt{-63}$ **27.** $\sqrt{-125}$ **28.** $\sqrt{-108}$

29. $4 + \sqrt{-49}$ **30.** $6 + \sqrt{-64}$ **31.** $-5 - \sqrt{-81}$ **32.** $-9 - \sqrt{-144}$

Find the sums or differences of the complex numbers. Express answers in standard form a + bi. (See Example 3.)

33. $2i + (3 - 5i)$ **34.** $6i + (5 - 2i)$ **35.** $-4i - (3 - 3i)$ **36.** $8i - (5 - 4i)$

37. $(4 + 3i) + (5 + 6i)$ **38.** $(2 + 9i) + (1 + 4i)$ **39.** $(6 - 3i) + (-5 - 5i)$ **40.** $(-4 - 3i) + (4 - 7i)$

41. $(4 + 2i) - (3 + 7i)$ **42.** $(8 + 4i) - (3 + i)$ **43.** $(7 - 4i) - (-3 - 6i)$ **44.** $(-7 - 5i) - (3 - 8i)$

Find the products or quotients of the complex numbers. Express answers in standard form a + bi. (See Examples 4 and 5.)

45. $3i(4 + 2i)$ **46.** $4i(4 + 3i)$ **47.** $5i(4 - 2i)$ **48.** $2i(3 - 6i)$

49. $(2 + i)(3 + 2i)$ **50.** $(5 + 2i)(1 + i)$ **51.** $(2 - 3i)(4 + 2i)$ **52.** $(5 - 3i)(3 + 2i)$

53. $(4 - 2i)(2 - 3i)$ **54.** $(3 - 4i)(1 - 4i)$ **55.** $(-2 + 5i)(4 + 2i)$ **56.** $(-3 - 2i)(2 - i)$

57. $\dfrac{1 - 2i}{2 + 2i}$ **58.** $\dfrac{2 + 3i}{2 + 4i}$ **59.** $\dfrac{3 - i}{1 - 3i}$ **60.** $\dfrac{4 + 2i}{2 - i}$

Find the solutions of the quadratic equations. Express answers in standard form. (See Example 6.)

61. $(x + 2)^2 = -25$ **62.** $(x + 4)^2 = -4$ **63.** $(x - 5)^2 = -64$ **64.** $(x - 3)^2 = -100$

65. $x^2 - 4x + 5 = 0$ **66.** $x^2 - 6x + 10 = 0$ **67.** $x^2 - 6x + 13 = 0$ **68.** $x^2 - 4x + 8 = 0$

69. $x^2 - 4x + 6 = 0$ **70.** $x^2 + 6x + 11 = 0$ **71.** $x^2 - 6x + 12 = 0$ **72.** $x^2 - 2x + 6 = 0$

73. $x^2 + 3x + 6 = 0$ **74.** $x^2 - 5x + 7 = 0$ **75.** $x^2 + 7x + 15 = 0$ **76.** $x^2 + x + 2 = 0$

Section 11.5   Applications Involving Quadratic Equations

OBJECTIVES *When you complete this section, you will be able to:*

Solve the following types of application problems involving quadratic equations:

A Numbers.
B Geometric figures.
C Pythagorean theorem.
D Distance, rate, and time.
E Work.
F Applications from science.

PREREQUISITE SKILLS *Before beginning this section, you should be able to:*

a. Solve linear equations. (Sections 3.1–3.3)
b. Solve quadratic equations. (Sections 5.8 and 11.1–11.3)
c. Use the Pythagorean theorem. (Section 10.7)
d. Represent consecutive integers, including consecutive even and odd. (Section 3.5)
e. Solve equations containing rational expressions. (Section 7.6)

GETTING READY FOR SECTION 11.5

Solve the following equations:

1. $5x + 16 = 0$ **2.** $4x^2 + 20x + 24 = 80$

3. $\dfrac{200}{x} = \dfrac{200}{x + 1} + 10$

4. The longer leg of a right triangle is one foot more than the shorter leg, and the hypotenuse is 5 feet. Find the lengths of the two legs.

5. Represent three consecutive even integers, using x as the variable.

Introduction

Application problems that resulted in quadratic equations were first introduced in Section 5.8. In that section, we were limited to integer problems and areas of squares and rectangles. In later chapters, we learned to solve other types of equations. Consequently, in this section, we will concentrate on application problems that involve solving these other types of equations and equations from the sciences.

As in Section 5.8, care must be taken when solving application problems that involve quadratic equations. Quite often, a number may solve the equation, but not satisfy the physical conditions of the problem. For example, the length of the side of a rectangle cannot be negative. Consequently, it is very important that all solutions of the equation be checked against the wording of the original problem. Also, as in Section 5.8, most of the equations in this section can be solved by factoring, though you have the option of using the quadratic formula if you wish.

Answers: Getting Ready: 1. $-\dfrac{16}{5}$ **2.** $x = -7, 2$ **3.** $x = -5, 4$ **4.** 3 ft, 4 ft **5.** $x, x + 2, x + 4$

OBJECTIVE A Solving number problems involving quadratic equations

Example 1

The sum of the reciprocals of two consecutive even integers is $\frac{7}{24}$. Find the integers.

Solution

Understand: We are given that the numbers are consecutive even integers and the sum of their reciprocals. The unknowns are the consecutive even integers.

> Relationship 1: There are two consecutive even integers.
> Relationship 2: The sum of the reciprocals is $\frac{7}{24}$.

Plan: Use relationship 1 to represent the unknowns in terms of a variable, and relationship 2 to write the equation.

Translate: Using relationship 1, we represent the unknowns in terms of a variable.

Let x represent the smaller of the two consecutive even integers. Then, $x + 2$ represents the larger of the two consecutive even integers.

$\dfrac{1}{x}$ = the reciprocal of the smaller integer.

$\dfrac{1}{x + 2}$ = the reciprocal of the larger integer.

From relationship 2, we know that the sum of the reciprocals is $\frac{7}{24}$, so the equation is
$\dfrac{1}{x} + \dfrac{1}{x + 2} = \dfrac{7}{24}.$

Solve:

$\dfrac{1}{x} + \dfrac{1}{x + 2} = \dfrac{7}{24}$	Multiply both sides by the LCD $24x(x + 2)$.
$24x(x + 2)\left(\dfrac{1}{x} + \dfrac{1}{x + 2}\right) = 24x(x + 2)\left(\dfrac{7}{24}\right)$	Simplify both sides.
$24x(x + 2)\left(\dfrac{1}{x}\right) + 24x(x + 2)\left(\dfrac{1}{x + 2}\right) = x(x + 2)(7)$	Multiply.
$24(x + 2) + 24x = 7x(x + 2)$	Distribute.
$24x + 48 + 24x = 7x^2 + 14x$	Add like terms.
$48x + 48 = 7x^2 + 14x$	Subtract $48x$ and 48 from both sides.
$0 = 7x^2 - 34x - 48$	Factor $7x^2 - 34x - 48$.
$0 = (7x + 8)(x - 6)$	Set each factor equal to 0.
$7x + 8 = 0, x - 6 = 0$	Solve each equation.
$7x = -8, \qquad x = 6$	Solution.
$x = -\dfrac{8}{7}$	Solution.

State: Since $-\frac{8}{7}$ is not an integer, the only possible answer is 6. Since x represents the smaller integer, the smaller integer is 6 and the larger is $x + 2 = 6 + 2 = 8$.

Check: Is the sum of the reciprocals of 6 and 8 equal to $\frac{7}{24}$? Since $\frac{1}{6} + \frac{1}{8} = \frac{4}{24} + \frac{3}{24} = \frac{7}{24}$, our solutions are correct.

PRACTICE EXERCISE 1

The product of the reciprocals of two consecutive integers is $\frac{1}{12}$. Find the integers.

OBJECTIVE B Solving geometric problems involving quadratic equations

Example 2 A rectangular picture, 4 inches by 6 inches, is enclosed by a frame of uniform width. If the area of the picture and the frame is 80 square inches, find the width of the frame.

Solution

Understand: We are given the dimensions of the picture and the area of the picture and the frame. The unknown is the width of the frame.

> Relationship 1: The picture is 4 inches by 6 inches and is surrounded by a frame of uniform width.
>
> Relationship 2: The area of a rectangle is $A = LW$.

Plan: Use relationship 1 to represent the length and width of the picture plus frame in terms of a variable, and use relationship 2 to write the equation.

Translate:

Let x represent the width of the frame.

Draw and label a figure, using relationship 1.

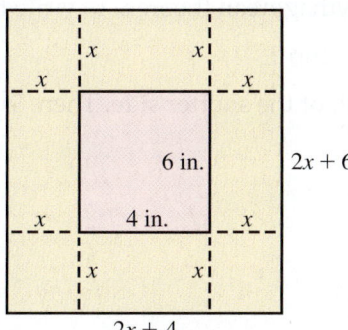

Since there are x inches of frame to the left and the right of the picture and the picture is 4 inches wide, the outside dimension is $2x + 4$ inches. Since there are x inches of the frame above and below the picture and the picture is 6 inches high, the outside dimension is $2x + 6$ inches.

Use relationship 2 to write the equation. Since the area of the picture and the frame is 80 square inches, substitute 80 for A, $2x + 6$ for L, and $2x + 4$ for W into the formula.

$A = LW$ Substitute for A, L, and W.
$80 = (2x + 6)(2x + 4)$

Solve:

$80 = (2x + 6)(2x + 4)$ Multiply the right side.
$80 = 4x^2 + 20x + 24$ Subtract 80 from both sides.
$0 = 4x^2 + 20x - 56$ Remove the common factor of 4.
$0 = 4(x^2 + 5x - 14)$ Factor $x^2 + 5x - 14$.
$0 = 4(x + 7)(x - 2)$ Set each factor, except 4, equal to 0.
$4 \neq 0, x + 7 = 0, x - 2 = 0$ Solve each equation.
$x = -7, \qquad\qquad x = 2$ Solutions of the equation.

State: Since a frame cannot have a width of -7 inches, 2 inches is the only answer. Therefore, the width of the frame is 2 inches.

Check: The check is left as an exercise for the student.

PRACTICE EXERCISE 2

The length of a rectangular flower garden is 2 feet more than its width. The flower garden is surrounded by a uniform border of mulch that is 2 feet wide. If the area of the flower garden and the border is 80 square feet, find the length and width of the flower bed.

OBJECTIVE **Solving Pythagorean theorem problems involving quadratic equations**

Example 3 | **The longer side of a rectangular mural is 3 feet less than three times the shorter side. If the mural is 13 feet diagonally across, find the lengths of the sides.**

Solution

Understand: We are given the relationship between the length and width of a rectangle and the length of the diagonal. The unknowns are the lengths of the sides.

    Relationship 1: The longer side is 3 feet less than three times the shorter side.
    Relationship 2: The diagonal, which is 13 feet, is the hypotenuse of a right triangle.

Plan: Use relationship 1 to represent the length and width in terms of a variable. Use relationship 2 and the Pythagorean theorem to write the equation.

Translate: Using relationship 1,

Let x represent the length of the shorter side. Then, $3x - 3$ represents the length of the longer side.

Draw and label a figure.

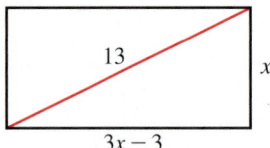

Use relationship 2 and the Pythagorean theorem to write the equation.

$a^2 + b^2 = c^2$ Substitute for a, b, and c.
$x^2 + (3x - 3)^2 = 13^2$

Solve:

$x^2 + (3x - 3)^2 = 13^2$ Raise to powers.

$x^2 + 9x^2 - 18x + 9 = 169$ Add like terms.

$10x^2 - 18x + 9 = 169$ Subtract 169 from both sides.

$10x^2 - 18x - 160 = 0$ Remove the common factor of 2.

$2(5x^2 - 9x - 80) = 0$ Factor the trinomial.

$2(5x + 16)(x - 5) = 0$ Set each factor, except 2, equal to 0.

$5x + 16 = 0, x - 5 = 0$ Solve each equation.

$5x = -16, \qquad x = 5$ Solution.

$x = -\dfrac{16}{5}$ Solution.

State: Since a side of a rectangle cannot be $-\frac{16}{5}$, the only possible answer is 5 feet. Therefore, the shorter side is 5 feet. The longer side is $3x - 3 = 3(5) - 3 = 15 - 3 = 12$ feet.

Check: The check is left as an exercise for the student.

Answers: Practice Exercise 2: length = 6 ft, width = 4 ft

PRACTICE EXERCISE 3

The distance diagonally across a rectangular vacant lot is 10 feet more than the length of the longer side. If the shorter side is 30 feet long, find the length of the longer side.

OBJECTIVE Solving distance, rate, and time problems involving quadratic equations

Example 4 The average speed of a car is 10 miles per hour more than the average speed of a bus. The time required for the bus to travel 200 miles is one hour more than the time required by the car. Find how long it takes the car to travel 200 miles.

Solution

Understand: We are given the relationship between the speed of the car and the speed of the bus. We are also given the relationship between the times of the car and bus. The unknown is the time it takes the car.

  Relationship 1: The time for the bus is one hour more than the time for the car.

  Relationship 2: The speed of the car is 10 miles per hour more than the speed of the bus.

Plan: Use a chart. Use relationship 1 to represent the time of the car and the time of the bus in terms of variables. Use relationship 2 to write the equation.

Translate: Remember from Section 3.5 that we use a chart to solve distance, time, and rate problems.

	d	r	t
Bus			
Car			

Using relationship 1, we let x represent the time for the car to travel 200 miles. Then, $x + 1$ represents the time for the bus to travel 200 miles. Fill in the time column. Since both distances equal 200 miles, fill in the distance column also.

	d	r	t
Bus	200		$x + 1$
Car	200		x

Since $r = \frac{d}{t}$, the distance traveled by the bus is $\frac{200}{x + 1}$ and the distance traveled by the car is $\frac{200}{x}$. Put these in the rate column.

	d	r	t
Bus	200	$\frac{200}{x + 1}$	$x + 1$
Car	200	$\frac{200}{x}$	x

From relationship 2, we have
(the rate of the car) = (the rate of the bus) + 10. Consequently, the equation is

$$\frac{200}{x} = \frac{200}{x + 1} + 10.$$

Answers: Practice Exercise 3: longer side = 40 ft

Solve:

$$\frac{200}{x} = \frac{200}{x+1} + 10$$ Multiply both sides by the LCD $x(x+1)$.

$$x(x+1)\left(\frac{200}{x}\right) = x(x+1)\left(\frac{200}{x+1} + 10\right)$$ Simplify both sides.

$$(x+1)(200) = x(x+1)\left(\frac{200}{x+1}\right) + x(x+1)(10)$$ Continue simplifying.

$200x + 200 = 200x + 10x^2 + 10x$ Subtract $200x$ and 200 from both sides.

$0 = 10x^2 + 10x - 200$ Remove the common factor of 10.

$0 = 10(x^2 + x - 20)$ Factor $x^2 + x - 20$.

$0 = 10(x+5)(x-4)$ Set each factor, except 10, equal to 0.

$x + 5 = 0, x - 4 = 0$ Solve each equation.

$x = -5, \qquad x = \mathbf{4}\ \text{hours}$ Solutions of the equation.

State: Since time cannot be negative, 4 is the only possible answer. Since x represents the time of the car, it takes the car 4 hours to go 200 miles.

Check: The rate of the car is $\frac{200}{x} = \frac{200}{4} = 50$ miles per hour. So, in 4 hours the car can travel $50(4) = 200$ miles. The time necessary for the bus to travel 200 miles is $x + 1 = 4 + 1 = 5$ hours. The rate of the bus is $\frac{200}{x+1}$. Therefore, the rate of the bus is $\frac{200}{4+1} = \frac{200}{5} = 40$ miles per hour. So, in 5 hours the bus can travel $5(40) = 200$ miles. Therefore, our answers are correct.

PRACTICE EXERCISE 4

The average speed of a passenger train is 25 miles per hour more than the average speed of a car. The time required for the car to travel 300 miles is 2 hours more than the time required for the train. Find the average speed of the car.

OBJECTIVE E Solving work problems involving quadratic equations

In Section 7.7, we did work problems that resulted in linear equations, but they often result in quadratic equations as well.

Example 5   It would take Ramon 4 hours longer to plant a garden than it would take his wife Aurea. If they can plant the garden in $4\frac{4}{5}$ hours working together, how long would it take each to plant the garden alone?

Solution

Understand: We are given the relationship between the times that it takes Ramon and Aurea to plant the garden and the amount of time it takes them to plant the garden, working together. The unknowns are the amount of time it takes each to plant the garden alone.

> Relationship 1: It takes Ramon 4 hours longer to plant the garden than it does Aurea.
> Relationship 2: Together it takes them $4\frac{4}{5}$ hours to plant the garden.

Plan: Use relationship 1 to represent, in terms of a variable, the times it takes each working alone. Represent the part of the job that each can do in one hour and the part they can do together in one hour, and use these to write the equation.

Answers: Practice Exercise 4: 50 mph

Translate: Use relationship 1 to represent each unknown in terms of variable. Let x represent the number of hours for Aurea to plant the garden working alone. Then, $x + 4$ represents the number of hours for Ramon to plant the garden working alone.

$\dfrac{1}{x}$ represents the part of the garden Aurea can plant in 1 hour.

$\dfrac{1}{x + 4}$ represents the part of the garden Ramon can plant in 1 hour.

$\dfrac{1}{4\frac{4}{5}} = \dfrac{1}{\frac{24}{5}} = \dfrac{5}{24}$ represents the part of the garden they can plant in 1 hour together.

The word equation is
(the part done by Aurea in 1 hr) + (the part done by Ramon in 1 hr) =
    (the part done in 1 hr together).

Hence, the equation is

$$\frac{1}{x} + \frac{1}{x + 4} = \frac{5}{24}$$

Solve:

$\dfrac{1}{x} + \dfrac{1}{x + 4} = \dfrac{5}{24}$	Multiply both sides by the LCD $24x(x + 4)$.
$24x(x + 4)\left(\dfrac{1}{x} + \dfrac{1}{x + 4}\right) = 24x(x + 4)\left(\dfrac{5}{24}\right)$	Simplify both sides.
$24x(x + 4) \cdot \dfrac{1}{x} + 24x(x + 4) \cdot \dfrac{1}{x + 4} = 5x(x + 4)$	Continue simplifying.
$24(x + 4) + 24x = 5x^2 + 20x$	Distribute.
$24x + 96 + 24x = 5x^2 + 20x$	Add like terms.
$48x + 96 = 5x^2 + 20x$	Subtract $48x$ and 96 from both sides.
$0 = 5x^2 - 28x - 96$	Factor $5x^2 - 28x - 96$.
$0 = (5x + 12)(x - 8)$	Set each factor equal to 0.
$5x + 12 = 0, x - 8 = 0$	Solve each equation.
$5x = -12, \qquad x = 8$	Solution.
$x = -\dfrac{12}{5}$	Solution.

State: Since the garden cannot be planted in $-\frac{12}{5}$ hours, 8 is the only possible answer. So, it takes Aurea 8 hours working alone and Ramon $x + 4 = 8 + 4 = 12$ hours working alone.

Check: The check is left as an exercise for the student.

PRACTICE EXERCISE 5

Terri and Tommy run a flea market. It takes Terri 2 hours longer to put the merchandise out than it does Tommy. If they can put the merchandise out together in $1\frac{7}{8}$ hours, how long would it take each working alone?

Answer: **Practice Exercise 5:** Tommy = 3 hours, Terri = 5 hours

OBJECTIVE Solving science problems involving quadratic equations

Example 6

A particle moves along a straight line according to the formula $s = 2t^2 + t - 6$, where s represents the distance in inches from the beginning point and t represents the time in seconds. Find the time(s), to the nearest tenth of a second, when the particle is 3 inches from the beginning point.

Solution

Since s represents the distance from the beginning point, we need to find t when $s = 3$.

$$s = 2t^2 + t - 6 \qquad \text{Substitute 3 for } s.$$

$$3 = 2t^2 + t - 6 \qquad \text{Subtract 3 from both sides of the equation.}$$

$$0 = 2t^2 + t - 9 \qquad \text{Solve, using the quadratic formula.}$$

$$t = \frac{-1 \pm \sqrt{1^2 - 4(2)(-9)}}{2(2)} \qquad \text{Simplify the radical.}$$

$$t = \frac{-1 \pm \sqrt{1 + 72}}{4} \qquad \text{Add 1 and 72.}$$

$$t = \frac{-1 \pm \sqrt{73}}{4} \qquad \text{Write as two separate answers.}$$

$$t = \frac{-1 + \sqrt{73}}{4}, \qquad t = \frac{-1 - \sqrt{73}}{4} \qquad \sqrt{73} \approx 8.5440.$$

$$t \approx \frac{-1 + 8.5440}{4}, \qquad t \approx \frac{-1 - 8.5440}{4} \qquad \text{Simplify.}$$

$$t \approx 1.886, \qquad t \approx -2.386 \qquad \text{Round to the nearest tenth. Since we cannot have negative time, the only answer is 1.9 seconds.}$$

$$t \approx 1.9, \qquad t \approx -2.4$$

Check:

Substitute 3 for s and 1.9 for t in the formula $s = 2t^2 + t - 6$.

$$3 \approx 2(1.9)^2 + 1.9 - 6$$

$$3 \approx 2(3.61) + 1.9 - 6$$

$$3 \approx 7.22 + 1.9 - 6$$

$$3 \approx 3.12$$

These are not exactly equal, due to the round-off error. Since they are approximately equal, our answer is correct.

PRACTICE EXERCISE 6

A ball is thrown upward with an initial velocity of 32 feet per second from the top of a building that is 64 feet high. The distance, s, above the ground t seconds after the ball is thrown is given by the formula $s = -16t^2 + 32t + 64$, where s is measured in feet. Find the time(s) when the ball is 32 feet above the ground.

Answer: Practice Exercise 6: $t \approx 2.7$ seconds

Exercise Set 11.5

For Extra Help

MyMathLab®

Solve the following number problems. (See Example 1.)

1. The sum of the reciprocals of two consecutive integers is $\frac{9}{20}$. Find the integers.

2. The sum of the reciprocals of two consecutive odd integers is $\frac{12}{35}$. Find the integers.

3. One integer is 3 more than another. If the sum of the squares of the integers is 117, find the integers.

4. The larger of two integers is 1 less than twice the smaller. If the difference of the squares of the integers is 56, find the integers.

5. The product of the reciprocals of two consecutive odd integers is $\frac{1}{35}$. Find the integers.

6. The product of the reciprocals of two consecutive even integers is $\frac{1}{48}$. Find the integers.

Solve the following geometric problems. (See Example 2.)

7. The length of a small rectangular field is 10 yards more than its width. A road 5 yards wide completely encircles the field. If the area of the road and the field together is 4200 square yards, find the dimensions of the field.

8. A rectangular slab whose length is 2 feet more than its width is poured for a utility shed. The shed is placed on the slab so that 1 foot of the slab is exposed on all sides of the shed. If the area of the floor of the shed is 120 square feet, find the dimensions of the slab.

9. A rectangular sheet of paper is 3 inches longer than it is wide. Material is printed on the page with 1-inch margins at the top and bottom of the page and $\frac{1}{2}$-inch margins on the left and right. If the area of the printed material is 63 square inches, find the length and width of the page.

10. The length of a rectangular picture is 2 inches more than its width. It is to be mounted with 1 inch of matting on each side and $1\frac{1}{2}$ inches each at the top and bottom. If the area of the picture and the matting is 130 square inches, find the length and width of the picture.

Solve the following Pythagorean Theorem problems. (See Example 3.)

11. The sides of a right triangle are three consecutive integers. Find the lengths of the sides.

12. The sides of a right triangle are three consecutive even integers. Find the lengths of the sides.

13. A guy wire is attached to a vertical pole at a point 24 feet above the ground. The length of the wire is 6 feet more than twice the distance from the pole to the point where the wire is attached to the ground. Find the length of the wire.

14. A telephone pole is supported by two guy wires. The distance from the pole to the points on the ground where the wires are attached is 15 feet. The length of each wire is 5 feet more than the distance from the ground to the point on the pole where the wires are attached. Find the total length of the wires.

15. Two cars are approaching the same intersection at right angles to each other. The distance from the intersection to one of the cars is 10 feet less than twice the distance from the intersection to the other car. If the distance between the cars is 170 feet, find the distance from each car to the intersection.

16. Two ships are approaching the same port and are traveling on courses that are at right angles to each other. One ship is 8 miles from port. The distance of the other ship from port is 2 miles less than the distance between the ships. If the maximum range of their radios is 20 miles, can the ships communicate with each other?

Solve the following distance, rate, time problems. (See Example 4.)

17. The average rate of a bus is 15 miles per hour more than the average rate of a truck. It takes the truck 1 hour longer to go 180 miles than it takes the bus. How long does it take the bus to travel 180 miles?

18. The time required for a train to travel 200 miles is 2 hours more than the time required for a car to travel 180 miles. The average rate of the car is 20 miles per hour more than the average rate of the train. Find the rate of the train.

19. The time required for a bus to travel 360 miles is 3 hours more than the time required for a motorcycle to travel 300 miles. The average rate of the motorcycle is 15 miles per hour more than the average rate of the bus. Find the average rate of the bus.

20. The time required for a truck to travel 400 miles is 3 hours more than the time required for a car to travel 300 miles. The average rate of the car is 10 miles per hour more than the average rate of the truck. Find the average rate of the truck.

Solve the following work problems. (See Example 5.)

21. Billy and Jody are commercial fisherman. Working alone, it takes Billy 2 hours longer to run the hoop nets than it takes Jody working alone. Together, they can run the hoop nets in $2\frac{2}{5}$ hours. How long does it take each when working alone?

22. Using a riding lawn mower, Fran can mow the grass at the campground in 4 hours less time than it takes Donnie using a push-type mower. Together, they can mow the grass in $2\frac{2}{3}$ hours. How long does it take each when working alone?

23. It takes Buster 1 hour longer to stack a load of hay than it takes Ronnie. Together, they can stack a load of hay in $1\frac{1}{5}$ hours. How long does it take each when working alone?

24. It takes John, a novice mechanic, 5 hours longer to repair a transmission than it takes Ramon, an experienced mechanic. Together, they can repair the transmission in $3\frac{1}{3}$ hours. How long does it take each when working alone?

Solve the following science problems. (See Example 6.)

25. A particle is moving along a straight line, according to the formula $s = t^2 - 2t + 3$, where s is the distance in inches and t is the time in seconds. Find the time(s) when the particle is 18 inches from the beginning point.

26. A particle is moving along a straight line, according to the formula $s = t^2 - 6t + 12$, where s is the distance in inches and t is the time in seconds. Find the time(s) when the particle is 4 inches from the beginning point.

27. The distance a free-falling object falls is given by $s = 16t^2$, where s is in feet and t is in seconds. Find the time(s) to the nearest tenth of a second it takes a free-falling object to fall 48 feet.

28. The distance a free-falling object falls is given by $s = 16t^2$, where s is in feet and t is in seconds. Find the time(s) to the nearest tenth of a second it takes a free-falling object to fall 80 feet.

29. An object is thrown upward with an initial velocity of 48 feet per second from a height of 80 feet. The distance, s, above the ground after t seconds is given by $s = -16t^2 + 48t + 80$, where s is measured in feet. Find the time(s) to the nearest tenth of a second when the object is 40 feet above the ground.

30. An object is thrown upward with an initial velocity of 52 feet per second from a height of 100 feet. The distance, s, above the ground after t seconds is given by $s = -16t^2 + 52t + 100$, where s is measured in feet. Find the time(s) to the nearest tenth of a second when the object is 120 feet above the ground.

Writing Exercises (31 and 32)

31. Write an application problem that results in a quadratic equation and involves integers.

32. Write an application problem that results in a quadratic equation and involves geometric figures.

Chapter 11 Summary

Concept/Procedure	Example

Solving Incomplete Quadratic Equations:
[Section 11.1]

1. Solve incomplete quadratic equations that can be put in the form of $ax^2 + bx = 0$ by factoring.

Example 1:

Solve $5x^2 - 3x = 0$.

Solution:

$5x^2 - 3x = 0$ Remove the common factor.

$x(5x - 3) = 0$ Set each factor equal to 0.

$x = 0, 5x - 3 = 0$ Solve the linear equation.

$$5x = 3$$

$$x = \frac{3}{5}$$ Therefore, the solutions are $x = 0, \frac{3}{5}$.

2. Solve incomplete quadratics that can be put in the form $ax^2 + c = 0$ by extraction of roots. This involves the following:

a. Solve the equation for x^2. This gives an equation of the form $x^2 = k$.

b. Find the positive and negative square roots of k.

c. Write the solutions in simplest form.

Example 2:

Solve $4x^2 - 8 = 0$.

Solution:

$4x^2 - 8 = 0$ Add 8 to both sides.

$4x^2 = 8$ Divide both sides by 4.

$x^2 = 2$ Take the positive and negative square roots of 2.

$x = \pm\sqrt{2}$ Therefore, the solutions are $\sqrt{2}$ and $-\sqrt{2}$.

3. Solve equations of the form $(ax + b)^2 = c$ by doing the following:

a. Set $ax + b = \pm\sqrt{c}$.

b. Rewrite as two equations.

c. Solve each equation.

Example 3:

Solve $(3x + 4)^2 = 49$.

Solution:

$(3x + 4)^2 = 49$ Take the positive and negative square roots of 49.

$3x + 4 = \pm 7$ Write as two equations.

$3x + 4 = 7, 3x + 4 = -7$ Solve each equation.

$3x = 3 \qquad 3x = -11$

$x = 1 \qquad x = -\dfrac{11}{3}$ Therefore, the solutions are 1 and $-\dfrac{11}{3}$.

Completing the Square: [Section 11.2]

- To determine the number to be added to $x^2 + bx$ in order to get a perfect-square trinomial, find one-half of b and square it. The resulting trinomial is the square of a binomial whose first term is the square root of the first term of the trinomial, whose second term is the square root of the last term of the trinomial, and whose sign is the same as the sign of the middle term of the trinomial.

Concept/Procedure	Example

Solving Quadratic Equations by Completing the Square: [Section 11.2]

1. Write the equation in the form $x^2 + bx = c$. This may require several steps, depending upon the form of the given equation.
2. Add $\left(\frac{1}{2} \cdot b\right)^2$ to both sides of the equation to make the left side a perfect square trinomial.
3. Write one side of the equation as the square of a binomial, and add the numbers on the other side of the equation.
4. Solve the equation resulting from step 3 by extraction of roots.
5. Solve each of the resulting equations.
6. Check each solution in the original equation.

Example 4:
Solve $x^2 - 4x - 24 = 0$ by completing the square.

Solution:

$x^2 - 4x - 24 = 0$	Add 24 to both sides.
$x^2 - 4x = 24$	$\frac{1}{2} \cdot 4 = 2$ and $2^2 = 4$, so add 4 to both sides.
$x^2 - 4x + 4 = 24 + 4$	Write the left side as the square of a binomial.
$(x - 2)^2 = 28$	Solve, using extraction of roots.
$x - 2 = \pm\sqrt{28}$	Simplify $\sqrt{28}$.
$x - 2 = \pm 2\sqrt{7}$	Write as two equations.
$x - 2 = 2\sqrt{7}, x - 2 = -2\sqrt{7}$	Solve each equation.
$x = 2 + 2\sqrt{7}, x = 2 - 2\sqrt{7}$	Therefore, the solutions are $2 + 2\sqrt{7}$ and $2 - 2\sqrt{7}$.

Solving Quadratic Equations, Using the Quadratic Formula: [Section 11.3]

1. If necessary, rewrite the equation in the form of $ax^2 + bx + c = 0$.
2. Find the values of a, b, and c.
3. Substitute the values of a, b, and c into the quadratic formula, which is
$$x = \frac{-b \pm \sqrt{b^2 - 4ac}}{2a}.$$
4. Simplify the results.
5. Check the solutions in the original equation.

Example 5:
Solve $2x^2 - 7x - 3 = 0$, using the quadratic formula.

Solution:
$a = 2, b = -7$, and $c = -3$.

$x = \dfrac{-b \pm \sqrt{b^2 - 4ac}}{2a}$	Substitute for a, b, and c.
$x = \dfrac{-(-7) \pm \sqrt{(-7)^2 - 4(2)(-3)}}{2(2)}$	Simplify.
$x = \dfrac{7 \pm \sqrt{49 + 24}}{4}$	Continue simplifying.
$x = \dfrac{7 \pm \sqrt{73}}{4}$	The solutions are $\dfrac{4 + \sqrt{73}}{4}$ and $\dfrac{4 - \sqrt{73}}{4}$.

Imaginary Unit: [Section 11.4]

- The imaginary unit, i, is the number whose square is -1. In symbols, $\sqrt{-1} = i$ and $i^2 = -1$.

Writing a Radical in i Form: [Section 11.4]

- To write a radical in i form, remove a factor of -1 and apply $\sqrt[n]{ab} = \sqrt[n]{a} \cdot \sqrt[n]{b}$, as in $\sqrt{-a} = \sqrt{-1 \cdot a} = \sqrt{-1} \cdot \sqrt{a} = i\sqrt{a}$.

Example 6:
Write $\sqrt{-98}$ in i form.

Solution:
$\sqrt{-98} = \sqrt{-49 \cdot 2} = \sqrt{-49} \cdot \sqrt{2} = \sqrt{-1 \cdot 49 \cdot 2} = \sqrt{-1} \cdot \sqrt{49} \cdot \sqrt{2} = 7i\sqrt{2}$

Concept/Procedure	Example
Standard Form of a Complex Number: [Section 11.4] • The standard form of a complex number is $a + bi$, where a and b are real numbers and $i = \sqrt{-1}$. If $a = 0$, then $a + bi = bi$, which is an imaginary number. If $b = 0$, then $a + bi = a$, which is a real number.	
Finding Sums or Differences of Complex Numbers: [Section 11.4] • To add or subtract complex numbers, add or subtract as though they were polynomials and treat i as though it were a variable.	Example 7: Simplify $(4 + 6i) - (2 - 3i)$. Solution: $(4 + 6i) - (2 - 3i) =$ Remove parentheses. $4 + 6i - 2 + 3i =$ Add the real parts and the imaginary parts. $2 + 9i$ Difference.
Finding Products of Complex Numbers: [Section 11.4] • To find the product of complex numbers, use the distributive property or FOIL. Remember to use $i^2 = -1$ when necessary.	Example 8: Simplify $(2 + 6i)(5 - 3i)$. Solution: $(2 + 6i)(5 - 3i) =$ Apply FOIL. $10 - 6i + 30i - 18i^2 =$ Add $-6i$ and $30i$, and replace i^2 with -1. $10 + 24i - 18(-1) =$ $-18(-1) = 18$ $10 + 24i + 18 =$ Add 10 and 18. $28 + 24i$ Product.
Finding Quotients of Complex Numbers: [Section 11.4] • To find the quotient of complex numbers, multiply the numerator and the denominator by the complex conjugate of the denominator.	Example 9: Simplify $\dfrac{2 + 3i}{3 - i}$. Solution: Multiply both numerator and denominator by $3 + i$, the conjugate of the denominator. $\dfrac{2 + 3i}{3 - i} = \dfrac{2 + 3i}{3 - i} \cdot \dfrac{3 + i}{3 + i}$ Multiply. $= \dfrac{6 + 11i + 3i^2}{9 - i^2}$ Replace i^2 with -1 and multiply. $= \dfrac{6 + 11i - 3}{9 - (-1)}$ Add. $= \dfrac{3 + 11i}{10}$ Write in standard form. $= \dfrac{3}{10} + \dfrac{11}{10}i$ Quotient.

Concept/Procedure	Example
Solving Quadratic Equations with Complex Solutions: [Section 11.4] • To solve equations of the form $(ax + b)^2 = -d$, with $d > 0$, use the extraction of roots method. To solve equations of the form $ax^2 + bx + c = 0$, use the quadratic formula.	**Example 10:** Solve $2x^2 + 5x + 4 = 0$, using the quadratic formula. Solution: $a = 2, b = 5, c = 4$ $x = \dfrac{-b \pm \sqrt{b^2 - 4ac}}{2a}$ Substitute for a, b, and c. $x = \dfrac{-5 \pm \sqrt{5^2 - 4(2)(4)}}{2(2)}$ Simplify. $x = \dfrac{-5 \pm \sqrt{25 - 32}}{4}$ Continue simplifying. $x = \dfrac{-5 \pm \sqrt{-7}}{4}$ $\sqrt{-7} = i\sqrt{7}$. $x = \dfrac{-5 \pm i\sqrt{7}}{4}$ Solution.
Solving Application Problems Involving Quadratic Equations: [Section 11.5] • Solve the application problems, following the same procedure you used earlier. Be sure you answer the question(s) asked. Check the solution(s) against the wording of the problem. Be aware that some solutions of the equation may not satisfy the conditions of the original problem.	

Chapter 11   Review Exercises

Solve the incomplete quadratic equations. If the solutions are not real numbers, write "no real-number solution." [Section 11.1]

1. $6x^2 + 18x = 0$ **2.** $5x^2 - 80 = 0$

3. $5a^2 = -13a$ **4.** $2c^2 = 13c$

5. $n(n + 3) = 5n$ **6.** $a(a - 5) = 8a$

7. $r^2 = 144$ **8.** $c^2 = 400$

9. $x^2 - 28 = 0$ **10.** $c^2 - 45 = 0$

11. $b^2 + 63 = 0$ **12.** $k^2 + 48 = 0$

13. $3x^2 = 75$ **14.** $2q^2 = 98$

15. $4x^2 - 80 = 0$ **16.** $5w^2 - 90 = 0$

17. $(x + 6)^2 = 49$ **18.** $(a - 3)^2 = 64$

19. $(3x + 4)^2 = 80$ **20.** $(2x - 5)^2 = 125$

Solve the quadratic equations by completing the square. If the solutions are not real numbers, write "no real-number solutions." [Section 11.2]

21. $x^2 + 10x + 24 = 0$ **22.** $x^2 - 10x + 16 = 0$

23. $c^2 = 5c + 14$ **24.** $r^2 - 18 = -3r$

25. $2w^2 + 6 = -13w$ **26.** $3n^2 = 4n + 4$

27. $y^2 + 8y - 2 = 0$ **28.** $c^2 - 10c + 4 = 0$

29. $a^2 + 7a = 3$ **30.** $n^2 - 9n + 2 = 0$

31. $x^2 + 3x + 5 = 0$ **32.** $y^2 + 7y + 9 = 0$

33. $2x^2 + 12x - 5 = 0$ **34.** $2x^2 - 14x + 3 = 0$

Solve the quadratic equations, using the quadratic formula. If the solutions are not real numbers, write "no real-number solutions." [Section 11.3]

35. $a^2 + 3a + 2 = 0$ **36.** $b^2 + 5b + 4 = 0$

37. $a^2 = 8a - 12$ **38.** $b^2 - 3b = 10$

39. $2x^2 + x - 10 = 0$ **40.** $3t^2 + t - 2 = 0$

41. $6v^2 = 8v + 3$ **42.** $6m^2 = -5m + 4$

43. $x^2 + 6x - 4 = 0$ **44.** $q^2 + 8q - 2 = 0$

45. $2x^2 + 1 = 8x$ **46.** $4n^2 = -6n + 5$

47. $6r^2 + 2r = 3$ **48.** $6x^2 = 4x + 1$

49. $3d^2 + 2d + 4 = 0$ **50.** $4z^2 - 3z + 5 = 0$

51. $2x^2 + 5x - 2 = 0$ **52.** $3x^2 + 7x = -1$

Write each of the following as the product of a real number and i: [Section 11.4]

53. $\sqrt{-81}$ **54.** $\sqrt{-98}$

Write each of the following in standard form $a + bi$: [Section 11.4]

55. 12 **56.** $11i$

57. $\sqrt{-90}$ **58.** $4 + \sqrt{-121}$

Find the sums or differences of the given complex numbers. Express answers in standard form $a + bi$. [Section 11.4]

59. $7i + (-4 - 6i)$ **60.** $(7 - 8i) + (-10 - 5i)$

61. $-4i - (-4 + 5i)$ **62.** $(-3 + 7i) - (-2 - 4i)$

Find the following products or quotients of complex numbers: [Section 11.4]

63. $5i(4 - 8i)$ **64.** $(4 + 2i)(3 - 2i)$

65. $(5 + 5i)(5 - 5i)$ **66.** $\dfrac{3 - i}{4 - i}$

Solve the quadratic equations. Express answers in standard form $a + bi$. [Section 11.4]

67. $(x + 6)^2 = -64$ **68.** $x^2 - 6x + 25 = 0$

69. $x^2 - 8x + 18 = 0$ **70.** $x^2 + 3x + 7 = 0$

Solve the following application problems: [Section 11.5]

71. One integer is 3 more than the other. The sum of the reciprocals of the integers is $\frac{7}{10}$. Find the integers.

72. One integer is 4 more than the other. The sum of the squares of the integers is 40. Find the integers.

73. The sum of the squares of two consecutive integers is 61. Find the integers.

74. One integer is twice the other. The product of the reciprocals of the integers is $\frac{1}{32}$. Find the integers.

75. A rectangular room is 4 feet longer than it is wide. A rectangular rug is placed on the floor, and there is a uniform border of exposed floor 1 foot wide all the way around the rug. If the area of the rug is 140 square feet, find the dimensions of the room.

76. A rectangular piece of plywood is 3 feet longer than it is wide. The whole piece of plywood is to be painted with a black rectangle surrounded by a uniform border of red that is 1 foot wide. If the area of the red border is 30 square feet, find the length and width of the piece of plywood.

77. Two airplanes are approaching the same airport and are flying on courses that are perpendicular to each other. One airplane is 1 mile less than twice as far from the airport as the other. If the distance between the airplanes is 17 miles, how far is each from the airport?

78. A diver's flag is rectangular in shape and is red with a black stripe going diagonally across it. If the width of the flag is 5 inches less than the length and the length of the black stripe is 25 inches, find the length and width of the flag.

79. The average rate of a recreational vehicle is 20 miles per hour less than the average rate of a car. It takes the recreational vehicle 1 hour more to travel 200 miles than it takes the car to travel 240 miles. Find the average rate of the recreational vehicle.

80. It takes a bus 2 hours longer to travel 200 miles than it takes a motorcycle to travel 120 miles. The average rate of the motorcycle is 10 miles per hour more than the average rate of the bus. Find how long it takes the motorcycle to travel 120 miles.

81. Horace is planning to build a house. The lot he has chosen has to be built up with fill dirt. It would take Horace 4 hours longer to spread the dirt than it would take his worker, Ahmad. If they can spread the dirt in $3\frac{3}{4}$ hours working together, how long would it take each when working alone?

82. After Horace built his house in Exercise 81, he and Ahmad planned to paint it. It would take Horace 2 days longer to paint the house than it would take Ahmad. Together, they could paint the house in $3\frac{3}{7}$ days. How long would it take each when working alone?

Chapter 11 Test

Solve the following incomplete quadratic equations:

1. $6x^2 - 12x = 0$

2. $4x^2 = 20x$

3. $x^2 - 121 = 0$

4. $3x^2 = 81$

5. $(x - 6)^2 = 36$

6. $(3x + 2)^2 = 18$

Find the number needed to make each polynomial a perfect-square trinomial. Then write each as the square of a binomial.

7. $a^2 - 20a +$ ____

8. $b^2 + \dfrac{7}{3}b +$ ____

Solve the following quadratic equations by completing the square:

9. $x^2 - 6x - 16 = 0$

10. $2x^2 - 6x - 5 = 0$

Solve the quadratic equations, using the quadratic formula. If the solutions are not real numbers, write "no real-number solutions."

11. $a^2 - 5a + 6 = 0$

12. $2c^2 - 3c = 20$

13. $r^2 - 2r + 6 = 0$

14. $3w^2 = -8w + 3$

Solve the quadratic equations, using the method that seems most appropriate. If the solutions are not real numbers, write "no real solutions."

15. $x^2 = 2x + 35$

16. $a^2 + \dfrac{7}{2}a = 3$

17. $3d^2 + 8 = 5d$

18. $x^2 - 8x - 109 = 0$

19. Write $-3 + \sqrt{-36}$ as a complex number in standard form $a + bi$.

Find the sums or differences of the complex numbers. Express answers in standard form $a + bi$.

20. $(3 - 8i) + (-4 - 5i)$

21. $(8 - 4i) - (5 - 2i)$

Find the products or quotients of the complex numbers. Express answers in standard form $a + bi$.

22. $(2 + 6i)(3 - 2i)$

23. $\dfrac{4 - 2i}{2 + 5i}$

Solve the following application problems:

24. Find two consecutive even integers the sum of whose squares is 164.

25. The length of a picture is 2 inches more than the width. The picture is in a frame that is $1\frac{1}{2}$ inches wide. If the area of the picture and the frame is 99 square inches, find the length and width of the picture.

26. In mountainous country, it takes a recreational vehicle (RV) 1 hour longer to travel 120 miles than it takes a van. If the rate of the van is 10 miles per hour more than the rate of the RV, find how long it takes the van to travel 120 miles.

Answers to Odd-Numbered Exercises, Chapter Reviews and Chapter Tests

Exercise Set R.1 **1.** hundred **3.** million **5.** one or unit
7. $6 \cdot 10,000 = 60,000$ **9.** $0 \cdot 10,000 = 0$ **11.** $4 \cdot 1,000,000 = 4,000,000$
13. $300 + 90 + 7$ **15.** $3000 + 30 + 3$ **17.** $400,000 + 9000 + 100 + 30 + 5$
19. twelve **21.** nine hundred five **23.** seven thousand, one hundred
forty-nine **25.** eight thousand, nine hundred two **27.** thirty thousand,
two hundred nine **29.** four million, seventy-eight thousand, seventy-
four **31.** twenty-nine million, seven hundred fifty-six thousand, eleven
33. 407 **35.** 14,073 **37.** 902,460 **39.** ⋒ IIIII

Exercise Set R.2 **1.** 868 **3.** 930 **5.** 9741 **7.** 79,962 **9.** 87
11. 1724 **13.** 6529 **15.** 211 **17.** 1051 **19.** 38 **21.** 453 **23.** 876
25. 38 **27.** 370 **29.** 1873 **31.** $1270 **33.** $3000 **35.** 13_{five}
37. $12,000

Exercise Set R.3 **1.** 68 **3.** 230 **5.** 158 **7.** 1794 **9.** 520
11. 1026 **13.** 11,860 **15.** 8772 **17.** 103,632 **19.** 217,136
21. 690 chairs **23.** $33,600 **25.** 96 windows **27.** 525 entries
29. 8 **31.** 7 R 2 **33.** 9 R 3 **35.** 9 R 6 **37.** 18 **39.** 118 R 4
41. 21 R 4 **43.** 14 R 41 **45.** 22 R 216 **47.** 16 children **49.** $60
51. 24 bundles **53.** $5 **55.** 5 days **57.** no **59.** $520
61. 10 in by 8 in

Exercise Set R.4 **1.** $\dfrac{9}{12}$ **3.** $\dfrac{1}{3}$ **5.** $\dfrac{36}{60}$

7. 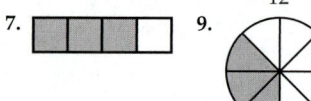 **9.**

11. $2\dfrac{3}{5}$ **13.** $5\dfrac{1}{7}$ **15.** $3\dfrac{8}{13}$ **17.** $2\dfrac{5}{29}$ **19.** $\dfrac{22}{5}$ **21.** $\dfrac{49}{9}$ **23.** $\dfrac{77}{12}$ **25.** $\dfrac{77}{6}$

27. $\dfrac{343}{23}$ **29.** $\dfrac{3}{4}$ **31.** $\dfrac{4}{5}$ **33.** $\dfrac{2}{3}$ **35.** $\dfrac{1}{15}$ **37.** $\dfrac{8}{45}$ **39.** $\dfrac{3}{14}$ **41.** $\dfrac{7}{36}$

43. 2 **45.** 6 **47.** 6 **49.** $\dfrac{12}{35}$ **51.** $\dfrac{9}{10}$ **53.** $\dfrac{7}{9}$ **55.** $\dfrac{2}{15}$ **57.** 8 **59.** $\dfrac{5}{7}$

61. $\dfrac{15}{17}$ **63.** $\dfrac{2}{5}$ **65.** $\dfrac{5}{13}$ **67.** $\dfrac{6}{7}$ **69.** $\dfrac{9}{13}$ **71.** 5 cups **73.** $\dfrac{5}{8}$

75. $\dfrac{14}{3}$ ft **77.** 18 pieces **79.** $\dfrac{4}{15}$ hr

Exercise Set R.5 **1.** six tenths **3.** eighty-four hundredths
5. seven hundred sixty-three thousandths **7.** twenty-seven and forty-
five hundredths **9.** sixty-eight and six hundred seven thousandths
11. tenths **13.** thousandths **15.** tenths **17.** 52.0 **19.** 18,043
21. 298.20 **23.** .66499 **25.** 1.1022 **27.** 1.4 **29.** .71 **31.** 6.06
33. 119.497 **35.** 1190.61 **37.** 206.27 **39.** .6 **41.** .58 **43.** 29.57
45. 35.26 **47.** 113.956 **49.** 19.1631 **51.** 32.34 **53.** 220.5
55. 1158.353 **57.** seven hundred six and $\frac{45}{100}$ dollars **59.** three hundred
five and $\frac{2}{100}$ dollars **61.** $8.29 million **63.** .8 ft **65.** 298.1 mi
67. $.16 **69.** $3.03 **71.** 28.971 gal **73.** 225.9 cents

Exercise Set R.6 **1.** 56.7 **3.** 14.08 **5.** .936 **7.** 23.4 **9.** 42.84
11. .267 **13.** .3885 **15.** 3.17781 **17.** .82 **19.** 8.4 **21.** 65 **23.** .007
25. 5.0 **27.** .7 **29.** .08 **31.** .7 **33.** 22.67 **35.** 1476.21 **37.** 36.57
39. 0.25 **41.** 0.88 **43.** 0.56 **45.** 0.67 **47.** 2.25 **49.** 4.4 **51.** 1.32
53. 2.21 **55.** 96 km/hr **57.** 4.17 pages per day **59.** $32.67
61. $1142.86 **63.** $504

Exercise Set R.7 **1.** 24 in. **3.** 3 yd **5.** 26,400 ft **7.** 3520 yd
9. 5 ft **11.** 8 yd **13.** 8 yd **15.** 765 ft **17.** 9240 yd **19.** 6 mi
21. 300 m **23.** 4 dm **25.** 8000 cm **27.** 1400 m **29.** 700 mm
31. 4.58 dam **33.** 10 km **35.** .84 m **37.** 9 m **39.** .0162 hm
41. 15 yd **43.** 35,376 ft **45.** .00528 km **47.** .0001 mm

Chapter R Review Exercises **1.** six thousand fifty-one
2. thirty thousand, four hundred fifty-nine **3.** 37,204 **4.** ten
thousand **5.** one million **6.** $4 \times 100,000 = 400,000$
7. $0 \times 10 = 0$ **8.** $300,000 + 90,000 + 6000 + 70 + 1$ **9.** 99 **10.** 85
11. 628 **12.** 6156 **13.** 729,351 **14.** 14,383 **15.** 42 points **16.** 61 ft
17. 45,000 trees **18.** 20 ft **19.** 688 **20.** 79 **21.** 75,156 **22.** 182 mi
23. 136 books **24.** 34 cousins **25.** $506 **26.** 455 **27.** 2964
28. 12,852 **29.** 20,130 **30.** 44,820 **31.** 360 test tubes **32.** 195 band
members **33.** $316 **34.** 37 **35.** 89 R 2 **36.** 33 **37.** 454 R 8

38. $5250 **39.** 35 classes **40.** $2\dfrac{1}{8}$ **41.** $5\dfrac{5}{6}$ **42.** $2\dfrac{11}{21}$ **43.** $2\dfrac{5}{11}$

44. $\dfrac{39}{7}$ **45.** $\dfrac{61}{9}$ **46.** $\dfrac{61}{18}$ **47.** $\dfrac{189}{23}$ **48.** $\dfrac{15}{8}$ **49.** $\dfrac{54}{77}$ **50.** $\dfrac{7}{11}$ **51.** $\dfrac{4}{9}$

52. 1.129 **53.** 197.149 **54.** 2.20 **55.** 146.81 **56.** 503.74 **57.** 467.942
58. 3.81 pounds **59.** $62.75 **60.** .088 **61.** .034 **62.** 38.646
63. 38.69 **64.** 73.5 **65.** 3 **66.** 31.13 **67.** 30.80 **68.** 0.31 **69.** 0.78
70. 0.85 **71.** 0.63 **72.** 2.2 **73.** 4.25 **74.** $450.48 **75.** 29.58 ml.
76. $34.30 **77.** 60 in. **78.** 3.6 ft **79.** 7.5 ft **80.** 3700 mm
81. 1,800,000 cm **82.** 890 dm **83.** 9 in. **84.** 3.2 mi **85.** 9.215 m
86. 43,600 dm

Chapter R Test **1.** ten thousand [R.1]
2. $80,000 + 2,000 + 60 + 3$ [R.1] **3.** 1402 [R.2] **4.** 96,021 [R.2]
5. 20,183 [R.2] **6.** 533 cards [R.2] **7.** 5981 [R.2] **8.** 517,370 [R.2]
9. 86 seeds [R.2] **10.** $2\dfrac{3}{13}$ [R.4] **11.** $\dfrac{46}{7}$ [R.4] **12.** 5992 [R.3]
13. 53,935 [R.3] **14.** 646 pencils [R.3] **15.** 95 [R.3] **16.** 37.15 [R.3]
17. $\dfrac{18}{5}$ [R.4] **18.** $\dfrac{5}{12}$ [R.4] **19.** 80.967 [R.5] **20.** 538.034 [R.5]
21. 14.654 [R.5] **22.** 6.2 pounds [R.5] **23.** 3.8775 [R.6] **24.** 1.90938 [R.6]
25. 3.66 [R.6] **26.** 65.00 [R.6] **27.** 2.10 [R.6] **28.** 936.67 [R.6]
29. 0.31 [R.6] **30.** 14 yd [R.7] **31.** 66 in. [R.7] **32.** 76,000 cm [R.7]
33. 90 mm [R.7]

Exercise Set 1.1 **1.** 8 squared, 64 **3.** 5 cubed, 125 **5.** 9 to the
fourth power, 6,561 **7.** 2 to the fifth power, 32 **9.** 10 to the sixth
power, 1,000,000 **11.** $\frac{4}{5}$ squared, $\frac{16}{25}$ **13.** 1.3 cubed, 2.197 **15.** $a \cdot a \cdot a \cdot a$
17. $a \cdot a \cdot b \cdot b \cdot b$ **19.** $6 \cdot b \cdot b \cdot b \cdot b$ **21.** $(4a)(4a)(4a)(4a)$ **23.** p^4
25. $b^4 d^3$ **27.** $3x^4$ **29.** $(4r)^6$ **31.** 9 **33.** 2 **35.** 19 **37.** 56 **39.** 29
41. 6 **43.** 12 **45.** 288 **47.** 4 **49.** $\dfrac{25}{72}$ **51.** 56 **53.** 24 **55.** 76
57. 16 **59.** 21 **61.** 5 **63.** 8 **65.** 32 **67.** 4 **69.** 10 **71.** 24 **73.** 4
75. 61 **77.** 2 **79.** 3 **81.** 4 **83.** 2,187 **85.** 53.5824 **87.** 43.7235
89. 430.204708 **91.** 18 **93.** 48 **95.** 12 **97.** 180 **99.** 48 **101.** 4
103. 2 **105.** $\dfrac{3}{25}$ **107.** $\dfrac{9}{25}$ **109.** 13.44 **111.** 181.8432 **113.** 99.5328
115. $150 + 28(18) = \$654$ **117.** $600 + 190(24) = \$5,160$
119. $3000(4)(12) + 3500(6)(12) = \$396,000$
121. $[3(10.99) + 5(1.29)] \div 5 = \$7.88$
123. $800 + 299(3)(12) + .15(58,000 - 45,000) = 13,514$
125. 70 **127.** 1096

Exercise Set 1.2 **1.** 14 m **3.** 33 in. **5.** 72 in. **7.** 42 m
9. 180 ft **11.** 36 ft **13.** 2.1 dm **15.** 4.8 yd **17.** 56 cm **19.** 26.2 in.
21. 30 mm **23.** 22 in. **25.** 10.2 m **27.** 14π m, 43.96 m **29.** 5.5π ft,
17.27 ft **31.** $\dfrac{8}{3}\pi$ in., $\dfrac{176}{21}$ in. **33.** 40 ft **35.** 32.28 ft **37.** 208 in. **39.** 69 ft
41. $82.50 **43.** 21 timbers, $103.95 **45.** 52 lights **47.** 376.8 in.

Exercise Set 1.3 **1.** 54 m^2 **3.** 1400 ft^2 **5.** 45 ft^2 **7.** 1.44 yd^2
9. $\dfrac{25}{49}$ cm^2 **11.** 19.32 m^2 **13.** 30 mm^2 **15.** 216 in.^{2} **17.** 20.4 in.^{2}

19. 49π m^2, 153.9 m^2 **21.** 42.25π ft^2, 132.7 ft^2 **23.** $\dfrac{550}{63}$ in^2 **25.** 112 mm^2
27. 1.56 mi^2 **29.** 7.13 cm^2 **31.** 42 mi^2 **33.** 107.8 in.^{2} **35.** 28.75 dm^2
37. 52 ft^2 **39.** 7.74 cm^2 **41.** 18,000 yd^2 **43.** 30,000 plants, \$88,500
45. 260 pieces of tile, \$673.40 **47.** \$1050 **49.** \$454.80 **51.** \$1,560
53. 13,600 ft^2 **55.** 12.36375 in.^{2}

Exercise Set 1.4 **1.** $V = 72$ ft^3, $SA = 108$ ft^2 **3.** $V = 9$ yd^3,
$SA = 27$ yd^2 **5.** $V = 270.4$ m^3, $SA = 254.8$ m^2 **7.** $V = 125$ mi^3,
$SA = 150$ mi^2 **9.** $V = \dfrac{27}{125}$ cm^3, $SA = \dfrac{54}{25}$ cm^2 **11.** $V = 300\pi$ cm^3,
942 cm^3 $SA = 170\pi$ cm^2, 533.8 cm^2 **13.** $V = 93.492\pi$ m^3, 293.6 m^3,
$SA = 79.8\pi$ m^2, 250.6 m^2 **15.** $V = 14.52\pi$ m^3, 45.6 m^3
17. 0.5π yd^3, 1.6 yd^3 **19.** $V = 972\pi$ cm^3, 3052.08 cm^3 **21.** $V = \dfrac{1100}{147}$ in.^{3}
23. $V = 250\pi$ in.^{3} **25.** $V = 108\pi$ ft^3 **27.** 14.22 yd^3 **29.** 31,752π in.^{3}
31. 390.5 in.^{3} **33.** \$720 **35.** 1,272,000 in.^{2} **37.** \$432 **39.** \$8.40
41. 50.24 in.^{2} **43.** \$1,256

Exercise Set 1.5 **1.** -8 **3.** losing 10 pounds **5.** 250 mi west
7.
9.
11. 0 or any negative integer **13.** 1, 2, 3, . . . **15.** a) 5 b) 0, 5
c) $-10, 0, 5$ d) $-6.2, \dfrac{7}{10}$ **17.** sometimes **19.** always **21.** always
23. -17 **25.** 15 **27.** -16 **29.** -8 **31.** 15 **33.** 12 **35.** $>$ **37.** $<$
39. $>$ **41.** $=$ **43.** $=$ **45.** $<$ **47.** $>$ **49.** $=$ **51.** 8 **53.** 5 **55.** 6
57. 10 **59.** 2 **61.** 16 **63.** 8 **65.** 5 **67.** 78

Exercise Set 1.6 **1.** 18 **3.** 11 **5.** -7 **7.** 9 **9.** -11 **11.** -11
13. 48 **15.** 11 **17.** -6 **19.** 29 **21.** -1.4 **23.** -4.7 **25.** 7 **27.** -5
29. -25 **31.** -70 **33.** 3 **35.** -11 **37.** -17 **39.** -5 **41.** $\dfrac{4}{11}$
43. $-\dfrac{4}{17}$ **45.** $-\dfrac{7}{11}$ **47.** -18.59 **49.** -942.2249 **51.** $-1,086.75$
53. commutative of addition **55.** additive inverse **57.** commutative
of addition **59.** additive identity **61.** commutative of addition
63. associative of addition **65.** $6 + (-5)$ **67.** 34 **69.** 0
71. $8 + (3 + 5)$ **73.** $-9 + 5 = -4$ **75.** $-3 + 6 + (-7) = -4$
77. $6 + 15 = 21$ **79.** $-18 + 14 = -4$ **81.** $36 + 23 = 59$
83. $-30 + 42 = 12$ **85.** $-14 + 6 = -8$ **87.** $-11 + (-6) = -17$
89. $78° + (-15°), 63°$ **91.** $18° + (-28°), -10°$ **93.** $-5° + 15°, 10°$
95. $-14° + (-12°), -26°$ **97.** $-43 + 27, -16$ ft
99. $-53 + (-22), -75$ ft **101.** 2nd parking level below ground
103. 20th floor **105.** \$360 **107.** \$101 **109.** 10,997.35 **111.** -650 ft or
650 ft below rim **113.** 1,832 feet **115.** 64 **117.** 324

Exercise Set 1.7 **1.** 6 **3.** -6 **5.** 12 **7.** 4 **9.** -8 **11.** -1.7
13. -9 **15.** -12 **17.** 25 **19.** 8.8 **21.** 4 **23.** 12 **25.** 1 **27.** -14
29. 39 **31.** -4 **33.** 17 **35.** -26 **37.** $-\dfrac{3}{4}$ **39.** $-\dfrac{23}{20}$ **41.** $-\dfrac{7}{17}$
43. $\dfrac{29}{24}$ **45.** $-\dfrac{6}{13}$ **47.** 16,138.3227 **49.** -8.93 **51.** $11x$ **53.** $6y$
55. $7ab$ **57.** $-7x^2$ **59.** $4z + 3$ **61.** $-10y + 3$ **63.** $-11x$ **65.** $-5r$
67. $6xy + 7x$ **69.** $-8x^2 - 2y^2$ **71.** $12z + x - 12$ **73.** $7b^2 - 10b$
75. $m^2r + 6r$ **77.** $7 - (-5) = 12$ **79.** $-10a - 6a = -16a$
81. $8 - 5 = 3$ **83.** $-5m - 8m = -13m$ **85.** $6 - 9 = -3$
87. $9 - (-13) = 22$ **89.** $26r - 32r = -6r$ **91.** $-3x - 5x = -8x$
93. $12a - 19a = -7a$ **95.** $14 + [6 - (-5)] = 25$
97. $[4ab - (-6ab)] + 15ab = 25ab$
99. $[9 - (-4)] - (-5 + 6) = 12$
101. $[7y - (-3y)] + [-5y - (-2y)] = 7y$ **103.** $-62 - (-18), -44$ ft
105. $-15 - (-34), 19$ ft **107.** $350 - (-145) = 495$ ft
109. $86,000 - (-28,000) = 114,000$ **111.** $-125 - 250 = -375$ ft
113. $2500 - (-187) = 2687$ ft **115.** $100° - (-173°) = 273°$C
117. $+157$ profit **119.** 100 **121.** 100 **123.** $\dfrac{2}{5}x^2y + \dfrac{5}{7}xy^2$
125. $-\dfrac{4}{13}m^3n^2 - \dfrac{7}{9}mn^2$ **127.** $-\dfrac{11}{23}x^3y - \dfrac{23}{29}xy^2$

Exercise Set 1.8 **1.** 1, monomial **3.** 3, trinomial
5. 2, binomial **7.** 3, trinomial **9.** not a polynomial **11.** 1, monomial
13. 4 **15.** 3; x; 1 **17.** 1; x; 6 **19.** -5; x; 1 **21.** 4; x; 3 **23.** -1; x; 3
25. 10; x and y; 6 **27.** -22; a, b; 11 **29.** $-5x + 4$; 1
31. $2x^2 + 3x - 4$; 2 **33.** $-x^4 + x^3 - 4x^2 + 3x$; 4 **35.** $6b^4 + b$; 4
37. 0 **39.** -5 **41.** -1 **43.** 45 **45.** 19 **47.** 100 ft **49.** 440.48 m
51. 27 units **53.** \$9,800 **55.** $9x - 3y$ **57.** $2x + 2$ **59.** $6y^2 + 3y - 9$
61. $5m^2c^2 + 4mc - 9c^2 + 3$ **63.** $-5r^2 + 6r - 10$ **65.** $rp^2 - 5rp - r^3$
67. $h^2 - 9$ **69.** $11x^2 + 6x + 3$ **71.** $-3x^2 - 10x + 4$ **73.** $8x + 5$
75. $3a + 7$ **77.** $3x^2 + x - 2$ **79.** $12z - 11$ **81.** $x^2 - 11x + 6$
83. $4x - 18$ **85.** $8x$ cm **87.** $(8x + 1)$ in. **89.** $16x$ in. **91.** $(7x + 9)$ ft
93. $(3x - 2)$ ft **95.** $(x - 2)$ ft **97.** $(5x - 2)$ m **99.** $\dfrac{6}{7}x - \dfrac{1}{11}y$
101. $\dfrac{2}{13}r - \dfrac{16}{17}m$ **103.** $\dfrac{1}{3}a^2 - \dfrac{2}{5}a + \dfrac{7}{15}$ **105.** $-\dfrac{4}{23}y^2 - \dfrac{1}{19}y + \dfrac{4}{13}$

Chapter 1 Review Exercises **1.** 9^5 **2.** $(-2)^4$ **3.** $3^3 \cdot 5^4$
4. $12 \cdot 9^3 \cdot 2^2$ **5.** $3^3 \cdot a^3$ **6.** $7^3 \cdot r^4 \cdot s^2$ **7.** 72 **8.** 80 **9.** 900
10. 50,000 **11.** 22 **12.** 36 **13.** 384 **14.** 14 **15.** 36 **16.** 1 **17.** 34
18. 18 **19.** 9 **20.** 23 **21.** 58 **22.** 4 **23.** 31 **24.** 27 **25.** 20 in.
26. 36 ft **27.** 252 in. **28.** 10πm, 31.4 m **29.** 21 in.^{2} **30.** 81 ft^2
31. 18 ft^2 **32.** 25πm^2, 78.5 m^2 **33.** 84 yd^2 **34.** 81 cm^2
35. $V = 240$ ft^3, $SA = 236$ ft^2 **36.** $V = 216$ in.^{3}, $SA = 216$ in.^{2}
37. $V = 175\pi$ m^3, $SA = 120\pi$ m^2 **38.** 32π m^3 **39.** 1,000 ft, 62, 500 ft^2
40. 942 ft **41.** \$17.50 **42.** \$3290 **43.** 2704 in.^{2} **44.** 64 in.^{2}
45. 70,650 ft^2 **46.** 120 in.^{2} **47.** a) 48 in.^{3} b) 124 in.^{2} **48.** a) 35.325 in.^{3}
b) 61.23 in.^{2} **49.** -23 **50.** 35 **51.** -7 **52.** -32 **53.** 64 **54.** 86
55. $<$ **56.** $<$ **57.** $>$ **58.** $=$ **59.** $<$ **60.** $>$ **61.** 11 **62.** 10 **63.** 4
64. 0 **65.** 6 **66.** 15 **67.** -7 **68.** -21 **69.** 23 **70.** -7 **71.** -95
72. -30 **73.** associative of addition **74.** commutative of addition
75. additive inverse **76.** additive identity **77.** $-9 + 6 = -3$
78. $-15 - (-8) = -7$ **79.** $-5 + 16 = 11$ **80.** $3 - (-6) = 9$
81. $-9 + 8 = -1$ **82.** $-3 - (-7) = 4$
83. $(-5 + 13) + [18 + (-8)] = 18$
84. $[-9 - (-2)] + (-5 - 9) = -21$
85. $[14 + (-8)] - [4 - (-5)] = -3$
86. $5(-6 + 10) = 20$ **87.** $-3x$ **88.** $2x + y$ **89.** $-5w^3 - 6u^2 + 3$
90. $-5x^2y + 6xy^2 + 4$ **91.** 9, x, 2 **92.** -9, x and y, 6
93. $4x^3 + 6x$, binomial, 3 **94.** $7y^2 - 8y + 2$, trinomial, 2
95. $5x^7$, monomial, 7 **96.** $-7x^5 + 9x^4 + 2x$, trinomial, 5
97. $14a^6 + 2a^3 - 5a^2 + 19$, 6 **98.** not a polynomial **99.** -3
100. 253 **101.** $10x - 5$ **102.** $5a + 11b$ **103.** $5u^4 - 5u^3$
104. $2a^2 - 14a + 5$ **105.** $z^4 + z^3 + 2z + 2$ **106.** $z^2 + 6z - 8$
107. $11x^2 - 11x + 4$ **108.** $3x^3 + 6x^2 - 4x - 11$ **109.** $-2x^2 + 13x - 5$
110. $-2u^4 + 3u^3 - 3u^2 - 3u + 8$ **111.** $-2x^3 - 13x^2 + 2x + 5$
112. $(9x - 6)$ ft

Chapter 1 Test **1.** $6^4 5^2$ [1.1] **2.** $4^2 x^4 y^2$ [1.1] **3.** $3x^5$ [1.1]
4. $(4a)^3$ [1.1] **5.** 33 [1.1] **6.** 23 [1.1] **7.** 70 [1.1] **8.** 2 [1.1]
9. 256 [1.1] **10.** 4 [1.1] **11.** $P = 33$ cm [1.2], $A = 42$ cm^2 [1.3]
12. $P = 26$ ft [1.2], $A = 36$ ft^2 [1.3] **13.** 40 m^2 [1.3] **14.** 38 ft^2 [1.3]
15. $C = 14\pi$m, 43.96 m [1.2] $A = 49\pi$ m^2, 153.86 m^2 [1.3]
16. $V = 90$ in.^{3}, $SA = 126$ in.^{2} [1.4] **17.** $V = 1,538.6$ m^3,
$SA = 747.32$ m^2 [1.4] **18.** \$36 [1.2] **19.** \$324 [1.3] **20.** 4,500 lbs [1.4]
21. $<$ [1.5] **22.** $<$ [1.5] **23.** -2 [1.5] **24.** -1 [1.5] **25.** -11 [1.6]
26. -7 [1.6] **27.** -1 [1.7] **28.** -44 [1.7] **29.** commutative of
addition [1.6] **30.** additive inverse [1.6] **31.** $-7 + 11 = 4$ [1.6]
32. $-12 + [9 - (-6)] = 3$ [1.7]
33. $[4 + (-13)] - (-3 + 9) = -15$ [1.7]
34. $(x^2 - 6x + 2) - (2x^2 + 5x - 3) = -x^2 - 11x + 5$ [1.8]
35. $(4x + 5) - [(3x + 6) + (2x - 3)] = -x + 2$ [1.8]
36. a) 4, b) 3, c) $4x^3 + 3x^2 + 6x$, d) trinomial [1.8] **37.** 16 [1.8]
38. $5xy - 4yz$ [1.7] **39.** $5x^2 - 3x - 4$ [1.8] **40.** $6n^3 - 2n^2 + 7n - 2$
[1.8] **41.** No, $-x$ is positive if x is negative. [1.5]

Exercise Set 2.1 **1.** -24 **3.** -24 **5.** 24 **7.** -21 **9.** 44
11. 3.5 **13.** -9.12 **15.** 15 **17.** -48 **19.** -90 **21.** -120 **23.** 120
25. -81 **27.** 121 **29.** -625 **31.** -72 **33.** 108 **35.** 125 **37.** $-\dfrac{8}{15}$
39. $\dfrac{56}{75}$ **41.** $\dfrac{9}{40}$ **43.** $-\dfrac{45}{64}$ **45.** -49.651 **47.** 6029.6236 **49.** -6

51. 16 **53.** 24 **55.** 40 **57.** -256 **59.** 64 **61.** $5(-3) + 6 = -9$
63. $-5 - 4(-7) = 23$ **65.** $3(-4)^2 = 48$ **67.** $(-6)(-3) - 8 = 10$
69. $-5(4) + 4(-2) = -28$ **71.** $4(-3) - (-5)(-7) = -47$

73. $-8(4) + 2(3)(-2) = -44$ **75.** $\frac{1}{2}(-8 + 6) = -1$ **77.** associative
of multiplication **79.** commutative of multiplication **81.** identity for
multiplication **83.** commutative of multiplication **85.** inverse for
multiplication **87.** commutative of multiplication **89.** multiplication
property of 0 **91.** $[3(-9)](-7)$ **93.** $(-7)(3)$ **95.** $4(-9) + 4(2)$
97. $(-9 + 2)(4)$ **99.** -2 **101.** 1 **103.** 0 **105.** 14 **107.** -11
109. $-12°$ **111.** -70 ft **113.** -6 points

Exercise Set 2.2
1. r^8 **3.** 4^7 **5.** x^9 **7.** $20x^9$ **9.** $21a^8$
11. $x^8 y^7$ **13.** $8a^{10}b^{11}$ **15.** $a^5 b^5$ **17.** $2^4 y^4$ or $16y^4$ **19.** $16x^2$ **21.** $-25x^2$
23. $27c^3$ **25.** $-27b^3$ **27.** y^{18} **29.** 4^{15} **31.** $x^7 y^7$ **33.** $4^8 x^8$ **35.** $c^{24}d^{18}$
37. $4^{20}a^{12}$ **39.** $-4^{20}b^{25}$ **41.** $6^8 b^{12}$ **43.** $108a^{17}b^{12}$ **45.** $4^{20}b^{28}$
47. $a^{14}b^{16}c^{22}$ **49.** $16a^2$ **51.** $3x^3$ **53.** $89x^6$ **55.** $-24a^4$ **57.** $-2m^4 n^6$
59. a^{8x} **61.** a^{5b-3} **63.** $2^{7x}a^{5y}b^{3z}$ **65.** $2^c x^{ac} y^{bc}$ **67.** b^{6n-3}

Exercise Set 2.3
1. $-7a^3 x^3$ **3.** $84f^7 k^2 b^5$ **5.** $12p^5 q^5 r^8$
7. $-53.7654x^9 y^{13}$ **9.** $-34.56x^{11}y^9$ **11.** $2x + 6$ **13.** $-8x + 12$
15. $12a - 8b$ **17.** $-20s + 30t$ **19.** $6x^2 - 18x + 15$
21. $-12b^2 + 4b - 8$ **23.** $4b^3 - 8b^2 + 10b$ **25.** $2h^5 + 16h^4 - 10h^3$
27. $2b^6 + 14b^5 - 12b^4$ **29.** $-6r^6 + 8r^5 - 4r^4 + 6r^3$
31. $10x^4 y^3 - 20x^3 y^4$ **33.** $-12p^6 q^4 + 4p^4 q^5 - 4p^2 q^4$
35. $22.79x^5 - 33.54x^4 - 30.96x^3$ **37.** $7x + 6$ **39.** $12a - 16$
41. $8y^2 - 8y - 6$ **43.** $21a^2 + 8ab - 4b^2$ **45.** $-8x^2 + 2x - 15$
47. $24c^2 - c - 6$ **49.** $15r^3 + 16r^2 - 21r - 10$
51. $8a^3 - 18a^2 + 11a - 3$ **53.** $11a^2 + 4a + 14$ **55.** $a^2 + 9a + 20$
57. $z^2 - 3z - 40$ **59.** $x^2 - 16$ **61.** $4x^2 - y^2$ **63.** $x^2 - xy - 2y^2$
65. $20w^2 - 17w - 24$ **67.** $6z^2 - 26z + 24$ **69.** $6a^2 + 5ab - 4b^2$
71. $x^3 + 2x - 3$ **73.** $2x^3 + 2x^2 - 9x + 9$ **75.** $x^3 + 8$
77. $6x^3 - x^2 - 18x + 9$ **79.** $37.63x^2 - 15x - 17.28$
81. $12a^4 - 7a^2 b^2 - 10b^4$ **83.** $4a^4 + 8a^3 - 4a^2 + 12a - 15$
85. $a^4 - a^3 - 7a^2 + a + 6$ **87.** $x^{2n} - x^n - 6$ **89.** $2x^2 + 10x + 12$
91. $6x^2 - 5x - 25$ **93.** $2a^3 - 6a^2 - 8a + 24$
95. $8y^3 + 8y^2 - 14y + 4$ **97.** $2x^2$ **99.** $4x^2 + 24x$ **101.** $2x^2 + x - 6$
103. $3x^3 - x^2 + 4x + 4$ **105.** $4y^2$ **107.** $9z^2$ **109.** $(2y^2 - 8y)$ ft^2
111. $(2x^2 + 3x)$ ft^2 **113.** $(12.5x^2 + 37.5x)$ dollars

Exercise Set 2.4
1. $x^2 + 5x + 6$ **3.** $a^2 - a - 12$
5. $c^2 - 7c + 10$ **7.** $a^2 + 7ab + 12b^2$ **9.** $r^2 - 2rs - 15s^2$
11. $6a^2 + 13ab + 6b^2$ **13.** $6x^2 - xy - 15y^2$ **15.** $6x^2 - 23xy + 20y^2$
17. $25.42x^2 + 35.54xy - 32.76y^2$ **19.** $x^2 - 9$ **21.** $p^2 - q^2$ **23.** $a^2 - 16b^2$
25. $36a^2 - b^2$ **27.** $16a^2 - 25$ **29.** $4x^2 - 25y^2$ **31.** $51.84a^2 - 44.89$
33. $x^2 + 4x + 4$ **35.** $a^2 - 8a + 16$ **37.** $r^2 + 2rs + s^2$ **39.** $t^2 - 2st + s^2$
41. $4x^2 + 12x + 9$ **43.** $9a^2 - 6a + 1$ **45.** $4x^2 + 20xy + 25y^2$
47. $16x^2 - 24xy + 9y^2$ **49.** $25t^4 - 10t^2 w + w^2$
51. $22.09x^2 + 34.78xy + 13.69y^2$ **53.** $6a^2 + 11ab - 10b^2$ **55.** $16a^2 - 9b^2$
57. $9a^2 + 30a + 25$ **59.** $8 - 18b + 9b^2$ **61.** $36 - y^2$ **63.** $16 + 24w + 9w^2$
65. $16c^2 - 49d^2$ **67.** $(3 + x)^2 = 9 + 6x + x^2$ **69.** $3 - x^2$
71. $(x - 5)^2 = 25$ **73.** $x^2 - 3x - 40$ **75.** $12a^2 + 21a - 45$
77. $x^2 + 16x + 64$ **79.** $9a^2 - 30ab + 25b^2$ **81.** $p^3 + q^3$ **83.** $x^3 + 27$
85. $b^3 - 64$ **87.** $27a^3 + 8b^3$

Exercise Set 2.5
1. -4 **3.** -6 **5.** 6 **7.** Undefined **9.** -8
11. $-\frac{9}{20}$ **13.** $-\frac{12}{25}$ **15.** $\frac{12}{-4} + 8 = 5$ **17.** $\frac{-18}{-9} - (-6) = 8$
19. $12 - \frac{36}{-9} = 16$ **21.** $7 - \frac{48}{-16} = 10$ **23.** $\frac{14}{-7} + \frac{-9}{3} = -5$
25. $\frac{48}{-8} - \frac{-36}{18} = -4$ **27.** $-7°$ per hour **29.** $-\frac{9}{150}$ or $\frac{3}{50}$ or
-0.06 ft per foot **31.** -9 ft/sec **33.** -2 **35.** 22 **37.** -16 **39.** 23
41. 2 **43.** 12 **45.** -22 **47.** 10 **49.** -6 **51.** 20 **53.** -5 **55.** -16
57. 27 **59.** -4 **61.** 10 **63.** 25 **65.** 60 **67.** 2 **69.** -2 **71.** $\frac{17}{2}$
73. 11 **75.** $-\frac{14}{29}$ **77.** 60.96 **79.** 5.8 **81.** 9948.2 **83.** -2665.18
85. -2.74 **87.** 10 **89.** -89 **91.** -7 **93.** 36 **95.** -60 **97.** -324
99. 15 **101.** 60 **103.** -26 **105.** -92 **107.** 1 **109.** $-\frac{5}{18}$ **111.** $\frac{43}{42}$

113. $55(.15) + 22(-.08) = \$6.49$ profit
115. $30(40) + 60(-25) = -\$300$ loss
117. $30(45) + (-12)(20) + 8(36) = \$1398$ profit

Exercise Set 2.6
1. 2^4 **3.** c^4 **5.** z^4 **7.** $\frac{1}{2^4}$ **9.** $\frac{1}{a^6}$ **11.** 5^3
13. y^5 **15.** 1 **17.** 3 **19.** 1 **21.** 2 **23.** -1 **25.** $\frac{2}{x^3}$ **27.** $-\frac{7}{b^4}$
29. $3^4 x^4$ **31.** $\frac{1}{2^3 x^3}$ **33.** 8^6 **35.** 2^2 **37.** $\frac{1}{5^2}$ **39.** a^{10} **41.** $\frac{1}{x^5}$ **43.** $\frac{1}{z^2}$
45. a^4 **47.** $\frac{1}{x^3}$ **49.** $\frac{1}{5^3}$ **51.** $\frac{1}{3^4}$ **53.** $\frac{1}{y^9}$ **55.** 6^7 **57.** r^7 **59.** 6
61. $\frac{1}{z^4}$ **63.** 8^{18} **65.** r^{20} **67.** $\frac{1}{6^{15}}$ **69.** $\frac{1}{y^{21}}$ **71.** $\frac{1}{8^{18}}$ **73.** $\frac{1}{z^{15}}$ **75.** 9^{10}
77. $-\frac{108x^3}{y^5}$ **79.** $-\frac{2m^{14}}{n^{16}}$ **81.** $\frac{8b^{18}}{a^{21}}$ **83.** x^{2a} **85.** a^{m+1}

Exercise Set 2.7
1. $\frac{4}{9}$ **3.** $\frac{125}{8}$ **5.** $\frac{4}{9}$ **7.** $\frac{2401}{81}$ **9.** $\frac{a^3}{b^3}$ **11.** $\frac{q^4}{p^4}$
13. $\frac{8x^3}{y^3}$ **15.** $-\frac{27q^3}{p^3}$ **17.** $\frac{a^{12}}{b^{20}}$ **19.** $\frac{y^{12}}{x^{24}}$ **21.** $\frac{1}{64x^6}$ **23.** $\frac{y^{10}}{4}$ **25.** $\frac{25}{a^4 b^{10}}$
27. $\frac{z^4}{4w^6}$ **29.** $\frac{w^9}{z^{18}}$ **31.** $\frac{z^{30}}{x^{20}}$ **33.** $24w^3$ **35.** $-\frac{3}{10x}$ **37.** x^2 **39.** $\frac{1}{a^{15}}$
41. a^{14} **43.** $\frac{16}{y^8}$ **45.** x^4 **47.** $\frac{1}{4x^4}$ **49.** $\frac{8^4}{16}$ **51.** $\frac{1}{x^{15}}$ **53.** $\frac{1}{q}$ **55.** $\frac{1}{a^{12}}$
57. $\frac{8}{x^{12}}$ **59.** $81a^3$ **61.** $\frac{216a^7}{b^{19}}$ **63.** $a^{mq}b^{nq}$ **65.** $x^{ac-md}y^{bc-nd}$

Exercise Set 2.8
1. $5x$ **3.** $-3z^2$ **5.** $2x^2$ **7.** $-\frac{5}{a^3}$ **9.** $-6n^2$
11. 2 **13.** $x^2 y^2$ **15.** $-\frac{1}{r^2}$ **17.** $3x^2 y^2$ **19.** $-\frac{4y^3}{x^3}$ **21.** $\frac{no^3}{m}$ **23.** $-\frac{4y^3}{x^2 z^2}$
25. $\frac{2n^2}{p^3}$ **27.** $9x^2 y^2$ **29.** $-\frac{5}{y^2}$ **31.** $x + 3$ **33.** $3a - 1$ **35.** $3 + \frac{7}{y}$
37. $2z - 1$ **39.** $3y + \frac{2}{y}$ **41.** $4a^3 b - 2ab^4$ **43.** $\frac{2n}{m} + \frac{3m^2}{n}$
45. $2x^2 + 3x - 1$ **47.** $y^3 - y^2 - 1$ **49.** $3m^3 - m^2 + 2m$
51. $3y^2 - 4y + \frac{2}{y}$ **53.** $3m - 2m^2 n + 1$ **55.** $3b - \frac{3}{4b}$ **57.** $\frac{4n}{m^2} - \frac{3m}{n^2}$
59. $4r^2 - 2 + \frac{3}{r^2}$ **61.** $5r^2 - 2 + \frac{5}{6r}$ **63.** $5r^5 - \frac{2}{r} - \frac{2}{9r^2}$
65. $\frac{8}{pq} - \frac{3p^2}{4q^2} + 5q^2$ **67.** $\frac{14m^4 n^8}{7m^3 n^6} = 2mn^2$ **69.** $\frac{-32m^4 n^2}{-16m^7 n} = \frac{2n}{m^3}$
71. $\frac{16a^7 - 24a^5}{8a^3} = 2a^4 - 3a^2$ **73.** $\frac{12y^5 - 16y^3 - 8}{4y^3} = 3y^2 - 4 - \frac{2}{y^3}$
75. $\frac{8x^3 y^3 + (-2x^3 y^3)}{3xy} = 2x^2 y^2$
77. $\frac{(4y^5 - 6y^3 + y^2) + (2y^5 - 4y^3 + 7y^2)}{2y^2} = 3y^3 - 5y + 4$

Exercise Set 2.9
1. 35,000 **3.** .0095 **5.** 4,790,000
7. .00000924 **9.** 400 **11.** .00000001 **13.** 7.6×10^3 **15.** 3.5×10^{-4}
17. 8.57×10^5 **19.** 4.98×10^{-7} **21.** 6×10^5 **23.** 1×10^{-8}
25. 3×10^8; 300,000,000 **27.** 1.376×10^3; 1376
29. 1.31976×10^{-2}; .0131976 **31.** 2.64388×10^{-7}; .000000264388
33. 2×10^3; 2000 **35.** 1.5×10^{-7}; .00000015 **37.** 1.2×10^6; 1,200,000
39. 5×10^1; 50 **41.** 2.5×10^4; 25,000 **43.** 9×10^6; 9,000,000
45. 9.9×10^7; 99,000,000 **47.** 2.52×10^{-1}; .252
49. 9×10^{-14}; .00000000000009 **51.** 2.3×10^4; 23,000
53. 2.5×10^{-5}; .000025 **55.** 2.2×10^6; 22,000,000 **57.** 1.8×10^{-1}; .18
59. 6×10^{-2}; .06 **61.** 8×10^0; 8 **63.** 1.5×10^8 km
65. 1.36×10^4 kg per cubic meter **67.** 1.3×10^{-6} meters
69. 299,792,500 meters per second **71.** 19,300 kg per cubic meter
73. .0000000000000001673 grams **75.** 4.2 light years **77.** 1.5 moles
79. 5,977,286,312,010,000,000,000,000 atoms **81.** 601,503,759,399,
000,000,000,000 atoms

Chapter 2 Review Exercises

1. -36 **2.** 15 **3.** -72
4. -120 **5.** 144 **6.** -16 **7.** 15 **8.** -12 **9.** -72 **10.** 96
11. $(-4)(8) + 7 = -25$ **12.** $-6 - (-5)(5) = 19$
13. $(-6)(3) + 6(-4) = -42$ **14.** commutative property of multiplication   **15.** associative property of multiplication
16. commutative property of multiplication   **17.** distributive property
18. multiplicative inverse   **19.** identity for multiplication   **20.** w^{13}
21. 4^{14} **22.** $-40x^{10}$ **23.** $-30a^6 b^{12}$ **24.** $x^7 y^7$ **25.** $125d^3$ **26.** $16a^4 b^4$
27. $x^9 y^9$ **28.** $-2048x^8$ **29.** x^{20} **30.** 4^{24} **31.** $a^{16}b^{24}$ **32.** $4^4 a^{12} b^{28}$
33. $-a^{37}b^{19}$ **34.** $72m^{13}n^{19}$ **35.** $15x^{10}y^5$ **36.** $18y^3 - 12y$
37. $-20x^4 + 25x^3$ **38.** $21c^5 - 49c^4 + 14c^3$
39. $-6x^6 y^4 + 16x^3 y^8 - 8x^5 y^3$ **40.** $12x + 13$ **41.** $3x + 9$
42. $5a^2 - 23a + 13$ **43.** $-b^2 + 19b + 13$ **44.** $x^2 + 7x + 10$
45. $12m^2 - 39m - 30$ **46.** $20m^2 - 9mn - 18n^2$ **47.** $x^2 - 81$
48. $25x^2 - 36$ **49.** $4s^2 - 25t^2$ **50.** $m^2 + 2mn + n^2$
51. $9a^2 - 12a + 4$ **52.** $36a^2 + 24ab + 4b^2$ **53.** $a^3 - a^2 - 16a + 16$
54. $6m^3 - 11m^2 - 18m + 20$ **55.** $12x^3 - 26x^2 y + 18xy^2 - 4y^3$
56. $2(3x + 7) + 5(-2x + 1) = -4x + 19$
57. $4(4m - 2) - 3(2m + 5) = 10m - 23$ **58.** $x^3 - 64$
59. $27x^3 + 125$ **60.** $15a^2 - 11a - 14$ **61.** $10x^2$ **62.** $9z^2 + 48z + 64$
63. $(6x - 4)^2 = 36x^2 - 48x + 16$ **64.** $(3z + 2)^2 = 9z^2 + 12z + 4$
65. -8 **66.** 2 **67.** -9 **68.** -4 **69.** 2 **70.** 2 **71.** 4 **72.** -2
73. -12 **74.** 7 **75.** -61 **76.** 4 **77.** -20 **78.** -37 **79.** 16 **80.** 3
81. -2 **82.** -6 **83.** -3 **84.** -1 **85.** $\dfrac{36}{19}$ **86.** 35 **87.** 432 **88.** 72
89. -41 **90.** -10 **91.** -2 **92.** $\dfrac{34}{57}$ **93.** 2^5 **94.** m^2 **95.** $\dfrac{1}{3^4}$ **96.** $\dfrac{1}{x^6}$
97. $\dfrac{1}{5^2}$ **98.** 4 **99.** $\dfrac{1}{5}$ **100.** z^2 **101.** $\dfrac{1}{x^6}$ **102.** $\dfrac{1}{5^6}$ **103.** a^3 **104.** 7^{10}
105. $\dfrac{9}{25}$ **106.** $\dfrac{4}{9}$ **107.** $\dfrac{1}{3^{10}}$ **108.** r^{20} **109.** $\dfrac{x^{12}}{y^{12}}$ **110.** $\dfrac{x^6}{64}$ **111.** $\dfrac{1}{5^3 a^{12}}$
112. $\dfrac{x^{12}}{8y^{18}}$ **113.** $\dfrac{n^6}{m^8}$ **114.** m^{32} **115.** $8n^3$ **116.** $\dfrac{n^4}{m^2}$ **117.** $\dfrac{6}{m^2}$
118. $-5x^2$ **119.** $\dfrac{3}{m^3 n^6}$ **120.** m^7 **121.** $x - 2$ **122.** $5y^2 + 3$
123. $7a - 4$ **124.** $\dfrac{4m^2}{n^2} + \dfrac{2n^2}{m}$ **125.** $4x^2 - 3x - \dfrac{7}{x}$
126. $\dfrac{1}{mn^2} - \dfrac{4n^2}{m^4} + \dfrac{3}{n}$ **127.** $\dfrac{48m^6 n^3}{-16m^4 n^5} = \dfrac{-3m^2}{n^2}$
128. $\dfrac{24a^6 - 16a^4}{8a^3} = 3a^3 - 2a$
129. $\dfrac{(4x^4 - 2x^3 - 3x^2) + (2x^4 - 7x^3 + 2x^2)}{3x^2} = 2x^2 - 3x - \dfrac{1}{3}$
130. $\dfrac{28x^6 - 40x^3}{4x^4} = 7x^2 - \dfrac{10}{x}$
131. $\dfrac{(5x^2 - 7x + 5) + (3x^2 + 3x - 9)}{2x} = 4x - 2 - \dfrac{2}{x}$ **132.** $460{,}000$
133. $3{,}060{,}000$ **134.** 0.0000703 **135.** 0.007123 **136.** 9.7×10^{10}
137. 4.67×10^4 **138.** 4.78×10^{-6} **139.** 3.07×10^{-4}
140. 1.395×10^{-2}; $.01395$ **141.** 2.05842×10^{-6}; $.00000205842$
142. 3×10^{-2}; $.03$ **143.** 3×10^{-8}; $.00000003$ **144.** 5.52×10^{-3}; $.00552$
145. 1.598×10^{-11}; $.00000000001598$ **146.** 5×10^7; $50{,}000{,}000$
147. 1.5×10^3; $1{,}500$

Chapter 2 Test

1. 60 [2.1]   **2.** -72 [2.1]   **3.** $-6x^5 y^5$ [2.2]
4. $48x^6 y^3$ [2.2]   **5.** $6x^3 - 8x^2$ [2.3]   **6.** $2p^3 q^4 - 4p^4 q^2 - 2p^2 q$ [2.3]
7. $3x + 8$ [2.3]   **8.** $6p^2 + 5p - 4$ [2.3], [2.4]   **9.** $6a^2 - 11ab + 4b^2$ [2.3], [2.4]   **10.** $3y^3 + 5y^2 + 13y - 5$ [2.3]   **11.** $25x^2 - 9y^2$ [2.4]
12. $4a^2 - 28a + 49$ [2.4]   **13.** -4 [2.5]   **14.** 3 [2.5]   **15.** $\dfrac{-24n^7}{m^3}$ [2.2]
16. 7 [2.6]   **17.** $\dfrac{-2z^3}{x^3}$ [2.6]   **18.** $\dfrac{x^{15}}{27y^{12}}$ [2.7]   **19.** $\dfrac{2a^3}{b^{18}}$ [2.7]   **20.** $\dfrac{2b^6}{a^{10}}$ [2.7]
21. $2m^3 n - 8mn^3 + 5m^2$ [2.8]   **22.** $4x - 5 + \dfrac{3}{x}$ [2.8]   **23.** -14 [2.5]
24. -3 [2.5]   **25.** -5 [2.5]   **26.** $\dfrac{12}{-4} - 7(-3) = 18$ [2.1]
27. $(3x^2 - 7x + 5) + (x + 3)(2x - 5) = 5x^2 - 6x - 10$ [2.4]

28. a) 4.79×10^9 b) 7.49×10^{-8} [2.9]   **29.** 1075 [2.9]
30. 4.2×10^{10} [2.9]

Exercise Set 3.1

1. $t = 8$ **3.** $r = -10$ **5.** $a = 8$ **7.** $x = 10.5$
9. $m = -1.2$ **11.** $w = 9$ **13.** $x = 7$ **15.** $z = 3$ **17.** $r = -2$
19. $x = 8.1$ **21.** $r = 4$ **23.** $x = 0$ **25.** $x = -2$ **27.** $x = 6$
29. $x = 2$ **31.** $x = -2.6$ **33.** $y = -7.7$ **35.** $t = 12$ **37.** $x = 0$
39. $x = 10$ **41.** $x = -.77$ **43.** Answers vary   **45.** Answers vary
47. Answers vary   **49.** $x - 2 = 9, x = 11$ **51.** $x - 14 = 8, x = 22$
53. $x - 2 = 15, x = 17$ **55.** $5 + x = 13, x = 8$
57. $x - 8 = -10, x = -2$ **59.** $x - (-2) = -8, x = -10$
61. $275 + x = 550, x = 275$ miles   **63.** $x + 75 = 160, x = \$85$
65. $x - 8 = 24, x = 32$ points   **67.** $632 + x = 800, x = 168$ points
69. $x + 46 = 326, x = \$280$ **71.** $x - 23 = 118, x = 141$ pounds
73. $x + 18{,}000 = 40{,}000, x = \$22{,}000$ **75.** $18 = 6 + 8 + x, x = 4$ in.

Exercise Set 3.2

1. $u = 8$ **3.** $n = -12$ **5.** $x = -3$ **7.** $p = 4$
9. $u = .8$ **11.** $x = 50$ **13.** $t = 110$ **15.** $n = 16$ **17.** $q = -18$
19. $z = -49$ **21.** $w = 9$ **23.** $m = -75$ **25.** $x = 9$ **27.** $a = -2$
29. $y = -7$ **31.** $x = 3$ **33.** $x = -4$ **35.** $x = -3$ **37.** $u = 9$
39. $x = 3$ **41.** Answers vary   **43.** Answers vary   **45.** Answers vary
47. $5x = 35, x = 7$ **49.** $4x = -28, x = -7$ **51.** $.4x = 2, x = 5$
53. $\dfrac{x}{4} = 5, x = 20$ **55.** $\dfrac{x}{-3} = 8, x = -24$ **57.** $\dfrac{3}{7}x = 9, x = 21$
59. $\dfrac{2}{9}x = -6, x = -27$ **61.** 8 ft   **63.** 30 yd   **65.** 150 calories
67. $\$8.50$ **69.** 115 people   **71.** 4 miles   **73.** 36 months
75. $\$17{,}500$ **77.** 6 in.

Exercise Set 3.3

1. $x = 2$ **3.** $x = -2$ **5.** $a = -3$ **7.** $x = 3$
9. $u = 2$ **11.** $v = 1$ **13.** $n = 0$ **15.** $p = -2$ **17.** $v = -2$
19. $k = 1$ **21.** $a = 2$ **23.** $y = 4$ **25.** $n = -8$ **27.** $r = 5$
29. $u = 13$ **31.** $w = -6$ **33.** $y = \dfrac{5}{2}$ or 2.5 **35.** $x = 10$ **37.** $n = 1$
39. $p = 2$ **41.** $n = -2$ **43.** $u = 1$ **45.** $b = 1$ **47.** $a = -9$
49. $v = \dfrac{13}{2}$ or 6.5 **51.** Identity, all real numbers   **53.** Identity, all real numbers   **55.** Contradiction, no solutions or $\varnothing$ **57.** $x = 2$
59. Answers vary   **61.** Answers vary   **63.** Answers vary
65. Answers vary   **67.** $3x - 8 = 13, x = 7$ **69.** $3x + 7 = 4, x = -1$
71. $0 = 9(x + 3), x = -3$ **73.** $4(3x - 2) = 16, x = 2$ **75.** 4 hrs
77. 5 hrs   **79.** $\$160$ **81.** $\$15.25$ **83.** 121 copies   **85.** $\$12.24$
87. 100 minutes   **89.** 12 oz salt, 96 oz water   **91.** 550 yd   **93.** 8 ft

Exercise Set 3.4

1. $A = 8$ mi^2 **3.** $h = 4$ ft   **5.** $t = 1.5$ hrs
7. $r = 4$ in.   **9.** $P = 42$ in.   **11.** $A = 24$ **13.** $V = 15$ **15.** $P = 60$
17. $W = 2624$ **19.** $P = 12$ **21.** $y = -6$ **23.** $x = 6$
25. $C = K - 273$ **27.** $T = \dfrac{P}{k}$ **29.** $m = \dfrac{F}{a}$ **31.** $t = \dfrac{v}{g}$ **33.** $V = \dfrac{kT}{P}$
35. $m = \dfrac{E}{c^2}$ **37.** $r = \dfrac{I}{pt}$ **39.** $a^2 = c^2 - b^2$ **41.** $g = \dfrac{2d}{t^2}$
43. $M = DV$ **45.** $T = \dfrac{A - P}{Pr}$ **47.** $a = \dfrac{2s - ln}{n}$ or $\dfrac{2s}{n} - l$
49. $T = \dfrac{v - v_0}{-32}$ **51.** $y = -2x + 3$ **53.** $y = \dfrac{-3x + 6}{2}$
55. $y = \dfrac{-6x + 7}{-3}$ **57.** $B = \dfrac{2A - hb}{h}$ or $\dfrac{2A}{h} - b$ **59.** $q_1 = \dfrac{Fr^2}{kq_2}$

Exercise Set 3.5

1. 346 **3.** 1600 **5.** Bud Light, $\$1434.1$ million; Coors Light, $\$708.1$ million   **7.** grapefruit $= 22$, orange juice $= 27$
9. U.S., 104; France, 59; Japan, 55 **11.** death of spouse $= 100$, divorce $= 73$, separation $= 65$ **13.** $\$35$ per day   **15.** $\$1500$
17. $\$34{,}596.15$ **19.** $11, 13$ **21.** $24, 25, 26$ **23.** $123, 125, 127$
25. $23, 24, 25$ **27.** $13, 15$ **29.** $18, 20, 22$ **31.** $12, 14, 16$ **33.** 7 hrs
35. 8 hrs   **37.** 9 hrs   **39.** 6.5 hrs   **41.** 6 hrs   **43.** 5 hrs   **45.** 9 ft, 12 ft, 15 ft, 18 ft

Exercise Set 3.6

1. $(400 - x)$ miles   **3.** $(24 - x)$ dimes
5. $(150 - x)$ camellias   **7.** 16 at $\$25$, 20 at $\$32$ **9.** 80 of 36 in., 25 of 54 in.   **11.** 22 at $36¢$, 14 at $43¢$ **13.** 6 roses, 8 azaleas
15. four $\$5$ bills, forty-four $\$20$ bills   **17.** 25 dimes, 30 nickels

19. 20 dimes, 35 quarters   **21.** $10,000 at 6%, $26,000 at 5%
23. $40,000   **25.** $18,000 at 6%, $25,000 at 8%   **27.** $30,000   **29.** 4 oz
31. $6.00 per pound   **33.** 4 lb peanuts, 8 lb cashews   **35.** 18 ml
37. 30 tons   **39.** 10 liters   **41.** 40 gallons   **43.** 25 at $.49, 15 at $1.19
45. 6.25 gallons   **47.** 12 nickels, 15 dimes, 8 quarters

Exercise Set 3.7
1. 220 ft   **3.** $L = 75$ yd, $W = 35$ yd
5. $L = 855$ m, $W = 530$ m   **7.** 32 in., 34 in., 64 in.   **9.** 27 ft, 27 ft, 40 ft
11. 45°, 45°, 90°   **13.** 75°, 25°, 80°   **15.** 50°, 40°   **17.** 40°, 140°
19. $x = 20$, $\angle A$ and $\angle B$ are 78°   **21.** $x = 15$, $\angle A$ and $\angle B$ are 57°
23. $x = 20$, $\angle A$ is 70°, $\angle B$ is 110°   **25.** $x = 125°$ $\angle A$ is 70°
and $\angle B$ is 110°

Exercise Set 3.8

1.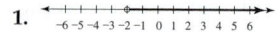
3.
5.
7.
9.
11.

13. $q < -2$
15. $p \le 2$

17. $p < -2$
19. $s \ge 1$

21. $x > 2$
23. $u > 41$

25. $t \le -3$

27. $y < 5$ **29.** $u < -3$ **31.** $a > 7$ **33.** $x > -15$ **35.** $r < 2$
37. $q > 12$ **39.** $a \ge 3$ **41.** $p \ge 5$ **43.** $w > -8$ **45.** $a \le 1.2$
47. $t > .3$ **49.** $s > -1$ **51.** $p \le 3$ **53.** $w \ge -1$ **55.** $m > 9$
57. $x \le -1$ **59.** $u \le -4$ **61.** $u > -1$ **63.** $n > 0$
65. $z \ge -3$ **67.** $r > -5$ **69.** $-8 < x < -2$ **71.** $-3 \le y \le 4$
73. $-2 \le x < 3$ **75.** $1 > y \ge -1$ or $-1 \le y < 1$ **77.** $-1 < t < 5$
79. $y \ge 5$ **81.** Answers vary   **83.** Answers vary   **85.** Answers vary
87. $6x < 18, x < 3$ **89.** $19 - 4x > 18, x < \frac{1}{4}$
91. $3(2x + 7) \ge -3, x \ge -4$ **93.** $18 < 4x - 6, x > 6$
95. $4(x - 3) \ge 8, x \ge 5$ **97.** $2(x - 1) + 3 \le 5, x \le 2$

Chapter 3 Review Exercises
1. $x = 4$ **2.** $x = 1$
3. $z = -3$ **4.** $a = 8$ **5.** $z = 40$ **6.** $x = -2.6$ **7.** $u = -3$
8. $b = 7$ **9.** $a = 7$ **10.** $z = 5$ **11.** Answers vary   **12.** Answers
vary   **13.** $x + 12 = -2, x = -14$ **14.** $x - 4 = 6, x = 10$ **15.** $140
16. $280   **17.** $-\frac{7}{2}$ **18.** $\frac{1}{4}$ **19.** $x = -8$ **20.** $y = -5$ **21.** $t = -12$
22. $x = -12$ **23.** Answers vary   **24.** Answers vary   **25.** $4x = -12$,
$x = -3$ **26.** $-\frac{2}{3}x = 6, x = -9$ **27.** $156   **28.** 6 m   **29.** $w = -14$
30. $z = -10$ **31.** $329   **32.** 24   **33.** 75   **34.** $180,000   **35.** 9 cm
36. 8 in.   **37.** $t = -2$ **38.** $u = -2$ **39.** $r = 3$ **40.** $s = 9$ **41.** $v = 1$
42. $p = 13$ **43.** $q = 5$ **44.** $a = 4$ **45.** Contradiction, no solutions
46. Identity, all real numbers   **47.** Answers vary   **48.** Answers vary
49. $4x + 3 = 11, x = 2$ **50.** $54 = 6(3x - 9), x = 6$ **51.** $45   **52.** $35
53. 18.84   **54.** $x = \frac{c - b}{a}$ **55.** 138 women   **56.** 7 ft
57. heart = 32.6%, cancer = 23.4%   **58.** 80% chuck = 15 g,
93% turkey = 8 g   **59.** 25 watt = 215 lumens, 60 watt = 880 lumens
60. pee wee = 15 oz, small = 18 oz, medium = 21 oz
61. feta = 4 g, mozzarella = 8 g, provolone = 7 g   **62.** 10, 11, 12
63. 5, 7, 9   **64.** 30 light, 20 heavy   **65.** 20 nickels, 12 dimes
66. $3500 at 6%, $3000 at 5%   **67.** 200 ml at 25%, 300 ml at 50%
68. 8 lbs   **69.** $L = 100$ ft, $W = 75$ ft   **70.** 40°, 95°   **71.** 53°, 37°
72. $x = 25$, $\angle A$ and $\angle B$ are 135°.   **73.**
74.
75.
76.
77. $p < 15$ **78.** $q \le 12$ **79.** $x > 16$

80. $y \le 2.2$ **81.** $z > -12$ **82.** $n \ge -5$ **83.** Answers vary
84. $11 \ge x + 6$ **85.** $t \le -7$ **86.** $u > -6$ **87.** $x < -1$ **88.** $y \ge 3$
89. $z > -5$ **90.** $r < 1$ **91.** $u \le 6$ **92.** $v > 1$ **93.** $s \ge -1$
94. $t \le -5$ **95.** $-4 < x < 1$ **96.** $5 \le x \le 7$ **97.** $-2 \le x \le 2$
98. $-3 \le x < 3$

Chapter 3 Test
1. $u = -13$ [3.1]   **2.** $v = -3$ [3.1]
3. $k = -6$ [3.2]   **4.** $m = -.675$ [3.2]   **5.** $v = -10$ [3.3]   **6.** $t = 5$ [3.3]
7. $y = 5$ [3.3]   **8.** $a = -2$ [3.3]   **9.** $W = \frac{P - 2L}{2}$ or $\frac{P}{2} - L$ [3.4]
10. $B = \frac{2A - bh}{h}$ or $\frac{2A}{h} - b$ [3.4]   **11.** Answers vary [3.2]
12. Answers vary [3.3]   **13.** $3(x - 6) = x + 1$ [3.3]   **14.** 8 ft [3.2]
15. 10 in. [3.3]   **16.** $5250 [3.2]   **17.** $26 [3.3]   **18.** 750 ml [3.2]
19. 2532 [3.5]   **20.** $-1, 0, 1$ [3.5]   **21.** 15 incandescent, 10 fluorescent [3.6]
22. $\frac{1}{2}$ hr [3.5]   **23.** $W = 31$ cm, $L = 40$ cm [3.7]   **24.** $p > 4$ [3.8]
25. $q \ge -2$ [3.8]   **26.** $u \ge 17$ [3.8]   **27.** $x > 7$ [3.8]   **28.** $y \ge -6$ [3.8]
29. $z > -14$ [3.8]   **30.** $-2 < x < 3$ [3.8]

Exercise Set 4.1
1. a) Wolf, 2.8% b) 3 c) 31,623   **3.** a) France, 76%
b) France, Lithuania, Slovakia, Belgium c) 47% d) 31%   **5.** a) 2008–2009,
2004–2005 b) rising sharply c) $2.30 d) 2003   **7.** Yes   **9.** No   **11.** Yes
13. No   **15.** Yes   **17.** Yes   **19.** No   **21.** Yes   **23.** $-3, \frac{3}{2}, 3, -2$
25. $-4, -2, 2, 2$ **27.** $-4, 6, -8, -3$ **29.** 4, 5, 12, -5 **31.** $-2, -3, -8, 3$
33. 4, -3, 12, -12 **35.** 4, 4, 4, 4   **37.** 4, 4, 4, 4
39.   **41.**   **43.**

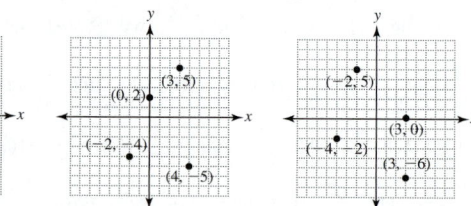

45. $A(3, 1), B(-4, 2), C(-2, -3), D(4, -5)$
47. $A(3, 0), B(0, 4), C(-5, 0), D(0, -6)$
49. $A(3, 4), B(0, 6), C(-4, -7), D(4, 0)$
51. a) (2002, 2077) (2003, 2001) (2004, 1907) (2005, 1818) (2006, 1780)
(2007, 1676) (2008, 1585) (2009, 1506) b) In 2007, there were 1676 national banks in the U.S.
c)   d) decreasing

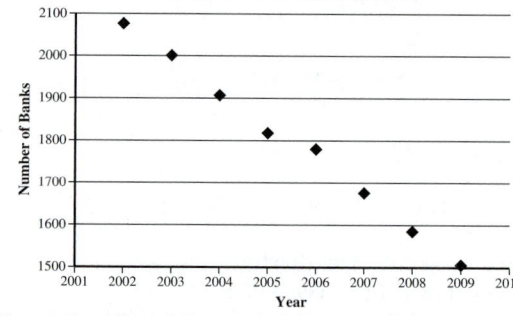

Exercise Set 4.2
1. $-3, 1, -7$ **3.** 5, 5, 3   **5.** 6, 3, 1

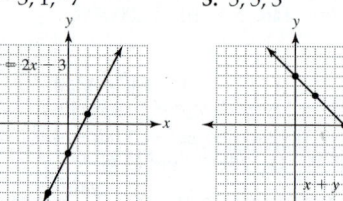
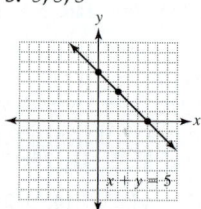
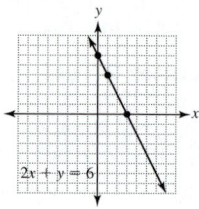

7. $(0, 2)$ $(-2, 0)$ $(2, 4)$ **9.** $(0, 4)$ $(-2, 0)$ $(1, 6)$

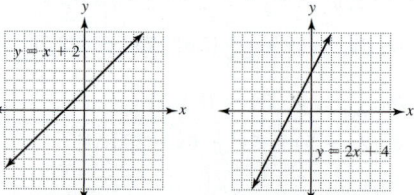
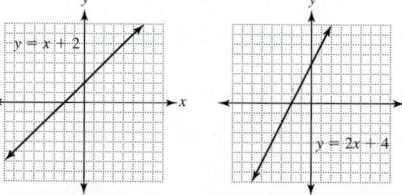

11. $(0, 1)$ $(2, -3)$ $(-2, 5)$

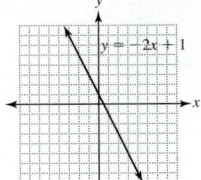

13. $(0, 3)$ $(3, 0)$ $(5, -2)$

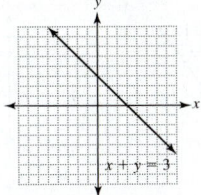

39. $(0, 0)$ $(2, -4)$ $(-2, 4)$

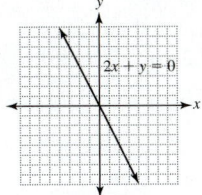

41. $(0, 0)$ $(2, 4)$ $(-2, -4)$

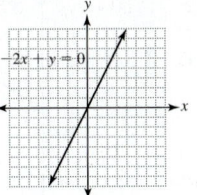

15. $(9, 0)$ $(0, 3)$ $(6, 1)$

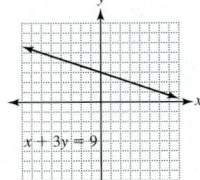

17. $(0, -4)$ $(8, 0)$ $(4, -2)$

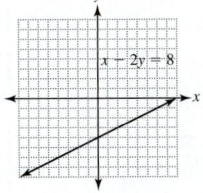

43. $(0, 0)$ $(4, 1)$ $(-4, -1)$

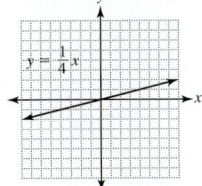

45. $(0, 0)$ $(6, -2)$ $(-6, 2)$

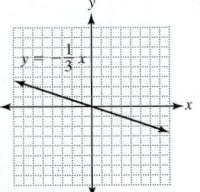

19. $(0, -2)$ $(2, 6)$ $(-1, -6)$

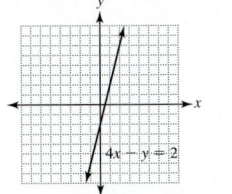

21. $(0, 2)$ $(3, 0)$ $(6, -2)$

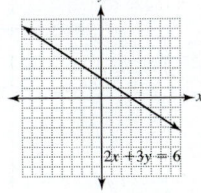

47. $(0, 0)$ $(3, 2)$ $(-3, -2)$

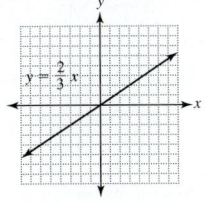

49. $(2, 0)$ $(2, 1)$ $(2, 2)$

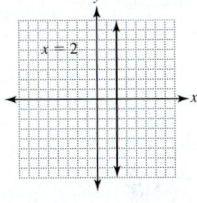

23. $(0, -2)$ $(5, 0)$ $(-5, -4)$

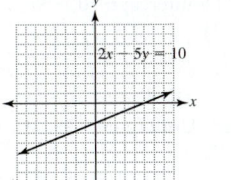

25. $(0, 2)$ $(6, 0)$ $(3, 1)$

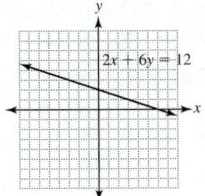

51. $(0, 3)$ $(3, 3)$ $(-4, 3)$

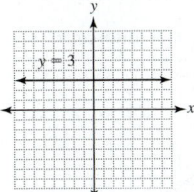

53. $(-4, 0)$ $(-4, 2)$ $(-4, -5)$

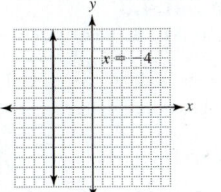

27. $(0, -3)$ $(4, 0)$ $(-4, -6)$

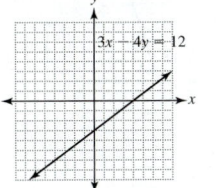

29. $(0, 1)$ $(-5, 3)$ $(5, -1)$

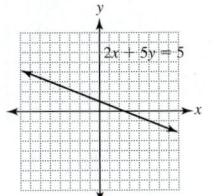

55. $(0, -4)$ $(-2, -4)$ $(5, -4)$

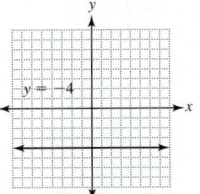

57. $(0, -9)$ $(3, -9)$ $(-4, -9)$

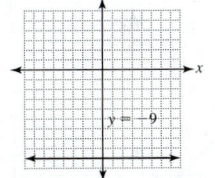

31. $(0, -2)$ $(5, 1)$ $(-5, -5)$

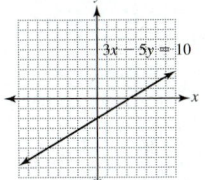

33. $(1, -2)$ $(4, 2)$ $(-2, -6)$

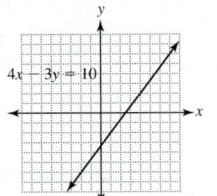

59. $(6, 0)$ $(6, -2)$ $(6, 4)$

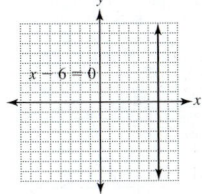

61. $(0, -1)$ $(2, -1)$ $(-2, -1)$

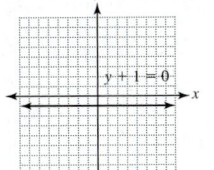

35. $(0, 0)$ $(3, 3)$ $(-3, -3)$

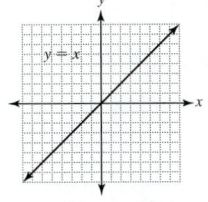

37. $(0, 0)$ $(1, -3)$ $(-1, 3)$

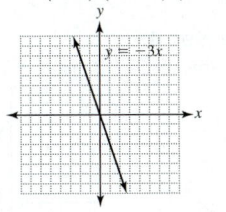

63. a) $59.45 b) 85 miles c)   d) 90 miles

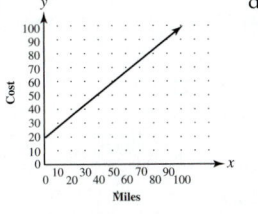

65. a) $9.20 b) 7 units c) 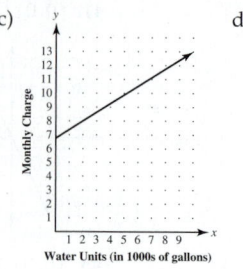 d) 9 units

67. a) 7.5 b) 8.5 c) 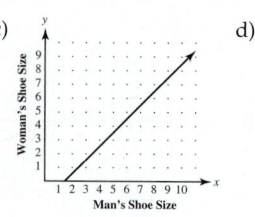 d) 8.5

69. a) $(0,0)$ $(2,-64)$ $(4,-128)$

b) 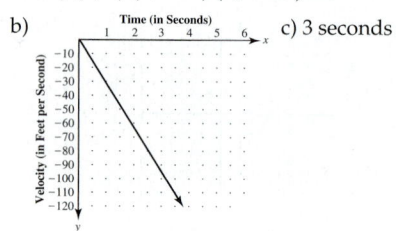 c) 3 seconds

Exercise Set 4.3

1. $(-4,0),(0,3)$ **3.** $(-4,0),(0,-3)$ **5.** $(6,0);$ No y − intercept.

7. 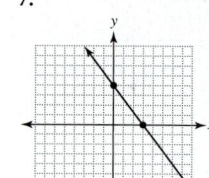 **9.** **11.**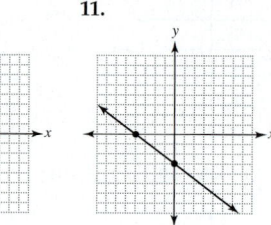

13. **15.** **17.**
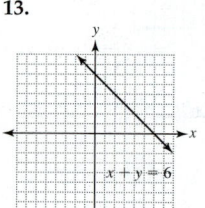 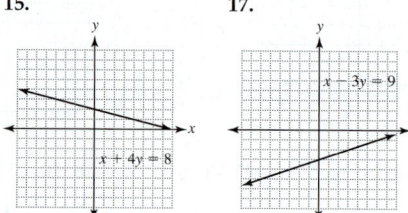

19. **21.** **23.**
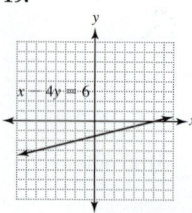 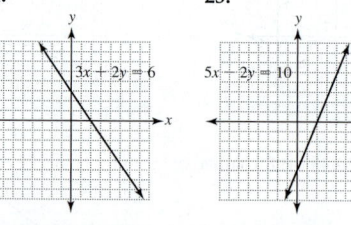

25. **27.** **29.**
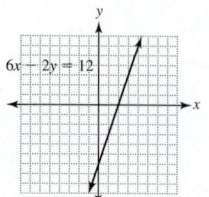 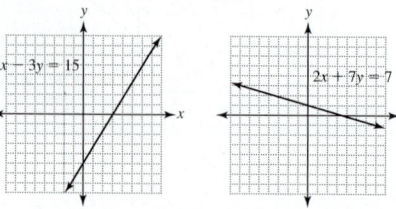

31. **33.** **35.**
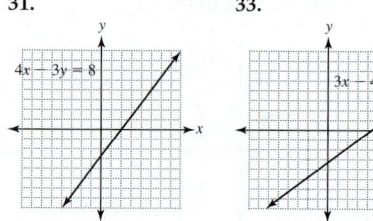 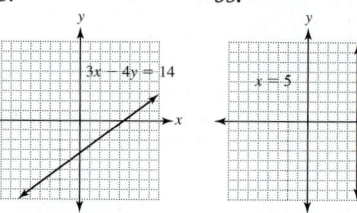

37. **39.** **41.**
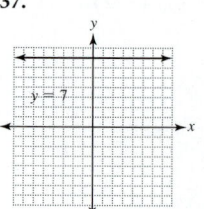 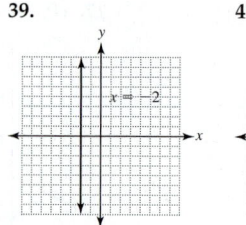 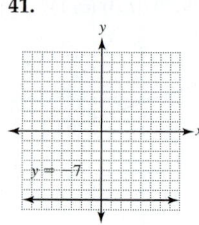

43. **45.**
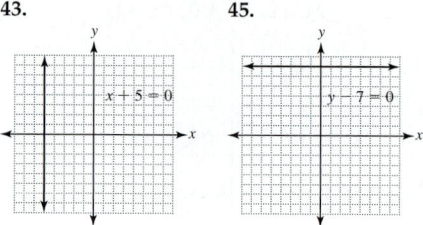

47. $4x + 6y = 36$ **49.** $5x + 10y = 450$ **51.** $50x + 60y = 500$
53. $a = 3, b = 5$ **55.** x-intercepts: $(-3,0),(2,0)$ y-intercept: $(0,-5)$
57. x-intercept: $(-3,0)$, y-intercepts: $(0,-1),(0,3)$

Exercise Set 4.4

1. $\frac{1}{2}$ **3.** -2 **5.** -2 **7.** $-\frac{2}{3}$ **9.** -1 **11.** $\frac{3}{4}$ **13.** $\frac{1}{2}$ **15.** 0
17. Undefined **19.** 0 **21.** 1 **23.** $-\frac{3}{4}$ **25.** $\frac{3}{4}$ **27.** Undefined **29.** 0

31. **33.** **35.**
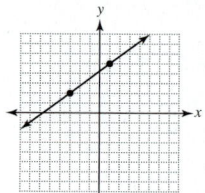 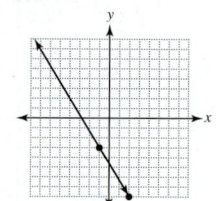 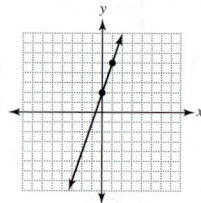

37. **39.** **41.**
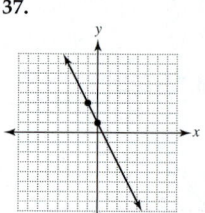 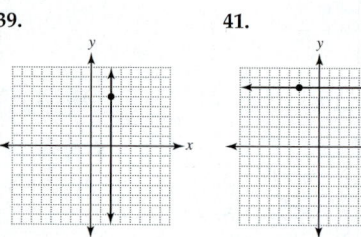

43. Undefined **45.** Undefined **47.** 0 **49.** 0 **51.** a) $\dfrac{30}{500} = \dfrac{3}{50}$ b) 6%

53. $\dfrac{24}{9} = \dfrac{8}{3}$

Exercise Set 4.5

1. $m = 2, y$ int. $= 3$ **3.** $m = -2, y$ int. $= 9$ **5.** $m = \frac{3}{4}, y$ int. $= -1$

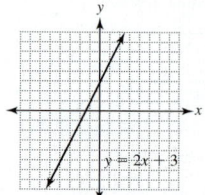 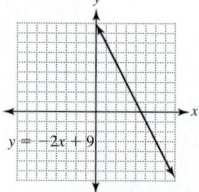 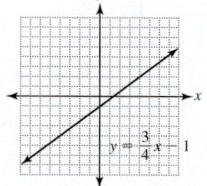

7. $m = -\frac{4}{3}$, y int. $= 2$ **9.** $m = 3$, y int. $= -7$ **11.** $m = -\frac{2}{3}$, y int. $= 3$

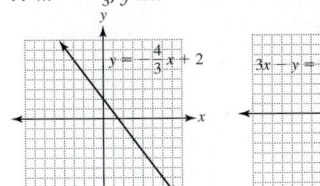
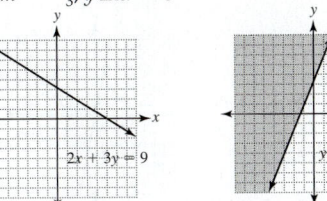
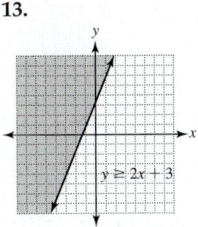

13.
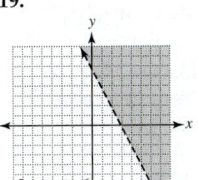

15.
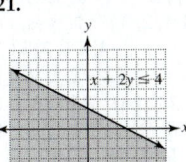

17.
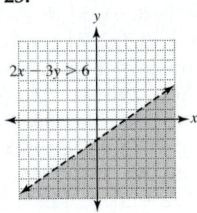

13. $m = -\frac{4}{5}$, y int. $= 4$ **15.** $m = \frac{5}{3}$, y int. $= -5$ **17.** $m = \frac{2}{5}$, y int. $= 2$

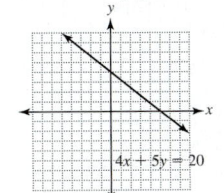
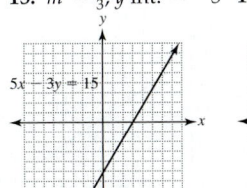
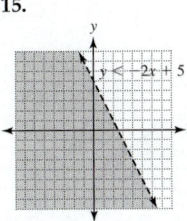

19.
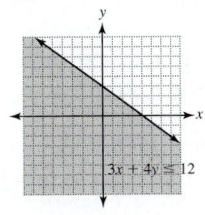

21.
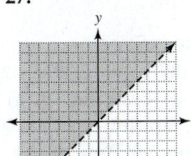

23.
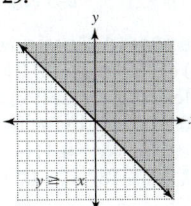

19. $m = -2$, y int. $= 0$ **21.** $m = -\frac{3}{2}$, y int. $= 0$ **23.** $m = -\frac{3}{2}$, y int. $= \frac{5}{2}$

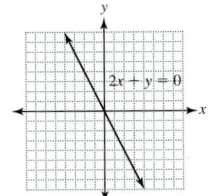
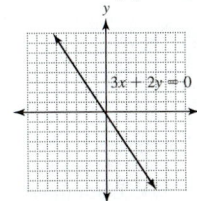
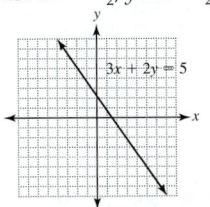

25.
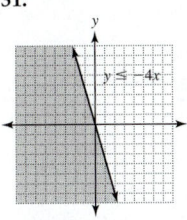

27.
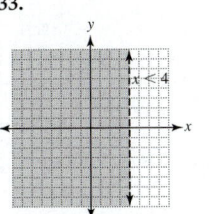

29.
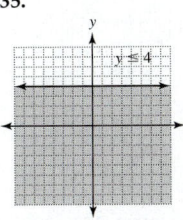

25. Undefined   **27.** Undefined   **29.** 0   **31.** 0   **33.** $y = 2x + 4$
35. $y = \frac{3}{5}x - 1$ **37.** $y = -\frac{2}{3}x + \frac{4}{5}$ **39.** $x = -4$ **41.** $x = 6$ **43.** $x = 4$
45. Parallel   **47.** Perpendicular   **49.** Parallel   **51.** Neither
53. Parallel   **55.** Perpendicular   **57.** Parallel   **59.** Perpendicular
61. Neither   **63.** Parallel   **65.** Perpendicular
67. a) $y = 6x$ b)   c) Same

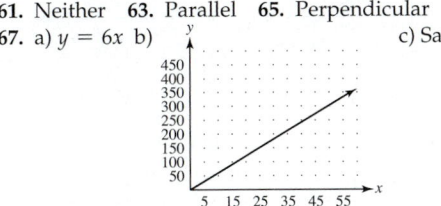

69. a) $y = 2x + 100$ b)   c) Same

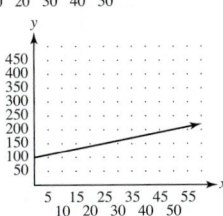

31.
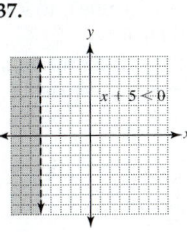

33.
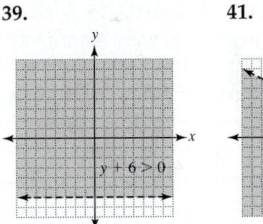

35.
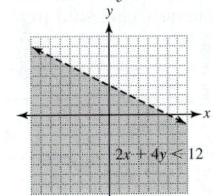

37.
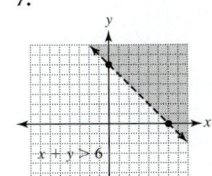

39.
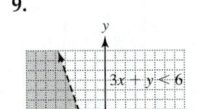

41. $2x + 4y < 12$
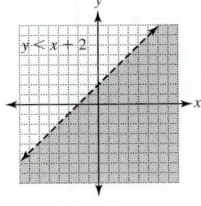

Exercise Set 4.6   **1.** $2x - y = 5$ **3.** $2x + y = 0$
5. $x - 4y = -23$ **7.** $x + 3y = -10$ **9.** $4x - 3y = 17$
11. $2x + 3y = -19$ **13.** $x = 3$ **15.** $y = 3$ **17.** $x = 5$ **19.** $y = 4$
21. $y = -2x + 5$ **23.** $y = x + 4$ **25.** $y = -\frac{3}{2}x - 5$
27. $y = -\frac{2}{3}x + 4$ **29.** $x = 5$ **31.** $y = -4$ **33.** $x + 2y = 1$
35. $2x - 3y = 6$ **37.** $4x - y = 17$ **39.** $3x - y = -18$ **41.** $x = 6$
43. $y = -4$ **45.** a) $y = .15x + 25$ b) $100 c) 250 miles
47. a) $y = .8x + 1.2$ b) $7.60 c) 5 miles   **49.** a) $y = -5000x + 50,000$
b) $10,000 c) 10 years   **51.** a) $y = .05x + 200$ b) $12,700 c) $175,000

43. $x - 2y \geq 8$
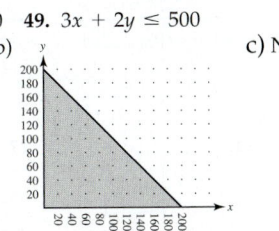

45. $y < 3x + 5$

Exercise Set 4.7   **1.** No   **3.** No   **5.** Yes
7.
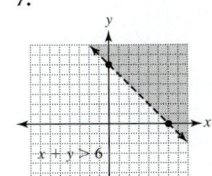

9.

11.

47. $20x + 25y \leq 150$ **49.** $3x + 2y \leq 500$
51. a) $x + y \leq 200$ b)   c) No

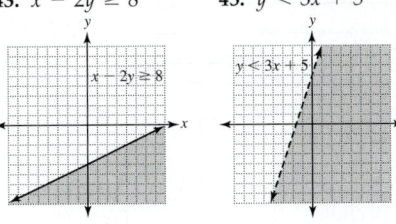

53. a) $2x + 3y \le 84$ b) 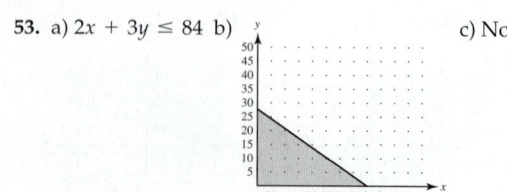 c) No

Exercise Set 4.8 **1.** yes, $D = \{-6, -3, 5, 7\}$, $R = \{2, -2, -7, 9\}$
3. yes, $D = \{-5, -2, 0, 3, 6\}$, $R = \{-2, 1, 3, 4, 7\}$ **5.** no,
$D = \{-8, -6, 5\}$, $R = \{-3, 1, 2, 3\}$ **7.** yes, $D = \{-7, -3, 3, 4, 6\}$,
$R = \{-3, 1, 2, 7\}$ **9.** yes, $D = \{-4, -1, 3, 5\}$, $R = \{1, 2, 3\}$
11. no, $D = \{-5, 2, 3, 5\}$, $R = \{-2, -1, 1, 2, 4\}$ **13.** Yes **15.** No
17. Yes **19.** No **21.** Yes **23.** No **25.** Yes **27.** Yes **29.** Yes
31. No **33.** No **35.** No **37.** Yes **39.** Yes **41.** No **43.** Yes
45. 5, (0, 5) **47.** $3a + 5$, $(a, 3a + 5)$ **49.** $-1, (-1, -1)$ **51.** 16, (1, 16)
53. $16a^2$, $(a, 16a^2)$ **55.** $-6, (-2, -6)$ **57.** $7, (-2, 7)$
59. $|2z - 3|$, $(z, |2z - 3|)$ **61.** a) $34 b) $64 c) The cost of driving
250 miles is $52. **63.** a) $1675 b) $1175 c) Her salary for $20,000 of
sales is $2175. **65.** No **67.** Yes

Chapter 4 Review Exercises
1. a) 65 and older, 19.2% b) 189 c) 48,929,024 **2.** a) Finland, 83%
b) Mexico, Germany c) 71% d) Australia **3.** a) 2005 and 2006,
$8,500,000,000 b) Decreasing rapidly c) $3,500,000,000 d) 2003
4. Yes **5.** No **6.** Yes **7.** Yes **8.** Yes **9.** No **10.** 3, 1, −6, 2
11. 2, $-\frac{1}{2}$, −18, −2 **12.** 6, 4, 12, 2 **13.** −5, 3, 5, 6
14.   **15.**

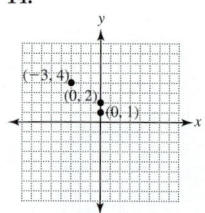   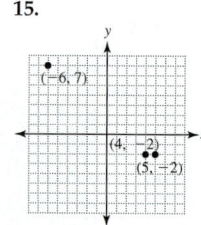

16. a) (1999, 8.7), (2000, 9.8), (2001, 9.9), (2002, 11.5), (2003, 10.7), (2004,
10.9), (2005, 12.0), (2006, 14.8), (2007, 15.5), (2008, 16.8) b) In 2003, 10.9%
of the new cars sold in the U.S. were manufactured in Japan.
c) 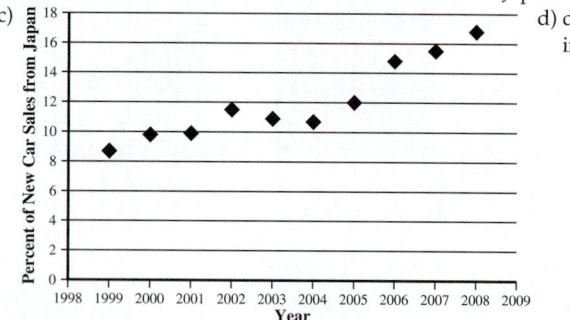   d) decreasing,
increasing

17.   **18.**   **19.**

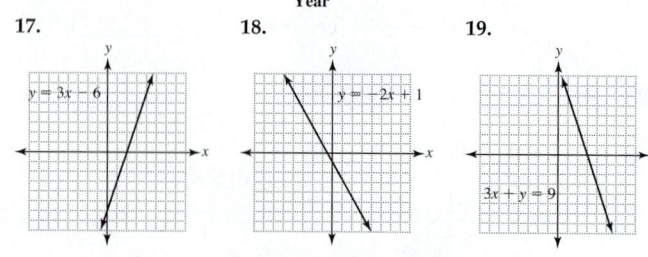

20.   **21.**   **22.**

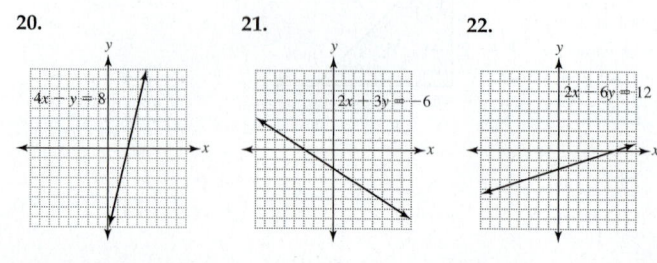

23.   **24.**   **25.**

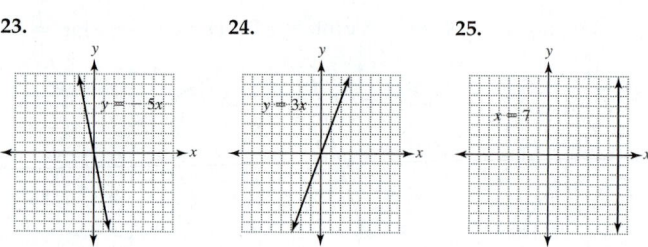

26.   **27.** a) 20 hrs b) 14 hrs

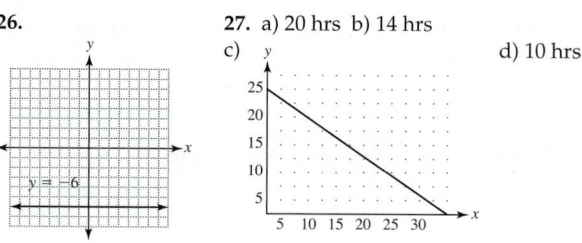   d) 10 hrs

28. a) $156 b) 45 items c)   d) $140

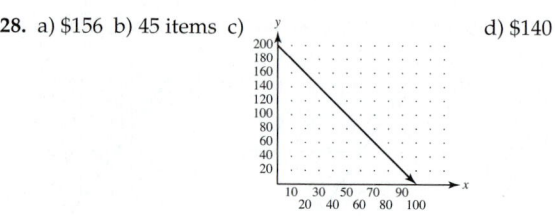

29. 2 **30.** $-\frac{4}{3}$ **31.** $-\frac{1}{2}$ **32.** $\frac{5}{2}$ **33.** Undefined **34.** 0
35.   **36.**   **37.**

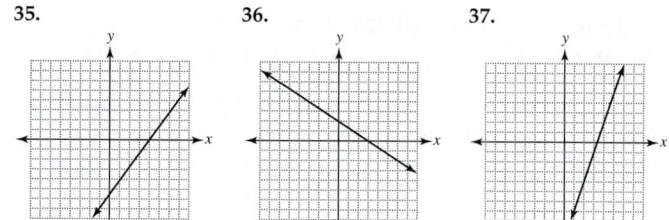

38.   **39.**   **40.**

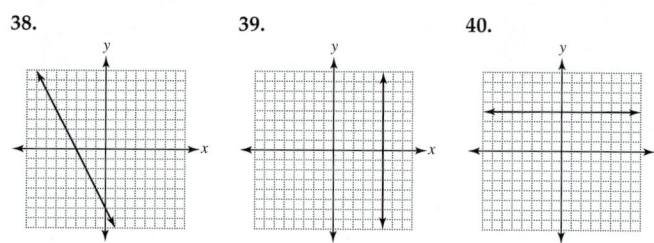

41. $\frac{20}{30} = \frac{2}{3}$ **42.** $-\frac{2}{25}$ **43.** $-\frac{1}{2}, (0, 4)$ **44.** $\frac{5}{2}, \left(0, \frac{-15}{2}\right)$ **45.** Undefined,
none **46.** 0, (0, −3) **47.** $x = 6$ **48.** Parallel **49.** Perpendicular
50. Neither **51.** Perpendicular **52.** Parallel **53.** Neither
54. Parallel **55.** Perpendicular **56.** $y = -3x + \frac{5}{6}$ **57.** $y = \frac{6}{5}x + 4$
58. $y = -3$ **59.** $3x - y = 0$ **60.** $2x + y = 1$ **61.** $5x + 2y = 18$
62. $4x - 3y = -32$ **63.** $x = -7$ **64.** $y = -5$ **65.** $y = 2$
66. $x = -8$ **67.** $x + 2y = 5$ **68.** $x - y = 4$ **69.** $3x + 2y = 12$
70. $3x - y = 12$ **71.** $y = 5$ **72.** $x = 2$ **73.** $3x + 5y = -2$
74. $3x + y = -9$ **75.** $2x - 3y = -22$ **76.** $x + 4y = -5$
77. $x = 2$ **78.** $y = -1$ **79.** $y = 3$ **80.** a) $y = 1.50x + 250$
b) $700 c) 225 sandwiches
81.   **82.**   **83.**

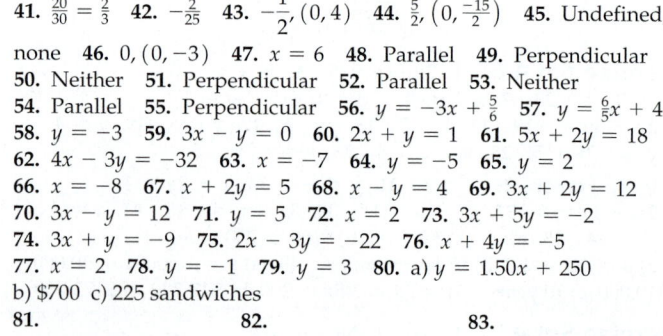

84.   **85.**   **86.**

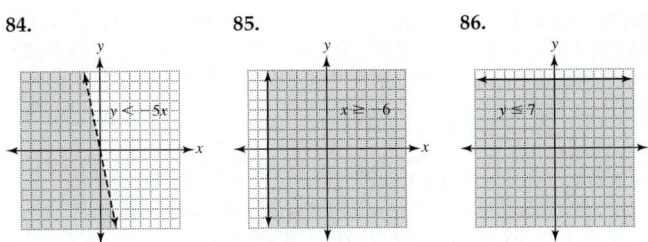

87. $3y - 2x \geq 8$ **88.** $28x + 35y \leq 180$
89. No, $D = \{-3, -1, 0\}$, $R = \{2, 4, 5, 7\}$
90. Yes, $D = \{-4, -1, 3, 6\}$, $R = \{2, 4, 5\}$ **91.** Yes   **92.** No
93. No   **94.** Yes   **95.** Yes   **96.** Yes   **97.** No   **98.** Yes   **99.** a) 11, (2, 11)
b) 21, $(-3, 21)$ c) $2a^2 + 3$, $(a, 2a^2 + 3)$ **100.** a) -18, $(-2, -18)$
b) 2, (0, 2)   c) $b^3 - 3b^2 + 2$, $(b, b^3 - 3b^2 + 2)$ **101.** a) \$22,800
b) \$12,000   c) After 8 years, the machinery is worth \$15,600.

Chapter 4 Test

1. a) Superior b) Huron and Erie c) 190 miles d) 65 miles [4.1]
2. Yes [4.1]   **3.** $-6, -8, -3, -12$ [4.1]   **4.** a) 9 million b) 8 million
c) The number of milk cows decreased rapidly from 1960–1975 and then
leveled off somewhat though still decreased until 2005. [4.1]
5. [4.2] [4.3]   **6.** [4.2] [4.3]   **7.** [4.2] [4.3]

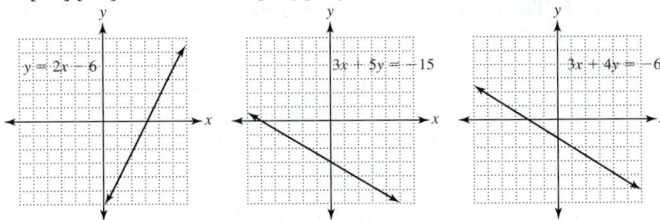

8. [4.2] [4.3]   **9.** a) $9x + 20y = 480$ b) 6 shovels
c)   d) 20 hammers [4.2]

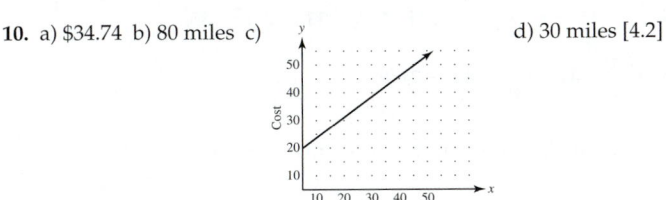

10. a) \$34.74 b) 80 miles c)   d) 30 miles [4.2]

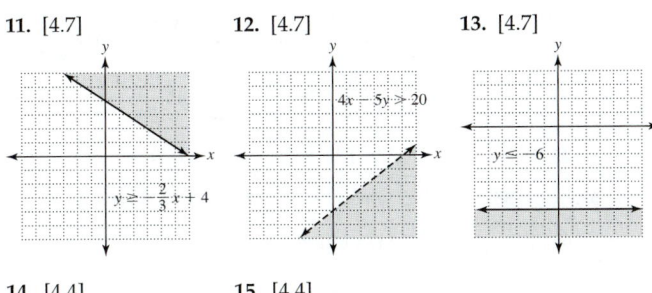

11. [4.7]   **12.** [4.7]   **13.** [4.7]

14. [4.4]   **15.** [4.4]

16. -1 [4.4]   **17.** 0 [4.4]   **18.** $\frac{1}{4}$, $(0, -2)$ [4.5]   **19.** $\frac{5}{2}$, $\left(0, -\frac{3}{2}\right)$ [4.5]

20. Undefined, None [4.5]   **21.** $x = -4$ [4.5]   **22.** Perpendicular [4.5]
23. Parallel [4.5]   **24.** $5x + 2y = 16$ [4.6]   **25.** $x - 2y = 8$ [4.6]
26. $x - 3y = -11$ [4.6]   **27.** $y = -6$ [4.6]   **28.** The line with slope of
-2 slants downward and drops 2 units vertically for each horizontal
change of 1 to the right. The line with slope of 2 slants upward and rises
2 units vertically for each horizontal change of 1 to the right. [4.4]
29. No, $D = \{-4, 8, 6\}$, $R = \{-4, -2, 3, 7\}$ [4.8]   **30.** No [4.8]
31. Yes [4.8]   **32.** a) 0, (2, 0) b) 6, $(-1, 6)$ [4.8]   **33.** a) \$4500 b) It costs
\$7000 to operate the business for 10 days. [4.8]

Exercise Set 5.1   **1.** composite   **3.** prime   **5.** composite
7. composite   **9.** prime   **11.** 1, 2, 3, 6, 9, 18   **13.** 1, 2, 3, 4, 6, 8, 12, 24
15. 1, 2, 3, 4, 6, 9, 12, 18, 36   **17.** 1, 2, 3, 6, 9, 18, 27, 54   **19.** 2, 3, 5
21. 3   **23.** 2, 3   **25.** none   **27.** 5   **29.** 2, 3   **31.** 2, 3, 5   **33.** $2 \cdot 7$
35. $2 \cdot 23$ **37.** $2^2 \cdot 7$ **39.** prime   **41.** $3^2 \cdot 5$ **43.** $2^2 \cdot 5^2$ **45.** $2 \cdot 3 \cdot 7$
47. $2 \cdot 7 \cdot 11$ **49.** $2^2 \cdot 3^2 \cdot 5$ **51.** $3^2 \cdot 5 \cdot 7$ **53.** $2^2 \cdot 3^3 \cdot 7$ **55.** 1, 2, 3, 4, 6,
9, 12, 18, 27, 36, 54, 108   **57.** $2^3 \cdot 3^3 \cdot 7$ **59.** 3   **61.** 5   **63.** 1   **65.** 3
67. 10   **69.** 90   **71.** 1   **73.** 8   **75.** 8   **77.** 24   **79.** mn **81.** cd^2
83. a^3b^2 **85.** 1   **87.** a^2bc^2 **89.** pr^3 **91.** $6pq$ **93.** $9s^2t^3$ **95.** $3abc^2$
97. $4a^2b^3c^2$ **99.** 3   **101.** 36   **103.** $x - 3$ **105.** $2(c - 2d)$
107. $16a^3b(x + 2)^2$

Exercise Set 5.2   **1.** $c(d + f)$ **3.** $r(s - t)$ **5.** $3(x + 3)$
7. $4(2x - 3)$ **9.** $x(x + 1)$ **11.** $r^2(r^2 - 1)$ **13.** $7a^2(a + 2)$
15. $9p^3(2 - p^2)$ **17.** $5x(3x^2 + 1)$ **19.** $11r^2(2r^3 - 1)$
21. $4c^2(3c + 2)$ **23.** $6z^2(3 - 2z^4)$ **25.** $x^2y^2(y + 1)$ **27.** $u^3v^2(u - v)$
29. $cd^2(c^2d^2 + 1)$ **31.** $x^2y^2(1 - xy^2)$ **33.** $6a^2b^2(3b^2 + 2a)$
35. $3xy^2(3y - 4x)$ **37.** $3xy^3(3xy + 1)$ **39.** $3(3x^2 - 4x + 2)$
41. $5x^2(3x^2 - 2x - 4)$ **43.** $6(a^4 + 2a^2 - 4)$ **45.** $11x(2x^2 - 3x - 1)$
47. $4cd^2(4c^2 - 6cd^2 + 9)$ **49.** $7xy^2(2x^2y - 3xy^2 - 1)$
51. $-3(m - 2n)$ **53.** $-8(2c - d)$ **55.** $-7(2x - 1)$
57. $-4(x^2 - 2x + 4)$ **59.** $-5x(2x^2 + 3x - 5)$ **61.** $-(x - y)$
63. $(m + n)(a + b)$ **65.** $(c + 2)(a - b)$ **67.** $(t - 3)(t + 6)$
69. $(a - 6)(a - 7)$ **71.** $2x + 3, 3x - 5$ **73.** $180x^6y^4(2y - 3x^2)$
75. $3(a + b)^2(a + b + 3)$ **77.** $(x + y)(z + w)$ **79.** $(n - 4)(m + 3)$
81. $(b + 5)(a + 3)$ **83.** $(x - y)(x + 5)$ **85.** $(b - 3)(a + 1)$
87. $(m + n)(x - y)$ **89.** $(n - 3)(m - 6)$ **91.** $(d - e)(c - 3)$
93. $(x + 3)(2x - y)$ **95.** $(s + 4)(r - 1)$ **97.** $(d - 3)(c - 1)$
99. $(2b + 5)(2a + 3)$ **101.** $(2z - w)(4x - y)$
103. $(3c + 2d)(2a - 3b)$ **105.** $(2x + 3)(4x - y)$

Exercise Set 5.3   **1.** $(x + 1)(x + 2)$ **3.** $(x - 3)(x - 1)$
5. $(a + 2)(a + 3)$ **7.** $(b - 4)(b - 2)$ **9.** $(y - 5)(y + 1)$
11. $(a - 5)(a + 7)$ **13.** $(x + 2)(x + 8)$ **15.** $(r - 8)(r + 9)$
17. $(a - 9)(a - 3)$ **19.** prime   **21.** $(x - 7)(x - 5)$
23. $(y + 7)(y - 6)$ **25.** $(z + 3)(z + 12)$ **27.** $(a + b)(a + 3b)$
29. $(r - 7s)(r - s)$ **31.** $(c - 2d)(c + 9d)$ **33.** $(r - 11s)(r + 4s)$
35. $(y + 8z)(y - 5z)$ **37.** $(a - 9b)(a + 6b)$ **39.** $(r - 12s)(r - 3s)$
41. $3(a + 2)(a + 3)$ **43.** $5(z - 3)(z - 2)$ **45.** $x(x - 7)(x + 4)$
47. $y^2(y - 15)(y - 5)$ **49.** $3a(a - 7)(a + 6)$
51. $xy(x - 8)(x + 10)$ **53.** $y^2(x + 4)(x + 5)$ **55.** $4s(r - 2)(r + 7)$
57. $3b(a - 7)(a + 4)$ **59.** $(x - 12)(x + 9)$ **61.** $(y + 8)(y + 16)$
63. $(4 + x)(6 - x)$ **65.** $(6 + a)(3 + a)$ **67.** $(8 + b)(2 - b)$
69. $k = \pm 7, \pm 8, \pm 13$ **71.** $(x + 4), (x + 8)$ **73.** $x, (x + 10), (x - 4)$

Exercise Set 5.4   **1.** $(a + 1)(2a + 5)$ **3.** $(a + 3)(2a - 1)$
5. $(7m + 1)(m - 5)$ **7.** $(3x - 7)(x + 1)$ **9.** prime
11. $(11x + 5y)(x + y)$ **13.** prime   **15.** $(5x - 4)(x + 2)$
17. $(2x + 1)(2x + 3)$ **19.** $(2x - 7)(5x + 1)$ **21.** $(2x + 3)(x + 4)$
23. $(3a - 8)(a - 2)$ **25.** $(5c - 4)(c + 3)$ **27.** prime
29. $(2x + 5)(3x + 1)$ **31.** $(2x - 5)(4x - 1)$ **33.** $(4a - 3)(6a + 1)$
35. $(2n + 1)(8n - 5)$ **37.** $(2x + 3)(3x + 5)$ **39.** $(3c - 2)(4c - 3)$
41. $(3a + 5)(6a - 7)$ **43.** $(2w + 3)(4w - 9)$ **45.** prime
47. $(3y + 4)(6y - 7)$ **49.** $(2a + 5b)(3a + 7b)$
51. $(3c - 4d)(6c + 5d)$ **53.** $(3a + 4b)(4a - 3b)$
55. $(2f - 3g)(6f - 7g)$ **57.** $3(2a - 3)(a + 2)$ **59.** $y(4x - 1)(x - 3)$
61. $3x(x - 1)(2x - 5)$ **63.** $3x(2x - 3)(3x + 4)$
65. $-1(x - 7)(x - 6)$ **67.** $-1(3x + 5)(x - 4)$
69. $-1(3x - 4)(4x + 7)$ **71.** $(4x + 5)(9x - 4)$
73. $[(x + 2) - 3][(x + 2) + 2] = (x - 1)(x + 4)$
75. $(3 + 2a)(2 - a)$ **77.** $(3 - 2b)(4 + 3b)$ **79.** $2(3 + 4n)(3 - n)$
81. $\pm 7, \pm 11$

Exercise Set 5.5

1. $(m + n)(m - n)$ **3.** $(x + 2)(x - 2)$
5. $(r + 8)(r - 8)$ **7.** prime **9.** $(c + 5d)(c - 5d)$
11. $(a + 10b)(a - 10b)$ **13.** $(2x + 5y)(2x - 5y)$
15. $(7p - 9q)(7p + 9q)$ **17.** prime **19.** $(11a + 7b)(11a - 7b)$
21. $(a^2 + b^2)(a + b)(a - b)$ **23.** $(x^2 + 1)(x + 1)(x - 1)$
25. $(4x^2 + 9)(2x + 3)(2x - 3)$ **27.** $(x + y + 3)(x + y - 3)$
29. $(6 + x - y)(6 - x + y)$ **31.** $(2x - 2y + 3)(2x - 2y - 3)$
33. $3(x + 5)(x - 5)$ **35.** $4(2x + 3y)(2x - 3y)$
37. $x^2(3x + 4)(3x - 4)$ **39.** $9x^2(x + 3)(x - 3)$
41. $3(x^2 + 4)(x + 2)(x - 2)$ **43.** $L = 3x + 7, W = 3x - 7$
45. $(14x - 25)(14x + 25)$ **47.** $(16c - 19d)(16c + 19d)$
49. $2(13x - 12)(13x + 12)$ **51.** $(x^4 + y^4)(x^2 + y^2)(x + y)(x - y)$
53. $[9(x + 2) + 4y][9(x + 2) - 4y] = (9x + 18 + 4y)(9x + 18 - 4y)$
55. $(x^n + y^n)(x^n - y^n)$ **57.** $(x + y)(x^2 - xy + y^2)$
59. $(c + 2)(c^2 - 2c + 4)$ **61.** $(n - 4)(n^2 + 4n + 16)$
63. $(m + 1)(m^2 - m + 1)$ **65.** $(2x + y)(4x^2 - 2xy + y^2)$
67. $(3a - 2)(9a^2 + 6a + 4)$ **69.** $(2x + 5y)(4x^2 - 10xy + 25y^2)$

Exercise Set 5.6

1. $(x + 2)^2$ **3.** $(x + 1)^2$ **5.** $(x + 6)^2$
7. $(y + 10)^2$ **9.** $(5x - 1)^2$ **11.** $(a - b)^2$ **13.** $(2a + b)^2$
15. $(4c - 3)^2$ **17.** $(3x + 5)^2$ **19.** $(4x - y)^2$ **21.** $(5x + 4y)^2$
23. not a perfect square **25.** $2(2x - 3)^2$ **27.** $2(x + 7)^2$
29. $4(x + 10)^2$ **31.** $x^2(5x - 3)^2$ **33.** $2xy(5x + 4y)^2$ **35.** $x - 7$
37. $(9x - 5y)^2$ **39.** $25(2a^2 + 3b^2)^2$ **41.** 49 **43.** 25

Exercise Set 5.7

1. $8(2 - x^2)$ **3.** $(c + 3)(c + 9)$
5. $3a(a + 3)(a - 6)$ **7.** $(x + 10)(x - 10)$ **9.** prime
11. $(a + b)(x + 2)$ **13.** $6(a - 2)(a - 6)$ **15.** $(r + 9)(r - 8)$
17. $m^2n^2(mn^2 + 1)$ **19.** $3a^2(2a - 5)(2a + 5)$ **21.** $(3a + 5b)^2$
23. $8(r + 4)(2r - 3)$ **25.** $(ab + 8)(ab - 8)$ **27.** $(5n + 2)(n + 4)$
29. prime **31.** $(y + z)(x + w)$ **33.** $(a - b + 6)(a - b - 6)$
35. $(2y - 3)(5x - 2)$ **37.** $(a^2 + 2b)(a^2 - 2b)$ **39.** prime
41. $(4b^2 + 1)(2b + 1)(2b - 1)$ **43.** $(6x + 1)(x + 5)$
45. $(2x + 3z)(3x - y)$ **47.** $(b + 3)(a - 1)$ **49.** $2u(8u - v)(u + 3v)$
51. $ab(2a - 5ab + 7b)$ **53.** $(8c + d)(c + 4d)$ **55.** $3(3a + b)(x - 1)$
57. $(3x - 5)(2x + 3)$ **59.** $(3a + 2b)(4c - 3d)$
61. $36(x + 2y)(x - 2y)$ **63.** $(x - 10)(x + 12)$
65. $3a^2b^3(4a + b)(3a - 4b)$ **67.** $(2z + 5)(4z^2 - 10z + 25)$
69. $(3c - 2d)(9c^2 + 6cd + 4d^2)$ **71.** $3(2c - d)(4c^2 + 2cd + d^2)$
73. $(3x - 8)(4x + 3)$ **75.** $(9a^2 + 25b^2)(3a + 5b)(3a - 5b)$
77. $(x + 5 - y)(x + 5 + y)$ **79.** $(x + 2)(x - 2)(x + 3)(x - 3)$
81. $(x - 2)(xy + 2y - 3)$

Exercise Set 5.8

1. $6, -4$ **3.** $5, 9$ **5.** $0, \frac{2}{5}$ **7.** $0, -\frac{11}{7}$
9. $3, -2, 5$ **11.** $-7, \frac{5}{3}, \frac{1}{4}$ **13.** 2 **15.** $-2, 3$ **17.** $-2, -5$ **19.** $4, -\frac{1}{3}$
21. $\frac{3}{4}, 5$ **23.** $-\frac{5}{2}, \frac{2}{3}$ **25.** $0, 2$ **27.** $0, \frac{2}{3}$ **29.** $2, -2$ **31.** $5, -5$
33. $3, 4$ **35.** $-\frac{2}{3}, 4$ **37.** 6 **39.** $-\frac{5}{2}$ **41.** $0, -4, 6$ **43.** $0, \frac{1}{2}, -5$
45. $12, 14$ **47.** $-6, -4$ **49.** -4 or -8 **51.** 3 **53.** $W = 4$ ft, $L = 8$ ft
55. $W = 8$ in., $L = 11$ in. **57.** $b = 6$ mi, $h = 2$ mi **59.** a) 7 sec,
b) 3, 4 sec **61.** 4 sec **63.** 7 positive integers **65.** 6 logs
67. 9 vertices **69.** a) -8 b) 0 c) -5 **71.** $0, 4, 9$ **73.** $0, -6$

Chapter 5 Review Exercises

1. composite **2.** prime
3. $2^2 \cdot 3 \cdot 5$ **4.** $2^2 \cdot 7 \cdot 11$ **5.** 42 **6.** 70 **7.** 14 **8.** 6 **9.** m^2n
10. b^2 **11.** $6r^2s^2t^2$ **12.** $5ab^2$ **13.** $a(x + y)$ **14.** $3(2x^2 + 1)$
15. $6xy(3x - 4)$ **16.** $r^4s^3(1 + r^2s)$ **17.** $6(3x^2 - 4x + 2)$
18. $2ab^3(4ab - 6a^2b^2 + 1)$ **19.** $(s - 2)(r - 5)$ **20.** $(n - p)(m - q)$
21. $-5x(3x^2 - x + 5)$ **22.** $(a + b)(x + y)$ **23.** $(s - 8)(r + 3)$
24. $(a - 3)(b - 6)$ **25.** $(x - 7)(y - 4)$ **26.** $(b - 1)(3a - 1)$
27. $(2x + 5)(3x - 2y)$ **28.** $(x + 11)(x + 1)$ **29.** $(x - 5)(x - 7)$
30. $(x + 7)(x - 1)$ **31.** $(m - 5)(m + 2)$ **32.** $(x - 8)(x - 3)$
33. $(r + 9s)(r - 8s)$ **34.** $(a + 27b)(a - 2b)$
35. $(m - 21n)(m + 2n)$ **36.** $2(x + 7)(x + 5)$
37. $3y(y + 11z)(y - 4z)$ **38.** $(x - 1)(5x - 2)$ **39.** $(z + 1)(3z - 5)$
40. $(x + 4)(5x - 2)$ **41.** $(3a + 2)(a + 8)$ **42.** prime
43. $(9a + 5)(2a - 1)$ **44.** $(5x + 2y)(7x + 3y)$
45. $(5a + 6b)(4a - 3b)$ **46.** $2(2a - 7)(5a + 1)$

47. $3c(3c - 2d)(4c - 3d)$ **48.** $(g + h)(g - h)$ **49.** $(r + 11)(r - 11)$
50. $(5x - 4)(5x + 4)$ **51.** $(7a + 8b)(7a - 8b)$ **52.** $6(x + 2)(x - 2)$
53. $16(2c - d)(2c + d)$ **54.** $(m + n + 9)(m + n - 9)$
55. $(8 + c - d)(8 - c + d)$ **56.** prime
57. $(m^2 + n^2)(m + n)(m - n)$ **58.** $(m + n)(m^2 - mn + n^2)$
59. $(r - 1)(r^2 + r + 1)$ **60.** $(a + 5)(a^2 - 5a + 25)$
61. $(4x + 3y)(16x^2 - 12xy + 9y^2)$ **62.** $(r + 9)^2$ **63.** $(x - 12)^2$
64. $(x + 8)^2$ **65.** $(7a - 1)^2$ **66.** $(7a + 2b)^2$ **67.** $(5c + 3d)^2$
68. $2(3p + 4)^2$ **69.** $4(5a - 3b)^2$ **70.** $13a(2a + 1)$
71. $(9x + y)(9x - y)$ **72.** $2(x - 8)(x + 5)$ **73.** $(x - 9)(x + 7)$
74. $11a^2b(2 - ab + 3b)$ **75.** $3(3m - 4)(m - 4)$
76. $(x - 9)(3y - 4)$ **77.** $(2a + 2b + 7)(2a + 2b - 7)$
78. $4(2x^2 + 7x - 14)$ **79.** $(7a - 2b)^2$ **80.** $3(x + 5)(2x - 3)$
81. $2(y - 5)(3x + 1)$ **82.** $3y(3x - 5)(3x + 5)$
83. $(11a + 13)(11a - 13)$ **84.** $(9x - 4y)^2$ **85.** $9(x^2 + 9)$
86. $4(x + 3)(2x - y)$ **87.** $(r + 10)(r - 4)$ **88.** prime
89. $(y^2 + 25)(y + 5)(y - 5)$ **90.** $3xy^2(x - 9y)(x + 5y)$

91. $(8x - 7)(x - 1)$ **92.** $0, -5$ **93.** $0, 6$ **94.** $\frac{9}{2}, -\frac{9}{2}$ **95.** $\frac{8}{3}, -\frac{8}{3}$
96. $7, -5$ **97.** $7, -3$ **98.** $4, -\frac{1}{2}$ **99.** $\frac{2}{3}, -4$ **100.** $W = 5$ cm, $L = 5$ cm
101. -9 and -7 or 7 and 9 **102.** 6 yds **103.** -7 or 4

Chapter 5 Test

1. a) composite b) prime [5.1] **2.** $2^2 \cdot 3^2 \cdot 7$ [5.1]
3. 24 [5.1] **4.** $12x^2y$ [5.1] **5.** $3x^2(2x + 1)$ [5.2] **6.** $(m - 9)^2$ [5.6]
7. $(m - 8n)(m + 5n)$ [5.3] **8.** $(7a + 3b)(7a - 3b)$ [5.5]
9. $7x^2y(2x - 4y + 5x^2y^2)$ [5.2] **10.** $(3y - 4)^2$ [5.6] **11.** prime [5.5]
12. $(s + 5)(r - 7)$ [5.2] **13.** $(x - y - 10)(x - y + 10)$ [5.5]
14. $3(2q - 5)(q + 7)$ [5.3] **15.** $(a - 6)(3a + 5)$ [5.4]
16. $(2a - 5b)(4a - 7b)$ [5.4] **17.** $3(3x - 2y)(3x + 2y)$ [5.5]
18. $(3b - 2)(2a - 5)$ [5.2] **19.** $5(a + 5b)^2$ [5.6]
20. $(x^2 + 25)(x + 5)(x - 5)$ [5.5] **21.** $(4a - 3b)(5a + 2b)$ [5.4]
22. $(x - 6)(x + 4)$ [5.3] **23.** $0, -\frac{2}{3}$ [5.8] **24.** $\frac{2}{3}, \frac{5}{2}$ [5.8]
25. $W = 4$ in., $L = 8$ in. [5.8]

Exercise Set 6.1

1. ⟵ ●─●─●─┼─┼─┼─┼─┼─┼ ⟶   **3.** ⟵ ●─┼─●─┼─●─┼─┼─┼─┼ ⟶   **5.** $\frac{1}{6}$
$\quad$ $-4\,-3\,-2\,-1\ \ 0\ \ 1\ \ 2\ \ 3\ \ 4$ $\qquad$ $-4\,-3\,-2\,-1\ \ 0\ \ 1\ \ 2\ \ 3\ \ 4$
7. $\frac{3}{2}$ **9.** $\frac{10}{21}$ **11.** $\frac{2}{3}$ **13.** $\frac{3}{2}$ **15.** $\frac{4}{3}$ **17.** $\frac{2}{3}$ **19.** $\frac{5}{4}$ **21.** $\frac{21}{11}$ **23.** $\frac{2}{7}$
25. $\frac{b^2}{a^2}$ **27.** $-r$ **29.** $\frac{1}{ab^2}$ **31.** $4x^3y$ **33.** $-2xy^2$ **35.** $\frac{-7}{4m^4n}$ **37.** $\frac{a^2c^2}{b^3}$
39. $\frac{16n^2}{9p}$ **41.** $-\frac{3}{5rs}$ **43.** $-2x^2y$

Exercise Set 6.2

1. $x \neq -5$ **3.** $r \neq 8$ **5.** $x \neq -4$ **7.** $a \neq \frac{4}{3}$
9. $x \neq 7$ or -3 **11.** $x \neq -3, 3$ **13.** $x \neq -2, 5$ **15.** $r \neq -\frac{3}{2}, 5$
17. $\frac{4}{9}$ **19.** $-\frac{5}{6}$ **21.** $\frac{2}{3}$ **23.** $\frac{2}{3}$ **25.** $\frac{2}{5}$ **27.** $-\frac{3}{4}$ **29.** $\frac{x + 2}{x - 3}$ **31.** $\frac{a - 2}{a + 3}$
33. $\frac{1}{x + 2}$ **35.** $\frac{x - 3}{x - 5}$ **37.** $\frac{2x - 3y}{3x - 5y}$ **39.** $\frac{2a - 3b}{3a - 2b}$ **41.** $\frac{c - d}{x + y}$
43. $\frac{y - 4}{y - 6}$ **45.** $-1(x - y)$ **47.** $-1(c - 2d)$ **49.** $-1(2s - 3r)$
51. -1 **53.** -1 **55.** $-(a + 5)$ or $-a - 5$ **57.** $-(x + 2)$ or $-x - 2$
59. $\frac{-1}{x - 5}$ **61.** a) $-\frac{1}{2}$ b) $-\frac{1}{3}$ c) Undefined

Exercise Set 6.3

1. $\frac{6}{35}$ **3.** $-\frac{45}{88}$ **5.** 14 **7.** $\frac{4}{3}$ **9.** $\frac{4}{3}$ **11.** 4
13. $-\frac{10}{9}$ **15.** $\frac{15}{7}$ **17.** $\frac{12}{5}$ **19.** -8 **21.** $\frac{48}{5}$ **23.** $-\frac{25}{4}$ **25.** 68 **27.** 39
29. $-\frac{6}{5}$ **31.** 15 **33.** 3 **35.** 27 **37.** $-\frac{7}{8}$ **39.** $-\frac{1}{2}$ **41.** 13 **43.** 45
45. $\frac{s^2}{r^3}$ **47.** x^2y^2 **49.** $\frac{1}{m^2}$ **51.** $\frac{a^2x}{b^2y^2}$ **53.** $\frac{x}{y^2}$ **55.** $\frac{1}{rs}$ **57.** $\frac{3xy^2}{2}$

59. $\dfrac{10y^2z^2}{3x^3w}$ **61.** $\dfrac{4xy^2}{9}$ **63.** $\dfrac{8c}{3b^3}$ **65.** 104 sq. yd **67.** 1452 yd

69. $\dfrac{2}{9}$ of a beaker **71.** $\dfrac{9}{4}$ ft **73.** $\dfrac{75}{8}$ cm^2 or $9\dfrac{3}{8}$cm^2 **75.** 38 ft^2

77. $\dfrac{63}{160}$ m^2 **79.** $\dfrac{169}{48}$ ft^2 or $3\dfrac{25}{48}$ ft^2 **81.** 11 cm^3 **83.** 72 yd^3 **85.** 10 in.^{2}

Exercise Set 6.4
1. $\dfrac{4a(x+y)}{3}$ **3.** $\dfrac{xy^2(r+s)}{8}$ **5.** $\dfrac{6}{35}$ **7.** $\dfrac{2}{5}$

9. $\dfrac{4(a+1)}{3(a+4)}$ **11.** $(2x-5)(3x+4)$ **13.** $-\dfrac{9}{7}$ **15.** $-\dfrac{3}{2}$ **17.** $\dfrac{-4y^2}{3x}$

19. $\dfrac{1}{(x-1)(x+4)}$ **21.** $\dfrac{x-2}{x+3}$ **23.** $\dfrac{x-1}{x+4}$ **25.** $\dfrac{x+2}{x-2}$ **27.** 1

29. $\dfrac{x+2}{x-3}$ **31.** $-(b+5)$ or $-b-5$ **33.** $-(3x-1)$ or $1-3x$

35. $\dfrac{(d-5)(a+2)}{(b-4)(d+1)}$ **37.** $\dfrac{c-d}{m+n}$ **39.** $2a-b$

Exercise Set 6.5
1. $\dfrac{7}{16}$ **3.** $-\dfrac{3}{5}$ **5.** $\dfrac{6}{5}$ **7.** $\dfrac{20}{11}$ **9.** $-\dfrac{8}{5}$ **11.** $\dfrac{4}{3}$

13. 9 **15.** -27 **17.** $-\dfrac{56}{5}$ **19.** $\dfrac{2}{9}$ **21.** $\dfrac{7}{45}$ **23.** $\dfrac{9}{5}$ **25.** -6 **27.** $\dfrac{28}{27}$

29. $-\dfrac{9}{20}$ **31.** $\dfrac{21}{2}$ **33.** $-\dfrac{5}{6}$ **35.** $\dfrac{4}{5}$ **37.** \$24 **39.** 14 lots **41.** $\dfrac{1}{6}$ of a bag

43. $\dfrac{1}{a^2b^2}$ **45.** $\dfrac{2}{3ab}$ **47.** $\dfrac{a^2}{b^3c^2d^2}$ **49.** $\dfrac{32x^6}{45}$ **51.** $\dfrac{15}{28}$ **53.** $\dfrac{4a}{9}$ **55.** $\dfrac{8}{5}$

57. $\dfrac{1}{4}$ **59.** $-\dfrac{2}{3}$ **61.** $\dfrac{1}{(h+7)^2}$ **63.** $\dfrac{d-4}{2d(3d+4)}$ **65.** $\dfrac{(m-5)(m+6)}{6}$

67. $\dfrac{x-8}{x+8}$ **69.** $\dfrac{a-5}{a-6}$ **71.** $\dfrac{x-2}{x+3}$ **73.** $\dfrac{b-4}{c-3}$

Exercise Set 6.6
1. $x-5$ **3.** $x+3$ **5.** $x+2$ **7.** $2x+5$

9. $2x-5$ **11.** $x+5+\dfrac{1}{x-3}$ **13.** $3x+1+\dfrac{9}{4x-3}$

15. $3x-4+\dfrac{12}{2x+7}$ **17.** $2x^2-6x+3$ **19.** $2x^2+4x-6$

21. $3x^2-2x+1+\dfrac{4}{2x+5}$ **23.** $3x^2+15x+2$

25. $2x^2+2x+3+\dfrac{27}{2x-3}$ **27.** $4x^2-3x+2+\dfrac{4}{4x+3}$

29. x^2+2x+4 **31.** x^3+2x^2+4x+8 **33.** $x^2-3x+9+\dfrac{-18}{x+3}$

35. 18 **37.** -5

Chapter 6 Review Exercises
1. **2.**

3. $\dfrac{6}{7}$ **4.** $\dfrac{8}{9}$ **5.** $\dfrac{30}{31}$ **6.** $\dfrac{7}{8}$ **7.** $\dfrac{x}{y^2}$ **8.** $\dfrac{1}{m^2n}$ **9.** r^2s^2 **10.** $4xy^2$

11. $-\dfrac{14p^3}{9q^2}$ **12.** $-\dfrac{1}{3m^2n^3}$ **13.** $x\neq 9$ **14.** $x\neq\dfrac{-5}{2}$

15. $x\neq -7, -2$ **16.** $a\neq -6, 2$ **17.** $\dfrac{7}{5}$ **18.** $\dfrac{4}{7}$ **19.** $\dfrac{a-3}{a+2}$

20. $\dfrac{1}{3x+4}$ **21.** $\dfrac{5x-3}{2x-1}$ **22.** $\dfrac{1}{4c+7d}$ **23.** -1 **24.** $\dfrac{-1}{2x+5}$

25. $\dfrac{3}{2}$ **26.** $-\dfrac{9}{4}$ **27.** $-\dfrac{7}{12}$ **28.** 16 **29.** 76 **30.** $\dfrac{3}{2}$ **31.** \$44.20 **33.** $\dfrac{7}{2}$ ft

33. 3 sq ft **34.** $\dfrac{42}{5}$ in.^{2} **35.** $\dfrac{x^3}{y}$ **36.** $\dfrac{10mn^2p^2}{3q^3}$ **37.** $\dfrac{9n^2}{8m^2(m-n)}$

38. $\dfrac{7}{8}$ **39.** $\dfrac{3(5x+4)}{4}$ **40.** $-\dfrac{5}{6}$ **41.** $\dfrac{x-4}{x+5}$ **42.** $\dfrac{4x-5}{3x+2}$ **43.** $-y-2$

44. 1 **45.** $-\dfrac{8}{5}$ **46.** -256 **47.** $\dfrac{4}{57}$ **48.** $-\dfrac{42}{5}$ **49.** $-\dfrac{32}{9}$ **50.** \$360

51. 32.5 mpg **52.** $\dfrac{14b^3y^2}{9ax^2}$ **53.** $\dfrac{40}{27}$ **54.** $\dfrac{3}{4}$ **55.** $-\dfrac{3}{2}$ **56.** $\dfrac{1}{3z(z-4)}$

57. $\dfrac{(2p+3)(6p-7)}{20}$ **58.** $\dfrac{5p+3q}{2p+5q}$ **59.** $4x+5$

60. $3x-7+\dfrac{-2}{5x+2}$ **61.** $2a^2-3a+1+\dfrac{3}{3a+4}$

62. $4z-1+\dfrac{9}{4z-3}$ **63.** $4x^2+5x-6$ **64.** $3x^2+9x+3+\dfrac{2}{x-3}$

Chapter 6 Test
1. [6.1]

2. $a\neq 4$ [6.2] **3.** $b\neq -2$ or 4 [6.2] **4.** $\dfrac{6}{7}$ [6.1] **5.** $-\dfrac{x^3}{y^4}$ [6.1]

6. $\dfrac{21a^2}{8b^2}$ [6.1] **7.** $\dfrac{4}{3}$ [6.2] **8.** $\dfrac{3x-2}{2x-3}$ [6.2] **9.** $10\dfrac{1}{2}$ gal [6.3]

10. \$4.50 [6.5] **11.** $\dfrac{21}{20}$ [6.3] **12.** $\dfrac{25}{2}$ [6.3] **13.** $\dfrac{2}{21}$ [6.5] **14.** $\dfrac{b}{a}$ [6.3]

15. $\dfrac{15b^2y^2}{4a^3x^3}$ [6.3] **16.** 1 [6.5] **17.** $\dfrac{-13}{21}$ [6.4] **18.** $\dfrac{a+2}{a-4}$ [6.5]

19. $\dfrac{y}{(2y-3)}$ [6.4] **20.** $4x+3+\dfrac{3}{3x-5}$ [6.6]

Exercise Set 7.1
1. $\dfrac{5}{7}$ **3.** $\dfrac{2}{3}$ **5.** $\dfrac{9}{16}$ **7.** $\dfrac{1}{4}$ **9.** $\dfrac{15}{17}$ **11.** $-\dfrac{2}{11}$

13. 1 **15.** $\dfrac{3}{5}$ **17.** $\dfrac{1}{3}$ **19.** $\dfrac{1}{3}$ **21.** $\dfrac{89}{7}$ or $12\dfrac{5}{7}$ **23.** $\dfrac{16}{3}$ or $5\dfrac{1}{3}$

25. $\dfrac{61}{9}$ or $6\dfrac{7}{9}$ **27.** $\dfrac{10}{b}$ **29.** $\dfrac{u-5}{v}$ **31.** $\dfrac{5m}{7u}$ **33.** $\dfrac{5}{2y}$ **35.** $-\dfrac{x}{7y^2}$ **37.** $\dfrac{4}{x}$

39. $\dfrac{b}{3}$ **41.** $a+2$ **43.** $\dfrac{9}{n+5}$ **45.** $\dfrac{2}{t-8}$ **47.** $\dfrac{3c}{c+2}$ **49.** $\dfrac{6a}{a+2}$

51. 2 **53.** 5 **55.** $\dfrac{x+2}{x-4}$ **57.** $\dfrac{t-1}{t+7}$ **59.** $\dfrac{u+5}{u+3}$ **61.** $\dfrac{x-1}{x-6}$

63. $\dfrac{x-3}{x+5}$ **65.** $-\dfrac{1}{7}$ **67.** $-\dfrac{1}{17}$ **69.** $\dfrac{8}{4-t}$ or $\dfrac{-8}{t-4}$ **71.** $\dfrac{-4v}{v-u}$ or $\dfrac{4v}{u-v}$

73. $\dfrac{3m-n-3}{m-n}$ or $\dfrac{-3m+n+3}{n-m}$ **75.** $\dfrac{-a-5b}{3a-b}$ or $\dfrac{a+5b}{b-3a}$

77. $10\dfrac{1}{4}$ ft **79.** 22 in. **81.** $\dfrac{23}{24a}$ ft **83.** $\dfrac{7a+7b}{2a+3b}$ dm **85.** $\dfrac{x-4}{x-5}$ cm

87. $\dfrac{20}{x}$ m **89.** $\dfrac{6a+2b}{2a-3b}$ ft **91.** $\dfrac{4x+6}{2x-1}$ m **93.** $\dfrac{1}{x}$ **95.** $\dfrac{t-3}{t-1}$

Exercise Set 7.2
1. 15 **3.** 99 **5.** 150 **7.** 180 **9.** 770 **11.** 252
13. x^3y^3 **15.** u^6v^4 **17.** $a^3b^6c^4$ **19.** $x^4y^6z^3$ **21.** $9u^3v^5$ **23.** $72y^5z^4$
25. $45m^5np^9$ **27.** $60r^4s^8t^3$ **29.** $108u^4v^6$ **31.** $84r^4s^9t^6$ **33.** $15(a+5)$
35. $5r(r-7)$ **37.** $(v-5)(v+3)$ **39.** $(x-4)(x+3)(x+5)$
41. $(y+4)(y+1)$ **43.** $w(w+3)(w-6)$ **45.** $(z+1)^2(z-5)$
47. $(t-3)(t+3)(t-5)(t-4)$ **49.** 132 **51.** $6(y+5)(y-5)^2$
53. 21 **55.** 35 **57.** 5 **59.** 60 **61.** 243 **63.** cf **65.** 24 **67.** $35tz^2$
69. $99rs^3$ **71.** $63pq^2$ **73.** $108y^2z^5$ **75.** $27tx^3y^2$ **77.** $5y$ **79.** $9t$
81. $16c$ **83.** $11r+33$ **85.** $5v^2+10v$ **87.** y^2+2y-8
89. w^2+2w-3 **91.** $n^2-12n+27$ **93.** $(x+2)t$ or $xt+2t$

Exercise Set 7.3
1. $14, \dfrac{7}{14}, \dfrac{2}{14}$ **3.** $15, \dfrac{10}{15}, \dfrac{9}{15}$ **5.** $30, \dfrac{25}{30}, \dfrac{24}{30}$

7. $15, \dfrac{6}{15}, \dfrac{7}{15}$ **9.** $72, \dfrac{27}{72}, \dfrac{32}{72}$ **11.** $30, \dfrac{3}{30}, \dfrac{2}{30}$ **13.** $120, \dfrac{33}{120}, \dfrac{52}{120}$

15. $336, \dfrac{189}{336}, \dfrac{176}{336}$ **17.** $120, \dfrac{20}{120}, \dfrac{45}{120}, \dfrac{108}{120}$ **19.** $72, \dfrac{66}{72}, \dfrac{20}{72}, \dfrac{51}{72}$

21. $mn, \dfrac{2m}{mn}, \dfrac{3n}{mn}$ **23.** $bd, \dfrac{ad}{bd}, \dfrac{bc}{bd}$ **25.** $abc, \dfrac{6c}{abc}, \dfrac{5a}{abc}$

27. $mstu, \dfrac{mr}{mstu}, \dfrac{ns}{mstu}$ **29.** $swv^2, \dfrac{sx}{swv^2}, \dfrac{vwy}{swv^2}$ **31.** $a^3b^5, \dfrac{2at}{a^3b^5}, \dfrac{3b^2z}{a^3b^5}$

33. $x^2y^5z^4, \dfrac{5ay^4}{x^2y^5z^4}, \dfrac{7dxz}{x^2y^5z^4}$ **35.** $24u^2v^4, \dfrac{28}{24u^2v^4}, \dfrac{9uv}{24u^2v^4}$

37. $60p^6q^8, \dfrac{16}{60p^6q^8}, \dfrac{27p^2q^3}{60p^6q^8}$ **39.** $14x^2y^2, \dfrac{2ry}{14x^2y^2}, \dfrac{tx}{14x^2y^2}$

41. $50w^6z^7, \dfrac{15a}{50w^6z^7}, \dfrac{12bw^4z}{50w^6z^7}$

43. $(u+8)(u-10), \dfrac{3u-30}{(u+8)(u-10)}, \dfrac{9u+72}{(u+8)(u-10)}$

45. $(x+7)(x-5), \dfrac{x^2-5x}{(x+7)(x-5)}, \dfrac{x^2+7x}{(x+7)(x-5)}$

47. $12(t-3), \dfrac{4t}{12(t-3)}, \dfrac{15t}{12(t-3)}$

49. $45(y+2), \dfrac{30y}{45(y+2)}, \dfrac{24y}{45(y+2)}$

51. $(a+1)(a-1)(a+4), \dfrac{3a^2+12a}{(a+1)(a-1)(a+4)},$
$\dfrac{a^2+a}{(a+1)(a-1)(a+4)}$

53. $(v+1)(v-3)(v-2), \dfrac{v^2-7v+10}{(v+1)(v-3)(v-2)},$
$\dfrac{v^2+4v+3}{(v+1)(v-3)(v-2)}$

55. $(m+4)^2(m+1), \dfrac{m^2+2m+1}{(m+4)^2(m+1)}, \dfrac{m^2-16}{(m+4)^2(m+1)}$

57. $x^2(x+3)(x-2)(x+4), \dfrac{x^3+9x^2+20x}{x^2(x+3)(x-2)(x+4)},$
$\dfrac{x^2-5x+6}{x^2(x+3)(x-2)(x+4)}$

Exercise Set 7.4   **1.** $\dfrac{19}{21}$ **3.** $\dfrac{7}{10}$ **5.** $\dfrac{1}{2}$ **7.** $-\dfrac{5}{24}$ **9.** $\dfrac{23}{9}$

11. $-\dfrac{11}{4}$ **13.** $\dfrac{13}{12}$ **15.** $-\dfrac{1}{36}$ **17.** $-\dfrac{5}{36}$ **19.** $\dfrac{11}{36}$ **21.** $\dfrac{53}{8}$ or $6\dfrac{5}{8}$

23. $\dfrac{283}{24}$ or $11\dfrac{19}{24}$ **25.** $\dfrac{205}{36}$ or $5\dfrac{25}{36}$ **27.** $\dfrac{49}{16}$ or $3\dfrac{1}{16}$ **29.** $-\dfrac{19}{20}$

31. $\dfrac{a^2+2b}{ab}$ **33.** $\dfrac{1}{24u}$ **35.** $\dfrac{31z}{20x}$ **37.** $\dfrac{m+38}{6m}$

39. $\dfrac{45p^2q-10pq+9p+3}{30p^3q^2}$ **41.** $\dfrac{-2k+8}{k(k+2)}$ **43.** $\dfrac{9w+13}{(w-3)(w+7)}$

45. $\dfrac{3a^2+8a+12}{(a-3)(a+4)}$ **47.** $\dfrac{2t^2+11t+8}{(t+4)(t+6)}$ **49.** $\dfrac{3a^2-5a-13}{(2a+3)(a+4)}$

51. $\dfrac{x-13}{6(x+3)}$ **53.** $\dfrac{8t-24}{(t-4)^2}$ **55.** $\dfrac{1}{a-1}$ **57.** $\dfrac{3}{3x-2}$

59. $\dfrac{2x^2-3x+19}{(x-3)(x+4)(x+2)}$ **61.** $\dfrac{2z^2-z+9}{(z+3)^2(z-2)}$

63. $\dfrac{5}{(x+3)(x-2)}$ **65.** $\dfrac{x-2}{(x+4)(x+2)}$ **67.** $\dfrac{3x-1}{x(x-2)}$

69. $\dfrac{3x-5}{(x-3)(x+2)}$ **71.** 17 mi   **73.** \$7.10   **75.** $\dfrac{3a+1}{a^2}$ ft

77. $\dfrac{2yz+3xz+4xy}{xyz}$ in.   **79.** $\dfrac{3a+3b+10c+10d}{(a+b)(c+d)}$ ft   **81.** $\dfrac{8x+6y}{xy}$ ft

83. $\dfrac{4x^2+2y^2}{(x+y)(2x-y)}$ cm   **85.** $\dfrac{4a^2-2a+44}{(a+4)(a-6)(a+2)}$ in.

Exercise Set 7.5   **1.** $\dfrac{4}{5}$ **3.** $\dfrac{14}{9}$ **5.** $\dfrac{9}{2}$ **7.** 2   **9.** $\dfrac{1}{16}$ **11.** $\dfrac{3}{35}$

13. 4   **15.** $\dfrac{4}{11}$ **17.** $\dfrac{2}{13}$ **19.** $\dfrac{230}{71}$ **21.** $-\dfrac{5}{3}$ **23.** $\dfrac{187}{30}$ **25.** $\dfrac{88}{15}$ **27.** $\dfrac{3m}{2n}$

29. $\dfrac{5f}{3d}$ **31.** $\dfrac{tx}{ry}$ **33.** $\dfrac{g}{4h}$ **35.** $\dfrac{ac}{b}$ **37.** $\dfrac{a}{xb}$ **39.** $\dfrac{uv}{w}$ **41.** $\dfrac{1}{b}$ **43.** $\dfrac{u^4}{v^2w^4}$

45. $\dfrac{4}{2z-z^2}$ **47.** $\dfrac{2+x}{2-x}$ **49.** $\dfrac{t}{1+t}$ **51.** $\dfrac{v+2}{v-3}$ **53.** $\dfrac{1}{u+6}$

55. $\dfrac{x-6}{x(x+3)}$ **57.** $\dfrac{5r+44}{(r+8)^2}$ **59.** $\dfrac{t-4}{2}$ **61.** 2   **63.** $3-p$

65. $\dfrac{x-1}{x}$ **67.** $\dfrac{2a^2+a+6}{9a^3+54a^2-3a}$

Exercise Set 7.6   **1.** $x=2$ **3.** $t=14$ **5.** $u=6$

7. $m=-\dfrac{15}{2}$ **9.** $x=2$ **11.** $t=5$ **13.** $x=2$ **15.** $y=\dfrac{1}{3}$

17. $x=-2$ **19.** $v=1$ **21.** $q=1$ **23.** $m=24$ **25.** $r=\dfrac{2}{3}$

27. $u=1$ **29.** $w=-2$ **31.** $x=4$ **33.** $t=0$ **35.** $a=11$

37. $c=2$ or -5 **39.** $y=\dfrac{1}{3}$ or $y=-2$ **41.** $m=0$ **43.** $y=-3$

45. $a=2\,(-2$ is extraneous$)$ **47.** $x=3$, $(2$ is extraneous$)$ **49.** $t=1$

51. $w=-\dfrac{35}{4}$ **53.** $x=2\,(0$ is extraneous$)$ **55.** $\varnothing$, $(-1$ and 2 are

extraneous$)$ **57.** $x=\dfrac{9}{4}$

Exercise Set 7.7   **1.** 5   **3.** $\dfrac{10}{7}$ **5.** 9, 12   **7.** 6, 12   **9.** 4, 6

11. \$18,000   **13.** $\dfrac{12}{5}$ hr $=2\dfrac{2}{5}$ hr $=2$ hr 24 min   **15.** $\dfrac{175}{12}=14\dfrac{7}{12}$ days

17. 12 hr   **19.** $22\dfrac{1}{2}$ min   **21.** 3 P.M.   **23.** 75 mph   **25.** 60 mph

27. 441 mph   **29.** $\dfrac{8}{16}$ **31.** catcher $=51$, pitcher $=34$

33. $\dfrac{20}{3}$ hr $=6\dfrac{2}{3}$ hr $=6$ hr 40 min   **35.** 3   **37.** 10 mi

Chapter 7 Review Exercises   **1.** $\dfrac{2}{7}$ **2.** 1   **3.** $\dfrac{12}{13}$ **4.** $\dfrac{7}{11}$

5. $15\dfrac{4}{5}$ **6.** $\dfrac{79}{12}$ or $6\dfrac{7}{12}$ **7.** 7   **8.** $\dfrac{178}{17}$ or $10\dfrac{8}{17}$ **9.** $\dfrac{9r+2p}{14x}$ **10.** $\dfrac{1}{y+2}$

11. $\dfrac{x-7}{x+5}$ **12.** $\dfrac{x-4}{x-6}$ **13.** $\dfrac{3p^2+3p-6}{p^2-16}$ **14.** $\dfrac{6x-5y}{x-y}$ **15.** 36

16. \$4.44   **17.** $13\dfrac{8}{9}$ m   **18.** $1\dfrac{1}{4}$ km   **19.** 42   **20.** 200   **21.** 96   **22.** 108
23. a^2y^4 **24.** $60t^5v^7$ **25.** $42x^5w^4$ **26.** $(z-4)(z+3)$ **27.** $(a+2)^2$
28. $(y+4)^2(y-3)$ **29.** 24   **30.** 45   **31.** $20p^2r$ **32.** $32t^2x^3z^3$
33. $8a$ **34.** $7b+21$ **35.** $p^2-11p+24$ **36.** $2qy+6q$

37. $14, \dfrac{8}{14}, \dfrac{35}{14}$ **38.** $18, \dfrac{15}{18}, \dfrac{7}{18}$ **39.** $36, \dfrac{32}{36}, \dfrac{33}{36}$ **40.** $60, \dfrac{24}{60}, \dfrac{36}{60}, \dfrac{39}{60}$

41. $t^2v^5, \dfrac{4}{t^2v^5}, \dfrac{6tv^2}{t^2v^5}$ **42.** $36p^3q^5, \dfrac{32bq}{36p^3q^5}, \dfrac{33cp}{36p^3q^5}$

43. $(t-4)(t+2), \dfrac{7t+14}{(t-4)(t+2)}, \dfrac{9t-36}{(t-4)(t+2)}$

44. $(y-11)(y-3), \dfrac{y^2-3y}{(y-11)(y-3)}, \dfrac{y^2-11y}{(y-11)(y-3)}$

45. $28(2u+3), \dfrac{28u}{28(2u+3)}, \dfrac{36u}{28(2u+3)}$

46. $(w+4)(w-4)(w+1),$
$\dfrac{2w^2+2w}{(w+4)(w-4)(w+1)}, \dfrac{6w^2-24w}{(w+4)(w-4)(w+1)}$

47. $(a+3)^2(a-5), \dfrac{3a^2-15a}{(a+3)^2(a-5)}, \dfrac{10a^2+30a}{(a+3)^2(a-5)}$

48. $(m-4)(m+2)(m+1),$
$\dfrac{m^2+2m+1}{(m-4)(m+2)(m+1)}, \dfrac{m^2+m-2}{(m-4)(m+2)(m+1)}$ **49.** $\dfrac{41}{48}$

50. $\dfrac{7}{20}$ **51.** $-\dfrac{3}{5}$ **52.** $\dfrac{31}{42}$ **53.** $\dfrac{5}{2}$ **54.** $\dfrac{33}{8}$ **55.** $\dfrac{7n+bm}{mn}$ **56.** $\dfrac{4}{9x}$

57. $\dfrac{9y-7}{20y}$ **58.** $\dfrac{4tu^3-6u^3+15t^2+3t}{36t^4u^5}$ **59.** $\dfrac{11w-9}{w(w-3)}$

60. $\dfrac{6r-2}{r-5}$ **61.** $\dfrac{-v^2+v-4}{4(v+2)(v-3)}$ **62.** $\dfrac{2}{4x-5}$ **63.** $\dfrac{x+1}{x(x-3)}$

64. $\dfrac{5u+8}{5(u-3)}$ **65.** $39\dfrac{3}{4}$ in.   **66.** \$1.00   **67.** $\dfrac{8}{3}$ **68.** $\dfrac{25}{33}$ **69.** $\dfrac{1}{18}$

70. $\dfrac{13}{2}$ **71.** $\dfrac{8}{9}$ **72.** $\dfrac{12}{7}$ **73.** $\dfrac{3x}{y}$ **74.** $\dfrac{ps}{r}$ **75.** $\dfrac{t}{5}$ **76.** $\dfrac{u^3w}{v}$ **77.** $\dfrac{x^4y}{t^3}$

78. $\dfrac{7w}{24-w}$ **79.** $\dfrac{y+3}{y-2}$ **80.** $x+5$ **81.** $y=3$ **82.** $p=2$

83. $t=\dfrac{12}{55}$ **84.** $w=\dfrac{22}{23}$ **85.** $v=\dfrac{17}{6}$ **86.** $z=\dfrac{-17}{14}$ **87.** $t=\dfrac{7}{4}$

88. $m=3$ **89.** $q=-2$ or 1   **90.** $x=4$ $(-2$ is extraneous$)$ **91.** $\dfrac{16}{19}$
92. 5 or -2 **93.** 18 days   **94.** 23 mph   **95.** 9:06 A.M.

Chapter 7 Test   **1.** 48 [7.2]   **2.** $45x^2y^4$ [7.2]   **3.** $15tu^2w$ [7.2]

4. $6p(q-4)$ [7.2]   **5.** $54, \dfrac{12}{54}, \dfrac{21}{54}, \dfrac{8}{54}$ [7.3]   **6.** $(a+3)(a-6),$

$\dfrac{7a}{(a+3)(a-6)}$, $\dfrac{b(a-6)}{(a+3)(a-6)}$ [7.3] **7.** $\dfrac{5}{7}$ [7.1] **8.** $\dfrac{r}{r^2+16}$ [7.1]

9. $\dfrac{1}{4}$ [7.4] **10.** $\dfrac{11}{30}$ [7.4] **11.** $\dfrac{37}{12}$ or $3\dfrac{1}{12}$ [7.4] **12.** $\dfrac{3v^2+26v+6}{9v^2}$ [7.4]

13. $\dfrac{46w-3}{15(w-4)}$ [7.4] **14.** $\dfrac{x-2}{(x+4)(x+2)}$ [7.4] **15.** $10\dfrac{7}{10}$ in. [7.4]

16. $\dfrac{24a-13}{(2a+3)(a-2)}$ m [7.4] **17.** $\dfrac{4x+12y}{xy}$ ft [7.4]

18. $\dfrac{16m^2+42mn-6n^2}{(3m-n)(2m+3n)}$ [7.4] **19.** \$4.67 [7.4] **20.** $\dfrac{3}{10}$ [7.5] **21.** 4 [7.5]

22. $\dfrac{4m}{3}$ [7.5] **23.** $\dfrac{12-u}{8u}$ [7.5] **24.** $v=3$ [7.6] **25.** $x=6$ [7.6]

26. $t=-7$ [7.6] **27.** $n=2$ (-1 is extraneous) [7.6]
28. 0 and 4 [7.7] **29.** 30 days [7.7] **30.** 2 hr [7.7]

Exercise Set 8.1
1. 2 to 3 **3.** $3:5$ **5.** 5 to 2 **7.** $3:1$ **9.** $\dfrac{3}{7}$

11. $\dfrac{4}{3}$ **13.** $\dfrac{2}{5}$ **15.** $\dfrac{13}{7}$ **17.** $100:49$ **19.** $\dfrac{10}{13}$ **21.** $\dfrac{4}{1}$ **23.** $\dfrac{4}{1},\dfrac{1}{4}$ **25.** 280 mg

sodium : 1 oz mix **27.** $\dfrac{101}{500}$ **29.** no **31.** 50 mi per hr **33.** 6 cans per
student **35.** \$.04 **37.** \$.38 **39.** \$.37 **41.** \$62.50 **43.** 3.2¢ per oz.
45. 2.8¢ per oz. **47.** 1.0¢ per oz. **49.** $10:4:2:3:2:4$

Exercise Set 8.2
1. yes **3.** yes **5.** yes **7.** yes **9.** $x=15$

11. $t=12$ **13.** $y=\dfrac{32}{7}$ **15.** $t=91$ **17.** $p=28$ **19.** $w=.6$

21. $y=3$ **23.** $x=3$ **25.** $t=.96$ **27.** $w=20$ **29.** $x=4$
31. $x=9$ **33.** $x=5$ **35.** $x=6$ **37.** $x=7$ **39.** $x=4,-7$

41. $x=-\dfrac{5}{2},6$ **43.** $W=Pt$ **45.** $P=\dfrac{KT}{V}$ **47.** 280 mi **49.** 7.5 novels

51. 77 innings **53.** $7\dfrac{1}{2}$ in. **55.** 400 trout **57.** 240 in. = 20 ft **59.** \$248
61. 65 min **63.** 154 lbs, \$98.50

Exercise Set 8.3
1. $\dfrac{4}{25}$ **3.** $\dfrac{9}{10,000}$ **5.** $\dfrac{17}{125}$ **7.** $\dfrac{13}{4}$ **9.** $\dfrac{4}{75}$

11. $\dfrac{17}{140}$ **13.** .42 **15.** .006 **17.** .375 **19.** 1.20 **21.** .0808 **23.** .128

25. 20% **27.** 38% **29.** .8% **31.** 321% **33.** $55\dfrac{5}{9}$% **35.** 500%

37. 1000% **39.** 40% **41.** 45% **43.** 76% **45.** 180% **47.** 37.5%

49. $133\dfrac{1}{3}$% or $133.\overline{3}$% **51.** $122\dfrac{2}{9}$% or $122.\overline{2}$% **53.** $53\dfrac{11}{13}$% or 53.8%

55. 39% **57.** 20.9% **59.** 14.23% **61.** 6.4% **63.** 9.5% **65.** 27%

67. 17.65% **69.** $\dfrac{103}{1100}$

Exercise Set 8.4
1. What number is 25% of 36? **3.** 40% of 60 is
what number? **5.** 20% of what number is 8? **7.** 52 is 65% of what
number? **9.** What percent of 30 is 21? **11.** 4.8 is what percent of 80?
13. 10 **15.** 28 **17.** 80 **19.** 600 **21.** 30% **23.** 12.5% **25.** 5
27. 45 **29.** 88 **31.** 1.75 **33.** 72 **35.** 80 **37.** 4% **39.** 23.5% **41.** 22
43. 180 **45.** 81% **47.** 945 accidents **49.** 105 ml **51.** 40%
53. \$85,000 **55.** 190 lbs **57.** 40 **59.** 100 ml

Exercise Set 8.5
1. \$6.30 **3.** 4.5% **5.** \$340 **7.** \$591.25 **9.** \$27
11. 20% **13.** \$150, 30% **15.** \$45 **17.** \$125 **19.** 10% **21.** \$48
23. \$3466.67 **25.** $t=3$ **27.** $r=8\%$ **29.** \$600 **31.** \$680
33. \$28.33 **35.** \$384 **37.** \$192 **39.** \$3785.60 **41.** 8% **43.** $8\dfrac{1}{3}$% or
8.3% to the nearest tenth **45.** 36% **47.** \$6831.14 **49.** 41.2
51. \$9043.63

Chapter 8 Review Exercises
1. $3:5$ **2.** 7 to 6 **3.** $\dfrac{9}{5}$

4. $\dfrac{3}{4}$ **5.** $25:14$ **6.** $2:1:1$ **7.** $4:2:1$ **8.** $\dfrac{227}{500}$ **9.** 4 pieces **10.** 20 lbs

11. \$.08 **12.** 4¢ or \$0.04 **13.** Yes **14.** No **15.** $x=2.5$
16. $y=96$ **17.** $t=10$ **18.** $u=0$ **19.** $y=5.6$ **20.** $x=2.2$

21. $v=2$ **22.** $v=\dfrac{14}{3}$ **23.** $t=5$ **24.** $w=3$ **25.** $x=7$
26. $x=3.5$ **27.** $y=-3$ **28.** $t=10$ **29.** $x=-4,7$ **30.** $x=\dfrac{5}{2},-3$

31. 3 acres **32.** 10,080 **33.** 540 mi **34.** 1890 **35.** 25% **36.** 87.5%

37. 250% **38.** 400% **39.** 42.9% **40.** 83.3% **41.** $\dfrac{2}{5}$ **42.** $\dfrac{109}{100}$

43. $\dfrac{41}{500}$ **44.** $\dfrac{1}{3}$ **45.** 2.5% **46.** 657% **47.** 30% **48.** 4.61% **49.** 60.5%

50. $50\dfrac{1}{3}$% **51.** $\dfrac{2}{25}$ **52.** $\dfrac{1}{8}$ **53.** $\dfrac{37}{400}$ **54.** $\dfrac{1}{2500}$ **55.** .18 **56.** .0675

57. .026 **58.** .00005 **59.** 3.27 **60.** .097 **61.** 35 **62.** 80% **63.** 150%
64. 40 **65.** 64% **66.** 400 **67.** 11.88% or 12% **68.** 840 **69.** 6.5%
70. \$13,250 **71.** \$1150 **72.** \$830 **73.** 12% **74.** \$4950 **75.** \$3000
76. 8% **77.** \$8.40 **78.** \$2679.65 **79.** \$185 billion

Chapter 8 Test
1. $4:3:5$ [8.1] **2.** faculty to students,
$1:100$ [8.1] **3.** $8:1:5$ [8.1] **4.** \$800,000 : 1 [8.1] **5.** \$.05 [8.1]
6. $y=10$ [8.2] **7.** $x=4$ [8.2] **8.** $x=-4,-6$ [8.2] **9.** 175 min [8.2]
10. 54 hogs [8.2] **11.** 160% [8.3] **12.** 57.1% [8.3] **13.** 30.8% [8.3]

14. 766.7% [8.3] **15.** $\dfrac{3}{8}$ [8.3] **16.** $\dfrac{37}{400}$ [8.3] **17.** 950 [8.4] **18.** 4375 [8.4]

19. $33\dfrac{1}{3}$% [8.4] **20.** \$51 [8.4] **21.** 6% [8.5] **22.** \$297.50 [8.5]
23. \$35,000 [8.5] **24.** \$72,250 [8.5] **25.** \$12,730.80 [8.5] **26.** \$19,795 [8.5]

Exercise Set 9.1
1. yes **3.** no **5.** yes **7.** no **9.** a) Consistent
b) 1 c) $(3,-4)$ **11.** a) Inconsistent b) 0 **13.** a) Dependent
b) Infinite **15.** $(0,1)$ **17.** $(-2,3)$ **19.** $(-3,-2)$ **21.** $(5,1)$
23. $(3,-4)$ **25.** $(3,3)$ **27.** $(-2,-4)$ **29.** $(-3,4)$ **31.** $(3,3)$ **33.** $(0,0)$
35. Inconsistent **37.** Dependent **39.** Inconsistent **41.** Consistent
$(1,-6)$ **43.** yes

Exercise Set 9.2
1. $(2,0)$ **3.** $(1,2)$ **5.** $(3,-4)$ **7.** $(-3,-2)$
9. $(-1,4)$ **11.** $(3,0)$ **13.** $(4,6)$ **15.** $(-1,2)$ **17.** $(4,-3)$
19. $(-4,3)$ **21.** $(4,-4)$ **23.** $(2,5)$ **25.** $(3,-1)$ **27.** $(2,-2)$
29. $(-3,-2)$ **31.** $\left(\dfrac{3}{4},-5\right)$ **33.** $\left(3,-\dfrac{1}{2}\right)$ **35.** $\left(\dfrac{2}{3},-\dfrac{2}{3}\right)$ **37.** $\left(\dfrac{37}{41},-\dfrac{7}{41}\right)$
39. $(3,-4)$ **41.** $(4,-5)$ **43.** $(1,-1)$ **45.** $(0,4)$ **47.** $(9,2)$
49. $(3,-1)$ **51.** Dependent **53.** Consistent, $(6,4)$ **55.** Inconsistent
57. $(1,-1)$ **59.** $\left(2\dfrac{1}{3},-\dfrac{1}{2}\right)$

Exercise Set 9.3
1. $(2,-1)$ **3.** $(1,5)$ **5.** $(2,-3)$
7. $(1,-3)$ **9.** $(-2,4)$ **11.** $(-4,-1)$ **13.** $(3,4)$ **15.** $(3,3)$
17. $(-5,2)$ **19.** $(-3,-2)$ **21.** $(4,-4)$ **23.** $(-1,4)$ **25.** $\left(\dfrac{1}{3},1\right)$
27. $(2,2)$ **29.** $(-3,-2)$ **31.** Inconsistent **33.** $(7,-9)$
35. Dependent **37.** Inconsistent **39.** $(4,-1)$ **41.** $(6,8)$ **43.** $(4,0)$

Exercise Set 9.4
1. U.S. = 332, Russia = 106 **3.** TV = 45.9%,
newspapers = 17.7% **5.** Burj Khalifa = 828 m, KVLY-TV Mast = 629 m
7. Ceres = 584 mi, Pallas = 365 mi **9.** 16 at \$25, 20 at \$32
11. 80 36-in fans, 25 54-in fans **13.** 27 dimes, 25 quarters **15.** 22 at 18
cents, 14 at 25 cents **17.** Cutter = 15 mph, current = 3 mph
19. ship = 12 knots, current = 3 knots **21.** 60 mph-passenger train,
45 mph-freight train **23.** \$2000 at 6%, \$18000 at 8.5% **25.** \$150,000
at 12%, \$100,000 at 8% **27.** 8 lb peanuts, 4 lb cashews **29.** 15 L of 4%,
15 L of 12% **31.** 10, 15 **33.** 15 cars, 12 trucks **35.** 700 quarters, 200
silver dollars **37.** 40 kph in still water, 5 kph current **39.** \$1600 at
18%, \$1400 at 9% **41.** 8 cups rice, 12 cups mixture **43.** w = 6 ft,
l = 10 ft **45.** 7 cm, 5 cm **47.** 75°, 105°

Exercise Set 9.5
1.

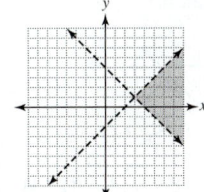

3. **5.**

7. **9.**

11. **13.**

15. **17.**

19. **21.**

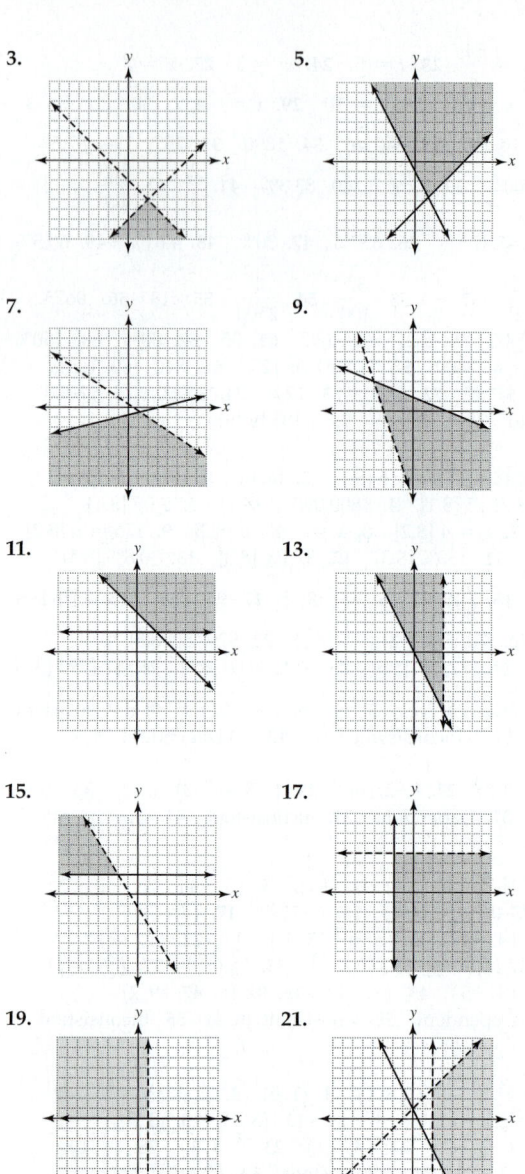

Chapter 9 Review Exercises **1.** No **2.** Yes **3.** Yes

4. No **5.** $(-1, 2)$ **6.** $(3, -2)$

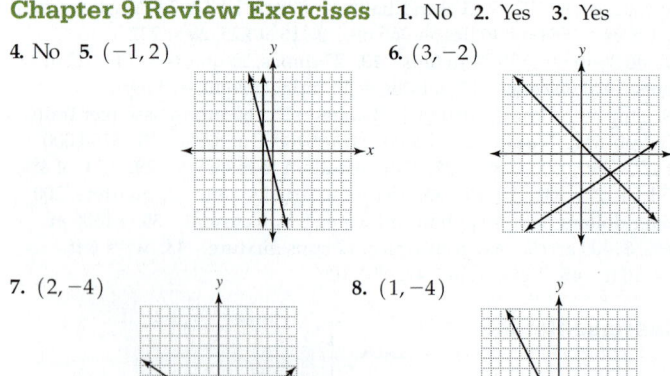

7. $(2, -4)$ **8.** $(1, -4)$

9. Dependent **10.** Consistent, $(4, -6)$ **11.** Inconsistent
12. Dependent **13.** $(13, 1)$ **14.** $(6, 14)$ **15.** $(-4, 6)$ **16.** $(-1, 1)$
17. $\left(-3, \frac{1}{4}\right)$ **18.** $\left(\frac{5}{3}, -\frac{1}{2}\right)$ **19.** $(-3, -5)$ **20.** $(-6, 1)$ **21.** Inconsistent

22. $(-4, 3)$ **23.** Dependent **24.** Inconsistent **25.** $(-2, 1)$ **26.** $(3, 5)$
27. $(4, 6)$ **28.** $(-2, 9)$ **29.** $(-5, -9)$ **30.** Inconsistent **31.** $(-3, -2)$
32. $(4, 7)$ **33.** Dependent **34.** $(-2, -5)$ **35.** $(-1, 4)$
36. Inconsistent **37.** $\left(\frac{1}{2}, 2\right)$ **38.** $\left(\frac{3}{4}, \frac{2}{5}\right)$ **39.** Men = 19.19 sec,
women = 21.34 sec **40.** Basic = \$28, executive = \$35
41. 25 nickels, 40 dimes **42.** plane = 270 mph, wind = 20 mph
43. \$10,000 at 12%, \$40,000 at 7.5% **44.** 4 quarts of each

45. **46.**

47. **48.**

49. **50.**

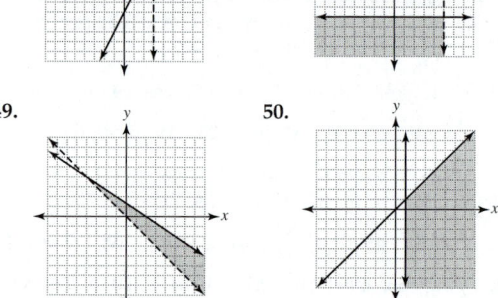

Chapter 9 Test **1.** No [9.1] **2.** $(4, -4)$ [9.1] **3.** $(-3, 2)$ [9.1]
4. Dependent [9.1] **5.** $(6, 4)$ [9.2] **6.** $(5, 7)$ [9.2] **7.** $(-3, 8)$ [9.2]
8. $(2, -4)$ [9.3] **9.** $(4, 1)$ [9.3] **10.** $(-2, 5)$ [9.3] **11.** $(-4, -7)$ [9.1–9.3]
12. $\left(\frac{4}{5}, -2\right)$ [9.1–9.3] **13.** $(3, 4)$ [9.1–9.3] **14.** $(-7, -2)$ [9.1–9.3]
15. Panama = 8065, Liberia = 2306 [9.4] **16.** 60 dimes,
30 quarters [9.4] **17.** Canoeist = 6 kph, current = 2 kph [9.4]
18. \$12,000 at 22% and \$4000 at 7% [9.4]
19. [9.5] **20.** [9.5]

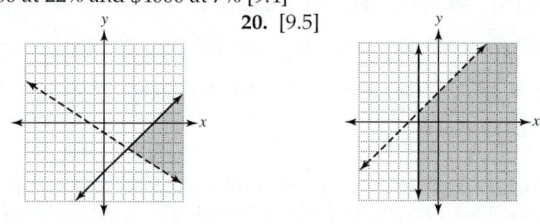

Exercise Set 10.1 **1.** 5 **3.** -5 **5.** -10 **7.** does not exist as a
real number **9.** 25 **11.** 13 **13.** 3 **15.** -3 **17.** -6 **19.** 6
21. does not exist as a real number **23.** -3 **25.** irrational, 4.899
27. rational, 12 **29.** irrational, 3.684 **31.** rational, 4 **33.** rational, 1.3
35. irrational, 28.460 **37.** rational, .5 **39.** irrational, 4.642
41. rational, 46 **43.** rational, 21 **45.** b **47.** n^2 **49.** r^4 **51.** n **53.** r^2
55. c^3 **57.** s^2 **59.** a **61.** $x - 5$ **63.** $2a + b$ **65.** $x + 5$ **67.** $2a - 3b$

Exercise Set 10.2 **1.** $2\sqrt{3}$ **3.** $7\sqrt{2}$ **5.** $8\sqrt{2}$ **7.** $8\sqrt{5}$
9. $15\sqrt{6}$ **11.** $24\sqrt{5}$ **13.** $20\sqrt{7}$ **15.** $63\sqrt{3}$ **17.** $a\sqrt{a}$ **19.** $b^6\sqrt{b}$
21. xy^2 **23.** $x^3y^4z^5$ **25.** $x^2y^4\sqrt{x}$ **27.** $m^8n^9\sqrt{mn}$ **29.** $c^2d^3\sqrt{c}$
31. $r^5s^3\sqrt{rs}$ **33.** $2x^3\sqrt{5}$ **35.** $18x^2\sqrt{2x}$ **37.** $25x^4y^7\sqrt{6x}$
39. $24a^9c^3\sqrt{11a}$ **41.** $2\sqrt[3]{4}$ **43.** $8\sqrt[3]{5}$ **45.** $18\sqrt[3]{7}$ **47.** $x^2\sqrt[3]{x}$
49. $x^2y\sqrt[3]{y^2}$ **51.** $4z^2\sqrt[3]{2z^2}$ **53.** $4\sqrt[3]{3}$ **55.** $12a^3\sqrt[3]{6a}$ **57.** 3 sec.
59. 10.53 knots **61.** 6 in.^{2} **63.** $4\sqrt{6}$ ft.^{2} **65.** $3\sqrt[4]{2}$ **67.** $ab\sqrt[4]{a^2b^3}$
69. $3\sqrt[5]{4}$ **71.** $ab^3\sqrt[5]{a^4b}$

Exercise Set 10.3

1. $\sqrt{30}$ **3.** $\sqrt{35}$ **5.** 6 **7.** 15 **9.** $\sqrt{5x}$
11. $\sqrt{15y}$ **13.** $\sqrt{xy}$ **15.** k **17.** x^2 **19.** w **21.** x^3 **23.** w **25.** $2ab$
27. $2\sqrt{5}$ **29.** $4\sqrt{3}$ **31.** $60\sqrt{3}$ **33.** $24\sqrt{10}$ **35.** $30\sqrt{14}$ **37.** a^2
39. $y^2\sqrt{y}$ **41.** $m^4n^3\sqrt{m}$ **43.** $x^4y^3\sqrt{y}$ **45.** $30m^4\sqrt{15}$ **47.** $24c^4\sqrt{15c}$
49. $\dfrac{1}{2}$ **51.** $\dfrac{7}{8}$ **53.** 2 **55.** 7 **57.** $6\sqrt{2}$ **59.** $6\sqrt{5}$ **61.** x **63.** $c^2d\sqrt{c}$
65. $3x$ **67.** $12a^2$ **69.** $6c^3\sqrt{3c}$ **71.** $36x^2y^3\sqrt{2y}$ **73.** $\dfrac{2\sqrt{6}}{7}$ **75.** $\dfrac{a^4}{2}$
77. $\dfrac{3x^5\sqrt{5}}{4}$

Exercise Set 10.4

1. $7\sqrt{7}$ **3.** $-6\sqrt{6}$ **5.** $9\sqrt{a}$ **7.** $-3\sqrt{n}$
9. $3x\sqrt[3]{9}$ **11.** $18x^2\sqrt[4]{5x}$ **13.** $5\sqrt{3}$ **15.** $-\sqrt{3}$ **17.** $-\sqrt{5y}$
19. $6n\sqrt{6}$ **21.** $9y^2\sqrt{2y}$ **23.** $15\sqrt{5}$ **25.** $-8\sqrt{5}$ **27.** $6\sqrt{6}$
29. $14a\sqrt{3a}$ **31.** $4\sqrt{6}$ **33.** $18\sqrt{2}-17\sqrt{3}$ **35.** $28x\sqrt{2y}$ **37.** $7\sqrt{5}$
39. $-15\sqrt{6}$ **41.** $7\sqrt{2}$ **43.** $-14\sqrt{5}$ **45.** $3\sqrt{2}+2$ **47.** $3-3\sqrt{5}$
49. $\sqrt{15}+10\sqrt{3}$ **51.** $24-48\sqrt{2}$ **53.** $12-3\sqrt{2}+4\sqrt{5}-\sqrt{10}$
55. $6+5\sqrt{x}+x$ **57.** $68-24\sqrt{6}$ **59.** $6+10\sqrt{2}+9\sqrt{3}+15\sqrt{6}$
61. $\sqrt{6}+\sqrt{10}+3+\sqrt{15}$ **63.** $x-\sqrt{xy}-2y$ **65.** $-13-2\sqrt{6}$
67. $-70+24\sqrt{10}$ **69.** $12\sqrt{14}-12\sqrt{6}+6\sqrt{35}-6\sqrt{15}$
71. $8a+10\sqrt{ab}-3b$ **73.** $22+8\sqrt{6}$ **75.** $39+12\sqrt{3}$
77. $16+8\sqrt{3}$ **79.** $36-12\sqrt{x}+x$ **81.** $7-6\sqrt{x-2}+x$
83. 1 **85.** 4 **87.** $9-x$ **89.** -14 **91.** 25 **93.** 1 **95.** $x-y$
97. 50 **99.** 21 **101.** $2+\sqrt{2}$ **103.** $2+4\sqrt{6}$ **105.** $1+2\sqrt{5}$
107. $3-2\sqrt{6}$ **109.** $\dfrac{2+3\sqrt{3}}{4}$ **111.** $\dfrac{4-2\sqrt{6}}{3}$ **113.** $5\sqrt[3]{2}$
115. $9xy\sqrt[3]{2xy^2}$ **117.** $5\sqrt[4]{3}$ **119.** $2\sqrt[3]{4}-2-6\sqrt[3]{2}$

Exercise Set 10.5

1. $\dfrac{3\sqrt{7}}{7}$ **3.** $\sqrt{5}$ **5.** $3\sqrt{2}$ **7.** $2\sqrt{10}$
9. $\dfrac{4\sqrt{x}}{x}$ **11.** $\dfrac{m\sqrt{n}}{n}$ **13.** $\dfrac{\sqrt{10}}{2}$ **15.** $\dfrac{\sqrt{6}}{4}$ **17.** $\dfrac{\sqrt{10}}{8}$ **19.** $\dfrac{\sqrt{ab}}{b}$
21. $\dfrac{\sqrt{15}}{3}$ **23.** $\dfrac{\sqrt{21}}{6}$ **25.** $\dfrac{\sqrt{pq}}{q}$ **27.** $\dfrac{\sqrt{7x}}{x^3}$ **29.** $\dfrac{\sqrt{35y}}{5y}$ **31.** $\dfrac{3\sqrt{7a}}{7a}$
33. $\dfrac{2\sqrt[3]{3}}{3}$ **35.** $2\sqrt[3]{5}$ **37.** $\dfrac{\sqrt[3]{rs^2}}{s}$ **39.** $\dfrac{\sqrt[3]{xy}}{y}$ **41.** $\dfrac{\sqrt[3]{10b^2}}{2b}$ **43.** $\dfrac{\sqrt[3]{28d}}{2d}$
45. $2-\sqrt{5}$ **47.** $\dfrac{21+7\sqrt{3}}{6}$ **49.** $\dfrac{7a-a\sqrt{a}}{49-a}$ **51.** $\dfrac{15+7\sqrt{3}}{6}$
53. $\dfrac{29-10\sqrt{5}}{-11}$ **55.** $\dfrac{20+5\sqrt{5}+4\sqrt{3}+\sqrt{15}}{11}$ **57.** $\dfrac{14+9\sqrt{a}+a}{4-a}$
59. $4-\sqrt{15}$ **61.** $\dfrac{x+2\sqrt{xy}+y}{x-y}$ **63.** $\dfrac{5\sqrt{2}-2\sqrt{3}}{2}$
65. $-\dfrac{2\sqrt{10}+2\sqrt{35}+6\sqrt{3}+3\sqrt{42}}{-15}$ **67.** $\sqrt[4]{9}$ or $\sqrt{3}$ **69.** $y\sqrt[4]{y}$ **71.** $3\sqrt[5]{4}$

Exercise Set 10.6

1. 4 **3.** 49 **5.** $\varnothing$ **7.** 16 **9.** 64 **11.** 25
13. 6 **15.** 6 **17.** 3, 5 **19.** 6 **21.** 7 **23.** 14 **25.** 8 **27.** 4, 5 **29.** 6
31. 5 **33.** 3 **35.** 0, 4 **37.** 5 **39.** $\varnothing$ **41.** 4 **43.** a) 66.6 mph b) 400 ft
45. a) 20 amps b) 27,000 watts **47.** $\dfrac{\pi}{2}$ or approximately 1.57 sec
b) 1.82 ft **49.** 6 **51.** 8 **53.** -11 **55.** 16 **57.** 41

Exercise Set 10.7

1. 15 in. **3.** 4 ft **5.** $4\sqrt{5}$ m **7.** $\sqrt{14}$ in.
9. $\sqrt{31}$ in. **11.** $3\sqrt{13}$ ft **13.** $2\sqrt{5}$ m **15.** $2\sqrt{13}$ in. **17.** $2\sqrt{2}$ cm
19. 10 in. **21.** $6\sqrt{2}$ ft **23.** $8\sqrt{3}$ mm **25.** $4\sqrt{5}$ m **27.** $\sqrt{30}$ m
29. $\sqrt{89}$ cm **31.** 8 in. **33.** $7\sqrt{2}$ ft **35.** $3\sqrt{6}$ cm **37.** $4\sqrt{2}$
in. **39.** 17 ft **41.** $\sqrt{69}$ in. **43.** 50 in. **45.** 75 ft **47.** $2\sqrt{7}$ ft

Chapter 10 Review Exercises

1. 4 **2.** does not exist as a
real number **3.** -4 **4.** 5 **5.** -5 **6.** -5 **7.** 3 **8.** does not exist as a
real number **9.** -3 **10.** b^2 **11.** $n^2\sqrt{n}$ **12.** $m\sqrt[3]{m^2}$ **13.** $z\sqrt[5]{z^3}$
14. $m^2n^3\sqrt{m}$ **15.** $ab^2\sqrt[3]{a}$ **16.** irrational, 5.657 **17.** rational, 13
18. irrational, 3.979 **19.** rational, 11 **20.** rational, 1.2 **21.** rational, .3
22. $3\sqrt{7}$ **23.** $30\sqrt{3}$ **24.** $3\sqrt[3]{2}$ **25.** $12\sqrt[4]{4}$ **26.** $9x^3\sqrt{2x}$

27. $10a^4\sqrt{2}$ **28.** $6z^2y\sqrt{10zy}$ **29.** $4a\sqrt[3]{2a}$ **30.** $15nm^2\sqrt[3]{4m}$
31. r **32.** x^3 **33.** $\sqrt{35}$ **34.** $\sqrt{mn}$ **35.** 7 **36.** $5\sqrt{7}$ **37.** 96
38. $126\sqrt{3}$ **39.** c^2 **40.** s^4 **41.** $v^2\sqrt{v}$ **42.** $60n^3\sqrt{3}$ **43.** $\dfrac{9x}{2y^2}$
44. $\dfrac{\sqrt{13}}{5}$ **45.** 6 **46.** x^2 **47.** $cd\sqrt{c}$ **48.** $6a^2n\sqrt{7n}$ **49.** $\dfrac{\sqrt{15}}{4}$
50. $\dfrac{n^3\sqrt{n}}{m^2}$ **51.** $\dfrac{a\sqrt{21}}{6b}$ **52.** $-2\sqrt{5}$ **53.** $-2a\sqrt[3]{x}$ **54.** $3\sqrt{3}$
55. $2\sqrt{3}$ **56.** $-11a\sqrt{5}$ **57.** $12a\sqrt{7a}+48a\sqrt{a}$ **58.** $6\sqrt{3}$ **59.** $\sqrt{6}$
60. $3\sqrt{5}-5\sqrt{6}$ **61.** $3+6\sqrt{2}$ **62.** $15+\sqrt{5}$ **63.** $-14-7\sqrt{15}$
64. 21 **65.** $11-4\sqrt{7}$ **66.** $\dfrac{3-\sqrt{2}}{6}$ **67.** $2+2\sqrt{3}$ **68.** $\dfrac{6\sqrt{7}}{7}$
69. $\dfrac{4\sqrt{2}}{3}$ **70.** $\dfrac{\sqrt{15}}{3}$ **71.** $\dfrac{a\sqrt{b}}{b}$ **72.** $\dfrac{\sqrt{5n}}{n}$ **73.** $\dfrac{4\sqrt[3]{3}}{3}$ **74.** $2\sqrt[3]{4a^2}$
75. $-8-4\sqrt{5}$ **76.** $-3\sqrt{3}-3\sqrt{5}$ **77.** $2\sqrt{2}+2\sqrt{3}$
78. $-9-4\sqrt{5}$ **79.** 81 **80.** 36 **81.** 21 **82.** $\varnothing$ **83.** $\varnothing$ **84.** 3, 7
85. 7 **86.** 15 in. **87.** $\sqrt{17}$ ft **88.** $3\sqrt{2}$ cm **89.** $\sqrt{89}$ ft
90. $\sqrt{181}$ in. **91.** $8\sqrt{2}$ m **92.** $3\sqrt{7}$ yd **93.** $3\sqrt{5}$ ft **94.** $5\sqrt{10}$ ft
95. 54 ft.

Chapter 10 Test

1. 9 [10.1] **2.** -4 [10.1] **3.** x^4 [10.1]
4. c^3d^2 [10.1] **5.** $4\sqrt{6}$ [10.2] **6.** $21x^2\sqrt{2x}$ [10.2] **7.** $8\sqrt[3]{5}$ [10.2]
8. $12y^2\sqrt[3]{y}$ [10.2] **9.** $30\sqrt{7}$ [10.3] **10.** $60a^3\sqrt{6a}$ [10.3] **11.** 4 [10.3]
12. $16c^2\sqrt{2c}$ [10.3] **13.** $\dfrac{3c^2\sqrt{7}}{5}$ [10.3] **14.** $-6\sqrt{6}$ [10.4]
15. $22\sqrt{5}$ [10.4] **16.** $18a\sqrt{5a}$ [10.4] **17.** $3\sqrt{2}+12\sqrt{6}$ [10.4]
18. $3+5\sqrt{30}$ [10.4] **19.** -12 [10.4] **20.** $25-10\sqrt{x}+x$ [10.4]
21. $\dfrac{4\sqrt{3}}{3}$ [10.5] **22.** $2\sqrt[3]{4}$ [10.5] **23.** $\dfrac{\sqrt{mn}}{n}$ [10.5] **24.** $\dfrac{21+3\sqrt{5}}{11}$ [10.5]
25. $-\dfrac{16+7\sqrt{6}}{2}$ [10.5] **26.** 25 [10.6] **27.** 9 [10.6] **28.** -2 [10.6]
29. $3\sqrt{2}$ ft [10.7] **30.** 75 ft [10.7]

Exercise Set 11.1

1. $0, -2$ **3.** $0, -3$ **5.** $0, \dfrac{2}{3}$ **7.** $0, 2$ **9.** ± 5
11. no real-number solutions **13.** ± 11 **15.** $\pm\sqrt{7}$ **17.** no real-
number solutions **19.** $\pm 4\sqrt{3}$ **21.** $\pm 2\sqrt{7}$ **23.** ± 2 **25.** $\pm\sqrt{5}$
27. $\pm 2\sqrt{5}$ **29.** $\pm 2\sqrt{2}$ **31.** no real-number solutions **33.** $\pm\dfrac{\sqrt{14}}{2}$
35. $\pm\dfrac{\sqrt{55}}{5}$ **37.** $\pm\dfrac{2\sqrt{3}}{3}$ **39.** $1, -9$ **41.** $-5\pm 2\sqrt{11}$ **43.** $1, -\dfrac{11}{3}$
45. $-4, 7$ **47.** $3\pm\sqrt{7}$ **49.** $-5\pm\sqrt{11}$ **51.** $\dfrac{-4\pm\sqrt{2}}{2}$ **53.** $2\pm 7\sqrt{2}$
55. $\dfrac{-2\pm 2\sqrt{13}}{3}$ **57.** 2 and -8 **59.** 8 in. **61.** 3 cm
63. $\dfrac{-15\pm\sqrt{21}}{3}$ or $-5\pm\dfrac{\sqrt{21}}{3}$ **65.** $\dfrac{-15\pm\sqrt{55}}{10}$ or $-\dfrac{3}{2}\pm\dfrac{\sqrt{55}}{10}$

Exercise Set 11.2

1. $9, (a+3)^2$ **3.** $81, (x-9)^2$
5. $\dfrac{25}{4}, \left(n+\dfrac{5}{2}\right)^2$ **7.** $\dfrac{49}{4}, \left(b-\dfrac{7}{2}\right)^2$ **9.** $\dfrac{25}{16}, \left(x+\dfrac{5}{4}\right)^2$
11. $\dfrac{9}{196}, \left(a-\dfrac{3}{14}\right)^2$ **13.** 3, 5 **15.** $-2, -6$ **17.** $-2, 6$ **19.** 2, -5
21. $-4, -5$ **23.** $2, \dfrac{3}{2}$ **25.** $6, -\dfrac{4}{3}$ **27.** $4, -\dfrac{5}{2}$ **29.** $3, \dfrac{2}{3}$ **31.** no real-
number solutions **33.** $-2\pm\sqrt{5}$ **35.** $4\pm\sqrt{11}$ **37.** $7\pm\sqrt{51}$
39. $-4\pm\sqrt{13}$ **41.** $5\pm\sqrt{30}$ **43.** $\dfrac{-5\pm\sqrt{37}}{2}$ **45.** $\dfrac{2\pm\sqrt{10}}{2}$
47. $\dfrac{-3\pm\sqrt{19}}{2}$ **49.** $\dfrac{-3\pm 2\sqrt{3}}{3}$ **51.** no real-number solutions
53. $\dfrac{1}{3}, -\dfrac{2}{3}$

Exercise Set 11.3 **1.** $1, 5$ **3.** $2, 4$ **5.** $-4, 6$ **7.** $-1, -6$
9. $-2, 7$ **11.** $6, \dfrac{3}{2}$ **13.** $6, -\dfrac{4}{3}$ **15.** no real-number solutions
17. $\dfrac{5}{2}, -\dfrac{3}{4}$ **19.** $2, -\dfrac{1}{15}$ **21.** $\dfrac{-3 \pm \sqrt{13}}{2}$ **23.** $\dfrac{-1 \pm \sqrt{13}}{2}$
25. $\dfrac{3 \pm \sqrt{41}}{4}$ **27.** $\dfrac{1 \pm \sqrt{61}}{10}$ **29.** no real-number solutions
31. $\dfrac{-7 \pm 3\sqrt{5}}{2}$ **33.** $\dfrac{9 \pm \sqrt{105}}{4}$ **35.** $1 \pm \sqrt{3}$ **37.** $5 \pm \sqrt{5}$
39. $4 \pm 2\sqrt{2}$ **41.** $\dfrac{2 \pm \sqrt{14}}{2}$ **43.** $\dfrac{-5 \pm \sqrt{5}}{4}$ **45.** $\dfrac{-1 \pm \sqrt{13}}{3}$
47. no real-number solutions **49.** 3 and 5 or -3 and -5 **51.** $W = 5$
in., $L = 11$ in. **53.** $W = 4$ yds, $L = 7$ yd **55.** a) -3 b) 5 c) 0
57. a) 1 b) 2 c) 17 **59.** $2, \dfrac{3}{2}$ **61.** $3, -\dfrac{2}{3}$

Exercise Set 11.4 **1.** $2i$ **3.** $8i$ **5.** $i\sqrt{6}$ **7.** $i\sqrt{15}$ **9.** $2i\sqrt{6}$
11. $5i\sqrt{3}$ **13.** $6 + 0i$ **15.** $-7 + 0i$ **17.** $0 + 6i$ **19.** $0 - 4i$
21. $3\sqrt{3} + 0i$ **23.** $4\sqrt{5} + 0i$ **25.** $0 + 3i\sqrt{2}$ **27.** $0 + 5i\sqrt{5}$
29. $4 + 7i$ **31.** $-5 - 9i$ **33.** $3 - 3i$ **35.** $-3 - i$ **37.** $9 + 9i$
39. $1 - 8i$ **41.** $1 - 5i$ **43.** $10 + 2i$ **45.** $-6 + 12i$ **47.** $10 + 20i$
49. $4 + 7i$ **51.** $14 - 8i$ **53.** $2 - 16i$ **55.** $-18 + 16i$ **57.** $-\dfrac{1}{4} - \dfrac{3}{4}i$
59. $\dfrac{3}{5} + \dfrac{4}{5}i$ **61.** $-2 + 5i, -2 - 5i$ **63.** $5 + 8i, 5 - 8i$
65. $2 - i, 2 + i$ **67.** $3 + 2i, 3 - 2i$ **69.** $2 + i\sqrt{2}, 2 - i\sqrt{2}$
71. $3 + i\sqrt{3}, 3 - i\sqrt{3}$ **73.** $\dfrac{-3 + i\sqrt{15}}{2}, \dfrac{-3 - i\sqrt{15}}{2}$
75. $\dfrac{-7 + i\sqrt{11}}{2}, \dfrac{-7 - i\sqrt{11}}{2}$

Exercise Set 11.5 **1.** $4, 5$ **3.** 6 and 9 or -6 and -9 **5.** 5 and 7
or -5 and -7 **7.** $W = 50$ yd, $L = 60$ yd **9.** $W = 8$ in., $L = 11$
in. **11.** 3, 4, and 5 **13.** 26 ft **15.** 80 ft and 150 ft **17.** 3 hrs
19. 45 mph **21.** Jody $= 4$ hrs, Billy $= 6$ hrs
23. Ronnie $= 2$ hrs, Buster $= 3$ hrs
25. 5 sec **27.** $\sqrt{3}$ or 1.7 sec **29.** 3.7 sec

Chapter 11 Review Exercises **1.** $0, -3$ **2.** ± 4 **3.** $0, \dfrac{-13}{5}$
4. $c = 0, \dfrac{13}{2}$ **5.** $0, 2$ **6.** $a = 0, 13$ **7.** ± 12 **8.** ± 20 **9.** $\pm 2\sqrt{7}$
10. $\pm 3\sqrt{5}$ **11.** $\varnothing$ **12.** $\varnothing$ **13.** ± 5 **14.** ± 7 **15.** $\pm 2\sqrt{5}$
16. $\pm 3\sqrt{2}$ **17.** $1, -13$ **18.** $-5, 11$ **19.** $\dfrac{-4 \pm 4\sqrt{5}}{3}$
20. $\dfrac{5 \pm 5\sqrt{5}}{2}$ **21.** $-4, -6$ **22.** $2, 8$ **23.** $-2, 7$ **24.** $3, -6$

25. $-6, -\dfrac{1}{2}$ **26.** $2, -\dfrac{2}{3}$ **27.** $-4 \pm 3\sqrt{2}$ **28.** $5 \pm \sqrt{21}$
29. $\dfrac{-7 \pm \sqrt{61}}{2}$ **30.** $\dfrac{9 \pm \sqrt{73}}{2}$ **31.** no real-number solutions
32. $\dfrac{-7 \pm \sqrt{13}}{2}$ **33.** $\dfrac{-6 \pm \sqrt{46}}{2}$ **34.** $\dfrac{7 \pm \sqrt{43}}{2}$ **35.** $-1, -2$
36. $-1, -4$ **37.** $2, 6$ **38.** $5, -2$ **39.** $2, -\dfrac{5}{2}$ **40.** $\dfrac{2}{3}, -1$
41. $\dfrac{4 \pm \sqrt{34}}{6}$ **42.** $\dfrac{1}{2}, -\dfrac{4}{3}$ **43.** $-3 \pm \sqrt{13}$ **44.** $-4 \pm 3\sqrt{2}$
45. $\dfrac{4 \pm \sqrt{14}}{2}$ **46.** $\dfrac{-3 \pm \sqrt{29}}{4}$ **47.** $\dfrac{-1 \pm \sqrt{19}}{6}$ **48.** $\dfrac{2 \pm \sqrt{10}}{6}$ **49.** no
real-number solutions **50.** no real-number solutions **51.** $\dfrac{-5 \pm \sqrt{41}}{4}$
52. $\dfrac{-7 \pm \sqrt{37}}{6}$ **53.** $9i$ **54.** $7i\sqrt{2}$ **55.** $12 + 0i$ **56.** $0 + 11i$
57. $0 + 3i\sqrt{10}$ **58.** $4 + 11i$ **59.** $-4 + i$ **60.** $-3 - 13i$
61. $4 - 9i$ **62.** $-1 + 11i$ **63.** $40 + 20i$ **64.** $16 - 2i$ **65.** 50
66. $\dfrac{13}{17} - \dfrac{1}{17}i$ **67.** $-6 + 8i, -6 - 8i$ **68.** $3 + 4i, 3 - 4i$
69. $4 + i\sqrt{2}, 4 - i\sqrt{2}$ **70.** $\dfrac{-3 + i\sqrt{19}}{2}, \dfrac{-3 - i\sqrt{19}}{2}$ **71.** 2 and 5
72. 2 and 6 or -2 and -6 **73.** 5 and 6 or -5 and -6 **74.** 4 and 8 or
-4 and -8 **75.** $W = 12$ ft, $L = 16$ ft **76.** $L = 10$ ft, $W = 7$ ft
77. 8 mi and 15 mi **78.** $L = 20$ in., $W = 15$ in. **79.** 40 mph **80.** 2 hrs
81. Ahmad $= 6$ hrs and Horace $= 10$ hrs **82.** Ahmad $= 6$ days,
Horace $= 8$ days

Chapter 11 Test **1.** $0, 2$ [11.1] **2.** $0, 5$ [11.1] **3.** ± 11 [11.1]
4. $\pm 3\sqrt{3}$ [11.1] **5.** $0, 12$ [11.1] **6.** $\dfrac{-2 \pm 3\sqrt{2}}{3}$ [11.1]
7. $100, (a - 10)^2$ [11.2] **8.** $\dfrac{49}{36}, \left(b + \dfrac{7}{6}\right)^2$ [11.2] **9.** $-2, 8$ [11.2]
10. $\dfrac{3 \pm \sqrt{19}}{2}$ [11.2] **11.** $2, 3$ [11.3] **12.** $4, -\dfrac{5}{2}$ [11.3] **13.** no real-
number solutions [11.3] **14.** $-3, \dfrac{1}{3}$ [11.3] **15.** $-5, 7$ [11.1–11.3]
16. $\dfrac{-7 \pm \sqrt{97}}{4}$ [11.1–11.3] **17.** no real-number solutions [11.1–11.3]
18. $4 \pm 5\sqrt{5}$ [11.1–11.3] **19.** $-3 + 6i$ [11.4] **20.** $-1 - 13i$ [11.4]
21. $3 - 2i$ [11.4] **22.** $18 + 14i$ [11.4] **23.** $-\dfrac{2}{29} - \dfrac{24}{29}i$ [11.4] **24.** 8 and
10 or -8 and -10 [11.5] **25.** $W = 6$ in., $L = 8$ in. [11.5] **26.** 3 hrs [11.5]

TABLE 1: POWERS AND ROOTS

n	n^2	n^3	$\sqrt{n}$	$\sqrt[3]{n}$	$\sqrt{10n}$
1	1	1	1.000	1.000	3.162
2	4	8	1.414	1.260	4.472
3	9	27	1.732	1.442	5.477
4	16	64	2.000	1.587	6.325
5	25	125	2.236	1.710	7.071
6	36	216	2.449	1.817	7.746
7	49	343	2.646	1.913	8.367
8	64	512	2.828	2.000	8.944
9	81	729	3.000	2.080	9.487
10	100	1,000	3.162	2.154	10.000
11	121	1,331	3.317	2.224	10.488
12	144	1,728	3.464	2.289	10.954
13	169	2,197	3.606	2.351	11.402
14	196	2,744	3.742	2.410	11.832
15	225	3,375	3.873	2.466	12.247
16	256	4,096	4.000	2.520	12.649
17	289	4,913	4.123	2.571	13.038
18	324	5,832	4.243	2.621	13.416
19	361	6,859	4.359	2.688	13.784
20	400	8,000	4.472	2.714	14.142
21	441	9,261	4.583	2.759	14.491
22	484	10,648	4.690	2.802	14.832
23	529	12,167	4.796	2.844	15.166
24	576	13,824	4.899	2.884	15.492
25	625	15,625	5.000	2.924	15.811
26	676	17,576	5.099	2.962	16.125
27	729	19,683	5.196	3.000	16.432
28	784	21,952	5.292	3.037	16.733
29	841	24,389	5.385	3.072	17.029
30	900	27.000	5.477	3.107	17.321
31	961	29,791	5.568	3.141	17.607
32	1,024	32,768	5.657	3.175	17.889
33	1,089	35,937	5.745	3.208	18.166
34	1,156	39,304	5.831	3.240	18.439
35	1,225	42,875	5.916	3.271	18.708
36	1,296	46,656	6.000	3.302	18.974
37	1,369	50,653	6.083	3.332	19.235
38	1,444	54,872	6.164	3.362	19.494
39	1,521	59,319	6.245	3.391	19.748
40	1,600	64,000	6.325	3.420	20.000
41	1,681	68,921	6.403	3.448	20.248
42	1,764	74,088	6.481	3.476	20.494
43	1,849	79,507	6.557	3.503	20.736
44	1,936	85,184	6.633	3.530	20.976
45	2,025	91,125	6.708	3.557	21.213
46	2,116	97,336	6.782	3.583	21.148
47	2,209	103,823	6.856	3.609	21.679
48	2,304	110,592	6.928	3.534	21.909
49	2,401	117,649	7.000	3.659	22.136
50	2,500	125,000	7.071	3.684	22.361

TABLE 1: POWERS AND ROOTS

n	n^2	n^3	$\sqrt{n}$	$\sqrt[3]{n}$	$\sqrt{10n}$
51	2,601	132,651	7.141	3.708	22.583
52	2,704	140,608	7.211	3.733	22.804
53	2,809	148,877	7.280	3.756	23.022
54	2,916	157,464	7.348	3.780	23.238
55	3,025	166,375	7.416	3.803	23.452
56	3,136	175,616	7.483	3.826	23.664
57	3,249	185,193	7.550	3.849	13.875
58	3,364	195,112	7.616	3.871	24.083
59	3,481	205,379	7.681	3.893	24.290
60	3,600	216,000	7.746	3.915	24.495
61	3,721	226,981	7.810	3.936	24.698
62	3,844	238,328	7.874	3.958	24.900
63	3,969	250,047	7.937	3.979	25.100
64	4,096	262,144	8.000	4.000	25.298
65	4,225	274,625	8.062	4.021	25.495
66	4,356	287,496	8.124	4.041	25.690
67	4,489	300,763	8.185	4.062	25.884
68	4,624	314,432	8.246	4.082	26.077
69	4,761	328,509	8.307	4.102	26.268
70	4,900	343,000	8.367	4.121	26.458
71	5,041	357,911	8.426	4.141	26.646
72	5,184	373,248	8.485	4.160	26.833
73	5,329	389,017	8.544	4.179	27.019
74	5,476	405,224	8.602	4.198	27.203
75	5,625	421,875	8.660	4.217	27.386
76	5,776	438,976	8.718	4.236	27.568
77	5,929	456,533	8.775	4.254	27.749
78	6,084	474,552	8.832	4.273	27.928
79	6,241	493,039	8.888	4.291	28.107
80	6,400	512,000	8.944	4.309	28.284
81	6,561	531,441	9.000	4.327	28.460
82	6,724	551,368	9.055	4.344	28.636
83	6,889	571,787	9.110	4.362	28.810
84	7,056	592,704	9.165	4.380	28.983
85	7,225	614,125	9.220	4.397	29.155
86	7,396	636,056	9.274	4.414	29.326
87	7,569	658,503	9.327	4.431	29.496
88	7,744	981,472	9.381	4.448	29.665
89	7,921	704,969	9.434	4.465	29.833
90	8,100	729,000	9.487	4.481	30.000
91	8,281	753,571	9.539	4.498	30.166
92	8,464	778,688	9.592	4.514	30.332
93	8,649	804,357	9.644	4.531	30.496
94	8,836	830,584	9.695	4.547	30.659
95	9,025	857,375	9.747	4.563	30.882
96	9,216	884,736	9.798	4.579	30.984
97	9,409	912,673	9.849	4.595	31.145
98	9,604	941,192	9.899	4.610	31.305
99	9,801	970,299	9.950	4.626	31.464
100	10,000	1,000,000	10.000	4.642	21.623

Index